ORGANIC SYNTHESES

ORGANIC SYNTHESES

Collective Volume 6

A REVISED EDITION OF ANNUAL VOLUMES 50–59

JOHN WILEY & SONS
NEW YORK • CHICHESTER • BRISBANE • TORONTO • SINGAPORE

Published by John Wiley & Sons, Inc.

Library of Congress Catalog Card Number: 21-17747
ISBN 0-471-85243-0

Printed in the United States of America

10 9 8 7 6 5 4 3

Ralph L. and the Late Rachel H. Shriner
50th Wedding Anniversary, 1979,
Taken at Interlaken Camp, Deer River, Minnesota

This volume is dedicated to Ralph L. and the Late Rachel H. Shriner. Ralph Shriner served Organic Syntheses as the editor of annual volume 27 (1947), as a long-time member of the Board of Directors (1951–1977), and as its Vice President (1974–1976) and a senior statesman. Ralph, together with his wife, Rachel, in their crowning achievement for Organic Syntheses, edited the *Cumulative Indices to Collective Volumes I–V* (which also contains the contents of annual volumes 50–54).

PREFACE

Continuing the tradition of its predecessor collective volumes, Collective Volume 6 contains the procedures previously published in ten annual volumes, 50–59 (1970–79), but revised and the discussion sections frequently updated and expanded. To achieve this objective the Editor-in-Chief wrote to each senior submitter asking for suggestions for corrections of any errors, typographical or procedural, changes in writeup of the Discussion section, and important new references that should be included. Some did not reply, others saw no need for changes, but many generously provided updating improvements in their discussion sections for which the Editor-in-Chief is very grateful.

Errors found in the original printings have been corrected, calculations have been rechecked, and journal abbreviations in the references have been made to conform to those recommended by CASSI *(Chemical Abstracts Service Source Index)*. Efforts have been made to standardize the style of the experimental procedures where variations have arisen from differences in style between the ten editors of annual volumes and the larger number of checkers of individual procedures. A particular effort has been made to standardize the presentation of spectroscopic data and make them more compact in accord with current usage. The period covered in the annual volumes (1970–79) has seen a rapid increase in the use of spectroscopic data and an evolution in the styles of presentation, so that it is quite understandable that there have been differences from volume to volume.

Increasing emphasis on the toxicology of organic compounds has led to the inclusion of additional hazard warnings, particularly for benzene, hexamethylphosphoric triamide (HMPA), and methyl iodide at the many places where they appear. The practice of organic chemistry frequently involves the use or preparation of hazardous compounds. *Organic Syntheses* would be remiss and incomplete if it did not include such examples. It is better that it should deal with such situations and teach, by example, methods of keeping risks as low as possible in cases where it is necessary to deal with hazardous compounds.

Where experience indicated, some changes have been made in the experimental procedures from the annual volumes. Thus, the experimental procedure for benzyl chloromethyl ether has been improved and rechecked. Two procedures have been replaced by improved procedures for related compounds, though the titles of the original procedures are still listed in alphabetical order by compound name in the Contents section, in the indices, and in the text where the details would otherwise have appeared. Thus, the procedure for ethyl (*E*)-3-nitroacrylate has been replaced by an improved procedure for methyl (*E*)-3-nitroacrylate and the procedure for hydroboration of olefins: (+)-isopinocampheol has been replaced by an improved procedure for (−)-isopinocampheol. The procedure for preparation of alkenes by reaction of lithium dipropenylcuprates with alkyl halides: (*E*)-2-undecene has been deleted because the required starting material, lithium dispersion containing 1% sodium, is no longer commercially available in the quality needed.

Although many titles of procedures begin with a description of the method used, followed by the name of the compound, procedures are arranged by compound common name in alphabetical order (by letter, not by first word) in the Contents section and in the text. While desirable from the indexing standpoint, this means that procedures that are part of a longer synthetic sequence will be separated and found in different parts of the volume. In the titles in the text (but not in the Contents), the *Chemical Abstracts* name (if different) is given below the main title name in brackets in smaller type. Following the title in the Contents section, reference is made in parentheses to the annual volume (in boldface type) and page number where each procedure first appeared. This concordance listing makes it possible, if the title of the procedure is known, to quickly locate the revised procedure in this collective volume. In the few cases where the compound name is known but has been changed, it is possible to scan the Contents section to find the correct procedure. If only a literature citation to the annual volume is available, it would usually be most convenient to use the concordance index at the back of this volume.

In a departure from the practice in previous collective volumes, there is a greatly expanded, consolidated general index containing the features of several separate indices from previous volumes (type of reaction, type of compound, preparation or purification of solvents and reagents, and special or unique apparatus) and a number of additional features. The general index includes common and *Chemical Abstracts* names for title compounds, isolated intermediates, reagents, and Chemical Abstracts Service Registry numbers (italicized and in brackets) and hazard warnings for such compounds (also repeated in a separate hazard index). Other compounds, such as those mentioned only in the discussion section or in related tables, are indexed only by commonly used but systematic names. In addition, there are separate author (submitter), formula, and hazard warning indices. The formula index has been expanded to include all compounds covered in the general index. Finally, there is a concordance index indicating in which annual volume each procedure first appeared. This should make it possible to take a literature citation to a procedure in annual volumes 50–59 and quickly locate the revised procedure in this collective volume.

Attention is called to the very useful volume, *Cumulative Indices to Collective Volumes I–V*, published in 1976. In addition to its thorough coverage of the collective volumes, this volume contains in Appendix B an interesting history and discussion of the origin, development, organization, and operations of *Organic Syntheses*. Also, a brief history entitled "Fifty Years of Organic Syntheses" by Roger Adams, one of the founders, appeared in the Golden Anniversary Volume, annual Volume 50 (p. vii). Further information about the historical development of *Organic Syntheses* can be obtained from reading the prefaces of the 63 annual volumes published to date and from the obituary biographies that appear in them from time to time.

The editors of *Organic Syntheses* welcome corrections, suggestions, and procedures being submitted for consideration by the Board of Editors. Prospective submitters should consult the section entitled "Submission of Preparations" at the front of one of the latest annual volumes for guidance. Correspondence should be addressed to the current Secretary of the Board of Editors of Organic Syntheses, Dr. Jeremiah P. Freeman, Department of Chemistry, University of Notre Dame, Notre Dame, IN 46556. In addition to his duties as Secretary, Dr. Freeman has the responsibility of serving as Editor-in-Chief of the first five-year collective volume, *Collective Volume 7*, which will cover the revised contents of annual volumes 60–64.

The Editor-in-Chief is grateful to the submitters, checkers, and editors of annual volumes 50–59, who made this collective volume possible, to the Editors-in-Chief of Collective Volumes I–V for their example and to Henry E. Baumgarten and Norman Rabjohn for their advice, and to Dr. Richard E. Benson who initiated the preparation of the large-scale version, subsequently updated by fellow members of the Board of Editors, of the useful *Style Guide for Organic Syntheses,* now in its latest version, for *Volumes 60–69.* It is also a pleasure to acknowledge the helpful suggestions and encouragement provided by the officers, Jeremiah P. Freeman (Secretary), Carl R. Johnson (Treasurer), Nelson J. Leonard (President), and my remaining fellow members of the Board of Directors. Thanks are due to the secretaries who served with me during my ten-year term as Secretary to the Board of Editors (during which the procedures for annual Volumes 50–59 were collected and processed): Dorothy McDougall, Arliss T. Piri, Doris E. Berg, Roberta Andrich, and Marion P. Gorman; and during the preparation of this manuscript: Sheryl Denise Bortot (who handled the major wave of correspondence and procedure revisions), and Marcy Scherbel, Brenda Ford, Ruth Kay Granheim, Joyce Garcia, and Susan J. Wrayge. An undergraduate student, Evan David Goldstein, assisted with library work on nomenclature and registry numbers. Finally, special thanks are due to my editorial assistants: Rodney D. DeKruif, who, in addition to the efforts by the Editor-in-Chief, reviewed each procedure at least twice for accuracy, consistency of style, and brevity, and prepared the index file cards; Garrett H. Schure, who has been very helpful in organizing the final copy of the manuscript for the publisher and in correcting proofs; Todd S. Gac, who did invaluable work in assigning page numbers to all the index entries and in preparing the printed indexes, to Keith W. Palmer and Mark E. Brigham, who assisted him; and to Randy Röhl, a word processor who did valuable work in putting the data from the index file cards on disks and in printing out hard copies of the index.

Wayland E. Noland

Minneapolis, Minnesota
January 13, 1988

CONTENTS

CONTENTS

ORGANIC SYNTHESES

ACYLAMIDOALKYL ACETOPHENONES FROM SUBSTITUTED PHENETHYLAMINES: 2-(2-ACETAMIDOETHYL)-4,5-DIMETHOXYACETOPHENONE

[Acetamide, *N*-[2-(2-acetyl-4,5-dimethoxyphenyl)ethyl]-]

Submitted by A. Brossi, L. A. Dolan, and S. Teitel[1]
Checked by Hiroshi Itazaki and Wataru Nagata

1. Procedure

Caution! Part A should be conducted in a hood to avoid inhalation of hydrogen chloride fumes.

A. 6,7-*Dimethoxy*-1-*methyl*-3,4-*dihydroisoquinoline*. A 2-l., three-necked, round-bottomed flask equipped with a mechanical stirrer, a reflux condenser protected by a calcium chloride tube, and a pressure-equalizing dropping funnel is charged with 54.0 g. (0.243 mole) of *N*-acetylhomoveratrylamine (Note 1) and 275 ml. of dry toluene (Note 2). The mixture is stirred, warmed to 40°, and treated with 86.4 g. (52.5 ml., 0.572 mole) (Note 3) of phosphorus oxychloride (Note 4), which is added over 1 hour (Note 5). After addition, the reaction mixture is stirred at reflux for 2 hours, then cooled with an ice bath for 4 hours. The resulting crystals are collected by filtration and dried overnight at 50° in a vacuum oven, giving 79.0–79.5 g. (Note 6) of 6,7-dimethoxy-1-methyl-3,4-dihydroiso-quinoline dichlorophosphate, m.p. 148–152° (Note 7). This material is dissolved in 150 ml. of water (Note 8), and the solution is treated with 100 ml. of aqueous 40% sodium hydroxide (Note 9). The oil which separates is drawn off, and the aqueous solution is washed with three 20-ml. portions of chloroform. These extracts and the oil are combined, washed with 15 ml. of water, and dried over anhydrous sodium sulfate. Removal of chloroform with a rotary evaporator provides 47.0–48.0 g. (95–96%) of product, which is used without purification in Part B (Note 10).

B. 2-*Acetyl*-6,7-*dimethoxy*-1-*methylene*-1,2,3,4-*tetrahydroisoquinoline*. A 1-l., three-necked, round-bottomed flask equipped with a mechanical stirrer, a reflux condenser topped with a calcium chloride drying tube, and a thermometer is charged with 110 ml. of acetic anhydride, 110 ml. of pyridine, and 45.0 g. (0.220 mole) of the dihydroisoquinoline prepared in Part A. The reaction mixture is stirred and heated at 90–95° for 30 minutes, stored at room temperature overnight, and concentrated by distillation at 50° using a rotary evaporator. The residue is diluted with 20 ml. of ethyl acetate, and another evaporation under reduced pressure gives material that can be crystallized from 75 ml. of ethyl acetate to yield 38.5–41.0 g. (72–77%) of product m.p., 106–107° (Note 11).

C. 2-(2-*Acetamidoethyl*)-4,5-*dimethoxyacetophenone*. A slurry of 31.0 g. (0.125 mole) of the methylene derivative obtained in Part B and 75 ml. of 5% hydrochloric acid is stirred and warmed on a steam bath to 60–65°. As soon as all the solid has dissolved, the solution is cooled with an ice bath to 30° and basified by slowly adding a solution of 6.25 g. of potassium carbonate in 12.5 ml. of water (Note 12). The crystalline precipitate is collected by filtration, washed with three 12.5-ml. portions of water, and air-dried, yielding 30.5–32.0 g. (91–93%) of the acetophenone, m.p. 123–125° (Note 13).

2. Notes

1. *N*-Acetylhomoveratrylamine[2] was prepared by adding 190 ml. of acetic anhydride to a stirred solution of 300 g. (1.80 moles) of β-(3,4-dimethoxyphenyl)ethylamine (Aldrich Chemical Company, Inc.) in 150 ml. of pyridine at such a rate that the temperature is maintained at 90–95° (*ca.* 1.5 hours is required). After the solution had been stored at room temperature overnight, the volatile material was evaporated under reduced pressure, and the residue was crystallized from ethyl acetate to give 286–306 g. (78–83%) of acetylated product, m.p. 99–100°.

2. The checkers used reagent-grade toluene dried over Linde-type 5A molecular sieves.

3. The checkers obtained an identical result when the molar ratio of phosphorous oxychloride to substrate was reduced from 2.35 to 1.5.

4. The checkers obtained phosphorus oxychloride from Wako Pure Chemical Industries, Ltd., Japan and distilled it prior to use.

5. The reaction temperature increased gradually to reflux, at which time the rate of addition was adjusted to maintain reflux.

6. This weight varies with the amount of solvent remaining.

7. Analysis calculated for $C_{12}H_{15}O_2N \cdot HOPOCl_2$: C, 42.37; H, 4.74; N, 4.12; Cl, 20.85; P, 9.10. Found: C, 42.30; H, 4.92; N, 4.21; Cl, 19.08; P, 8.51. IR (KBr) cm.$^{-1}$: 2800, 1665, 1602, 1565, 1105; ^{1}H NMR (D_2O), δ (multiplicity, number of protons, assignment): 3.17 (s, 3H, N=CCH_3), 4.25 (s, 3H, OCH_3), 4.30 (s, 3H, OCH_3), 7.37 (s, 1H, aryl CH), 7.62 (s, 1H, aryl CH).

The submitters, working on a kilogram scale without purification of reagent or solvent and with no precaution against moisture, obtained 6,7-dimethoxy-1-methyl-3,4-dihydroisoquinoline hydrochloride,[2] m.p. 202–203°, instead of the dichlorophosphate at this stage. The checkers obtained this hydrochloride by either treating the free base with hydrochloric acid or recrystallizing the dichlorophosphate from methanol–ethyl acetate.

8. The crystals dissolve gradually in water and, since dissolution is exothermic due to decomposition of the dichlorophosphoric acid, ice cooling is desirable.

9. Ice is added during neutralization to keep the temperature below 30°.

10. A pure sample may be prepared by crystallization from ether: m.p. 105–107°; UV (C$_2$H$_5$OH) nm. max. (ϵ): 227 (24,000), 270 (7360), 307 (6640); UV (0.01 N hydrochloric acid) nm. max. (ϵ): 244 (17,250), 302 (8740), 352 (8440); IR (KBr) cm.$^{-1}$: 1650 (C=N).

11. UV (C$_2$H$_5$OH) nm. max. (ϵ): 220 (30,750), 267 (13,200), 304 (7080); UV (0.01 N hydrochloric acid) nm. max. (ϵ): 232 (23,500), 276 (9345), 305 (6120); IR (KBr) cm.$^{-1}$: 880–910 (C=CH$_2$). A dimorphic form melts at 100–102°. A mixture of this material and the 6,7-dimethoxy-1-methyl-3,4-dihydroisoquinoline described in Note 10 melted below 90°.

12. The rate of addition is dependent on the amount of foaming. Ice is added periodically to keep the temperature below 35°.

13. The product can be used without further purification. Recrystallization from water gave an analytical specimen, m.p. 126–127°; UV (95% C$_2$H$_5$OH) nm. max. (ϵ): 231 (24,100), 274 (8750), 304 (5500); infrared (KBr) cm.$^{-1}$: 1673, 1633; ^{1}H NMR (CDCl$_3$), δ (multiplicity, number of protons, assignment): 1.92 (s, 3H, Ha), 2.60 (s, 3H, Hb), 2.68–3.17 (m, 2H, Hg), 3.30–3.68 (m, 2H, Hh), 3.93 (s, 6H, Hc, and Hd), 6.68–7.07 (broad, 1H, Hi), 6.80 (s, 1H, He), 7.22 (s, 1H, Hf).

TABLE I

SUBSTITUTED ACETAMIDOETHYL ACETOPHENONES[5]

	Melting Point	Ultraviolet C$_2$H$_2$OH nm. max (ϵ)
R$_1$ = OCH$_3$, R$_2$ = R$_3$ = H	86°	221 (14,700, 269 (13,600)
R$_1$, R$_2$ = OCH$_2$O, R$_3$ = H	120°	230 (36,500), 273 (5200), 307 (5460)
R$_1$ = R$_2$ = R$_3$ = OCH$_3$	58°	222 (14,700), 263 (16,250)

3. Discussion

This procedure provides a facile method for converting substituted 1-methyl-3,4-dihydroisoquinolines into the corresponding 2-(2-acetamidoethyl)acetophenones, which are useful intermediates in the synthesis of 1-(substituted phenethyl)-2-methyl-1,2,3,4-tetrahydroisoquinolines.[3,4] The sequence is uncomplicated and affords, in excellent yield, a product that requires no further purification. In addition to the examples in Table I, this method has been utilized for the synthesis of other substituted acetophenones,[3,5,6] as well as related benzophenones and a heptanophenone.[7] The latter two classes of compounds have also been obtained by ring opening of 2-ethyl-1-phenyl- or 2-ethyl-1-hexyl-6,7-dialkoxy-3,4-dihydroisoquinolinium iodides with benzoyl chloride.[8]

1. Chemical Research Department, Hoffmann-La Roche Inc., Nutley, New Jersey 07110.
2. E. Späth and N. Polgar, *Monatsh. Chem.,* **51**, 190 (1929).
3. A. Brossi, H. Besendorf, B. Pellmont, M. Walter, and O. Schnider, *Helv. Chim. Acta,* **43**, 1459 (1960).
4. A. Brossi, H. Besendorf, L. A. Pirk, and A. Rheiner, Jr., in G. DeStevens, Ed., "Analgetics," Academic Press, New York, 1965, pp. 287–289.
5. A. Brossi, J. Würsch, and O. Schnider, *Chimia,* **12**, 114 (1958).
6. G. Dietz, G. Faust, and W. Fiedler, *Pharmazie,* **26**, 586 (1971).
7. J. Gardent, *C. R. Hebd. Seances Acad. Sci.,* **247**, 2010 (1958).
8. J. Gardent, *Bull. Soc. Chim. Fr.,* **10**, 1260 (1957).

THIOL PROTECTION WITH THE ACETAMIDOMETHYL GROUP:
S-ACETAMIDOMETHYL-L-CYSTEINE HYDROCHLORIDE

(L-Cysteine, S-[(acetylamino)methyl]-, monohydrochloride)

$$CH_3C \overset{O}{\diagup}{}_{NH_2} \quad + \quad \underset{H}{\overset{O}{\diagdown}}C\overset{\diagup}{}{}_{H} \quad \xrightarrow[\text{water}]{K_2CO_3} \quad CH_3C\overset{O}{\diagup}{}_{NHCH_2OH}$$

$$CH_3C\overset{O}{\diagup}{}_{NHCH_2OH} \quad + \quad \underset{{}^+NH_3Cl^-}{HSCH_2CHCO_2H} \quad \xrightarrow[\substack{1\text{-}2 \text{ days}}]{\substack{HCl \\ \text{water, } 25°,}}$$

$$\underset{{}^+NH_3Cl^-}{CH_3\overset{O}{\overset{\|}{C}}NHCH_2SCH_2CHCO_2H}$$

Submitted by JOHN D. MILKOWSKI,[1] DANIEL F. VEBER,[2]
and RALPH HIRSCHMANN[1]
Checked by A. PEARCE and G. BÜCHI

1. Procedure

A. N-(*Hydroxymethyl*)*acetamide*.   In a 2-l., round-bottomed flask, 100 g. (1.70 moles) of acetamide (Note 1) is added to a solution of 10 g. (0.072 mole) of anhydrous potassium carbonate in 137 g. (1.7 moles) of an aqueous 36–38% solution of formaldehyde (Note 2). The mixture is swirled, heated on a steam bath for 3 minutes, and allowed to stand overnight at room temperature. Several pieces of crushed dry ice are added (Note 3), after which the mixture is evaporated under reduced pressure with a heating bath kept below 40° (Note 4). A 128-g. portion of anhydrous sodium sulfate is added to the remaining colorless oil, which may have some precipitated salt suspended in it. After several hours the oil is dissolved in 1 l. of acetone, the suspended drying agent and salts are filtered, and the filtrate (Note 5) is dried further with additional anhydrous sodium sulfate. The suspension is filtered, and the clear filtrate is evaporated under reduced pressure. The yield of N-(hydroxymethyl)acetamide, a colorless hygroscopic oil at this point, is 148–151 g. (98–100%) (Note 6). The oily product, which may solidify (Note 7) on standing for several days, is used directly in step B.

B. S-*Acetamidomethyl*-L-*cysteine hydrochloride*.   A 1-l., round-bottomed flask is charged with 127 g. (1.43 moles) of N-(hydroxymethyl)acetamide, 228 g. (1.30 moles) of

L-cysteine hydrochloride monohydrate (Note 8), and 350 ml. of water. The resulting solution is swirled and cooled in an ice bath as 50 ml. of concentrated hydrochloric acid is slowly added (Note 9). The flask is flushed with nitrogen, capped with a nitrogen-filled balloon, and allowed to stand for 1–2 days at room temperature. The progress of the reaction is monitored by TLC (Note 10). When L-cysteine hydrochloride is no longer detectable, the solution is evaporated under reduced pressure at a bath temperature of *ca.* 40°. The remaining solid is suspended in a small amount of absolute ethanol, and the mixture is again carefully evaporated to avoid bumping. This entrainment procedure with absolute ethanol is repeated several times to remove traces of water. The dry solid is dissolved in the minimum amount of methanol (Note 11), and anhydrous diethyl ether is added until the cloud point is reached. The cloudy solution is allowed to stand in a refrigerator at *ca.* 4–5° for 1 week, during which the crystalline mass is broken up several times. The white crystalline product is collected, washed with ether, and dried under reduced pressure, yielding 152–190 g. (51–64%) of S-acetamidomethyl-L-cysteine hydrochloride, dec. 159–163°, $[\alpha]_D^{25}$ −30.7° ($c = 1$, water) (Notes 12–14).

2. Notes

1. The submitters obtained acetamide from Merck & Company, Inc. Acetamide was purchased by the checkers from the Fisher Scientific Company.

2. A 37% solution of formaldehyde in water is available from the Aldrich Chemical Company, Inc.

3. The submitters state that the failure to add dry ice at this point may result in greatly reduced yields. The purpose of the dry ice is presumably to lower the pH of the solution by converting potassium carbonate to potassium bicarbonate.

4. When the mixture was heated above 40° by the submitters, it became discolored and an insoluble precipitate was formed.

5. The filtrate may be cloudy.

6. Apparent yields in excess of the theoretical amount may be observed, owing to the presence of a small portion of water. The oily product may be dried at high vacuum over phosphorous pentoxide for several days.

7. A melting point of 50–52° is reported for N-(hydroxymethyl)acetamide.[3]

8. The submitters purchased L-cysteine hydrochloride monohydrate from Schwartz/Mann Division, Becton, Dickinson, and Company, Mountain View Avenue, Orangeburg, New York 10962. The checkers used material supplied by Aldrich Chemical Company, Inc.

9. The pH of the solution is *ca.* 0.5.

10. The balloon was removed briefly while aliquots were taken. The flask was flushed again with nitrogen and the balloon was then replaced. TLC analyses were carried out on glass plates coated with silica gel G purchased from Analtech, Newark, Delaware. With a 10:2:3 (v/v/v) solution of 1-butanol, acetic acid, and water as developing solvent, the R_f values for the product and L-cysteine hydrochloride are 0.19 and 0.25, respectively.

11. The submitters dissolved the solid in methanol at room temperature; however, the solid obtained by the checkers was not very soluble under these conditions. Consequently, the material was dissolved in approximately 2–3 l. of methanol by gentle warming on a steam bath.

12. The product obtained by the checkers had $[\alpha]_D^{25}$ $-28°$ ($c = 1$, water); IR (Nujol) cm.$^{-1}$: 1715 (C=O), 1580 (C=O); ^{1}H NMR (DCl in D$_2$O), δ (multiplicity, number of protons, assignment): 2.05 (s, 3H, CH$_3$), 3.2–3.4 (m, 2H, CH$_2$CH), 4.3–4.5 (m, 1H, CH$_2$CH), 4.39 (s, 2H, NCH$_2$S).

13. On several occasions the product isolated by the submitters was contaminated with L-cystine dihydrochloride, which was not easily removed by recrystallization. In this event the product was converted to the zwitterionic form and recrystallized in the following manner: The pH of a solution of the product in water was adjusted to 6 with aqueous 2.5 N potassium hydroxide, and the solution was evaporated to dryness under reduced pressure at $ca.$ 40°. The residue was dissolved in a minimum amount of hot water, and two volumes of 95% ethanol were added to precipitate S-acetamidomethyl-L-cysteine monohydrate, dec. 187°, $[\alpha]_{589}^{25}$ $-42.5°$ ($c = 1$, water).

14. The following unchecked procedure for liberating L-cysteine from S-acetamido-methyl-L-cysteine was provided by the submitters as a model for removing the S-acet-amidomethyl group from peptides. The pH of a solution of 96.1 mg. (0.500 mmole) of S-acetamidomethyl-L-cysteine in 10.0 ml. of water is adjusted to 4.0 with aqueous 0.25 N hydrochloric acid. The solution is stirred, 159.3 mg. (0.5000 mmole) of mercury(II) acetate is added, and the pH is readjusted to 4.0 by adding more 0.25 N hydrochloric acid. The resulting suspension is stirred for 1 hour at room temperature then diluted with an equal volume of water. Hydrogen sulfide gas is introduced to complete the precipitation of mercury from solution, the mixture is filtered, and the aqueous filtrate is evaporated to dryness under reduced pressure. TLC analysis on the residue as described in Note 10 showed the presence of L-cysteine and the absence of S-acetamidomethyl-L-cysteine.

3. Discussion

The present procedure provides a convenient method for preparing S-acetamido-methyl-L-cysteine hydrochloride.[4] The zwitterionic form may be obtained readily from the hydrochloride by the procedure described in Note 13, by ion-exchange chromatography,[5] or by precipitation from 2-propanol with pyridine.[6] S-Acetamido-methyl-L-cysteine has also been prepared from N-(hydroxymethyl)acetamide under anhydrous conditions in liquid hydrogen fluoride[4] and in trifluoroacetic acid.[7] The preparation of N-(hydroxymethyl)acetamide described in Part A is based on the procedure of Einhorn.[3]

The acetamidomethyl group serves as a useful thiol-protecting group for cysteine during peptide synthesis.[4,7–9] The protecting group is stable to the conditions generally prevailing in peptide synthesis, including not only typical solution and solid-phase procedures but also reactions carried out in liquid hydrogen fluoride.[4] Peptides containing S-acetamidomethyl-L-cysteine generally have good water solubility[8] and are not prone to racemization at the α-position of the cysteine residue.[4] The acetamidomethyl protecting group may be easily removed by reaction with either mercury(II) acetate at pH 4, as described in Note 14, or iodine.[7,8] S-Acetamidomethylation of the cysteine residues of proteins may be accomplished in liquid hydrogen fluoride, and the group may be removed by reaction with mercury(II) ion.[4] The thiol group of β-mercaptopropionic acid has also been protected by formation of the S-acetamidomethyl derivative.[7]

1. Merck Sharp & Dohme Research Laboratories, P.O. Box 2000, Rahway, New Jersey 07065.
2. Merck Sharp & Dohme Research Laboratories, West Point, Pennsylvania 19486.
3. A. Einhorn, *Justus Liebigs Ann. Chem.*, **343**, 207 (1905).
4. D. F. Veber, J. D. Milkowski, S. L. Varga, R. G. Denkewalter, and R. Hirschmann, *J. Am. Chem. Soc.*, **94**, 5456 (1972).
5. P. Hermann and E. Schreier, in H. Hanson and H. D. Jakubke, Eds., "Peptides 1972," 12th European Symposium on Peptides, North Holland, Amsterdam, 1973, pp. 126–127.
6. H. Romovacek, S. Drabarek, K. Kawasaki, S. R. Dowd, R. Obermeier, and K. Hofmann, *Int. J. Pept. Protein Res.*, **6**, 435 (1974).
7. P. Marbach and J. Rudinger, *Helv. Chim. Acta*, **57**, 403 (1974).
8. B. Kamber, *Helv. Chim. Acta*, **54**, 927 (1971).
9. R. Hirschmann, R. F. Nutt, D. F. Veber, R. A. Vitali, S. L. Varga, T. A. Jacob, F. W. Holly, and R. G. Denkewalter, *J. Am. Chem. Soc.*, **91**, 507 (1969).

ACETIC FORMIC ANHYDRIDE

(Acetic acid, anhydride with formic acid)

$$CH_3COCl + HCOONa \rightarrow CH_3COOCHO + NaCl$$

Submitted by Lewis I. Krimen[1]
Checked by James Savage and Peter Yates

1. Procedure

A dry, 2-l., three-necked, round-bottomed flask equipped with a stirrer, a thermometer, a reflux condenser fitted with a calcium chloride tube, and a dropping funnel is charged with 300 g. (4.41 moles) of sodium formate (Note 1) and 250 ml. of anhydrous diethyl ether (Note 2). To this stirred mixture is added 294 g. (266 ml., 3.75 moles) of acetyl chloride (Note 3) as rapidly as possible, while the temperature is maintained at 23–27° (Note 4). After the addition is complete, the mixture is stirred for 5.5 hours at 23–27° to ensure complete reaction. The mixture is then filtered with suction, the solid residue is rinsed with 100 ml. of ether, and the washings are added to the original filtrate (Note 5). The ether is removed by distillation at reduced pressure, and the residue is distilled, yielding 212 g. (64%) of colorless acetic formic anhydride, b.p. 27–28° (10 mm.), 38–38.5° (39 mm.); n_D^{20} 1.388 (Note 6).

2. Notes

1. Reagent grade sodium formate from J. T. Baker Chemical Co. was used; it was finely ground to ensure better contact. It is imperative that extreme care be taken to ensure anhydrous conditions throughout the procedure, since hydrolysis produces formic and acetic acids, which are very difficult to remove from the product. A slight excess of sodium formate ensures a product free of acetyl chloride.

2. Mallinckrodt AR grade ether was used without further drying by the submitter. The checkers, working at half scale, found it essential to dry the ether over sodium.

3. Acetyl chloride from Matheson, Coleman and Bell was used without further purification.

4. The addition of acetyl chloride is mildly exothermic; the exotherm can be controlled by slower addition or by the use of a cooling bath (20–24°). The addition is completed in *ca*. 5 minutes.

5. The filtration and subsequent ether rinse should be carried out quickly in order to keep the filtrate dry.

6. The acetic formic anhydride may be stored at 4° in a standard-taper, round-bottomed flask fitted with a polyethylene stopper. Moisture catalyzes the decomposition of the product to acetic acid, with the evolution of carbon monoxide. *The material must not be stored in sealed containers!*

7. The IR spectrum of neat acetic formic anhydride shows two bands in the carbonyl region at 1765 and 1791 cm.$^{-1}$ and carbon-oxygen-carbon stretching absorption at 1050 cm.$^{-1}$ (a band at 1180 cm.$^{-1}$ could also be due to C—O—C). The ^{1}H NMR spectrum (neat) shows a singlet at δ 2.25 (acetyl protons) and a singlet at δ 9.05 (formyl proton). If the product is not pure, the following peaks may also be observed: δ 2.05 (CH_3CO_2H), 2.20 [$(CH_3CO)_2O$], 2.68 (CH_3COCl), 8.05 ($HCOOH$), 8.85 [$(HCO)_2O$]. The spectrum of the product obtained by the checkers showed slight contamination with acetic anhydride and formic anhydride.

3. Discussion

Acetic formic anhydride has been prepared by the reaction of formic acid with acetic anhydride[2,3] and ketene,[4,5] and of acetyl chloride with sodium formate.[6] The present procedure is essentially that of Muramatsu.[6] It is simpler than others previously described, gives better yields, and is easily adapted to the preparation of large quantities, usually with an increase in yield. Acetic formic anhydride is a useful intermediate for the formylation of amines,[3,7] amino acids,[8,9] and alcohols,[2,10] for the synthesis of aldehydes from Grignard reagents,[11] and for the preparation of formyl fluoride.[12]

1. Abbott Laboratories, Abbott Park, North Chicago, Illinois 60064.
2. A. Behal, *C. R. Hebd. Seances Acad. Sci.*, **128**, 1460 (1899).
3. C. W. Huffman, *J. Org. Chem.*, **23**, 727 (1958).
4. R. E. Dunbar and F. C. Garven, *J. Am. Chem. Soc.*, **77**, 4161 (1955).
5. C. D. Hurd and A. S. Roe, *J. Am. Chem. Soc.*, **61**, 3355 (1939).
6. I. Muramatsu, M. Murakami, T. Yoneda, and A. Hagitani, *Bull Chem. Soc. Jpn.*, **38**, 244 (1965).
7. G. R. Clemo and G. A. Swan, *J. Chem. Soc.*, 603 (1945).
8. S. G. Waley, *Chem. Ind. (London)*, 107 (1953).
9. J. C. Sheehan and D. D. H. Yang, *J. Am. Chem. Soc.*, **80**, 1154 (1958).
10. C. D. Hurd, S. S. Drake, and O. Fancher, *J. Am. Chem. Soc.*, **68**, 789 (1946).
11. W. R. Edwards, Jr., and K. P. Kammann, Jr., *J. Org. Chem.*, **29**, 913 (1964).
12. G. A. Olah and S. J. Kuhn, *J. Am. Chem. Soc.*, **82**, 2380 (1960).

ACETONE HYDRAZONE

(2-Propanone, hydrazone)

$$CH_3COCH_3 \xrightarrow{N_2H_4 \cdot H_2O} (CH_3)_2C=NN=C(CH_3)_2 \xrightarrow{N_2H_4}$$
$$(CH_3)_2C=NNH_2$$

Submitted by A. C. Day[1] and M. C. Whiting[2]
Checked by G. Swift and W. D. Emmons

1. Procedure

Caution! Hydrazine is toxic and should be handled in a hood. Anhydrous hydrazine is extremely reactive with oxidizing agents (including air) and should always be prepared and used behind a protective screen.

A. *Acetone azine.* A 500-ml., round-bottomed flask containing 145 g. (183 ml., 2.50 moles) of acetone (Note 1) is fitted with a mechanical stirrer (Note 2) and a dropping funnel and cooled in an ice bath. With vigorous stirring, 65.5 g. (1.31 moles) of 100% hydrazine hydrate (Note 1) is added at such a rate that the temperature is maintained below 35°. The addition takes 20–30 minutes. The mixture is stirred for an additional 10–15 minutes before 50 g. of potassium hydroxide pellets is added with vigorous stirring and continued cooling (Note 3). The upper liquid layer is separated and allowed to stand over 25 g. of potassium hydroxide pellets for 30 minutes, with occasional swirling (Note 4). After filtration, the liquid is further dried with two successive 12.5-g. portions of potassium hydroxide. Distillation gives 120–126 g. (86–90%) of almost colorless acetone azine, b.p. 128–131°, n_D^{22} 1.4538 (Note 5).

B. *Acetone hydrazone.* Anhydrous hydrazine is prepared by heating under reflux 100% hydrazine hydrate with an equal weight of sodium hydroxide pellets for 2 hours, followed by distillation in a slow stream of nitrogen introduced through a capillary leak. *(Caution! Distillation in air can lead to explosion.)* The distillate boils at 114–116° and the yield is 95–97% (Note 6).

A mixture of 112 g. (1.00 mole) of acetone azine and 32 g. (1.0 mole) of anhydrous hydrazine is placed in a 300-ml. round-bottomed flask fitted with a reflux air condenser and drying tube, and kept at 100° for 12–16 hours. *(Caution! This reaction and the subsequent distillation should be carried out behind a protective screen.)* The crude product is then distilled rapidly through a water-cooled condenser, and the colorless fraction boiling at 122–126° is collected, n_D^{22} 1.4607 (Note 7), yielding 111–127 g. (77–88%, Notes 7 and 8) of essentially pure acetone hydrazone (Note 9).

2. Notes

1. The acetone and hydrazine hydrate were good-quality commercial products purified before use.

2. A Hershberg stirrer made of Nichrome wire is most efficient for aiding dissolution of the potassium hydroxide added after azine formation is complete.

3. The dissolution of the potassium hydroxide is strongly exothermic. A small proportion may remain undissolved.

4. A lower, aqueous phase may form at this stage, but the product is easily decanted from it.

5. The distillation gives a small forerun, b.p. 120–128°, containing hydrazine and acetone hydrazone. There is virtually no distillation residue. The submitters carried out the preparation of both acetone azine and acetone hydrazone on a fourfold scale with comparable results.

6. The purity is 95–98% by this method.[3] The purity is lower (85–95%) by an alternative procedure[4] which requires separation of the hydrazine and alkaline phases above 60°; with the latter method the submitters found that a frequent problem was the solidification of the lower phase in the separating funnel, and in one case *a very serious fire* occurred during the transfer of the hot (*ca.* 100°) mixture to the separating funnel.

7. The forerun contains hydrazine; material boiling above 126° contains much acetone azine. With a slow rate of distillation, disproportionation occurs and the yield of acetone hydrazone is reduced. If the forerun and material boiling above 126° are combined and reheated at 100° for 12–16 hours, they give more acetone hydrazone on redistillation. With further repetitions of this procedure, the yield is almost quantitative.

8. The highest yields were obtained in cases where the anhydrous hydrazine was treated with barium oxide for several hours before use.

9. The hydrazone should be used as soon as possible. If it is stored, care must be taken to exclude moisture, which catalyzes disproportionation to hydrazine and acetone azine.[5-7] Even in the absence of moisture it disproportionates slowly at room temperature and so should be redistilled immediately before use. Old samples can be regenerated fairly satisfactorily by reheating them for 12–16 hours at 100° before redistillation, but there is always some irreversible decomposition to high-boiling products during storage.

3. Discussion

The procedure for acetone azine is essentially that of Curtius and Thun.[5] The method for acetone hydrazone is adapted from that of Staudinger and Gaule.[8] The hydrazone has been prepared directly from acetone and hydrazine, but this is much less satisfactory.[6]

Acetone hydrazone is produced in good yield by the method described, but an inferior product is obtained without the precautions noted. The compound is used for the preparation of 2-diazopropane.[8-10]

1. Dyson Perrins Laboratory, Oxford University, England.
2. Department of Organic Chemistry, University of Bristol, England.
3. F. Raschig, *Ber. Dtsch. Chem. Ges.*, **43**, 1927 (1910).
4. R. A. Pennman and L. F. Audrieth, *J. Am. Chem. Soc.*, **71**, 1644 (1949).
5. T. Curtius and K. Thun, *J. Prakt. Chem.*, [2] **44**, 161 (1891).
6. T. Curtius and L. Pflug, *J. Prakt. Chem.*, [2] **44**, 535 (1891).
7. *Cf.* K. Heyns and A. Heins, *Justus Liebigs Ann. Chem.*, **604**, 133 (1957).
8. H. Staudinger and A. Gaule, *Ber. Dtsch. Chem. Ges.*, **49**, 1897 (1916).

9. A. C. Day, P. Raymond, R. M. Southam, and M. C. Whiting, *J. Chem. Soc. C*, 467 (1966); D.
   E. Applequist and H. Babad, *J. Org. Chem.*, **27**, 288 (1962).
10. S. D. Andrews, A. C. Day, P. Raymond, and M. C. Whiting, *Org. Synth.*, **Coll. Vol. 6**, 392
   (1988).

PREPARATION OF HYDRAZONES:
ACETOPHENONE HYDRAZONE

(Benzeneethanimidamide)

$$C_6H_5COCH_3 + (CH_3)_2NNH_2 \xrightarrow{CH_3CO_2H} C_6H_5\overset{\overset{\displaystyle CH_3}{|}}{C}{=}NN(CH_3)_2$$

$$C_6H_5\overset{\overset{\displaystyle CH_3}{|}}{C}{=}NN(CH_3)_2 + N_2H_4 \longrightarrow C_6H_5\overset{\overset{\displaystyle CH_3}{|}}{C}{=}NNH_2$$

Submitted by G. R. Newkome[1] and D. L. Fishel[2]
Checked by G. Swift and W. D. Emmons

1. Procedure

*Caution! Hydrazines are toxic and should be handled in a hood. Anhydrous hydrazine
is extremely reactive with oxidizing agents (including air) and should always be used
behind a protective screen.*

A.   *Acetophenone N,N-dimethylhydrazone.*   A mixture of acetophenone (12.0 g.,
0.100 mole), anhydrous *N,N*-dimethylhydrazine (18.0 g., 0.300 mole) (Note 1), absolute
ethanol (25 ml.), and glacial acetic acid (1 ml.) (Note 2) is heated at reflux for 24 hours.
During this period the colorless solution becomes bright yellow. The volatile reactants and
solvent are removed under reduced pressure and the residual oil is fractionally distilled
through a 10-cm. Vigreux column, giving a small forerun of unreacted acetophenone,
b.p. 30–40° (0.15 mm.), followed by 14.6–15.2 g. (90–94%) of acetophenone *N,N*-
dimethylhydrazone, b.p. 55–56° (0.15 mm.), n_D^{25} 1.5443 (Notes 3 and 4).

B.   *Acetophenone hydrazone.*   A mixture of acetophenone *N,N*-dimethylhydrazone
(8.1 g., 0.050 mole) and anhydrous hydrazine (6.4 g., 0.20 mole) (Note 5) in absolute
ethanol (15 ml.) is heated at reflux until the reaction mixture turns colorless (Note 6). The
volatile materials are removed on a rotary evaporator without allowing the flask temperature to rise above 20° (Note 7). The colorless residual acetophenone hydrazone, which
solidifies as the last traces of solvent are removed, weighs 6.5–6.6 g. (97–99%) and is
sufficiently pure for most purposes, m.p. 24–25° (Notes 8 and 9).

2. Notes

1. Anhydrous N,N-dimethylhydrazine obtained from Matheson, Coleman and Bell is used directly. It can also be prepared by the method in *Org. Synth.*, **Coll. Vol. 2,** 213 (1943).

2. It is not necessary to use glacial acetic acid as a catalyst, but without it the reaction time required for completion is prolonged.

3. Physical constants previously reported:[3] b.p. 100.5–102° (10 mm.), n_D^{25} 1.5455.

4. This method has been used to prepare various N,N-dimethylhydrazones in 70–99% yield.[4]

5. A good commercial grade of anhydrous hydrazine (Eastman Organic Chemicals) is satisfactory.

6. The reaction time for complete conversion is usually less than 24 hours. A convenient "end point" is the visual color change from bright yellow to colorless or very pale yellow.

7. It is of utmost importance that the flask temperature during the removal of the volatile materials be kept below 20° to minimize possible azine formation by decomposition of the hydrazone.

8. The reported melting points are 16–20°,[5] 22°,[6] 24–25°,[4,7] and 26°.[7]

9. Acetophenone hydrazone can be stored at temperatures below 0° for indefinite periods of time.

3. Discussion

The formation of acetophenone hydrazone has been accomplished by heating acetophenone with hydrazine or hydrazine hydrate,[7–10] by heating acetophenone azine with anhydrous hydrazine,[6,11] by the reaction of α-dimethylaminoacetophenone with hydrazine,[5] and by the present method.[4]

This synthetic process is applicable to the preparation of most hydrazones from aldehydes and ketones. The two-step preparation offers several distinct advantages over the one-step method:[6,9] (1) the yield of both steps is high; (2) the product is not contaminated with azine; (3) the isolated product is pure enough to be used in subsequent reactions without further purification. This method excels in the preparation of unstable liquid or low-melting hydrazones over the common methods of preparation.

1. Department of Chemistry, Louisiana State University, Baton Rouge, Louisiana 70803.
2. Department of Chemistry, Kent State University, Kent, Ohio 44240.
3. P. A. S. Smith and E. E. Most, Jr., *J. Org. Chem.*, **22,** 358 (1957).
4. G. R. Newkome and D. L. Fishel, *J. Org. Chem.*, **31,** 677 (1966).
5. R. L. Letsinger and R. Collat, *J. Am. Chem. Soc.*, **74,** 621 (1952).
6. H. Staudinger and A. Gaule, *Ber. Dtsch. Chem. Ges.*, **49,** 1897 (1916).
7. G. Lock and K. Stach, *Ber. Dtsch. Chem. Ges.*, **77,** 293 (1944).
8. D. E. Pearson, K. N. Carter, and C. M. Greer, *J. Am. Chem. Soc.*, **75,** 5905 (1953).
9. T. Curtius and L. Pflug, *J. Prakt. Chem.*, [2] **44,** 535 (1891).
10. J. Stanek, *Collect. Czech. Chem. Commun.*, **12,** 671 (1947).
11. H. Staudinger and L. Hammet, *Helv. Chim. Acta*, **4,** 217 (1921).

3β-ACETOXY-5α-CYANOCHOLESTAN-7-ONE

[Cholestane-5-carbonitrile, 3-(acetyloxy)-7-oxo-,(3β, 5α)-]

Submitted by W. NAGATA and M. YOSHIOKA[1]
Checked by ROBERT E. IRELAND, ROBERT CZARNY,
and CONRAD J. KOWALSKI

1. Procedure

Caution! This preparation should be carried out in a good hood. Also, great care should be taken in handling neat triethylaluminum because it is pyrophoric—that is, it ignites spontaneously upon contact with air (Note 2).

Benzene has been identified as a carcinogen; OSHA has issued emergency standards on its use. All procedures involving benzene should be carried out in a well-ventilated hood, and glove protection is required.

A dry, 50-ml., three-necked, round-bottomed flask equipped with a gas-inlet tube for nitrogen, a magnetic stirring bar, and a serum stopper for the introduction of reagents is flushed with nitrogen, stoppered with a glass stopper, charged with 17 ml. of anhydrous tetrahydrofuran (Note 1), then immersed in an ice bath. Stirring is begun and 3.3 g. (3.9 ml., 0.028 mole) of triethylaluminum is introduced into the flask with a dry hypodermic syringe (Note 2, Note 3). After 5–10 minutes, 4.8 ml. of a 3.57 M solution of hydrogen cyanide (0.017 mole) in anhydrous tetrahydrofuran (Note 4) is added with a dry hypodermic syringe. The stirring is continued for about 5–10 minutes.

A dry, 100-ml., three-necked, round-bottomed flask equipped with a gas-inlet tube for nitrogen, a magnetic stirring bar, and a serum stopper, as described above, is flushed with nitrogen. While the flask is being flushed, 2.50 g. (0.00565 mole) of 3β-acetoxy-cholest-5-en-7-one (Note 5) and 0.0521 ml. (0.00289 mole) of water (Note 6) are added to the reaction flask. The flask is stoppered with a glass stopper and charged with 17 ml. of anhydrous tetrahydrofuran. After the starting material has dissolved, the cold triethylaluminum-hydrogen cyanide solution is transferred to the reaction flask with a dry hypodermic syringe. The resulting pale yellow solution is stirred at room temperature under a positive nitrogen pressure. After 3 hours, a solution of 0.044 ml. (0.0024 mole) of water in 0.87 ml. of anhydrous tetrahydrofuran is added, and the solution is allowed to stir for an additional 4 hours.

The reaction mixture is poured slowly into a vigorously stirred solution of 28 ml. (0.28 mole) of concentrated hydrochloric acid and 350 ml. of ice water in a 1-l., three-necked, round-bottomed flask fitted with an efficient stirrer and immersed in an ice bath (Note 7, Note 8). The mixture is stirred for 20 minutes with ice cooling and extracted with three 200-ml. portions of a 3:1 (v/v) mixture of diethyl ether and dichloromethane. The extracts are washed with three 200-ml. portions of aqueous 2 M sodium hydroxide, two 200-ml. portions of water, and one 200-ml. portion of saturated aqueous sodium chloride, dried over anhydrous sodium sulfate, and evaporated under reduced pressure (Notes 9 and 10). The crystalline residue, weighing 2.70 g., is recrystallized by dissolving it in 7.5–8 ml. of hot (almost boiling) benzene and adding 25 ml. of n-pentane (distilled) to the hot solution (Note 11). 3β-Acetoxy-5α-cyanocholestan-7-one is obtained as white crystals, m.p. 192.5–193.5°; the yield is 2.27–2.41 g. (86–91%). A second crop (50–170 mg.) can be obtained, m.p. 188.5–190° (Note 12); the total yield is 92–93% (Notes 13 and 14).

2. Notes

1. Prior to use the tetrahydrofuran was distilled from lithium aluminum hydride into a dry flask flushed with nitrogen and sealed with a serum stopper.

2. *Caution! Triethylaluminum is pyrophoric. Use safety glasses, gloves, and an apron. Use dry sand to extinguish fires.* The submitters note that a description of the properties and handling procedures for triethylaluminum are available from the Ethyl Corporation, Louisiana. The checkers used triethylaluminum in lecture bottles from Alpha Inorganics, Inc., and suggest the handling procedure described below. Since this procedure was submitted and checked, standardized solutions of triethylaluminum in various hydrocarbon solvents, which may be substituted for pure triethylaluminum, have become available from Texas Alkyls, Inc., a division of Stauffer Chemical Company.

a. *Checkers handling procedure.* Figures 1 and 2 suggest the equipment to be used, and how to assemble it. The hose end fitting is connected to the stopcock with a piece of Teflon-lined tubing and fastened with copper wire. The stopcock should be well greased and held secure with a taut rubber band since it does have a tendency to pop out. The triethylaluminum may now be removed from the lecture bottle by the following procedure. A dry, three-necked flask equipped with serum stopper and gas-inlet tube is flushed thoroughly with nitrogen from a nitrogen bubbler. The triethylaluminum transfer apparatus is put into the open neck of the flask, the joint being well greased (the gas-inlet tube joint should also be greased). With the stopcock of the transfer apparatus *open* and well secured as suggested above, the tank valve is opened (usually one or two full turns) with a wrench. The flow of triethylaluminum may now be adjusted with the stopcock. When one obtains as much triethylaluminum as desired, the tank valve is closed, the transfer apparatus is allowed to drain, and the stopcock is closed. The transfer apparatus is removed from the flask which is quickly stoppered with a glass stopper while the flask is

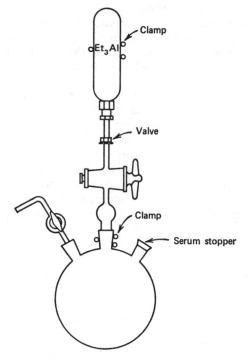

Figure 1.   Apparatus for collecting triethylaluminum from a lecture bottle.

flushed with nitrogen. This flask of triethylaluminum can be stored in this manner for many weeks since this reagent is quite stable. The transfer apparatus, which may still contain some triethylaluminum, should be carefully removed, rinsed quickly with acetone, and cleaned with dilute hydrochloric acid.

b. *Submitters handling procedure.* The submitters suggest another handling procedure especially useful for removing triethylaluminum from a lecture bottle having a clogged valve outlet. Figure 3 shows the apparatus and how to transfer the material. With the cylinder clamped in upright position, remove the valve unit so that only the bottle remains. Quickly attach the adaptor to the opening of the bottle and apply a slow stream of nitrogen. Transfer the triethylaluminum into dry, nitrogen-flushed, 50-ml. ampoules using a 100- or 200-ml. Luer-lock hypodermic syringe with a needle, 43 cm. long and 2 mm. in diameter. Sweep the opening of the ampoule with nitrogen during the transfer. The syringe should be slightly greased, and the ampoules should be strong with a long, thick stem so that they can be resealed. The ampoules are sealed as soon as possible. The use of rubber caps is effective for temporary protection of the ampoules from air. The material in a 50-ml. ampoule can be divided in smaller ampoules using the apparatus shown in Figure 4, which can be used also for transferring triethylaluminum from an ampoule to a reaction flask.

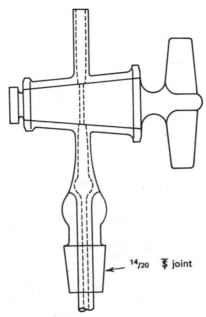

Figure 2.   Detailed construction of the stopcock for the apparatus shown in Figure 1.

This procedure was not tested by the checkers. It has the advantage that one does not have to contend with a clogged lecture bottle, which in the checkers experience is best discarded. It has the disadvantage that one must handle large syringe-fulls of triethylaluminum and make several transfers without exposing the liquid to air. Caution must be exercised with either procedure.

3. Neat triethylaluminum may be replaced with a 10–25% stock solution of it in anhydrous tetrahydrofuran, decreasing the amount of solvent in the reaction flask. The stock solution is prepared using a graduated flask to measure the volume of the triethylaluminum and solvent, added appropriately. The stock solution is very stable and not pyrophoric.

4. The submitters have prepared hydrogen cyanide as directed in *Org. Synth.*, **Coll. Vol. 1,** 314, (1944). The checkers used a similar procedure described by Brauer.[3] The hydrogen cyanide was collected in a tared flask and diluted with anhydrous tetrahydrofuran to a previously marked volume. The flask was capped with two serum stoppers (the second put on in an inverted position) to ensure against leakage and stored in a freezer. Solutions such as these seem to be stable for several months when kept cold.

5. Checkers obtained the steroid from K & K Laboratories.

6. A small amount of water has been found to accelerate the reaction. In the absence of water, the reaction was about 80% complete after 7.5 hours. Despite the rate accelera-

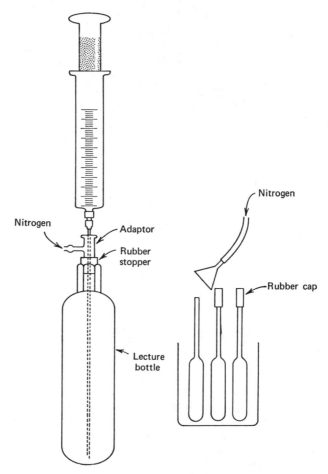

Figure 3. Apparatus for removing triethylaluminum from the lecture bottle.

tion by water, the reaction mixture should be protected from moisture, because a larger amount of water than that specified retards the reaction, owing to decomposition of the triethylaluminum.

7. This treatment is an exothermic reaction with evolution of gaseous ethane. The reaction mixture should be added in a slow stream, with good stirring, at such a rate that the content in the flask does not overflow. When the ice has melted, additional ice should be added.

8. The reaction mixture remaining on the wall of the reaction flask is treated with a small amount of cold, dilute hydrochloric acid and combined with the extraction mixture.

9. Acid treatment followed by alkaline washing of the extracts prevents possible hydrolysis of the acetoxyl and the cyano groups.

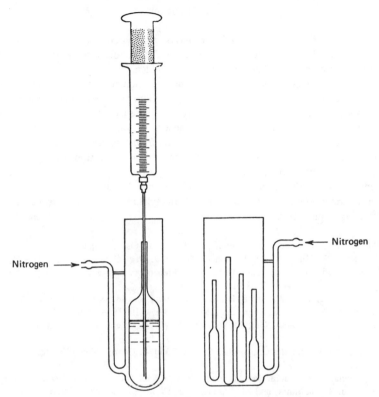

Figure 4. Apparatus for collecting aliquots of triethylaluminum.

10. The submitters have performed the extraction and washing in a countercurrent manner using three 2-l. separatory funnels.

11. The recrystallization was carried out in a 40-ml. centrifuge tube. The recrystallization mixture, after cooling to room temperature, was cooled in a freezer and washed twice with 20-ml. portions of cold recrystallization solvent.

12. When the second crop is contaminated with a polar material (α-cyanohydrin of the product) as evidenced by a lower melting point (m.p. 160–170°) and TLC (Kiesel Gel GF, benzene-ethyl acetate, 4:1, R_f = 0.2), the residue from the mother liquor must be treated again with alkali before crystallization.

13. This procedure is applicable to smaller or larger scale preparations with some modification. In a smaller scale experiment, the submitters suggest using a stock solution of triethylaluminum (Note 3). In a larger scale preparation, to avoid using large syringes, it is possible to run the triethylaluminum into a dry, graduated, pressure-equalizing dropping funnel, in which case the triethylaluminum-hydrogen cyanide solution may be added to the reaction flask through a funnel instead of using a syringe.

14. The submitters used 5.76 g. of 3β-acetoxycholest-5-en-7-one and obtained 5.65 g. (92%) of the cyano ketone.

3. Discussion

3β-Hydroxy-5α-cyanocholestan-7-one has been prepared in 43% yield by the action of potassium cyanide and ammonium chloride[2] on 7-ketocholesterol at 100° for 33 hours.[4] The present method was developed by the submitters.[4]

This process is superior to classical hydrocyanation methods using an alkali metal cyanide[5] and to the improved method using potassium cyanide and ammonium chloride[2] with respect to reactivity, stereospecificity, and absence of side reactions. Also, the process is applicable to conjugate hydrocyanation of various α,β-unsaturated carbonyl and imino compounds,[4,6] cyanohydrin formation from less-reactive ketones,[4] and cleavage of epoxides to produce β-cyanohydrins,[7] the reaction conditions being varied depending on the substrate to be used. The present procedure is typical of hydrocyanation procedures using other organoaluminum compounds. Cyanotrimethylsilane reacts with α,β-unsaturated ketones in the presence of triethylaluminum to give trimethylsilyl enol ethers of β-cyano ketones.[8]

1. Shionogi Research Laboratory, Shionogi & Co., Ltd., Osaka, Japan.
2. W. Nagata, S. Hirai, H. Itazaki, and K. Takeda, *J. Org. Chem.*, **26**, 2413 (1961).
3. O. Glemser, in G. Brauer, Ed., "Handbook of Preparatory Inorganic Chemistry," 2nd ed., Vol. 1, Academic Press, New York, 1963, p. 658.
4. W. Nagata, M. Yoshioka, and S. Hirai, *Tetrahedron Lett.*, 48 (1962); W. Nagata, M. Yoshioka, and M. Murakami, *J. Am. Chem. Soc.*, **94**, 4644, 4654 (1972); W. Nagata, M. Yoshioka, and T. Terasawa, *J. Am. Chem. Soc.*, **94**, 4672 (1972); W. Nagata and M. Yoshioka, *Org. React.*, **25**, 255 (1977).
5. P. Kurtz, "Methoden der Organischen Chemie," Vol. 8, Georg Thieme Verlag, Stuttgart, 1952, p. 265.
6. W. Nagata, T. Okumura, and M. Yoshioka, *J. Chem. Soc. C*, 2347 (1970); W. Nagata, M. Yoshioka, T. Okumura, and M. Murakami, *J. Chem. Soc. C*, 2355 (1970).
7. W. Nagata, M. Yoshioka, and T. Okumura, *Tetrahedron Lett.*, 847 (1966); *J. Chem. Soc. C*, 2365 (1970).
8. K. Utimoto, M. Obayashi, Y. Shishiyama, M. Inoue, and H. Nozaki, *Tetrahedron Lett.*, **21**, 3389 (1980).

p-ACETYL-α-BROMOHYDROCINNAMIC ACID

(Benzenepropanoic acid, 4-acetyl-α-bromo-)

$$p\text{-}CH_3COC_6H_4NH_2 \xrightarrow[\text{HBr}]{\text{NaNO}_2} p\text{-}CH_3COC_6H_4 \overset{\oplus}{-}N\equiv N \ Br^-$$

$$p\text{-}CH_3COC_6H_4 \overset{\oplus}{-}N\equiv N \ Br^- + CH_2=CHCO_2H$$

$$\xrightarrow[\text{HBr}]{\text{CuBr}} p\text{-}CH_3COC_6H_4CH_2\underset{\underset{Br}{|}}{C}HCO_2H$$

Submitted by GEORGE H. CLELAND[1]
Checked by MICHAEL J. UMEN and HERBERT O. HOUSE

1. Procedure

Caution! Since bromoacetone, a powerful lachrymator, is produced as a by-product in this preparation, the reaction should be performed in a hood.

A tared, 500-ml., two-necked, round-bottomed flask is equipped with a magnetic stirring bar, a thermometer, and an ice-filled cooling bath. A solution of 13.5 g. (0.100 mole) of 4-aminoacetophenone (Note 1) in 200 ml. of acetone is placed in the flask and stirred while 32 ml. (about 0.3 mole) of aqueous 48% hydrobromic acid is added. After the resulting solution has been cooled to 5–7°, it is stirred continuously while 20 ml. of an aqueous solution containing 6.90 g. (0.100 mole) of sodium nitrite is added rapidly (30 seconds) beneath the surface of the reaction solution with a hypodermic syringe or a long-stemmed dropping funnel. Stirring and cooling are continued until the exothermic reaction subsides (Note 2) and the reaction solution has cooled to 14–15°. After 106 g. (100 ml., 1.47 moles) of acrylic acid (Note 3) is added, the solution is again cooled to 14–15°, with stirring, before 0.10–0.11 g. (0.00069–0.00077 mole) of copper(I) bromide (Note 4) is added. Stirring is continued, during which time the solution darkens, and nitrogen evolution is observed; the temperature of the reaction mixture is kept below 33° by use of the external cooling bath. As soon as the evolution of nitrogen has ceased (usually 20 minutes is sufficient), the reaction solution is concentrated with a rotary evaporator, giving a mixture weighing about 120–130 g. The residual brown suspension is mixed with 5 g. of decolorizing charcoal and 200 ml. of water, and the resulting mixture is boiled for 3 minutes, and filtered while hot through a Büchner funnel containing Celite filter aid. The residue on the filter is washed with 100 ml. of boiling water, and the combined filtrates are diluted with 300 ml. of water. The resulting aqueous solution, from which the product begins to crystallize, is cooled in a water bath and then allowed to stand in a refrigerator (0–3°) for 24 hours to complete the crystallization of the crude product. The crystalline solid is collected on a filter, washed with two 100-ml. portions of cold water, and dried in the air. The crude product, a pale yellow solid (19.1–22.2 g.), is recrystallized from 40 ml. of a 2:3 (v/v) formic acid-water mixture. The resulting crystals are collected on a filter, washed with 20 ml. of 2:3 (v/v) cold formic acid-water, and dried in the air, yielding 16.6–18.2 g. (61–67%) of white needles, m.p. 158–160°, which are sufficiently pure for most purposes. Three additional crystallizations from 20-ml. portions of 2:3 (v/v) formic acid-water give 15.2–16.0 g. (56–59%) (Note 5) of the pure p-acetyl-α-bromohydrocinnamic acid, m.p. 159–161° (Note 6).

2. Notes

1. Commercial grades of acetone and 4-aminoacetophenone (Matheson, Coleman and Bell or Aldrich Chemical Company, Inc.) were used without further purification.

2. The temperature of the reaction mixture rises to about 30° and then falls to 15° as stirring and cooling are continued. If this preparation were performed on a larger scale, it would probably be necessary to add the sodium nitrite solution over a longer period of time in order to control the temperature.

3. A freshly opened bottle of acrylic acid, obtained from Eastman Organic Chemicals, was used without further purification. The checkers encountered difficulty in attempting to use samples of acrylic acid that had been stored in partially filled bottles for long periods of time.

4. A reagent grade of copper(I) bromide, obtained from Fisher Scientific Company, was washed with acetone until the washings were colorless and then dried.

5. The combined filtrates from these recrystallizations can be concentrated to obtain an additional 1–2 g. of product.

6. The product has IR absorption (KBr) at 1735, 1645, and 1607 cm.$^{-1}$ with a UV maximum (95% C_2H_5OH solution) at 252.5 nm (ϵ 17,000). The sample has ^{1}H NMR peaks (CF_3CO_2H) at δ 2.78 (s, 3H, CH_3CO), 3.2–3.9 (m, 2H, benzylic CH_2), 4.60 (t, $J = 7.5$ Hz., 1H, CHBr), 7.47 (d, $J = 9$ Hz., 2H, aryl CH), and 8.10 (d, $J = 9$ Hz., 2H, aryl CH). The mass spectrum has weak molecular peaks at m/e 270 and 272 with the following relatively abundant fragment peaks: m/e (rel. int.), 191 (73), 175 (100), 131 (52), 103 (40), 77 (55), and 51 (43). The product gives a deep red color when treated with sodium nitroprusside and aqueous base; this color changes to dark blue upon acidification with acetic acid.

3. Discussion

This procedure has been used to prepare a variety of substituted α-bromohydrocinnamic acids;[2] p-acetyl-α-bromohydrocinnamic acid was prepared for the first time by this method. The method illustrates a typical application of the Meerwein reaction for the arylation of unsaturated substrates.[3] In this reaction a catalytic amount of a copper(I) salt is used to reduce an aryl diazonium salt forming an aryl radical and a copper(II) halide. Addition of the aryl radical to an unsaturated substrate forms an alkyl radical that is reoxidized by the copper(II) halide present forming an alkyl halide and regenerating the copper(I) salt catalyst. In this preparation, the product, an α-bromo acid, is formed in an acidic reaction mixture and dehydrohalogenation does not occur. However, de-hydrohalogenation of the intermediate halide is often observed in analogous reactions performed under neutral or basic reaction conditions.[3] The use of the Meerwein reaction to form ultimately 1-(4-nitrophenyl)-1,3-butadiene by the addition of an intermediate aryl radical to 1,3-butadiene followed by dehydrohalogenation of the initially formed alkyl halide is illustrated in *Organic Syntheses.*[4]

1. Department of Chemistry, Occidental College, Los Angeles, California 90041.
2. G. H. Cleland, *J. Org. Chem.*, **26**, 3362 (1961); *J. Org. Chem.*, **34**, 744 (1969).
3. C. S. Rondestvedt, Jr., *Org. React.*, **11**, 189 (1960); *Org. React.*, **24**, 225 (1976).
4. G. A. Ropp and E. C. Coyner, *Org. Synth.*, **Coll. Vol. 4**, 727 (1963).

α,β-DEHYDROGENATION OF β-DICARBONYL COMPOUNDS BY SELENOXIDE ELIMINATION: 2-ACETYL-2-CYCLOHEXEN-1-ONE

(2-Cyclohexen-1-one, 2-acetyl-)

Submitted by JAMES M. RENGA and HANS J. REICH[1]
Checked by ALBERT W. M. LEE and ROBERT V. STEVENS

1. Procedure

Caution! Most selenium compounds are toxic; consequently care should be exercised in handling them. The hydrogen peroxide oxidation of selenides is highly exothermic, acid-catalyzed, and autocatalytic. The procedure given for adding the hydrogen peroxide solution should be carefully followed.

A. *2-Acetyl-2-phenylselenocyclohexanone.* A 500-ml., three-necked, round-bottomed flask is fitted with a mechanical stirrer, a pressure-equalizing dropping funnel, and a combined inlet-outlet assembly connected to a nitrogen source and a bubbler. The flask is charged with 3.36 g. (0.140 mole) of sodium hydride (Note 1), the apparatus is flushed with nitrogen, and 100 ml. of tetrahydrofuran (Note 2) is added. The suspension is stirred and cooled in an ice bath under a static nitrogen atmosphere as a solution of 14.02 g. (0.100 mole) of 2-acetylcyclohexanone (Note 3) in 15 ml. of tetrahydrofuran is added over a 15-minute period. The formation of the sodium enolate is complete when hydrogen evolution ceases and a thick suspension has developed. Stirring and cooling are continued for 20 minutes, after which a solution of 20.1 g. (0.105 mole) of benzeneselenenyl chloride (Note 4) in 20 ml. of tetrahydrofuran is rapidly added. The contents of the flask are stirred at 0° for 15 minutes and poured into a beaker containing a magnetically stirred mixture of 200 ml. of 1 : 1 (v/v) diethyl ether-pentane, 50 ml. of aqueous 7% sodium hydrogen carbonate, and 50 g. of ice. The layers are separated, and the aqueous layer is extracted with 50 ml. of 1 : 1 (v/v) ether–pentane. The combined organic extracts are washed with 50 ml. of saturated aqueous sodium chloride and dried by filtration through a cone of anhydrous sodium sulfate. Evaporation of the solvents under reduced pressure gives 29.2–30 g. of crude, solid 2-acetyl-2-phenylselenocyclohexanone which is used in Part B without purification (Note 5).

B. *2-Acetyl-2-cyclohexen-1-one*. A 500-ml., three-necked, round-bottomed flask equipped with a pressure-equalizing dropping funnel, a reflux condenser, a thermometer, and a magnetic stirring bar is charged with a solution of 29.2–30 g. (*ca.* 0.1 mole) of crude 2-acetyl-2-phenylselenocyclohexanone in 100 ml. of dichloromethane (Note 6). The solution is stirred at room temperature, and a 2–3 ml. portion from a solution of 23.8 g. of 30% hydrogen peroxide (7.14 g., 0.21 mole) (Note 7) in 20 ml. of water is added to initiate the oxidation (*Caution!* Note 8). After the exothermic reaction begins, the mixture is stirred and cooled in an ice–salt bath as necessary to keep the temperature between 30 and 35° while the remainder of the hydrogen peroxide solution is added. When the oxidation is complete (Note 9), the ice–salt bath is removed, and vigorous stirring is continued for 15 minutes at room temperature and 15 minutes at 0°. The chilled suspension of benzeneseleninic acid is filtered, and the filter cake is washed with 50 ml. of dichloromethane (Note 10). The dichloromethane layer from the filtrate is washed with 50 ml. of aqueous 7% sodium hydrogen carbonate, dried by filtration through a cone of anhydrous sodium sulfate, and evaporated, providing 12.8–13.7 g. of crude product (Note 11). Distillation in carefully washed glassware (Note 12) at 0.1 mm. using a Kügelrohr apparatus (Note 13) with an oven temperature of 50–55° gives 11.0–11.9 g. (79–85%) of 2-acetyl-2-cyclohexen-1-one (Note 14).

2. Notes

1. A 57% dispersion of sodium hydride in mineral oil was purchased from Alfa Division, Ventron Corporation. A 5.90-g. portion of the dispersion was placed in the reaction vessel and washed free of mineral oil with three 50-ml. portions of pentane by decanting the supernatant pentane after each washing. The pentane that remains in the flask is evaporated as the assembled apparatus is purged with nitrogen prior to adding the tetrahydrofuran.

2. Tetrahydrofuran was purified by the submitters by distillation from the sodium ketyl of benzophenone.

3. 2-Acetylcyclohexanone was used as supplied by Aldrich Chemical Company, Inc.

4. Benzeneselenenyl chloride was prepared by the procedure in *Org. Synth.*, **Coll. Vol. 6,** 533 (1988). A freshly prepared solution of 24.8 g. (0.105 mole) of benzeneselenenyl bromide[2] in 25 ml. of tetrahydrofuran may also be used.

5. The crude selenide is contaminated by volatile impurities including some 2-acetylcyclohexanone which may be removed by sublimation at 50–60° to a cold finger cooled with dry ice, or by recrystallization from ether–pentane. The purified product melts at 72–73°; IR (CCl₄) cm⁻¹: 1693 strong, 1579 weak; ¹H NMR (CCl₄) δ (multiplic-

ity, number of protons, assignment): 1.3–2.3 (m, 7H, ring protons), 2.30 (s, 3H, CH_3), 2.5–2.8 (m, 1H, $CH_AH_BC=O$), 7.28 (m, 5H, C_6H_5Se). The product was analyzed by the submitters. Analysis calculated for $C_{14}H_{16}O_2Se$: C, 56.96; H, 5.46. Found: C, 57.12; H, 5.48.

6. Dichloromethane was used without purification.

7. A 30% solution of hydrogen peroxide in water was purchased from Mallinckrodt Chemical Works. The reaction requires 2 molar equivalents of hydrogen peroxide, the first to oxidize the selenide to the selenoxide and the second to oxidize the elimination product, benzeneselenenic acid, to benzeneseleninic acid. The submitters recommend that the hydrogen peroxide solution be taken from a recently opened bottle, or titrated to verify its concentration.

8. The oxidation is autocatalytic, being catalyzed by the product, benzeneseleninic acid.[3] If the temperature drops significantly below 30°, the addition of hydrogen peroxide should be stopped, and the ice–salt bath should be removed to maintain the rate of oxidation and avoid an accumulation of hydrogen peroxide in the flask.

9. The yellow dichloromethane solution turns colorless, and a precipitate of benzeneseleninic acid appears.

10. The benzeneseleninic acid weighs 14.4–16 g. (73–82%) and melts at 123–124°. It may be reconverted to diphenyl diselenide by reduction with sodium thiosulfate[2] or sodium bisulfite.[4]

11. The enol content of the product at this point is less than 2%. If the unenolized enedione is desired, the following distillation should be omitted and the product used without purification to avoid further isomerization.

12. The glassware was cleaned in a sodium dichromate–sulfuric acid bath, washed with aqueous 10% ammonium hydroxide, and rinsed with water. The extent of enolization apparently depends on the care taken in washing the glassware and conducting the distillation.

13. Kügelrohr distillation ovens manufactured by Büchi Glasapparatfabrik are available from Brinckmann Instruments, Inc., Westbury, New York.

14. The product is contaminated by 5–15% of 2-acetylcyclohexanone, which was present in the crude selenide. This impurity may be avoided by purifying the selenide as described in Note 5. The enol content of the product obtained by the submitters varied from 5 to 50%. At equilibrium the enol content is 84%. The spectral properties of the enedione are as follows: IR (CCl_4) cm.$^{-1}$: 1694 strong, 1602 weak; ^1H NMR (CCl_4), δ (multiplicity, coupling constant J in Hz., number of protons, assignment): 1.9–2.2 (m, 2H, $CH_2CH_2CH_2$), 2.35 (s, 3H, CH_3), 2.3–2.7 (m, 4H, $CH_2CH_2CH_2$), 7.56 (t, $J = 4.3$, H, $CH_2CH=C$). The enol form of the product exhibits the following ^1H NMR (CCl_4), δ (multiplicity, coupling constant J in Hz., number of protons, assignment): 2.07 (s, 3H, CH_3), 2.1–2.7 (m, 4H, CH_2CH_2), 5.55 (d of t, $J = 4.5$ and 10, 1H, $CH_2CH=CH$), 6.19 (d of t, $J = 1.5$ and 10, 1H, $CH_2CH=CH$), 15.8 (s, 1H, OH).

3. Discussion

The procedure described here serves to illustrate a new, general method for effecting the α,β-dehydrogenation of ketones,[2,5–7] aldehydes,[7] esters,[2,5,7] lactones,[7,8] nitriles,[9] sulfones, and related compounds.[2,10] The individual steps in the process are formation of an α-carbanion or enol derivative, phenylselenenylation with diphenyl diselenide or benzene selenenyl halides, oxidation of the resulting α-phenylseleno compound to the selenoxide, and thermal *syn*-elimination of benzeneselenenic acid. The advantages of this method include (*a*) the ease of introducing the α-phenylseleno group; (*b*) the rapid stoichiometric oxidation of the selenide with aqueous hydrogen peroxide at $25-35°$, sodium metaperiodate in aqueous media, or ozone in dichloromethane at $-78°$; and (*c*) the fact that the elimination occurs at about room temperature under essentially neutral conditions.

The mild character of the reaction conditions is exemplified effectively here by the preparation of 2-acetyl-2-cyclohexen-1-one from 2-acetylcyclohexanone.[2] The crude product is initially isolated entirely in the less stable enedione form which is partially converted to the more stable enol form, 2-acetyl-1,3-cyclohexadien-1-ol,[11,12] during distillation at $45-55°$. A series of α,β-unsaturated β-keto esters, β-diketones, and a β-keto sulfoxide have also been prepared in the unenolized form by this procedure (Table I).[2,5] In the case of the highly sensitive ethyl 5-oxo-1-cyclopentene-1-carboxylate, the

TABLE I

α,β-Unsaturated β-Keto Esters, β-Diketones, and a β-Keto Sulfoxide Prepared by Selenoxide Elimination

α,β-Unsaturated Product	Yield (%)[a]	α,β-Unsaturated Product	Yield (%)[a]
	81[b]		74
	89		89
	84		55
	89		

[a]Overall yield from β-keto ester, β-diketone, or β-keto sulfoxide. The scale was $0.01-0.005$ mole.
[b]The starting β-keto ester and the product were $2:1$ mixtures of ethyl and methyl esters.

hydrogen carbonate extraction must be omitted to avoid base-catalyzed decomposition during isolation.

The enolized form of 2-acetyl-2-cyclohexen-1-one has been synthesized in low yield by dehydrochlorination of 2-acetyl-2-chlorocyclohexanone in collidine at 180°[11] and by elimination of acetamide from N-(2-acetyl-3-oxo-1-cyclohexyl)acetamide at 120–140°.[12] The preparation of other α,β-unsaturated β-dicarbonyl compounds has been attempted with varying degrees of success. The dehydrogenation of 2-hydroxymethylene-3-keto steroids to 2-formyl-Δ¹-3-keto compounds with 2,3-dichloro-5,6-dicyano-1,4-benzoquinone has been reported.[13] Ethyl 5-oxo-1-cyclopentene-1-carboxylate has been prepared by selenium dioxide oxidation of the parent β-keto ester.[14] α-Acetoxylation of 3-methyl- and 3-isopropyl-2,4-pentanedione with lead tetraacetate followed by acetate pyrolysis provided the α,β-unsaturated β-diketones.[15] Chlorination and dehydrochlorination of 2-acetylcycloheptanone gave an enolic tautomer of 2-acetyl-2-cyclohepten-1-one.[11b] Numerous failures in attempts to synthesize these and other α,β-unsaturated β-dicarbonyl compounds by halogenation and dehydrohalogenation have been recorded as a consequence of competing Favorskii rearrangement, migration of halogen to the α'-position, and decomposition of the products from a combination of the high temperatures and basic conditions employed.[11,13–16] A number of α,β-unsaturated β-keto esters and β-diketones have been prepared by intermolecular aldol condensations under Knoevenagel conditions,[17] aldol cyclization,[16,18] and Robinson annelation.[19] All these procedures lead to equilibrium mixtures of keto and enol forms.

1. Department of Chemistry, University of Wisconsin, Madison, Wisconsin 53706.
2. H. J. Reich, J. M. Renga, and I. L. Reich, J. Am. Chem. Soc., 97, 5434 (1975).
3. H. J. Reich, F. Chow, and S. L. Peake, Synthesis, 299 (1978).
4. K. B. Sharpless and R. F. Lauer, J. Org. Chem., 39, 429 (1974).
5. (a) H. J. Reich, I. L. Reich, and J. M. Renga, J. Am. Chem. Soc., 95, 5813 (1973); (b) H. J. Reich, J. M. Renga, and I. L. Reich, J. Org. Chem., 39, 2133 (1974); (c) H. J. Reich and J. M. Renga, J. Org. Chem., 40, 3313 (1975).
6. D. L. J. Clive, J. Chem. Soc. Chem. Commun., 695 (1973).
7. K. B. Sharpless, R. F. Lauer, and A. Y. Teranishi, J. Am. Chem. Soc., 95, 6137 (1973).
8. P. A. Grieco and M. Miyashita, J. Org. Chem., 39, 120 (1974).
9. D. N. Brattesani and C. H. Heathcock, Tetrahedron Lett., 2279 (1974).
10. For recent reviews, see H. J. Reich, Acc. Chem. Res., 12, 22 (1979); H. J. Reich, "Organoselenium Oxidations," in W. Trahanovsky, Ed., "Oxidation in Organic Chemistry," Part C, Academic Press, New York, 1978, pp. 1–130; D. L. J. Clive, Tetrahedron, 34, 1049 (1978).
11. (a) M. E. McEntee and A. R. Pinder, J. Chem. Soc., 4419 (1957); (b) C. W. T. Hussey and A. R. Pinder, J. Chem. Soc., 3525 (1961).
12. A. A. Akhrem, A. M. Moiseenkov, and F. A. Lakhvich, Izv. Akad. Nauk SSSR, 407 (1972); Bull. Acad. Sci. USSR, Div. Chem. Sci., 355 (1972).
13. J. A. Edwards, M. C. Calzada, L. C. Ibanez, M. E. Cabezas Rivera, R. Urquiza, L. Cardona, J. C. Orr, and A. Bowers, J. Org. Chem., 29, 3481 (1964).
14. J. N. Marx, J. H. Cox, and L. R. Norman, J. Org. Chem., 37, 4489 (1972).
15. D. Gorenstein and F. H. Westheimer, J. Am. Chem. Soc., 92, 634 (1970); C. W. T. Hussey and A. R. Pinder, J. Chem. Soc., 1517 (1962).
16. J. A. Brenner, J. Org. Chem., 26, 22 (1962).
17. G. B. Payne, J. Org. Chem., 24, 1830 (1959); F. Tiemann and P. Krüger, Ber. Dtsch. Chem. Ges., 28, 2121 (1895).

18. S. N. Huckin and L. Weiler, *J. Am. Chem. Soc.,* **96,** 1082 (1974).
19. G. Stork and R. N. Guthikonda, *J. Am. Chem. Soc.,* **94,** 5109 (1972).

2-ACETYL-1,3-CYCLOPENTANEDIONE

(2-Acetyl-3-hydroxy-2-cyclopenten-1-one, Note 1)

Submitted by FERENC MERÉNYI and MARTIN NILSSON[1]
Checked by E. J. COREY, JOEL I. SHULMAN, and LAWRENCE LIBIT

1. Procedure

Caution! Since hydrogen chloride is evolved in this reaction, it should be conducted in a hood.

A dry, 3-l., three-necked round-bottomed flask equipped with a sealed mechanical stirrer (all-glass or glass-Teflon), a reflux condenser fitted with a calcium chloride drying tube and a 100-ml. dropping funnel is charged with 50.0 g. (0.500 mole) of finely powdered succinic anhydride (Note 2), 133.4 g. (1.000 mole) of freshly crushed anhydrous aluminum chloride (Note 3), and 500 ml. of anhydrous 1,2-dichloroethane. The mixture is stirred vigorously at room temperature for about 2 hours to dissolve as much of the solid reactants as possible before 50.0 g. (0.500 mole) of isopropenyl acetate (Note 4) is added rapidly through the dropping funnel; the reaction starts immediately as indicated by a rise in temperature to about 60° to 70°. The mixture is refluxed for 15 minutes with continuous stirring. The hot reaction mixture, which contains a sticky oil, is poured into a stirred mixture of 200 ml. of 12 *M* hydrochloric acid and 1000 g. of crushed ice, and the reaction flask is rinsed with part of the acidic aqueous phase. When the dark mass has dissolved, 200 ml. of 12 *M* hydrochloric acid is added, and the mixture is stirred vigorously for about 3 hours (Note 5). The dichloroethane phase is separated, and the aqueous phase is extracted with eight 600-ml. portions of dichloromethane. The extracts are combined with the dichloroethane phase (Note 6) and extracted first with 600 ml. and then twice with 200-ml. portions of aqueous saturated sodium hydrogen carbonate solution. The combined sodium hydrogen carbonate extracts are washed with 200 ml. of dichloromethane and then cautiously acidified in a 3-l. beaker with 150 ml. of 12 *M* hydrochloric acid, with vigorous stirring. The acidic solution (Note 7) is extracted first with 600 ml. and then with four 400-ml. portions of dichloromethane. The bulk of the dichloromethane is removed by distillation at atmospheric pressure and water bath temperatures; however, the last 150 ml. of solvent is removed below room temperature under reduced pressure (about 20 mm.), leaving 24–27 g. of crude 2-acetyl-1,3-cyclopentanedione as a light brown solid, m.p. 68–71° (Note 8).

The crude product is purified by decolorization with charcoal and recrystallization from 100–150 ml. of diisopropyl ether, giving 19–21 g. (27–30%) of colorless needles, m.p. 70–72°. This material is sufficiently pure for most purposes. Further purification may be achieved by recrystallization from diisopropyl ether and/or sublimation at 60° (0.1 mm.) onto a cold-finger condenser, giving material melting at 73–74° (Notes 9, 10).

2. Notes

1. 2-Acetyl-1,3-cyclopentanedione is completely enolized in the solid state as well as in solution.[2] Indirect evidence indicates that the carbon-carbon double bond of the enol is within the ring.

2. The submitters used a practical grade of succinic anhydride obtained from Matheson, Coleman and Bell.

3. The submitters used anhydrous, sublimed aluminum chloride obtained from E. Merck AG, Darmstadt, Germany. The checkers used analytical reagent grade material obtained from Mallinckrodt Chemical Works.

4. The submitters used either a pure grade (more than 98.5%) of isopropenyl acetate or a practical grade (95–98%) from Fluka AG, Buchs SG, Switzerland. Generally, it was not essential to distill the material before use.

5. This treatment increases the amount of product which can be extracted from the aqueous solution.

6. Evaporation of the solvents at this stage gives an oily product which may be decolorized with charcoal and then recrystallized from diisopropyl ether to give 2-acetyl-1,3-cyclopentanedione, m.p. 68–71°. However, the extraction with sodium hydrogen carbonate is preferred.

7. Decolorization of the acidic aqueous solution with charcoal at this stage improves the quality of the product.

8. 2-Acetyl-1,3-cyclopentanedione is quite volatile and appreciable losses may occur if the evaporation of the concentrated solution is continued at elevated temperature.

9. Further amounts of 2-acetyl-1,3-cyclopentanedione can be obtained by continuous extraction of the original aqueous phase with dichloromethane. The submitters have obtained total yields of product as high as 41–43%. 2-Acetyl-1,3-cyclopentanedione is

sparingly soluble in diethyl ether and is not extracted very efficiently from aqueous solution with ether.

10. 2-Acetyl-1,3-cyclopentanedione can be hydrolyzed to 1,3-cyclopentanedione.[3,4] The preferred method[4] is to keep a 0.1 M solution of 2-acetyl-1,3-cyclopentanedione in 0.1 M aqueous acetic acid at 100° for 24 hours. The solvent is then evaporated at 40° with a water pump and the yellow solid is recrystallized from butanone (with charcoal) to give 1,3-cyclopentanedione, m.p. 148–150°, in 65–80% yield.

3. Discussion

2-Acetyl-1,3-cyclopentanedione has been obtained in small amounts from the aluminum chloride-catalyzed reactions of vinyl acetate and succinyl chloride in 1,1,2,2-tetrachloroethane.[5] 2-Acetyl-1,3-cyclopentanediones have been prepared via Dieckmann condensation of 1,4-bis(ethoxycarbonyl)-3,5-hexanediones.[6] The present procedure is essentially that of Merényi and Nilsson[7] with some modifications.

The diacylation of isopropenyl acetate with anhydrides of dicarboxylic acids is applicable to the synthesis of several other cyclic β-triketones in moderate yield.[3,7] It has been used for the synthesis of 2-acetyl-1,3-cyclohexanedione (40% yield), 2-acetyl-4-methyl-1,3-cyclopentanedione (10% yield), 2-acetyl-4,4-dimethyl-1,3-cyclopentanedione (10% yield), 2-acetyl-5,5-dimethyl-1,3-cyclohexanedione (10% yield), 2-acetyl-1,3-cycloheptanedione (12% yield), and 2-acety-1,3-indanedione (25% yield). Maleic anhydrides under more drastic conditions give acetyl-1,3-cyclopent-4-enediones in yields from 5 to 12%.[8] The corresponding acylation of the enol acetate of 2-butanone with succinic anhydride has been used to prepare 2-methyl-1,3-cyclopentanedione, an important intermediate in steroid synthesis.[9,10]

1. Department of Organic Chemistry, Chalmers University of Technology and the University of Göteborg, S-412 96 Göteborg 5, Sweden.
2. S. Forsén, F. Merényi, and M. Nilsson, *Acta Chem. Scand.*, **18**, 1208 (1964).
3. F. Merényi and M. Nilsson, *Acta Chem. Scand.*, **18**, 1368 (1964).
4. M. Nilsson, *Acta Chem. Scand. Ser. B*, **35**, 667 (1981).
5. A. Sieglitz and O. Horn, *Chem. Ber.*, **84**, 607 (1951).
6. M. Vandewalle, *Bull. Soc. Chim. Belg.*, **73**, 628 (1964).
7. F. Merényi and M. Nilsson, *Acta Chem. Scand.*, **17**, 1801 (1963).
8. M. Nilsson, *Acta Chem. Scand.*, **18**, 441 (1964).
9. H. Scheck, G. Lehman, and G. Hilgetag, *J. Prakt. Chem.*, **35**, 28 (1967); *Angew. Chem.*, **79**, 97 (1967).
10. V. J. Grenda, G. W. Lindberg, N. L. Wendler, and S. H. Pines, *J. Org. Chem.*, **32**, 1236 (1967).

3-ACETYL-2,4-DIMETHYLFURAN

[Ethanone, 1-(2,4-dimethyl-3-furanyl)-]

A. $(CH_3)_2S + BrCH_2C{\equiv}CH \longrightarrow (CH_3)_2\overset{+}{S}CH_2C{\equiv}CH\ Br^-$

B. $(CH_3)_2\overset{+}{S}CH_2C{\equiv}CH\ Br^-$
   $+$
   $CH_3COCH_2COCH_3$
   $\xrightarrow[\text{ethanol, }\Delta]{C_2H_5ONa}$

   (furan ring with substituents CH$_3$, COCH$_3$, CH$_3$, O) $+ (CH_3)_2S$

Submitted by P. D. Howes and C. J. M. Stirling[1]
Checked by C. Reese, M. Uskoković, and A. Brossi

1. Procedure

Caution! These reactions should be performed in a hood because of the noxious odors.

A. *Dimethyl-2-propynylsulfonium bromide.* A mixture of 6.2 g. (0.10 mole) of dimethyl sulfide (Note 1), 11.9 g. (0.100 mole) of 3-bromopropyne (Note 2), and 10 ml. of acetonitrile (Note 3) is stirred magnetically for 20 hours (Note 4) in a darkened, 100-ml., round-bottomed flask (Note 5) fitted with a calcium chloride drying tube. The resulting white, crystalline mass is filtered with suction and washed with three 50-ml. portions of dry diethyl ether (Note 6), giving 16.4 g. (90%) of the sulfonium salt, m.p. 105–106°. This material may be used in the next step without purification but, if desired, it may be recrystallized from ethanol-ether (Note 7) with minimal loss, m.p. 109–110°.

B. *3-Acetyl-2,4-dimethylfuran.* To a solution of 8.7 g. (0.087 mole) of acetylace-tone (Note 8) in 175 ml. of 0.5 *M* ethanolic sodium ethoxide (0.087 mole), contained in a 500-ml., round-bottomed flask, fitted with a condenser topped with a calcium chloride drying tube, is added a solution of 15.75 g. (0.0870 mole) of dimethyl-2-propynylsulfonium bromide in 150 ml. of ethanol (Note 9). The mixture is refluxed until the odor of dimethyl sulfide is no longer appreciable (Note 10). The reaction flask is then fitted with a 30-cm., helix-packed column, and by heating the flask with a water bath, ethanol is distilled through the column (Note 11). The residue is treated with 200 ml. of ether, and the suspension is filtered. Ether is distilled from the filtrate at atmospheric pressure, and the residue is distilled, giving 9.7 g. (81%) of 3-acetyl-2,4-dimethylfuran (Notes 12 and 13), b.p. 90–95° (12 mm.), n_D^{24} 1.4965.

2. Notes

1. Dimethyl sulfide was used as supplied by British Drug Houses.
2. 3-Bromopropyne, supplied by British Drug Houses, was distilled before use (b.p. 84–86°).

TABLE I
Furans Prepared Via Acetylenic Sulfonium Salts

Sulfonium Salt	Addend	R_1	R_2	R_3	Yield,%[2]
$(CH_3)_2\overset{+}{S}CH_2C\equiv CH\ Br^-$	$CH_3COCH_2CO_2CO_2C_2H_5$	CH_3	$CO_2C_2H_5$	CH_3	86
$(CH_3)_2\overset{+}{S}CH_2C\equiv CH\ Br^-$	$CH_3COCH_2SO_2C_6H_4\text{-}4\text{-}CH_3$	CH_3	$SO_2C_6H_4\text{-}4\text{-}CH_3$	CH_3	78
$(CH_3)_2\overset{+}{S}CH_2C\equiv CH\ Br^-$	$C_6H_5COCH_2COC_6H_5$	C_6H_5	COC_6H_5	CH_3	72
$(CH_3)_2\overset{+}{S}CH_2C\equiv CC_6H_5Br^-$	$CH_3COCH_2CO_2C_2H_5$	CH_3	$CO_2C_2H_5$	$CH_2C_6H_5$	63
$(CH_3)_2\overset{+}{S}CHC\equiv CH\ Br^-$ $\quad\mid$ $\quad CH_3$	$CH_3COCH_2CO_2H_5$	CH_3	$CO_2C_2H_5$	C_2H_5	50

3. Acetonitrile (Matheson, Coleman and Bell, spectral grade) was used without further treatment.

4. The maximum yield was obtained after 20 hours. Shorter reaction times give slightly lower yields.

5. If a brown glass flask is unavailable, an ordinary flask wrapped with aluminum foil may be used.

6. The ether was dried over sodium.

7. The salt was dissolved in 10 ml. of ethanol, 75 ml. of ether was added portionwise, and the mixture was allowed to stand overnight at room temperature.

8. Acetylacetone, supplied by British Drug Houses, was distilled before use (b.p. 137°).

9. The ethanol was dried with magnesium ethoxide.[2]

10. About 6 hours is required on this scale.

11. Distillation through the packed column is essential to prevent loss of furan by co-distillation with ethanol.

12. The product has IR absorption (neat) at 1690 cm.$^{-1}$ (ketone C=O and ^{1}H NMR peaks (CCl$_4$) at δ 2.20 (s, 3H, COCH_3), 2.40 (s, 3H, CH_3), 3.60 (s, 3H, CH_3), and 7.40 (s, 1H, furyl).

13. A convenient modification of this procedure gives the furan in 70–75% yield; the sulfonium salt is preformed in acetonitrile and, without isolation, the other reagents are added.

3. Discussion

This procedure illustrates a recently published,[3] simple, general method for the synthesis of substituted furans. The scope of the reaction is shown in Table I. Many variations of this procedure are clearly possible.

The method described has some features in common with the well-known, but apparently little-used, Feist-Benary furan synthesis,[4] which uses an α-haloketone in place of the sulfonium salt. Acetylenic bromides suitable for preparing the sulfonium salts are readily available by well-documented procedures involving acetylenic organometallic compounds.

The mechanism of furan formation by this route is determined by the structure of the sulfonium salt; the course, hence the end product, is governed by whether an α-substituent is present. This must be considered when syntheses based on this procedure are being planned. Plausible mechanisms for the reaction have been suggested.[3]

Direct treatment of propargyl halides with β-dicarbonyl compounds and subsequent treatment of the products with zinc carbonate yields 2,3,5-trisubstituted furans.[5]

1. School of Physical and Molecular Sciences, University College of North Wales, Bangor, Caernarvonshire, U.K.
2. D. D. Perrin, W. L. F. Armarego, and D. R. Perrin, "Purification of Laboratory Chemicals," 1st ed., Pergamon Press, New York, 1966, p. 157.
3. J. W. Batty, P. D. Howes, and C. J. M. Stirling, *J. Chem. Soc., Perkin Trans. 1*, 65 (1973).
4. A. T. Blomquist and H. B. Stevenson, *J. Am. Chem. Soc.*, **56**, 146 (1934).
5. K. E. Schulte, J. Reisch, and A. Mock, *Arch. Pharm. Ber. Dtsch. Pharm. Ges.*, **295**, 627 (1962).

2-ACETYL-6-METHOXYNAPHTHALENE

[Ethanone, 1-(6-methoxy-2-naphthalenyl)-]

Submitted by L. ARSENIJEVIC,[1] V. ARSENIJEVIC,[1] A. HOREAU,[2] and J. JACQUES[2]
Checked by DAVID WALBA and ROBERT E. IRELAND

1. Procedure

A 1-l., three-necked, round-bottomed flask is fitted with a mechanical stirrer and a thermometer; the third neck of the flask is fitted with a 50-ml., pressure-equalizing addition funnel, carrying a drying tube attached to a gas trap. The flask is charged with 200 ml. of dry nitrobenzene (Note 1), followed by 43 g. (0.32 mole) of anhydrous aluminum chloride. After the aluminum chloride has dissolved, 39.5 g. (0.250 mole) of finely ground 2-methoxynaphthalene (nerolin, Note 2) is added. An ice bath is used to cool the stirred solution to about 5° before 25 g. (23 ml., 0.32 mole) of redistilled acetyl chloride (Note 3) is added dropwise over a 15–20 minute period, with stirring and at a rate which holds the temperature between 10.5 and 13° (Note 4). After addition of the acetyl chloride is complete, the flask is kept immersed in the ice water while stirring is continued for 2 hours. The mixture is then allowed to stand at room temperature for at least 12 hours.

The reaction mixture is cooled in an ice bath, poured with stirring into a 600-ml. beaker containing 200 g. of crushed ice, and treated with 100 ml. of concentrated hydrochloric acid. The resulting two-phase mixture and 50 ml. of chloroform are transferred to a 1-l. separatory funnel (Note 5); the chloroform-nitrobenzene layer is separated and washed with three 100-ml. portions of water. The organic layer is transferred to a 2-l., round-bottomed flask, and steam-distilled. A fairly rapid flow of steam is used, and the distillation flask is heated in an oil bath at about 120°. After about 3 hours (3–4 l. of water) the distillation is stopped, and the residue in the flask is allowed to cool. Residual water in the flask is decanted from the solid organic material and extracted with chloroform. The solid residue in the flask is dissolved in 100 ml. of chloroform and separated from any water left in the flask, and the chloroform layers are combined and dried over anhydrous magnesium sulfate. The chloroform is removed on a rotary evaporator, and the solid residue, weighing 50–65 g. (still slightly wet with chloroform), is distilled under vacuum (Note 6). The receiving flask should be immersed in ice water, and the fraction boiling about 150–165° (0.02 mm.) is collected (Note 7).

The yellow distillate (ca. 40 g., m.p. 85–95°) is recrystallized from 75 ml. of methanol, cooled in an ice bath (Note 8) and filtered, yielding 22.5–24 g. (45–48%) of white, crystalline 2-acetyl-6-methoxynaphthalene (Note 9), m.p. 106.5–108° (lit. 104–105°).[3]

2. Notes

1. The nitrobenzene may be dried by distilling the first 10% and using the residue directly, or standing over anhydrous calcium chloride overnight and filtering.

2. 2-Methoxynaphthalene (Matheson, Coleman and Bell), m.p. 71.5–73°, was used without further purification.

3. Acetic anhydride may be used instead of acetyl chloride. However, it is then necessary to use two molecular equivalents of aluminum chloride per mole of anhydride and increase the amount of nitrobenzene by about 30%. About the same yield of ketone is obtained.

4. Temperature control is very important (see discussion).

5. The addition of chloroform is not always indispensable, but very useful to prevent emulsification and facilitate separation of the nitrobenzene layer. The reaction vessel and beaker are rinsed with this chloroform before it is added to the nitrobenzene layer. If an emulsion does form and phase separation becomes inconveniently slow, as much nitrobenzene as possible is withdrawn, and the emulsion and water layers are filtered by suction through a Celite cake wet with chloroform. The phases should then separate easily.

6. The material may be distilled from a two-bulb distillation flask as described in *Org. Synth.*, **Coll. Vol. 3,** 133 (1955), or from a small Claisen flask.

7. Care must be taken to prevent solidification and possible blocking in the condenser. A small burner may be used to keep the adapter just hot enough to melt the distillate.

8. If the methanol is cooled below 0°, the 1-acetyl isomer that is formed during the reaction will also crystallize with the product.

9. ^{1}H NMR (CDCl$_3$): δ 2.65 (s, 3H, COCH_3), 3.92 (s, 3H, OCH_3), 7.20 (m, 4H, ArH), 7.80 (m, 1H, ArH), 8.30 (m, 1H, ArH).

3. Discussion

The procedure described herein is a modification of that of Haworth and Sheldrick,[3] the efficiency of which has been confirmed by many authors.[4–10] 2-Acetyl-6-methoxynaphthalene has also been prepared by the action of methylzinc iodide on 6-methoxy-2-naphthoyl chloride.[11]

In this reaction, nitrobenzene has an important function because it causes acylation to occur predominantly at the 6-position, whereas 1-acetyl-2-methoxynaphthalene is the principal product when carbon disulfide is used. The main feature of this procedure is the particular attention given to temperature control in order to obtain reliable results. It has been observed that the ratio of 6-acetylated to 1-acetylated nerolin is dependent on the temperature, the lower temperatures favoring 1-acetylation. Below 0° the yield of 6-acetylated product is only 3–10%; at higher temperatures the 6-acetylated product predominates, but an increased amount of tarry material is formed.

1. Department of Organic Chemistry, Faculty of Pharmacy, Belgrade, Yugoslavia.
2. Organic Chemistry of Hormones Laboratory, College of France, 75231 Paris 5, France.
3. R. D. Haworth and G. Sheldrick, *J. Chem. Soc.,* 864 (1934).

4. R. Robinson and H. N. Rydon, *J. Chem. Soc.*, 1394 (1939).
5. L. Novak and M. Protiva, *Collect. Czech. Chem. Commun.*, **22**, 1637 (1957).
6. N. P. Buu Hoi, D. Lavit, and J. Collard, *Croat. Chem. Acta*, **29**, 291 (1957) [*Chem. Abstr.*, **53**, 2170b (1959)].
7. J. Cason and H. Rapoport, "Laboratory Text in Organic Chemistry," 2nd ed., Prentice-Hall, Englewood Cliffs, N.J., 1962, p. 439.
8. J. Fried and I. Harrison, S. African Pat. 67,07,597 (1969) [*Chem. Abstr.*, **71**, 91162 (1979)].
9. F. Alvarez, Ger. Pat. 1,934,460 (1970) [*Chem. Abstr.*, **72**, 100364 (1970)].
10. J. Fried and I. Harrison, U.S. Pat. 3,978,124 (1976) [*Chem. Abstr.*, **86**, 43446 (1977)].
11. K. Fries and K. Schimmelschmidt, *Ber. Dtsch. Chem. Ges.*, **58**, 2835 (1925).

COPPER CATALYZED ARYLATION OF β-DICARBONYL COMPOUNDS: 2-(1-ACETYL-2-OXOPROPYL)BENZOIC ACID

[Benzoic acid, 2-(1-acetyl-2-oxopropyl)-]

$$+ CH_3COCH_2COCH_3 \xrightarrow[\text{CuBr, 80–85°}]{\text{2 NaH}}$$

Submitted by Alle Bruggink, Stephen J. Ray, and Alexander McKillop[1]
Checked by Sadao Hayashi and Wataru Nagata

1. Procedure

A 250-ml., three-necked, round-bottomed flask equipped with a sealed-mechanical stirrer, a Claisen adapter fitted with a thermometer and a gas-inlet adapter, and a calcium chloride drying tube (Note 1) is charged with 150 ml. of acetylacetone (Notes 2 and 3), 25 g. (0.12 mole) of 2-bromobenzoic acid (Note 4), and 1.0 g. of copper(I) bromide (Note 5). The mixture is thoroughly purged with dry nitrogen and stirred rapidly while 9.0 g. (0.47 mole) of an 80% dispersion of sodium hydride in mineral oil (Note 6) is slowly added portionwise through the inlet protected by the calcium chloride drying tube. Addition of the first portions of the sodium hydride results in an immediate exothermic reaction, and the temperature of the mixture rises rapidly to 50–55°. The remainder of the sodium hydride is added at such a rate that the temperature remains within the 50–55° range; external cooling with a cold-water bath may be necessary (Note 7). After addition of the sodium hydride is complete (30–35 minutes), the flask is placed in a hot-water bath heated to 80–85°, and the reaction mixture is stirred and heated for 5 hours, during which time a slow stream of dry nitrogen is passed through the apparatus.

When the reaction mixture has cooled, the contents of the flask are poured into a 1-l. Erlenmeyer flask containing 400 ml. of distilled water. The reaction flask and stirrer are thoroughly washed with an additional 100 ml. of distilled water, with the washings added to the bulk of the reaction mixture. The aqueous mixture is allowed to stand at room temperature for 15 minutes to ensure completion of hydrolysis and precipitation of inorganic salts. After the salts are removed by filtration under reduced pressure (Note 8) and discarded, the filtrate is transferred to a 1-l. separatory funnel, and the excess acetylacetone is separated (Note 9). The aqueous phase is washed with five 100-ml. portions of diethyl ether (Note 9) and transferred to a 1-l. conical flask. Nitrogen is blown through the solution for 15 minutes to remove traces of ether (Note 10). The aqueous solution is then acidified to pH 3 with concentrated hydrochloric acid, with the flask being constantly swirled during addition of the acid, and the mixture is allowed to stand at room temperature for 30 minutes to ensure complete precipitation of the product. The crude material is collected by filtration under reduced pressure, washed with 25 ml. of distilled water, and dried under reduced pressure over phosphorus pentoxide, giving 21–22 g. (76–80%) of crude product, m.p. 138–145° (Note 11). Recrystallization from a mixture of 50 ml. of methanol and 100 ml. of water gives 19.5–21 g. (71–76%) of pure material (Note 12) as colorless prisms, m.p. 142–144.5° (Note 11).

2. Notes

1. The checkers inserted a condenser between the flask and the drying tube to prevent acetylacetone from being carried away by the nitrogen stream.

2. The acetylacetone was washed with aqueous sodium hydrogen carbonate solution, dried over sodium sulfate, and distilled prior to use. The checkers used acetylacetone with a purity greater than 95%, purchased from Wako Pure Chemical Industries, Ltd., Japan, and purified by distillation prior to use.

3. The use of a large amount of acetylacetone as both reagent and solvent is essential to prevent precipitation of the sodium salts of acetylacetone and 2-bromobenzoic acid. If this happens, the whole reaction mixture rapidly solidifies, resulting in incomplete mixing and poor, irreproducible yields of the product.

4. Commercial 2-bromobenzoic acid, purchased from Aldrich Chemical Company, Inc., is a gray powder, m.p. 144–147°, and was purified as follows: the crude acid was dissolved in warm 2 N sodium hydroxide solution; the mixture was heated to reflux, treated with activated carbon, filtered, and cooled, and the filtrate was acidified with concentrated hydrochloric acid. The colorless solid that precipitated was collected by filtration under reduced pressure and recrystallized from aqueous methanol, giving pure 2-bromobenzoic acid as colorless needles, m.p. 148–150°. The checkers used 2-bromobenzoic acid of m.p. 151–151.5°, obtained from Wako Pure Chemical Industries, Ltd., Japan, without further purification. Their results were comparable to those of the submitters.

5. Copper(I) bromide was prepared according to *Org. Synth., Coll. Vol. 3,* 186 (1955); the salt was dried under reduced pressure over phosphorus pentoxide prior to use. The checkers observed that good results could be realized by using a commercially available copper(I) bromide with a purity greater than 95% from Wako Pure Chemical Industries, Ltd., Japan, without purification.

6. The checkers used a 50% dispersion of sodium hydride in mineral oil, available from Wako Pure Chemical Industries, Ltd., Japan, and obtained results comparable to those of the submitters.

7. Cooling is normally required. It is important, however, to ensure that the temperature of the reaction mixture does not fall below 50°; otherwise efficient stirring becomes impossible.

8. The inorganic salts that precipitate at this stage are generally in a very finely divided form. A medium-sized Büchner funnel (11–15 cm.) or a small Büchner funnel (5–7.5 cm.) fitted with a Celite pad should be used to avoid slow filtration.

9. Acetylacetone can readily be recovered and recycled. The bulk of it, which is obtained at the separation step, is combined with the ether washings and dried over anhydrous sodium sulfate. The solvent is then removed by evaporation under reduced pressure and the residual crude acetylacetone is purified by distillation, giving 85–95 g. of pure acetylacetone.

10. Acidification without prior removal of the small amount of ether present in the aqueous solution results in precipitation of the product as an oily, semisolid mass that is difficult to filter.

11. Evolution of water is noticeable from *ca.* 130° upward; this is a result of lactone formation between the enolic hydroxyl group of the β-dicarbonyl unit and the aromatic carboxylic acid group.

12. The product, which exists entirely in the enolic form, has the following spectral data: UV (CH_3OH) nm. max. (ϵ): 227 (7760) and 287 (8190); IR (Nujol) cm^{-1}: 3300–2400, 1700–1690, 1305, 1270, 810, 770, 720, and 700; ^{1}H NMR (CDCl$_3$), δ (multiplicity, number of protons, assignment): 1.82 (sharp s, 6H, 2CH_3), 7.2–8.2 (complex m, 4H, C$_6H_4$), 10.1 (broad s, 1H, COOH), and 16.3 (broad s, 1H, enolic OH).

3. Discussion

2-(1-Acetyl-2-oxopropyl)benzoic acid has been prepared by the copper-catalyzed arylation of acetylacetone with 2-bromobenzoic acid. Facile condensation of β-dicarbonyl compounds with 2-bromobenzoic acid was first demonstrated by Hurtley in 1929,[2] and reactions of this type have occasionally been employed with limited success in a number of natural product syntheses.[3] The reaction conditions originally employed for these condensations and subsequently adopted by all other workers involve the use of sodium ethoxide as base, ethanol as solvent, and copper powder as catalyst. Under these conditions, however, 2-ethoxybenzoic acid is obtained as a by-product in substantial amounts (25–35%), together with smaller amounts of unchanged 2-bromobenzoic acid (5–10%); furthermore, separation and purification of the desired α-(2-carboxyphenyl)-β-dicarbonyl product is often difficult and tedious.

The present method of preparation, described by Bruggink and McKillop,[4] has the particular advantages of high yield and manipulative simplicity, and avoids the problem inherent in Hurtley's procedure of separation of mixtures of carboxylic acids by fractional crystallization or column chromatography. The method has wide applicability with respect to both the β-dicarbonyl compound and the 2-bromobenzoic acid. The synthetic scope and limitations of this procedure for the direct arylation of β-dicarbonyl compounds have been fully defined with respect to a wide range of substituted 2-bromobenzoic acids and to certain other bromoaromatic and heteroaromatic carboxylic acids.[4]

1. School of Chemical Sciences, University of East Anglia, Norwich, England.
2. W. R. H. Hurtley, *J. Chem. Soc.*, 1870 (1929).

3. K. A. Cirigottis, E. Ritchie, and W. C. Taylor, *Aust. J. Chem.*, **27**, 2209 (1974).
4. A. Bruggink and A. McKillop, *Tetrahedron*, **31**, 2607 (1975).

5-ACETYL-1,2,3,4,5-PENTAMETHYLCYCLOPENTADIENE

[Ethanone, 1-(1,2,3,4,5-pentamethyl-2,4-cyclopentadien-1-yl)-]

Submitted by R. B. KING, W. M. DOUGLAS, and A. EFRATY[1]
Checked by J. X. McDERMOTT, G. M. WHITESIDES, and G. BÜCHI

1. Procedure

Caution! Explosive mixtures often result when peracids other than m-*chloroperbenzoic acid are used in this reaction. No such problem has ever been encountered using* m-*chloroperbenzoic acid.*

A 3-1., three-necked, round-bottomed flask is fitted with a nitrogen-inlet tube, a pressure-equalizing addition funnel, and an air-driven stirring apparatus. After the system has been thoroughly purged with nitrogen, it is charged with a solution of 100 g. (0.615 mole) of 1,2,3,4,5,6-hexamethylbicyclo[2.2.0]hexa-2,5-diene (Hexamethyldewarbenzene Note 1) in 200 ml. of toluene. The reaction mixture is cooled to 0° with a bath of ice water and stirred rapidly while a solution of 130 g. (0.640 mole) of 85% *m*-chloroperbenzoic acid (Note 2) in 1.5 l. of chloroform is added dropwise over 3–4 hours. Throughout the addition and for 4 hours following its completion the reaction mixture is kept at 0°, after which it is stirred at room temperature for 36 hours. The white precipitate that forms during the reaction (mainly *m*-chlorobenzoic acid) is removed by filtration through a 650-ml., sintered-glass funnel (porosity 10–15 μm), and the filter cake is washed with two 100-ml. portions of chloroform. Combination of the filtrates gives a chloroform–toluene solution, which is condensed on a rotary evaporator with the water bath at

35–40°. When all of the chloroform has been removed, the residual toluene solution is diluted with 500 ml. of pentane and refiltered. The filtrate is washed with four 200-ml. portions of 10% aqueous sodium hydroxide and two 250-ml. portions of water, dried over anhydrous magnesium sulfate, and evaporated to dryness on a rotary evaporator with the water bath at 50°. Traces of toluene are removed under high vacuum (0.02 mm.) at 25°, and the residue is distilled (Note 3). Collection of material boiling between 72° and 95° (4 mm.) gives about 50 g. of crude product, which is further purified by dissolving it in 50 ml. of pentane and cooling the pentane solution at −78° for several hours. The resulting white crystals are collected by filtration and dried, giving 31.3–37.5 g. (29–34%) of pure 5-acetyl-1,2,3,4,5-pentamethylcyclopentadiene, m.p. 54–56° (Notes 4 and 5).

2. Notes

1. 1,2,3,4,5,6-Hexamethylbicyclo[2.2.0]hexa-2,5-diene may be purchased from Columbia Organic Chemicals, Columbia, South Carolina, or Henley Chemical Company, New York, New York.

2. *m*-Chloroperbenzoic acid may be purchased from Columbia Organic Chemicals or the Aldrich Chemical Company, Inc., Milwaukee, Wisconsin.

3. Distillation was performed using a 30-cm., vacuum-jacketed column wrapped with heating tape and containing a spiral of wire (Nichrome or Chromel) in the center bore. The receiver portion of the assembly contained two flasks, one to collect the material that distilled below 72° and another for the desired fraction. An air-cooled condenser was used with this apparatus, since water cooling often causes the product to solidify and clog the condenser tube.

4. If the product is to be used for the preparation of metal carbonyl derivatives, further usable material may be obtained from the mother liquors. Removal of pentane on a rotary evaporator leaves a yellow–orange, viscous oil that is suitable for most preparative purposes.

If an extremely pure product is required, the crystalline material can be sublimed at 35–40° (0.01 mm.).

5. IR (cyclohexane) cm.$^{-1}$: 1703, 1340, 1190, 1090, 960, 760; ^{1}H NMR (CCl$_4$): δ (number of protons, multiplicity): 1.0 (s, 3H), 1.5 (s, 3H), 1.7 (s, 6H), 1.9 (s, 6H).

3. Discussion

The present preparation of 5-acetyl-1,2,3,4,5-pentamethylcyclopentadiene is more reliable and convenient than that previously available.[2] This compound has been used to prepare many pentamethylcyclopentadienyl metal carbonyl derivatives[3] and is also a convenient source of pentamethylcyclopentadiene for use in preparing other (Me$_5$Cp)$_m$ML$_n$ derivatives.[4,5]

1. Department of Chemistry, University of Georgia, Athens, Georgia 30602.
2. H. N. Junker, W. Schafer, and H. Niedenbruck, *Chem. Ber.,* **100,** 2508 (1967).
3. R. B. King and A. Efraty, *J. Am. Chem. Soc.,* **94,** 3773 (1972).
4. R. B. King and M. Bisnette, *J. Organomet. Chem.,* **8,** 287 (1967).
5. U. Burger, A. Delay, and F. Mazenod, *Helv. Chim. Acta,* **57,** 2106 (1974); D. Feitler and G. M. Whitesides, *Inorg. Chem.,* **15,** 466 (1976).

2-ADAMANTANECARBONITRILE

[Tricyclo[3.3.1.1^{3,7}]decane-2-carbonitrile]

$$\text{(adamantanone)} + CH_3 -\!\!\left\langle\bigcirc\right\rangle\!\!- SO_2CH_2N\!=\!C + KOC(CH_3)_3 + C_2H_5OH \xrightarrow[5-35°]{\text{dimethoxy-ethane}}$$

$$\text{(2-adamantanecarbonitrile)} + CH_3 -\!\!\left\langle\bigcirc\right\rangle\!\!- SO_2K + H\overset{\overset{\displaystyle O}{\|}}{C}OC_2H_5 + HOC(CH_3)_3$$

Submitted by O. H. OLDENZIEL, J. WILDEMAN, and A. M. VAN LEUSEN[1]
Checked by TERESA Y. L. CHAN and S. MASAMUNE

1. Procedure

A 500-ml., three-necked, round-bottomed flask equipped with a mechanical stirrer, a thermometer, and a calcium chloride drying tube is charged with 15.0 g. (0.100 mole) of adamantanone (Tricyclo[3.3.1.1^{3,7}]decan-2-one), (Note 1), 25.4 g. (0.130 mole) of p-tolylsulfonylmethyl isocyanide (Notes 2, 3), 10 ml. (0.17 mole) of absolute ethanol (Note 4), and 350 ml. of 1,2-dimethoxyethane (Note 5). The stirred solution is cooled in an ice bath to 5°, and 28 g. (0.25 mole) of potassium *tert*-butoxide is added in portions at such a rate that the temperature is kept between 5° and 10° (Notes 6, 7). After the addition is complete, the ice bath is removed and stirring is continued for 30 minutes. The reaction mixture is heated for 30 minutes at 35–40°, the stirred suspension is cooled to room temperature, and the precipitated potassium p-toluenesulfinate is removed by filtration. The precipitate is extracted with three 50-ml. portions of 1,2-dimethoxyethane, and the combined 1,2-dimethoxyethane solutions are concentrated to a volume of 25–35 ml. on a rotary evaporator. The concentrated solution is chromatographed (Note 8) through a short column of alumina using distilled petroleum ether (b.p. 40–60°) as the eluent. The combined fractions are refluxed for 15 minutes with 1 g. of activated carbon (Note 9). After removal of the carbon, the solution is concentrated to dryness in a rotary evaporator. The white solid residue is dried overnight in a vacuum desiccator over silica gel, yielding 13.5–14.5 g. (84–90%) of analytically pure 2-adamantanecarbonitrile, m.p. 170–177° (Note 10).

2. Notes

1. Commercial adamantanone (Aldrich Chemical Company, Inc.) was used; the synthesis of adamantanone is described in *Org. Synth.*, **Coll. Vol. 6**, 48 (1988).

2. The synthesis of p-tolylsulfonylmethyl isocyanide is described in *Org. Synth.*, **Coll. Vol. 6**, 987 (1988).[2] The light-brown compound, m.p. 111–114°, was used without further purification.

3. p-Tolylsulfonylmethylisocyanide was used in slight excess to effect complete

conversion of the adamantanone, which otherwise is difficult to remove from the final product.

4. Commercial absolute ethanol was used.

5. Commercial 1,2-dimethoxyethane, "zur Synthese" quality, was purchased from E. Merck, Darmstadt.

6. Scoops of solid potassium tert-butoxide (purchased from E. Merck, Darmstadt, and specified to be at least 95% pure) were added over 20–30 minutes by temporarily removing the drying tube. At the beginning of the reaction much heat is evolved; therefore, the base should be added in small portions in order to keep the temperature below 10°. During the addition of the base, a precipitate is formed.

7. The reaction has also been carried out successfully with sodium ethoxide.[3]

8. The submitters recommend using a 5 cm. by 10 cm. column packed with 200 g. of neutral alumina (activity I) in petroleum ether (b.p. 40–60°), and eluting with 250-ml. of this solvent. The checkers have found that the elution may require more solvent depending on the amount of residual 1,2-dimethoxyethane: they recommend following the chromatography by GC analysis (see Note 9).

9. Treatment with activated carbon (purchased from J. T. Baker Chemical Company) can be omitted. In that case, removal of the solvent will provide 14–15 g. (87–93%) of a near-white product with a melting range of 160–180° (see Note 10). Despite this wide range, this material is over 99.8% pure, according to a GC analysis carried out at 190° on a 2-m. SE-30 column. This high degree of purity was confirmed on three different types of column.

10. Melting points of 2-adamantanecarbonitrile were determined in sealed tubes to prevent sublimation. Varying values were found for the melting point which apparently is not a very reliable indication of the purity of this compound. Occasionally, a value as high as 184–187° has been found by following the same procedure. Spectral properties of this product: IR (CHCl$_3$) cm.$^{-1}$: 2240 (CN); ^{1}H NMR (CDCl$_3$), δ (multiplicity, number of protons): 1.4–2.4 (m, 14H), 2.9 (m, 1H).

3. Discussion

The procedure described is an example of a more general synthetic method for the direct conversion of ketones into cyanides.[3-6] The reaction has been carried out successfully with acyclic and cyclic aliphatic ketones, including numerous steroidal ketones and aryl-alkyl ketones. The conversion of diaryl or highly hindered ketones such as camphor and β,β-dimethyl-α-tetralone requires the use of a more polar solvent. In those cases, the dimethoxyethane used in the present procedure should be replaced by dimethyl sulfoxide.[6] By introduction of a slight modification, the method applies to aldehydes also.[7]

2-Adamantanecarbonitrile was prepared previously by a more laborious method,[8] also starting from adamantanone, in 46% overall yield.

The hydrolysis of 2-adamantanecarbonitrile with hydrogen bromide in acetic acid provides a useful route to 2-adamantanecarboxylic acid (m.p. 143–144°),[9] which the submitters obtained in 95% yield. Stetter and Tillmans[8] reported a yield of 62% starting with impure 2-adamantanecarbonitrile.

1. Department of Organic Chemistry, Groningen University, Nijenborgh 16, 9747 RG Groningen, The Netherlands.
2. See also A. M. van Leusen, G. J. M. Boerma, R. B. Helmholdt, H. Siderius, and J. Strating, *Tetrahedron Lett.*, 2367 (1972); and A. M. van Leusen, R. J. Bouma, and O. Possel, *Tetrahedron Lett.*, 3487 (1975).
3. O. H. Oldenziel and A. M. van Leusen, *Synth. Commun.*, **2**, 281 (1972).
4. O. H. Oldenziel and A. M. van Leusen, *Tetrahedron Lett.*, 1357 (1973).
5. J. R. Bull and A. Tuinman, *Tetrahedron*, **31**, 2151 (1975).
6. O. H. Oldenziel, D. van Leusen, and A. M. van Leusen, *J. Org. Chem*, **42.**, 3114 (1977).
7. A. M. van Leusen and P. G. Oomkes, *Synth. Commun.*, **10**, 399 (1980).
8. H. Stetter and V. Tillmans, *Chem. Ber.*, **105**, 735 (1972).
9. For other syntheses, see A. H. Alberts, H. Wynberg, and J. Strating, *Synth. Commun.*, **2**, 79 (1972) and D. Fǎrcaşiu, *Synthesis*, 615 (1972).

TERTIARY ALCOHOLS FROM HYDROCARBONS BY OZONATION ON SILICA GEL: 1-ADAMANTANOL

(Tricyclo[3.3.1.1^{3,7}]decan-1-ol)

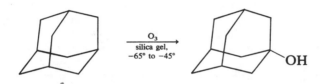

Submitted by Zvi Cohen, Haim Varkony, Ehud Keinan, and Yehuda Mazur[1]
Checked by Frank E. Blaney and Robert M. Coates

1. Procedure

Caution! Ozone is toxic and potentially explosive. This procedure should be carried out in an efficient hood and behind a suitable protective shield.

A solution of 6 g. (0.044 mole) of adamantane (Note 1) in 100 ml. of pentane and 500 g. of silica gel (Note 2) are placed in a 2-l., round-bottomed flask (Note 3). The pentane is removed by rotary evaporation at room temperature under reduced pressure (20 mm.), and the resulting dry silica gel is allowed to rotate for an additional 2 hours (Note 4). The adamantane–silica gel dispersion is poured through a powder funnel into the ozonation vessel (Note 5), which is then immersed in a 2-propanol–dry ice bath at $-78°$. A flow of oxygen is passed through the vessel at a rate of 1 l. per minute for 2 hours, after which the internal temperature reaches -60 to $-65°$ (Note 6). The ozone generator (Note 7) is turned on, and the ozone–oxygen mixture is passed through the vessel for *ca.* 2 hours, causing the silica gel to become dark blue (Notes 8 and 9). The cooling bath is removed, and the vessel is allowed to warm to room temperature in the hood over a 3-hour period.

The silica gel is transferred to a chromatography column, and the organic material is eluted with 3 l. of ethyl acetate. Evaporation of the solvent affords 6.1–6.4 g. of crude adamantanol (Note 10), which is dissolved in 200 ml. of 1:1 (v/v) dichloromethane–hexane by heating on a steam bath. The solution is filtered, concentrated to incipient crystallization and placed in a freezer at −20°. After a crop of fine, white needles (3.0–3.2 g.), m.p. 280–282° (sealed capillary), is collected, the mother liquor is concentrated and cooled to separate two additional crops, which give 2.2–2.6 g. and have melting point ranges of 270–274° to 275–280° (sealed capillary) (Note 11). The total yield of 1-adamantanol is 5.4–5.6 g. (81–84%) (Note 12).

2. Notes

1. Adamantane is available from Aldrich Chemical Company, Inc., and Fluka AG, Buchs, Switzerland.

2. Silica gel 60, with particle sizes ranging from 0.063 to 0.200 mm. (70–230 mesh), is suitable and may be purchased from Brinkmann Instruments, Inc., or E. Merck, Darmstadt, Germany. The submitters report that silica gel of this type normally contains *ca.* 5% water, which may be removed by drying at 300° for several hours, and that somewhat better yields are obtained when the silica gel is dried in this manner before use.

3. The submitters have found that the adsorption of adamantane on silica gel may also be accomplished by mixing the dry solids in a closed flask for a few hours.

4. Heating should be avoided to prevent loss of some of the adamantane through sublimation.

5. The submitters have used both a tightly closed, 1-l. gas-washing bottle and the apparatus shown in Figure 1 for ozonation vessels. They recommend that the glass joints not be greased. The apparatus used by the checkers consisted of a cylindrical, two-necked vessel having dimensions given in Figure 1. One neck of the vessel was fitted with a Claisen distillation head and the other with a thermometer with its bulb positioned in the middle of the vessel. A bent gas-dispersion tube with an extra-coarse sintered-glass frit extending through the vertical branch of the Claisen head to within 2–3 mm. of the center of the bottom of the flask served as the gas inlet. The curved branch of the Claisen head was fitted with a drying tube and this functioned as the gas exit.

6. The checkers found that the maintenance of a flow of oxygen during the cooling period prevented clogging of the glass frit and a building up of pressure in the gas-inlet tube in their apparatus.

7. A Welsbach T-816 Ozonator purchased from the Welsbach Corporation, Philadelphia, Pennsylvania, was used. The oxygen stream was dried by passage through dry silica gel and molecular sieves and introduced into the ozonator with the operating voltage set at 115 V., the gas pressure at 8 p.s.i.g., and the gas flow rate at 1 l. per minute. The resulting ozone flow rate was 0.00245 mole per minute, as determined by titration of a potassium iodide trap. *Org. Synth.,* **Coll. Vol. 5,** 489 (1973)][2].

8. In the apparatus used by the checkers, the internal temperature was between −45° and −65° while ozone was being passed through the silica gel. The use of lower bath temperatures results in the adsorption of a greater quantity of ozone on the silica gel; consequently, shorter reaction times and higher conversions were realized. *However, since ozone liquifies at −112°, there is a serious danger of explosion.*

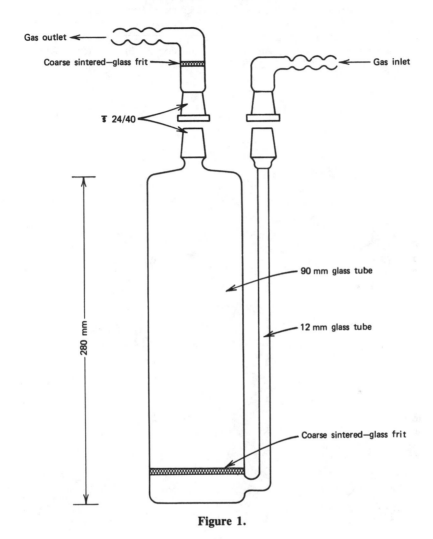

Figure 1.

9. The ozone flow is stopped when the silica gel reaches a constant, dark blue color. The time required for saturating the silica gel with ozone depends on the type of silica gel used and on whether it has been dried (Note 2).

10. A GC analysis on the crude adamantanol was carried out by the checkers using a 1.8 m. × 3 mm. column packed with 5% silicone oil (SE-30) supported on Chromosorb W and the following column temperature program: hold at 120° for 6 minutes and then increase at *ca*. 8° per minute. The chromatogram of the product from one run showed a major peak at retention time of 10 minutes and three minor peaks with retention times of 11.2, 12, and 13.7 minutes and relative areas amounting to 1.5, 1.6, and 4% of the major peak, respectively. A GC analysis by the submitters with 5% diethylene glycol succinate

TABLE I
PREPARATION OF TERTIARY ALCOHOLS FROM HYDROCARBONS
WITH OZONE ON SILICA GEL

Hydrocarbon	Tertiary Alcohol	Conversion (%)	Yield (%)[a]
		>99.5	65[b]
		72	79[c]
		92	76[d]
		>99.5	99
		88	72[e]
		>99.5	90

[a]Based on the amount of hydrocarbon consumed, as determined by GC.
[b]A mixture of the three methyl cyclohexanones was also formed to the extent of 34%.
[c]The epimeric alcohol was also present to the extent of 0.6%.
[d]The epimeric alcohol was also present to the extent of 3.5%.
[e]trans-1-Decalone (10%) and trans-2-decalone (16%) were also formed.

supported on Chromosorb W as a stationary phase at 110–160° showed peaks for adamantan-1,3-diol and adamantanone, as by-products totaling 7%, in addition to the peak for 1-adamantanol.

11. A GC analysis by the checkers (see Note 10) on the material in the third crop from one run showed a major peak for 1-adamantanol and a second minor peak having an area *ca.* 12% of that of the major peak. In another run the area of the peak from this by-product in the third crop was less than 2% relative to that of 1-adamantanol.

12. A yield of 5.8 g. (87%), m.p. 280–282°, was obtained by the submitters. The IR, ^{1}H NMR, ^{13}C NMR, and mass spectra of the product were identical to those of an authentic sample of 1-adamantanol. A mixed melting point with an authentic sample of 1-adamantanol showed no depression.

The spectral characteristics of the product are as follows: IR (KBr) cm.$^{-1}$: 3350(OH), 1455, 1352, 1302, 1118, 1088; ^{1}H NMR (CDCl$_3$), δ (multiplicity, number of protons, assignment): 1.53 (s, 1H, O*H*), 1.55–1.80 (m, 12H, 6 C*H*$_2$), 2.17 (broad s, 3H, 3C*H*); ^{13}C NMR (CDCl$_3$), δ (assignment): 30.7 (3*C*H), 36.1 (3*C*H), 36.1 (3*C*H$_2$), 45.4 (3*C*H$_2$), 68.2 (*C*OH).

3. Discussion

This "dry ozonation" procedure is a general method for hydroxylation of tertiary carbon atoms in saturated compounds (Table I).[3] The substitution reaction occurs predominantly with retention of configuration. Thus, *cis*-decalin gives *cis*-1-decalol, whereas *cis*- and *trans*-1,4-dimethylcyclohexane afford *cis*- and *trans*-1,4-dimethylcyclohexanol, respectively. The amount of epimeric alcohol formed in these ozonation reactions is usually less than 1%. The tertiary alcohols may be further oxidized to diols by repeating the ozonation; however, the yields in these reactions are poorer. For instance, 1-adamantanol is oxidized to 1,3-adamantanediol in 43% yield. Secondary alcohols are converted to the corresponding ketone. This method has been employed for the hydroxylation of tertiary positions in saturated acetates and bromides.

Dry ozonation may be carried out according to the following alternative procedure: The ozone–oxygen mixture is passed through the silica gel at −45°C followed by removal of the excess ozone at the same temperature (−45°C) by passing an inert gas (nitrogen or argon) through the sample.

1-Adamantanol has been prepared by oxidation of adamantane with peroxyacetic acetic[4] and by hydrolysis of 1-bromoadamantane with silver nitrate[5] or hydrochloric acid.[6]

1. Department of Organic Chemistry, Weizmann Institute of Science, Rehovot, Israel.
2. P. S. Bailey, *Chem. Rev., 58,* 986 (1958).
3. (a) Z. Cohen, E. Keinan, Y. Mazur, and T. H. Varkony, *J. Org. Chem., 40,* 2141 (1975); (b) Z. Cohen, E. Keinan, Y. Mazur, and A. Ulman, *J. Org. Chem., 41,* 2651 (1976); (c) E. Keinan and Y. Mazur, *Synthesis,* 523 (1976); (d) E. Keinan and T. H. Varkony, "Ozonization of Single Bonds," in S. Patai, Ed., "The Chemistry of Peroxides," Wiley-Interscience, New York, in press.
4. P. von R. Schleyer and R. D. Nicholas, *J. Am. Chem. Soc., 83,* 182 (1961).
5. H. Stetter, M. Schwarz, and A. Hirschhorn, *Chem. Ber., 92,* 1629 (1959).
6. H. W. Geluk and J. L. M. A. Schlatmann, *Tetrahedron, 24,* 5361 (1968).

ADAMANTANONE

(Tricyclo[3.3.1.1^{3,7}]decanone)

98% H$_2$SO$_4$ | 80°

Submitted by H. W. Geluk and V. G. Keizer[1]
Checked by L. Foley, W. Jackson, and A. Brossi

1. Procedure

Caution! This procedure should be carried out in an efficient hood to avoid exposure to sulfur dioxide.

Benzene has been identified as a carcinogen; OSHA has issued emergency standards on its use. All procedures involving benzene should be carried out in a well-ventilated hood, and glove protection is required.

A 1-l., three-necked, round-bottomed flask equipped with an efficient mechanical stirrer (Note 1), a thermometer, and a vent (Note 2) is placed in a water bath and charged with 600 ml. of 98% sulfuric acid (Note 3). Powdered adamantane [*Org. Synth.*, **Coll. Vol. 5,** 16 (1973)], (100g., 0.735 mole), is added in one portion to the stirred acid, and the mixture is heated rapidly (with the water bath) to an internal temperature of 70°. The internal temperature is then raised gradually to 80° over a 2-hour period (Note 4) while vigorous stirring is maintained (Note 5). After stirring at 80° is continued for an additional 2 hours, the temperature is raised to 82°. When almost all the adamantane is dissolved, the residual sublimed material is scraped and rinsed from the walls of the flask (Notes 1 and 6). When GC analysis indicates that 2–3% of adamantanol is present (Note 6), the hot reaction mixture is poured immediately onto 800 g. of crushed ice, giving a 1500-ml. suspension containing crude adamantanone.

A 750-ml. portion of this suspension of crude adamantanone is transferred to a 2-l., round-bottomed flask, equipped for steam distillation (Notes 7 and 8), which is placed in a heating mantle. The contents of the distillation flask are heated to 70°; the external heating is turned off (Note 9) and steam is introduced carefully through both inlet tubes (Note 10). The two layers of distillate are separated, and the aqueous layer is extracted with two 75-ml. portions of dichloromethane. This steam distillation procedure is then repeated with the second half of the suspension of crude adamantanone. The organic extracts are combined, washed with 100 ml. of aqueous, saturated sodium chloride, dried over anhydrous sodium sulfate, and evaporated under reduced pressure, yielding 52−53 g. (47−48%) of adamantanone (Notes 11, 12, and 13).

2. Notes

1. The stirrer should be placed just under the surface of the sulfuric acid. The flask should be filled to at least three-quarters of its volume so the sublimed material can be rinsed from the walls with vigorous stirring.

2. Sulfur dioxide generated during the reaction can escape through the third neck of the flask. This neck should be wide enough to prevent clogging by subliming adamantane.

3. The required amount of 98% sulfuric acid is prepared by adding 53 ml. of fuming sulfuric acid (65% free sulfur trioxide) to 530 ml. of 96% sulfuric acid, or by adding 120 ml. of fuming sulfuric acid (30−33% free sulfur trioxide) to 480 ml. of 96% sulfuric acid.

4. During this part of the reaction a vigorous evolution of sulfur dioxide takes place. Care should be taken to ensure that the internal temperature does not rise too fast to prevent an uncontrollable increase in the evolution of the sulfur dioxide.

5. An adamantane layer floating on the sulfuric acid, caused by ineffective stirring or too rapid evolution of sulfur dioxide, can lead to heavy foaming and subsequent losses. This layer, if formed, may be brought into contact with the reaction mixture by increasing the speed of the stirrer or by lowering the bath temperature to 65−70°.

6. At this stage the reaction should be monitored carefully by GC. For this purpose samples are taken periodically from the reaction mixture, poured onto ice, extracted with dichloromethane, washed with water, and subjected to GC. The submitters used an F and M Model 700 gas chromatograph equipped with a 2 m. by 3 mm. glass column, filled with 80/100 mesh Chromosorb-W-Hp impregnated with 9.5% Apiezon-L and 0.5% Carbowax-20 M, at 120° with a flame ionization detector and nitrogen as carrier gas. To obtain a good yield of a fairly pure product, the reaction should be stopped when the amount of 1-adamantanol remaining is between 2% and 3%. Prolonged heating will give a further reduction in the 1-adamantanol content but it should be emphasized that concurrently the yield of the adamantanone will diminish rapidly.

7. The flask is fitted with two inlet tubes, one narrow and adjustable positioned above the surface of the adamantanone suspension, and the other reaching half-way between the bottom of the flask and the surface. Both tubes are connected to the steam supply. The flask is connected through a splash head and a short adapter to a 1-l., three-necked, round-bottomed flask equipped with two very efficient reflux condensers each stoppered with a wad of cotton. The receiving flask is charged with 100 ml. of benzene, which, during the course of the steam distillation, begins to reflux and rinses the

adamantanone from the condensers. *Added in proof.* Because of the toxicity of benzene the submitters advise replacement of this solvent with ethyl acetate at this point and in Note 8.

8. The checkers found that it is advisable to use superheated steam and also to distill each portion twice. Thus when the first receiving flask fills with distillate, it is replaced with a second 1-l., round-bottom flask charged with 100 ml. of benzene and distillation is resumed.

9. Prolonged external heating will stimulate the foaming. Insulation by the heating mantle is sufficient to prevent extensive steam condensation in the distilling flask.

10. The short inlet tube above the surface (Note 7) will break the foam and enable smooth removal of the adamantanone. When foaming is very heavy, steam is introduced only through this short inlet tube.

11. The checkers found that an additional 3 g. of adamantanone could be obtained by extracting the combined contents of the two distillation flasks with dichloromethane, removing the solvent under reduced pressure and steam-distilling the residue, resulting in a total yield of 55–56 g. (50–51%).

12. The product is 97–98% pure by GC (Note 6) and is satisfactory for most purposes. If desired, the adamantanone may be purified by either column chromatography (alumina, activity grade IV; eluent: ether) or by treatment with fuming sulfuric acid (20% free sulfur trioxide). For example, 8.0 g. of adamantanone is added portionwise to 40 ml. of ice-cold, fuming sulfuric acid, and the solution is heated to 40° and maintained at this temperature for one hour. After pouring the mixture onto ice, the adamantanone is recovered by extraction with dichloromethane.

13. The IR spectrum (KBr) shows a strong band at 1717 with minor peaks or shoulders at 1670, 1690, 1725, and 1740 cm.$^{-1}$; ^{1}H NMR (CDCl$_3$), δ 2.04 (broad s, 12H, 2CH, 5CH_2) and 2.55 (broad s, 2H, 2CHC$=$O).

3. Discussion

Adamantanone is a very versatile starting material for the preparation of adamantane derivatives, especially those substituted at secondary carbon atoms.

The preparative method presented is a slight modification of that reported by Geluk and co-workers,[2,3] who also give a detailed account of the several reactions of adamantane, 1-adamantanol, and 2-adamantanol that take place in sulfuric acid.[2,4] Adamantanone can be prepared essentially as described herein starting with 1-adamantanol instead of adamantane[2]—the yield is better (70%) and the reaction time is shorter, but adamantane is a more suitable starting material.

Adamantane can also be oxidized to adamantanone with ozone,[5] and the oxime can be made directly from adamantane by photoöximation.[6]

1. Philips-Duphar B. V. Research Laboratories, Weesp, The Netherlands.
2. H. W. Geluk and J. L. M. A. Schlatmann, *Tetrahedron,* **24,** 5361 (1968).
3. H. W. Geluk and V. G. Keizer, *Synth. Commun.,* **2,** 201 (1972). This preparation appears in *Org. Synth., Coll. Vol. 6,* 48 (1988), by courtesy of Marcel Dekker, Inc., New York.
4. H. W. Geluk and J. L. M. A. Schlatmann, *Tetrahedron,* **24,** 5369 (1968).
5. S. Landa and L. Vodička, Czech. Pat. 119,348 (1966) [*Chem. Abstr.,* **67,** P21508v (1967)].
6. E. Müller and G. Fiedler, *Chem. Ber.,* **98,** 3493 (1965).

ENONE REDUCTION–ENOLATE ALKYLATION SEQUENCE: 2-ALLYL-3-METHYLCYCLOHEXANONE

[Cyclohexanone, 3-methyl-2-(3-propenyl)-]

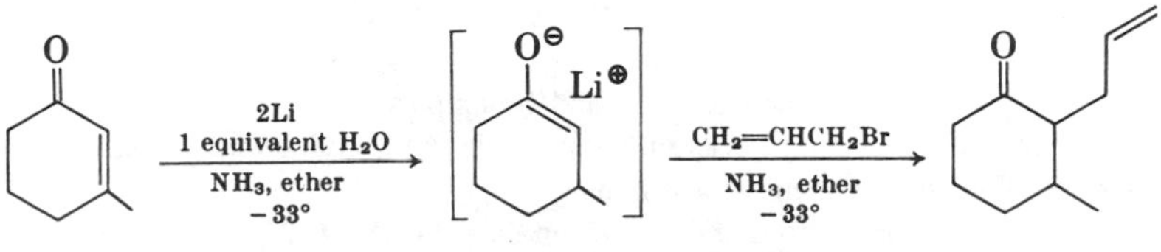

Submitted by Drury Caine,[1] Sam T. Chao,[1]
and Homer A. Smith[2]
Checked by Mark E. Jason and Kenneth B. Wiberg

1. Procedure

Caution! Since liquid ammonia is to be used, this preparation should be carried out in a well-ventilated hood.

A 2-l., three-necked, round-bottomed flask is fitted with a mechanical stirrer (Note 1), a pressure-equalizing dropping funnel, and a two-necked Claisen adapter holding a Dewar condenser (in the offset neck) and an adapter with stopcock for introduction of ammonia (Note 2). The entire apparatus is flame-dried under a stream of nitrogen, which enters at the top of the Dewar condenser, and a positive pressure of nitrogen is maintained throughout the experiment (Note 3). After anhydrous ammonia (1 l.) is introduced into the flask in liquid form (Note 4), the condenser is filled with an acetone–dry ice slurry, and the inlet adapter is removed and replaced with a glass stopper. Stirring is begun, and the ammonia is dried by adding very small pieces of lithium (*ca.* 5 mg. each) until the blue color persists (Note 5). Stirring is continued throughout the following reaction sequence.

Freshly-cut lithium wire (2.77 g., 0.400 g.-atom) (Note 6) is introduced into the flask and allowed to dissolve for 20 minutes, before a solution of 20.0 g. (0.182 mole) of 3-methyl-2-cyclohexen-1-one (Note 7) and 3.27 g. (0.182 mole) of water in 400 ml. of anhydrous diethyl ether (Notes 8 and 9) is added dropwise over 60 minutes (Note 10). Ten minutes after completion of the addition, a solution of 65 g. (0.54 mole) of allyl bromide (Note 11) in 150 ml. of anhydrous ether is added from the dropping funnel in a stream over 60 seconds (Note 12). Five minutes later 30 g. of ammonium chloride is added as rapidly as possible. Stirring is discontinued, the Dewar condenser is removed, and the ammonia is allowed to evaporate. A mixture of 300 ml. of ether and 300 ml. of water is added, and the reaction mixture is transferred to a separatory funnel and shaken. After separation of the ether layer, the aqueous layer is saturated with sodium chloride and extracted with two 100-ml. portions of ether. The combined ether extracts are washed with 100 ml. of 5% hydrochloric acid and 100 ml. of aqueous saturated sodium chloride and dried over anhydrous magnesium sulfate. After filtration, ether is removed with a rotary evaporator. The residual oil is distilled through a 37-cm. column packed with 6-mm. Raschig rings and equipped with a resistance heater for thermal balance. There is a forerun, b.p. 74–80° (12 mm.), consisting principally of 3-methylcyclohexanone, before 12.3–13.6 g. (45–49%) (Note 13) of 2-allyl-3-methylcyclohexanone distils as a 20:1 mixture of *trans*- and *cis*-isomers, b.p. 99–102° (12 mm.), n_D^{24} 1.4680–1.4683 (Note 14).

2. Notes

1. A glass stirring shaft and blade should be employed.

2. An apparatus with four necks makes possible the introduction of solid reactants without removal of the condenser or dropping funnel.

3. An apparatus similar to that described in *Org. Synth.*, **Coll. Vol. 4,** 132 (1963). was used for maintaining a nitrogen atmosphere in the system. A positive nitrogen flow was maintained whenever the system was open.

4. Some commercial cylinders are provided with an eductor tube, which allows the extraction of liquid ammonia when the cylinder is upright. Cylinders not so equipped may be secured in a suitable wooden or metal cradle, constructed so that the outlet valve is inclined below the body of the cylinder. The checkers dried commercial ammonia with sodium in a separate flask, then distilled the ammonia into the reaction flask.

5. Although water was used as a proton donor in this preparation, drying of the liquid ammonia and ether was carried out so that the lithium–ammonia solution would not be exposed to water during preparation and so that the amount of water added could be accurately controlled.

6. Lithium wire, purchased from the Lithium Corporation of America, Inc., was cut in 2.5–5.0-cm. pieces, washed free of the protective oil with ether, and weighed by transferring quickly to a tared beaker containing mineral oil. The lengths of wire were again washed in a beaker of ether and then held with forceps over an open neck of the reaction flask (positive nitrogen flow), cut with scissors into 3–4-mm. pieces, and allowed to fall directly into the flask.

7. 3-Methyl-2-cyclohexen-1-one, b.p. 76–78° (11 mm.), was purchased from Aldrich Chemical Company, Inc., or prepared according to the procedure of Cronyn and Riesser.[3]

8. Ether was heated to reflux over lithium aluminum hydride and distilled prior to use.

9. The three-component mixture required a few minutes of shaking to ensure the formation of a homogeneous solution. It is essential that the water be completely dissolved before the solution is used.

10. The deep blue solution of lithium in liquid ammonia generally turned pale blue at the end of the addition, and a white precipitate of the lithium enolate and/or lithium hydroxide could be seen.

11. Allyl bromide purchased from Eastman Organic Chemicals, Inc., was dried over calcium chloride and distilled, b.p. 69–70° (760 mm.), prior to use.

12. Since the ammonia boils vigorously during this addition, it is advisable to open one neck of the flask to avoid excessive pressure buildup.

13. The checkers obtained an additional 2–5 g. of 2-allyl-3-methylcyclohexanone by redistilling the column wash and the pot residue through a short-path distillation column, raising the total yield to 14.9–18.1 g. (54–66%).

14. IR (neat) cm.$^{-1}$: 1714 (C=O), 1641 (C=C); ^{1}H NMR (CCl$_4$), δ (number of

protons, multiplicity): 0.82 and 1.10 (two d's, total of 3H, CH_3 of *cis*- and *trans*-isomers, respectively), 1.4–2.7 (m, 10H), 5.1 (two t's, total of 2H, $-CH=CH_2$), 5.8 (m, 1H, $-CH=CH_2$). The distillate was analyzed by GC on a 3 mm. by 2 m. column containing 10% Carbowax 20M on HMDS Chromosorb W (60–80 mesh). Using a column temperature of 104° and a carrier gas flow rate of 30 ml./minute, the retention times for the *trans*- and *cis*-isomers were 9.0 minutes and 11.2 minutes, respectively. The product contained greater than 98% of the mixture of *trans*- and *cis*-2-allyl-3-methylcyclohexanone.

A portion of the product was heated to reflux with methanolic sodium methoxide to convert it into the thermodynamic mixture of *trans*- (*ca.* 65%) and *cis*- (*ca.* 35%) isomers. Small amounts of the isomers were collected by preparative GC using an 8 mm. by 1.7 m. column containing 15% Carbowax 20M on Chromosorb W, and each isomer exhibited the expected spectral and analytical properties. The same thermodynamic mixture of isomers was prepared independently by lithium–ammonia reduction[4] of 2-allyl-3-methylcyclo-2-hexen-1-one,[5] followed by equilibration with methanolic sodium methoxide.

Using the 3 mm. by 2m. column described above, a mixture of stereoisomers of 2-allyl-5-methylcyclohexanone, prepared by allylation of the enamine of 3-methylcyclohexanone,[6a] showed peaks at retention times of 8.4 minutes (more stable isomer) and 9.6 minutes. A mixture of the two isomeric 2-allyl-3-methylcyclohexanones and the two isomeric 2-allyl-5-methylcyclohexanones clearly exhibited four distinct peaks.

3. Discussion

Alternative preparations of 2-allyl-3-methylcyclohexanone include (a) lithium–ammonia reduction of 2-allyl-3-methyl-2-cyclohexen-1-one (see Note 13), which can be prepared by alkylation of 3-methyl-2-cyclohexen-1-one or by alkylation of ethyl 2-methyl-4-oxo-2-cyclohexene-1-carboxylate (Hagemann's ester), followed by hydrolysis and decarboxylation; and (b) conjugate addition of lithium dimethylcuprate to 2-cyclohexen-1-one followed by trapping of the enolate with allyl iodide[6b] or allyl bromide[6c] in an appropriate solvent.

The present procedure utilizes the enone reduction–enolate alkylation procedure discovered by Stork and co-workers[7] and provides a means of directing alkylation to relatively inaccessible α-positions of unsymmetrical ketones. This procedure has been applied to the synthesis of specifically alkylated steroidal ketones,[8] decalones,[7] hydrindanones,[7] and monocyclic[4] and acyclic ketones.[9] The method is successful because, unlike enolates with other alkali metal counterions, lithium enolates of unsymmetrical ketones undergo alkylation with relatively reactive alkylating agents more rapidly than they equilibrate to isomeric enolates by proton transfer.

The direct conversion of 3-methyl-2-cyclohexen-1-one into 2-allyl-3-methylcyclohexanone provides an interesting example of the utility of the reduction–alkylation procedure. Synthesis of this compound from 3-methylcyclohexanone would be difficult because the latter is converted mainly into 2-alkyl-5-methylcyclohexanones either by direct base-catalyzed alkylation[10] or by indirect methods such as alkylation of its enamine (see Note 13) or alkylation of the magnesium salt derived from its cyclohexylimine.[11]

The procedure may also be extended to other alkylating agents and enones. We have reported[4] that 2,3-dimethylcyclohexanone can be prepared by reduction–methylation of 3-methyl-2-cyclohexen-1-one and have found that 2-benzyl-3-methylcyclohexanone is obtained by substituting benzyl chloride for allyl bromide in a procedure similar to that described above, but employing a longer alkylation time.[12] In a related case, Conia and Berlin[13] have prepared 3-(3-butenyl)-2-methylcyclohexanone by reduction–methylation of 3-(3-butenyl)-2-cyclohexen-1-one.

High-boiling products found in this procedure and in similar experiments involving cyclohex-2-enone derivatives[4] probably result from bimolecular reduction processes.[14] 3-Methylcyclohexanone, which arises by protonation rather than alkylation of the enolate (and which made up *ca.* 12% of the volatile products), is probably the result of reaction of allyl bromide with liquid ammonia to form the acidic species allyl ammonium bromide.[4,9]

Unless a proton donor is added, the lithium–ammonia reduction of an enone leads to the lithium enolate and lithium amide. The latter is a sufficiently strong base to rapidly convert the mono-alkylated ketone into its enolate (see Figure 1), which can be further alkylated. The function of the added proton donor is to buffer the system by reaction with the lithium amide. *tert*-Butyl alcohol has been widely used as a proton donor in lithium–ammonia reductions,[4,7,9] but lithium *tert*-butoxide is also a sufficiently strong base to cause a significant amount of polyalkylation in easily ionizable systems. In a procedure similar to that above but in which *tert*-butyl alcohol was substituted for water, 2,2- and/or 2,6-diallyl-3-methylcyclohexanone made up 15–20% of the volatile products. However, when water was used as the proton donor so that lithium hydroxide was the strongest base present during the alkylation step, the amount of diallyl product was 2% or less.

It is interesting that *trans*-2-allyl-3-methylcyclohexanone is by far the major product of this reduction–alkylation sequence, being formed in greater than the equilibrium ratio (see Note 13). The lithium enolate would be expected to exist in the two conformations shown in Figure 2. The conformation having the methyl group quasiaxial may be quite important, because $A^{1,2}$-strain may be significant when the methyl group is

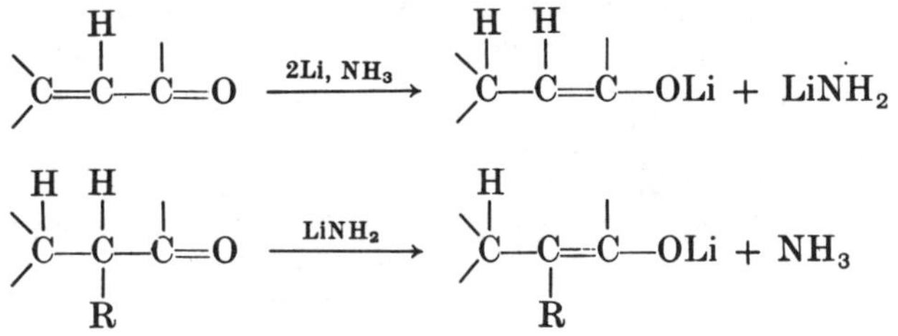

Figure 1.

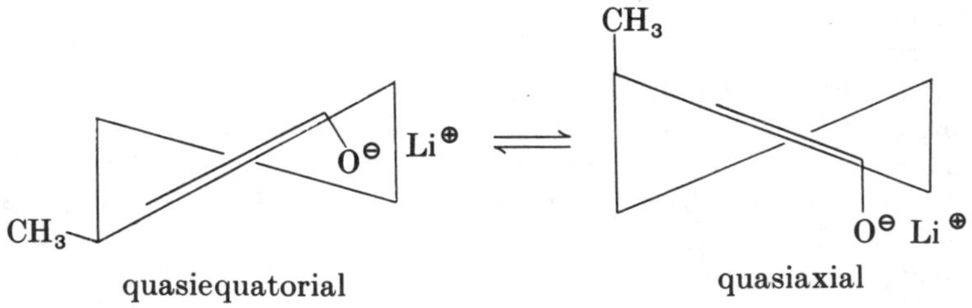

quasiequatorial   quasiaxial

Figure 2.

quasiequatorial.[15] Recent work on the alkylation of lithium enolates of cyclohexanone derivatives has revealed that the stereochemistry of the reaction is governed largely by steric factors within the reactants.[16] From examination of models of the two conformations shown above, it appears that steric interactions between the methyl group and the approaching alkylating agent would be minimized in either of the two possible transition states which lead to the *trans* product.

1. School of Chemistry, Georgia Institute of Technology, Atlanta, Georgia 30332.
2. Present Address: Department of Chemistry, Hampden-Sydney College, Hampden-Sydney, Virginia 23943.
3. M. W. Cronyn and G. H. Reisser, *J. Am. Chem. Soc.*, **75**, 1664 (1953).
4. H. A. Smith, B. J. L. Huff, W. J. Powers III, and D. Caine, *J. Org. Chem.*, **32**, 2851 (1967).
5. J. M. Conia and F. Rouessac, *Bull. Soc. Chim. Fr.*, 1925 (1963).
6. (a) S. K. Malhotra, D. F. Moakley, and F. Johnson, *Chem. Commun.*, 448 (1967); (b) R. K. Boeckman, Jr., *J. Org. Chem.*, **38**, 4450 (1973); (c) R. M. Coates and L. O. Sandefur, *J. Org. Chem.*, **39**, 277 (1974).
7. G. Stork, P. Rosen, N. L. Goldman, R. V. Coombs, and J. Tsuji, *J. Am. Chem. Soc.*, **87**, 275 (1965).
8. C. Djerassi, Ed., "Steroid Reactions," Holden-Day, San Francisco, 1963, p. 322.
9. L. E. Hightower, L. R. Glasgow, K. M. Stone, D. A. Albertson, and H. A. Smith, *J. Org. Chem.*, **35**, 1881 (1970).
10. R. Cornubert and R. Humeau, *Bull. Soc. Chim. Fr.*, **49**, 1238 (1931).
11. G. Stork and S. R. Dowd, *J. Am. Chem. Soc.*, **85**, 2178 (1963).
12. H. A. Smith, unpublished work.
13. J. M. Conia and P. Berlin, *Bull. Soc. Chim. Fr.*, 483 (1969).
14. (a) K. W. Bowers, R. W. Giese, J. Grimshaw, H. O. House, N. H. Kolodny, K. Kronberger, and D. K. Roe, *J. Am. Chem. Soc.*, **92**, 2783 (1970); (b) H. O. House, R. W. Giese, K. Kronberger, J. P. Kaplan, and J. P. Simeone, *J. Am. Chem. Soc.*, **92**, 2800 (1970).
15. F. Johnson, *Chem. Rev.*, **68**, 375 (1968).
16. (a) H. O. House, B. A. Tefertiller, and H. O. Olmstead, *J. Org. Chem.*, **33**, 935 (1968); (b) B. J. L. Huff, F. N. Tuller, and D. Caine, *J. Org. Chem.*, **34**, 3070 (1969).

PREPARATION OF *N*-AMINOAZIRIDINES:
trans-1-AMINO-2,3-DIPHENYLAZIRIDINE, 1-AMINO-2-PHENYLAZIRIDINE, AND 1-AMINO-2-PHENYLAZIRIDINIUM ACETATE

(1-Aziridinamine, *trans*-(±)-2,3-diphenyl-,
1-aziridinamine, (±)-2-phenyl-, and
1-aziridinamine, monoacetate, (±)-2-phenyl-)

Submitted by Robert K. Müller, Renato Joos, Dorothee Felix,
Jakob Schreiber, Claude Wintner, and A. Eschenmoser[1]
Checked by Robert E. Ireland, J. Kleckner, and David M. Walba

1. Procedure

Caution! Steps B *and* D *should be carried out in a hood behind a safety screen.*

A. trans-2,3-*Diphenyl*-1-*phthalimidoaziridine* (Note 1). A 1-l., three-necked, round-bottomed flask equipped with an efficient mechanical stirrer is charged with a mixture of 19.5 g (0.120 mole) of *N*-aminophthalimide (Note 2), 108 g. (0.600 mole) of (*E*)-stilbene (Note 3), and 300 ml. of dichloromethane (Note 4). To the resulting suspension is added 60 g. (*ca.* 0.12 mole) of lead tetraäcetate (Note 5) over a period of 10 minutes, at room temperature with *vigorous* stirring. Stirring is continued for an additional 30 minutes, after which time the mixture is filtered through Celite®, which is washed twice with 50-ml. portions of dichloromethane. The combined filtrates are transferred to a 4-l. beaker, and 1.5 l. of pentane is added with gentle stirring and cooling in an ice bath. The yellow precipitate that forms after 15 minutes is suction filtered and redissolved in 200 ml. of dichloromethane. The solution obtained is swirled for 5 minutes with 20 g. of silica gel, filtered through Celite®, which is washed with two 100-ml. portions of dichloromethane. To this dichloromethane solution is added 1.5 l. of pentane, with cooling. The precipitate that forms after 0.5 hour is dried under vacuum (10 mm.) at room temperature for 5 hours, yielding 16–21 g. (39–51%) of product, m.p. 175°, which is of sufficient purity for use in the next step (Note 6).

B. trans-1-*Amino*-2,3-*diphenylaziridine* (Note 7). To a magnetically stirred suspension of 20.4 g. (0.0600 mole) of *trans*-2,3-diphenyl-1-phthalimidoaziridine in 150 ml. of 95% ethanol in a 500-ml., round-bottomed flask at room temperature is added 150 ml. (3 moles) (Note 8) of hydrazine hydrate (Note 9). The mixture is stirred for 40 minutes while maintaining the temperature at 43–45° (Note 10) with a thermostatted oil bath. The resulting cloudy yellow solution (Note 11) is cooled and filtered through Celite®. The filtrate is poured into a 2-l. separatory funnel containing 400 ml. of diethyl ether and 200 g. of ice, and shaken vigorously. The organic phase is separated and washed with three 200-ml. portions of ice-cold water, and the aqueous washings are reextracted with a 250-ml. portion of ether (Note 12). The combined ethereal extracts are dried over anhydrous potassium carbonate, filtered through Celite® if necessary, and concentrated to approximately 300 ml. on a rotary evaporator at room temperature. Addition of 400 ml. of pentane and overnight storage at −20° leads to crystallization of 7.9–9.5 g. (63–75%) of *trans*-1-amino-2,3-diphenylaziridine (Note 13) as colorless crystals, m.p. 93–94° (dec.) (Note 14). An additional 1.2–2.2 g. (10–17%) is obtained by concentration of the mother liquor to 50–80 ml. and addition of approximately 50 ml. of pentane (Note 7).

C. *Styrene glycol dimesylate* (Note 15). A 300-ml., three-necked, round-bottomed flask equipped with a thermometer, an efficient stirrer (Note 16), and a dropping funnel is charged with a solution of 34.5 g. (0.250 mole) of styrene glycol (Note 17) in 90 ml. of pyridine (Note 18). The solution is cooled to −5° with an ice–salt bath, and 64.7 g. (43.7 ml., 0.560 mole) of methanesulfonyl chloride (Note 19) is added dropwise over a 1-hour period, while maintaining the temperature at or below 0° (Note 20). Stirring is continued for 4 hours at 2–4°, with the flask cooled with an ice-water bath. The reaction mixture is mixed thoroughly with 600 g. of ice, and the dimesylate precipitates. After careful acidification of the mixture with 6 N hydrochloric acid to approximately pH 3 (Note 21), the dimesylate is suction filtered, washed twice with 100-ml. portions of ice water, and pressed as dry as possible. This product, which is still moist, is transferred to a separatory funnel and is shaken well with 200 ml. of dichloromethane. The dichloromethane is separated, and the aqueous layer is extracted further with two 20-ml. portions of dichloromethane. The combined dichloromethane layers are dried over anhydrous magnesium sulfate, and 250–300 ml. of pentane is added to the solution, until crystallization just begins. After 2 hours in a deep freeze at −25°, the crystals are collected, washed with two 30-ml. portions of pentane precooled to 0°, and dried to constant weight in a vacuum desiccator (10 mm.) at room temperature, yielding 62–64 g. (84–86%) of white, crystalline dimesylate, m.p. 93–94° (Note 15).

D. *Hydrazinolysis of styrene glycol dimesylate.* A 1-l., round-bottomed flask equipped with a magnetic stirring bar is charged with 50 ml. (1 mole) of hydrazine hydrate (Note 9). Finely powdered styrene glycol dimesylate (20 g., 0.068 mole), m.p. 93–94°, is then added with gentle stirring at room temperature. To the resulting slurry is slowly added 600 ml. of pentane. The stirring speed should be adjusted in such a manner that the two phases mix somewhat, but the dimesylate–hydrazine hydrate layer is not deposited on the upper walls of the flask. After 20–24 hours of stirring at room temperature two entirely clear layers can be observed upon the cessation of stirring (Note 22), indicating that the reaction is complete. The hydrazine hydrate is separated from the pentane and extracted with two 30-ml. portions of pentane. The combined pentane layers are filtered through cotton, which holds back any remaining droplets of hydrazine hydrate, into a 1-l., round-bottomed flask. At this point *Step E* is followed for 1-amino-2-phenylaziridine, and *Step F* for 1-amino-2-phenylaziridinium acetate.

E. 1-*Amino-2-phenylaziridine* (Note 23). If the pentane solution from *Step D* is removed on a rotary evaporator at room temperature, 7.5–7.7 g. (82–85%) of 1-amino-2-phenylaziridine, suitable for preparative use, is obtained. Kügelrohr distillation of this material on a 1–2 g. scale (0.01 mm./60–65° oven temperature) (Note 24) gives a recovery of over 90% (Notes 23 and 25).

F. 1-*Amino-2-phenylaziridinium acetate* (Note 23). The pentane solution from *Step D* is stirred with a magnetic stirrer and cooled to 0°, and 3.9 ml. (0.068 mole) of acetic acid is measured. Three drops of acetic acid are added at first, and stirring is continued at 0° until the precipitation of white 1-amino-2-phenylaziridinium acetate begins. If necessary, crystallization is initiated by scratching with a glass rod or by addition of a seed crystal from a previous run. The remainder of the acetic acid is added over a 10-minute period, and stirring is continued for an additional 20 minutes, while maintaining the temperature at 0°. The salt is filtered, washed with 30 ml. of pentane precooled to 0°, and dried in a vacuum desiccator (10 mm.) at room temperature, yielding 10.0–10.5 g. (76–79%) (Note 26) of product, m.p. 69–70° (Note 27) which is suitable for preparative purposes. Recrystallization is possible; however, it must be carefully carried out to avoid the formation of a yellow product, whose melting point is lower than that of the crude product. A solution of 10 g. of 1-amino-2-phenylaziridinium acetate in 40 ml. of dichloromethane is prepared at a maximum temperature of 20–22°. The turbid solution is immediately filtered through Celite®, which is washed with two 10-ml.

portions of dichloromethane. The resulting clear solution is treated with 200–250 ml. of pentane until crystallization just begins, and placed in a deep freeze at −25° for 2 hours. Filtration, washing with 30 ml. of pentane precooled to 0°, and drying as before, afford 9.2–9.3 g. of 1-amino-2-phenylaziridinium acetate, m.p. 70–72° (Notes 23 and 28).

2. Notes

1. *trans*-2,3-Diphenyl-1-phthalimidoaziridine is available from Fluka AG, CH-9470 Buchs.

2. *N*-Aminophthalimide is available from Fluka AG or may be prepared from phthalimide and hydrazine.[2] The quality is important; the m.p. should be 199–202° with subsequent resolidification of the melt due to thermal reaction. Recrystallization, if necessary, can be carried out in ethanol. The checkers observed that with one batch of recrystallized material, the solid never really did melt, but seemed to sinter at ~200°.

3. Technical grade (*E*)-stilbene obtained from Fluka AG gives satisfactory results, although a better grade is preferable. The checkers used reagent grade material obtained from Aldrich Chemical Company, Inc.

4. This was distilled over phosphorus pentoxide.

5. "Purum" grade lead tetraäcetate, 85–90%, moistened with acetic acid, obtained from Fluka AG was used. The checkers used reagent grade material, moistened with acetic acid, purchased from Matheson, Coleman and Bell.

6. The yield of the reaction, while always at least 39%, is subject to fluctuation. The product may be contaminated with small amounts of (*E*)-stilbene and/or lead salts. The presence of (*E*)-stilbene can easily be monitored by TLC, using ready-prepared Silica Gel F_{254} plates available from E. Merck & Company, Darmstadt, Germany. The plates are developed with dichloromethane, and the spots detected under UV light. In runs of smaller scale, or if product of higher purity is desired, column chromatography on silica gel may replace the work-up described; the reaction mixture is filtered through Celite®, and the resulting solution is concentrated on a rotary evaporator. The residue is then chromatographed on 570 g. of silica gel. Eluting with dichloromethane gives the stilbene first, then small amounts of unidentified impurities, and finally the desired adduct in 70–73% yield.

7. *trans*-1-Amino-2,3-diphenylaziridine decomposes thermally, largely to (*E*)-stilbene and nitrogen. In the crystalline state it is stable for several hours at room temperature, and for several weeks (probably for several months) at −20°. It decomposes within 3 days at room temperature in aprotic solvents and much more rapidly in protic solvents.[3] Decomposition is still faster in the presence of traces of acids. The work-up described should be completed as quickly and as *precisely* as possible, and the product should be stored in a deep freeze if it is not to be used immediately. For the use of this reagent in the α,β-epoxyketone fragmentation, see references 3 and 4.

8. Use of lesser amounts of hydrazine hydrate or ethanol causes precipitation of the product as a pasty mass that dissolves only slowly in ether, thereby making the work-up difficult.

9. "Purum" grade hydrazine hydrate obtained from Fluka AG was used. The checkers used 99–100% hydrazine hydrate purchased from Matheson, Coleman and Bell.

10. The reaction temperature must be carefully controlled. At temperatures above 48°

the yield is markedly reduced by decomposition of the product, and below 43° the reaction time is greatly lengthened.

11. The submitters report that if very pure trans-2,3-diphenyl-1-phthalimidoaziridine is used, no insoluble matter is present at this point, and filtration through Celite® is not necessary.

12. Before working up the reaction mixture, it is recommended to test whether the reaction is complete by TLC (Note 6). If (E)-stilbene is observed, the reaction should be interrupted immediately. The checkers found that the work-up was complicated by formation of emulsions; small quantities of brine were used to aid separation of the phases.

13. IR (CHCl$_3$) cm.$^{-1}$: 3340, 1603, 1495, 1450, 1085, 1070, 1030 (a weak band at 960 cm.$^{-1}$ is due to (E)-stilbene, as there is always some decomposition of the trans-1-amino-2,3-diphenylaziridine during the recording of the spectrum) (Note 7); ^{1}H NMR (CDCl$_3$), δ (multiplicity, coupling constant J in Hz., number of protons, assignment): 3.10 (broad m, 2H, NH$_2$), 3.22 and 3.36 (AB q, 2H, $J = 5$), 7.1–7.6 (m, 10H). The nonequivalence of the two methine protons is due to slow inversion at nitrogen, and confirms the trans-2,3-substitution.

14. The melting-point tube is placed in the bath at 85° and heated rapidly.

15. A second crystallization from dichloromethane–pentane is sometimes necessary to achieve material having this melting point. Styrene glycol dimesylate must be stored in a refrigerator, since slow decomposition takes place at room temperature.

16. Toward the end of the reaction, the mixture becomes quite viscous. Unless the stirring assembly is capable of mixing material at the flask walls, homogeneous temperature control cannot be guaranteed.

17. Styrene glycol is available from Aldrich Chemical Company, Inc., or from Eastman Organic Chemicals. Alternatively, it may be prepared by hydrolysis of styrene oxide.[3] If the glycol melts at lower than 63°, it should be recrystallized before use.

18. Dried over potassium hydroxide and distilled.

19. "Purum" grade methanesulfonyl chloride supplied by Fluka AG or 98% pure material supplied by Eastman Organic Chemicals was used.

20. The addition rate is about 30 drops/minute. After about half of the methanesulfonyl chloride has been added, white crystals of pyridine hydrochloride begin to precipitate, and the solution becomes viscous (Note 16).

21. About 110–120 ml. of 6 N hydrochloric acid is needed. The temperature should not be allowed to rise above 5°.

22. The rate of the reaction is influenced by the speed of stirring. At slow speeds, 30 hours may be required for completion without any decrease in the final yield. The role of pentane is to continuously remove newly formed product from the hydrazine solvent.

23. 1-Amino-2-phenylaziridine decomposes at temperatures over 0°, and must therefore be stored in a deep freeze at −25°, at which temperature it is crystalline. 1-Amino-2-phenylaziridinium acetate is somewhat more stable, but it too decomposes within 2 days at room temperature. It can be kept unchanged in a deep freeze for months. For the use of these two reagents in the α,β-epoxyketone fragmentation, see references 3 and 4.

24. In order to minimize decomposition, the distillation should be carried out on small portions at the lowest possible temperature and pressure.

25. The distilled product has the following ^{1}H NMR spectrum (CDCl$_3$), δ (number of

protons, assignment): 1.91, 1.98, and 2.06 (2H, AB part of ABX), 2.50, 2.58, 2.64, and 2.72 (1H, X part of ABX), 3.60 (2H, NH_2), 7.25 (5H). Undistilled material shows substantially the same spectrum.

26. By the addition of two further drops of acetic acid to the mother liquor, and overnight cooling at $-25°$, an additional 0.3–0.4 g. (2–3%) of product, m.p. 60–62°, can be isolated.

27. For the melting point determination the capillary is placed in the apparatus at 60°, and the temperature is raised 4°/minute.

28. Recrystallized 1-amino-2-phenylaziridinium acetate has the following ^{1}H NMR spectrum (CDCl$_3$), δ (number of protons, assignment): 1.95–2.20 (5H, AB part of ABX and CH_3 at δ 2.02), 2.67–2.95 (1H, X part of ABX), 6.50–6.70 (3H, NH_3) 7.0–7.4 (5H).

3. Discussion

trans-1-Amino-2,3-diphenylaziridine and 1-amino-2-phenylaziridine are reagents for the α,β-epoxyketone → alkynone fragmentation,[3] an example of which is given in this volume.[4] An alternative preparation of 1-amino-2-phenylaziridine is the hydrazinolysis of 2-phenyl-1-phthalimidoaziridine.[3]

The lead tetraäcetate reaction between *N*-aminophthalimide and (*E*)-stilbene was first described by Rees,[5] and the hydrazinolysis of the addition product by Carpino.[6] The procedures described here incorporate their methods, with improvements. The dimesylate–hydrazine reaction was first described by Paulsen[7] in the carbohydrate series.

1. Laboratorium für Organische Chemie, Eidgenössische Technische Hochschule, CH-8006 Zürich, Switzerland.
2. H. D. K. Drew and H. H. Hatt, *J. Chem. Soc.*, 16 (1937).
3. D. Felix, R. K. Müller, U. Horn, R. Joos, J. Schreiber, and A. Eschenmoser, *Helv. Chim. Acta*, **55**, 1276 (1972).
4. D. Felix, C. Wintner, and A. Eschenmoser, *Org. Synth.*, **Coll. Vol. 6**, 679 (1988).
5. D. J. Anderson, T. L. Gilchrist, D. C. Horwell, and C. W. Rees, *J. Chem. Soc. C*, 576 (1970).
6. L. A. Carpino and R. K. Kirkley, *J. Am. Chem. Soc.*, **92**, 1784 (1970).
7. H. Paulsen and D. Stoye, *Angew. Chem.*, **80**, 120 (1968) [*Angew. Chem. Int. Ed. Engl.*, **7**, 134 (1968)]; *Chem. Ber.*, **102**, 820 (1969).

REDUCTION OF KETONES BY USE
OF THE TOSYLHYDRAZONE DERIVATIVES:
ANDROSTAN-17β-OL

[Androstan-17-ol, (5α, 17β)-]

Submitted by L. Caglioti[1]
Checked by J. F. Moser and A. Eschenmoser

1. Procedure

A 100-ml., round-bottomed flask equipped with a reflux condenser is charged with 1.00 g. (0.00345 mole) of 5α-androstan-17β-ol-3-one (Note 1), 0.90 g. (0.0048 mole) of tosylhydrazide (Note 2), and 70 ml. of methanol (Note 3). The mixture is heated under gentle reflux for 3 hours, then cooled to room temperature. To the solution is added 2.5 g. (0.075 mole) of sodium borohydride in small portions over one hour (Note 4) and the resulting mixture is heated under reflux for an additional 8 hours. The reaction mixture is cooled to room temperature before the solvent is removed under reduced pressure. The residue is dissolved in diethyl ether, transferred to a separatory funnel, and washed successively with water, dilute aqueous sodium carbonate, 2 M hydrochloric acid, and water. The ethereal solution is dried over anhydrous sodium sulfate and evaporated under reduced pressure. The residue, 0.95 g. of white crystals (Note 5), is dissolved in about 20 ml. of a 7:3 (v/v) mixture of cyclohexane-ethyl acetate and applied to a column packed with 60 g. of silica gel (Merck, 0.05–0.2 mm.). The column is eluted with the 7:3 (v/v) cyclohexane-ethyl acetate mixture and a 200-ml. fraction is collected. Evaporation of this fraction under reduced pressure affords 0.70–0.73 g. (73–76%) of pure 5α-androstan-17β-ol. Recrystallization from aqueous methanol provides 0.64 g. of analytically pure product, m.p. 161–163°.

2. Notes

1. 5α-Androstan-17β-ol-3-one was supplied by Aldrich Chemical Co., Inc.

2. Tosylhydrazide was supplied by Aldrich Chemical Co., Inc. Alternately, it may be prepared by a procedure described in *Org. Synth., Coll. Vol. 5,* 1055 (1973).

3. Tetrahydrofuran serves equally well as a solvent. However, the quantity of sodium borohydride should be reduced to 1.0 g. and the isolation procedure modified in the following way. After the solution has been refluxed for 8 hours, the reaction mixture is cooled and the excess sodium borohydride is decomposed by the slow addition of dilute hydrochloric acid. The resulting mixture is extracted with ether and the ethereal solution is washed as described.

4. Because of the ready decomposition of sodium borohydride in methanol, the solution is maintained at room temperature during the addition of the metal hydride.

5. The crude product is contaminated with a small amount of a more polar substance which is subsequently removed by chromatography.

3. Discussion

The preparation of 5α-androstan-17β-ol from 5α-androstan-17β-ol-3-one may be realized by classical methods such as the Wolff-Kishner or Clemmensen reduction.

This procedure illustrates a general method for the reduction of aldehyde and ketone functions to methylene groups under very mild conditions. Since strong acids and bases are not employed, this procedure is of particular importance for the reduction of ketones possessing an adjacent chiral center.[2,3] Moreover, the use of deuterated metal hydrides permits the preparation of labeled compounds.[4]

The reduction of the preformed tosylhydrazones with sodium borohydride may be effected in aprotic solvents, such as tetrahydrofuran or dioxane.[5] The use of lithium aluminium hydride in nonhydroxylic solvents permits the reduction of aromatic aldehydes and ketones.

1. Istituto di Chimica Organica, Universita di Roma, Rome, Italy.
2. The reduction of the tosylhydrazone of (+)(S)-4-methyl-3-hexanone affords (+)(S)-3-methylhexane, optical purity 85%; L. Lardicci and C. Botteghi, private communication.
3. A. N. De Belder and R. Weigel, *Chem. Ind. (London),* 1689 (1964).
4. E. J. Corey and S. K. Gros, *J. Am. Chem. Soc.,* **89,** 4561 (1967); M. Fischer, Z. Pelah, D. H. Williams, and C. Djerassi, *Chem. Ber.,* **98,** 3236 (1965).
5. L. Caglioti, *Tetrahedron,* **22,** 487 (1966); L. Caglioti and P. Grasselli, *Chim. Ind. (Milan),* **46,** 1492 (1964); S. Cacchi, L. Caglioti, and G. Paolucci, *Bull. Chem. Soc. Jpn.,* **47,** 2323 (1974).

ALDEHYDES FROM 4,4-DIMETHYL-2-OXAZOLINE
AND GRIGNARD REAGENTS: o-ANISALDEHYDE

(Benzaldehyde, 2-methoxy-)

Submitted by R. S. BRINKMEYER, E. W. COLLINGTON, and A. I. MEYERS[1]
Checked by R. E. IRELAND and R. R. SCHMIDT III

1. Procedure

Caution! Hexamethylphosphoric triamide (HMPA) vapors have been reported to cause cancer in rats.[2] All operations with hexamethylphosphoric triamide should be performed in a good hood, and care should be taken to keep the liquid off the skin.

Methyl iodide, in high concentrations for short periods or in low concentrations for long periods, can cause serious toxic effects in the central nervous system. Accordingly, the American Conference of Governmental Industrial Hygienists[3] has set 5 p.p.m., a level which cannot be detected by smell, as the highest average concentration in air to which workers should be exposed for long periods. The preparation and use of methyl iodide should always be performed in a well-ventilated fume hood. Since the liquid can be absorbed through the skin, care should be taken to prevent contact.

A 1-l., three-necked, round-bottomed flask equipped with a 500-ml. dropping funnel (Note 1), a mechanical stirrer, and an argon inlet tube is charged with 80 g. (0.33 mole) of N,4,4-trimethyl-2-oxazolinium iodide (Note 2). The system is flushed with argon; 150 ml. of dry tetrahydrofuran (Note 3) is added, and the stirred suspension is cooled in an ice bath. Meanwhile, to a cooled solution of freshly prepared 2-methoxyphenylmagnesium bromide (0.414 mole) (Note 4) is added 150 g. (146 ml., 0.828 mole) of dry hexamethylphosphoric triamide (Note 5). This solution is then transferred under an argon atmosphere to the 500-ml. dropping funnel with the aid of an argon-pressurized siphon. The solution is slowly run into the cooled suspension, whereupon the methiodide salt dissolves. When the addition is complete, the reaction mixture is stirred at room temperature overnight.

The suspension is slowly poured into 600 ml. of saturated ammonium chloride solution which has previously been cooled to 10–15°, and the mixture is extracted with three 250-ml. portions of diethyl ether. Concentration of the combined extracts gives the crude addition product (Note 6), which is added to 200 ml. of ice-cold water and quickly acidified with cold 3 *N* hydrochloric acid. The acidic solution is rapidly extracted with 300 ml. of cold hexane, and the extract is discarded. The aqueous solution is then made basic by the addition of aqueous 20% sodium hydroxide solution (Note 7), and the mixture is extracted with three 250-ml. portions of ether. Concentration of the combined ethereal extracts affords 70–75 g. of crude oxazolidine (Note 8).

A 1-l., round-bottomed flask is charged with 72 g. of the crude oxazolidine in 600 ml. of water, and 201.6 g. (1.6 moles) of hydrated oxalic acid is added. The mixture is heated under reflux for 1 hour, cooled, treated with 600 ml. of water to dissolve precipitated oxalic acid, and extracted with three 100-ml. portions of ether. The combined ethereal extracts are washed with 50 ml. saturated sodium hydrogen carbonate, dried over anhydrous potassium carbonate, and concentrated, giving 30–35 g. of crude aldehyde. Distillation of this material at 70–75° (1.5 mm.) gives pure *o*-anisaldehyde (22.8–26.3 g.; 51–59%), m.p. 35.5–38° (Note 9).

2. Notes

1. The dropping funnel should be equipped so that the transfer of the Grignard reagent to it can be carried out under a positive nitrogen pressure.

2. 4,4-Dimethyl-2-oxazoline is commercially available from Columbia Organic Chemicals, 912 Drake Street, Columbia, South Carolina, or may be prepared as follows. A 250-ml., three-necked flask is charged with 89.14 g. (1.001 mole) of 2-amino-2-methylpropanol and cooled in an ice bath. The amine is carefully neutralized with 52.3 g. (1.25 mole) of 90.6% formic acid over a 1-hour period. A magnetic stirring bar is added, the flask is fitted with a short path distillation head, and the reaction mixture is placed in a silicon oil bath which is rapidly heated to 220–250°. The azeotropic mixture of water and oxazoline distills over a period of 2–4 hours and is collected in an ice-cooled flask containing ether. The aqueous layer is separated, saturated with sodium chloride, and extracted with three 50-ml. portions of ether. The combined ethereal extracts are dried over potassium carbonate and filtered. The ether is removed at 35–40° at atmospheric pressure, and 4,4-dimethyl-2-oxazoline is collected as the temperature rises above 85°, yielding 56.7–62.7 g. (57–63%) of a colorless mobile liquid, b.p. 99–100° (758 mm.).

The checkers found that if the azeotropic mixture is distilled more slowly from the reaction mixture at a pot temperature of 175–195°, the yield is greatly reduced and large amounts of polymeric material are formed.

N,4,4-Trimethyl-2-oxazolinium iodide is prepared by adding 49.5 g. (0.500 mole) of 4,4-dimethyl-2-oxazoline to an excess of cold methyl iodide (182 g., 78.2 ml., 1.28 moles) in a 500-ml. flask and stirring at room temperature under argon for 20 hours. The light brown solid is suction filtered and dissolved in 350 ml. of dry acetonitrile. The methiodide salt is precipitated by the addition of 750 ml. of dry ether to this acetonitrile solution. The purified salt is again suction filtered, and the white solid is washed with 250 ml. of dry ether and finally dried under vacuum, giving 96 g. (80%) of the methiodide, m.p. 215° (dec.). The salt can be stored in an inert atmosphere without deterioration.

3. Tetrahydrofuran was dried by distillation from lithium aluminium hydride. See *Org. Synth.,* **Coll. Vol. 5,** 976, 1973, for a warning concerning the purification of tetrahydrofuran.

4. 2-Methoxyphenylmagnesium bromide was prepared from 77.5 g. (0.414 mole, distilled from calcium hydride) of *o*-bromoanisole and 11 g. (0.46 g.-atom) of magnesium turnings in 100 ml. of dry tetrahydrofuran. The solution of this Grignard reagent is heated to reflux for 30 minutes prior to use.

5. Hexamethylphosphoric triamide was dried by distillation from calcium hydride.

6. If pure oxazoline is required, residual amounts of hexamethylphosphoric triamide can be removed by elution of the ethereal solution through silica gel (20–40 mesh).

7. Ice may be added to keep the mixture cool during the neutralization.

8. If this step is omitted, the *o*-anisaldehyde obtained after hydrolysis of the oxazolidine is contaminated with 5–10% *o*-bromoanisole.

9. *o*-Anisaldehyde is commercially available from Aldrich Chemical Co. and Eastman Organic Chemicals, Eastman Kodak Co.

3. Discussion

The conversion of a Grignard or an organolithium reagent to an aldehyde has been accomplished by a variety of reagents. The methods utilize such reagents as *N*-ethoxymethyleneaniline;[4a] ethyl orthoformate;[4a] 4-dimethylaminobenzaldehyde;[4b] *N,N*-dimethylformamide;[5] a dihydroquinazolinium salt;[6] and, more recently, a *tert*-alkyl isonitrile.[7]

This procedure illustrates a general method for the preparation of aryl, benzyl, alkynyl, and vinyl aldehydes.[8] Table I gives the aldehydes which have been prepared from the corresponding Grignard reagents by conditions similar to those described here.

This method does not allow the formylation of aliphatic Grignard or organolithium reagents, since in these cases, the enhancement in base strength in the presence of hexamethylphosphoric triamide produces side reactions due to proton abstraction.

The present method is simple, proceeds easily and in good yield. The starting materials are readily available. The method is of particular value for the ready preparation of C-1 deuterated aldehydes using the 2-deuterio-*N*,4,4-trimethyl-2-oxazolinium iodide.[8] Also,

TABLE I
ALDEHYDES FROM *N*,4,4-TRIMETHYL-2-OXAZOLINIUM IODINE

Grignard Reagent	Aldehyde	Yield, %
C_6H_5MgBr	C_6H_5CHO	69
$C_6H_5CH_2MgCl$	$C_6H_5CH_2CHO$	87
$C_6H_5CH=CHMgBr$	$C_6H_5CH=CHCHO$	64
$C_6H_5C\equiv CMgBr$	$C_6H_5C\equiv CCHO$	51
$2\text{-}CH_3OC_6H_4MgBr$	$2\text{-}CH_3OC_6H_4CDO$	70[a]

[a]From 2-deuterio-*N*,4,4-trimethyl-2-oxazolinium iodide.

since ^{14}C-labeled formic acid is routinely available, this provides easy access to isotopically labeled aldehydes.

1. Department of Chemistry, Colorado State University, Fort Collins, Colorado 80523.
2. J. A. Zapp, *Science,* **190,** 422 (1975).
3. American Conference of Governmental Industrial Hygienists (ACGIH), "Documentation of Threshold Limit Values," 3rd ed., Cincinnati, Ohio, 1971, p. 166.
4. (a) L. I. Smith and J. Nichols, *J. Org. Chem.,* **6,** 489 (1941); (b) A. J. Sisti, *Org. Synth.,* **Coll. Vol. 5,** 46 (1973).
5. R. A. Barnes and W. M. Bush, *J. Am. Chem. Soc.,* **81,** 4705 (1959).
6. H. M. Fales, *J. Am. Chem. Soc.,* **77,** 5118 (1955).
7. H. M. Walborsky, W. M. Morrison III, and G. E. Niznik, *J. Am. Chem. Soc.,* **92,** 6675 (1970).
8. A. I. Meyers and E. W. Collington, *J. Am. Chem. Soc.,* **92,** 6676 (1970).

[18]ANNULENE

(1,3,5,7,9,11,13,15,17-Cyclooctadecanonaene)

A.

3

$$\begin{array}{c} C\!\!\equiv\!\!CH \\ | \\ CH_2 \\ | \\ CH_2 \\ | \\ C\!\!\equiv\!\!CH \end{array} \xrightarrow[\text{pyridine, 55°}]{Cu(OCOCH_3)_2 \cdot H_2O}$$

B.

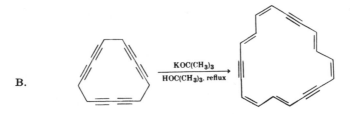

$$\xrightarrow[\text{HOC(CH}_3)_3, \text{ reflux}]{KOC(CH_3)_3}$$

C.

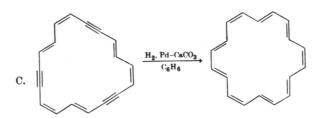

$$\xrightarrow[\text{C}_6\text{H}_6]{\text{H}_2, \text{ Pd–CaCO}_3}$$

Submitted by K. Stöckel and F. Sondheimer[1]
Checked by R. C. Wheland, R. E. Benson, H. Ona, and S. Masamune

1. Procedure

Caution! Benzene has been identified as a carcinogen; OSHA has issued emergency standards on its use. All procedures involving benzene should be carried out in a well-ventilated hood, and glove protection is required.

A. *Hexadehydro[18]annulene.* A 12-l., four-necked, round-bottomed flask provided with a stopcock at the bottom (Figure 1) (Note 1) is fitted with a stirrer (Note 2), a reflux condenser, and a 500-ml. pressure-equalizing dropping funnel. The flask is placed in a large metal vessel equipped with a hot–cold water inlet and a drain. The flask is charged with 600 g. (3.00 moles) of copper(II) acetate monohydrate (Note 3) and 3.8 l. of pyridine (Note 4). Stirring is begun and warm water is added to the metal vessel to heat the contents of the flask to $55 \pm 1°$, and the mixture is slowly stirred at this temperature for 1 hour. With vigorous stirring (approximately 600 r.p.m., Note 4), a solution of 50 g. (0.64 mole) of 1,5-hexadiyne (Note 5) in 400 ml. of pyridine (Note 6) is added over 30 minutes to the 55°, green suspension. The mixture is stirred vigorously at this temperature for an additional 2 hours before the warm water in the metal vessel is allowed to drain and is replaced with a mixture of ice and water. When the contents of the flask have cooled, the mixture is drawn off through the stopcock at the bottom of the flask and filtered through a 25-cm. Büchner funnel (Filtrate A). The green residue is transferred to a 5-l. beaker and 2.5 l. of 1:1 (v/v) benzene–diethyl ether is added. The mixture is stirred well, avoiding emulsion formation, and filtered (Filtrate B). The residue is similarly extracted with 2.5 l. of 1:1 (v/v) benzene–ether and filtered (Filtrate C). Each filtrate is kept separate.

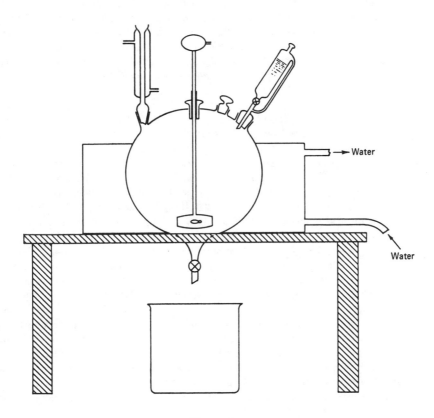

Figure 1.

Filtrate A is concentrated to approximately 200 ml. (Note 7) and added to Filtrate C. The resulting mixture is well agitated and filtered through a 25-cm. Büchner funnel. The residue is extracted twice in a similar manner with 500 ml. of 1:1 (v/v) benzene–ether. All the filtrates are now combined in the original 12-l. reaction vessel (Note 8) and washed successively with two 500-ml. portions of water, two 500-ml. portions of cold (0°), 2 N hydrochloric acid, and three 500 ml. portions of water. The organic layer is dried for 2 hours over *ca.* 110 g. of magnesium sulfate (Note 9), filtered, and evaporated to dryness (Note 7), yielding approximately 16 g. of crude product.

B. *Tridehydro[18]annulene.* The dark brown residue from Part A and 800 ml. of benzene (Note 10) are added to a 2-l., round-bottomed flask fitted with a reflux condenser and a calcium chloride drying tube. The solid dissolves on heating to reflux, then the solution is allowed to cool briefly. The condenser is removed, and 800 ml. of a saturated solution of potassium *tert*-butoxide in *tert*-butyl alcohol (Notes 11 and 12) is rapidly added (Note 13). After the resulting solution is heated to reflux for 30 minutes, the hot mixture is transferred to the original 12-l. reaction vessel (Note 14), which contains approximately 2 l. of ice, and 1.5 l. of ether is added. The resulting mixture is stirred well then allowed to stand for a few minutes. The incompletely separated phases, in which is suspended a large amount of black solid, are filtered through a 25-cm. Büchner funnel. Whenever the filtration becomes slow, the black material on the filter paper is washed with ether, and the paper is replaced. The organic layer of the filtrate is separated, washed three times with 500-ml. portions of water, dried for 2 hours over *ca.* 100 g. of anhydrous magnesium sulfate (Note 9), filtered, and evaporated to dryness (Notes 7 and 15). The red-brown viscous residue is dissolved in 60 ml. of benzene (Note 10) and chromatographed on alumina.

One liter of 95:5 (v/v) pentane–ether (Note 16) is poured into a closed chromatography column (100 × 4.5 cm.), the bottom of which is protected by a small plug of cotton wool. One kilogram of alumina (Note 17) is added in portions, with slow rotation by hand. The stopcock is opened, and the level of the supernatant liquid is allowed to fall to the top of the alumina. The benzene solution is now added with a long pipette and is allowed to seep in. The column is developed first with 1 l. of 95:5 (v/v) pentane–ether in order to wash out the benzene, then with 80:20 (v/v) pentane–ether. Two bands are observable on the column. The faster moving, light brown band consists of tridehydro[18]annulene (Note 18) and the slower moving, dark brown band consists of tetradehydro[24]annulene. When the first band is approximately 15 cm. from the bottom of the column, after approximately 4–6 l. of solvent have been eluted, 150-ml. fractions are collected, and the electronic spectrum of each fraction is determined. As soon as the maxima at 385 and 400 nm. characteristic of tridehydro[18]annulene appear (Note 19), the fractions are combined ("mixed fractions") until the first band is approximately 1 cm. from the bottom of the column. Essentially "pure" tridehydro[18]annulene (Note 18) is now eluted, and this material is collected until the second band is approximately 5 cm. from the bottom of the column. Smaller fractions (150 ml. each) are now collected again; the electronic spectrum of each fraction is determined, and those still showing maxima at 385 and 400 nm. are combined with the previously mentioned "mixed fractions." If required, the tetradehydro[24]annulene suitable for conversion to [24]annulene[2] can then be eluted with pentane–ether (70:30 to 50:50).

The electronic spectrum of the fractions containing the "pure" tridehydro[18]annulene exhibits the strongest absorption maximum (in benzene) at 342 nm. (ϵ 155,000)[3] and the spectroscopic yield, based on the molar extinction coefficient, is 1.17 g. (2.40% from 1,5-hexadiyne). The yield of tridehydro[18]annulene in the "mixed fractions," based on the 342 nm. maximum, is 0.27 g. (0.55%). The tridehydro[18]annulene is best stored in solution in the refrigerator.

C. *[18]Annulene*.   In a 300-ml., conical flask fitted with a side arm (with a closed stopcock) and a 3.5-cm. Teflon®-coated magnetic stirring bar is placed 1 g. of a 10% palladium-on-calcium carbonate catalyst (Note 20) and 30 ml. of benzene (Note 10). The flask is attached to an atmospheric pressure hydrogenation apparatus, and the air in the system is replaced by hydrogen by evacuating the flask and refilling with hydrogen three times. The catalyst suspension is stirred until no more gas is absorbed. One-third (390 mg., determined spectroscopically) of the "pure" tridehydro[18]annulene solution described in Part B is evaporated to dryness (Note 7), and the resulting brown crystalline residue is dissolved in 30 ml. of benzene (Note 10). This solution is added to the hydrogenation flask through the side arm without stirring. The mixture is now stirred under a hydrogen atmosphere as rapidly as possible (*ca.* 600 r.p.m.) (Note 21) until 207 ml. (4.9 molar equivalents) of gas (22°, 740 mm.) are absorbed over *ca.* 5 minutes. The mixture is filtered through a sintered glass funnel, and the catalyst is washed with three 20-ml. portions of benzene.

This hydrogenation experiment is repeated twice with the remaining two-thirds of the "pure" tridehydro[18]annulene solution, and the combined filtrates from the three hydrogenations are evaporated to approximately 30 ml. (Note 7). The solution is then transferred with a pipette to a test tube (10 × 3 cm.) and concentrated to approximately 15 ml. at 40° with a stream of nitrogen. The dark green solution is diluted with 20 ml. of ether and cooled in an ice bath for 30 minutes. The resulting red-brown crystals of [18]annulene are collected by filtration on a sintered glass funnel and washed with approximately 3 ml. of cold (−20°) ether. After drying in air for a few minutes this material weighs 114 mg. A second crop weighing 42 mg. is obtained by evaporation of the mother liquors to dryness (Note 7), followed by crystallization from 6 ml. of benzene and 20 ml. of ether. Both crops give a single spot on TLC (Note 22).

The [18]annulene mother liquors contain 181 mg. of unchanged tridehydro[18]annulene, as determined by the electronic spectrum (see Part B). They are combined with the "mixed fractions" described in Part B (containing 270 mg. of tridehydro[18]annulene), and the resulting solution is evaporated to dryness (Note 7). The residue is dissolved in 30 ml. of benzene and stirred with hydrogen and 1.0 g. of a 10% palladium-on-calcium carbonate catalyst (Note 20), as before, until 216 ml. (4.4 molar equivalents) of hydrogen (20°, 740 mm.) are absorbed. The catalyst is removed, and the filtrate is concentrated to approximately 6 ml., as described previously. After the addition of 20 ml. of ether and cooling in an ice bath, 112 mg. of [18]annulene is collected by filtration (Note 23). The purity of the product is confirmed by TLC (Note 22). The total yield of [18]annulene is 268 mg. (0.54% overall from 1,5-hexadiyne, Note 24).

[18]Annulene is best stored in benzene–ether solution in the refrigerator. The crystals can also be kept for some time in the refrigerator, small crystals being less stable than large ones.[2] Material that has decomposed to some extent may be purified as follows. The

substance (300 mg.) is broken up with a spatula and heated with 30 ml. of benzene, and the insoluble material is removed by filtration. The filtrate is poured onto a chromatography column (20 × 1.8 cm.), prepared with 35 g. of alumina (Note 17) and benzene. The column is washed with benzene until the eluate is colorless. Polymeric material is retained at the top of the column. The eluent is concentrated to approximately 6 ml., diluted with 20 ml. of ether, and cooled in an ice–salt bath. The resulting crystals of [18]annulene are collected on a sintered glass funnel, washed with 3 ml. of cold (−20°) ether, and dried in air for a few minutes.

2. Notes

1. The submitters used an ordinary 10-l., three-necked, round-bottomed flask.

2. A stirrer with one 11-cm., Teflon® blade was employed by the checkers. The submitters used a stirrer with two 7.5-cm. paddles, 9 cm. apart.

3. Copper(II) acetate monohydrate A. R. available from Fisons Scientific Apparatus Ltd. (Loughborough, England) was employed by the submitters. The checkers used both Baker Analytical Reagent and Fisher A.C.S. certified copper(II) acetate monohydrate. Small portions were finely ground with a mortar and pestle.

4. Vigorous stirring is important, since the yield of product appears to be reduced if the mixture is not stirred well.[5] The checkers used rates of 530–560 r.p.m.

5. 1,5-Hexadiyne was obtained from Farchan Research Laboratories and was distilled at atmospheric pressure (b.p. 84–85°) before use. It can be prepared as described by Raphael and Sondheimer.[4]

6. The submitters employed pyridine from BDH Chemicals Ltd. (Poole, England) that was dried over solid potassium hydroxide for 24 hours, then distilled under slightly reduced pressure. The checkers used a similarly treated product.

7. All evaporations, unless otherwise stated, were carried out with a rotary evaporator under reduced pressure (water pump) in a water bath kept at about 40°. The checkers' water bath was maintained at 44–50°, and the evaporation required 4–5 hours rather than the 2 hours found by the submitters.

8. The submitters used a 10-l. separatory funnel.

9. Magnesium sulfate ("dried") from BDH Chemicals Ltd. (Poole, England) was used by the submitters. The checkers used anhydrous magnesium sulfate available from Fisher Scientific Company.

10. Benzene A. R. from Fisons Scientific Apparatus Ltd. (Loughborough, England) was employed by the submitters, and Fisher reagent grade benzene was used by the checkers.

11. The solution was obtained from 44 g. of potassium (Fisons Scientific Apparatus Ltd., Loughborough, England, submitters; Fisher Scientific Company, checkers) and 1 l. of tert-butyl alcohol (Note 12) by boiling under reflux under nitrogen, in the apparatus described by Vogel,[6] until all the metal had dissolved. The solution is kept well stoppered. If crystals of potassium tert-butoxide separate, they are dissolved by gentle heating before use.

12. tert-Butyl alcohol from May & Baker Ltd. of Dagenham, England, was dried by refluxing 1 l. of the alcohol with approximately 40 g. of sodium wire until about two-thirds of the metal had dissolved, and recovering the alcohol by distillation (see Note

2 in Reference 5). The checkers used *tert*-butyl alcohol available from Fisher Scientific Company and recommend conducting this purification on twofold scale.

13. An exothermic reaction was observed at this point. Too rapid addition of the *tert*-butoxide solution to the benzene solution will lead to loss of product by frothing.

14. The submitters used a 10-l. separatory funnel and shook the mixture after addition to the ice and ether.

15. It is important that all volatile solvents are removed, particularly *tert*-butyl alcohol, since its presence will interfere with the subsequent chromatography procedure. The checkers obtained approximately 9.0 g. of crude material.

16. The submitters used "pentane" petroleum spirit (b.p. 25–40°) available from Hopkin & Williams Ltd., Hadwell Heath, England, and anhydrous ether available from May & Baker Ltd., Dagenham, England. Both solvents were distilled over solid potassium hydroxide before use. The checkers purchased pentane from Phillips Petroleum Company and distilled it over solid potassium hydroxide. Anhydrous ether supplied by Mallinckrodt Chemical Works was used without further purification.

17. Aluminum oxide (neutral, activity grade I) available from M. Woelm, Eschwege, Germany, was deactivated by the addition of 7 ml. of water to 1 kg. of the adsorbent before use.

18. "Tridehydro[18]annulene" here and in the sequel refers to the symmetrical isomer shown in the formula, admixed with smaller quantities of an unsymmetrical isomer and tetradehydro[18]annulene.[7] These can be separated by chromatography on alumina coated with 20% silver nitrate, but this is unnecessary for the synthesis of [18]annulene, since all three substances give this annulene on catalytic hydrogenation.[7]

19. For the full electronic spectra of tridehydro[18]annulene and tetradehydro[24]annulene in isoöctane, see reference 3.

20. The submitters used 10% palladium-on-calcium carbonate available from Fluka AG, Buchs, Switzerland. The checkers prepared the catalyst in a manner similar to that described by Busch and Stöve.[8] To a solution of calcium chloride (5.1 g.) in 60 ml. of water was added 4.8 g. of sodium carbonate. The precipitate was filtered, washed thoroughly with water, and then suspended in 30 ml. of water. To this calcium carbonate suspension was added with stirring a solution of 0.833 g. of palladium chloride in a few drops of 6 N hydrochloric acid. The catalyst mixture was then filtered through a sintered glass funnel, and the solid was washed with water until the chloride test (silver nitrate) became negative. After being washed with ethanol and ether, the catalyst was dried under reduced pressure.

21. Vigorous stirring is essential, since the yield of [18]annulene is reduced considerably if the mixture is stirred more slowly.[9]

22. TLC was performed on silica gel GF$_{254}$ plates supplied by E. Merck AG, Darmstadt, Germany, using (92:4:4) pentane–cyclohexane–benzene. The electronic spectra in benzene were essentially identical to those previously reported,[2] and exhibit absorption maxima at 378 (ϵ 270,000), 414 (ϵ 8,600), 428 (6,700), and 455 nm. (26,200). ^{1}H NMR spectra of [18]annulene (tetrahydrofuran-d_8, vacuum-sealed) are temperature-dependent and show a singlet at 5.45 p.p.m. at 120° and two multiplets at 9.25 (12H) and −2.9 p.p.m. (6H) at −60°. The latter two signals merge just above room temperature.[10] The checkers observed the similar temperature dependence of ^{13}C NMR (tetrahydrofuran-d_8, 22.6 MHz) of the [18]annulene. Thus, proton-decoupled spectra

show a singlet at 126 p.p.m. at 60° and two singlets at 128 and 121 p.p.m. at −70°. The rapid processes exchanging the inner and outer nuclei (both proton and ^{13}C) in solution are responsible for the above spectral behavior.

23. The submitters reported in their original procedure that some additional amounts of [18]annulene and tridehydro[18]annulene were obtained by chromatography of the mother liquors.

24. The yields described in Parts B and C are the average values of two runs performed by the checkers, and both runs proceeded with virtually identical results at every stage. The submitters obtained a 0.63% overall yield of [18]annulene from 1,5-hexadiyne.

3. Discussion

Only two general methods have been developed for the synthesis of the macrocyclic annulenes.[11] The first of these, developed by Sondheimer and co-workers, involves the oxidative coupling of a suitable terminal diacetylene to a macrocyclic polyacetylene of required ring size, using, typically, copper(II) acetate in pyridine. The cyclic compound is then transformed to a dehydroannulene, usually by prototropic rearrangement effected by potassium *tert*-butoxide. Finally, partial catalytic hydrogenation of the triple bonds to double bonds leads to the annulene.

The presently described procedure for the synthesis of [18]annulene, although the overall yield is low by the standard normally set for *Organic Syntheses*, illustrates the above general route leading to the theoretically important macrocyclic annulenes, and in this way [14]-, [16]-, [18]-, [20]-, [22]-, and [24]annulenes have been prepared in pure crystalline form.

[18]Annulene was the first macrocyclic annulene containing $(4n + 2)$ π-electrons to be synthesized. The compound is of considerable interest, since it is the type of annulene that was predicted to be aromatic by Hückel.[12] It proved to be aromatic in practice, as evidenced from the ^1H NMR spectrum,[10,13] the X-ray crystallographic analysis,[14] and the fact that electrophilic substitution reactions could be effected.[15]

The method of synthesis is essentially that described by Sondheimer and Wolovsky[3] (preparation of tridehydro[18]annulene) and by Sondheimer, Wolovsky, and Amiel[2] (hydrogenation of tridehydro[18]annulene to [18]annulene). It has been simplified, since [18]annulene is now obtained from tridehydro[18]annulene without chromatography, and the whole procedure involves only one chromatographic separation. [18]Annulene has also been obtained by Figeys and Gelbcke[16] in 0.42% overall yield by a six-step sequence from propargyl alcohol via propargyl aldehyde, *meso*-1,5-hexadiyne-3,4-diol, *meso*-1,5-hexadiyne-3,4-diol ditosylate, *cis*-3-hexene-1,5-diyne, and 1,3,7,9,13,15-hexadehydro[18]annulene.

1. Deceased Feb. 11, 1981; formerly at Chemistry Department, University College, London WC1H OAJ, England.
2. F. Sondheimer, R. Wolovsky, and Y. Amiel, *J. Am. Chem. Soc.*, **84**, 274 (1962).
3. F. Sondheimer and R. Wolovsky, *J. Am. Chem. Soc.*, **84**, 260 (1962).
4. R. A. Raphael and F. Sondheimer, *J. Chem. Soc.*, 120 (1950).
5. J. E. Fox, unpublished observation.

6. A. I. Vogel, "Practical Organic Chemistry," 3rd ed., Longmans, Green, London, 1967, p. 921.
7. R. Wolovsky, *J. Am. Chem. Soc.*, **87**, 3638 (1965).
8. M. Busch and H. Stöve, *Ber. Dtsch. Chem. Ges.*, **49**, 1063 (1916).
9. R. E. Wolovsky, unpublished observation.
10. J.-M. Gilles, J. F. M. Oth, F. Sondheimer, and E. P. Woo, *J. Chem. Soc. B*, 2177 (1971).
11. For a review of the annulenes, see F. Sondheimer, *Acc. Chem. Res.*, **5**, 81 (1972).
12. E. Hückel, *Z. Physik*, **70**, 204 (1931); "Grundzüge der Theorie Ungesättigter und Aromatischer Verbindungen," Verlag Chemie, Berlin, 1938.
13. L. M. Jackman, F. Sondheimer, Y. Amiel, D. A. Ben-Efraim, Y. Gaoni, R. Wolovsky, and A. A. Bothner-By, *J. Am. Chem. Soc.*, **84**, 4307 (1962).
14. J. Bregman, F. L. Hirschfeld, D. Rabinovich and G. M. J. Schmidt, *Acta Crystallogr.*, **19**, 227 (1965); F. L. Hirschfeld and D. Rabinovich, *Acta Crystallogr.*, **19**, 235 (1965).
15. I. C. Calder, P. J. Garratt, H. C. Longuet-Higgins, F. Sondheimer, and R. Wolovsky, *J. Chem. Soc. C*, 1041 (1967); E. P. Woo and F. Sondheimer, *Tetrahedron*, **26**, 3933 (1970).
16. H. P. Figeys and M. Gelbcke, *Tetrahedron Lett.*, 5139 (1970).

AZETIDINE

$$HO(CH_2)_3NH_2 + 2CH_2{=}CHCO_2C_2H_5 \longrightarrow$$

$$HO(CH_2)_3N(CH_2CH_2CO_2C_2H_5)_2$$

$$HO(CH_2)_3N(CH_2CH_2CO_2C_2H_5)_2 + SOCl_2 \longrightarrow$$

$$Cl(CH_2)_3N(CH_2CH_2CO_2C_2H_5)_2 + HCl + SO_2$$

$$Cl(CH_2)_3N(CH_2CH_2CO_2C_2H_5)_2 + Na_2CO_3 \longrightarrow$$

$$N(CH_2)_2CO_2C_2H_5 + CH_2{=}CHCO_2C_2H_5 + NaCl + NaHCO_3$$

$$N(CH_2)_2CO_2C_2H_5 + KOH \longrightarrow$$

$$NH + CH_2{=}CHCOOK + C_2H_5OH$$

Submitted by DONALD H. WADSWORTH[1]
Checked by F. THOENEN, E. VOGEL, R. HOBI, and A. ESCHENMOSER

1. Procedure

A. *Ethyl 3-(1-azetidinyl)propionate.* A solution of 150 g. (2.00 moles) of 3-amino-1-propanol in 500 g. (5.00 moles) of ethyl acrylate (Note 1) is refluxed for 2 hours in a 1-l., round-bottomed flask. Subsequent vacuum removal of the excess ethyl acrylate at steam temperature yields 548 g. (99%) of crude diethyl 3-*N*-(3-hydroxypropyl)iminodi-propionate. A stirred, cooled solution of this material (548 g.) in 1 l. of chloroform and 10 ml. of *N,N*-dimethylformamide is treated dropwise with 262 g. (2.20 moles) of thionyl chloride. By cooling with an ice bath and controlling the addition rate, the reaction temperature is maintained below 40° (Note 2). After the addition is complete, the reaction mixture is stirred for 30 minutes at room temperature and poured slowly into a slurry of 340 g. of sodium hydrogen carbonate in 1 l. of water (Note 3). The organic layer is separated (Note 4) and dried over sodium sulfate, and the solvent is removed under reduced pressure, below 50°, yielding 570 g. (97%) of crude diethyl 3-*N*-(3-chloropropyl)iminodipropionate. A mixture of 100 g. of this crude material, 200 g. of anhydrous, powdered sodium carbonate (Note 5), and 10.0 g. of pentaerythritol (Note 6) in 200 ml. of diethyl phthalate is placed in a 500-ml., round-bottomed flask fitted with a vacuum-distillation head and an effective stirrer. The system is evacuated through a trap of sufficient capacity to contain 50 ml. of liquid, and the product is distilled by heating the stirred suspension with a heating mantle at 10–15 mm. By proper adjustment of the heat source, the distillation temperature is maintained below 150°, minimizing codistillation of diethyl phthalate. The distillate is collected until the head temperature cannot be kept below 150°. Redistillation of the resulting crude product through a 4-in. Vigreux column yields 34.0 g. (57–68%) of ethyl 3-(1-azetidinyl)propionate, b.p. 86–87° (12 mm.), 99% pure by GC (Note 7).

B. *Azetidine*. A stirred mixture of 38 g. (0.68 mole) of potassium hydroxide pellets in 100 ml. of white mineral oil (Note 8) is heated to 140–150° in a four-necked, 500-ml., round-bottomed flask, fitted with an air-driven Hershberg stirrer, a thermometer, a dropping funnel, and a 6-in. Vigreux column fitted with a vacuum-distillation head. The flask is removed from the heat source, and 50 g. (0.32 mole) of purified ethyl 3-(1-azetidinyl)propionate is added dropwise at a rate sufficient to maintain the reaction temperature at 150° (Note 9). After addition is complete, the reaction mixture is heated to 200° at 50 mm. to remove all traces of ethanol (Note 10). The flask is fitted with a distillation head and a nitrogen bubbler, and the distillation is resumed at atmospheric pressure until azetidine distills (210° maximum pot temperature, Note 11). The resulting product (19.6 g., 85% purity) is dried over potassium hydroxide and redistilled through a short Vigreux column yielding 14.5–15.8 g. (80–87%) of purified azetidine, b.p. 62–63° (Note 12).

2. Notes

1. Eastman Organic Chemicals, practical grade, 3-amino-1-propanol, and ethyl acrylate were used. The checkers used 3-amino-1-propanol, Fluka purum.
2. If the reaction temperature is not controlled, a tarry by-product is formed. By dissolving the crude aminochloride in petroleum ether, the impurity is separated as an insoluble tar.
3. Considerable foaming occurs during neutralization, which is best accomplished in a 4-l. beaker with rapid stirring.
4. Any excess of insoluble salt should be filtered from the reaction mixture before the chloroform layer is separated. The checkers observed no insoluble salt at this point.
5. Baker and Adamson reagent grade sodium carbonate powder was used. The checkers found that the use of crystalline anhydrous sodium carbonate lowered the yield.

6. The presence of some hydroxylic material appears to be necessary to ensure reproducibility of this step, and the submitters have found empirically that pentaerythritol is very effective for this purpose.

7. The crude distillate can be used directly without intermediate isolation of ethyl 3-(1-azetidinyl)propionate to give a 50% yield of azetidine, assuming that all of the crude material was ethyl 3-(1-azetidinyl)propionate and running the reaction as described for the pure material.

8. Fisher Scientific Company Paraffin Oil, N.F., Saybolt Viscosity 125/135 and Mobil Oil Corp. S/V Industrial White Oil Number 320, Saybolt Viscosity 200/210 gave comparable results.

9. If the reaction does not start immediately, 5 ml. of ethanol may be added as an initiator.

10. Any ethanol that is not removed will contaminate the product. Although careful distillation will separate azetidine and ethanol, a considerable yield loss is encountered.

11. The submitters used a higher pot temperature (230–275°). The checkers, however, recommend the lower temperature to minimize losses of azetidine.

12. ^{1}H NMR (CDCl$_3$), δ 1.85 (s, 1H, NH), 2.0–2.6 (m, 2H, CH_2), 3.68 (t, J = 8 Hz., 4H, 2CH_2).

3. Discussion

Azetidine has been prepared by the following procedures: cyclization of 3-bromopropylamine with potassium hydroxide (low yield);[2] cleavage of 1-p-toluenesulfonylazetidine with sodium and refluxing 1-pentanol (85–100% yield)[3,4] or sodium and liquid ammonia (30% yield);[5] hydrogenolysis of 1-benzylazetidine (50%);[6] cyclization of diethyl 3-N-(3-chloropropyl)iminodipropenoate with sodium carbonate without solvent (60–70% yield).[7] A review of methods for preparing azetidine has been published.[8]

All other preparations of azetidine suffer either from low yields, arduous preparative procedures, or cumbersome purification operations. In this procedure, cyclization is accomplished in relatively concentrated solutions, and azetidine is obtained directly from the cleavage reaction in a state sufficiently pure for most applications. In addition, stable ethyl 3-(1-azetidinyl)propionate can be prepared in advance, and the air-sensitive azetidine can be formed readily as needed with a one-step procedure.

1. Eastman Kodak Company, Rochester, New York 14650.
2. S. Gabriel and J. Weiner, *Ber. Dtsch. Chem. Ges.*, **21**, 2669 (1888).
3. C. C. Howard and W. Markwald, *Ber. Dtsch. Chem. Ges.*, **32**, 2031 (1899).
4. W. F. Vaughn, R. S. Klonowski, R. S. McElhinney, and B. B. Millward, *J. Org. Chem.*, **26**, 138 (1961). Many investigators have had difficulty in repeating this procedure. Apparently the azetidine tends to be swept out of the reaction mixture with the hydrogen.
5. A. B. Burg and C. D. Good, *J. Inorg. Nucl. Chem.*, **2**, 237 (1956).
6. R. S. Klonowski, Ph.D. Dissertation, University of Michigan, 1959.
7. D. H. Wadsworth, *J. Org. Chem.*, **32**, 1184 (1967). Although this procedure will furnish good yields on a small scale, scale-up causes mechanical difficulties and lower yields.
8. James A. Moore, in "The Chemistry of Heterocyclic Compounds," Vol. 19, Part II, A. Weissberger, Ed., Interscience, New York, 1964, Chap. VII.

AZOETHANE

(Hydrazone, diethyl)

$$C_2H_5NH_2 + SO_2Cl_2 \xrightarrow[-15°]{\text{pyridine, petroleum ether}} (C_2H_5NH)_2SO_2 \xrightarrow[H_2O]{\text{1. NaOCl,NaOH}}$$

$$\begin{matrix} NaO_3S-N-C_2H_5 \\ | \\ NH-C_2H_5 \end{matrix} \longrightarrow \begin{matrix} HN-C_2H_5 \\ | \\ HN-C_2H_5 \end{matrix} \xrightarrow[H_2O,25°]{NaOCl,NaOH} \begin{matrix} N-C_2H_5 \\ \| \\ N-C_2H_5 \end{matrix}$$

Submitted by ROLAND OHME, HELMUT PREUSCHHOF, and HANS-ULRICH HEYNE[1]
Checked by HARVEY W. TAYLOR and HENRY E. BAUMGARTEN

1. Procedure

Caution! Azoalkanes have been reported to have carcinogenic properties.[2,3] Care should be taken to avoid inhalation of these substances and contact of them with the skin. It is advisable to prepare and handle these compounds in a good fume hood.

A. *N,N'-Diethylsulfamide.* A dry, 2-l., three-necked, round-bottomed flask fitted with a mechanical stirrer, a reflux condenser, a thermometer, and a dropping funnel, and protected from atmospheric moisture with calcium chloride drying tubes is charged with 500 ml. of petroleum ether, 100 g. (2.20 moles) of ethylamine, and 140 g. (1.76 moles) of pyridine (Note 1). The stirred mixture is cooled in an acetone–dry ice bath to −30° to −15°; a solution of 120 g. (0.889 mole) of sulfuryl chloride in 220 ml. of petroleum ether is added, dropwise and with stirring, to the reaction flask at such a rate that the temperature remains below −15°. After addition is complete, the reaction mixture is stirred at room temperature for one hour. The petroleum ether layer is separated and discarded. The dark semisolid residue is made acidic by addition of 6 M hydrochloric acid, and the acidic mixture is heated under reflux for 2 hours (Note 2). The resulting solution is extracted with diethyl ether in a continuous extractor (Note 3) until all of the diethylsulfamide has dissolved. The ether is evaporated using a rotary evaporator, yielding 58−61 g. (44−45%) of crude *N,N'*-diethylsulfamide, m.p. 65−67°, (Note 4) which is of sufficient purity for use in the next step.

B. *Azoethane.* A 3-l., three-necked, round-bottomed flask fitted with a mechanical stirrer, a reflux condenser, a thermometer, and a dropping funnel is charged with 500 ml. of an aqueous 2 M solution of sodium hydroxide and 152 g. (1.00 mole) of *N,N'*-diethylsulfamide, which is brought into solution by warming the reaction flask. The reaction flask is cooled in a cold water bath before 715 ml. (1.0 mole) of aqueous 1.4 M sodium hypochlorite (Note 5) is added dropwise with stirring. After addition is complete, the reaction mixture is stirred for 15 minutes at room temperature. The mixture is brought to pH 1 by addition of 6 M hydrochloric acid and stirred for an additional 30 minutes at 60° (Note 6). The mixture is cooled to room temperature, then is brought to pH 14 by addition of aqueous 2 M sodium hydroxide (Note 7). Addition of 715 ml. (1.0 mole) of aqueous 1.4 M sodium hypochlorite solution causes the separation of azoethane as an oil, having a fruitlike odor. The mixture is extracted with three 100-ml. portions of toluene (Note 8). The combined extracts are dried over anhydrous sodium sulfate and distilled through a 50-cm. packed column, yielding 44–46 g. (51–54%) of azoethane, b.p. 58–59°, n_D^{20} 1.3861 (Note 9).

2. Notes

1. The submitters dried the ethylamine and pyridine by distillation over potassium hydroxide pellets and used 600 ml. of petroleum ether, 113 g. (2.50 moles) of ethylamine, and 158 g. (2.00 moles) of pyridine to which was added 135 g. (1.00 mole) of sulfuryl chloride in 250 ml. of petroleum ether. In the United States ethylamine is sold in 100-g. quantities in sealed-glass vials (Eastman Organic Chemicals) or as the compressed gas in cylinders (Matheson Gas Products). The checkers used the contents of a freshly opened vial (without distillation) for each run, as a matter of convenience, and used either pentane or petroleum ether (b.p. 38–51°). Note that step B requires the product from at least two [submitters' scale *and yield* (Note 4)] or three (checkers' scale) step A runs. Step B can be run at half scale with the same percentage yield.

2. The purpose of this step is to hydrolyze any alkyl imido compound that may have formed from the further reaction of the sulfamide.[4]

3. A convenient continuous extractor has been described earlier in this series.[5]

4. After purification by dissolving the crude product in ether and precipitating with petroleum ether, N,N'-diethylsulfamide is obtained as shiny, white leaflets, m.p. 67°. The submitters reported a 54% yield of the purified product.

Under identical conditions using 78 g. (2.5 mole) of methylamine, 71 g. (57%) of N,N'-dimethylsulfamide, m.p. 76°, may be obtained as fine, white needles after recrystallization from benzene.

For sulfamides with larger alkyl groups (C_3 to C_6) the following procedure is preferred. A solution of 316 g. (4.00 moles) of pyridine in 400 ml. of chloroform is added dropwise, with cooling to −10° to −5°, to a stirred mixture of 135 g. (1.00 mole) of sulfuryl chloride and 500 ml. of chloroform. Maintaining a temperature of −5° to 0°, a solution of 2.5 moles of alkylamine in 600 ml. of chloroform is added to the reaction. After addition is complete the mixture is stirred for 30 minutes at room temperature, evaporated under reduced pressure to a thick brown liquid, and treated with 2 M hydrochloric acid until the pyridine dissolves. Cooling the acidic solution, the crystalline sulfamide precipitates and is filtered. Any dissolved sulfamide may be recovered by extraction of the filtrate with ether. The crude product may be purified by recrystallization from 50% ethanol.

The pyridine used in the submitters' procedures apparently reacts with the sulfuryl chloride to form an intermediate, quaternary pyridinium complex which undergoes aminolysis, yielding the sulfamide.[6] However, in many instances the pyridine may be replaced by an equivalent quantity of the primary alkylamine being used.[4,7] Using this variation the checkers obtained a 78% yield of N,N'-dicyclohexylsulfamide (compare with Table I). Moreover, in the reaction of 4-aminospiro(cyclohexane-1,9'-fluorene) with sulfuryl chloride no sulfamide could be isolated from reactions run in the presence of pyridine (or triethylamine); however, a 54% (purified) yield of N,N'-dispiro(cyclohexane-1,9'-fluorene)-4-ylsulfamide was obtained when 2.7 *equivalents* of the amine (relative to sulfuryl chloride) were used. Probably the failure of the mixed pyridine-alkylamine technique was the result of combined bulk of the pyridinium complex and the amine.

5. The sodium hypochlorite solution was prepared by passing chlorine, with stirring and cooling to 0–5°, into 1.5 l. of aqueous 1.4 M sodium hydroxide solution.

In some small-scale preparations of this type the checkers used commercial household bleach (Chlorox®, 5.25% NaOCl) and followed the course of the reaction by TLC. The yields appear to be somewhat lower than those obtained with sodium hypochlorite prepared as described above. The obvious attractive alternative, preparation of potassium hypochlorite as described elsewhere in this series,[8] apparently has not been tried.

6. In the preparation of 2,2'-azoisobutane and azocyclohexane the acid hydrolysis step is not necessary and the two moles of sodium hypochlorite may be added in one step.

7. In the preparation of azomethane a gas-inlet tube is used to pass nitrogen slowly through the reaction mixture, during the second oxidation stage while the temperature is raised to 60°. The reflux condenser is fitted with a drying tube filled with potassium hydroxide pellets connected via rubber hose to two dry ice-cooled cold traps connected in series and terminated with a second drying tube filled with potassium hydroxide pellets. The azomethane collects in the cold traps. Redistillation gives a 39% yield of azomethane, b.p. 1°.

8. For the homologous azoalkanes ether, pentane, or petroleum ether may be used for extraction. The extraction solvent may be added before the addition of hypochlorite.[6]

9. The checkers used a 60-cm. Vigreux column. Their product gave the following ^{1}H NMR spectrum (CDCl₃): δ 1.17 (q, J = 7 Hz., 2H, CH_2), 3.77 (t, J = 7Hz., 3H, CH_3).

3. Discussion

Azoalkanes have been prepared by oxidation of N,N'-dialkylhydrazines with copper-(II) chloride[9] or with yellow mercury(II) oxide.[10,11] The dialkyl hydrazines are obtained by alkylation of N,N'-diformylhydrazine and subsequent hydrolysis,[9] by reduction of the

TABLE I

PREPARATION OF AZOALKANES

R	RNHSO₂NHR		R—N=N—R	
	m.p.	Yield, %	b.p.	Yield, %
n-C₃H₇	118°	69	113–115°	54
n-C₄H₉	126°	66	59–60° (18 mm.)	54
tert-C₄H₉	140–142°	68	109–110°	84
cyclo-C₆H₁₁	154°	59	m.p. 33–34°	80
4-NO₂-C₆H₄	197°	58	m.p. 216°	31

corresponding azine with lithium aluminum hydride,[11] or by catalytic hydrogenation of the azine over a platinum catalyst.[10]

The present procedure may be used for the preparation of azoalkanes with alkyl, cycloalkyl, or aromatic substituents (Table I). Azoalkanes have been used as radical sources for inducing of radical reactions (*e.g.*, polymerization). The present procedure may also be used for the preparation of *N,N'*-dialkylhydrazines.[6] For this purpose only one equivalent of sodium hypochlorite solution is employed and the reaction mixture is worked up after its addition (yields: 60–95%).

1. Institut für chemische Technologie, Akademie der Wissenschaften der DDR, 1199 Berlin-Adlershof, Rudower Chaussee 5.
2. H. Druckrey, R. Preussmann, S. Ivanković, C. H. Schmidt, B. T. So, and C. Thomas, *Z. Krebsforsch.*, **67**, 31 (1965).
3. R. Preussmann, H. Druckrey, S. Ivankovic, and A. von Hondenberg, *Ann. N.Y. Acad. Sci.*, 163 (1969).
4. R. Sowada, *J. Prakt. Chem.*, [4] **20**, 310 (1963).
5. See Note 10 in G. Billek, *Org. Synth.*, **Coll. Vol. 5**, 630 (1973).
6. R. Ohme and H. Preuschhof, *Justus Liebigs Ann. Chem.*, **713**, 74 (1968).
7. J. C. Stowell, *J. Org. Chem.*, **32**, 2360 (1967).
8. M. S. Newman and H. L. Holmes, *Org. Synth.*, **Coll. Vol. 2**, 428 (1943).
9. J. L. Weininger and O. K. Rice, *J. Am. Chem. Soc.*, **74**, 6216 (1952).
10. A. U. Blackham and N. L. Eatough, *J. Am. Chem. Soc.*, **84**, 2922 (1962).
11. R. Renaud and L. C. Leitch, *Can. J. Chem.*, **32**, 545 (1954).

DIELS-ALDER ADDITION OF PERCHLOROBENZYNE: BENZOBARRELENE

(1,4-Ethenonaphthalene, 1,4-dihydro)

Submitted by Neil J. Hales, Harry Heaney,[1] John H. Hollinshead, and Pritpal Singh
Checked by G. Crass, M. Pohmakotr, and D. Seebach

1. Procedure

Caution! Benzene has been identified as a carcinogen; OSHA has issued emergency standards on its use. All procedures involving benzene should be carried out in a well-ventilated hood, and glove protection is required.

A. *Tetrachlorobenzobarrelene.* A carefully dried, 5-l., three-necked, round-bottomed flask equipped with a large magnetic stirring bar, a low-temperature thermometer, a 500-ml., pressure-equalizing dropping funnel bearing a gas-inlet tube, and a Nujol bubbler (Note 1) is charged with 28.5 g. (0.100 mole) of hexachlorobenzene (Note 2). The apparatus is flushed with argon or nitrogen (Note 3) before 600 ml. of dry diethyl ether (Note 4) is added. The resulting suspension is stirred and cooled to −72° to −78° in a 4-l. acetone–dry ice bath. A solution of *n*-butyllithium (0.110 mole) in hexane (Note 5) is added over a 30-minute period such that the temperature does not exceed −70° (Note 6). The mixture is then allowed to warm to −60° over an additional 1.5 hours. Four liters of dry, thiophene-free benzene (Note 4) is added to the pentachlorophenyllithium solution

(Note 7) over a 1-hour interval, during which the temperature rises to *ca.* +10° (Note 8). The mixture is allowed to warm slowly to room temperature over a period of at least 14 hours and then is heated at +30° for another 2 hours to ensure complete reaction (Note 9). A 10-g. portion of solid ammonium chloride is added, and 15 minutes later the contents of the flask are filtered through 20 g. of Celite. The volume of the filtrate is reduced to 75 ml. with a rotary evaporator, and 100 g. of alumina (Note 10) is added to the concentrate in a 250-ml. flask. The rotary evaporation is continued until the weight remains constant and a free-flowing consistency is attained. The material is placed on a column packed with 800 g. of alumina and eluted with low-boiling petroleum ether (Note 11). Fractions of *ca.* 200 ml. are collected and analyzed by GC or TLC (Note 6), with the appropriate fractions combined and evaporated, providing 16.9–17.5 g. (58–60%, Note 12) of essentially pure tetrachlorobenzobarrelene, m.p. 127–129° (Note 13).

B. *Benzobarrelene.* A dry, 1-l., three-necked, round-bottomed flask equipped with a magnetic stirring bar, a combined gas-inlet tube and rubber septum, a 500-ml., pressure-equalizing dropping funnel, and a reflux condenser connected to a Nujol bubbler is charged with 500 ml. of dry tetrahydrofuran (Note 14) and 17 g. (0.74 mole) of sodium wire having a diameter of *ca.* 0.5 mm. The mixture is stirred and heated at reflux under an atmosphere of argon or nitrogen while 50 ml. of freshly distilled *tert*-butyl alcohol is added. Immediately afterward a solution of 15 g. (0.051 mole) of tetrachlorobenzobarrelene in 200 ml. of tetrahydrofuran is added over a 15-minute period. After 4 hours under reflux (Note 15) the contents of the flask are cooled to room temperature and filtered through a plug of glass wool (Note 16) into a 2-l. beaker containing 50 ml. of methanol. After any remaining pieces of sodium have reacted with the methanol, 400 ml. of water is added, and the mixture is extracted with six 150-ml. portions of ether. The combined ether layers are washed with two 200-ml. portions of aqueous saturated sodium chloride, dried over magnesium sulfate, and evaporated with a rotary evaporator operated at water aspirator pressure and room temperature. The semicrystalline residue (7.3–8.1 g.) is mixed with 40 g. of alumina (Note 10) and swirled at room temperature under reduced pressure until it attains a free-flowing consistency. The material is then placed on a column packed with 600 g. of alumina and eluted with low-boiling petroleum ether (Note 11). Fractions of *ca.* 200 ml. are collected, evaporated, and assayed by GC (Notes 15 and 17). Combination of the appropriate fractions yields 5.9–6.8 g. (75–86%) of benzobarrelene, m.p. 62–64.5° (Note 18).

2. Notes

1. The dropping funnel must be arranged so that the drops fall directly into the solution and not onto the side of the flask. The checkers used a 4-l., four-necked flask equipped with a mechanical stirrer and a ground-glass stirring assembly, and carried out the reaction on four-fifths scale.

2. Technical-grade hexachlorobenzene was purchased by the submitters from BDH Chemicals, Ltd., and recrystallized twice from benzene: m.p. 227°. The submitters found that if the technical-grade material is used without purification, some insoluble material remains after the reaction with *n*-butyllithium, though the yield of tetrachlorobenzobarrelene is only slightly reduced. The checkers used 22.8 g. (0.0800 mole) of hexachloro-

benzene of 98% purity, purchased as a fine powder from EGA-Chemie K. G., an affiliate of Aldrich Chemical Company, Inc., without further purification.

3. The flushing operation was accomplished by replacing the bubbler with a stopcock and alternately evacuating and filling the apparatus with inert gas three times. A slight outflow of inert gas should be maintained during all subsequent operations. When the flask is being cooled, it is necessary to increase the gas flow.

4. Dry ether and dry, thiophene-free benzene were prepared by the submitters according to procedures presented in reference 2.

5. n-Butyllithium as $1.5-3.0\ M$ solutions in hexane is available from the following firms: Pfizer, Ltd., Sandwich, England; Metallgesellschaft, Frankfurt, Germany; Alfa Division, Ventron Corporation. The appropriate volume of the solution is transferred with a 50-ml. syringe to the dropping funnel with care being taken to exclude air. An excess of n-butyllithium above the 10% recommended here may lead to the formation of dilithiotetrachlorobenzene.

6. The reaction of n-butyllithium and hexachlorobenzene and, later, the formation of tetrachlorobenzobarrelene may be monitored by GC or TLC. Samples withdrawn from the reaction mixture with a syringe are injected into a small amount of water, and the organic layer is analyzed. GC was carried out by the submitters with flame ionization detection and a 1.5 m. × 4 mm. (inside diameter) glass column packed with 3% silicone rubber (SE-30) supported on $80-100$ mesh Gaschrom Q. With a column temperature of 150° and a nitrogen carrier gas flow rate of 45 ml. per minute, the retention times of pentachlorobenzene, hexachlorobenzene, and tetrachlorobenzobarrelene are $ca.$ 2, 4, and 18 minutes, respectively. Normally trace amounts of hexachlorobenzene are still detectable at the end of the reaction with n-butyllithium. TLC was performed on silica gel with 5% ether in pentane as developing solvent. The R_f value of tetrachlorobenzobarrelene is less than that of chlorobenzenes.

7. A clear yellow solution is usually obtained at this stage; however, some suspended material may be present, particularly when technical grade hexachlorobenzene is used.

8. For proper temperature control the cooling bath should be free from excess amounts of dry ice. The benzene should be added in the following manner (volume of benzene added, period of addition, final temperature reached): 0.5 l., 15 minutes, $ca.$ $-20°$; 0.5 l., 15 minutes, $ca.$ $-10°$; 3.0 l., 30 minutes, $ca.$ $+10°$.

9. The reaction is relatively slow at a laboratory temperature of $18-20°$ and may require as much as 40 hours to reach completion.

10. The submitters used Activity I (Brockmann) Camag alumina, which was purchased from Hopkins and Williams. The checkers used comparable material obtained from E. Merck, Darmstadt, Germany.

11. Low-boiling petroleum ether (b.p. $30-50°$ or $40-60°$) was distilled from calcium chloride prior to use.

12. The submitters usually combined fractions $5-14$ and obtained $18-19.5$ g. ($62-67\%$) of tetrachlorobenzobarrelene, m.p. $127-131°$. The checkers, using a 3.5 cm. × 1 m. column for the chromatography, isolated $13.5-14$ g. ($58-60\%$) of product from fractions $10-25$.

13. ^{1}H NMR (CDCl$_3$): δ (multiplicity, number of protons, assignment): 5.45 (m, 2H, bridgehead H), 6.95 (m, 4H, vinyl H). A melting point of 125° is reported in the literature.[3]

14. The tetrahydrofuran was freshly distilled from lithium aluminum hydride. For a warning regarding this method of purifying tetrahydrofuran, see *Org. Synth.*, **Coll. Vol. 5**, 976 (1973).

15. An aliquot may be removed at this stage and analyzed by either GC or TLC to ensure that the reaction is complete. Benzobarrelene has a retention time of *ca.* 5 minutes under the conditions stated in Note 6, but with a column temperature of 104°. The completion of the reaction is also indicated by a purple coloration of the precipitated sodium chloride.

16. The glass wool removes the larger pieces of unreacted sodium and much of the purple sodium chloride.

17. The dimensions of the column used by the checkers were the same as those specified in Note 12, and the product was obtained from fractions 10–20. The submitters evaporated the fractions with a rotary evaporator operated at water aspirator pressure and room temperature (*ca.* 20°); however, the checkers caution that the product sublimes very readily.

18. IR cm.$^{-1}$, strong peaks: 1460, 1325, 790, 750, 690, 660; ^{1}H NMR (CDCl$_3$): δ (multiplicity, number of protons, assignment): 4.9 (m, 2H, bridgehead *H*), 6.8–7.3 (m, 8H, aryl and vinyl *H*). The reported melting point is 65.5–66°.[4] From 20 g. of tetrachlorobenzobarrelene the submitters obtained 8.3–8.8 g. (79–83%) of benzobarrelene, m.p. 64–65°.

3. Discussion

Although benzobarrelene has been used in a number of recent studies, the best available published synthesis[4] starts with the Diels-Alder reaction of β-naphthol and maleic anhydride, affording benzobarrelene in *ca.* 1% yield after five additional steps. Minor improvements allow small quantities of benzobarrelene to be prepared in an overall

TABLE I

PREPARATION OF SUBSTITUTED BENZOBARRELENES

Tetrachlorobenzobarrelenes	Yield of Benzobarrelenes (%)
R^1 = OMe; R^2 = R^3 = R^4 = R^5 = H	100
R^1 = OMe; R^2 = R^5 = Me; R^3 = R^4 = H	90
R^1 = OMe; R^2 = R^4 = Me; R^3 = R^5 = H	95
R^1 = OMe; R^2 = Me; R^3 = R^4 = R^5 = H	95

yield of *ca.* 10%.[5] The reaction of benzyne with benzene is relatively inefficient, giving benzobarrelene in *ca.* 2% yield.[6] When benzyne is generated by decomposition of benzenediazonium-2-carboxylate at high dilution in benzene, the yield of benzobarrelene is raised to 14%.[7] The reactions of benzyne with other aromatic substrates are equally inefficient.

Tetrahalobenzynes, however, react with a variety of aromatic compounds to afford tetrahalobenzobarrelene derivatives in good yields, frequently in the range of 55 to 75%.[8] The dehalogenation of a variety of alkenyl chlorides with alkali metals in tetrahydrofuran containing *tert*-butyl alcohol[9] suggested this approach to the dechlorination of tetrachlorobenzobarrelenes.

The generation of pentachlorophenyllithium by the reaction of *n*-butyllithium with hexachlorobenzene has been reported previously by Rausch, Tibbetts, and Gordon.[10] The present procedure for the preparation of benzobarrelene is based on the submitters' previously published note.[11] By this method 10-g. quantities of benzobarrelene may be obtained in *ca.* 3 working days without the use of large-scale apparatus. The generality of the procedure is shown by the examples given in Table I.

1. Department of Chemistry, The University of Technology, Loughborough, Leicestershire, England, LE11 3 TU.
2. J. A. Riddick and E. Toops, Jr., "Organic Solvents," Vol. VII of "Technique of Organic Chemistry," A. Weissburger, Ed., Interscience, New York, 1955; A. I. Vogel, "A Text-Book of Practical Organic Chemistry," 3rd ed., Longmans, Green, London, 1956, pp. 163–165, 172–173.
3. H. Heaney and J. M. Jablonski, *J. Chem. Soc. C,* 1895 (1968).
4. H. E. Zimmerman, R. S. Givens, and R. M. Pagni, *J. Am. Chem. Soc.,* **90,** 6096 (1968).
5. H. Hart and G. M. Love, personal communication; L. A. Paquette, personal communication.
6. R. G. Miller and M. Stiles, *J. Am. Chem. Soc.,* **85,** 1798 (1963); L. Friedman, *J. Am. Chem. Soc.,* **89,** 3071 (1967).
7. H. E. Zimmerman, R. J. Boettcher, N. E. Buehler, G. E. Keck, and M. G. Steinmetz, *J. Am. Chem. Soc.,* **98,** 7680 (1976).
8. H. Heaney, *Fortschr. Chem. Forsch.,* **16,** 35 (1970).
9. P. Bruck, D. Thompson, and S. Winstein, *Chem. Ind. (London),* 405 (1960); P. G. Gassman and P. G. Pape, *J. Org. Chem.,* **29,** 160 (1964); G. W. Griffin and A. K. Price, *J. Org. Chem.,* **29,** 3192 (1964); P. G. Gassman and J. L. Marshall, *Org. Synth.,* **Coll. Vol. 5,** 424 (1973).
10. M. D. Rausch, F. E. Tibbetts, and H. B. Gordon, *J. Organomet. Chem.,* **5,** 493 (1966).
11. N. J. Hales, H. Heaney, and J. H. Hollinshead, *Synthesis,* 707 (1975).

BENZOCYCLOPROPENE

(Bicyclo [4.1.0]hepta-1,3,5-triene)

Submitted by W. E. BILLUPS,[1] A. J. BLAKENEY, and W. Y. CHOW[1]
Checked by NOBUO NAKAMURA and S. MASAMUNE

1. Procedure

Caution! Benzocyclopropene is characterized by an extremely unpleasant (foul) odor, and use of a good hood is recommended for the preparation.

A. *7,7-Dichlorobicyclo[4.1.0]hept-3-ene.*[2] A 2-l., three-necked, round-bottomed flask is equipped with a sealed mechanical stirrer, a reflux condenser, and a pressure-equalizing dropping funnel. The system is flushed with nitrogen with a gas-inlet tube attached to the top of the condenser before 126 g. (1.123 moles) of potassium *tert*-butoxide (Note 1) and 1.2 l. of pentane are added. The stirred suspension is cooled to 0–5° with an ice bath and 90 g. (1.12 moles) of 1,4-cyclohexadiene (Note 2) is introduced rapidly through the dropping funnel; 135 g. (1.131 moles) of chloroform (Note 3) is then added dropwise over a period of 1.5–2 hours. The resulting mixture is stirred for an additional 30 minutes before 300 ml. of cold water is added to dissolve all of the precipitated salts. The organic phase is separated, and the aqueous phase is extracted with one 50-ml. portion of pentane. The extract is combined with the original pentane solution and dried over approximately 20 g. of anhydrous sodium sulfate. The solvent is removed on a rotary evaporator, and the product is distilled through a 15-cm. Vigreux column, giving 69–72 g. (38–39%) of 7,7-dichlorobicyclo[4.1.0]hept-3-ene, b.p. 50–51° (0.8 mm.) (Note 4).

B. *Benzocyclopropene.* A dry, three-necked, round-bottomed flask fitted with a sealed mechanical stirrer, a reflux condenser, and a pressure-equalizing dropping funnel is flushed with nitrogen and charged with 35.0 g (0.312 mole) of potassium *tert*-butoxide (Note 1), followed by 200 ml. of dimethyl sulfoxide (Note 5). The stirred mixture is cooled to 15–20° (Note 6) with an ice bath before 24.5 g. (0.154 mole) of 7,7-dichlorobicyclo[4.1.0]hept-3-ene is added over a 7-minute period. The bath is removed, the mixture stirred an additional 25 minutes, and the reaction quenched by first cooling the flask with an ice bath and then adding 180 ml. of ice water. The crude product is pumped directly into an acetone–dry ice cold trap through a glass vacuum take-off adapter. The *tert*-butyl alcohol and dimethyl sulfoxide are removed by washing the distillate once with 400 ml. of ice water (Notes 7 and 8). The benzocyclopropene that separates as the lower layer is distilled from 1 g. of anhydrous sodium sulfate, using the apparatus shown in Figure 1. This procedure gives 4.35–5.48 g. (32–41%) of almost pure benzocyclopropene (Note 9).

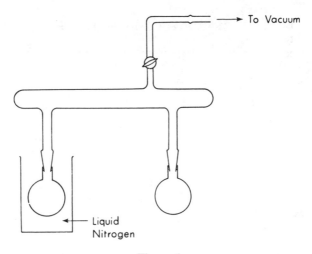

Figure 1.

2. Notes

1. Potassium *tert*-butoxide was used from a freshly opened bottle supplied by the MSA Research Corporation.

2. The checkers purchased 1,4-cyclohexadiene from Aldrich Chemical Company, Inc., and distilled it prior to use.

3. Reagent grade chloroform was used without removal of stabilizer.

4. The product was shown to be approximately 95% pure by GC analysis, using a 180 cm. × 0.24 cm. column packed with UCW-98 and heated to 130°.

5. Dimethyl sulfoxide (supplied by the Aldrich Chemical Company, Inc.) was dried by distilling over calcium hydride at 5 mm.

6. Care should be taken not to freeze the dimethyl sulfoxide.

7. On one occasion the checkers observed that the mixture formed an emulsion and that centrifugation facilitated the separation of the layers.

8. An alternative procedure is to extract the product into pentane.

9. The ^{1}H NMR spectrum shows that the product is approximately 95% pure. The major impurities are toluene and styrene. The product has the following spectral properties; IR (neat) cm.$^{-1}$: 1666, 1380, 1060, 735; ^{1}H NMR (CDCl$_3$), δ (number of protons): 3.17 (2H), 7.21 (4H); ^{13}C NMR (CDCl$_3$) δ: 18.56, 115.07, 125.86, 129.15; UV (cyclohexane) nm. max. (log ε): 263 (3.88), 268 (3.96), 276 (3.90).

3. Discussion

Fusion of the smallest cycloalkene, cyclopropene, to benzene would be expected to result in accommodation of the aromatic sextet with consequent bond length alteration in the aromatic ring. Benzocyclopropene thus arouses theoretical interest, and the high strain energy (approximately 68 kcal./mole)[3] associated with the compound suggests unusual chemical reactivity. Two review articles have appeared.[4,5]

The first successful synthesis of benzocyclopropene was reported by Vogel and coworkers[6] and is illustrated below. Though elegant, this method does require the prior, lengthy synthesis of the commercially unavailable 1,6-methano[10]annulene.[7]

The procedure described here[8] is a convenient two-step reaction which relies on the base-induced elimination-isomerization reactions of *gem*-dichlorocyclopropanes.[9-16] The reaction mechanism has been studied.[17] The principal advantage of this method is the availability of reagents.

1. Department of Chemistry, Rice University, Houston, Texas 77001.
2. B. S. Farah and E. E. Gilbert, *J. Chem. Eng. Data, 7,* 568 (1962).
3. W. E. Billups, W. Y. Chow, K. H. Leavell, E. S. Lewis, J. L. Margrave, R. L. Sass, J. J. Shieh, P. J. Werness, and J. L. Wood, *J. Am. Chem. Soc., 95,* 7878 (1973).
4. B. Halton, *Chem. Rev., 73,* 113 (1973).
5. W. E. Billups, *Acc. Chem. Res., 11,* 245 (1978).
6. E. Vogel, W. Grimme, and S. Korte, *Tetrahedron Lett.,* 3625 (1965).
7. E. Vogel and H. D. Roth, *Angew. Chem., 76,* 145 (1964) [*Angew. Chem. Int. Ed. Engl., 3,* 228 (1964)]; E. Vogel, W. Klug, and A. Breuer, *Org. Synth.,* **Coll. Vol. 6,** 731 (1988).
8. W. E. Billups, A. J. Blakeney, and W. Y. Chow. *J. Chem. Soc. D,* 1461 (1971).
9. C. L. Osborn, T. C. Shields, B. A. Shoulders, J. F. Krause, H. V. Cortez, and P. D. Gardner, *J. Am. Chem. Soc., 87,* 3158 (1965).
10. T. C. Shields, B. A. Shoulders, J. F. Krause, C. L. Osborn, and P. D. Gardner, *J. Am. Chem. Soc., 87,* 3026 (1965).
11. T. C. Shields and P. D. Gardner, *J. Am. Chem. Soc., 89,* 5425 (1967).
12. T. C. Shields and W. E. Billups, *Chem. Ind. (London),* 1999 (1967).
13. T. C. Shields, W. E. Billups, and A. R. Lepley, *J. Am. Chem. Soc., 90,* 4749 (1968).
14. T. C. Shields and W. E. Billups, *Chem. Ind. (London),* 619 (1969).
15. W. E. Billups, K. H. Leavell, W. Y. Chow, and E. S. Lewis, *J. Am. Chem. Soc., 94,* 1770 (1972).
16. W. E. Billups, T. C. Shields, W. Y. Chow, and N. C. Deno, *J. Org. Chem., 37,* 3676 (1972).
17. J. Prestien and H. Gunther, *Angew. Chem., 86,* 278 (1974) [*Angew. Chem. Int. Ed. Engl., 13,* 276 (1974)].

α-CHLORINATION OF CARBOXYLIC ACIDS MEDIATED BY CHLOROSULFONIC ACID: ε-BENZOYLAMINO-α-CHLOROCAPROIC ACID

(Hexanoic acid, 6-benzoylamino-2-chloro-)

$$C_6H_5\overset{\overset{\displaystyle O}{\|}}{C}NH(CH_2)_5-C\overset{\displaystyle O}{\underset{\displaystyle OH}{\diagdown}} \xrightarrow[\text{1,2-dichloroethane}]{\overset{\displaystyle Cl_2,\ O_2}{ClSO_3H}} C_6H_5\overset{\overset{\displaystyle O}{\|}}{C}NH(CH_2)_4-\underset{\displaystyle Cl}{C}HC\overset{\displaystyle O}{\underset{\displaystyle OH}{\diagdown}}$$

Submitted by Yoshiro Ogata, Toshiyuki Sugimoto, and Morio Inaishi[1]
Checked by Angela Hoppmann and George Büchi

1. Procedure

Caution! Since chlorine is poisonous, this procedure should be conducted in an efficient hood. Chlorosulfonic acid is a strong skin irritant and should be handled with gloves and a protective face shield.

A 500-ml., four-necked, round-bottomed flask is equipped with an air-tight mechanical stirrer (Note 1), a gas dispersion tube with a porous glass frit, a Dimroth reflux condenser (Note 2), and a thermometer, making sure all joints are greased with silicone grease. The top of the condenser is connected to a series of three traps with polyvinyl chloride tubing (Figure 1). The first trap is empty, and the other two contain aqueous $10\,N$ sodium hydroxide. The gas dispersion tube extends to near the bottom of the flask, just above the stirrer blade, and is connected to a gas-mixing chamber having two inlet tubes, one for oxygen and the other for chlorine. The flask is charged with 47.1 g. (0.200 mole) of ε-benzoylaminocaproic acid (Note 3) and 200 ml. of 1,2-dichloroethane. The solution is stirred and heated to 60–70°, before 25.5 g. (0.219 mole) of chlorosulfonic acid (Note 4) is added gradually. A 2:1 (v/v) mixture of gaseous chlorine and oxygen (Note 5) is bubbled into the flask for 3 hours while the contents are stirred and heated at reflux. The chlorine-oxygen gas flow is discontinued, and nitrogen is passed through the reaction mixture for 1 hour at 60–70° to remove chlorine remaining in solution. The flask is stoppered, allowed to stand for 1 hour at room temperature, and stored in a refrigerator for 12 hours. The supernatant liquid is removed, and *ca.* 800 ml. of aqueous 1 *N* sodium hydroxide is added to the solid remaining in the flask with ice cooling. Nitrogen is bubbled through the alkaline solution for 30 minutes to expel 1,2-dichloroethane. The solution is decolorized with 5 g. of activated carbon, mixed with *ca.* 400 g. of ice, and acidified to a pH of *ca.* 6 with 6 *N* hydrochloric acid. If available, a few seed crystals of ε-benzoylamino-α-chlorocaproic acid are added to the solution to facilitate crystallization. After 1 hour, more 6 *N* hydrochloric acid (Note 6) is added gradually until the pH is lowered to 1. An hour later the precipitate is filtered and washed thoroughly with 300 ml. of cold water until sulfate ion in the aqueous wash is no longer detectable with a test solution of barium chloride.

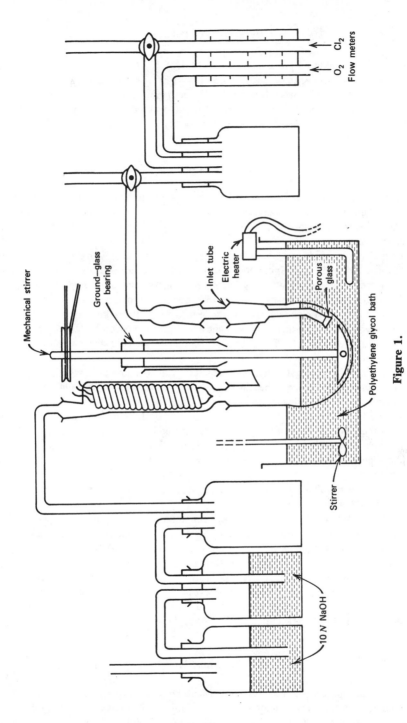

Figure 1.

91

Drying under reduced pressure yields 39.1–43.1 g. (72–80%) of crude, crystalline ε-benzoylamino-α-chlorocaproic acid, m.p. 138–140°. The product is dissolved in 320 ml. of hot 95% ethanol, 480 ml. of boiling water is added, and the resulting solution is allowed to cool slowly. The crystals are collected, washed with cold water, and dried, yielding 26.1–28.2 g. (48–52%) of pure ε-benzoylamino-α-chlorocaproic acid, m.p. 143–144° (Note 7).

2. Notes

1. Vigorous stirring action is necessary to disperse the heavy, viscous mixture. The use of a magnetic stirrer is not advisable since the mixture may separate into two layers. A mechanical stirrer with ground-glass shaft and bearing lubricated with 1,2-dichloroethane is recommended.

2. A Dimroth condenser has an internal, spiral cooling tube with the inlet and outlet both connected at the top. Spiral condensers of this type are available from Ace Glass Incorporated, Vineland, New Jersey 08360. A Dimroth condenser is recommended for use with refluxing liquids that boil up to 160°.[2] Since the points of sealing are situated above the zone with a high temperature gradient, the risk of cracking from thermal stress is minimized. The α-chlorination of aliphatic acids by this procedure is usually carried out at 110–140° (see Table I). The submitters circulated ice-cold water through the condenser.

3. ε-Benzoylaminocaproic acid was prepared by the reaction of benzoyl chloride with ε-aminocaproic acid, as described in *Org. Synth.*, **Coll. Vol. 2,** 76 (1943).

4. Chlorosulfonic acid was purified by distillation before use, b.p. 86–88° (33 mm.).

5. The flow rates of the two gases are regulated by flow meters inserted in parallel between the gas-mixing chamber and the chlorine and oxygen tanks. Appropriate flow rates for chlorine and oxygen are 80–100 and 40–50 ml. per minute, respectively. The checkers purchased gas flow meters from Arthur H. Thomas Company, Philadelphia, Pennsylvania.

6. If the warm alkaline solution is acidified rapidly with 6 *N* hydrochloric acid, the product is likely to separate as an oil.

7. A melting point of 145–147° has been reported.[3] The submitters performed a high-pressure liquid chromatographic analysis on a 25 × 0.2 cm. column packed with porous, dichlorodimethylsilane-treated silica gel (Yanapak DMS). With 40:60 (v/v) methanol–water as carrier liquid and a flow rate of 80–100 ml. per hour, the product appeared as a single peak. IR (KBr) cm.$^{-1}$: 3360, 3040, 2920, 1700, 1600, 1550, 820, 770, 720, 690; ^{1}H NMR (dimethyl sulfoxide-d_6), δ (multiplicity, coupling constant *J* in Hz., number of protons, assignment): 1.2–2.2 (m, 6H, $CH_2CH_2CH_2$), 3.0–3.4 (m, 2H, CH_2N), 4.40 (t, *J* = 7, 1H, C*H*Cl), 7.2–7.9 (m, 5H, C_6H_5), 8.40 (broad t, *J* = 6, 1H, N*H*).

3. Discussion

The present procedure, a modification of one reported earlier by the submitters,[4] has been applied to the α-chlorination of a series of aliphatic carboxylic acids (Table I).[5] In these reactions solvent (1,2-dichloroethane) was unnecessary, 0.25 molar equivalents of

TABLE I
α-Chloro Carboxylic Acids Prepared by Chlorination in the Presence
of Chlorosulfonic Acid and Oxygen[a]

Entry	α-Chloro Acid	Scale (mole)	Temperature (°)	Yield[b] (%)
1	$(CH_3)_2CCO_2H$ Cl	0.45	120	75
2	$CH_3CH_2CHCO_2H$ Cl	0.45	120	82[c]
3	$(CH_3)_2CHCHCO_2H$ Cl	0.60	140	73
4	CH_3 $CH_3(CH_2)_2CCO_2H$ Cl	0.35	120	81
5	$CH_3CH_2CHCHCO_2H$ H_3C Cl	0.18	120	78
6	$(CH_3)_2CHCH_2CHCO_2H$ Cl	0.16	120	79[d]
7		0.23	110	73

[a] A 4:1:0.04 molar ratio of carboxylic acid, chlorosulfonic acid, and chloranil was used. A 2:1 mixture of chlorine and oxygen was passed into the reaction for 3 hours.
[b] The yields were determined by gas chromatographic analysis after esterification of aliquots with sulfuric acid and methanol in 1,2-dichloroethane.
[c] β-Chloro acid was also formed in 1.6% yield.
[d] β-Chloro acid was also formed in 6.4% yield.

chlorosulfonic acid was sufficient, and higher temperatures in the range of 110–140° were employed. The α-chloro acids were converted efficiently to the corresponding methyl esters, for characterization, by reaction with methanol and a catalytic amount of concentrated sulfuric acid in 1,2-dichloroethane at reflux for 10 hours.[6] The methyl esters of the α-chloro acids shown in entries 3–6 have not been previously prepared.

Chlorosulfonic acid is particularly effective at mediating the α-chlorination of carboxylic acids, evidently owing to both its high acidity and its ability to render the reaction mixture more nearly homogeneous than other acidic catalysts. The function of oxygen is to scavenge free radicals, thereby suppressing the free radical chlorination at other positions of the carboxylic acid.[7] The chlorination of isovaleric acid (entry 3) in the absence of oxygen gives an appreciable amount of β-chloro acid. In the presence of oxygen only trace amounts (0–6.4%) of the β-chloro, or other isomers, were formed in the chlorinations shown in the table despite the tertiary hydrogens present in entries 3, 5, and 6. This method, which uses chlorosulfonic acid and 1,2-dichloroethane, can be applied to α-bromination[8] and α-iodination[9] of carboxylic acids, where no radical trapper such as molecular oxygen is necessary.

ϵ-Benzoylamino-α-chlorocaproic acid has been previously prepared by chlorination of ϵ-benzoylaminocaproic acid with sulfuryl chloride in the presence of iodine.[3] The corresponding bromo analog has been obtained by reaction with bromine and red phosphorous and subsequent hydrolysis.[10,11] ϵ-Benzoylamino-α-halocaproic acid is an intermediate in the synthesis of d,l-lysine dihydrochloride.[3,12]

1. Department of Applied Chemistry, Faculty of Engineering, Nagoya University, Chikusa-ku, Nagoya, Japan.
2. B. J. Hazzard, "Organicum: Practical Handbook of Organic Chemistry," Addison-Wesley, Reading, Massachusetts, 1973, pp. 6–9.
3. A. Galat, *J. Am. Chem. Soc.*, **69**, 86 (1947).
4. Y. Ogata and T. Sugimoto, *Chem. Ind. (London)*, 538 (1977).
5. Y. Ogata, T. Harada, K. Matsuyama, and T. Ikejiri, *J. Org. Chem.*, **40**, 2960 (1975).
6. R. O. Clinton and S. C. Laskowski, *J. Am. Chem. Soc.*, **70**, 3135 (1948).
7. J. C. Little, A. R. Sexton, Y.-L. Tong, and T. E. Zurawic, *J. Am. Chem. Soc.*, **91**, 7098 (1969).
8. Y. Ogata and T. Sugimoto, *J. Org. Chem.*, **43**, 3684 (1978).
9. Y. Ogata and S. Watanabe, *J. Org. Chem.*, **44**, 2768 (1979); *J. Org. Chem.*, **45**, 2831 (1980); *Bull. Chem. Soc. Jpn.*, **53**, 2417 (1980).
10. J. C. Eck and C. S. Marvel, *Org. Synth.*, **Coll. Vol. 2**, 74 (1943).
11. E. E. Howe and E. W. Pietrusza, *J. Am. Chem. Soc.*, **71**, 2581 (1949).
12. J. C. Eck and C. S. Marvel, *Org. Synth.*, **Coll. Vol. 2**, 374 (1943).

1-*N*-ACYLAMINO-1,3-DIENES FROM 2,4-PENTADIENOIC ACIDS BY THE CURTIUS REARRANGEMENT: BENZYL *trans*-1,3-BUTADIENE-1-CARBAMATE

[Carbamic acid, 1,3-butadienyl-, (*E*)-, phenylmethyl ester]

Submitted by PETER J. JESSUP, C. BRUCE PETTY, JAN ROOS, and LARRY E. OVERMAN[1]
Checked by PAUL H. LIANG and G. BÜCHI

1. Procedure

A. trans-2,4-*Pentadienoic acid*. A 1-l., three-necked, round-bottomed flask is equipped with a mechanical stirrer, a condenser cooled with ice-cold water (Note 1) bearing a calcium chloride drying tube, and a powder funnel. The flask is charged with 206 g. (210 ml., 2.61 moles) of pyridine (Note 2), and with vigorous stirring 208 g. (2.00 moles) of powdered malonic acid (Note 3) is added in portions. The powder funnel is replaced by a 250-ml., pressure-equalizing dropping funnel containing 126 g. (150 ml., 2.25 moles) of acrolein (Note 4), which is added, with vigorous stirring, over a 30-minute period. The exothermic reaction begins immediately with evolution of carbon dioxide, and the gently refluxing mixture becomes homogeneous. After 1 hour, as carbon dioxide evolution ceases, the solution is poured onto 1 l. of ice in a 3-l. Erlenmeyer flask, and carefully acidified with 130 ml. of concentrated sulfuric acid. The aqueous layer is extracted with four 250-ml. portions of dichloromethane, and the organic extracts are dried over magnesium sulfate for *ca.* 10 minutes. The dichloromethane solution is concentrated to *ca.* 300 ml. on a rotary evaporator with the water bath at 20–30° and allowed to crystallize in a freezer at −10° for several hours. After a first crop, yielding 40–50 g. of product, m.p. 72–73°, is collected, three additional crops are taken following successive concentration of the mother liquor to 150, 70, and 30 ml. Dried under reduced pressure in the presence of phosphorous pentoxide, the four crops of off-white crystals total 82–90 g. (42–46%), m.p. 69–71° (Note 5).

B. *Benzyl* trans-1,3-*butadiene-1-carbamate. Caution! The following reaction should be carried out in a fume hood to avoid accidental exposure to toxic hydrazoic acid.*

A dry, 1-l., three-necked, round-bottomed flask is equipped with a magnetic stirring bar, a thermometer, and a 250-ml., pressure-equalizing dropping funnel bearing a nitrogen inlet. The flask is flushed with nitrogen and charged with 49 g. (0.50 mole) of *trans*-2,4-pentadienoic acid, 80 g. (0.62 mole) of *N,N*-diisopropylethylamine, and 300 ml. of acetone (Note 6). The resulting solution is stirred and cooled to 0° in an ice–salt bath. A solution of 55 g. (0.51 mole) of ethyl chloroformate in 150 ml. of acetone is added over 30 minutes while the temperature is maintained below 0° (Note 7). Stirring is continued for an additional 30 minutes at 0°, after which a chilled solution of 65 g. (1.0 mole) of sodium azide (Note 8) in 170 ml. of water is added over a 20-minute interval, keeping the temperature below 0°. The contents of the flask are stirred for an additional 10–15 minutes at 0° (Note 9) and poured into a 2-l. separatory funnel containing 500 ml. of ice-water. The acyl azide is isolated by extraction with six 250-ml. portions of toluene. The combined toluene extracts are dried over anhydrous magnesium sulfate for 20 minutes and concentrated to a volume of *ca.* 300 ml. on a rotary evaporator at a water bath temperature of 40–50° (Note 10). *Caution! The acyl azide is potentially explosive. The solution should not be evaporated to dryness.* While the toluene solution is being concentrated, a dry, 2-l., three-necked, round-bottomed flask equipped with a mechanical stirrer, a 500-ml. pressure-equalizing dropping funnel, a simple distillation head, and a heating mantle is charged with 43 g. (0.40 mole) of benzyl alcohol, 250 mg. of 4-*tert*-butylcatechol (Note 11), and 200 ml. of toluene. About 30 ml. of toluene is distilled from the flask to remove trace amounts of water, and the distillation head is replaced with a condenser fitted with a nitrogen inlet. The toluene solution is stirred and heated at a rapid reflux under a nitrogen atmosphere as the toluene solution of the acyl azide is added over 30 minutes. The disappearance of the acyl azide and isocyanate is followed by IR analysis (Note 12). Conversion to the carbamate is complete in 10–30 minutes, after which the solution is cooled rapidly to room temperature by immersing the flask in an ice bath. The toluene is rapidly removed on a rotary evaporator with the water bath at 40–50°, producing a yellow solid residue (Note 13) which is dissolved in 50 ml. of 95% ethanol and allowed to crystallize in a freezer at −25° for several hours. Two crops of pale yellow crystals, m.p. 69–72°, are isolated which total 39–46 g. after drying under reduced pressure. Concentration of the mother liquor affords an oily residue that is placed on a 6 × 80-cm. column packed with 500 g. of silica gel (Note 14) and eluted with 1:9

(v/v) ethyl acetate–hexane. An additional 11–12 g. of crystalline product is obtained from
the chromatography, raising the total yield to 50–58 g. (49–57%) of nearly pure benzyl *trans*-1,3-butadiene-1-carbamate, a pale yellow solid, m.p. 70–73° (Note 15).

2. Notes

1. A water pump purchased from Little Giant Pump Company, Oklahoma City, Oklahoma, was used by the submitters to circulate ice-water through the condenser. The checkers used a dry ice condenser.

2. Reagent grade pyridine was stored over Linde type 4A molecular sieves for at least 24 hours before use.

3. Malonic acid was purchased from Aldrich Chemical Company, Inc. Most, but not all, of the malonic acid dissolves after 30 minutes. If pyridine is added to the malonic acid, a hard cake results.

4. Acrolein contaminated by 3% water was obtained by the submitters from Aldrich Chemical Company, Inc., and was purified by distillation from anhydrous calcium sulfate, b.p. 53°. The checkers used acrolein purchased from MC and B Manufacturing Chemists.

5. This material is of satisfactory purity for use in Part B. A thin-layer chromatogram on silica gel developed with 1:1 (v/v) ethyl acetate–hexane containing 1% acetic acid and visualized with 10% phosphomolybdic acid in ethanol as a spray reagent showed a major spot at R_f = 0.4 and a faint spot at 0.2. The crystalline acid may be stored for several months without significant decomposition. A melting point of 72° is reported[2] for *trans*-2,4-pentadienoic acid. The spectral properties of the product are as follows: IR (CHCl$_3$) cm.$^{-1}$: 3200–2700 (OH), 1696 (C=O), 1640, 1600, 1280, 1010; ^{1}H NMR (CDCl$_3$), δ (multiplicity, coupling constant J in Hz., number of protons, assignment): 5.1–5.8 (m, 2H, two H at C-5), 5.92 (d, J = 15, 1H, H at C-2), 6.2–6.8 (m, 1H, H at C-4), 7.37 (d of d, $J_{2,3}$ = 15, $J_{3,4}$ = 11, 1H, H at C-3), 12.0 (s, 1H, CO$_2$$H$).

6. *N,N*-Diisopropylethylamine was supplied by Aldrich Chemical Company, Inc., and purified by distillation from sodium hydride, b.p. 127°. Reagent grade acetone was stored over Linde type 4A molecular sieves for at least 24 hours before use.

7. Ethyl chloroformate obtained from Aldrich Chemical Company, Inc., was distilled, b.p. 93°. The progress of the reaction may be followed by ^{1}H NMR spectroscopy. Aliquots are partitioned between dichloromethane and water, the organic layer is concentrated, and the spectrum is recorded. A quartet from the ethoxy group of the mixed anhydride appears at δ 4.2. Ethyl chloroformate, which exhibits a quartet at δ 4.3, is removed in the concentration step.

8. Analytical reagent grade sodium azide purchased from Alfa Division, Ventron Corporation, or J. T. Baker Chemical Company was used as supplied.

9. The formation of the acyl azide may be followed by the growth of the 2130-cm.$^{-1}$ (—N=N=N) IR absorption of concentrated dichloromethane extracts of aliquots removed from the reaction.

10. The solution is concentrated for the purpose of removing residual ethanol. If this step is omitted, ethyl *trans*-1,3-butadiene-1-carbamate will be formed, contaminating the final product.

11. Reagent grade benzyl alcohol was purified by distillation, b.p. 205°. 4-*tert*-Butylcatechol purchased from Aldrich Chemical Company was sublimed at 50° (0.1 mm.) and recrystallized from hexane.

12. The acyl azide IR absorption occurs at 2130 cm.$^{-1}$ and the isocyanate at 2270 cm.$^{-1}$.

13. Impure samples of the product are particularly prone to decomposition. The purification steps should be carried out immediately.

14. Grade 60, activity III silica gel, purchased from W. R. Grace and Company, was used.

15. Material of this purity is suitable for most applications and may be stored in a freezer for several months with only slight decomposition. A thin-layer chromatogram on silica gel developed with 1 : 3 (v/v) ethyl acetate–hexane and visualized by UV fluorescence showed a single spot. Benzyl *trans*-1,3-butadiene-1-carbamate, like all *N*-acylamino-1,3-dienes, is quite acid sensitive and can, for example, be decomposed by traces of deuterium chloride in chloroform-*d*. A pure sample may be obtained by recrystallization from 1 : 20 (v/v) ethyl acetate–hexane, m.p. 74–75°; IR (Nujol) cm.$^{-1}$: 3300, 1692, 1625, 1515, 1230, 690; ^{1}H NMR (CDCl$_3$), δ (multiplicity, coupling constant *J* in Hz., number of protons, assignment): 4.8–5.0 (m, 2H, two *H* at C-4), 5.15 (s, 2H, C*H*$_2$C$_6$H$_5$), 5.4–5.8 (m, 2H, *H* at C-2 and N*H*), 6.26 (apparent d of t, *J* = 10 and *J* = 17, 1H, *H* at C-3), 6.71 (broadened d, *J* = 9, 1H, *H* at C-1), 7.33 (s, 5H, C$_6$H$_5$); ^{13}C NMR (CDCl$_3$), δ (assignment): 67.5 (*C*H$_2$C$_6$H$_5$), 112.5 (*C*-2), 113.5 (*C*-4), 127.2 (*C*-1), 128.3 (*para*-C$_6$H$_5$), 128.4 (*meta*-C$_6$H$_5$), 128.7 (*ortho*-C$_6$H$_5$), 134.6 (*C*-3), 136.0 (*peri*-C$_6$H$_5$), 153.7 (*C*=O).

3. Discussion

This procedure illustrates a general method[3] for the preparation of *trans*-1-*N*-acylamino-1,3-butadienes from 2,4-pentadienoic acids. A number of 1,3-butadienyl and 1,3-pentadienyl carbamates, thiocarbamates, and ureas have been prepared by this procedure (Table I).[3] These dienamides are reasonably stable crystalline solids which, when pure, may be stored in a freezer for several months with little decomposition. Since a variety of conjugated dienoic acids are readily accessible from Knoevenagel, Wittig, and related reactions,[4] 1-*N*-acylamino-1,3-butadienes with a diversity of carbon skeletons and heteroatom acyl substituents are potentially available by this method. Although this procedure describes the only known preparation of benzyl *trans*-1,3-butadiene-1-carbamate, a number of similar 1-*N*-acylamino-1,3-butadiene derivatives have been prepared by Curtius rearrangement of 2,4-hexadienoyl azide,[5] by Hofmann rearrangement of 2,4-hexadienamide,[6] by *N*-acylation of 1-*N*-alkylamino-1,3-butadiene anions generated from crotonaldehyde imines,[7] and by thermal rearrangement of trichloroacetimidic esters of acetylenic alcohols.[8]

The Curtius rearrangement procedure described here is a modification of one reported by Weinstock.[9] The submitters have found this procedure to be considerably more reproducible when *N,N*-diisopropylethylamine is substituted for triethylamine. The procedure described for the preparation of *trans*-2,4-pentadienoic acid is a modification of an earlier one by Doebner.[10] The submitters have found this method to give reproducibly higher yields, and to be more convenient, than other commonly used procedures for preparing this material.[2,11] The use of dichloromethane as the extracting and crystallizing solvent greatly simplifies the isolation of polymer-free samples of the crystalline acid.

trans-1-*N*-Acylamino-1,3-butadienes are useful dienes in Diels-Alder reactions. They are the most convenient synthetic equivalents currently available for the parent 1-amino-1,3-butadienes. These electron-rich dienamides undergo Diels-Alder cycloaddition with remarkable ease and high regio- and stereoselectivity.[12,13] As illustrated below, they react readily with even relatively unreactive dienophiles such as methyl acrylate and *trans*-crotonaldehyde. A recent quantitative study[12] has confirmed the expectation of an increase in reactivity with increasing electron-donating ability of the acyl substituent of an acylaminobutadiene, although the effects observed are not large.

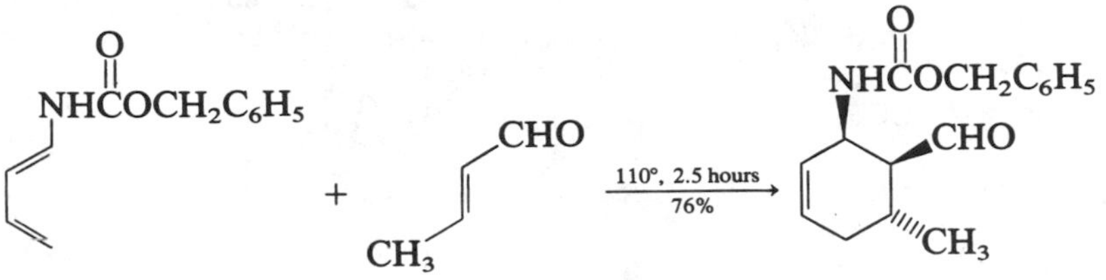

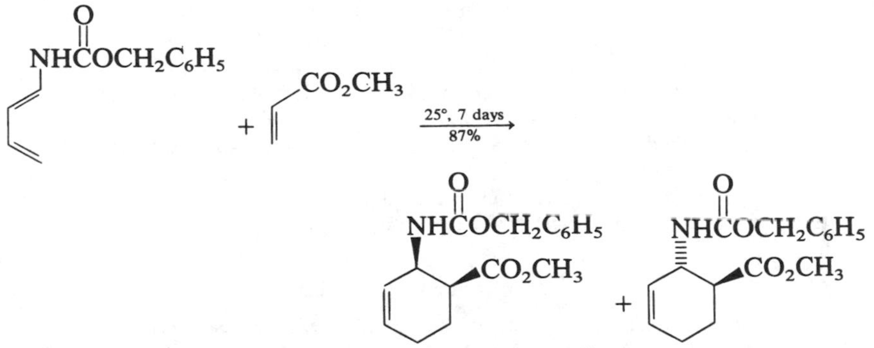

The choice of the acyl substituent X for Diels-Alder reactions of 1-*N*-acylamino-1,3-butadienes depends on the particular synthetic problem. The acyl substituent has a moderate effect on the cycloaddition reactivity of these dienes,[12] and also determines what amine unmasking procedures are required. As a result of their stability[12,13] and the variety of amine deprotection procedures available,[14] the diene carbamates are the components of choice in most cases. A particularly attractive aspect of the diene synthesis detailed here is the ability to "tailor" the amino-protecting group ($\overset{\overset{\text{O}}{\parallel}}{\text{C}}$—X) for the application at hand. Benzyl *trans*-1,3-butadiene-1-carbamate has the amino group protected with the benzyloxycarbonyl group. The ability to remove this functionality by catalytic hydrogenation was an important design feature in a recent alkaloid synthesis[13] that utilized this diene.

TABLE I
PREPARATION OF *trans*-1-*N*-ACYLAMINO-1,3-BUTADIENES[3]

R	X	Procedure[a]	M.p. (°)	Yield (%)
H	$OCH_2C_6H_5$	A	74–75	53[b]
H	$OC(CH_3)_3$	A	67–68	59
CH_3	OCH_2CH_3	A	91–92	80
H	OCH_2CH_3	A	44–45	71[b]
CH_3	OC_6H_5	A	118–120	72
H	OC_6H_5	B[c]	118–119	66
		A		45
CH_3	$N(CH_2)_4$	B	164–165	77
		A		(10)[d]
H	$N(CH_2)_4$	B	163–164	44
CH_3	SC_6H_5	B	116–118	78
H	SC_6H_5	B	92–93	47

[a] Isocyanate trapped as formed at 100° (Procedure A) or isocyanate preformed at 110° and trapped at 25° (Procedure B).
[b] Mean yield of four runs. All other entries are non-optimized yields from one run.
[c] A few drops of triethylamine were added.
[d] Estimated from the ^{1}H NMR spectrum.

1. Department of Chemistry, University of California, Irvine, California 92717.
2. I. E. Muskat, B. C. Becker, and J. S. Lowenstein, *J. Am. Chem. Soc.*, **52**, 326 (1930).
3. L. E. Overman, G. F. Taylor, and P. J. Jessup, *Tetrahedron Lett.*, 3089 (1976); L. E. Overman, G. F. Taylor, C. B. Petty, and P. J. Jessup, *J. Org. Chem.*, **43**, 2164 (1978).
4. H. von Brachel and U. Bahr, in E. Müller, Ed., "Methoden der Organischen Chemie" (Houben-Weyl), 4th ed., Vol. 5/1C, Georg Thieme Verlag, Stuttgart, Germany, 1970, pp. 519–555.
5. F. Tanimoto, T. Tanaka, H. Kinato, and K. Fukui, *Bull. Chem. Soc. Jpn.*, **39**, 1922 (1966).
6. V. B. Mochalin, I. S. Varpakhovskaya, and O. P. Beletskaya, *J. Org. Chem. USSR*, **10**, 1556 (1974).
7. W. Oppolzer and W. Fröstl, *Helv. Chim. Acta*, **58**, 587 (1975).
8. L. E. Overman and L. A. Clizbe, *J. Am. Chem. Soc.*, **98**, 2352, 8295 (1976).
9. J. Weinstock, *J. Org. Chem.*, **26**, 3511 (1961).
10. O. Doebner, *Ber. Dtsch. Chem. Ges.*, **35**, 1136 (1902).
11. E. P. Kohler and F. R. Butler, *J. Am. Chem. Soc.*, **48**, 1036 (1926); K. Alder and W. Vogt, *Justus Liebigs Ann. Chem.*, **570**, 190 (1950).
12. L. E. Overman, G. F. Taylor, K. N. Houk, and L. N. Domelsmith, *J. Am. Chem. Soc.*, **100**, 3182 (1978).

13. L. E. Overman and P. J. Jessup, *Tetrahedron Lett.*, 1253 (1977); L. E. Overman and P. J. Jessup, *J. Am. Chem. Soc.*, **100**, 5179 (1978); L. E. Overman and C. Fukaya, *J. Am. Chem. Soc.*, **102**, 1454 (1980).
14. J. F. W. McOmie, "Protective Groups in Organic Chemistry," Plenum Press, London, 1973, Chapter 2.

BENZYL CHLOROMETHYL ETHER

[Benzene, (chloromethyoxy)methyl-]

$$C_6H_5CH_2OH + HCl + HCHO \xrightarrow{25°} C_6H_5CH_2OCH_2Cl$$

Submitted by D. S. CONNOR,[1,2] G. W. KLEIN,[1,3] G. N. TAYLOR,[1,4] R. K. Boeckman, Jr.,[5] and J. B. MEDWID[6]
Checked by E. M. CARREIRA, S. E. DENMARK, and ROBERT M. COATES

1. Procedure

Caution! Benzyl chloromethyl ether is a powerful alkylating agent and a potential carcinogen. Furthermore, it is a mild lachrymator and reacts with water and alcohols, forming hydrogen chloride. The procedure should be conducted in a hood, and inhalation and skin contact should be avoided.

A 1-l, three-necked flask equipped with an overhead mechanical stirrer with a Teflon paddle, gas-inlet tube, thermometer, and a calcium chloride drying tube (Note 1) is charged with 216 g. (2.00 moles) of benzyl alcohol (Note 2) and 66 g. (2.20 moles as CH_2O) of paraformaldehyde (Note 3). The resulting mixture is maintained at 20–25° with a water bath during addition of anhydrous hydrogen chloride (Note 4) at a moderate rate, with stirring (Note 5). After approximately 2 hours the reaction is complete, as judged by the appearance of two clear homogeneous phases (Note 6). The layers are separated, and the upper layer is diluted with 800 ml. of pentane and dried over anhydrous magnesium sulfate for 3 hours at 0°, with stirring. The drying agent is removed by filtration, 2–3 g. of anhydrous calcium chloride is added to the filtrate, and the solution is concentrated on a rotary evaporator (Note 7). The residual liquid, which is nearly pure benzyl chloromethyl ether, is decanted, affording 260 g. (83%) of crude product (Note 8). This crude benzyl chloromethyl ether, which is suitable for use in some applications, is stored over anhydrous calcium chloride at 0° under an inert atmosphere (Notes 9 and 10).

If further purification is desired, just prior to use the crude material (40 g.) may be distilled at approximately 3 mm from anhydrous calcium chloride (Note 11), affording very pure benzyl chloromethyl ether (35 g.), b.p. 70–71° (3 mm.) (Notes 12 and 13).

2. Notes

1. A Claisen adapter is utilized to accommodate both the thermometer and calcium chloride drying tube.

2. Fisher Scientific reagent grade benzyl alcohol was freshly distilled prior to use.

3. Fisher Scientific reagent grade paraformaldehyde was used.

4. Anhydrous hydrogen chloride was obtained from Matheson Gas Products and dried by passing it through concentrated sulfuric acid.

5. The gas-inlet tube utilized was a Pasteur pipet; however, a fritted glass gas-dispersion tube could be utilized. Hydrogen chloride is introduced as a stream of fine bubbles; the rate of addition controls the reaction temperature.

6. To judge whether the reaction is complete, stirring is stopped and the phases are permitted to separate. ^1H NMR analysis of the upper phase (CDCl$_3$) showed that the reaction is complete and devoid of major side-products.

7. Decomposition was noted during concentration and distillation in the absence of anhydrous calcium chloride.

8. The checkers obtained 316.3–316.6 g. (101%).

9. The crude material is satisfactory for the C-alkylation of an ester enolate; little difference was noted when the crude material was substituted for distilled material.

10. The crude product exhibits singlets in the ^1H NMR (CCl$_4$) at δ 4.68 (2H), 5.41 (2H), and 7.29 (5H). Both ^1H NMR and GC analyses indicate a purity of greater than 90%. GC analysis was carried out at 155° with a 2 m. × 0.7 cm. column packed with silicone fluid No. 710 suspended on 60–80 mesh finebrick. The major impurities appear to be varying amounts of benzyl chloride and dibenzyl formal, by ^1H NMR analysis.

11. Complete decomposition occurs if distillation is attempted at atmospheric pressure. Minor to occasionally major decomposition occurs upon attempted distillation at reduced pressure in the absence of anhydrous calcium chloride, which retards the decomposition significantly.

12. The product gave satisfactory microanalytical data after one distillation. The reported physical constants for benzyl chloromethyl ether are b.p. 96–98° (9.5 mm.),[7] n_D^{20} 1.5264–1.5292.[8,9]

13. The checkers used a procedure identical to that described above at one-eighth scale to prepare benzyl bromomethyl ether, using hydrogen bromide. A quantitative yield of crude material was obtained and distilled, giving a 97% yield of pure benzyl bromomethyl ether, b.p. 55–57° (1 mm.), n_D^{20} 1.5547: ^1H NMR (CDCl$_3$): δ 4.67 (s, 2H), 5.66 (s, 2H), 7.30 (s, 5H). Analysis calculated for C$_8$H$_9$BrO (201.09): C, 47.79; H, 4.51; Br, 39.74. Found: C, 48.05; H, 4.68; Br, 40.05. This was found to be a superior alkylating agent.

3. Discussion

Benzyl chloromethyl ether is useful for introduction of a potential hydroxymethyl group in alkylation reactions. Hill and Keach[10] first used this method and found it convenient in barbiturate syntheses. Graham and McQuillin,[11] and Graham, McQuillin, and Simpson[12] have extended the scope of the alkylation reaction to various ketone derivatives. They also have investigated the conditions for obtaining maximum C-alkylation and the stereochemistry of alkylation in various octalone systems.[11] Alkylation of ketones followed by sodium borohydride reduction and catalytic hydrogenolysis represents a convenient method for obtaining 1,3-diols.[11] Similarly, Wolff-Kishner reduction and catalytic hydrogenolysis give primary alcohols.[11] A procedure of this type has been used for obtaining bridgehead methanol derivatives of bicyclic compounds.[13] Alkylation of ester enolates, generated by lithium diisopropylamide, has been reported.[14]

Several other alkylations with benzyl chloromethyl ether using phosphorus compounds as nucleophiles have been reported.[8] Hydrolysis and alcoholysis reactions of the reagent[15] have been investigated, along with the addition of the chloroether to propylene in the presence of zinc chloride.[16] Alkylation of enamines with benzyl bromomethyl ether has been reported.[17]

Benzyl chloromethyl ether has been prepared from benzyl alcohol, aqueous formaldehyde solution, and hydrogen chloride.[9,10,18] Gaseous formaldehyde[9] and 1,3,5-trioxane[19] have also been used. This chloromethyl ether has also been prepared by the chlorination of benzyl methyl ether.[16] The present procedure is based on the first method, but avoids the use of a large excess of formaldehyde and provides a considerably simplified isolation procedure.

1. Work done at Department of Chemistry, Yale University, New Haven, Connecticut 06520.
2. Present address: The Procter and Gamble Company, Miami Valley Laboratories, P. O. Box 39175, Cincinnati, Ohio 45247.
3. Present address: Eastman Kodak Research Laboratories, 1667 Lake Ave., Rochester, New York 14650.
4. Present address: Bell Telephone Laboratories, 600 Mountain Avenue, Murray Hill, New Jersey 07974.
5. Department of Chemistry, University of Rochester, Rochester, New York 14627.
6. Department of Chemistry, Wayne State University, Detroit, Michigan 48202.
7. A. Rieche and H. Gross, *Chem. Ber.*, **93**, 259 (1960).
8. V. S. Abramov, E. V. Sergeeva, and I. V. Chelpanova, *Zh. Obshch. Khim.*, **14**, 1030 (1944); *Chem. Abstr.*, **41**, 700 (1947).
9. Sh. Mamedov, M. A. Avanesyan, and B. M. Alieva, *Zh. Obshch. Khim.*, **32**, 3635 (1962); *Chem. Abstr.*, **58**, 12444 (1963).
10. A. J. Hill and D. T. Keach, *J. Am. Chem. Soc.*, **48**, 257 (1926).
11. C. L. Graham and F. J. McQuillin, *J. Chem. Soc.*, 4634 (1963).
12. C. L. Graham, F. J. McQuillin, and P. L. Simpson, *Proc. Chem. Soc. (London)*, 136 (1963).
13. K. B. Wiberg and G. W. Klein, *Tetrahedron Lett.*, 1043 (1963).
14. R. K. Boeckman, Jr., M. Ramaiah, and J. B. Medwid, *Tetrahedron Lett.*, 4485 (1977).
15. H. Böhme and A. Dörries, *Chem. Ber.*, **89**, 719 (1956).
16. H. Böhme and A. Dörries, *Chem. Ber.*, **89**, 723 (1956).
17. A. T. Blomquist and E. J. Mariconi, *J. Org. Chem.*, **26**, 3761 (1961).
18. P. Carré, *C. R. Hebd. Seances Acad. Sci.*, **186**, 1629 (1928); *Bull. Soc. Chim. Fr.*, **43**, 767 (1928).
19. S. Sabetay and P. Schving, *Bull. Soc. Chim. Fr.*, **43**, 1341 (1928).

1-BENZYLINDOLE

[1*H*-Indole, 1-(phenylmethyl)-]

Submitted by HARRY HEANEY[1] and STEVEN V. LEY
Checked by A. BROSSI, E. E. GARCIA, and R. P. SCHWARTZ

1. Procedure

A 500-ml. Erlenmeyer flask equipped with a magnetic stirring bar is charged with 200 ml. of dimethyl sulfoxide (Note 1) and 26.0 g. (0.399 mole) of potassium hydroxide (Note 2). The mixture is stirred at room temperature for 5 minutes before 11.7 g. (0.100 mole) of indole (Note 3) is added. Stirring is continued for 45 minutes before 34.2 g. (0.200 mole) of benzyl bromide (Note 4) is added (Note 5). After being stirred for an additional 45 minutes the mixture is diluted with 200 ml. of water. The mixture is extracted with three 100-ml. portions of diethyl ether, and each ether layer is washed with three 50-ml. portions of water. The combined ether layers are dried over calcium chloride, and the solvent is removed at slightly reduced pressure (Note 6). The excess benzyl bromide is removed by distillation at approximately 15 mm., and the residue is distilled, yielding 17.5–18.4 g. (85–89%) of 1-benzylindole, b.p. 133–138° (0.3 mm.). The distillate crystallizes upon cooling and scratching; recrystallization from ethanol gives material melting at 42–43° (Notes 7 and 8).

2. Notes

1. The dimethyl sulfoxide used was not rigorously dried but should not contain an appreciable amount of water.
2. Freshly crushed 86% potassium hydroxide pellets were used.
3. A commercial grade of indole is satisfactory.
4. Reagent grade benzyl bromide was used without further purification.
5. Cooling with an ice–water bath moderates the exothermic reaction.
6. The submitters used a Büchi rotary evaporator (water aspirator).
7. The submitters have obtained yields as high as 20 g. (97%).
8. ^{1}H NMR (CDCl$_3$): 5.21 (s, 2H), 6.52 (d, J = 3.4 Hz., 1H), 7.0–7.4 (m, 9H), and 7.5–7.7 (m, 1H).

3. Discussion

Although the *N*-alkylation of pyrrole[2] and indole[3] has been reported on many occasions, a generally applicable, simple, high yield procedure was not available. Many simple procedures give mixtures of products because of the ambident nature of the anions; however, alkylation at nitrogen is usually predominant in strongly ionizing solvents. Recent methods include alkylations of indole in liquid ammonia,[4] N,N-dimethylformamide,[5] and hexamethylphosphoric triamide.[6]

The use of dimethyl sulfoxide as a dipolar aprotic solvent is well known;[7] the present method can be regarded as a model procedure and has been applied to the preparation of a number of N-n-alkyl pyrroles and N-n-alkyl indoles.[8] The yield of N-benzylindole is considerably higher than in previously reported preparations and is as good as that reported for the preparations of N-methylindole in liquid ammonia.[4] The present method is, however, less laborious and quicker. Very high yields are obtained using n-alkyl halides and moderately good yields with secondary alkyl halides. The reactions may be compared with those recently reported for pyrrylthallium.[9]

The procedure has been much used for the alkylation of a variety of indoles,[10] including a modification using sodium methoxide in the methylation of tetrahydroalstonine.[11] The wide applicability of the method is indicated by the methylation of diarylamines such as 10,11-dihydro-5H-dibenzo[b,f]azepine.[12] An extension to the acylation of indoles has been reported using, for example, acetic anhydride-KOH in dimethyl sulfoxide.[13]

1. Department of Chemistry, The University of Technology, Loughborough, Leicestershire, LE11 3TU, England.
2. K. Schofield, "Heteroaromatic Nitrogen Compounds, Pyrroles and Pyridines," Butterworth, London, 1967.
3. L. R. Smith, "Indoles," Part Two, in W. J. Houlihan, "The Chemistry of Heterocyclic Compounds," Wiley-Interscience, New York, 1972.
4. K. T. Potts and J. E. Saxton, *Org. Synth.*, **Coll. Vol. 5,** 769 (1973).
5. J. Szmuzkovicz, Belg. Pat. 621,047 (1963).
6. H. Normant and T. Cuvigny, *Bull. Soc. Chim. Fr.,* 1866 (1965).
7. L. F. Fieser and M. Fieser, "Reagents for Organic Synthesis," Vol. 1, Wiley, New York, 1967, pp. 296–318; Vol. 2, 1969, pp. 157–173; Vol. 3, 1972, pp. 119–123; D. J. Cram, B. Rickborn, and G. R. Knox, *J. Am. Chem. Soc.,* **82,** 6412 (1960); D. J. Cram, B. Rickborn, C. A. Kingsbury, and P. Haberfield, *J. Am. Chem. Soc.,* **83,** 3678 (1961).
8. H. Heaney and S. V. Ley, *J. Chem. Soc. Perkin Trans. 1,* 499 (1973).
9. C. F. Candy and R. A. Jones, *J. Org. Chem.,* **36,** 3993 (1971).
10. G. W. Gribble and J. H. Hoffman, *Synthesis,* 859 (1977); R. A. W. Johnstone and M. E. Rose, *Tetrahedron,* **35,** 2169 (1979); R. A. W. Johnstone, D. Tuli, and M. E. Rose, *J. Chem. Res. Synop.,* 283 (1980).
11. R. T. Brown, C. L. Chapple, R. Platt, and H. Spencer, *J. Chem. Soc. Chem. Commun.,* 929 (1974).
12. P. Beresford, D. H. Iles, L. J. Kricka, and A. Ledwith, *J. Chem. Soc. Perkin Trans. 1,* 276 (1974).
13. G. W. Gribble, L. W. Reilly, and J. L. Johnson, *Org. Prep. Proced. Int.,* **9,** 271 (1977).

N-ALKYLINDOLES FROM THE ALKYLATION OF SODIUM INDOLIDE IN HEXAMETHYLPHOSPHORIC TRIAMIDE: 1-BENZYLINDOLE

[1H-Indole, 1-(phenylmethyl)-]

Submitted by GEORGE M. RUBOTTOM[1] and JOHN C. CHABALA[2]
Checked by R. E. IRELAND and JAMES E. KLECKNER

1. Procedure

Caution! Hexamethylphosphoric triamide (HMPA) vapors have been reported to cause cancer in rats.[3] All operations with hexamethylphosphoric triamide should be performed in a good hood, and care should be taken to keep the liquid off the skin.

A 100-ml., three-necked flask fitted with a reflux condenser, a magnetic stirring bar, and a gas-inlet tube is charged with 2.34 g. (0.0200 mole) of indole (Note 1) and 15 ml. of hexamethylphosphoric triamide (Note 2) under a static atmosphere of argon. The flask is cooled to 0° with an ice bath, and 0.53 g. (0.022 mole) of sodium hydride is added to the stirred solution over a period of 10 minutes (Note 3). The resulting slurry is stirred for 5 hours at room temperature (Note 4) then cooled to 0° (ice bath) before 2.53 g. (2.30 ml., 0.0200 mole) of benzyl chloride (Note 5) is added as rapidly as possible to the stirred mixture. The mixture is stirred for 8–15 hours (overnight), during which time the ice in the ice bath melts, and the temperature of the reaction flask gradually rises to room temperature. The mixture is then diluted with 15 ml. of water and extracted with three 25-ml. portions of diethyl ether. The combined ethereal extracts are washed with two 40-ml. portions of water and dried with anhydrous magnesium sulfate. After filtration the solvent is removed at reduced pressure, and 4.4 g. of crude 1-benzylindole is obtained as a liquid. After bulb-to-bulb distillation of this material in a Kügelrohr oven [120–130° (0.0025 mm.)], crystallization of the distillate from 15 ml. of hot ethanol affords 3.46–3.61 g. (83–87%) of 1-benzylindole. A second crop of 0.17–0.26 g. (4–6%) is obtained on concentration of the mother liquors to 6 ml. The total yield of 1-benzylindole, m.p. 43–44°, is 3.72–3.78 g. (90–91%) (Notes 6 and 7).

2. Notes

1. Commercial indole (Matheson, Coleman and Bell) was used with no further purification.

2. Commercial HMPA (Aldrich) was stored over Linde 4 A Molecular Sieves and used without further purification.

3. A batch of 0.93 g. of a 57% sodium hydride dispersion in mineral oil is washed with hexane to remove the mineral oil immediately prior to use. The slow addition in the cold minimizes the small amount of foaming caused by hydrogen evolution.

4. This stirring time insures complete formation of sodium indolide.

5. Commercial benzyl chloride (Matheson, Coleman and Bell) was used without further purification.

6. The recrystallized product has ^{1}H NMR absorptions (CDCl$_3$) at δ 5.21 (s, 2H), 6.52 (d, J = 3.4 Hz., 1H), 7.0–7.4 (m, 9H), and 7.5–7.7 (m, 1H).

7. 1-Benzylindole colors significantly in contact with air at room temperature (*ca.* 1 week), but keeps indefinitely under argon.

3. Discussion

Generally, the alkylation of sodium indolide, generated from indole and sodium amide in liquid ammonia, has been used for the preparation of *N*-alkylindoles.[4–12] The drawback to this method is the use of liquid ammonia. The procedure outlined here[13] overcomes this problem and affords pure *N*-alkylindoles in excellent yields. Further, the use of the hexamethylphosphoric triamide–sodium hydride system affords conditions leading to the formation of the *N*-alkylindoles with little or no side reaction leading to *C*-alkylated products.[13,14] Table I illustrates the generality of the method.

TABLE I
ALKYLATION OF INDOLE SODIUM WITH R-X IN HMPA SOLVENT

R-X	Yield of N-Alkylindole, %	b.p. or m.p.
CH$_3$I	94	b.p. 73–75°/2.4 mm. (Ref. 15, b.p. 70–75/2 mm.)
C$_2$H$_5$I	92	b.p. 83–86°/0.6 mm. (Ref. 16, b.p. 82–85°/0.7 mm.)
CH$_2$=CHCH$_2$Br	84	b.p. 72–73°/0.12 mm. (Ref. 7, b.p. 114–116°/6 mm.)
C$_6$H$_5$CH$_2$Cl	90–91	m.p. 43–44° (Ref. 6, m.p. 44°)

1. Department of Chemistry, University of Puerto Rico, Rio Piedras, Puerto Rico 00931. [Present address: Department of Chemistry, University of Idaho, Moscow, Idaho 83843.]
2. Merck Sharp & Dohme Research Laboratories, Rahway, New Jersey 07065.

3. J. A. Zapp, *Science*, **190**, 422 (1975).
4. K. T. Potts and J. E. Saxton, *J. Chem. Soc.*, 2641 (1954).
5. K. T. Potts and J. E. Saxton, *Org. Synth.*, **Coll. Vol. 5**, 769 (1973).
6. H. Plieninger, *Chem. Ber.*, **87**, 127 (1954).
7. M. Nakazaki and S. Isoe, *Nippon Kagaku Zasshi*, **76**, 1159 (1955); *Chem. Abstr.*, **51**, 17877.
8. M. Nakazaki, *Bull. Chem. Soc. Jpn.*, **32**, 838 (1959).
9. N. I. Grineva, V. L. Sadovskaya and V. N. Ufimtsev, *Zh. Obshch. Khim.*, **33**, 552 (1963).
10. M. Julia and P. Manoury, *Bull. Soc. Chim. Fr.*, 1946 (1964).
11. S. Yamada, *Chem. Pharm. Bull. (Tokyo)*, **13**, 88 (1965).
12. A. H. Jackson and A. E. Smith, *Tetrahedron*, **21**, 989 (1965).
13. Essentially the procedure outlined in G. M. Rubottom and J. C. Chabala, *Synthesis*, 566 (1972).
14. B. Cardillo, G. Casnati, A. Pochini, and R. Ricca, *Tetrahedron*, **23**, 3771 (1967).
15. L. Marion and C. W. Oldfield, *Can. J. Res. B*, **25**, 1 (1947).
16. A. P. Gray, H. Kraus, and D. E. Heitmeier, *J. Org. Chem.*, **25**, 1939 (1960).

3-ALKYLATED AND 3-ACYLATED INDOLES FROM A COMMON PRECURSOR: 3-BENZYLINDOLE AND 3-BENZOYLINDOLE

[1H-Indole, 3-(phenylmethyl)- and Methanone, 1H-indole-3-ylphenyl-]

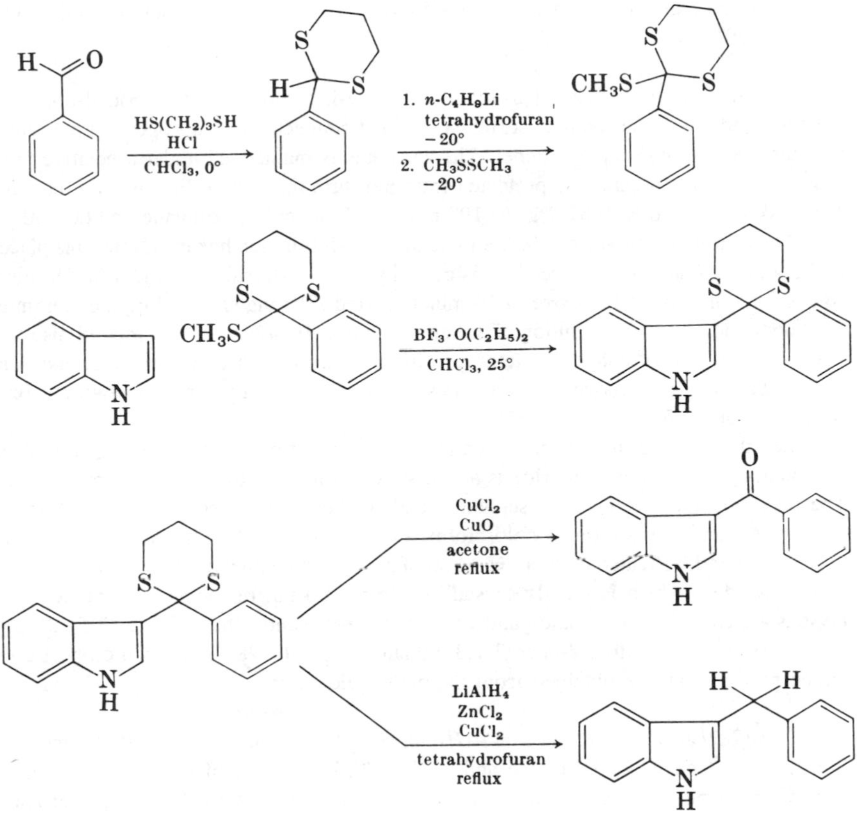

Submitted by P. Stütz and P. A. Stadler[1]
Checked by M. Cushman and G. Büchi

1. Procedure

Caution! The following reactions should be conducted in a well-ventilated hood, since in each step evil-smelling sulfur compounds are either used as starting materials or generated during the reaction.

A. *2-Phenyl-1,3-dithiane.* A 100 ml., two-necked, round-bottomed flask fitted with a gas-inlet tube and a drying tube containing glass wool is charged with 5.52 g. (5.28

ml., 0.0521 mole) of benzaldehyde, 5.6 g. (5.2 ml., 0.052 mole) of 1,3-propanedithiol, and 30 ml. of chloroform (Note 1). The flask is immersed in an ice bath, and a slow stream of hydrogen chloride gas is bubbled through the solution. When saturated (after *ca.* 5 minutes), the reaction mixture is left at room temperature for 30 minutes, then evaporated at reduced pressure. The oily residue (Note 2) is taken up in 50 ml. of methanol and vigorously agitated to induce crystallization. The crystals are collected by filtration, washed with ligroin, and air dried. By condensing the mother liquors, two more crops may be obtained, giving a total of 9.5–9.7 g. (93–95%) of 2-phenyl-1,3-dithiane, m.p. 69–70° (Note 3).

B. *2-Methylthio-2-phenyl-*1,3-*dithiane.* A 2-l., three-necked, round-bottomed flask is fitted with a mechanical stirrer, a gas inlet connected to a nitrogen source, and a pressure-equalizing dropping funnel. The apparatus is maintained under a positive pressure of nitrogen and carefully protected from moisture throughout the ensuing reaction. The flask is charged with 37.2 g. (0.190 mole) of 2-phenyl-1,3-dithiane and 600 ml. of anhydrous tetrahydrofuran, and 0.228 mole of *n*-butyllithium in hexane (Note 4) is placed in the funnel. With stirring, the flask is cooled with a −20° bath, and the *n*-butyllithium solution is run into the flask over a 10-minute period. Stirring and cooling are continued for 2 hours, after which a solution of 32.2 g. (30.4 ml., 0.342 mole) of dimethyl disulfide (Note 5) in 50 ml. of anhydrous tetrahydrofuran is added over a 10-minute period. The cooling bath is then removed, and as the solution is stirred at room temperature a fine, white precipitate forms.

When the reaction mixture has been at room temperature for 90 minutes, 300 ml. of saturated aqueous sodium chloride is added slowly, followed by 500 ml. of chloroform. The resulting aqueous layer is separated and washed with three 100-ml. portions of chloroform, and the combined chloroform solutions are dried with sodium sulfate and decolorized with activated carbon. Removal of chloroform with a rotary evaporator leaves an oily residue, which is readily crystallized from 100 ml. of methanol. The colorless crystals are collected by filtration and washed with petroleum ether, giving 33.5–35.3 g. (72–77%) of 2-methylthio-2-phenyl-1,3-dithiane, m.p. 76–78°. A second crop of crystalline product may be obtained from the mother liquors.

C. *3-(2-Phenyl-*1,3-*dithian-*2-*yl)-*1*H-indole.* A solution of 24.2 g. (0.102 mole) of 2-methylthio-2-phenyl-1,3-dithiane and 5.86 g. (0.0501 mole) of indole in 600 ml. of chloroform is placed in a 2-l., three-necked flask fitted with a mechanical stirrer and a 100-ml., pressure-equalizing dropping funnel. The reaction mixture is stirred vigorously as a solution of 25 ml. (about 0.2 mole) of boron trifluoride diethyl etherate (Note 6) in 50 ml. of chloroform is added over 10 minutes. An orange–brown, resinous precipitate forms as the slightly exothermic reaction proceeds. After an additional 10 minutes of stirring, a solution of 5.86 g. (0.0501 mole) of indole in 50 ml. of chloroform is added in one portion.

Stirring is continued for 2 hours at room temperature, and methanol is added until a clear solution is obtained (*ca.* 10 ml. of methanol is required, and some heat is generated). When the solution has cooled, it is washed successively with 200 ml. of aqueous 2 *N* potassium carbonate and 200 ml. of water. The aqueous phases are combined, washed with three 100-ml. portions of chloroform, and discarded. The organic phases are combined, dried over sodium sulfate, and decolorized with activated carbon. Concentration of the chloroform solution obtained provides three crops of pale yellow crystals, which are washed with 30% hexane in chloroform and dried for 2 hours at 80°/0.1 mm, yielding 22.3 – 25.4 g. (72 – 81%) of 3-(2-phenyl-1,3-dithian-2-yl)-1*H*-indole, m.p. 167 – 169° (Note 7). This material requires no further purification for use in Parts D or E.

D.   *3-Benzoylindole*.   A 100-ml., three-necked flask fitted with a magnetic stirring bar, a condenser, and a pressure-equalizing dropping funnel is charged with 0.48 g. (0.0060 mole) of copper(II) oxide, 1.61 g. (0.0120 mole) of anhydrous copper(II) chloride (Note 8), and 40 ml. of acetone. The resulting suspension is brought to reflux with vigorous stirring, and a solution of 1.55 g. (0.00508 mole) of 3-(2-phenyl-1,3-dithian-2-yl)-1*H*-indole in 9 ml. of acetone and 1 ml. of *N,N*-dimethylformamide is added over 5 minutes. Reflux temperatures are maintained for 90 minutes, during which time the reaction mixture gradually turns yellow, then the mixture is cooled and filtered. The insoluble material is washed with three 20-ml. portions of hot 10% ethanol in dichloromethane, and the combined organic solutions are washed with 50 ml. of aqueous 2 *N* sodium carbonate (Note 9), dried over sodium sulfate, and filtered. Concentration of the filtrate to a small volume under reduced pressure gives a residue which gradually deposits crystals. Filtration provides 0.94 – 0.97 g. (85 – 88%) of pure 3-benzoylindole, m.p. 238 – 240°.

E.   *3-Benzylindole*.   A suspension of 1.34 g. (0.0100 mole) of anhydrous copper(II) chloride (Note 8) and 2.72 g. (0.0200 mole) of anhydrous zinc chloride (Note 10) in 150 ml. of anhydrous tetrahydrofuran is prepared in a 500-ml., three-necked, round-bottomed flask fitted with a mechanical stirrer and a condenser connected to a nitrogen source. This mixture is stirred at room temperature and maintained under a nitrogen atmosphere while 3.04 g. (0.0800 mole) of lithium aluminum hydride is added cautiously in small portions (Note 11). The resulting exothermic reaction gives a black precipitate, which is stirred at room temperature for 45 minutes before 1.55 g. (0.00508 mole) of 3-(2-phenyl-1,3-dithian-2-yl)-1*H*-indole is added. Stirring is continued as the mixture is refluxed for 1 hour, cooled to room temperature, and quenched by careful dropwise addition of 10 ml. of water. The resulting slurry is diluted with 100 ml. of dichloromethane and filtered to remove inorganic salts, which are washed by digesting with three 50-ml. portions of refluxing dichloromethane. The combined filtrates are then washed with a solution of 4.0 g. (0.012 mole) of mercury(II) acetate in 100 ml. of water, dried over sodium sulfate, filtered, and concentrated on a rotary evaporator, leaving about 1.1 g. of a yellow crystalline residue. Bulb-to-bulb distillation at 145 – 155° (0.1 mm.) yields 0.74 – 0.85 g. (72 – 82%) of practically pure 3-benzylindole as pale yellow crystals, m.p. 102 – 105° (Note 12).

2. Notes

1. Benzaldehyde and 1,3-propanedithiol were purchased from the Aldrich Chemical Company, Inc., and used without further purification. Exclusion of moisture during the reaction is advantageous but not essential.

2. The checkers obtained a crystalline residue, which was triturated with 10 ml. of methanol and filtered.

3. The method used is quite general for substituted 1,3-dithianes.[2] A different

method is required for the preparation of 1,3-dithiane itself [*Org. Synth.*, **Coll. Vol. 6,** 556 (1988)].

4. The volume of solution required will depend on the concentration of *n*-butyllithium, which should be determined by titration prior to use. A convenient procedure utilizes 1,10-phenanthroline as an indicator; titration to the colorless end point with 2-butanol[3] in xylene gives organolithium concentration directly.[4] A method for transferring large volumes of *n*-butyllithium solution is outlined in *Org. Synth.*, **Coll. Vol. 6,** 979 (1988).

5. Dimethyl disulfide was obtained from the Aldrich Chemical Company, Inc., and used without further purification. The molar ratio used is the same as that in the original reference.[5]

6. Boron trifluoride etherate was purchased from either Fluka AG, Switzerland or Aldrich Chemical Company, Inc. The checkers distilled the commercial material from calcium hydride immediately prior to use.[6]

7. ^{1}H NMR (CDCl$_3$), δ (multiplicity, coupling constant J in Hz., number of protons): 1.5–2.2 (m, 2H), 2.7–3.0 (m, 4H), 7.0 (d, $J = 2.5$, 1H), 7.1–7.5 (m, 6H), 7.7–8.2 (m, 4H). A sample recrystallized from methanol–chloroform melted at 183–185°. The submitters also obtained pure product, m.p. 181–183°, after chromatography on basic alumina with 20% petroleum ether in dichloromethane as eluent.

8. Anhydrous copper(II) chloride was prepared by heating the dihydrate at 110° overnight.

9. This extraction must be performed gently, since violent agitation will give an emulsion.

10. Reagent grade anhydrous zinc chloride was obtained from Merck & Company, Inc.

11. Lithium aluminum hydride was obtained from Fluka AG or from Ventron Corporation. One convenient technique for the addition is to weigh the reagent into an Erlenmeyer flask, which is then connected to the reaction flask by a short piece of Gooch tubing. In this way the solid can be added in portions without exposing it to the atmosphere.

12. Various melting points are reported in the literature for 3-benzylindole: 96–98°,[7] 103°, and 107°.[8]

3. Discussion

There are other convenient methods for the preparation of 3-benzylindole[7,8] and 3-benzoylindole.[9] The present procedure, however, has two useful elements of flexibility: it produces both 3-alkyl- and 3-acylindoles from a single precursor, and it tolerates the presence of a wide variety of substituents.

The pivotal step in this sequence is an electrophilic substitution on indole. Although the use of 1,3-dithian-2-yl carbanions is well documented, it has been shown only recently that 1,3-dithian-2-yl carbenium ions can be used in a Friedel-Crafts type reaction. This was accomplished initially using 2-methoxy-1,3-dithiane or 2-methoxy-1,3-dithiolane and titanium tetrachloride as the Lewis acid catalyst.[10] 2-Substituted lysergic acid derivatives and 3-substituted indoles have been prepared under these conditions, but the method is limited in scope by the difficulties of preparing substituted 2-methoxy-1,3-dithianes. 1,3-Dithian-2-yl carbenium ions have also been prepared by protonation of ketene dithioacetals with trifluroacetic acid,[11] but this reaction cannot be used to introduce 1,3-dithiane moieties into indole.

The procedure described herein is fairly general for indoles, and since 2-methylthio-1,3-dithianes are readily available,[3,5] it should prove versatile. Two further examples are as follows:

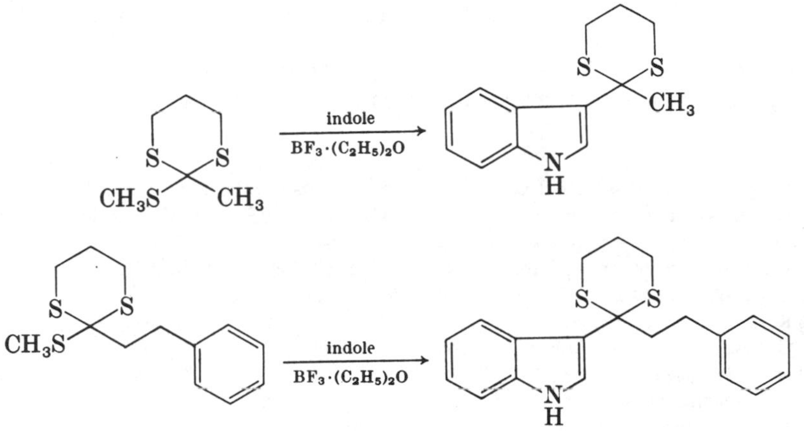

In attempting to extend the method to other activated aromatics, it was found that pyrroles give mixtures of 2- and 3-substituted products, and that naphthol ethers and benzo[b]thiophene fail to react.

The hydrolytic step (Part D) uses conditions described by Narasaka, Sakashita, and Mukaiyama.[12] It was necessary to modify the original stoichiometry, since the recommended molar ratio of substrate : copper(II) chloride : copper(II) oxide (1 : 2 : 4) gave only a 57% yield of 3-benzoylindole. The more generally known mercuric oxide–mercuric chloride hydrolysis[2] may also be used, and in the present case it gives a yield of about 90%. The reductive desulfurization of Part E, also based on the work of Mukaiyama,[13] is clearly superior to Raney nickel desulfurization, which gives only 35–45% of 3-benzylindole.

Some new reagents of the same general type, leading to intermediate carbocations of dithians, have been reported in the literature recently. Hiratani, Nakai, and Okawara synthesized 1,3-dithian-2-yltrimethylammonium iodide.[14] Corey and Walinsky[15] applied 1,3-dithian-2-yl fluoroborate, prepared by hydride ion exchange from 1,3-dithian and trityl fluoroborate,[16] to a new kind of electrophilic reaction for the preparation of cyclopentane derivatives.

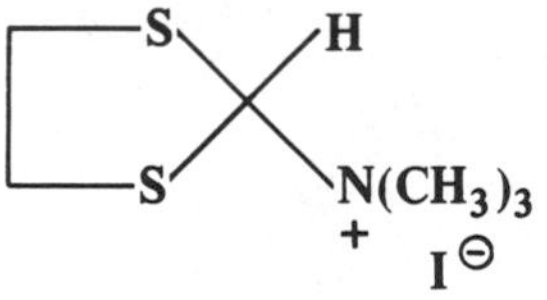

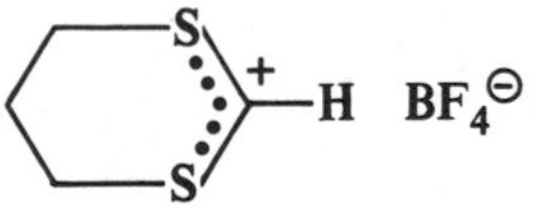

Further, substantial progress leading to a generally applicable method is shown by the preparation of 2-chloro-1,3-dithiane and its application in electrophilic substitution reactions with reactive aromatic molecules like phenol and N,N-dimethylaniline.[17]

So far, however, no reagent of the dithianylcarbocation type has been found which allows electrophilic substitution reactions with unactivated aromatic molecules such as benzene.

1. Sandoz Ltd., Pharmaceutical Division, Preclinical Research, Basel, Switzerland.
2. D. Seebach, *Angew. Chem. Int. Ed. Engl.*, **6**, 442 (1967); D. Seebach, B. W. Erickson, and G. Singh, *J. Org. Chem.*, **31**, 4303 (1966).
3. For a warning concerning 2-butanol, see *Chem. Eng. News*, **59**, (19), 3 (1981).
4. S. C. Watson and J. F. Eastham, *J. Organomet. Chem.*, **9**, 165 (1967).
5. R. A. Ellison, W. D. Woessner, and C. C. Williams, *J. Org. Chem.*, **37**, 2757 (1972).
6. G. Zweifel and H. C. Brown, *Org. React.*, **13**, 28 (1963).
7. H. Plieninger, *Chem. Ber.*, **87**, 127 (1954).
8. R. Cornforth and R. Robinson, *J. Chem. Soc.*, 680 (1942).
9. W. C. Anthony, *J. Org. Chem.*, **25**, 2049 (1960).
10. P. Stütz and P. A. Stadler, *Helv. Chim. Acta*, **55**, 75 (1972).
11. F. A. Carey and A. S. Court, *J. Org. Chem.*, **37**, 1926 (1972).
12. K. Narasaka, T. Sakashita, and T. Mukaiyama, *Bull. Chem. Soc. Jpn.*, **45**, 37 (1972).
13. T. Mukaiyama, *Int. J. Sulfur Chem.*, **7**, 173 (1972).
14. K. Hiratani, T. Nakai, and M. Okawara, *Bull. Chem. Soc. Jpn.*, **46**, 3510 (1973).
15. E. J. Corey and S. W. Walinsky, *J. Am. Chem. Soc.*, **94**, 8932 (1972).
16. H. J. Dauben, L. R. Honnen, and K. M. Harmon, *J. Org. Chem.*, **25**, 1442 (1960).
17. K. Arai and M. Oki, *Tetrahedron Lett.*, **1975**, 2183; K. Arai and M. Oki, *Bull. Chem. Soc. Jpn.*, **49**, 553 (1976); C. G. Kruse, N. L. J. M. Broekhof, A. Wijsman, and A. van der Gen, *Tetrahedron Lett.*, **1977**, 885; C. G. Kruse, A. Wijsman, and A. van der Gen, *J. Org. Chem.*, **44**, 1847 (1979).

ALKYLATION OF ISOQUINOLINES
via 2-BENZOYL-1,2-DIHYDROISOQUINALDONITRILES:
1-BENZYLISOQUINOLINE

[Isoquinoline, 1-(phenylmethyl)-]

Submitted by BARRIE C. UFF,[1] JOHN R. KERSHAW,[1] and JOHN L. NEUMEYER[2]
Checked by A. BROSSI, H. BRUDERER, and J. METZGER

1. Procedure

Caution! This reaction involves highly toxic cyanide salts. It may be carried out safely, however, if prudent laboratory procedures are practiced. In particular, cyanide residues should be collected and disposed of separately (Note 1), and the entire sequence should be performed in an efficient hood.

A. *2-Benzoyl-1,2-dihydroisoquinaldonitrile.* A 1-l., three-necked, round-bottomed flask fitted with a mechanical stirrer, a thermometer, and a 250-ml., pressure-equalizing dropping funnel is charged with 64.6 g. (0.499 mole) of freshly distilled isoquinoline (Note 2) in 400 ml. of dichloromethane and 97.7 g. (1.50 moles) of potassium cyanide (Note 2) in 200 ml. of water. The mixture is stirred vigorously as 126.5

g. (121 ml., 0.900 mole) of freshly distilled benzoyl chloride is added from the dropping funnel over 1 hour. As the addition proceeds, the temperature rises to 38°, and the dichloromethane comes to reflux. Stirring is continued for an additional 3 hours, and the resulting brown reaction mixture is filtered through 20 g. of Dicalite Speedex. After the insoluble material has been washed with 200 ml. of water and 200 ml. of dichloromethane, the filtrate and washings are transferred to a separatory funnel. The layers are separated (Notes 1 and 3), and the dichloromethane layer is washed successively with 300 ml. of water, three 200-ml. portions of 2 N hydrochloric acid, 200 ml. of water, three 200-ml. portions of aqueous 2 N sodium hydroxide, and 200 ml. of water (Note 1). After drying over anhydrous potassium carbonate, the dichloromethane solution is filtered and evaporated under reduced pressure, giving 108–110 g. of a pale brown solid. This crude product is dissolved in 200 ml. of boiling ethyl acetate, filtered, and set aside to cool. On standing, the Reissert compound crystallizes as cream rhombs, which are collected on a Büchner funnel and dried in a vacuum desiccator, giving 89.9 g. (69%) of 2-benzoyl-1,2-dihydroisoquinaldonitrile, m.p. 125–127° (Notes 4 and 5).

B.   1-*Benzylisoquinoline*.   An 11.4-g. portion of 55% sodium hydride–mineral oil dispersion (Note 6) is washed free of oil by slurrying with two 40-ml. portions of dry hexane (Note 7) and decanting the liquid. The sodium hydride is transferred to a dry, 2-l., three-necked, round-bottomed flask fitted with a mechanical stirrer, a 500-ml., pressure-equalizing dropping funnel, and a nitrogen-inlet tube (Note 8). A slurry formed by adding 200 ml. of N,N-dimethylformamide (Note 9) is cooled to −10° with a methanol–ice bath. Stirring is begun, and a solution of 65.0 g. (0.250 mole) of 2-benzoyl-1,2-dihydroisoquinaldonitrile and 32 g. (29 ml., 0.25 mole) of benzyl chloride in 400 ml. of dry N,N-dimethylformamide is added dropwise over 1 hour. During addition the reaction mixture becomes dark, then fades to light brown. Ice is added as required to hold the bath temperature near −10° (Note 10).

When the addition is complete, the reaction mixture is stirred overnight at room temperature, maintaining a nitrogen atmosphere. Excess sodium hydride is destroyed by slow addition of 10 ml. of water. The N,N-dimethylformamide is evaporated at 40° (0.01 mm.), and the residue is diluted with 800 ml. of toluene and 800 ml. of water. After thorough shaking, the mixture is transferred to a separatory funnel. The aqueous layer is discarded, and the toluene layer is washed with two 200-ml. portions of water, dried over anhydrous potassium carbonate, and filtered. Removal of toluene under reduced pressure leaves a yellow oil, which crystallizes on standing (Note 11).

This material is dissolved in 500 ml. of ethanol and transferred to a 2-l., round-bottomed flask. A solution of 200 g. of sodium hydroxide in 200 ml. of water is added, and the mixtured is refluxed for 2 hours. Ethanol is then removed by distillation, and the residue is shaken with 500 ml. of water and 800 ml. of toluene. The toluene layer is separated, washed with two 200-ml. portions of water, and then vigorously shaken with 600 ml. of 2 N hydrochloric acid. A portion of the 1-benzylisoquinoline hydrochloride precipitates at this point and is collected on a Büchner funnel and washed with 200-ml. portions of water and toluene. The filtrate is then transferred to a separatory funnel, the acidic layer is separated, and the crystals from the Büchner funnel are suspended in this layer. After basifying the suspension with 50% aqueous sodium hydroxide, the oil that separates is extracted with three 200-ml. portions of dichloromethane. The combined dichloromethane layers are dried over anhydrous potassium carbonate, filtered, and evaporated under reduced pressure, leaving the crude product as a yellow oil. Distillation under reduced pressure yields 49.8 g. (91%) of pure 1-benzylisoquinoline as a pale yellow oil, b.p. 145–150° (0.01 mm.). The product solidifies on standing and may be crystallized from chloroform–hexane, giving colorless prisms, m.p. 54–55° (Note 12).

2. Notes

1. The original aqueous layer and the aqueous washings of the dichloromethane layer contain cyanide residues. These should be destroyed prior to disposal by making the solution strongly basic with sodium hydroxide and then adding, with stirring, a large excess of iron(II) sulfate. The resulting suspension should be boiled for several hours in the hood before disposal. This process converts cyanide to the nontoxic Prussian Blue [iron(III) ferrocyanide], which precipitates.

2. This product is supplied by Fluka AG, Buchs, Switzerland.

3. Any precipitate occurring during the washing procedure was removed by filtration through a small amount of Dicalite Speedex, obtained from Chemische Fabrik Schweizerhalle, Switzerland.

4. A further fraction of less pure material can be obtained by evaporating the filtrate to approximately 50 ml. and refrigerating the solution overnight.

5. The literature[3] gives m.p. 124–126°. IR (KBr)cm.$^{-1}$: 2240 very weak (—C≡N), 1658 strong (C=O), 1632 strong (C=C—N); ^{1}H NMR (8% w/w in CDCl$_3$), δ (multiplicity, coupling constant J in Hz., number of protons, assignment): 6.06 (d, $J_{3,4}$ = 7.8, 1H, H_4), 6.57 (broad s, 1H, H_1), 6.60 (d of d, $J_{3,4}$ = 7.8 and $J_{1,3}$ = 1, 1H, H_3), 7.0–7.7 (m, 9H, aromatic CH).

6. This product is supplied by Fluka AG, Buchs, Switzerland. The amount of dispersion used should provide 6.25 g. (0.260 mole) of pure sodium hydride. A small molar excess is used to allow for variation in the sodium hydride:oil ratio of commercial material.

7. The hexane was dried by filtration through alumina.

8. The apparatus was dried in an oven and assembled hot under a stream of dry, oxygen-free nitrogen. A nitrogen atmosphere was maintained throughout the reaction, since oxygen reacts with the Reissert anion, giving 1-cyanoisoquinoline.[4] Commercial nitrogen was dried by bubbling through concentrated sulfuric acid and found to be sufficiently oxygen-free, requiring no special treatment.

9. N,N-Dimethylformamide was purchased from Merck, Darmstadt and dried over molecular sieves (Union Carbide, type 4A).

10. It is necessary to hold the reaction temperature below $-5°$ in order to prevent 1,2-rearrangement of the Reissert anion to 1-benzoylisoquinoline.[5]

11. The intermediate 1-benzyl-2-benzoyl-1,2-dihydroisoquinaldonitrile can be crystallized from ethyl acetate–hexane, m.p. 130–131°. The compound has been described as an oil[5,6] and as a crystalline product, m.p. 129°.[7] IR (KBr) cm.$^{-1}$: 2236 weak (C≡N), 1669 strong (C=O), 1641 strong (C=C—N); ^{1}H NMR 8% w/w in CDCl$_3$), δ (multiplicity, coupling constant J in Hz., number of protons, assignment): 3.48 and 3.78 (AB qtr, $J = 13$, 2H, C$_6$H$_5$CH$_2$), 5.54 (d, $J = 8$, 1H, H_4), 6.37 (d, $J = 8$, 1H, H_3), 6.75–7.8 (m, 14H, aromatic CH). Analysis calculated for C$_{24}$H$_{18}$N$_2$O: C, 82.26; H, 5.18; N, 7.99. Found: C, 82.08; H, 5.29; N, 7.89.

12. The literature gives m.p. 56°.[8] IR (KBr) cm.$^{-1}$: 1621, 1601, 1585, 1560, 1500, 1493; ^{1}H (8% w/w in CDCl$_3$), δ (multiplicity, coupling constant J in Hz., number of protons, assignment): 4.66 (s, 2H, CH_2), 7.1–8.25 (m, 10H, aromatic CH), 8.5 (d, $J = 6$, 1H, H_3).

3. Discussion

Other methods for the synthesis of 1-benzylisoquinolines include: (a) dehydrogenation of 1-benzyl-3,4-dihydroisoquinolines,[9,10] which in turn are produced in the Bischler-Napieralski reaction by heating N-phenylacetyl-β-phenylethylamines with a dehydrating agent such as phosphorus pentoxide in xylene,[9,10] (b) the Pictet-Gams modification of (a), in which N-phenylacetyl-2-hydroxy-2-phenylethylamines are dehydrated with phorphorus pentoxide,[8,10,11,12] (c) thermal rearrangements of N-benzylisoquinolinium chlorides in the presence of copper,[13] and (d) addition of benzylmagnesium chloride to isoquinolines.[14] The first two methods are limited in scope by the accessibility of starting materials and the requirement that the aromatic ring carry an electron-donating substituent *para* to the point of closure for reasonable yields. The last two methods often lead to mixtures and have not been shown to be of general applicability.

The Reissert method[15]—conversion of an isoquinoline to a 2-benzoyl-1,2-dihydroisoquinaldonitrile (Reissert compound), alkylation, and hydrolysis—has enjoyed wide success in the synthesis of benzylisoquinoline and related alkaloids.[16,17] In particular, aporphines are prepared conveniently by converting isoquinolines to 1-(*o*-nitrobenzyl)-isoquinolines *via* a Reissert sequence, followed by N-alkylation, reduction, and Pschorr cyclization.[17]

The present procedure illustrates two recent and highly useful modifications of the Reissert method. First, the Reissert compound is formed by the two-phase method of Popp and Blount.[18] This modification generally gives much higher yields for isoquinolines[19] and quinolines[20] than does the single (aqueous) phase method used previously,[21] succeeding in many cases where the aqueous method fails altogether. The aqueous method is generally less clean and has the disadvantage that both starting material and product are insoluble in water. A nonaqueous benzene–hydrogen cyanide method[22] has also been used for Reissert compound formation, but it has the obvious drawbacks associated with the use of hydrogen cyanide. Second, the Reissert anion is formed with sodium hydride–N,N-dimethylformamide. This modification, developed by the submitters[23] and independently by Popp and Wefer,[6,24] has several advantages over the earlier reagent, phenyllithium in ether,[5,21] in that the sodium hydride does not have to be specially prepared, and its strength is known without titration; cessation of hydrogen evolution indicates that carbanion generation is complete; and the use of N,N-dimethylformamide overcomes solubility problems often encountered because of the ether used in the earlier method.

It has been observed[25,26] that in lower yielding syntheses of Reissert compounds improvements can be obtained by the inclusion of a phase transfer catalyst such as benzyltrimethylammonium chloride, in $1-3\%$ quantities by weight with respect to the weight of potassium cyanide. For example, methyl 1-cyano-1,2-dihydoisoquinoline-2-carboxylate forms in 69% yield in the presence of benzyltrimethylammonium chloride, but in only 24% in its absence.[25]

1. Department of Chemistry, Loughborough University of Technology, Leicestershire, LE11 3TU, England.
2. Graduate School of Pharmacy and Allied Health Professions, Northeastern University, Boston, Massachusetts 02115.
3. J. J. Padbury and H. G. Lindwall, *J. Am. Chem. Soc.*, **67**, 1268 (1945).
4. G. W. Kirby, S. L. Tan, and B. C. Uff, IUPAC Congress, Boston, 1971, Abstract 270, p. 113.
5. V. Boekelheide and J. Weinstock, *J. Am. Chem. Soc.*, **74**, 660 (1952).
6. F. D. Popp and J. M. Wefer, *J. Heterocycl. Chem.*, **4**, 183 (1967).
7. M. Markosza, *Tetrahedron Lett.*, 677 (1969).
8. R. Forsyth, I. Kelly, and F. L. Pyman, *J. Chem. Soc.*, **127**, 1659 (1925).
9. A. Pictet and F. W. Kay, *Ber. Dtsch. Chem. Ges.*, **42**, 1973 (1909); C. I. Brodrick and W. F. Short, *J. Chem. Soc.*, 2587 (1949).
10. W. M. Whaley and T. R. Govindachari, *Org. React.*, **6**, 74 (1951).
11. A. Pictet and A. Gams, *Ber. Dtsch. Chem. Ges.*, **43**, 2384 (1910); R. Robinson, *J. Chem. Soc.*, **95**, 2167 (1909).
12. A. Brossi and S. Teitel, *Helv. Chim. Acta*, **49**, 1757 (1966).
13. J. v. Braun, J. Nelles, and A. May, *Ber. Dtsch. Chem. Ges.*, **70B**, 1767 (1937).
14. E. Bergmann and W. Rosenthal, *J. Prakt. Chem.*, [2] **135**, 267 (1932).
15. A. Reissert, *Ber. Dtsch. Chem. Ges.*, **38**, 3415 (1905); F. D. Popp, *Adv. Heterocycl. Chem.*, **9**, 1 (1968).
16. F. D. Popp, *Heterocycles*, **1**, 165 (1973); A. H. Jackson and G. W. Stewart, *J. Chem. Soc. D*, 149 (1971); S. F. Dyke and A. C. Ellis, *Tetrahedron*, **27**, 3083 (1971); S. F. Dyke and A. C. Ellis, *Tetrahedron*, **28**, 3999 (1972); B. C. Uff, J. R. Kershaw, and S. R. Chhabra, *J. Chem. Soc. Perkin Trans. 1*, 479 (1972); A. J. Birch, A. H. Jackson, P. V. R. Shannon, and P. S. P. Varma, *Tetrahedron Lett.*, 4789 (1972); F. R. Stermitz and D. K. Williams, *J. Org. Chem.*, **38**, 1761 (1973).
17. J. L. Neumeyer, B. R. Neustadt, and J. Weintraub, *Tetrahedron Lett.*, 3107 (1967); J. L. Neumeyer, K. H. Oh, K. K. Weinhardt, and B. R. Neustadt, *J. Org. Chem.*, **34**, 3786 (1969); M. P. Cava and M. Srinivasan, *Tetrahedron*, **26**, 4649 (1970); M. P. Cava and M. V. Lakshmikantham, *J. Org. Chem.*, **35**, 1867 (1970); M. P. Cava and I. Noguchi, *J. Org. Chem.*, **37**, 2936 (1972); M. P. Cava and I. Noguchi, *J. Org. Chem.*, **38**, 60 (1973); J. L. Neumeyer, B. R. Neustadt, K. H. Oh, K. K. Weinhardt, and C. B. Boyce, *J. Med. Chem.*, **16**, 1223 (1973); J. L. Neumeyer, U.S. Pat. 3,717,639 (1973).
18. F. D. Popp and W. Blount, *Chem. Ind. (London)*, 550 (1961).
19. F. D. Popp and W. Blount, *J. Org. Chem.*, **27**, 297 (1962).
20. F. D. Popp, W. Blount, and P. Melvin, *J. Org. Chem.*, **26**, 4930 (1961).
21. J. Weinstock and V. Boekelheide, *Org. Synth.*, **Coll. Vol. 4**, 641 (1963).

22. J. M. Grosheintz and H. O. L. Fischer, *J. Am. Chem. Soc.*, **63**, 2021 (1941).
23. J. R. Kershaw and B. C. Uff, *Chem. Commun.*, 331 (1966); B. C. Uff and J. R. Kershaw, *J. Chem. Soc. C*, 666 (1969).
24. F. D. Popp and J. M. Wefer, *Chem. Commun.*, 207 (1966).
25. B. C. Uff and R. S. Budhram, *Heterocycles*, **6**, 1789 (1977).
26. D. Bhattacharjee and F. D. Popp, *Heterocycles*, **6**, 1905 (1977).

THE FORMATION AND ALKYLATION OF SPECIFIC ENOLATE ANIONS FROM AN UNSYMMETRICAL KETONE: 2-BENZYL-2-METHYLCYCLOHEXANONE AND 2-BENZYL-6-METHYLCYCLOHEXANONE

[Cyclohexanone, 2-methyl-2-(phenylmethyl)- and cyclohexanone, 2-methyl-6-(phenylmethyl)-]

Submitted by Martin Gall and Herbert O. House[1]
Checked by K. E. Wilson and S. Masamune

1. Procedure

A.   *2-Methyl-1-cyclohexen-1-yl acetate.   Caution! Since mixtures of perchloric acid with small amounts of organic material can explode violently, the perchloric acid should always be the last component added to the reaction mixture.*

To a 1-l. flask are added 600 ml. of carbon tetrachloride, 270 g. (250 ml., 2.65 moles) of acetic anhydride, 56 g. (0.50 mole) of 2-methylcyclohexanone, and 0.34 ml. (0.002 mole) of 70% perchloric acid. The reaction flask is stoppered and allowed to stand at room temperature for 3 hours during which time the reaction solution becomes first yellow-orange and finally red in color. The reaction mixture is poured into a cold (0–5°) mixture of 400 ml. of saturated aqueous sodium hydrogen carbonate and 400 ml. of pentane contained in a 4-l. Erlenmeyer flask equipped with a mechanical stirrer. While the mixture is stirred vigorously at 0–5°, solid sodium hydrogen carbonate is added in 3–5 g. portions as rapidly as foaming of the reaction mixture will permit. The addition of solid sodium hydrogen carbonate is continued until the acetic acid has been neutralized and the aqueous phase remains slightly basic (pH 8). This neutralization requires approximately 400 g. of solid sodium hydrogen carbonate added in portions over a period of *ca.* 3 hours. As soon as the neutralization is complete (Note 1), the organic layer (the lower layer) is separated and the aqueous phase is extracted with three 200-ml. portions of pentane. The combined organic solutions are dried over anhydrous magnesium sulfate and concentrated by distilling the bulk of the pentane through a 30-cm. Vigreux column. The remaining solvents are removed with a rotary evaporator and the residual liquid is distilled (Note 2) under reduced pressure, yielding 66.6–70.9 g. (87–92%) of 2-methyl-1-cyclohexen-1-yl acetate, as a colorless liquid, b.p. 81–86° (18 mm.), n_D^{25} 1.4562–1.4572 (Note 3).

B.   *2-Benzyl-2-methylcyclohexanone.   Caution! Ethereal solutions of methyllithium in contact with atmospheric oxygen may catch fire spontaneously. Therefore any manipulations with this reagent must be carried out with the utmost care to avoid accidental spillage. Benzyl bromide is a powerful lachrymator. Steps B and C should be performed in an efficient fume hood.*

A 1-l., three-necked flask is equipped with a nitrogen-inlet tube fitted with a stopcock, a glass joint fitted with a rubber septum, a 125-ml., pressure-equalizing dropping funnel, a thermometer, and a glass-covered magnetic stirring bar. After the apparatus has been dried in an oven, 20 mg. of 2,2'-bipyridyl is added to the flask and the apparatus is thoroughly flushed with anhydrous, oxygen-free nitrogen (Note 4). A static nitrogen atmosphere is maintained in the reaction vessel throughout subsequent operations involving organometallic reagents (Note 5). An ethereal solution containing 0.40 mole of methyllithium (Note 6) is added to the reaction vessel with a hypodermic syringe. The diethyl ether is removed by evacuating the apparatus while the solution is stirred and the flask is warmed with a water bath (40°) (Note 7). The reaction vessel is refilled with nitrogen and 400 ml. of 1,2-dimethoxyethane (b.p. 83°, freshly distilled from lithium aluminum hydride) is transferred to the reaction vessel with a hypodermic syringe or a stainless steel cannula. The resulting purple solution of methyllithium and the methyllithium bipyridyl charge-transfer complex is cooled to 0–10° before 29.3 g. (0.190 mole) of 2-methyl-1-cyclohexen-1-yl acetate is added, dropwise and with stirring, over a period of 35–45 minutes (Note 8) while the temperature of the reaction mixture is maintained at 0–10° with an ice bath. After the addition of the enol acetate, the reaction solution must still retain a light red-orange color indicating the presence of a small amount of excess methyllithium (Note 9). To this cold (10°) solution is added rapidly (10–15 seconds) and with stirring, 68.4 g. (0.400 mole) of freshly distilled benzyl bromide [b.p. 78–79° (12 mm.), n_D^{25} 1.5738]. The resulting yellow solution is stirred for 2–2.5 minutes (during which time the temperature of the reaction mixture rises from 10° to about 30°), poured into 500 ml. of cold (0–10°), saturated aqueous sodium hydrogen carbonate, and extracted with three 150-ml. portions of pentane. The combined organic extracts are dried over anhydrous magnesium sulfate and concentrated with a rotary evaporator. The residual liquid is fractionally distilled under reduced pressure, separating 31–41 g. of forerun fractions, b.p. 71–89° (20 mm.) and 41–87° (0.3 mm.) (Note 10), and 20.7–22.2 g. (54–58%) of 2-benzyl-2-methylcyclohexanone as a colorless to pale yellow liquid, b.p. 87–93° (0.3 mm.), n_D^{25} 1.5322–1.5344 (Notes 11 and 12).

C.   *2-Benzyl-6-methylcyclohexanone. Caution! The same precaution as that described in part B should be exercised in this step.*

A 1-l., three-necked flask is equipped as described in part B. After the assembled apparatus has been dried in an oven, 45 mg. of 2,2′-bipyridyl is added to the flask and the apparatus is thoroughly flushed with anhydrous, oxygen-free nitrogen (Note 4). A static nitrogen atmosphere is maintained in the reaction vessel throughout subsequent operations involving organometallic reagents (Note 5). An ethereal solution containing 0.20 mole of methyllithium (Note 6) is added to the reaction flask with a hypodermic syringe. After the ether is removed under reduced pressure as described in part B (Note 7), the reaction vessel is refilled with nitrogen and 400 ml. of 1,2-dimethoxyethane (b.p. 83°, freshly distilled from lithium aluminum hydride) is added to the vessel with a hypodermic syringe or a stainless steel cannula. The resulting purple solution of methyllithium and the methyllithium-bipyridyl charge-transfer complex is cooled to −50° with a dry ice–methanol bath before 21.0 g. (29.2 ml., 0.208 mole) of diisopropylamine (b.p. 84–85°, freshly distilled from calcium hydride) is added with a hypodermic syringe, dropwise and with stirring. During this addition, which requires 2–3 minutes, the temperature of the reaction solution should not be allowed to rise above −20° (Note 12). The resulting reddish-purple solution of lithium diisopropylamide and the bipyridyl charge-transfer complex is stirred at −20° for 2–3 minutes before 50 ml. of a 1,2-dimethoxyethane solution containing 21.3 g. (0.190 mole) of 2-methylcyclohexanone is added, dropwise and with stirring. During this addition the temperature of the reaction solution should not be allowed to rise above 0° (Note 12). After the addition of the ketone, the solution of the lithium enolate must still retain a pale reddish-purple color indicating the presence of a slight excess of lithium diisopropylamide (Notes 9 and 13). The enolate solution is stirred and warmed to 30° with a water bath before 68.4 g. (0.400 mole) of freshly distilled benzyl bromide [b.p. 78–79° (12 mm.), n_D^{25} 1.5738] is added, rapidly and with vigorous stirring, from a hypodermic syringe. The temperature of the reaction mixture rises to about 50° within 2 minutes and then begins to fall. After a total reaction period of 6 minutes, the reaction mixture is poured into 500 ml. of cold (0–10°), saturated aqueous sodium hydrogen carbonate and extracted with three 150-ml. portions of pentane. The combined organic extracts are washed successively with two 100-ml. portions of 5% hydrochloric acid and 100 ml. of saturated aqueous sodium hydrogen carbonate, dried over anhydrous magnesium sulfate, and concentrated with a rotary evaporator. The residual yellow liquid is fractionally distilled under reduced pressure (Note 14), separating 31–32 g. of forerun fractions, b.p. 67–92° (20 mm.) and 40–91° (0.3 mm.) (Note 10), and 21.3–23.3 g. (58–61%) of crude 2-benzyl-6-methylcyclohexanone as a colorless liquid, b.p. 91–97° (0.3 mm.), n_D^{25} 1.5282–1.5360. The residue (10–11 g.) contains dibenzylated products. The crude reaction product contains (Notes 11 and 15) 2-benzyl-6-methylcyclohexanone (86–90%) and 2-benzyl-2-methylcyclohexanone (10–14%) accompanied in some cases by small amounts of *trans*-stilbene (Note 13).

To obtain the pure 2,6-isomer, the following procedure may be followed. A 200-ml. flask, equipped with a Teflon-covered magnetic stirring bar, dried in an oven, and flushed with nitrogen, is charged with 2.59 g. (0.0479 mole) of sodium methoxide (Note 16) and stoppered with a rubber septum. A static nitrogen atmosphere is maintained in the reaction vessel throughout the remainder of the reaction. Using a hypodermic syringe, 90 ml. of ether (freshly distilled from lithium aluminum hydride) is added to the flask. The resulting suspension is cooled with an ice bath before a mixture of the crude, distilled, alkylated product (about 21–23 g.) and 3.74 g. (0.0505 mole) of ethyl formate (Note 17) is added with a hypodermic syringe. The mixture is stirred for 10 minutes with ice-bath cooling. The bath is then removed and stirring is continued for an additional 50 minutes. The resulting yellow suspension is treated with 300 ml. of water and extracted with 250 ml. of ether. The ethereal extract is washed with 100 ml. of aqueous 1 *M* sodium hydroxide, dried over anhydrous magnesium sulfate, and concentrated with a rotary evaporator. The residual yellow liquid is distilled under reduced pressure, yielding 16.2–17.3 g. (overall yield 42–45%) of pure (Note 11) 2-benzyl-6-methylcyclohexanone as a colorless liquid, b.p. 95–100° (0.3 mm.), n_D^{25} 1.5299–1.5328.

2. Notes

1. Because the enol acetate is slowly hydrolyzed, even by neutral aqueous solutions, the reaction mixture should be neutralized and the organic product separated and dried as rapidly as is practical.

2. The glassware employed in the distillation should be washed first with ammonium hydroxide, then water, and dried in an oven before use to avoid the possibility of acid-catalyzed hydrolysis or rearrangement of the enol acetate during the distillation.

3. The submitters have been unsuccessful in finding a convenient GC column which will separate 2-methyl-1-cyclohexen-1-yl acetate from its double-bond isomer, 6-methyl-1-cyclohexen-1-yl acetate. However, the ^1H NMR spectrum (CCl₄) of the product exhibits a peak at δ 2.02 (singlet, CH_3CO) superimposed on a multiplet at δ 1.3–2.2 (vinyl CH_3 and aliphatic CH_2) and lacks absorption δ at 0.98 where 6-methyl-1-cyclohexen-1-yl acetate exhibits a doublet ($J = 7$ Hz.) attributable to the aliphatic methyl group.[2] Consequently, the product contains less than 5% of the unwanted double-bond isomer. The product exhibits IR absorption (CCl₄) at 1755 cm.$^{-1}$ (enol ester C=O) and 1705 cm.$^{-1}$ (C=C).

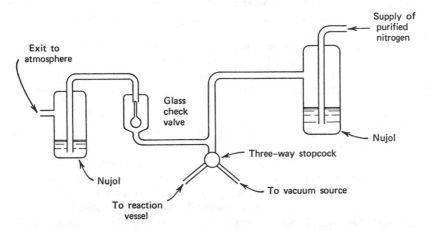

Figure 1. Apparatus for either evacuating or supplying a nitrogen atmosphere to the reaction vessel.

4. A good grade of commercial, prepurified nitrogen can be used without further purification. A suitable method for the purification of nitrogen is described in the literature.[3]

5. The apparatus illustrated in Figure 1 is convenient both for evacuating the reaction vessel, refilling it with nitrogen, and also for maintaining a static atmosphere of nitrogen at slightly above atmospheric pressure in the reaction vessel.

6. A solution of methyllithium in ether was purchased from Alfa Inorganics, Inc. Directions for the preparation of methyllithium from methyl bromide are also available.[4] Solutions of methyllithium should be standardized immediately before use by the titration procedure of Watson and Eastham.[5] A standard 0.500 M solution of 2-butanol (b.p. 99–100°, freshly distilled from calcium hydride)[6] in p-xylene (b.p. 137–138°, freshly distilled from sodium) is prepared in a volumetric flask. A 25-ml. round-bottom flask, fitted with a rubber septum and a glass-covered magnetic stirring bar, is dried in an oven. After 1–2 mg. of 2,2′-bipyridyl has been added to the flask, it is flushed with anhydrous, oxygen-free nitrogen by inserting hypodermic needles through the rubber septum to allow gas to enter and escape. The tip of a 10-ml. burette is forced through the rubber septum and a measured volume of the standard 2-butanol solution is added to the flask, followed by 2.50 ml. of methyllithium solution. The mixture is stirred and additional standard 2-butanol solution is added to the flask from the burette until the purple color of the methyllithium-bipyridyl complex is just discharged. For a 1.66 M solution of methyllithium, 8.30 ml. of the standard 2-butanol solution is required in this titration.

7. Since some lithium enolates are significantly less soluble and less reactive in ether than in 1,2-dimethoxyethane, the submitters recommend the general use of this simple procedure to remove the ether before the lithium enolate is generated.

8. Lithium enolates react relatively slowly with enol acetates to form C-acetylat-

ed products. Consequently, the enol acetate should be added slowly with efficient stirring so that high local concentrations of both the enolate anion and the enol acetate are avoided.

9. It is important that the indicator color, showing a small excess of strong base, not be discharged completely since the presence of any excess enol acetate or ketone will permit equilibration of the isomeric metal enolates. Consequently, the addition of this reactant is complete if further additions will discharge completely the color of the indicator.

10. The various fractions of the forerun were analyzed employing a GC column packed with silicone gum, No. XE-60, suspended on Chromosorb P and heated to 248°. The components found (retention times) were: benzyl bromide (9.0 minutes), 2-methylcyclohexanone (5.3 minutes), and, in some cases, bibenzyl (22.6 minutes). The bibenzyl, formed by reaction of the benzyl bromide with the excess methyllithium[7] was identified from the IR spectrum of a sample collected from the GC.

11. Using a 6-m. GC column packed with silicone gum, No. XE-60, suspended on Chromosorb P and heated to 240°, 2-benzyl-2-methylcyclohexanone (retention time 35.0 minutes) and 2-benzyl-6-methylcyclohexanone (retention time 33.2 minutes, cis- and trans-isomers not resolved) are partially resolved. However, the use of this analytical method to detect small amounts of one structural isomer in the presence of the other is not reliable. GC, however, can be used to determine the presence of any trans-stilbene (retention time 39.0 minutes) in the crude product.

The proportions of structurally isomeric benzylmethylcyclohexanones can be more accurately measured from the ^{1}H NMR spectra of the distilled monoalkylated products. Pure 2-benzyl-6-methylcyclohexanone (principally the more stable cis-isomer in which both substituents are equatorial) exhibits the following ^{1}H NMR (C_6D_6): δ 0.97 (d, $J =$ 6.0 Hz., 3H, CH_3), 1.1–2.6 (m, 9H, aliphatic CH and one of the two nonequivalent benzylic protons), 2.9–3.5 (m with at least 5 lines, 1H, the second of the nonequivalent benzylic protons), and 7.0–7.3 (m, 5H, aryl CH). In CCl_4 the corresponding peaks are found at δ 0.97 (d, $J = 6.0$ Hz.), 1.1–2.7, 2.9–3.5, and 7.0–7.3; in this solvent a second weak doublet ($J = 6.5$ Hz.) is present at δ 1.04 and is attributable to the small amount of the less stable trans-2-benzyl-6-methylcyclohexanone (one equatorial and one axial substituent) present. The 2,6-isomer exhibits IR absorption (CCl_4) at 1710 cm.$^{-1}$ ($C{=}O$) and a series of weak (ϵ 202 to 335) UV maxima (95% C_2H_5OH) in the region 240–270 nm. The mass spectrum exhibits a molecular ion at m/e 202 with relatively abundant fragment peaks at m/e 159, 145, 117, 111, and 91 (base peak). Pure 2-benzyl-2-methylcyclohexanone has the following ^{1}H NMR absorptions (C_6D_6): δ 0.91 (s, 3H, CH_3), 1.2–1.7 (m, 6H, aliphatic CH), 2.1–2.4 (m, 2H, CH_2CO), 2.78 (s, 2H, benzylic CH_2), and 6.9–7.3 (m, 5H, aryl CH). In CCl_4 the corresponding peaks are found at 0.95, 1.4–2.0, 2.2–2.6, 2.78, and 6.9–7.3. This ketone has IR absorption (CCl_4) at 1710 cm.$^{-1}$ ($C{=}O$) and shows a series of weak (ϵ 140 to 284) UV maxima (95% C_2H_5OH) in the region 240–270 nm. The mass spectrum exhibits a molecular ion at m/e 202 with relatively abundant fragment peaks at m/e 159, 117, 92, 91 (base peak), 55, 44, 43, and 41. Mixtures of the 2,6- and 2,2-isomers could be analyzed by measuring their ^{1}H NMR spectra in C_6D_6 and integrating the region δ 2.6–3.5. The peak at δ 2.78, attributable to both benzylic hydrogen atoms of the 2,2-isomer, is well resolved from the multiplet at δ 2.9–3.5, attributable to one of the two benzylic hydrogen atoms of the 2,6-isomer.

Utilizing this method (which is in agreement with the less reliable value obtained by GC analysis), no 2,6-isomer is detected in the 2-benzyl-2-methylcyclohexanone prepared by the present procedure. The 2-benzyl-6-methylcyclohexanone product contains 10–14% of the 2,2-isomer.

12. At temperatures above 0°, 1,2-dimethoxyethane is slowly attacked by lithium diisopropylamide resulting in the protonation of the strong base.

13. If this precaution is not followed, partial or complete equilibration of the enolates will occur because of proton transfers between the enolates and the excess un-ionized ketone. In an experiment where a slight excess of ketone was added, the distilled, monoalkylated product (40% yield) contained 77% of the undesired 2,2-isomer and only 23% of the desired 2,6-isomer. However, it is also important in this preparation not to allow a large excess of lithium diisopropylamide to remain in the reaction mixture; this base reacts with benzyl bromide, forming *trans*-stilbene[8] which is difficult to separate from the reaction product.

14. During the early part of the distillation when a substantial amount of benzyl bromide is present, serious discoloration can be avoided by not heating the still pot above 140°. When the bulk of the benzyl bromide has been removed, the temperature of the still pot may be raised to 150–160° to facilitate distillation of the product.

15. The proportions of the desired 2,6-isomer and the unwanted 2,2-isomer in the alkylated product will vary depending on the rate and efficiency of mixing of the benzyl bromide with the lithium enolate. If the alkylation of the initially formed enolate could be effected without any enolate equilibration, less than 2% of the unwanted 2,2-isomer would be expected.[9]

16. Sodium methoxide was purchased from Matheson, Coleman and Bell. Material from a freshly opened bottle was used without further purification.

17. Commercial ethyl formate (Eastman Organic Chemicals) was purified by stirring it successively over anhydrous sodium carbonate and over anhydrous magnesium sulfate. The material was distilled to separate pure ethyl formate, b.p. 54–54.5°.

3. Discussion

2-Benzyl-6-methylcyclohexanone has been prepared by the hydrogenation of 2-benzylidene-6-methylcyclohexanone over a platinum or nickel catalyst,[10] and by the alkylation of the sodium enolate of 2-formyl-6-methylcyclohexanone with benzyl iodide followed by cleavage of the formyl group with aqueous base.[11] The 2,6-isomer was also obtained as a minor product (about 10% of the monoalkylated product) along with the major product, 2-benzyl-2-methylcyclohexanone by successive treatment of 2-methylcyclohexanone with sodium amide and then with benzyl chloride or benzyl bromide.[12,13] Reaction of the sodium enolate of 2-formyl-6-methylcyclohexanone with potassium amide in liquid ammonia formed the corresponding dianion which when first treated with 1 equiv. of benzyl chloride, then deformylated with aqueous base gave 2-benzyl-2-methylcyclohexanone.[14]

These synthetic routes illustrate the classical methods which have been used for the alkylation of unsymmetrical ketones. Reaction of the ketone with a strong base such as sodium amide under conditions which permit equilibration of the enolates affords an equilibrium mixture of enolates, and subsequent reaction with an alkylating agent yields a mixture of monoalkylated products, as well as polyalkylated products. In the present case, the equilibrium mixture of metal enolates from 2-methylcyclohexanone contains 10–35% of the less highly substituted double bond isomer.[2,15,16] Consequently, the major alkylation product from this mixture is the 2,2-isomer. If the methylene group is protected with a blocking group, the resulting ketone is alkylated solely at the more highly substituted alpha carbon. Removal of the blocking group affords pure 2,2-isomer. Alternatively, an activating group such as a formyl group or a carboalkoxyl group can be introduced at the less highly substituted alpha carbon to permit selective alkylation at this position; the activating group is then removed.

An alternative solution to the problem of effecting the selective alkylation of an unsymmetrical ketone consists of generating a specific enolate under conditions where the enolate isomers do not equilibrate.[17] The methods which have been used to generate specific enolate anions include the reduction of enones[18,19] or α-haloketones[20,21] with metals, the reaction of organolithium reagents with enol silyl ethers[9,22,23] or enol esters,[17,23] and the kinetically controlled abstraction of the least hindered alpha proton from a ketone with a hindered base such as lithium diisopropylamide.[9] The present procedures illustrate the last two methods. To prevent the equilibration of lithium enolates during their formation, care is taken that no proton-donor BH (such as an alcohol or the un-ionized ketone) is present. Although with attention to this precaution, either of the structurally isomeric enolate ions can be prepared and maintained in solution, this fact does not ensure a structurally specific alkylation. As the accompanying equations illustrate, once reaction of the enolate with the alkyl halide is initiated, the reaction mixture will necessarily contain an un-ionized ketone, namely the alkylated product, and equilibration of the enolate ions can occur. Consequently, a structurally specific alkylation of an enolate anion can be successful only if the alkylation reaction is more rapid than equilibration so that the starting enolate **2** is consumed by the alkylating agent before significant amounts of the unwanted enolate **1** have been formed. In practice, this criterion normally is fulfilled with very reactive alkylating agents such as methyl iodide. With less reactive alkylating agents such as benzyl bromide and *n*-alkyl iodides, some equilibration is usually observed.[23] The problem is aggravated when the alkylation involves the less stable and/or the less reactive enolate isomer (*e.g.*, **2**,). In the present procedures, relatively high concentrations of the enolate and the alkyl halide are employed to increase the alkylation rate and, consequently, decrease the proportion of the unwanted monoalkylation product which results from equilibration prior to alkylation. As is to be expected from the foregoing discussion, the alkylation of the enolate **1**, forming 2-benzyl-2-methylcyclohexanone, exhibits more structural specificity than the alkylation of the enolate **2**, forming 2-benzyl-6-methylcyclohexanone. The unwanted 2,2-isomer (10–14%) in this alkylation is removed by a well-known chemical separation procedure in which the 2,2-isomer is converted to its formyl derivative.[24,25]

1. School of Chemistry, Georgia Institute of Technology, Atlanta, Georgia 30332.
2. H. O. House and V. Kramer, *J. Org. Chem.*, **28**, 3362 (1963).
3. H. Metzger and E. Müller, in E. Müller, Ed., "Methoden der Organischen Chemie" (Houben-Weyl), 4th ed., Vol. 1/2, Georg Thieme Verlag, Stuttgart, 1959, p. 327.
4. G. Wittig and A. Hesse, *Org. Synth.*, **Coll. Vol. 6**, 901 (1988).
5. S. C. Watson and J. F. Eastham, *J. Organomet. Chem.*, **9**, 165 (1967).
6. For a warning concerning 2-butanol, see *Chem. Eng. News*, **59**, (19), 3 (1981).
7. H. Gilman and F. K. Cartledge, *J. Organomet. Chem.*, **2**, 447 (1964); H. Gilman and A. H. Haubein, *J. Am. Chem. Soc.*, **66**, 1515 (1944).
8. C. R. Hauser, W. R. Brasen, P. S. Skell, S. W. Kantor, and A. E. Brodhag, *J. Am. Chem. Soc.*, **78**, 1653 (1956).
9. H. O. House, L. J. Czuba, M. Gall, and H. D. Olmstead, *J. Org. Chem.*, **34**, 2324 (1969).
10. R. Cornubert and C. Borrel, *C.R. Hebd. Seances Acad. Sci.*, **183**, 294 (1926); *Bull. Soc. Chim. Fr.*, **46**, 1148 (1929); P. Anziani, A. Aubrey, P. Bourguignon, and R. Cornubert, *Bull. Soc. Chim. Fr.*, 1202 (1950).
11. H. K. Sen and K. Mondal, *J. Indian Chem. Soc.*, **5**, 609 (1928).
12. R. Cornubert and H. Le Bihan, *C.R. Hebd. Seances Acad. Sci.*, **186**, 1126 (1928).
13. R. Cornubert, C. Borrel, and H. Le Bihan, *Bull. Soc. Chim. Fr.*, **49**, 1381 (1931).
14. S. Boatman, T. M. Harris, and C. R. Hauser, *J. Am. Chem. Soc.*, **87**, 82 (1965).
15. H. O. House, W. L. Roelofs, and B. M. Trost, *J. Org. Chem.*, **31**, 646 (1966).
16. D. Caine, *J. Org. Chem.*, **29**, 1868 (1964); D. Caine and B. J. L. Huff, *Tetrahedron Lett.*, 4695 (1966).
17. For a brief review, see H. O. House, *Rec. Chem. Prog.*, **28**, 99 (1967).
18. G. Stork, P. Rosen, N. Goldman, R. V. Coombs, and J. Tsuji, *J. Am. Chem. Soc.*, **87**, 275 (1965).
19. H. A. Smith, B. J. L. Huff, W. J. Powers, III, and D. Caine, *J. Org. Chem.*, **32**, 2851 (1967); L. E. Hightower, L. R. Glasgow, K. M. Stone, D. A. Albertson, and H. A. Smith, *J. Org. Chem.*, **35**, 1881 (1970).

20. M. J. Weiss, R. E. Schaub, G. R. Allen, J. F. Poletto, C. Pidacks, R. B. Conrow, and C. J. Coscia, *Tetrahedron,* **20,** 357 (1964).

21. T. A. Spencer, R. W. Britton, and D. S. Watt, *J. Am. Chem. Soc.,* **89,** 5727 (1967).

22. G. Stork and P. F. Hudrlik, *J. Am. Chem. Soc.,* **90,** 4462, 4464 (1968).

23. H. O. House and B. M. Trost, *J. Org. Chem.,* **30,** 2402 (1965); H. O. House, M. Gall, and H. D. Olmstead, *J. Org. Chem.,* **36,** 2361 (1971).

24. W. J. Bailey and M. Madoff, *J. Am. Chem. Soc.,* **76,** 2707 (1954); F. E. King, T. J. King, and J. G. Topliss, *J. Chem. Soc.,* 919 (1957).

25. The application of this procedure to the separation of the benzylmethylcyclohexanone isomers was developed by Michael J. Umen.

SULFIDE SYNTHESIS: BENZYL SULFIDE

(Benzene, 1,1-[thiobis(methylene)]bis-)

$$(C_6H_5CH_2)_2S_2 + [(CH_3)_2N]_3P \xrightarrow[\text{reflux}]{\text{benzene}} (C_6H_5CH_2)_2S + [(CH_3)_2N]_3PS$$

Submitted by David N. Harpp[1] and Roger A. Smith
Checked by Susan A. Vladuchick and William A. Sheppard

1. Procedure

Caution! Benzene has been identified as a carcinogen; OSHA has issued emergency standards on its use. All procedures involving benzene should be carried out in a well-ventilated hood, and glove protection is required.

A 100-ml., one-necked, round-bottomed flask equipped with a magnetic stirring bar and reflux condenser topped with a drying tube is charged with 3.92 g. (0.0240 mole) of freshly-distilled hexamethylphosphorous triamide (Note 2), 15 ml. of dry benzene (Note 3), and 4.93 g. (0.0200 mole) of benzyl disulfide (Note 4). The mixture is refluxed for 1 hour (Notes 5 and 6), during which time it may develop a slight yellow color. After cooling to room temperature, the solution is concentrated on a rotary evaporator (Note 7), and the residue is chromatographed over 75 g. of silica gel (Note 8). After eluting with 8:2 (v/v) hexane–chloroform at 3–4 ml. per minute (Note 9) and monitoring the eluant by GC (Note 10), an initial fraction containing no product is collected (typically 100 ml.). Successive fractions (*ca.* 1.45 l.) contain the product. These are combined and evaporated (Note 11); 20 ml. of absolute ethanol is added to the residue, and the mixture is heated on a steam bath for *ca.* 30 seconds to form a homogeneous solution. The solution is filtered hot through a Büchner funnel (medium-porosity fritted disk), which is washed with three 3-ml. portions of absolute ethanol. The solvent is then removed to constant weight of residue by rotary evaporation, yielding 4.15–4.27 g. (96.7–99.5%) of benzyl sulfide as a white solid or colorless oil that crystallizes on standing (cooling on dry ice for *ca.* 1 minute may be required), m.p. 46.5–47.5°. This product is of sufficient purity for most purposes (Note 12). Further elution of the column with 1.1 l. of chloroform (same flow rate as above) and removal of solvent by rotary evaporation afford 3.94–4.06 g. (101–104%) of crude (*ca.* 95% pure) tris(dimethylamino)phosphine sulfide (Note 13).

2. Notes

1. To avoid benzene, dry toluene or acetonitrile can be used as solvents following the same procedure.

2. The hexamethylphosphorous triamide was obtained from the Eastman Kodak Company, distilled under reduced pressure (water aspirator), yielding a clear, colorless oil, b.p. 46–48° (7 mm.) and stored under nitrogen (rubber septums should be avoided as they tend to deteriorate on prolonged contact with phosphine vapors). Hexamethylphosphorous triamide not freshly distilled often requires much longer reaction times. Since this chemical is an irritant, all operations with it should be performed in a well-ventilated hood. Hexamethylphosphoric triamide is classified as a carcinogen.[2] It should not be confused with hexamethylphosphorous triamide used in this preparation.

3. The benzene was dried over type 3Å molecular sieves. Use of a greater amount of solvent slows the reaction rate. The checker flamed the reaction flask under nitrogen and ran the reaction under a nitrogen atmosphere.

4. The benzyl disulfide was obtained from the Eastman Kodak Company. Some batches, not homogeneous by GC, were recrystallized from absolute ethanol (ca. 2.5 ml. per g.), yielding colorless crystals, m.p. 69–71° (lit.,[3] m.p. 70–71.5°).

5. The temperature of the oil bath is 100–105°; the reaction mixture is ca. 88°.

6. After 1 hour, the reaction is complete as monitored by GC on an Hewlett-Packard F&M 5751 research chromatograph (1.85 m. × 0.313 cm. stainless-steel column of 10% Apiezon L on Chromosorb W AW/DMCS; column temperature 250°). TLC was found to be of little use in monitoring the reaction, as the R_f values for benzyl disulfide and benzyl sulfide are virtually identical for a variety of solvent systems tried. ^{1}H NMR (C_6H_6) shows that these two compounds have coincidental chemical shifts for the benzylic protons: δ 3.4 (singlet).

A 20% excess of hexamethylphosphorous triamide was utilized to increase the reaction rate. When a 10% excess of the phosphine was used, the reaction was not quite complete (as monitored by GC) even after 3 hours. If the reaction is not allowed to go to completion, the chromatographed product will contain benzyl disulfide. Note that the desulfurization rate for hexaethylphosphorous triamide is comparable to that for hexamethylphosphorous triamide. However, the chromatographic separation following the reaction is much more efficient when hexamethylphosphorous triamide is used.

7. The pressure should be reduced gradually and cautiously to avoid the "bumping" often characteristic of benzene distillations.

8. To 75 g. of silica gel (see below) that has been freshly heated at 120–130° for 18 hours and cooled in a desiccator (see below) is added ca. 200 ml. of a 8 : 2 (v/v) hexane–chloroform mixture. The resulting slurry is stirred to remove air bubbles and added to a 26-mm. column (in which a flat base of glass wool and sand has been prepared) in one portion with tapping of the column. Any remaining silica gel is also added to the column by washing with more of the solvent mixture, resulting in a ca. 28-cm. column of silica gel. A sand layer is then carefully added to the top of the column, and the solvent is drained to the sand level. The reaction residue is placed on the column as is, and the reaction flask is rinsed with small portions of hexane–chloroform which are also loaded onto the column.

The volume of column packing used represents a length–width ratio of *ca.* 10:1. Since column efficiency improves as the length–width ratio is increased (for a given volume of packing), these approximate dimensions should be maintained. For example, when a length–width ratio of *ca.* 9:1 was applied (with the same amount of silica gel), there was not complete separation, and the yield of pure benzyl sulfide was less than optimum (93%).

The packing material is Silica Gel 60, Cat 7734 (70–230 mesh ASTM), for column chromatography, an EM Reagent of E. Merck, Darmstadt, available from Brinkmann Instruments Ltd. (Canada). Lower grades of silica gel such as the 60–120 mesh grade of BDH Chemicals Ltd. were found to be much less efficient, not effecting a complete separation of the compounds. The amount of Silica Gel 60 utilized and found to be most efficient was 75 g., or 8.5 g. of silica gel per g. of reaction mixture. The checkers used Mallinckrodt CC-7 Special (100–200 mesh) column packing, a flow rate of 8.5 ml. per minute, a 9:1 ratio of hexane–chloroform eluant and obtained a lower product yield (77%).

Eighteen hours at 120–130° was found to dry various batches of Silica Gel 60 to the same activity. The silica gel was cooled in a desiccator over 8-mesh Drierite available from Anachemia Chemicals Ltd. This procedure should be performed so that the silica gel is used immediately after cooling; otherwise, the activity will decrease with time even if the silica is left in the desiccator.

The solvent mixture of 8:2 (v/v) was found to be the most effective, with increased concentrations of chloroform not effecting a complete separation. The two solvents should therefore be mixed thoroughly. Chloroform (A.C.S. grade) and an isomeric mixture of hexanes (A.C.S. grade) supplied by Fisher Scientific Company are quite suitable.

9. Increased flow rate resulted in a poorer chromatographic resolution.

10. The chromatography may be monitored by collection of 50-ml. fractions analyzed by GC (same conditions as in Note 6).

11. The residue will be a white solid or colorless oil.

12. The solidified residue is homogeneous by GC, and its IR and ^{1}H NMR spectra and GC are identical to those of recrystallized authentic benzyl sulfide. The product may be recrystallized from *ca.* 25 ml. of absolute ethanol to yield colorless plates, m.p. 47–48° (lit.,[4] m.p. 50°), which give a satisfactory combustion analysis.

13. This material has IR and ^{1}H NMR spectra identical to those of an authentic sample prepared by the procedure of Stuebe and Lankelma.[5]

3. Discussion

The disulfide linkage is found in a large number of natural products,[6] and chemical manipulations of this functionality are of considerable interest.[7] In contrast to desulfurization methods utilizing various phosphines[8] and phosphites,[9] aminophosphines have proved to be efficient desulfurizing agents for a variety of alkyl, aralkyl, alicyclic, and certain diaryl disulfides.[10] In addition, the reaction conditions are mild enough to be compatible with a variety of common functional groups. Hence, the desulfurization procedure given here for benzyl disulfide merely demonstrates the practibility of this method for which the synthetic scope and mechanism have already been developed.[10]

1. Department of Chemistry, McGill University, P.O. Box 6070, Montreal, Quebec, Canada H3A 2K6.
2. J. A. Zapp, Jr., *Science,* **190,** 422 (1975).
3. F. H. McMillan and J. A. King, *J. Am. Chem. Soc.,* **70,** 4143 (1948).
4. C. Forst, *Justus Liebigs Ann. Chem.,* **178,** 370 (1875).
5. C. Stuebe and H. P. Lankelma, *J. Am. Chem. Soc.,* **78,** 976 (1956).
6. (a) L. Young and G. A. Maw, "The Metabolism of Sulfur Compounds," Wiley, New York, 1958; (b) R. F. Steiner, "The Chemical Foundations of Molecular Biology," Van Nostrand, Princeton, 1965; (c) K. G. Stern and A. White, *J. Biol. Chem.,* **117,** 95 (1937); (d) K. Jost, V. Debabov, H. Nesvadba, and J. Rudinger, *Collect. Czech. Chem. Commun.,* **29,** 419 (1964).
7. (a) W. A. Pryor and K. Smith, *J. Am. Chem. Soc.,* **92,** 2731 (1970); (b) J. L. Kice, "Mechanisms of Reactions of Sulfur Compounds," Vol. 3, Wiley, New York, 1968, p. 91; (c) N. Kharasch and A. J. Parker, *Q. Rep. Sulfur Chem.,* **1,** 285 (1966); (d) N. Kharasch, Ed., "Organic Sulfur Compounds," Vol. 1, Pergamon Press, New York, 1961; (e) A. J. Parker and N. Kharasch, *Chem. Rev.,* **59,** 583, 589 (1959).
8. (a) A. Schönberg, *Ber. Dtsch. Chem. Ges.,* **68,** 163 (1935); (b) A. Schönberg and M. Barakat, *J. Chem. Soc.,* 892 (1949); (c) F. Challenger and D. Greenwood, *J. Chem. Soc.,* 26 (1950); (d) C. Moore and B. Trego, *Tetrahedron,* **18,** 205 (1962).
9. (a) H. Jacobson, R. Harvey, and E. V. Jensen, *J. Am. Chem. Soc.,* **77,** 6064 (1955); (b) A. Poshkus and J. Herweh, *J. Am. Chem. Soc.,* **79,** 4245 (1957); (c) C. Walling and R. Rabinowitz, *J. Am. Chem. Soc.,* **79,** 5326 (1957); (d) C. Walling and R. Rabinowitz, *J. Am. Chem. Soc.,* **81,** 1243 (1959); (e) C. Moore and B. Trego, *J. Chem. Soc.,* 4205 (1962); (f) R. Harvey, H. Jacobson, and E. Jensen, *J. Am. Chem. Soc.,* **85,** 1618 (1963); (g) K. Pilgram, D. Phillips, and F. Korte, *J. Org. Chem.,* **29,** 1844 (1964); (h) K. Pilgram and F. Korte, *Tetrahedron,* **21,** 203 (1965); (i) A. J. Kirby, *Tetrahedron,* **22,** 3001 (1966); (j) R. S. Davidson, *J. Chem. Soc. C,* 2131 (1967); (k) D. N. Harpp and R. A. Smith, *J. Org. Chem.,* **44,** 4140 (1979).
10. D. N. Harpp and J. G. Gleason, *J. Am. Chem. Soc.,* **93,** 2437 (1971); D. N. Harpp, J. Adams, J. G. Gleason, D. Mullins, and K. Steliou, *Tetrahedron Lett.,* 3989 (1978).

BICYCLO[1.1.0]BUTANE

(Bicyclo[1.1.0]butane)

$$Cl-\langle\rangle-Br + 2Na \longrightarrow \langle\rangle + NaCl + NaBr$$

Submitted by Gary M. Lampman and James C. Aumiller[1]
Checked by R. A. Fenoglio and K. B. Wiberg

1. Procedure

A 300-ml., three-necked, round-bottomed flask is equipped with a mechanical stirrer, a reflux condenser, and a pressure-equalizing addition funnel. The condenser is connected in series with two traps, immersed in liquid nitrogen, with the exit leading to a drying tube (Note 1). A line for dry nitrogen that has a T-tube joined to a U-tube containing mercury is connected to the top of the addition funnel (Note 2). To the flask are added 150 ml. of purified dioxane (Note 3) and 13.6 g. (0.591 g.-atom) of freshly cut sodium (Note 4). The

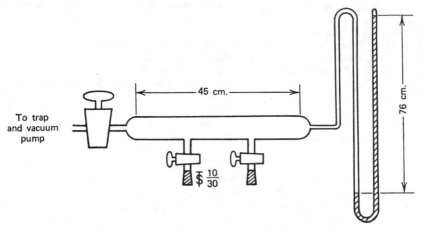

Figure 1. Vacuum manifold.

mixture is heated to reflux, and the molten sodium is broken up with the stirrer. A solution of 20.0 g. (0.118 mole) of 1-bromo-3-chlorocyclobutane (Note 5) in 20 ml. of dioxane (Note 3) is added to the refluxing dioxane over a 1-hour period; refluxing is maintained for an additional 2 hours (Notes 6 and 7). The product in the traps is separated from any dioxane with the vacuum manifold system shown in Figure 1 (Note 8). The two traps containing the product are cooled in liquid nitrogen and connected to one of the stopcocks on the manifold. A gas storage bulb (Figure 2) is attached to the other stopcock. All the stopcocks are opened, and the system is evacuated. The stopcock to the pump is then closed, and the liquid nitrogen bath is removed from the traps and used to cool the gas storage bulb. The traps are warmed slightly, and the bicyclobutane condenses in the storage bulb, leaving the dioxane behind, yielding 5–6 g. (78–94%) (Note 9) of bicyclobutane.

2. Notes

1. Although the entire gaseous product is caught in the first trap, this trap tends to plug during the reaction. Therefore, the second trap is used as a safety measure to collect the bicyclobutane as the first trap is thawed to open the system.

2. It is essential that a U-tube containing mercury is connected by a T-connector to the nitrogen inlet. The U-tube monitors the pressure on the system and acts as a safety valve in the event of a plugged trap. A slight positive pressure is maintained.

3. Reagent grade dioxane (2 1.) was heated to reflux with the sodium ketyl of benzophenone, prepared from 10 g. of benzophenone and 1 g. of sodium, until a deep blue solution results. If the color is not developed, another portion of benzophenone and sodium is added and heating continued until the color persists. The peroxide-free dioxane was distilled from the flask and used immediately.

4. The excess sodium allows the reaction to proceed at a greater rate and decomposes any peroxides remaining in the dioxane.

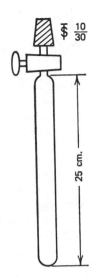

Figure 2.   Gas storage bulb.

5. 1-Bromo-3-chlorocyclobutane was prepared as described in *Org. Synth.,* **Coll. Vol. 6,** 179 (1988).

6. The reflux rate and flow of nitrogen gas must be kept at a minimum to assure that the amount of dioxane carried over to the traps in liquid nitrogen is kept as small as possible.

7. This reflux time is a minimum since decreased yields were observed when the reflux time was shortened. There was no increase in yield when the refluxing time was increased to 4 hours.

8. The basic apparatus consists of a large diameter (about 25 mm.) glass tube to which are attached at least two stopcocks, a closed-end manometer, and a large stopcock, used to isolate the manifold from the vacuum pump.

9. The product, which is about 90% bicyclobutane and 10% cyclobutene, is sufficiently pure for most purposes. The purity of the product can be determined by GC analysis at room temperature, using a 275-cm.-long column containing 20% β,β'-oxydipropionitrile on Chromosorb W (45/60). The retention times are 2.7 and 3.8 minutes for cyclobutene and bicyclobutane, respectively. Bicyclobutane (b.p. 8°) can be stored temporarily in the gas storage bulb as a liquid in an acetone–dry ice bath or for longer periods of time in an ampoule, sealed under vacuum, and stored in a freezer.

3. Discussion

Bicyclobutane has been prepared by intramolecular addition of divalent carbon to an olefinic double bond,[2] irradiation of butadiene,[3] decomposition of cyclopropanecarboxaldehyde tosylhydrazone,[4] and deamination of cyclobutylamine and cyclopropylcarbinylamine.[5] The present procedure, based upon a published method,[6] gives the highest yield of the known methods and provides a process for making moderate quantities of material.

The procedure provides a good example of a high-yield intramolecular Wurtz reaction. Intermolecular Wurtz reactions normally do not give high yields of coupled products and are accompanied by formation of alkenes and alkanes corresponding to the alkyl halide.[7] In contrast, intramolecular reactions of 1,3-dihalides with metals such as sodium are important synthetic methods for making cyclopropane derivatives. Examples are the reactions of sodium with pentaerythrityl tetrabromide, giving spiropentane[8] and of sodium-potassium alloy with 1,3-dibromohexamethylcyclobutane, giving hexamethylbicyclo[1.1.0]butane.[9] If 1,3-dihalides are not used, the yields of cyclic compounds may be considerably reduced. Thus, 1,4-dibromobutane[10] and 3-(bromomethyl)cyclobutyl bromide[11] give very little cyclobutane and bicyclo[1.1.1]pentane, respectively.

Metals other than sodium may be considered for the reduction in intramolecular Wurtz reactions.[11] One of the most common of these is zinc under various reaction conditions. Examples of use of this reagent that have resulted in high yields of cyclopropane and derivatives of cyclopropane include cyclopropane from 1,3-dichloropropane,[12] spiropentane from pentaerythrityl tetrabromide in the presence of a chelating agent,[13] spiro[2.5]octane from 1,1-bis-(bromomethyl)cyclohexane,[14] and 1,1-dialkylcyclopropanes from 1,3-dibromo-2,2-dialkylpropanes.[14] However, 1-bromo-3-chlorocyclobutane yields no bicyclobutane on reaction with zinc,[6] even in the presence of a chelating agent.

1. Department of Chemistry, Western Washington University, Bellingham, Washington 98225.
2. D. M. Lemal, F. Menger, and G. W. Clark, *J. Am. Chem. Soc.*, **85**, 2529 (1963).
3. R. Srinivasan, *J. Am. Chem. Soc.*, **85**, 4045 (1963).
4. H. M. Frey and I. D. R. Stevens, *Proc. Chem. Soc. (London)*, 144 (1964).
5. J. Bayless, L. Friedman, J. A. Smith, F. B. Cook, and H. Schechter, *J. Am. Chem. Soc.*, **87**, 661 (1965).
6. K. B. Wiberg, G. M. Lampman, R. P. Ciula, D. S. Conner, P. Schertler, and J. Lavanish, *Tetrahedron*, **21**, 2749 (1965).
7. A. A. Morton, J. B. Davidson, and B. L. Hakan, *J. Am. Chem. Soc.*, **64**, 2242 (1942).
8. H. O. House, R. C. Lord, and H. S. Rao, *J. Org. Chem.*, **21**, 1487 (1956).
9. D. P. G. Hamon, *J. Am. Chem. Soc.*, **90**, 4513 (1968).
10. J. Cason and R. L. Way, *J. Org. Chem.*, **14**, 31 (1949).
11. K. B. Wiberg and D. S. Connor, *J. Am. Chem. Soc.*, **88**, 4437 (1966).
12. H. B. Hass, E. T. McBee, G. E. Hinds, and E. W. Gluesenkamp, *Ind. Eng. Chem.*, **28**, 1178 (1936).
13. D. E. Applequist, G. F. Fanta, and B. W. Henrikson, *J. Org. Chem.*, **23**, 1715 (1953).
14. R. W. Shortridge, R. A. Craig, K. W. Greenlee, J. M. Derfer, and C. E. Boord, *J. Am. Chem. Soc.*, **70**, 946 (1948).

BORANES IN FUNCTIONALIZATION OF DIENES TO CYCLIC KETONES: BICYCLO[3.3.1]NONAN-9-ONE

A. $(C_2H_5)_3COH \xrightarrow[\text{hexane, } 0°]{C_4H_9Li} (C_2H_5)_3COLi$

1

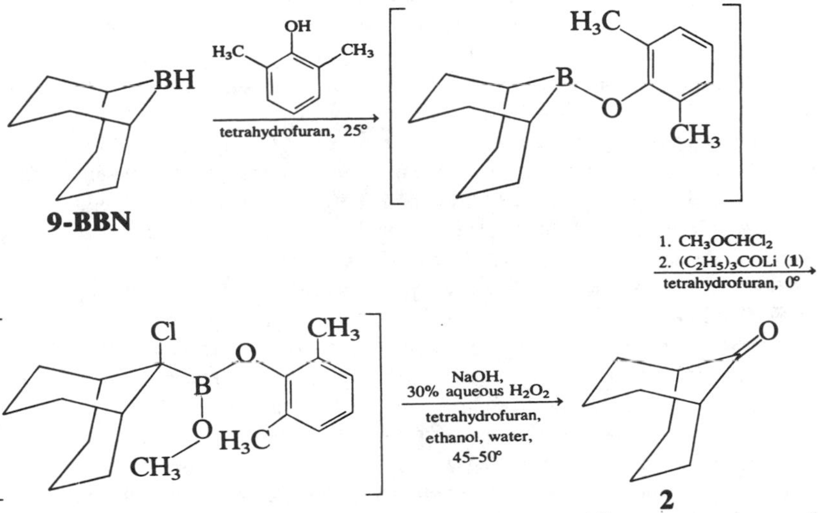

9-BBN

1. CH_3OCHCl_2
2. $(C_2H_5)_3COLi$ **(1)**, tetrahydrofuran, 0°

NaOH, 30% aqueous H_2O_2

tetrahydrofuran, ethanol, water, 45–50°

2

Submitted by Bruce A. Carlson[1] and Herbert C. Brown[2]
Checked by J. C. Bottaro and G. Büchi

1. Procedure

Caution! The oxidation with 30% hydrogen peroxide in the last step of Part B may become vigorous and exothermic. No difficulties have been encountered under the conditions described; however, the oxidation should be carried out in a hood behind a protective shield.

A.   *Lithium triethylcarboxide* (Solution **1**).   A thoroughly dried, 1-l., three-necked, round-bottomed flask fitted with a septum inlet with serum cap, a reflux condenser, and a magnetic stirrer is purged with and maintained under an atmosphere of dry nitrogen. After a solution of 300 ml. of 1.67 *M* n-butyllithium (0.500 mole) in hexane (Note 1) is introduced into the flask by syringe (Note 2) and cooled to 0° in an ice bath, 58 g. (0.50 mole) of 3-ethyl-3-pentanol (Note 3) is added slowly, but constantly, by syringe (Note 4). At the end of addition the yellow tint of the n-butyllithium solution disappears. The alkoxide solution is standardized by hydrolysis of aliquots in water and by titration of the resulting lithium hydroxide with standard acid to a phenolphthalein end point (Note 5).

B.   *Bicyclo[3.3.1]nonan-9-one* (**2**).   A thoroughly dried, 3-l., three-necked, round-bottomed flask fitted with a serum cap and a magnetic stirrer under nitrogen is charged with 42.3 g. (0.347 mole) of 9-borabicyclo[3.3.1]nonane (9-BBN) (Note 6) before 500 ml. of anhydrous tetrahydrofuran (Note 7) is added using a double-ended syringe needle (Note 2). The flask is fitted with a dry, 500-ml., pressure-equalizing dropping funnel while the flask is purged with a rapid stream of dry nitrogen. The apparatus is maintained under a static pressure of nitrogen throughout the reaction. A solution of 42.3 g. (0.347 mole) of 2,6-dimethylphenol (Note 8) in 75 ml. of anhydrous tetrahydrofuran is added slowly by syringe, and the mixture is stirred at room temperature for 3 hours. Once hydrogen evolution is complete (Note 9), 350 ml. of a 1.46 M solution of 1 in hexane is introduced into the dropping funnel after purging it with dry nitrogen. The clear solution in the flask is cooled to 0°, and 44 g. (35 ml., 0.38 mole) of dichloromethyl methyl ether (Note 10) is added. Solution 1 is then added slowly over approximately 30 minutes. The ice bath is removed, and the flask is warmed to room temperature for 90 minutes (Note 11). A heavy white precipitate of lithium chloride forms (Note 12), and the nitrogen atmosphere is no longer required. A solution of 300 ml. of 95% ethanol, 70 ml. of water, and 42 g. of sodium hydroxide is added, and the mixture is cooled to 0° with efficient stirring. Oxidation is carried out over a 40–45 minute period by the slow, dropwise addition of 70 ml. of 30% hydrogen peroxide (Note 13), while the temperature is maintained below 50° with a cooling bath. After addition the mixture is heated with stirring to 45–50° for 2 hours (Note 14) and then cooled to room temperature. Water (300 ml.) is added, and the aqueous phase is saturated with sodium chloride. The organic phase is separated and washed with 100 ml. of saturated aqueous sodium chloride. The solvents are removed on a rotary evaporator, and the resulting orange liquid is diluted with 500 ml. of pentane, which is then extracted with 250-ml. and 100-ml. portions of aqueous 3 M sodium hydroxide to remove 2,6-dimethylphenol. After washing with 100 ml. of saturated aqueous sodium chloride, the pentane is removed on a rotary evaporator. The 3-ethyl-3-pentanol is removed by distillation under a water aspirator vacuum, b.p. 54–56° (16 mm.) (Note 15). The resulting semisolid residue is dissolved in 200 ml. of pentane and filtered to remove the impurities. Ketone 2 is crystallized by cooling the filtrate to −78°. Ketone 2 is collected by suction filtration, washed with 50 ml. of −78° pentane (Note 16), and dried (Note 17), yielding 35–40 g. (78–83%) of pure ketone 2, m.p. 154–156° (Note 18). Evaporation of the filtrate to *ca.* 50 ml. and cooling to −78° gives another 3–4 g. of ketone 2, m.p. 149–154°. The total yield of bicyclo[3.3.1]nonan-9-one (2) is 38–44 g. (79–91%).

2. Notes

1. *n*-Butyllithium in hexane was obtained from Foote Mineral Company. The solution can be titrated for total alkyllithium by the procedure of Watson and East-ham,[5] and for total base by hydrolysis of aliquots and titration against standard acid prior to use. *n*-Butyllithium solution of good purity is essential for success of the reaction.

2. A double-ended syringe needle is most convenient for transfer of reagents and handling of air-sensitive solutions. Transfer is completed by applying nitrogen pressure. See references 3 and 4 for descriptions of this procedure and for general techniques for handling air-sensitive reagents.

3. 3-Ethyl-3-pentanol (triethylcarbinol) (98%) was purchased from Chemical Sam-

ples Company, 469 Kenny Road, Columbus, Ohio 43221, and used without further purification. It is also available from Aldrich Chemical Company, Inc.

4. Addition causes vigorous reaction with evolution of butane and heat. Redissolution of the butane in the solution may create a partial vacuum, causing air to be sucked back into the flask; the addition should be kept at such a rate to prevent this. The solution may reflux, but this causes no harm.

5. Alternatively, if only one use of the solution of lithium triethylcarboxide (1) is planned, the required amounts of butyllithium solution and 3-ethyl-3-pentanol may be reacted, and the resulting solution of lithium triethylcarboxide used without standardization.

6. 9-Borabicyclo[3.3.1]nonane (9-BBN) was purchased from Aldrich Chemical Company, Inc., and used directly. Alternatively, 9-BBN may be prepared by hydroboration of 1,5-cyclooctadiene.[4] *Caution! 9-BBN is air sensitive and can be pyrophoric.*

7. Tetrahydrofuran, containing less than 0.002% peroxides, supplied by the William A. Mosher, Ph.D. Company, 101 Townsend Road, Newark, Delaware 19711, was used directly. Tetrahydrofuran from other sources may require purification prior to use. Stirring with lithium aluminum hydride or sodium benzophenone ketyl, followed by distillation under an inert atmosphere, is the recommended procedure for removal of water and peroxides (see reference 4, p. 256).

8. 2,6-Dimethylphenol (99%), from Chemical Samples Company, was used without purification. It is also available from Aldrich Chemical Company, Inc.

9. Addition causes evolution of hydrogen and is carried out over *ca.* 10 minutes to prevent foaming. The undissolved 9-BBN rapidly goes into solution. Completion of the reaction may be measured by monitoring hydrogen evolution with a gas meter. The theoretical amount of hydrogen is 7.8 l. at STP.

10. Dichloromethyl methyl ether (98%) was purchased from Aldrich Chemical Company, Inc. (listed as α,α-dichloromethyl methyl ether) and used without purification. Better results can be obtained if the material is distilled under nitrogen prior to use. Unlike chloromethyl methyl ether and bis(chloromethyl) ether, dichloromethyl methyl ether is reported to have no significant carcinogenic activity;[6] however, as a precaution it should be handled carefully in a well-ventilated hood.

11. Longer reaction times (*e.g.*, allowing the reaction to stir overnight) will not affect the product.

12. The checkers observed no precipitate in their two runs. On occasion the submitters have also noted no lithium chloride precipitate during a run or sudden precipitation of the salt. Apparently the lithium chloride supersaturates on these occasions.

13. The 30% hydrogen peroxide was obtained from Fisher Scientific Company.

14. Initial heating should be done cautiously in case an additional exotherm is experienced.

15. 3-Ethyl-3-pentanol may be removed rapidly by condensation into an acetone–dry ice cold trap.

16. The temperature of the solution should be kept as close as possible to $-78°$ during filtration to prevent losses of the first crop of product. On humid days crystallization and filtration are best accomplished in a closed system to prevent excessive condensation of water.

17. The product has a high vapor pressure and sublimes readily. Care should be taken in drying so that losses of product are not incurred. Drying briefly in a vacuum desiccator over Drierite is recommended.

18. Literature,[7] m.p. 155–158.5°; the 2,4-dinitrophenylhydrazone was prepared by the standard procedure and recrystallized from ethyl acetate–ethanol to provide orange prisms, m.p. 190–191.2°; lit.,[7] m.p. 191.8–192.3°.

3. Discussion

Dialkylborinic acids are conveniently prepared by the hydroboration of olefins with monochloroborane diethyl etherate[4c] or, in some cases, by controlled hydroboration of two equivalents of olefin[4c] with borane–tetrahydrofuran or borane–methyl sulfide complex followed by alcoholysis. The base-induced reaction of dichloromethyl methyl ether with borinic acid esters provides a route to a variety of ketones[8] and olefins.[9,10] The reaction presumably proceeds *via* generation of a haloalkoxy carbanion or carbene which reacts rapidly with the borinic acid ester, resulting in transfer of the alkyl groups from boron to carbon. The α-chloroboronic ester intermediate, thus generated, is exceptionally versatile. Oxidation results in the formation of the corresponding ketone in high yield,[8,11] whereas pyrolysis or solvolysis with aqueous silver nitrate[10] provides the internal olefin:

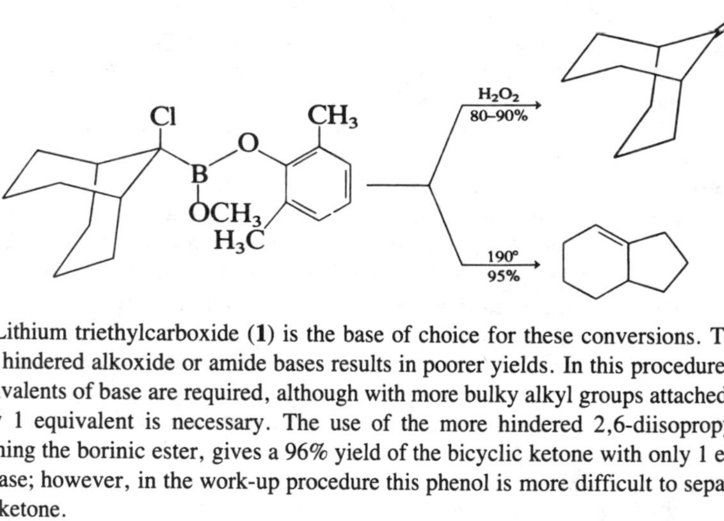

Lithium triethylcarboxide (**1**) is the base of choice for these conversions. The use of less hindered alkoxide or amide bases results in poorer yields. In this procedure 1.5–2.0 equivalents of base are required, although with more bulky alkyl groups attached to boron only 1 equivalent is necessary. The use of the more hindered 2,6-diisopropylphenol, forming the borinic ester, gives a 96% yield of the bicyclic ketone with only 1 equivalent of base; however, in the work-up procedure this phenol is more difficult to separate from the ketone.

Carbonylation of trialkylboranes in the presence of water[4e] and the cyanation of thexyldialkylboranes[12] offer alternative routes to ketones from organoboranes. Neither procedure can utilize dialkylborinic esters in the synthesis and, thus, result in loss of one alkyl group. Carbonylation requires a moderate pressure of carbon monoxide in some cases, and the cyanation reaction involves the use of strongly electrophilic reagents; neither route has been used successfully in the preparation of bicyclo[3.3.1]nonan-9-one (**2**). Previous synthetic routes to this interesting bicyclic ketone[13] have generally required numerous steps and resulted in low overall yields.[7,14] The best alternative procedure, which gives a 60% yield of the bicyclic ketone, involves reaction of nickel carbonyl with 1,5-cyclooctadiene.[15]

The base-induced reaction of dichloromethyl methyl ether with trialkyl- and triarylboranes also provides a powerful method for the preparation of the corresponding tertiary carbinols.[16,17] In this case, all three groups transfer readily from boron to carbon under mild conditions, and oxidation with alkaline hydrogen peroxide provides the tertiary alcohol.

$$R_3B + CHCl_2OCH_3 \xrightarrow{LiOC(C_2H_5)_3} R_3C-B \begin{smallmatrix} Cl \\ \\ OCH_3 \end{smallmatrix} \xrightarrow{[O]} R_3COH$$

70–90%

1. Central Research & Development Department, E. I. du Pont de Nemours & Co., Wilmington, Delaware 19898.
2. Richard B. Wetherill Professor of Chemistry, Purdue University, West Lafayette, Indiana 46391.
3. Aldrich Chemical Company, Inc., Bulletin No. A74, "Handling Air-Sensitive Solutions."
4. H. C. Brown, "Organic Syntheses via Boranes," Wiley, New York, 1975; (a) pp. 191–261; (b) pp. 32–34; (c) pp. 45–47; (d) pp. 3–9; (e) pp. 127–130.
5. S. C. Watson and J. F. Eastham, *J. Organomet. Chem.*, **9**, 165 (1967). The procedure is described in detail in reference 4, pp. 248–249.
6. B. L. Van Duuren, C. Katz, B. M. Goldschmidt, K. Frenkel, and A. Sivak, *J. Nat. Cancer Inst.*, **48**, 1431 (1972).
7. C. S. Foote and R. B. Woodward, *Tetrahedron*, **20**, 687 (1964).
8. B. A. Carlson and H. C. Brown, *J. Am. Chem. Soc.*, **95**, 6876 (1973).
9. J. J. Katz, B. A. Carlson, and H. C. Brown, *J. Org. Chem.*, **39**, 2817 (1974).
10. H. C. Brown, J. J. Katz, and B. A. Carlson, *J. Org. Chem.*, **40**, 813 (1975).
11. B. A. Carlson and H. C. Brown, *Synthesis*, 776 (1973).
12. A. Pelter, K. Smith, M. G. Hutchings, and K. Rowe, *J. Chem. Soc., Perkin Trans.* 1, 129 (1975).
13. A review of structure and reactivities in the bicyclo[3.3.1]nonane system is provided by H. Kato in *J. Synth. Org. Chem. Jpn.*, **28**, 682 (1966).
14. R. D. Allan, B. G. Cordiner, and R. J. Wells, *Tetrahedron Lett.*, 6055 (1968).
15. B. Fell, W. Seide, and F. Asinger, *Tetrahedron Lett.*, 1003 (1968).
16. H. C. Brown and B. A. Carlson, *J. Org. Chem.*, **38**, 2422 (1973).
17. H. C. Brown, J. J. Katz, and B. A. Carlson, *J. Org. Chem.*, **38**, 3968 (1973).

BICYCLO[3.2.1]OCTAN-3-ONE

(Bicyclo[3.2.1]octan-3-one)

$$\text{Cl}_3\text{CCO}_2\text{C}_2\text{H}_5 \quad / \quad \text{NaOCH}_3$$

$$\text{LiAlH}_4$$

$$\text{H}_2\text{O} \quad / \quad \text{H}_2\text{SO}_4$$

Submitted by C. W. Jefford,[1,2] J. Gunsher,[1] D. T. Hill,[1]
P. Brun,[3] J. Le Gras,[3] and B. Waegell[3]
Checked by R. W. Begland and R. E. Benson

1. Procedure

A. exo-3,4-*Dichlorobicyclo*[3.2.1]*oct-2-ene*. A 1-l., four-necked, round bottomed flask is fitted with an efficient stirrer, a thermometer, a reflux condenser protected by a calcium chloride tube, and a 500-ml., stoppered, pressure-equalizing addition funnel. After the addition of a solution of 52.5 g. (0.558 mole) of norbornene (Note 1) in 400 ml. of petroleum ether (Note 2, b.p. 45–60°) to the flask, 112 g. (2.06 moles) of sodium methoxide (Note 3) is added, and stirring is begun. The flask is immersed in an ice–salt mixture (Note 4) before 349 g. (1.82 moles) of ethyl trichloroacetate (Note 5) is placed in the addition funnel and allowed to drip slowly into the stirred mixture at a rate such that the temperature does not rise above 0° (Note 6). The addition requires about 4 hours, and the originally white reaction mixture becomes increasingly yellow in color. The mixture is stirred at a temperature below 0° for 4 hours (Note 6) before the temperature is allowed to rise gradually to room temperature overnight. The reaction mixture is poured onto 500 g. of crushed ice in 300 ml. of water. After the ice has melted, the organic layer is separated, and the aqueous layer is shaken with four 200-ml. portions of diethyl ether. The aqueous layer is neutralized with 10% hydrochloric acid and shaken again with two 200-ml. portions of ether. The original organic layer and the ether extractions are combined, washed with 300 ml. of a saturated solution of sodium chloride, dried for 6 hours over 20 g. of anhydrous magnesium sulfate, filtered, and concentrated to about 200 ml. by distillation. The resulting product is distilled through a 20-cm. Vigreux column, giving 72.5–87.0 g. (74–88%) of exo-3,4-dichlorobicyclo[3.2.1]oct-2-ene as a colorless liquid, b.p. 72–73° (0.9 mm.), n_D^{25} 1.5333 (Notes 7 and 8).

B.   *3-Chlorobicyclo[3.2.1]oct-2-ene.*   A 2-l., three-necked, round-bottomed flask equipped with a stirrer, a reflux condenser protected by a calcium chloride tube, and a 500-ml., stoppered, pressure-equalizing addition funnel is charged with 350 ml. of dry ether and 15 g. (0.40 mole) of powdered lithium aluminum hydride (Note 9). The flask is placed in a mixture of ice and water, and 1050 ml. of dry tetrahydrofuran (Note 10) is added. Stirring is begun, and a solution of 39.5 g. (0.224 mole) of *exo*-3,4-dichlorobicyclo[3.2.1]oct-2-ene in 50 ml. of dry tetrahydrofuran (Note 10) is added dropwise from the addition funnel over a 30-minute period. After the addition is complete, the reaction mixture is heated under gentle reflux for 18 hours. The mixture is cooled to 0°, and the remaining lithium aluminum hydride is decomposed by the careful addition of wet ether, followed by the cautious dropwise addition of 10 ml. of water. The resulting mixture is poured onto 500 g. of crushed ice in 200 ml. of water. After the ice has melted, the organic layer is separated, and the aqueous layer is acidified to a pH of 5−6 with 10% hydrochloric acid, dissolving the lithium and aluminum salts present (Note 11). The aqueous solution is shaken five times with 200-ml. portions of ether. The organic layers are combined, washed with 200 ml. of a saturated solution of sodium chloride, and dried overnight over 20 g. of anhydrous magnesium sulfate. The ether and tetrahydrofuran are removed by distillation at atmospheric pressure, and the product is distilled through a 20-cm. Vigreux column, yielding 23.5−23.9 g. (74−75%) of 3-chlorobicyclo[3.2.1]oct-2-ene as a colorless oil, b.p. 76−77° (21 mm.), n_D^{20} 1.5072 (Note 12).

C.   *Bicyclo[3.2.1]octan-3-one.*   A magnetic stirring bar is placed in a 300-ml., round-bottomed flask and 100 ml. of concentrated sulfuric acid (Note 13) is added. Stirring is begun, the flask is cooled in an ice bath, and 9.0 g. (0.63 mole) of 3-chlorobicyclo[3.2.1]oct-2-ene is added in one portion. The mixture is stirred, allowed to warm slowly over 4 hours to room temperature, then stirred overnight. The resulting solution is poured onto 200 g. of ice with stirring. After the ice has melted, the mixture is shaken with three 100-ml. portions of ether. The ether layers are combined, washed once with 50 ml. of water, and dried over 10 g. of magnesium sulfate. The ether is removed by careful distillation, and the crude product is sublimed at a bath temperature of 70° (15 mm.) directly onto a cold finger inserted into the flask. The crude product is twice sublimed, yielding 5.9−6.3 g. (75−81%) of bicyclo[3.2.1]octan-3-one, m.p. 137°, 98% pure by analysis (Notes 14, 15, and 16).

2. Notes

1. Norbornene can be prepared by the Diels-Alder reaction of ethylene with di-cyclopentadiene, as described in *Org. Synth.,* **Coll. Vol. 4,** 738 (1963). It can be purchased from Matheson, Coleman and Bell or the Aldrich Chemical Company, Inc.

2. The petroleum ether used must be olefin-free.

3. The submitters used sodium methoxide available from Schuchardt, Ainmiller-strasse 25, 8-Munich 13, Germany. The checkers used product available from Matheson, Coleman and Bell.

4. An amount sufficient to fill a 5-l. container is recommended.

5. It is used as purchased from either Schuchardt or Eastman Organic Chemicals. It appears that when equimolar amounts of carbene precursor and olefin are used, the adduct is obtained in only 25% yield.[4]

6. Efficient stirring and maintenance of a low temperature are required if high yields are to be obtained.

7. For the final distillation the pressure is regulated with a manostat.

8. The IR spectrum (neat) shows absorption at 1645, 1450, 1305, 1223, 1052, 974, 959, 866, 796, and 750 cm.$^{-1}$. The ^{1}H NMR spectrum (CCl$_4$) shows absorption at δ 4.1 (d, $J \sim$ 3 Hz.), and 6.08 (d, $J \sim$ 4 Hz.) attributable to the vinyl and allyl protons, respectively.

9. This was purchased from Metal Hydrides, Inc., and is also obtainable from Schuchardt.

10. The checkers used reagent grade tetrahydrofuran (available from Fisher Scientific Company) from a freshly opened bottle. The submitters used tetrahydrofuran purified as described in the literature.[5] [See *Org. Synth.*, **Coll. Vol. 5,** 976 (1973) for a warning regarding the purification of tetrahydrofuran.]

11. A small amount of insoluble material may remain at this point. If it interferes with the extraction procedure it may be removed by filtration.

12. The IR spectrum (neat) shows absorption at 1635, 1440, 1038, 952, 841, 833, and 685 cm.$^{-1}$. The ^{1}H NMR spectrum (CCl$_4$) shows three major peaks between δ 2.0–2.8 and a doublet at 5.91 ($J \sim$ 7 Hz., 1H, CH).

13. Sulfuric acid of specific gravity 1.84 is used.

14. The IR spectrum (Nujol) has an intense band at 1710 cm.$^{-1}$. The ^{1}H NMR spectrum (CDCl$_3$) has a broad peak at δ 1.7 (6H), a sharp peak at 2.35 (4H) and a broad peak at 2.55 (2H).

15. The GC data were obtained using 3 m. × 7 mm. column packed with nonacid-washed Chromosorb 45/60 W containing 15% 200M Apiezon silicone oil as the immobile phase.

16. The oxime has m.p. 96°. The ketone can be further purified *via* the semicarbazone derivative, which can be purified by crystallization and subsequently hydrolyzed by dilute hydrochloric acid.

3. Discussion

Studies by the submitters have indicated that the procedure reported here is the preferred method for the preparation of bicyclo[3.2.1]octan-3-one. It employs readily available, inexpensive reagents, and the overall yield is good. In addition, the method can be used for the synthesis of the relatively inaccessible, higher homologues of bicyclo[2.2.2]oct-2-ene, as well as for derivatives of norbornene. Bicyclo[3.2.2]nonan-3-one and 1-methylbicyclo[3.2.1]octan-3-one have been prepared by a similar route[6] in 60% and 47% yields, respectively (based on adduct). However, the preferred procedure for the formation of the dichlorocarbene adduct of bicyclo[2.2.2]oct-2-ene is that of Seyferth, using phenyltrichloromethylmercury. Even in this case the overall yield is moderate (37%).

The present procedure is a refinement of existing methods,[6–9] an adaption of the method of Parham and Schweizer[10] that furnishes the carbene adduct in higher yield than other methods. The submitters' studies indicate that the procedure of Doering,[11] involving the interaction of chloroform with potassium *tert*-butoxide, is unsatisfactory since traces of *tert*-butyl alcohol present react with the rearranged adduct during the reduction step. In addition, the yields of adduct are poor (4–15%).[4] The method of Wagner,[12] utilizing the pyrolysis of sodium trichloroacetate, is easy to perform, but the yields of initial adduct are poor (13%). That of Seyferth,[13] involving the pyrolysis of phenyltrichloromethylmercury, gives the adduct in 45% yield. However, the higher cost and additional steps entailed in the preparation of the reagent, together with the hazards associated with mercury, detract from its use.

Other preparations leading to bicyclo[3.2.1]octan-3-one include the oxidation of the mixture of alcohols obtained by the action of nitrous acid on 2-aminomethylnorbornane,[14] the chromic acid oxidation of bicyclo[3.2.1]octane,[15] and the oxidative hydroboration of bicyclo[3.2.1]oct-2-ene.[16] These reactions lack preparative utility, and all have a common disadvantage of being accompanied by isomer formation.

1. Chemistry Department, Temple University, Philadelphia, Pennsylvania 19122.
2. Present address: Département de Chimie Organique, Section de Chimie, 1211 Genève 4, Suisse.
3. Département de Chimie Organique, Faculté des Sciences (St. Charles), Université d'Aix-Marseille, France.
4. R. C. DeSelms and C. M. Combs, *J. Org. Chem.*, **28**, 2206 (1963).
5. L. F. Fieser, "Experiments in Organic Chemistry," 3rd ed., rev., Heath, Boston, 1957, p. 292.
6. C. W. Jefford, S. Mahajan, J. Waslyn, and B. Waegell, *J. Am. Chem. Soc.*, **87**, 2183 (1965).
7. C. W. Jefford, *Proc. Chem. Soc. (London)*, 64 (1963).
8. B. Waegell and C. W. Jefford, *Bull. Soc. Chim. Fr.*, 844 (1964).
9. W. R. Moore, W. R. Moser, and J. E. LaPrade, *J. Org. Chem.*, **28**, 2200 (1963).
10. W. E. Parham and E. E. Schweizer, *J. Org. Chem.*, **24**, 1733 (1959).
11. W. von E. Doering and A. K. Hoffmann, *J. Am. Chem. Soc.*, **76**, 6162 (1954).
12. W. M. Wagner, H. Kloosterziel, and S. van der Ven, *Recl. Trav. Chim. Pays-Bas*, **80**, 740 (1961).
13. T. J. Logan, *Org. Synth.*, **Coll. Vol. 5**, 969 (1973).
14. K. Alder and R. Reubke, *Chem. Ber.*, **91**, 1525 (1958).
15. P. von R. Schleyer and R. D. Nicholas, Abstracts of Papers, 75Q, Division of Organic Chemistry, Am. Chem. Soc. 140th National Meeting, Chicago, Ill., September 1961.
16. R. R. Sauers and R. J. Tucker, *J. Org. Chem.*, **28**, 876 (1963).

BICYCLO[2.1.0]PENT-2-ENE

Submitted by A. Harry Andrist, John E. Baldwin,[1] and Robert K. Pinschmidt, Jr.
Checked by Keith C. Bishop, III and Kenneth B. Wiberg

1. Procedure

Caution! The photochemical reaction should be carried out under a light absorbent cover. The operator should wear goggles affording protection from ultraviolet light.

While still slightly warm from the drying oven, a photolysis vessel with a water-jacketed quartz immersion well (Note 1) (section *A* of Figure 1) is charged with 500 ml. of anhydrous tetrahydrofuran (Note 2) and 10.0 ml. (8.05 g., 0.122 mole) of cyclopentadiene (Note 3). The solution is cooled in an ice bath and purged with dry nitrogen for 2

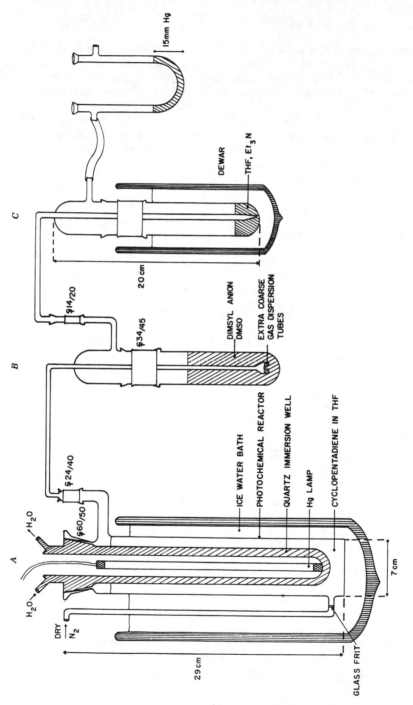

Figure 1.

minutes. The vessel is then sealed, the lamp inserted, and the solution irradiated at 0° for 30 minutes. During this period, sections B and C (see diagram) are removed from the oven to cool slightly. While still warm, the bottom part of C is filled with 0.5 ml. of anhydrous tetrahydrofuran and approximately 25 μl. of triethylamine (Note 4), greased, and stoppered. The top parts of B and C are thoroughly greased and connected to the 24/40 ⊤ joint of A. The bottom of part B is filled rapidly with 200 ml. of an approximately 1 M solution of dimsyl anion (Note 5) in dimethyl sulfoxide. The bottom part of C is then connected and cooled in a 2-propanol–dry ice bath. The system is irradiated and subjected to a steady flow of nitrogen (approximately 400–500 ml./minute). After the first hour, this flow rate is reduced to 200–250 ml./minute for the remaining 4.5 hours. At the end of a total irradiation period of 6 hours, trap C is stoppered and stored at dry ice temperature. This trap contains 15–25 ml. of a 2–5% solution of bicyclopentene in tetrahydrofuran, (Note 6) contaminated with minor amounts (<15%) of tricyclo[2.1.0.0^{2,5}]pentane,[2] which is highly labile and difficult to separate.

2. Notes

1. The submitters used an unfiltered, Hanovia, 450-watt, medium-pressure mercury vapor lamp (catalog No. 679A, Hanovia Lamp Division, Canrad Precision Industries, 100 Chestnut Street, Newark, N.J. 07105) and a reactor equipped with a nitrogen-inlet containing a glass frit, which enters the bottom of the reactor. All connections (Note 9) were glass with the exception of the rubber tubing between the second trap and the mercury bubbler. All glassware was soaked for 30 minutes in aqueous 30% ammonium hydroxide and oven-dried prior to use without further rinsing. Any polymer formed on the quartz immersion well during this preparation may be removed with nitric acid or with warm water and cleansing powder; if it is not removed, yields in subsequent preparations are significantly diminished.

2. Tetrahydrofuran (purchased from Mallinckrodt Chemical Works) was dried by distillation from lithium aluminum hydride, and stored over Linde 4A Molecular Sieves.

3. Cyclopentadiene was prepared [compare with *Org. Synth.*, **Coll. Vol. 4**, 473, Note 4 (1963) and **Coll. Vol. 5**, 414 (1973)] by heating dicyclopentadiene (purchased from Eastman Organic Chemicals) and a pinch of hydroquinone under a column of glass helices or a Vigreux column at 175° and collecting the distillate in a receiver cooled with a 2-propanol–dry ice bath. The monomer was dried over Linde 4A Molecular Sieves at −20° and could be stored at this temperature for several weeks without excessive dimerization.

4. Triethylamine (purchased from Matheson, Coleman and Bell) was dried over Linde 4A Molecular Sieves.

5. Dimsyl anion[3] was prepared from 10.2 g. (0.241 mole) (Note 10) of 56.8% sodium hydride, which was washed with pentane and vacuum dried, and 200 ml. of anhydrous dimethyl sulfoxide. The mixture was heated at 65–70° for about 50 minutes, until hydrogen evolution ceased. *Caution! This mixture should not be heated above 80° because of the possibility of explosive decomposition.*

6. Analysis and purification of the product solution is best accomplished by GC. The submitters used a 500 cm. by 0.6 cm. aluminum or polyethylene column packed with 21% β,β'-oxydipropionitrile on Chromosorb P with column, injector, and detector operated at

25° and a flow rate of 50 ml./minute. Under these conditions the retention times of bicyclopentene and cyclopentadiene were 3 and 5 minutes, respectively, beyond that of coinjected air. Tricyclopentane elutes in low yield with the bicyclopentene. Since bicyclopentene is extremely labile with respect to acid catalysis, any contact with water, hydroxylic solvents, and aprotic acids should be avoided (Note 11). Bicyclopentene stored at −78° in anhydrous tetrahydrofuran is stable indefinitely.

7. Decalin, benzene, 1,4-dioxane, and ethanol may be used as solvents for the photolysis. In an alternative procedure, volatile materials swept from the photolysis vessel are condensed in a dry ice trap. This cold mixture is added to a flask containing a magnetically stirred solution of dimsyl anion in dimethyl sulfoxide, and fractionation at reduced pressure provides a solution of bicyclopentene in tetrahydrofuran. In both variants additional bicyclopentene remains in the dilute reactor and trap solutions. It may be more fully recovered by using a modified distillation assembly composed of a gently warmed flask, a distillation head with a dry ice condenser filled with ice, an ice-cooled receiver, and a 2-propanol–dry ice vapor trap prior to a vacuum pump. (A rotary evaporator with an ice-filled dry ice condenser also works.) The vacuum is adjusted to rapidly distill the high boiling solvent into the receiver and bicyclopentene past the condenser into the −78° trap. The final distillation should be from the dimsyl anion solution (mainly tetrahydrofuran distills here) to remove codistilled cyclopentadiene.

8. A purified, undiluted sample of bicyclopentene has been reported to explode.[4]

9. All of the connections must be well secured by sturdy rubber bands to avoid leakage caused by substantial back pressure that develops in the course of the reaction.

10. A higher base concentration or substitution of one or two gas-dispersion tubes leads to clogging of the inlet to the trap.

11. Two injections of 10 ml. of ammonia vapors greatly helps to eliminate decomposition on the column during GC collection.

3. Discussion

The present procedure, a modification of that previously reported,[5–8] permits the ready preparation of a cyclopentadiene-free solution of bicyclopentene on a synthetic scale. Until now, synthetic chemistry involving the use of bicyclopentene has been limited to preparations of bicyclopentane[5,8] and bicyclopentane-2,3-d_2,[8] which is more a reflection of the extraordinary care required in handling this unusually sensitive bicyclic alkene than a lack of interesting potential or useful reactivity.

Most cyclic and acyclic 1,3-dienes, such as cyclopentadiene, undergo photochemical ring-closure to cyclobutenes.[9] Bicyclo[2.1.0]pent-2-ene derivatives have also been prepared via photolysis of cyclopentadiene-5-d-, -d_6, and -1,5-$^{13}C_2$;[2,10] 1-, 2-, and 5-methylcyclopentadiene;[2,11] 5,5-dimethylcyclopentadiene;[12] 1,5-dimethyl- and 2,5-dimethyl-1,3-cyclopentadien-5-yl acetic acid;[13] 2,2-dimethylisoindene[14] and 2,3,5-tri-tert-butylcyclopentadienone.[15] Substituted thiophenes have been converted photochemically to 5-thia derivatives of bicyclopentene.[16]

1. Department of Chemistry, University of Oregon, Eugene, Oregon 97403.
2. G. D. Andrews and J. E. Baldwin, J. Am. Chem. Soc., 99, 4851, 4853 (1977); 94, 1775 (1972) and references therein.

3. E. J. Corey and M. Chaykovsky, *J. Am. Chem. Soc.*, **84**, 866 (1962); **87**, 1345 (1965).
4. T. J. Katz, private communication.
5. J. I. Brauman, L. E. Ellis, and E. E. van Tamelen, *J. Am. Chem. Soc.*, **88**, 846 (1966).
6. J. I. Brauman and D. M. Golden, *J. Am. Chem. Soc.*, **90**, 1920 (1968).
7. D. M. Golden and J. I. Brauman, *Trans. Faraday Soc.*, **65**, 464 (1969).
8. P. G. Gassman, K. T. Mansfield, and T. J. Murphy, *J. Am. Chem. Soc.*, **91**, 1684, (1969); **90**, 4746 (1968).
9. G. B. Gill, *Q. Rev. Chem. Soc.*, **22**, 338 (1968); G. J. Fonken, *Org. Photochem.*, **1**, 197 (1967).
10. J. E. Baldwin, R. K. Pinschmidt, Jr., and A. H. Andrist, *J. Am. Chem. Soc.*, **92**, 5249 (1970).
11. J. E. Baldwin, A. H. Andrist, *J. Chem. Soc. D*, 1561 (1970).
12. W. R. Roth, F.-G. Klärner, and H. W. Lennartz, *Chem. Ber.*, **113**, 1818 (1980).
13. F.-G. Klärner, and F. Adamsky, *Angew. Chem. Int. Ed. Engl.*, **18**, 674 (1979).
14. W. R. Dolbier, K. Matsui, H. J. Dewey, D. V. Horak, and J. Michl, *J. Am. Chem. Soc.*, **101**, 2136 (1979).
15. G. Maier, U. Schäfer, W. Sauer, H. Hartan, and J. F. M. Oth, *Tetrahedron Lett.*, 1837 (1978).
16. J. A. Barltrop, A. C. Day, and E. Irving, *J. Chem. Soc. Chem. Commun.*, 966, 881 (1979); H. A. Wiebe, S. Braslavsky, and J. Heicklen, *Can. J. Chem.*, **50**, 2721 (1972).

DEHYDROXYLATION OF PHENOLS;
HYDROGENOLYSIS OF PHENOLIC ETHERS:
BIPHENYL

Submitted by WALTER J. MUSLINER and JOHN W. GATES, JR.[1]
Checked by D. ROBERT COULSON and RICHARD E. BENSON

1. Procedure

Caution! Benzene has been identified as a carcinogen; OSHA has issued emergency standards on its use. All procedures involving benzene should be carried out in a well-ventilated hood, and glove protection is required.

A. *4-(1-Phenyl-5-tetrazolyloxy)biphenyl*. A 1-l., round-bottomed flask fitted with an efficient condenser and a magnetic stirring bar is charged with 17 g. (0.10 mole) of 4-phenylphenol, 18.1 g. (0.100 mole) of 1-phenyl-5-chlorotetrazole (Note 1), 27.6 g. (0.200 mole) of anhydrous potassium carbonate, and 250 ml. of acetone. The mixture is stirred and heated under reflux for 18 hours (Note 2). Water (250 ml.) is added to the hot mixture producing a clear solution that is chilled in ice. After 1 hour, the solid is collected by filtration and dried in air, giving 32–33 g. of the crude product, m.p. 151–153°, which is then dissolved in 250 ml. of hot ethyl acetate. The solution is filtered while hot to remove a small amount of insoluble material and cooled on ice, yielding 25 g. of 4-(1-phenyl-5-tetrazolyloxy)biphenyl, as white crystals, m.p. 150–153°. An additional 2–3 g. of product is recovered from the filtrate by concentration to 125 ml. bringing the total yield to 27–28 g. (86–89%).

B. *Biphenyl.* Added to a solution of 10 g. (0.032 mole) of the product from Part A in 200 ml. of benzene is 2 g. of 5% palladium-on-charcoal, and the mixture is shaken with hydrogen in a Parr apparatus at 40 p.s.i. and 35–40° for 8 hours (Note 3). The mixture is filtered, and the insoluble residue is washed with three 100-ml. portions of hot ethanol (Note 4). The filtrates are combined, and the solvent is removed with a rotary evaporator at 60° (12 mm.), leaving a solid residue, which is dissolved in 100 ml. of benzene. After adding 100 ml. of 10% aqueous sodium hydroxide the mixture is shaken, and the layers separated. The aqueous layer is extracted with 100 ml. of benzene, and the original benzene layer is washed with 100 ml. of water (Note 5). The benzene solutions are combined and dried over magnesium sulfate. Removal of the benzene by distillation yields 4.0–4.7 g. (82–96%) of biphenyl as a white powder, m.p. 68–70° (Note 6). The IR spectrum is identical with that of an authentic sample, and a purity of at least 99.5% was indicated by GC analysis.

2. Notes

1. 4-Phenylphenol and 1-phenyl-5-chlorotetrazole were obtained from Eastman Organic Chemicals.
2. A reflux period of 18 hours was chosen because it represents an overnight reaction time; the reaction is essentially completed in 8 to 10 hours.
3. The hydrogenolysis can also be carried out in ethanol or tetrahydrofuran. An amount of catalyst equivalent to 10–20% by weight of tetrazolyl ethers is most satisfactory for this reaction. Platinum oxide also catalyzes this hydrogenolysis.
4. A large portion of 1-phenyl-5-tetrazolone (and a small amount of biphenyl) remains mixed with and adsorbed to the catalyst and is removed by the ethanol treatment.
5. 1-Phenyl-5-tetrazolone can be recovered from the combined aqueous solutions by acidification with dilute hydrochloric acid. The yield is 4.2–4.7 g. (82–92%), m.p. 190–191°.
6. Benzoxazolyl ethers can also be used in this reaction sequence but an amount of catalyst equivalent to 20–40% by weight of ether is necessary in the hydrogenolysis step. 2-Chlorobenzoxazole is available from Eastman Organic Chemicals.

3. Discussion

The preparation is essentially that described by the submitters[2] and is cited as an example of this general procedure for replacement of phenolic hydroxyl groups by hydrogen.

The reaction sequence, which involves the conversion of the phenolic hydroxyl groups to a phenyltetrazolyl ether (see Note 6) followed by reduction to effect removal of the phenolic hydroxyl group, illustrates a mild, efficient, general, and convenient procedure.

TABLE I
HYDROGENOLYSIS OF PHENOLIC ETHERS

Substituted Phenol	Yield of Tetrazolyl Ether, %	Hydrogenolysis Time, hours	Hydrogenolysis	
			Product	Yield, %
Guaiacol	94	15	Anisole	86[a]
3-Methoxyphenol	95	16	Anisole	85[a]
4-Methoxyphenol	97	6	Anisole	83[a]
2-Phenylphenol	98	8	Biphenyl	82
4-Aminophenol	86	9	Aniline	46[b]
4-Carbethoxyphenol	91	16	Ethyl benzoate	89[a]
Thymol	93	15	p-Cymene	72[a]
1-Naphthol	88	7	Naphthalene	50
2-Naphthol	94	17	Naphthalene	65
4-Chlorophenol	92	18	Benzene[c]	70[a]

[a] Filtered solution analyzed directly by gas chromatography with toluene as internal standard.
[b] Isolated as the hydrochloride salt.
[c] From hydrogenolysis of carbon-chlorine bond.

It has been applied successfully by the submitters[2] to a variety of substituted phenols, as shown in Table I.

Phenols having a variety of substituents including alkyl, alkoxyl, aryl, amino, and carbalkoxyl have been successfully converted to the desired product in good yield. The only limitation yet found is in the hydrogenolysis of the halogen–carbon bond. Thus 4-chlorophenol was converted to benzene using this procedure.

Other procedures include zinc-dust distillation, not generally useful except for exhaustive degradation of phenols to hydrocarbons, and various sodium and liquid ammonia cleavages of phenol ethers.[3-7] These latter reactions lack generality and are often unpredictable. They require conditions too harsh for certain aromatic substituents, and the yields are frequently low.

1. Research Laboratories, Eastman Kodak Company, Rochester, New York 14650.
2. W. J. Musliner and J. W. Gates, Jr., *J. Am. Chem. Soc.*, **88**, 4271 (1966).
3. W. H. Pirkle and J. L. Zabriskie, *J. Org. Chem.*, **29**, 3124 (1964) and references cited therein.
4. Y. K. Sawa, N. Tsuji, and S. Maeda, *Tetrahedron*, **15**, 144, 154 (1961); Y. K. Sawa, N. Tsuji, K. Okabe, and T. Miyamoto, *Tetrahedron*, **21**, 1121 (1965); Y. K. Sawa and J. Irisawa, *Tetrahedron*, **21**, 1129 (1965); Y. K. Sawa, M. Horiuchi, and K. Tanaka, *Tetrahedron*, **21**, 1133 (1965).
5. P. A. Sartoretto and F. J. Sowa, *J. Am. Chem. Soc.*, **59**, 603 (1937); A. L. Kranzfelder, J. J. Verbanc, and F. J. Sowa, *J. Am. Chem. Soc.*, **59**, 1488 (1937); F. C. Weber and F. J. Sowa, *J. Am. Chem. Soc.*, **60**, 94 (1938).
6. M. Tomita, H. Furukawa, S.-T. Lu, and S. M. Kupchan, *Tetrahedron Lett.*, 4309 (1965).
7. E. J. Strojny, *J. Org. Chem.*, **31**, 1662 (1966).

CONTROLLED-POTENTIAL ELECTROLYTIC REDUCTION:
1,1-BIS(BROMOMETHYL)CYCLOPROPANE

[Cyclopropane, 1,1-bis(bromomethyl)-]

$$C(CH_2Br)_4 + 2e^- \xrightarrow[\substack{(n\text{-}C_4H_9)_4NBr, \\ (CH_3)_2NCHO}]{\substack{Hg\ cathode \\ (-1.8\ volt\ vs.\ sce)}} \triangleright(CH_2Br)_2 + 2Br^-$$

Submitted by M. R. Rifi[1]
Checked by Edith Feng and Herbert O. House

1. Procedure

Caution! Since bromine is liberated during the electrolysis, the reaction should be conducted in a hood.

The electrolysis cell (Note 1) is assembled in a 1000-ml., flat-bottomed Pyrex reaction kettle. The Pyrex cover contains four standard taper outer joints in which are mounted: (1) a 11.5 × 2-cm.-diameter carbon rod (Note 2) surrounded by a 15 × 5.5-cm.-diameter, porous, porcelain cup (Note 3); (2) a 31-cm. × 8-mm.-diameter length of soda-lime glass tubing (Note 4) with a short length of 0.6-mm.-diameter platinum wire fused into the bottom: (3) a tee-tube fitted to hold a thermometer and to allow nitrogen to be passed into the reaction vessel: and (4) a saturated calomel reference electrode fitted with successive salt bridges containing aqueous 1 *M* sodium nitrate and 1.5 *M* tetraethylammonium tetrafluoroborate in *N,N*-dimethylformamide (Note 5). The cover is also equipped with a suitable clamp so that it may be fastened to the reaction kettle during the electrolysis.

A sufficient quantity of mercury (about 700 g.) is added to the reaction kettle to form a cathode pool 1-cm. deep, and a Teflon-covered magnetic stirring bar is placed on this mercury pool. A solution of 25.0 g. (0.0651 mole) of pentaerythrityl tetrabromide (Note 6) in 250 ml. of 0.2 *M* tetra-*n*-butylammonium bromide (Note 7) in *N,N*-dimethylformamide (Note 8) is added to the reaction vessel. An additional 175 ml. (Note 9) of 0.2 *M* tetra-*n*-butylammonium bromide (Note 7) in *N,N*-dimethylformamide (Note 8) is added to the porous porcelain cup surrounding the carbon-rod anode, and the cover is clamped to the reaction kettle. The glass tubing with a platinum wire contact sealed in the bottom is adjusted so that all of the exposed platinum is below the mercury surface (Note 10). The bottom of the porous porcelain cup should be a sufficient distance above the mercury pool so as not to interfere with the magnetic stirring bar. The salt bridge associated with the calomel reference electrode (Note 5) should be adjusted so that the lower, porous Vycor plug is between 5 and 10 mm. above the surface of the mercury pool. Before the electrolysis is begun a 0.5-ml. aliquot of the reaction solution should be removed from the cathode compartment and analyzed polarographically (Note 11), and the electrical resistances between the cathode and anode and between the cathode and the reference electrode should be measured with a suitable resistance bridge (Note 12). If all electrical connections are satisfactory the cathode-anode resistance should be in the range 20–30 ohms and the cathode-reference resistance should be in the range 5000–15,000 ohms. The electrical leads from the anode and cathode should be connected with a direct current source whose potential may be conveniently adjusted from 0 to 40 volts with a continuous current output of at least 1 amp. A suitable d.c. voltmeter is mounted in parallel with the anode and cathode leads, and a d.c. ammeter, capable of measuring currents of 0 to 3 amps is placed in series in either of the two leads. Finally, a vacuum-tube d.c. voltmeter or some equivalent *high-impedance* (Note 13) potential measuring device is connected to measure the potential difference between the cathode and the reference electrode. The reaction kettle should be placed in a nonmagnetic bath, to which cooling water may be added if necessary and magnetic stirring started. A very slow stream of nitrogen (1–2 ml. per minute) is passed through the apparatus throughout the electrolysis (Note 14). The potential of the direct current source is adjusted, giving a potential difference of 1.7 to 1.8 volts between the cathode and the reference electrode (Note 15), and the current source is adjusted at 10–15 minute intervals to maintain this potential difference throughout the electrolysis. It is convenient to keep a record of time and the current passing through the cell so that the time when the theoretical amount of electricity (12,500 coulombs or 3.5 amp.-hours) has been passed through the cell can be estimated (Note 16). During the electrolysis the temperature of the catholyte solution is kept below 40° (Note 17) by the use of external cooling if necessary. When approximately the theoretical amount of electricity has been passed through the electrolysis cell (Note 16, typically 4–6 hours), a 0.5-ml. aliquot is removed and analyzed polarographically (Note 11). The electrolysis is continued until the polarographic analysis (Note 11) indicates the consumption of practically all the pentaerythrityl tetrabromide.

The solutions are removed from the cathode and anode (Note 18) compartments and added to 200 ml. of water. The resulting mixture is extracted with four 150-ml. portions of pentane, and the combined pentane extracts are washed with water, dried over anhydrous sodium sulfate, and concentrated by distillation through a short Vigreux column (Note 19). The remaining pentane is removed by distillation and the residual yellow liquid is distilled under reduced pressure, yielding 6.9–8.5 g. (47–58%) of 1,1-bis-(bromomethyl)cyclopropane as a colorless liquid, b.p. 65–67° (5 mm.), n_D^{25} 1.5341–1.5347 (Note 20).

2. Notes

1. The electrolysis cell designed by the checkers is shown in Figure 1. The adapters, which hold the reference electrode salt bridge (Note 5), the glass tube with the platinum contact, the tee tube for the nitrogen-inlet, and the thermometer are commercially available from Ace Glass, Inc., Vineland, New Jersey. The adapter which supports the carbon rod anode and the surrounding porous porcelain cup was machined from a Teflon rod; the dimensions of the adapter used by the checkers are indicated in Figure 2. Holes were drilled in the porcelain cup to permit fastening to the Teflon adapter with three stainless-steel machine screws. Electrical contact between the carbon anode and the wire to the external circuitry was achieved by drilling and tapping the end of the carbon rod for a small machine screw. The arrangement used for the remainder of the electrical circuit is indicated in Figure 1.

2. Carbon rods of approximately the dimensions indicated are commercially available from welding supply companies.

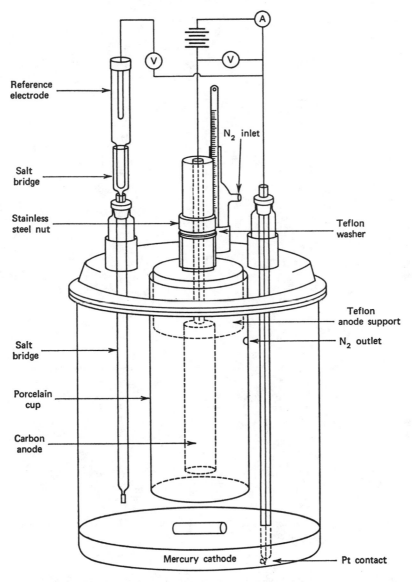

Figure 1. Cell for controlled-potential electrolytic reduction.

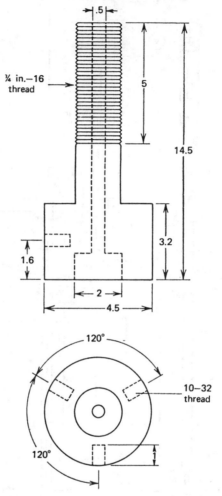

Figure 2.   Teflon anode support for the electrolysis cell. Unless otherwise stated, the dimensions are in centimeters.

3. A suitable porous porcelain cup (Coors 700, unglazed) may be purchased from the Arthur H. Thomas Company.

4. To obtain a satisfactory seal between the platinum and the glass, soda-lime glass rather than Pyrex glass should be employed. Before the platinum wire is sealed into the glass a length of copper wire should be silver soldered to the platinum to provide an accessible electrical lead at the top of the glass tubing.

5. Any commercial saturated calomel electrode of convenient dimensions may be employed. The arrangement of the salt bridges between the calomel electrode and the reaction solution is illustrated in Figure 3. The Teflon tubing is available from Bolab,

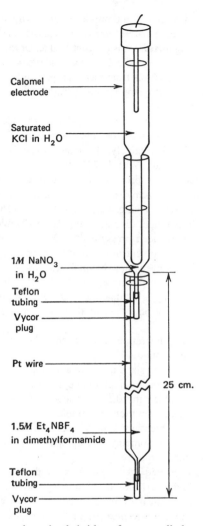

Figure 3. Reference electrode and salt bridges for controlled potential electrolysis.

Inc., 359 Main Street, Reading, Massachusetts 01867, and the porous Vycor plugs are cut from lengths of ⅛-in.-diameter porous Vycor rod ("thirsty" glass or Corning Vycor No. 7930), available from the Electronic Parts Department, Corning Glass Works, Houghton Park, Corning, New York, 14830. The intermediate salt bridge containing aqueous sodium nitrate is used to prevent precipitation of the insoluble potassium tetrafluoroborate at the small fiber in the tip of the calomel electrode.[2] To minimize resistance in the long, lower, nonaqueous salt bridge, a concentrated (1.5 M) solution of tetraethylammonium tetrafluoroborate[3] in N,N-dimethylformamide is used with a length of platinum wire inside the tube. Tetraethylammonium tetrafluoroborate was prepared by mixing 5.3 g.

(0.025 mole) of tetraethylammonium bromide (Eastman Organic Chemicals) and 3.6 ml. (*ca.* 0.026 mole) of aqueous 48–50% fluoroboric acid (Allied Chemical Corporation) in 8 ml. of water. The resulting mixture was concentrated under reduced pressure, diluted with ether and filtered, yielding 4.6 g. (85%) of the crude tetrafluoroborate salt, m.p. 375–378° (dec.). Two recrystallizations from methanol-petroleum ether mixtures afforded 3.7 g. (69%) of the pure tetraethylammonium tetrafluoroborate as white needles, m.p. 377–378° (dec.) (after drying).

6. Crude pentaerythrityl tetrabromide, purchased from Columbia Organic Chemicals Company, Inc., was recrystallized from chloroform, yielding the tetrabromide as tan needles, m.p. 158–160°. Alternatively, this material may be obtained by the procedure described in *Org. Synth.*, **Coll. Vol. 4**, 753 (1963).

7. Tetra-*n*-butylammonium bromide, obtained from Eastman Organic Chemicals, was recrystallized from chloroform. The white prisms that separated were pulverized and dried under reduced pressure, giving the pure salt, m.p. 116–117.5°.

8. *N*,*N*-Dimethylformamide, obtained from Allied Chemical Corporation, was purified by distillation under reduced pressure, b.p. 39–41° (6 mm.).

9. Since the amount of solvent in the anode compartment is slowly depleted as the electrolysis proceeds, it is convenient to begin the electrolysis with the level of solution in the anode compartment about 9 cm. above the level of the solution in the cathode compartment. Alternatively, additional 0.2 M tetra-*n*-butylammonium bromide in *N*,*N*-dimethylformamide may be added to the anode compartment as the electrolysis proceeds.

10. To avoid competing liberation of hydrogen at the cathode, no platinum should be exposed to the catholyte solution.

11. The polarographic analysis is obtained at a dropping mercury electrode with any conventional three-electrode polarograph employing a saturated calomel electrode as the reference. The checkers added the 0.5-ml. aliquots of the reaction mixture to 10-ml. portions of 0.2 M tetra-*n*-butylammonium bromide (Note 7) in *N*,*N*-dimethylformamide (Note 8). The half-wave potentials ($E_{1/2}$ vs. sce) for the reduction of pentaerythrityl tetrabromide and 1,1-bis-(bromomethyl)cyclopropane are -1.71 ($\alpha n = 0.44$) volts and -2.18 ($\alpha n = 0.31$) volts, respectively.

12. The checkers measured these resistances with a Serfass Conductivity Bridge, Model RCM 15, employing a 1000-Hz. alternating current.

13. Since the resistance in the circuit containing the reference electrode is approximately 10,000 ohms, an accurate measure of the cathode-reference electrode potential can only be obtained by the use of a potential measuring device with an input impedance of at least 100,000 ohms. Although a vacuum-tube voltmeter (VTVM, typical imput impedance 11×10^6 ohms) is suitable for this purpose, a common multimeter (VOM, typical imput impedance 20,000 ohms per volt) *is not a satisfactory alternative*.

14. This slow stream of nitrogen passes through the hole about 2.5 cm. from the top of the porous cup surrounding the anode compartment and sweeps the bromine formed at the anode out the top of the apparatus through the hole in the center of the Teflon adapter.

15. If the cathode-reference potential is allowed to rise significantly above -1.8 volts, further reduction of the 1,1-bis-(bromomethyl)cyclopropane will occur. The submitter reports that the dibromide may be reduced to spiropentane in 39% yield by employing a cathode-reference potential of -2.0 volts to -2.2 volts.

16. The checkers found that about 1.3 times the theoretical amount of electricity was passed through the cell during the time required to consume the starting material. Presumably the excess electricity is consumed partially in the reduction of impurities (*e.g.*, oxygen) and partially in the further reduction of the dibromide to spiropentane (Note 15).

17. This temperature limit was selected to avoid the possible loss of solvent and the

TABLE I

ELECTROLYTIC REDUCTIVE RING CLOSURE OF DIHALIDES

Dihalide	Product(s)	Yield, %	Ref.
$Br(CH_2)_3Br$	[cyclopropane] (only product)	60–80	9, 12
$Br(CH_2)_4Br$	25% [cyclobutane] + 75% $CH_3CH_2CH_2CH_3$	—	9
$Br(CH_2)_5CH_3$	$CH_3(CH_2)_3CH_3$ (only product)	80	9

$C_6H_5CHCH_2CH_2Br$
   $\vert$
   Br
	C_6H_5—[cyclopropane]	70	11

[cyclopropane]$(CH_2Br)_2$	[spiropentane]	39	12
CH_3\ /CH_3 [diamond] Br/ \Br	CH_3—[cyclobutane (rhombus)]—CH_3	55–94	9
Cl—[cyclobutane]—Br	60% [cyclobutene] + 20% [cyclobutane] + 10% [cyclobutane]	—	9

$CH_3CHCH_2CHCH_3$
   $\vert$ $\vert$
   Br Br
 (either meso or racemic isomer)
	CH_3—[cyclopropane]—CH_3 +
	 (both *cis* and *trans* isomers)
	 $n\text{-}C_5H_{12}$ + isomeric pentenes
		—	13

volatile product in the slow stream of nitrogen passed through the cell. Since the rate of the electrolytic reduction is increased with an increase in the reaction temperature, it is advantageous to maintain the temperature of the reaction solution in the range $35-40°$.

18. Since some diffusion of the product from the cathode compartment to the anode compartment is probable, both solutions are subjected to the isolation procedure.

19. To avoid the possible loss of the volatile product, the pentane solvent should be removed by distillation through a $20-30$-cm. Vigreux column rather than by distillation in a rotary evaporator.

20. On a GC column, packed with Apiezon M suspended on Chromosorb P and heated to 70°, the product exhibits a single peak with a retention time of 23.2 minutes. The sample exhibits IR absorption (CCl_4) at 3100, 3030, 2985, 1440, 1340, and 1235 cm.$^{-1}$ with ^{1}H NMR (CCl_4) singlets at δ 3.45 (CH_2Br) and 0.90 (cyclopropyl CH_2). The mass spectrum of the sample has abundant fragment peaks at m/e 149, 147, 67, 41, and 39.

3. Discussion

1,1-Bis-(bromomethyl)cyclopropane has been obtained as one component in a mixture of isomeric bromides by reaction of methylenecyclobutane with bromine[4] and by reaction of cyclopropane-1,1-dimethanol with phosphorus tribromide.[5-7] The present method illustrates a general procedure for the preparation of cyclopropane and cyclobutane derivatives by the electrolytic reduction of 1,3- and 1,4-dihalides.[8-15] In at least some cases, this method is clearly superior to the reductive cyclizations of dihalides effected with metals or with chromium(II) salts. Examples of this reductive ring closure are provided in Table I. The reaction appears to give the best results when dibromides rather than dichlorides are used as starting materials and when an aprotic solvent such as N,N-dimethylformamide or acetonitrile is used. These electrolytic ring closures proceed by way of a stepwise reaction mechanism.[12]

1. Chemicals and Plastics Division, Union Carbide Corporation, Bound Brook, New Jersey 08805.
2. B. McDuffie, L. B. Anderson, and C. N. Reilley, *Anal. Chem.,* **38,** 883 (1966).
3. (a) H. O. House, E. Feng, and N. P. Peet, *J. Org. Chem.,* **36,** 2371 (1971); (b) C. W. Wheeler, Jr., and R. A. Sandstedt, *J. Am. Chem. Soc.,* **77,** 2025 (1955).
4. D. E. Applequist and J. D. Roberts, *J. Am. Chem. Soc.,* **78,** 874 (1956).
5. Ya. M. Slobodin and I. N. Shokhor, *J. Gen. Chem. USSR Engl. Transl.,* **21,** 2231 (1951); *Zh. Obshch. Khim.,* **21,** 2005 (1951).
6. N. Zelinsky, *Ber. Dtsch. Chem. Ges.,* **46,** 160 (1913).
7. W. M. Schubert and S. M. Leahy, Jr., *J. Am. Chem. Soc.,* **79,** 381 (1957).
8. M. R. Rifi, *J. Am. Chem. Soc.,* **89,** 4442 (1967).
9. M. R. Rifi, *Tetrahedron Lett.,* 1043 (1969).
10. R. Gerdil, *Helv. Chim. Acta,* **53,** 2100 (1970).
11. M. R. Rifi, unpublished work.
12. A. J. Fry and W. E. Britton, *Tetrahedron Lett.,* 4363 (1971).
13. M. R. Rifi in M. M. Baizer, Ed., "Organic Electrochemistry," Dekker, New York, 1973, p. 279.
14. M. R. Rifi and F. H. Covitz, "Introduction to Organic Electrochemistry," Dekker, New York, 1974, p. 194.

15. M. R. Rifi in N. L. Weinberg, Ed., "Technique of Electroörganic Synthesis," Vol. V, Part II of "Techniques of Chemistry," A. Weissberger, Ed., Wiley, New York, 1975, Chap. VIII, p. 83.

BIS(TRIFLUOROMETHYL)DIAZOMETHANE

(Propane, 2-diazo-1,1,1,3,3,3-hexafluoro-)

$$(CF_3)_2C{=}NH + H_2NNH_2 \longrightarrow (CF_3)_2\overset{\displaystyle NH_2}{\underset{|}{C}}{-}NHNH_2$$

$$(CF_3)_2\overset{\displaystyle NH_2}{\underset{|}{C}}{-}NHNH_2 \xrightarrow{P_2O_5} (CF_3)_2C{=}NNH_2$$

$$(CF_3)_2C{=}NNH_2 + Pb(OCOCH_3)_4 \longrightarrow$$
$$(CF_3)_2CN_2 + Pb(OCOCH_3)_2 + 2CH_3CO_2H$$

Submitted by W. J. MIDDLETON[1] and D. M. GALE
Checked by L. SCERBO and W. D. EMMONS

1. Procedure

Caution! This procedure should be carried out in a good hood and behind a shield to avoid exposure to the toxic compounds, hexafluoroacetone imine and hydrazine.

A. *Hexafluoroacetone hydrazone.* A 100-ml., three-necked, round-bottomed flask fitted with a thermometer, a gas-inlet tube, a Dewar condenser, and a magnetic stirring bar is charged with 16 g. (0.50 mole) of anhydrous hydrazine [*Org. Synth.*, **Coll. Vol. 6**, 10 (1988)] and cooled in an ice-methanol bath. The condenser is filled with a cooling mixture of ice and methanol, the magnetic stirrer is started, and 82.5 g. (0.500 mole) (Note 1) of hexafluoroacetone imine [*Org. Synth.*, **Coll. Vol. 6**, 664 (1988)] is slowly distilled into the flask through the gas-inlet tube, while the temperature is maintained below 10°. After the addition, which requires about 1 hour, the cooling bath is removed and the reaction mixture is allowed to warm to room temperature, then poured rapidly into a 500-ml., single-necked, round-bottomed flask containing 150 g. of phosphorus pentoxide. A spatula is used to partially mix the liquid with the phosphorus pentoxide (Note 2); the flask is quickly fitted with a simple still head with condenser and heated with a heating mantle until no further distillation occurs (Note 3). The crude distillate is redistilled through a 45-cm. spinning-band column, giving 48–53 g. (53–59% conversion) of hexafluoroacetone hydrazone as a colorless liquid, b.p. 95.5–96°, n_D^{25} 1.3298.

B. *Bis(trifluoromethyl)diazomethane.* A 1-l., three-necked, round-bottomed flask is equipped with a sealed mechanical stirrer and a 100-ml., pressure-equalizing dropping funnel. The third neck is connected with pressure tubing to a 50-ml. capacity cold trap immersed in an acetone-dry ice cooling bath and protected from the atmosphere with a calcium chloride drying tube. The system is purged with nitrogen, and 375 ml. of benzonitrile and 120 g. (0.271 mole) of lead tetraäcetate (Note 4) are placed in the flask. Stirring is started and the flask is cooled with an ice bath before a solution of 45 g. (0.25 mole) of hexafluoroacetone hydrazone in 50 ml. of benzonitrile is added with the dropping funnel over a period of 1 hour. The ice bath is removed and the reaction mixture is stirred for an additional hour at room temperature. During this time some product collects in the cold trap. The remaining product is distilled into the trap at 10 mm. pressure by removing the drying tube and connecting the trap to a pump or aspirator (a small amount of acetic acid and other impurities can be removed by trap-to-trap distillation), yielding 20–21 ml. or 34–35 g. (76–79%) (Note 5) of bis(trifluoromethyl)-diazomethane as a yellow liquid, b.p. 12–13°. The product is neither impact- nor static-sensitive (Note 6) and is stable to long storage at −78°; however, it partly decomposes with loss of nitrogen after several weeks at room temperature.

2. Notes

1. If it is inconvenient to distill the hexafluoroacetone imine directly from a cylinder, it may be condensed first in a calibrated cold trap cooled to −10°, then distilled from the trap into the reaction mixture. About 55 ml. of the imine, measured at −10°, corresponds to 0.5 mole.

2. *Caution! A spontaneous reaction may result if mixing is prolonged beyond 1 minute. Rubber gloves should be worn to avoid exposure to hydrazine vapors.*

3. About 14 g. (17%) of hexafluoroacetone imine can be recovered during the redistillation step if the receivers for both distillations are cooled to −10° or lower. The yield of the hydrazone is about 67% if the recovered imine is taken into account.

4. Since acetic acid does not interfere with this reaction, a grade of lead tetraäcetate that has been stabilized with acetic acid may be used.

5. The submitters have obtained yields as high as 90% on runs two times this scale.

6. *Caution! Toxicity data on this material are not available; therefore, it should be assumed to be as toxic as other diazo compounds and handled only in an efficient hood. Although it has been stored in stainless-steel cylinders at autogenous pressure and handled without incident, contact with reagents likely to initiate rapid evolution of nitrogen should be carried out with caution.*

3. Discussion

Bis(trifluoromethyl)diazomethane has been prepared by the nitrosation of 1,1,1,3,3,3-hexafluoroisopropylamine[2] and by the present procedure,[3] which gives higher yields. Bis(perfluoroethyl)diazomethane has also been prepared by an extension of this method.[2]

Bis(trifluoromethyl)diazomethane is a reactive, electrophilic compound, in that it forms adducts with nucleophiles such as amines and phosphines[3] and adds to olefins, acetylenes,[3] and thiocarbonyl compounds, forming heterocycles. It has been used as a source of bis(trifluoromethyl)carbene in reactions with benzene,[3] saturated hydrocarbons,[5] carbon disulfide,[6] and transition metal compounds,[7] and it undergoes a unique radical chain reaction with saturated hydrocarbons, giving hydrazone and azine adducts.[5]

1. Central Research and Development Department, Experimental Station, E. I. du Pont de Nemours and Company, Wilmington, Delaware 19898.
2. E. P. Mochalina and B. L. Dayatkin, *Izv. Akad. Nauk SSSR Ser. Khim.*, **5**, 926 (1965).
3. D. M. Gale, W. J. Middleton, and C. G. Krespan, *J. Am. Chem. Soc.*, **88**, 3617 (1966).
4. W. J. Middleton, *J. Org. Chem.*, **34**, 3201 (1969).
5. W. J. Middleton, D. M. Gale, and C. G. Krespan, *J. Am. Chem. Soc.*, **90**, 6813 (1968).
6. M. S. Raasch, *Chem. Commun.*, 577 (1966).
7. J. Cooke, W. R. Cullen, M. Green, and F. G. A. Stone, *Chem. Commun.*, 170 (1968).

BIS[2,2,2-TRIFLUORO-1-PHENYL-1-(TRIFLUOROMETHYL) ETHOXY] DIPHENYL SULFURANE

[Sulfur, bis[α,α-bis(trifluoromethyl)benzenemethanolato]diphenyl-]

$$C_6H_5C(CF_3)_2OH + KOH \xrightarrow{\text{water}} C_6H_5C(CF_3)_2OK$$

$$2C_6H_5C(CF_3)_2OK + (C_6H_5)_2S + Br_2 \xrightarrow[25°]{CCl_4} \begin{array}{c} O-C(CF_3)_2C_6H_5 \\ C_6H_5 \diagdown | \\ S: \\ C_6H_5 \diagup | \\ O-C(CF_3)_2C_6H_5 \end{array} + 2KBr$$

Submitted by J. C. Martin,[1] R. J. Arhart, J. A. Franz, E. F. Perozzi, and L. J. Kaplan
Checked by H. G. Corkins, C. J. Stark, and C. R. Johnson

1. Procedure

A. *Potassium 1,1,1,3,3,3-hexafluoro-2-phenyl-2-propanolate.*   A 500-ml., round-bottomed flask equipped for simple vacuum distillation is charged with a solution of 25.4 g. (0.390 mole) of 86% potassium hydroxide (Notes 1 and 2) in 50 ml. of water, to which is added 100.0 g. (0.4098 mole) of 1,1,1,3,3,3-hexafluoro-2-phenyl-2-propanol (Note 3). The colorless solution obtained is concentrated to a syrup by vacuum distillation at aspirator pressure. Further evacuation with a vacuum pump with heating to 140° results in a white solid; under the greater vacuum volatile substances can be collected in a trap immersed in an acetone-dry ice bath. The trap must be cleaned or be replaced several times during the first hour. After drying the white solid at 140° for 12 hours, the flask is transferred to a glove bag (Note 4) equipped with a mortar and pestle, a vacuum adapter, a powder funnel, a spatula, and a tared, 500-ml., round-bottomed flask. Under a dry nitrogen atmosphere the white solid is ground to a fine powder, transferred to the 500-ml. flask, and dried on a vacuum pump (10^{-2} mm.) to constant weight, yielding $107-109$ g. ($97-99\%$) of potassium 1,1,1,3,3,3-hexafluoro-2-phenyl-2-propanolate as a white powder.

B. *Bis[2,2,2-trifluoro-1-phenyl-1-(trifluoromethyl)ethoxy]   diphenyl   sulfurane.* Carbon tetrachloride is distilled directly from phosphorous pentoxide into a dry, 2-l., three-necked flask, fitted with stoppers, until 700 ml. is collected. The flask is quickly fitted with an adapter for use as a nitrogen-inlet, a mechanical stirrer, and an adapter for solid addition (Note 5) which is attached to the 500-ml. flask containing the potassium alkoxide from Part A. A positive pressure of dry nitrogen is used to maintain inert atmosphere conditions. A white slurry is obtained when the powdered alkoxide is added at room temperature to the stirred carbon tetrachloride. Since all of the alkoxide does not transfer, the tared, 500-ml. flask is reweighed, and the amount added to the reaction vessel is determined by difference (Note 6). The adaptor used for solid addition is quickly exchanged for a septum.

To a stirred suspension of 105 g. (0.372 mole) of the alkoxide, 34.7 g. (31.0 ml., 0.186 mole) of diphenyl sulfide is added by syringe. Bromine, 29.9 g. (9.6 ml., 0.186 mole), is then added by syringe over a 5-minute period, giving a red-brown mixture which gradually fades to a pale yellow within 30 minutes. Stirring is continued at room temperature for 2.5 hours, leaving a pale-yellow solution and a copious precipitate of potassium bromide containing some potassium alkoxide.

A glove bag is equipped with a spatula, a tared, 1-l., single-necked round-bottomed flask, a 350-ml. Buchner funnel with filter paper, a 1-l. flask, a vacuum adapter, an aspirator hose (Note 7), and a flask containing 100 ml. of dry carbon tetrachloride. Filtration of the reaction mixture under a nitrogen atmosphere can be achieved by pouring the solution into the filter funnel *via* one neck of the three-necked flask, which is inserted through a hole in the glove bag. The potassium bromide is removed and washed with two 50-ml. portions of dry carbon tetrachloride. The filtrate is then transferred to the 1-l. flask, fitted with the vacuum adapter. After removal of the sulfurane solution from the glove bag (Note 8), the flask is quickly placed on a rotary evaporator (Note 9) and concentrated to a semisolid. Drying under reduced pressure (10^{-2} mm.) for 24 hours results in crude sulfurane (115–119 g., 93–96%) as slightly yellow crystals.

The flask containing the crude sulfurane is transferred to the nitrogen atmosphere of the glove bag which contains a powder funnel, a fluted filter paper, a 1-l., single-necked, round-bottomed flask, a 250-ml. graduated cylinder, a 1-l. flask containing 700 ml. of dry pentane (Note 10), and a 500-ml. flask containing 250 ml. of dry diethyl ether (Note 10). After dissolving the sulfurane in 150 ml. of ether, 500 ml. of pentane is added, giving a cloudy solution that clarifies when filtered directly into the 1-l. flask. The flask is stoppered with the vacuum adapter and removed from the bag. At this time everything but the pentane is removed from the glove bag and the following items are added: a 600-ml., medium-frit sintered-glass funnel, a 1-l., single-necked, round-bottomed flask, a tared, 500-ml., single-necked round-bottomed flask, a vacuum adaptor, two spatulas, and a powder funnel. The sulfurane solution is concentrated to *ca.* 350 ml. at reduced pressure and a temperature not exceeding 40° (Note 9), using a magnetic stirrer to prevent bumping as the solvent is evaporated through the vacuum adapter. After a drying tube is connected to the adapter, the flask is cooled in an acetone–dry ice bath to induce crystallization; swirling the flask during the crystallization prevents crystals from adhering to the sides. When crystallization is complete, the flask is exchanged for the three-necked flask inserted into the side of the glove bag. Cooling in the acetone–dry ice bath is continued while the neck of the flask penetrates the glove bag. The crystals, collected in the sintered-glass funnel by vacuum filtration, are washed with one 50-ml. portion of cold (−78°) pentane and transferred to the 500-ml. flask. After fitting with the vacuum adapter, the flask is removed from the bag and the crystals dried under reduced pressure (10^{-2} mm.) until the powdered, white sulfurane is at constant weight (76–79 g., 61–64%, m.p. 103–108°). Further crystallization from the mother liquors, after concentration by rotary evaporation to one-half the original volume, gives up to 7 g. of crude sulfurane (m.p. < 95°) for a total yield of 79–83 g. (63–67%) (Note 11). Recrystallization by the same method as above gives analytically pure material, m.p. 109.5–110.5°.

2. Notes

1. The submitters report no problems in running the entire procedure on four times the scale described here.

2. The use of potassium hydroxide rather than sodium hydroxide is dictated by solubility characteristics which make purification of the sodium alkoxide difficult.

3. The alcohol was obtained from PCR, Incorporated, or was prepared from hexafluoroacetone (E. I. du Pont de Nemours and Company), benzene, and aluminum chloride by the published[2] procedure.

4. Glove bags may be purchased from Instruments for Research and Industry, 108 Franklin Avenue, Cheltenham, Pennsylvania 19012. The 27 × 27 × 15 inch bag was found to be a convenient size for this procedure. The submitters performed all inert atmosphere operations in a dry box. All apparatus must be oven or flame dried prior to use.

5. This adapter consisted of two 24/40 joints connected with Gooch tubing.

6. The quantities of bromine and diphenyl sulfide must be adjusted according to the amount of alkoxide added.

7. A stopcock and a drying tube were inserted into the hose between the glove bag and the aspirator.

8. The other materials in the bag may be removed at this time, but the three-necked flask inserted through the side of the glove bag must not be removed.

9. During solvent removal or recrystallization, temperatures should be kept below 50° to avoid degradation of product quality.

10. Dry ether was obtained by distillation of reagent grade dry ether from sodium dispersion or by drying over three portions of sodium wire over a 48-hour period. Dry pentane was prepared by adding sodium wire directly to reagent grade solvent at least twice or by distilling it from a sodium dispersion.

11. Hexafluoro-2-phenyl-2-propanol may be recovered from mother liquors, recovered solvent, and the KBr salt cake by extracting the mixture with aqueous base. Neutralization of the aqueous phase gives the alcohol (13–23 g.), which is purified by distillation.

3. Discussion

This dialkoxy diphenyl sulfurane has been prepared by the reaction of diphenyl sulfide, 2,2,2-trifluoro-1-phenyl-1-(trifluoromethyl)ethyl hypochlorite, and potassium 1,1,1,3,3,3-hexafluoro-2-phenyl-2-propanolate[3] and by the reaction of diphenyl sulfide with 1 equivalent of chlorine and 2 equivalents of potassium 1,1,1,3,3,3-hexafluoro-2-phenyl-2-propanolate in ether.[4]

The present method offers several advantages over earlier methods. The use of carbon tetrachloride instead of ether as solvent avoids the intrusion of certain radical-chain reactions with the solvent which are observed with bromine and to a lesser degree with chlorine. In addition, the potassium bromide has a reduced solubility in carbon tetrachloride compared to ether, thus providing additional driving force for the reaction and ease of purification of product. The selection of bromine over chlorine as the oxidizing agent is made in consideration of the ease of handling the bromine by syringe compared to the greater number of operations and more complex apparatus required for chlorine.

Another dialkoxy diaryl sulfurane has been prepared[5] under conditions similar to those reported here, indicating that this reaction for sulfurane formation may have wide applicability.

The great reactivity of the sulfurane prepared by this procedure toward active hydrogen compounds, coupled with an indefinite shelf life in the absence of moisture, makes this compound a useful reagent for dehydrations,[6,7] amide cleavage reactions,[8] epoxide formation,[9] sulfilimine syntheses,[10] and certain oxidations and coupling reactions.

1. Department of Chemistry, University of Illinois, Urbana, Illinois 61801. This research was supported by the National Science Foundation (GP 30491X) and the National Cancer Institute (CA 13963).
2. B. S. Farah, E. E. Gilbert, and J. P. Sibilia, *J. Org. Chem.*, **30**, 998 (1965).
3. J. C. Martin and R. J. Arhart, *J. Am. Chem. Soc.*, **93**, 2339 (1971).
4. J. C. Martin and R. J. Arhart, *J. Am. Chem. Soc.*, **93**, 2341 (1971).
5. E. F. Perozzi and J. C. Martin, *J. Am. Chem. Soc.*, **94**, 5519 (1972).
6. J. C. Martin and R. J. Arhart, *J. Am. Chem. Soc.*, **93**, 4327 (1971).
7. R. J. Arhart and J. C. Martin, *J. Am. Chem. Soc.*, **94**, 5003 (1972).
8. J. A. Franz and J. C. Martin, *J. Am. Chem. Soc.*, **95**, 2017 (1972).
9. J. C. Martin, J. A. Franz, and R. J. Arhart, *J. Am. Chem. Soc.*, **96**, 4604 (1974).
10. J. A. Franz and J. C. Martin, *J. Am. Chem. Soc.*, **97**, 583 (1975).

ACYLOIN CONDENSATION IN WHICH CHLOROTRIMETHYLSILANE IS USED AS A TRAPPING AGENT: 1,2-BIS(TRIMETHYLSILYLOXY)-CYCLOBUTENE AND 2-HYDROXYCYCLOBUTANONE

[Trimethylsilane, 1-cyclobuten-1,2-ylenedioxybis- and Cyclobutanone, 2-hydroxy-]

$$\begin{array}{l} CH_2CO_2C_2H_5 \\ | \\ CH_2CO_2C_2H_5 \end{array} + 4Na + 4ClSi(CH_3)_3 \longrightarrow$$

$$+ 2C_2H_5OSi(CH_3)_3 + 4NaCl$$

$$\xrightarrow{CH_3OH} \quad + 2CH_3OSi(CH_3)_3$$

Submitted by Jordan J. Bloomfield[1] and Janice M. Nelke
Checked by Ž. Stojanac and Z. Valenta

1. Procedure

Caution! See warnings concerning the use of impure chlorotrimethylsilane (Note 8) and finely dispersed alkali metals (Notes 1, 9, 11, and 19).

A. *1,2-Bis(trimethylsilyloxy)cyclobutene. Method* 1. A 1-l., three-necked, creased flask is fitted with a stirrer capable of forming a fine dispersion of molten sodium (Note 1), a reflux condenser, and a Hershberg addition funnel and maintained under an oxygen-free, nitrogen atmosphere. The flask is charged with 250–300 ml. of dry solvent (Note 2) and 9.6–9.8 g. (~0.4 g.-atom) of freshly cut sodium (Notes 3, 4). The solvent is brought to gentle reflux, and the stirrer is operated at full speed until the sodium is fully dispersed (Note 5). The stirrer speed is then reduced (Note 6), and a mixture of 17.4 g. (0.100 mole) of diethyl succinate (Note 7) and 45–50 g. (~0.4 mole) of chlorotrimethylsilane (Note 8) in 125 ml. of solvent is added over 1–3 hours. The reaction is exothermic, and a dark purple precipitate appears within a few minutes (Note 9). The solvent is maintained at reflux during and after the addition (Note 10). After five hours of additional stirring, the contents of the flask are cooled and filtered through a 75-mm., coarse sintered-disk funnel in a nitrogen dry-box (Note 11). The precipitate is washed several times with anhydrous diethyl ether or petroleum ether.

The colorless to pale yellow filtrate is transferred to a distilling flask, the solvent is evaporated, and the residue is distilled under reduced pressure (Note 12). After a small forerun (0.5–1.0 g.), 18 g. (78%) of the product is obtained at 82–86° (10 mm.) as a colorless liquid, n_D^{25} 1.4331 (Note 13).

Method 2. The apparatus described above is charged with 4.8–5.0 g. (~0.2 g.-atom) of clean sodium and 8.0–8.2 g. (~0.2 g.-atom) of clean potassium (Notes 3, 4). The flask is heated with a heat gun, forming the low-melting alloy, and 300–350 ml. of anhydrous ether is added from a freshly opened can. The stirrer is operated at full speed until the alloy is dispersed, then at a slower speed for the remainder of the reaction (Notes 5, 6). A mixture of 17.4 g. (0.100 mole) of diethylsuccinate (Note 7), 44 g. (0.41 mole) of chlorotrimethylsilane (Note 8), and 125 ml. of anhydrous ether is added at a rate sufficient to keep the reaction under control (Note 14). The purple mixture is stirred for another 4–6 hours (Note 9), then filtered and washed as above in a nitrogen dry-box (Note 11). The product is distilled as above, collecting a forerun, 0.5–2 g. to 80° (10 mm.), and then the product at 82–86° (10 mm.) as a colorless liquid, 13.8–16.1 g. (60–70%), n_D^{25} 1.4323–1.4330 (Notes 15–19).

B. *2-Hydroxycyclobutanone*. A 1-l., three-necked flask fitted with a magnetic stirring bar, a sintered-disk gas-inlet tube, a dropping funnel, and a reflux condenser is charged with 450 ml. of reagent grade methanol (Note 20). Dry, oxygen-free nitrogen is bubbled vigorously through the methanol for about 1 hour before 23 g. (0.10 mole) of freshly distilled (Note 21) 1,2-bis(trimethylsilyloxy)cyclobutene is transferred under nitrogen to the addition funnel and added dropwise to the stirred methanol. Stirring under a reduced nitrogen flow is continued for 24–30 hours (Note 22). The methanol and methoxytrimethylsilane are removed under reduced pressure, and the residual 2-hydroxycyclobutanone is distilled through a short-path still as a colorless liquid, b.p. 52–57° (0.1 mm.), 6.1–7.4 g. (71–86%), n_D^{25} 1.4613–1.4685 (Notes 23–25).

2. Notes

1. Generally a "Stir-O-Vac" stirrer (available from Labline Instruments, Inc., Melrose Park, Ill.) is used. A Vibromixer type of stirrer is also satisfactory, especially when high-dilution conditions are required. Both were used in the submitter's laboratory in conjunction with a short condenser surrounding the stirrer shaft to prevent loss of solvent or reactants. The condenser is required in high-dilution procedures with the Vibromixer to protect the diaphragm from solvent vapor. A plain sleeve stirrer with a Teflon paddle was used by the checkers, but the submitters warn that a Teflon paddle with highly dispersed metal, especially sodium-potassium alloy, is dangerous; explosions have been reported when Teflon and molten sodium are in contact. The submitters also emphasize that, because the large particle size can cause or lead to side reactions, a fine dispersion of metal is desirable, even with chlorotrimethylsilane.

2. Toluene is commonly used and can be dried with molecular sieves or directly distilled from calcium hydride into the reaction flask. Solvent stored over calcium hydride for several days is usually sufficiently dry to decant directly into the reaction flask, but distillation gives more consistent results. Any solvent with a boiling point sufficiently high to melt sodium is satisfactory. The submitters have also used methylcyclohexane and xylene in acyloin condensations. After the sodium is dispersed, the high-boiling solvent

can be removed and replaced with anhydrous ether (as noted by the submitters) or can be retained and used in combination with ether (checkers).

3. The submitters routinely used a nitrogen dry-box to clean, cut, and weigh alkali metals. The checkers cut and weighed these metals under dry toluene.

4. An excess of metal is used because aromatic solvents are reduced to some extent, and it is easier than weighing out the exact amount.

5. One or two minutes of a "Stir-O-Vac" operated through a variable transformer at full voltage is required.

6. A setting of 30–40 volts on the variable transformer is used.

7. Diethyl succinate was obtained by the submitters from Eastman Organic Chemicals and used without purification. The checkers obtained the ester from British Drug Houses, Ltd., and distilled it at 100° (11 mm.). In general, it is preferable to distill or crystallize and dry all esters before attempting acyloin condensations.

8. Chlorotrimethylsilane, obtained from Eastman Organic Chemicals (submitters) and Aldrich Chemical Co. (checkers), was distilled from calcium hydride under nitrogen, then stored and weighed in a nitrogen dry-box. *Caution! It is particularly important that the chlorotrimethylsilane be distilled, preferably from calcium hydride, under nitrogen.* In at least one laboratory[2] the use of this reagent without prior purification led to explosions. Chlorotrimethylsilane may contain some dichlorodimethylsilane as an impurity, which hydrolyzes more readily than the monochlorosilane. Cautious treatment with a small amount of water, followed by distillation from calcium hydride, under nitrogen, removes this impurity. A further cautionary note concerning these reactions is also necessary. The explosions occurred in reactions run on a scale larger than 0.1 mole, using undistilled chlorotrimethylsilane and following a published procedure.[3] This procedure requires mixing all the reagents at 20–30° and gradually warming the mixture. When this procedure was applied to diethyl pentanedioate on a large scale, the reaction became uncontrollably exothermic at about 50°.[2] It is recommended that the ester and chlorosilane be added together, dropwise, at a rate sufficient to maintain the exothermic reaction. It is often unsafe with many esters to have a large amount of unreacted ester in the reaction mixture at any time.

9. The purple color seems to be indicative of a satisfactory reduction. When the color is light or no color develops, the yield is usually poor. Sometimes no reaction occurs. In this instance it is best to discard all residues *(pyrophoric)* carefully and start over with scrupulous attention to the dryness of all apparatus and reagents.

10. This is the usual procedure. The submitters report that equally good results are obtained if all the dispersing solvent is replaced by ether and the reaction is run at room temperature. The checkers have found a slightly modified procedure in which refluxing toluene (90 ml.) is used for dispersion, anhydrous ether (250 ml.) is added without removal of toluene, additional ether (120 ml.) is used for addition of diethyl succinate and chlorotrimethylsilane, and the mixture is heated under reflux for 14 hours, to be particularly convenient and to give consistently high yields (77–86%). The checkers have also found that prior removal of toluene does not affect the yields, but simplifies final purification. With toluene as solvent, better results are usually obtained at or near reflux. The amount of time following completion of addition of the ester is not critical. (It may vary with the compound being reduced.) The submitters generally use 4–6 hours or overnight, whichever happens to be more convenient.

11. Since a slight excess of metal is used, some may be left over. The excess chlorosilane and the product are sensitive to moisture. To avoid unpleasantness due to the pyrophoric nature of finely divided alkali metal residues, to hydrolysis of product, or production of free acid from the excess silane the submitters *always* filtered the reaction mixture in a nitrogen dry-box. The checkers used simple sintered-glass funnel filtration under a stream of dry nitrogen.

12. The submitters used a 250-mm. vacuum-jacketed Vigreux column fitted with a variable take-off head. Any good column should be as satisfactory.

13. The yield varies from 65 to 86%, n_D^{25} 1.4322–1.4338; b.p. 58–59° (2 mm.); 68–70° (6 mm.); 82–86° (10 mm.); 88–92° (13–14 mm.). In twelve separate runs (in toluene, toluene-ether, or in ether), the checkers did not obtain a yield below 76%.

14. The reaction is exothermic. Two hours is more than enough time. Too vigorous a reaction can be controlled with an acetone–dry ice bath.

15. The yields given are those obtained by the checkers. For this particular reaction the submitters have found the product to be cleaner and the yields higher (78–93%) with sodium-potassium reduction. This is not necessarily a general observation inasmuch as other reactions can occur with the alloy.[4] The checkers found that a modification involving formation of the alloy in hot toluene (10 ml.), removal of most of the toluene with a stream of dry nitrogen, and dispersion of the alloy in ether led to somewhat better yields (81–85%). Because of convenience, safety, and yield reproducibility, they strongly favor Method 1.

16. The product can be examined for purity by ^{1}H NMR or by GC. The submitters have used XF-1150 columns successfully. Columns with polar sites will strip silyloxy groups from the bis(silyloxy) compounds and are unsatisfactory.

17. Both Methods 1 and 2 have been successfully applied to a wide range of 1,2-diesters[5] and to a variety of other esters.[6] The use of a high-dilution cycle permits this procedure to be applied to medium- and large-ring acyloins with good to excellent results.

18. The product is stable if stored in a tightly screw-capped bottle. Prolonged exposure to moist air leads to decomposition. The submitters have stored samples for several years with no change in physical properties.

19. *Caution! Disposal of residues must be made with care. When excess metal, especially sodium-potassium alloy, is used, the residues can be pyrophoric!*

20. The submitters used a freshly opened bottle for each hydrolysis.

21. Use of freshly distilled bis(silyloxy) compound is critical in many cases, especially in this example. The yield and, more particularly, the quality of the product deteriorate with the age of the sample. Traces of acid should be avoided because even as little as one drop of chlorotrimethylsilane added to the reaction mixture produces a different product. The longer the reaction time in the presence of acid, the greater is the number of other products formed.

22. The reaction time can be reduced considerably by gentle reflux. It is advantageous to follow the reaction by GC (see Note 16) if heating is used, because prolonged reflux can lead to side reactions.

23. The wide range of refractive indices is related to the time interval between distillation and measurement. The longer one waits, the higher the refractive index. This is apparently due to rapid formation of dimer.

24. In the IR (CHCl$_3$), 2-hydroxycyclobutanone has a carbonyl band at 1780 cm^{-1}.

Kept in nitrogen-filled screw-capped vials in the freezing compartment of a refrigerator, 2-hydroxycyclobutanone slowly but completely solidifies as its dimer. The IR spectrum (KBr) of the solid shows no carbonyl. However, a CHCl₃ solution of the solid does show the characteristic 1780 cm^{-1} band, indicating rapid equilibration with the monomer.

25. Air readily oxidizes 2-hydroxycyclobutanone; quantitative conversion to succinic acid occurs on standing in the open for several days.

3. Discussion

The discovery that chlorotrimethylsilane will react *in situ* with alcoholates and the acyloin enediolates[3,7] provided the opportunity to prepare a wide range of four-membered acyloins for the first time.[5] In addition, most acyloin reaction yields are improved, and some diesters, which were found to give Dieckmann condensation products as a result of the base formed concomitantly with the acyloin, can now be reductively cyclized in good yield.[6] The general reaction conditions given for the cyclization of diethyl succinate have been applied to synthesis of four- to eight-membered rings with very good results.[3,7] Likewise, when a high-dilution cycle is used, good to excellent cyclizations of eight- to fourteen-membered rings can be obtained. (Yields are 8, 72–85%; 9, 68%; 10, 58–69%; 11, 48%; 12, 68%; 13, 84%; 14, 67%.)[8]

For many reasons, use of the trapping agent is recommended as the most efficient method for running acyloin condensations. Among them are: (a) the work-up is very simple: filter and distill; (b) the bis(silyloxy)olefin is usually easier to store than the free acyloin and is readily purified by redistillation; (c) unwanted base-catalyzed side reactions during reduction are completely avoided; and (d) the bis(silyloxy)olefin can be easily converted directly into the diketone by treatment with 1 mole of bromine in carbon tetrachloride.[9,10] Other reactions are described in Rühlmann's review and in *Organic Reactions*.[11]

Bis(silyloxy)cyclobutenes are also subject to a variety of special reactions. Probably the most interesting is the observation that they readily undergo a ring-opening reaction leading to butadiene derivatives.[5] This reaction has already been used to prepare large-ring diketones from cyclic 1,2-diesters.[12]

The synthesis of 2-hydroxycyclobutanone was chosen as a model for the use of a trapping agent because diethyl succinate is the most accessible of 1,2-diesters and the hydrolysis step for this compound is more difficult than most. Procedures developed for succinoin have been found broadly applicable in preparation of other sensitive acyloins.

1. Corporate Research Department, Monsanto Company, St. Louis, Missouri 63166.
2. Private communication from Professor P. E. Eaton, Department of Chemistry, University of Chicago, Chicago, Ill. 60637.
3. See review by K. Rühlmann, *Synthesis,* 236 (1971).
4. J. J. Bloomfield, R. A. Martin, and J. M. Nelke, *J. Chem. Soc. Chem. Commun.,* 96 (1972).
5. J. J. Bloomfield, *Tetrahedron Lett.,* 587 (1968).
6. J. J. Bloomfield, *Tetrahedron Lett.,* 591 (1968).
7. (a) K. Rühlmann and S. Poredda, *J. Prakt. Chem.,* [4] **12,** 18 (1960); (b) U. Schräpler and K. Rühlmann, *Chem. Ber.,* **96,** 2780 (1963); *Chem. Ber.,* **97,** 1383 (1964); (c) K. Rühlmann, H. Seefluth, and H. Becker, *Chem. Ber.,* **100,** 3820 (1967).
8. J. J. Bloomfield, unpublished studies.

9. (a) H. Wynberg, S. Rieffers, and J. Strating, *Recl. Trav. Chim. Pays-Bas.*, **89**, 982 (1970); (b) J. Strating, S. Rieffers, and H. Wynberg, *Synthesis*, 209, 211 (1971).
10. J. M. Conia and J. M. Denis, *Tetrahedron Lett.*, 2845 (1971).
11. J. J. Bloomfield, D. C. Owsley, and J. M. Nelke, *Org. React.*, **23**, 259 (1976).
12. T. Mori, T. Nakahara, and H. Nozaki, *Can. J. Chem.*, **47**, 3266 (1969).

2-BORNENE

(1,7,7-Trimethylbicyclo[2.2.1]hept-2-ene)

Submitted by ROBERT H. SHAPIRO[1] and J. H. DUNCAN
Checked by ROBERT CZARNY and ROBERT E. IRELAND

1. Procedure

A dry, 1-l., three-necked flask equipped with a reflux condenser protected with a calcium sulfate drying tube, a 250-ml., pressure-equalizing dropping funnel, and a magnetic stirring bar is charged with 32 g. (0.10 mole) of camphor tosylhydrazone (Note 1) and 400 ml. of dry diethyl ether (Note 2). The flask is immersed in a cold-water bath (20–25°), and the contents are stirred magnetically. About 50 ml. of dry ether is placed in the dropping funnel, and the addition rate is set at 2–3 ml./minute (Note 3). After this addition 150 ml. of 1.6 N (0.24 mole) methyllithium (Note 4) in ether is added to the dropping funnel and dropped into the reaction flask over 1 hour (Note 5), while the cooling bath temperature is maintained at 20–25°. The yellow-orange solution is stirred for 8–9 hours, during which time lithium p-toluenesulfinate precipitates and the solution develops a deep red-orange color. A small amount of water is carefully added to destroy excess methyllithium before an additional 200 ml. is added. The layers are separated, the organic phase is washed with four 250-ml. portions of water, and the combined aqueous phases are shaken twice with 100-ml. portions of ether. After drying the combined ethereal extracts over anhydrous sodium sulfate, the volume of the solution is reduced to 50–60 ml. by distillation of the ether through a 25-cm. Vigreux column with gentle boiling on a steam bath. After 100 ml. of distilled pentane (Note 6) is added to the orange solution, the solvent is again gently boiled away, reducing the volume to 30–50 ml. The addition and removal of pentane is repeated two additional times to assure the complete removal of ether, and the final volume of the solution of 2-bornene is reduced to 30–40 ml. The solution is added to an 80 × 5 cm. chromatography column packed with 500 g. of alumina (Note 7), and the product is eluted with 750 ml. of pentane. After concentration of the eluate by distillation of the solvent through a Vigreux column, the residue is transferred to a 50-ml. flask and distilled through a U-tube with the aid of an oil bath and a heating lamp (Note 8). After collecting a forerun (pentane and 2-bornene), 8.5–8.8 g. (63–65%) of 2-bornene, m.p. 110–111° (lit.[2], 109–110°), is collected as colorless crystals in a cooled flask (Note 9). GC analysis[3] shows this product to be 98–99% pure, containing no camphene or tricyclene (Note 10).

2. Notes

1. Camphor tosylhydrazone[4] is prepared in the following manner. To a 1-l., one-necked, round-bottomed flask are added 44 g. (0.24 mole) of p-toluenesulfonylhydrazide [Org. Synth., Coll. Vol. 5, 1055 (1973)], 31.6 g. (0.208 mole) of camphor, and 300 ml. of 95% ethanol. One milliliter of concentrated hydrochloric acid is added, the flask is fitted with a reflux condenser, and the solution is heated under reflux for 2 hours. The resulting solution is cooled in an ice bath; colorless needles are collected by suction filtration and dried in air (Note 11). Recrystallization from ethanol yields 50 g. (73%) of pure camphor tosylhydrazone, m.p. 163–164°.

2. Anhydrous ether available from Mallinckrodt Chemical Works can be used without further drying.

3. This prevents clogging of the funnel during the subsequent addition of methyl-lithium solution.

4. Methyllithium available from either Foote Mineral Company or Alfa Inorganics, Inc., can be used without further purification. The checkers used 137 ml. of a 1.85 M solution.

5. During the first half of the addition each drop of methyllithium solution produces a yellow color that quickly disappears. The solution turns yellow during the second half of the addition and slowly becomes more intensely colored until it reaches red-orange near the end of the reaction period.

6. Since the solvent is never completely removed at any time prior to final distillation of the product, the accumulation of higher boiling hydrocarbons results if petroleum ether is used. As a result the forerun of the final distillation will be larger, and the yield of 2-bornene will be reduced. The pentane was distilled to assure the removal of any higher boiling impurities.

7. Neutral, reagent grade aluminum oxide available from Merck & Co., Inc., was used by the checkers.

8. When the temperature of the bath reaches 140–143°, the heating lamp is turned directly on the U-tube, and the receiver is changed to collect the 2-bornene.

9. The yield of the product is greatly reduced if the receiver is not cooled. A 2-propanol–dry ice bath was used.

10. On a 180 cm. × 32 mm. GC column containing 10% Apiezon L on Chromosorb P at 60° a synthetic mixture of 2-bornene (10.3 minutes), tricyclene (12.3 minutes), and camphene (15.5 minutes) was readily resolvable.

11. The product has m.p. 161–163° and can be used without further purification.

3. Discussion

2-Bornene has been prepared from the reaction of 2-bromobornane-3-carboxylic acid with aqueous sodium hydrogen carbonate,[5] by pyrolysis of isoborneol methyl xanthate,[6] and by the β-elimination of hydrogen chloride from bornyl chloride with sodium alkoxides in various solvents.[2]

This procedure appears to be general for the preparation, without rearrangement, of lesser substituted olefins. 2-Methylcyclohexanone tosylhydrazone gives 3-methylcyclohexene (98% yield by GC analysis).[7] Cholestan-6-one tosylhydrazone gives cholest-6-ene (95% isolated yield), androstan-17-one tosylhydrazone gives androst-16-ene (91% isolated yield),[7] and phenylacetone tosylhydrazone gives allylbenzene (70% yield by GC analysis, accompanied by the substitution product isobutylbenzene in 30% yield). Advantages of this procedure include its simplicity and the availability of carbonyl compounds as precursors. The reaction proceeds smoothly but more slowly at −25° and can be employed with heat-sensitive or volatile compounds in ordinary laboratory equipment.

1. Department of Chemistry, University of Colorado, Boulder, Colorado 80302. [Present address: Department of Chemistry, James Madison University, Harrisonburg, Virginia 22807.]
2. (a) H. Meerwein and J. Joussen, *Ber. Dtsch. Chem. Ges.,* **55,** 2529 (1922); (b) M. Hanack and R. Hähnle, *Chem. Ber.,* **95,** 191 (1962); (c) L. Borowiecki and Y. Chrétien-Bessière, *Bull. Soc. Chim. Fr.,* 2364 (1967).
3. W. J. Zubyk and A. Z. Conner, *Anal. Chem.,* **32,** 912 (1960).
4. W. R. Bamford and T. S. Stevens, *J. Chem. Soc.* 4735 (1952).
5. J. Bredt and H. Sandkuhl, *Justus Liebigs Ann. Chem.,* **366,** 11 (1909).
6. L. Tschugaeff, *Ber. Dtsch. Chem. Ges.,* **32,** 3332 (1899).
7. R. H. Shapiro and M. J. Heath, *J. Am. Chem. Soc.,* **89,** 5734 (1967).

SELECTIVE α-BROMINATION OF AN ARALKYL KETONE WITH PHENYLTRIMETHYLAMMONIUM TRIBROMIDE: 2-BROMOACETYL-6-METHOXYNAPHTHALENE AND 2,2-DIBROMOACETYL-6-METHOXYNAPHTHALENE

[Ethanone, 2-bromo-1-(6-methoxy-2-naphthalenyl)- and
Ethanone, 2,2-dibromo-1-(6-methoxy-2-naphthalenyl)-]

$$C_6H_5N(CH_3)_2 \xrightarrow[\text{toluene}]{(CH_3)_2SO_4} C_6H_5\overset{\oplus}{N}(CH_3)_3\,CH_3SO_4^{\ominus} \xrightarrow[\text{HBr}]{Br_2}$$

$$C_6H_5\overset{\oplus}{N}(CH_3)_3Br_3^{\ominus}$$
$$(PTT)$$

Submitted by J. JACQUES and A. MARQUET[1]
Checked by DAVID WALBA and ROBERT E. IRELAND

1. Procedure

Caution! All operations should be carried out in a well-ventilated hood because dimethyl sulfate is highly toxic and the bromoketones are lachrymators and skin irritants.

A. *Phenyltrimethylammonium sulfomethylate.* A solution of 24.8 g. (25.9 ml., 0.205 mole) of freshly distilled *N,N*-dimethylaniline (Note 1) in 100 ml. of toluene (Note 2) is prepared in a 250-ml. Erlenmeyer flask equipped with a thermometer and a magnetic stirrer. The solution is stirred and heated to about 40°. Heating is stopped and 25 g. (19 ml., 0.20 mole) of distilled dimethyl sulfate (Note 3) is added with an addition funnel over 20 minutes. Within minutes, the colorless sulfomethylate starts to crystallize. The temperature which varies very little during the previous addition, rises slowly for one hour thereafter and approaches 50°. The reaction is allowed to proceed at ambient temperature for 1.5 hours after the addition is complete before it is heated on a steam bath for one hour. After cooling, the phenyltrimethylammonium sulfomethylate is filtered, washed with 20 ml. of dry toluene, and dried under vacuum, yielding 44–46.5 g. (89–94%) (Note 4).

B. *Phenyltrimethylammonium tribromide*. A solution of 10 g. (0.040 mole) of phenyltrimethylammonium sulfomethylate in 10 ml. of 48% hydrobromic acid diluted with 10 ml. of water is prepared in a 125-ml. Erlenmeyer flask equipped with a magnetic stirrer. Bromine (7.8 g., 2.5 ml., 0.049 mole) (Note 5) is added to the stirred solution from a dropping funnel over 20 minutes. An orange-yellow precipitate forms immediately, and the slurry is stirred at room temperature for 5–6 hours. The product, phenyltrimethylammonium tribromide (PTT), is filtered, washed with about 10 ml. of water, and air-dried under an efficient hood. The crude PTT, *ca.* 15 g., is recrystallized from 25 ml. of acetic acid, giving, after filtration and air-drying, 12.9–14.0 g. (86–93%) (Note 6) of orange crystals, m.p. 113–115°.

C. *2-Bromoacetyl-6-methoxynaphthalene*. To a solution of 1 g. (0.005 mole) of 2-acetyl-6-methoxynaphthalene [*Org. Synth.*, **Coll. Vol. 6**, 34 (1988)] in 10 ml. of anhydrous tetrahydrofuran (Note 7) contained in a 125-ml. Erlenmeyer flask is added 1.88 g. (0.00500 mole) of PTT in small portions over a 10 minute period. A white precipitate forms immediately and the solution becomes pale yellow. After 20 minutes, 50 ml. of cold water is added, and the crystalline precipitate (Note 8) is filtered and washed with 10 ml. of water. The crude, white 2-bromoacetyl-6-methoxynaphthalene (*ca.* 1.3 g., m.p. 100–105°) is recrystallized from 32 ml. of cyclohexane, yielding 1.1 g. (79%) of crystalline product, m.p. 107–109° (lit. 107–108°)[4] (Notes 9 and 10).

D. *2,2-Dibromoacetyl-6-methoxynaphthalene*. To a solution of 1 g. (0.005 mole) of 2-acetyl-6-methoxynaphthalene [*Org. Synth.*, **Coll. Vol. 6**, 34 (1988)] in 10 ml. of anhydrous tetrahydrofuran (Note 7) contained in a 125-ml. Erlenmeyer flask is added 3.76 g. (0.0100 mole) of PTT in small portions over 10 minutes. A white precipitate forms and the solution becomes yellow over one hour. Cold water (50 ml.) is added, and the crystalline product (Note 8) is filtered and washed with 10 ml. of water. The crude 2,2-dibromoacetyl-6-methoxynaphthalene (*ca.* 1.7 g., m.p. 110–117°) is recrystallized from 15 ml. of ethanol, filtered, and washed with 2 ml. of ethanol, yielding 1.40–1.55 g. (78–87%) of slightly yellow product, m.p. 116.5–118° (lit. 118–119°)[3] (Notes 9 and 11).

2. Notes

1. Commercial *N,N*-dimethylaniline was redistilled, b.p. 78° (13 mm.).

2. Benzene can also be used, but toluene is preferable because of its lower toxicity.

3. Commercial dimethyl sulfate was distilled, b.p. 70° (13 mm.). A slight deficiency of dimethyl sulfate ensures the complete utilization of this toxic product.

4. This product is slightly hygroscopic, but no special precautions are required for handling.

5. Bromine (B & A, ACS Reagent Grade) was used without further purification.

6. The "active bromine" can be titrated according to the following procedure: about 300 mg. of PTT is dissolved in 50 ml. of acetic acid, 10 ml. of a 5% solution of potassium iodide in ethanol is added, and the liberated iodine is titrated with a 0.1 N solution of sodium thiosulfate. Percent "active bromine": calculated 42.5%; found 42.1–42.5%. The molecular weight of PTT is 375.96.

7. Tetrahydrofuran was purified and dried as previously described [*Org. Synth., Coll. Vol. 5*, 976 (1973)]. PTT is remarkably soluble in tetrahydrofuran (630 g. per l. at 20°). Under the same conditions, the solubility of the resulting phenyltrimethylammonium bromide is only 0.09 g. per l.

8. If the product precipitates as an oil, mere standing at room temperature may cause it to crystallize. If not, the addition of *ca.* 3 ml. of tetrahydrofuran, followed by swirling will usually induce crystallization.

9. *This product, like other bromoketones, can be very irritating to exposed skin.*

10. ^{1}H NMR (CDCl$_3$): δ 3.94 (s, 3H, OCH_3), 4.54 (s, 2H, COCH_2Br), 7.20 (m, 4H, ArH), 7.90 (m, 1H, ArH), 8.21 (m, 1H, ArH).

11. ^{1}H NMR (CDCl$_3$): δ 3.93 (s, 3H, OCH_3), 6.86 (s, 1H, COCHBr$_2$), 7.20 (m, 4H ArH), 7.90 (m, 1H, ArH), 8.50 (m, 1H, ArH).

3. Discussion

Quaternary ammonium perhalogenides, being solid compounds, constitute convenient halogen sources. Of the different compounds studied and reported,[4] pyridine hydrobromide perbromide[5] is the most popular. Phenyltrimethylammonium tribromide (PTT), the utility of which was recognized by Marquet and Jacques,[6] has the advantage of high stability and ease of preparation. The procedure described herein is a modification of that of Vorländer and Siebert.[7]

When dissolved in tetrahydrofuran, PTT (like pyridine hydrobromide perbromide) is a source of Br$_3^\ominus$ ions, the properties of which are different from those of molecular bromine. In particular, it is much less electrophilic and less reactive toward aromatic rings and double bonds,[8] and is thus a selective brominating reagent for ketones[3] or ketals[3,9] when the molecule has double bonds or activated aromatic nuclei which would be attacked by bromine. The two examples of use of this reagent clearly differentiate between its reactivity from that of bromine. Reaction of bromine with 2-acetyl-6-methoxynaphthalene (in diethyl ether solution) gives a mixture of products, the main constituent of which results from ring bromination (2-acetyl-5-bromo-6-methoxynaphthalene).[2]

Many other examples have been described that illustrate the selectivity of PTT, not possible with bromine. Steroid and terpene α-bromoketones have been selectively obtained in molecules containing double bonds[3] or cyclopropane rings;[10,11,12] anisyl cyclohexyl ketone gives the α-bromoketone in very good yield with the aromatic ring remaining unattacked;[3] 5,7-dimethoxyflavanone can be brominated in good yield at the position alpha to the keto group, although the aromatic ring is activated by two methoxy groups.[13] A similar selectivity has been observed in the analogous case of 2-methyl-2-(2-benzyloxy-5-methoxyphenyl)cyclopentanone.[14] PTT has also been used for the selective bromination of 4-oxo-4,5,6,7-tetrahydrobenzofuran.[15]

Anhydrous tetrahydrofuran contributes to the selectivity of the reagent because of the stability of Br$_3^\ominus$ in this solvent. Moreover, tetrahydrofuran acts as a buffer by reaction with the liberated hydrobromic acid, which is why PTT in tetrahydrofuran can also be very useful if the substrate bears acid-sensitive functions. Acid-catalyzed epimerization can also be avoided: 2-bromobenzo[6.7]bicyclo[3.2.1]oct-6-en-3-one gives the diaxial 2,4-dibromoketone[16] and 3,4-dihydro-1-(p-tolylsulfonyl)benz[c,d]indol-5(1H)-one leads to the expected bromoketone, in spite of the tendency of this type of molecule to isomerize into the naphthalenoid system.[17] The importance of the solvent appears in the bromination of 1,5-cyclooctanedione: in tetrahydrofuran, the pure *cis-trans-cis* isomer is isolated,[18] whereas in dichloromethane, a mixture of two tetrabromo ketones is obtained.[19] It must be emphasized that anhydrous tetrahydrofuran must be used because small amounts of water can greatly retard the rate of bromination of ketones, with resulting decreased selectivity.

Other uses of this reagent have also been described. Tosylhydrazones undergo oxidation to tosylazoalkenes, using PTT followed by treatment with base; this reaction fails with molecular bromine, dioxane dibromide, or *N*-bromosuccinimide in a range of solvents.[21] PTT has also been recommended for the aromatization of 2-substituted-6-benzoyl-4,5-dihydro-6*H*-pyrrolo[3,2-*e*]-benzothiazoles.[21]

Recently, other bromination reagents containing $Br_3^{\ominus}$ have been described: pyrrolidone hydrotribromide,[22] 2-carboxyethyltriphenylphosphonium perbromide,[23] and Amberlyst A 26-$Br_3^{\ominus}$.[24]

1. Organic Chemistry of Hormones Laboratory, College of France, 75231 Paris 5, France.
2. A. Marquet, A. Horeau, J. Jacques, L. Novak, and M. Protiva, *Collect. Czech. Chem. Commun.*, **26**, 1475 (1961).
3. A. Marquet and J. Jacques, *Bull. Soc. Chim. Fr.*, 90 (1962).
4. A. Marquet, M. Dvolaitzky, H. B. Kagan, L. Mamlok, C. Ouannes, and J. Jacques, *Bull. Soc. Chim. Fr.*, 1822 (1961).
5. L. F. Fieser and M. Fieser, "Reagents for Organic Synthesis," Wiley, New York, 1967, p. 967.
6. A. Marquet and J. Jacques, *Tetrahedron Lett.* (9) 24 (1959).
7. D. Vorländer and E. Siebert, *Ber. Dtsch. Chem. Ges.*, **52**, 283 (1919).
8. A. Marquet, J. Jacques, and B. Tchoubar, *Bull. Soc. Chim. Fr.*, 511 (1965).
9. W. J. Johnson, J. Dolf Bass, and K. L. Williamson, *Tetrahedron*, **19**, 861 (1963).
10. C. Berger, M. Franck-Neumann, and G. Ourisson, *Tetrahedron Lett.*, 3451 (1968).
11. W. J. Gensler and P. H. Solomon, *J. Org. Chem.*, **38**, 1726 (1973).
12. V. Cerny, *Collect. Czech. Chem. Commun.*, **38**, 1563 (1973).
13. D. Brulé and C. Mentzer, *C. R. Hebd. Seances Acad. Sci. Paris*, **250**, 365 (1960).
14. W. K. Anderson, E. J. LaVoie, and G. E. Lee, *J. Org. Chem.*, **42**, 1045 (1977).
15. W. A. Remers and G. S. Jones, Jr., *J. Heterocycl. Chem.*, **12**, 421 (1975).
16. J. W. Wilt and R. R. Rasmussen, *J. Org. Chem.*, **40**, 1031 (1975).
17. R. E. Bowman, D. D. Evans, J. Guyett, H. Nagy, J. Weale, D. J. Weyel, and A. C. White, *J. Chem. Soc. Perkin Trans. 1*, 1926 (1972).
18. J. E. Heller and A. S. Dreiding, *Helv. Chim. Acta*, **56**, 413 (1973).
19. J. Heller, A. Yogev, and A. S. Dreiding, *Helv. Chim. Acta*, **55**, 1003 (1979).
20. G. Rosini and G. Baccolini, *J. Org. Chem.*, **39**, 826 (1974).
21. W. A. Remers, R. H. Roth, and M. J. Weiss, *J. Med. Chem.*, **14**, 860 (1971).
22. D. V. C. Awang and S. Wolfe, *Can. J. Chem.*, **47**, 706 (1969).
23. V. W. Armstrong, N. H. Chishti, and R. Ramage, *Tetrahedron Lett.*, 373 (1975).
24. S. Cacchi, L. Caglioti, and E. Cernia, *Synthesis*, 64 (1979).

MERCURY(II) OXIDE-MODIFIED HUNSDIECKER REACTION: 1-BROMO-3-CHLOROCYCLOBUTANE

$$2\,Cl-\!\!\!\bigcirc\!\!\!-CO_2H + HgO + 2Br_2$$

$$\downarrow$$

$$2\,Cl-\!\!\!\bigcirc\!\!\!-Br + H_2O + 2CO_2 + HgBr_2$$

Submitted by Gary M. Lampman and James C. Aumiller[1]
Checked by G. Nelson and K. B. Wiberg

1. Procedure

In a 1-l., three-necked, round-bottomed flask, wrapped with aluminum foil to exclude light, and equipped with a mechanical stirrer, a reflux condenser, and an addition funnel, is suspended 37 g. (0.17 mole) of red mercury(II) oxide (Note 1) in 330 ml. of carbon tetrachloride (Note 2). To the flask is added 30.0 g. (0.227 mole) of 3-chlorocyclobutane-carboxylic acid (Note 3). With stirring, the mixture is heated to reflux before a solution of 40 g. (0.25 mole) of bromine in 180 ml. of carbon tetrachloride is added dropwise, but as fast as possible (4–7 minutes) without loss of bromine from the condenser (Note 4). After a short induction period, carbon dioxide is evolved at a rate of 150–200 bubbles per minute (Note 5). The solution is allowed to reflux for 25–30 minutes, until the rate of carbon dioxide evolution slows to about 5 bubbles per minute (Note 5). The mixture is cooled in an ice bath, and the precipitate is removed by filtration. The residue is washed with carbon tetrachloride, and the filtrates are combined. The solvent is removed by distillation using a modified Claisen distillation apparatus with a 6-cm. Vigreux column; vacuum distillation of the residual oil gives 13–17 g. (35–46%) of 1-bromo-3-chlorocyclobutane, b.p. 67–72° (45 mm.), n_D^{23} 1.5065 (Notes 7 and 8).

2. Notes

1. Purified product is available from J. T. Baker Chemical Company.

2. Reagent grade carbon tetrachloride was used.

3. 3-Chlorocyclobutanecarboxylic acid was prepared as described in *Org. Synth.*, **Coll. Vol. 6,** 271 (1988).

4. The heating bath should be maintained at about 120° to ensure that the solution continues to reflux while the bromine solution is added.

5. The gas evolution can be monitored by conducting the gas through rubber tubing from the condenser into a small amount of water where the bubbling can be observed. A small amount of bromine is lost because of entrainment by the gas.

6. There is no increase in yield on heating the mixture under reflux for 3 hours.

7. The submitters reported a 48–52% yield (18–20 g.) using the indicated scale, and a 35% yield when the reaction was carried out using twice the scale. The checkers obtained the product in 28–29% yield when the reaction was conducted on a scale 10 times that indicated.

8. The product was analyzed by GC at 130° on a Beckman GC-2 chromatograph equipped with a 180 cm. × 6 mm. column (Beckman 17449) containing 42/60 Johns-Manville C-22 firebrick coated with Dow-Corning 550 silicone oil. The retention times are 12 and 14 minutes for the *trans* and *cis* compounds, respectively.

3. Discussion

This procedure, a modified Hunsdiecker reaction based upon the method of Cristol and Firth,[2] results in moderate to high yields of bromides and iodides from aliphatic[2,3] and alicyclic carboxylic acids.[4–6] Carbon tetrachloride is most frequently used as the solvent, but others can be employed.[3,6] Attempts to prepare chlorides by the method have proved to be unsuccessful.[7]

The main advantage of this procedure over that of the standard method[8] is one of convenience. For example, the present method is a one-step reaction while the usual method is a two-step sequence involving an intermediate silver salt. In addition, the presence of water produced in the reaction apparently does not reduce the yield in the present method while water markedly reduces the yield in that involving the silver intermediate.

Some variations of the method have been used to prepare cyclopropyl and cyclobutyl halides. Simultaneous addition of bromine and 3-bromocyclobutanecarboxylic acid to the suspension of mercury(II) oxide gives 1,3-dibromocyclobutane in good yield.[7] Similarly, cyclopropanecarboxylic acid gives bromocyclopropane,[9] and 3-(bromomethyl)cyclobutanecarboxylic acid gives 1-bromo-3-(bromomethyl)cyclobutane.[10] In the latter reaction, it was found desirable to remove the water from the reaction as it is formed in order to obtain high yields. Another variation is the addition of a mixture of the acid and mercury (II) oxide to excess bromine in bromotrichloromethane.[6]

The conversion of 1-bromo-3-chlorocyclobutane to bicyclo[1.1.0]butane is described in *Organic Syntheses*.[11]

1. Department of Chemistry, Western Washington University, Bellingham, Washington 98225.
2. S. J. Cristol and W. C. Firth, Jr., *J. Org. Chem.*, **26**, 280 (1961).
3. J. A. Davis, J. Herynk, S. Carroll, J. Bunds, and D. Johnson, *J. Org. Chem.*, **30**, 415 (1965).
4. S. J. Cristol, J. R. Douglass, W. C. Firth, Jr., and R. E. Krall, *J. Org. Chem.*, **27**, 2711 (1962).
5. S. J. Cristol, L. K. Gaston, and T. Tiedeman, *J. Org. Chem.*, **29**, 1279 (1964).
6. F. W. Baker, H. D. Holtz, and L. M. Stock, *J. Org. Chem.*, **28**, 514 (1963).
7. K. B. Wiberg and G. M. Lampman, *J. Am. Chem. Soc.*, **88**, 4429 (1966).
8. C. V. Wilson, *Org. React.*, **9**, 332 (1957).
9. J. S. Meek and D. T. Osuga, *Org. Synth.*, **Coll. Vol. 5**, 126 (1973).
10. K. B. Wiberg and D. S. Connor, *J. Am. Chem. Soc.*, **88**, 4437 (1966).
11. G. M. Lampman and J. C. Aumiller, *Org. Synth.*, **Coll. Vol. 6**, 133 (1988).

para-BROMINATION OF AROMATIC AMINES:
4-BROMO-*N*,*N*-DIMETHYL-3-(TRIFLUOROMETHYL)ANILINE

[Benzenamine, 4-bromo-*N*,*N*-dimethyl-3-(trifluoromethyl)]

Submitted by G. J. Fox,[1] G. Hallas,[2]
J. D. Hepworth,[1] and K. N. Paskins[2]
Checked by J. D. Lock, Jr. and S. Masamune

1. Procedure

Caution! The reaction should be conducted in a hood to avoid inhalation of bromine vapor.

A. *2,4,4,6-Tetrabromo-2,5-cyclohexadien-1-one.* A mixture of 66.2 g. (0.200 mole) of 2,4,6-tribromophenol (Note 1), 27.2 g. (0.197 mole) of sodium acetate trihydrate, and 400 ml. of glacial acetic acid is placed in a 1-l. Erlenmeyer flask and warmed (*ca.* 70°) until a clear solution is obtained. The solution is magnetically stirred and cooled to room temperature to produce a finely divided suspension of the phenol, to which a solution of 32 g. (0.20 mole) of bromine in 200 ml. of glacial acetic acid is added dropwise over 1 hour (Note 2). The resulting mixture is kept at room temperature for 30 minutes, then poured onto 2 kg. of crushed ice. The yellow solid which separates is removed by suction filtration after the ice has melted, and the damp crystals are dissolved in the minimum of warm chloroform (Note 3). The upper aqueous layer is removed with a pipet fitted with a suction bulb. The dienone crystallizes from the chloroform solution upon cooling, yielding 50–55 g. (61–67%) of crystals, m.p. 125–130° (dec.), sufficiently pure for use in the next step (Notes 4 and 5).

B. *4-Bromo-N,N-dimethyl-3-(trifluoromethyl)aniline.* A solution of 9.45 g. (0.0500 mole) of *N,N*-dimethyl-3-(trifluoromethyl)aniline (Note 6) in 200 ml. of dichloromethane is placed in a 500-ml. Erlenmeyer flask, cooled to −10°, and stirred magnetically as 20.5 g. (0.0500 mole) of finely powdered 2,4,4,6-tetrabromo-2,5-

cyclohexadien-1-one is added in 0.5-g. portions. During this addition the temperature of the mixture should be maintained between −10° and 0° (Note 7). The cooling bath is removed and the reaction mixture is allowed to warm to room temperature over a 30-minute period and extracted twice with 50 ml. of aqueous 2 N sodium hydroxide to remove 2,4,6-tribromophenol (Note 8). The organic layer is washed with 25 ml. of water and dried over anhydrous magnesium sulfate. Removal of the solvent yields 12–12.5 g. of crude 4-bromo-N,N-dimethyl-3-(trifluoromethyl)aniline. Distillation through an 8-cm. Vigreux column provides 11–12 g. (82–90%) of pure bromoamine, b.p. 134–136° (15 mm.), which solidifies, giving colorless crystals, m.p. 29–30° (Notes 9 and 10).

2. Notes

1. The submitters used reagent grade 2,4,6-tribromophenol. The checkers recrystallized the practical grade reagent purchased from Fisher Scientific Company. The solvent used was Skelly B, and the melting point of the phenol, after recrystallization, was 93–95°.

2. It is essential to maintain the temperature of the solution below 25° during the addition of the bromine solution. If external cooling is applied, initially with an ice-water bath, the addition can be completed within 20 minutes.

3. Some decomposition of the dienone is observed if the chloroform solution is vigorously refluxed for any length of time; bromine is evolved and there is a reduction in yield. Approximately 400 ml. of chloroform is needed to dissolve the dienone at approximately 60°. The checkers used 450 ml. of the solvent.

4. The submitters used 0.5-molar quantities with no reduction in yield.

5. ^{1}H NMR (dioxane-d_8), δ (multiplicity): 7.98 (singlet).

6. The amine was prepared according to the procedure described in *Org. Synth.*, **Coll. Vol. 5,** 1085 (1973).

7. The reaction proceeds satisfactorily over a range between −30° and +20°. At lower temperatures, the reaction proceeds rather slowly.

8. 2,4,6-Tribromophenol may be recovered by acidification of the aqueous alkaline extracts and reused in the preparation of the tetrabromo-compound after crystallization from petroleum ether (b.p. 80–100°).

9. The product can be crystallized from petroleum ether (b.p. 30–40°).

10. ^{1}H NMR (CDCl$_3$), δ (multiplicity, number of protons, assignment): 2.94 (s, 6H, 2CH_3), 6.7 (approximate d of d, 1H), 7.0 (approximate d, 1H), 7.5 (approximate d, 1H).

3. Discussion

4-Bromo-N,N-dimethyl-3-(trifluoromethyl)aniline has been prepared by the methylation of 4-bromo-3-(trifluoromethyl)aniline with trimethyl phosphate in 70–80% yield.[3] The present method, which effectively uses 3-(trifluoromethyl)aniline as starting material, offers advantages in cost, yield, and ease of purification.

Aromatic amines are usually polybrominated on treatment with bromine. Several mild brominating agents have been introduced in attempts to achieve partial bromination without the necessity of protecting and deprotecting the amino group, but these give variable results when applied to a large variety of amines. Dioxane dibromide[4] monobrominates tertiary aromatic amines, but gives poor yields with primary and secondary aryl amines. The use of *N*-bromosuccinimide[5,6] (1-bromo-2,5-pyrrolidinedione) leads to monobrominated compounds frequently contaminated with decomposition products.

The dienone, which is prepared essentially as described by Benedikt[7] and Caló,[8] monobrominates a wide range of primary, secondary, and tertiary aromatic amines almost exclusively in the *para*-position. The procedure described is of general synthetic utility for the preparation of *para*-brominated aromatic and heteroaromatic amines in high yields and frequently in a high state of purity. The submitters have used this technique to *para*-brominate many compounds in quantities ranging from 0.01–0.1 mole, including the following (yields after one crystallization): aniline (92), *N*-methylaniline (94), *N,N*-dimethylaniline (91), *N,N*-diethylaniline (94), *o*-toluidine (88), *N*,2-dimethylaniline (91), *N,N*,2-trimethylaniline (90), *m*-toluidine (90), *N,N*,3-trimethylaniline (86), 2,3-dimethylaniline (91), 2,5-dimethylaniline (91), 3,5-dimethylaniline (81), 2-chloroaniline (86), 3-chloroaniline (86), 2-bromoaniline (78), 3-bromoaniline (82), 2-nitroaniline (91), 3-nitroaniline (85), *m*-phenylenediamine (82), *o*-anisidine (85), 3-methoxyaniline (58, 4-bromo; 30, 6-bromo), diphenylamine (90), 1-aminonaphthalene (86), 1-dimethylaminonaphthalene (84), 2-(trifluoromethyl)aniline (85), 2-aminopyridine (75), 2-dimethylaminopyridine (70), 3-dimethylaminopyridine (60),[9] 2-amino-6-methylpyridine (76). Where solubility of the amine in dichloromethane is low, chloroform may be used as solvent. For example, 2-aminopyrimidine gave 2-amino-5-bromopyrimidine (82%) in this manner, compared with 41% when the amine is brominated conventionally in aqueous solution.[10] In the case of anthranilic acid, 2-amino-5-bromobenzoic acid (82%) precipitated from the chloroform reaction medium. In addition to its use with amines, the dienone reagent monobrominates a variety of phenols,[11] and behaves as an oxidizing agent toward sulfides, converting them to sulfoxides.[12]

1. Department of Chemistry, College of Art and Technology, Derby, England. [Present address: Department of Chemistry and Biology, Preston Polytechnic, Preston, England.]
2. Department of Colour Chemistry, The University, Leeds, England.
3. D. E. Grocock, G. Hallas, and J. D. Hepworth, *J. Chem. Soc. Perkin Trans. 2*, 1792 (1973).
4. G. M. Kosolapoff, *J. Am. Chem. Soc.*, **75**, 3596 (1953).
5. L. Horner, E. Winkelmann, K. H. Knapp, and W. Ludwig, *Chem. Ber.*, **92**, 288 (1959).
6. J. B. Wommack, T. G. Barbee, Jr., D. J. Thoennes, M. A. McDonald, and D. E. Pearson, *J. Heterocycl. Chem.*, **6**, 243 (1969).
7. R. Benedikt, *Justus Liebigs Ann. Chem.*, **199**, 127 (1879).
8. V. Caló, F. Ciminale, L. Lopez, and P. E. Todesco, *J. Chem. Soc. C*, 3652 (1971).
9. G. J. Fox, J. D. Hepworth, and G. Hallas, *J. Chem. Soc. Perkin Trans. 1*, 68 (1973).
10. J. P. English, J. H. Clark, J. W. Clapp, D. Seeger, and R. H. Ebel, *J. Am. Chem. Soc.*, **68**, 453 (1946).
11. V. Caló, F. Ciminale, L. Lopez, G. Pesce, and P. E. Todesco, *Chim. Ind. (Milan)*, **53**, 467 (1971).
12. V. Caló, F. Ciminale, G. Lopez, and P. E. Todesco, *Int. J. Sulfur Chem., Part A*, **1**, 130 (1971).

BROMOHYDRINS FROM ALKENES AND *N*-BROMO-
SUCCINIMIDE IN DIMETHYL SULFOXIDE:
erythro-2-BROMO-1,2-DIPHENYLETHANOL

(Ethanol, 2-bromo-1,2-diphenyl-, *erythro*-)

Submitted by A. W. LANGMAN and D. R. DALTON[1]
Checked by I. DAVID REINGOLD and S. MASAMUNE

1. Procedure

A 500-ml., round-bottomed flask equipped with a magnetic stirring bar and a thermometer is charged with 18.0 g. (0.100 mole) of (*E*)-stilbene (Note 1), 5.0 ml. (0.28 mole) of water, and 300 ml. (4.23 moles) of dimethyl sulfoxide (Note 2). The resulting suspension is stirred for 5 minutes at room temperature (20–25°) (Note 3). Stirring is continued as 35.6 g. (0.200 mole) of *N*-bromosuccinimide (Note 4) is added in small portions over *ca.* 10 minutes. A yellow color appears when the first portion of *N*-bromosuccinimide is added, and by the time the addition is complete, the solution is bright orange. During the addition the temperature of the mixture rises to 50–55°, and all the (*E*)-stilbene dissolves. The contents of the flask are stirred for another 15 minutes and poured into 1 l. of ice water; the product separates immediately as a white solid (Note 5). The aqueous slurry is transferred to a separatory funnel with the aid of 50-ml. portions of water and diethyl ether, and extracted with four 200-ml. portions of ether. The combined ethereal extracts are washed with 250 ml. of water and 250 ml. of sodium chloride solution, dried over anhydrous magnesium sulfate, and evaporated with a rotary evaporator at a water bath temperature of *ca.* 30°. The pale yellow, crystalline residue is dissolved, to the extent possible, in 600 ml. of hot hexane, and the resulting suspension is filtered while hot, removing a small amount of an insoluble impurity. Cooling the filtrate provides colorless fibers of analytically pure *erythro*-2-bromo-1,2-diphenylethanol, m.p. 83–84° (Note 6); a second crop of crystals is obtained by concentrating the mother liquor to 200 ml. (Note 7). The combined yield is 22.0–24.9 g. (80–90%) (Note 8).

2. Notes

1. (*E*)-Stilbene, m.p. 123–126°, was purchased from Aldrich Chemical Company, Inc., and used as received. It is also available from J. T. Baker Chemical Company and from Eastman Organic Chemicals.

2. Reagent grade dimethyl sulfoxide was used without purification. The amount of dimethyl sulfoxide can be varied. A large excess is employed in this case to facilitate dissolution of the stilbene.

3. The suspension may be warmed to dissolve the alkene more rapidly; (E)-stilbene dissolves completely at ca. 65°. If the suspension is warmed, it *must* be cooled below 30° before proceeding further to prevent a vigorous reaction when the N-bromosuccinimide is added. The submitters recommend that the warm suspension be cooled under an atmosphere of nitrogen. If a volatile alkene is used, the mixture should be cooled prior to and during the addition of N-bromosuccinimide to prevent losses by evaporation.

4. N-Bromosuccinimide purchased from Arapahoe Chemical Company was used without purification. If the purity of the N-bromosuccinimide is in doubt, it should be titrated before use by the standard iodide–thiosulfate method and purified, if necessary, by recrystallization from 10 times its weight of water.[2] Solutions of N-bromosuccinimide in dimethyl sulfoxide cannot be stored, since the solvent is oxidized by the brominating reagent.

5. The product does not appear to deteriorate if allowed to stand at this point.

6. The submitters recrystallized the product from 600 ml. of petroleum ether (b.p. 30–60°) and reported a melting point of 84–84.5° (lit., m.p. 84.5–85.5°[3] and 86°[4]).

7. The submitters found that the residue (2.8 g.) obtained upon evaporation of the mother liquor was largely *erythro*-2-bromo-1,2-diphenylethanol contaminated with a small amount of succinimide. Absorptions for the *threo* isomer could not be detected in the IR and ^{1}H NMR spectra of this material.

8. The product obtained by the checkers was analyzed. Analysis calculated for $C_{14}H_{13}BrO$: C, 60.67; H, 4.73. Found: C, 60.74; H, 4.77. The spectral properties of the product are as follows: IR (CCl_4) cm.$^{-1}$: 3610, 1500, 1460, 700; ^{1}H NMR ($CDCl_3$), δ (multiplicity, coupling constant J in Hz., number of protons, assignment): 5.06 and 5.16 (AB doublet, $J = 6.5$, 2H, C*H*BrC*H*OH), 7.35 (m, 10H, aryl *H*).

3. Discussion

The present procedure affords a simple, general method for preparing bromohydrins from alkenes and avoids the heterogeneous solvent systems often used in such reactions. Labeling experiments have demonstrated that the oxygen from the dimethyl sulfoxide appears in the hydroxyl group of the bromohydrin;[5] therefore, the role of the water is to hydrolyze the intermediate β-bromodimethylsulfoxonium ion.

Many alkenes have been converted into their respective bromohydrins by this procedure, usually with high regio- and stereoselectivity (Table I).[5,6] Although the regioselectivity of the addition generally follows Markovnikov's rule, the opposite orientation is observed with alkenes bearing the bulky *tert*-butyl substituent (entries 7–9). The reaction of conjugated dienes with N-bromosuccinimide in aqueous dimethyl sulfoxide also occurs in a regio- and stereoselective manner, leading exclusively to vicinal bromohydrins in high yield.[7]

When electron-withdrawing groups are attached to the double bond, the reaction is strongly inhibited and may fail completely. In such cases, the bromide anion, produced by the reaction of dimethyl sulfoxide with N-bromosuccinimide, competes with the dimethyl sulfoxide for the bromonium (or bromo carbonium) ion intermediate. Thus, dibromide may accompany recovered alkene or any bromohydrin formed. Similarly, exogenous anions often compete with dimethyl sulfoxide for the cation.[6]

erythro-2-Bromo-1,2-diphenylethanol has been prepared by reaction of (E)-stilbene

TABLE I

BROMOHYDRINS FROM ALKENES WITH N-BROMOSUCCINIMIDE IN AQUEOUS
DIMETHYL SULFOXIDE

Entry	Alkene	Bromohydrin	Yield (%)[a]
1	(Z)-C_6H_5CH=CHC_6H_5	threo-$C_6H_5CH(OH)CH(Br)C_6H_5$	82
2	(E)-C_6H_5CH=$CHCH_3$	erythro-$C_6H_5CH(OH)CH(Br)CH_3$	92
3	(Z)-C_6H_5CH=$CHCH_3$	threo-$C_6H_5CH(OH)CH(Br)CH_3$	95
4	C_6H_5CH=CH_2	$C_6H_5CH(OH)CH_2Br$	76
5	C_6H_5—$C(CH_3)$=CH_2	$C_6H_5C(CH_3)(OH)CH_2Br$	89
6	$C_6H_5CH_2CH$=CH_2	$C_6H_5CH_2CH(OH)CH_2Br$	83
7	(E)-$(CH_3)_3CCH$=$CHCH_3$	erythro-$(CH_3)_3CCH(Br)CH(OH)CH_3$	90
8	(Z)-$(CH_3)_3CCH$=$CHCH_3$	threo-$(CH_3)_3CCH(Br)CH(OH)CH_3$	90
9	$(CH_3)_3CCH$=CH_2	$(CH_3)_3CCH(Br)CH_2OH$	89
10	$(CH_3)_3CC(CH_3)$=CH_2	$(CH_3)_3CC(CH_3)(OH)CH_2Br$	60[b]

[a]Average yield from two or more runs.
[b]Accompanied by 24% dibromide.

with N-bromoacetamide in buffered aqueous acetone,[3] by addition of hydrogen bromide
to (E)-stilbene oxide,[4] and by reaction of (E)-stilbene with bromotrinitromethane in
dimethyl sulfoxide followed by hydrolysis.[8]

1. Department of Chemistry, Temple University, Philadelphia, Pennsylvania 19122.
2. L. F. Fieser and M. Fieser, "Reagents for Organic Synthesis," Vol. 1, Wiley, New York, 1967,
   p. 78.
3. H. O. House, J. Am. Chem. Soc., 77, 3070 (1955).
4. D. Reulos, C.R. Hebd. Seances Acad. Sci., 216, 774 (1943).
5. D. R. Dalton, V. P. Dutta, and D. C. Jones, J. Am. Chem. Soc., 90, 5498 (1968).
6. D. R. Dalton and V. P. Dutta, J. Chem. Soc. B, 85 (1971).
7. D. R. Dalton and R. M. Davis, Tetrahedron Lett., 1057 (1972).
8. K. Torssell, Ark. Kemi, 23, 543 (1965).

CHAIN ELONGATION OF ALKENES
via gem-DIHALOCYCLOPROPANES:
2-BROMO-3,3-DIPHENYL-2-PROPEN-1-YL ACETATE

[2-Propen-1-ol, 2-bromo-3,3-diphenyl-, acetate]

Submitted by Stanely R. Sandler[1]
Checked by D. W. Brooks and S. Masamune

1. Procedure

A. *1,1-Dibromo-2,2-diphenylcyclopropane.* A 500-ml., three-necked, round-bottomed flask equipped with a mechanical stirrer, a dropping funnel, and a condenser fitted with a drying tube is flushed with dry nitrogen, then charged with 25.0 g. (0.139 mole) of 1,1-diphenylethylene (Note 1), 100 ml. of pentane, and 28 g. (0.25 mole) of potassium *tert*-butoxide (Note 2). The mixture is stirred and cooled to 0° before 66.0 g. (0.261 mole) of bromoform (Note 3) is added dropwise over 30–45 minutes. Stirring is continued for an additional 2–3 hours at room temperature, and 200 ml. of water is added. The yellowish insoluble product is filtered, dried, and digested with 300 ml. of refluxing 2-propanol for 30 minutes. After cooling, the product is filtered and washed with 100 ml. of 2-propanol, yielding 31–38 g. (63–78%) of colorless crystals, m.p. 151–152°.

B. *2-Bromo-3,3-diphenyl-2-propen-1-yl acetate.* A 250-ml. flask equipped with a condenser is charged with 17.6 g. (0.0500 mole) of 1,1-dibromo-2,2-diphenylcyclopropane, 12.5 g. (0.0748 mole) of silver acetate (Note 4), and 50 ml. of glacial acetic acid, then immersed in an oil bath at 100–120° for 24 hours (Note 5). After cooling, the mixture is diluted with 200 ml. of diethyl ether and filtered. The ethereal filtrate is washed with two 100-ml. portions of water, two 100-ml. portions of aqueous saturated sodium carbonate, and finally with two 100-ml. portions of water. After drying over anhydrous sodium sulfate, the ether is removed on a rotary evaporator. Distillation of the resulting residue under reduced pressure yields 12.0 g. (72%) of the product, b.p. 142–145° (0.15 mm.), n_D^{22} 1.6020–1.6023 (Note 6).

TABLE I
CHAIN ELONGATION OF ALKENES *via* gem-DIBROMOCYCLOPROPANES

Alkene	Conditions for Dibromocyclopropane Opening	Product
(cyclopentene)	$AgNO_3$, H_2O	(cyclohexene ring) OH, Br
(cyclopentene)	Heat	(cyclohexene ring) Br, Br
(cyclohexene)	CH_3CO_2Ag, CH_3CO_2H	(cycloheptene ring) O_2CCH_3, Br
C_6H_5 (vinyl)	CH_3CO_2Ag, CH_3CO_2H	C_6H_5 Br O_2CCH_3
(2-butene) *(Z)* or *(E)*	CH_3CO_2Ag, CH_3CO_2H	Br O_2CCH_3
(2-methyl-2-butene)	Heat or CH_3CO_2Ag, CH_3CO_2H	Br
(2-methyl-2-butene)	CH_3CO_2Ag, CH_3CO_2H	Br O_2CCH_3 + Br
(2-methyl-1-propene)	CH_3CO_2Ag, CH_3CO_2H	Br O_2CCH_3 + Br CH_3CO_2

188

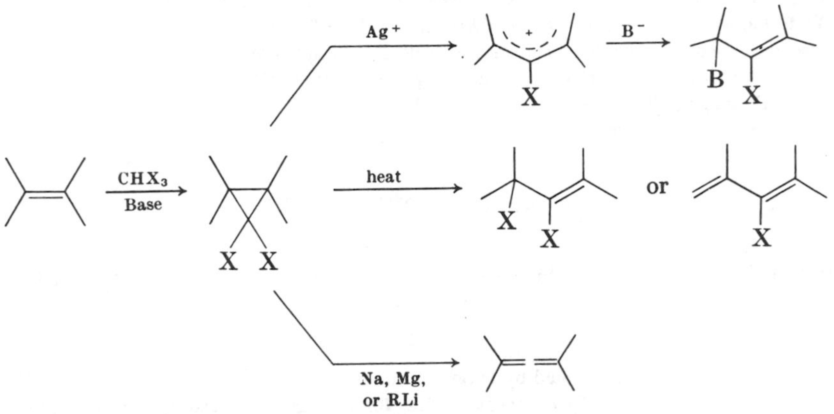

Figure 1.

2. Notes

1. 1,1-Diphenylethylene was purchased from Eastman Organic Chemicals.

2. Potassium *tert*-butoxide was supplied by Mine Safety Appliances (MSA) Research Corporation. See end of Discussion section below.

3. Bromoform was supplied by the Dow Chemical Company and used without further purification.

4. The silver acetate can be replaced by a mixture of sodium acetate and silver nitrate.

5. A 24-hour period may not be required but was found to be convenient.

6. UV (CH$_3$OH) nm. max. (log ϵ): 260 (3.94); ^{1}H NMR (CDCl$_3$), δ (multiplicity, number of protons): 2.08 (s, 3H), 4.87 (s, 2H), 7.3 (m, 10H).

3. Discussion

The present procedure is that of the submitter[2] and illustrates a general method for the chain extension of alkenes *via gem*-dihalocyclopropanes, earlier described by Skell and Sandler.[3] The reaction of dihalocyclopropanes with electrophilic reagents yields haloallylic derivatives,[2] the thermal reaction yields haloallylic halides or halodienes, and the reaction with magnesium, sodium, or lithium alkyl reagents yields allenes.[2] These reactions are summarized in Figure 1, and examples are given in Table I.

This general method has been used by Parham and co-workers[4] to transform indenes into β-halonaphthalenes. The method is also useful for the conversion of pyrroles to β-substituted pyridines and of indoles to β-haloquinolines.[3] More recently, phase transfer agents have been used to aid the preparation of *gem*-dihalocyclopropanes by the reaction of olefins with haloforms, using aqueous sodium hydroxide.[5,6,7]

1. Borden, Inc., Chemical Division, Central Research Laboratory, Philadelphia, Pennsylvania 19124 [Present address: Pennwalt Corp., King of Prussia, Pennsylvania 19406].

2. S. R. Sandler, *J. Org. Chem.*, **32**, 3876 (1967) and references cited therein.

3. P. S. Skell and S. R. Sandler, *J. Am. Chem. Soc.*, **80**, 2024 (1958).
4. W. E. Parham and H. E. Reiff, *J. Am. Chem. Soc.*, **77**, 1177 (1955) and subsequent papers on the reaction of indenes with dihalocarbenes to yield β-halonaphthalenes.
5. E. V. Dehmlow, *Angew. Chem. Int. Ed. Engl.*, **13**, 170 (1974).
6. E. V. Dehmlow, *Chem. Tech.*, 210 (April 1975).
7. I. Crossland, *Org. Synth.*, **60**, 6 (1981).

2-BROMOHEXANOYL CHLORIDE

$$CH_3CH_2CH_2CH_2CH_2COOH \xrightarrow[\text{(2) } (CH_2CO)_2NBr, \text{ HBr, } 85°]{\text{(1) } SOCl_2, \text{ } 65°} CH_3CH_2CH_2CH_2\underset{\underset{Br}{|}}{C}HCOCl$$

Submitted by DAVID N. HARPP, L. Q. BAO,
CHRISTOPHER COYLE, JOHN G. GLEASON, and SHARON HOROVITCH[1]
Checked by JAMES E. KLECKNER and ROBERT E. IRELAND

1. Procedure

Caution! This reaction should be conducted in a good hood since hydrogen chloride and bromine vapors are evolved.

A 200-ml., round-bottomed flask equipped with a magnetic stirring bar is charged with 11.6 g. (0.100 mole) of hexanoic acid (Note 1) and 10 ml. of carbon tetrachloride. After 46.9 g. (28.8 ml., 0.394 mole) of thionyl chloride (Note 2) is added to the solution, an efficient reflux condenser with an attached drying tube is fitted to the flask. The solution is stirred and heated with an oil bath at 65° for 30 minutes (Note 3). The flask is removed from the oil bath and cooled to room temperature. To the reaction mixture are added successively 21.4 g. (0.120 mole) of finely powdered *N*-bromosuccinimide (Note 4), 50 ml. of carbon tetrachloride, and 7 drops of 48% hydrogen bromide (Note 5). The flask is heated at 70° for 10 minutes (Note 6), before the temperature of the bath is increased to 85°, until the color of the reaction becomes light yellow (*ca.* 1.5 hours; Note 7). The reaction mixture is cooled to room temperature, and the carbon tetrachloride and excess thionyl chloride are removed under reduced pressure (Note 8). The residue is suction filtered, the solid (Note 9) is washed several times with carbon tetrachloride (total 20 ml.) and the combined filtrate collected in a 50-ml. flask. The solvent is removed from the solution as before, and the residue is distilled into a dry ice-cooled receiver (short-path column), giving, after a small forerun, 16.1–17.1 g. (76–80%) of 2-bromohexanoyl chloride, b.p. 44–47° (1.5 mm.) as a clear, slightly yellow oil, n_D^{22} 1.4707. This material is of sufficient purity for most synthetic purposes (Note 10).

The yellow product (Note 11) is decolorized by dissolving it in an equal volume of carbon tetrachloride (*ca.* 12 ml.) and vigorously shaking the solution thus obtained with 1.5 ml. of freshly prepared aqueous 35% sodium thiosulfate. The two layers are completely separated after 5 minutes. The colorless bottom layer is drawn off into a 50-ml. Erlenmeyer flask. The top layer is extracted three times with 1.5 ml. of carbon tetrachloride. The combined carbon tetrachloride extracts are dried over 0.5 g. (Note 12) of anhydrous magnesium sulfate for 30 minutes. The solution is filtered into a 50-ml.

anhydrous magnesium sulfate for 30 minutes. The solution is filtered into a 50-ml. distilling flask, and the magnesium sulfate is washed several times with carbon tetrachloride (total 5 ml.). The solvent is removed, and the colorless product is distilled as described above, affording 14.7–15.8 g. (69–74% overall, based on hexanoic acid; 88–92% for the decolorization step) of colorless 2-bromohexanoyl chloride, b.p. 45–47° (1.5 mm.), n_D^{22} 1.4706 (Note 13), d_4^{24} 1.4017 (Notes 14 and 15).

2. Notes

1. Practical grade hexanoic acid is obtainable from Matheson, Coleman and Bell. The submitters report slightly higher yields using purified grade hexanoic acid obtained from Fisher Scientific Company.

2. Thionyl chloride was obtained from Anachemia Chemicals Ltd., Fisher Scientific Company (reagent grade), or Matheson, Coleman and Bell. The first two were slightly yellow, and the latter was colorless; however, the yields of final product were identical with each brand. The excess thionyl chloride serves as a drying agent for the hexanoic acid and as a solvent for the N-bromosuccinimide, which is not very soluble in carbon tetrachloride.

3. ^{1}H NMR analysis indicates complete conversion to the acid chloride. This may be monitored by following the disappearance of the triplet (CH_2CO_2H) at δ 2.40 and the emergence of a new triplet (CH_2COCl) at δ 2.87.

4. N-Bromosuccinimide was obtained from Matheson, Coleman and Bell or Aldrich Chemical Company, Inc. Product yields were optimized using 20% excess, although only 5–10% yield reductions were noted using 5% excess reagent. Recrystallizing the reagent prior to use had no noticeable effect on the overall yield of product.

5. Aqueous 48% hydrogen bromide was obtained from Baker and Adamson. Without added hydrogen bromide, the reaction was much slower.

6. If the reaction was heated too rapidly to 85°, vigorous foaming resulted.

7. Initially the reaction mixture is dark red, and there is bromine vapor in the condenser. Toward the end of the reaction the color lightens considerably and after a short period (ca. 15 minutes) begins to darken again. The heat should be removed when this darkening commences. On standing, the yellow solution may also turn black, but the yield of the product is not noticeably affected. When the stirring is stopped, succinimide floats to the top of the solution. The reaction may be conveniently monitored by following the disappearance of the triplet (CH_2COCl) at δ 2.87 and appearance of a triplet ($CHBrCOCl$) at δ 4.54 in the ^{1}H NMR spectrum.

8. The evaporation of solvents under reduced pressure should be performed carefully with vigorous stirring at room temperature. An oil pump protected with a dry ice trap and equipped with a manometer is used. Initially the pressure should be adjusted to prevent excessive foaming; it is reduced progressively to approximately 5 mm.

9. About 10 g. of the solid (succinimide) is collected.

10. The IR and ^{1}H NMR spectra are identical with those of colorless, doubly distilled material; n_D^{22} 1.4706.

11. When the decolorization procedure was carried out before the first distillation, inconsistent yields were obtained. About 2.5 ml. of a dark viscous liquid (giving a violet solution on dilution in carbon tetrachloride) remained in the distillation flask.

12. When more drying agent was employed, the product yield was lower.

13. A central fraction had n_D^{22} 1.4704.

14. Analysis calculated for $C_6H_{10}BrClO$: C, 33.75; H, 4.72; Br, 37.42; Cl, 16.60. Found: C, 33.42; H, 4.77; Br, 37.29; Cl, 16.74. IR (NaCl) cm.$^{-1}$: 2955, 2925, 1785, 1470; ^1H NMR δ (multiplicity, number of protons): 0.94 (m, 3H), 1.43 (m, 4H), 2.10 (m, 2H), 4.54 (t, 1H), mass spectrum m/e: 179, 177 (M−Cl).

15. The corresponding α-bromo acid is prepared by the following procedure. A 500-ml., round-bottomed flask is charged with 10.28 g. (0.04826 mole) of 2-bromohexanoyl chloride and 92 ml. of acetone. The flask is fitted with a magnetic stirring bar, a thermometer, and a 200-ml. dropping funnel in which is placed 115 ml. of aqueous saturated sodium hydrogen carbonate (ca. 0.115 mole). The flask is cooled to approximately 10°, while the base is added over a period of about 45 minutes. The mixture is acidified with concentrated hydrochloric acid. An organic layer forms at the top and is separated from the aqueous layer, which is extracted with three 30-ml. portions of chloroform. The combined organic extracts are dried over anhydrous magnesium sulfate and the solvent is removed under reduced pressure, giving 9.36 g. of crude 2-bromohexanoic acid as a colorless liquid. This product is 96% pure by GC analysis, using a Hewlett-Packard 5750 Research Chromatograph with a 1.8 m. 4% SE-30 column at 130°, and having a flow rate of 60 ml./minute. This product can be distilled through a short-path column, yielding, after an 11% forerun, 7.76 g. (83%) of 2-bromohexanoic acid, b.p. 64−66° (0.075 mm.), which shows one peak by GC analysis (as above). IR and ^1H NMR spectra are consistent with the structure.

3. Discussion

The α-bromination of acids (via the acid chloride) has been achieved by the Hell−Volhard−Zelinsky reaction or its variances,[2] however, this technique can involve reaction times of up to 2−3 days,[3] high reaction temperatures (>100°), copious evolution of hydrogen bromide, and variable yields. A recent procedure,[4] while affording good overall yields, involves several steps to achieve the transformation (alkylation, proton abstraction, bromination, deacylation and deësterification).

The submitters have found the N-bromosuccinimide procedure to be a very general reaction. Alkyl, alicyclic, aryl, and heterocyclic acetic acids have been brominated in 50−80% yield.[5] The reaction may be applied in the presence of labile benzylic hydrogens; for example, 3-phenylpropanoic acid gives exclusively 2-bromo-3-phenylpropanoyl chloride.[6] The procedure has several significant advantages; it is considerably faster than the known methods (overall reaction times of 2 hours are common),[7] the use of bromine is circumvented, and work-up is simplified considerably.

1. Department of Chemistry, McGill University, 801 Sherbrooke Street West, Montreal, PQ, Canada H3A 2K6.
2. H. O. House, "Modern Synthetic Reactions," 2nd ed., W. A. Benjamin, New York, 1972, pp. 477−478, and references cited therein.
3. L. A. Carpino and L. V. McAdams III, Org. Synth., Coll. Vol. 6, 403 (1988).
4. P. L. Stotter and K. A. Hill, Tetrahedron Lett., 4067, (1972).
5. J. G. Gleason and D. N. Harpp, Tetrahedron Lett., 3431, (1970); J. G. Gleason, unpublished results.

6. D. N. Harpp, L. Q. Bao, C. J. Black, R. A. Smith, and J. G. Gleason, *Tetrahedron Lett.*, 3235 (1974).
7. D. N. Harpp, L. Q. Bao, C. J. Black, J. G. Gleason, and R. A. Smith, *J. Org. Chem.*, **40**, 3420 (1975).

1-BROMO-3-METHYL-2-BUTANONE

$$(CH_3)_2CH-\underset{\overset{\|}{O}}{C}-CH_3 + Br_2 \xrightarrow[0° \text{ to } 5°]{CH_3OH} (CH_3)_2CH-\underset{\overset{\|}{O}}{C}-CH_2Br + HBr$$

Submitted by M. Gaudry and A. Marquet[1]
Checked by Diana Metzger and Richard E. Benson

1. Procedure

Caution! This preparation must be carried out in an efficient hood. Bromomethyl ketones are highly lachrymatory and are skin irritants.

A 2-l., four-necked, round-bottomed flask equipped with a sealed mechanical stirrer, a thermometer, a reflux condenser fitted with a calcium chloride drying tube, and a 100-ml., pressure-equalizing dropping funnel is charged with 86.0 g. (105 ml., 1.00 mole) of 3-methyl-2-butanone (Note 1) and 600 ml. of anhydrous methanol (Note 2). The solution is stirred and cooled in an ice–salt bath to 0–5°, and 160 g. (54.6 ml., 1.00 mole) of bromine (Note 3) is added in a rapid, steady stream from the dropping funnel (Note 4). During this time, the temperature is allowed to rise but not permitted to exceed 10°. The reaction temperature is maintained at 10° during the remaining reaction time (Note 5). The red color of the solution fades gradually in about 45 minutes (Note 6), 300 ml. of water is then added (Note 7), and the mixture is stirred at room temperature overnight (Note 8).

To the solution is added 900 ml. of water, and the resulting mixture is washed with four 500-ml. portions of diethyl ether. The ether layers are combined, washed with 200 ml. of aqueous 10% potassium carbonate and then twice with 200-ml. portions of water (Note 9), and dried for 1 hour over 200 g. of anhydrous calcium chloride (Note 10). The solvent is removed on a rotary evaporator at room temperature, yielding 145–158 g. of crude product (Note 11). Distillation under reduced pressure through a Vigreux column gives 115–128 g. of a fraction, b.p. 83–86° (54 mm.), n_D^{22} 1.4620–1.4640, containing 95% of 1-bromo-3-methyl-2-butanone as established by ^{1}H NMR measurements (Note 11).

2. Notes

1. The checkers used 3-methyl-2-butanone purchased from Eastman Organic Chemicals. One sample that gave a positive test for peroxides was purified by passage through a column of alumina before distillation. The material was distilled routinely before use.
2. The methanol was distilled twice from magnesium turnings. Alternately, it was

dried overnight over molecular sieves then distilled. The checkers also found freshly opened reagent grade methanol (purchased from Fisher Scientific Company) to be satisfactory.

3. The submitters used R.P. bromine obtained from Prolabo, Paris, without further purification. The checkers used bromine available from Fisher Scientific Company.

4. It is very important to add the bromine in a single portion. When it is added dropwise, a mixture containing significant amounts of 3-bromo-3-methyl-2-butanone is obtained.

5. The temperature must be controlled carefully, especially at the end of the addition when the reaction becomes more exothermic. If the solution becomes warm, a mixture of the two isomeric bromoketones is obtained.

6. If a slight excess of bromine has been added, a light yellow color remains after reaction of one equivalent since dibromination is very slow under these conditions.

7. The quantity of water added is such that the brominated products do not separate from the aqueous methanol.

8. The water is added in order to hydrolyze the α-bromodimethyl ketals produced during the reaction. The ease of hydrolysis of these bromoketals depends on the structure of the ketone. With acetylcyclohexane or acetylcyclopentane, stirring with water for 10 minutes is sufficient for complete hydrolysis. In contrast, with phenylacetone or methyl ethyl ketone, after dilution with water, the addition of 10 equivalents of concentrated sulfuric acid with respect to ketone and stirring for 15 hours at room temperature are necessary for complete hydrolysis.

9. The submitters state that the hydrobromic acid can also be neutralized before extraction by adding 75 g. of potassium carbonate (6 g. excess) in small portions.

10. Under these extraction conditions, the ether solution contains significant amounts of water and methanol which cannot be removed efficiently with anhydrous sodium sulfate.

11. In the crude product the ratio of 1-bromo-3-methyl-2-butanone to 3-bromo-3-methyl-2-butanone is estimated by ^{1}H NMR to be 95:5. The ^{1}H NMR properties of the two isomers are as follows; 1-bromo-3-methyl-2-butanone: (CDCl$_3$), δ (multiplicity, coupling constant J in Hz., number of protons, assignment): 1.17 (d, J = 6.9, 6H, 2CH_3), 3.02 (m, 1H, CH), 4.10 (s, 2H, CH_2); 3-bromo-3-methyl-2-butanone. (CDCl$_3$): δ (multiplicity, number of protons, assignment): 1.89 (s, 6H, 2CH_3), 2.46 (s, 3H, COCH_3).

3. Discussion

Pure isomeric, monobrominated ketones substituted at the less substituted or at the more substituted α-carbon are not readily accessible by direct bromination of unsymmetrical ketones since the reaction often leads to a mixture of products, with the more substituted isomer usually predominating.[2] Radical bromination of unsymmetrical ketones in the presence of epoxides yields exclusively the monobromo ketone corresponding to bromination at the more substituted α-position.[3] The action of hydrobromic acid on diazo ketones has been, for a long time, the only method of preparing bromomethyl ketones.[4] Recently, some indirect routes involving halogenation of preformed isomeric enol silyl ethers[5] or enamines[6] have been described. The bromination of unsymmetrical ketals (e.g., dioxolanes or dimethyl ketals) occurs to a greater extent on the less substituted carbon atom, and this constitutes an efficient route to the corresponding α-bromo ketones.[7-9] Direct bromination of 2-substituted cyclohexanones[8] and various methyl ketones[10] in methanol leads to the same result.

This procedure, in contrast to methods mentioned above, has only one step and is readily adapted to large-scale preparative work. Furthermore, because dibromination is very slow in methanol, the crude reaction products contain only traces of dibromo ketones. This contrasts with the behavior in other solvents, such as ether or carbon tetrachloride, where larger amounts of dibromo ketones are always present, even when one equivalent of bromine is used. Methanol is thus recommended as a brominating solvent even when no orientation problem is involved. It should be noted that α-bromomethyl ketals are formed along with α-bromoketones and must be hydrolyzed during the workup (Note 8).[10]

The regiospecificity of bromination depends on the structure of the ketone.[10] This regiospecificity is very high for methyl ketones when the α'-position is tertiary, and not as high when it is secondary. For example, cyclohexyl methyl ketone and cyclopentyl methyl ketone lead to crude products containing 100 and 85%, respectively, of bromomethyl ketone, while 2-methylcyclohexanone, methyl ethyl ketone, and phenylacetone give 65,[8] 70,[10] and 40%,[10] respectively, of ketone brominated at the less substituted carbon. In these latter cases, bromination of the corresponding dimethyl ketal in methanol affords better yields of these bromo ketones.[10]

Errata for
ORGANIC SYNTHESIS
COLLECTIVE VOLUME 6
Wayland E. Noland, Editor-in-Chief

The following reference list should appear on page 195.

1. Organic Chemistry of Hormones Laboratory, College of France, Paris: Cedex 05. [Present address: Laboratoire de Chimie Organique Biologique, Université Pierre et Marie Curie, Tour 44–45, 4 Place Jussieu, 75230 Paris Cedex 05, France.]
2. J. R. Catch, D. F. Elliott, D. H. Hey, and E. R. H. Jones, *J. Chem. Soc.*, 272 (1948); J. R. Catch, D. H. Hey, E. R. H. Jones, and W. Wilson, *J. Chem. Soc.*, 276 (1948); H. M. E. Cardwell and A. E. H. Kilner, *J. Chem. Soc.*, 2430 (1951).
3. V. Calo, L. Lopez, and G. Pesce, *J. Chem. Soc. Perkin Trans. 1*, 501 (1977).
4. D. A. Clibbens and M. Nierenstein, *J. Chem. Soc.*, 107, 1491 (1915); J. R. Catch, D. F. Elliott, D. H. Hey, and E. R. H. Jones, *J. Chem. Soc.*, 278 (1948).
5. R. H. Reuss and A. Hassner, *J. Org. Chem.*, 39, 1785 (1974); L. Bianco, P. Amice, and J. M. Conia, *Synthesis*, 194 (1976); P. Amice, L. Blanco, and J. M. Conia, *Synthesis*, 196 (1976).
6. R. Carlson and C. Rappe, *Acta Chem. Scand.*, B31, 485 (1977); R. Carlson, *Acta Chem. Scand.*, B32, 646 (1978).
7. A. Marquet, M. Dvolaitzky, H. B. Kagan, L. Mamlok, C. Ouannes, and J. Jacques, *Bull. Soc. Chim. Fr.*, 1822 (1961).
8. E. W. Garbisch, Jr., *J. Org. Chem.*, 30, 2109 (1965).
9. M. Gaudry and A. Marquet, *Bull. Soc. Chim. Fr.*, 1849 (1967); 4169 (1969).
10. M. Gaudry and A. Marquet, *Tetrahedron*, 26, 5611, 5617 (1970).

trans,trans-1,3-BUTADIENE-1,4-DIYL DIACETATE

(1,3-Butadiene-1,4-diol, diacetate, *E,E*-)

Submitted by Robert M. Carlson[1] and Richard K. Hill[2]
Checked by Jack M. Pal and Peter Yates

1. Procedure

Caution! Benzene has been identified as a carcinogen; OSHA has issued emergency standards on its use. All procedures involving benzene should be carried out in a well-ventilated hood, and glove protection is required.

A. trans-7,8-*Diacetoxybicyclo*[4.2.0]*octa-2,4-diene.* A 1-l., three-necked flask fitted with a reflux condenser, an efficient stirrer, and a thermometer dipping well into the solution is charged with a suspension of mercury(II) acetate (Note 1) (160 g., 0.502 mole) in glacial acetic acid (400 ml). While the suspension is stirred, 52.0 g. of cycloöctatetraene (0.500 mole) (Note 2) is added rapidly. The white addition compound that separates after 10–15 minutes is decomposed by careful heating of the reaction mixture at 70–75° for 2 hours (Note 3). While still warm, the mixture is poured through funnels containing glass wool plugs into two, 4-l. beakers, each containing 2 l. of water (Note 4). The mixture is allowed to stand for several hours, and the solid that separates is collected on a Büchner funnel and pressed as dry as possible on the funnel. The moist, yellow solid is spread out on a large piece of filter paper and allowed to dry overnight, yielding 83–86 g. (75–77.5%) of trans-7,8-diacetoxybicyclo[4.2.0]octa-2,4-diene, m.p. 52–55° (Note 5); it may be used in the next step without further purification (Note 6).

B.   trans,trans-1,3-*Butadiene*-1,4-*diyl diacetate*.   A solution of the diacetate (83.0 g., 0.373 mole) and dimethyl acetylenedicarboxylate [*Org. Synth.*, **Coll. Vol. 4,** 329 (1963); 54.0 g., 0.380 mole] in benzene (250 ml.) is placed in a 500-ml. flask and refluxed for 6 hours (Note 7). The solution is filtered to remove the remaining mercury and mercury salts, and the benzene is distilled under reduced pressure. The residual viscous yellow oil is distilled under reduced pressure (Note 8). A mixture of 1,3-butadiene-1,4-diyl diacetate and dimethyl phthalate is collected at 140–155° (18–20 mm.), bath temperature 170–200°, from which the diene crystallizes as colorless needles in the cooled receiver. The solid is broken up, washed onto a Büchner funnel with petroleum ether (b.p. 60–70°), pressed between sheets of filter paper to remove excess dimethyl phthalate, and recrystallized from *ca.* 1:2 acetone–petroleum ether (b.p. 60–70°) (Note 9), yielding 26–31 g. (41–49%) of *trans,trans*-1,3-butadiene-1,4-diyl diacetate, as colorless needles, m.p. 102–104° (Notes 10, 11, and 12).

2. Notes

1. "Baker analyzed" mercury(II) acetate was used as obtained from J. T. Baker Chemical Company.

2. Cycloöctatetraene was used as obtained from Badische Anilin- und Soda-Fabrik, 67 Ludwigshafen, Rhein, Germany.

3. Temperatures in excess of 75° cause darkening of the reaction mixture. The submitters found that adequate temperature control could be maintained with a Bunsen flame; the checkers used a heating mantle.

4. This operation should be carried out in a well-ventilated hood. Scratching the sides of the beakers at the surface of the water promotes crystallization. The beakers are stirred occasionally to promote rapid crystallization and to minimize the formation of a solid cake on the bottom.

5. The checkers obtained 92–93 g. (83–84%); m.p. 58–61°.

6. Melting points of 61–62°, 61.4–62.5°, 64–65°, and 66° have been reported.[3–5] The diacetate may be recrystallized from aqueous acetic acid,[3–5] ligroin,[3] or ethanol.[4]

7. A longer reflux period, *e.g.*, overnight, did not affect the yield.

8. The submitters used a 500-ml. distilling flask with a 250-ml. distilling flask as receiver. The side arm of the distilling flask was extended, if necessary, with a Tygon joint into the bulb of the receiver. The receiver was cooled with a stream of cold water. The distillation was continued until about three-fourths of the material in the flask had distilled. The checkers used standard-taper glassware for the distillation and found it necessary to heat the distillation adapter with a microburner from time to time in order to prevent plugging by solidified product. They continued the distillation until no further solid product distilled.

9. The checkers found that it was essential to conduct the recrystallization rapidly in order to obtain maximum yields.

10. The checkers obtained a yield of 40.8 g. (64%) when distillation of the product mixture was continued until no additional solid product distilled (*cf.* Notes 8 and 9).

11. A satisfactory alternative to recrystallization of the diene is the following. The crude solid is placed in a 100-ml. beaker containing 40–50 ml. of a 15% solution of acetone in petroleum ether (b.p. 60–70°). The lumps are broken up, and the colorless

solid is filtered from the pale-yellow solution. A final wash with 20 ml. of ice-cold acetone leaves 31–32 g. of diene, m.p. 99–102°, pure enough for many purposes.

12. The IR and UV spectra of the product have been reported.[6]

3. Discussion

This method is essentially that described by Reppe, Schlichting, Klager, and Toepel,[3] although the correct structures were assigned by others.[4,7] 7,8-Diacetoxybicyclo[4.2.0]octa-2,4-diene has also been prepared by oxidation of cyclooctatetraene with lead tetraäcetate,[5] and by chlorination of cyclooctatetraene with sulfuryl chloride followed by displacement with potassium acetate.[3,7] The two other geometric isomers of the diene have been prepared by another method.[6] *trans,trans*-1,3-Butadiene-1,4-diyl diacetate is a reactive diene in the Diels-Alder reaction. It has been used as the starting material in stereospecific syntheses of conduritol-D[8] and shikimic acid,[9,10] and in simple, general methods of preparation of benzene derivatives, especially unsymmetrical biphenyls.[11,12]

1. Present address: Department of Chemistry, University of Minnesota Duluth, Duluth, Minnesota 55812.
2. Frick Chemical Laboratory, Princeton University, Princeton, New Jersey. [Present address: Department of Chemistry, The University of Georgia, Athens, Georgia 30602.]
3. W. Reppe, O. Schlichting, K. Klager, and T. Toepel, Justus Liebigs Ann. Chem., **560**, 1 (1948).
4. A. C. Cope, N. A. Nelson, and D. S. Smith, J. Am. Chem. Soc., **76**, 1100 (1954).
5. M. Finkelstein, Chem. Ber., **90**, 2097 (1957).
6. H. H. Inhoffen, J. Heimann-Trosien, H. Muxfeldt, and H. Krämer, Chem. Ber., **90**, 187 (1957).
7. R. Criegee, W. Hörauf, and W. D. Schellenberg, Chem. Ber., **86**, 126 (1953).
8. R. Criegee and P. Becher, Chem. Ber., **90**, 2516 (1957).
9. R. McCrindle, K. H. Overton, and R. A. Raphael, J. Chem. Soc., 1560 (1960).
10. E. E. Smissman, J. T. Suh, M. Oxman, and R. Daniels, J. Am. Chem. Soc., **84**, 1040 (1962).
11. R. K. Hill and R. M. Carlson, J. Org. Chem., **30**, 2414 (1965).
12. R. K. Hill and R. M. Carlson, J. Org. Chem., **30**, 1571 (1965).

A NEW REAGENT FOR *tert*-BUTOXYCARBONYLATION: 2-*tert*-BUTOXYCARBONYLOXYIMINO- 2-PHENYLACETONITRILE

(Benzeneacetonitrile, α-[[[(1,1-dimethylethoxy)carbonyl]carbonyl]oxy]imino]-)

$$C_6H_5CH_2CN + CH_3ONO \xrightarrow[\substack{methanol, \\ 0-15°}]{NaOH} C_6H_5-C\begin{smallmatrix}NOH \\ \\ CN\end{smallmatrix}$$

$$C_6H_5-C\begin{smallmatrix}NOH \\ \\ CN\end{smallmatrix} + \underset{Cl}{\overset{O}{\underset{\|}{C}}}Cl \xrightarrow[\substack{benzene, dioxane, \\ 5-6°}]{C_6H_5N(CH_3)_2}$$

$$C_6H_5-C\begin{smallmatrix}NOCCl \\ \\ CN\end{smallmatrix} \xrightarrow[\substack{pyridine, \\ 5-10°}]{(CH_3)_3COH} C_6H_5-C\begin{smallmatrix}NOCOC(CH_3)_3 \\ \\ CN\end{smallmatrix}$$

Submitted by MASUMI ITOH, DAIJIRO HAGIWARA, and TAKASHI KAMIYA[1]
Checked by HIROYUKI ISHITOBI, TERUJI TSUJI, and WATARU NAGATA

1. Procedure

Caution! Phosgene is highly toxic. Part B should be performed in an efficient hood. Benzene has been identified as a carcinogen; OSHA has issued emergency standards on its use. All procedures involving benzene should be carried out in a well-ventilated hood, and glove protection is required.

A. *2-Hydroxyimino-2-phenylacetonitrile.* A 1-l., round-bottomed flask fitted with a mechanical stirrer, a calcium chloride drying tube, a thermometer, and a gas-inlet tube is charged with 117 g. (1.00 mole) of benzyl cyanide and a solution of 40.0 g. (1.00 mole) of sodium hydroxide in 300 ml. of methanol (Note 1). The resulting solution is stirred and cooled at 0° as methyl nitrite is introduced through the gas-inlet tube, which extends below the surface of the liquid. The methyl nitrite is generated by dropwise addition of a cold solution of 32 ml. of concentrated sulfuric acid in 65 ml. of water from a 100-ml., pressure-equalizing dropping funnel into a 300-ml. Erlenmeyer flask containing a suspension of 83 g. (1.2 moles) of sodium nitrite in 53 ml. of methanol and 50 ml. of water (Note 2). The rate of generation of methyl nitrite is adjusted so that the reaction temperature does not exceed 15°. After the addition is complete (Note 3), stirring is continued for another 2 hours, and the solvent is removed under reduced pressure with a rotary evaporator. The residue is dissolved in 500 ml. of water, and the resulting solution is washed with two 100-ml. portions of toluene. The aqueous layer is acidified with concentrated hydrochloric acid and cooled in an ice bath. The resulting precipitate is filtered, washed thoroughly with cold water, and dried, yielding 111–120 g. (76–82%) of 2-hydroxyimino-2-phenylacetonitrile, m.p. 119–124° (Note 4). This material is used in Part B without further purification.

B.   2-tert-*Butoxycarbonyloxyimino-2-phenylacetonitrile*.   A 200-ml., three-necked, round-bottomed flask equipped with a dropping funnel, a mechanical stirrer, a thermometer, and a calcium chloride drying tube is charged with a solution of 10.9 g. (0.110 mole) of phosgene (Note 5) in 30 ml. of benzene. The contents of the flask are stirred and cooled in an ice bath while a solution of 14.6 g. (0.100 mole) of 2-hydroxyimino-2-phenylacetonitrile and 13.2 g. (0.113 mole) of *N,N*-dimethylaniline in 5 ml. of dioxane and 80 ml. of benzene (Note 6) is added dropwise over 1 hour at 5–6°. Stirring is continued for 6 hours at the same temperature, after which the mixture is allowed to stand overnight in an ice bath. A solution of 11.1 g. (0.150 mole) of *tert*-butyl alcohol and 12.0 ml. (0.150 mole) of pyridine (Note 7) in 30 ml. of benzene (Note 6) is added over 1 hour as the mixture is stirred and cooled at 5–10°. Stirring is continued for an additional 6 hours while the reaction temperature is allowed to rise to room temperature. The reaction mixture is allowed to stand overnight (Note 8) and is then mixed with 50 ml. of water and 50 ml. of benzene. The organic layer is separated and washed successively with three 30-ml. portions of cold 1 *N* hydrochloric acid, 30 ml. of water, two 30-ml. portions of 5% sodium hydrogen carbonate solution, and two 30-ml. portions of water. Each of the aqueous washings is extracted with 30 ml. of benzene. The organic layers are combined, dried with magnesium sulfate, and concentrated to dryness under reduced pressure at a temperature lower than 35°. The crystalline residue is triturated with 20 ml. of aqueous 90% methanol. The solid is filtered, washed with 30 ml. of aqueous 90% methanol, and dried, giving 15.8–17.0 g. of crude product, m.p. 84–86° (Note 9). Recrystallization from methanol (Note 10) affords 14.6–15.7 g. (59–64%) of 2-*tert*-butoxycarbonyloxy-imino-2-phenylacetonitrile as white needles or plates, m.p. 84–86° (Note 11).

2. Notes

1. The submitters used reagent grade solvents and reagents without further purification. The yield of 2-hydroxyimino-2-phenylacetonitrile was 76% when the checkers used technical grade benzyl cyanide purchased from Wako Pure Chemical Industries, Ltd., Osaka, Japan. The yield was improved to 81% with distilled material, b.p. 75–77° (3 mm.). Benzyl cyanide is also available from Aldrich Chemical Company, Inc.

2. This method for the preparation of methyl nitrite is described in *Org. Synth.*, **Coll. Vol. 2,** 363 (1943).

3. The addition of sulfuric acid requires *ca.* 1 hour. Occasional swirling of the Erlenmeyer flask is recommended for smooth generation of methyl nitrite.

4. Material of this quality is satisfactory for most purposes; however, if further purification is necessary, it may be recrystallized from hot water, giving a solid that melts at 126–128°. The checkers obtained product melting at 104–116° and 104–117° in the first and second runs, respectively. Evidently the product is a mixture of *syn* and *anti* isomers, the ratio of which was different in the material obtained by the submitters and the checkers. This difference in the isomer ratio might be attributed to a slight variation of experimental conditions. The submitters later informed the checkers that the methanol

was evaporated at 70–80°; the checkers removed the solvent at 35–40°. On partial recrystallization from hot water, the checkers isolated both the less soluble *anti* isomer, m.p. 127.5–129°, and the more soluble *syn* isomer, m.p. 97–99°. The melting points given in the literature for the *syn* and *anti* isomers are 129° and 99°, respectively.[2] The UV spectra (95% C_2H_5OH) of the *syn* and *anti* isomers show maxima at 274 nm. (log ϵ, 3.99) and 260 nm. (log ϵ, 4.05), respectively.

5. Phosgene may be replaced by a 0.5 molar equivalent of trichloromethyl chloroformate. This reagent may be purchased from Hodogaya Chemical Company, Ltd., Tokyo, Japan, or prepared by the procedure in *Org. Synth.*, **Coll. Vol. 6,** 715 (1988).

6. *N,N*-Dimethylaniline, pyridine, *tert*-butyl alcohol, and the solvents were dried with Linde type 3A molecular sieves.

7. The use of a 0.5 molar excess of pyridine and *tert*-butyl alcohol is necessary in this case to obtain a satisfactory yield. However, when this procedure is applied to the preparation of other alkoxycarbonates (Table II), excess alcohol should be avoided since it may contaminate the product.

8. The yield was reduced to 46% in a run in which the product was isolated without the additional overnight reaction time.

9. The checkers obtained 12.8–13.0 g. (52–53%), m.p. 84–86°, in the first crop and 2.7–3.4 g. (11–14%), m.p. 52–62°, in the second crop. Recrystallization of the former from methanol gave 11.5 g. of crystals, m.p. 84–86°, suggesting that the first crop is a pure single isomer. A ^{1}H NMR spectrum ($CDCl_3$) of the second crop shows two singlets at δ 1.62 and 1.64 for the *tert*-butyl groups. Thus, this material is a mixture of *syn* and *anti* isomers. Both the first and second crops proved equally useful for *tert*-butoxycarbonylation of an amino acid.

10. Recrystallization from boiling methanol should be avoided owing to the thermal instability of the product.

11. IR (Nujol) cm.$^{-1}$: 1785 (C=O); ^{1}H NMR ($CDCl_3$), δ (multiplicity, number of protons, assignment): 1.62 (s, 9H, 3CH_3), 7.2–8.2 (m, 5H, C_6H_5). A TLC on silica gel (Merck precoated plate, 60 F_{254}) using UV detection and 10% methanol in chloroform as the developing solvent showed a major and a minor spot at an R_f value of 0.74 and 0.50, respectively. The minor spot arises from 2-hydroxyimino-2-phenylacetonitrile formed by partial hydrolysis of the product on the silica gel.

The submitters recommend that the product be stored in a stoppered brown bottle in a refrigerator. Although the material can be kept at room temperature for several weeks without noticeable decomposition, gradual evolution of carbon dioxide occurs over a period of several months, with the attendant risk of explosion. However, storage in the presence of a small amount of silica gel as a drying agent extends the shelf life of the material to more than a year.

3. Discussion

2-Hydroxyimino-2-phenylacetonitrile has been prepared from benzyl cyanide by reaction with nitrous acid,[3] with isoamyl nitrite and sodium ethoxide,[4] and with butyl nitrite and hydrogen chloride.[2]

The *tert*-butoxycarbonyl group is one of the most important amino protecting groups in peptide synthesis. Many *tert*-butoxycarbonylating reagents[5,6] have been prepared as substitutes for *tert*-butyl azidoformate,[7] which is toxic, shock-sensitive, and relatively unreactive.[8] 2-*tert*-Butoxycarbonyloxyimino-2-phenylacetonitrile,[9] one such reagent, possesses the following advantages: (1) it is stable, highly reactive, and ready for use; (2) *tert*-butoxycarbonylation of an amino acid is usually complete within 4–5 hours at room temperature in the presence of a 0.5 molar excess of triethylamine in 50% aqueous dioxane (Table I); and (3) the by-product, 2-hydroxyimino-2-phenylacetonitrile, is easily and completely removed by extraction into an organic solvent, leaving the *tert*-butoxycarbonylamino acid salt in the aqueous phase. The present procedure is also applicable to preparation of other amino-protecting reagents (Table II).

TABLE I

PREPARATION OF *N-tert*-BUTOXYCARBONYL-PROTECTED AMINO ACIDS
WITH 2-*tert*-BUTOXYCARBONYLOXYIMINO-2-PHENYLACETONITRILE[a]

Amino Acid	Solvent[b]	Time (hours)	Yield (%)
Glycine	A	2	87
Alanine	B	4	80
S-Benzyl cysteine	A	3	94
Glutamic acid	A	3	78
Leucine	A	3	72
Methione	A	3	82
Phenylalanine	A	2	65
Proline	C	1.5	88
Threonine	A	3	100
Asparagine	A	20	86

[a] The reactions were carried out with 0.010 mole of the amino acid, 0.011 mole of 2-*tert*-butoxycarbonyloxyimino-2-phenyl-acetonitrile, and 0.015 mole of triethylamine at 20–25°.
[b] The solvents were as follows: A, aqueous dioxane; B, aqueous acetone; C, methanol–dioxane–water, 15:5:10.

TABLE II

OTHER ALKOXYCARBONYLATING REAGENTS PREPARED FROM
2-HYDROXYIMINO-2-PHENYLACETONITRILE

$$C_6H_5C \underset{CN}{\overset{NOCOR}{\diagdown}} \quad \overset{O}{\underset{}{\parallel}}$$

R	Solvent for Recrystallization	M.p. (°)	Yield (%)
$C_6H_5CH_2-$	ethyl acetate-hexane	73–75	62
$4-CH_3OC_6H_4CH_2-$	ethyl acetate-hexane	112–113	36
Cl_3CCH_2-	methanol	82–84	87

1. Research Laboratories, Fujisawa Pharmaceutical Company, Ltd., Yodogawa-Ku, Osaka 532, Japan.
2. T. E. Stevens, *J. Org. Chem.*, **32**, 670 (1967).
3. A. Meyer, *Ber. Dtsch. Chem. Ges.*, **21**, 1306 (1888).
4. M. Murakami, R. Kawai, and K. Suzuki, *Nippon Kagaku Zasshi*, **84**, 669 (1963) [*Chem. Abstr.*, **60**, 4053b (1964)].
5. E. Wünsch, Ed., "Synthese von Peptiden Teil I," in E. Müller, Ed., "Methoden der Organischen Chemie" (Houben-Weyl), 4th ed., Vol. 15/1, George Thieme Verlag, Stuttgart, 1974, pp. 117–125.
6. U. Ragnarsson, S. M. Karlsson, B. E. Sandberg, and L.-E. Larsson, *Org. Synth.*, **Coll. Vol. 6**, 203 (1988).
7. R. Schwyzer, P. Sieber, and H. Kappeler, *Helv. Chim. Acta*, **42**, 2622 (1959); L. A. Carpino, B. A. Carpino, P. J. Crowley, C. A. Giza, and P. A. Terry, *Org. Synth.*, **Coll. Vol. 5**, 157 (1973); M. A. Insalaco and D. S. Tarbell, *Org. Synth.*, **Coll. Vol. 6**, 207 (1988).
8. For warnings regarding the use of *tert*-butyl azidoformate, see *Org. Synth.*, **Coll. Vol. 6**, 207 (1988). P. Feyen, *Angew, Chem. Int. Ed. Engl.*, **16**, 115 (1977).
9. (a) M. Itoh, D. Hagiwara, and T. Kamiya, *Tetrahedron Lett.*, 4393 (1975); (b) M. Itoh, D. Hagiwara, and T. Kamiya, *Bull. Chem. Soc. Jpn.*, **50**, 718 (1977).

tert-BUTOXYCARBONYL-L-PROLINE

[1,2-Pyrrolidinedicarboxylic acid, 1-(1,1-dimethylethyl) ester, (S)-]

$$H_2C - CH_2$$
$$H_2C\diagdown_N\diagup CHCO_2H \quad \xrightarrow[\substack{(CH_3)_2NC(=NH)N(CH_3)_2 \\ \text{dimethyl sulfoxide}}]{(CH_3)_3COCO_2C_6H_5}$$
$$\underset{H}{}$$

$$H_2C - CH_2$$
$$H_2C\diagdown_N\diagup CHCO_2H + C_6H_5OH$$
$$\underset{COOC(CH_3)_3}{|}$$

Submitted by ULF RAGNARSSON,[1] SUNE M. KARLSSON,
BENGT E. SANDBERG, and LARS-ERIC LARSSON
Checked by S. WANG and A. BROSSI

1. Procedure

A 1-l. Erlenmeyer flask (Note 1) equipped with a magnetic stirrer, and a thermometer is charged with 115 g. (1.00 mole) of L-proline (Note 2) and 500 ml. of dimethyl sulfoxide (Note 3). To the stirred suspension are added simultaneously, over 5 minutes, 115 g. (1.00 mole) of 1,1,3,3-tetramethylguanidine (Note 4) and 214 g. (1.10 moles) of *tert*-butyl phenyl carbonate (Note 5). The proline dissolves completely within a few minutes in an exothermic reaction, the temperature of which reaches a maximum of 50–52° after 10–15 minutes. After stirring for 3 hours, the clear reaction mixture is transferred to a 6-l. separatory funnel and shaken with 2.2 l. of water and 1.8 l. of diethyl ether (Note 6). The aqueous layer, after being washed with 500 ml. of ether, is acidified to pH 3.0 by the addition of 10% sulfuric acid (Note 7), which generally causes partial crystallization of the product. The acidic solution, including the solid, is extracted with three 600-ml. portions of a mixture of equal volumes of ethyl acetate and ether. The combined extracts are washed with three 25-ml. portions of water, dried over magnesium sulfate, filtered, and evaporated with a rotary evaporator at a bath temperature not exceeding 40°. After drying in a vacuum oven at 50°, the crude product weighs 202 g., m.p. 129–132°. It is recrystallized from 300 ml. of hot ethyl acetate, and clarified by filtration and the addition of 1 l. of petroleum ether (40–60°), yielding, after drying under vacuum at 50°, 179–193 g. (83–90%) of tert-butoxycarbonyl-L-proline, m.p. 132–134°, $[\alpha]_D^{25}$ −59.84° to −61.6° (c = 1, glacial acetic acid) (Notes 8, 9, and 10).

2. Notes

1. A three-necked flask equipped with a U-tube may also be used for the reaction.

2. The submitters used L-proline obtained from Tanabe Seiyaku Company, Ltd., Osaka, Japan, and checked its purity by the method of Manning and Moore.[2] The L-proline used by the checkers was obtained from Ajinomoto Company, New York.

3. Dimethyl sulfoxide, Fisher Scientific Company, was used without further purification.

4. 1,1,3,3-Tetramethylguanidine, b.p. 159–160°, was used without further purification. The submitters obtained their material from Schuchardt, Munich, Germany, and the checkers obtained theirs from Pfaltz and Bauer, New York.

5. tert-Butyl phenyl carbonate furnished by Ega-Chemie KG, Steinheim, Germany, was used by the submitters. The checkers used material obtained from Aldrich Chemical Company, Inc.

6. The pH of the aqueous layer was 7.2. If the pH, as measured with a pH meter, is not between 7 and 8, it should be adjusted to within these limits by the addition of either 10% sulfuric acid or 1,1,3,3-tetramethylguanidine. The submitters worked up the reaction by the following alternate, but less convenient, method. The reaction mixture was poured into 1.25 l. of 1 M sodium hydrogen carbonate solution, 500 ml. of ether, and sufficient water (ca. 1 l.) to give two clear phases. The pH, which was 8.9, was adjusted to 8.0 by the addition, with stirring, of solid potassium hydrogen sulfate.

7. 10% Sulfuric acid (1.1 M) was prepared by diluting 25 ml. of concentrated sulfuric acid with 398 ml. of water. The submitters used solid potassium hydrogen sulfate for acidification to pH 3.0.

8. An additional 5 g. of product, m.p. 127–129°, may be obtained from the mother liquor.

9. Some reported yields, melting points, and rotations are: 55%, 136–137°, and $[\alpha]_D^{25}$ −60.2° (2.011 in acetic acid)[3]; 96%, 134–136°, $[\alpha]_{578}^{18-25}$ −68.5° (c = 1, acetic acid)[4]; 95%, 132–134°, $[\alpha]_{578}$ −62.5° (acetic acid)[5]; 90%, 133–135°, $[\alpha]_D^{25}$ −60.4° (c = 2.2 in acetic acid)[6]; 93%, 134–135°, no rotation reported.[7]

TABLE I
Other Boc[a] Amino Acids Synthesized by This Procedure

Compound	Solvent	Temperature, °	Time, Hr.	Yield, %	Remarks
Boc-Ala[b]	DMSO[c]	25	40	58	
Boc-Ala[b]	DMSO[c]	40	40	79	
Boc-Asn[d]	DMSO[c]	25	66	70	2 equiv. TMG[e]
Boc-Asp[f]	DMSO[c]	25	18	89	2 equiv. TMG
Boc-Cys (Bzl)[g]	DMSO[c]	25	40	62	CHA salt[h]
Boc-Cys (Bzl)[g]	DMSO[c]	40	40	78	CHA salt
Boc-Gln[i]	DMSO[c]	50	48	62	2 equiv. TMG, DCHA salt,[j] continuous extraction with ethyl acetate
Boc-Glu[k]	DMSO[c]	25	2.5	80	2 equiv. TMG
Boc-Ile[l]	DMSO[c]	40	72	72	
Boc-Leu[m]	DMSO[c]	25	48	73	Calc. as hemihydrate
Boc-Met[n]	DMSO[c]	25	24	86	81% solid + 5% DCHA salt
Boc-Phe[o]	DMSO[c]	25	40	81	DCHA salt
Boc-Phe[o]	DMSO[c]	25	96	88	DCHA salt
Boc-Phe[o]	DMSO[c]	40	18	81	DCHA salt
Boc-Phe[o]	DMF[p]	25	48	59	DCHA salt
Boc-Phe[o]	D—W[q]	25	48	5	DCHA salt
Boc-Pro[r]	DMSO[c]	25	2.5	90	
Boc-Pro[r]	DMF[p]	25	2.5	92	
Boc-Pro[r]	D—W[q]	25	21	84	
Boc-Thr[s]	DMSO[c]	25	67	66	DCHA salt
Boc-Val[t]	DMSO[c]	25	71	77	2 equiv. TMG, 65% solid + 12% DCHA salt

[a]tert-Butoxycarbonyl. [b]L-Alanine. [c]Dimethylsulfoxide. [d]L-Asparagine. [e]1,1,3,3-Tetramethylguanidine. [f]L-Aspartic acid. [g]S-Benzyl-L-cysteine. [h]Cyclohexylamine. [i]L-Glutamine. [j]Dicyclohexylamine. [k]L-Glutamic acid. [l]L-Isoleucine. [m]L-Leucine. [n]L-Methionine. [o]L-Phenylalanine. [p]Dimethylformamide. [q]Dioxane-water 1:1. [r]L-Proline. [s]L-Threonine. [t]L-Valine.

205

10. ^{1}H NMR (dimethyl sulfoxide-d_6): δ 1.38 (s, 9H, 3CH_3), 1.92 (m, 4H, 2CH_2), 3.31 (t, 2H, CH_2N), 4.05 (t, 1H, CHN), 12.3 (s, 1H, CO$_2H$). Analysis calculated for $C_{10}H_{17}NO_4$: C, 55.80; H, 7.96; N, 6.51. Found: C, 55.81; H, 7.95; N, 6.44.

3. Discussion

Since their introduction by McKay and Albertson,[8] and Anderson and McGregor,[3] tert-butoxycarbonyl amino acids have been prepared by several different methods. The simplest procedure[6] requires working with large quantities of phosgene. Another very good method,[5] but one that has not found wide application, involves the use of tert-butoxycarbonylfluoride, which is not commercially available. At the present the most useful reagent has been tert-butoxycarbonylazide, for which good procedures[9,10] are available; the excellent method of Schnabel[4] and one more recently reported[7] are based on this reagent. Of the procedures for the preparation of tert-butoxycarbonylazide, one,[9] which is readily adaptable for large-scale operations, involves three steps and the other,[10] a two-step process, is more suitable for small-scale work.

Our procedure[11] represents a simplification in that tert-butyl phenyl carbonate, which is used as a starting material, is the first intermediate in the three-step synthesis[9] of tert-butoxycarbonylazide. This reagent is easy to prepare in quantity and is commercially available in bulk (Note 5). Further, 1,1,3,3-tetramethylguanidine is inexpensive and the experimental operations are extremely simple.

Proline dissolves readily in dimethyl sulfoxide. Some other amino acids which are less soluble require longer reaction times and, in some instances, other solvents.[11] These details and the scope of the reaction are illustrated in Table I.

More recently, a few additional reagents have been found useful for the synthesis of Boc-amino acids. Among these are tert-butyl-4,6-dimethylpyrimidyl-2-thiol carbonate,[12] di-tert-butyl dicarbonate,[13] 2-tert-butoxycarbonyloxyimino-2-phenylacetonitrile,[14] and tert-butyl α-methoxyvinyl carbonate.[15] With the advent of a relatively simple method for the preparation of carbonyl chloride fluoride[16], tert-butoxycarbonylfluoride[17] is now more readily available.

1. Biokemiska Institutionen, Uppsala Universitet, Box 531, S-751 23 Uppsala, Sweden.
2. J. M. Manning and S. Moore, *J. Biol. Chem.*, **243**, 5591 (1968).
3. G. W. Anderson and A. C. McGregor, *J. Am. Chem. Soc.*, **79**, 6180 (1957).
4. E. Schnabel, *Justus Liebigs Ann. Chem.*, **702**, 188 (1967).
5. E. Schnabel, H. Herzog, P. Hoffmann, E. Klauke, and I. Ugi, in E. Bricas, Ed., "Peptides 1968," North-Holland, Amsterdam, 1968, p. 91.
6. S. Sakakibara, I. Honda, K. Takada, M. Miyoshi, T. Ohnishi, and K. Okumura, *Bull. Chem. Soc. Jpn.*, **42**, 809 (1969).
7. A. Ali, F. Fahrenholz, and B. Weinstein, *Angew. Chem.*, **84**, 259 (1972) [*Angew. Chem. Int. Ed. Engl.*, **11**, 289 (1972)].
8. F. C. McKay and N. F. Albertson, *J. Am. Chem. Soc.*, **79**, 4686 (1957).
9. L. A. Carpino, B. A. Carpino, P. J. Crowley, C. A. Giza, and P. H. Terry, *Org. Synth., Coll. Vol. 5*, 157 (1973).
10. M. A. Insalaco and D. S. Tarbell, *Org. Synth., Coll. Vol. 6*, 207 (1988).
11. U. Ragnarsson, S. M. Karlsson, and B. E. Sandberg, *Acta Chem. Scand.*, **26**, 2550 (1972).
12. T. Nagasawa, K. Kuroiwa, K. Narita, and Y. Isowa, *Bull. Soc. Chem. Jpn.*, **46**, 1269 (1973).

13. B. M. Pope, Y. Yamamoto, and D. S. Tarbell, *Org. Synth.*, **Coll. Vol. 6,** 418 (1988).
14. M. Itoh, D. Hagiwara, and T. Kamiya, *Org. Synth.*, **Coll. Vol. 6,** 199 (1988).
15. Y. Kita, J. Haruta, H. Yasuda, K. Fukunaga, Y. Shirouchi, and Y. Tamura, *J. Org. Chem.*, **47,** 2697 (1982).
16. G. Siegemund, *Angew. Chem.*, **85,** 982 (1973) [*Angew. Chem. Int. Ed. Engl.*, **12,** 918 (1973)].
17. L. Wackerle and I. Ugi, *Synthesis,* 598 (1975).

tert-BUTYL AZIDOFORMATE

(Carbonazidic acid, 1,1-dimethylethyl ester)

$$(CH_3)_3COK + CO_2 \rightarrow (CH_3)_3COCO_2K$$

$$(CH_3)_3COCO_2K + ClP(O)(OC_2H_5)_2 \rightarrow (CH_3)_3COCO_2P(O)(OC_2H_5)_2$$

$$(CH_3)_3COCO_2P(O)(OC_2H_5)_2 + KN_3 \rightarrow$$

$$(CH_3)_3COCON_3 + (C_2H_5O)_2PO_2K$$

Submitted by MICHAEL A. INSALACO and D. STANLEY TARBELL[1]
Checked by FREDERICK J. SAUTER, WALTER J. CAMPBELL, and HERBERT O. HOUSE

Caution! Tests conducted by the Eastman Kodak Company have shown that tert-*butyl azidoformate, also known as* tert-*butoxy carbonyl azide and* t-BOC *azide, is a thermally unstable, shock-sensitive compound (TNT equivalence: 45%).*

A number of less-hazardous reagents that can be substituted for tert-*butyl azidoformate in* tert-*butoxycarbonylation reactions are available, including 2-(tert-butoxycarbonyloxyimino)-2-phenylacetonitrile (Aldrich Chemical Company), O-*tert-*butyl S-phenyl thiocarbonate (Eastman Organic Chemicals), di-*tert-*butyl dicarbonate[2] and* tert-*butyl phenyl carbonate.[3]*

1. Procedure

A. tert-*Butylcarbonic diethylphosphoric anhydride.* A 500-ml., three-necked flask fitted with a mechanical stirrer, a pressure-equalizing dropping funnel, and a gas-inlet tube is dried in an oven, flushed with nitrogen, and allowed to cool while an atmosphere of nitrogen is maintained in the reaction vessel. To the flask are added successively 250 ml. of anhydrous tetrahydrofuran (Note 1) and 13.9 g. (0.124 mole) of alcohol-free potassium *tert*-butoxide (Note 2). After the mixture has been stirred for 10 min. under a nitrogen atmosphere to complete dissolution of the salt, the solution is cooled in an ice bath, and a slow stream of anhydrous carbon dioxide (Note 3) is bubbled through the cold solution for 1.5 hours with continuous stirring. While cooling, stirring, and the flow of carbon dioxide are maintained, a solution of 20.6 g. (0.120 mole) of pure diethyl phosphorochloridate (Note 4) in 50 ml. of anhydrous tetrahydrofuran (Note 1) is added dropwise to the reaction mixture, and the cold reaction mixture is stirred for an additional

30 minutes under a carbon dioxide atmosphere. After the solvent has been removed from the reaction mixture by concentration with a rotary evaporator at room temperature, the residual mixture of anhydride and potassium chloride is diluted with 300 ml. of anhydrous diethyl ether and centrifuged, separating the insoluble salt. The salt is successively suspended in two 150-ml. portions of anhydrous ether and centrifuged to remove any remaining ether soluble product. The combined ether solutions are concentrated at room temperature with a rotary evaporator, leaving 25–28 g. (84–91%) of the crude anhydride, as a colorless to pale yellow liquid (Note 5), sufficiently pure for use in the next step.

B. *tert-Butyl azidoformate*. A 300-ml., three-necked flask fitted with a mechanical stirrer, a pressure-equalizing dropping funnel, and a gas-inlet tube is dried in an oven, flushed with nitrogen, and allowed to cool while a nitrogen atmosphere is maintained in the reaction vessel. To the flask are added 120 ml. of anhydrous dimethyl sulfoxide (Note 6) and 8.1 g (0.10 mole) of powdered potassium azide (Note 7). While a nitrogen atmosphere is maintained in the reaction vessel, the mixture is stirred for 30 minutes to dissolve the bulk of the potassium azide before 25.4 g. (0.100 mole) of *tert*-butylcarbonic diethylphosphoric anhydride is added dropwise with stirring. During this addition the temperature of the reaction mixture is maintained at approximately 25° with a water bath. The resulting solution is stirred at 20–25° for 1 hour; 120 ml. of water is then added dropwise while stirring and external cooling are maintained. The resulting mixture is extracted with three 120-ml. portions of pentane, and the combined pentane extracts are washed successively with 20 ml. of water and 20 ml. of saturated aqueous sodium chloride. After the pentane solution has been dried over magnesium sulfate, the pentane is removed at room temperature with a rotary evaporator. The residual crude product, 11–14 g. of a pale yellow liquid, is distilled under reduced pressure from a distilling flask heated in a water bath. *Caution! This distillation should be conducted in a hood behind a safety shield.* (Note 8). The product is collected as 7.2–9.3 g. (50–65%) of colorless liquid, b.p. 57–61° (40 mm.), n_D^{24} 1.4224–1.4230 (Note 9).

2. Notes

1. A reagent grade of tetrahydrofuran should be redistilled from lithium aluminum hydride immediately before use. In this distillation discontinue heating when the residue in the stillpot has reached a volume of 50–100 ml.

2. Alcohol-free potassium *tert*-butoxide, obtained from the MSA Research Corporation, Callery, Pennsylvania, should be weighed and transferred under anhydrous conditions.

3. The carbon dioxide, obtained from a cylinder or from a container packed with dry ice, should be passed through a drying tube packed with silica gel or anhydrous calcium sulfate.

4. The diethyl phosphorochloridate, available from Stauffer Chemical Company, New York, should be redistilled before use: b.p. 60° (2 mm.), n_D^{25} 1.4153.

5. IR (CCl₄), 1770 (C=O), 1255 and 1292 cm.$^{-1}$ (P=O); ¹H NMR (CCl₄), δ 1.38 (pair of t, J_{H-H} = 6.6 Hz., J_{P-H} = 2.5 Hz., 6H, 2CH₂C*H₃*), 1.53 [s, 9H, C(C*H₃*)₃], and 4.17 (pair of quadruplets, J_{H-H} = 6.6 Hz., J_{P-H} = 8.5 Hz., 4H, 2C*H₂*CH₃).

6. An analytical grade of dimethyl sulfoxide obtained from J. T. Baker Co. or Matheson, Coleman and Bell was used. The checkers redistilled this solvent from calcium hydride before use; b.p. 83–85° (17 mm.).

7. A commercial grade of potassium azide, obtained from either Eastman Organic Chemicals or Alfa Inorganics, Inc., was purified by recrystallization from aqueous ethanol (1:2, v/v) as previously described.[4] The pure azide, obtained as white solid, m.p. 343° (dec.), was dried in an oven at 110° for 2 hours before use. In this preparation it is preferable to use the more soluble potassium azide rather than sodium azide.

8. Although this distillation has been conducted repeatedly without incident, azides are potentially explosive.[5] Consequently, the distillation should be conducted behind a safety shield and the operator should wear protective equipment. The distilling flask should be heated with a water bath to avoid the possibility of overheating the distillation residue.

9. IR (CCl$_4$), 2170, 2120 (azide), 1760, and 1735 cm.$^{-1}$ (C=O); ^{1}H NMR (CCl$_4$), δ 1.51, singlet, (CH_3)$_3$C; mass spectrum, m/e (rel. int.), 115 (17), 100 (16), 59 (85), 57 (20), 56 (41), 44 (18), 43 (84), 41 (21), 39 (19), 29 (30), 28 (100), 27 (21).

3. Discussion

Although *tert*-butyl azidoformate, a useful reagent for the protection of primary and secondary amino groups[6-8] has been prepared previously from *tert*-butyl carbazate,[5] the present procedure[9] is more convenient. The intermediate carbonic phosphoric anhydride reacts with nucleophiles other than the azide ion; for example, reaction of this anhydride with amines yields the *tert*-butoxycarbonyl derivatives of amines.[9] Other carbonic phosphoric anhydrides have been prepared by procedures analogous to the method described here.[10]

1. Department of Chemistry, University of Rochester, Rochester, New York. [Present address: Department of Chemistry, Vanderbilt University, Nashville, Tennessee 37203.]
2. B. M. Pope, Y. Yamamoto, and D. S. Tarbell, *Org. Synth., Coll. Vol. 6*, 418 (1988).
3. U. Ragnarsson, S. M. Karlsson, B. E. Sandberg, and L. E. Larsson, *Org. Synth., Coll. Vol. 6*, 205 (1988).
4. A. W. Browne, *Inorg. Synth.*, 1, 79 (1939).
5. L. A. Carpino, B. A. Carpino, P. J. Crowley, C. A. Giza, and P. H. Terry, *Org. Synth., Coll. Vol. 5*, 157 (1973).
6. (a) L. A. Carpino, *J. Am. Chem. Soc.*, 79, 4427 (1957); (b) L. A. Carpino, C. A. Giza, and B. A. Carpino, *J. Am. Chem. Soc.*, 81, 955 (1959).
7. T. Wieland and H. Determann, *Angew. Chem., Int. Ed. Engl.*, 2, 358 (1963).
8. R. A. Boissonnas, *Advan. Org. Chem.*, 3, 159 (1963).
9. D. S. Tarbell and M. A. Insalaco, *Proc. Nat. Acad. Sci. U.S.*, 57, 233 (1967).
10. A. A. Shamshurin, O. E. Krivoshchekova, and M. Z. Krimer, *J. Gen. Chem. USSR*, 35, 1871 (1965).

tert-BUTYLCYANOKETENE

(Butanenitrile, 2-carbonyl-3,3-dimethyl-)

Submitted by WALTER WEYLER, JR., WARREN G. DUNCAN,
MARGO BETH LIEWEN, and HAROLD W. MOORE[1]
Checked by B. E. SMART and R. E. BENSON

1. Procedure

Caution! Benzene has been identified as a carcinogen; OSHA has issued emergency standards on its use. All procedures involving benzene should be carried out in a well-ventilated hood, and glove protection is required.

A. 2,5-*Di*-tert-*butyl*-5,6-*dichloro*-2-*cyclohexene*-1,4-*dione*. A 2-l. Erlenmeyer flask is charged with 110 g. (0.500 mole) of 2,5-di-*tert*-butyl-1,4-benzoquinone (Note 1) and 500 ml. of glacial acetic acid. The reaction flask is equipped with an efficient magnetic stirring bar, and chlorine gas is introduced through a safety trap into the well-stirred mixture (Note 2). During the course of the reaction, which should be completed in 35–40 minutes (Note 3), the mixture becomes homogeneous and warms to about 60°. The mixture is flushed with nitrogen to expel excess chlorine. During this process the product crystallizes from the solution; precipitation is completed by cooling the reaction mixture to 20°. The mixture is filtered, and the product is washed with 1 l. of water and air dried, yielding 112 g. (77%) of 2,5-di-*tert*-butyl-5,6-dichloro-2-cyclohexene-1,4-dione, m.p. 125–129° (Note 4).

B. *3-Chloro-2,5-di-*tert-*butyl-1,4-benzoquinone*. A 2-l. Erlenmeyer flask is charged with a solution of 112 g. (0.385 mole) of 2,5-di-*tert*-butyl-5,6-dichloro-2-cyclohexene-1,4-dione in 800 ml. of diethyl ether. A solution of 28.4 g. (0.383 mole) of diethylamine in 50 ml. of ether is added in one portion to the vigorously swirled flask (Note 5). The reaction is instantaneous, resulting in a voluminous precipitate. The mixture is washed with two 1-l. portions of water, then with 500 ml. of aqueous saturated sodium chloride. The yellow ether solution is dried over anhydrous magnesium sulfate, filtered, and concentrated on a rotary evaporator, yielding 96–97 g. (98–99%) of 3-chloro-2,5-di-*tert*-butyl-1,4-benzoquinone as a yellow oil which is used without further purification (Note 6).

C. *2,5-Di-*tert-*butyl-3,5,6-trichloro-2-cyclohexene-1,4-dione*. A 2-l. Erlenmeyer flask is charged with a solution of 96.6 g. (0.379 mole) of crude 3-chloro-2,5-di-*tert*-butyl-1,4-benzoquinone in 500 ml. of glacial acetic acid. The reaction flask is equipped with an efficient magnetic stirring bar, and chlorine gas is introduced (Note 7). The reaction is complete in 4–5 hours (Note 8), at which time the solution is flushed with nitrogen to expel excess chlorine. Approximately 1 l. of water is added, and the resulting mixture is extracted with 300 ml. of dichloromethane. The dichloromethane solution is washed three times with water, dried over anhydrous magnesium sulfate, filtered, and concentrated on a rotary evaporator, yielding 116–117 g. (94%) of 2,5-di-*tert*-butyl-3,5,6-trichloro-2-cyclohexene-1,4-dione as a lemon yellow oil which is used directly in the next step (Note 9).

D. *2,5-Di-*tert-*butyl-3,6-dichloro-1,4-benzoquinone*. A 2-l. Erlenmeyer flask is charged with a solution of 116.6 g. (0.3582 mole) of 2,5-di-*tert*-butyl-3,5,6-trichloro-2-cyclohexene-1,4-dione in 800 ml. of ether. To the vigorously swirled solution is added, in one portion, 26.2 g. (0.359 mole) of diethylamine dissolved in *ca.* 50 ml. of ether. The reaction, which is instantaneous, results in a voluminous precipitate (Note 10). The reaction mixture is washed with two 1-l. portions of water and then with 500 ml. of aqueous saturated sodium chloride (Note 11). The ether solution is dried over anhydrous magnesium sulfate, filtered, and concentrated on a rotary evaporator. The crude product, a yellow semisolid (109 g.), is dissolved in 300 ml. of hot ethanol, then cooled first to room temperature and finally to 0°. After crystallization has begun, the flask is left at −5° to −10° overnight. The product is filtered, washed with 85% ethanol, and air dried, yielding 62–70 g. (60–67%) of yellow crystalline 2,5-di-*tert*-butyl-3,6-dichloro-1,4-benzoquinone, m.p. 68–69° (Notes 12 and 13).

E. *3,6-Diazido-2,5-di-*tert-*butyl-1,4-benzoquinone*. A solution of 10 g. (0.035 mole) of 2,5-di-*tert*-butyl-3,6-dichloro-1,4-benzoquinone in 375 ml. of methanol is cooled to 5–15°. To the solution is added, over 1–2 minutes, a solution of 5 g. (0.08 mole) of sodium azide in 15 ml. of water. The initial yellow solution becomes orange during addition of the azide. The flask is then cooled to $-5°$ to $-10°$ for at least 4 hours. The product precipitates from the solution and is collected by filtration, yielding $8.3-8.8$ g. ($80-85\%$) of 3,6-diazido-2,5-di-*tert*-butyl-1,4-benzoquinone, m.p. $88.9-90°$ (dec.), which is recrystallized at room temperature by dissolving it in a minimum amount of chloroform, filtering, and adding 2 parts of 95% ethanol to the chloroform solution. The resulting solution is cooled to $-5°$ to $-10°$; the crystalline precipitate is isolated (86% recovery) by filtration (Note 14), with no appreciable change in melting point (Note 15).

F. tert-*Butylcyanoketene*. Typically, the ketene is prepared by dissolving 1 g. (0.003 mole) of 3,6-diazido-2,5-di-*tert*-butyl-1,4-benzoquinone in $10-25$ ml. of anhydrous benzene (Note 16). The solution is refluxed, and the disappearance of the starting material as well as the intermediate cyclopentenedione is followed by TLC (Note 17). When the cyclopentenedione is no longer detectable, after approximately 90 minutes, the heating is stopped. The solution contains *tert*-butylcyanoketene in amounts equivalent to at least a 95% yield (Note 18).

2. Notes

1. Practical grade 2,5-di-*tert*-butyl-1,4-benzoquinone of m.p. $151-154°$ obtained from Eastman Organic Chemicals was used. Chlorine available from Air Products and Chemicals, Inc., was used by the checkers.

2. A satisfactory way to introduce chlorine with minimal loss of the gas is to seal the reaction flask with a two-holed stopper equipped with a gas-inlet tube, reaching just above the surface of the reaction mixture, and an exit tube, connected to a U-tube filled with mineral oil which is used as a gas-flow indicator. Chlorine is introduced from the cylinder through a safety trap at such a rate as to maintain a small positive pressure in the reaction flask.

3. The reaction can be followed by ^1H NMR spectroscopy. The original absorption for the vinyl proton disappears and two new absorption peaks appear, one in the vinyl region ($ca.$ $\delta 6.5$, CDCl$_3$) and the other in the methine region of the spectrum. There are two products formed, presumably the *cis*- and *trans*-isomers, in the ratio of $95:5$, respectively. The checkers also obtained the same yield when the reaction quantities were doubled.

4. An analytical sample has a m.p. of $127-129°$. Additional product can be recovered from the mother liquor by addition of approximately 1 l. of water followed by filtration. The yield of this product is about 31 g. (21%). However, it contains about 20% of the minor isomer (Note 3) that is not dehydrohalogenated under the reaction conditions employed in the next step. The second crop can be recrystallized from hot methanol, giving predominantly the desired isomer. In some preparations the submitters did not separate the minor product and observed no significant loss in yield in the subsequent steps. The spectral properties of the product are as follows; IR (Nujol) cm.$^{-1}$: 1700 (C=O), 1600 (C=C); ^1H NMR (CDCl$_3$), δ (multiplicity, number of protons, assignment): 1.28 [s, 9H, C(CH$_3$)$_3$], 1.37 (s, 9H, C(CH$_3$)$_3$], 4.75 (s, 1H, CH), 6.47 (s, 1H, =CH).

5. The reaction mixture warms slightly, resulting in the boiling of the ether. The

large amount of diethylamine hydrochloride formed transforms the reaction mixture into a thick paste.

6. The spectral properties of the product are as follows; IR (Nujol) cm.$^{-1}$: 1680 (C=O), 1660 (C=C); ^{1}H NMR (CDCl$_3$), δ (multiplicity, number of protons, assignment): 1.30 [s, 9H, C(CH$_3$)$_3$], 1.46 [s, 9H, C(CH$_3$)$_3$], 6.59 (s, 1H, =CH).

7. The experimental setup in this reaction is exactly as that described in Note 2.

8. The progress of the reaction is followed by ^{1}H NMR spectroscopy. When the absorption for the vinyl proton (*ca.* δ 6.6, CDCl$_3$) is completely absent, the reaction is stopped. Several minor products that were not identified are also formed in this step.

9. IR (Nujol) cm.$^{-1}$: 1710 (C=O); ^{1}H NMR (CDCl$_3$), δ (multiplicity, number of protons, assignment): 1.1–1.4 [m, 18H, 2C(CH$_3$)$_3$], 4.68 (s, 1H, CH), 4.87 (s, 1H, CH).

10. Comments given in Note 5 apply here also.

11. The solution may have a brown tint, partially masking the yellow color of the quinone. The dark color is probably due to reaction of the diethylamine with the 2,5-di-*tert*-butyl-3,6-dichloro-1,4-benzoquinone.

12. The mother liquor and wash solution are combined and concentrated to 200 ml. on a rotary evaporator. Upon cooling, a second crop (15–18 g.) of product is obtained. This second crop was a semisolid material. The spectral properties of the crystalline product are as follows; IR (Nujol) cm.$^{-1}$: 1660 (C=O); ^{1}H NMR (CDCl$_3$), δ (multiplicity, assignment): 1.45 [s, C(CH$_3$)$_3$].

13. Theoretically there remains about 22% of product to be isolated. Some of this material can be recovered indirectly by converting it to the diazide. The submitters diluted the mother liquor, which contains at most 23 g. (0.079 mole) of 2,5-di-*tert*-butyl-3,6-dichloro-1,4-benzoquinone, with 500 ml. of methanol, and then added, with swirling, a solution of 10.4 g. (0.0160 mole) of sodium azide in 30 ml. of water over a 2-minute period, turning the yellow solution orange. It is cooled to −5° to −10°, and the resulting orange precipitate is collected, yielding 12 g. of the diazide. The minimum yield is thus 88%.

14. Water can be added to the mother liquor, and the mixture extracted with chloroform to increase the diazide yield to 95–98%. During the course of any purification method that might be employed the diazide should not be heated above 50° since decomposition occurs quite noticeably at that temperature. It is best to store the pure product below −5° in the dark since it undergoes a facile photochemical rearrangement to the cyclopentenedione.

15. IR (Nujol) cm.$^{-1}$: 2110 (N$_3$), 1640 (C=O); ^{1}H NMR (CDCl$_3$), δ (multiplicity, assignment): 1.31 [s, C(CH$_3$)$_3$].

16. The Discussion contains comments on the stability of *tert*-butylcyanoketene in various solvents.

17. TLC is carried out on silica gel using 1:1 (v/v) petroleum ether–chloroform as eluent. The cyclopentenedione has an R_f value about half that of the diazide, and can be detected with an UV lamp when silica gel containing fluorescent indicator is used. The ketene undoubtedly reacts with the hydroxyl groups of the silica gel and remains at the origin. The checkers found the reaction to be complete in 1.5–2 hours. The yield was established by the checkers to be ⩾ 95% by ^{1}H NMR, by integration studies in the presence of an internal standard.

18. The submitters have not been successful in isolating *tert*-butylcyanoketene by any method. If the solvent is removed, the ketene polymerizes. The spectral properties of the product are as follows; IR (C_6H_6) cm.$^{-1}$: 2220 (C≡N), 2130 (C=C=O); ^{1}H NMR (C_6H_6), δ (multiplicity, assignment): 0.75 [s, C(CH$_3$)$_3$].

3. Discussion

In benzene at room temperature, *tert*-butylcyanoketene (TBCK) does not undergo rapid self-condensation;[2] however, it is quite reactive toward cycloaddition reactions with alkenes, allenes, ketenes, imines, and formimidates.[3] The mechanisms of a number of these cycloaddition reactions have been investigated and in some cases have been shown to involve the formation of a zwitterionic intermediate. Typical examples of TBCK cycloadditions are summarized in the formula below.

The ketene is less stable in nonaromatic hydrocarbon solvents than in aromatic solvents. For example, it has a half-life of greater than 7 days in benzene at 25°. On the other hand, in cyclohexane at the same temperature its half-life is only a few hours. All

Reactions of *tert*-Butylcyanoketene (TBCK)

attempts to isolate *tert*-butylcyanoketene have failed. Either removal of the solvent or cooling the solution to low temperature ($-70°$) causes polymerization of the ketene, a very efficient process, giving a white solid polymer which appears to have repeating ketenimine units. This assignment is consistent with the very strong absorption at 2140 cm^{-1} in the IR spectrum.[4]

The method described here for the synthesis of *tert*-butylcyanoketene has marked advantages over other possible classical routes, *e.g.*, dehydrohalogenation of the corresponding acid chloride. The only other product formed is molecular nitrogen and no external catalyst, *e.g.*, triethylamine, is necessary. In fact, when *tert*-butylcyanoketene reacts with triethylamine, or when α-*tert*-butyl-α-cyanoacetyl chloride is subjected to dehydrohalogenation conditions, 1,3-di-*tert*-butyl-1,3-dicyanoallene is immediately formed and no ketene can be detected.

tert-Pentylcyanoketene can be prepared in an analogous fashion starting from the commercially available 2,5-di-*tert*-pentylbenzoquinone. This ketene seems to be very similar in stability and reactivity to the *tert*-butyl homolog. Other cyanoketenes which have been prepared from azidoquinones or related compounds include dicyano-, chlorocyano-, bromocyano-, iodocyano-, methylcyano-, and isopropylcyanoketene.[3] However, all of these must be generated *in situ* since they undergo rapid self-condensation in the absence of a ketenophile.

1. Department of Chemistry, University of California, Irvine, California 92717.
2. W. Weyler, W. G. Duncan, and H. W. Moore, *J. Am. Chem. Soc.*, **97**, 6187 (1975).
3. H. W. Moore, *Acc. Chem. Res.*, **12**, 125 (1979).
4. H. K. Hall, Jr., E. P. Blanchard, Jr., S. C. Cherkofsky, J. B. Sieja, and W. A. Sheppard, *J. Am. Chem. Soc.*, **93**, 110 (1971).

cis-4-*tert*-BUTYLCYCLOHEXANOL

[Cyclohexanol, 4-(1,1-dimethylethyl)-, *cis*-]

Submitted by E. L. ELIEL,[1] T. W. DOYLE, R. O. HUTCHINS,[2] and E. C. Gilbert
Checked by MITCHELL WINNIK and RONALD BRESLOW

1. Procedure

To a solution of 4.0 g. (0.012 mole) of iridium tetrachloride (Note 1) in 4.5 ml. of concentrated hydrochloric acid is added 180 ml. of water followed by 52 g. (50 ml., 0.42 mole) of trimethyl phosphite (Note 2). This solution then is added to a solution of 30.8 g. (0.200 mole) of 4-*tert*-butylcyclohexanone (Note 3) in 635 ml. of 2-propanol contained in a 2-l. flask equipped with a reflux condenser. The solution is heated at reflux for 48 hours

(Note 4), at which time the 2-propanol is removed with a rotary evaporator. The remaining solution is diluted with 250 ml. of water and extracted with four 150-ml. portions of diethyl ether. The combined ether extracts are washed with two 100-ml. portions of water, which are combined with the aqueous residue (Note 5), dried over magnesium sulfate or potassium carbonate, and concentrated on a rotary evaporator, yielding 29–31 g. (93–99%) of *cis*-4-*tert*-butylcyclohexanol as a white solid. Analysis of the crude product by GC shows it contains 95.8–96.2% *cis*-alcohol and 4.2–3.8% of the *trans* isomer with essentially no ketone remaining (Note 6). Recrystallization from 40% aqueous ethanol affords greater than 99% pure *cis*-alcohol, m.p. 82–83.5° after sublimation (Note 7).

2. Notes

1. Iridium tetrachloride was originally obtained from Platinum Chemicals, Inc., Box 565, Asbury Park, New Jersey 07712, or from Alfa Products, Thiokol/Ventron Division, P.O. Box 299, 152 Andover St., Danvers, Massachusetts 01923. More recently, the procedure has been repeated successfully with material obtained from Pfaltz and Bauer, Inc., a subsidiary of Aceto Chemical Co., Inc., 375 Fairfield Ave., Stamford, Connecticut 06902.

2. The order of mixing the catalyst components is required for good results, and the sequence described should be followed. Particular care should be taken *not* to add the trimethyl phosphite before the water, as the reaction between it and concentrated hydrochloric acid is extremely violent.

3. 4-*tert*-Butylcyclohexanone was obtained from Dow Chemical Company or Aldrich Chemical Co. Inc.

4. The reaction solution is often dark-colored at the beginning but lightens as reflux continues. The reflux time may be varied with the amount of ketone to be reduced. The completeness of the reaction may be followed by removing small aliquots, working up these samples as described in the text, and analyzing the product mixture by GC (see Note 6).

5. The iridium catalyst used in this preparation may be regenerated by reducing the volume of the aqueous residue to about 200 ml. at diminished pressure. This solution is then used instead of the iridium tetrachloride and water called for in the procedure.

6. The product was analyzed by GC using a 9-ft. 20% Carbowax 20M on 45/60 Chromosorb W column at 150°. The order of increasing retention times is: ketone, *cis*-alcohol, *trans*-alcohol.

7. Recrystallization is best accomplished by dissolving the crude product in hot ethanol (approx. 35 ml. per 10 g.) followed by adding water (approx. 25 ml. per 10 g.) and allowing the solution to cool slowly to 0°. The fluffy white needles are filtered using a sintered-glass funnel and dried over P_2O_5 at atmospheric pressure. Recooling the filtrate affords a second crop of product, for an overall yield of 75–87%.

3. Discussion

4-*tert*-Butylcyclohexanol has been prepared from *p-tert*-butylphenol by reduction under a variety of conditions.[4,5] Eliel and Ro[3] obtained *cis*-rich 4-*tert*-butylcyclohexanol by the reduction of 4-*tert*-butylcyclohexanone with hydrogen on platinum oxide in glacial acetic acid containing some hydrogen chloride; Sommerville and Theimer have similarly reduced *p-tert*-butylphenol with 5% rhodium on carbon.[6,7] They have also prepared *cis*-rich alcohol by fractional distillation of the *cis–trans* mixture over an equilibrating catalyst, such as aluminum isopropoxide in the presence of 4-*tert*-butylcyclohexanone.[6] Eliel and Nasipuri[8] have also obtained 4-*tert*-butylcyclohexanol containing 80–92% of the *cis* isomer by the reduction of 4-*tert*-butylcyclohexanone with isobornyloxyaluminum dichloride.

The present[9] procedure employs a readily available starting material and produces essentially pure *cis* isomer in good yield. In view of the fact that the catalyst may be reused several times with little loss in stereoselectivity, the expense of the iridium tetrachloride is not a serious impediment.

The method is useful in the preparation of other axial alcohols. Henbest[9] has reported the reductions of 3-*tert*-butylcyclohexanone, 3,3,5-trimethylcyclohexanone, and cholestanone to the axial alcohols by this procedure, although for the preparation of cholestan-3α-ol the procedure of Edward[10] is preferred by the checkers. Recently[11] 2,4,4-trimethylcyclohexanone has been reduced to the pure axial alcohol by this method in 90% yield.

Since this preparation was submitted, a number of reductions of 4-*tert*-butylcyclohexanone to the *cis* alcohol, with 93–100% selectivity, using various bulky, complex metal hydrides have been described: lithium tri-*sec*-butylborohydride (L-Selectride),[12,13] lithium trisiamylborohydride,[13–15] lithium tri-*trans*-2-methylcyclopentyl borohydride,[14] lithium dimesitylborohydride,[16] lithium 2,6-di-*tert*-butylphenoxyneopentoxyaluminumhydride;[17] high (94–99%) selectivity is also attained by catalytic hydrogenation with various rhodium catalysts.[18,19] Hydrogenation over rhodium-on-carbon[18] (94% *cis*) followed by purification with liquid chromatography appears to be an attractive method, one which avoids the need for special reagents.

1. Department of Chemistry, University of Notre Dame, Notre Dame, Indiana 46556. [Present address: Department of Chemistry, University of North Carolina, Chapel Hill, North Carolina 27514.]
2. Present address: Department of Chemistry, Drexel University, Philadelphia, Pennsylvania 19104.
3. E. L. Eliel and R. S. Ro, *J. Am. Chem. Soc.*, **79**, 5992 (1957).
4. G. Vavon and M. Barbier, *Bull. Soc. Chim. Fr.*, [4] **49**, 567 (1931).
5. H. Pines and V. Ipatieff, *J. Am. Chem. Soc.*, **61**, 2728 (1939).
6. W. T. Sommerville and E. T. Theimer, U.S. Pat. 2,840,599 (1958) [*Chem. Abstr.*, **52**, 18265 (1958)].
7. W. T. Sommerville and E. T. Theimer, U.S. Pat. 2,927,127 (1960) [*Chem. Abstr.*, **54**, 13027g (1960)].
8. E. L. Eliel and D. Nasipuri, *J. Org. Chem.*, **30**, 3809 (1965).
9. Y. M. Y. Haddad, H. B. Henbest, J. Husbands, and T. R. B. Mitchell, *Proc. Chem. Soc.*, 361 (1964).
10. J. T. Edward and J-M. Ferland, *Can. J. Chem.*, **44**, 1311 (1966).
11. D. J. Pasto and F. M. Klein, *J. Org. Chem.*, **33**, 1468 (1968).
12. H. C. Brown and S. Krishnamurthy, *J. Am. Chem. Soc.*, **94**, 7159 (1972).
13. H. C. Brown, J. L. Hubbard, and B. Singaram, *J. Org. Chem.*, **44**, 5004 (1979).
14. S. Krishnamurthy and H. C. Brown, *J. Am. Chem. Soc.*, **98**, 3383 (1976).
15. H. C. Brown, S. Krishnamurthy, and J. L. Hubbard, *J. Organomet. Chem.*, **166**, 271 (1979).

16. J. Hooz, S. Akiyama, F. J. Cedar, M. J. Bennett, and R. M. Tuggle, *J. Am. Chem. Soc.*, **96**, 274 (1974).
17. H. Haubenstock, *J. Org. Chem.*, **40**, 926 (1975).
18. S. Mitsui, H. Saito, Y. Yamashita, M. Kaminaga, and Y. Senda, *Tetrahedron*, **29**, 1531 (1973).
19. S. Nishimura, M. Ishige, and M. Shiota, *Chem. Lett.*, 963 (1977).

POLYMERIC CARBODIIMIDE. MOFFAT OXIDATION: 4-*tert*-BUTYLCYCLOHEXANONE

[Cyclohexanone, 4-(1,1-dimethylethyl)-]

$\widehat{P}$ = styrene–divinylbenzene copolymer

Submitted by NED M. WEINSHENKER,[1] CHAH M. SHEN, and JACK Y. WONG
Checked by A. FUKUZAWA and S. MASAMUNE

1. Procedure

Caution! Benzene has been identified as a carcinogen; OSHA has issued emergency standards on its use. All procedures involving benzene should be carried out in a well-ventilated hood, and glove protection is required.

A 250-ml., three-necked, round-bottomed flask equipped with a mechanical stirrer, a gas-inlet, and a stopper is charged with 540 mg. (0.00346 mole) of a mixture of *cis*- and *trans*-4-*tert*-butylcyclohexanols (Note 1), 50 ml. of anhydrous benzene (Note 2), and 25 ml. of anhydrous dimethyl sulfoxide (Note 3). While a slight positive pressure of argon is maintained in the system, 13.19 g. of carbodiimide resin (Note 4) is added, followed by 0.2 ml. of dimethyl sulfoxide (Note 3) containing 98 mg. (0.0010 mole) of anhydrous orthophosphoric acid (Note 5). The resulting mixture is stirred at room temperature for 3.5 days. The beads are then separated by filtration and washed with three 100-ml. portions of diethyl ether, and the combined filtrates are washed with five 100-ml. portions of water. After evaporation of the organic phase to dryness, the residue crystallizes, providing 446–450 mg. (83–84%) of crude 4-*tert*-butylcyclohexanone, m.p. 42–45° (Note 6). The deactivated carbodiimide resin can be regenerated by treatment with triethylamine and 4-toluenesulfonyl chloride (Note 4).

2. Notes

1. This mixture is available from Aldrich Chemical Company, Inc. The checkers used a 7:93 mixture of the *cis*- and *trans*-isomers, prepared by lithium aluminum hydride reduction of 4-*tert*-butylcyclohexanone and recrystallization of the crude product. The ketone was purchased from Aldrich Chemical Company, Inc.

2. Benzene was dried by distillation from sodium.

3. The submitters dried dimethyl sulfoxide over Linde-type 3A molecular sieves. The checkers distilled this reagent from calcium hydride at 10 mm. prior to use.

4. The amount of resin used contains about 0.012 mole of active carbodiimide. Methods for preparing this reagent, determining its carbodiimide content, and regenerating spent resin are described in *Org. Synth.*, **Coll. Vol. 6,** 951 (1988).

5. Anhydrous orthophosphoric acid was prepared according to the equation:

$$P_2O_5 + 3H_2O \rightarrow 2H_3PO_4$$

The submitters added 5.88 ml. of 85% phosphoric acid to 3.98 g. of phosphorous pentoxide and heated the mixture for 15 minutes or until all of the solid had dissolved. The checkers placed 71.0 g. of phosphorous pentoxide in a flask, cooled it in ice, and cautiously added 27 ml. of water.

6. IR (CHCl$_3$) cm.$^{-1}$: 1712 (C=O). GC analysis (10% Carbowax 20M, 3 mm. by 1.8 m., 180°) showed the crude product to be 97% pure. 4-*tert*-Butylcyclohexanone has been reported to melt at 49.5–51°.[2]

3. Discussion

The general procedure described here was originally published by the submitters.[3] Both ketones and aldehydes may be prepared, and this method is particularly effective when the mild conditions of the Moffat oxidation are required, but the dicyclohexylurea by-product formed with the usual reagents causes purification problems.

1. Dynapol, 1454 Page Mill Road, Palo Alto, California 94304.
2. E. L. Eliel and M. N. Rerick, *J. Am. Chem. Soc.,* **82,** 1367 (1960).
3. N. M. Weinshenker and C. M. Shen, *Tetrahedron Lett.,* 3285 (1972).

OXIDATION OF ALCOHOLS BY METHYL SULFIDE-
N-CHLOROSUCCINIMIDE- TRIETHYLAMINE:
4-*tert*-BUTYLCYCLOHEXANONE

[Cyclohexanone, 4-(1,1-dimethylethyl)-]

Submitted by E. J. Corey,[1] C. U. Kim,[2] and P. F. Misco[2]
Checked by S. Yamamoto and W. Nagata

1. Procedure

Caution! Benzene has been identified as a carcinogen; OSHA has issued emergency standards on its use. All procedures involving benzene should be carried out in a well-ventilated hood, and glove protection is required.

A 1-l., three-necked, round-bottomed flask equipped with a mechanical stirrer (Note 1), a thermometer, a dropping funnel, and an argon-inlet tube is charged with 8.0 g. (0.060 mole) of N-chlorosuccinimide (Notes 2 and 3) and 200 ml. of toluene (Note 4). While a continuous positive argon pressure is maintained, the solution is stirred and cooled to 0°, and 6.0 ml. (0.10 mole) of methyl sulfide (Note 2) is added. A white precipitate appears immediately after addition of the sulfide. The mixture is cooled to −25° using a carbon tetrachloride–dry ice bath (Note 5), and a solution of 6.24 g. (0.0400 mole) of 4-*tert*-butylcyclohexanol (mixture of *cis*- and *trans*-) (Note 2) in 40 ml. of toluene is added dropwise over 5 minutes (Note 6). The stirring is continued for 2 hours at −25° (Note 5) before a solution of 6.0 g. (0.59 mole) of triethylamine (Note 2) in 10 ml. of toluene is added dropwise over 3 minutes (Note 7). The cooling bath is removed; after 5 minutes 400 ml. of diethyl ether is added (Note 8). The organic layer is washed with 100 ml. of 1% hydrochloric acid, then with two 100-ml. portions of water (Note 9), dried over anhydrous magnesium sulfate, and evaporated under reduced pressure. The residue is transferred to a 50-ml., round-bottomed flask and distilled bulb to bulb at 120° (25 mm.), yielding 5.54–5.72 g. (90–93%) of 4-*tert*-butylcyclohexanone, m.p. 41–45° (Note 10).

2. Notes

1. Efficient magnetic stirring could be used as well.
2. The submitters used N-chlorosuccinimide, 4-*tert*-butylcyclohexanol, methyl sulfide, and triethylamine, available from Aldrich Chemical Company, Inc., without further purification.

The checkers used *cis*- and *trans*-4-*tert*-butylcyclohexanol (36:64) obtained from E. Merck A G, Darmstadt, methyl sulfide obtained from Tokyo Kasei Kogyo Co., Ltd., and triethylamine obtained from Wako Pure Chemical Industries, Ltd., without further purification.

3. The checkers observed that, when N-chlorosuccinimide of 93–95% purity, m.p. 128–132° (checked by iodometry), obtained from E. Merck A G, Darmstadt, was used without purification, the oxidation was incomplete, resulting in 93–94% yields of a product containing a 12–15% amount of the starting alcohol (a mixture of the *cis*- and *trans*-). The use of 98% pure N-chlorosuccinimide, m.p. 150–151° (recrystallized from benzene), resulted again in recovery of the alcohol in a considerable amount, mainly because of low solubility of the pure reagent in toluene. Therefore, the checkers modified slightly the earlier part of the procedure as follows: the 98% pure N-chlorosuccinimide (8.0 g., 0.06 mole) m.p. 150–151°, is dissolved in 400 ml. of toluene (twice as much as the volume used by the submitters) at 40°, and the solution is cooled to room temperature, stirred under nitrogen atmosphere, and cooled to 0–5°. Methyl sulfide (6 ml., 0.10 mole) is added at this temperature and, after addition, the reaction mixture is stirred for an additional 20 minutes at 0–3°, during which time white precipitate appeared.

4. Reagent grade toluene employed by the submitters was obtained from Mallinckrodt Chemical Works. The checkers used reagent grade toluene purchased from Wako Pure Chemical Industries, Ltd., and dried over molecular sieves (4 Å) before use. Regarding the volume of the toluene used by the checkers, see Note 3.

5. The checkers observed that the internal temperature was −20° with this cooling mixture.

6. The checkers observed that the reaction was slightly exothermic as judged from an internal temperature rise of *ca.* 5°.

7. The checkers needed a little longer time (3–5 minutes) for this operation to maintain the internal temperature at −15 to −20°.

8. The checkers successively added 100 ml. 1% hydrochloric acid to the reaction mixture.

9. Vigorous shaking is necessary for complete removal of succinimide. The triethylamine hydrochloride is also removed in this step. The methyl sulfide codistils with the ether.

10. An authentic sample from Aldrich Chemical Company, Inc., had m.p. 45–50°. The product was analyzed by GC at 80° on F & M Research chromatograph model 1810 with 3% OV-17 column, which indicated the contamination of the product by 4-*tert*-butylcyclohexanol (<2%) and 4-*tert*-butylcyclohexylmethyl methylthiomethyl ether (<2%).[3] The submitters reported a yield of 6.0 g. (96%) of purity greater than 96%.

The checkers recrystallized the product from petroleum ether (dissolved at room temperature and cooled to −20°) and obtained a pure product, m.p. 45–46° (88.5% recovery); IR (CHCl$_3$) cm^{-1}: 1712 (C=O); ^{1}H NMR (CDCl$_3$), δ (multiplicity, number of protons, assignment): 0.93 [s, 9H, C(CH$_3$)$_3$], 1.3–2.2 (m, 5H, CH$_2$CHCH$_2$), and 2.2–2.5 (m, 4H, CH$_2$COCH$_2$).

3. Discussion

The procedure described here is general for the oxidation of primary and secondary alcohols to carbonyl compounds,[3] but not for allylic and dibenzylic alcohols, which give halides in high yields.[4] The yields of carbonyl compounds are usually high, and the

TABLE I
OXIDATION OF PRIMARY AND SECONDARY ALCOHOLS[3]

Alcohol	Product	Yield (%)
$C_6H_5CH_2OH$	C_6H_5CHO	90
		96
		94
		91
		86[5]

formation of methyl thiomethyl ether can be minimized in nonpolar media. Some examples are listed in Table I. The quantitative conversion of catechols to o-quinones using this oxidation procedure has been reported.[6] For oxidation of allylic and dibenzylic alcohols to the corresponding carbonyl compounds, a dimethyl sulfoxide–chlorine reagent[7] is suitable.

1. Department of Chemistry, Harvard University, Cambridge, Massachusetts 02138.
2. Medicinal Chemical Research Department, Bristol Laboratories, Syracuse, New York 13201.
3. E. J. Corey and C. U. Kim, *J. Am. Chem. Soc.*, **94**, 7586 (1972); E. J. Corey and C. U. Kim, *J. Org. Chem.*, **38**, 1233 (1973).
4. E. J. Corey, C. U. Kim, and M. Takeda, *Tetrahedron Lett.*, 4339 (1972).
5. E. J. Corey and C. U. Kim, *Tetrahedron Lett.*, 287 (1974).
6. J. P. Marino and A. Schwatz, *J. Chem. Soc., Chem. Commun.*, 812 (1974).
7. E. J. Corey and C. U. Kim, *Tetrahedron Lett.*, 919 (1973).

2-*tert*-BUTYL-1,3-DIAMINOPROPANE

[1,3-Propanediamine, 2-(1,1-dimethylethyl)-]

$$CH_2(CN)_2 + (CH_3)_3CCl \xrightarrow[\text{nitromethane}]{AlCl_3} (CH_3)_3CCH(CN)_2$$

$$\xrightarrow[\text{tetrahydrofuran}]{B_2H_6} \xrightarrow{HCl} (CH_3)_3CCH(CH_2\overset{+}{N}H_3\overset{-}{Cl})_2$$

$$\xrightarrow{NaOH} (CH_3)_3CCH(CH_2NH_2)_2$$

Submitted by R. O. HUTCHINS[1] and B. E. MARYANOFF
Checked by ABRAHAM PINTER and RONALD BRESLOW

1. Procedure

A.   tert-*Butylmalononitrile.*   A dry, 2-l., three-necked flask equipped with a thermometer, mechanical stirrer, and a Claisen adapter fitted with a dropping funnel and condenser protected with a drying tube is charged with 200 ml. of nitromethane (Note 1). The flask is cooled to 0° in an ice–salt bath (Note 2), and anhydrous powdered aluminum chloride (90.0 g., 0.674 mole) is added, with slow stirring, through a powder funnel that temporarily replaces the thermometer. The temperature may rise to *ca.* 50° but quickly drops to 0°. A solution of 45.0 g. (0.682 mole) of malononitrile (Note 3) in 50 ml. of nitromethane is added through the dropping funnel at a rate such that the temperature is kept below 10° (approximately one hour). This is followed by slow, dropwise addition of a solution of 150 g. (1.62 moles) of *tert*-butyl chloride in 50 ml. of nitromethane at a rate such that the temperature is maintained below 10° (approximately 3–4 hours). The reaction mixture is stirred at 0–5° for 10 hours before 1 l. of saturated sodium hydrogen carbonate (*ca.* 80 g. in 1000 ml. of water) is added slowly and cautiously (still in the cold), keeping the temperature below 10°. The mixture is poured into a 3 or 4 l. beaker and solid sodium hydrogen carbonate (*ca.* 100 g.) is added in *small* portions, with stirring. The organic phase is separated, and the aqueous layer extracted with three half-volume portions of dichloromethane, which are concentrated on a rotary evaporator, combined with the organic phase, and concentrated further. The resultant brown oil is subjected to steam distillation. The first fraction is collected using a cold water condenser until solidification is observed in the condenser, at which time warm water is passed through the condenser and the receiver is changed (Note 5). The product is collected until occasional passage of cold water through the condenser no longer causes apparent solidification. At this point, the receiver is changed again, and a third fraction (*ca.* 500 ml.) is collected. The middle fraction is cooled in ice and filtered with vacuum through a medium fritted glass funnel, yielding 48–50 g. of light-yellow product. Extraction of the filtrate with two half-volumes of diethyl ether followed by evaporation gives an additional small amount (*ca.* 2 g.) of product. Another small crop of product may be gleaned from the first and third steam distillation fractions by separating any organic phase, removing distillable material on a rotary evaporator, cooling, and filtering the resulting solid. The total combined yield of crude product is 54–58 g. (65–70%). Further purification is accomplished by careful sublimation at 80–90 mm. (*ca.* 85°), giving 52–56.5 g. (63–68%) of white, waxy dinitrile, m.p. 76–79° with softening at 71°.

B.   2-tert-*Butyl*-1,3-*diaminopropane*.   A dry, 500-ml., three-necked flask fitted with a magnetic stirrer, nitrogen inlet, dropping funnel, and condenser attached to an acetone gas trap (Note 6) is flushed with dry nitrogen for 30 minutes and charged with purified *tert*-butylmalononitrile (30.5 g., 0.250 mole), sodium borohydride (17.5 g., 98% assay, 0.453 mole), and 150 ml. of dry tetrahydrofuran (Note 1). A dry nitrogen atmosphere is maintained while boron trifluoride diethyl etherate (85.3 g., 0.600 mole) (Note 7) in 50 ml. of dry tetrahydrofuran is added dropwise, with slow magnetic stirring, at a rate that permits gentle reflux. The addition takes about 4 hours (Note 8). The mixture is stirred for an additional 90 minutes, hydrolyzed by the very cautious dropwise addition of 30 ml. of concentrated hydrochloric acid, and transferred to a 1-l. flask, with rinsing by 100 ml. of tetrahydrofuran. The solution is evaporated to dryness on a rotary evaporator, yielding a dry, white, solid mass, to which is added a small portion (*ca.* 10 ml.) of 125 ml. of aqueous 40% (w/w) sodium hydroxide. The mixture is triturated with a glass rod and warmed on a steam bath until a reaction begins; heat is generated, and white smoke is evolved. The reaction is controlled by cooling in an ice bath. When the reaction appears to have subsided at room temperature, the trituration is repeated cautiously until all 125 ml. of the hydroxide solution is added. The resulting mixture is warmed for 30 minutes on a steam bath, with occasional stirring, cooled, and filtered with vacuum. The solid material is washed with ten 20-ml. portions of ether. The filtrate is separated, and the aqueous phase is extracted with three 100-ml. portions of ether. The combined ethereal extracts and organic phase are dried over anhydrous sodium sulfate. The drying agent is filtered and washed with ether, and the filtrate concentrated on a rotary evaporator. The residue is fractionally distilled at reduced pressure through a short Vigreux column. After removal of residual solvent and collection of a small forerun, 11.5–15.5 g. (36–48%) of product is obtained, b.p. 92.5–95° (21 mm.), 96–98° (27 mm.), n_D^{28} 1.4570–1.4585. Analysis of the product by GC (OV-1 column) indicated the purity to be *ca.* 95% (Note 9).

2. Notes

1. Fisher Scientific Company certified reagent grade was used without further purification from a freshly opened bottle.
2. As a convenience, the reaction may be conducted in a refrigerated room, in which case it will be unnecessary to replenish the ice during the course of the reaction.
3. Malononitrile was obtained from Eastman Organic Chemicals and distilled prior to use, b.p. 80–82° (3 mm.).
4. The commercial material was distilled prior to use, b.p. 50–51°.
5. Passing warm water through the condenser prevents blockage by solid product; otherwise, pressure may build and force the joints apart. As an alternative, steam may be passed through the condenser periodically.

6. Impure diborane is a hazardous material and may combust explosively on contact with air. Therefore, precautions must be taken to prevent escape from the reaction.

7. The commercial material was distilled prior to use, b.p. 59–60° (20 mm.). The material employed should not be more than a few days old.

8. The duration of the reaction is important and should not be curtailed. For example, if one is operating on a one-tenth scale, the reaction should be heated at gentle reflux for an additional 3.5 hours after the addition (25 minutes).

9. A sample of the compound, collected by GC, gave the correct analyses for carbon, hydrogen, and nitrogen. Two solid derivatives were prepared in good yield and both gave correct analyses.

3. Discussion

This preparation illustrates the alkylation of malononitrile under acid-catalyzed conditions, and the use of diborane for the reduction of a dinitrile to a diamine. The procedure for the preparation of *tert*-butylmalononitrile has been outlined briefly by Boldt and co-workers.[2] The generation of diborane *in situ* and the general method for nitrile reduction is that described by Brown and co-workers.[3] Attempts to reduce the dinitrile to the diamine by other methods including catalytic hydrogenation (5% rhodium on alumina, 5 atm.), lithium aluminum hydride, and lithium aluminum hydride–aluminum chloride were singularly unsuccessful.

1. Department of Chemistry, Drexel University, Philadelphia, Pennsylvania, 19104.
2. P. Boldt and W. Thielecke, *Angew. Chem. Int. Ed. Engl.*, **5**, 1044 (1966); P. Boldt and L. Schulz, Ger. Pat. 1,200,298 (1965) [*Chem. Abstr.*, **63**, 16221e (1965)]; P. Boldt and L. Schulz, *Naturwissenschaften*, **51**, 288 (1964).
3. H. C. Brown and B. C. Subba Rao, *J. Am. Chem. Soc.*, **82**, 681 (1960); and more recently, H. C. Brown, P. Heim, and N. M. Yoon, *J. Am. Chem. Soc.*, **92**, 1637 (1970).

PHOTOCHEMICAL RING CONTRACTION OF
2-ETHOXYPYRROLIN-5-ONES TO
CYCLOPROPANONE DERIVATIVES:
tert-BUTYL *N*-(1-ETHOXYCYCLOPROPYL)CARBAMATE

[Carbamic acid, (1-ethoxycyclopropyl)-, 1,1-dimethylethyl ester]

Submitted by GEOFFREY C. CROCKETT and TAD H. KOCH[1]
Checked by S. BOETTGER and M. F. SEMMELHACK

1. Procedure

Caution! The photochemical reaction in Part C should be carried out behind a light-absorbent cover or shield. Protective goggles should be worn to avoid exposure of the eyes to ultraviolet light.

A. *Succinimide silver salt (Note 1).* A 3-l., two-necked, round-bottomed flask equipped with a mechanical stirrer and a pressure-equalizing dropping funnel is charged with a solution of 28.9 g. (0.292 mole) of succinimide (Note 2) in 1.2 l. of absolute ethanol (Note 3). A solution of 48.58 g. (0.2860 mole) of silver nitrate in 200 ml. of dimethyl sulfoxide (Note 4) is added in one portion. The resulting solution is stirred as 700 ml. (0.280 mole) of 0.400 *M* sodium ethoxide in ethanol (Notes 5 and 6) is added dropwise over 1.5 hours. The off-white silver salt begins to precipitate after *ca.* 140 ml. of the sodium ethoxide solution has been added. Stirring is continued for 45 minutes after the addition is completed, and the reaction mixture is then stored in a refrigerator at *ca.* 5° overnight to complete the precipitation and aggregation of the product.

The precipitate is collected on filter paper (Note 7) in a Büchner funnel by vacuum filtration and washed with 100 ml. of absolute ethanol. The solid is slurried in three 75-ml. portions of distilled water (Note 8), 100 ml. of absolute ethanol, two 100-ml. portions of reagent grade acetone, and two 100-ml. portions of anhydrous ethyl ether. The filter cake is pressed dry in the funnel with suction using a piece of rubber dam, transferred to a tared, 500-ml., round-bottomed flask, and dried under reduced pressure (0.01 mm.) at room temperature for 24 hours (Note 9), yielding 51–54 g. (88–94%) of the silver salt of succinimide.

B.   2-*Ethoxypyrrolin-5-one*.   The flask containing 51–54 g. (0.25–0.26 mole) of succinimide silver salt is equipped with a magnetic stirring bar (Note 10), a heating mantle, and reflux condenser bearing a silica gel drying tube. The solid is suspended in 295 ml. of dry chloroform (Note 11), and 51.4 g. (26.4 ml., 0.330 mole) of ethyl iodide is added in one portion. The flask is covered with aluminum foil, and the mixture is stirred vigorously and heated under reflux for 48 hours. The mixture is cooled, the silver iodide is removed by vacuum filtration through Celite, and the filter cake is washed well with dry chloroform. The filtrate is concentrated by rotary evaporation to a mixture of a dark liquid and a white solid identified as succinimide. Anhydrous diethyl ether is added to dissolve the liquid, and the resulting suspension is filtered through a plug of glass wool, separating 7.7–11.5 g. of succinimide. The ether is removed from the filtrate by rotary evaporation at aspirator pressure, and the residual liquid is distilled under reduced pressure with a short-path distillation apparatus. The product is collected at 74–82° (0.01 mm.) as a faintly yellow oil which crystallizes in the freezer. The yield is 11.5–16.7 g. (32–46% based on sodium ethoxide in Part A) (Notes 12 and 13).

C.   tert-*Butyl* N-(1-*ethoxycyclopropyl*)*carbamate*.   A three-necked, cylindrical irradiation vessel is equipped with a magnetic stirring bar, a water-jacketed quartz immersion well, an inert gas-inlet, and a gas-exit tube connected to a bubbler (Note 14). The vessel is charged with 6.26 g. (0.0493 mole) of redistilled 2-ethoxypyrrolin-5-one (Note 20) and 180 ml. of dry *tert*-butyl alcohol (Note 15). The solution is stirred and degassed by bubbling nitrogen or argon through the gas-inlet tube for 15 minutes. The degassed solution is stirred and irradiated with ultraviolet light from a 450-watt, Hanovia, medium-pressure, mercury lamp filtered through a Vycor glass sleeve. During the irradiation an atmosphere of nitrogen or argon is maintained, and the lamp is cooled with warm water (35–40°) circulated through the cooling jacket of the immersion well. The progress of the irradiation is monitored by GC (Note 16). When 90% of the 2-ethoxypyrrolin-5-one has reacted, the irradiation is stopped. The solution (Note 17) is transferred to a 250-ml., round-bottomed flask equipped with a magnetic stirring bar and an air-cooled reflux condenser mounted with a T-shaped nitrogen inlet. Nitrogen is passed through the apparatus for 30 minutes, after which the solution is stirred and heated at reflux under a nitrogen atmosphere for 20 hours (Note 18). The solvent is removed by rotary evaporation, and the residual orange oil is refrigerated to induce crystallization. Sublimation of the solid at 35–40° (0.05 mm.) affords 5.5–6.3 g. (56–64%) of the carbamate as white needles, m.p. 38–40° (Notes 19 and 20).

2. Notes

1. In parts A and B care should be taken to minimize the exposure of silver-containing reactants and products to light.

2. Succinimide purchased from MC and B Manufacturing Chemists was used without purification.

3. Absolute ethanol from a commercial supplier was used.

4. A mixture of silver nitrate and dimethyl sulfoxide was stirred vigorously for *ca*. 1 hour to dissolve all of the salt. Reagent grade dimethyl sulfoxide was used without purification.

5. The sodium ethoxide solution was prepared from the reaction of 9.2 g. (0.40 mole) of sodium with 1 l. of absolute ethanol and is standardized by titration with aqueous 0.1 *N* hydrochloric acid. The appropriate volume of the solution to give 0.280 mole of base was used.

6. Slightly less than equivalent amounts of both silver nitrate and sodium ethoxide are used to minimize the formation of silver oxide which imparts a brown color to the product.

7. The submitters state that the use of a sintered-glass funnel may cause discoloration of the product. However, the checkers used a sintered glass funnel in one run with no adverse effect on the yield or purity of the product.

8. A considerable amount of sodium nitrate is present in the precipitate. Although the presence of sodium nitrate did not hinder small-scale alkylation reactions (*ca*. 250 mg.), the submitters recommend that it be removed in larger runs to facilitate the isolation and drying of the silver salt.

These washings are most easily done without removing the material from the filter. However, the solid must be slurried thoroughly in each portion of solvent, particularly with acetone and ether. Care must be taken to ensure that the filter paper is not lifted from the bottom of the funnel. The submitters accomplished this by holding the filter paper in the funnel with a ring of flexible polyvinyl chloride (inside diameter 0.64 cm.), the ends of which were joined by a small piece of rigid polyethylene tubing. The ring was expanded to fit snugly in the bottom of the funnel over the paper.

9. Silver salts may be unstable when heated. An explosion occurred while the silver salt of isatin was drying under reduced pressure at *ca*. 100°.

10. The checkers found that the heavy suspension could not be stirred effectively with a magnetic stirring bar and recommend that a mechanical stirrer be used.

11. Reagent grade chloroform was dried by filtering through alumina (50 g. per l. of solvent).

12. The checkers' data are given. The submitters recovered 4.7–11.4 g. of succinimide and collected 15.6–21.2 g. (44–60% based on sodium ethoxide in Part A) of product, b.p. 65–70° (0.05 mm.). Based on the amount of unrecovered succinimide, the yield of product obtained by the checkers and submitters was 49–61% and 68–69%, respectively. The product is best stored in a freezer. If sufficient care is taken to exclude moisture, 2-ethoxypyrrolin-5-one is stable indefinitely.

13. The product obtained by the submitters was contaminated with impurities amounting to *ca*. 10% which were primarily succinimide and N-ethylsuccinimide. Although this material was considered to be of satisfactory purity for use in Part C, further purification

can be accomplished, if desired, by redistillation, giving product boiling at 72–74° (0.05 mm.). The checkers found it necessary to redistil the 2-ethoxypyrrolin-5-one to obtain product with the reported melting point in Part C (Note 20); IR (neat) cm.$^{-1}$: 2940, 1748, 1562; ^{1}H NMR ($CDCl_3$), δ (multiplicity, coupling constant J in Hz., number of protons, assignment): 1.4 (t, J = 7, 3H, OCH_2CH_3), 2.4–3.0 (m, 4H, CH_2CH_2), 4.45 (q, J = 7, 2H, OCH_2CH_3); UV (cylcohexane) nm. max. (ϵ): 273 (55).

14. The irradiation apparatus was similar to one depicted in the procedure for bicyclo[2.1.0]pent-2-ene in *Org. Synth.*, **Coll. Vol. 6,** 145 (1988), Figure 2, Section A. The height and inside diameter of the irradiation vessel used by the submitters were approximately 35 cm. and 6.2 cm., respectively. Two short necks with ₸ 14/20 outer joints were located on the shoulder of the vessel just below the ₸ 60/50 center joint. One neck was capped with a rubber septum and the other was connected to the exit bubbler. The nitrogen inlet was a syringe needle passing through the septum and connected to a section of Teflon tubing that extended to the bottom of the vessel. The checkers used a similar 23 × 7.5 cm. irradiation vessel that had a fritted-glass inlet for argon situated at the base as shown in the figure referred to above. The solution was agitated during the irradiation by a continual flow of argon rather than by magnetic stirring.

The apparatus is dried in an oven at 140° overnight and cooled under nitrogen or argon prior to the irradiation. A Vycor filter sleeve and a 450-watt, medium-pressure mercury lamp are placed in the immersion well. The Vycor filter, the quartz immersion well (catalog No. 19434), the 450-watt mercury lamp (catalog No. 679A36), and the requisite transformer are all available from Hanovia Lamp Division, Canrad-Hanovia Inc., 100 Chestnut Street, Newark, New Jersey 07105.

15. Reagent grade *tert*-butyl alcohol was distilled from calcium hydride prior to use. The scale described is that used by the checkers. The submitters irradiated 4.0 g. (0.032 mole) of 2-ethoxypyrrolin-5-one in 115 ml. of dry *tert*-butyl alcohol.

16. The submitters used a 2.1 m. × 0.64 cm. column with 5% fluorosilicone (FS-1265) supported on Diatoport S as stationary phase. With a column temperature of 170° and a helium flow rate of 60 ml. per minute, 2-ethoxypyrrolin-5-one has a retention time of 2.2 minutes. The analysis was carried out at 160° by the checkers, using a column of 5% diethylene glycol succinate–Bentone^{34} supported on Diatoport S. The starting material had a retention time of 3.9 minutes under these conditions. Bentone^{34} is available from Applied Sciences Laboratory, Box 440, State College, Pennsylvania 16801.

17. At this point the product consisted mostly of the isocyanate, since the reaction with *tert*-butyl alcohol is relatively slow at 35–40°. If the photolysis is carried out in an aprotic solvent such as tetrahydrofuran, the isocyanate may be isolated.[2] However, care must be exercised to avoid losses of this rather volatile and moisture-sensitive compound.

18. The isocyanate is completely consumed at this time, as evidenced by the disappearance of the absorption band at 2250 cm.$^{-1}$ in the IR spectrum.

19. Alternatively, the product may be distilled at 40° (0.05 mm.). However, the distillate tends to crystallize in the condenser and plug the apparatus.

20. The checkers found that product with the reported melting point was obtained only when the starting 2-ethoxypyrrolin-5-one was redistilled carefully and was largely free of the *N*-ethyl isomer and succinimide. With 2-ethoxypyrrolin-5-one purified by a single distillation, the product was obtained as a gummy solid that was difficult to purify. Nevertheless, the IR and ^{1}H NMR spectra of this material were essentially identical to

TABLE I
PREPARATION AND IRRADIATION OF 2-ETHOXYPYRROLIN-5-ONES

2-Ethoxy-pyrrolin-5-one	Yield (%)	Photoproduct or Its Derivative	Yield (%)
(bicyclic imide structure) OC_2H_5	46^a	(cyclobutane) $NHCN(CH_3)_2$, OC_2H_5	43^b
(cyclopentane-fused imide) OC_2H_5	59	(bicyclic) $N=C=O$, OC_2H_5	76^b
(cyclohexane-fused imide) OC_2H_5	89	(bicyclic) $N=C=O$, OC_2H_5	64^b
(cyclohexene-fused imide) OC_2H_5	74	(bicyclic) $N=C=O$, OC_2H_5	53^b
(dimethyl pyrrolinone, cis) OC_2H_5	88^c	(cyclopropane) $NHCOCH_3$, OC_2H_5	58^d
(dimethyl pyrrolinone, trans) OC_2H_5		(cyclopropane) $NHCOCH_3$, OC_2H_5	60

[a] The yield was 84% based on unrecovered imide.
[b] The product is a mixture of *endo*- and *exo*- isomers.
[c] The isomers were separated by preparative GC.
[d] The product is a mixture of *cis*- and *trans*- isomers.

those of pure *tert*-butyl *N*-(1-ethoxycyclopropyl)carbamate. The submitters obtained 4.4–4.8 g. (70–76%) of carbamate, m.p. 40–42°, from 4.0 g. of 2-ethoxypyrrolin-5-one. The product has the following spectral characteristics: IR (neat) cm.$^{-1}$: 3333, 2940, 1754 (C=O); ^{1}H NMR (CDCl$_3$), δ (multiplicity, coupling constant *J* in Hz., number of protons, assignment): 0.80–1.15 (m, 4H, cyclopropyl *H*), 1.13 (t, *J* = 7, 3H, OCH$_2$C*H*$_3$), 1.47 [s, 9H, C(C*H*$_3$)$_3$], 3.68 (q, *J* = 7, 2H, OC*H*$_2$CH$_3$), 5.75 (broad, 1H, N*H*).

3. Discussion

The procedure described here for the preparation of succinimide silver salt is a modification of one reported for the formation of the silver derivative of maleimide.[3] The alkylation step is modeled after the procedure of Comstock and Wheeler,[4] who prepared 2-ethoxypyrrolin-5-one in unspecified yield, and is an improvement over a later procedure developed in the laboratories of the submitters.[2] The general scheme has been successfully applied to the preparation of a variety of 2-ethoxypyrrolin-5-ones (Table I)[5–7] as well as 6-ethoxy- and 6-propoxy-4,5-dihydro-2(3*H*)-pyridone from the corresponding five- and six-membered cyclic imides.[2]

The photochemical rearrangement of substituted 2-ethoxypyrrolin-5-ones is a general reaction of synthetic utility and high stereoselectivity, which affords the corresponding 1-ethoxycyclopropyl isocyanates and their derivatives in useful yields (Table I).[6,7] The procedure reported here is the only known preparation of *tert*-butyl *N*-(1-ethoxycyclopropyl)carbamate, a precursor of 1-aminocyclopropanol and 1-ethoxycyclopropylamine.[8] 1-Aminocyclopropanol has previously been prepared in low yield by the addition of ammonia to cyclopropanone.[8] The photorearrangement of 2-ethoxypyrrolin-5-one to *tert*-butyl *N*-(1-ethoxycyclopropyl)carbamate followed by hydrolysis to 1-aminocyclopropanol is a key step in the synthesis of the alkaloid coprine.[8] Cyclopropanone derivatives have been used as precursors for a variety of compounds[9] such as β-lactams,[10] cyclobutanones,[11] and cyclopropanols.[12]

2-Ethoxypyrrolin-5-one reacts with secondary amines, giving 2-aminopyrrolin-5-ones which photochemically rearrange to aminocyclopropyl isocyanates in 80–90% yield.[13] Furthermore, 1,4-bis[(pyrrolin-3-onyl)methylamino]-2-butyne photorearranges to 1,4-bis[(isocyanatocyclopropyl)methylamino]-2-butyne. This reaction is of potential use for photocrosslinking polyurethanes and polyureas.

R = Me

1. Chemistry Department, University of Colorado, Boulder, Colorado 80309.
2. T. H. Koch, R. J. Sluski, and R. H. Moseley, *J. Am. Chem. Soc.*, **95**, 3957 (1973).

3. A. L. Schwartz and L. M. Lerner, *J. Org. Chem.*, **39**, 21 (1974). For an alternative procedure for the preparation of succinimide silver salt, see W. R. Benson, E. T. McBee, and L. Rand, *Org. Synth.*, **Coll. Vol. 5**, 664 (1973).
4. W. J. Comstock and H. L. Wheeler, *Am. Chem. J.*, **13**, 522 (1891) [*Beilstein*, 4th ed., **21**, 575 (1935)].
5. K. A. Howard and T. H. Koch, *J. Am. Chem. Soc.*, **97**, 7288 (1975).
6. G. C. Crockett and T. H. Koch, *J. Org. Chem.*, **42**, 2721 (1977).
7. J. M. Burns, M. Ashley, G. C. Crockett, and T. H. Koch, *J. Am. Chem. Soc.*, **99**, 6924 (1977).
8. P. Lindberg, R. Bergman, and B. Wickberg, *J. Chem. Soc. Chem. Commun.*, 946 (1975).
9. For a discussion of cyclopropanone chemistry see: H. H. Wasserman, G. M. Clark, and P. C. Turley, *Top. Curr. Chem.*, **47**, 73 (1974): W. J. M. van Tilborg, H. Steinberg, and T. H. deBoer, *Recl. Trav. Chim. Pays-Bas*, **93**, 294 (1974); W. J. M. van Tilborg, Ph.D. Thesis, University of Amsterdam, 1971; N. J. Turro, *Acc. Chem. Res.*, **2**, 25 (1969).
10. H. H. Wasserman, H. W. Adickes, and O. E. de Ochoa, *J. Am. Chem. Soc.*, **93**, 5586 (1971); H. H. Wasserman and M. S. Baird, *Tetrahedron Lett.*, 3721 (1971); H. H. Wasserman, E. A. Glazer, and M. J. Hearn, *Tetrahedron Lett.*, 4855 (1973).
11. H. H. Wasserman, M. J. Hearn, B. Haveaux, and M. Thyes, *J. Org. Chem.*, **41**, 153 (1976).
12. J. Szmuszkovicz, D. J. Duchamp, E. Cerda, and C. G. Chidester, *Tetrahedron Lett.*, 1309 (1969); H. H. Wasserman, R. E. Cochoy, and M. S. Baird, *J. Am. Chem. Soc.*, **91**, 2375 (1969).
13. B. J. Swanson, G. C. Crockett, and T. H. Koch, *J. Org. Chem.*, **46**, 1082 (1981).

PHASE-TRANSFER
HOFMANN CARBYLAMINE REACTION: *tert*-BUTYL ISOCYANIDE

(Propane, 2-isocyano-2-methyl-)

$$(CH_3)_3CNH_2 + CHCl_3 + 3\ NaOH \xrightarrow[H_2O,\ \text{dichloromethane, reflux}]{[(C_2H_5)_3NCH_2C_6H_5]^+Cl^-}$$

$$(CH_3)_3CN=C + 3\ NaCl$$

Submitted by GEORGE W. GOKEL,[1,2] RONALD P. WIDERA, and WILLIAM P. WEBER[1]

Checked by F. A. SOUTO-BACHILLER, S. MASAMUNE, CHARLES J. TALKOWSKI, and WILLIAM A. SHEPPARD

1. Procedure

Caution! This preparation should be conducted in an efficient hood because of the evolution of carbon monoxide and the obnoxious odor of the isocyanide.[3]

A 2-l., round-bottomed flask equipped with a magnetic stirring bar, a reflux condenser, and a pressure-equalizing dropping funnel is charged with 300 ml. of water. Stirring is begun and 300 g. (7.50 moles) of sodium hydroxide are added in portions in order to maintain efficient stirring (Note 1). A mixture of 141.5 g. (203.3 ml., 1.938 moles) of *tert*-butylamine, 117.5 g. (78.86 ml., 0.9833 mole) of chloroform (Note 2), and

2 g. (0.009 mole) of benzyltriethylammonium chloride (Note 3) in 300 ml. of dichloromethane is added dropwise to the stirred, warm (*ca.* 45°) solution over a 30-minute period. The reaction mixture begins to reflux immediately (Note 4) and subsides within 2 hours; stirring is continued for an additional hour (Note 5). After the reaction mixture is diluted with 800 ml. of ice and water, the organic layer is separated and retained, and the aqueous layer is extracted with 100 ml. of dichloromethane. The dichloromethane solutions are combined and washed successively with 100 ml. of water and 100 ml. of aqueous 5% sodium chloride, and dried over anhydrous magnesium sulfate.

The drying agent is removed by filtration, and the filtrate is distilled under nitrogen through a spinning band column (Notes 6 and 7). The fraction, boiling at 92–93° (725 mm.), is collected, yielding 54.2–60.0 g. (66–73%, based on chloroform) of *tert*-butyl isocyanide (Notes 8 and 9).

2. Notes

1. Efficient stirring is required. A solution of 225 g. (5.62 moles) of sodium hydroxide in 225 ml. of water can be added to the stirred mixture of the organic substrates in dichloromethane if a more efficient mechanical stirrer is used. In the original procedure, the submitters noted an induction period of about 20 minutes which was found to vary somewhat with the stirring rate, stirring-bar size, and relative amount of phase-transfer catalyst. Three moles of base are required for the reaction: one to generate the carbene and two to react with the additional two moles of hydrochloric acid lost by the amine–carbene adduct in the isonitrile formation step. If less base is used, the excess hydrochloric acid reacts with the isonitrile by α-addition, and the yield is substantially reduced.

2. Chloroform, commercially available, normally contains 0.75% ethanol and was used as supplied.

3. Benzyltriethylammonium chloride is available from Eastman Organic Chemicals. The checkers prepared the salt in a state of high purity by a modification of a reported procedure.[4] A solution of 33.7 g. (0.334 mole) of triethylamine and 50.0 g. (0.395 mole) of benzyl chloride (both from Eastman Organic Chemicals) in 60 ml. of absolute ethanol was refluxed for 64 hours, cooled to room temperature, and treated with 300 ml. of ether. The precipitated ammonium salt was removed by filtration, redissolved in the minimum amount of hot acetone, and reprecipitated with ether.

4. The volatilities of both *tert*-butylamine and dichloromethane necessitate the use of an efficient condenser as a precaution, although the rate of reflux is generally not vigorous. In preparations where higher boiling amines are used, this precaution is less critical.

5. The submitters noted that a longer stirring period did not seem to affect the yield appreciably.

6. The bulk of the residual *tert*-butylamine is recovered.

7. A 60-cm., annular, Teflon®, spinning band, distillation column is recommended to achieve clean separation of solvent and unreacted reagent from product rather than a column packed with glass helices. For higher boiling isocyanides, separation of solvent and unreacted reagents may be effected by the use of a rotary evaporator, although the thermal instability of the isocyanides should be taken into consideration.

8. Yields higher than about 70% for any of these isonitrile preparations generally

TABLE I
Preparation of Isocyanides (RN=C) by the Carbylamine Reaction[a]

$$\text{CHCl}_3 + \text{NaOH} \xrightarrow{\text{PTC}} [:\text{CCl}_2] \xrightarrow{\text{RNH}_2} \text{RN}{=}\text{C}$$

R	Yield (%)	b.p. of RN=C
$CH_3(CH_2)_2CH_2-$	60	40–42° (11 mm.)
$C_6H_5CH_2-$	45	92–93° (11 mm.)
$CH_3(CH_2)_{10}CH_2-$	41	115–118° (0.1 mm.)
cyclo-$C_6H_{11}-$	48	67–72° (13 mm.)
C_6H_5-	57	50–52° (11 mm.)
CH_3-[b]	24	59–60° (760 mm.)
CH_3CH_2-[b]	47	78–79° (760 mm.)

[a]Prepared by the phase-transfer method using chloroform and aqueous sodium hydroxide with the corresponding amines.[7]
[b]Bromoform substituted for chloroform for ease of fractionation.[8]

indicate incomplete fractionation. The purity of the product may be conveniently checked by ^{1}H NMR (CDCl_3) spectroscopy. The characteristic 1:1:1 triplet for β-hydrogen: [14]N coupling in *tert*-butyl isocyanide appears at δ 1.45. A small upfield peak usually indicates the presence of unreacted amine. Other common contaminants are dichloromethane and chloroform. The purity may be determined more accurately by GC analysis on a 230 cm. by 0.6 cm. column packed with 10% SE-30 on Chromosorb G, 60–80 mesh, at 80°.

9. Glassware may be freed from the isocyanide odor by rinsing with a 1:10 mixture of concentrated hydrochloric acid and methanol.

3. Discussion

The present method utilizes dichlorocarbene generated by the phase-transfer method of Makosza[5] and Starks.[6] The submitters have routinely realized yields of pure distilled isocyanides in excess of 40%.[7] With less sterically hindered primary amines a 1:1 ratio of amine to chloroform gives satisfactory results. Furthermore, by modifying the procedure, methyl and ethyl isocyanides may be prepared directly from the corresponding aqueous amine solutions and bromoform.[8] These results are summarized in Table I.

Various synthetic routes to isocyanides have been reported since their identification over 100 years ago.[9] Until now,[9,11] the useful synthetic procedures required a dehydration reaction.[10-12] Although the carbylamine reaction involving the dichlorocarbene intermediate is one of the early methods,[9] it had not been preparatively useful until the innovation of phase-transfer catalysis (PTC).[5,6]

The phase-transfer catalysis method has also been utilized effectively for addition of dichlorocarbene to olefins,[5] as well as for substitution and elimination reactions, oxidations, and reductions.[13] A later procedure in this volume is another example.[14]

1. Department of Chemistry, University of Southern California, University Park, Los Angeles, California 90007. This work was supported in part by a grant from the National Science Foundation, grant number GP 40331X.

2. Present address: Department of Chemistry, University of Maryland, College Park, Maryland 20742.
3. Many isocyanides are reported to exhibit no appreciable toxicity to mammals. See J. A. Green II and P. T. Hoffmann in "Isonitrile Chemistry," I. Ugi, Ed., Academic Press, New York, 1971, p. 2. However, since certain isocyanides are highly toxic (*e.g.*, 1,4-diisocyanobutane), the checkers recommend that all isocyanides be handled with due caution.
4. R. A. Moss and W. L. Sunshine, *J. Org. Chem.*, **35**, 3581 (1970).
5. M. Makosza and M. Wawrzyniewicz, *Tetrahedron Lett.*, 4659 (1969).
6. C. M. Starks, *J. Am. Chem. Soc.*, **93**, 195 (1971).
7. W. P. Weber and G. W. Gokel, *Tetrahedron Lett.*, 1637 (1972).
8. W. P. Weber, G. W. Gokel, and I. Ugi, *Angew. Chem. Int. Ed. Engl.*, **11**, 530 (1972).
9. For a review, see P. T. Hoffmann, G. Gokel, D. Marquarding, and I. Ugi in "Isonitrile Chemistry," I. Ugi, Ed., Academic Press, New York, 1971, Chapter II.
10. I. Ugi, R. Meyr, M. Lipinski, F. Bodesheim, and F. Rosendahl, *Org. Synth.*, **Coll. Vol. 5**, 300 (1973); I. Ugi and R. Meyr, *Org. Synth.*, **Coll. Vol. 5**, 1060 (1973).
11. G. E. Niznik, W. H. Morrison III, and H. M. Walborsky, *Org. Synth.*, **Coll. Vol. 6**, 751 (1988).
12. R. E. Schuster, J. E. Scott, and J. Casanova, Jr., *Org. Synth.*, **Coll. Vol. 5**, 772 (1973).
13. J. Dockx, *Synthesis*, 441 (1973); E. V. Dehmlow, *Angew. Chem. Int. Ed. Engl.*, **13**, 170 (1974); A. W. Herriott and D. Picker, *Tetrahedron Lett.*, 1511 (1974).
14. M. Makosza and A. Jónczyk, *Org. Synth.*, **Coll. Vol. 6**, 897 (1988).

SULFIDE SYNTHESIS IN PREPARATION OF UNSYMMETRICAL DIALKYL DISULFIDES: *sec*-BUTYL ISOPROPYL DISULFIDE

(Disulfide, 1-methylethyl 1-methylpropyl)

A. $(CH_3)_2CHBr + Na_2S_2O_3 \cdot 5H_2O \xrightarrow[\text{reflux}]{CH_3OH-H_2O} (CH_3)_2CH-S-SO_3Na + NaBr$

1

B. $CH_3CH_2CH(CH_3)-SH + NaOH \xrightarrow[25°]{H_2O} CH_3CH_2CH(CH_3)-SNa$

2

C. $1 + 2 \xrightarrow[0-5°]{H_2O} (CH_3)_2CH-S-S-CH(CH_3)CH_2CH_3$

3

Submitted by M. E. ALONSO[1] and H. ARAGONA
Checked by LINDA D.-L. LU and S. MASAMUNE

1. Procedure

Caution! This procedure should be carried out in an efficient hood to prevent exposure to alkane thiols.

A. *Sodium isopropyl thiosulfate* (1). A 5-l., three-necked, round-bottomed flask equipped with a condenser, a 300-ml. dropping funnel, and a mechanical stirrer is charged with 123 g. (1.00 mole) of freshly distilled 2-bromopropane (Note 1) in 1.2 l. of methanol. Water is added slowly with stirring until a slight turbidity develops (Note 2). The stirred mixture is heated to reflux, and 310 g. (1.25 mole) of sodium thiosulfate pentahydrate (Note 3) in 250 ml. of water is added over a period of 30 minutes. The slightly yellow-tinted solution is heated for an additional 3.5–4.0 hours and allowed to cool to room temperature. The alcohol is removed with a rotary evaporator; the remaining milky solution is diluted with water to a total volume of about 1.2 l and extracted twice with hexane (Note 4). Discarding the organic layers, the aqueous solution of crude thiosulfate is cooled to 0° and stored.

B. *Sodium 2-butanethiolate* (2). A 500-ml., three-necked, round-bottomed flask equipped with a dropping funnel, a mechanical stirrer, and a gas-inlet is charged with 40 g. (1.0 mole) of sodium hydroxide in 100 ml. of water. Under an atmosphere of argon (Note 5), 90 g. (1.0 mole) of 2-butanethiol (Note 6) is added dropwise over a 2-hour period, with rapid stirring at room temperature (Note 7). Thiolate solution 2 becomes very viscous toward the end of the addition; it is diluted with 30 ml. of water and cooled to 0°.

C. sec-*Butyl isopropyl disulfide* (3). A 3-l., three-necked, round-bottomed flask equipped with a dropping funnel, a thermometer, and a mechanical stirrer is charged with the crude thiosulfate solution 1 and cooled to 0° with the aid of an ice–salt bath. The cold thiolate solution 2 is added rapidly, with vigorous stirring for 3 minutes, followed by 200 ml. of aqueous saturated sodium chloride (Note 8), and the mixture is warmed to 5°. Stirring is stopped after 10 minutes, counted from the start of the addition of the aqueous sodium chloride. The crude disulfide 3, which separates as an oil, is removed, and the aqueous layer is extracted twice with 250-ml. portions of diethyl ether. The extracts are combined with the oil, washed twice with 150-ml. portions of water, dried briefly over granular calcium sulfate, and filtered through a glass-wool plug. Removal of the solvent leaves 125–133 g. of crude disulfide 3 (Note 9). Distillation of this material at 44.5–45° (1.25 mm.) gives 118–123 g. (73–75%) of pure disulfide 3 (Note 10).

2. Notes

1. The submitters used 2-bromopropane available from Aldrich Chemical Company, Inc. The checkers purchased the reagent from J. T. Baker Chemical Company.
2. The turbidity indicates saturation of alkyl halide. In this way both sodium thiosulfate and 2-bromopropane are nearly in a one-phase system, thus shortening significantly the heating period. Furthermore, competing hydrogen bromide elimination and acid-promoted decomposition of thiosulfate into sulfur and sulfur dioxide are minimized. The checkers added 300 ml. of water over a period of 90 minutes.

3. The submitters used sodium thiosulfate pentahydrate supplied by E. Merck A G, Darmstadt, and the checkers used A.C.S. reagent grade material available from Fisher Scientific Company.

4. This extraction is intended to remove traces of unreacted alkyl halide that might compete for the thiolate in the nucleophilic substitution (Step B).

5. The submitters performed this step in air, but the checkers found that use of an inert atmosphere resulted in a somewhat improved yield.

6. This reagent was purchased from Aldrich Chemical Company, Inc.

7. The submitters observed the separation of a pasty solid at this stage, added four 10-ml. portions of water during the addition of the thiol, and dissolved the entire solid with approximately 240 ml. of water. Sometimes, at this point, as much as 10% of the added thiol separated out as a floating oil. The presence of this thiol affects the course of the reaction, yielding symmetrical disulfides.[2] In this case the organic layer should be separated, added dropwise to an equivalent amount of sodium hydroxide dissolved in a minimum amount of water, and mixed with the original thiolate solution.

8. The submitters found that the addition of sodium chloride facilitated the separation of the insoluble disulfide.

9. This material contained no less than 90% of disulfide 3 according to GC analysis (1.5-m. 5% SE-30 column).

10. The distilled disulfide 3 has the following ^{1}H NMR spectrum (CDCl$_3$), δ (multiplicity, coupling constant J in Hz., number of protons, assignment): 1.00 (t, $J = 7.0$, 3H, CH_3CH$_2$), 1.31 (d, $J = 7.0$, 9H, (CH_3)$_2$CH and CH_3CH), 1.62 (m, 2H, CH_2), 2.86 (sextuplet, $J = 7.0$, 1H, CH$_3$CHCH$_2$), 3.02 (septuplet, $J = 7.0$, 1H, CH$_3$CHCH$_3$).

3. Discussion

Unsymmetrical, dialkyl disulfides[3] can be prepared by several methods; three procedures appear to be generally applicable. First, the reaction of an N-(alkylthio)- or N-(arylthio)phthalimide with thiols[4] gives unsymmetrical disulfides in good yield; however, the synthesis of the thiophthalimide[5] requires the corresponding sulfenyl chloride, which is rather unstable and undergoes undesirable side reactions when α-protons are available.[6] Second, the adduct of a thiol and diethyl azodicarboxylate reacts with a thiol, giving unsymmetrical disulfides in high yield.[7] The adduct formation, however, is severely suppressed by steric hindrance in the alkyl portion of the thiol; secondary and tertiary thiols are normally unreactive.[8a] Third, the reaction of sodium alkylthiosulfates[9] with thiolates provides unhindered, mixed disulfides in low to moderate yields,[10] and hindered compounds[8] in yields of 6–10%. A general and satisfactory synthetic procedure for hindered, unsymmetrical disulfides was not available at the time the present study was undertaken,[5b] the probable reason being that the known methods utilize a sterically sensitive bimolecular attack by a nucleophilic form of sulfur onto a sulfur atom bearing a suitable leaving group. Forcing conditions usually lead to disproportionation[11] and polysulfide formation.[10] The availability and low cost of starting materials and the expeditious process involved in the "Bunte Salt" approach[9] provided a reasonable basis for modifying the existing procedure,[10] extending its applicability to bulky, unsymmetrical disulfides. Table I shows boiling points and distillate composition of a number of mixed disulfides

TABLE I

Unsymmetrical Dialkyl Disulfides Prepared by the Thiosulfate Procedure. Distillation Conditions and Composition[a,b]

R	R'	bp, °C/torr	yield (%)	RSSR' (%)	RSSR (%)	R'SSR' (%)
$(CH_3)_2CHCH_2-$	$(CH_3)_2CH-$	33–34/0.15	60	92	6	2
$(CH_3)_2CHCH_2-$	CH_3CH_2-	51/3	62	98		1
$(CH_3)_2CHCH_2-$	CH_3-	52/8.5	65	97		1.5
$(CH_3)_2CHCH_2-$	$CH_2=CHCH_2-$	57.5/1.5	81	96	2	1.5
$(CH_3)_2CH-$	CH_3CH_2-	49/9	60	98	<1	c
$(CH_3)_2CH-$	CH_3-	37–38/11	72	98.5	<1	
$(CH_3)_2CH-$	$CH_2=CHCH_2-$	55.5/5	69	98.5		<1
$C_2H_5CH(CH_3)-$	$(CH_3)_2CH-$	40–41/0.2	73	96.5	3	c
$C_2H_5CH(CH_3)-$	CH_2CH_2-	47/1.5	64	98.5	<1	<1
$C_2H_5CH(CH_3)-$	CH_3-	46.5/6	71	99	<1	
$C_2H_5CH(CH_3)-$	$CH_2=CHCH_2-$	51/1	71	98.5	<1	
$(CH_3)_3C-$	$(CH_3)_2CH$	30–32/2	53	96.5	1	<1
$(CH_3)_3C-$	CH_3CH_2-	43–44/5	52	97	<1	2
$(CH_3)_3C-$	CH_3-	46/12	49	98.5	<1	<1
$(CH_3)_3C-$	$CH_2=CHCH_2-$	53/2	54	99	<1	<1

[a]Determined by GC analysis on a 12-ft 5% SE-30 on Chromosorb G column.
[b]Satisfactory analytical data (±0.52% for C, H, S) were obtained for all compounds.
[c]Could not be separated by GC.

238

prepared in up to 80% yield by the method presented here;[12] base-catalyzed disproportionation[11] and polymerization[10] appear to be minimized.

1. Centro de Química, Instituto Venezolano de Investigaciones Científicas, IVIC, Apartado 1827, Caracas 1010 A, Venezuela.
2. A. J. Parker and N. Kharasch, *Chem. Rev.*, **59**, 583 (1959).
3. A Schöberl and A. Wagner, in E. Müller, Ed., "Methoden der Organischen Chemie" (Houben-Weyl), 4th ed., Vol. 9, Georg Thieme Verlag, Stuttgart, 1955, pp. 72–73; C. G. Moore and M. Porter, *J. Chem. Soc.*, 2890 (1958); L. Field, T. C. Owen, R. R. Crenshaw, and A. W. Bryan, *J. Am. Chem. Soc.*, **83**, 4414 (1961); T. F. Parsons, J. D. Buckman, D. E. Pearson, and L. Field, *J. Org. Chem.*, **30**, 1923 (1965), and references cited therein; V. I. Dronov and N. V. Pokaneshchikova, *Zh. Org. Khim.*, **6**, 2225 (1970); R. H. Cragg, J. P. N. Husband, and A. F. Weston, *J. Chem. Soc. D*, 1701 (1970); L. Field, *Synthesis*, 101 (1972); D. A. Armitage, M. J. Clark, and C. C. Tso, *J. Chem. Soc., Perkin Trans. 1*, 680 (1972); J. Meijer and P. Vermees, *Recl. Trav. Chim. Pays-Bas*, **93**, 242 (1974); O. Abe, M. F. Lukacovic, and C. Ressler, *J. Org. Chem.*, **39**, 253 (1974); P. Dubs and R. Stüssi, *Helv. Chim. Acta*, **59**, 1307 (1976); K. C. Mates, O. L. Chapman, and J. A. Klun, *J. Org. Chem.*, **42**, 1814 (1977).
4. K. S. Boustany and A. B. Sullivan, *Tetrahedron Lett.*, 3547 (1970); D. N. Harpp, D. K. Ash, T. G. Back, J. G. Gleason, B. A. Orwig, W. F. Van Horn, and J. P. Snyder, *Tetrahedron Lett.*, 3551 (1970).
5. (a) E. Kühle, *Synthesis*, 617 (1971); (b) One notable exception appeared; see M. Fukurawa, T. Suda, A. Tsukamoto, and S. Hayashi, *Synthesis*, 165 (1975).
6. M. Behforouz and J. E. Kerwood, *J. Org. Chem.*, **34**, 51 (1969).
7. T. Mukaiyama and K. Takahashi, *Tetrahedron Lett.*, 5907 (1968).
8. (a) M. E. Alonso, unpublished observations; (b) H. E. Wijiers, H. Boelens, A. Van der Gen, and L. Brandsma, *Recl. Trav. Chim. Pays-Bas*, **88**, 519 (1969).
9. H. B. Footner and S. Smiles, *J. Chem. Soc.*, **127**, 2887 (1925).
10. B. Milligan and J. M. Swan, *J. Chem. Soc.*, 6008 (1963); B. Milligan, B. Saville, and J. M. Swan, *J. Chem. Soc.*, 4850 (1961).
11. M. Calvin, "Mercaptans and Disulfides," Oak Ridge, Tenn., 1954, U.S. Atomic Energy Report UCRL-2438; A. P. Ryle and F. Sanger, *Biochem. J.*, **60**, 535 (1955); J. L. Kice and G. E. Ekman, *J. Org. Chem.*, **40**, 711 (1975); J. P. Danehy, in N. Kharasch and C. Y. Meyers, Eds., "The Chemistry of Organic Sulfur Compounds," Vol. 2, Pergamon Press, New York, 1966, p. 337.
12. M. E. Alonso, H. Aragona, A. W. Chitty, R. Compagnone, and G. Martin, *J. Org. Chem.*, **43**, 4491 (1978).

3-BUTYL-2-METHYLHEPT-1-EN-3-OL
[5-Nonanol, 5-(2-propenyl)-]

$$n\text{-}C_4H_9Br + 2\ Li \rightarrow n\text{-}C_4H_9Li + LiBr$$

$$2\ n\text{-}C_4H_9Li + CH_2{=}C{-}\overset{\displaystyle O}{\overset{\|}{C}}OCH_3 \rightarrow CH_2{=}C{-}\overset{\displaystyle n\text{-}C_4H_9}{\underset{\displaystyle n\text{-}C_4H_9}{\overset{|}{\underset{|}{C}}}}{-}OLi + CH_3OLi$$

$$\underset{\displaystyle CH_3}{\overset{\displaystyle CH_3}{}}$$

$$CH_2{=}\overset{\displaystyle n\text{-}C_4H_9}{\underset{\displaystyle CH_3}{C}}{-}\overset{|}{\underset{\displaystyle n\text{-}C_4H_9}{C}}{-}OLi \xrightarrow[\text{H}_2\text{O}]{\text{HCl}} CH_2{=}\overset{\displaystyle n\text{-}C_4H_9}{\underset{\displaystyle CH_3}{C}}{-}\overset{|}{\underset{\displaystyle n\text{-}C_4H_9}{C}}{-}OH + LiCl$$

Submitted by P. J. Pearce,[1] D. H. Richards, and N. F. Scilly
Checked by N. Cohen, R. Lopresti, and A. Brossi

1. Procedure

A 2-l., four-necked flask equipped with a sealed, Teflon-paddle stirrer, a mercury thermometer, a gas-inlet tube, and a dropping funnel is charged with 1.2 l. of anhydrous tetrahydrofuran (Note 1) and 50 g. (7.1 g.-atoms) of lithium pieces (Note 2) under an atmosphere of prepurified nitrogen. The stirred mixture is cooled to −20° with an acetone dry-ice bath and a mixture of 100 g. (1.00 mole) of methyl methacrylate (Note 3), and 411 g. (3.00 moles) of n-butyl bromide (Note 4) is added dropwise over a period of 3−4 hours. During this addition, an exothermic reaction ensues and is controlled at −20° (Note 5), and on completion of the addition, the vessel is maintained at this temperature, with stirring, for an additional 30 minutes. The contents of the flask are filtered with suction through a 70-mm.-diameter, slit-sieve Buchner funnel, without filter paper, to remove the excess lithium metal. The filtrate is concentrated on a rotary evaporator at aspirator pressure. The residual lithium alcoholate is hydrolyzed by the addition of 1 l. of 10% hydrochloric acid, with ice bath cooling. The liberated alcohol is extracted with two 400-ml. portions of diethyl ether, and the combined ether extracts are washed with two 400-ml. portions of water and dried over 100 g. of anhydrous magnesium sulfate. After suction filtration and removal of the ether on a rotary evaporator at aspirator pressure, the crude alcohol is distilled under reduced pressure through a 40-cm. Vigreux column, yielding 147−158 g. (80−86%) of 3-butyl-2-methyl-1-hepten-3-ol, b.p. 80° (1mm.). The purity of the product, determined by GC analysis, is greater than 99%.

2. Notes

1. Reagent grade (stabilized) tetrahydrofuran was allowed to stand over molecular sieves for 24 hours, refluxed for 2 hours with sodium wire, and finally distilled and used

within 48 hours. The checkers found that it was convenient simply to percolate the tetrahydrofuran, after preliminary drying over molecular sieves, through a column of grade I, neutral aluminum oxide, under nitrogen, directly into the reaction flask, until the required volume of solvent was collected.

2. A convenient form of lithium metal can be purchased from Associated Lead Manufacturers Ltd., 14 Gresham Street, London. A typical analysis shows a purity of 99.6%, and it can be obtained as 1.3-cm.-diameter rod coated with petroleum jelly. A comparable form of lithium metal can be purchased from Ventron Corporation, Chemicals Division, Beverly, Massachusetts. Preparation for use involves weighing, washing with petroleum ether (b.p. 40–60°), and cutting the rod by scissors so that the pieces fall into the reaction vessel. The rod is cut into pieces about 0.5 cm. long that have an average weight of 0.3 g. per piece. Since excess lithium is employed in this reaction, accurate weighing is unnecessary.

3. Reagent grade methyl methacrylate monomer was dried over powdered calcium hydride and freshly distilled before use. The checkers found that identical yields could be obtained when Matheson, Coleman and Bell Chromatoquality methyl methacrylate monomer was used as received with no purification.

4. Reagent grade n-butyl bromide (greater than 98% pure) was used after drying over molecular sieves.

5. The reaction is highly exothermic and the submitters have found that isothermal conditions are best maintained by using cooling equipment consisting of a cooling bath seated on a pneumatically operated labjack and controlled by a temperature sensor which is attached to the thermometer dipping into the reaction vessel. This equipment, known as Jack-o-matic, is supplied by Instruments for Research and Industry, Cheltenham, Pennsylvania.

3. Discussion

This method has general applicability in that the carbonyl compound may be an aldehyde, a ketone, or an ester.[2] Similarly, the halide may be chloride, bromide, or iodide, although yields are generally lower with iodides. Alkyl and aryl halides react with equal facility, and the alkyl halide may be primary, secondary, or tertiary. A few examples of the yields obtained with a variety of reagents are given in Table I (the yields quoted are obtained by GC analysis of the reaction mixture using an internal standard).

TABLE I

Carbonyl Compound	Halide	Product	Yield, %
Propionaldehyde	Ethyl bromide	3-Pentanol	90
Benzaldehyde	Chlorobenzene	Benzhydrol	100
Di-n-butyl ketone	n-Butyl bromide	Tri-n-butyl carbinol	91
Ethyl formate	n-Butyl bromide	5-Nonanol	91
Acrolein	Ethyl bromide	1-Penten-3-ol	90
Butyraldehyde	sec-Butyl bromide	3-Methyl-4-heptanol	89

For maximum yield, care must be taken to ensure that the rate of addition of the reagents is not excessive. If this occurs, the alkyllithium is generated in the presence of significant amounts of unchanged alkyl halide, and Wurtz condensation may be favored. The rate of formation of the alkyllithium is proportional to the surface area of the lithium metal; therefore, at a constant rate of addition, an increase in the lithium surface available for reaction will reduce the probability of Wurtz condensation.

Excess alkyl halide is required to compensate for these side reactions; commonly, only a 10–20% excess is used, rather than the 50% quoted in the method above. The yields given in the table are those obtained with 20% excess halide. The submitters have scaled up the reaction by a factor of 40 with no lowering of yield.

The technique is more efficient than the conventional Grignard reaction for three main reasons: (1) it is a one-stage process; (2) the yields are generally higher; and (3) the final product isolation is cleaner and more convenient.

1. Explosives Research and Development Establishment, Ministry of Defense, Waltham Abbey, Essex, U.K.

2. P. J. Pearce, D. H. Richards, and N. F. Scilly, *J. Chem. Soc. D,* 1160 (1970); Brit. Pat. Appl. 61956 (1969).

MONOALKYLATION OF α,β-UNSATURATED KETONES
via METALLOENAMINES:
1-BUTYL-10-METHYL-$\Delta^{1(9)}$-2-OCTALONE

[2(3*H*)-Naphthalenone, 1-butyl-4,4a,5,6,7,8-hexahydro-4a-methyl-]

Submitted by G. STORK[1] and J. BENAIM[2]
Checked by K. J. BRUZA, R. K. BOECKMAN, and C. R. JOHNSON

1. Procedure

Caution! Hexamethylphosphoric triamide vapors have been reported to cause cancer in rats.[3] All operations with hexamethylphosphoric triamide should be performed in a good hood, and care should be taken to keep the liquid off the skin.

Benzene has been identified as a carcinogen; OSHA has issued emergency standards on its use. All procedures involving benzene should be carried out in a well-ventilated hood, and glove protection is required.

A. 10-*Methyl*-$\Delta^{1(9)}$-2-*octalone* N,N-*dimethylhydrazone*. A 250-ml., round-bottomed flask equipped with a magnetic stirring bar and a Dean-Stark trap is maintained under a dry nitrogen atmosphere (Note 1) and charged with 7.4 g. (0.045 mole) of 10-methyl-$\Delta^{1(9)}$-2-octalone (Note 2), 9.0 g. (0.15 mole) of N,N-dimethylhydrazine, 150 ml. of dry benzene, and 0.02 g. of p-toluenesulfonic acid. This mixture is refluxed for 10–14 hours, after which time no additional water separates. Benzene and excess N,N-dimethylhydrazine are removed by simple distillation, and the residue is distilled under reduced pressure, giving 8.1 g. (87%) of the dimethylhydrazone as a pale-yellow liquid, b.p. 94–98° (0.2 mm.) (Notes 3, 4).

B. 1-*Butyl*-10-*methyl*-$\Delta^{1(9)}$-2-*octalone*. A 250-ml., three-necked flask equipped with a magnetic stirring bar, a reflux condenser, a 50-ml., pressure-equalizing funnel, and a rubber septum is charged with 1.4 g. (0.032–0.035 mole) of 55–60% sodium hydride dispersion in mineral oil and put under a dry nitrogen atmosphere. The mineral oil is removed by washing the sodium hydride three or four times with 5-ml. portions of dry toluene (Note 5). A solution of 6.12 g. (0.0297 mole) of 10-methyl-$\Delta^{1(9)}$-2-octalone N,N-dimethylhydrazone in 100 ml. of dry toluene is placed in the flask, and 10 ml. of dry hexamethylphosphoric triamide is added. The rubber septum is replaced with a glass stopper. The solution is warmed with an oil bath to reflux the toluene; hydrogen evolution is observed. Reflux is maintained for 14–16 hours, during which time the solution becomes dark brown. The solution is then cooled to −10° and 4.5 g. (0.033 mole) of 1-bromobutane in 10 ml. of dry toluene is slowly added. The solution is warmed to 60° and maintained at that temperature for 4–5 hours. An abundant precipitate of sodium bromide is formed. The solution is cooled to 0°, and an acetate-buffer solution (Note 6) is added. The mixture is refluxed for 4–5 hours to complete the hydrolysis before it is cooled and decanted. The aqueous phase is extracted three times with 25-ml. portions of diethyl ether. The combined organic layers are successively washed with three 80- to 100-ml. portions of 10% hydrochloric acid, three 50-ml. portions of saturated sodium hydrogen carbonate, two 50-ml. portions of saturated sodium chloride, and then dried over sodium sulfate. The solvents are removed by rotary evaporation (Note 7). The residue is distilled under high vacuum using a short column. After a small forerun, 4.3–4.7 g. (65–72%) of pure 1-butyl-10-methyl-$\Delta^{1(9)}$-2-octalone, b.p. 84–92° (0.2 mm.), is obtained (Note 8).

2. Notes

1. A positive pressure of nitrogen is maintained using a mercury bubbler.
2. 10-Methyl-$\Delta^{1(9)}$-2-octalone, which can be prepared from 4-(diethylamino)-2-butanone, 2-methylcyclohexanone, and sodium,[4] has b.p. 65–70° (0.1 mm.) and n_D^{20} 1.523. IR (neat) cm.$^{-1}$: 1610, 1670; ^{1}H NMR (CCl$_4$), δ (assignment): 1.25 (CH_3), 5.6 (CH).
3. The hydrazone should be stored under dry nitrogen at −10°.
4. ^{1}H NMR (CCl$_4$), δ (multiplicity, assignment): 1.15 (s, CCH_3), 2.35 [s, N(CH_3)$_2$], 6.35 (40%) and 5.70 (60%) (s, CH of diastereomeric hydrazones); IR (neat) cm.$^{-1}$: 1620, 1580; n_D^{20} 1.505.
5. A 5-ml. portion of dry toluene is introduced into the flask with a syringe, and

sodium hydride dispersion is stirred for 1 minute before 4 ml. of the supernatant toluene is carefully removed from the flask with a syringe.

6. The buffer solution is prepared by dissolving 20 g. of anhydrous sodium acetate in a mixture of 40 ml. of acetic acid and 40 ml. of water.

7. At this point the submitters reported 7.07 g. of crude product; GC analysis on an SE-30 column at 200° showed 1–3% of 10-methyl-$\Delta^{1(9)}$-2-octalone and 85% of the desired alkylated product.

8. ^{1}H NMR (CCl$_4$), δ (multiplicity, assignment): 0.9 (t, CH$_2$CH_3), 1.2 (s, CCH_3); IR (neat) cm.$^{-1}$: 1660, 1600; $n_D^{25} = 1.511$.

3. Discussion

Alkylations of enamines of α,β-unsaturated ketones with alkyl halides often give very poor yields of C-alkylated products because of competing N-alkylation.[5,6] In the type of transformation illustrated here, direct alkylations of enamines are completely unsuccessful, even in cases where hindered enamines[7] are used. Generally, the metalloenamine method[8] can be applied with good success to the problem of monoalkylation of α,β-unsaturated ketones.[9]

Metalloenamines can be formed from N,N-dimethylhydrazones, as illustrated here, or from N-cyclohexylimines. Various strong bases have been used, including n-butyllithium, lithium diisopropylamide, sodium hydride, and lithium bis(trimethylsilyl)amide. The nature and sometimes the stoichiometry of the strong base used can be important. Poor yields of alkylated compounds are obtained with Grignard reagents, and in the case of n-butyllithium, excess base can result in the formation of significant amounts of kinetically controlled alkylation products (e.g., alkylation at C-3 of 10-methyl-$\Delta^{1(9)}$-2-octalone). In the cases of octalones and steroid compounds (cholestenone, testosterone benzoate) it has been found that sodium hydride and lithium diisopropylamide gave the best yields of the desired alkylated compounds.[9]

1. Department of Chemistry, Columbia University, New York, NY 10027.
2. U. E. R. Sciences et Techniques, Centre Universitaire de Toulon et du Var, 83130 La Garde, France.
3. J. A. Zapp, Jr., *Science*, **190**, 422 (1975).
4. M. Yaganita, M. Hirakura, and F. Seki, *J. Org. Chem.*, **23**, 841 (1958). (A procedure adapted from that described by these authors for their compound IV may be used.)
5. G. Stork, A. Brizzolara, H. Landesman, J. Szmuskovicz, and R. Terrel, *J. Am. Chem. Soc.*, **85**, 207 (1963).
6. G. Stork and G. Birnbaum, *Tetrahedron Lett.*, 313 (1961).
7. T. J. Curphey and J. C. Yu-Hung, *Chem. Commun.*, 510 (1967).
8. G. Stork and S. R. Dowd, *J. Am. Chem. Soc.*, **85**, 2178 (1963).
9. G. Stork and J. Benaim, *J. Am. Chem. Soc.*, **93**, 5938 (1971).

β-DIKETONES FROM METHYL ALKYL KETONES:
3-*n*-BUTYL-2,4-PENTANEDIONE

$$n\text{-}C_4H_9CH_2COCH_3 + (CH_3CO)_2O \xrightarrow{4\text{-}CH_3C_6H_4SO_3H}$$

$$\overset{\displaystyle OCOCH_3}{\underset{\displaystyle n\text{-}C_4H_9CH=C-CH_3}{|}}$$

$$BF_3 \cdot CH_3COOH \bigg| (CH_3CO)_2O$$

$$n\text{-}C_4H_9CH(COCH_3)_2 \xleftarrow[H_2O, \Delta]{CH_3CO_2Na} \quad n\text{-}C_4H_9-C\overset{\displaystyle \overset{CH_3}{|}}{\underset{\displaystyle \underset{CH_3}{|}}{C=O}}\overset{C-O}{\underset{C=O}{}}BF_2$$

Submitted by CHUNG-LING MAO[1] and CHARLES R. HAUSER[1,2]
Checked by DAVID G. MELILLO and HERBERT O. HOUSE

1. Procedure

A mixture of 28.6 g. (0.251 mole) of 2-heptanone (Note 1), 51.0 g. (0.500 mole) of acetic anhydride (Note 2), and 1.9 g. (0.010 mole) of *p*-toluenesulfonic acid monohydrate (Note 3) contained in a stoppered 500-ml., round-bottomed flask, equipped with a magnetic stirrer, is stirred at room temperature for 30 minutes before 55 g. (0.43 mole) of the solid 1:1 boron trifluoride-acetic acid complex (Note 4) is added, resulting in the evolution of heat. The amber-colored solution is stirred in the stoppered flask at room temperature for 16–20 hours (Note 5), and a solution of 136 g. (1.00 mole) of sodium acetate trihydrate (Note 6) in 250 ml. of water is added. After the flask has been fitted with a reflux condenser, the reaction mixture is heated at reflux for 3 hours and cooled, and the product is extracted with three 100-ml. portions of petroleum ether (b.p. 30–60°). The combined organic extracts are washed successively with aqueous 5% sodium hydrogen carbonate and saturated aqueous sodium chloride and dried over anhydrous calcium sulfate (Drierite). The solvent is removed with a rotary evaporator, and the residual oil is distilled, yielding 25–30 g. (64–77%) of 3-*n*-butyl-2,4-pentanedione as a colorless liquid, b.p; 84–86° (6 mm.), n_D^{25} 1.4422–1.4462 (Note 7).

2. Notes

1. 2-Heptanone, obtained from Eastman Organic Chemicals, was distilled before use, b.p. 145–147°.

2. Acetic anhydride purchased from Merck & Co., Inc., was fractionally distilled and the fraction, b.p. 139–141°, was used.

3. p-Toluenesulfonic acid monohydrate was obtained from Eastman Organic Chemicals and used without purification.

4. The submitters employed 75 g. (0.50 mole) of the liquid 1:2 boron trifluoride-acetic acid complex obtained from Harshaw Chemical Company. Since the checkers were unable to obtain this complex from a commercial source, a solid 1:1 complex was prepared according to the literature.[3,4] A 2-l., three-necked flask fitted with a mechanical stirrer, a gas-outlet tube, and a gas-inlet tube extending to the bottom of the flask is charged with a solution of 230 ml. of acetic acid in 750 ml. of 1,2-dichloroethane. A stream of boron trifluoride gas is passed through the reaction flask while the solution is stirred and cooled with an ice bath. After approximately 1 hour, when the mixture is saturated, the addition of boron trifluoride is stopped and the insoluble 1:1 boron trifluoride-acetic acid complex is rapidly collected on a filter, washed with 200 ml. of 1,2-dichloroethane, and transferred to a dry, stoppered container. Since this solid complex tends to liquefy partially on storage, portions to be used in this preparation should be washed with 1,2-dichloroethane immediately prior to use. The amount of catalyst obtained is sufficient to perform this preparation several times.

5. A longer reaction time gives similar results.

6. Sodium acetate trihydrate was obtained from Eastman Organic Chemicals.

7. On a GC column packed with SE-30 silicone gum on Chromosorb P and heated to 150°, the product exhibits a single peak with a retention time of 12.3 minutes; under the same conditions the peak for 2-heptanone has a retention time of 4.4 minutes. The product, which is partially enolic, has IR bands (CCl_4) at 1725(sh), 1695, and 1605 cm.$^{-1}$ with a UV maximum (95% C_2H_5OH) at 288 nm. (ϵ 2560) and ^1H NMR peaks (CCl_4) at δ 0.7–2.0 (m, 9H, aliphatic CH), 2.10 (s, 6H, 2COCH_3), 3.57 (t, J = 7 Hz., 0.7H, COCHCO), and 16.50 (s, 0.3H, enolic OH). The mass spectrum exhibits a molecular ion at m/e 156 with abundant fragment peaks at m/e 100, 71, 58, 44, and 43 (base peak).

3. Discussion

This procedure for the acetylation of methyl alkyl ketones to β-diketones is a modification[5] of an earlier method, which used boron trifluoride gas as the catalyst.[6] 3-n-Butyl-2,4-pentanedione has also been prepared by the acetylation of 2-heptanone catalyzed with boron trifluoride gas,[7] by the thermal rearrangement of the enol acetate of 2-heptanone,[7] and by the alkylation of the potassium enolate of 2,4-pentanedione with n-butyl bromide.[8]

In this procedure, the ketone is first converted to its enol acetate by reaction with acetic anhydride in the presence of a protic acid. Since enol acetylation is performed under equilibrating conditions, the more stable enol acetate (usually the more highly substituted isomer) is produced. Acetylation of this enol acetate, catalyzed by boron trifluoride, usually leads to the formation of the enol acetate of a β-diketone, which is cleaved by boron trifluoride, forming acetyl fluoride and the borofluoride complex of the β-diketone. Thus, this procedure offers a convenient and general synthetic route to 3-substituted-2,4-pentanediones.[5] The acylation of 2-butanone to 3-methyl-2,4-pentanedione (48%); 2-pentanone to 3-ethyl-2,4-pentanedione (57%); phenylacetone to 3-phenyl-2,4-pentanedione (68%); and 3-methyl-2-butanone to 3,3-dimethyl-2,4-pentanedione (40–48%) have been reported by the submitters.

A similar acetylation procedure (without *p*-toluenesulfonic acid) has been employed to prepare other β-diketones.[5] For example, cyclohexanone was converted to 2-acetylcyclohexanone (73%); cyclopentanone yielded 2-acetylcyclopentanone (80%); 3-pentanone yielded 3-methyl-2,4-hexanedione (81%); dibenzyl ketone yielded 1,3-diphenyl-2,4-pentanedione (72%); and acetophenone gave benzoylacetone (70%).

1. Department of Chemistry, Duke University, Durham, North Carolina 27706.
2. Deceased January 6, 1970.
3. R. M. Manyik, F. C. Frostick, Jr., J. J. Sanderson, and C. R. Hauser, *J. Am. Chem. Soc.,* **75,** 5030 (1953).
4. L. F. Fieser and M. Fieser, "Reagents for Organic Synthesis," Vol. 1, Wiley, New York, 1967, p. 69.
5. C.-L. Mao, F. C. Frostick, Jr., E. H. Man, R. M. Manyik, R. L. Wells, and C. R. Hauser, *J. Org. Chem.,* **34,** 1425 (1969).
6. C. R. Hauser, F. W. Swamer, and J. T. Adams, *Org. React.,* **8,** 59 (1954).
7. F. G. Young, F. C. Frostick, Jr., J. J. Sanderson, and C. R. Hauser, *J. Am. Chem. Soc.,* **72,** 3635 (1950).
8. D. F. Martin, W. C. Fernelius, and M. Shamma, *J. Am. Chem. Soc.,* **81,** 130 (1959).

SECONDARY AND TERTIARY ALKYL KETONES FROM CARBOXYLIC ACID CHLORIDES AND LITHIUM PHENYLTHIO(ALKYL)CUPRATE REAGENTS: *tert*-BUTYL PHENYL KETONE

(1-Propanone, 2,2-dimethyl-1-phenyl)

$$C_6H_5SLi + CuI \xrightarrow[\text{tetrahydrofuran}]{25°} C_6H_5SCu + LiI$$

$$C_6H_5SCu + (CH_3)_3CLi \xrightarrow[\text{tetrahydrofuran, pentane}]{-60° \text{ to } -65°} C_6H_5S[(CH_3)_3C]CuLi$$

$$C_6H_5S[(CH_3)_3C]CuLi + C_6H_5COCl \xrightarrow[\text{tetrahydrofuran}]{-60° \text{ to } -65°} (CH_3)_3CCOC_6H_5$$

Submitted by GARY H. POSNER[1] and CHARLES E. WHITTEN
Checked by JOYCE M. WILKINS and HERBERT O. HOUSE

1. Procedure

Caution! Since the odor of the thiophenol (benzenethiol) used in this preparation is unpleasant, both steps of this preparation should be conducted in a hood and the glassware used should be washed before it is removed from the hood.

A.   *Lithium phenylthio*(tert-*butyl*)*cuprate.*   A dry, 200-ml., round-bottomed flask is fitted with a magnetic stirring bar and a 100-ml., pressure-equalizing dropping funnel, the top of which is connected to a nitrogen inlet. After the apparatus has been flushed with nitrogen, 50 ml. of 1.60 M (0.080 mole) n-butyllithium (Note 1) solution is placed in the flask and cooled with an ice bath. Under a nitrogen atmosphere, a solution of 8.81 g. (0.0801 mole) of freshly distilled thiophenol (Note 2) in 30 ml. of anhydrous tetrahydrofuran (Note 3) is added dropwise to the cooled, stirred solution. An aliquot of the resulting solution (Note 4) is standardized by quenching in water, followed by titration with 0.10 N hydrochloric acid to a green end point with a bromocresol indicator. The concentration of lithium thiophenoxide prepared in this manner is typically 1.0 M.

A dry, 250-ml., three-necked, round-bottomed flask is equipped with a sealed mechanical stirrer (Note 5), a glass stopper, and a rubber septum through which are inserted hypodermic needles used to evacuate the flask and to admit nitrogen. After the apparatus has been flushed with nitrogen, 4.19 g. (0.0220 mole) of purified copper(I) iodide (Note 6) is added, and while warming with a flame, the apparatus is evacuated, then refilled with nitrogen. After this procedure has been performed twice, the flask is allowed to cool, the stopper is replaced with a thermometer, and 45 ml. of anhydrous tetrahydrofuran is added (Note 3) with a hypodermic syringe. With continuous stirring, 22 ml. of 1.0 M (0.022 mole) lithium thiophenoxide solution is added with a syringe to the slurry of copper(I) iodide. After 5 minutes, the resulting yellow solution is cooled, with continuous stirring, to −65° with an acetone – dry ice cooling bath. Some copper(I) thiophenoxide usually separates from solution at *ca.* −45°. When the temperature of the mixture has reached *ca.* −65°, 13.6 ml. (0.0218 mole) of 1.60 *M tert*-butyllithium (Note 7) solution is added with a syringe to the stirred mixture at such a rate that the temperature of the mixture remains at −60° to −65°. The resulting cloudy yellow-orange solution of the cuprate reagent is stirred at −60° to −65° for 5 minutes (Note 8).

B.   tert-*Butyl phenyl ketone.*   With a syringe a solution of 2.81 g. (0.0200 mole) of freshly distilled benzoyl chloride (Note 9) in 15 ml. of anhydrous tetrahydrofuran (Note 3) is added dropwise, with stirring, to the cold solution (−60° to −65°) of the cuprate reagent. The resulting yellow-brown solution is stirred for 20 minutes (at −60° to −65°) and quenched by the addition, with a syringe, of 5 ml. of anhydrous methanol. The red-orange reaction mixture is allowed to warm to room temperature and then poured into 100 ml. of aqueous saturated ammonium chloride. The copious precipitate of copper(I) thiophenoxide is separated by suction filtration and washed thoroughly with several 50-ml. portions of diethyl ether. The combined filtrate is extracted with three 100-ml. portions of ether. The combined ethereal solution is washed with two 50-ml. portions of aqueous 1 *N* sodium hydroxide and with one 50-ml. portion of aqueous 2% sodium thiosulfate. Each of the aqueous washes is extracted in turn with a fresh 50-ml. portion of ether. The combined ethereal solution is dried with anhydrous magnesium sulfate, filtered, and concentrated by distillation through a short Vigreux column. The residual pale yellow liquid (Note 10) is distilled through a short column under reduced pressure, yielding 2.73 – 2.82 g. (84 – 87%) of *tert*-butyl phenyl ketone as a colorless liquid b.p. 105 – 106° (15 mm.), 114 – 115° (44 mm.), n_D^{20} 1.5092, n_D^{25} 1.5066 (Note 11).

2. Notes

1. Solutions containing approximately 1.6 *M n*-butyllithium in hexane were purchased either from Alfa Inorganics, Inc., or from Foote Mineral Company. The concentration of *n*-butyllithium in these solutions can be determined either by a double titration procedure[2] or by dilution with anhydrous tetrahydrofuran, followed by titration with 2-butanol[3] in the presence of a 2,2′-bipyridyl indicator [*Org. Synth.*, **Coll. Vol. 6,** 121 (1988)]. In either case the total base concentration in the reagent is determined by titration with standard aqueous acid.

2. Thiophenol, purchased from Aldrich Chemical Company, Inc., was redistilled before use; b.p. 65 – 66° (42 mm.).

3. Commercial anhydrous tetrahydrofuran was distilled from lithium aluminum hydride and stored under nitrogen.

4. The submitters report that this solution may be stored under nitrogen at 0° for several days without deterioration. Phenylthiocopper is now commercially available from the Alpha Division of the Ventron Corp., 152 Andover St., Danvers, Massachusetts 01923.

5. Although the submitters had recommended use of a magnetic stirring bar, the checkers encountered considerable difficulty in maintaining adequate stirring of the cold reaction mixture with a magnetic stirrer and recommend use of a sealed mechanical stirrer such as a Truebore® stirrer.

6. Copper(I) iodide, purchased from Fisher Scientific Company, was purified by continuous extraction with anhydrous tetrahydrofuran in a Soxhlet extractor for approximately 12 hours, to remove colored impurities. The residual copper(I) iodide was then dried under reduced pressure at 25° and stored under nitrogen in a desiccator.

7. A pentane solution of *tert*-butyllithium (purchased from either Alfa Inorganics, Inc., or Lithium Corporation of America, Inc.) was standardized by one of the previously described titration procedures (Note 1). If possible, it is desirable to use a freshly opened bottle of *tert*-butyllithium since previously used bottles of this reagent often contain lithium *tert*-butoxide, which will lead to formation of a contaminant in the final product (Note 10).

8. Although the submitters report that this reagent is stable at 0° (*i.e.*, still reactive toward benzoyl chloride) for periods of at least one hour under a nitrogen atmosphere,[3,4] the checkers repeatedly observed evidence of thermal decomposition when the solution was allowed to warm above −40°. This decomposition was indicated by the appearance of a red-brown coloration as the reagent was warmed to −40°; as the temperature was raised further to −25° and to 0°, the mixture progressively exhibited a darker brown color.

9. Benzoyl chloride (purchased from Eastman Organic Chemicals) was redistilled before use; b.p. 35 – 36° (0.5 mm.).

10. The checkers found that with previously opened bottles of *tert*-butyllithium, the crude product was often contaminated with *tert*-butyl benzoate (from lithium *tert*-butoxide; see Note 7). The presence of this impurity in the crude product may be detected either by the presence of an extra IR peak at 1720 cm.$^{-1}$ (conjugated ester), or by GC analysis. On a 1.3-m. GC column, packed with silicone fluid, No. SE-52, suspended on Chromosorb P and operated at 155°, the retention time of *tert*-butyl phenyl ketone was 4.4 minutes, and the retention times of potential impurities, methyl benzoate and *tert*-butyl benzoate were 2.4 minutes and 7.8 minutes, respectively. If a small amount of *tert*-butyl benzoate is present in the crude product, it is most easily removed by heating a mixture of the crude product with 1% by weight of *p*-toluenesulfonic acid on a steam bath for 10 minutes followed by partitioning the product between ether and aqueous sodium hydrogen carbonate. After the resulting ether solution has been dried and distilled, pure *tert*-butyl phenyl ketone is obtained.

11. The product exhibits a single GC peak (see Note 10). The spectral properties of the product are as follows; IR (CCl$_4$) cm.$^{-1}$: 1680 (conjugated ketone), 1395 and 1370 [C(CH$_3$)$_3$]; ^{1}H NMR (CCl$_4$), δ (multiplicity, number of protons, assignment): 1.30 [s, 9H, C(CH$_3$)$_3$], 7.2 – 7.9 (m, 5H, C$_6$H$_5$); UV (95% C$_2$H$_5$OH) nm. max. (ε): 237 (7350) and 272 (620); mass spectrum *m/e* (relative intensity): 162 (M, 45), 106 (28), 105 (100), 77 (63), 57 (40), 51 (23), and 41 (30). The physical constants reported for the product are: b.p. 103 – 104° (13 mm.),[5] n_D^{20} 1.5090.[6]

3. Discussion

tert-Butyl phenyl ketone has been prepared by the reactions of benzoic acid with *tert*-butyllithium,[7,8] of acetophenone with methyl iodide and base,[5,9] of benzaldehyde with *tert*-butylmagnesium chloride followed by oxidation,[10] and of 2,2-dimethylpropanoyl chloride with phenylmagnesium bromide.[11]

The procedure described here illustrates the preparation of mixed lithium arylhetero-

(alkyl)cuprate reagents and their reactions with carboxylic acid chlorides.[4] These mixed cuprate reagents also react with α,α'-dibromoketones,[12] primary alkyl halides,[4] and α,β-unsaturated ketones,[4] with selective transfer of the alkyl group.

Two limitations on the utility of organocopper reagents have often been the difficulty in using thermally unstable lithium *sec*- and especially *tert*-alkylcuprates[13] and the need for a large (*e.g.*, 300 – 500%) excess of an organocuprate to achieve complete conversion of substrate to product. Both of these limitations are circumvented by using lithium phenylthio(*tert*-alkyl)cuprates, which react with approximately equimolar amounts of carboxylic acid chlorides, forming the corresponding *tert*-alkyl ketones in high yield, even with the yield based on the transferred alkyl group. Furthermore, this alkyl group transfer can be achieved in the presence of other functional groups (*e.g.*, remote halogen or ester functionalities) in the carboxylic acid chloride substrate (Equation 1). Transfer of secondary alkyl groups can also be accomplished efficiently in this way (Equation 2).

$$C_2H_5O_2CCH_2CH_2COCl \xrightarrow[\text{tetrahydrofuran, }-78°, \text{ 15 minutes}]{1.2 \text{ equivalent } C_6H_5S(\text{tert-}C_4H_9)CuLi}$$

$$C_2H_5O_2CCH_2CH_2COC_4H_9\text{-}\textit{tert} \quad (1)$$

$$(65\%)$$

$$C_6H_5COCl \xrightarrow[\text{tetrahydrofuran, }-78°, \text{ 15 minutes}]{1.3 \text{ equivalent } C_6H_5S(\text{sec-}C_4H_9)CuLi} C_6H_5COC_4H_9\text{-}\textit{sec} \quad (2)$$

$$(80\%)$$

The reaction of *tert*-alkyl Grignard reagents with carboxylic acid chlorides in the presence of a copper catalyst provides *tert*-alkyl ketones in substantially lower yields than those reported here.[4,14] The simplicity and mildness of experimental conditions and isolation procedure, the diversity of substrate structural type, and the functional group selectivity of these mixed organocuprate reagents render them very useful for conversion of carboxylic acid chlorides to the corresponding secondary and tertiary alkyl ketones.[15]

1. Department of Chemistry, the Johns Hopkins University, Baltimore, Maryland 21218; this work was supported by the National Science Foundation (GP-33667).
2. G. M. Whitesides, C. P. Casey, and J. K. Krieger, *J. Am. Chem. Soc.*, **93**, 1379 (1971).
3. For warning concerning 2-butanol, see *Chem. Eng. News*, **59**, (19), 3 (1981).
4. G. H. Posner, C. E. Whitten, and J. J. Sterling, *J. Am. Chem. Soc.*, **95**, 7788 (1973).
5. A. Haller and E. Bauer, *C.R. Hebd. Seances Acad. Sci.*, **148**, 73 (1909).
6. C. Cherrier and J. Metzger, *C.R. Hebd. Seances Acad. Sci.*, **226**, 797 (1948).
7. Unpublished results of C. H. Heathcock and R. Radcliff as reported in Ref. 8.
8. For a general discussion of ketone formation from carboxylic acids and organolithium reagents, see M. J. Jorgenson, *Org. React.*, **18**, 1 (1970).
9. J. U. Nef, *Justus Liebigs Ann. Chem.*, **310**, 316 (1900).
10. A. Favorskii, *Bull. Soc. Chim. Fr.*, **3**, 239 (1936).
11. J. Thiec, *Ann. Chim. (Paris)*, **9**, 51 (1954).
12. G. H. Posner and J. J. Sterling, *J. Am. Chem. Soc.*, **95**, 3076 (1973).
13. G. M. Whitesides, W. F. Fischer, Jr., J. San Filippo, Jr., R. W. Bashe, and H. O. House, *J. Am. Chem. Soc.*, **91**, 4871 (1969).

14. J. E. Dubois, M. Boussu, and C. Lion, *Tetrahedron Lett.*, 829 (1971) and references cited therein; J. A. MacPhee and J. E. Dubois, *Tetrahedron Lett.*, 467 (1972).

15. For use of other organocopper reagents in converting carboxylic acid chlorides to ketones, see G. H. Posner and C. E. Whitten, *Tetrahedron Lett.*, 1815 (1973); G. H. Posner, C. E. Whitten, and P. E. McFarland, *J. Am. Chem. Soc.*, **94**, 5106 (1972). For a recent report on direct and convenient preparation of lithium phenylthio(alkyl)cuprate reagents, see G. H. Posner, D. J. Brunelle, and L. Sinoway, *Synthesis*, 662 (1974); G. H. Posner, "An Introduction to Synthesis Using Organocopper Reagents," Wiley, New York, 1980.

REMOVAL OF N^{α}-BENZYLOXYCARBONYL GROUPS FROM SULFUR-CONTAINING PEPTIDES BY CATALYTIC HYDROGENATION IN LIQUID AMMONIA: *O-tert*-BUTYL-L-SERYL-*S-tert*-BUTYL-L-CYSTEINE *tert*-BUTYL ESTER

Submitted by ARTHUR M. FELIX, MANUEL H. JIMENEZ, and JOHANNES MEIENHOFER[1,2]
Checked by LÁSZLÓ RÉVÉSZ and G. BÜCHI

1. Procedure

Caution! All operations described in these procedures must be carried out in a well-ventilated hood, since ammonia is highly toxic, hydrogen is extremely flammable, and palladium black is pyrophoric.

A. L-*Methionine*. A dry, 1-l., three-necked, round-bottomed flask is equipped with a dry ice reflux condenser (Note 1), a gas-inlet tube, and a magnetic stirring bar as illustrated in Figure 1. The reaction vessel is immersed in an acetone – dry ice bath, and a total of 300 ml. of ammonia (Note 2) is passed through a drying tower containing potassium hydroxide pellets and collected in the flask. The bath is removed to permit the reaction to proceed at the boiling point of ammonia ($-33°$), and a gentle stream of dry nitrogen (Note 2) is bubbled into the flask. A solution of 0.708 g. (0.80250 mole) of N-benzyloxycarbonyl-L-methionine (Note 3) in 10 ml. of N,N-dimethylacetamide (Note 4), 1.02 g. (1.40 ml., 0.0101 mole) of triethylamine (Note 5), and 1.25 g. of freshly prepared palladium black (Note 6) are added. The nitrogen stream is discontinued and replaced by a stream of hydrogen (Note 2) that has been passed through a concentrated sulfuric acid scrubber. The mixture is stirred under reflux for 5.5 hours to effect hydrogenolysis (Note 1). The hydrogen stream is discontinued, a flow of nitrogen is resumed, and the dry ice is removed from the reflux condenser, permitting rapid evaporation of ammonia (Note 7). The flask is attached to a rotary evaporator (Note 8), and the mixture is evaporated to dryness under reduced pressure. The residue is dissolved in water and filtered through a sintered funnel of medium porosity to remove the catalyst. The filtrate is evaporated to dryness, and the residue (354 mg., 95%) is crystallized from water – ethanol. The white crystalline product, after drying under reduced pressure at 25°, weighs 272 – 305 mg. (73 – 82%), m.p. 280 – 282° (dec.) (Note 9), $[\alpha]_D^{25}$ +23.1° ($c = 1$, aqueous 5 N hydrochloric acid) (Note 10).

B. O-tert-*Butyl*-L-*seryl*-S-tert-*butyl*-L-*cysteine* tert-*butyl ester*. A dry, 1-l., three-necked, round-bottomed flask is equipped with a dry ice reflux condenser (Note 1), a gas-inlet tube, and a magnetic stirring bar, as illustrated in Figure 1. The reaction vessel is immersed in an acetone – dry ice bath, and a total of 300 ml. of ammonia (Note 2) is

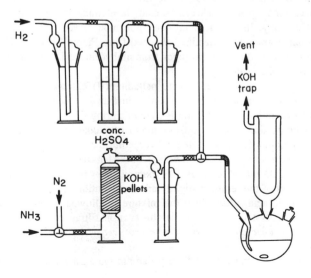

Figure 1.

passed through a drying tower containing potassium hydroxide pellets and then collected in the flask. The bath is removed, and a gentle stream of nitrogen (Note 2) is bubbled into the flask. A solution of 200 mg. (0.392 mmole) of N^α-benzyloxycarbonyl-O-$tert$-butyl-L-seryl-S-$tert$-butyl-L-cysteine $tert$-butyl ester (Note 11) in 4 ml. of N,N-dimethylacetamide (Note 4), 0.160 g. (0.220 ml., 0.00158 mole) of triethylamine (Note 5), and 200 mg. of palladium black freshly prepared from 333 mg. (0.00188 mole) of palladium(II) chloride (Note 6) are added. The nitrogen stream is discontinued and replaced by a stream of hydrogen (Note 2) that has been passed through a concentrated sulfuric acid scrubber. The mixture is stirred and hydrogenated at reflux temperature for 6 hours (Note 1). The hydrogen stream is discontinued, a stream of nitrogen is again passed into the flask, and the dry ice is removed from the reflux condenser to permit rapid evaporation of ammonia (Note 7). The flask is attached to a rotary evaporator (Note 8) and evaporated to dryness under reduced pressure. The residue is dissolved in 50 ml. of methanol (Note 12), and the suspension is filtered through a 5 × 25-mm. bed of Celite (Note 13) to remove the catalyst. The Celite bed is washed thoroughly with three 20-ml. portions of methanol. The filtrate is evaporated to dryness, and the residue is recrystallized from petroleum ether (b.p. 60 – 90°). The white crystalline product, after drying under reduced pressure at 25°, weighs 121 – 127 mg. (82 – 86%), m.p. 71 – 73° (Note 14), $[\alpha]_D^{25}$ −5.8° (c = 1, methanol) (Note 15).

2. Notes

1. The condenser is filled with crushed dry ice (no solvent). More dry ice is added periodically as necessary throughout Parts A and B.

2. Anhydrous ammonia, prepurified nitrogen, and prepurified hydrogen were all purchased from Matheson Gas Products.

3. N-Benzyloxycarbonyl-L-methionine was obtained from Bachem, Inc., 3132 Kashiwa Street, Torrance, California 90505.

4. Spectrophotometric-grade N,N-dimethylacetamide was purchased from Aldrich Chemical Company, Inc., and stored over molecular sieves.

5. Sequanal-grade triethylamine, obtained from Pierce Chemical Company, Rockford, Illinois, was distilled under nitrogen from ninhydrin, which was also purchased from Pierce Chemical Company.

6. The catalyst is prepared[3,4] with palladium(II) chloride and 97 – 100% formic acid, which were purchased from Engelhard Industries Division (Engelhard Minerals and Chemicals Corporation) and MC and B Manufacturing Chemists, respectively. A 10.4-ml. aliquot containing 2.08 g. (0.0117 mole) of palladium(II) chloride from a stock solution of palladium(II) chloride in 2 N hydrochloric acid (10 g. per 50 ml.) is added to 104 ml. of boiling water in a 600-ml. beaker. A 0.51-g. (0.42 ml., 0.011 mole) portion of formic acid and 33 ml. of aqueous 10% potassium hydroxide are added to the boiling solution. The pH of the resulting slightly alkaline solution (pH ∼ 8) is adjusted to 6 – 7 by adding formic acid, after which the mixture is allowed to boil for an additional 5 minutes. The catalyst is isolated by careful suction filtration. *Caution! The palladium catalyst is pyrophoric and must always be kept wet with water or methanol to prevent contact with air.*

To minimize the danger in handling palladium black, the submitters recommend that filtration, washing, and transfer of the catalyst be performed with a "syringe filter."[5] This device was fashioned from a 10-ml. Plastipak syringe, purchased from Becton-Dickinson and Company, Rutherford, New Jersey, by cutting off the tip at the end of the cylindrical barrel and forcing a tight-fitting, porous disk of polypropylene into its place. The use of this "syringe filter" permits the removal of most of the solvent and the safe transfer of the catalyst to the flask with little danger of ignition or moisture absorption.

The catalyst is washed thoroughly with 100 ml. of water and with 200 ml. of absolute methanol to remove all traces of water, after which it is transferred to the flask under nitrogen with a minimal amount of absolute methanol. To be effective the catalyst must be pyrophoric, and extreme care must be taken during this operation to prevent ignition of the methanol or ammonia. The catalyst must not be allowed to become dry or to collect on the wall of the flask above the surface of the liquid ammonia.

7. Evaporation of the ammonia generally requires several hours. Toward the end of the evaporation, it is advantageous to immerse the flask in an acetone bath, taking care to avoid bumping.

8. N,N-Dimethylacetamide remains in the flask and is removed by rotary evaporation under reduced pressure with a water bath kept at a temperature lower than 35°. The submitters recommend that the evaporation be carried out directly in the same three-necked flask by stoppering the two side arms and adjusting the angle of the rotary evaporator.

9. The melting point is corrected [lit.,[6] m.p. $280-281°$ (dec.)].

10. The literature[7] reports $[\alpha]_D^{25}$ +23.2° ($c = 1$, 5 N hydrochloric acid). The product was analyzed by the submitters. Analysis calculated for $C_5H_{11}NO_2S$: C, 40.25; H, 7.43; N, 9.39; S, 21.49. Found: C, 40.14; H, 7.42; N, 9.50; S, 21.52. The product was homogeneous according to TLC on precoated silica gel G plates purchased from Analtech, Inc., Newark, Delaware, and developed with the following two solvent systems (solvents, volume ratios of solvents in the same order): 1-butanol – acetic acid – ethyl acetate – water, 1:1:1:1, R_f 0.49; 1-butanol-acetic acid-pyridine-water, 15:3:10:12, R_f 0.51.

11. The protected dipetide was prepared by the procedure described in the following paragraph using N-benzyloxycarbonyl-O-tert-butyl-L-serine purchased from Chemical Dynamics Corporation (P. O. Box 395, South Plainfield, New Jersey 07080), tetrahydrofuran distilled from lithium aluminum hydride [*Caution!* For a warning regarding this method for purifying tetrahydrofuran, see *Org. Synth., Coll. Vol. 5*, 976 (1973)], and N-methylmorpholine distilled from ninhydrin. S-tert-Butyl-L-cysteine tert-butyl ester[8] was prepared as follows: To a suspension of 10 g. (0.082 mole) of L-cysteine in 75 ml. of dry dioxane in a 200-ml. pressure bottle cooled in an ice bath are added 10.0 ml. of concentrated sulfuric acid and 56 g. (95 ml., 1.0 mole) of isobutylene. The pressure bottle is stoppered and shaken at room temperature for 18 hours. The mixture is cooled to 0°, the pH is adjusted to 10 by adding 160 ml. of aqueous 2 N sodium hydroxide, and the product is extracted with three 100-ml. portions of diethyl ether. The ethereal solution is washed with three 80-ml. portions of aqueous 5% sodium hydrogen carbonate and three 80-ml. portions of water, dried over anhydrous magnesium sulfate, and concentrated to a volume of *ca.* 200 ml. The concentrate is stirred and cooled at 0° as 90.8 ml. (0.0908 mole) of 1 M hydrogen chloride in ether is added. The resulting mixture is stirred for several minutes, the precipitated hydrochloride is filtered, and the filter cake is washed with ether. The

white crystalline product weighs 18.1 g. (81%) and is used without further purification. The submitters caution that the hydrochloride sublimes under reduced pressure. Recrystallization from chloroform – petroleum ether affords an analytical sample, double m.p. 187° and 219 – 222°, $[\alpha]_D^{25}$ +5.85° (c = 1, methanol). Analysis calculated for $C_{11}H_{23}NO_2S \cdot HCl$: C, 48.96; H, 8.96; N, 5.19; S, 11.88; Cl, 13.14. Found: C, 49.07; H, 9.25; N, 5.13; S, 12.06; Cl, 13.07. The free base is obtained by dissolving the hydrochloride in aqueous 10% sodium carbonate, extracting the mixture with ether, and evaporating the ethereal solution under reduced pressure.

A dry, three-necked, round-bottomed flask is equipped with a mechanical stirrer, a rubber septum, and a 200-ml., pressure-equalizing dropping funnel mounted with a T-shaped gas-inlet connected to both a nitrogen source and a bubbler serving as a gas exit. The flask is charged with a solution of 14.9 g. (0.0506 mole) of N^α-benzyloxycarbonyl-O-$tert$-butyl-L-serine in 100 ml. of tetrahydrofuran and purged with nitrogen. The solution is stirred and cooled at −15° as 5.12 g. (5.67 ml., 0.0506 mole) of N-methylmorpholine and 6.91 g. (6.61 ml., 0.0506 mole) of isobutyl chloroformate are added rapidly through the septum with syringes. One minute after the addition is completed, a precooled (−20°) solution of 11.8 g. (0.0506 mole) of S-$tert$-butyl-L-cysteine $tert$-butyl ester in 100 ml. of tetrahydrofuran is added dropwise at −15°. The contents of the flask are stirred for 1 hour at −15° and for 3 hours at room temperature. The mixture is then evaporated to dryness under reduced pressure and dissolved in 150 ml. of ethyl acetate. The solution is washed with three 50-ml. portions of each of the following: 5% sodium hydrogen carbonate in water, water, 1 M citric acid in water, and water. The ethyl acetate solution is then dried over anhydrous magnesium sulfate, the solvent is evaporated, and the remaining solid is recrystallized from ethyl acetate – petroleum ether (b.p. 35 – 60°). The white crystalline product, after drying under reduced pressure at 25°, weighs 25.5 g. (98.6%), m.p. 94.5 – 95°, $[\alpha]_D^{25}$ −2.47° (c = 1, methanol). TLC of the product on plates precoated with silica gel G and purchased from Analtech, Inc., Newark, Delaware, each showed a single spot when developed with the following three solvent systems (solvents, volume ratio of solvents in the same order): 1-butanol – acetic acid – ethyl acetate – water, 1 : 1 : 1 : 1, R_f 0.90; 1-butanol – acetic acid – water, 4 : 1 : 1, R_f 0.81; 1-butanol – acetic acid – pyridine – water, 15 : 3 : 10 : 12, R_f 0.81. Analysis calculated for $C_{26}H_{42}N_2O_6S$: C, 61.15; H, 8.29; N, 5.49; S, 6.28. Found: C, 61.15; H, 8.52; N, 5.42; S, 6.44.

12. Other solvents including N,N-dimethylformamide and water may also be used to dissolve the products.

13. In some cases the catalyst was not entirely removed, and the filtrate contained trace amounts of palladium. In those instances the submitters evaporated the filtrate to a small volume and repeated the filtration using a large bed of Celite.

14. The melting point was taken on a Reichert hot stage microscope. This instrument is available from William J. Hacker and Company, Inc., P.O. Box 646, West Caldwell, New Jersey 07006.

15. The recrystallized product was analyzed by the submitters. Analysis calculated for $C_{18}H_{36}N_2O_4S$: C, 57.41; H, 9.63; N, 7.43; S, 8.51. Found: C, 57.60; H, 9.66; N, 7.37; S, 8.25. TLC (Note 10) run by the submitters showed a single spot for the product in each of three following solvent systems (solvents, volume ratio of solvents in the same order): chloroform – methanol – acetic acid, 85 : 10 : 5, R_f 0.60; 1-butanol – acetic acid – water, 4 : 1 : 1, R_f 0.58; 1-butanol – acetic acid – pyridine – water, 15 : 3 : 10 : 12, R_f 0.71.

TABLE I

REMOVAL OF N^α-BENZYLOXYCARBONYL GROUPS FROM SULFUR-CONTAINING PEPTIDES

Peptide[a]	Amount (g.) (mmole.)	Dimethylacetamide (ml.)	Ammonia (ml.)	Triethylamine (mmole.)	Palladium(II) Chloride (g.)	Yield[b] (%)
R = Lys-Thr-Phe-Thr-Ser-Cys-OBu^t Boc Bu^t | Bu^t Bu^t						
Z-R	3.05 (2.54)	10.0	300	40	2.08	98
Z-Phe-Trp-R	0.418 (0.272)	3.0	35	4.0	0.227	81
Z-Phe-Phe-Trp-R	0.263 (0.160)	3.5	30	4.0	0.133	91
Z-Asn-Phe-Phe-Trp-R	0.305 (0.170)	3.0	60	4.7	0.532	100
Z-Lys-Asn-Phe-Phe-Trp-R | Boc	0.368 (0.182)	3.0	70	4.7	0.596	89
Z-Cys-Lys-Asn-Phe-Phe-Trp-R | Bu^t Boc	0.281 (0.129)	2.0	40	10.0	0.596	86
Z-Gly-Cys-Lys-Asn-Phe-Phe-Trp-R | Bu^t Boc	0.178 (0.080)	1.3	40	10.0	0.380	85

[a] Protecting group abbreviations are as follows: N-benzyloxycarbonyl (Z), N-tert-butyloxycarbonyl (Boc), tert-butyl ether (Bu^t), and S-tert-butyl (Bu^t).

[b] The reaction time was 5.0–6.0 hours in all entries.

257

3. Discussion

The protection of the amino terminus of a peptide with the benzyloxycarbonyl group combined with protection of the carboxyl terminus and all side-chain functions with tert-butyl-derived groups[9] enables totally selective liberation of the terminal amino function by catalytic hydrogenolysis. This combination of protecting groups is currently considered "ideal" for peptide synthesis,[10] except for the rather serious limitation that catalyst poisoning has prevented its application to the preparation of peptides that contain cysteine, methionine, or other residues bearing divalent sulfur groups. The submitters have recently discovered[11] that catalyst poisoning is greatly diminished when liquid ammonia is used as solvent for palladium-catalyzed hydrogenation. This solvent enables quantitative cleavage of N^{α}-benzyloxycarbonyl groups on many protected peptides bearing S-protected cysteine residues. The method has been used successfully in syntheses of oxytocin[12] and somatostatin.[13]

The present procedures illustrate this method with the regeneration of L-methionine and the preparation of the tert-butyl ester of O-tert-butyl-L-seryl-S-tert-butyl-L-cysteine[13] from their respective N^{α}-benzyloxycarbonyl derivatives. No other procedures for the preparation of this protected dipeptide have been reported. These preparations and other peptide syntheses have served to establish the complete stability of the S-methyl substituent and various protecting groups including the tert-butyl ester (OBu^t), tert-butyl ether (Bu^t), N-tert-butyloxycarbonyl (Boc), S-tert-butyl (Bu^t), S-benzyl (Bzl), and S-acetamidomethyl (Acm) groups.[14] Table I summarizes the results of hydrogenations carried out with sulfur-containing peptides having chain lengths varying from 6 to 13 amino acid residues, demonstrating the stability of various protecting groups to this procedure for hydrogenolysis. In each case the N^{α}-benzyloxycarbonyl (Z) group was removed quantitatively and the tert-butyl based protecting groups were unaffected. As a result of these findings, sulfur-containing amino acids may now be used in peptide synthesis by the "ideal" combination of aminoterminal benzyloxycarbonyl protection with tert-butyl-type blocking groups on all other functions.

Since the use of N,N-dimethylacetamide and triethylamine improved the rate and extent of cleavage of the N-benzyloxycarbonyl group in several difficult cases, these additives have been incorporated into the submitters' standard procedure and are included in the present procedures. Deprotection with this method has been carried out with as much as 25 g. of the protected peptide.

1. Chemical Research Department, Hoffmann-La Roche Inc., Nutley, New Jersey 07110.
2. Paper VI in the series "Reactions in Liquid NH$_3$"; for paper V, see ref. 13.
3. H. Willstätter and E. Waldschmidt-Leitz, Ber. Dtsch. Chem. Ges., 54, 113 (1921).
4. H. Wieland, Ber. Dtsch. Chem. Ges., 45, 484 (1912).
5. M. H. Jimenez, Org. Prep. Proced. Int., 10, 295 (1978).
6. "The Merck Index," 8th ed., Merck & Co., Inc., Rahway, New Jersey, 1968, p. 675.
7. J. P. Greenstein and M. Winitz, "Chemistry of the Amino Acids," Vol. 3, Wiley, New York, 1961, p. 2125.
8. A. Chimiak, T. Kolasa, and J. F. Biernat, Z. Chem., 12, 264 (1972).
9. R. Schwyzer and P. Sieber, Nature (London), 199, 172 (1963).
10. R. Schwyzer, Naturwissenschaften, 53, 189 (1966); F. M. Finn and K. Hofmann, "The Synthesis of Peptides by Solution Methods with Emphasis on Peptide Hormones," in H. Neurath

and R. L. Hill, Eds., "The Proteins," 3rd ed., Vol. 2, Academic Press, New York, 1976, pp. 105 – 253.
11. J. Meienhofer and K. Kuromizu, *Tetrahedron Lett.*, 3259 (1974).
12. K. Kuromizu and J. Meienhofer, *J. Am. Chem. Soc.*, **96,** 4978 (1974).
13. A. M. Felix, M. H. Jimenez, T. Mowles, and J. Meienhofer, *Int. J. Pept. Protein Res.*, **11,** 329 (1978).
14. A procedure for *S*-acetamidomethylation is described in *Org. Synth.*, **Coll. Vol. 6,** 5 (1988).

ESTERIFICATION OF HINDERED ALCOHOLS:
tert-BUTYL *p*-TOLUATE

(Benzoic acid, 4-methyl-, 1,1-dimethylethyl ester)

$$n\text{-}C_4H_9Li + HOC(CH_3)_3 \longrightarrow LiOC(CH_3)_3 + n\text{-}C_4H_{10}$$

Submitted by G. P. Crowther,[1] E. M. Kaiser,[2]
R. A. Woodruff,[2] and C. R. Hauser[1,3]
Checked by A. Brossi, R. A. LeMahieu, and P. LaSalle

1. Procedure

A 200-ml., one-necked, round-bottomed flask equipped with a Claisen adapter, a condenser, an addition funnel, and a magnetic stirring bar is charged with 50 ml. of *tert*-butyl alcohol (Note 1). Under nitrogen, 22.6 ml. of a 1.55 M solution (0.0350 mole) of *n*-butyllithium in hexane (Note 2) is added slowly from a syringe (Note 3), giving a turbid reaction mixture. A water bath is used to keep the mixture near room temperature. After stirring for 15 minutes, a solution of 5.42 g. (0.0351 mole) of *p*-toluoyl chloride (Note 4) in 25 ml. of anhydrous diethyl ether (Note 5) is added dropwise to the stirred mixture. The resulting yellow slurry is stirred at room temperature for 15 hours (Note 6). The yellow suspension (Note 7) is transferred with 100 ml. of ether to a separatory funnel and washed with three 25-ml. portions of saturated sodium chloride, and dried over magnesium sulfate. The ether is removed by distillation, and the residual oil distilled under reduced pressure, yielding a small forerun (0.10 g.) and 5.31 – 5.51 g. (79 – 82%) of *tert*-butyl *p*-toluate, b.p. 98 – 101° (4.2 mm.) (Note 8).

2. Notes

1. *tert*-Butyl alcohol (Eastman Organic Chemicals white label) was dried by distillation from calcium hydride.
2. The solution of 1.55 M *n*-butyllithium in hexane was obtained from Foote Mineral Company.

TABLE I
ESTERS PREPARED BY ALKOXIDE METHODS

Ester	Yield, %	Ester	Yield, %
$C_6H_5CH_2CO_2C(CH_3)_3$	47[a]	$C_6H_5CO_2CH_2CH_2CH(C_6H_5)_2$	70[b]
		$C_6H_5CO_2C(CH_3)_3$	89[c]
		$C_6H_5CO_2CH_3(C_2H_5)_2$	87[c]
		$C_6H_5CO_2C(C_2H_5)_3$ CH_3	94[c]
		$C_6H_5CO_2CH(t\text{-}C_4H_9)$	76[c]
		$C_6H_5CO_2CH_2(t\text{-}C_4H_9)$	78[c]
	72[b]		70[c]
			69[c]

260

$C_6H_5CO_2CH_2$—◁   91[c]

$C_6H_5CO_2$—⬡—CH_3 94[c]

H_3C—⬡—$SO_3CH_2CH(C_6H_5)_2$ 85[d]

$(CH_3)_3CCO_2C(CH_3)_3$	64[c]
$(CH_3)_3CCO_2C(C_2H_5)_3$	75[c]
$C_6H_5CH=CHCO_2C(CH_3)_3$	88[c]
$C_6H_5CH_2CH_2CO_2C(CH_3)_3$	72[c]

$(i\text{-}C_3H_7)_2CHCO_2C(C_2H_5)_3$	88[e]
$(t\text{-}C_4H_9)CH_2CO_2C(C_2H_5)_3$	86[e]
$(t\text{-}C_4H_9)_2CHCO_2C(C_2H_5)_3$	30[e]

[a] As described in the accompanying procedure.

[b] As described in the accompanying procedure except ether used as solvent (see Note 8).

[c] Prepared in refluxing tetrahydrofuran with 1.0 equivalent of appropriate alcohol, 1.1 equivalents of n-butyllithium, and 1.1 equivalents of acid chloride.[4]

[d] Prepared by adding an equivalent amount of p-toluenesulfonyl chloride to a suspension of sodium 2,2-diphenylethoxide in ether.[6]

[e] Prepared by adding 0.50 equivalent of acid chloride to an ether suspension of sodium triethylmethoxide, which was obtained from 0.52 equivalent of sodium amide and 0.55 equivalent of triethylcarbinol.[7]

3. Formation of the lithium *tert*-butoxide in this manner is very exothermic and causes the hexane to boil during addition.

4. *p*-Toluoyl chloride was prepared by treating *p*-toluic acid (Eastman Organic Chemicals white label) with thionyl chloride (Eastman Organic Chemicals white label). The *p*-toluoyl chloride used was distilled, b.p. 48 – 49° (0.1 mm.).

5. Anhydrous ether was distilled from lithium aluminum hydride and stored over sodium ribbon prior to use.

6. In one instance an additional 75 ml. of anhydrous ether was added to make the slurry less viscous. The ester was obtained in the same yield in another run after stirring only 30 minutes.

7. Alternatively, the reaction mixture may be concentrated with a rotary evaporator, removing excess *tert*-butyl alcohol. Ether and water are added, and the mixture transferred to the separatory funnel; the yield of ester is unchanged.

8. With the same procedure *tert*-butyl phenylacetate has been prepared in 47% yield.[4] When esters of less common alcohols were prepared, anhydrous ether was used as a solvent instead of excess alcohol, with equivalent amounts of alcohol, *n*-butyllithium, and acid chloride employed. Thus, the triethylcarbinol ester of *p*-toluic acid and the 2,2-diphenylethanol ester of benzoic acid have been prepared in 72 and 70% yields, respectively.

3. Discussion

The present procedure[4] is an especially effective method for the synthesis of esters of aromatic acids and hindered tertiary alcohols or of acid-labile alcohols such as 2,2-diphenylethanol. The yields are excellent, and the reaction procedure is simple. The method is illustrated by the preparation of *tert*-butyl *p*-toluate, a compound that could not be prepared by a conventional method[5] of esterification involving the acid chloride and *tert*-butyl alcohol in the presence of N,N-dimethylaniline. Examples of esters prepared by this method are illustrated in Table I.

1. Chemistry Department, Duke University, Durham, North Carolina 27706. This work was supported at Duke University by the Army Research Office (Durham).
2. Chemistry Department, University of Missouri, Columbia, Missouri 65211.
3. Deceased January 6, 1970.
4. E. M. Kaiser and R. A. Woodruff, *J. Org. Chem.*, **35**, 1198 (1970).
5. C. R. Hauser, B. E. Hudson, B. Abramovitch, and J. C. Shivers, *Org. Synth.*, **Coll. Vol. 3**, 142 (1955).
6. P. J. Hamrick, Jr., and C. R. Hauser, *J. Org. Chem.*, **26**, 4199 (1961).
7. M. S. Newman and T. Fukunaga, *J. Am. Chem. Soc.*, **85**, 1176 (1963).

PEPTIDE SYNTHESES USING *N*-ETHYL-5-PHENYLISOXAZOLIUM-3'-SULFONATE: CARBOBENZOXY-L-ASPARAGINYL-L-LEUCINE METHYL ESTER AND *N*-CARBOBENZOXY-3-HYDROXY-L-PROLYLGLYCYLGLYCINE ETHYL ESTER

[L-Leucine, *N*-[*N*²-[(phenylmethoxy)carbonyl]-L-asparginyl]-, methyl ester and Glycine, *N*-[3-hydroxy-1-[(phenylmethoxy)carbonyl]-L-prolyl]-, ethyl ester]

Submitted by R. B. Woodward[1] and R. A. Olofson[2]
Checked by David L. Carroll, James C. Powers, and Herbert O. House

1. Procedure

A. *Carbobenzyloxy-L-asparaginyl-L-leucine methyl ester.* A mixture of 2.024 g. (0.00800 mole) of *N*-ethyl-5-phenylisoxazolium-3'-sulfonate (Note 1) and 20 ml. of nitromethane (Note 2) is prepared in a 50-ml., glass-stoppered Erlenmeyer flask and stirred vigorously, at room temperature, with a magnetic stirrer (Note 3). A solution of 2.128 g. (0.00800 mole) of carbobenzyloxy-L-asparagine (Note 4) and 810 mg. (0.00802 mole) of triethylamine (Notes 5 and 6) in 15 ml. of nitromethane (Note 2) is added. Stirring is continued until dissolution of the isoxazolium salt is practically complete, giving a pale yellow solution (*ca.* 8 minutes is required; Note 3), before 1.452 g. (0.00800 mole) of L-leucine methyl ester hydrochloride (Note 7) is added, followed by a solution of 810 mg. (0.00802 mole) of triethylamine (Note 5) in 5 ml. of nitromethane (Note 2). The resulting mixture is stirred overnight at room temperature, during which time some solid may separate from the solution. The mixture is then transferred to a 100-ml., round-bottomed flask and concentrated with a rotary evaporator. The residue is triturated with warm (ca. 60°) aqueous 0.5% sodium hydrogen carbonate, and the resulting suspension is cooled and allowed to stand at 5° for at least 4 hours. The precipitate is collected on a filter, washed thoroughly with a total of 25 ml. of cold water, and dried, leaving 2.36 – 2.65 g. (75 – 84%) of the crude peptide, which melts within the range of 168 – 178°. Recrystallization from acetone – water provides 2.09 – 2.21 g. (66 – 70%) of pure carbobenzyloxy-L-asparaginyl-L-leucine methyl ester as fine, colorless crystals, m.p. 175 – 176.3°, $[\alpha]_D^{24}$ −27.7° (c 1.0, methanol) (Note 8).

B.   N-*Carbobenzyloxy-3-hydroxy-L-prolylglycylglycine ethyl ester*.   A mixture of 760 mg. (0.00300 mole) of N-ethyl-5-phenylisoxazolium-3'-sulfonate (Note 1) and 5 ml. of acetonitrile (Note 9) is prepared in a 25-ml., glass-stoppered Erlenmeyer flask. The mixture is cooled to 0° with an ice bath and stirred vigorously with a magnetic stirrer (Note 3). A solution of 796 mg. (0.00300 mole) of N-carbobenzyloxy-3-hydroxy-L-proline (Note 10) and 304 mg. (0.00301 mole) of triethylamine (Notes 5 and 6) in 5 ml. of acetonitrile (Note 9) is added. The cold (0 – 5°) mixture is stirred vigorously until almost all of the isoxazolium salt has dissolved (ca. 1 – 1.5 hours; Note 3), then treated with 590 mg. (0.00300 mole) of glycylglycine ethyl ester hydrochloride (Note 11), followed by addition of a solution of 304 mg. (0.00301 mole) of triethylamine (Note 5) in 6 ml. of acetonitrile (Note 9). After this mixture has been stirred for 1 hour at 0 – 5°, the resulting pale yellow solution is allowed to warm to room temperature and stirred overnight. The reaction solution is then transferred to a 100-ml., round-bottomed flask and concentrated with a rotary evaporator. The residue is partitioned between 10 ml. of aqueous 1% sodium hydrogen carbonate and 50 ml. of ethyl acetate. After the phases are separated, the aqueous phase is extracted with three 10-ml. portions of ethyl acetate, and the combined organic extracts are dried over anhydrous sodium sulfate. The organic solution is concentrated with a rotary evaporator, the residual white crystalline solid is triturated with 20 ml. of water, and the resulting slurry is cooled to 5° and allowed to stand overnight in a refrigerator. Filtration of the cold mixture provides a solid, which is washed with a small amount of cold water and dried. The combined aqueous filtrates are concentrated with a rotary evaporator, and the residual solid is again triturated with 10 ml. of water, allowed to stand overnight at 5°, and filtered. The solids collected total 958 – 980 mg. (78 – 80%) of N-carbobenzyloxy-3-hydroxy-L-prolylglycylglycine ethyl ester, m.p. 145 – 146°, $[\alpha]_D^{24}$ −18.5° (c 1.0, ethanol) (Note 12).

2. Notes

1. This isoxazolium salt reagent is commonly known and sold as Woodward's Reagent K (see, for example: The Merck Index, 9th ed., 1976, entry 9715). This salt (10 g.) (obtained from the Aldrich Chemical Company, Inc.) was dissolved in 45 ml. of 1 M hydrochloric acid and reprecipitated by the slow addition with swirling of 400 ml. of acetone. The salt was collected, washed with 300 ml. of acetone, and dried overnight at 25° under reduced pressure (<1 mm.) to give a fluffy product, m.p. 206 – 208° (dec.). An

isomeric salt, N-ethyl-5-phenylisoxazolium-4'-sulfonate, which may be obtained by the usual synthetic procedure,[3] is also useful in peptide synthesis.

2. An anhydrous spectral grade of nitromethane, obtained from Fisher Scientific Company, was used without purification.

3. Since the initially formed enol ester rearranges slowly to an imide,[4] the yield depends on the rate at which the isoxazolium salt reacts, and that rate is increased by vigorous stirring. The reaction time for the activation step is approximately 8 minutes in nitromethane at 25° and approximately 1 hour in acetonitrile at 0°. In reactions performed in acetonitrile, the checkers did not obtain complete solution. The reaction flask should be kept in a water bath to minimize heat transfer from the magnetic stirrer to the reaction mixture.

4. Carbobenzyloxy-L-asparagine (obtained from the Aldrich Chemical Company, Inc.) was recrystallized from acetone – water to give the pure acid, m.p. 162 – 163°.

5. Commercial triethylamine (obtained from Eastman Organic Chemicals), was distilled (b.p. 89 – 90°) from phosphorus pentoxide and stored under a nitrogen atmosphere. Since the presence of even a small excess of triethylamine is deleterious in these reactions, the quantities of this amine used should be measured by weight rather than volume.

6. The triethylamine salts of peptide acids are often relatively insoluble in acetonitrile or nitromethane; therefore, the supersaturated solution formed on mixing the amine and the acid should be added to this reaction mixture immediately, before crystallization occurs. If crystallization does occur, the mixture should be heated to dissolve the salt, cooled rapidly, and added to the reaction mixture immediately. If it is impossible to obtain a solution of the salt, the peptide acid and then the triethylamine solution may be added separately to the reaction mixture with only a small sacrifice in yield.

7. L-Leucine methyl ester hydrochloride (obtained from the Mann Research Laboratories, Inc.) was recrystallized from methanol – ether, m.p. 149 – 150°.

8. The reported[5] rotation for this product is $[\alpha]_D^{23}$ $-26.3°$ (c 2.0, methanol). IR (KBr) cm.$^{-1}$: 3455, 3400, 3295 (amide NH stretching), 1737 (urethane and ester $C{=}O$), 1692 (amide $C{=}O$), 1647, 1535 (amide NH bending); ^{1}H NMR (dimethyl sulfoxide-d_6), δ (multiplicity, coupling constant J in Hz., number of protons, assignment): 0.7 – 1.1 (m, 6H, 2CH_3), 1.4 – 1.8 (m, 3H, aliphatic CH), 2.3 – 2.7 (m, 2H, CH_2CO), 3.62 (s, 3H, OCH_3), 4.1 – 4.7 (m, 2H, 2 NCHCO), 5.03 (s, 2H, benzylic CH_2), 6.92 (broad, 1H, NH), 7.25 (broad, 2H, NH_2), 7.35 (s, 5H, aromatic H), 8.17 (broad d, $J = 7$, 1H, NH); mass spectrum m/e (relative intensity): 393 (M$^+$, 1), 316 (5), 210 (12), 177 (15), 108 (37), 91 (100), 86 (50), 43 (28).

·9. Reagent grade acetonitrile (obtained from Eastman Organic Chemicals) was dried over Linde type 4A molecular sieves, decanted, and used without further purification.

10. N-Carbobenzyloxy-3-hydroxy-L-proline (purchased from either Mann Research Laboratories or Sigma Chemical Company) was recrystallized prior to use. The submitters recommend recrystallization from water, but the checkers found it easier to recrystallize the material from a mixture of ethyl acetate and petroleum ether (b.p. 30 – 60°). In either case, the checkers added a seed crystal to induce crystallization. The recrystallized product melted at 106 – 107°.

11. Glycylglycine ethyl ester hydrochloride (obtained from Nutritional Biochemicals

Corporation) was recrystallized twice from mixtures of ethanol and ether to separate the pure salt, m.p. 181 – 182°.

12. The reported[6] rotation for this product (m.p. 144 – 145°) is $[\alpha]_D^{21}$ −11.1° (c 1.0, ethanol). The submitters report that the melting point of the product is not changed by recrystallization. IR (CHCl$_3$) cm.$^{-1}$: 3390, 3295 (OH and NH stretching), 1730 (ester and urethane C=O), 1620 (broad, amide C=O), 1520 (broad, amide NH bending); ^{1}H NMR (CDCl$_3$), δ (multiplicity, coupling constant J in Hz., number of protons, assignment): 1.2 (t, $J = 7$, 3H, OCH$_2$CH$_3$), 2.0 – 2.4 (m, 3H, aliphatic CH), 3.5 – 4.7 (m, 10H, OH and aliphatic CH), 5.08 (s, 2H, benzylic CH$_2$), 7.30 (s, 5H, aromatic H), 7.6 (broad, 2H, 2NH); mass spectrum, m/e (relative intensity): 207 (6), 149 (17), 108 (93), 107 (64), 91 (18), 79 (100), 77 (55), 65 (13).

3. Discussion

These procedures illustrate the use of N-ethyl-5-phenylisoxazolium-3′-sulfonate as a reagent for peptide synthesis.[3,4] Procedure A is recommended for peptides not soluble in either organic solvents or water. Procedure B illustrates the formation of a peptide that is soluble both in organic solvents and in water. For peptides soluble in organic solvents and insoluble in water, the submitters recommend the use of Procedure B, except that the peptide product may be recovered directly from its solution in ethyl acetate after this organic solution has been washed successively with aqueous 5% sodium hydrogen carbonate, water, 1 M hydrochloric acid, and water. Table I summarizes the preparation of various peptides by these procedures. Additional complex examples from other laboratories are listed elsewhere.[3b]

Since there are a number of excellent and extensive reviews of peptide chemistry,[7 – 10] no attempt will be made here to describe the known methods of peptide synthesis. Absolute comparisons of the procedure presented herein with other methods are impossible due to a number of factors: (a) the practice of many authors of reporting only yields of crude materials of unknown purity, (b) the necessities that prompt many experimenters to use a large excess of either the carboxyl component or the amine component in peptide synthesis, and (c) the difficulty of comparing this one-step synthesis of the amide bond with the ordinary two- and three-step syntheses (carboxyl activation, isolation of the free amine component from its hydrohalide, and aminolysis of the activated carboxyl group).

The method illustrated here does have, however, several excellent features.[11] The yields are very good even in the synthesis of asparaginyl and glutaminyl peptides, which are ordinarily very difficult to prepare in reasonable yield. Furthermore, all by-products are water-soluble and, therefore, easily removed from the product peptide derivative. One recrystallization, even under conditions of almost complete precipitation, usually suffices to yield pure material. A stringent test of this statement is the synthesis of carbobenzyl-oxyhydroxy-L-prolylglycylglycine ethyl ester, a peptide which is itself very soluble in water. The water solubility of the starting isoxazolium salt and of the by-products from coupling has also been useful in studies of the reaction of proteins with the isoxazolium salt in aqueous solution[12,13] and to effect intermolecular cross linking of polypeptides.[14] Finally, use of the isoxazolium salt procedure for activation of the carboxyl groups of serine, tyrosine, and threonine offers the advantage that protection of the hydroxyl groups

TABLE I[3]

Peptide	Reaction Procedure	Yield
Z-(N^ϵ-Z)-Lys·Gly-OEt[Glycine, N-[N^2,N^6-bis-[(phenylmethoxy)carbonyl]-L-lysyl]-, ethyl ester]	B	95%
Z-Phe·Gly-OEt[Glycine, N-[N-[(phenylmethoxy)carbonyl]-L-phenylalanyl]-, ethyl ester]	B	93%
Z-Phe·Leu-OMe[L-Leucine, N-[N-[(phenylmethoxy)carbonyl]-L-phenylalanyl]-, methyl ester]	B	90%
Z-Met·Gly·Gly-OEt[Glycine, N-[N-[N-(phenylmethoxy)carbonyl]-L-methionyl]glycyl-, ethyl ester]	B	86%
Phth-Gly·Gly-OEt[Glycine, N-[(1,3-dihydro-1,3-dioxo-2H-isoindol-2-yl)acetyl]-, ethyl ester]	B	88%
Z-Gly·Gly·Tyr-OMe[L-Tyrosine, N-[N-[N-[(phenylmethoxy)carbonyl]glycyl]glycyl]-, methyl ester]	B	84%
Z-Gly·DL-Phe·Gly-OEt[Glycine, N-[N-[N-[(phenylmethoxy)carbonyl]glycyl]-DL-phenylalanyl]-, ethyl ester]	A	89%
Z-Gly·Gly·Gly-OEt[Glycine, N-[N-[N-[(phenylmethoxy)carbonyl]glycyl]glycyl]-, ethyl ester]	A	91%
Z-Gly-NHBz[Carbamic acid, [2-oxo-2-[(phenylmethyl)amino]ethyl]-, phenylmethyl ester]	A	94%
Z-Asp·Gly-OEt [Glycine, N-[N^2-[(phenylmethoxy)carbonyl]-L-asparaginyl]-, ethyl ester]	A	80%
Z-Gln·Val-OMe[L-Valine, N-[N^2-[(phenylmethoxy)carbonyl]-L-glutaminyl]-, methyl ester]	A	77%
Z-Gln·Tyr-OMe[L-Tyrosine, N-[N^2-[(phenylmethoxy)carbonyl]-L-glutaminyl]-, methyl ester]	A	75%

is often unnecessary.[10b] Among the disadvantages of this method of peptide synthesis are the high cost of the isoxazolium salt and the limitations in the choice of solvent.

In tests devised to determine the amount of racemization to be expected in peptide syntheses in which the carboxyl component is a di- or higher peptide, this method ranks below the racemization-resistant azide procedure but above almost all other standard methods. Using a very sensitive and accurate isotope dilution assay, 1% racemization was observed[15] in the formation of the Anderson test peptide (Z-Gly·Phe·Gly-Oet)[16] and 7% racemization was observed[15] in the formation of the Young test peptide (Bz-Leu·Gly·OEt)[17] under optimized conditions. (The Young test was designed to exaggerate racemization problems, thus permitting more accurate studies of the effects of reaction condition variations.)

1. 1965 Nobel Laureate in Chemistry; deceased July 8, 1979; formerly at the Department of Chemistry, Harvard University, Cambridge, Massachusetts 02138.
2. Present address: Department of Chemistry, The Pennsylvania State University, University Park, Pennsylvania 16802.

3. (a) R. B. Woodward, R. A. Olofson, and H. Mayer, *J. Am. Chem. Soc.*, **83**, 1010 (1961); (b) R. B. Woodward, R. A. Olofson, and H. Mayer, *Tetrahedron*, **Suppl. 8**, Pt. I, 321 (1967).

4. R. B. Woodward and R. A. Olofson, *J. Am. Chem. Soc.*, **83**, 1007 (1961); R. B. Woodward and R. A. Olofson, *Tetrahedron*, **Suppl. 7**, 415 (1966).

5. W. Koenig and R. Geiger, *Chem. Ber.*, **103**, 788 (1970).

6. N. C. Davis and E. L. Smith, *J. Biol. Chem.*, **200**, 373 (1953).

7. G. W. Kenner and M. Goodman, *Adv. Protein Chem.*, **12**, 465 (1957).

8. J. P. Greenstein and M. Winitz, "Chemistry of Amino Acids," Wiley, New York, 1961, p. 763.

9. E. Schröder and K. Lübke, "The Peptides," Vol. 1, Academic Press, New York, 1965.

10. (a) M. Bodanszky and M. Ondetti, "Peptide Synthesis," Wiley, New York, 1966; (b) Y. S. Klausner and M. Bodanszky, *Synthesis*, 453 (1972).

11. In a survey of the methods used for peptide synthesis in 1968, the isoxazolium salt method was used in 7% of the examples surveyed: J. H. Jones in "Amino Acids, Peptides, and Proteins," Vol. 2, The Chemical Society, London, 1970, p. 145.

12. P. Bodlaender, G. Feinstein, and E. Shaw, *Biochemistry*, **8**, 4941 (1969).

13. G. Feinstein, P. Bodlaender, and E. Shaw, *Biochemistry*, **8**, 4949 (1969).

14. P. S. Marfey, T. J. Gill, and H. W. Kunz, *Biopolymers*, **3**, 27 (1965).

15. D. S. Kemp, S. W. Wang, G. Busby, and G. Hugel, *J. Am. Chem. Soc.*, **92**, 1043 (1970).

16. G. W. Anderson and F. M. Callahan, *J. Am. Chem. Soc.*, **80**, 2902 (1958).

17. M. W. Williams and G. T. Young, *J. Chem. Soc.*, 881 (1963).

CARBONYL CYANIDE

(Propanedinitrile, oxo-)

Submitted by E. L. Martin[1]
Checked by R. Kottke and W. D. Emmons

1. Procedure

Caution! Carbonyl cyanide and water react with explosive violence, forming hydrogen cyanide and carbon dioxide. This preparation should be carried out in a good hood with shielding, and rubber gloves should be worn.

A 500-ml., three-necked flask equipped with a magnetic stirring bar, a pressure-equalizing dropping funnel, a thermometer, and a 25-cm. Vigreux column attached to a trap cooled in an acetone – dry ice mixture is charged with 100 ml. of diethyl phthalate (Note 1) and 43 g. (0.30 mole) of tetracyanoethylene oxide (Note 2). The dropping funnel is charged with 44 g. (0.30 mole) of distilled *n*-butyl sulfide, the pressure is reduced to 5 – 20 mm. with a water aspirator, and the reaction flask is warmed with a 50° water bath. The *n*-butyl sulfide is added dropwise with stirring over 20 – 25 minutes, and the internal temperature is maintained at 50 ± 2° by controlling the temperature of the water bath. The reaction is exothermic at the start, but it is necessary to supply heat toward the end of the addition. The internal temperature is increased to 80° over 10 – 15 minutes after the sulfide has been added. The vacuum on the system is released by the introduction of nitrogen. The solid carbonyl cyanide that has collected in the trap is allowed to warm to room temperature under nitrogen, and 2 g. of tetracyanoethylene oxide (Note 3) and a boiling stone are added. The mixture is then warmed to 50°, and the product is distilled under reduced pressure (5 – 20 mm.) into a second trap cooled in an acetone – dry ice mixture. Another portion of tetracyanoethylene oxide (1 g.) is added to the distillate, and the carbonyl cyanide is distilled, again under reduced pressure (5 – 20 mm.), into a distillation flask cooled in acetone – dry ice mixture. Fractionation of the distillate through a 20-cm. column packed with glass helices gives 20.8 – 21.8 g. (86 – 91%) (Note 4) of faintly yellow carbonyl cyanide, b.p. 65 – 66° (Notes 5, 6, and 7).

2. Notes

1. The diethyl phthalate is freed of traces of water and ethanol by distilling about 5% of it under reduced pressure, b.p. 185° (20 mm.) and then cooling the residue with protection from moisture. Moisture must be excluded because carbonyl cyanide reacts vigorously with water.

2. Tetracyanoethylene oxide was prepared as described in *Org. Synth.*, **Coll. Vol. 5,** 1007 (1973).

3. The distillate is treated with tetracyanoethylene oxide to remove the small amount of *n*-butyl sulfide that codistills with carbonyl cyanide.

4. The yield is based on the amount of tetracyanoethylene oxide initially charged.

5. There is very little if any low-boiling material when the reaction is carried out as described and care is taken to have the apparatus dry and exclude moisture.

6. GC analyses of various cuts indicate essentially pure carbonyl cyanide with traces of hydrogen cyanide and carbon dioxide. Both are hydrolysis products of carbonyl cyanide and probably are formed because of traces of moisture on the column.

7. Carbonyl cyanide has m.p. −38°, n_D^{19} 1.3923.[2]

3. Discussion

The procedure described is that of Linn, Webster, and Benson.[3] Carbonyl cyanide has previously been prepared by the pyrolysis of the diacetyl derivative of diisonitrosoacetone, a multistep process that suffers from low yield, lack of reproducibility, and risk of explosion.[2] The present procedure provides a convenient high-yield synthesis of carbonyl cyanide.

Carbonyl cyanide reacts with alcohols and phenols, giving cyanoformate esters;[4] with primary and secondary amines, giving cyanoformamides; with N,N-dimethylaniline, giving bis[4-(dimethylamino)phenyl]malononitrile; and with pyrrole, giving 2-(cyanoformyl)pyrrole.[5] With olefins of the type $C{=}C{-}CH$, products with structures $C{=}CC{-}C(CN)_2OH$, $C{=}CC{-}COCN$, and $C{=}CC{-}C(CN)_2OCOCN$ are obtained, depending on the nature of the olefin and the reaction conditions.[6] Carbonyl cyanide also undergoes Diels-Alder reactions with some conjugated dienes, yielding dicyanodihydropyrans.[7]

1. Central Research Department, Experimental Station, E. I. du Pont de Nemours & Co. (Inc.), Wilmington, Delaware 19898.
2. R. Malachowski, L. Jurkiewicz, and J. Wojtowicz, *Ber. Dtsch. Chem. Ges.*, **70**, 1012 (1937).
3. W. J. Linn, O. W. Webster, and R. E. Benson, *J. Am. Chem. Soc.*, **87**, 3651 (1965); W. J. Linn, U.S. Pat. 3,115,517 (1963) [*Chem. Abstr.*, **60**, 7919a (1964)].
4. O. Achmatowicz, K. Belniak, C. Borecki, and M. Leplawy, *Rocz. Chem.*, **39**, 1443 (1965) [*Chem. Abstr.*, **64**, 17457f (1966)].
5. R. Malachowski and J. Jankiewicz-Wasowska, *Rocz. Chem.*, **25**, 35 (1951) [*Chem. Abstr.*, **47**, 10483f (1953)].
6. O. Achmatowicz and F. Werner-Zamojska, *Bull. Acad. Pol. Sci. Cl. 3*, **5**, 923 (1957) [*Chem. Abstr.*, **52**, 6333a (1958)].
7. O. Achmatowicz and A. Zamojski, *Rocz. Chem.*, **35**, 799 (1961) [*Chem. Abstr.*, **56**, 7257c (1962)].

3-CHLOROCYCLOBUTANECARBOXYLIC ACID

(Cyclobutanecarboxylic acid, 3-chloro-)

Submitted by GARY M. LAMPMAN and JAMES C. AUMILLER[1]
Checked by G. NELSON and K. B. WIBERG

1. Procedure

Caution! Benzene has been identified as a carcinogen; OSHA has issued emergency standards on its use. All procedures involving benzene should be carried out in a well-ventilated hood, and glove protection is required.

A 2-l., three-necked, round-bottomed flask equipped with a Trubore stirrer and paddle is charged with 172.8 g. (1.200 moles) of 1,1-cyclobutanedicarboxylic acid (Note 1) and 1500 ml. of benzene. The mixture is stirred and heated at reflux, and 200 ml. of benzene and benzene-water azeotrope is removed by distillation to ensure anhydrous conditions. The flask is then fitted with an addition funnel and a reflux condenser attached to a drying tube. Stirring and heating are continued, and over a 40-minute period, 170 g. (102 ml., 1.26 moles) of sulfuryl chloride (Note 2) is added from the funnel, while 4.0 g. of benzoyl peroxide (Note 3) is added simultaneously in small portions through the top of the condenser. After a short induction period, hydrogen chloride and sulfur dioxide are evolved. After the addition is complete, heating at reflux is maintained for 22 hours. The solid is dissolved after 1 hour, leaving a light brown solution. After the heating period is complete, the benzene is removed by distillation, and the residue is heated to 190–210° for 45 minutes to effect decarboxylation. The black residue is transferred to a small flask and distilled under vacuum through a 6-cm. Vigreux column. After a forerun of about 25–30 g. (Note 4), 65–79 g. (40–49%) of *cis-* and *trans*-3-chlorocyclobutanecarboxylic acid is collected as a light yellow liquid, b.p. 131–137° (15 mm.) n_D^{24} 1.4790 (Note 5). A black residue remains in the distillation flask.

2. Notes

1. Diethyl 1,1-cyclobutanedicarboxylate is prepared by the method in *Org. Synth., Coll. Vol. 4*, 288 (1963). The diester is isolated in 55% yield, b.p. 111–114° (16 mm.). The diester can be saponified by the method in *Org. Synth., Coll. Vol. 3*, 213 (1955), but omitting the barium chloride step, to give the diacid. This material upon recrystallization from ethyl acetate gives the diacid in high purity. The diacid may also be purchased from Aldrich Chemical Company, Inc.

2. Eastman Organic Chemicals or Matheson, Coleman and Bell practical grade material was distilled before use. Since hydrogen chloride and sulfur dioxide are evolved, the preparation should be carried out in an efficient hood.

3. Eastman Organic Chemicals white label material was used.

4. The forerun contains cyclobutanecarboxylic acid and 3-chlorocyclobutanecarboxylic acid, b.p. 100–130° (15 mm.) n_D^{24} 1.4623. The presence of cyclobutanecarboxylic acid indicates that some of the diacid was not chlorinated. Attempts were made to reduce the amount of unchlorinated product by increasing the amount of sulfuryl chloride. Instead, this increased the amount of a dichlorinated impurity which is difficult to separate from the desired product.

5. The product is analyzed by GC at 190° on a Beckman GC-2 chromatograph equipped with a 180 cm. × 6 mm. column (Beckman 17449) containing 42/60 Johns-Manville C-22 firebrick coated with Dow-Corning 550 silicone oil. The retention times are 8 and 9 minutes for the *trans-* and *cis-* compounds, respectively.

3. Discussion

3-Chlorocyclobutanecarboxylic acid has been prepared from the rather inaccessible 3-hydroxy-1,1-cyclobutanedicarboxylic acid.[2] The related 3-bromocyclobutanecarboxylic acid has also been prepared by an eight-step synthetic scheme.[3] The present method, based upon the procedure of Nevill, Frank, and Trepka,[4] affords the 3-chloro acid in high yield in one step. Thus this method provides a compound which cannot be easily made by other methods.

The use of sulfuryl chloride for free radical chlorination of aliphatic carboxylic acids gives mixtures of positional isomers;[5] however, with the cyclobutane ring, the attack is much more selective. The present method provides a procedure for free radical halogenation of a cyclobutane ring.

The conversion of 3-chlorocyclobutanecarboxylic acid to 1-bromo-3-chlorocyclobutane is described in *Organic Syntheses.*[6]

1. Department of Chemistry, Western Washington University, Bellingham, Washington 98225.
2. R. C. Jones, Ph.D. Thesis, Harvard University, 1941.
3. K. B. Wiberg and G. M. Lampman, *J. Am. Chem. Soc.*, **88**, 4429 (1966).
4. W. A. Nevill, D. S. Frank, and R. D. Trepka, *J. Org. Chem.*, **27**, 422 (1962).
5. M. S. Kharasch and H. C. Brown, *J. Am. Chem. Soc.*, **62**, 925 (1940).
6. G. M. Lampman and J. C. Aumiller, *Org. Synth., Coll. Vol. 6*, 179 (1988).

(Z)-4-CHLORO-4-HEXENYL TRIFLUOROACETATE

[Acetic acid, trifluoro-, (Z)-4-chloro-4-hexenyl ester]

$$CH_3C\equiv CH \xrightarrow[\text{2. BrCH}_2\text{CH}_2\text{CH}_2\text{Cl}]{\text{1. NaNH}_2, \text{ liquid ammonia}} CH_3C\equiv CCH_2CH_2CH_2Cl \xrightarrow[\text{reflux}]{\text{CF}_3\text{COOH}}$$

$$
\begin{array}{c}
H \\
\diagdown \\
C = C \\
\diagup \quad \diagdown \\
CH_3 \qquad Cl
\end{array}
\qquad
\begin{array}{c}
CH_2CH_2CH_2O\overset{\displaystyle O}{\overset{\|}{C}}CF_3 \\
\end{array}
$$

Submitted by P. E. Peterson and M. Dunham[1]
Checked by K.-C. Luk and G. Büchi

1. Procedure

A. *6-Chloro-2-hexyne*. A 2-l., three-necked, round-bottomed flask is equipped with a low-temperature condenser (Note 1), a gas-inlet, and a magnetic stirrer. Sodium hydroxide drying tubes are placed to precede the inlet and on the condenser. The system is purged with nitrogen, and approximately 650 ml. of anhydrous ammonia (Note 2) is condensed. Freshly cut sodium metal (20.2 g., 0.878 g.-atom) is added to the refluxing ammonia. After dissolution of the sodium (about 20 minutes is required), 0.2 g. of iron(III) nitrate is added. Two hours of stirring are allowed for conversion of the deep-blue solution of sodium in ammonia to sodium amide. The reaction mixture is cooled to $-73°$ using an acetone–dry ice bath and 35.3 g. (50.0 ml., 0.882 mole) of precondensed propyne (Notes 3, 4) is added in portions over a 1-minute period through a glass funnel precooled in dry ice (Note 4). The mixture is stirred, with continued cooling for 15 minutes, and 153.5 g. (0.975 mole) of 1-bromo-3-chloropropane (Note 5) is added from an addition funnel over 20 minutes (Note 6). After 30 minutes of additional stirring, 250 ml. of diethyl ether is added to the flask, the dry ice bath is removed, and the ammonia is allowed to evaporate (Note 7). Water (200 ml.) is added to the reaction vessel, and the resulting solution is transferred to a 1-l. separatory funnel. The water layer is removed and extracted once with 100 ml. of ether. The combined ether extracts are treated with 6 M hydrochloric acid until the aqueous layer is acidic (approximately 20 ml. is required). The ether layer is separated and dried in three stages over magnesium sulfate. Removal of the solvent and distillation of the crude product through a 20-cm. Widmer column (Note 8) yields 49.4–56.0 g. of 6-chloro-2-hexyne, b.p. 58–64° (20 mm.), which contained some 1-bromo-3-chloropropane (to be removed in the next step). The corrected yield is 29–31% (Note 9).

B. *4-Chloro-4-hexenyl trifluoroacetate*. A 500-ml., one-necked flask equipped with a magnetic stirrer is charged with 200 ml. of redistilled trifluoroacetic acid (Note 10) and 6-chloro-2-hexyne (0.16 mole, calculated from GC analysis) containing 1-bromo-3-chloropropane. The flask is fitted with a Friedrichs condenser, and the mixture is refluxed for 3 hours. The flask is transferred to a vacuum distillation apparatus; with the aid of an aspirator and a controlled leak, excess trifluoroacetic acid is removed under slightly reduced pressure, maintaining the pot temperature below 65° to prevent reactions of the acid with the double bond. The remaining trifluoroacetic acid is removed by pouring the product into 100 ml. of ice water and extracting with 100- and 50-ml. portions of dichloromethane. The organic extracts are brought to pH 7 with saturated sodium hydrogen carbonate (*ca.* 40 ml.). The resulting aqueous layer is extracted once with an additional 25 ml. of dichloromethane, and the combined dichloromethane extracts are dried in three stages by stirring over magnesium sulfate. The final stage is allowed to stand in a refrigerator for 24 hours. The solvent is removed by distillation through a 20-cm. Widmer column (Note 8) at atmospheric pressure. Distillation at reduced pressure gives 16.0–16.4 g. of 4-chloro-4-hexenyl trifluoroacetate, b.p. 86–89° (22 mm.) (Note 11), which contains trace amounts of 1-bromo-3-chloropropane (Note 12). On the basis of the alkyne present in the reactant the yield is 50–52%.

2. Notes

1. A cold finger condenser packed with dry ice and 2-propanol may be used. A Friedrichs condenser in combination with a circulating low-temperature bath (−70°) is more convenient.

2. Ammonia was purchased from Matheson Gas Products.

3. Propyne can be purchased from Linde Specialty Gases or Farchan Research Laboratories. Matheson Gas Products sells propyne also, but only in 100-lb. quantities.

4. The propyne (b.p. −23.2°) is precondensed to the mark in a volumetric flask cooled by acetone–dry ice. Evaporation of some propyne during addition will lead to a moderate molar excess of 1-bromo-3-chloropropane, regarded as desirable in preventing formation of diyne product.

5. 1-Bromo-3-chloropropane was purchased from Aldrich Chemical Company, Inc.

6. Sudden foaming occurred in a run involving insufficient cooling or overly rapid additions. Slow addition could lead to diyne product.

7. The checkers maintained the condenser at −5° during the evaporation to minimize loss of alkyne.

8. The submitters use a platinum spinning band apparatus for the distillation.

9. GC analysis at 79° using a flame detector in conjunction with a 183×0.32 cm. stainless-steel column containing Dow-Corning 550 fluid on silanized support gave peaks for 1-bromo-3-chloropropane (6.5 minutes) and 6-chloro-2-hexyne (9.3 minutes) whose areas were shown to be proportional to the mole fractions. The latter were determined by integration of the expanded (50 Hz. sweep width) 100 MHz. ^1H NMR spectrum in the region of overlapping triplets near δ 3.6.

10. The checkers purchased trifluoroacetic acid from Aldrich Chemical Company, Inc., and distilled it from phosphorous pentoxide. The submitters point out that some trifluoroacetic anhydride, whose effects have not been fully investigated, is obtained under these conditions. The submitters prefer to use trifluoroacetic acid which has been distilled through a glass packed column without the use of a drying agent.

11. n_D^{25} 1.4025; ^1H NMR (CCl$_4$), δ (multiplicity, coupling constant J in Hz., number of protons): 1.70 (d, $J = 7$, 3H), 1.8–2.6 (m, 4H), 4.35 (t, $J = 6$, 2H), 5.58 (q, $J = 7$, 1H).

12. 1-Bromo-3-chloropropane, b.p. 46–55° (15 mm.) is well separated in the early fractions. GC analysis at 120° (*cf.* Note 9) gives peaks of proportional areas for 1-bromo-3-chloropropane (1.2 minutes), 4-chloro-4-hexenyl trifluoroacetate (2.3 minutes), and traces of 4-chloro-4-hexen-1-ol (2.5 minutes).

3. Discussion

The presence, in the 4-chloro-4-hexenyl trifluoroacetate, of small amounts of two *cis–trans* pairs of products of addition of trifluoroacetic acid to the triple bond without concomitant halogen shift remains speculative. In any event, these compounds would be removed as ketones upon hydrolysis[2] of the trifluoroacetate. Both the 4-chloro-4-hexenyl trifluoroacetate and the alcohol resulting from its hydrolysis have been shown to contain 9% of the (*E*) isomer.[2] In the present study, the hydrogen decoupled ^{13}C magnetic resonance spectra of the ester and alcohol were shown to contain peaks attributable to approximately 9% of (*E*) isomer.

Unsymmetrical *trans*-vinyl halides have been prepared from acetylenic alcohols by Corey and co-workers[3] (as illustrated in the accompanying formulation) in connection with their synthesis of farnesol and *Cecropia* juvenile hormone. Several syntheses of vinyl halides (with identical R groups, *trans*) have been reported, including procedures involving halogenation and elimination,[4] addition of hydrochloric acid to alkynes,[5] and preparation from alkynes *via* vinyl alanes.[6] The synthesis of unsymmetrical *trans*-vinyl halides by 1,4-halogen shift reactions is exemplified by the procedure given here. Such compounds are potentially useful in the synthesis of compounds containing trisubstituted double bonds.[7]

$$R-C{\equiv}C-CH_2OH \xrightarrow[\text{2. } I_2\,(-78°)]{\substack{\text{1. LiAlH}_4\text{-AlCl}_3, \\ \text{tetrahydrofuran, reflux}}} \begin{array}{c} H \qquad CH_2OH \\ \diagdown \diagup \\ \diagup \diagdown \\ R \qquad I \end{array}$$

1. Department of Chemistry, University of South Carolina, Columbia, South Carolina 29208.
2. P. E. Peterson, R. J. Bopp, and M. M. Ajo, *J. Am. Chem. Soc.*, **92**, 2834 (1970).
3. (a) E. J. Corey, J. A. Katzenellenbogen, and G. H. Posner, *J. Am. Chem. Soc.*, **89**, 4245 (1967); (b) E. J. Corey, J. A. Katzenellenbogen, N. W. Gilman, S. A. Roman, and B. W. Erickson, *J. Am. Chem. Soc.*, **90**, 5618 (1968).
4. M. C. Hoff, K. W. Greenlee, and C. E. Boord, *J. Am. Chem. Soc.*, **73**, 3329 (1951).
5. R. C. Fahey and D. L. Lee, *J. Am. Chem. Soc.*, **89**, 2780 (1967).
6. G. Zweifel and C. C. Whitney, *J. Am. Chem. Soc.*, **89**, 2753 (1967).
7. For a review of synthetic methods applicable to trisubstituted alkenes see J. Faulkner, *Synthesis*, 175 (1971).

m-CHLOROPERBENZOIC ACID

(Benzenecarboperoxoic acid, 3-chloro-)

$$m\text{-ClC}_6\text{H}_4\text{COCl} + \text{H}_2\text{O}_2 + \text{NaOH} \xrightarrow[\text{dioxane}]{\substack{\text{MgSO}_4 \\ \text{H}_2\text{O}}} \xrightarrow{\text{H}_2\text{SO}_4} m\text{-ClC}_6\text{H}_4\text{CO}_3\text{H}$$

Submitted by RICHARD N. McDONALD,[1] RICHARD N. STEPPEL, and JAMES E. DORSEY
Checked by WILLIAM N. WASHBURN and RONALD BRESLOW

1. Procedure

A 1-l. beaker (Note 1) equipped with a magnetic stirrer is charged with 1.5 g. of magnesium sulfate heptahydrate, 36 g. of sodium hydroxide, 360 ml. of water, 90 ml. of 30% hydrogen peroxide, and 450 ml. of dioxane. This mixture is cooled to 15° with an ice-water bath and by the addition of a small amount of ice to the mixture before 52.5 g. (0.300 mole) of *m*-chlorobenzoyl chloride (Note 2) is added in one portion while vigorous stirring is maintained. Small portions of ice are added to maintain the temperature below 25°. The reaction mixture is stirred at this temperature for 15 minutes, then transferred to a 3-l. separatory funnel. Cold, 20% sulfuric acid (900 ml.) (Note 3) is added to the separatory funnel, the mixture is shaken and separated, and the aqueous layer is extracted with four 200 ml. portions of cold dichloromethane (Note 4). The combined extracts are dried over anhydrous magnesium sulfate, and the dichloromethane is removed under reduced pressure *via* a short-path distillation (Note 5). After most of the solvent has been removed, a white pasty solid remains and the residual solvent is removed under full vacuum for an additional 2 hours or until a white, flaky powder remains. The product weighs approximately 51 g. Sodium thiosulfate analysis indicates 80–85% active oxygen present (Note 6).

2. Notes

1. The checkers used a Pyrex vessel; the submitters utilized a Nalgene beaker, because contact with the glass surface may catalyze the decomposition of the peracid.

2. The *m*-chlorobenzoyl chloride is prepared by refluxing *m*-chlorobenzoic acid, either commercial or recovered, with excess thionyl chloride. Distillation gives the *m*-chlorobenzoyl chloride in high yield, b.p. 112° (11 mm.).[2]

The recovery of the *m*-chlorobenzoic acid, the by-product from *m*-chloroperbenzoic acid oxidations, is facilitated by using dichloromethane as the solvent since the peracid is very soluble whereas the acid is quite insoluble.

3. The cold 20% sulfuric acid solution is made by adding concentrated sulfuric acid (180 ml.) to crushed ice followed by the addition of more ice until the required volume is reached, *i.e.*, 900 ml. The acid solution is kept cold in an ice bath until required. The dichloromethane is also precooled before use.

4. Iodometric titration of the moist extracts indicates approximately 51 g. of *m*-chloroperbenzoic acid present. A 4.0-ml. aliquot of the solution requires approximately

20 ml. of a 0.1000 *N* solution of sodium thiosulfate. To prepare the sample, 10 ml. of 10% sodium iodide and 5 ml. of acetic acid are added to the 5-ml. aliquot, and the mixture is diluted to 50 ml. The dark red solution is titrated to a pale yellow. At this point 1 ml. of starch solution is added and the titration continued to the end point, *i.e.*, a change from dark blue to clear.

$$\text{Total weight of peracid} = \frac{N_{Na_2S_2O_3} \cdot V_{Na_2S_2O_3(ml.)} \cdot V_{total\ extract\ (l.)} \cdot 86.29}{V_{aliquot\ (ml.)}}$$

5. The dichloromethane solvent should be removed as rapidly as possible because contact with glass and heat cause decomposition of the peracid. A convenient method for the removal of the solvent involves a short-path vacuum distillation using a 2-l. distillation flask and trapping the solvent at dry ice or liquid nitrogen temperatures. The pressure should be reduced slowly at first and at least three traps used to minimize the amount of solvent introduced into the vacuum pump. Adjustment of a stopcock located between the first and second traps will help to control this problem. The distillation flask is placed in a water bath maintained between 25° and 35° so that the rate of solvent evaporation is quite rapid. Removal of the solvent over prolonged periods and drying of the solid peracid by excessive heating results in drastic losses of active oxygen. *m*-Chloroperbenzoic, once isolated, is stable over long periods of time when stored in polyethylene containers and refrigerated.

6. Iodometric titration of the solid product involves the use of 0.2-g. samples of the peracid and the procedure in Note 4.

$$\text{Percent peracid} = \frac{N_{Na_2S_2O_3} \cdot V_{Na_2S_2O_3} \cdot 0.8629}{\text{Weight of sample}}$$

3. Discussion

This method is an extension of the reported perhydrolysis of certain other acid chlorides and anhydrides.[3] Although *m*-chloroperbenzoic acid is commercially available, this preparation requires only a short reaction time and simple equipment, and it affords high yields of this relatively stable and useful peracid. The dichloromethane–dioxane extracts can be stored and used directly for many peroxidations, in which case the total preparation time should not exceed 2 hours.

1. Department of Chemistry, Kansas State University, Manhattan, Kansas 66502.
2. R. M. Herbst and R. R. Wilson, *J. Org. Chem.*, **22**, 1142 (1957).
3. Y. Ogata and Y. Swaki, *Tetrahedron*, **23**, 3327 (1967).

3-(4-CHLOROPHENYL)-5-(4-METHOXYPHENYL)ISOXAZOLE

[Isoxazole, 3-(4-chlorophenyl)-5-(4-methoxyphenyl)-]

Submitted by MATILDA PERKINS, CHARLES F. BEAM, JR.,[1]
MORGAN C. D. DYER,[2] and CHARLES R. HAUSER[3]
Checked by H. W. JACOBSON and R. E. BENSON

1. Procedure

Caution! This preparation should be carried out in an efficient hood.

A 2-l., three-necked, round-bottomed flask equipped with a magnetic stirring bar, a nitrogen-inlet tube, and a 250-ml. pressure-equalizing dropping funnel attached to a calcium chloride tube is charged with 16.96 g. (0.1000 mole) of 4-chloroacetophenone oxime (Note 1) and 500 ml. of anhydrous tetrahydrofuran (Note 2). The flask is stoppered (Note 3) and cooled in an ice-water bath (Note 4). In the dropping funnel is placed 140 ml. (0.22 mole) of 1.6 M n-butyllithium in hexane (Note 5), which is rapidly added dropwise to the stirred solution over a 12–15-minute period. The solution is stirred for 30 minutes after the addition is complete, and the addition funnel is replaced by a clean one (125-ml. capacity), fitted with a calcium chloride drying tube (Note 6). Cooling is continued, a solution of 8.31 g. (0.0501 mole) of methyl anisate (Notes 7 and 8) in 100 ml. of anhydrous tetrahydrofuran is added to the stirred mixture over a 6–10-minute period, and the resulting mixture is stirred for an additional 30 minutes (Note 9). At the end of this period, 300 ml. of 3 N hydrochloric acid is added. The nitrogen-inlet tube is removed and replaced by a reflux condenser, and the dropping funnel is replaced by a ground-glass stopper. The ice bath is removed, and the mixture is heated under reflux for 1 hour. The flask is then cooled, and its contents are poured into a 2-l. Erlenmeyer flask, and solid sodium hydrogen carbonate is added to the mixture until neutralization is complete (Note 10).

The resulting mixture consists of an organic phase and a lower aqueous phase containing a small amount of insoluble material. The mixture is transferred to a 2-l. separatory funnel and the phases are separated. The aqueous phase is extracted with 100 ml. of tetrahydrofuran, which is combined with the original organic phase and concentrated to dryness on a rotary evaporator. Approximately 150 ml. of xylene is added to the flask and the contents of the flask are heated to reflux to remove any water present as the azeotrope. The resulting hot solution is filtered rapidly through a large Büchner funnel with light suction. The volume is reduced to approximately 100 ml., and the solution is cooled in an ice bath. The tan crystals which separate are collected in a Büchner funnel and washed with 10 ml. of ice-cold xylene. The crude product is recrystallized from 150 ml. of xylene (Note 11), yielding, after drying, 7.4–7.6 g. (52–53%) of 3-(4-chlorophenyl)-5-(4-methoxyphenyl)isoxazole, m.p. 175–176° (Note 12).

2. Notes

1. 4-Chloroacetophenone oxime was prepared by a modification of the method described by Shriner, Fuson, and Curtin.[4] A mixture of 100 g. (0.647 mole) of reagent grade 4-chloroacetophenone, 300 ml. of water, 200 ml. of aqueous 10% sodium hydroxide, 50 g. (0.72 mole) of hydroxylamine hydrochloride, and 500 ml. of ethanol is heated at reflux in a 2-1., round-bottomed flask for 2 hours. The crystals that separate on cooling in an ice bath are recovered by filtration and air dried. The product is added to approximately 1 l. of hexane, and the mixture is heated to reflux to remove any remaining water as the azeotrope. The resulting solution is cooled, yielding 70–74 g. (64–68%) of 4-chloroacetophenone oxime as white crystals, m.p. 96–97°.

2. Tetrahydrofuran was obtained from E. I. du Pont de Nemours and Company and distilled from lithium aluminum hydride or sodium–benzophenone immediately before use. The submitters used reagent grade tetrahydrofuran available from Matheson, Coleman and Bell.

3. Ground-glass stoppers proved most convenient.

4. The initial reaction of n-butyllithium (or lithium diisopropylamide) with the oxime is exothermic, and if the bath is not used, a slightly lower yield of colored product is obtained.

5. The concentration of the n-butyllithium obtained from Foote Mineral Company is generally close to the 1.6 M as quoted. An exact measurement of the volume (hypodermic syringe recommended) is not necessary, but a slight excess above the stoichiometrically required amount (0.20 mole) is needed. The submitters used n-butyllithium available from Lithium Corporation of America, Inc. A recent modification utilizes lithium diisopropylamide (0.33 mole) instead of n-butyllithium (0.22 mole).[5] A 0.33-mole sample of n-butyllithium is cooled to 0° and blanketed with nitrogen. To this base is added 0.33 mole of diisopropylamine dissolved in 150 ml. of tetrahydrofuran.

6. The purpose of the exchange is to provide a clean funnel for the addition of the ester solution. If the funnel is not changed, the yield is slightly lower. When lithium diisopropylamide is used the ratio of the reagents is 2 oxime : 6 base : 2 ester.

7. The ratio of the reagents is 2 oxime : 4 base : 1 ester and is consistent with similar procedures used for a modified Claisen condensation.[6] The yield is based on the ester. When a ratio of reagents of 1 oxime : 2 base : 1 ester was used, a yield of 21% based on the ester was obtained.

8. Methyl anisate was obtained from Eastman Organic Chemicals.

9. At least 30 minutes is required for an optimum yield of the isoxazole.

10. Care should be taken to add the sodium hydrogen carbonate in small amounts initially, in order to avoid excessive frothing. The mixture was tested with pH paper to establish complete neutralization.

11. The product can also be recrystallized from ethanol, but a substantially larger volume of solvent is required.

12. The product has the following spectral properties; IR (KBr) cm.$^{-1}$: 3103 and 3006 (aromatic C—H), 2955, 2925, and 2830 (aliphatic C—H stretching), 1257 and 1032 (aromatic methyl ether), 841 and 812 (C—H out-of-plane bending of isoxazole C_4—H and 4-substituted phenyl); ^{1}H NMR (trifluoroacetic acid), δ (multiplicity, number of protons, assignment): 3.98 (s, 3H, OCH_3), 7.00–7.27 (m, 1H, isoxazole C_4—H; and 2H, aryl H), and 7.42–7.97 (m, 6H, aryl H).

3. Discussion

This procedure has several advantages over previous methods. It provides a simple, direct route to unsymmetrically substituted isoxazoles in which the location of substituents is unequivocal. The method uses readily available starting materials and can be used for the synthesis of a variety of substituted isoxazoles in which the substituents are stable to n-butyllithium. Examples of products synthesized by this method[7] are given in Table I.

3-(4-Chlorophenyl)-5-(4-methoxyphenyl)isoxazole has been prepared from the dilithio derivative of 4-chloroacetophenone oxime by three other methods: (a) reaction with anisonitrile (4-methoxybenzonitrile) followed by acid-catalyzed cyclization,[8] (b) reaction with N,N-dimethylbenzamide followed by acid-catalyzed cyclization,[9] and (c) condensation of the dilithio intermediate, prepared in an excess of lithium diisopropylamide, with methyl 4-methoxybenzoate, followed by acid-catalyzed cyclization.[6] The condensation of dilithioöxime with aroyl chlorides, followed by acid-catalyzed cyclization, could result in 4-acylisoxazoles.[10]

The use of dilithio reagents for the preparation of heterocyclic systems has been extended to the synthesis of 2-isoxazolin-5-ones[11] by carboxylation of a dilithioöxime, followed by cyclization, and the synthesis of pyrazoles from dilithiophenylhydrazones

TABLE I
ISOXAZOLES DERIVED FROM OXIMES[7]

R_1	R_2	Yield[a] (%)
C_6H_5—	C_6H_5—	59
C_6H_5—	$4\text{-}CH_3OC_6H_4$—	51
C_6H_5—	$4\text{-}ClC_6H_4$—	59
$4\text{-}ClC_6H_4$—	C_6H_5—	65
$4\text{-}CH_3OC_6H_4$—	$4\text{-}ClC_6H_4$—	66[b]

[a]Yield obtained using ~2.25 M n-butyllithium reagent.
[b]100% yield obtained using lithium diisopropylamide reagent.

and trilithiothiohydrazones.[12,13] If the dilithiophenylhydrazones are formed in an excess of lithium diisopropylamide, they can be treated with diethyl carbonate, followed by cyclization to give 2-pyrazolin-5-ones.[14] Dilithiooximes formed in an excess of lithium diisopropylamide can be condensed with electrophilic-nucleophilic reagents such as methyl anthranilate.[15]

1. Submitted from William Chandler Chemistry Laboratory, Lehigh University, Bethlehem, Pennsylvania 18015. [Present address: Department of Chemistry, Newberry College, Newberry, South Carolina 29108. Technical assistance by David C. Reames is acknowledged.]

2. National Aeronautics and Space Administration Trainee, 1967–1969.

3. Deceased January 6, 1970.

4. R. L. Shriner, R. C. Fuson, and D. Y. Curtin, "The Systematic Identification of Organic Compounds," 5th ed., Wiley, New York, 1964, p. 289 (4th ed., 1956, p. 255).

5. T. D. Fulmer, L. P. Dasher, B. L. Bobb, J. D. Wilson, K. L. Sides, and C. F. Beam, *J. Heterocycl. Chem.*, **17**, 799 (1980).

6. C. R. Hauser, F. W. Swamer, and J. T. Adams, *Org. React.*, **8**, 113 (1954).

7. C. F. Beam, M. C. D. Dyer, R. A. Schwarz, and C. R. Hauser, *J. Org. Chem.*, **35**, 1806 (1970).

8. C. F. Beam, R. S. Foote, and C. R. Hauser, *J. Heterocycl. Chem.*, **9**, 183 (1972).

9. G. W. Barber and R. A. Olofson, *J. Org. Chem.*, **43**, 3015 (1978).

10. R. M. Sandifer, L. M. Shaffer, W. M. Hollinger, D. C. Reames, and C. F. Beam, *J. Heterocycl. Chem.*, **13**, 607 (1976).

11. J. S. Griffiths, C. F. Beam, and C. R. Hauser, *J. Chem. Soc. C*, 974 (1971).

12. R. S. Foote, C. F. Beam, and C. R. Hauser, *J. Heterocycl. Chem.*, **7**, 589 (1970).

13. C. F. Beam, R. S. Foote, and C. R. Hauser, *J. Chem. Soc. C*, 1658 (1971).

14. J. D. Wilson, T. D. Fulmer, L. P. Dasher, and C. F. Beam, *J. Heterocycl. Chem.*, **17**, 389 (1980).

15. J. Brown, K. L. Sides, T. D. Fulmer, and C. F. Beam, *J. Heterocycl. Chem.*, **16**, 1669 (1979).

α-CHLORO ENAMINES, REACTIVE
INTERMEDIATES FOR SYNTHESIS:
1-CHLORO-*N,N*,2-TRIMETHYLPROPENYLAMINE

(Propenylamine, 1-chloro-*N,N*,2-trimethyl-)

Submitted by B. Haveaux, A. Dekoker, M. Rens,
A. R. Sidani, J. Toye, and L. Ghosez[1]
Checked by Masayuki Murakami, Mitsuru Yoshioka,
and Wataru Nagata

1. Procedure

Caution! Phosgene is highly toxic. This preparation should be carried out in a well-ventilated hood.

A.   1-*Chloro*-N,N,2-*trimethylpropylideniminium chloride*.   A 1-l., three-necked, round-bottomed flask is equipped with a magnetic stirring bar, a pressure-equalizing dropping funnel, an inlet tube connected to a graduated trap with a flexible polyethylene tube, and a dry-ice condenser connected to a series of three traps. The first and last traps in the series contain sulfuric acid and 10% potassium hydroxide, respectively, and the middle trap is left empty, as shown in Figure 1. Phosgene (85–100 ml., 1.2–1.4 moles) (Note 1) is condensed in the graduated trap which is cooled in an acetone–dry ice bath. The flask is charged with 200 ml. of anhydrous dichloromethane (Note 2) and cooled in an ice–salt bath. The liquid phosgene is slowly poured into the flask (Note 3), the inlet tube is replaced with a thermometer, and a solution of 115 g. (1.00 mole) of freshly distilled *N,N*-dimethylisobutyramide (Note 4) in 150 ml. of anhydrous dichloromethane is added dropwise from the dropping funnel over 20 minutes. The temperature is maintained at 0° during this time and gradually raised to room temperature within *ca.* 1 hour. The gas evolution becomes vigorous, and the phosgene begins to boil. The reaction mixture, containing a white precipitate, is left overnight at room temperature. The flask is prepared for distillation and connected to a water pump (Note 5), maintaining a slightly reduced pressure in the system. The excess phosgene and most of the solvent are removed by

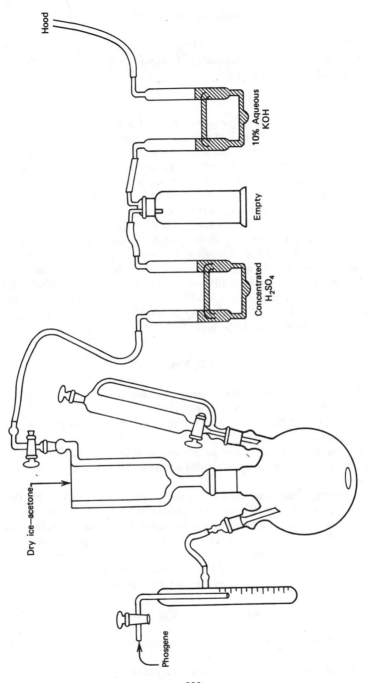

Figure 1.

Hood

10% Aqueous KOH

Empty

Concentrated H₂SO₄

Dry ice—acetone

Phosgene

warming the flask in a water bath at *ca.* 50° and collected in an ice-cold receiver (Note 6). The white or pale-yellow solid remaining, 1-chloro-*N,N*,2-trimethylpropylideniminium chloride, is used directly in Part B.

B. 1-*Chloro*-N,N,2-*trimethylpropenylamine*. The flask is equipped with a dropping funnel, a mechanical stirrer, and a reflux condenser protected from moisture with a sulfuric acid trap. The iminium chloride is suspended in 200 ml. of anhydrous dichloromethane, and 140 g. (1.39 moles) of triethylamine (Note 7) is slowly added to the mixture from the dropping funnel, with vigorous stirring, over 1 hour (Note 8). The temperature rises to 45°, and the solvent begins to reflux. The resulting suspension is stirred at room temperature for an additional 2 hours, after which 150 ml. of dry, low-boiling petroleum ether (Note 9) is added to complete the precipitation of triethylamine hydrochloride (Note 10). The mixture is quickly filtered under nitrogen (Note 11) into a 1-l., round-bottomed flask through an Iena sintered-glass filter. A 300-ml. portion of petroleum ether is used to wash the flask and the triethylamine hydrochloride on the filter. The solvent is removed by distillation under nitrogen. Further distillation through a Vigreux column under nitrogen gives 93–103 g. (69–77%) of 1-chloro-*N,N*,2-trimethylpropenylamine, b.p. 125–130° (760 mm.) (Notes 12 and 13). The compound is very sensitive to humidity and should be immediately stored in ampoules (Note 14).

2. Notes

1. The submitters used technical grade phosgene purchased from Gardner Cryogenics Europe N. V., 1800 Vilvoorde, Belgium.

2. Technical grade dichloromethane was dried by refluxing over phosphorus pentoxide for 24 hours and distilled.

3. Alternatively phosgene may be allowed to distill into the flask cooled in an acetone–dry ice bath. For this operation the inlet tube should extend to the bottom of the flask.

4. Following a procedure reported in the literature,[2] the checkers prepared *N,N*-dimethylisobutyramide, b.p. 67–68° (15 mm.), in 85% yield by treating reagent grade isobutyryl chloride with 2 molar equivalents of reagent grade dimethylamine in anhydrous ethyl ether at 0°. The reported boiling point for *N,N*-dimethylisobutyramide is 175–176° (744 mm.).[2] Isobutyryl chloride and dimethylamine were both purchased from Tokyo Kasei Kogyo Company Ltd., Tokyo, Japan. These two reagents are also available from Aldrich Chemical Company, Inc., and the Specialty Gas Division, J. T. Baker Chemical Company, respectively.

5. A phosphorus pentoxide tube is placed between the water pump and the distillation apparatus.

6. The submitters found that filling the flask with argon helped to reduce exposure to moisture in the air.

7. The submitters used Baker-grade triethylamine purchased from J. T. Baker Chemicals N.V., P.O. Box 1, Deventer, Holland, after distillation from potassium hydroxide.

8. The reaction is exothermic. The checkers found that the yield of the final product was raised from 57% in the first run to 71% in the second and third runs when the triethylamine was added at 30–34° with slight cooling. Another procedural change made by the checkers in Part A in these last two runs was that the dry ice condenser was kept in place for more than 8 hours after the addition of N,N-dimethylisobutyramide was completed. In the first run the condenser was removed after 20 minutes.

9. The submitters used technical petroleum ether, b.p. <70°, which was distilled from sodium wire.

10. Leaving the reaction mixture overnight before filtration did not affect the yield.

11. A slow stream of dry nitrogen was passed through an inverted funnel placed over the filtration apparatus.

12. The checkers collected several fractions during the distillation. Early fractions boiling at 100–125° (760 mm.) were shown to be a mixture of the product and triethylamine. The product from two runs carried out at one-half scale was collected in two main fractions amounting to 2.2–5.8 g., b.p. 125–130° (760 mm.), and 41.7–45.4 g., 130–134° (760 mm.). The total yield was 47.5–47.6 g. (71%), b.p. 125–134°. The submitters obtained 105–110 g. (78–82%), b.p. 129–130° (760 mm.).

13. The checkers obtained an analysis on the distilled product. Analysis calculated for $C_6H_{12}NCl$: C, 53.93; H, 9.05; N, 10.48; Cl, 26.54. Found: C, 54.51; H, 9.21; N, 10.69; Cl, 26.49. The spectral properties of the product are as follows: IR (CCl$_4$) cm.$^{-1}$: 1653, 1470, 1451, 1295, 1124, 1013; ^{1}H NMR (ca. 15% w/v in CCl$_4$), δ (multiplicity, number of protons, assignment): 1.73 (s, 3H, allylic CH$_3$), 1.79 (s, 3H, allylic CH$_3$), 2.37 (s, 6H, 2NCH$_3$); ^{1}H NMR (about 15% w/v in CDCl$_3$), δ (multiplicity, number of protons, assignment): 1.77 (s, 6H, two allylic CH$_3$) and 2.38 (s, 6H, 2NCH$_3$); mass spectrum (225°, 70 e.v.) m/e (relative intensity): 135 (M + 2, 24), 133 (M, 77), 98 (100), 83 (56), 82 (23), 72 (31), 44 (36), 42 (60).

14. The tubes should be sealed immediately to avoid hydrolysis. In spite of this precaution, a light precipitate is always formed.

3. Discussion

1-Chloro-N,N,2-trimethylpropenylamine has been prepared by reaction of 2-methylpropenylidenebis(dimethylamine) with phosphorous trichloride or dichlorophenyl-phosphine.[3] The present method[4] is far more convenient and general. The reagents are inexpensive, the amide reactants are readily available, and the procedure is applicable to the synthesis of various α-chloro enamines on a large scale with only minor modifications (Table I).[5]

The reaction of the more basic amides with phosgene is exothermic (Caution!); consequently the reaction mixture must be cooled in an ice bath. With the less reactive amides (e.g., entries 2, 4, 5, and 8–10), however, the reaction often requires several days. It can be accelerated by the addition of catalytic amounts of N,N-dimethylformamide. With the monosubstituted acetamides or acetanilides, the solution must be saturated with gaseous hydrogen chloride before adding phosgene to avoid the formation of β-chlorocarbonyl α-chloro enamines resulting from elimination of hydrogen chloride and acylation of the α-chloro enamine with phosgene. The subsequent elimination reaction must be conducted with an excess of triethylamine in ethyl ether, carbon

TABLE I
SYNTHESIS OF α-CHLORO ENAMINES FROM AMIDES

Entry	Amide	α-Chloro Enamine	Yield (%)
1	$(CH_3)_2CHCON$⟨piperidine⟩	$(CH_3)_2C=C$ with Cl and N(piperidine)	85
2	$(CH_3)_2CHCON(CH_3)C_6H_5$	$(CH_3)_2C=C$ with Cl and $N(CH_3)C_6H_5$	55–79
3	cyclohexyl–H, $CON(C_2H_5)_2$	cyclohexylidene=C with Cl and $N(C_2H_5)_2$	40
4	$C_6H_5CH(CH_3)CON(CH_3)_2$	$C_6H_5C(CH_3)=C$ with Cl and $N(CH_3)_2$	76[a]
5	$CH_3CH(Cl)CON$⟨pyrrolidine⟩	$CH_3C(Cl)=C$ with Cl and N(pyrrolidine)	60–70[a]
6	azepanone with CH₃, N–CH₃, =O	azepine ring with CH₃, Cl, N–CH₃	75
7	$(CH_3)_3CCH_2CON(CH_3)_2$	$(CH_3)_3CCH=C$ with Cl and $N(CH_3)_2$	65[a]
8	$CH_3CH_2CON(CH_3)C_6H_5$	$CH_3CH=C$ with Cl and $N(CH_3)C_6H_5$	45–62[a]

TABLE I (*continued*)

SYNTHESIS OF α-CHLORO ENAMINES FROM AMIDES

Entry	Amide	α-Chloro Enamine	Yield (%)
9	$C_6H_5CH_2CON(CH_3)C_6H_5$	$C_6H_5CH{=}C{\overset{\displaystyle Cl}{\underset{\displaystyle N(CH_3)C_6H_5}{}}}$	45
10	$CH_3CON(CH_3)C_6H_5$	$CH_2{=}C{\overset{\displaystyle Cl}{\underset{\displaystyle N(CH_3)C_6H_5}{}}}$	42

[a] The product is a mixture of *cis* and *trans* isomers.

tetrachloride, or petroleum ether. Dichloromethane and chloroform are not suitable since these solvents promote the formation of condensation products to a considerable extent. With the exceptions of entries 7–10, α-chloroenamines derived from monosubstituted acetamides are unstable and should be kept in solution at concentrations below 1 M.

Most α-chloro enamines can be readily converted into the corresponding α-fluoro enamines by reaction with potassium or cesium fluoride.[6] The less stable α-iodo enamines are more conveniently prepared *in situ* from α-chloro enamines and potassium iodide. 1-Bromo-*N,N*,2-trimethylpropenylamine is easily obtained from the corresponding chloro compound and refluxing dibromomethane.[7] All α-haloenamines are highly hygroscopic and must be stored in sealed tubes.

α-Halo enamines are useful organic reagents that show versatile chemical behavior and have great synthetic potential.[5] As enamines derived from carboxylic acid halides, they react with a variety of electrophilic reagents on C-2 to give, after hydrolysis, a carbox-amide substituted at the α-position. Moreover, spontaneous or catalyzed ionization leads to keteniminium ions that are strongly electrophilic and add various nucleophilic reagents at C-1.[5,8–10] Keteniminium ions are also capable of undergoing [2 + 2]-cycloaddition reactions with olefins,[11] acetylenes,[12] and imines[13] extremely readily. 1-Halo-*N,N*,2-trimethylpropenylamines are also highly effective reagents for the replacement of hydroxyl groups by chlorine, bromine, and iodine under neutral conditions.[5,7] A summary of some of the reactions of α-chloro enamines follows.

1. Laboratoire de Chimie Organique de Synthèse, Université de Louvain, Place L. Pasteur, 1, B - 1348 Louvain-La-Neuve, Belgium.
2. N. Gavrilov and A. V. Koperina, *Zh. Obshch. Khim*, **9**, 1394 (1939) [*Chem. Abstr.*, **34**, 1615[5] (1940)]; N. Gavrilov, A. V. Koperina, and M. Klyuchareva, *Bull. Soc. Chim. Fr.*, **12**, 773 (1945) [*Beilstein*, 4th ed., 3rd suppl., **4**, 127 (1962)].
3. H. Weingarten, *J. Org. Chem.*, **35**, 3970 (1970).
4. L. Ghosez, B. Haveaux, and H. G. Viehe, *Angew. Chem. Int. Ed. Engl.*, **8**, 454 (1969).
5. For a review see L. Ghosez and J. Marchand-Brynaert, "α-Halo Enamines and Keteniminium Salts," in H. Böhme and H. G. Viehe, Eds., "Iminium Salts in Organic Chemistry," Part 1, Vol.

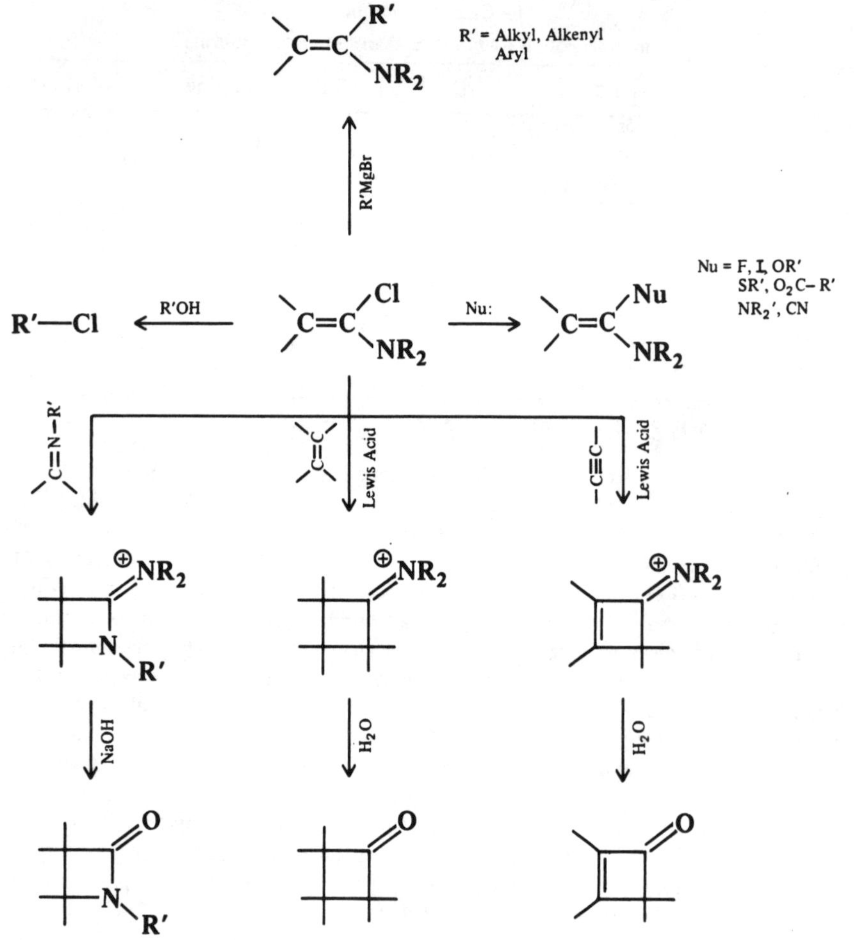

R' = Alkyl, Alkenyl
Aryl

R'MgBr

R'—Cl ←R'OH— —Nu:→

Nu = F, I, OR'
SR', O₂C–R'
NR₂', CN

Cl

NR₂

Nu

NR₂

C≡N–R'

C=C
Lewis Acid

–C≡C–
Lewis Acid

⊕NR₂

N
R'

⊕NR₂

⊕NR₂

NaOH

H₂O

H₂O

O

N
R'

O

O

288

9, in "Advances in Organic Chemistry," E. C. Taylor, Ed., Wiley-Interscience, New York, 1976, p. 421.

6. A. Colens, M. Demuylder, B. Téchy, and L. Ghosez, *Nouveau J. Chim.*, **1**, 369 (1977).

7. A. Devos, J. Remion, A. M. Frisque-Hesbain, A. Colens, and L. Ghosez, *J. Chem. Soc., Chem. Commun.*, 1180 (1979).

8. M. Rens and L. Ghosez, *Tetrahedron Lett.*, 3765 (1970).

9. J. Marchand-Brynaert and L. Ghosez, *J. Am. Chem. Soc.*, **94**, 2869 (1972).

10. J. Toye and L. Ghosez, *J. Am. Chem. Soc.*, **97**, 2276 (1975).

11. J. Marchand-Brynaert and L. Ghosez, *J. Am. Chem. Soc.*, **94**, 2870 (1972); A. Sidani, J. Marchand-Brynaert, and L. Ghosez, *Angew. Chem. Int. Ed. Engl.*, **13**, 267 (1972).

12. C. Hoornaert, A. M. Hesbain-Frisque, and L. Ghosez, *Angew. Chem. Int. Ed. Engl.*, **14**, 569 (1975).

13. M. De Poortere, J. Marchand-Brynaert, and L. Ghosez, *Angew. Chem. Int. Ed. Engl.*, **13**, 268 (1974).

MODIFIED CLEMMENSEN REDUCTION: CHOLESTANE

Submitted by Shosuke Yamamura,[1] Masaaki Toda,[2] and Yoshimasa Hirata[2]
Checked by A. Laurenzano, L. A. Dolan, and A. Brossi

1. Procedure

A 500-ml., four-necked, round-bottomed flask (Note 1) equipped with a sealed mechanical stirrer (Note 2), a gas-inlet tube, a low-temperature thermometer, and a calcium chloride tube is charged with 250 ml. of dry diethyl ether. With an acetone–dry ice bath the temperature of the ether is lowered to -10 to $-15°$ and maintained within this range while a slow stream (Note 3) of hydrogen chloride is introduced, with slow stirring, for about 45 minutes. The gas-inlet tube is replaced with a glass stopper, and 10.0 g. (0.0259 mole) of cholestan-3-one (Note 4) is added while the temperature of the stirred solution (Note 5) is kept below $-15°$. The reaction mixture is cooled to $-20°$, and 12.3 g.

(0.188 g.-atom) of activated zinc (Note 6) is added over a 2–3 minute period. The temperature of the reaction mixture is allowed to rise to −5° (Note 7), and it is maintained between −4° and 0° (Note 8) for 2 hours. Stirring is not interrupted for the duration of the reaction. The mixture is finally cooled to −15° and poured slowly onto about 130 g. of crushed ice. The ethereal layer is separated, and the aqueous layer is extracted with 100 ml. of ether that had been used to rinse the reaction vessel. The ethereal solutions are combined, washed with saturated aqueous sodium chloride, dried over anhydrous magnesium sulfate, and filtered. The ether is distilled under reduced pressure with a 50° water bath, leaving 9.3–9.5 g. of a colorless, liquid residue that solidifies on cooling. This solid is dissolved in 30–40 ml. of *n*-hexane (Note 9). The solution is poured onto a 3.5 cm. by 17 cm. column of silica gel (Note 10) and eluted with 80–90 ml. of *n*-hexane. Distillation of the solvent under reduced pressure with a 50° water bath leaves 8.0–8.2 g. (82–84%) of cholestane (Note 11), which, after recrystallization from ethanol-ether (Note 12), yields 7.3–7.5 g. (76–77%) of product as plates, m.p. 78–79° (lit., m.p. 80°)[3] (Note 13).

2. Notes

1. A standard, three-necked flask fitted with a Y-tube may be used.

2. An efficient magnetic stirrer may be substituted.

3. Approximately one bubble per second can be spot-checked periodically by connecting the calcium chloride tube to an oil-filled bubble counter.

4. Cholestan-3-one was prepared according to *Org. Synth., Coll. Vol. 2,* 139 (1943); single spot on TLC with the system described in Note 11.

5. The cholestanone does not dissolve completely at this low temperature, but the reaction is not affected.

6. The submitters prepared activated zinc by the following procedure. Commercial zinc powder (16 g.), special grade, *ca.* 300 mesh, obtained from either Kishida Chemical Company Ltd. or Hayashi Pure Chemical Company Ltd., is added with stirring to a 300-ml., round-bottomed flask containing 100 ml. of 2% hydrochloric acid. Vigorous stirring is continued until the surface of the zinc becomes bright (*ca.* 4 minutes). The aqueous solution is decanted, and the zinc powder in the flask is washed by decantation with four 200-ml. portions of distilled water. The activated zinc powder is transferred to a suction filter with 200 ml. of distilled water and washed successively with 50 ml. of ethanol, 100 ml. of acetone, and 50 ml. of dry ether. Filtration and washing should be done as rapidly as possible to minimize exposure of the activated zinc to air. The zinc is finally dried at 85–90° for 10 minutes in a vacuum oven (*ca.* 15 mm.), cooled, and used immediately; the yield is 13–14 g.

The checkers used this procedure with certified zinc powder, 325 mesh, obtained from Fisher Scientific Company.

7. This requires *ca.* 20 minutes.

8. The temperature is regulated by adding pieces of dry ice to the cooling bath as required. As the reduction proceeds, the solution separates into two phases.

9. The solution is decanted from any insoluble matter.

10. Silicic acid, 100 mesh (Mallinckrodt), was used.

11. This material melts at 78–79°. On TLC [silica, development with *n*-hexane,

visualization with sulfuric acid-methanol $(1:1)$ and heating] the product had $R_f = 0.74$. An impurity, $R_f = 0.65$, was present.

12. The cholestane is dissolved in 50 ml. of ether. Ether is distilled until the volume is 25 ml., 200 ml. of ethanol is added, and the mixture is refrigerated.

13. Recovery is 92%. Recrystallization has no effect on the quality of the product as judged by m.p. and TLC (Note 11).

3. Discussion

The well-known Clemmensen reduction[4] is a general method by which aralkyl ketones are readily converted to the corresponding hydrocarbons with amalgamated zinc and hydrochloric acid. It is not particularly effective, however, with alicyclic and aliphatic ketones. The procedure described herein provides a simple method of reducing a variety of ketones to their desoxy derivatives in high yields under much milder conditions $(0°,$ $1-2$ hours) than those normally used in the Clemmensen reaction.[4] This permits selective deoxygenation of ketones in polyfunctional molecules[5] containing groups such as cyano, amido, acetoxy, and carboalkoxy, which are stable under the mild reaction conditions. For example, the following reduction[6] has been carried out successfully by the modification of our procedure, using acetic anhydride as the solvent.[5]

Wide latitude is permitted in choosing the solvent for the reaction. Several organic solvents (tetrahydrofuran, benzene, n-hexane)[7] and particularly acetic anhydride[5,8] may be used instead of dry ether. α-Halo- and α-acetoxycholestanone[5] are converted to cholestane with Zn-HCl-Et_2O and also with Zn-HCl-Ac_2O.[7,8] These reduction systems, however, have given different results with α,β-unsaturated ketones.[9] With Zn-HCl-Et_2O, cholest-1-en-3-one gave cholestane in 88% yield, while cholest-4-en-3-one gave an 88% yield of a mixture of 1.2 parts of cholestane and 1 part of coprostane. By contrast, reaction of Zn-HCl-Ac_2O with cholest-1-en-3-one afforded a mixture of three compounds: cholestane $(30-32\%)$, 3-acetoxycholest-2-ene $(10-24\%)$, and cholestan-3-one $(30-40\%)$. Cholestan-3-one appears to be formed from the corresponding cyclopropanol acetate[10] during the work up. The mechanism of this reduction is probably similar to that of the Clemmensen reaction.[11]

1. Faculty of Pharmacy, Meijo University, Showa-ku, Nagoya, Japan. [Present address: Department of Chemistry, Faculty of Engineering, Keio University, Hiyoshi, Yokohama, Japan.]
2. Chemical Institute, Nagoya University, Chikusa-ku, Nagoya, Japan.
3. O. Diels and K. Linn, *Ber. Dtsch. Chem. Ges.*, **41**, 548 (1908); A. Windaus, *Ber. Dtsch. Chem. Ges.*, **50**, 133 (1917).
4. E. L. Martin, *Org. React.*, **1**, 155 (1942).

5. S. Yamamura and Y. Hirata, *Chem. Commun.*, 2887 (1968).
6. Private communication from H. Kakisawa, Tokyo Kyoiku University, Tokyo, Japan.
7. M. Toda, Y. Hirata, and S. Yamamura, *J. Chem. Soc. D*, 919 (1969); E. Vedejs, *Org. React.*, **22**, 401 (1975).
8. S. Yamamura, *Chem. Commun.*, 1494 (1968).
9. M. Toda, M. Hayashi, Y. Hirata, and S. Yamamura, *Bull. Chem. Soc. Jpn.*, **45**, 264 (1972).
10. M. I. Elphimoff-Felkin and P. Sarda, *Tetrahedron Lett.*, 3045 (1969).
11. H. O. House, "Modern Synthetic Reactions," W. A. Benjamin, New York, 1965, p. 58; J. G. St. C. Buchanan and P. D. Woodgate, *Q. Rev. (London)*, **23**, 522 (1969) and references cited therein.

CONJUGATE REDUCTION OF α,β-UNSATURATED *p*-TOLUENESULFONYLHYDRAZONES TO ALKENES WITH CATECHOLBORANE: 5β-CHOLEST-3-ENE

[Cholest-3-ene, (5β)-]

Submitted by GEORGE W. KABALKA,[1] ROBERT HUTCHINS,[2] NICHOLAS R. NATALE,[2] DOMINIC T. C. YANG,[3] and VICKY BROACH[3]

Checked by STEVEN J. BRICKNER and MARTIN F. SEMMELHACK

1. Procedure

A. *Cholest-4-en-3-one* p-*toluenesulfonylhydrazone*. A 100-ml., round-bottomed flask equipped with a magnetic stirring bar and a reflux condenser is charged with 10.19 g. (0.0265 mole) of cholest-4-en-3-one (Note 1), 5.53 g. (0.0297 mole) of *p*-toluenesulfo-nylhydrazide (Note 1), and 17 ml. of 95% ethanol. The solution is stirred and heated at reflux for 10 minutes and allowed to cool to room temperature. The precipitated solid is collected by filtration and recrystallized from 95% ethanol, affording 13.1–13.3 g (89–91%) of cholest-4-en-3-one p-toluenesulfonylhydrazone, m.p. 139–141° (Note 2), in two crops.

B. 5β-Cholest-3-ene. A dry, 100-ml., two-necked, round-bottomed flask equipped with a magnetic stirring bar, a rubber septum, and a reflux condenser connected to a mercury bubbler (Note 3) is charged with 4.98 g. (0.00950 mole) of cholest-4-en-3-one p-toluenesulfonylhydrazone and 20 ml. of chloroform, and the apparatus is evacuated with an aspirator and filled with nitrogen three times. The solution is stirred and cooled at 0° as 1.29 g. (1.21 ml., 0.0108 mole) of catecholborane (Note 4) is injected through the septum into the flask. Stirring and cooling are continued for 2 hours, after which 2.5 g. (0.018 mole) of sodium acetate trihydrate and 20 ml. of chloroform are added. The mixture is allowed to warm to room temperature over *ca.* 30 minutes, heated under reflux for 1 hour, cooled to room temperature, and filtered. The solid material is washed with 50 ml. of chloroform, and the combined filtrates are evaporated under reduced pressure. The remaining oil is purified by chromatography on a 5 × 50 cm. column packed with 200 g. of alumina (Note 5). The column is eluted with hexane and 200-ml. fractions are collected. Evaporation of the second 200-ml. fraction affords 2.76–2.95 g. (83–88%) of 5β-cholest-3-ene as a colorless oil which eventually crystallizes on standing, m.p. 48–50°, $[\alpha]_D^{24} = 19.6°$ (c = 63, chloroform) (Note 6).

2. Notes

1. Cholest-4-en-3-one and p-toluenesulfonylhydrazide are available from Aldrich Chemical Company, Inc. Procedures for the preparation of cholest-4-en-3-one and p-toluenesulfonylhydrazide are described in *Org. Synth.*, **Coll. Vol. 4,** 192, 195 (1963) and **Coll. Vol. 5,** 1055 (1973). The checkers used 5.70 g. of p-toluenesulfonylhydrazide, the purity of which was 97%.

2. The reported[4] melting point is 139–142°.

3. Nitrogen is introduced *via* a syringe needle that pierces the septum. A positive pressure of nitrogen is maintained in the apparatus during the following operations.

4. Catecholborane with a purity of 95% was purchased from Aldrich Chemical Company, Inc.

5. Activity grade I, neutral alumina was supplied by Brinckmann Instruments, Inc., Westbury, New York. The checkers used a 3 × 30 cm. column.

6. A TLC analysis was carried out by the submitters on a precoated silica gel plate (type Q6) purchased from Quantum Industries, 341 Kaplan Drive, Fairfield, New Jersey 07006. The chromatogram was developed with cyclohexane and showed a single spot for the product after visualization by charring with concentrated sulfuric acid. 5β-Cholest-3-ene is reported[5] to melt at 48–49°. The spectral properties of the product are as follows: IR (CHCl₃) cm.⁻¹: 2926, 1658, 1465, 831, 758, 678; ¹H NMR (CDCl₃), δ (multiplicity, number of protons, assignment): 0.66 (s, 3H, C-18 CH₃), 0.82 (s, 3H, CH₃), 0.92 (s, 3H, CH₃), 0.94 (s, 3H, C-19 CH₃), 5.2–5.7 (m, 2H, vinyl H); mass spectrum m/e: 370 (M+).

The submitters prepared the dibromide derivative, 3α,4β-dibromo-5β-cholestane, m.p. 98–99°. The melting point of the dibromide is reported as 98–100°.[6] The

TABLE I

CONJUGATE REDUCTION OF α,β-UNSATURATED p-TOLUENESULFONYLHYDRAZONES TO ALKENES

p-Toluenesulfonylhydrazone[a,b]	Alkene[b]	Procedure[c]	Yield (%)
$C_6H_5CH=CH-\underset{\underset{\displaystyle NNHTs}{\|\|}}{C}-CH_3$	$C_6H_5CH_2CH=CHCH_3$	A B C	72[d] 54 54
$C_6H_5CH=CH-\underset{\underset{\displaystyle NNHTs}{\|\|}}{C}-H$	$C_6H_5CH_2CH=CH_2$	A B C	53[d] 98[d] 42–56
$(CH_3)_2C=CH-\underset{\underset{\displaystyle NNHTs}{\|\|}}{C}-CH_3$	$(CH_3)_2CH-CH=CHCH_3$	A	65[d]
		A B C	77[d] 79 61–72
		A B C	66[e] 4[d,f] 18

TABLE I (continued)

CONJUGATE REDUCTION OF α,β-UNSATURATED p-TOLUENESULFONYLHYDRAZONES TO ALKENES

p-Toluenesulfonylhydrazone[a,b]	Alkene[b]	Procedure[c]	Yield (%)
(structure with TsNHN, CH₃, CH₃)	(cyclohexene structure with CH₃)	B C	70 51
CH₃(CH₂)₃—C≡C—C(NNHTs)—CH₃	CH₃(CH₂)₃—CH=C=CH—CH₃	A	64
CH₃—C(NNHTs)—C≡C—C₆H₅	CH₃—CH=C=CH—C₆H₅	A	75

[a]The abbreviation Ts stands for p-toluenesulfonyl.
[b]The p-toluenesulfonylhydrazones and alkenes with acyclic disubstituted double bonds are the E isomers.
[c]See text for descriptions of the procedures.
[d]Yield determined by GC.
[e]Yield determined by ¹H NMR spectroscopy.
[f]The cycloalkane was also formed in 32% yield.

296

mass spectrum of the dibromide exhibits three molecular ions at m/e (relative intensity, assignment): 532 (25%,$C_{27}H_{46}{}^{81}Br^{81}Br$), 530 (50%,$C_{27}H_{46}{}^{79}Br^{81}Br$), 528 (25%, $C_{27}H_{46}{}^{79}Br^{79}Br$).

3. Discussion

The reduction of p-toluenesulfonylhydrazone derivatives of α,β-unsaturated ketones and aldehydes with aluminum[7] or boron hydride reagents[8-11] effects a formal "conjugate" hydride transfer and produces alkenes in which the double bond has migrated to the position between the α-carbon and the carbonyl carbon. The mechanism of the reaction is presumed to involve initial reduction of the $C=N$ double bond, elimination of p-toluenesufinate, forming an allyl diazene, and concerted fragmentation of the diazene with 1,5-hydrogen transfer. One or both of the last two steps may take place during a subsequent hydrolysis. The reductions have been carried out with excess lithium aluminum hydride in tetrahydrofuran,[7] with catecholborane in chloroform at 0° followed by hydrolysis at $ca.$ 60° (Procedure A),[8] with sodium cyanoborohydride in 1:1 (v/v) N,N-dimethylformamide–sulfolane acidified with concentrated hydrochloric acid at 100–105° (Procedure B),[9,10] and with sodium borohydride in acetic acid at 70° (Procedure C).[11] A selection of examples of these reductions is given in Table I.

This method provides a convenient synthesis of alkenes with the double bond in a relatively unstable position. Thus, reduction of the p-toluenesulfonylhydrazones of α,β-unsaturated aryl ketones and conjugated dienones gives rise to nonconjugated olefins. Unsaturated ketones with endocyclic double bonds produce olefins with double bonds in the exocyclic position. The reduction of p-toluenesulfonylhydrazones of conjugated alkynones furnishes a simple synthesis of 1,3-disubstituted allenes.[12,13]

The present procedure illustrates this method with the preparation of 5β-cholest-3-ene by reduction of cholest-4-en-3-one p-toluenesulfonylhydrazone, using catecholborane as the reducing agent.[8,14] The advantages of catecholborane include its high solubility in common aprotic and nonpolar solvents, the low temperatures required for the reduction (0–25°), and the generally mild conditions used. Although the sodium cyanoborohydride and sodium borohydride procedures require higher temperatures, the use of polar solvents and protic conditions offers a valuable complement to the nonpolar, aprotic medium employed in the catecholborane procedure. However, the reduction of cholest-4-en-3-one p-toluenesulfonylhydrazone with sodium cyanoborohydride (Procedure B) gave a 71% yield of a mixture consisting of 5β-cholest-3-ene (32.5%), 5β-cholestane (30.5%), 5α-cholestane (30.5%), and 5α-cholest-3-ene (6.5%).[15]

5β-Cholest-3-ene has been prepared previously by deamination of 5β-cholestan-3β-yl amine,[16] by reduction of a mixture of 4β-bromo-5β-cholestan-3α-ol and its 3β epimer with zinc in acetic acid,[5] and as component of a mixture of cholestenes by Wolff-Kishner reduction of cholest-4-en-3-one.[6]

1. Department of Chemistry, University of Tennessee, Knoxville, Tennessee 37916.
2. Department of Chemistry, Drexel University, Philadelphia, Pennsylvania 19104.
3. Department of Chemistry, University of Arkansas at Little Rock, Little Rock, Arkansas 72204.
4. Y. Inouye and K. Nakanishi, *Steroids*, **3**, 487 (1964).
5. G. Bellucci, F. Macchia, and V. Malaguzzi, *Tetrahedron Lett.*, 4973 (1966).

6. A. Nickon, N. Schwartz, J. DiGiorgio, and D. Widdowson, *J. Org. Chem.*, **30**, 1711 (1965).
7. I. Elphimoff-Felkin and M. Verrier, *Tetrahedron Lett.*, 1515 (1968).
8. G. W. Kabalka, D. T. C. Yang, and J. D. Baker, Jr., *J. Org. Chem.*, **41**, 574 (1976).
9. R. O. Hutchins, M. Kacher, and L. Rua, *J. Org. Chem.*, **40**, 923 (1975); for a review of cyanoborohydride chemistry including deoxygenations, see R. O. Hutchins and N. R. Natale, *Org. Prep. Proceed. Int.*, **11**, 201 (1979).
10. R. O. Hutchins, C. A. Milewski, and B. Maryanoff, *J. Am. Chem. Soc.*, **95**, 3662 (1973).
11. R. O. Hutchins and N. R. Natale, *J. Org. Chem.*, **43**, 2299 (1978).
12. G. W. Kabalka, R. J. Newton, Jr., J. H. Chandler, and D. T. C. Yang, *J. Chem. Soc. Chem. Commun.*, 726 (1978).
13. G. W. Kabalka and J. H. Chandler, *Synth. Commun.*, **9**, 275 (1979).
14. G. W. Kabalka, J. D. Baker, Jr., and G. W. Neal, *J. Org. Chem.*, **42**, 512 (1977).
15. E. J. Taylor and C. Djerassi, *J. Am. Chem. Soc.*, **98**, 2275 (1976).
16. C. W. Shoppee, D. E. Evans, and G. H. R. Summers, *J. Chem. Soc.*, 97 (1957).

5β-CHOLEST-3-ENE-5-ACETALDEHYDE

[Cholest-3-ene-5-acetaldehyde, (5β)-]

Submitted by R. E. IRELAND[1] and D. J. DAWSON
Checked by W. PAWLAK and G. BÜCHI

1. Procedure

A 50-ml., round-bottomed flask equipped with a magnetic stirring bar and a 20-ml. calibration mark (Note 1) is charged with 970 mg. (2.51 mmoles) of cholest-4-en-3β-ol (Note 2). Ethyl vinyl ether is distilled into the flask to the 20-ml. mark (Note 3). The mixture is stirred to effect solution before 820 mg. (2.55 mmoles) of mercury(II) acetate

(Note 4) is added to the reaction mixture. The flask is fitted with a reflux condenser connected to a gas-inlet tube and flushed with argon. The reaction mixture is then stirred and heated (Note 5) at reflux under a positive argon pressure for 17 hours. After the solution has cooled to room temperature, 0.062 ml. (1.1 mmoles) of glacial acetic acid (Note 6) is added, and stirring is continued for 3 hours. The reaction mixture is poured into a preshaken mixture of 150 ml. of petroleum ether (Note 7) and 50 ml. of 5% aqueous potassium hydroxide. The aqueous phase is extracted with 50 ml. of petroleum ether, and the combined extracts are washed with three 50-ml. portions of a 20% aqueous sodium chloride, dried over anhydrous sodium carbonate, filtered and evaporated at reduced pressure (Note 8), giving 1.11 g. of an oil which, upon filtration through 5 g. of silica gel (Note 9) with 200 ml. of petroleum ether, affords 0.81 g. of the cholestenyl vinyl ether as a clear, colorless oil. If desired, crystallization of this oil from 10 ml. of acetone will give 0.74 g. (71%) of the vinyl ether as colorless prisms, m.p. 55–56.5° (Note 10).

Alternatively, the crude vinyl ether (0.81 g.) is transferred with petroleum ether into a 50-ml., round-bottomed flask fitted with a long gas-inlet tube. After the petroleum ether is removed at reduced pressure (Note 8), the flask is filled with argon and heated (Note 11) under a positive argon pressure at 220–225° for 5 hours; little or no bubbling should occur. After cooling, the oil is chromatographed on 75 g. of silica gel using 10% diethyl ether in petroleum ether as the elution solvent (Notes 7, 9, 12). The first 175 ml. of eluant contains side products and is discarded; elution with another 175 ml. of the solvent gives 0.45–0.55 g. (50–53% overall yield from cholest-4-en-3β-ol) of 5β-cholest-3-ene-5-acetaldehyde as white prisms, m.p. 66.5–68° (Note 10).

2. Notes

1. This flask must be cleaned with hot chromic acid solution and then, along with *all* other glassware used in this preparation, soaked in a base solution, rinsed with distilled water, and oven dried. Thermal rearrangement of the intermediate vinyl ether in a new (untreated) flask resulted in elimination.

2. Cholest-4-en-3β-ol can be prepared by the procedure of Burgstahler and Nordin.[2] A melting point below 130° indicates that the material is contaminated with some of the 3α-hydroxy isomer. The material used above melted at 130.5–131° (from ethanol).

3. Eastman practical grade ethyl vinyl ether was dried over anhydrous sodium carbonate, distilled (b.p. 36°) from sodium wire, and then redistilled from calcium hydride (b.p. 36°) into the reaction flask after a 5-ml. forerun is discarded.

4. Matheson, Coleman and Bell mercury(II) acetate was partially dissolved in hot absolute ethanol containing 0.02% glacial acetic acid (Note 6) and filtered by suction. The filtrate was cooled, and the white plates of mercury(II) acetate were collected by suction filtration and stored under vacuum.

5. An oil bath at 50–55° was found to be satisfactory.

6. DuPont 99.7% acetic acid was used without purification.

7. Baker petroleum ether (b.p. 30–60°) was used.

8. The solvent was removed by rotary evaporation followed by vacuum (0.01 mm.) drying for 1 hour.

9. Merck silica gel (0.05–0.2 mm., 70–325 mesh ASTM) was used. The filtration column (1.4 × 7 cm.) is prepared in the same way as one used for chromatography, only one (200-ml.) fraction is collected. Use of alumina for the filtration gives variable results.

10. Burgstahler and Nordin report the melting point for the vinyl ether as 56–57°, and for the aldehyde, 66–69°.[2]

11. A Kügelrohr oven was used.

12. Mallinckrodt anhydrous ether was used. The chromatography column was 2.7 × 27 cm.

3. Discussion

The Claisen rearrangement[3] has been adapted in recent years to provide a viable synthetic sequence for the preparation of functional groups other than aldehydes and ketones. Ester[4] and amide[5] syntheses have been reported which proceed through the Claisen intermediate (**A**). The Claisen rearrangement has also been used to generate *trans*-trisubstituted double bonds stereoselectively,[4,6–9] angularly-functionalized derivatives,[10] substituted cyclohexenes,[11] acids,[12] and furans.[7]

$$R = {-}H, {-}Alkyl, {-}Aryl, {-}OR', {-}NR'_2, {-}OSiR'_3$$

A

The procedure given above is an excellent example of the utilization of the Claisen rearrangement to generate an angularly functionalized steroid. The vinyl ether and aldehyde were originally prepared by Burgstahler and Nordin.[2] This procedure combines variations employed by Ireland and co-workers and, in addition, introduces the use of silica gel for the purification of the vinyl ether, thereby improving the reproducibility of the procedure.

1. Division of Chemistry and Chemical Engineering, Gates and Crellin Laboratories of Chemistry, California Institute of Technology, Pasadena, California 91109.
2. A. W. Burgstahler and I. C. Nordin, *J. Am. Chem. Soc.,* **83,** 198 (1961).
3. P. de Mayo, "Molecular Rearrangements," Part One, Wiley, New York, 1963, pp. 660–684.
4. W. S. Johnson, L. Werthemann, W. R. Bartlett, T. J. Brocksom, T.-t. Li, D. J. Faulkner, and M. R. Peterson, *J. Am. Chem. Soc.,* **92,** 741 (1970).
5. A. E. Wick, D. Felix, K. Steen, and A. Eschenmoser, *Helv. Chim. Acta,* **47,** 2425 (1964); D. Felix, K. Gschwend-Steen, A. E. Wick, and A. Eschenmoser, *Helv. Chim. Acta,* **52,** 1030 (1969).
6. C. L. Perrin and D. J. Faulkner, *Tetrahedron Lett.,* 2783 (1969).
7. D. J. Faulkner and M. R. Petersen, *J. Am. Chem. Soc.,* **95,** 553 (1973).
8. R. Marbet and G. Saucy, *Helv. Chim. Acta,* **50,** 2095 (1967).
9. R. I. Trust and R. E. Ireland, *Org. Synth., * **Coll. Vol. 6,** 606 (1988).
10. R. F. Church, R. E. Ireland, and J. A. Marshall, *J. Org. Chem.,* **31,** 2526 (1966).

11. G. Büchi and J. E. Powell, Jr., *J. Am. Chem. Soc.*, **92**, 3126 (1970).
12. R. E. Ireland and R. H. Mueller, *J. Am. Chem. Soc.*, **94**, 5897 (1972).

18-CROWN-6

(1,4,7,10,13,16-Hexaoxacyclooctadecane)

$$HOCH_2CH_2(OCH_2CH_2)_2OH + Cl(CH_2CH_2O)_2CH_2CH_2Cl \xrightarrow[\text{tetrahydrofuran}]{\text{KOH} \atop \text{aqueous}}$$

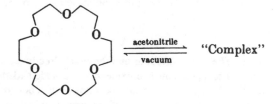

$$\xrightleftharpoons[\text{vacuum}]{\text{acetonitrile}} \quad \text{"Complex"}$$

Submitted by George W. Gokel,[1] Donald J. Cram,[2] Charles L. Liotta,[3]
Henry P. Harris,[3] and Fred L. Cook[3]
Checked by E. A. Noe, M. Raban, and C. R. Johnson

1. Procedure

Caution! Crown ethers may be toxic.[4] *Due care should be exercised in the preparation and handling of* 18-*crown*-6. *An explosion has been reported during the thermal decomposition of the crude* 18-*crown*-6-*potassium salt complex; see Note 8.*

A 3-l., three-necked flask equipped with a mechanical stirrer, a reflux condenser, and an addition funnel is charged with 112.5 g. (100.0 ml., 0.7492 mole) of triethylene glycol and 600 ml. of tetrahydrofuran (Note 1). Stirring is begun and a 60% potassium hydroxide solution, prepared by dissolving 109 g. (1.65 moles) of 85% potassium hydroxide in 70 ml. water, is added (Note 2). The solution warms slightly. After about 15 minutes of vigorous stirring (the solution begins to develop color and gradually becomes rust brown; Note 3), a solution of 140.3 g. (0.7503 mole) of 1,2-bis(2-chloroethoxy)ethane (Note 4) in 100 ml. of tetrahydrofuran is added in a stream. After the addition is complete, the solution is heated at reflux and stirred vigorously for 18–24 hours. The solution is allowed to cool and the bulk of the tetrahydrofuran is evaporated under reduced pressure (Note 5). The resulting thick, brown slurry is diluted with 500 ml. of dichloromethane and filtered through a glass frit. The salts removed by filtration are washed with more dichloromethane to remove absorbed crown and the combined organic solution is dried over anhydrous magnesium sulfate (Note 6), filtered, evaporated to minimum volume (aspirator vacuum), and distilled under high vacuum using a simple distillation head. The distillation should be carried out at the lowest possible pressure; a typical fraction contains 76–87 g. (38–44%) of crude 18-crown-6 and is collected over 100–167° (0.2 mm.) (Notes 7–9).

To 50 g. of the crude 18-crown-6 in a 250-ml. Erlenmeyer flask is added 100 ml. of acetonitrile. A magnetic stirring bar is added, and the flask is equipped with a calcium chloride drying tube. The resulting slurry is heated on a hot plate to effect solution. The solution is stirred vigorously as it is allowed to cool to ambient temperature; fine white crystals of crown-acetonitrile complex are deposited. The flask is allowed to stand in a freezer for 24–48 hours and is finally cooled in a −30° bath to precipitate as much of the complex as possible. The solid is collected by rapid filtration (Note 10) and washed once with a small amount of cold acetonitrile. The hygroscopic crystals are transferred to a 200-ml., round-bottomed flask equipped with a magnetic stirring bar and a vacuum takeoff. The acetonitrile is removed from the complex under high vacuum (0.1–0.5 mm.), with gentle heating (~35°), over 2–3 hours. The pure colorless crown (28–33 g., 56–66%) (Note 11) crystallizes on standing, m.p. 38–39.5° (Note 12).

2. Notes

1. The tetrahydrofuran may be used directly without drying or purification.

2. The potassium hydroxide may be added to the water in one portion, but the resulting base solution should be allowed to cool to nearly room temperature before adding to the reaction mixture. If the potassium hydroxide solution is cooled much below ambient temperature, the potassium hydroxide begins to separate; hot potassium hydroxide solution could cause the tetrahydrofuran solution to boil.

3. The rate of darkening is related to the temperature of the solution and, if warm potassium hydroxide solution is used, the color will develop somewhat more rapidly. Differences in the rate of darkening do not appear to affect the yield or purity of product.

4. 1,2-Bis(2-chloroethoxy)ethane is available from Eastman Organic Chemicals.

5. As much water as possible should be removed during evaporation so that the salts will filter more readily and the solution can be dried more easily.

6. Drying agents containing complexable cations, such as K^+ or Na^+, should be avoided.

7. There is generally a forerun (room temperature to *ca.* 100°) the size of which varies according to the vigor of the previous evaporation steps (see Note 5).

8. In a large batch preparation of 18-crown-6 an explosion has been reported as a result of difficulties occurring during this distillation step.[5] In this instance the head temperature rose to near 200°. When the system was vented to the atmosphere at this temperature an explosion occurred, apparently the result of autoignition of 1,4-dioxane vapors. Dioxane is reported to undergo autoignition in air at temperatures in excess of 180°. It is recommended that the head and pot be allowed to cool and then be vented with a nitrogen atmosphere.

9. The material obtained in the distillation cut contains both alcoholic and vinylic impurities. The crown may be purified by a second, more careful distillation followed by recrystallization, sublimation or by chromatography in addition to the method described here (see Discussion).

10. The filtration should be conducted in a dry-box or by using an inverted funnel-nitrogen flow, whichever is more convenient.

11. The yield of pure crown depends somewhat on the purity of the crude material

used. Additional crown may be obtained by combining mother liquors and repeating the distillation and complex formation process.

12. The ^{1}H NMR spectrum (CCl_4) exhibits a singlet at δ 3.56.

3. Discussion

The compound known as 18-crown-6 is one of the simplest and most useful of the macrocyclic polyethers. Its synthesis in low yield was first reported by Pedersen.[5] Greene[6] and Dale and Kristiansen[7] have reported syntheses of the title compound from triethylene glycol and triethylene glycol di-p-toluenesulfonate. Both of these procedures use strong base and anhydrous conditions and achieve purification by more or less classical methods. The combination of distillation and formation of the acetonitrile complex affords crown of high purity without lengthy chromatography or sublimation.[8,9]

1. Department of Chemistry, Pennsylvania State University, University Park, PA 16802. [Present address: University of Maryland, College Park, MD 20742.]
2. Department of Chemistry, University of California at Los Angeles, CA 90024.
3. School of Chemistry, Georgia Institute of Technology, Atlanta, GA 30332.
4. C. J. Pedersen, *J. Am. Chem. Soc.,* **89,** 7017 (1967).
5. P. E. Stott, *Chem. Eng. News,* **54** (37), 5 (1976).
6. R. N. Greene, *Tetrahedron Lett.,* 1793 (1972).
7. J. A. Dale and P. O. Kristiansen, *Acta Chem. Scand.,* **26,** 1471 (1972).
8. G. W. Gokel, D. J. Cram, C. L. Liotta, H. P. Harris, and F. L. Cook, *J. Org. Chem.,* **39,** 2445 (1974).
9. Acknowledgment is made to E. P. Kyba (University of Texas) for noting that this crown can be distilled and to E. R. Wonchoba (Du Pont Co.) for helpful comments.

CINNAMONITRILE

(2-Propenenitrile, 3-phenyl-)

$$CH{=}CH{-}COOH$$

$$\xrightarrow[-CO_2]{+ClSO_2NCO}$$

$$CH{=}CH{-}CONHSO_2Cl$$

$$\xrightarrow{HCON(CH_3)_2}$$

$$CH{=}CH{-}CN$$

$$+ SO_3 + HCl$$

Submitted by G. Lohaus[1]
Checked by Frank J. Weigert and Richard E. Benson

1. Procedure

Caution! Chlorosulfonyl isocyanate is a highly corrosive, irritating compound. This reaction should be carried out in an efficient hood.

A 2-l., four-necked flask equipped with a stirrer, thermometer, dropping funnel, and reflux condenser is charged with 296 g. (2.00 moles) of cinnamic acid and 600 ml. of dichloromethane. The mixture is heated to reflux and a solution of 290 g. (2.05 moles) of chlorosulfonyl isocyanate (Note 1) in 100 ml. of dichloromethane is added dropwise, with stirring, over a period of 45 minutes. After a few minutes the solution becomes clear and, after about one-half of the isocyanate has been added, the carboxylic acid amide *N*-sulfonyl chloride begins to precipitate. At the end of the addition, the reaction mixture is heated for an hour to complete the evolution of carbon dioxide (Note 2). *N,N*-Dimethylformamide (300 g., 4.11 moles) (Note 3) is added, with stirring, over a 15-minute period, while cooling with ice to an internal temperature of 15–20°. The reaction mixture is stirred for an additional 15 minutes and then poured onto *ca.* 800 g. of ice. After the ice has melted, the resulting layers are separated and the aqueous phase is extracted once with 100 ml. of dichloromethane. The organic phases are combined and extracted six times with 100-ml. portions of water to remove most of the *N,N*-dimethylformamide (Note 4). The resulting organic solution is dried for 2 hours with 50 g. of potassium carbonate, decanted from the drying agent, and concentrated by distillation at atmospheric pressure. The resulting oil is distilled through a 10-cm. Vigreux column (Note 5) to give 197–225 g. (78–87%) of cinnamonitrile, b.p. 92–94° (1 mm.), n_D^{25} 1.5998 (Note 6). Its ^{1}H NMR spectrum (60 MHz., CCl$_4$, 36°) δ, (multiplicity, coupling constant *J* in Hz., number of protons): 5.79 (d, *J* = 17, 1H), 7.29 (d, *J* = 17, 1H), 7.35 (s, 5H).

2. Notes

1. Chlorosulfonyl isocyanate, *Org. Synth.*, **Coll. Vol. 5,** 226 (1973), is available from Farbwerke Hoechst AG. The checkers found it necessary to distill the product before use.

2. The rate of evolution of carbon dioxide can be followed easily with a bubble counter attached to the reflux condenser.

3. Other amides also can be used, but *N,N*-dimethylformamide generally is preferred because of its volatility, high solvating ability, and miscibility with water.

4. The checkers found that, if the mixture at this point was allowed to stand overnight, a crystalline product separated that was identified as the *N,N*-dimethylformamide – sulfur trioxide complex.

5. The presence of a crystalline residue at the end of the distillation prevents the use of a spinning band column.

6. GC indicates the purity to be greater than 99.9%. A column containing Chromosorb W/DMCS/AW with 10% Triton X 305 as the stationary phase is used.

3. Discussion

This reaction illustrates a broadly applicable method for converting carboxylic acids to the corresponding nitriles.[2] It avoids the necessity for conversion of the acid to the amide, followed by dehydration with vigorous reagents such as phosphorus pentachloride or phosphorus oxychloride. The reaction is characterized by easy workup, generally good yields, and by mild reaction conditions that permit certain functional groups that may be present to remain unchanged. For example, the half ethyl ester of succinic acid is converted to the corresponding nitrile in 72% yield with this procedure. Aliphatic

TABLE I
NITRILES DERIVED FROM CORRESPONDING ACID

Nitrile	Yield, %
cyclo-$C_6H_{11}CN$	78
$ClCH_2CH_2CN$	66
$(CH_3)_3CCN$	68
$CH_3CH{=}CHCH{=}CHCN$	76
$C_6H_5CH_2CN$	84
$C_2H_5OCOCH_2CH_2CN$	72
$NC\text{-}(CH_2)_8CN$	86

| | 63 |

unsaturation may be present; thus, 2,4-hexadienenitrile is obtained from 2,4-hexadienoic acid in 76% yield. The reaction is operable with chlorine-containing acids, and an aromatic acid has been converted to the nitrile with this procedure. The results are summarized in Table I.[2]

Other specific procedures for the synthesis of cinnamonitrile include the dehydration of cinnamamide with phosphorus pentachloride[3] or phosphorus oxychloride,[4] the dehydration of cinnamaldehyde oxime with acetic anhydride,[5] and the dehydrochlorination of α-chloro-β-phenylpropionitrile with quinoline,[6,7] N,N-diethylaniline,[8] or triethylamine.[9]

1. Hoechst AG., previously Meister Lucius & Brüning, Frankfurt/Main-Höchst, Germany.
2. G. Lohaus, *Chem. Ber.,* **100,** 2719 (1967).
3. J. v. Rossum, *Z. Chem.,* 362 (1866) [*Beilstein,* 4th ed., **9,** 589 (1926)].
4. K. v. Auwers and M. Seyfried, *Justus Liebigs Ann. Chem.,* **484,** 212 (1930).
5. T. Posner, *Justus Liebigs Ann. Chem.,* **389,** 117 (1912).
6. A. H. Cook, J. Downer, and B. Hornung, *J. Chem. Soc.,* 502 (1941).
7. W. H. Brunner and H. Perger, *Monatsh. Chem.,* **79,** 187 (1948).
8. C. F. Koelsch, *J. Am. Chem. Soc.,* **65,** 57 (1943).
9. N. O. Pastushak, N. F. Stadniichuk and A. V. Dombrovskii, *Zh. Obshch. Khim.,* **33,** 2950 (1963).

PREPARATION OF CYANO COMPOUNDS USING ALKYLALUMINUM INTERMEDIATES: 1-CYANO-6-METHOXY-3,4-DIHYDRONAPHTHALENE

(Naphthalenecarbonitrile, 3,4-dihydro-6-methoxy-)

Submitted by W. Nagata,[1] M. Yoshioka, and M. Murakami
Checked by R. Wong, C. Kowalski, R. Czarny, and R. E. Ireland

1. Procedure

Caution! Benzene has been identified as a carcinogen; OSHA has issued emergency standards on its use. All procedures involving benzene should be carried out in a well-ventilated hood, and glove protection is required.

A 200-ml., two-necked, round-bottomed flask charged with 6.15 g. (0.0347 mole) of 6-methoxy-1-tetralone (Note 1) and a 100-ml., round-bottomed flask are flushed with nitrogen, and each of the flasks is fitted with an adaptor with a side arm connected to a nitrogen bubbler system and then charged with 30 ml. of anhydrous toluene. The 200-ml. flask is cooled to −20° to −25° (bath temperature) (Note 2). Into the 100-ml. flask is introduced 60 ml. (0.07 mole) of a 13% solution of diethylaluminum cyanide in benzene (Note 3) with a hypodermic syringe, and this flask is cooled with ice water. The cooled diethylaluminum cyanide solution is added to the cold solution of 6-methoxytetralone with a hypodermic syringe and the resulting mixture, after being swirled, is kept at −15° for 80 minutes under nitrogen. The stopper of the flask is replaced by a glass tube which has one end extending to the bottom of the reaction flask and the other end mounted in a neck of a 2-l., three-necked flask, equipped with an efficient stirrer and containing a cold (−70°) mixture of 250 ml. of methanol and 150 ml. of concentrated hydrochloric acid as shown in Figure 1. The reaction mixture is added through the glass tube to the vigorously stirred acid mixture by applying a positive nitrogen pressure to the reaction flask (Note 4). After the bulk of the reaction mixture is added, about 50 ml. of a cold mixture of methanol and hydrochloric acid is added to the reaction flask and this mixture is transferred to the

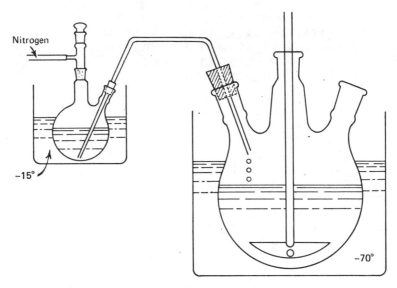

Figure 1. Apparatus for acid treatment of the reaction mixture.

2-1. flask in the same way, as described above. Stirring is continued for one hour, and the resulting mixture is poured into a mixture of 200 ml. of concentrated hydrochloric acid and 1 l. of ice water (Note 5) and extracted with three 500-ml. portions of dichloromethane. The combined organic phases are washed once with 1.5 l. of water, dried over anhydrous sodium sulfate, and evaporated from a flask containing 55 mg. of p-toluenesulfonic acid monohydrate (Note 6), using a rotary evaporator at a temperature below 40°.

The residue, obtained as a pale yellow oil, weighs approximately 7.4 g. and consists of 1-cyano-1-hydroxy-6-methoxytetralin and a small amount of unchanged 6-methoxy-1-tetralone. The oil is transferred to a 10-ml. Claisen flask, a small amount of a mixture of dichloromethane and diethyl ether being used to complete the transfer. Two hundred milligrams of powdered potassium hydrogen sulfate is added, and the flask is heated at 130° under reduced pressure (5 mm.) for 30 minutes. The pressure is then reduced to 0.01 mm. and the temperature is raised to about 150°, collecting all the distillate [b.p. 113–117° (0.01 mm.)] in a 50-ml. flask. The viscous distillate (including material adhering to the distillation apparatus), weighs 6.0–6.2 g. and yields 4.91–5.05 g. (76–78%) of product, m.p. 50–51.5°, after two or three crystallizations from methanol. The residue from the mother liquors (1.0–1.3 g.) is adsorbed on a column of 100 times its weight of silica gel (70–325 mesh), and the column is eluted with approximately 1 l. of 40% ether in petroleum ether (b.p. 30–60°). The first 200 ml. of eluent is discarded, and 510–550 mg. of the product is eluted in the next 250 ml. of eluent. Crystallization of this material from an ether–petroleum ether (b.p. 30–60°) mixture affords an additional 460–500 mg. (7.0–7.8%) of pure product, m.p. 50.5–51.5°. The total yield of the unsaturated nitrile is 5.41–5.51 g. (83.8–85.5%). (Note 7). The final 500 ml. fraction from chromatography contains 330–660 mg. (5.4–10.7%) of the starting material, m.p. 77–78°.

2. Notes

1. 6-Methoxy-1-tetralone is available from K & K Laboratories, New York, although the submitters used a material, m.p. 77–80°, produced by Osaka Yuki Gosei K. K., Nishinomiya-shi, Japan.

2. Crystals of 6-methoxy-1-tetralone may separate from the solution on cooling, but redissolve upon addition of the cooled diethylaluminum cyanide solution.

3. For the preparation of diethylaluminum cyanide, see *Org. Synth.*, **Coll. Vol. 6,** 436 (1988). Both the submitters and the checkers employed a crude reagent solution rather than a solution prepared from distilled diethylaluminum cyanide. A 1–2 M solution of diethylaluminum cyanide in benzene is commercially available from Alfa Products, Ventron Corporation, Danvers, Massachusetts.

4. Application of the nitrogen pressure may be made conveniently by capping the outlet of the mercury bubbler.

5. The two-step decomposition is effective for preventing reconversion of the cyanohydrin into the starting ketone.

6. The cyanohydrin initially formed is unstable and readily reconverted to the starting 6-methoxy-1-tetralone on evaporation of the extracts unless the solution is kept slightly acidic by addition of a trace amount of *p*-toluenesulfonic acid monohydrate. As this acid is relatively insoluble in dichloromethane, it should be added directly to the flask used for evaporation of the solvent.

7. Preferably, the product should be stored in an oxygen-free atmosphere. Samples not stored in an inert atmosphere have deteriorated to dark-brown masses within several months, whereas no appreciable change has been observed in a sample stored for 2 years in an ampoule filled with argon.

3. Discussion

The present method developed by the submitters[2] is the only practical process for the preparation of 1-cyano-6-methoxy-3,4-dihydronaphthalene. Birch and Robinson[3] have reported that 6-methoxy-1-tetralone did not react with hydrogen cyanide or sodium acetylide.

This process presents a procedure applicable to the preparation of cyanohydrins from relatively unreactive ketones and aldehydes. 1-Cyano-6-methoxy-3,4-dihydronaphthalene is useful as an intermediate in the synthesis of polycyclic compounds.

Cyanotrimethylsilane[4] is useful for the preparation of trimethylsilyl ethers of cyanohydrins, obtained from ketones, aldehydes, α,β-unsaturated ketones, and quinones.[5]

1. Shionogi Research Laboratory, Shionogi & Co., Ltd., Osaka, Japan.
2. W. Nagata and M. Yoshioka, *Tetrahedron Lett.*, 1913 (1966); W. Nagata, M. Yoshioka, and M. Murakami, *J. Am. Chem. Soc.,* **94,** 4654 (1972).
3. A. J. Birch and R. Robinson, *J. Chem. Soc.,* **503,** (1944).

4. S. Hünig and G. Wehner, *Synthesis*, 522 (1979); J. K. Rasmussen and S. M. Heilmann, *Synthesis*, 523 (1979).
5. D. A. Evans, L. K. Truesdale, and G. L. Carroll, *J. Chem. Soc. Chem. Commun.*, 55 (1973); D. A. Evans, J. M. Hoffman, and L. K. Truesdale, *J. Am. Chem. Soc.*, **95**, 5822 (1973); D. A. Evans, G. L. Carroll, and L. K. Truesdale, *J. Org. Chem.*, **39**, 914 (1974); K. Deuchert, U. Hertenstein, and S. Hünig, *Synthesis*, 777 (1973).

CYCLOBUTADIENEIRON TRICARBONYL

[Iron, tricarbonyl (η^4-1,3-cyclobutadiene)-]

Submitted by R. Pettit[1] and J. Henery
Checked by J. Napierski and R. Breslow

1. Procedure

Caution! Benzene has been identified as a carcinogen; OSHA has issued emergency standards on its use. All procedures involving benzene should be carried out in a well-ventilated hood, and glove protection is required.

In a well-ventilated hood a 500-ml., three-necked flask is immersed in an oil bath and fitted with a condenser and a mechanical stirrer; a T-piece is inserted through a rubber stopper placed in the top of the condenser. One lead of the T-piece is connected to a nitrogen supply and the other to a gas bubbler. *cis*-3,4-Dichlorocyclobutene [*Org. Synth.*, **Coll. Vol. 6**, 422 (1988)], 20 g. (0.16 mole), and 125 ml. of anhydrous benzene are added to the flask, and the apparatus is flushed with nitrogen. Diiron nonacarbonyl, 25 g. (Note 1) is then added, the flow of N_2 is stopped, and the mixture is heated to 50–55°, with stirring. After about 15 minutes the initial rapid evolution of carbon monoxide becomes greatly diminished and an additional 8 g. of the nonacarbonyl is added; additional 8-g. quantities are added at intervals (approximately 15 minutes), governed by the rate of carbon monoxide evolution. The addition is continued until no more carbon monoxide is liberated (Note 2), and the reaction mixture is stirred at 50° for an additional hour. Approximately 140 g. of diiron nonacarbonyl is required for the complete conversion of the dichlorocyclobutene, the total reaction time being about 6 hours.

The contents of the flask are then filtered with suction through Filtercel and the residue, while kept in the Buchner funnel, is thoroughly washed with pentane until the washings are colorless (Note 3). The pentane and much of the benzene are evaporated from the combined filtrates with a water aspirator.

The residual liquid is transferred to a flask equipped with an efficient fractionating column and distilled under reduced pressure. Benzene is removed first, followed by considerable quantities of iron pentacarbonyl (b.p. 20°, 30 mm.); when the diiron nonacarbonyl has been removed, the pressure is reduced further and cyclobutadieneiron tricarbonyl[3] is collected as a pale yellow oil, b.p. 47° (3 mm.), yielding 13.8–14.4 g. (45–46% based on dichlorocyclobutene), (Note 4).

2. Notes

1. Diiron nonacarbonyl is readily available through photolysis of iron pentacarbonyl.[2]

2. The conversion of the dichlorocyclobutene to cyclobutadieneiron tricarbonyl can be conveniently monitored by GC. On a 5 ft. × ⅛ in. column of 20% Carbowax on Chromosorb W, under conditions where the retention time of dichlorocyclobutene is 2.6 minutes, the retention time of cyclobutadieneiron tricarbonyl is 2.4 minutes.

3. The brown insoluble residue is frequently pyrophoric if it is allowed to dry; it should be immediately wetted with water before it is disposed of.

4. In some preparations the last portion of the distillate of the complex may be dark green in color. This color is due to trace amounts of $Fe_3(CO)_{12}$. If desired, this can be readily removed by chromatography over alumina. The submitters report a similar yield on three times the scale.

3. Discussion

Cyclobutadieneiron tricarbonyl may also be produced by the reaction of 3,4-dichlorocyclobutene with disodium irontetracarbonyl[5] and by irradiation of α-pyrone followed by treatment with diiron nonacarbonyl.[5] The method outlined here is the most convenient, especially when considerable quantities (10 g. or more) of cyclobutadieneiron tricarbonyl are required. The analogous reaction of derivatives of 3,4-dihalocyclobutenes with diiron nonacarbonyl affords the corresponding cyclobutadieneiron tricarbonyl complexes. Cyclobutadieneiron tricarbonyl can be oxidized to generate cyclobutadiene *in situ*.[6]

1. Deceased, December 10, 1981; work done at the Department of Chemistry, The University of Texas at Austin, Austin, Texas 78712.
2. E. H. Braye and W. Huebel, *Inorg. Syn.*, **8**, 178 (1966).
3. G. F. Emerson, L. Watts, and R. Pettit, *J. Am. Chem. Soc.*, **87**, 131 (1965).
4. R. G. Amiet, P. C. Reeves, and R. Pettit, *Chem. Commun.*, 1208 (1967).
5. M. Rosenblum and C. Gatsonis, *J. Am. Chem. Soc.*, **89**, 5074 (1967).
6. L. Watts and R. Pettit, *Advan. Chem. Series*, "Werner Centennial," **62**, 549 (1966).

ALDEHYDES FROM ACID CHLORIDES BY REDUCTION OF
ESTER-MESYLATES WITH SODIUM BOROHYDRIDE:
CYCLOBUTANECARBOXALDEHYDE

Submitted by M. Ross Johnson and Bruce Rickborn[1]
Checked by Saul C. Cherkofsky and Richard E. Benson

1. Procedure

A. erythro-2,3-*Butanediol monomesylate*. A 2-l., round-bottomed flask is equipped with a 1-l. dropping funnel attached to a calcium chloride drying tube. A magnetic stirring bar is placed in the flask and a solution of 48.0 g. (0.500 mole) of methanesulfonic acid (Note 1) in 500 ml. of anhydrous diethyl ether is added. Stirring is begun, and the flask is cooled in an ice-water bath while a solution of 37 g. (0.52 mole) of trans-2-butene oxide (Notes 2 and 3) in 500 ml. of anhydrous ether is added over a period of 3–4 hours (Note 4). After 6 hours the cooling bath is removed and the mixture is stirred an additional 12 hours. The ether and any excess epoxide are removed with a rotary evaporator at 25° and water aspirator pressure, giving 83–84 g. (99–100%) of erythro-2,3-butanediol monomesylate as a clear, colorless, somewhat viscous liquid (Note 5).

B. erythro-3-*Methanesulfonyloxy-2-butyl cyclobutanecarboxylate*. A 500-ml., round-bottomed flask, cooled in an ice-water bath, equipped with a 50-ml. dropping funnel and a magnetic stirring bar is charged with 35.3 g. (0.210 mole) of erytho-2,3-butanediol monomesylate and 150 ml. of dry pyridine. Stirring is begun, and 23.7 g. (0.200 mole) of cyclobutanecarboxylic acid chloride (Note 6) is added over a period of 1 hour. The cooling bath is removed, and stirring is continued for 8 hours. The mixture is added to 500 ml. of ether, and the resulting solution washed with three 250-ml. portions of 3 N sulfuric acid. The pyridine-free solution is washed with 250 ml. of a saturated sodium hydrogen carbonate solution and then with 250 ml. of water. The ether solution is dried over 2 g. of anhydrous magnesium sulfate. The solvent is removed with a rotary evaporator at 25°, giving 45.1–48.0 g. (90–96%) of *erythro*-3-methanesulfonyloxy-2-butyl cyclobutanecarboxylate as a pale yellow, viscous liquid (Note 7).

C. 2-*Cyclobutyl*-cis-4-trans-5-*dimethyl*-1,3-*dioxolane*. A 2-l., three-necked, round-bottomed flask is equipped with a mechanical stirrer, a 125-ml. dropping funnel, and a condenser attached to a nitrogen line with a bubbler device to permit maintenance of a positive pressure of nitrogen. Anhydrous pyridine (650 ml.) and 5 g. (0.1 mole) of sodium borohydride (Note 8) are added to the flask, stirring is begun, and the mixture is heated at reflux. A solution of 25 g. (0.10 mole) of *erythro*-3-methanesulfonyloxy-2-butyl cyclobutanecarboxylate in 50 ml. of anhydrous pyridine is added from the dropping funnel over a period of 30 minutes, and heating at reflux is continued for 8 hours. After cooling, 50 ml. of water is added (some heat is evolved), and the mixture is transferred to a 4-l. separatory funnel with 1 l. of pentane (Note 9), and 700 ml. of cold 3 *N* sulfuric acid saturated with sodium chloride. The aqueous layer is separated and washed with two 250-ml. portions of pentane. The pentane extractions are combined and washed with three 500-ml. portions of 3 *N* sulfuric acid saturated with sodium chloride and finally with 500 ml. of saturated sodium hydrogen carbonate. The pentane solution is dried over 1 g. of anhydrous potassium carbonate and evaporated on a steam bath. The product is distilled through a short Vigreux column, yielding 6.7–7.6 g. (43–49%, Note 10) of 2-cyclobutyl-*cis*-4-*trans*-5-dimethyl-1,3-dioxolane, b.p. 79–83° (22 mm.) (Note 11).

D. *Cyclobutanecarboxaldehyde*. A 1-l., round-bottomed flask equipped with a magnetic stirring bar and a 60-cm. glass helix-packed column is charged with 600 ml. of 3 *N* sulfuric acid, 200 ml. of *N,N*-dimethylformamide (Note 12), and 20 g. (0.13 mole) of 2-cyclobutyl-*cis*-4-*trans*-5-dimethyl-1,3-dioxolane. The mixture is heated to gentle reflux, and cyclobutanecarboxaldehyde is collected as a steam distillate, b.p. 86°. After the distillation of the oil has ceased, the product is transferred to a separatory funnel, and the lower layer of water is discarded. The oil is dissolved in 25 ml. of ether and dried over anhydrous sodium sulfate. The product is distilled through a small Vigreux column. After removal of the ether, 6.2–6.7 g. (58–63%) of cyclobutanecarboxaldehyde is collected, b.p. 56–59° (120 mm.) (Note 13).

2. Notes

1. Methanesulfonic acid was obtained from Aldrich Chemical Company, Inc., and distilled prior to use. The fraction collected at 140° (0.2 mm.) was used.

2. *trans*-2-Butene oxide was prepared by appropriate modification of the procedure in *Org. Synth.*, **Coll. Vol. 4**, 860 (1963). A 2-l., four-necked, round-bottomed flask fitted with a mechanical stirrer, a 1-l. dropping funnel, an acetone–dry ice condenser, and a thermometer is charged with 1 l. of 1,1,2,2-tetrachloroethane. The condenser is packed with dry ice and acetone, and the flask is cooled in a methanol-ice bath to −15°. *trans*-2-Butene (153 g., 2.73 moles) (Phillips Petroleum Company, 99%) is distilled into the flask from a tared, chilled trap. Six hundred milliliters of 40% peracetic acid (FMC Corporation), to which has been added 30 g. sodium acetate to neutralize the sulfuric acid present, is added to the stirred solution from the dropping funnel over a period of 2 hours. The mixture is stirred at −15° for another hour, then allowed to warm to room tempera-

ture. The mixture is poured into 1 l. of ice-cold water. The organic layer is separated, washed first with 10% sodium carbonate solution, then with water, dried over magnesium sulfate, and filtered. Distillation of the filtrate through a 75-cm. spinning-band column gives 133 g. (68%) of *trans*-2-butene oxide as a colorless oil, b.p. 52.5–55°.

3. A slight excess of *trans*-2-butene oxide is used to assure complete utilization of methanesulfonic acid. The checkers' experiments indicated that a 15% excess of the epoxide substantially reduced the amount of unreacted methane–sulfonic acid present in the product and did not appear to interfere with the succeeding steps of this procedure.

4. This order of addition and dilution is required to avoid dimerization or polymerization of the epoxide.

5. No attempt was made to purify this compound further. It had a very characteristic ^{1}H NMR spectrum (CDCl$_3$, external tetramethylsilane reference): δ 1.22 (d, $J = 7.5$ Hz., 3H), 1.37 (d, $J = 7.5$ Hz., 3H), 3.1 (s, 3H), 3.4 (s, OH, position variable), 4.0 (d of q, $J = 4.0$, 7.5 Hz., 1H), and 4.78 (d of q, $J = 4.0$, 7.5 Hz., 1H). A sample stored for several weeks at room temperature showed no change in its spectrum.

6. Cyclobutanecarboxylic acid chloride was obtained from Aldrich Chemical Company, Inc., and distilled prior to use. The acid chloride can be prepared by the reaction of thionyl chloride with the corresponding acid (available from Aldrich) by the general procedure in *Org. Synth.*, **Coll. Vol. 1,** 147 (1941). The preparation of cyclobutanecarboxylic acid has been described in *Org. Synth.*, **Coll. Vol. 3,** 213 (1955) and elsewhere.[2]

7. The ^{1}H NMR spectrum (CDCl$_3$, external tetramethylsilane reference): δ 1.27 (d, $J = 6.5$ Hz., 3H), 1.41 (d, $J = 6.5$ Hz., 3H), 2.18 (m, 6H), 3.10 (s, 3H), superimposed on 3.2 (m, 1H), and 5.0 (m, 2H). IR (CDCl$_3$): 1725 cm.$^{-1}$.

8. Commercial material from Matheson, Coleman and Bell and recrystallized reagent gave comparable results. The yield is decreased by use of less than 1 mole of sodium borohydride per mole of mesylate.

9. Either purified pentane or Spectranalyzed pentane available from Fisher Scientific Company was used.

10. The submitter reports yields of 10–11 g. (64–71%). The checker obtained the dioxolane in 57% yield on conducting the experiment on a sixfold scale.

11. The ^{1}H NMR spectrum (neat, external tetramethylsilane reference) δ 1.1 (two overlapping d, $J = 6$ Hz., 6H), 1.7–2.0 (m, 6H), 2.1–2.6 (m, 1H), 3.2–3.8 (m, 2H), and 4.94 (d, $J = 5$ Hz., 1H).

12. This proportion of water to N,N-dimethylformamide is needed to assure solubility (hence facile reaction) of the acetal on heating at reflux.

13. The ^{1}H NMR spectrum (neat, external tetramethylsilane reference): δ 1.4–2.4 (m, 6H), 2.6–3.2 (m, 1H), and 9.8 (d, $J \approx 1.5$ Hz., 1H); IR (CCl$_4$); 1730 cm.$^{-1}$ (C=O).

3. Discussion

Cyclobutanecarboxaldehyde has been prepared in very low yield by the Rosenmund reduction procedure.[3] A 46% yield of the 2,4-dinitrophenylhydrazone derivative has also been reported, with the aldehyde formed as an intermediate, in the reaction of the acid chloride and lithium tri-t-butoxyaluminum hydride at $-78°$ in diglyme.[4]

Methods now available for the reduction of carboxylic acid derivatives to aldehydes require careful control of conditions to avoid overreduction or underreduction. The

TABLE I
ALDEHYDES FROM ESTER-MESYLATES

$$R-\overset{\overset{O}{\|}}{C}-O \quad OSO_2CH_3 \longrightarrow R-\overset{O-CHCH_3}{\underset{O-CHCH_3}{\overset{|}{CH}}} \longrightarrow RCHO$$

$$CH_3-CH-CH-CH_3$$

(A)	(B) ·	(C)

Ester-Mesylates (A), R =	Acetal (B) Yield, %[a]	Aldehyde (C) Yield, %[a]
n-C_5H_{11}-	63	78
C_6H_5-	64	89
cyclo-C_6H_{11}-	75	93
$(CH_3)_3C$-	77	50[b]
$C_6H_5CH{=}CH$-	62	87[c,d]
(cyclohexenyl)	67	81[c]
cyclo-C_3H_5-	72	81

[a]Yield of distilled product; in several instances numerous runs were made, and the lowest yield is given.
[b]This acetal hydrolyzes quite slowly, and the relatively low yield of pivalaldehyde appears to be associated with this observation.
[c]Yields determined by GC analysis only.
[d]Fifteen percent of this product is the dihydro derivative, that is, the acetal of 3-phenylpropanal.

procedure described here is particularly convenient in that the acetal, not subject to further reduction, is formed directly in the reducing medium.

The scope of the reaction is indicated in Table I. An interesting aspect of the reaction is that the rate of the borohydride reduction step appears to be relatively insensitive to the substitutent R. It is suggested that the reaction occurs with formation of an intermediate acyloxonium ion, which is rapidly converted to acetal by reaction with the borohydride ion. Pyridine–borane has been shown to be the other product of this reaction; yield studies also indicate that only one hydride per borohydride ion is used efficiently in the formation of acetal.

$$\underset{CH_3-HC-CH-CH_3}{\overset{R}{O \overset{+}{\triangle} O}} \xrightarrow[\text{pyridine}]{BH_4^-} \underset{CH_3-HC-CH-CH_3}{\overset{R \quad H}{O \quad O}} + \quad \overset{+}{N}-\overset{-}{B}H_3$$

1. Department of Chemistry, University of California, Santa Barbara, California 93106.
2. J. Cason and H. Rapoport, "Laboratory Text in Organic Chemistry," 2nd ed., Prentice-Hall, Englewood Cliffs, N.J., 1962, p. 407.

3. E. D. Venus-Danilova, *Zh. Obshch. Khim.*, **8**, 1179 (1938) [*Chem. Abstr.*, **33**, 4203 (1939)].
4. H. C. Brown and B. C. Subba Rao, *J. Am. Chem. Soc.*, **80**, 5377 (1958).

CYCLIC KETONES FROM
1,3-DITHIANE: CYCLOBUTANONE

Submitted by D. Seebach[1] and A. K. Beck
Checked by Jose F. Pazos and Richard E. Benson

1. Procedure

A.   *5,9-Dithiaspiro[3.5]nonane.*   A dry, 2-l., one-necked, round-bottomed flask containing a magnetic stirring bar (Note 1) is flushed with dry nitrogen (Note 2), and 1.25 l. of dry tetrahydrofuran (Note 3) and 50 g. (0.42 mole) of 1,3-dithiane (Note 4) are added. The flask is quickly equipped with a three-way stopcock bearing a standard tapered joint, a rubber septum, and a nitrogen inlet, as shown in Figure 1 (Note 5). The solution is stirred with an efficient magnetic stirrer and cooled to an external temperature of −20° with a methanol–dry ice bath. A 3% excess (total of 0.43 mole) of 1.5–2.5 M *n*-butyllithium in *n*-hexane (Note 6) is added through the rubber septum with a syringe. The bath temperature is kept between −10° and −20° for 2 hours before the temperature of the bath is reduced to −75° (Note 7) and 65.5 g. (44.5 ml., 0.417 mole) of 1-bromo-3-chloropropane (Note 8) is added with a syringe over 10 minutes. The temperature of the bath is raised to −30° over a 2-hour period by gradually replacing the cold methanol with warm methanol. The bath is removed and stirring is continued until the reaction flask is at room temperature (Note 9). The flask is again cooled to −75° (Note 7) and 0.44 mole of *n*-butyllithium in *n*-hexane (Note 6) is added with a syringe over 10 minutes. After the addition is complete, the temperature of the reaction flask is allowed to rise to room temperature overnight (Note 10). The solvent is removed from the product with a rotary evaporator at 50° and aspirator pressure. Water (300 ml.) and diethyl ether (500 ml.) are added to the product in the flask, the ether layer is separated, and the aqueous layer is washed again with 500 ml. of ether. The organic layers are combined, washed with 200 ml. of water, and dried over 10 g. of anhydrous potassium carbonate. The ether is removed by distillation to yield about 75 g. of crude product. Distillation through a packed column gives 44−57 g. (65−84%) of 5,9-dithiaspiro[3.5]nonane, b.p. 65−75° (1 mm.), n_D^{20} 1.5700. This product is of sufficient purity for use in Part B (Note 11).

B.   *Cyclobutanone.*   A 2-l., three-necked flask is fitted with an efficient mechanical stirrer and a water-cooled condenser assembled for downward distillation to which is attached a 250-ml. receiver with a side arm. Two cold traps are attached consecutively to the distillation apparatus as shown in Figure 2 (Note 12). The receiver is immersed in an ice-water bath, and the traps are immersed in dry ice and acetone. To the flask is added 45

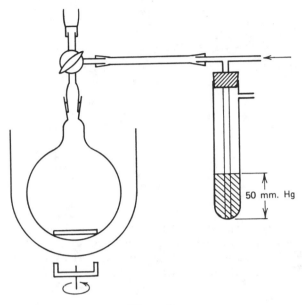

Figure 1.

g. (0.28 mole) of 5,9-dithiaspiro[3.5]nonane, 900 ml. of triethylene glycol, and 150 ml. of water. Stirring is begun, and 163 g. (0.600 mole) of mercury(II) chloride (Note 13) and 51.5 g. (0.300 mole) of cadmium carbonate (Note 13) are added. A nitrogen-inlet tube reaching to the bottom of the flask is inserted into the third neck of the flask, and nitrogen is introduced at approximately 50 ml. per minute. The reaction flask is heated to 90° in an oil bath, and the temperature is slowly increased to 110° over a 2–3 hour period. Water and cyclobutanone are carried into the receivers. The water in the receiving flask is saturated with sodium chloride, and the resulting solution is transferred to a separatory funnel. The flask is rinsed with 25 ml. of dichloromethane, and this solution is used for an initial washing of the aqueous solution. The aqueous solution is shaken three additional times with 25-ml. portions of 12chloromethane. The dichloromethane solutions are combined and added to the traps to dissolve cyclobutanone. The resulting solution is transferred to a 250-ml. flask, the traps are rinsed with a small amount of di-chloromethane, and the rinse is combined with the original solution. The organic solution is dried over 5 g. of anhydrous sodium sulfate, filtered, and the solvent is removed by distillation through a 20-cm., helix-packed, vacuum-insulated column. The product is transferred to a 25-ml. flask, 5 ml. of mesitylene is added, and the product is distilled through a spinning-band column. The fraction boiling at 95–100° is collected, yielding 12–15.8 g. (60–81%) (Note 14) of cyclobutanone.

2. Notes

1. Efficient stirring is required throughout the reaction.
2. If available, argon is preferable to nitrogen because of its higher density.

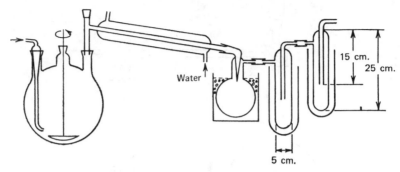

Figure 2.

3. The checkers used reagent grade tetrahydrofuran (available from Fisher Scientific Company) from a freshly opened bottle. The submitters used tetrahydrofuran purified by distillation from lithium aluminum hydride. See *Org. Synth.*, **Coll. Vol. 5**, 976 (1973), for warning regarding purification of tetrahydrofuran.

4. 1,3-Dithiane was prepared as described in *Org. Synth.*, **Coll. Vol. 6**, 556 (1988), and sublimed prior to use.

5. During the entire reaction sequence a positive pressure of approximately 50 mm. of nitrogen is maintained against the atmosphere with a mercury bubbler.

6. The titer of the solution should be determined prior to use. The checkers used product available from Foote Mineral Company.

7. Only a slight excess of dry ice should be added.

8. Available from Eastman Organic Chemicals. The product was distilled prior to use, b.p. 140–142°.

9. Approximately 2 hours is required.

10. The submitters have found that the ring closure reaction is essentially complete by the time the temperature reaches −20°.

11. GC analysis using a column containing 20% silicone DC 200 on Gas Chrom Z at 160° showed the product to be 96% pure.

12. The traps are attached in a reverse manner. The diameter of the inner tubing is 1.5 cm.

13. Anhydrous practical grade reagents were used.

14. The IR spectrum (neat) shows strong absorption at 1775 cm.$^{-1}$. The purity of the product is greater than 95% as established by GC on a 4-ft. column containing 20% silicone DC 200 on Gas Chrom Z at 50°. The ^{1}H NMR spectrum (CCl$_4$) shows a pentet at δ 1.83 (*J* = 8 Hz.) and a triplet at 3.01 (*J* = 8 Hz.) in a ratio 1:2, respectively.

3. Discussion

The best large-scale preparation of cyclobutanone is the reaction of diazomethane with ketene.[2] It requires a ketene generator and necessitates handling large quantities of the potentially hazardous diazo compound. A more frequently used method for the preparation of cyclobutanone starts with pentaerythritol, the final step being the oxidative degradation of methylenecyclobutane,[3,4] which can also be prepared from other precursors.[5] A general survey of all methods used to obtain cyclobutanone has been published.[6,7]

The procedure described here is an example of the use of the dithiane method[8] for the preparation of ketones. The reactions appear to be general, and the yields are satisfactory. The dithiane method can be successfully applied to the synthesis of rings containing up to seven carbon atoms with a slight modification[9,10] of the procedure above. The synthesis of larger rings may require high dilution methods.[9]

Aldehydes and open-chain ketones are also available from dithiane.[9] Carbonyl compounds with high optical activity[9] have been synthesized, including those that undergo facile racemization. An extensive review covering all applications of this reaction up to June 1969 has been authored by one of the submitters.[9] Methods and limitations for the preparation of silyl ketones[11] (R_3SiCOR) and germanyl ketones (R_3GeCOR) have been described.[12]

1. Institut für Organische Chemie der Universität (TH) Karlsruhe, West Germany. [Present address: Laboratorium für Organische Chemie, ETH-Zentrum, CH-8092 Zurich, Switzerland.]
2. P. Lipp and R. Köster, *Ber. Dtsch. Chem. Ges.*, **64**, 2823 (1931); P. Lipp, J. Buchkremer, and H. Seeles, *Justus Liebigs Ann. Chem.*, **499**, 1 (1932); S. Kaarsemaker and J. Coops, *Recl. Trav. Chim. Pays-Bas*, **70**, 1033 (1951); cf. ref. 4.
3. J. D. Roberts and C. W. Sauer, *J. Am. Chem. Soc.*, **71**, 3925 (1949).
4. J.-M. Conia, P. Leriverend, and J.-L. Ripoll, *Bull. Soc. Chim. Fr.*, 1803 (1961).
5. J.-M. Conia and J. Gore, *Bull. Soc. Chim. Fr.*, 735 (1963).
6. J.-M. Conia and J. Gore, *Bull. Soc. Chim. Fr.*, 726 (1963).
7. See also D. Seebach, in E. Müller, Ed., "Methoden der Organischen Chemie," (Houben-Weyl), Vol. 4/4, Georg. Thieme Verlag, Stuttgart, 1971.
8. E. J. Corey and D. Seebach, *Angew. Chem.*, **77**, 1134, 1135 (1965) [*Angew. Chem. Int. Ed. Engl.*, **4**, 1075, 1077 (1965)].
9. D. Seebach, *Synthesis*, **17**, (1969), and references cited therein.
10. D. Seebach, N. R. Jones, and E. J. Corey, *J. Org. Chem.*, **33**, 300 (1968).
11. E. J. Corey, D. Seebach, and R. Freedman, *J. Am. Chem. Soc.*, **89**, 434 (1967).
12. A. G. Brook, *Adv. Organomet. Chem.*, **7**, 95 (1968).

CYCLOBUTANONE FROM METHYLENECYCLOPROPANE
via OXASPIROPENTANE

Submitted by J. R. Salaun, J. Champion, and J. M. Conia[1]
Checked by Ž. Stojanac and Z. Valenta

1. Procedure

Caution! The preparation of methylenecyclopropane must be carried out in an efficient hood because ammonia is evolved. Oxaspiropentanes have been widely used as useful synthetic intermediates and appear to be quite stable; nevertheless, in view of the nature of the compounds, it is recommended that the preparation and handling of oxaspiropentane be carried out behind a safety screen.

A. *Methylenecyclopropane (Note 1).* A dry, 3-l., three-necked, round-bottomed flask with ground-glass fittings is equipped with a sealed stirrer (Note 2) driven by a heavy-duty motor, an efficient condenser fitted with a silica gel drying tube, and a 500-ml., pressure-equalizing dropping funnel connected to a nitrogen inlet. The flask is charged with 450 g. (11.5 moles) of sodium amide (Note 3) and 750 ml. of anhydrous tetrahydrofuran (Note 4), and the dropping funnel with a solution of 283.5 g. (3.831 moles) of anhydrous *tert*-butyl alcohol (Note 5) in 300 ml. of anhydrous tetrahydrofuran. While the sodium amide suspension is stirred vigorously under a nitrogen atmosphere, the solution of *tert*-butyl alcohol is added dropwise at room temperature over 3 hours. The stirred mixture is heated to 45°, with an oil bath, for 2 hours, at which point it may be necessary to add additional tetrahydrofuran (Note 6). The outlet of the condenser is connected with an adapter to a 250-ml. gas washing bottle containing 100 ml. of 5 N sulfuric acid, to eliminate evolved ammonia (Note 7). A silica gel drying tube (15 cm. long) joins the gas washing bottle to a 300-ml. cold trap protected from the atmosphere with a calcium chloride drying tube and cooled in a methanol–dry-ice bath maintained at −80° (Note 8). A solution of 228 g. (2.52 moles) of 3-chloro-2-methyl-1-propene (Note 9) in 500 ml. of dry tetrahydrofuran is added to the stirred basic mixture, which is then heated to 65° over a period of approximately 8 hours; a light nitrogen stream is used to carry the methylenecyclopropane into the cold trap. After the addition is complete, the reaction mixture is stirred and heated to 65° for 3 more hours (Note 10). The trap flask contains 58 g. (43%) of methylenecyclopropane (Note 11).

B. *Oxaspiropentane.* A 3-l., three-necked, round-bottomed flask equipped with a sealed stirrer, a thermometer, and an efficient condenser cooled by methanol–dry ice

(Note 12) is charged with 450 ml. of dichloromethane and 200 g. (1.09 moles) of 4-nitroperbenzoic acid (Note 13). The mixture is stirred and cooled to −50° by immersion of the flask in a methanol–dry-ice bath before 58 g. (1.1 moles) of methylenecyclopropane is distilled directly into the flask with a gas-inlet tube reaching to the bottom of the flask. The cooling bath is removed so that the temperature gradually rises; at about 0° the exothermic reaction starts. The temperature is maintained below 20° by occasional immersion of the flask in an ice–water bath; the methylenecyclopropane is allowed to reflux slowly (Note 14). After refluxing stops, the mixture is stirred overnight at room temperature. The 4-nitrobenzoic acid is removed by filtration and washed twice with 100-ml. portions of dichloromethane. The combined organic layers, which still contain about 10% of the total amount of 4-nitrobenzoic acid, are distilled at room temperature under reduced pressure (15 mm.) to eliminate the acid completely (Note 15). The distillate is concentrated to *ca.* 200 ml. of dichloromethane by distillation through a 15-cm., helix-packed, vacuum-insulated column, at a maximum oil bath temperature of 60° (Note 16).

C. *Cyclobutanone (Note 16)*. The resulting solution of oxaspiropentane (35%) in 200 ml. dichloromethane is added dropwise at room temperature to a magnetically stirred solution containing 5–10 mg. of lithium iodide in 50 ml. of dichloromethane (Notes 17, 18), at such a rate as to maintain gentle reflux of the solvent. After the addition, when the reaction mixture returns to room temperature, the transformation into cyclobutanone is complete. The dichloromethane solution is washed with 20 ml. of saturated aqueous sodium thiosulfate and 20 ml. of water, dried over magnesium sulfate, and concentrated by distillation of the solvent through a 15-cm., helix-packed, vacuum-insulated column. The residual liquid consists of cyclobutanone (95%) and 3-buten-2-one and 2-methylpropenal (5%).[5] A final distillation at 760 mm. through a 50-cm., stainless-steel spinning band column yields 41 g. (64% from methylenecyclopropane) of pure cyclobutanone, b.p. 100–101° (Notes 19, 20).

2. Notes

1. The procedure described for the synthesis of methylenecyclopropane is patterned after the method reported by Caubere and Coudert.[2] Methylenecyclopropane is also available from the stepwise method described by Köster and co-workers.[3]

2. The checkers used a stirrer for vacuum work (Teflon bearing, Fisher Scientific Company). The submitters used a mercury-sealed stirrer.

3. The submitters used sodium amide (obtained from Fluka A G as small lumps under kerosene) which was washed with anhydrous tetrahydrofuran and ground with a mill. The checkers used freshly opened and recently purchased cans of sodium amide powder (Fisher Scientific Company); older reagent gave unsatisfactory results.

4. Tetrahydrofuran is purified by distillation from lithium aluminium hydride after 48 hours of refluxing over potassium hydroxide (see *Org. Synth.*, **Coll. Vol. 5,** 976 (1973).

5. *tert*-Butyl alcohol was refluxed overnight over calcium hydride and distilled.

6. The checkers obtained a heavy slurry at this stage which became heavier during addition of 3-chloro-2-methyl-1-propene. They found it necessary to dilute with more

tetrahydrofuran (about 450 ml. for the scale given in the procedure) before the allyl chloride was added.

7. It is advisable to insert a safety bottle to avoid any run-back of sulfuric acid into the reaction flask. The gas washing bottle must be cooled by immersion in a large water bath (15°); the sulfuric acid solution is replaced by a fresh 5 N solution when neutralized by evolved ammonia (checked by phenolphthalein).

8. Methylenecyclopropane, b.p. 11° (760 mm.), is volatile at room temperature; all adapter fittings must be carefully checked. The checkers recommend the use of two cold traps in series.

9. 3-Chloro-2-methyl-1-propene (methallyl chloride) is available from Fluka A G and Eastman Organic Chemicals. The chloride, b.p. 72° (760 mm.), was distilled before use.

10. In the checkers' hands, at least 24 hours was needed to produce the bulk of methylenecyclopropane; small amounts of the product condensed during an additional 24-hour period.

11. The yield is determined by weighing the cold trap before and after distillation of methylenecyclopropane. Any small amounts of tetrahydrofuran carried into the methylenecyclopropane trap are eliminated in a subsequent distillation. By ^{1}H NMR analysis the checkers found that no tetrahydrofuran reached the cold traps; the spectrum (CD_2Cl_2) shows a triplet at δ 1.00 and a quintuplet at δ 5.35 in the ratio 4:2.

12. *Caution! The yield isolated from this reaction depends on the efficiency of this condenser; the epoxidation is exothermic and methylenecyclopropane is volatile.*

13. The 4-nitroperoxybenzoic acid (technical, 77–85%) may be obtained from the Aldrich Chemical Co., Inc., or Fluka A. G. (or from its U.S. representative, Tridom Chemical Inc.), or may be prepared from 4-nitrobenzoic acid.[4] The oxidation of methylenecyclopropane to oxaspiropentane has been reported to proceed in the same manner, in 48% yield, with the less expensive reagent *m*-chloroperbenzoic acid (MCPBA).[5]

14. Cooling below 0° stops the reaction.

15. A short-path distillation apparatus is used, the distillate (oxaspiropentane plus dichloromethane) being trapped in a receiver placed in a methanol–dry-ice bath cooled to −80°. The checkers found it useful to drive out last traces of product by adding several milliliters of dichloromethane to the residual thick paste and distilling. The ^{1}H NMR spectrum (CD_2Cl_2) shows an octet at δ 0.85 and a singlet at δ 3.00 in the ratio 4:2.

16. If the oil bath temperature reaches 80°, the residue consists of cyclobutanone (75%) and oxaspiropentane (25%). Distillation of this residue at 97–103° (760 mm.) yields cyclobutanone and oxaspiropentane; however, as reported[5] oxaspiropentane can be distilled at low temperature (36°) in vacuum (20 mm) without rearrangement.

17. *Caution! Addition of lithium iodide (catalytic amount) to a dichloromethane solution containing more than 30% oxaspiropentane leads to a very vigorous reaction.*

18. Dichloromethane from the previous distillation is used.

19. The purity of cyclobutanone was checked by GC on a 3.6-m. column containing 20% silicone SE-30 on chromosorb W at 65°. The IR spectrum (neat) shows carbonyl absorption at 1779 cm.$^{-1}$; the ^{1}H NMR spectrum (CCl_4) shows a multiplet at δ 2.00 and a triplet at δ 3.05 in the ratio 1:2.

20. The checkers obtained yields of 61–64% on smaller-scale runs (~ 10 g. of cyclobutanone).

3. Discussion

This method for the preparation of cyclobutanone *via* oxaspiropentane is an adaptation of that reported by Salaün and Conia.[6] Although cyclobutanone has been known now for over 75 years,[7] renewed interest in the potential of this small ring compound has led to several recently reported syntheses. The earlier syntheses of this compound consist of the reaction of the hazardous diazomethane with ketene[8] and the oxidative degradation[9] or the ozonization of methylenecyclobutane.[10] Recent syntheses involve the dithiane method of Corey and Seebach,[11,12] the solvolytic cyclization of 3-butyn-1-yl trifluoromethanesulfonate,[13] the ring enlargement of protected cyclopropanone cyanohydrin[14] and of 1-(phenylthio)cyclopropanemethanol,[15] and finally, the cyclodialkylation of tosylmethyl isocyanide.[16] These methods are more or less laborious, time-consuming, and based on expensive starting materials. Most of them present the disadvantage of producing an aqueous solution of the highly water-miscible and volatile cyclobutanone (b.p. 100°C).

The availability of methylenecyclopropane, its facile oxidation to oxaspiropentane, and the lithium iodide-induced ring enlargement to cyclobutanone described here remains the best laboratory scale preparation of cyclobutanone. Furthermore, this rearrangement of oxaspiropentanes appears to be general for the preparation of substituted cyclobutanones.[5,17]

1. Laboratoire des Carbocycles, Université de Paris-Sud, 91405 ORSAY, France.
2. P. Caubere and G. Coudert, *Bull. Soc. Chim. Fr.*, 2234 (1971).
3. R. Köster, S. Arora, and P. Binger, *Synthesis*, 322 (1971).
4. M. Vilkas, *Bull. Soc. Chim. Fr.*, 1401 (1959).
5. D. H. Aue, M. J. Meshishnek, and D. F. Shellhamer, *Tetrahedron Lett.*, 4799 (1973).
6. J. R. Salaün and J. M. Conia, *J. Chem. Soc. D*, 1579 (1971).
7. N. Kishner, *Zh. Russ. Fiz.-Khim. O-va.*, **37**, 106 (1905) [*Chem. Zentralbl.*, *I*, 1220 (1905)].
8. P. Lipp and R. Köster, *Ber. Dtsch. Chem. Ges.*, **64**, 2823 (1931).
9. J. D. Roberts and C. W. Sauer, *J. Am. Chem. Soc.*, **71**, 3925 (1949).
10. J. M. Conia, P. Leriverend, and J. L. Ripoll, *Bull. Soc. Chim. Fr.*, 1803 (1961).
11. E. J. Corey and D. Seebach, *Angew. Chem.*, **77**, 1134, 1135 (1965) [*Angew. Chem. Int. Ed. Engl.*, **4**, 1075, 1077 (1965)]; D. Seebach, N. R. Jones, and E. J. Corey, *J. Org. Chem.*, **33**, 300 (1968); D. Seebach and A. K. Beck, *Org. Synth.*, **Coll. Vol. 6**, 316 (1988).
12. K. Ogura, Y. Yamashita, M. Suzuki, and E. Tsuchihashi, *Tetrahedron Lett.*, 3653 (1974).
13. M. Hanack, T. Demesch, K. Hummel, and A. Nierth, *Org. Synth.*, **Coll. Vol. 6**, 324 (1988).
14. G. Stork, J. C. Depezay, and J. D'Angelo, *Tetrahedron Lett.*, 389 (1975).
15. B. M. Trost and W. C. Vladuchick, *Synthesis*, 821 (1978).
16. D. van Leusen and A. M. van Leusen, *Synthesis*, 325 (1980).
17. M. J. Bogdanowicz and B. M. Trost, *Tetrahedron Lett.*, 887 (1972).

CYCLOBUTANONE *via* SOLVOLYTIC CYCLIZATION

$$HC\equiv C-CH_2-CH_2OH + (CF_3SO_2)_2O \xrightarrow{Na_2CO_3}$$

$$HC\equiv C-CH_2-CH_2OSO_2CF_3$$

$$HC\equiv C-CH_2-CH_2OSO_2CF_3 \xrightarrow[\text{2. } H_2O]{\text{1. } CF_3CO_2H + CF_3CO_2Na}$$

Submitted by M. HANACK,[1] T. DEHESCH, K. HUMMEL, and A. NIERTH
Checked by H. ONA, B. A. BOIRE, and S. MASAMUNE

1. Procedure

A. *3-Butyn-1-yl trifluoromethanesulfonate.* A 500-ml., three-necked flask is fitted with a mechanical stirrer, a pressure-equalizing dropping funnel, and a stopper. The system is flushed with nitrogen through a gas-inlet tube attached to the top of the funnel. To 150 ml. of dry dichloromethane (Note 1) in the flask is added 75 g. (0.27 mole) of trifluoromethanesulfonic anhydride (Note 2), and the solution is cooled to −40°. After addition of 14.5 g. (0.175 mole) of finely powdered anhydrous sodium carbonate (Note 3), 15 g. (0.21 mole) of 3-butyn-1-ol (Note 4) is added dropwise over a 20-minute period to the well-stirred reaction mixture maintained at −40° to −55°. Stirring is continued at −30° for 2 hours, at 0° for another hour, and finally the reaction is quenched by dropwise addition of 50 ml. of water. The organic layer is separated, dried over anhydrous sodium sulfate, filtered, and concentrated with a rotary evaporator, the temperature of the water bath not exceeding 25°. The resulting residue is placed in a flask directly connected with a liquid nitrogen trap and distilled at 1 mm. The fractions boiling in the range from 40° to 50° are sufficiently pure for use in the next step. The yield of the sulfonate is 38.94 g. (90%) (Notes 5 and 6).

B. *Cyclobutanone.* A 500-ml., thick-walled ampoule is charged with 210 g. (137 ml.) of trifluoroacetic acid, 11.5 g. (0.0846 mole) of sodium trifluoroacetate (Note 7), and 17 g. (0.084 mole) of 3-butyn-1-yl trifluoromethanesulfonate, in this order. A magnetic stirring bar is added, and the ampoule is sealed. The stirred reaction mixture is immersed in a constant temperature bath kept at 65° (±2°) for 1 week. The ampoule is cooled slowly to −50° with a methanol–dry-ice bath (Note 8) and opened. With the aid of 200 ml. of diethyl ether the reaction mixture is transferred to a 1-l. Erlenmeyer flask to which 74 g. (1.83 mole) of sodium hydroxide in 150 ml. of water is added carefully. During the addition the flask is immersed in the bath, maintained at approximately −50° (Note 8). After the ethereal layer is separated, the aqueous layer is saturated with sodium chloride and extracted twice with ether. The original organic layer and ethereal extracts are combined, dried over anhydrous sodium sulfate, and directly distilled into a liquid nitrogen trap. The total condensate in the trap is placed in a distillation flask attached to a 40-cm. Vigreux column and a condenser cooled to −40° with a circulating cold bath

(Note 9). After the ether is distilled, all volatile materials are collected by raising the bath to 130°, yielding 1.84–2.05 g. (31–36%) of cyclobutanone. The purity of the product is greater than 95% by ^{1}H NMR, the only impurity being diethyl ether (Notes 10, 11).

2. Notes

1. The submitters treated dichloromethane first with sulfuric acid, then with sodium hydroxide, and distilled it before use. The checkers used reagent grade solvent, supplied by Fisher Scientific Company, stored overnight over 4A Molecular Sieves.

2. The submitters prepared the anhydride, following basically the procedure described by Burdon, Farazmand, Stacey, and Tatlow.[2] Trifluoromethanesulfonic acid (32.1 g., 0.214 mole) (supplied by Minnesota Mining and Manufacturing Company) maintained at 0°, is treated with 25 g. of phosphorus pentoxide in three portions. The resulting anhydride is distilled by gradually heating the reaction mixture to a bath temperature of 110° over a 1-hour period. The fractions boiling at 80–100° (760 mm.) are collected and redistilled from approximately 8 g. of phosphorus pentoxide until the distillate no longer fumes on exposure to air. Normally three distillations are necessary. The presence of the trifluoromethanesulfonic acid can be detected by dipping a glass rod into the distillate and waving the wet rod in the air. The anhydride, in contrast, does not fume. The final yield of the product is 25 g. (83%), b.p. 84°. The checkers purchased the anhydride from Pierce Chemical Company and used it without further purification.

3. Anhydrous sodium carbonate was ground into fine powder and dried in vacuum at 200° for 4 hours.

4. In one experiment the checkers used 3-butyn-1-ol available from Aldrich Chemical Company, Inc., and found that its purity was satisfactory. In other experiments, both the submitters and the checkers prepared the hydroxy compound from sodium acetylide and ethylene oxide in liquid ammonia according to the procedure described by Schulte and Reiss[3] and further attempted to maximize the yield by varying the ratio of sodium:ethylene oxide:liquid ammonia used in the reaction. While the submitters obtained 3-butyn-1-ol in a yield of 60%, the checkers failed to obtain consistent results in repeated experiments and consequently could not define the optimum conditions for the reaction. Thus, the yield of 3-butyn-1-ol varied from 15 to 45% and 15 to 31% on the basis of sodium and ethylene oxide, respectively. Unknown and apparently subtle experimental factors affect the yield significantly.

5. When 3-butyn-1-ol was added to a solution of the anhydride, cooled to 0°, and the mixture was allowed to react at room temperature for 3 hours, the yield of the sulfonate dropped to 70–75%.

6. ^{1}H NMR of 3-butyn-1-ol trifluoromethanesulfonate (CCl$_4$) δ, (multiplicity, coupling constant J in Hz., number of protons): 2.05 (t, $J_{2,4}$ = 2.6, 1H), 2.76 (d of t, $J_{1,2}$ = 6.7, $J_{2,4}$ = 2.6, 2H), 4.57 (t, $J_{1,2}$ = 6.7, 2H).

7. The salt is available from Aldrich Chemical Company Inc. However, the checkers readily prepared it in the following way. To a stirred solution of 8.8 g. (0.22 mole) of sodium hydroxide in 400 ml. of 98% ethanol was added dropwise 25 g. (0.22 mole) of trifluoroacetic acid. After the addition was completed, the ethanol was removed under reduced pressure, and the residue was suspended in approximately 100 ml. of ethcr,

filtered, and washed several times with ether. The yield was 25 g. (83.8%), and the salt was dried to a constant weight in a vacuum desiccator containing calcium sulfate (2 days).

8. Due to the high volatility of cyclobutanone, a substantial amount of the product is lost unless the mixture is sufficiently cooled during the process of neutralization.

9. The cooling to −40° was necessary to prevent the loss of highly volatile cyclobutanone.

10. The checkers redistilled this product through a 3-cm. column and determined its b.p. to be 96.5–97.5° (710 mm.); IR (CHCl₃), cm.⁻¹: 1780; ¹H NMR (CCl₄), δ (multiplicity, number of protons): 3.05 (m, 4H), 1.96 (m, 2H).

11. Subsequent to the publication of this procedure in the annual volume, the submitter has suggested the following changes, which have not been checked: After neutralization of the cyclobutanone with 1.83 moles of sodium hydroxide, the aqueous layer is made neutral or weakly alkaline with additional sodium hydroxide solution. Often the neutralization point can be observed as a weak color change (to a lighter color) of the dark brown reaction mixture. After separation of the ethereal layer and saturation of the aqueous layer with sodium chloride, the aqueous layer is extracted repeatedly (more than twice) with ether. The submitters obtained cyclobutanone in 66% yield.

3. Discussion

Cyclobutanone has been prepared by (1) reaction of diazomethane with ketene,[4] (2) treatment of methylenecyclobutane with performic acid, followed by cleavage of the resulting glycol with lead tetraacetate,[5] (3) ozonolysis of methylenecyclobutane,[6] (4) epoxidation of methylenecyclopropane followed by acid-catalyzed ring expansion,[7] (5) oxidative cleavage of cyclobutane trimethylene thioketal, which is prepared from 2-(ω-chloropropyl)-1,3-dithiane,[8] and (6) by hydrolytic conversion of 1-tosylcyclobutyl isocyanide, which is obtained by a one-step cyclodialkylation of tosylmethyl isocyanide with 1,3-dibromopropane.[9]

The present procedure[10] represents another synthesis of cyclobutanone through the unique acetylenic bond participation in solvolysis. Cyclobutane derivatives prepared in this way include 2-methyl-, 2-ethyl-, 2-isopropyl-, and 2-trifluoromethylcyclobutanone from the corresponding acetylenic compounds.[11] Condensed cyclobutanones are also easily accessible in good preparative yields by homopropargylic rearrangement: 7-methylbicyclo[3.2.0]heptan-6-one and 9-methylbicyclo[5.2.0]nonan-8-one are prepared by solvolyzing the corresponding trans-2-(1-propynyl)cycloalkyl-4-dimethylaminobenzenesulfonates in 67% sulfuric acid.[12]

1. Institut für Organische Chemie der Universität Tübingen, Lehrstuhl für Organische Chemie II, Auf er Morgenstelle 18, D-7400 Tübingen, Germany.
2. J. Burdon, I. Farazmand, M. Stacey, and J. C. Tatlow, J. Chem. Soc., 2574 (1957).
3. K. E. Schulte and K. P. Reiss, Chem. Ber., 86, 777 (1953).
4. P. Lipp and R. Köster, Ber. Dtsch. Chem. Ges., 64, 2823 (1931).
5. J. D. Roberts and C. W. Sauer, J. Am. Chem. Soc., 71, 3925 (1949).
6. J. M. Conia, P. Leriverend, and J. L. Ripoll, Bull. Soc. Chim. Fr., 1803 (1961).
7. J. R. Salaün and J. M. Conia, J. Chem. Soc. D, 1579 (1971).
8. D. Seebach, N. R. Jones, and E. J. Corey, J. Org. Chem., 33, 300 (1968); D. Seebach and A. K. Beck, Org. Synth., Coll. Vol. 6, 316 (1988).

9. D. van Leusen and A. M. van Leusen, *Synthesis*, 325 (1980).
10. K. Hummel and M. Hanack, *Justus Liebigs Ann. Chem.*, **746**, 211 (1971).
11. M. Hanack, S. Bocher, I. Herterich, K. Hummel, and V. Vott, *Justus Liebigs Ann. Chem.*, **733**, 5 (1970). See also references cited therein.
12. W. Schumacher and M. Hanack, *Synthesis*, **1981**, in press.

ONE-CARBON RING EXPANSION OF CYCLOALKANONES TO CONJUGATED CYCLOALKENONES: 2-CYCLOHEPTEN-1-ONE

Submitted by Yoshihiko Ito, Shotaro Fujii, Masashi Nakatsuka, Fumio Kawamoto, and Takeo Saegusa[1]
Checked by Peter Senter, William F. Burgoyne, and Robert M. Coates

1. Procedure

Caution! Diethylzinc, which is used in Part B of this procedure, is highly pyrophoric. Accordingly, this reagent must be kept under a nitrogen atmosphere; exposure to air must be avoided during transfers.

A. *1-Trimethylsilyloxycyclohexene.* A 500-ml., three-necked, round-bottomed flask fitted with a mechanical stirrer, a reflux condenser protected with a calcium chloride tube, and a rubber septum is charged with 100 ml. of *N,N*-dimethylformamide (Note 1) and 60.6 g. (0.600 mole) of triethylamine (Note 2). The solution is stirred while 32.6 g. (0.300 mole) of chlorotrimethylsilane (Note 3) and 24.5 g. (0.250 mole) of cyclohexanone are injected in succession through the septum into the flask. The resulting mixture is stirred and heated under reflux for 6 hours, cooled to room temperature, and diluted with 300 ml. of pentane. The triethylamine hydrochloride that precipitates is removed by filtering through a coarse, sintered-glass Büchner funnel, and the filter cake is washed with three 100-ml. portions of pentane. The filtrates are combined and washed with three 300-ml. portions of ice-cold sodium hydrogen carbonate solution. The organic layer is washed rapidly with 100 ml. of ice-cold 3% hydrochloric acid and 100 ml. of ice-cold sodium hydrogen carbonate in succession. The pentane solution is washed with 50 ml. of sodium chloride solution, dried over anhydrous sodium sulfate, and evaporated. The residual liquid is distilled at reduced pressure through a 10-cm. Vigreux column, affording, after separation of a small forerun, 33–35.5 g. (78–84%) of 1-trimethylsilyloxycyclohexene, b.p. 74–75° (20 mm.) (Note 4).

B. *1-Trimethylsilyloxybicyclo[4.1.0]heptane.* A 250 ml., three-necked, round-bottomed flask is equipped with a magnetic stirring bar, a pressure-equalizing dropping funnel, a reflux condenser bearing a nitrogen inlet at its top, and a rubber septum. The apparatus is purged with nitrogen, flamed dry, and allowed to cool (Note 5). The flask is charged with 130 ml. of diethyl ether (Note 6), 17.0 g. (0.100 mole) of 1-trimethylsilyloxycyclohexene, and 18.5 g. (0.146 mole) of diethylzinc (Note 7), each being added through the septum with a syringe. The solution is stirred and maintained at room temperature with a water bath while 40.2 g. (0.150 mole) of diiodomethane (Note 8) is added slowly from the dropping funnel over a 1-hour period (Note 9). The reaction mixture is stirred and heated under reflux for 8 hours (Note 10). After the reaction is complete (Note 11), the contents of the flask are stirred and cooled in an ice-water bath as 5.4 ml. of concentrated aqueous ammonium chloride is added over *ca.* 30 minutes. A large amount of gas is evolved, and a white solid is formed during the hydrolysis. The salts are separated by filtering through a sintered-glass Büchner funnel and washed with 100 ml. of a 1 : 1 (v/v) ether–pentane solution. The combined filtrates are washed with four 50-ml. portions of ice-cold saturated aqueous ammonium chloride and two 100-ml. portions of ice-cold aqueous sodium chloride. The solution is filtered through a pad of anhydrous sodium sulfate and evaporated. The residual liquid is distilled through a 17.5-cm. Vigreux column under reduced pressure, affording a forerun of 2.1–3.5 g., b.p. 65–80° (12 mm.), and 14.2–15.2 g. (77–83%) of 1-trimethylsilyloxybicyclo[4.1.0]heptane, b.p. 80–82° (12 mm.) (Note 12).

C. *2-Cyclohepten-1-one.* A 250-ml., three-necked, round-bottomed flask equipped with a mechanical stirrer, a pressure-equalizing dropping funnel bearing a nitrogen inlet at its top, and a thermometer is charged with 17.9 g. (0.110 mole) of anhydrous iron(III) chloride (Note 13). The flask is immersed in an ice-water bath, stirring is begun, and 70 ml. of N,N-dimethylformamide (Note 1) is added slowly (Note 14). When all the iron(III) chloride has dissolved, a solution of 9.2 g. (0.050 mole) of 1-trimethylsilyloxybicyclo[4.1.0]heptane in 20 ml. of N,N-dimethylformamide is added dropwise through the dropping funnel over 1 hour while the internal temperature is maintained at 0–5°. After the addition is complete, the brown solution is stirred at room temperature for 2 more hours, then poured into *ca.* 200 ml. of ice-cold 1 N hydrochloric acid. The aqueous solution is extracted with three 50-ml. portions of chloroform. The combined chloroform extracts are washed successively with 50-ml. portions of 1 N hydrochloric acid, saturated sodium hydrogen carbonate, and sodium chloride solution. The solution is dried by filtration through a pad of anhydrous sodium sulfate and evaporated. The remaining liquid (Note 15) is dissolved in 50 ml. of methanol saturated with sodium acetate and heated at reflux for 3 hours. The volume is reduced to *ca.* 25 ml. by evaporation under reduced pressure, 50 ml. of water is added, and the mixture is extracted with three 30-ml. portions of ether. The combined extracts are dried over anhydrous sodium sulfate, the ether is evaporated, and the residual liquid is distilled under reduced pressure through a 17.5-cm. Vigreux column, yielding after separation of a 0.4–1.0 g. forerun, 4.3–4.5 g. (78–82%) of 2-cyclohepten-1-one as a colorless liquid, b.p. 73–76° (18 mm.) (Note 16).

2. Notes

1. *N,N*-Dimethylformamide was purified by distillation from calcium hydride under a nitrogen atmosphere and stored over Linde type 4A molecular sieves.

2. Triethylamine was distilled from lithium aluminum hydride.

3. Chlorotrimethylsilane is available from Aldrich Chemical Company, Inc. The reagent was distilled before use.

4. The product has the following spectral properties: IR (neat) cm.$^{-1}$: 1675 (C=C); ^{1}H NMR, δ (multiplicity, number of protons, assignment): 0.16 (s, 9H, Si(CH$_3$)$_3$), 1.3–2.1 (m, 8H, 4CH$_2$), 4.78 (m, 1H, vinyl *H*).

5. A slight positive pressure of nitrogen is maintained in the apparatus throughout this procedure.

6. Anhydrous diethyl ether from Mallinckrodt Chemical Works was distilled from sodium and benzophenone before use.

7. Diethylzinc in a cylinder pressurized with nitrogen was purchased from Alfa Division, Ventron Corporation, and distilled at atmospheric pressure under a nitrogen atmosphere before use, b.p. 118°. The distillate was collected in a two-necked receiver fitted with a rubber septum and kept under a nitrogen atmosphere. Aliquots of diethylzinc were withdrawn with a gas-tight syringe. The checkers destroyed excess or waste reagent by injecting it cautiously beneath the surface of ice-cold water through which argon was vigorously bubbled.

8. Diiodomethane from both Eastman Organic Chemicals and Aldrich Chemical Company, Inc., was used by the checkers after distillation under reduced pressure, b.p. 68–70° (12 mm.).

9. The solution becomes somewhat cloudy as the diiodomethane is added.

10. The checkers found considerable variation in the rate of the reaction in different runs, the time required for its completion ranging from 3 to 10 hours. It is therefore advisable to monitor the progress of the reaction. For this purpose small aliquots (*ca.* 0.05 ml.) were withdrawn from the flask with a syringe and hydrolyzed by injection into a vial containing ether and saturated ammonium chloride. The relative amounts of enol silane and cyclopropoxy silane were determined by GC on an 0.6 cm. × 3.7 m. column of 3% OV-17 coated on 100–120 mesh Chromosorb W. With a column temperature of 120° and a carrier gas flow rate of 20 ml. per minute, the retention times for the enol silane and the cyclopropoxy silane are *ca.* 1.9 and 2.3 minutes, respectively.

11. In one run that was particularly slow, an additional 9.9 g. of diiodomethane was added. The reaction then proceeded quickly to completion.

12. The spectral properties of the product are as follows: IR (neat) cm.$^{-1}$: 1250, 1209, 1010, 900, 865, 840; 220-MHz. ^{1}H NMR (CDCl$_3$), δ (multiplicity, coupling constant J in Hz., number of protons, assignment): 0.13 [s, 9H, Si(CH$_3$)$_3$], 0.29 (t, $J = 5$, 1H, *endo* cyclopropyl H at C-7), 0.84 (d of d, $J = 5$ and $J = 11$, 1H, *exo* cyclopropyl H at C-7), 0.98–1.70 (m, 6H, cyclohexyl H), 1.82–2.18 (m, 3H, cyclohexyl H). A GC analysis as described in Note 10 indicated the purity of the product to be *ca.* 95–98%, the remainder being 3–5% of unreacted enol silane.

13. Anhydrous iron(III) chloride was purchased by the submitters from Merck & Company, Inc. The checkers obtained the reagent from Aldrich Chemical Company, Inc. The reagent was dried at 60–70° under reduced pressure for several hours before use.

14. The dissolution of iron(III) chloride in *N,N*-dimethylformamide is exothermic.

15. A GC analysis on the liquid by the submitters using a Carbowax 20 M (polyethylene glycol) column at 170° showed a major peak assigned to 3-chlorocycloheptanone and minor peak for 2-cycloheptenone. The spectral properties of 3-chlorocycloheptanone are as follows: IR (neat) cm.$^{-1}$: 1705 (C=O); ^{1}H NMR (CCl$_4$), δ (multiplicity, number of protons, assignment): 1.4–2.3 [m, 6H, (CH$_2$)$_3$-CHCl], 2.3–2.6 (m, 4H, CH$_2$COCH$_2$), 4.1–4.4 (m, 1H, CHCl); mass spectrum, *m/e* (intensity ratio): M+, 146 and 148 (3:1).

16. A GC analysis by the submitters as described in the preceding note indicated that the purity of the product was 98%. The purity of the product obtained by the checkers was estimated at 95% by a GC analysis at 140° as described in Note 10. 2-Cyclohepten-1-one has the following spectral properties: IR (neat) cm.$^{-1}$: 1700 (C=O), 1660 (C=C), 1445, 1090, 888; ^{1}H NMR, δ (multiplicity, coupling constant J in Hz., number of protons, assignment): 1.75 [m, 4H, CH$_2$(CH$_2$)$_2$CH$_2$], 2.45 [m, 4H, CH$_2$(CH$_2$)$_2$CH$_2$], 5.90 (d, $J = 13$, 1H, CH=CHCO), 6.52 (d of t, $J = 5$ and $J = 13$, 1H, CH=CHCO).

3. Discussion

This procedure illustrates a new three-step reaction sequence for the one-carbon ring expansion of cyclic ketones to the homologous α,β-unsaturated ketones.[2] The key step in the sequence is the iron(III) chloride-induced cleavage of the central bond of trimethyl-silyloxycyclopropanes which are obtained by cyclopropanation of trimethylsilyl enol ethers. The procedure for the preparation of 1-trimethylsilyloxycyclohexene from cyclohexanone described in Part A is that of House, Czuba, Gall, and Olmstead.[3]

The cyclopropanation of 1-trimethylsilyloxycyclohexene in the present procedure is accomplished by reaction with diiodomethane and diethylzinc in ethyl ether.[4] This modification of the usual Simmons–Smith reaction[5,6] in which diiodomethane and activated zinc are used has the advantage of being homogeneous and is often more effective for the cyclopropanation of olefins such as enol ethers which polymerize readily. However, in the case of trimethylsilyl enol ethers, the heterogeneous procedures with either zinc–copper[7] or zinc–silver couple[8] are also successful. Attempts by the checkers to carry out Part B in benzene or toluene at reflux instead of ethyl ether afforded the trimethylsilyl ether of 2-methylenecyclohexanol, evidently owing to zinc iodide-catalyzed isomerization of the initially formed cyclopropyl ether.[9] The preparation of

TABLE I

Preparation of 2-Cycloalkenones and Cycloalkane-1,3-diones by Iron(III) Chloride-Induced Ring Opening of 1-Trimethylsilyloxy- and 1,2-Bis(trimethylsilyloxy)bicyclo[n.1.0]alkanes

Silyloxybicyclo-[n.1.0]alkane	2-Cycloalkenone or Cycloalkane-1,3-dione	Yield (%)[a]
		98
		80[b]
		83
		83
		92
		81[c]

[a] The scale was 0.002–0.005 mole except as noted.
[b] This reaction was conducted on a 0.05-mole scale.
[c] This compound was a mixture of cis and trans isomers.

331

TABLE 1 *(continued)*

PREPARATION OF 2-CYCLOALKENONES AND CYCLOALKANE-1,3-DIONES
BY IRON(III) CHLORIDE-INDUCED RING OPENING OF 1-TRIMETHYLSILYLOXY-
AND 1,2-BIS(TRIMETHYLSILYLOXY)BICYCLO[n.1.0]ALKANES

Silyloxybicyclo-[n.1.0]alkane	2-Cycloalkenone or Cycloalkane-1,3-dione	Yield (%)[a]
		68
		72

1-trimethylsilyloxybicyclo[4.1.0]heptane by cyclopropanation with diethylzinc and chloroiodomethane in the presence of oxygen has been reported.[10]

The ring-opening reaction with iron(III) chloride in N,N-dimethylformamide is effective with a series of 1-trimethylsilyloxybicyclo[n.1.0]alkanes, as shown by the examples presented in Table I.[2] The corresponding 3-chlorocycloalkanones are usually isolable intermediates which are separately subjected to dehydrochlorination with sodium acetate in methanol, as in the preparation of 2-cyclohepten-1-one described here. However, the reaction of 1-trimethylsilyloxybicyclo[3.1.0]hexane with iron(III) chloride at 0–5° afforded 2-cyclohexen-1-one directly. The slower ring opening of 1-trimethylsilyloxybicyclo[10.1.0]tridecane was carried out at 80°, conditions which also effected spontaneous dehydrochlorination to *trans*-2-cyclotridecenone. The regiospecific ring enlargement of the unsymmetrical ketones, 2-methylcyclohexanone and β-tetralone, are of particular interest in view of the diversity of synthetic routes to trimethysilyl enol ethers.[11]

The present procedure for ring expansion has also been applied to 1,2-bis(trimethylsilyloxy)bicyclo[n.1.0]alkanes,[2,12] which are prepared by cyclopropanation of 1,2-bis(silyloxy)cycloalkenes.[13] The latter are readily available from acyloin condensations in the presence of chlorotrimethylsilane.[14] This reaction provides a new route to cyclic 1,3-diketones and macrocyclic compounds containing two 1,3-diketone units in the ring. It has been reported[15] that methylcyclopropanation of 1,2-bis-(trimethylsilyloxy)cyclohexene by 1,1-diiodoethane with zinc–copper couple or diethylzinc and the subsequent iron(III) chloride-induced ring expansion afforded 2-methylcycloheptane-1,3-dione in moderate yield.

The regiospecificity of the iron(III) chloride-induced ring cleavage contrasts with that observed in reactions of 1-silyloxybicyclo[n.1.0]alkanes with bromine[7c] and potassium tert-butoxide.[8b] Although the mechanism of the reaction is not known with certainty, it is reasonable to suppose that an alkoxy radical is involved, that this radical undergoes homolytic scission of the more highly substituted carbon–carbon bond of the cyclopropane ring, and that the resulting carbon radical abstracts a chlorine atom from iron(III) chloride.[16]

2-Cyclohepten-1-one has been prepared from cycloheptanone by dehydrohalogenation of the ethylene ketals of 2-chloro- and 2-bromocycloheptanone and subsequent hydrolysis.[17] The α,β-dehydrogenation of cycloheptanone has also been effected via the α-phenylthio[18a] and α-phenylseleno[18b] ketones which were subjected to oxidation and thermal elimination. Another route to the title compound starts with cyclohexene, which is subjected to allylic bromination, hydrolysis, and chromic acid oxidation.[19]

1. Department of Synthetic Chemistry, Faculty of Engineering, Kyoto University, Kyoto 606, Japan.
2. Y. Ito, S. Fujii, and T. Saegusa, *J. Org. Chem.*, **41**, 2073 (1976).
3. H. O. House, L. J. Czuba, M. Gall, and H. D. Olmstead, *J. Org. Chem.*, **34**, 2324 (1969).
4. J. Furukawa, N. Kawabata, and J. Nishimura, *Tetrahedron*, **24**, 53 (1968).
5. R. D. Smith and H. E. Simmons, *Org. Synth.*, **Coll. Vol. 5**, 855 (1973).
6. For a review see H. E. Simmons, T. L. Cairns, S. A. Vladuchick, and C. M. Hoiness, *Org. React.*, **20**, 1 (1973).
7. (a) R. Le Goaller and J.-L. Pierre, *Bull. Soc. Chim. Fr.*, 1531 (1973); (b) G. M. Rubottom and M. I. Lopez, *J. Org. Chem.*, **38**, 2097 (1973); (c) S. Murai, T. Aya, and N. Sonoda, *J. Org. Chem.*, **38**, 4354 (1973).
8. (a) J. M. Denis, C. Girard, and J. M. Conia, *Synthesis*, 549 (1972); (b) J. M. Conia and C. Girard, *Tetrahedron Lett.*, 2767 (1973); (c) C. Girard, P. Amice, J. P. Barnier, and J. M. Conia, *Tetrahedron Lett.*, 3329 (1974).
9. S. Murai, T. Aya, T. Renge, I. Ryu, and N. Sonoda, *J. Org. Chem.*, **39**, 858 (1974); I. Ryu, S. Murai, S. Otani, and N. Sonoda, *Tetrahedron Lett.*, 1995 (1977).
10. S. Miyano, Y. Izumi, H. Fujii, and H. Hashimoto, *Synthesis*, 700 (1977).
11. J. K. Rasmussen, *Synthesis*, 91 (1977).
12. Y. Ito, T. Sugaya, M. Nakatsuka, and T. Saegusa, *J. Am. Chem. Soc.*, **99**, 8366 (1977); Y. Ito and T. Saegusa, *J. Org. Chem.*, **42**, 2326 (1977).
13. M. Audibrand, R. Le Goaller, and P. Arnaud, *C.R. Hebd. Seances Acad. Sci. Ser. C*, **268**, 2322 (1969).
14. K. Rühlmann, *Synthesis*, 236 (1971); J. J. Bloomfield, D. C. Owsley, and J. M. Nelke, *Org. React.*, **23**, 259 (1976); J. J. Bloomfield and J. M. Nelke, *Org. Synth.*, **Coll. Vol. 6**, 167 (1988).
15. S. Lewicka-Piekut and W. H. Okamura, *Synth. Commun.*, **10**, 415 (1980).
16. C. H. DePuy, W. C. Arney, Jr., and D. H. Gibson, *J. Am. Chem. Soc.*, **90**, 1830 (1968); C. H. DePuy and R. J. Van Lanen, *J. Org. Chem.*, **39**, 3360 (1974).
17. (a) W. Treibs and P. Grossman, *Chem. Ber.*, **92**, 267 (1959); (b) E. W. Garbisch, Jr., *J. Org. Chem.*, **30**, 2109 (1965).
18. (a) B. M. Trost, T. N. Salzmann, and K. Hiroi, *J. Am. Chem. Soc.*, **98**, 4887 (1976); (b) H. J. Reich, J. M. Renga, and I. L. Reich, *J. Am. Chem. Soc.*, **97**, 5434 (1975).
19. N. Heap and G. H. Whitham, *J. Chem. Soc. B*, 164 (1966).

NITRILES FROM KETONES: CYCLOHEXANECARBONITRILE

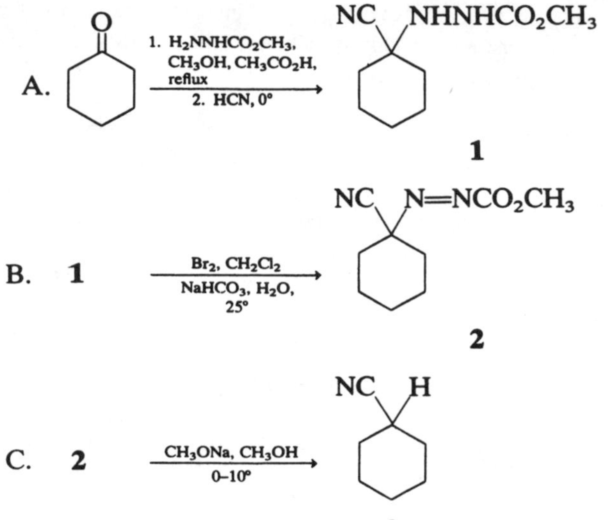

A.

1. $H_2NNHCO_2CH_3$,
   CH_3OH, CH_3CO_2H,
   reflux
2. HCN, 0°

NC $NHNHCO_2CH_3$

1

B. **1**

Br_2, CH_2Cl_2
$NaHCO_3$, H_2O,
25°

NC $N=NCO_2CH_3$

2

C. **2**

CH_3ONa, CH_3OH
0–10°

NC H

3

Submitted by P. A. WENDER,[1,2] M. A. EISSENSTAT,[1] N. SAPUPPO,[1] and F. E. ZIEGLER[3]
Checked by D. F. BUSHEY and G. BÜCHI

1. Procedure

Caution! Because of the toxicity of hydrogen cyanide, this procedure should be conducted in a well-ventilated hood, and rubber gloves should be worn. (See also Note 3.)

A. *Methyl 2-(1-cyanocyclohexyl)hydrazine carboxylate* (**1**). A 100-ml., three-necked flask equipped with a reflux condenser, thermometer, magnetic stirring bar, and gas-exhaust tube (Note 1) is charged with 9.0 g. (0.10 mole) of methyl carbazate (Note 2), 20 ml. of methanol, 2 drops of acetic acid, and 9.8 g. (0.10 mole) of cyclohexanone. The resulting mixture is refluxed for 30 minutes, then cooled to 0° and treated with 6 ml. (0.15 mole) of hydrogen cyanide (Note 3), added dropwise over a period of 3 minutes. After approximately 15 minutes the solution is allowed to warm to room temperature, during which time the hydrazine **1** crystallizes (Notes 4 and 5). After 2 hours the resulting mixture is vacuum filtered, and the crystalline residue is washed with 10 ml. of cold methanol (Note 6), yielding 17.7 g. of hydrazine **1**. Concentration of the filtrate provides an additional 1.4 g. of product. The total yield of crude hydrazine **1**, m.p. 130–133°, (Note 7) is 97%. Recrystallization (methanol–pentane) provides an analytically pure sample, m.p. 135–136°.

B. *Methyl 2-(1-cyanocyclohexyl)diazenecarboxylate* (**2**). A 5.5 *M* solution of bromine in dichloromethane is added dropwise to a vigorously stirred mixture of 19.1 g.

(0.0970 mole) of hydrazine **1**, 75 ml. of dichloromethane, 75 ml. of water, and 22 g. (0.26 mole) of sodium hydrogen carbonate in a 250-ml. flask, until a persistent, positive potassium iodide–starch paper test is obtained (Note 8). Excess bromine is discharged with aqueous sodium sulfite, the phases are separated, and the aqueous phase is extracted with two 50-ml. portions of dichloromethane. The combined organic extracts are washed with 30 ml. of water, dried over magnesium sulfate, filtered, concentrated using a rotary evaporator, and distilled, yielding 17.3 g. (93%) of the diazene **2** as a clear, bright yellow oil, b.p. 95–97° (0.2 mm.) (Note 7).

C. *Cyclohexanecarbonitrile* (**3**). A 100-ml., three-necked, round-bottomed flask fitted with a thermometer, a magnetic stirring bar, and an addition funnel is charged with 2.7 g. (0.050 mole) of sodium methoxide and 25 ml. of methanol. The solution is cooled in ice and stirred while a solution of 17.3 g. (0.0887 mole) of diazene **2** in 10 ml. of methanol is added dropwise at a rate that maintains the solution temperature at 0–10° (Note 9). Stirring is continued for an additional 30 minutes at ambient temperature, and the mixture is poured into 70 ml. of water. The resulting solution is extracted with five 20-ml. portions of pentane, and the combined organic extracts are dried over magnesium sulfate, filtered, concentrated, and distilled to provide 8.6–9.4 g. (78–86%) of carbonitrile **3**, b.p. 117–119° (90 mm.) (Note 10).

2. Notes

1. The tube is connected to the top of the reflux condenser and attached to a length of tygon tubing. The end of the tubing is positioned in the exhaust vent of the hood to remove any hydrogen cyanide vapor.

2. Methyl carbazate was prepared by the method of Diels.[4] This reagent is available from Aldrich Chemical Company, Inc., under the name methyl hydrazinocarboxylate. All other chemicals used in this sequence were of the highest purity commercially available and were not further purified before use.

3. Hydrogen cyanide can be purchased from Fumico Inc., Amarillo, Texas. When large amounts of HCN are used, it is recommended that amyl nitrite pearls and an oxygen cylinder with mask be available. These in combination are effective antidotes for HCN poisoning. The checkers prepared HCN according to the method in *Org. Synth.*, **Coll. Vol. 1,** 314 (1941). The use of smaller quantities of HCN results in slower reaction and reduced yield.

4. Spontaneous crystallization usually occurs within 30 minutes; if not, crystallization is induced by scratching the bottom of the reaction flask.

5. The solid formed at this point usually interferes with stirring. The yield of product, however, is not affected by the nature of mixing beyond this point.

6. The methanol is used to transfer any residual material from the reaction flask as well as to wash the residue.

7. Further purification of this intermediate is unnecessary for the preparation of nitriles.

8. The endpoint of this oxidation is indicated by the appearance in the reaction mixture of a red-orange color caused by excess bromine.

9. Nitrogen is vigorously and instantaneously evolved as each drop is added.

10. Physical and analytical characterizations of cyclohexanecarbonitrile (3) agree with literature reports:[5] ^{1}H NMR (CDCl$_3$), δ (multiplicity, number of protons, assignment): 1.30–2.00 (m, 10H, 5CH$_2$), 2.40–2.90 (m, 1H, CHCN). The average overall yield of cyclohexanecarbonitrile (3) from cyclohexanone is 85% when the diazene 2 is distilled and 86–90% when crude diazene is used directly. The submitter reported 8.9 g. (92%) of carbonitrile 3.

3. Discussion

The transformation of ketones to nitriles has been accomplished with varying success using several methods including cyanide displacement reactions of acetates[6] and halides[7] obtained from ketones; cyanohydrin formation, dehydration, and reduction of the unsaturated nitriles;[8] reaction of tosylhydrazones with potassium cyanide followed by pyrolysis of the cyanohydrazine;[9] and reaction of tosylmethylisocyanide with ketones.[10] Related transformations of ketones and ketone derivatives to nitrile derivatives have also been reported.[11]

This procedure uses readily available reagents and provides a simple, efficient method for nitrile synthesis. The entire sequence of four steps can be performed in a single day. Although product formation in the second step is presumably thermodynamically controlled, the cyanohydrazine is favored in all cases studied except with aryl ketones. A

TABLE I
KETONE TO NITRILE TRANSFORMATION

Ketone	Diazene Yield (%)[a]	Nitrile Yield (%)[b] (ratio of diastereomers)
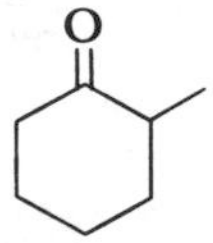	95	89 (50:50)
	94	97 (58:42)
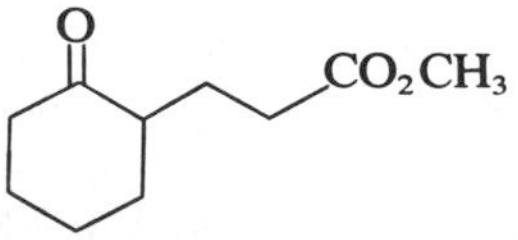	94	90 (50:50)

[a] Distilled yield based on starting ketone, obtained without purification of intermediate hydrazine.
[b] Yield based on purified diazene as determined by chromatographic analysis.

water–methanol solution of ammonium chloride and potassium cyanide may also be employed for cyanohydrazine formation, but lower yields (*ca.* 60%) are obtained. The third step, a conveniently performed titration procedure with bromine as oxidant, can be effected with other oxidizing reagents such as 4-phenyl-4*H*-1,2,4-triazole-3,5-dione, *tert*-butyl hypochlorite, and Jones reagent.[12] The final diazene decomposition step is induced with bases and nucleophiles such as methoxide, ethoxide, hydroxide, and iodide. The diazine decomposition can be extended to the preparation of α-methylnitriles and α-carboalkoxynitriles.[13]

This procedure is general for the preparation of secondary nitriles (Table I), and can be used in the presence of other functional groups by the appropriate choice of oxidation and decomposition reagents.[14]

1. Department of Chemistry, Harvard University, Cambridge, Massachusetts 02138.
2. Present address: Department of Chemistry, Stanford University, Stanford, California 94305.
3. Department of Chemistry, Yale University, New Haven, Connecticut 06520.
4. O. Diels, *Ber. Dtsch. Chem. Ges.*, **47**, 2183 (1914).
5. "Dictionary of Organic Compounds," Vol. 2, Oxford University Press, New York, 1965, p. 782.
6. B. A. Dadson and J. Harley-Mason, *J. Chem. Soc. D*, 665 (1969).
7. L. Friedman and H. Shechter, *J. Org. Chem.*, **25**, 877 (1960).
8. K. Meyer, *Helv. Chim. Acta*, **29**, 1580 (1946); N. Danieli, Y. Mazur, and F. Sondheimer, *J. Am. Chem. Soc.*, **84**, 875 (1962).
9. S. Cacchi, L. Caglioti, and G. Paolucci, *Chem. Ind. (London)*, 213 (1972); cf. J. Jiricny, D. M. Orere, and C. B. Reese, *J. Chem. Soc., Perkin Trans. 1*, 1487 (1980).
10. O. H. Oldenziel and A. M. van Leusen, *Synth. Commun.*, **2**, 281 (1972); O. H. Oldenziel, D. Van Leusen, and A. M. Van Leusen, *J. Org. Chem.*, **42**, 3114 (1977); O. H. Oldenziel, J. Wildeman, and A. M. Van Leusen, *Org. Synth.*, **Coll. Vol. 6**, 41 (1987).
11. B. M. Trost, M. J. Bogdanowicz, and (in part) J. Kern, *J. Am. Chem. Soc.*, **97**, 2218 (1975), and references contained therein.
12. L. F. Fieser and M. Fieser, "Reagents for Organic Synthesis," Wiley, New York, 1967, p. 142.
13. F. E. Ziegler and P. A. Wender, *J. Am. Chem. Soc.*, **93**, 4318 (1971); F. E. Ziegler and P. A. Wender, *J. Org. Chem.*, **42**, 2001 (1977).
14. K. C. Mattes, M. T. Hsia, C. R. Hutchinson, and S. A. Sisk, *Tetrahedron Lett.*, 3541 (1977).

ALDEHYDES FROM OLEFINS: CYCLOHEXANECARBOXALDEHYDE

$$\text{(cyclohexene)} + CO + H_2 \xrightarrow[\substack{\text{benzene} \\ \text{60-150 atm., 100°}}]{Rh_2O_3} \text{(cyclohexane-CHO)}$$

Submitted by P. Pino[1] and C. Botteghi[2]
Checked by Mary M. Borecki, Joseph J. Mrowca,
and Richard E. Benson

1. Procedure

Caution! Benzene has been identified as a carcinogen; OSHA has issued emergency standards on its use. All procedures involving benzene should be carried out in a well-ventilated hood, and glove protection is required.

To a stainless-steel, 0.5-l. pressure vessel (Note 1) equipped with a 450-atm. manometer and a temperature recorder is added 0.2 g. (0.8 mmole) of rhodium(III) oxide (Note 2). The vessel is sealed and evacuated to 0.1 mm. pressure. A solution of 82 g. (1.0 mole) of cyclohexene (Note 3) in 140 ml. of anhydrous benzene is introduced by suction into the vessel. The vessel is placed in a heatable shaking device and pressured to 75 atm. with carbon monoxide; the total pressure is then increased to 150 atm. with hydrogen (Note 4). Shaking is begun and the vessel is heated to an internal temperature of 100° (Note 5). When the internal temperature reaches 100°, the pressure begins to fall. Whenever the pressure falls to 60 atm., rocking is stopped and the pressure is first increased to 105 atm. with carbon monoxide, then to 150 atm. with hydrogen. Rocking is started again, and the process is continued until no appreciable pressure decrease occurs. Approximately 2 hours is required, and the pressure decrease corresponds to the consumption of 2 moles of gas. The vessel is rapidly cooled to room temperature (Note 6) and the residual gas is carefully vented.

The vessel is opened, and the slightly yellow reaction mixture is transferred immediately to a 2-l., round-bottomed flask containing a freshly prepared solution of 200 g. of sodium hydrogen sulfite in 400 ml. of water. The flask is fitted with a stopper and is occasionally shaken at room temperature for a period of 3 hours (Note 7). The resulting precipitate is collected by suction filtration on a sintered-glass funnel and washed with 500 ml. of diethyl ether (Note 8). After drying in air, the bisulfite derivative is transferred to a 2-l. distillation flask containing 1 l. of 20% aqueous potassium carbonate. The resulting mixture is distilled, and the azeotropic mixture of water and aldehyde (b.p. 94–95°) is collected under nitrogen (Note 9).

The aldehyde is separated from the lower aqueous layer as a colorless liquid, dried over 10 g. of anhydrous sodium sulfate, filtered, and distilled under reduced pressure using a Claisen distillation apparatus, yielding 92–94 g. (82–84%) of cyclohexanecarboxaldehyde, b.p. 52–53° (18 mm.), n_D^{25} 1.4484 (Notes 10, 11). A purity of about 98% was established by GC analysis (Note 12); the product is suitable for synthetic use without further purification (Note 13).

2. Notes

1. The pressure vessel was tested to a pressure of 700 atm. at 300°.

2. The submitters used rhodium(III) oxide available from Fluka A G without further purification. The checkers obtained rhodium(III) oxide from Alfa Inorganics.

3. The cyclohexene was purified by distillation over sodium metal before use (n_D^{25} 1.4452). The submitters used the product available from Fluka A G, and the checkers used the product available from Aldrich Chemical Company, Inc.

4. The purity of the gases used was greater than 99%.

5. During the course of the reaction the temperature was maintained at 100° ± 2°.

6. This procedure avoids secondary reactions of the aldehydes, which lead to high-boiling products. It is particularly advisable when linear aliphatic aldehydes are synthesized using cobalt catalysts.[3]

7. The formation of the bisulfite derivative is an exothermic reaction; the flask is cooled with a bath of cold water for the first 10–15 minutes.

8. It is impossible to obtain a completely white precipitate by this procedure.

9. In order to avoid oxidation of the product the submitters recommend use of a nitrogen atmosphere for all manipulations involving cyclohexanecarboxaldehyde.

10. Literature[4] values for cyclohexanecarboxaldehyde: b.p. 78.5–80° (57 mm.), n_D^{25} 1.4485.

11. Cyclohexanecarboxaldehyde is stable at room temperature under nitrogen; the submitters noted no appreciable variation in the refractive index after 30 days.

12. The submitters state that GC analysis was made using a 2-m. column packed with polypropylene glycol (LB-550-X available from Perkin-Elmer) on Chromosorb. The retention time at 140° is 5.2 minutes at a flow rate of 30 ml./minute of nitrogen. The 2,4-dinitrophenylhydrazone derivative[4] melts at 173–174°, and the semicarbazone derivative[5] melts at 172–173°.

13. In addition to rhodium(III) oxide, cobalt(II) acetylacetonate or dicobalt octacarbonyl has been used by the submitters as catalyst precursors for the hydroformylation of cyclohexene. The results are given in Table I.

3. Discussion

This preparation is an illustration of the hydroformylation of olefins (oxo synthesis). The reaction occurs in the presence of soluble catalytic complexes of Group VIII metals. Although the metal originally used by Roelen[6] and still largely used in industry for the production of aliphatic aldehydes and alcohols[7] is cobalt, the most active and selective catalysts are rhodium-containing compounds. The catalytic activity of the other Group VIII metals is, in general, much poorer. Although the hydroformylation of unsaturated substrates is a very general reaction,[7,8] some important limitations associated with the olefin structure may lead to the formation of isomeric aldehydes. In addition, especially in the presence of cobalt catalysts, further reactions of synthesized aldehydes may occur under hydroformylation conditions.

With regard to the structure of the olefins, tetrasubstituted olefins do not undergo hydroformylation under typical reaction conditions, and olefinic substrates containing

TABLE I
HYDROFORMYLATION OF CYCLOHEXENE
WITH COBALT CATALYSTS[a]

Catalyst Precursor (mole/l.)	Solvent[b]	Reaction Temperature	Reaction Time (hours)	Yield[c] (%)
[Bis(acetylacetonate) cobalt(II)] (0.08)	Benzene	150°[d]	1.5	70
[Bis(acetylacetonate)[e] cobalt(II)] (0.08)	Heptane	110°	12	74
Dicobalt octacarbonyl[f] (0.006)	Benzene	120°	8	80

[a] 4.15 mole/l. of cyclohexene; $CO:H_2 = 1:1$; 150 atm. initial pressure.

[b] 140 ml.

[c] The aldehyde was isolated from the reaction mixture through its bisulfite derivative as described in the procedure.

[d] Induction time, 40–60 minutes.

[e] This catalyst precursor (5 g.) in 140 ml. of heptane was heated in the autoclave at 160° with a mixture of $CO:H_2$ (1:1) at 150 atm. for 2 hours. The vessel was cooled, the gas released, 1 mole of cyclohexene was charged, and the reaction was carried out according to the usual procedure.

[f] The submitters used product available from Fluka A G that was dried under reduced pressure after recrystallization from heptane at −70°.

functional groups sometimes give poor yields and unexpected products.[7,8] If there is no plane of symmetry in the substrate across the double bond, at least two isomeric aldehydes are obtained.[9] Although methods for shifting the isomeric composition of the products have been proposed,[10-12] complete control of the isomeric composition has not been achieved despite the fact that the reaction mechanism is fairly well understood.[13] In addition, if the structure of the olefin is such that a double-bond shift is possible, isomers other than the two shown below can be formed.[14] Further reactions of the synthesized aldehydes may occur, especially when cobalt catalysts are used, leading to alcohols, aldol condensation products or acetal derivatives. Some of the secondary reactions can be avoided by carrying out the hydroformylation in the presence of orthoformic acid esters[15] or of other reagents protecting the aldehyde group.[16] However, care must be taken when ortho esters are used, since hydroformylation of ortho esters may occur and yield aldehydes or acetals.[17]

$$R—CH{=}CH_2 + CO + H_2 \longrightarrow \underset{\underset{CHO}{|}}{R—CH—CH_3} + R—CH_2CH_2—CHO$$

Although cobalt catalysts are the best known and the most commonly used, in recent years rhodium has been preferred for laboratory syntheses because of its higher activity and selectivity. As catalyst precursors Rh_2O_3,[18] $Rh_4(CO)_{12}$,[19] or $HRh(CO)(PPh_3)_3$[11] are commonly used. Rhodium complexes supported on polymers have also been used.[12] For typical organic syntheses the easily accessible Rh_2O_3 seems preferable, even if higher temperature and pressures are required to carry out the olefin hydroformylation. Using $HRh(CO)(PPh_3)_3$, olefin hydroformylation at room temperature and pressure is possible.[11] Carrying out the reaction in the presence of (-)DIOP [(4R,5R)-2,2-dimethyl-4,5-bis(5-dibenzophosphol-5-ylmethyl)-1,3-dioxolane], produced optically active aldehydes from monosubstituted ethylenes, as well as from 1,1- and 1,2-disubstituted ethylenes.[20]

The hydroformylation of cyclohexene has been extensively investigated.[11,12,15,21-23] The present procedure is an adaptation of the rhodium-catalyzed hydroformylation of 2-butene.[18]

Other methods for the preparation of cyclohexanecarboxaldehyde include the catalytic hydrogenation of 3-cyclohexene-1-carboxaldehyde, available from the Diels–Alder reaction of butadiene and acrolein,[24] the reduction of cyclohexanecarbonyl chloride by lithium tri-tert-butoxyaluminum hydride,[25] the reduction of N,N-dimethylcyclohexanecarboxamide with lithium diethoxyaluminum hydride,[26] and the oxidation of the methane sulfonate of cyclohexylmethanol with dimethyl sulfoxide.[27] The hydrolysis, with simultaneous decarboxylation and rearrangement, of glycidic esters derived from cyclohexanone gives cyclohexanecarboxaldehyde.[4,28]

1. Technisch-Chemisches Laboratorium, ETH Zürich, Switzerland.
2. Istituto di Chimica Applicata, Università di Sassari, Italy.
3. P. Pino, F. Piacenti, M. Bianchi, and R. Lazzaroni, Chim. Ind. (Milan), 50, 106 (1968) [Chem. Abstr., 68, 86796c (1968)].
4. E. P. Blanchard, Jr., and G. Büchi, J. Am. Chem. Soc., 85, 955 (1963).
5. M. S. Newman, J. Am. Chem. Soc., 71, 379 (1949).
6. O. Roelen, Ger. Pat., 849,548 (1938).
7. J. Falbe, "Carbon Monoxide in Organic Synthesis," Springer Verlag, Berlin-Heidelberg-New York, 1970, and references cited therein.
8. O. Bayer, in E. Müller, Ed., "Methoden der Organischen Chemie" (Houben-Weyl), Vol. 7/1, Georg Thieme Verlag, Stuttgart, 1954, p. 55.
9. A. I. M. Keulemans, A. Kwantes, and T. h. van Bavel, Recl. Trav. Chim. Pays-Bas, 67, 298 (1948); see also ref. 14.
10. F. Piacenti, P. Pino, R. Lazzaroni, and M. Bianchi, J. Chem. Soc. C, 488 (1966); L. H. Slaugh and R. D. Mullineaux, J. Organomet. Chem., 13, 469 (1968).
11. D. Evans, J. A. Osborn, and G. Wilkinson, J. Chem. Soc. A, 3133 (1968).
12. L. V. Pittman, Jr., R. M. Hanes, J. Am. Chem. Soc., 98, 5042 (1976).
13. R. F. Heck and D. S. Breslow, J. Am. Chem. Soc., 83, 4023 (1961).
14. M. Johnson, J. Chem. Soc., 4859 (1963); F. Piacenti, S. Pucci, M. Bianchi, R. Lazzaroni, and P. Pino, J. Am. Chem. Soc., 90, 6847 (1968); P. Pino, F. Piacenti, and G. Dell'Amico, J. Chem. Soc. C, 1640 (1971); see also ref. 18.
15. P. Pino, Gazz. Chim. Ital., 81, 625 (1951).
16. W. W. Prichard, U.S. Pat. 2,517,416 (1950) [Chem. Abstr., 45, 648g (1951)].
17. F. Piacenti, Gazz. Chim. Ital., 92, 225 (1962); F. Piacenti, C. Cioni, and P. Pino, Chem. Ind. (London), 1240 (1960).
18. B. Fell, W. Rupilius, and G. Asinger, Tetrahedron Lett., 3261 (1968), and references cited therein.
19. P. Chini, S. Martinengo, and G. Garlaschelli, J. Chem. Soc. Chem. Commun., 709 (1972).

20. P. Pino, G. Consiglio, C. Botteghi, C. Salomon, *Adv. Chem. Series*, **132**, 295 (1974).

21. H. Wakamatsu, *Nippon Kagaku Zasshi*, **85**, 227 (1964) [*Chem. Abstr.*, **61**, 13173 (1964)].

22. N. S. Imyanitov and D. M. Rudkovskii, *Zh. Prikl. Khim. (Leningrad)*, **40**, 2020 (1967) [*Chem. Abstr.*, **68**, 95367 (1968)].

23. Montecatini, Brit. Pat. 782,459 (1957) [*Chem. Abstr.*, **52**, 5460 (1958)].

24. I. A. Heilbron, E. R. H. Jones, R. W. Richardson, and F. Sondheimer, *J. Chem. Soc.*, 737 (1949).

25. H. C. Brown and B. C. Subba Rao, *J. Am. Chem. Soc.*, **80**, 5377 (1958).

26. H. C. Brown and A. Tsukamoto, *J. Am. Chem. Soc.*, **81**, 502 (1959).

27. H. R. Nace, U.S. Pat. 2,888,488 (1959) [*Chem. Abstr.*, **53**, 19979 (1959)].

28. G. Darzens and P. Lefébure, *C.R. Hebd. Seances Acad. Sci.*, **142**, 714 (1906).

CATALYTIC OSMIUM TETROXIDE OXIDATION OF OLEFINS:
cis-1,2-CYCLOHEXANEDIOL

A.
$$\xrightarrow[\text{75°, 22.5 hours}]{\text{30\% aqueous H}_2\text{O}_2}$$

1

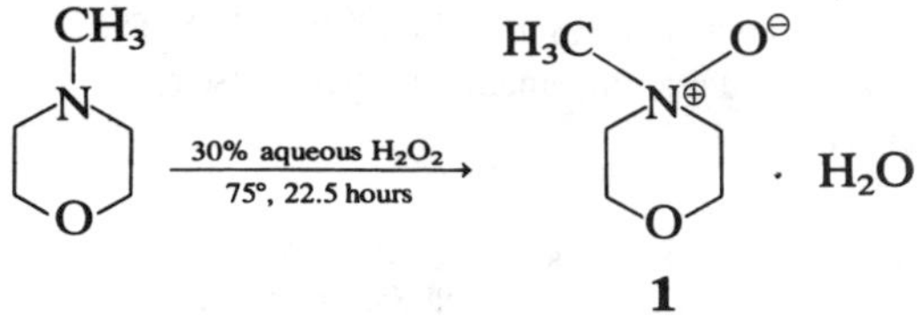

B.   **1** +
$$\xrightarrow[\text{H}_2\text{O, acetone}]{\text{OsO}_4}$$

2

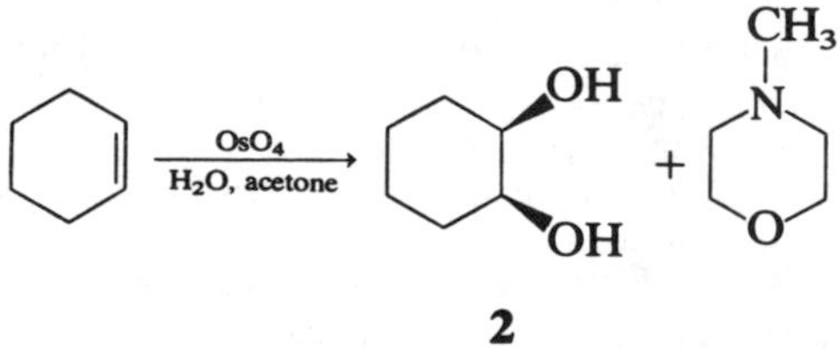

Submitted by V. VanRheenen, D. Y. Cha, and W. M. Hartley[1]
Checked by N. Meyer, W. Wykypiel, and D. Seebach

1. Procedure

Caution! Care should be taken in handling osmium tetroxide. The vapor is toxic, causing damage to eyes, respiratory tract, and skin.

A.   N-*Methylmorpholine* N-*oxide* (**1**) (Note 1).   A 100-ml., three-necked, round-bottomed flask equipped with a reflux condenser, a magnetic stirring bar, and a dropping funnel is flushed with nitrogen or argon and charged with 32.3 g. (35.1 ml., 0.320 mole) of *N*-methylmorpholine (Note 2). The flask is immersed in an oil bath maintained at 75°, and 32.4 g. (29.1 ml., 0.286 mole) of 30% aqueous hydrogen peroxide is added dropwise over a period of 2.5 hours (Note 3). The mixture is stirred for 20 hours at 75°, at which time a negative peroxide test (potassium iodide paper) is obtained (Note 4). The reaction mixture is cooled to 50°, and a slurry of 50 ml. of methanol, 0.5 g. of charcoal, and 0.5 g. of Celite (Note 5) is added. After being stirred for 1 hour, the mixture is filtered and the filter cake washed with three 15-ml. portions of methanol. The filtrate and combined washings are concentrated with a rotary evaporator (water aspirator vacuum), with the bath temperature finally reaching 95°, where it is held for 10 minutes. The flask is fitted with a reflux condenser, and the residual viscous oil is dissolved in 25 ml. of acetone at 60°. On cooling to 40° (with seeding, if crystals of the *N*-oxide are available) the product spontaneously crystallizes. The slurry is stored at room temperature overnight, cooled in an ice bath, and filtered. The crystals are washed with three 15-ml. portions of 0° acetone and dried overnight at 40° (0.01 mm.) The yield of colorless crystalline monohydrate **1** is 32.4–34.3 g. (83.8–88.7%), m.p. 75–76° (Notes 6 and 7).

B.   cis-1,2-*Cyclohexanediol* (**2**).   A 250-ml., three-necked, round-bottomed flask, with a magnetic stirrer and a nitrogen inlet, is charged with 14.81 g. (0.1097 mole) of monohydrate **1,** 40 ml. of water, and 20 ml. of acetone. To this solution is added *ca.* 70 mg. of osmium tetroxide (0.27 mmole) (Note 8) and 8.19 g. (10.1 ml., 0.100 mole) of cyclohexene (Note 9). This two-phase solution is stirred vigorously under nitrogen at room temperature. The reaction is slightly exothermic and is maintained at room temperature with a water bath. During the overnight stirring period, the reaction mixture becomes homogeneous and light brown in color. After 18 hours, TLC (Note 10) shows the reaction to be complete. Sodium hydrosulfite (0.5 g.) (Note 11) and 5 g. of Magnesol (Note 12) slurried in 20 ml. of water are added, the slurry is stirred for 10 minutes, and the mixture is filtered through a pad of 5 g. of Celite on a 150-ml. sintered-glass funnel. The Celite cake is washed with three 15-ml. portions of acetone. The filtrate, combined with acetone wash, is neutralized to pH 7 with 6.4 ml. of 12 *N* sulfuric acid. The acetone is evaporated under vacuum using a rotary evaporator. The pH of the resulting aqueous solution is adjusted to pH 2 with 2.3 ml. of 12 *N* sulfuric acid, and the *cis*-diol **2** is separated from *N*-methylmorpholine hydrosulfate by extraction with five 45-ml. portions of *n*-butanol (Note 13). The combined butanol extracts are extracted once with 25 ml. of 25% sodium chloride solution, and the aqueous phase is backwashed with 50 ml. of butanol. The butanol extracts are evaporated under vacuum, giving 12.1 g. of white solid. The *cis*-diol **2** is separated from a small amount of insoluble material (*ca.* 0.7 g.) by boiling the solid with a 200-ml., an 80-ml., and a 20-ml. portion of diisopropyl ether (Note 14), decanting the solvent each time. The combined ether fractions are evaporated to *ca.* 50 ml. under vacuum, and crystalline white plates precipitate. The mixture is cooled to *ca.* −15°. The crystals are filtered, washed with two 10-ml. portions of cold diisopropyl ether, and dried, yielding 10.18–10.32 g. (89–90%) of the *cis*-diol **2** (m.p. 96–97°).

2. Notes

1. *N*-Methylmorpholine *N*-oxide (**1**) can also be purchased from Eastman Organic Chemicals or Fluka A G.
2. Commercial material was used without purification (the purity was checked by refractive index and ¹H NMR).

3. The slow addition (2.5 hours) is required to avoid overheating of the reaction mixture. The potential danger of using hydrogen peroxide at an elevated temperature is minimized by using a 10% excess of N-methylmorpholine and by choosing reaction conditions that ensure rapid consumption, avoiding accumulation of peroxide in the mixture. A 50% aqueous hydrogen peroxide solution can also be used. The content of the commercial hydrogen peroxide (ca. 30 or 50%) must be determined by iodometric titration.

4. Very sensitive ether peroxide test strips (Merckoquant, Art. No. 10011), available from E. Merck, Darmstadt, are used. If the test is still positive at this point, an additional 0.2 ml. of N-methylmorpholine is added. Stirring and heating at 75° are continued for another 5 hours. Remaining peroxide renders the work-up and drying of the product potentially hazardous. N-Methylmorpholine N-oxide (1) and hydrogen peroxide form a strong 1:1 complex. In the reaction with osmium tetroxide, this complex produces conditions similar to those of the Milas reaction,[7] and some ketol formation may result.

5. Darco G 60, Aktivkohle, Fluka A G No. 05100, and Celite 512 Hyflosuper, Firma Schneider, Winterthur, Switzerland, No. 5100025, were used.

6. The procedure is designed to maintain the proper amount of water in the crystallization mixture so that the monohydrate 1 is obtained. It has the highest melting point and is the least hydroscopic. Other hydrated forms, such as the dihydrate (m.p. 35–60°) and mixed hydrates, may be isolated. A Karl Fischer assay of the water content is not necessary if the obtained material melts within the range given.

7. A second crop of 1.7–4.3 g. (4.3–11.1%) of the product can be obtained by evaporating the mother liquid, heating the residue at 97° for 20 minutes under reduced pressure, dissolving it in 55 ml. of acetone at 60°, and continuing as described.

8. Commercial osmium tetroxide was used without purification. It is not easy to accurately weigh this material because it rapidly sublimes.

9. Commercial cyclohexene was used (the purity was checked by refractive index). Addition of cyclohexene caused a darkening of the reaction mixture. This is caused by a finite concentration of the osmate ester. The reaction becomes lighter in color when complete.

10. The reaction may be followed by TLC. The ratio of the R_f values for cyclohexene and cis-diol 2 is 2:1 (commercial silica gel plates, ethyl acetate). The plates are best visualized by first spraying with 1% aqueous potassium permanganate, then with methanolic sulfuric acid, followed by charring with heat. The checkers found that, if the procedure is followed exactly, monitoring the reaction by TLC is unnecessary.

11. Sodium hydrosulfite reduces the osmium tetroxide to insoluble lower-valent osmium species.

12. The submitter used Magnesol, industrial grade, available from Reagent Chemical Research, Inc., Pilot Engineering Division. The checkers used Florisil TLC, available from E. Merck, Darmstadt, No. 12519.

13. Since the cis-diol 2 is very water soluble, a polar solvent such as n-butanol is required to extract it. n-Butanol forms an efficient water azeotrope. More conventional solvents may be used for less polar products.

14. Diisopropyl ether readily forms explosive peroxides. It should be tested for peroxides, and contact with air should be minimized.

TABLE I

PREPARATION OF *cis*-DIOLS BY CATALYTIC OXIDATION OF OLEFINS WITH OSMIUM TETROXIDE

Starting Material	Product	Procedure	(Isolated yields, %)	Reference
		NMO^a—OsO_4	(>95)	18
		NMO^a—OsO_4	(>95)	10^b
		NMO^a—OsO_4 $NaClO_3$—OsO_4 $KMnO_4$ H_2O_2—OsO_4	(79) (30) (50) (11.4)	10^b 11 12 11
		NMO^a—OsO_4 OsO_4, 1 mole $KMnO_4$	(31) (14) (3)	10^b 13 14

TABLE I (*continued*)

PREPARATION OF *cis*-DIOLS BY CATALYTIC OXIDATION OF OLEFINS WITH OSMIUM TETROXIDE

Starting Material	Product	Procedure	(Isolated yields, %)	Reference
		NMO^a—OsO_4 OsO_4, 1 mole $KMnO_4$	(25) (21) (28)	10[b] 15 15
		NMO^a—OsO_4 OsO_4, 1 mole $KClO_3$—OsO_4	(55) (53) (35)	10[b] 16 16
		NMO^a—OsO_4 $NaClO_3$—OsO_4	(78) (79)	17[c] 17[c]

[a] NMO = *N*-Methylmorpholine *N*-oxide.

[b] The reaction was carried out in aqueous acetone at room temperature using 0.2–1.0 mole % OsO_4 (see Experimental section).

[c] Solvent composition of 10:3:1 *tert*-butanol–tetrahydrofuran–water was preferred for this reaction.

346

3. Discussion

cis-Dihydroxylation of olefins may be effected with potassium permanganate, osmium tetroxide, or silver iodoacetate according to Woodward's procedure.[2] Oxidation of cyclohexene to cis-diol 2 with potassium permanganate is reported to proceed in only 30–40% yields.[3,4] A modification of Woodward's procedure, in which iodine, potassium iodate, and potassium acetate in acetic acid were used, has given cis-diol 2 in 86% yield.[5] This procedure is particularly useful for placement of cis-diols on the more hindered side of more complex substrates.

The reaction of an olefin with osmium tetroxide is the most reliable method for cis-dihydroxylation of a double bond, particularly for preparation of cis-diols on the least hindered side of the molecule. When used stoichiometrically, however, the high cost of osmium tetroxide can make a large-scale glycolization prohibitively expensive, and the work-up procedures can be cumbersome, particularly when pyridine is used. Also osmium tetroxide is volatile and toxic, resulting in handling problems. Catalytic osmylation using chlorate[6] or hydrogen peroxide (Milas reagent[7]) to regenerate osmium tetroxide avoids some of these problems, but overoxidation to an α-ketol commonly leads to losses in yield and separation problems. Preparation of cis-diol 2 with sodium chlorate and osmium tetroxide is reported to proceed in 46% yield,[3] and in 76% yield[8] when sodium chlorate, potassium osmate, and a detergent are used. A 62% yield of the cis-diol 2 from cyclohexene is reported in an interesting catalytic osmylation using tert-butyl hydroperoxide under alkaline conditions.[9] This method is particularly useful for oxidation of tri- and tetrasubstituted olefins.

In this report we describe the conversion of cyclohexene to cis-diol 2, in 90% yield, by catalytic osmylation using 1 mole equivalent of N-methylmorpholine N-oxide (1, NMO) to regenerate the osmium tetroxide catalyst. This procedure avoids the α-ketol by-products encountered with the currently available catalytic processes, and provides the high yields of the stoichiometric reaction without the expense and work-up problems.

The reaction is generally applicable to a variety of substrate types, as illustrated in Table I.[10] Compatible functionality includes hydroxyl, ester, lactone, acid, ketone, and electron-poor olefins such as those conjugated to α-ketones. Some selectivity between isolated double bonds is also found. The reaction generally gives nearly quantitative yields with simple olefins.

The reaction is usually run in aqueous acetone in either one- or two-phase systems, but substrate solubility may require the use of other solvents. Aqueous tert-butanol, tetrahydrofuran, and mixtures of these solvents have also been used successfully.

Other simple aliphatic amine oxides can be used as the oxidant in this reaction, but N-methylmorpholine N-oxide (1) is preferred because it generally gives a faster reaction rate and is easily prepared. The reaction can also be used to convert aliphatic amine oxides into amines.

1. The Upjohn Company, Kalamazoo, Michigan 49001.
2. F. D. Gunstone, in "Advances in Organic Chemistry," 17, Vol. 1, edited by R. A. Raphael, E. C. Taylor, and H. Wynberg, Interscience, New York, 1960, p. 110ff.
3. M. F. Clark and L. N. Owen, J. Chem. Soc., 315 (1949).
4. K. B. Wiberg and K. A. Saegebarth, J. Am. Chem. Soc., 79, 2822 (1957).

5. L. Mangoni, M. Adinolfi, G. Barone, and M. Parrilli, *Tetrahedron Lett.*, 4485 (1973).
6. K. A. Hofmann, *Ber. Dtsch. Chem. Ges.*, **45**, 3329 (1912).
7. N. A. Milas and S. Sussman, *J. Am. Chem. Soc.*, **58**, 1302 (1936); N. A. Milas, J. H. Trepagnier, J. T. Nolan, Jr., and M. I. Iliopulos, *J. Am. Chem. Soc.*, **81**, 4730 (1959); C. J. Norton and R. E. White, "Selective Oxidation Processes," *Adv. Chem. Ser.* **51**, 10–25 (1965).
8. W. D. Lloyd, B. J. Navarette, and M. F. Shaw, *Synthesis*, 610 (1972).
9. K. B. Sharpless and K. Akashi, *J. Am. Chem. Soc.*, **98**, 1986 (1976).
10. V. VanRheenen, R. C. Kelly, and D. F. Cha, *Tetrahedron Lett.*, 1973 (1976).
11. A. C. Cope, S. W. Fenton, and C. F. Spencer, *J. Am. Chem. Soc.*, **74**, 5884 (1952).
12. W. P. Weber and J. P. Shepherd, *Tetrahedron Lett.*, 4907 (1972).
13. K. Tanaka, *J. Biol. Chem.*, **247**, 7465 (1972).
14. J. L. Jernow, D. Gray, and W. D. Clossen, *J. Org. Chem.*, **36**, 3511 (1971).
15. Y. F. Shealy and J. D. Clayton, *J. Am. Chem. Soc.*, **91**, 3075 (1969).
16. R. C. Kelly and I. Schletter, *J. Am. Chem. Soc.*, **95**, 7156 (1973).
17. W. P. Schneider and A. V. McIntosh, U.S. Pat. 2,769,824 (1957) [*Chem. Abstr.*, **51**, 8822e (1957)]. The use of NMO in catalytic OsO_4 reactions was first disclosed in this patent during work to introduce the corticoid side chain (an α-ketol) in a steroid.
18. B. J. Magerlein, G. L. Bundy, F. H. Lincoln, and G. A. Youngdale, *Prostaglandins*, **9**(1), 5 (1975).

STEREOSELECTIVE HYDROXYLATION WITH THALLIUM(I) ACETATE AND IODINE: *trans*- AND *cis*-1,2-CYCLOHEXANEDIOLS

Submitted by R. C. Cambie and P. S. Rutledge[1]
Checked by D. Seebach, M. Liesner, and E.-M. Wilka

1. Procedure

Caution! Thallium salts are very toxic. These procedures should be carried out in a well-ventilated hood, and rubber gloves should be worn. For disposal of thallium wastes, see Org. Synth., **Coll. Vol. 6,** 791 (1988).

A.   trans-1,2-*Cyclohexanediol*.   In a 100-ml., round-bottomed flask equipped with a reflux condenser protected with a drying tube are placed a magnetic stirring bar, 17.56 g. (0.05447 mole) of thallium(I) acetate (Note 1), and 40 ml. of dried acetic acid (Note 2). The mixture is stirred and heated at reflux for 1 hour. The mixture is cooled before 2.84 g. (3.50 ml., 0.0346 mole) of cyclohexene (Note 3) and 8.46 g. (0.0333 mole) of iodine (Note 4) are added. The resulting suspension is stirred and heated at reflux for 9 hours (Note 5), and then cooled to room temperature. The yellow thallium(I) iodide precipitate is filtered and washed thoroughly with diethyl ether. The filtrates are combined, the solvents are removed with a rotary evaporator (Note 6), and the residual liquid is dissolved in dry ether. The turbid solution is dried with anhydrous potassium carbonate, and the solvent is again removed by rotary evaporation (Note 6), affording 5.4–6.3 g. of trans-1,2-cyclohexanediol diacetate as a mobile, brown liquid (Note 7).

The diacetate is dissolved in 25 ml. of 95% ethanol, a solution of 2.9 g. (0.073 mole) of sodium hydroxide in 11 ml. of water is added, and the resulting mixture is heatèd at reflux for 3 hours. The solution is concentrated by rotary evaporation and the remaining syrup is extracted with six 50-ml. portions of chloroform. The combined extracts are dried over anhydrous magnesium sulfate and evaporated, providing 3.1–3.3 g. of a pale brown crystalline solid, m.p. 97–103°. Recrystallization from carbon tetrachloride gives 2.5–2.7 g. (65–70% based on iodine) of trans-1,2-cyclohexanediol, m.p. 103–104° (Note 8).

B.   cis-1,2-*Cyclohexanediol*.   A 500-ml., round-bottomed flask equipped with a reflux condenser and a magnetic stirring bar is charged with 17.56 g. (0.05447 mole) of thallium(I) acetate (Note 1), 160 ml. of glacial acetic acid, 3.0 g. (3.7 ml., 0.036 mole) of cyclohexene (Note 3), and 8.46 g. (0.0333 mole) of iodine (Note 4) in the order given. The suspension is stirred and warmed in a heating bath at 80° for 30 minutes. An 80-ml. portion of water is added, stirring is continued, and the mixture is heated at reflux for 9 hours. The product is isolated and hydrolyzed as described in Part A, affording 3.2–4.9 g. of cis-1,2-cyclohexanediol, m.p. 88–95° (Note 9). Recrystallization from carbon tetrachloride gives 2.7–2.9 g. (70–75% based on iodine) of cis-1,2-cyclohexanediol, m.p. 97–98° (Note 10).

2. Notes

1. Thallium(I) acetate purchased from BDH Chemicals Ltd., Poole, England, or Fluka AG, Buchs, Switzerland, was used without further purification. This reagent is also available from Alfa Division, Ventron Corporation.

2. The submitters purchased glacial acetic acid from Showa Denko K. K., Tokyo, Japan, and acetic anhydride from Riedel de Haen AG, Seelze-Hannover, Germany. A solution prepared from 4 volumes of glacial acetic acid, 1 volume of acetic anhydride, and a catalytic amount of p-toluenesulfonic acid was heated under reflux for 24 hours and distilled. The distillate, which contained 5% water and 4% acetic anhydride according to ^{1}H NMR, analysis, was then used by the submitters. The water content was determined from the chemical shift of the hydroxyl proton.[2]

The checkers purchased analytical grade glacial acetic acid and acetic anhydride from E. Merck, Darmstadt, Germany. A 4:1 (v/v) solution of glacial acetic acid and acetic anhydride containing 500 mg. of *p*-toluenesulfonic acid per liter was heated at reflux for 24 hours and then distilled. A forerun corresponding to 25% of the solution was discarded before a main fraction of acetic acid corresponding to 60% of the solution was collected. The main fraction, containing 17% acetic anhydride and 1% or less water as determined from its ^{1}H NMR spectrum, was used by the checkers.

3. Cyclohexene was purchased from BDH Chemicals Ltd., Poole, England, by the submitters and used without purification. This reagent was purchased by the checkers from Fluka AG, Buchs, Switzerland, and distilled before use.

4. Iodine purchased from Riedel de Haen AG, Seelze-Hannover, Germany, was sublimed before use by the submitters. The checkers used iodine from Siegfried AG, Zofingen, Switzerland, without purification.

5. After *ca.* 30 minutes the initially black-green solid becomes yellow.

6. The checkers recommend that excessive heating and evacuation be avoided during rotary evaporation to minimize the loss of product during this operation. They kept the heating bath temperature below 80° and used a water aspirator.

7. The diacetate was judged to be virtually pure by the submitters on the basis of GC analysis carried out at 150° using a glass column packed with 3% OV 17 (1:1 methyl–phenyl silicone) supported on 70–80 mesh Chromosorb W.

8. The submitters obtained 2.9 g. (75%) of product that melted at 103–105° (lit.,[3a] m.p. 104°).

9. The unrecrystallized product obtained by the submitters melted at 91–95°. The purity of this material is estimated to be 96% on the basis of the melting point.[3a]

10. The submitters obtained 3.0 g. (78%) of product which melted at 99–100°. The melting point of 97–98° shown above was obtained by the checkers using a Tottoli melting point apparatus (Büchi) equipped with a 50° range *Anschütz* thermometer. A melting point of 98° has been reported.[3b]

3. Discussion

The present procedure offers a convenient alternative to the Prévost reaction[4] and the Woodward modification of the Prévost reaction[5] in which silver carboxylates are used instead of thallium(I) carboxylates. Thallium(I) salts have the advantages of being generally stable crystalline solids that can be readily prepared in high yield by neutralization of the appropriate carboxylic acid with thallium(I) ethoxide.[6] Silver salts, on the other hand, are frequently unstable and difficult to dry. Thallium and its compounds are, however, extremely toxic, and great care must be taken in the use and disposal of thallium salts.[7,8,9]

The mechanisms of these reactions are presumably analogous to those of the Prévost and Woodward–Prévost reactions.[4,5] In the first step of the reaction of iodine and thallium(I) acetate with cyclohexene, in both parts A and B of this procedure, produces *trans*-2-iodocyclohexyl acetate. The second equivalent of thallium(I) acetate scavenges iodide ion during formation of the 1,3-dioxolan-2-ylium ion intermediate. Under the anhydrous conditions in Part A, the carbonium ion reacts with acetate ion at a ring carbon with inversion to give the diacetate. In part B the ion is captured by water, and the resulting ortho ester undergoes ring opening to the *cis*-diol monoacetate. No appreciable reaction occurs unless thallium(I) acetate, iodine, and cyclohexene are all present. Thus, in contrast to the Prévost and Woodward–Prévost procedures, acetyl hypoiodite evidently cannot be prepared separately from thallium(I) acetate and iodine. The precise reasons for this difference are not clear.

trans-2-Iodocyclohexyl acetate can be isolated in essentially quantitative yield from the reaction of thallium(I) acetate, iodine, and cyclohexene in a 1:1:1 molar ratio in refluxing chloroform.[10] Similarly, iodo acetates from a representative series of alkenes, including cyclohexene, have been prepared in 80–98% yield[11] in glacial acetic acid which was not dried as described in this procedure. The corresponding iodo benzoates are obtained in comparable yields from reaction with thallium(I) benzoate and iodine in benzene. The deactivated olefin methyl cinnamate did not react under these conditions, and *o*-allylphenol underwent ring iodination to give 2-allyl-6-iodophenol.[12] The diterpenes, phyllocladene and isophyllocladene, upon reaction with thallium(I) benzoate and iodine,[13] afford the same mixture of allylic benzoates obtained from a Woodward–Prévost reaction. With the exception of 3-phenylpropene, the formation of iodo carboxylates from unsymmetrical alkenes occurs regioselectively in a Markovnikov sense.

Vicinal iodo carboxylates may also be prepared from the reaction of olefins either with iodine and potassium iodate in acetic acid,[14a] or with *N*-iodosuccinimide and a carboxylic acid in chloroform.[14b] A number of new procedures for effecting the hydroxylation or acyloxylation of olefins in a manner similar to the Prévost or Woodward–Prévost reactions include the following: iodo acetoxylation with iodine and potassium chlorate in acetic acid followed by acetolysis with potassium acetate;[14c] reaction with *N*-bromoacetamide and silver acetate in acetic acid;[15] reaction with thallium(III) acetate in acetic acid;[16] and reaction with iodine tris(trifluoroacetate) in pentane.[17]

The preparation of *trans*-1,2-cyclohexanediol by oxidation of cyclohexene with performic acid and subsequent hydrolysis of the diol monoformate has been described,[18] and other methods for the preparation of both *cis*- and *trans*-1,2-cyclohexanediols have been cited. Subsequently the *trans* diol has been prepared by oxidation of cyclohexene with various peracids,[19] with hydrogen peroxide and selenium dioxide,[20] and with iodine and silver acetate by the Prévost reaction.[21] Alternative methods for preparing the *trans* isomer are hydroboration of various enol derivatives of cyclohexanone[22] and reduction of *trans*-2-cyclohexen-1-ol epoxide with lithium aluminum hydride.[23] *cis*-1,2-Cyclohexanediol has been prepared by *cis* hydroxylation of cyclohexene with various reagents or catalysts derived from osmium tetroxide,[24] by solvolysis of *trans*-2-halocyclohexanol esters in a manner similar to that of the Woodward–Prévost reaction,[14c,15,17,21,25] by reduction of *cis*-2-cyclohexen-1-ol epoxide with lithium aluminum hydride,[23] and by oxymercuration of 2-cyclohexen-1-ol with mercury(II) trifluoroacetate in the presence of chloral and subsequent reduction.[26]

1. Department of Chemistry, University of Auckland, Auckland, New Zealand.
2. R. O. C. Norman and C. B. Thomas, *J. Chem. Soc. B*, 994 (1968); L. W. Reeves and W. G. Schneider, *Trans. Faraday Soc.*, **54**, 314 (1958).
3. (a) S. Winstein and R. E. Buckles, *J. Am. Chem. Soc.*, **64**, 2780 (1942); (b) S. Winstein and R. E. Buckles, *J. Am. Chem. Soc.*, **64**, 2787 (1942).
4. (a) C. V. Wilson, *Org. React.*, **9**, 332 (1957); (b) F. D. Gunstone, "Hydroxylation Methods," Chap. 4 in R. A. Raphael, E. C. Taylor, and H. Wynberg, Eds., "Advances in Organic Chemistry," Vol. 1, Wiley Interscience, New York, 1960, pp. 103–147.
5. R. B. Woodward and F. V. Brutcher, Jr., *J. Am. Chem. Soc.*, **80**, 209 (1958).
6. L. F. Fieser and M. Fieser, "Reagents for Organic Synthesis," Vol. 2, Wiley, New York, 1969, p. 407.
7. E. C. Taylor, R. L. Robey, D. K. Johnson, and A. McKillop, *Org. Synth.*, **Coll. Vol. 6**, 791 (1988).
8. E. C. Browning, "Toxicity of Industrial Metals," Butterworths, London, 1961, pp. 317–322; N. I. Sax, "Dangerous Properties of Industrial Materials," 3rd ed., Reinhold, New York, 1968, pp. 1154–1158.
9. E. C. Taylor and A. McKillop, *Acc. Chem. Res.*, **3**, 338 (1970).
10. A detailed procedure is available on request from the submitters.
11. R. C. Cambie, R. C. Hayward, J. L. Roberts, and P. S. Rutledge, *J. Chem. Soc., Perkin Trans. 1*, 1858 (1974).
12. R. C. Cambie, P. S. Rutledge, T. Smith-Palmer, and P. D. Woodgate, *J. Chem. Soc., Perkin Trans. 1*, 1161 (1976).
13. R. C. Cambie, R. C. Hayward, J. L. Roberts, and P. S. Rutledge, *J. Chem. Soc., Perkin Trans. 1*, 1120 (1974).
14. (a) M. Parrilli, G. Barone, M. Adinolfi, and L. Mangoni, *Gazz. Chim. Ital.*, **104**, 835 (1974), and references therein; (b) M. Adinolfi, M. Parrilli, G. Barone, G. Laonigro, and L. Mangoni, *Tetrahedron Lett.*, 3661 (1976); (c) L. Mangoni, M. Adinolfi, G. Barone, and M. Parrilli, *Gazz. Chim. Ital.*, **105**, 377 (1975).
15. D. Jasserand, J. P. Girard, J. C. Rossi, and R. Granger, *Tetrahedron Lett.*, 1581 (1976).
16. E. Glotter and A. Schwartz, *J. Chem. Soc., Perkin Trans. 1*, 1660 (1976).
17. J. Buddrus, *Angew. Chem. Int. Ed. Engl.*, **12**, 163 (1973).
18. A Roebuck and H. Adkins, *Org. Synth.*, **Coll. Vol. 3**, 217 (1955).
19. J. B. Brown, H. B. Henbest, and E. R. H. Jones, *J. Chem. Soc.*, 3634 (1950); W. D. Emmons, A. S. Pagano, and J. P. Freeman, *J. Am. Chem. Soc.*, **76**, 3472 (1954); R. J. Kennedy and A. M. Stock, *J. Org. Chem.*, **25**, 1901 (1960); R. Lombard and G. Schroeder, *Bull. Soc. Chim. Fr.*, 2800 (1963).
20. N. Sonoda and S. Tsutsumi, *Bull. Chem. Soc. Jpn.*, **38**, 958 (1965).

21. C. A. Bunton and M. D. Carr, *J. Chem. Soc.*, 770 (1963).
22. A. Hassner and B. H. Braun, *Univ. Colo. Stud. Ser. Chem. Pharm.*, No. 4, 48 (1962) [*Chem. Abstr.*, **58**, 11250e (1963)]; J. Klein, R. Levene, and E. Dunkelblum, *Tetrahedron Lett.*, 2845 (1972); H. Kono and Y. Nagai, *Org. Prep. Proceed. Int.*, **6**, 19 (1974).
23. H. B. Henbest and R. A. L. Wilson, *J. Chem. Soc.*, 1958 (1957).
24. W. D. Lloyd, B. J. Navarette, and M. F. Shaw, *Synthesis*, 610 (1972); K. B. Sharpless and K. Akashi, *J. Am. Chem. Soc.*, **98**, 1986 (1976); V. VanRheenen, D. Y. Cha, and W. M. Hartley, *Org. Synth.*, **Coll. Vol. 6**, 342 (1988).
25. F. V. Brutcher, Jr., and G. Evans, III, *J. Org. Chem.*, **23**, 618 (1958); S. J. Lapporte and L. L. Ferstandig, *J. Org. Chem.*, **26**, 3681 (1961); F. L. Scott and D. F. Fenton, *Tetrahedron Lett.*, 681 (1970).
26. L. E. Overman and C. B. Campbell, *J. Org. Chem.*, **39**, 1474 (1974).

CLEAVAGE OF METHYL ETHERS WITH IODOTRIMETHYLSILANE: CYCLOHEXANOL FROM CYCLOHEXYL METHYL ETHER

$$(CH_3)_3Si-O-Si(CH_3)_3 \xrightarrow[60-140°]{Al, I_2} (CH_3)_3SiI$$

Submitted by MICHAEL E. JUNG[1] and MARK A. LYSTER[1,2]
Checked by JOAN HUGUET, H. SHIBUYA, and S. MASAMUNE

1. Procedure

A. *Iodotrimethylsilane.* A 250-ml., two-necked, round-bottomed flask equipped with a magnetic stirring bar, an addition funnel for solids (Note 1), and a reflux condenser bearing a nitrogen inlet is charged with 5.6 g. (0.21 mole) of aluminum powder (Note 2) and 16.2 g. (0.100 mole) of hexamethyldisiloxane (Note 3) and purged with nitrogen. The mixture is stirred and heated with an oil bath at 60° as 50.8 g. (0.200 mole) of iodine is added slowly through the addition funnel over 55 minutes (Note 4). The bath temperature is raised to *ca.* 140°, and the mixture is heated under reflux for 1.5 hours. The reflux condenser is removed, and the flask is equipped for distillation at atmospheric pressure. The bath temperature is gradually raised from 140° to 210°, and the clear, colorless distillate is collected, yielding 32.6–35.3 g. (82–88%) of iodotrimethylsilane, b.p. 106–109° (Notes 5 and 6).

B. *Cyclohexanol.* A 25-ml., oven-dried, round-bottomed flask is charged with 1.722 g. (0.01524 mole) of cyclohexyl methyl ether (Note 7). The flask is purged with nitrogen and sealed with a rubber septum. With oven-dried syringes, 4 ml. of chloroform

(Note 8), 0.5 g. (0.5 ml., 0.006 mole) of pyridine (Notes 8 and 9), and 4.8 g. (3.5 ml., 0.024 mole) of freshly prepared iodotrimethylsilane are injected into the flask in the order specified. When the iodotrimethylsilane is added, the solution becomes slightly yellow and a precipitate appears. The mixture is heated without stirring at 60° for 64 hours, after which the reaction is normally complete (Note 10). Anhydrous methanol (2 ml.) is added, the mixture is cooled to room temperature, and the volatile components (Note 11) are removed on a rotary evaporator. Approximately 10 ml. of anhydrous diethyl ether (Note 12) is added, and the resulting suspension is filtered, removing pyridinium hydroiodide. The flask and the filter cake are washed thoroughly with *ca.* 50 ml. of anhydrous ether. The ether is evaporated, and the residual oil is purified by chromatography on 70 g. of silica gel packed with anhydrous ether in a 3 × 50 cm. glass column. The column is eluted with anhydrous ether, and 5–7 ml. fractions are collected and analyzed by TLC (Note 13). Fractions containing product are combined and evaporated, affording 1.26–1.35 g. (83–89%) of cyclohexanol (Note 14).

2. Notes

1. The submitters have used both an addition funnel with a worm gear delivery similar to those manufactured by Normag, and an Erlenmeyer flask attached to the neck of the reaction vessel with a piece of Gooch rubber tubing. Normag addition funnels are available from Lab Glass, Inc., P. O. Box 610, Vineland, New Jersey 08360.

2. The submitters purchased aluminum powder from MC and B Manufacturing Chemists. The metal used by the checkers was supplied by J. T. Baker Chemical Company.

3. Hexamethyldisiloxane is available from Aldrich Chemical Company, Inc. The reagent may also be prepared by the procedure described in the following paragraph. The submitters have used chlorotrimethylsilane purchased from Aldrich Chemical Company, Inc., and Silar Laboratories, Inc. (10 Alplaus Road, Scotia, New York 12302) either as supplied or after distillation from calcium hydride. No appreciable difference in yield was noted between preparations using undistilled and distilled reagent.

A 250-ml., three-necked, round-bottomed flask equipped with a magnetic stirring bar, a pressure-equalizing dropping funnel, and a reflux condenser bearing a nitrogen inlet is charged with a solution of 7 g. (0.4 mole) of water in 72.7 g. (76.0 ml., 0.601 mole) of N,N-dimethylaniline and flushed with nitrogen. Stirring is begun, and 62.49 g. (73.00 ml., 0.5754 mole) of chlorotrimethylsilane is added dropwise over 50 minutes. The mixture is heated under reflux in an oil bath at 125–130° for 1 hour. The reflux condenser is replaced by a distilling head, and the product is distilled at atmospheric pressure. The fraction boiling at 98–101° is collected, dried over anhydrous magnesium sulfate, and filtered, affording 43–44 g. (92–94%) of hexamethyldisiloxane as a clear colorless liquid.

4. In a similar procedure for the preparation of iodotrimethylsilane, aluminum, iodine, and hexamethyldisiloxane are combined, and the mixture is heated to reflux.[3] When this procedure was attempted by the submitters, violent exothermic reactions occurred at *ca.* 50–60°. The slow addition of iodine to the warm mixture described in the present procedure leads to a controlled, reproducible reaction.

5. The product is sometimes contaminated with a small amount of hexamethyldisiloxane. The amount of this contaminant is minimized by using longer reaction times and by careful handling to avoid contact with atmospheric moisture. The product may become discolored during storage, in which case it may be purified by distillation from copper powder. The ^{1}H NMR spectrum of iodotrimethylsilane (CDCl$_3$) exhibits a singlet at δ 0.71 in the presence of benzene as internal standard.

6. The submitters obtained 69.4 g. (87%) of product when the scale was doubled.

7. Cyclohexyl methyl ether was prepared by the method of Stoocknoff and Benoiton.[4] A 250-ml., two-necked, round-bottomed flask is equipped with a magnetic stirring bar, a rubber septum, and a reflux condenser mounted with a nitrogen inlet. The flask is purged with nitrogen and charged with 8 g. (0.2 mole) of a 60% dispersion of sodium hydride in mineral oil. The sodium hydride is washed free of mineral oil with pentane and suspended in 75 ml. of tetrahydrofuran. After 10 g. (0.10 mole) of cyclohexanol is added by syringe, the mixture is heated under reflux for 22 hours. A 28.4-g. (12.5 ml., 0.200 mole) portion of methyl iodide is injected into the flask, and the resulting mixture is heated under reflux for 18 hours. Water and chloroform are added to the cooled mixture. The aqueous layer is extracted with three 50-ml. portions of chloroform, the combined chloroform extracts are dried over anhydrous magnesium sulfate, and the solvents are evaporated. Distillation of the remaining liquid affords 7.1 g. (62%) of cyclohexyl methyl ether, b.p. 133–134°.

8. Chloroform and pyridine were dried over Linde type 4A molecular sieves.

9. Pyridine is added to neutralize small amounts of hydrogen iodide, often present in iodotrimethylsilane as a result of hydrolysis by contact with moisture. The amount of by-products, including cyclohexyl iodide, is reduced by the presence of pyridine. Hindered pyridine bases such as 2,6-di-*tert*-butyl-4-methylpyridine[5] have also been used for this purpose, by the submitters. The pyridine bases do not appear to react with iodotrimethylsilane.

10. The progress of the reaction may be conveniently monitored by ^{1}H NMR spectroscopy. After 64 hours the signal at δ 3.25 for the methoxyl group of cyclohexyl methyl ether had usually decreased to less than 1% of its original intensity, and peaks for cyclohexyl iodide could not be discerned. Although the submitters found that the reaction time was decreased by using larger amounts of iodotrimethylsilane, 5–10% of cyclohexyl iodide was also produced.

11. The volatile components are chloroform, methanol, methyl iodide, methyl trimethylsilyl ether, and hexamethyldisiloxane.

12. When the submitters used technical grade ether, the amount of iodine-containing by-products isolated from the chromatography was increased, and the yield of cyclohexanol was somewhat lower.

13. When an insufficient amount of iodotrimethylsilane was used by the submitters, cyclohexyl methyl ether remained at the end of the reaction and was eluted from the silica gel column before cyclohexanol. When present in the crude product, cyclohexyl iodide was also eluted from the column before cyclohexanol.

14. The identity and purity of the product were determined by GC, IR spectroscopy, and ^{1}H NMR spectroscopy by both the submitters and the checkers.

3. Discussion

This procedure describes a convenient method for the preparation of iodotrimethylsilane and illustrates the use of this reagent for ether cleavage, as in the regeneration of cyclohexanol from cyclohexyl methyl ether.[6] Iodotrimethylsilane was first prepared by Whitmore by the reaction of trimethylphenylsilane with iodine.[7] The reagent has also been generated *in situ* by halide exchange between magnesium iodide and chlorotrimethylsilane.[8] The present procedure is essentially that of Voronkov and Khudobin,[3] modified by slowly adding iodine to a mixture of aluminum and hexamethyldisiloxane heated at 60° (see Note 4).

The use of methyl ethers as protecting groups for aliphatic alcohols has been hampered by the difficulty of liberating the alcohol from this inert derivative.[9,10] The cleavage of methyl ethers has been previously accomplished with boron reagents such as boron trichloride,[11] boron trifluoride in acetic anhydride,[12] and diborane or sodium borohydride in the presence of iodine.[13] Two recent modifications of early methods for cleavage of aliphatic methyl ethers utilize hydrogen iodide generated *in situ*[10] and magnesium bromide in acetic anhydride.[14] Recently described methods for the hydrolysis of methyl ethers include the use of sodium cyanide in dimethyl sulfoxide,[15] anhydrous hydrogen bromide,[16] and thiotrimethylsilanes.[17]

The use of iodotrimethylsilane for this purpose provides an effective alternative to known methods. Thus, the reaction of primary and secondary methyl ethers with iodotrimethylsilane in chloroform or acetonitrile at 25–60° for 2–64 hours affords the corresponding trimethylsilyl ethers in high yield.[6] The alcohols may be liberated from the trimethylsilyl ethers by methanolysis. The mechanism of the ether cleavage is presumed to involve initial formation of a trimethylsilyl oxonium ion which is converted to the silyl ether by nucleophilic attack of iodide at the methyl group. *tert*-Butyl, trityl, and benzyl ethers of primary and secondary alcohols are rapidly converted to trimethylsilyl ethers by the action of iodotrimethylsilane, probably *via* heterolysis of silyl oxonium ion intermediates. The cleavage of aryl methyl ethers to aryl trimethylsilyl ethers may also be effected more slowly by reaction with iodotrimethylsilane at 25–50° in chloroform or sulfolane for 12–125 hours,[6] with iodotrimethylsilane at 100–110° in the absence of solvent,[18,19] and with iodotrimethylsilane generated *in situ* from iodine and trimethylphenylsilane at 100°.[19,20]

Alkyl esters are efficiently dealkylated to trimethylsilyl esters with high concentrations of iodotrimethylsilane either in chloroform or sulfolane at 25–80°[21] or without solvent at 100–110°.[18,20] Hydrolysis of the trimethylsilyl esters serves to release the carboxylic acid. Amines may be recovered from *O*-methyl, *O*-ethyl, and *O*-benzyl carbamates after reaction with iodotrimethylsilane in chloroform or sulfolane at 50–60° and subsequent methanolysis.[22] The conversion of dimethyl, diethyl, and ethylene acetals and ketals to the parent aldehydes and ketones under aprotic conditions has been accomplished with this reagent.[23] The reactions of alcohols (or the corresponding trimethylsilyl ethers) and aldehydes with iodotrimethylsilane give alkyl iodides[24] and α-iodosilyl ethers,[25] respectively. Iodomethyl methyl ether is obtained from cleavage of dimethoxymethane with iodotrimethylsilane.[26] A review by Schmidt[27] covers the applications of iodotrimethylsilane listed above along with many more recently published examples.

1. Contribution No. 3805 from the Department of Chemistry, University of California, Los Angeles, California 90024.
2. Present address: The Upjohn Company, Kalamazoo, Michigan 49001.
3. M. G. Voronkov and Yu. I. Khudobin, *Izv. Akad. Nauk SSSR, Ser. Khim.*, 713 (1956) [*Chem. Abstr.*, **51**, 1819e (1957)].
4. B. A. Stoochnoff and N. L. Benoiton, *Tetrahedron Lett.*, 21 (1973).
5. A. G. Anderson and P. J. Stang, *J. Org. Chem.*, **41**, 3034 (1976).
6. M. E. Jung and M. A. Lyster, *J. Org. Chem.*, **42**, 3761 (1977).
7. B. O. Pray, L. H. Sommer, G. M. Goldberg, G. T. Kerr, P. A. DiGiorgio, and F. C. Whitmore, *J. Am. Chem. Soc.*, **70**, 433 (1948).
8. U. Krüerke, *Chem. Ber.*, **95**, 174 (1962).
9. C. B. Reese, "Protection of Alcoholic Hydroxyl Groups and Glycol Systems," in J. F. W. McOmie, Ed., "Protective Groups in Organic Chemistry," Plenum Press, London, 1973, pp. 95–104.
10. For a compilation of recent methods, see C. A. Smith and J. B. Grutzner, *J. Org. Chem.*, **41**, 367 (1976).
11. (a) S. Allen, T. G. Bonner, E. J. Bourne, and N. M. Saville, *Chem. Ind. (London)*, 630 (1958); (b) W. Gerrard and M. F. Lappert, *Chem. Rev.*, **58**, 1081 (1958), and references therein; (c) A. B. Foster, D. Horton, N. Salim, M. Stacey, and J. M. Webber, *J. Chem. Soc.*, 2587 (1960); (d) T. G. Bonner, E. J. Bourne, and S. McNally, *J. Chem. Soc.*, 2929 (1960); (e) S. D. Géro, *Tetrahedron Lett.*, 591 (1966).
12. (a) R. D. Youssefyeh and Y. Mazur, *Tetrahedron Lett.*, 1287 (1962); (b) C. R. Narayanan and K. N. Iyer, *Tetrahedron Lett.*, 759 (1964).
13. (a) G. Odham and B. Samuelsen, *Acta Chem. Scand.*, **24**, 468 (1970); (b) L. H. Long and G. F. Freeguard, *Nature (London)*, **207**, 403 (1965).
14. D. J. Goldsmith, E. Kennedy, and R. G. Campbell, *J. Org. Chem.*, **40**, 3571 (1975), and pertinent references therein.
15. J. R. McCarthy, J. L. Moore, and R. J. Cregge, *Tetrahedron Lett.*, 5183 (1978).
16. D. Landin, F. Montanari, and F. Rolla, *Synthesis*, 771 (1978).
17. S. Hanessian and Y. Guidon, *Tetrahedron Lett.*, 2305 (1980).
18. T.-L. Ho and G. A. Olah, *Angew. Chem. Int. Ed. Engl.*, **15**, 774 (1976).
19. T.-L. Ho and G. A. Olah, *Proc. Natl. Acad. Sci. USA*, **75**, 4 (1978).
20. T.-L. Ho and G. A. Olah, *Synthesis*, 417 (1977).
21. M. E. Jung and M. A. Lyster, *J. Am. Chem. Soc.*, **99**, 968 (1977).
22. M. E. Jung and M. A. Lyster, *J. Chem. Soc. Chem. Commun.*, 315 (1978).
23. M. E. Jung, W. A. Andrus, and P. L. Ornstein, *Tetrahedron Lett.*, 2659 (1977).
24. M. E. Jung and P. L. Ornstein, *Tetrahedron Lett.*, 2659 (1977).
25. M. E. Jung, A. B. Mossman, and M. A. Lyster, *J. Org. Chem.*, **43**, 3698 (1978).
26. M. E. Jung, M. A. Mazurek, and R. M. Lim, *Synthesis*, 588 (1978).
27. A. H. Schmidt, *Chem. Ztg.*, **104**, 253 (1980).

PREPARATION OF α,β-UNSATURATED ALDEHYDES
via THE WITTIG REACTION:
CYCLOHEXYLIDENEACETALDEHYDE

(Acetaldehyde, cyclohexylidene-)

Submitted by Wataru Nagata,[1] Toshio Wakabayashi,[2] and Yoshio Hayase[1]
Checked by Kyo Abe and S. Masamune

1. Procedure

Caution! Benzene has been identified as a carcinogen; OSHA has issued emergency standards on its use. All procedures involving benzene should be carried out in a well-ventilated hood, and glove protection is required.

A 1-l., three-necked, round-bottomed flask, fitted with a magnetic stirrer, dropping funnel, and nitrogen inlet is charged with 5.45 g. (0.116 mole) of sodium hydride (51% oil dispersion) (Note 1) and 30 ml. of dry tetrahydrofuran (Note 2). The system is flushed with nitrogen and a solution of 30.2 g. (0.116 mole) of diethyl 2-(cyclohexylamino)vinyl-phosphonate [*Org. Synth.*, **Coll. Vol. 6,** 448 (1988)] in 90 ml. of dry tetrahydrofuran is added dropwise to the stirred mixture over a period of 15 minutes. During the addition the temperature is maintained at 0–5° with an ice bath. The mixture is stirred for an additional 15 minutes at 0–5° to ensure complete reaction. A solution of 10.3 g. (0.105 moles) of cyclohexanone (Note 3) in 70 ml. of dry tetrahydrofuran is added dropwise to the mixture over a period of 20 minutes, so that the temperature does not exceed 5°. The mixture is stirred for an additional 90 minutes at 20–25° in a water bath. During the stirring a gummy precipitate of sodium diethyl phosphate is observed. The mixture is poured into 500 ml. of cold water and extracted with three 300-ml. portions of diethyl ether. The combined ether extracts are washed twice with 200 ml. of saturated aqueous salt solution, dried over anhydrous sodium sulfate, and distilled under reduced pressure (35 mm.) at 25–30°. The residue is dissolved in 300 ml. of benzene and transferred to a 3-l., three-necked, round-bottomed flask equipped with a stirrer and a reflux condenser. To this solution is added a solution of 72 g. (0.57 mole) of oxalic acid dihydrate in 900 ml. of water (Note 4). The stirred mixture is refluxed for 2 hours under nitrogen, cooled, and transferred to a separatory funnel. The aqueous layer is extracted with two 300-ml. portions of ether. The combined organic extracts are washed with 200 ml. of water, then with 200 ml. of saturated aqueous salt solution, and dried over anhydrous sodium sulfate, and distilled under reduced pressure (35 mm.) at 25–30°. The residue is transferred to a 30-ml., round-bottomed flask and distilled under reduced pressure through a 5-cm. Vigreux column, yielding 10.8 g. (83%, Note 5) of cyclohexylideneacetaldehyde, b.p. 78–84° (12 mm.), containing *ca.* 15% of the isomeric cyclohexenylacetaldehyde (Note 6).

2. Notes

1. Sodium hydride (50–51% in mineral oil) was purchased from Metal Hydrides Inc. and used as 51%.

2. Reagent grade tetrahydrofuran was freshly distilled over sodium hydride before use. The checkers used lithium aluminum hydride to dry the solvent [see *Org. Synth.*, **Coll. Vol. 5,** 976 (1973) for warning note].

3. Reagent grade cyclohexanone was redistilled.

4. When a more concentrated solution (72 g. of oxalic acid in 450 ml. of water) was used, the product contained larger amounts of the β, γ-isomer, cyclohexenylacetaldehyde. To suppress this double bond isomerization, a 4–7% aqueous oxalic acid solution was used.

5. The yields were 80–85% in several runs.

6. The submitters found that analysis of the final product by GC indicated a 15% contaminant of the by-product, cyclohexenylacetaldehyde. The analysis was conducted on a column packed with 5% XE-60 on Chromosorb W at 120°. The retention times for cyclohexenylacetaldehyde and cyclohexylideneacetaldehyde were 1.3 and 3.3 minutes, respectively. The checkers found that the product contained 10–15% of cyclohexenylacetaldehyde by GC analysis and 12–16% by NMR spectral analysis ($CDCl_3$), using the relative intensity of two signals (δ 9.53 and 9.97) due to the aldehydic protons of the two compounds. Reported physical constants are b.p. 58–62° (16 mm.) for cyclohexenylacetaldehyde and b.p. 80–85° (16 mm.) for cyclohexylideneacetaldehyde.[3]

3. Discussion

For the conversion of ketones into α, β-unsaturated aldehydes containing two additional carbon atoms, several multistep processes *via* ethynyl or vinyl carbinol intermediates have been reported.[3-9] Although the overall yields obtained by these routes for the conversion of cyclohexanone into cyclohexylideneacetaldehyde have never exceeded 50%, they were the only useful methods for this type of conversion until the Wittig[10] method appeared. This process consists of the normal aldol condensations of ketones with the lithium salt of ethylidenecyclohexylamine, followed by dehydration and hydrolysis.

The present procedure also illustrates an excellent, general method for the conversion of ketones and aldehydes[11] into the corresponding α,β-unsaturated aldehydes, using diethyl 2-(cyclohexylamino)vinylphosphonate.[12] The yield is usually high, and the reaction proceeds stereoselectively, affording only the *trans* isomer. In the reaction of 3-ketosteroids with this reagent, no β,γ-isomers were formed.[11] Recently Meyers and co-workers[13] reported a new method for the synthesis of α,β-unsaturated aldehydes.

A remarkable improvement of the present procedure, involving the *in situ* preparation of a solution of the lithium salt of diethyl 2-(*tert*-butylamino)vinylphosphonate[12] followed by addition of an aldehyde or a ketone, has recently been reported.[14]

1. Shionogi Research Laboratories, Shionogi & Company, Ltd., Fukushima-ku, Osaka, 553 Japan.
2. Planning Department, Medicals & Pharmaceutical Division, Teijin Ltd., 1-1, Uchisaiwai-chō, 2-chōme, Chiyoda-ku, Tokyo 100, Japan.
3. J. B. Aldersley and G. N. Burkhardt, *J. Chem. Soc.*, 545 (1938).
4. E. A. Braude and O. H. Wheeler, *J. Chem. Soc.*, 320 (1955).
5. M. Julia and J.-M. Surzur, *Bull. Soc. Chim. Fr.*, 1615 (1956).
6. H. G. Viehe, *Chem. Ber.*, **92**, 1270 (1959).
7. G. Saucy, R. Marbet, H. Lindlar, and O. Isler, *Helv. Chim. Acta*, **42**, 1945 (1959), cf. V. T. Ramakrishnan, K. V. Narayanan, and S. Swaminathan, *Chem. Ind. (London)*, 2082 (1967).
8. M. C. Chaco and B. H. Iyer, *J. Org. Chem.*, **25**, 186 (1960).
9. A. Marcou and H. Normant, *Bull. Soc. Chim. Fr.*, 1400 (1966).
10. (a) G. Wittig, H.-D. Frommeld, and P. Suchanek, *Angew. Chem.*, **75**, 978 (1963); (b) G. Wittig and H. Reiff, *Angew. Chem. Int. Ed. Engl.*, **7**, 7 (1968).
11. W. Nagata and Y. Hayase, *Tetrahedron Lett.*, 4359 (1968); *J. Chem. Soc. C*, 460 (1969).
12. W. Nagata, T. Wakabayashi, and Y. Hayase, *Org. Synth.*, **Coll. Vol. 6**, 448 (1988).
13. A. I. Meyers, A. Nabeya, H. W. Adickes, J. M. Fitzpatrick, G. R. Malone, and I. R. Politzer, *J. Am. Chem. Soc.*, **91**, 764 (1969).
14. A. I. Meyers, K. Tomioka, and M. P. Fleming, *J. Org. Chem.*, **43**, 3788 (1978).

CYCLOPROPENONE

Submitted by R. BRESLOW,[1] J. PECORARO, and T. SUGIMOTO
Checked by R. LÜTHI, H. WÜEST, and G. BÜCHI

1. Procedure

Caution! Because liquid ammonia is used in Part B, this part of the procedure should be conducted in a well-ventilated hood.

A.   1-*Bromo-3-chloro-2,2-dimethoxypropane*.   In a good hood, a 1-l., three-necked, round-bottomed flask equipped with magnetic stirrer and reflux condenser is charged with 300 ml. of anhydrous methanol, 111 g. (1.00 mole) of 2,3-dichloro-1-propene (Note 1), and a few drops of concentrated sulfuric acid. With stirring, 178 g. (1.00 mole) of *N*-bromosuccinimide is added in small portions through the condenser. After the final addition, the reaction mixture is stirred for another hour at room temperature before 5 g. of anhydrous sodium carbonate is added to neutralize the catalyst. The solution is stirred for an additional 15 minutes and poured into a large separatory funnel containing 300 ml. of water. The lower, organic layer is removed, and the aqueous layer is extracted with two 500-ml. portions of pentane. The combined organic extracts are washed twice with an equal volume of water, dried over anhydrous magnesium sulfate, filtered, and evaporated, giving a white semicrystalline mass, which is dissolved in refluxing pentane (250 ml.). The solution is cooled in an acetone–dry-ice bath for 30 minutes, yielding 89–99 g. (41–45%) of the white crystalline ketal, m.p. 69.5–70.5° (Note 2).

B.   3,3-*Dimethoxycyclopropene*.   A 500-ml., three-necked, round-bottomed flask is equipped with a magnetic stirrer, a gas-inlet tube, a thermometer, and an acetone–dry-ice condenser topped with a drying tube containing sodium hydroxide pellets. An acetone–dry-ice bath is placed under the flask, and ammonia is condensed into the flask from a commercial cylinder. When 350–400 ml. of ammonia has condensed, the inlet tube is replaced by a stopper, and a small piece (0.5 g.) of potassium metal is added to the ammonia. The cooling bath is removed, and *ca.* 0.05 g. of anhydrous iron(III) chloride is added. When the ammonia reaches reflux temperature, the blue color of the dissolved potassium turns to gray, and the remainder of the potassium (11.7 g., 0.300-g.-atom total) is added in 0.5-g. pieces at such a rate that a gentle reflux is maintained. The stopper is then replaced by an addition funnel containing a solution of 1-bromo-3-chloro-2,2-dimethoxypropane (21.7 g., 0.100 mole) in 50 ml. of anhydrous diethyl ether, which is added to the gray potassium amide–ammonia suspension over a period of 15 minutes, during which time the mixture is maintained at $-50°$ to $-60°$ with the cooling bath (Note 3). After 3 hours at this temperature, solid ammonium chloride (10.8 g., 0.20 mole) is added with stirring. Ammonia is allowed to evaporate by removing the cooling bath, and during the course of the evaporation it is replaced with 350 ml. of anhydrous ether. When the reaction temperature reaches ca. $0°$, the resulting brown solution is filtered from inorganic salts and placed in a 500-ml., round-bottomed flask (Note 4). The ethereal solution is then subjected to a vacuum (50–80 mm.) applied through a carbon tetrachloride–dry-ice condenser (ca. $-25°$), while the flask is immersed in an ice bath. After 4–5 hours, when the quantity of residue seems to remain constant, the dry-ice condenser is replaced with a distilling head. The pressure is decreased to 1–2 mm., the receiver is maintained at $-78°$ with a cooling bath, and distillation yields 4.0–6.5 g. (40–65%) of 3,3-dimethoxycyclopropene as a clear liquid (Note 5). This material has been purified further,[2] but it can be used directly in the next step. If it is stored, it should be kept below $0°$.

C. *Cyclopropenone.* A stirred solution of 3.0 g. (0.030 mole) of 3,3-dimethoxycyclopropene in 30 ml. of dichloromethane, cooled to $0°$, is treated dropwise with 5 ml. of cold water containing 3 drops of concentrated sulfuric acid. The reaction mixture is stirred at $0°$ for an additional 3 hours before 30 g. of anhydrous sodium sulfate is added in portions, with stirring, to the $0°$ solution. The drying agent is removed by filtration, and the solvent is evaporated at 50–80 mm. with a water bath maintained at 0–10°. The brown, viscous residue is then distilled at 1–2 mm, at a water bath temperature of $10°$. The distillate, a mixture of methanol and dichloromethane, is collected in a receiver cooled to $-78°$. A new receiver is attached, and the bath temperature is gradually raised to $35°$ (Note 6), yielding 1.42–1.53 g. (88–94%) of cyclopropenone as a white solid, b.p. $26°$ (0.46 mm.), m.p. -29 to $-28°$ (Note 7).

Cyclopropenone prepared in this way is quite pure and suitable for most chemical purposes. It can be repurified by crystallization from 3 volumes of ethyl ether at $-60°$ using a cooled filtering apparatus. The residual ethyl ether is then removed by evaporation at 1–2 mm. and $0°$; very pure cyclopropenone is obtained in 60–70% recovery from the distilled material.

2. Notes

1. Commercial material was used without further purification. The reflux condenser is used to decrease evaporative losses of this material.

2. ^{1}H NMR (CCl_4), δ (multiplicity, number of protons): 3.27 (s, 6H), 3.48 (s, 2H), 3.63 (s, 2H); the IR and mass spectra are also as reported.[2]

3. Any crystals which may form at the tip of the addition funnel are scraped off and allowed to drop into the reaction flask.

4. The checkers found it inconvenient to complete Part B in one day and stored this ethereal solution overnight in the freezer compartment of a refrigerator.

5. The product usually contains small amounts of ether, as judged by its ^{1}H NMR spectrum. The yields given are based on pure cyclopropenone ketal. ^{1}H NMR (CDCl$_3$), δ (multiplicity, number of protons): 3.33 (s, 6H), 7.88 (s, 2H).

6. The bath temperature should be raised slowly to prevent decomposition of cyclopropenone.

7. IR (CHCl$_3$) cm.$^{-1}$: 1870, 1840, 1493; ^{1}H NMR (CDCl$_3$), δ: 9.11 (s).

3. Discussion

Cyclopropenone was first synthesized[3-5] by the hydrolysis of an equilibrating mixture of 3,3-dichlorocyclopropene and 1,3-dichlorocyclopropene (prepared by reduction of tetrachlorocyclopropene with tributyltin hydride), a procedure that has been adapted[5,4] for the preparation of labeled and deuterated cyclopropenones for use in physical studies. The current procedure is somewhat more convenient. It is closely based on the work of Baucom and Butler,[2] who have described this synthesis of dimethoxycyclopropene and shown that it can be hydrolyzed to cyclopropenone. The isolation of pure cyclopropenone by ketal hydrolysis parallels the method of Breslow and Oda,[5] which involves the hydrolysis of dichlorocyclopropenes.

Cyclopropenone is a molecule of considerable theoretical interest, since it combines remarkable stability with extreme strain. Various physical studies[6] suggest that much of its stability is derived from the special conjugative stabilization of the two-pi electron system, which is related to the cyclopropenyl cation. In addition, cyclopropenone has a number of interesting chemical properties[7,8] which suggest that it could be a useful synthetic intermediate. It has been used in the synthesis of cyclopropanone derivatives[7] and tropones,[7] the latter by rearrangement of products derived from Diels–Alder reactions. In addition, it undergoes a very interesting cyclization–rearrangement reaction with diazo compounds which leads to the overall insertion of a three-carbon unit between the diazo group and its original attachment point.[7] Perhaps the most remarkable reaction of cyclopropenone so far reported is its conversion with Grignard reagents into 2-substituted resorcinols.[8] This reaction seems to be of some generality, and it represents a simple way to elaborate a resorcinol ring (all six carbons of the resorcinol system are derived from two molecules of cyclopropenone) onto a variety of alkyl groups. The ready availability of this compound should lead to other synthetic applications.

1. Department of Chemistry, Columbia University, New York, N.Y. 10027.
2. K. B. Baucom and G. B. Butler, *J. Org. Chem.*, **37**, 1730 (1972).
3. R. Breslow and G. Ryan, *J. Am. Chem. Soc.*, **89**, 3073 (1967).
4. R. Breslow, G. Ryan, and J. T. Groves, *J. Am. Chem. Soc.*, **92**, 988 (1970).
5. R. Breslow and M. Oda, *J. Am. Chem. Soc.*, **94**, 4787 (1972).
6. R. C. Benson, W. H. Flygare, M. Oda, and R. Breslow, *J. Am. Chem. Soc.*, **95**, 2772 (1973), and references therein.
7. M. Oda, R. Breslow, and J. Pecoraro, *Tetrahedron Lett.*, 4419 (1972).
8. R. Breslow, M. Oda, and J. Pecoraro, *Tetrahedron Lett.*, 4415 (1972).

CYCLOPROPYLDIPHENYLSULFONIUM TETRAFLUOROBORATE

[Sulfonium, cyclopropyldiphenyl tetrafluoroborate(1-)]

A. $\quad ICH_2CH_2CH_2Cl + Ph_2S + AgBF_4 \xrightarrow[\text{nitromethane, r.t.} \sim 40°]{}$

$$Ph_2\overset{\oplus}{S}CH_2CH_2CH_2Cl \quad \overset{\ominus}{BF_4} + AgI$$

B. $\quad Ph_2\overset{\oplus}{S}CH_2CH_2CH_2Cl \quad \overset{\ominus}{BF_4} + NaH \xrightarrow[\text{tetrahydrofuran, r.t.}]{}$

$$\triangleright\!\!-\overset{\oplus}{S}Ph_2 \quad \overset{\ominus}{BF_4} + NaCl + H_2$$

Submitted by MITCHELL J. BOGDANOWICZ and BARRY M. TROST[1]
Checked by TSUTOMU AOKI and WATARU NAGATA

1. Procedure

A. *3-Chloropropyldiphenylsulfonium tetrafluoroborate*. A solution of 93.0 g. (0.500 mole) of diphenyl sulfide (Notes 1 and 2) and 347 g. (1.70 mole) of 1-chloro-3-iodopropane (Notes 2, 3, and 4) in 200 ml. of nitromethane (Note 5) in a 1-l., one-necked flask equipped with a magnetic stirring bar and a nitrogen inlet tube is stirred at room temperature under nitrogen. The flask is shielded from light (Note 6), and 78 g. (0.40 mole) of silver tetrafluoroborate (Note 7) is added in one portion. Initially the temperature rises to 40°, then gradually falls to room temperature. No external cooling is necessary. After 16 hours 200 ml. of dichloromethane is added, the mixture is filtered through a sintered glass funnel fitted with a pad of 35 g. of Florisil (Note 8), and the solid is washed with 100 ml. of dichloromethane. The dichloromethane portions are combined and evaporated at reduced pressure until a solid separates; 1 l. of diethyl ether is added to precipitate the product (Note 9). The off-white crystals are collected (Note 10), washed with ether, and dried under reduced pressure at 25°, yielding 122–140 g. (87–99%) of the sulfonium salt, m.p. 103–105° (Note 11).

B. *Cyclopropyldiphenylsulfonium tetrafluoroborate*. A suspension of 118.7 g. (0.3386 mole) of 3-chloropropyldiphenylsulfonium tetrafluoroborate (Note 2) in 500 ml. of dry tetrahydrofuran (Note 12) is placed in a 2-l., one-necked flask equipped with a magnetic stirring bar and nitrogen inlet tube under nitrogen before 5-g. portions of 55% sodium hydride–mineral oil dispersion (15.2 g., 0.350 mole) are added in 30-minute intervals. The resulting mixture is stirred (Note 13) at room temperature for 24 hours. An aqueous solution of 25 ml. of 48% fluoroboric acid (Note 14), 15 g. of sodium tetra-fluoroborate (Note 7, 15), and 400 ml. of water is added to the well-stirred reaction to destroy residual hydride and swamp out chloride ion (Note 16). After 5 minutes 300 ml. of dichloromethane is added, and the top organic layer is removed from the lower aqueous layer (Note 17). The dichloromethane solution is then extracted with 100 ml. of water.

The combined water layers are extracted with an additional 100 ml. of dichloromethane. The organic phases are combined, dried over anhydrous sodium sulfate, and evaporated at reduced pressure until precipitation occurs. Addition of 1 l. of ether completes the precipitation of the salt. The crystals are collected, washed with ether, recrystallized from hot absolute ethanol (approximately 400 ml.) (Note 18), and dried under reduced pressure, yielding 79.5–88.0 g. (75–83%) of cyclopropyldiphenylsulfonium tetrafluoroborate, m.p. 137–139° (Note 19).

2. Notes

1. Available from Matheson, Coleman and Bell and utilized without further purification. The checkers used reagent grade diphenyl sulfide obtained from Tokyo Kasei Kogyo Co. Ltd., Japan.

2. The checkers carried out the experiment on a half scale and obtained the same results as described by the submitters.

3. Available from K & K Laboratories or may be prepared in 89% yield by the following procedure. To a solution of 393 g. (2.63 moles) of sodium iodide in 1 l. of reagent grade acetone is added 394 g. (2.50 moles) of 1-bromo-3-chloropropane (Aldrich Chemical Co.). After stirring 2 hours at room temperature, the mixture is filtered, the sodium bromide is washed with acetone, and the acetone is evaporated at reduced pressure. A dark iodine color is present along with some solid sodium salts. The oil is dissolved in ether, and the solution is washed with a 10% aqueous sodium thiosulfate. The ethereal layer is separated, dried over anhydrous sodium sulfate, and evaporated at reduced pressure, yielding 454 g. of an oil that can be used without further purification.

4. An excess of 1-chloro-3-iodopropane must be employed to compete effectively with the diphenyl sulfide for complexation with silver fluoroborate.

5. Available from Aldrich Chemical Co. and used without further purification. Dichloromethane may be substituted for the nitromethane. The checkers used reagent grade nitromethane available from Tokyo Kasei Kogyo Co. Ltd., Japan.

6. The flask is wrapped with aluminium foil to prevent decomposition of the silver salts.

7. Available from Ozark Mahoning Corp.

8. A pad of Florisil is employed to facilitate removal of the suspended silver salts.

9. An oil separated initially. Vigorous shaking of the mixture to extract the excess starting material out of the oily sulfonium salt layer induces crystallization.

10. The crystals obtained by the checkers were light brown at this stage, but could be purified by the following procedure.

11. The material is normally utilized directly without further purification. If the solid is very gray, it may be recrystallized. For recrystallization the salt is dissolved in hot 95% ethanol (approximately 350 ml. per 100 g. of salt) containing decolorizing carbon and filtered rapidly. The clear supernatant liquid is allowed to cool in a freezer (−20°). In this way, white crystals, m.p. 106–107°, may be obtained with nearly quantitative recovery. The checkers obtained the purified material, m.p. 108–109°, with 95% recovery and used it for the next step. The purified material has the following spectral data; UV (95% C_2H_5OH) nm. max (ϵ): 236 shoulder (13,200), 262 (2200), 268 (2680), 275 (1010); IR

(Nujol) cm.$^{-1}$: 3090 weak, 3060 weak (aromatic CH), 1580 weak (C=C); ^{1}H NMR (CDCl$_3$), first-order analysis: δ 2–2.5 (m, J_{bc} = 8 Hz., J_{cd} = 6.5 Hz., 2H, CH_2^c). 3.75 (t, 2H, CH_2^d), 4.3 (poorly resolved t, 2H, CH_2^b), 7.5–8.2 (m, 10H, 2C$_6$H$_5^a$).

$$\begin{array}{c} C_6H_5^a \\ \diagdown \\ \overset{\oplus}{S}\cdot CH_2^b\cdot CH_2^c\cdot CH_2^d Cl \\ \diagup \\ C_6H_5 \end{array}$$

$$BF_4^{\ominus}$$

12. The tetrahydrofuran was dried by distilling from lithium aluminium hydride and then from sodium benzophenone ketyl (generated by adding small pieces of sodium metal and benzophenone) directly into the reaction flask. A blue-black color of the ketyl solution indicates dryness. The checkers purified tetrahydrofuran by distillation from sodium hydride dispersion under nitrogen, and used it immediately.

13. The checkers found that efficient stirring is essential for successful results.

14. Since 48% fluoroboric acid was not available in Japan, the checkers used 42% fluoroboric acid obtained from Wako Pure Chemicals Co. Ltd., and obtained the same result as described by the submitters.

15. The checkers prepared sodium tetrafluoroborate by neutralization of an ice-cold, aqueous, 42% solution of fluoroboric acid with an equivalent amount of sodium carbonate and addition of dry ethanol to the reaction mixture to effect complete crystallization of the product. The crystals were purified by washing with ethanol and obtained in 80% yield.

16. Normally no observable effect occurs upon this addition. Gas evolution with a slight exotherm indicates incomplete reaction.

17. The density of the dichloromethane and water layers are nearly equal. Thus, sometimes upon initial mixing, the dichloromethane starts out on the bottom, but the layers reverse on shaking. However, on occasion, the desired organic layer is found in fact to be the bottom one. It is therefore advisable to check the layers by addition of either water or dichloromethane. The checkers found that dichloromethane was on the bottom in the two experiments.

18. Ether may be added to the cold ethanol solution before filtration to insure complete precipitation.

19. The purified material has the following spectral data; UV (95% C$_2$H$_5$OH) nm. max (ϵ): 235 shoulder (12,200), 261 (1800), 267 (2200), 274 (1700); IR (Nujol) cm^{-1}: 3100 weak, 3045 weak (aromatic CH), 1582 weak (C=C); ^{1}H NMR (CDCl$_3$), first-order analysis: δ 1.4–1.75 (m, 4H, 2CH$_2^c$), 3.5–3.9 (m, 1H, CH^b), 7.5–8.1 (m, 10H, 2C$_6$H$_5^a$).

$$\begin{array}{c} C_6H_5^a \\ \diagdown \\ \overset{\oplus}{S}{\scriptstyle b}\!\!\triangleleft{\scriptstyle c} \\ \diagup \\ C_6H_5 \end{array} \qquad BF_4^{\ominus}$$

3. Discussion

The utility of sulfur ylides in organic synthesis demands methods for the efficient preparation of the precursor sulfonium salts.[2] Among the salts, the diphenylsulfonium moiety provides the ability to generate the higher alkylides unambiguously but the low nucleophilicity of the sulfur of diphenyl sulfide dictated the need for exceptionally reactive alkylating agents. Oxonium salts,[3] dialkoxycarbonium salts,[4] and fluorosulfate esters[5] are capable of achieving such alkylations; however, the unavailability of such alkylating agents except for the very simple alkyl groups (e.g., methyl and ethyl) does not allow generalization. On the other hand, alkyl halides complexed to silver salts form powerful alkylating agents and allow utilization of a wide range of alkyl halides susceptible to S_N2 displacement.[2,6] Although alkyl bromides may be employed, alkyl iodides are preferred. The latter are normally available in excellent yields from sulfonate esters, chlorides, or bromides by reaction with sodium iodide. Polyhalides may be employed without complications—reaction occurring preferably at a primary rather than secondary center. A nonsilver salt-based method for preparing 3-chloropropyldiphenylsulfonium tetrafluoroborate has also been reported recently.[7]

While sulfonium ylides do not normally undergo alkylations (except with reactive alkylating agents such as methyl iodide[8]), they do undergo intramolecular alkylation (cyclization) rather efficiently. The present procedure describes the synthesis of a particularly interesting reagent, cyclopropyldiphenylsulfonium tetrafluoroborate.[9] The ylide derived from this salt effects many different synthetic transformations which include facile syntheses of cyclobutanones,[9] γ-butyrolactones,[10] and specifically substituted cyclopentanones[11] from aldehydes and ketones and spiropentanes from α,β-unsaturated carbonyl partners.[9,12] Two reviews of the synthetic applications of the ylide have appeared.[13]

1. Department of Chemistry, University of Wisconsin, Madison, Wisconsin 53706.
2. E. J. Corey and M. Chaykovsky, *J. Am. Chem. Soc.*, **87**, 1353 (1965); E. J. Corey and W. Oppolzer, *J. Am. Chem. Soc.*, **86**, 1899 (1964); V. Franzen and H. E. Driesin, *Chem. Ber.*, **96**, 1881 (1963); V. Franzen, H. J. Schmidt, and C. Mertz, *Chem. Ber.*, **94**, 2942 (1961).
3. H. Meerwein, D. Delfs, and H. Morschel, *Angew. Chem.*, **72**, 927 (1960).
4. H. Meerwein, K. Bodenbenner, P. Borner, F. Kunert, and K. Wunderlich, *Justus Liebigs Ann. Chem.*, **632**, 38 (1960); H. Meerwein, P. Laasch, R. Mersch, and J. Spille, *Chem. Ber.*, **89**, 209 (1965); S. Kabuss, *Angew. Chem. Int. Ed. Engl.* **5**, 675 (1966).
5. M. G. Ahmed, R. W. Alder, G. H. James, M. L. Sinnott, and M. C. Whiting, *Chem. Commun.*, 1533 (1968); M. G. Ahmed and R. W. Alder, *J. Chem. Soc. D*, 1389 (1969).
6. Mercury salts (J. van der Veen, *Recl. Trav. Chim. Pays-Bas*, **84**, 540 (1965)) and antimony salts (G. A. Olah, J. R. DeMember, R. H. Schlosberg, and Y. Halpern, *J. Am. Chem. Soc.*, **94**, 156 (1972) and references therein) may also be employed.
7. B. Badet and M. Julia, *Tetrahedron Lett.*, 1101 (1979).
8. E. J. Corey, M. Jautelat, and W. Oppolzer, *Tetrahedron Lett.*, 2325 (1967).
9. B. M. Trost and M. J. Bogdanowicz, *J. Am. Chem. Soc.*, **93**, 3773 (1971); *J. Am. Chem. Soc.*, **95**, 5321 (1973).
10. M. J. Bogdanowicz and B. M. Trost, *Tetrahedron Lett.*, 923 (1973).
11. B. M. Trost and M. J. Bogdanowicz, *J. Am. Chem. Soc.*, **95**, 289, 5311 (1973).
12. For a related reagent see C. R. Johnson, G. F. Katekar, R. F. Huxol, and E. R. Janiga, *J. Am. Chem. Soc.*, **93**, 3771 (1971).
13. B. M. Trost, *Acc. Chem. Res.*, **7**, 85 (1974); *Pure Appl. Chem.*, **53**, 563 (1975).

RING CONTRACTION *via* A FAVORSKII-TYPE REARRANGEMENT: CYCLOUNDECANONE

Submitted by J. WOHLLEBE and E. W. GARBISCH, JR.[1]
Checked by J. M. DIAKUR and S. MASAMUNE

1. Procedure

Caution! Hydrazoic acid, which is used in Part C of this procedure, is very toxic. Consequently, the conversion of methyl 1-cycloundecenecarboxylate to cycloundecanone by the Schmidt degradation, including hydrolysis and subsequent steam distillation, should be conducted in a well-ventilated hood.

Pure hydrazoic acid in a condensed state has been reported in several instances to explode violently without apparent inducement, but explosions during Schmidt reactions do not seem to have been observed. Nevertheless, it is recommended that this reaction be carried out behind a safety shield.

Benzene has been identified as a carcinogen; OSHA has issued emergency standards on its use. All procedures involving benzene should be carried out in a well-ventilated hood, and glove protection is required.

A. 2,12-*Dibromocyclododecanone.* A 3-l., three-necked, round-bottomed flask is fitted with a magnetic stirrer (Note 1), a pressure-equalizing dropping funnel, a thermometer, and a gas-outlet tube. The outlet tube is connected by Tygon tubing to a calcium chloride drying tube, which is placed near an exhaust port of a hood. The flask is charged with 182 g. (1.00 mole) of cyclododecanone (Note 2), 1.4 l. of dry benzene (Note 3), and 150 ml. of anhydrous diethyl ether, and the funnel with 320 g. (2.00 moles) of bromine (Note 4). The reaction vessel is immersed in a water bath, stirring is initiated, and bromine is added at such a rate that the bromine in solution is consumed before the addition of each new drop; the addition requires 20–30 minutes. Ice is added to the water bath, as required, to hold the reaction temperature at 20–25°. The gas-outlet tube is then connected to a water aspirator, the dropping funnel is replaced with a stopper, and the water bath is filled with warm water. While stirring is continued, the pressure in the reaction flask is gradually decreased to evaporate the hydrobromic acid formed (together with most of the ether and some of the benzene) until the aspirator water is neutral to indicator paper (Note 5). Approximately 1 l. of solution remains in the flask, and this is used directly for the next step.

B. *Methyl 1-cycloundecenecarboxylate*. The benzene solution of 2,12-dibromocyclododecanone prepared in Part A is stirred and treated with 125 g. (2.31 moles) of powdered sodium methoxide (Note 6), which is added in portions over 30–40 minutes. Ice is added to the water bath, as required, to hold the reaction temperature at 25–30°. After being stirred at 25–30° for another 20 minutes, the reaction mixture is extracted successively with 500-ml. portions of water, 5% hydrochloric acid, and saturated aqueous sodium chloride. The aqueous phases are combined, extracted with 400 ml. of ether, and discarded. The combined organic phases are filtered through anhydrous sodium sulfate, the solvents are evaporated under reduced pressure, and the residual oil is distilled through a 7-cm., insulated Vigreux column (0.4 mm.). Collection of the fractions boiling below 104° provides 191–196 g. (91–93%) of methyl 1-cycloundecenecarboxylate (Note 7) as a pale yellow oil, most of which distils at 83–87° (0.4 mm.). This ester is sufficiently pure for use in the next step.

C. *Cycloundecanone*. In a well-ventilated hood a 3-l., three-necked, round-bottomed flask is fitted with a magnetic stirrer (Note 1), a thermometer, a reflux condenser protected by a calcium chloride tube, and a rubber stopper. Concentrated sulfuric acid (600 ml.) is placed in the flask, stirred slowly, and cooled to 5° with an ice bath. Methyl 1-cycloundecenecarboxylate (191–196 g., 0.91–0.92 mole) is added through a long-stemmed funnel, the rate of stirring is increased until the mixture becomes a homogeneous solution, and 500 ml. of chloroform is added. The resulting mixture is heated to 35° by immersion in a warm water bath. Vigorous stirring is continued while 78 g. (1.2 moles) of sodium azide (Note 8) is added in small portions over a 30–50-minute period, the reaction temperature being maintained at 40 ± 2° by adding ice to the water bath. *Caution! This operation should be performed behind a safety shield.* After an additional 10–15 minutes of stirring at 35–40°, the reaction mixture is cooled to 5°, poured onto 1 kg. of ice, and transferred together with 1.5–2 l. of water to a 5-l., three-necked flask set up for steam distillation. The chloroform is distilled off and saved, and the cycloundecanone is steam distilled with 3.5–4.0 kg. of steam. The steam distillate is extracted with the recovered chloroform, then with 500 ml. of ether. After the ethereal extract has been washed with concentrated aqueous sodium chloride, the organic phases are filtered through anhydrous sodium sulfate, combined, and concentrated under reduced pressure. Vacuum distillation of the residual oil affords 139–143 g. (83–85%) of cycloundecanone as a colorless or pale yellow oil, b.p. 84–85° (2 mm.), n_D^{25} 1.4794–1.4796 (Note 8).

2. Notes

1. A Thomas Magne-Matic Stirrer Model 15 (available from the Arthur H. Thomas Company, Philadelphia, Pennsylvania, was used in conjunction with a 5-cm., Teflon-coated, egg-shaped magnet.

2. Cyclododecanone was obtained from the Aldrich Chemical Company, Inc., and used without purification.

3. Solvent grade benzene was dried over sodium wire prior to use. If a voluminous sludge forms on drying, the solvent should be distilled from sodium.

4. Various brands and grades of bromine were used without noticeable difference (however, *cf.* reference 6). Bromine was added until a light orange color persisted for more than 2 minutes, which occasionally required the addition of a few drops in excess of the theoretical amount.

5. At this stage, the addition of a small amount of anhydrous potassium hydrogen carbonate powder to the residual reaction mixture should not cause evolution of carbon dioxide gas.

6. The submitters used the reagent as supplied by Mallinckrodt Chemical Works. The checkers have occasionally found that commercially available sodium methoxide has deteriorated on storage over an extended period, unless the reagent has been properly protected from moisture. Therefore, it was prepared in the following manner. A 2-l., three-necked, round-bottomed flask was equipped with a magnetic stirrer and a reflux condenser and flushed with dry nitrogen through a gas bubbler attached to the top of the condenser. Into this flask was distilled approximately 600 ml. of absolute methanol (dried with magnesium methoxide), and then 69 g. of sodium was added in 1–3-g. portions. After all of the sodium had dissolved, the methanol was distilled, first at atmospheric and then at reduced pressure. The resulting mass of sodium methoxide was powdered under nitrogen and dried under vacuum at 150° for 8 hours. Titration of the reagent against 0.1311 N hydrochloric acid showed it to be 97.5% pure.

7. GC analysis of the product showed two major peaks (relative intensity, 5 : 1), and the mass spectrum of each peak revealed a molecular ion at m/e 210. The ^{1}H NMR spectrum of the mixture showed that the two products were geometrically isomeric esters.

8. Practical grade sodium azide was obtained from Eastman Organic Chemicals, and lumps were broken up with a spatula. Care was taken to avoid contact of sodium azide with the skin.

9. The cycloundecanone solidified on cooling and melted at 16.2–16.6°. GC analysis of the product showed a single peak [1.5 m. by 3.2 mm. column, 5% SE-30 on Chromosorb W, at 125° (submitters); 1.5 m. by 3.2 mm. column, UC-W98, 150°, retention time 10.25 minutes (checkers)]. ^{1}H NMR (CDCl$_3$), δ (multiplicity, number of protons): 1.3–2.1 (m, 16H), 2.4–2.7 (m, 4H); IR (neat) cm^{-1}: 1700, very strong.

3. Discussion

Cycloundecanone has been prepared in several ways: (a) pyrolysis of the thorium salt of dodecanedioic acid,[2] (b) reduction of 2-hydroxycycloundecanone,[3,4] (c) ring expansion of several lower homologs of cycloundecanone,[5] (d) Curtius degradation of 1-cycloundecenecarboxylic acid,[6] and (e) hydrolysis of 1-methoxycycloundecene.[7]

The present method, a modification of a procedure described previously by the submitters,[8] gives higher yields and is less expensive and more expeditious than the previously published methods. This route, involving a Favorskii-type rearrangement, has the potential of being widely applicable for the preparation of the lower homolog of a ketone having at least one hydrogen atom at each α-position. For example, cyclodecanone has been prepared in 77% yield from cycloundecanone by essentially the same procedure.[8]

1. Department of Chemistry, University of Minnesota, Minneapolis, Minnesota 55455. [Present address: Environmental Concern, P.O. Box P, St. Michaels, Maryland 21663.]
2. L. Ruzicka, M. Stoll, and H. Schinz, *Helv. Chim. Acta*, **9**, 249 (1926).
3. K. Ziegler, H. Sauer, L. Bruns, H. Froitzheim-Kühlhorn, and J. Schneider, *Justus Liebigs Ann. Chem.*, **589**, 122 (1954).
4. R. W. Fawcett and J. O. Harris, *J. Chem. Soc.*, 2673 (1954).
5. E. Müller and M. Bauer, *Justus Liebigs Ann. Chem.*, **654**, 92 (1962).
6. K. Schank and B. Eistert, *Chem. Ber.*, **98**, 650 (1965).
7. Rhone-Poulenc S. A., Neth. Pat. 6,605,908 [*Chem. Abstr.*, **66**, 85538s (1967)].
8. E. W. Garbisch, Jr., and J. Wohllebe, *J. Org. Chem.*, **33**, 2157 (1968).

HYDROGENATION OF AROMATIC NUCLEI: 1-DECALOL

Submitted by A. I. MEYERS,[1] W. N. BEVERUNG, and R. GAULT[2]
Checked by P. FREIDENREICH and R. BRESLOW

1. Procedure

Caution! Benzene has been identified as a carcinogen; OSHA has issued emergency standards on its use. All procedures involving benzene should be carried out in a well-ventilated hood, and glove protection is required.

A 500 ml., Parr hydrogenation bottle is flushed with nitrogen, and 20.0 g. of 5% rhodium-on-alumina (Note 1) is weighed directly into the hydrogenation bottle. The catalyst is wet by cautiously adding 25 ml. of 95% ethanol, and a solution of 40.0 g. (0.278 mole) of 1-naphthol (Note 2) in 125 ml. of 95% ethanol is added to the bottle, along with 3 ml. of acetic acid. The mixture is shaken in a Parr apparatus (Note 3) at an initial pressure of 55–60 p.s.i. of hydrogen. The theoretical hydrogen absorption is reached in about 12 hours (Note 4). The catalyst is removed by suction filtration and washed twice with 50-ml. portions of ethanol (Note 5). The combined ethanol solutions are concentrated with a rotary evaporator, yielding a viscous residue (39–41 g.), which is dissolved in 150 ml. of benzene. The solution is washed with 75 ml. of 10% sodium hydroxide solution, then with 75 ml. of water, dried over magnesium sulfate for at least 3 hours, and concentrated with a rotary evaporator, giving 39–41 g. (94–97%) of a mixture[3] consisting of the geometrical isomers of 1-decalol. *cis,cis*-1-Decalol may be isolated as a crystalline solid from the mixture by the addition of 15–20 ml. of heptane, followed by cooling. The product is isolated by filtration and recrystallized from a minimum amount of n-heptane, yielding 13–14 g. (30–33%) of *cis,cis*-1-decalol, m.p.[4] 92–93°.

2. Notes

1. The catalyst is available from Engelhard Industries.

2. A purified grade of 1-naphthol should be used. Material available from Eastman Organic Chemicals, Aldrich Chemical Company, Inc., and Matheson, Coleman and Bell is satisfactory. Experiments with technical grade 1-naphthol have indicated that this material requires purification by sublimation in order to give satisfactory results.

3. It has been found that the rhodium catalyst is not nearly as sensitive to poisoning as platinum or palladium catalyst. The metal inlet tube to the reaction bottle was merely rinsed with acetone, followed by ethanol, and the rubber stopper was soaked in 30–40% sodium hydroxide solution overnight.

4. A variety of experiments have shown that for bicyclic aromatic nuclei the weight ratio of reactant to catalyst should be 2:1, whereas for monocyclic aromatic nuclei, the reactant to catalyst ratio should be 3:1. For the latter systems, hydrogen absorption is usually complete within 6–8 hours (see Discussion section).

5. The catalyst may be reused after washing thoroughly with ethanol and drying at 125° for 12–15 hours. The activity, however, is somewhat decreased. *Care should be exercised to never leave the catalyst exposed to air in the presence of a flammable solvent.*

3. Discussion

1-Naphthol has been reduced to 1-decalol using platinum,[4] Raney nickel,[5] and Raney copper.[6] The reactions catalyzed by nickel and copper required elevated temperatures and

TABLE I
HYDROGENATION OF AROMATIC NUCLEI[a]

Compound	g. Catalyst / g. Reactant	Product	Yield, %
2-Naphthol	0.50	2-Decalol[b]	88
2-Methylbenzofuran	0.33	*cis*-2-Methylhexahydro-benzofuran[c]	94
2,2-Dimethyl-2,3-dihydro-benzofuran	0.33	*cis*-2,2-Dimethylhexa-hydrobenzofuran[c]	91
3-Hydroxybenzoic acid	0.33	3-Hydroxycyclohexane-carboxylic acid[b]	81
4-Methoxyphenol	0.33	4-Methoxycyclohexanol[b]	88
Hydroquinone	0.33	1,4-Cyclohexanediol[b]	90
Resorcinol	0.33	1,3-Cyclohexanediol[b]	85

[a]From ref. 3.[7]
[b]Obtained as mixtures of geometric isomers.
[c]No detectable quantity of the *trans* isomer is obtained.

pressure. The present procedure allows the preparation of substantial quantities of 1-decalol under much more convenient conditions and shorter reaction times. Previous methods[4-6] require costly catalysts or high-pressure equipment and frequently result in a high degree of hydrogenolysis. The submitters have found that the present method is applicable to a wide variety of aromatic nuclei, some of which are listed in Table I.

1. Department of Chemistry, Louisiana State University in New Orleans, New Orleans, Louisiana 70122. [Present address: Department of Chemistry, Colorado State University, Fort Collins, Colorado 80523].
2. Present address: Wayne State University, Detroit, Michigan 48202.
3. A. I. Meyers, W. Beverung, and G. Garcia-Munoz, *J. Org. Chem.*, **29**, 3427 (1964). The discrepancy between the work reported earlier and the present work regarding isomer distribution may be due to variations in catalyst activity. The present reduction mixture consists of four decalol isomers of which the *cis-cis* product represents 50–55% as determined by gas chromatography analysis on a 250-cm. column containing 10% Carbowax 20M on Chromosorb P at 150–200°.
4. W. G. Dauben, R. C. Tweit, and C. Mannerskantz, *J. Am. Chem. Soc.*, **76**, 4424 (1954); C. D. Gutsche and H. H. Peter, *J. Am. Chem. Soc.*, **77**, 5974 (1955); H. E. Zimmerman and A. Mais, *J. Am. Chem. Soc.*, **81**, 3648 (1959).
5. D. M. Musser and H. Adkins, *J. Am. Chem. Soc.*, **60**, 665 (1938).
6. J. Jadot and R. Braine, *Bull. Soc. Roy. Sci. Liege*, **25**, 62 (1956) [*Chem. Abstr.*, **50**, 16651h (1956)].
7. Other examples may be found in: J. H. Stocker, *J. Org. Chem.*, **27**, 2288 (1962); M. Freifelder, R. M. Robinson, and G. R. Stone, *J. Org. Chem.*, **27**, 284 (1962); J. C. Sircar and A. I. Meyers, *J. Org. Chem.*, **30**, 3206 (1965); R. A. Finnegan and P. L. Bachman, *J. Org. Chem.*, **30**, 4145 (1965).

OXIDATION WITH THE CHROMIUM TRIOXIDE–PYRIDINE COMPLEX PREPARED *in situ:*
1-DECANAL

$$CH_3(CH_2)_8CH_2OH \xrightarrow[\text{dichloromethane, } 20°]{CrO_3 \cdot (\text{pyridine})_2} CH_3(CH_2)_8CHO$$

Submitted by R. W. RATCLIFFE[1]
Checked by ROBERT J. NEWLAND and CARL R. JOHNSON

1. Procedure

A 3-l., three-necked, round-bottomed flask equipped with a stirrer, a thermometer, and a drying tube is charged with 94.9 g. (1.20 moles) of pyridine (Note 1) and 1.5 l. of dichloromethane (Note 2). The solution is stirred with ice-bath cooling to an internal temperature of 5°, and 60.0 g. (0.600 mole) of chromium trioxide (Note 3) is added in one portion. The deep burgundy solution is stirred in the cold for an additional 5 minutes, then allowed to warm to 20° over a period of 60 minutes. A solution of 15.8 g. (0.100 mole) of 1-decanol (Note 4) in 100 ml. of dichloromethane is added rapidly, with immediate separation of a tarry, black deposit. The reaction mixture is stirred for 15 minutes and decanted from the tarry residue, which is washed with three 500-ml. portions of diethyl ether. The combined organic solution is washed successively with three 1-l. portions of ice-cold, aqueous 5% sodium hydroxide, 1 l. of ice-cold, 5% hydrochloric acid, 1 l. of aqueous 5% sodium hydrogen carbonate, and 1 l. of saturated brine. The solution is dried over anhydrous magnesium sulfate, filtered, and evaporated under reduced pressure. The resulting pale yellow liquid is distilled through a 15-cm., vacuum-jacketed Vigreux column (Note 5), yielding 9.8–10.2 g. (63–66%, Note 6) of 1-decanal, b.p. 96–98° (13 mm.) (Note 7).

2. Notes

1. Anhydrous pyridine was obtained by distillation of reagent grade material from barium oxide and storage over 4A molecular sieves.

2. Dichloromethane was purified by shaking with concentrated sulfuric acid, washing with aqueous sodium hydrogen carbonate and water, drying over anhydrous calcium chloride, and distilling. The purified solvent was stored in the dark over 4A molecular sieves.

3. Chromium trioxide (obtained from J. T. Baker Chemical Company) was stored in a vacuum desiccator over phosphorus pentoxide prior to use. Six-mole equivalents of oxidant is required for rapid, complete conversion to aldehyde. With less than the 6:1 molar ratio, a second, extremely slow oxidation step occurs (see reference 7).

4. 1-Decanol was obtained from Aldrich Chemical Company, Inc.

5. Vigorous magnetic stirring of the pot material prevents excessive foaming during the distillation.

6. The submitters obtained 12.9–13.0 g. (83%). The checkers obtained a yield of 66% when all solvent and wash volumes used in the procedure were reduced by 50%.

7. The product was identified through comparison of its IR, ^{1}H NMR, and mass spectra and GC mobility with authentic 1-decanal, available from Aldrich Chemical Company, Inc.

3. Discussion

Dipyridine–chromium(VI) oxide[2] was introduced as an oxidant for the conversion of acid-sensitive alcohols to carbonyl compounds by Poos, Arth, Beyler, and Sarett.[3] The complex, dispersed in pyridine, smoothly converts secondary alcohols to ketones, but oxidations of primary alcohols to aldehydes are capricious.[4] In 1968, Collins, Hess, and Frank found that anhydrous dipyridine–chromium(VI) oxide is moderately soluble in chlorinated hydrocarbons and chose dichloromethane as the solvent.[5] With this modification, primary and secondary alcohols were oxidized to aldehydes and ketones in yields of 87–98%. Subsequently, Dauben, Lorber, and Fullerton showed that dichloromethane solutions of the complex are also useful for accomplishing allylic oxidations.[6]

The chief drawbacks to using the Collins reagent are the nuisance involved in preparing pure dipyridine–chromium(VI) oxide,[6] its hygroscopic nature,[5] and its propensity to inflame during preparation.[2,3,6] The present method avoids these difficulties by simply preparing dichloromethane solutions of the complex directly.[7] In addition, as noted previously,[5] the use of dichloromethane as solvent facilitates isolation of the products. Other modifications of the Collins procedure include the use of a celite-supported reagent which facilitates the isolation of sensitive products[8] and the finding that a 1:1 pyridine-chromium trioxide ratio gives oxidation results[9] comparable with those obtained using the customary 2:1 ratio in the *in situ* procedure.

Several other reagents finding utility for the oxidation of primary and secondary alcohols to the corresponding carbonyl compounds under mild and nonaqueous conditions have been developed in recent years. Of the chromium(VI)-based oxidants, pyridinium chlorochromate[10] in dichloromethane and pyridinium dichromate[11] in *N,N*-dimethylformamide or dichloromethane deserve special mention. Both reagents are easily prepared and stored, allow the efficient oxidation of a variety of alcohols using only a modest excess of oxidant, and may be amenable to large-scale operations.

1. Merck Sharp and Dohme Research Laboratories, Division of Merck and Company, Inc., Rahway, New Jersey 07065.
2. H. H. Sisler, J. D. Bush, and O. E. Accountius, *J. Am. Chem. Soc.*, **70**, 3827 (1948).
3. G. I. Poos, G. E. Arth, R. E. Beyler, and L. H. Sarett, *J. Am. Chem. Soc.*, **75**, 422 (1953).
4. J. R. Holum, *J. Org. Chem.*, **26**, 4814 (1961).
5. J. C. Collins, W. W. Hess, and F. J. Frank, *Tetrahedron Lett.*, 3363 (1968); J. C. Collins and W. W. Hess, *Org. Synth.*, **Coll. Vol. 6**, 644 (1988).
6. W. G. Dauben, M. Lorber, and D. S. Fullerton, *J. Org. Chem.*, **34**, 3587 (1969).
7. R. Ratcliffe and R. Rodehorst, *J. Org. Chem.*, **35**, 4000 (1970).
8. N. H. Andersen and H. Uh, *Synth. Commun.*, **3**, 115 (1973).
9. E. Piers and P. M. Worster, *Can. J. Chem.*, **55**, 733 (1977).
10. E. J. Corey and J. W. Suggs, *Tetrahedron Lett.*, 2647 (1975).
11. E. J. Corey and G. Schmidt, *Tetrahedron Lett.*, 399 (1979).

REDUCTION OF ALKYL HALIDES AND TOSYLATES WITH SODIUM CYANOBOROHYDRIDE IN HEXAMETHYLPHOSPHORIC TRIAMIDE (HMPA):
A. 1-IODODECANE TO n-DECANE
B. 1-DODECYL TOSYLATE TO n-DODECANE

$$CH_3(CH_2)_8CH_2I \xrightarrow[\text{HMPA}]{\text{NaBH}_3\text{CN}} CH_3(CH_2)_8CH_3$$

$$CH_3(CH_2)_{10}CH_2OTs \xrightarrow[\text{HMPA}]{\text{NaBH}_3\text{CN}} CH_3(CH_2)_{10}CH_3$$

Submitted by ROBERT O. HUTCHINS,[1] CYNTHIA A. MILEWSKI,
and BRUCE E. MARYANOFF
Checked by RONALD I. TRUST and ROBERT E. IRELAND

1. Procedure

Caution! Hexamethylphosphoric triamide (HMPA) vapors have been reported to cause cancer in rats.[2] All operations with hexamethylphosphoric triamide should be performed in a good hood, and care should be taken to keep the liquid off the skin.

A. n-*Decane.* A dry, 100-ml., three-necked flask equipped with a stirring bar, a thermometer, and a condenser protected with a drying tube is charged with 25 ml. of hexamethylphosphoric triamide (HMPA) (Note 1), 1-iododecane (2.7 g., 0.010 mole) (Note 2) and sodium cyanoborohydride (0.943 g., 0.0157 mole) (Note 3). The solution is stirred at 70° for 2 hours, diluted with 25 ml. of water, and extracted with three 30-ml. portions of diethyl ether. The combined extracts are washed twice with water and dried over anhydrous magnesium sulfate. The solvent is removed by distillation on a steam bath through a 12-in. vacuum-jacketed Vigreux column (Note 4). The residue is distilled at reduced pressure in a short-path apparatus *(Caution! foaming)*, yielding 1.25–1.29 g. (88–90%) (Notes 5, 6) of n-decane, b.p. 68–70° (14 mm.); n_D^{20} 1.4122, n_D^{26} 1.4085 (lit.,[3] n_D^{25} 1.4097 (Note 7).

B. n-*Dodecane.* A dry, 200-ml., three-necked flask equipped exactly as described in Section A is charged with 50 ml. of hexamethylphosphoric triamide (HMPA), 1-dodecyl tosylate (6.80 g., 0.0201 mole) (Note 8), and sodium cyanoborohydride (5.02 g., 0.0797 mole) (Note 3). The solution is stirred at 80° for 12 hours (Note 9), diluted with 50 ml. of water, and extracted with three 60-ml. portions of hexane. The hexane solution is washed twice with water, dried over anhydrous magnesium sulfate, and concentrated with a rotary evaporator. Distillation of the residue through a short-path apparatus (Note 5) *(Caution! foaming)* affords 2.49–2.64 g. (73–78%) of n-dodecane, b.p. 79–81°; (3.75 mm.) n_D^{24} 1.4217 (lit.,[4] n_D^{20} 1.4219) (Note 7).

2. Notes

1. Commercial hexamethylphosphoric triamide was distilled from calcium hydride and stored over 13X molecular sieves (Linde).

2. Commercial 1-iododecane (Eastman Organic Chemicals) was filtered through activated charcoal and distilled before use.

3. Sodium cyanoborohydride was used as received from Aldrich Chemical Company, Inc., or Alfa Products, Thiokol/Ventron Division. No purification is necessary. Tan or brown material may be purified by the method of Purcell[5] or of Borch.[5]

4. If a hexane workup is used, and the solvent is removed with a rotary evaporator, considerable loss of product results from codistillation with the hexane. This should not present a significant problem when higher boiling materials are produced.

5. The condenser was cooled with an ethylene glycol–water mixture at $-5°$, and the receiver was cooled to $-10°$ in an ice–salt bath.

6. Considerable mechanical loss was observed because of inability to distill the last portions of the product at 14 mm. To avoid this problem, the pressure was reduced near the end of the distillation to *ca.* 5 mm. This did not affect the purity of the product (Note 7).

7. Both products showed IR and ^1H NMR spectra identical to those of authentic samples, and no side products were detected by GC or ^1H NMR.

8. Dodecyl tosylate was prepared from 1-dodecanol by the procedure in *Org. Synth., Coll. Vol. 3*, 366 (1955). Crystallization from a dried (magnesium sulfate) solution in light petroleum ether afforded white needles, m.p. 27.5–28.5°.

9. The large excess of sodium cyanoborohydride is recommended for the reduction of tosylates. Use of reduced molar excesses led to substantially lower yields. For example, a 3:1 cyanoborohydride to tosylate ratio afforded less than 60% yield of product at 80° for 5 hours, while a 1.5:1 excess gave only 52% yield at 70° for 8 hours.

3. Discussion

These preparations illustrate the use of sodium cyanoborohydride in hexamethylphosphoric triamide as an effective, selective, and convenient procedure for the reduction of alkyl halides and tosylates, and are essentially the same as previously described.[6] The very mild reducing ability of sodium cyanoborohydride makes the method particularly valuable when other functional groups are present in the molecule

$$\left(\text{i.e.,} \quad CO_2H, CO_2R, CN, NO_2, -\overset{\overset{\displaystyle O}{\|}}{C}N\overset{<}{\ }, \ \ \overset{>}{\ }C=O, -\overset{|}{C}\overset{O}{\diagup\diagdown}\overset{|}{C}-\right)^6$$

In addition, alkene side-products are seldom encountered, contrary to the situation with lithium aluminum hydride[7] or sodium borohydride in aqueous diglyme.[8] The combination of sodium borohydride in polar aprotic solvents is also effective for halide and tosylate removal,[9] although it is less selective.

1. Department of Chemistry, Drexel University, Philadelphia, Pennsylvania 19104.
2. J. A. Zapp, Jr., *Science*, **190**, 422 (1975).
3. A. F. Forziati, A. R. Glasgow, Jr., C. B. Willingham, and F. D. Rossini, *J. Res. Natl. Bur. Stand.*, **36**, 129 (1946) [*Chem. Abstr.*, **40**, 4341[9] (1946)].
4. A. F. Shepard, A. L. Henne, and T. Midgley, Jr., *J. Am. Chem. Soc.*, **53**, 1948 (1931).
5. R. C. Wade, E. A. Sullivan, J. R. Bershied, Jr., and K. F. Purcell, *Inorg. Chem.*, **9**, 2146 (1970); R. F. Borch, M. D. Bernstein, and H. D. Durst, *J. Am. Chem. Soc.*, **93**, 2897 (1971).
6. R. O. Hutchins, D. Kandasamy, C. A. Maryanoff, D. Masilamani, and B. E. Maryanoff, *J. Org. Chem.*, **42**, 82 (1977).
7. N. G. Gaylord, "Reductions with Complex Metal Hydrides," Interscience, New York, 1956.
8. H. M. Bell and H. C. Brown, *J. Am. Chem. Soc.*, **88**, 1473 (1966).
9. R. O. Hutchins, D. Kandasamy, F. Dux III, C. A. Maryanoff, D. Rotstein, B. Goldsmith, W. Burgoyne, F. Cistone, J. Dalessandro, and J. Puglis, *J. Org. Chem.*, **43**, 2259 (1978).

DIAMANTANE: PENTACYCLO[7.3.1.1^{4,12}.0^{2,7}.0^{6,11}]TETRADECANE

(Butanetetraylnaphthalene, 3,5,1,7-[1,2,3,4]-decahydro-)

Submitted by Tamara M. Gund, Wilfried Thielecke, and Paul v. R. Schleyer[1]
Checked by H. Gurien, R. Regenye, and A. Brossi

1. Procedure

A. *Binor-S.*[2] A 2-1., three-necked flask equipped with Teflon sleeves (Note 1), a thermometer, a condenser, a dropping funnel, and a mechanical stirrer is flushed with nitrogen and charged with 200 g. (2.18 moles) of freshly distilled norbornadiene (Note 2), 400 ml. of dry toluene, and 7.8 g. of cobalt bromide–triphenylphosphine catalyst (Note 3). While stirring at room temperature, 2.1 ml. of boron trifluoride diethyl etherate co-catalyst (Note 4) is added dropwise. The mixture is heated slowly to 105°, and the heating mantle is lowered. The ensuing exothermic reaction maintains the temperature at 105–110° for 15 minutes. When the temperature begins to fall, the mantle is raised, the mixture is brought to the reflux temperature, and stirring and refluxing are continued for 12 hours. The cooled mixture is diluted with 650 ml. of dichloromethane, transferred to a separatory funnel, and washed with three 650-ml. portions of water. The organic phase is dried over anhydrous magnesium sulfate, and the solvents are evaporated at reduced pressure. The residual crude material, 185–203 g., is distilled at 106–107° (1.5 mm.), giving 165–170 g. (82–85%) of Binor-S, which solidifies on cooling to a white solid, m.p. 59–63°.

B. *Tetrahydro-Binor-S.*   Binor-S (135.0 g., 0.734 mole) is dissolved in 670 ml. of glacial acetic acid containing 5.7 ml. of concentrated hydrochloric acid. To this solution is added 1.0 g. of platinum oxide catalyst. The reaction mixture is hydrogenated at 200 p.s.i. hydrogen pressure and 70° for 3 hours, using a 1200-ml., glass-lined autoclave (Note 5). After cooling to room temperature, the catalyst is removed with suction filtration, and water (*ca.* 1.5 l.) is added to the filtrate until two layers form. The bottom layer, containing only tetrahydro-Binor-S, is removed, and the top layer, consisting of a mixture of acetic acid and water, is extracted with one 400 ml. and two 100-ml. portions of dichloromethane. The combined dichloromethane–tetrahydro-Binor-S layers are washed twice with 100 ml. of water, dried over anhydrous magnesium sulfate and concentrated under reduced pressure. The residual tetrahydro-Binor-S is purified by distillation under reduced pressure, b.p. 105–110° (1.5 mm.), giving 125–130 g. (90–94%) of colorless liquid (Note 6).

C. *Diamantane.*   A 500-ml., three-necked flask, equipped with a reflux condenser, a drying tube, a magnetic stirring bar, and a dropping funnel is charged with 28 g. (0.11 mole) of fresh aluminum bromide and 100 ml. of cyclohexane (Note 7). The apparatus is flushed with hydrogen bromide gas (Note 8). When the aluminum bromide has dissolved, 100 g. (0.532 mole) of hydrogenated Binor-S is added dropwise to the rapidly stirred solution, and the reaction mixture refluxes for a short time without external heat. The course of the reaction is monitored by GC until no starting material remains (Note 9). Occasionally, an additional 5 g. portion of aluminum bromide, and application of external heat are needed to complete the reaction. The total reaction time is about 2–3 hours (Note 10). The hot cyclohexane layer is carefully decanted, and the aluminum bromide layer is extracted with five 200-ml. portions of hot cyclohexane. Diethyl ether (400 ml.) is added to the cooled cyclohexane extracts (Note 11), and the combined solvent fractions are washed with two 100-ml. portions of water and dried over anhydrous magnesium sulfate. Evaporation of the solvent leaves a semi-solid residue, which is partially dissolved in about 100 ml. of pentane. The undissolved white solid, diamantane, is collected by suction filtration. Additional diamantane is obtained by concentrating the pentane solution to a small volume and collecting the solid that precipitates. The total amount of diamantane obtained, after drying, is 60–62 g. (60–62%), m.p. 240–241° (closed tube) (Note 12). This product is sufficiently pure for most purposes, but it may be purified further by recrystallization from pentane, giving white crystals, m.p. 244.0–245.4°.

2. Notes

1. Teflon sleeves were used to keep the joints from freezing.

2. Once distilled, norbornadiene may be stored below 0°. Samples as old as 2 weeks were used successfully.

3. This catalyst[3] is prepared in quantitative yield by refluxing by 200 ml. benzene solution containing 10 g. (0.046 mole) of anhydrous cobalt dibromide and 24.4 g. (0.0931 mole) of triphenylphosphine. A color change is observed, and the blue-green solid that precipitates on cooling to room temperature is filtered and dried. This catalyst, stored in a dry atmosphere, appears to be active indefinitely.

4. Boron trifluoride etherate may be used without prior distillation only if fresh material is available. Care must be taken with this reagent because of fuming. The dimerization does not proceed without this co-catalyst.

5. At 70° Binor-S remains in solution, and the uptake of hydrogen is rapid. The checkers have observed that occasionally hydrogen uptake is incomplete, and an additional 1 g. of catalyst must be added to complete the absorption of hydrogen. The submitters carried out the hydrogenation in a large-scale Parr apparatus under three atmospheres of pressure with similar results.

6. After a small solvent-containing forefraction, which is discarded, essentially all of the material should distill in the indicated range, but occasionally material boiling as high as 130° (1.5 mm.) is obtained. This is included in the product.

7. Either carbon disulfide or cyclohexane may be used with comparable yields. The advantage of carbon disulfide is the greater solubility of diamantane. When extracting with cyclohexane, a boiling solution must be used to increase solubility. However, cyclohexane is less poisonous, does not have a foul odor, and gives a whiter product. Therefore, the use of carbon disulfide was not examined by the checkers. Use of dichloromethane has been reported[2] to lead to an 82% yield.

8. Flushing the apparatus with hydrogen bromide may not be necessary, especially for small-scale runs.

9. A Carbowax 20M or 1500 GC column at a temperature of 180° may be used. Diamantane has a shorter retention time than tetrahydro-Binor-S. The checkers used a 10% OV101 GCQ column, 100/120, at 200°; retention times are Binor-S, 11.5 minutes; tetrahydro-Binor-S, 7.6 minutes; diamantane, 6.2 minutes.

10. About 30 minutes after the addition of tetrahydro-Binor-S is complete, the reaction mixture begins to cool and external heat must be supplied to continue the refluxing and complete the reaction. If GC monitoring reveals that the rearrangement is proceeding slowly, an additional 0.5 g. of aluminum bromide is added and refluxing is continued until all the starting material is converted to product.

11. Addition of ether prevents crystallization of the diamantane from cooled cyclohexane.

12. The material obtained is pure by GC and ^{1}H NMR. The ^{1}H NMR spectrum of diamantane shows only a singlet at δ 1.68 (CDCl$_3$).[4] The pentane mother liquors contain a by-product. A comparable yield can be obtained using aluminum chloride in boiling dichloromethane, and the crude mixture at the end of the reaction need not, in some instances, be worked up before subsequent functionalization reactions are carried out.[2,5]

3. Discussion

Like adamantane,[6] diamantane (also known as congressane[4]), the second member of the diamond family, may also be prepared by aluminum halide-catalyzed isomerization. A variety of starting materials have been shown to give diamantane,[2,4,7,8] but the very best results are obtained by the present procedure[2,8] starting with Binor-S.[2,3,9] Diamantane may be converted to a variety of functionalized derivatives.[2,5,10,11] 4-Methyldiamantane may also be prepared by rearrangement.[10] The mechanisms of these transformations have been analyzed.[12]

1. Department of Chemistry, Princeton University, Princeton, New Jersey 08540. This work was supported by Grant GM-19134, National Institutes of Health, and by Hoffmann-La Roche Inc., Nutley, New Jersey 07110. [Present address: Institute für Organische Chemie, Universität Erlangen-Nürnberg, 8520 Erlangen, Federal Republic of Germany.]

2. T. Courtney, D. E. Johnston, M. A. McKervey, and J. J. Rooney, *J. Chem. Soc., Perkin Trans. I*, 2691 (1972).

3. G. N. Schrauzer, R. K. Y. Ho, and G. Schlesinger, *Tetrahedron Lett.*, 543 (1970).

4. C. Cupas, P. v. R. Schleyer, and D. J. Trecker, *J. Am. Chem. Soc.*, **87**, 917 (1965).

5. D. Faulkner, R. A. Glendinning, D. E. Johnston, and M. A. McKervey, *Tetrahedron Lett.*, 1671 (1971).

6. P. v. R. Schleyer, M. M. Donaldson, R. D. Nicholas, and C. Cupas, *Org. Synth.*, **Coll. Vol. 5**, 16 (1973).

7. V. Z. Williams, Jr., P. v. R. Schleyer, G. J. Gleicher, and L. B. Rodewald, *J. Am. Chem. Soc.*, **88**, 3862 (1966).

8. T. M. Gund, V. Z. Williams, Jr., E. Osawa, and P. v. R. Schleyer, *Tetrahedron Lett.*, 3877 (1970); T. M. Gund, E. Osawa, V. Z. Williams, Jr., and P. v. R. Schleyer, *J. Org. Chem.*, **39**, 2979 (1974).

9. G. N. Schrauzer, B. N. Bastian, and G. A. Fosselius, *J. Am. Chem. Soc.*, **88**, 4890 (1966).

10. T. M. Gund, M. Nomura, V. Z. Williams, Jr., and P. v. R. Schleyer, *Tetrahedron Lett.*, 4875 (1970); R. Hamilton, D. E. Johnston, M. A. McKervey, and J. J. Rooney, *J. Chem. Soc. Chem.*, 1209 (1972); T. M. Gund, M. Nomura, and P. v. R. Schleyer, *J. Org. Chem.*, **39**, 2987 (1974).

11. T. M. Gund and P. v. R. Schleyer, *Tetrahedron Lett.*, 1583 (1971). T. M. Gund, P. v. R. Schleyer, G. D. Unruh, and G. J. Gleicher, *J. Org. Chem.*, **39**, 2995 (1974); F. Blaney, D. E. Johnston, M. A. McKervey, and J. J. Rooney, *Tetrahedron Lett.*, 99 (1975).

12. T. M. Gund, P. v. R. Schleyer, P. H. Gund, and W. T. Wipke, *J. Am. Chem. Soc.*, **97**, 743 (1975).

MACROCYCLIC DIIMINES: 1,10-DIAZACYCLOÖCTADECANE

A. $\text{Cl}\overset{O}{\underset{\|}{C}}(\text{CH}_2)_6\overset{O}{\underset{\|}{C}}\text{Cl} + 2\ \text{H}_2\text{N}(\text{CH}_2)_8\text{NH}_2 \xrightarrow[25°]{\text{C}_6\text{H}_6}$

$$\begin{array}{c} \overset{O}{\underset{\|}{C}}(\text{CH}_2)_6\overset{O}{\underset{\|}{C}} \\ \diagup \qquad \diagdown \\ \text{NH} \qquad\qquad \text{NH} \\ \diagdown \qquad \diagup \\ (\text{CH}_2)_8 \end{array}$$

$$+\ \text{H}_2\text{N}(\text{CH}_2)_8\text{NH·2 HCl}$$

B. $$\begin{array}{c} \overset{O}{\underset{\|}{C}}(\text{CH}_2)_6\overset{O}{\underset{\|}{C}} \\ \diagup \qquad \diagdown \\ \text{NH} \qquad\qquad \text{NH} \\ \diagdown \qquad \diagup \\ (\text{CH}_2)_8 \end{array} \xrightarrow[\text{tetrahydrofuran, reflux}]{\text{LiAlH}_4}$$

$$\begin{array}{c} (\text{CH}_2)_8 \\ \diagup \qquad \diagdown \\ \text{NH} \qquad\qquad \text{NH} \\ \diagdown \qquad \diagup \\ (\text{CH}_2)_8 \end{array}$$

Submitted by CHUNG HO PARK and HOWARD E. SIMMONS[1]
Checked by STEPHEN R. WILSON and ROBERT E. IRELAND

1. Procedure

Caution! Benzene has been identified as a carcinogen; OSHA has issued emergency standards on its use. All procedures involving benzene should be carried out in a well-ventilated hood, and glove protection is required.

A. 1,10-*Diazacycloöctadecane-2,9-dione.* A 12-l., 4-necked, round-bottomed flask with four indents is fitted with a mechanical stirrer, two dropping funnels (Note 1), and an inlet tube to maintain a static nitrogen atmosphere throughout the reaction. Four liters of benzene (Note 2) are placed in the flask and stirred vigorously. Two solutions, one of 33.8 g. (0.160 mole) of suberyl dichloride (Note 3) in 2.0 l. of benzene and the other of 48.2 g. (0.335 mole) of 1,8-diaminoöctane (Note 4) in 2.0 l. of benzene, are added simultaneously over a 6–7 hour period at room temperature. After the addition is complete, the resulting suspension is stirred slowly overnight. The addition funnels are removed from the flask and replaced by stoppers fitted with tubes of suitable dimensions to permit the reaction mixture to be siphoned from the reaction flask when a slight positive nitrogen pressure is present in the flask. The fine suspension in the reaction flask is agitated and siphoned into a large, fritted, filter funnel. The white solid is washed three times with benzene and dried in a vacuum oven. The resulting white solid is pulverized, placed in a continuous extractor (Note 5), and extracted for three days with 1 l. of boiling benzene in a 2-l., round-bottomed flask equipped with a magnetic stirring bar and heating mantle. After the extractor is allowed to cool to room temperature, filtration of the white solid suspension gives 23.1–23.6 g. (51–52%) of crude 1,10-diazacycloöctadecane-2,9-dione, m.p. 198–201°, which is used in the subsequent step without further purification (Note 6).

B.   1,10-*Diazacyclooctadecane*.   A 500-ml., round-bottomed flask equipped with a
mechanical stirrer, condenser, and nitrogen bubbler is charged with 150 ml. of dry
tetrahydrofuran (Note 7) and 3.8 g. (0.10 mole; 33% excess) of lithium aluminum
hydride. While the suspension is stirred, 14.1 g. (0.0500 mole) of 1,10-diazacycloocta-
decane-2,9-dione is added in small portions through a Gooch tube (Note 8). When the
addition is complete and evolution of hydrogen subsides, the mixture is heated at reflux
under a nitrogen atmosphere for 48 hours. The mixture is cooled to 5–10° with an ice bath
and decomposed by cautious, dropwise addition of 3.8 ml. of water, followed by 3.8 ml.
of 15% sodium hydroxide, and finally by 11.8 ml. of water. The mixture is allowed to
come to room temperature, stirred for an additional hour, and filtered through a fritted-
glass funnel. The resulting cake is washed thoroughly with three 50-ml. portions of
tetrahydrofuran, followed by three 50-ml. portions of diethyl ether. The combined filtrate
is concentrated under reduced pressure. After purging with benzene to remove traces of
water and storing under vacuum overnight, 12.7 g. of crude diimine remains as a
colorless, waxy solid.

The crude product contains a small amount of unreduced dilactam, which can be
detected in the IR spectrum. Treatment of the crude diimine again under the identical
reduction conditions (Note 9) using the same amounts of the reagent and solvent gives
10.6 g. (83%) of product. Distillation under reduced pressure through a semimicro
Vigreux column (10-cm.) yields 9.5 g. (75%) of pure 1,10-diazacyclooctadecane, b.p.
120–130° (0.05 mm.), m.p. 59–61° (Note 10).

2. Notes

1. Two graduated, 500-ml., pressure-equalizing dropping funnels with small diameter tips to allow solutions to be added in fine streams were used. They were equipped
with screw-in type plungers with Teflon® fluorocarbon tips to allow fine adjustment of the
flow rates. The checkers utilized standard, 500-ml. dropping funnels and two needle valve
adapters, as shown in Figure 1.

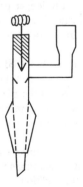

Figure 1.

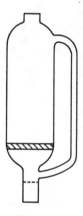

Figure 2.

It is important that the two solutions are added at the same slow rate. The submitters report that mixing of the two solutions in one portion halved the yield of the dilactam.

2. All of the benzene used was dried over sodium wire and distilled from sodium under a nitrogen atmosphere.

3. Suberyl dichloride purchased from Frinton Laboratories was used without further purification. The purity of the acid chloride was checked by GC after converting it to the diethyl ester.

4. 1,8-Diaminoöctane was purchased from Columbia Organic Chemicals Co., Inc. The purity was checked by GC and varied from batch to batch; only samples which were shown to be homogeneous were used. The diamine should be handled in an inert atmosphere, as it forms a carbonate rapidly.

5. A 2-l. extractor with a fritted disc with an 85 mm. diameter and coarse porosity was used (Figure 2).

The desired product is only slightly soluble in benzene. Depending on the efficiency of extraction, this step may be extended to longer periods with beneficial results. A lower yield of the dilactam generally indicates incomplete extraction. An attempt to isolate the product by washing away the dihydrochloride from the solid mixture with water resulted in the formation of an emulsion.

6. A sample of the crude dilactam dissolved in an ethanol–water mixture gave no precipitate with silver nitrate, indicating that the product is not contaminated with octamethylenediamine dihydrochloride. If the purity is questionable, the crude product can be conveniently purified at this point by recrystallization from N,N-dimethylformamide or 50% ethanol in water, giving better than 90% recovery of pure product, m.p. 207–209°.

7. Commercially available tetrahydrofuran was distilled from LiAlH$_4$ prior to use.

8. The submitters suggest that on ten times this scale, the reaction flask be purged with a stream of nitrogen during the addition of amide.

9. The repeated treatment with lithium aluminum hydride was necessary to obtain

complete reduction, since use of the reducing agent in manyfold excess and longer reaction times failed to accomplish complete reduction in one step. If the product was purified by distillation without the second reduction, the yield of pure diimine was 9.1 g. (71%). The submitters report that on ten times the scale the yield with one reduction was 81 g. (64%).

10. The product obtained is analytically pure, and the submitters claim it is suitable for polymerization. It is also shown to be homogeneous by GC.

3. Discussion

1,10-Diazacycloöctadecane has also been prepared by another general method which involves reaction of 1,8-dibromoöctane and N,N'-ditosyl-1,8-octanediamine with potassium carbonate as base, followed by cleavage of the sulfonamide groups;[2] the yield, however, is not as good as that by the present method. The present procedure is essentially a scale-up of the general method used by Stetter and Marx[3] for preparing a series of macrocyclic diimines of 10- to 21-membered rings, with a 25–78% yield for the crucial cyclization step. The procedure is adaptable for even larger scale preparations, using 5 moles of the acid chloride and a 100-gallon reaction kettle.

The reaction is generally applicable, to further extension, for making macrobicyclic diamines[4] with bridgehead nitrogen atoms by using monocyclic diimines and an appropriate acid chloride. The main difference in the procedures is that triethylamine is used as the hydrogen chloride acceptor in the cyclization step instead of using an extra mole of diimine. Triethylamine hydrochloride does not interfere with purification in this case since the bicyclic diamides are fairly soluble in benzene but insoluble in water. The bicyclic diamides are isolated by first removing triethylamine hydrochloride from the reaction mixture by filtration and water washings, followed by evaporation of the solvent. If triethylamine is used as the base in the preparation of monocyclic dilactams, the products are invariably contaminated with triethylamine hydrochloride, and the yields of the pure products are lower.

The same sequence of reactions are used to prepare macrobicyclic diamines with polyether linkages.[5,6] The reduction of macrobicyclic diamides can be accomplished with diborane using the procedure described by Brown and Heim,[7] but the present procedure is less involved and gives good yields.

$$\begin{array}{l}
\overset{\displaystyle \overbrace{}}{CH_2CH_2-O-CH_2CH_2-O-CH_2CH_2} \\
N-CH_2CH_2-O-CH_2CH_2-O-CH_2CH_2-N \\
\underset{\displaystyle \underbrace{}}{CH_2CH_2-O-CH_2CH_2-O-CH_2CH_2}
\end{array}$$

1. Central Research Department, E. I. Du Pont de Nemours and Co., Wilmington, Delaware 19898.
2. A. Müller and L. Kindlman, *Ber. Dtsch. Chem. Ges.*, **74B**, 416 (1941).
3. H. Stetter and J. Marx, *Justus Liebigs Ann. Chem.*, **607**, 59 (1957).
4. H. E. Simmons and C. H. Park, *J. Am. Chem. Soc.*, **90**, 2428 (1968).
5. H. E. Simmons and C. H. Park, unpublished work.
6. B. Dietrich, J. M. Lehn, and J. P. Sauvage, *Tetrahedron Lett.*, 2885 (1969).
7. H. C. Brown and P. Heim, *J. Am. Chem. Soc.*, **86**, 3566 (1964).

DIAZOACETOPHENONE

(Ethanone, 2-diazo-1-phenyl-)

$$C_6H_5COCl + CH_2N_2 + (C_2H_5)_3N \longrightarrow C_6H_5COCHN_2 + (C_2H_5)_3\overset{+}{N}H\overset{-}{C}l$$

Submitted by JOHN N. BRIDSON and JOHN HOOZ[1]
Checked by DENNIS R. MURAYAMA and RONALD BRESLOW

1. Procedure

Caution! All operations should be conducted in an efficient fume hood. Diazomethane is hazardous; directions for safe handling are given in Org. Synth., **Coll. Vol. 4,** 250 (1963); **Coll. Vol. 5,** 351 (1973). *Diazoacetophenone is a skin irritant and direct contact should be avoided.*

A solution of 0.375 mole of diazomethane in 1 l. of diethyl ether (Note 1) is placed in a 2-l. flask fitted with a large magnetic stirring bar, a two-necked adapter, equipped with a drying tube (containing potassium hydroxide pellets), and a pressure-equalizing dropping funnel. Triethylamine (37.9 g., 52.1 ml., 0.375 mole) (Note 2) is added, and the flask contents are cooled to *ca.* $-10°$ to $-5°$. A solution of 52.75 g. (43.56 ml., 0.3754 mole) of benzoyl chloride (Note 3) in 300 ml. of dry ether is added to the stirred mixture over a period of 0.5 hour (Note 4). An additional 50 ml. of ether is rinsed through the dropping funnel. Stirring is continued for one hour at approximately 0°, then overnight at room temperature.

The resulting triethylamine hydrochloride precipitate (41.4 g., 81%) is filtered and washed with 100 ml. of dry ether. The solvent is removed from the combined filtrate by rotary evaporation, and the semi-solid residue crystallizes to an orange-red solid after refrigeration for several hours at *ca.* 5°. Crystallization from a mixture of 150 ml. of pentane and 120 ml. of dry ether affords 38.8 g. of diazoacetophenone as yellow square plates, m.p. 44–48°. Concentration of the mother liquor and extraction of the residue with boiling pentane yields an additional 7.8 g. of pale yellow rods, m.p. 47.5–48.5°, bringing the total yield to 46.6 g. (85%) (Notes 5 and 6).

2. Notes

1. Diazomethane was prepared by the method in *Org. Synth:,* **Coll. Vol. 5,** 351 (1973), using 10% extra 2-(2-ethoxyethoxy)ethanol and an extra 100 ml. of water over that recommended, to prevent stirring difficulties in the later stages of the distillation. The ethereal diazomethane solution was dried at 0° over potassium hydroxide pellets, and the concentration was determined by reaction of an aliquot with benzoic acid and analyzing the resulting methyl benzoate by GC.

2. Triethylamine, purchased from J. T. Baker Chemical Company, was refluxed over calcium hydride, then fractionally distilled through a 40-cm. Vigreux column, b.p. 81–82° (700 mm.); b.p. 89.5–90° (760 mm.).

3. Benzoyl chloride, obtained from British Drug House (Canada) Ltd., was purified, as described in *Org. Synth., Coll.* **Vol. 3,** 112 (1955), by washing a benzene solution with 5% aqueous sodium hydrogen carbonate, drying over calcium chloride, and fractional distillation through a 40-cm. Vigreux column, b.p. 69–71° (12 mm.). The checkers used a fresh bottle, from Matheson, Coleman and Bell, without purification.

4. In the later stages of the addition a cake of crystals forms, preventing adequate stirring. This difficulty is overcome by temporarily interrupting the addition and swirling the flask manually—stirring then continues normally.

5. The submitters obtained a similar yield on twice the scale reported here.

6. Although crystallization from pentane gives better crystals, with an improved melting point range, recrystallization of the whole batch would require approximately 3 1. of solvent. Samples obtained from both ether-pentane and pentane evolve the theoretical amount of nitrogen on titration with 3 *N* hydrochloric acid.

3. Discussion

Apart from the reaction of diazomethane with benzoyl chloride,[2,3,4] diazoacetophenone has been prepared by the reaction of 2-aminoacetophenone hydrochloride with sodium nitrite,[5] from the mixed anhydride of benzoic acid and ethyl carbonate with diazomethane,[6] from benzoyl chloride and potassium methyldiazotate,[7] by treating the enamine formed from 2-formylacetophenone and *N*-methylaniline with *p*-toluenesulfonyl azide,[8] and from the reaction of the sodium enolate of 2-formylacetophenone with *p*-toluenesulfonyl azide.[9]

The reaction of an acid chloride with diazomethane illustrates a general method of preparing diazoketones. The acid chloride is slowly added to at least two equivalents of diazomethane; the hydrogen chloride liberated (Eq. 1) is then consumed according to Eq. 2. When the order of addition is reversed (*e.g.,* acid chloride is in excess) and only 1 mole of diazomethane is employed, the diazoketone reacts with hydrogen chloride, forming the α-chloroketone (Eq. 3).

$$\text{RCOCl} + \text{CH}_2\text{N}_2 \longrightarrow \text{RCOCHN}_2 + \text{HCl} \tag{1}$$

$$\text{CH}_2\text{N}_2 + \text{HCl} \longrightarrow \text{CH}_3\text{Cl} + \text{N}_2 \tag{2}$$

$$\text{RCOCHN}_2 + \text{HCl} \longrightarrow \text{RCOCH}_2\text{Cl} + \text{N}_2 \tag{3}$$

The method described here, discovered independently by Newman and Beal,[3] and Berenbom and Fones,[4] employs triethylamine (1 equivalent) to react with the hydrogen chloride; thus, only one equivalent of diazomethane is necessary. This modification was originally restricted to the use of either aromatic- or aliphatic acid chlorides lacking α-hydrogen atoms. Acid chlorides bearing α-hydrogens produce a mixture of products, presumably due to competing ketene formation and subsequent side reactions.

More recently it has been shown that, by operating at lower temperatures ($-78°\text{C}$), even simple aliphatic acid chlorides may also be successfully employed.[10] Some examples are the preparation of $\text{C}_6\text{H}_5\text{CH}_2\text{CH}_2\text{COCHN}_2$ (96%), $cyclo\text{-C}_6\text{H}_{11}\text{COCHN}_2$ (96%), $\text{CH}_3(\text{CH}_2)_7\text{COCHN}_2$ (96%), and $(\text{CH}_3)_2\text{CHCOCHN}_2$ (96%). However, this low temperature procedure is inapplicable to substrates with especially acidic α-hydrogens, such as phenylacetyl chloride, presumably due to competing ketene formation.

1. Department of Chemistry, University of Alberta, Edmonton, Alberta, Canada, T6G 2G2.
2. F. Arndt and J. Amende, *Ber.* Dtsch. Chem. Ges., **61**, 1122 (1928).
3. M. S. Newman and P. Beal III, *J. Am. Chem. Soc.*, **71**, 1506 (1949).
4. M. Berenbom and W. S. Fones, *J. Am. Chem. Soc.*, **71**, 1629 (1949).
5. H. E. Baumgarten and C. H. Anderson, *J. Am. Chem. Soc.*, **83**, 399 (1961).
6. D. S. Tarbell and J. A. Price, *J. Org. Chem.*, **22**, 245 (1957).
7. E. Mueller, W. Hoppe, H. Hagenmaier, H. Haiss, R. Huber, W. Rundel, and H. Suhr, *Chem. Ber.*, **96**, 1712 (1963).
8. R. Fusco, G. Bianchetti, D. Pocar, and R. Ugo, *Chem. Ber.*, **96**, 802 (1963).
9. M. Regitz, F. Menz, and J. Rueter, *Tetrahedron Lett.*, 739 (1967).
10. L. T. Scott and M. A. Minton, *J. Org. Chem.*, **42**, 3757 (1977).

2-DIAZOCYCLOALKANONES:
2-DIAZOCYCLOHEXANONE

(Cyclohexanone, 2-diazo-)

Submitted by Manfred Regitz,[1] Jörn Rüter,
and Annemarie Liedhegener
Checked by John D. Fenwick and Peter Yates

1. Procedure

Caution! 2-Diazocyclohexanone may explode, especially on being heated. The workup and distillation should be carried out in a fume hood behind a safety shield.

A 2-l., wide-necked Erlenmeyer flask is charged with 66.2 g. (0.525 mole) of 2-(hydroxymethylene)cyclohexanone (Note 1), 400 ml. of dichloromethane, and 106 g. (1.05 moles) of triethylamine (Note 2). The flask is cooled in an ice–salt bath at −12 to −15°, and 98.0 g. (0.500 mole) of *p*-toluenesulfonyl azide (Note 3) is added with vigorous mechanical stirring over a period of approximately 1 hour, at such a rate that the temperature of the reaction mixture does not rise above −5°. Stirring is continued for an additional 2 hours as the cooling bath melts. A solution of 30.8 g. (0.550 mole) of potassium hydroxide in 400 ml. of water is added, and the mixture is stirred for 15 minutes at room temperature. The resulting emulsion is placed in a 2-l. separatory funnel, the dichloromethane layer is separated after the emulsion has broken, and the aqueous, alcoholic layer is washed with two 100-ml. portions of dichloromethane. The combined dichloromethane solutions are washed with a solution of 2.8 g. of potassium hydroxide in 200 ml. of water, then with 200 ml. of water, and dried over anhydrous sodium sulfate (Note 4). The solvent is removed on a rotary evaporator at 35° (15 mm.) until the weight of the residue is constant, yielding 51.5–59.0 g. (83–95%) of yellow-orange 2-diazocyclohexanone (Note 5). Distillation with magnetic stirring (Note 6) of 20 g. of this crude product from a hot-water bath at 80° gave 17.0 g. of yellow-orange liquid, b.p. 46° (0.1 mm.) or 60° (0.4 mm.) (Note 7). The IR spectrum (liquid film) has a strong band at 2083 cm.$^{-1}$ attributable to the diazo function.

2. Notes

1. 2-(Hydroxymethylene)cyclohexanone was prepared from cyclohexanone [*Org. Synth.*, **Coll. Vol. 4,** 536 (1963)] and was freshly distilled before use.

2. Freshly distilled; b.p. 88.5–90.5°.

3. *p*-Toluenesulfonyl azide was prepared from *p*-toluenesulfonyl chloride and sodium azide [*Org. Synth.*, **Coll. Vol. 5,** 179 (1973)]. The submitters found that when somewhat less than the stoichiometric quantity of *p*-toluenesulfonyl azide is used, 2-diazocyclohexanone is obtained free of azide; the excess 2-(hydroxymethylene)cyclohexanone is readily removed in the alkaline workup. The crude product obtained by the checkers, however, contained *p*-toluenesulfonyl azide (Note 5).

4. The basic, aqueous solution on acidification with 6 *N* hydrochloric acid gives *N*-formyl-*p*-toluenesulfonamide in almost quantitative yield. Crystallization from benzene gave crystals, m.p. 101–102° (lit.[2] 102–103°).

5. The submitters found that this product gave a single spot by TLC (Note 8), and that it can be used for most preparative purposes without distillation. The checkers found by ^{1}H NMR spectroscopy that the product contained *p*-toluenesulfonyl azide (*ca.* 5%) and triethylamine.

6. If the distillation is carried out with a capillary leak, decomposition of the diazo compound occurs. The checkers found that distillation on a larger scale led to extensive decomposition.

7. The distillation is not carried to completion because of the danger of explosion. It is carried out with the usual safety precautions (safety shield), although no explosion has yet occurred.

8. TLC was carried out on DC-Fertigplatte Merck Kieselgel F_{254} supplied by Firma Merck AG, 61 Darmstadt, Germany. For the solvent system dichloromethane/methanol (97/3) the product has an R_f value of 0.45.

3. Discussion

In addition to previously described syntheses[2,3] by diazo group transfer with deformylation,[4] 2-diazocyclohexanone has been prepared by two variants of this method. In one, the reaction of 2-(hydroxymethylene)cyclohexanone with *p*-toluenesulfonyl azide is carried out in ether/diethylamine, and an enamine is assumed to be formed as an intermediate;[5] in the other, the sodium salt of the hydroxymethylene compound was treated with the lithium salt of *p*-carboxybenzenesulfonyl azide in ether/tetrahydrofuran.[6] Its preparation from 1,2-cyclohexanedione mono-*p*-toluenesulfonylhydrazone was described earlier.[7]

2-Diazocycloalkanones with five- to twelve-membered rings can be synthesized by the present procedure in good yields (Table I).[2] Diazo transfer with deformylation can also be used for the preparation of bicyclic α-diazo ketones.[8,9] A related procedure involving reaction of the sodium salt of an α-(hydroxymethylene) ketone with *p*-toluenesulfonyl azide in ethanol has been applied to the synthesis of diazoalkyl ketones, α-diazo aldehydes, and α-diazo carboxylic esters.[10]

TABLE I
Preparation of 2-Diazocycloalkanones

$(CH_2)_{n-2}$ $\begin{array}{c} C=O \\ | \\ C=N_2 \end{array}$ n	Boiling Point or [Melting Point], °C.	Yield, %
5	34–37 (0.8 mm.)	98
7	62 (0.4 mm.)	83
8	a	87
9	a	73
10	[54–55]	81
11	b	79
12	[42–43]	57

a Liquid, purified by crystallization from ether at $-60°$.
b Liquid, purified by crystallization from ethanol at $-20°$.

1. Institut für Organische Chemie der Universität des Saarlandes, 66 Saarbrucken 11, Germany. [Present address: Fachbereich Chemie, Universität Kaiserslautern, 675 Kaiserslautern, Germany.]
2. M. Regitz and J. Rüter, *Chem. Ber.*, **101**, 1263 (1968).
3. M. Regitz, F. Menz, and J. Rüter, *Tetrahedron Lett.*, 739 (1967).
4. M. Regitz, *Angew. Chem.*, **79**, 786 (1967); *Angew. Chem. Int. Ed. Engl.*, **6**, 733 (1967).
5. M. Rosenberger, P. Yates, J. B. Hendrickson, and W. Wolf, *Tetrahedron Lett.*, 2285 (1964).
6. J. B. Hendrickson and W. A. Wolf, *J. Org. Chem.*, **33**, 3610 (1968).
7. H. Stetter and K. Kiehs, *Chem. Ber.*, **98**, 1181 (1965).
8. T. Gibson and W. F. Erman, *J. Org. Chem.*, **31**, 3028 (1966).
9. K. B. Wiberg and A. de Meijere, *Tetrahedron Lett.*, 519 (1969).
10. M. Regitz and F. Menz, *Chem. Ber.*, **101**, 2622 (1968).

2-DIAZOPROPANE

(Propane, 2-diazo-)

$$(CH_3)_2C{=}NNH_2 \xrightarrow[\text{KOH}]{\text{HgO}} (CH_3)_2CN_2$$

Submitted by S. D. ANDREWS,[1] A. C. DAY,[1] P. RAYMOND,[1]
and M. C. WHITING[2]
Checked by G. SWIFT and W. D. EMMONS

1. Procedure

Caution! 2-Diazopropane is volatile and presumably toxic. All operations should be carried out in an efficient hood behind a protective screen.

A 250-ml., two-necked, round-bottomed flask, placed in a room temperature (*ca.* 20°) water bath, is equipped with a magnetic stirrer, a dropping funnel, and a distillation head carrying a thermometer, and connected *via* an acetone–dry-ice condenser to a receiver cooled to −78° in acetone and dry ice. In the distilling flask are placed 60 g. (0.27 mole) of yellow mercury(II) oxide (Note 1), 100 ml. of diethyl ether (Note 2), and 4.5 ml. of a 3 *M* solution of potassium hydroxide in ethanol (Note 3). The pressure throughout the system is reduced to 250 mm., and with vigorous stirring 15 g. (0.21 mole) of acetone hydrazone (Note 4) is added dropwise through the funnel (Note 5). With continued stirring, the pressure is reduced to 15 mm., and ether and 2-diazopropane co-distill and condense in the receiver (Note 6), yielding 70–90% (Notes 7, 8, and 9) of product.

2. Notes

1. Laboratory Reagent yellow mercury(II) oxide purchased from British Drug House was used for most runs. The preparation was not apparently improved by the use of freshly precipitated mercury(II) oxide.

2. The quantity of ether can be varied over a wide range (the submitters have successfully used as little as 60 ml.), and is adjusted to yield the desired concentration of 2-diazopropane in the distillate.

3. A stock solution of potassium hydroxide in ethanol was prepared and stored under nitrogen. Old stocks are brown and contain a dark sediment, but are apparently just as effective as the freshly prepared reagent. Methanolic potassium hydroxide has also been used by the submitters; this remains clear and colorless for long periods but offers no other advantage over the ethanolic solution. In absence of the basic solution, the acetone hydrazone is not oxidized by mercury(II) oxide.

4. *Org. Synth.,* **Coll. Vol. 6,** 161 (1988). Yields are lower if the hydrazone is not freshly redistilled.

5. No special precautions are necessary to keep the reaction mixture cool, since boiling of the ether provides adequate cooling.

6. It is usually unnecessary to dry the distillate, because the water produced in the

reaction is largely retained in the distilling flask. That which vaporizes is trapped as ice in the condenser.

7. The solution is *ca.* 2 *M*. Yields were determined by nitrogen evolution on adding acetic acid, or spectrometrically from the visible absorption band at 500 nm, which has ϵ ~ 2 as calculated from the nitrogen evolution. Yields estimated by addition of a standard solution of benzoic acid and titration with alkali were consistently much lower. Both methods underestimate the yield, since decomposition with acid gives tetramethylethylene and some acetone azine in addition to the 2-propyl ester.[3] The nitrogen evolution method (and therefore the spectrometric method) probably underestimates the yield by *ca.* 10–20%. the titration method by more than 50%.

8. The entire preparation is very rapid (*ca.* 30 minutes) and is easily adaptable to the preparation of larger amounts of 2-diazopropane. Without difficulty, 2–3 *M* solutions can be obtained (see Note 2). The solutions are essentially mercury-free.

9. 2-Diazopropane is an unstable material. The decay is first-order with a half-life of 3 hours at 0°.

3. Discussion

This method[4] is an adaptation of that given by Staudinger and Gaule.[5] Highly unstable solutions have been obtained by Applequist and Babad[5] by use of silver oxide in place of mercury(II) oxide.

Contrary to previous reports,[5–7] 2-diazopropane, as indicated by the present procedure, is neither difficult to prepare nor unduly unstable. The method may be extended to other secondary aliphatic diazo compounds, which have given difficulty in the past.[8] The success of the method depends on the use of a basic catalyst for the oxidation. The desirability of a basic catalyst has been recognized previously[9,10] and is well illustrated by the contrast between the two preparations of diphenyldiazomethane[10,11] (contrast also refs. 9 and 12). Miller has speculated on the role of the basic catalyst.[10]

2-Diazopropane is a potential source of *gem*-dimethyl groups. It undergoes 1,3-dipolar addition to acetylenes,[13,14] allenes,[15,16] and olefins,[17] and in all three classes the orientation of addition has been found to be sensitive to steric effects.[14,16,17] The adducts with acetylenes[14,18] and allenes[15,19] give cyclopropenes, and methylenecyclopropanes, respectively, upon photolysis. The adducts with certain acetylenes bearing an α-leaving group, however, are converted photochemically into allenes and conjugated dienes by an ionic mechanism.[13]

1. Dyson Perrins Laboratory, University of Oxford, Oxford, England.
2. Department of Chemistry, University of Bristol, Bristol, England.
3. D. E. Applequist and H. Babad, *J. Org. Chem.*, **27**, 288 (1962).
4. A. C. Day, P. Raymond, R. M. Southam, and M. C. Whiting, *J. Chem. Soc., C*, 467 (1966).
5. H. Staudinger and A. Gaule, *Ber. Dtsch. Chem. Ges.*, **49**, 1897 (1916).
6. G. M. Kaufman, J. A. Smith, G. G. Vander Stouw, and H. Shechter, *J. Am. Chem. Soc.*, **87**, 935 (1965).
7. J. R. Dyer, R. B. Randall, and H. M. Deutsch, *J. Org. Chem.*, **29**, 3423 (1964); D. W. Adamson and J. Kenner, *J. Chem. Soc.*, 286 (1935).
8. K. Heyns and A. Heins, *Justus Liebigs Ann. Chem.*, **604**, 133 (1957), and ref. 6.

9. C. D. Nenitzescu and E. Solomonica, *Org. Synth.*, **Coll. Vol. 2,** 496 (1943); cf. P. D. Bartlett and L. B. Gortler, *J. Am. Chem. Soc.*, **85,** 1864 (1963).
10. J. B. Miller, *J. Org. Chem.*, **24,** 560 (1959).
11. L. I. Smith and K. L. Howard, *Org. Synth.*, **Coll. Vol. 3,** 351 (1955).
12. L. I. Smith and H. H. Hoehn, *Org. Synth.*, **Coll. Vol. 3,** 356 (1955).
13. A. C. Day and M. C. Whiting, *J. Chem. Soc. B,* 991 (1967); *Chem. Commun.,* 292 (1965).
14. A. C. Day and R. N. Inwood, *J. Chem. Soc. C,* 1065 (1969).
15. A. C. Day and M. C. Whiting, *J. Chem. Soc. C,* 464 (1966); *Proc. Chem. Soc. London,* 368 (1964).
16. S. D. Andrews, A. C. Day, and R. N. Inwood, *J. Chem. Soc. C,* 2433 (1969); S. D. Andrews and A. C. Day, *Chem. Commun.,* 902 (1967).
17. S. D. Andrews, A. C. Day, and A. N. McDonald, *J. Chem. Soc. C,* 787 (1969).
18. A. C. Day and M. C. Whiting, *J. Chem. Soc. C.,* 1719 (1966).
19. S. D. Andrews and A. C. Day, *J. Chem. Soc. B,* 1271 (1968); *Chem. Commun.,* 667 (1966).

MACROCYCLIC POLYETHERS: DIBENZO-18-CROWN-6 POLYETHER AND DICYCLOHEXYL-18-CROWN-6 POLYETHER

(Dibenzo [*b,k*] [1,4,7,10,13,16] hexaoxacyclooctadecin, 6,7,9, 10,17,18,20,21-octahydro- and dibenzo [*b,k*] [1,4,7,10,13,16] hexaoxacyclooctadecin, eicosahydro-)

Submitted by CHARLES J. PEDERSEN[1]
Checked by EDITH FENG and HERBERT O. HOUSE

1. Procedure

Caution! The subsequently described macrocyclic polyethers are toxic (Note 1) and should be handled with care.

Benzene has been identified as a carcinogen. OSHA has issued emergency standards on its use. All procedures involving benzene should be carried out in a well-ventilated hood, and glove protection is required.

A. *Dibenzo-18-crown-6 polyether.* A dry, 5-l., three-necked flask is fitted with a reflux condenser, a 500-ml., pressure-equalizing dropping funnel, a thermometer, and a mechanical stirrer. An inlet tube at the top of the reflux condenser is used to maintain a static nitrogen atmosphere in the reaction vessel throughout the reaction. The flask is charged with 330 g. (3.00 moles) of catechol (Note 2) and 2 l. of commercial *n*-butanol before stirring is started, and 122 g. (3.05 moles) of sodium hydroxide pellets are added.

The mixture is heated rapidly to reflux (about 115°), and a solution of 222 g. (1.55 moles) of bis(2-chloroethyl) ether (Note 3) in 150 ml. of *n*-butanol is added, dropwise with continuous stirring and heating, over a 2-hour period. After the resulting mixture has been refluxed with stirring for an additional hour, it is cooled to 90° and an additional 122 g. (3.05 moles) of sodium hydroxide pellets are added. The mixture is refluxed, with stirring, for 30 minutes, and a solution of 222 g. (1.55 moles) of bis(2-chloroethyl) ether (Note 3) in 150 ml. of *n*-butanol is added, dropwise with stirring and heating, over a period of 2 hours. The final reaction mixture is refluxed, with stirring, for 16 hours (Note 4), then acidified by the dropwise addition of 21 ml. of concentrated hydrochloric acid. The reflux condenser is replaced with a distillation head and approximately 700 ml. of *n*-butanol is distilled from the mixture. As the distillation is continued, water is added to the flask from the dropping funnel at a sufficient rate to maintain a constant volume in the reaction flask. This distillation is continued until the temperature of the distilling vapor exceeds 99° (Note 5), and the resulting slurry is cooled to 30–40°, diluted with 500 ml. of acetone, stirred to coagulate the precipitate, and filtered with suction. The residual crude product is stirred with 2 l. of water, filtered with suction, stirred with 1 l. of acetone, and again filtered with suction. The residual product is washed with an additional 500 ml. of acetone and dried with suction, yielding 221–260 g. (39–48%) of tan, fibrous crystals, m.p. 161–162°, which are sufficiently pure for use in the next step. Dibenzo-18-crown-6 polyether may be recrystallized from benzene, giving white, fibrous needles, m.p. 162.5–163.5° (Note 6).

B. *Dicyclohexyl-18-crown-6 polyether.*   A 1-l., stainless-steel autoclave is charged with a mixture of 125 g. (0.347 mole) of dibenzo-18-crown-6 polyether, 500 ml. of redistilled *n*-butanol (Note 7), and 12.5 g. of 5% ruthenium-on-alumina catalyst (Note 8). After the autoclave has been closed, it is flushed with nitrogen and filled with hydrogen. The mixture is hydrogenated at 100° and a hydrogen pressure of about 70 atm. (1000 p.s.i.) until the theoretical amount of hydrogen (2.08 moles) has been absorbed. The autoclave is cooled to room temperature and vented, and the reaction mixture is filtered to remove the catalyst (Note 9). The filtrate is concentrated at 90–100° with a rotary evaporator (Note 10), and the residual crude product solidifies on standing (Note 11). To remove hydroxylic impurities, a solution of the crude product (about 130 g.) in 400 ml. of *n*-heptane is filtered through a 7-cm. by 20-cm. column of acid-washed alumina (80–100 mesh, activity I–II), and the column is eluted with additional *n*-heptane until the eluate exhibits hydroxyl absorption in the 3300–3400 cm^{-1} region of the IR. The solvent is removed from the combined eluates with a rotary evaporator, leaving 75–89 g. (58–69%) of mixture of diastereoisomeric dicyclohexyl-18-crown-6 polyethers as white prisms, melting within the range 38–54° (Note 12), which may be used to prepare complexes with various metal salts (Notes 13 and 14).

2. Notes

1. Dicyclohexyl-18-crown-6 polyether possesses unusual physiological properties which require care in its handling.[2] It is likely that other cyclic polyethers with similar complexing power are also toxic, and should be handled with equal care.

a. *Oral toxicity.* The approximate lethal dose of the dicyclohexyl-18-crown-6 polyether for ingestion by rats was 300 mg./kg. In a 10-day subacute oral test, the compound did not exhibit any cumulative oral toxicity when administered to male rats at a dose level of 60 mg./kg./day. It should be noted that dosage at the approximate lethal dose level caused death in 11 minutes, but that a dose of 200 mg./kg. was not lethal in 14 days.

b. *Eye irritation.* This dicyclohexyl polyether produced some generalized corneal injury, some iritic injury, and conjunctivitis when introduced as a 10% solution in propylene glycol. Although tests are not complete, there may be permanent injury to the eye even if the eye is washed after exposure.

c. *Skin absorption.* Dicyclohexyl-18-crown-6 polyether is very readily absorbed through the skin of test animals. It caused fatality when absorbed at the level of 130 mg./kg.

d. *Skin irritation.* Primary skin irritation tests run on this polyether indicate the material should be considered a very irritating substance.

2. Catechol of satisfactory purity may be purchased from Eastman Organic Chemicals or from Aldrich Chemical Company, Inc.

3. Bis(2-chloroethyl) ether may be obtained from Eastman Organic Chemicals. The checkers redistilled this material (b.p. 175–177°) before use.

4. A shorter period of refluxing may be sufficient.

5. The bulk of the material, a *n*-butanol–water azeotrope, distils at 92°.

6. The product has UV maxima (CH$_3$OH): 223 nm (ϵ 17,500) and 275 nm (ϵ 5500) with ^{1}H NMR peaks (CDCl$_3$), δ: 3.8–4.3 (m, 16H, 8CH_2O), 6.8–7.0 (m, 8H, aryl CH). The mass spectrum exhibits the following abundant peaks: *m/e* (rel. int.), 360 (M+, 29), 137 (29), 136 (74), 121 (100), 109 (23), 80 (31), 52 (21), 45 (27), and 43 (34).

7. It is advisable to use redistilled solvent to avoid the presence of catalyst poisons.

8. The 5% ruthenium-on-alumina catalyst is available from Engelhard Industries.

9. Since the catalyst, saturated with hydrogen, is pyrophoric, it should be kept wet with water after the filtration has been completed.

10. Since the product, a polyether, is apt to be oxidized by air, especially at elevated temperatures in the molten state, the product should be stored under a nitrogen atmosphere.

11. This residue is a mixture of stereoisomeric dicyclohexyl-18-crown-6 polyethers which may be contaminated with unchanged dibenzo-18-crown-6 polyether and alcohols, arising from hydrogenolysis of the polyether ring. The submitter reports that this residue is sufficiently pure for many purposes such as the preparation of complexes with potassium hydroxide which are soluble in aromatic hydrocarbons.

12. The submitter reports that the two major diastereoisomers present, designated isomer A, m.p. 61–62° and isomer B, as one of two crystalline forms, m.p. 69–70° or m.p. 83–84°, may be separated by chromatography on alumina.[3] An x-ray crystal structure determination for the complex of barium thiocyanate with isomer A of dicyclohexyl-18-crown-6 polyether has shown this polyether to have the *cis-syn-cis* stereochemistry.[4a] An x-ray crystal structure determination for the complex of sodium bromide with isomer B has shown this isomer to have the *cis-anti-cis* stereochemistry.[4b]

The mixture of isomers A and B has negligible UV absorption (95% C_2H_5OH) and exhibits ^1H NMR (C_6D_6) multiplets at δ 0.9–2.2 (16H, aliphatic CH) and 3.3–4.0 (20H, OCH). The mass spectrum of the mixture exhibits the following relatively abundant peaks: m/e (rel. int.), 372 (M+, 2), 187 (35), 143 (100), 141 (47), 99 (92), 98 (46), 97 (41), 89 (66), 87 (41), 83 (45), 82 (55), 81 (99), 73 (77), 72 (46), 69 (58), 67 (42), 57 (50), 55 (58), 45 (77), 43 (61), and 41 (58). Although the IR (CCl_4) and ^1H NMR (C_6D_6, 100 mHz.) spectra of the pure isomers A and B differ slightly from one another, the checkers were unable to use these spectra to determine quantitatively the composition of mixtures of the two isomers. The most notable difference in these spectra is the shape of the ^1H NMR multiplet in the region 3.3–4.0 p.p.m.; this multiplet is considerably broader in isomer A than in isomer B allowing a qualitative estimate of the purity of each isomer.

13. The submitter prepared a toluene solution of the complex of potassium hydroxide with dicyclohexyl-18-crown-6 polyether by the following procedure. A mixture of 14.9 g. (0.0402 mole) of dicyclohexyl-18-crown-6 polyether (mixture of isomers) and 2.64 g. (0.0400 mole) of 85% potassium hydroxide was dissolved in 50 ml. of methanol with gentle warming on a steam bath. The solution was diluted with 100 ml. of toluene and then concentrated with a rotary evaporator to a volume of 50 ml. An additional 100 ml. of toluene was added, and the solution was again concentrated to a volume of 50 ml. This solution was diluted with toluene to a volume of 100 ml., 1 g. of decolorizing charcoal was added, and the mixture was allowed to stand overnight under a nitrogen atmosphere. After gravity filtration, a clear toluene solution of the complex was obtained. Titration with standard hydrochloric acid indicated the solution to be approximately 0.3 M in base. This solution, which must be protected from atmospheric moisture and carbon dioxide, has been used for the saponification of sterically hindered esters.[2]

14. The checkers prepared a crystalline complex of potassium acetate with isomer B of dicyclohexyl-18-crown-6 polyether by the following procedure. To a stirred solution of 15.0 g. (0.0404 mole) of dicyclohexyl-18-crown-6 polyether (mixture of isomers) in 50 ml. of methanol was added a solution of 5.88 g. (0.0600 mole) of anhydrous potassium acetate (dried at 100° under reduced pressure) in 35 ml. of methanol. The resulting solution was concentrated with a rotary evaporator, and the residual white solid was extracted with 35 ml. of boiling dichloromethane. The resulting mixture was filtered, and the filtrate was cooled in an acetone–dry-ice bath and slowly diluted with petroleum ether (b.p. 30–60°, approximately 200 ml. was required) to initiate crystallization. The resulting suspension of the crystalline complex was allowed to warm to room temperature and filtered with suction. Recrystallization of this complex from a dichloromethane–petroleum ether (b.p. 30–60°) mixture separated 4.21–4.35 g. (22–23%) of the complex of potassium acetate with isomer B of dicyclohexyl-18-crown-6 polyether as white needles, m.p. 165–250° (dec.). This complex has IR absorption (CH_2Cl_2) at 1570 and 1385 cm.$^{-1}$ (COO$^-$) with ^1H NMR absorption ($CDCl_3$), δ 1.0–2.1 (m, 16H, aliphatic CH), 1.95 (s, 3H, CH_3CO), 3.3–4.0 (m, 20H, OCH). A 4.21-g. sample of this complex was partitioned between 75 ml. of water and three 25-ml. portions of ether. The combined ether solutions were dried over anhydrous magnesium sulfate and concentrated under reduced pressure, yielding 1.82 g. of isomer B of dicyclohexyl-18-crown-6 polyether as white prisms, m.p. 68–69°.

3. Discussion

The preparation of dibenzo-18-crown-6 polyether directly from catechol and bis(2-chloroethyl) ether has been reported previously.[2] The present procedure is an improvement of this method. Although dibenzo-18-crown-6 polyether can be obtained in 80% yield from bis-[2-(o-hydroxyphenoxy)-ethyl] ether and bis(2-chloroethyl) ether, the former intermediate has to be synthesized by a method involving several steps. One of the hydroxyl groups of catechol must be protected against alkali with dihydropyran or chloromethylmethyl ether. The intermediate is treated with bis(2-chloroethyl) ether in the presence of alkali and, finally, converted into the desired intermediate by acid hydrolysis.[2] The yield of bis[2-(o-hydroxyphenoxy)-ethyl] ether was less than 40% so that the overall yield of dibenzo-18-crown-6 polyether never approached 39–48%, the yield of the present, direct method.

Dibenzo-24-crown-8 and dibenzo-30-crown-10 polyethers can be prepared by this method with the substitution of the appropriate ω,ω'-dichloropolyether for bis(2-chloroethyl) ether. However, dibenzo-12-crown-4 and macrocyclic polyethers containing two or more benzo groups and an uneven number of oxygen atoms have to be prepared by the alternative method mentioned above, using the intermediate catechol monoethers. Macrocyclic polyethers containing one benzo group can be synthesized by the direct reaction between one molecule of catechol and one molecule of ω,ω'-dichloropolyethers in the presence of alkali. Certain substituted crown compounds can be obtained by using catechol derivatives, such as 4-(tert-butyl)-catechol and 4-chlorocatechol, which do not give side reactions in the presence of alkali.

It is unusual to form a ring of eighteen atoms in a single operation by the reaction of catechol with bis(2-chloroethyl) ether. It seems possible that the ring-closure step is facilitated by the presence of sodium ion, which is solvated by the intermediate acyclic polyether. Some experiments appear to support this hypothesis. The yields of dibenzo-18-crown-6 polyether are higher when it is prepared with sodium or potassium hydroxide than when lithium or tetramethylammonium hydroxide is used; lithium and quaternary ammonium ions are not strongly complexed by the polyethers. Furthermore, the best ligands for alkali metal cations, polyethers containing rings of 15 to 24 atoms including 5 to 8 oxygen atoms, are formed in higher yields than smaller or larger rings, or rings of equal sizes with only 4 oxygen atoms.

The physical properties of many macrocyclic polyethers and their salt complexes have been already described.[2,5] Dibenzo-18-crown-6 polyether is useful for the preparation of sharp-melting salt complexes. Dicyclohexyl-18-crown-6 polyether has the convenient property of solubilizing sodium and potassium salts in aprotic solvents, as exemplified by the formation of a toluene solution of the potassium hydroxide complex (Note 13). Crystals of potassium permanganate, potassium tert-butoxide, and potassium palladium(II) tetrachloride ($PdCl_2$ + KCl) are dissolved in liquid aromatic hydrocarbons merely by adding dicyclohexyl-18-crown-6 polyether.[2] The solubilizing power of the saturated macrocyclic polyethers permits ionic reactions to occur in aprotic media. It is expected that this property will find practical use in catalysis, enhancement of chemical reactivity, separation and recovery of salts, electrochemistry, and analytical chemistry. There are some limitations. Although salts with high lattice energy, such as fluorides, nitrates, sulfates, and carbonates, form complexes with macrocyclic polyethers in alcoholic solvents as readily as more polarizable (softer) salts, their complexes cannot be isolated in the solid state because one or the other uncomplexed component precipitates on concentrating the solutions. For the same reason, these salts cannot be rendered soluble in aprotic solvents by the polyethers.

1. Contribution No. 244 from Elastomer Chemicals Department, Research Division, Experimental Station, E. I. duPont de Nemours and Co., Wilmington, Delaware 19898. [Present address: 57 Market Street, Salem, New Jersey 08079.]
2. C. J. Pedersen, *J. Am. Chem. Soc.,* **89,** 7017 (1967); **92,** 386 (1970).
3. For further details, see H. K. Frensdorff, *J. Am. Chem. Soc.,* **93,** 4684 (1971).
4. (a) N. K. Dalley, D. E. Smith, R. M. Izatt, and J. J. Christensen, *J. Chem. Soc. Chem. Commun.,* 90 (1972); (b) D. E. Fenton, M. Mercer, and M. R. Truter, *Biochem. Biophys. Res. Comm.,* **48,** 10 (1972).
5. For reviews see (a) J. J. Christensen, J. O. Hill, and R. M. Izatt, *Science,* **174,** 459 (1971); (b) C. J. Pedersen and H. K. Frensdorff, *Angew. Chem., Int. Ed. Engl.,* **11,** 16 (1972).

1,2-DIAROYLCYCLOPROPANES:
trans-1,2-DIBENZOYLCYCLOPROPANE

(Methanone, 1,2-cyclopropanediylbis[phenyl]-, *trans*-)

$$C_6H_5COCH_2CH_2CH_2COC_6H_5 + I_2 + 2\ NaOH \xrightarrow[25-40°]{CH_3OH}$$

$$+ 2\ NaI + 2\ H_2O$$

Submitted by ISMAEL COLON, GARY W. GRIFFIN, and E. J. O'CONNELL, JR.[1]
Checked by STEPHEN P. PETERS and KENNETH B. WIBERG

1. Procedure

A 1-l., three-necked, round-bottomed flask equipped with a magnetic stirrer and a dropping funnel is charged with 35 g. (0.14 mole) of 1,3-dibenzoylpropane (Note 1) and a solution of 11.2 g. (0.280 mole) of sodium hydroxide in 400 ml. of methanol. The mixture is warmed to 45°, with stirring to dissolve the diketone. It is allowed to cool to 40°, and a solution of 35 g. (0.14 mole) of iodine in 200 ml. of methanol is slowly added to the stirred solution from the dropping funnel (Note 2). After the addition of iodine is completed, the resulting clear solution is stirred at room temperature for 1.5 hours. During this period, a white solid precipitates (Note 3). The mixture is filtered and the filtrate is placed in a round-bottomed flask. The white solid is washed with four, 100-ml. portions of water and dried at 1 mm. at 57° (1 mm.) for 15 hours, yielding 23–25 g. (66–72%) of *trans*-1,2-dibenzoylcyclopropane, m.p. 103–104° (Note 4). Additional material is obtained on evaporation of the filtrate using a rotary evaporator. The pale-red solid residue is treated with 25 ml. of aqueous 10% sodium bisulfite (Note 5), filtered, and dried as above giving 8–9 g. (23–26%) of the product, m.p. 95–98°. Recrystallization from methanol affords greater than 90% recovery of *trans*-1,2-dibenzoylcyclopropane, m.p. 102–103°. For the recrystallization of a 10-g. sample 200 ml. of methanol is used.

2. Notes

1. 1,3-Dibenzoylpropane was prepared by the method described for 1,4-dibenzoylbutane in *Org. Synth.*, **Coll. Vol. 2,** 169 (1943).
2. The iodine solution is added at a rate such that the color of the iodine is continually discharged by the rapidly stirred solution.
3. In some cases the precipitate may begin to form during the addition of the iodine solution. This has no effect on the yield.
4. Pertinent spectral data include a carbonyl stretching band in the IR at 1665 cm.$^{-1}$ and UV absorption maxima at 245 nm (ϵ 31,300), 278 nm (ϵ 2130), and 317 nm (ϵ 194).

The ^{1}H NMR spectrum (CCl$_4$) has two triplets of equal intensity at δ 3.32 and 1.68 in addition to the aromatic proton bands at δ 7.3–8.2. The integrated peak areas are in the ratio 1:1:5.

5. Sodium bisulfite solution is added to reduce unchanged iodine.

3. Discussion

1,3-Dibenzoylcyclopropane has been prepared by the method described here[2] and previously by Conant and Lutz.[3] Both procedures use 1,3-dibenzoylpropane as the reactant. The present procedure is accomplished in one step under very mild conditions and in nearly quantitative yield. The method of Conant and Lutz is a two-step process involving initial dibromination of the diketone followed by ring closure with zinc and sodium iodide, with an overall yield of trans-1,2-dibenzoylcyclopropane of approximately 15%. The submitters have extended the described method to the preparation of other trans-1,2-diaroylcyclopropanes, namely trans-1,2-bis(4-methoxybenzoyl)cyclopropane and trans-1,2-bis(2-methoxy-5-methylbenzoyl)cyclopropane. In theory the method is applicable to closure of all α,α'-disubstituted propanes having acidic hydrogens on both α-carbons. A probable mechanism for this transformation is as follows:

$$C_6H_5COCH_2CH_2CH_2COC_6H_5 \xrightarrow[CH_3OH]{NaOH} C_6H_5COCH_2CH_2\overset{Na^+}{\overline{C}H}COC_6H_5$$

$$\xrightarrow{I_2} C_6H_5COCH_2CH_2\underset{\overset{|}{I}}{C}HCOC_6H_5$$

$$\xrightarrow[CH_3OH]{NaOH} C_6H_5CO\overline{C}HCH_2\overset{}{C}HCOC_6H_5$$

$$\longrightarrow \quad \text{cyclopropane structure with } \begin{array}{c} H \quad COC_6H_5 \\ C_6H_5CO \quad H \end{array} \quad + I^-$$

1. Department of Chemistry, Fairfield University, Fairfield, Connecticut 06430.
2. G. W. Griffin, E. J. O'Connell, and H. A. Hammond, J. Am. Chem. Soc., 83, 1003 (1963).
3. J. B. Conant and R. E. Lutz, J. Am. Chem. Soc., 49, 1083 (1927).

α,α'-DIBROMODIBENZYL SULFONE

[Benzene, 1,1-[sulfonylbis(bromomethylene)]bis-]

Method 1

$$C_6H_5CH_2CO_2H \xrightarrow[PCl_3]{Br_2} \underset{\underset{Br}{|}}{C_6H_5CHCO_2H} \xrightarrow[Na_2CO_3]{Na_2S}$$

$$\underset{\underset{CO_2H}{|}}{(C_6H_5CH-)_2S} \xrightarrow[HOAc]{H_2O_2} \underset{\underset{CO_2H}{|}}{(C_6H_5CH-)_2SO_2} \xrightarrow[HOAc]{Br_2} \underset{\underset{Br}{|}}{(C_6H_5CH-)_2SO_2}$$

Submitted by Louis A. Carpino[1] and Louis V. McAdams, III
Checked by Timothy P. Higgs and Ronald Breslow

1. Procedure

Caution! Benzene has been identified as a carcinogen; OSHA has issued emergency standards on its use. All procedures involving benzene should be carried out in a well-ventilated hood, and glove protection is required.

A. *α-Bromophenylacetic acid.* A 3-l., round-bottomed flask fitted with a mechanical stirrer and an efficient reflux condenser (Note 1) is charged with 750 ml. of benzene, 230 g. (1.69 moles) of phenylacetic acid (Note 2), 15 g. (0.12 mole) of phosphorus trichloride and 288 g. (1.80 moles) of bromine (Note 3). The resulting solution is heated at gentle reflux for 2–3 days until the initial bromine color is discharged. The solution is allowed to cool to room temperature and after 1 hour is decanted from some polymeric material into a 2-l. distilling flask. Removal of the solvent by distillation at water bath temperatures with the aid of a water aspirator gives a black oil which is poured into 250–300 ml. of ligroin (b.p. 90–120°). The mixture is heated, dissolving the oil, and the solution is stored at −25° in a freezer for 12 hours. Filtration on a sintered-glass funnel followed by washing with 200 ml. of cold (10°) ligroin (b.p. 90–120°) gives 243 g. (67%) of α-bromophenylacetic acid as a white solid, m.p. 73–83°. Recrystallization from about 400 ml. of ligroin (b.p. 90–120°) with 15 g. of decolorizing carbon affords 217–233 g. (60–62%) of the purified acid, m.p. 80.5–84°.

B. *α,α'-Diphenylthiodiglycolic acid.* In a 4-l. Erlenmeyer flask a suspension of 223 g. (1.04 moles) of α-bromophenylacetic acid in 1.25 l. of water is brought into solution by addition of a solution of 157 g. (1.48 moles) of sodium carbonate in 700 ml. of water. A solution of 104 g. (0.8 mole) of sodium sulfide (60–62% pure fused flakes) (Note 4) in 700 ml. of water is added, and the resulting mixture is stirred mechanically at room temperature for 3 hours, heated to the boiling point, filtered while hot, cooled, and cautiously acidified (in a hood) with 3 *N* hydrochloric acid. Filtration followed by washing with 200 ml. of water yields 135–140 g. (86–89%) of crude acid, m.p. 130–140°, which is sufficiently pure for use in the next step (Note 5).

C. *α,α'-Dibromodibenzyl sulfone.* To a solution of 24.9 g. (0.0824 mole) of crude α,α'-diphenylthiodiglycolic acid in 250 ml. of glacial acetic acid contained in a 500-ml., three-necked, round-bottomed flask fitted with a bulb condenser is added 37.4 g. (0.33 mole) of 30% hydrogen peroxide over a 30-minute period, with ice bath cooling and magnetic stirring. The mixture is allowed to come to room temperature (Note 6), and after 3 days 30 g. (0.19 mole) of bromine is added in one portion, followed by 30 g. of potassium bromide in 150 ml. of water. A sunlamp (Note 7), focused on the reaction mixture from a distance of 1 in., causes the solution to warm to 80° (Note 8). After heating at 80° for 30 minutes the mixture is cooled, and the solid is filtered and washed with water and ethanol, yielding 9–10 g. (27–30%) of crude α,α'-dibromodibenzyl sulfone, m.p. 135–150°. The mixture of diastereomers is pure enough to be used directly in the synthesis of 2,3-diphenylvinylene sulfone (Note 9).

2. Notes

1. An efficient bulb condenser (Allihn) was used to prevent loss of bromine.
2. The phenylacetic acid was used as supplied by the Eastman Kodak Co.
3. The bromine was washed just before use in a separatory funnel with 200 ml. of concentrated sulfuric acid.
4. The sodium sulfide was dissolved in the aqueous solution by warming, but the solution was cooled to room temperature before addition. The checkers used 192 g. (0.800 mole) of sodium sulfide nonahydrate.
5. After repeated crystallization from nitromethane the *meso* isomer, m.p. 177–180°, was obtained in a pure state in low yield.
6. After a few hours a precipitate appeared but it usually redissolved after 1.5 days. If the precipitate had not dissolved, it could be brought into solution by heating. The solution was then cooled before addition of bromine.
7. A General Electric 275-watt sunlamp was used.
8. The heat of the sunlamp maintained the temperature near 80°. For larger runs a heating mantle must be used to keep the temperature near 80°.
9. By recrystallization from ethanol it was possible to separate two isomeric dibromides, m.p. 155–157.5° and 162–164°, in low yield.

Method II

$$C_6H_5CH_2Cl \xrightarrow{Na_2S} (C_6H_5CH_2)_2S \xrightarrow[\substack{2.}]{1.\ Br_2} (C_6H_5CH-)_2SO_2$$

1. Procedure

A. *Dibenzyl sulfide.* A solution of 25.8 g. (0.212 mole) of benzyl chloride in 75 ml. of 95% ethanol contained in a 250-ml., round-bottomed flask equipped with an efficient bulb condenser (Allihn) and a magnetic stirrer is brought to gentle reflux with a heating mantle. With stirring and heating, a solution of 36 g. (0.15 mole) of sodium sulfide nonahydrate in 50 ml. of water is added with a dropping funnel over a 4-hour period. The solution is heated at reflux for 3 days, after which the ethanol is removed by distillation at atmospheric pressure. The hot aqueous solution is poured with stirring into a 250-ml. beaker half-filled with chipped ice. After the ice has melted, the resulting yellow solid is filtered on a Büchner funnel and washed with 50 ml. of water. After air-drying, the solid is distilled from an ordinary 50-ml. Claisen flask, yielding 17.6 g. (80%) of dibenzyl sulfide, b.p. 120° (0.15 mm.) (Note 1), which is pure enough to use directly in the next step. Recrystallization from 70% ethanol gives a pure sample, m.p. 46–48°.

B. *α,α'-Dibromodibenzyl sulfone.* To a gently refluxing solution of 11.35 g. (0.05304 mole) of dibenzyl sulfide in 150 ml. of carbon tetrachloride (Note 2) contained in a three-necked, round-bottomed flask is added, dropwise over a period of 1.5 hours, a solution of 17.6 g. (0.0978 mole) of bromine in 50 ml. of carbon tetrachloride, while a sunlamp (Note 3) is focused on the reaction mixture from a distance of 1 in. The solution is refluxed with a heating mantle for 3 hours, and the carbon tetrachloride removed at a water bath temperature of 50° with the aid of a water aspirator. To the residual dark oil is added 25 ml. of anhydrous diethyl ether (Note 4), and after cooling in an ice bath a solution of 32.5 (0.189 mole) of *m*-chloroperbenzoic acid (Note 5) in 150 ml. of anhydrous ether is added dropwise over a 30-minute period. The mixture is allowed to warm to room temperature and stirred for 2 days.

The solvent is evaporated with an air jet at room temperature, and to the residual solid is added saturated sodium hydrogen carbonate solution until effervescence ceases. Filtration of the remaining solid followed by washing with water and cold ethanol gives 6.5 g. (30%) of crude α,α'-dibromodibenzyl sulfone, m.p. 142–158°. This mixture of diastereomers is pure enough for use in conversion to 2,3-diphenylvinylene sulfone (Note 6).

2. Notes

1. The submitters report a similar percent yield on twenty times the scale.
2. Carbon tetrachloride was freshly distilled over phosphorus pentoxide.
3. A General Electric 275-watt sunlamp was used.
4. The ether was dried over sodium metal.
5. *m*-Chloroperbenzoic acid (assay: 85%) was used either as supplied by FMC Corporation or prepared as described in *Org. Synth.*, **Coll. Vol. 6,** 276 (1988).
6. Repeated fractional crystallization from ethanol gave, in low yield, pure samples of the same two diastereomeric dibromides, m.p. 155–157.5° and 162–164°, obtained previously through application of Method I.

3. Discussion

α-Bromophenylacetic acid has been prepared by the bromination of phenylacetic acid with elemental bromine at high temperature[2] or under UV irradiation,[3] or with N-bromosuccinimide,[4] and by treatment of mandelic acid with phosphorus and bromine[5] or fuming hydrobromic acid.[6] α,α'-Diphenylthiodiglycolic acid has been prepared by reaction of α-bromophenylacetic acid with sodium sulfide.[7] Dibenzyl sulfide has been obtained from the reaction of benzyl chloride with potassium sulfide[8] or benzylmercaptide.[9] α,α'-Dibromodibenzyl sulfone has been obtained by oxidation and subsequent brominative decarboxylation of α,α'-diphenylthiodiglycolic acid[10] or by bromination and subsequent oxidation of dibenzylsulfide.[10]

Both methods of preparation represent general techniques for the synthesis of α,α'-dibromodibenzyl sulfones which are key intermediates in the synthesis of the vinylene sulfones.

1. Department of Chemistry, University of Massachusetts, Amherst, Massachusetts 01003.
2. B. Radziszewski, *Ber. Dtsch. Chem. Ges.*, **2**, 207 (1869).
3. Distillers Co., Ltd., Belg. Pat. 622,439 (March 13, 1963) [*Chem. Abstr.*, **59**, 11352 (1963)].
4. I. M. Panaiotov, *Izv. Khim. Inst. Bulg. Akad. Nauk.* **5**, 183 (1957) [*Chem. Abstr.*, **55**, 16500 (1961)].
5. C. Hell and S. Weinzweig, *Ber. Dtsch. Chem. Ges.*, **28**, 2445 (1895).
6. A. Darapsky and M. Prabhakar, *J. Prakt. Chem.*, **96**, 280 (1917).
7. T. Mazonski and B. Prajsnar, *Zesz. Nauk. Politech. Slask. Chem.*, No. 7, 17 (1961) [*Chem. Abstr.*, **62**, 13079d (1965)].
8. R. L. Shriner, H. C. Struck, and W. J. Jorison, *J. Am. Chem. Soc.*, **52**, 2060 (1930).
9. T. S. Price and D. F. Twiss, *J. Chem. Soc.*, **97**, 1179 (1910).
10. L. A. Carpino and L. V. McAdams III, *J. Am. Chem. Soc.*, **87**, 5804 (1965); L. V. McAdams III, Ph.D. Thesis, University of Massachusetts, 1966.

PHOSPHINE–NICKEL COMPLEX CATALYZED CROSS-COUPLING OF GRIGNARD REAGENTS WITH ARYL AND ALKENYL HALIDES: 1,2-DIBUTYLBENZENE

(Benzene, 1,2-di-n-butyl-)

$$CH_3(CH_2)_3Br + Mg \xrightarrow{\text{ethyl ether}} CH_3(CH_2)_3MgBr$$

$$+ 2CH_3(CH_2)_3MgBr \xrightarrow[\text{2. reflux}]{\begin{array}{c}\text{1. [NiCl}_2\text{(dppp)] (catalyst),}\\ \text{diethyl ether, 0°}\end{array}}$$

$$+ 2MgBrCl$$

$$dppp = (C_6H_5)_2P(CH_2)_3P(C_6H_5)_2$$

Submitted by Makoto Kumada, Kohei Tamao, and Koji Sumitani[1]
Checked by Teresa Y. L. Chan and S. Masamune

1. Procedure

A 500-ml., three-necked flask equipped with a mechanical stirrer, a pressure-equalizing dropping funnel, and a reflux condenser attached to a nitrogen inlet is charged with 12.2 g. (0.502 g.-atom) of magnesium turnings. The magnesium is dried under a rapid stream of nitrogen with a heat gun. After the flask has cooled to room temperature, the rate of nitrogen flow is reduced, and 200 ml. of anhydrous diethyl ether (Note 1) and approximately 5 ml. of a solution of 68.5 g. (0.500 mole) of 1-bromobutane (Note 2) in 50 ml. of anhydrous ether are added. The mixture is stirred at room temperature, and within a few minutes an exothermic reaction begins. The flask is immersed in an ice bath, and the remaining ether solution is added dropwise over *ca.* 1 hour. After addition is complete, the mixture is refluxed with stirring for 30 minutes and then cooled to room temperature.

A 1-l., three-necked flask, equipped in the same manner, is charged with *ca.* 0.25 g. (*ca.* 0.5 mmole) of dichloro[1,3-bis(diphenylphosphino)propane]nickel(II) (Note 3), 29.5 g. (0.201 mole) of 1,2-dichlorobenzene (Note 4), and 150 ml. of anhydrous ether (Note 5). The Grignard reagent prepared above is transferred to the dropping funnel and added over 10 minutes, with stirring, to the mixture cooled in an ice bath. The nickel complex reacts immediately with the Grignard reagent, and the resulting clear-tan reaction mixture is allowed to warm to room temperature, with stirring. An exothermic reaction starts within 30 minutes, and the ether begins to reflux gently. After stirring for 2 hours at room temperature, most of the magnesium bromochloride salt has deposited (Note 6). The mixture is refluxed with stirring for 6 hours, cooled in an ice bath, and cautiously hydrolyzed with 2 N hydrochloric acid (*ca.* 250 ml.) (Note 7). The nearly colorless organic layer is separated and the aqueous layer extracted with two 70-ml. portions of ether. The combined organic layer and extracts are washed successively with water, aqueous saturated sodium hydrogen carbonate, and again with water, dried over anhydrous calcium chloride, and filtered. After evaporation of the solvent the residue is distilled under reduced pressure through a 25-cm. column packed with glass helices, giving a forerun (*ca.* 4 g.) of 1-butyl-2-chlorobenzene, b.p. 52–54° (3.5 mm.), n_D^{20} 1.5110, followed by 30.0–31.5 g. (79–83%) of 1,2-dibutylbenzene, b.p. 76–81° (3.5 mm.), n_D^{20} 1.4920, as a colorless liquid (Note 8).

2. Notes

1. Ether was dried over sodium wire and freshly distilled before use.

2. Commercial 1-bromobutane was dried over anhydrous calcium chloride and distilled before use.

3. Dichloro[1,3-bis(diphenylphosphino)propane]nickel(II) can be easily prepared in an open reaction vessel.[2] To a hot solution of 9.5 g. (0.040 mole) of nickel(II) chloride hexahydrate in 175 ml. of a 5:2 (v/v) mixture of 2-propanol and methanol, is added, with stirring, a hot solution of 14.5 g. (0.0352 mole) of 1,3-bis(diphenylphosphino)propane in 200 ml. of 2-propanol. A reddish brown precipitate deposits immediately. The mixture is heated for 30 minutes and allowed to cool to room temperature. Filtration, washing with methanol, and drying under reduced pressure provide the red complex in almost quantitative yield. 1,3-Bis(diphenylphosphino)propane may be purchased from Strem Chemicals Inc., and nickel(II) chloride hexahydrate from Fisher Scientific Company.

4. Commercial G.R.-grade 1,2-dichlorobenzene can be used without further purification. At least a 20% excess of the Grignard reagent should be used to compensate for some loss through undesirable side reactions (see Discussion section).

5. The nickel complex is insoluble in the mixture.

6. If stirring becomes difficult, approximately 100 ml. of anhydrous ether may be added.

7. This hydrolysis is exothermic, and the acid should be added slowly, maintaining gentle refluxing of the ether.

8. GC analysis on a 3.5-m. column packed with Silicone DC 550 and operated at 200° showed that the product was at least 99.5% pure. The product has the following spectral properties: IR (neat): 750 cm^{-1} (1,2-disubstituted benzene); ^{1}H NMR (CCl$_4$), δ (multiplicity, number of protons, assignment): 0.7–1.7 (m, 14H, 2 CH_2CH$_2$CH_3), 2.60 (t, 4H, 2 benzylic CH_2), 7.0 (s, 4H, C$_6H_4$).

3. Discussion

1,2-Dibutylbenzene has been prepared from cyclohexanone by tedious, multistep procedures.[3,4] The present one-step method is based on the selective cross-coupling of a Grignard reagent with an organic halide in the presence of a phosphine–nickel catalyst.[5]

The phosphine–nickel complex catalyzes the cross-coupling of alkyl, alkenyl, aryl, and heteroaryl Grignard reagents with aryl, heteroaryl, and alkenyl halides. The method, thus, has wide application. Alkyl halides also exhibit considerable reactivity, but give a complex mixture of products. Some representative examples are listed in Table I. A labile diorganonickel complex involving the two organic groups originating from the Grignard reagent and from the organic halide, respectively, has been proposed as a reaction intermediate. The fact that simple alkyl Grignard reagents with hydrogen-bearing β-carbon atoms react with equal efficiency is one of the most remarkable features of the present method, considering the great tendency with which transition metal alkyls undergo a β-elimination reaction, forming alkenes and metal hydrides.[6] Although the β-elimination may be responsible for the side products formed during the coupling, an appropriate choice of phosphine ligand for the nickel catalyst minimizes this side reaction. The catalytic activity of the phosphine–nickel complex depends not only on the nature of the phosphine ligand but also on the combination of the Grignard reagent and organic halide. Of several catalysts, including [NiCl$_2$(dppp)], [NiCl$_2$(dppe)], [NiCl$_2$(dmpe)], and [NiCl$_2$\{P(C$_6$H$_5$)$_3$\}$_2$], where dppe = (C$_6$H$_5$)$_2$PCH$_2$CH$_2$P(C$_6$H$_5$)$_2$ and dmpe = (CH$_3$)$_2$PCH$_2$CH$_2$P(CH$_3$)$_2$, the first has been found most effective in almost all cases, except for alkenyl Grignard reagents. The halide may be chloride, bromide, or iodide, although chlorides usually give the most satisfactory results. Even fluorides react with comparable facility in some cases.[7] Vinyl chloride is one of the most reactive halocompounds, and the coupling (Table I, entry 10) can be conveniently conducted in an open system, similar to that described here, by the gradual addition of the Grignard reagent at 0° to a mixture of vinyl chloride, [NiCl$_2$(dppp)], and ether, followed by stirring at room temperature for 2 hours.

TABLE I

CROSS-COUPLING OF GRIGNARD REAGENTS WITH VARIOUS ARYL AND ALKENYL HALIDES IN THE PRESENCE OF [NiCl$_2$(dppp)] AS A CATALYST[a]

Entry	RMgX	Halides	Product (% yield)[b]
1	CH$_3$MgBr	cyclohexenyl—Cl	cyclohexenyl—CH$_3$ (98)
2	CH$_3$(CH$_2$)$_3$MgBr	C$_6$H$_5$Cl	C$_6$H$_5$C$_4$H$_9$ (96)
3	CH$_3$(CH$_2$)$_3$MgBr	2-bromopyridine	2-C$_4$H$_9$-pyridine (71)[c]
4	(CH$_3$)$_3$SiCH$_2$MgCl	1,3-Cl$_2$C$_6$H$_4$	1,3-[(CH$_3$)$_3$SiCH$_2$]$_2$C$_6$H$_4$ (83)[c]
5	(CH$_3$)$_2$CHMgCl	C$_6$H$_5$Cl	(CH$_3$)$_2$CH- and CH$_3$(CH$_2$)$_2$C$_6$H$_5$ (96:4) (89) (10:90) (84)[d]
6	C$_6$H$_5$CH(CH$_3$)MgCl	CH$_2$=CHBr[e,f]	(S)-(+)-C$_6$H$_5$CH(CH$_3$)CH=CH$_2$[g] (96)

TABLE I (*continued*)

CROSS-COUPLING OF GRIGNARD REAGENTS WITH VARIOUS ARYL AND ALKENYL HALIDES IN THE PRESENCE OF [NiCl$_2$(dppp)] AS A CATALYST[a]

Entry	RMgX	Halides	Product (% yield)[b]
7	BrMg(CH$_2$)$_{10}$MgBr	Cl[h,i] ... Cl	(CH$_2$)$_{10}$ (33)
8	C$_6$H$_5$MgBr	(Z)-C$_6$H$_5$CH=CHBr	(Z)-C$_6$H$_5$CH=CHC$_6$H$_5$ (100)
9	C$_6$H$_5$MgBr	ClCH=CHCl[d,i]	(Z)-C$_6$H$_5$CH=CHC$_6$H$_5$[k] (95–100)
10	4-ClC$_6$H$_4$MgBr	CH$_2$=CHCl[l]	4-ClC$_6$H$_4$CH=CH$_2$ (79)[c]
11	2,4,6-(CH$_3$)$_3$C$_6$H$_2$MgBr		(74)
12	CH$_2$=C(CH$_3$)MgBr	α-C$_{10}$H$_7$Br[d,h]	α-C$_{10}$H$_7$C(CH$_3$)=CH$_2$ (78)
13			(78)

[a]The reaction was carried out on a 0.01–0.02 mole scale in refluxing ethyl ether for 3–20 hours, unless otherwise noted.

[b]Determined by GC, unless otherwise noted.

[c]Isolated Yield.

[d]The catalyst is [NiCl$_2$(dmpe)], dmpe = (CH$_3$)$_2$PCH$_2$CH$_2$P(CH$_3$)$_2$.

[e]The catalyst is a mixture of NiCl$_2$ and (S)-Valphos. (S)-Valphos = (S)-(+)-(2-Dimethylamino-3-methylbutyl)diphenylphosphine (see ref. 8f).

[f]At 0° for 2 days.

[g]Configuration of the predominant enantiomer; 13.0% enantiomeric excess.

[h]Solvent is tetrahydrofuran.

[i]At 40° for 9 hours.

[j]At room temperature for 2 hours.

[k](Z)-Stilbene is formed stereoselectively regardless of whether (Z)- or (E)-dichloroethene is used (see text).

[l]Carried out on a 0.2 mole scale at 0° to room temperature over 2 hours.

Several other features deserve comment. The coupling of secondary alkyl Grignard reagents is accompanied by alkyl group isomerization from secondary to primary, the extent of which is strongly dependent on the electronic nature of the phosphine ligand in the catalyst (entry 5).[7] Asymmetric cross-coupling can be achieved by using optically active phosphine–nickel complexes as catalysts (entry 6).[8,9] Cyclocoupling of di-Grignard reagents with dihalides offers a new, one-step route to cyclophanes (entry 7).[10] Though the cross-coupling of monohaloalkenes proceeds stereospecifically (entry 8),[11] that of 1,2-dihaloethylenes proceeds nonstereospecifically, yielding a (Z)-alkene, the stereoselectivity being dependent on the nature of the phosphine ligand in the catalyst (entry 9).[11] Sterically hindered aryl Grignard reagents also react with ease (entry 11).[9]

The couplings are usually exothermic, and care must be taken *not* to add the phosphine–nickel catalyst to a mixture of a Grignard reagent and an organic halide, particularly in a large-scale preparation. For example the addition of a small amount of [$NiCl_2$(dppp)] to a mixture of vinyl chloride and 4-chlorophenylmagnesium bromide in ether at $0°$ led, after a few minutes' induction period, to an uncontrollable, violent reaction.

1. Department of Synthetic Chemistry, Faculty of Engineering, Kyoto University, Kyoto 606, Japan.

2. G. R. VanHecke and W. D. Horrocks, Jr., *Inorg. Chem.*, **5**, 1968 (1966).

3. M. Ogawa and G. Tanaka, *J. Chem. Soc. Jpn. Ind. Chem. Sect.*, **58**, 696 (1955).

4. B. B. Elsner and H. E. Strauss, *J. Chem. Soc.*, 583 (1957).

5. K. Tamao, K. Sumitani, and M. Kumada, *J. Am. Chem. Soc.*, **94**, 4374 (1972); R. J. P. Corriu and J. P. Masse, *J. Chem. Soc., Chem. Commun.*, 144 (1972); K. Tamao, K. Sumitani, Y. Kiso, M. Zembayashi, A. Fujioka, S. Kodama, I. Nakajima, A. Minato, and M. Kumada, *Bull. Chem. Soc. Jpn.*, **49**, 1958 (1976); see also D. G. Morell and J. K. Kochi, *J. Am. Chem. Soc.*, **97**, 7262 (1975).

6. For example, G. E. Coates, M. L. H. Green, and K. Wade, "Organometallic Compounds," Vol. 2, 3rd ed., Methuen and Co., London, England, 1968; W. Mowat, A. Shortland, G. Yagupsky, N. J. Hill, M. Yagupsky, and G. Wilkinson, *J. Chem. Soc., Dalton Trans.*, 533 (1972); P. S. Braterman and R. J. Cross, *Chem. Soc. Rev.*, **2**, 271 (1973).

7. K. Tamao, Y. Kiso, K. Sumitani, and M. Kumada, *J. Am. Chem. Soc.*, **94**, 9268 (1972); Y. Kiso, K. Tamao, and M. Kumada, *J. Organomet. Chem.*, **50**, C12 (1973).

8. (a) G. Consiglio and C. Botteghi, *Helv. Chim. Acta*, **56**, 460 (1973); (b) Y. Kiso, K. Tamao, N. Miyake, K. Yamamoto, and M. Kumada, *Tetrahedron Lett.*, 3 (1974); (c) T. Hayashi, M. Tajika, K. Tamao, and M. Kumada, *J. Am. Chem. Soc.*, **98**, 3718 (1978); (d) K. Tamao, H. Yamamoto, H. Matsumoto, N. Miyake, T. Hayashi, and M. Kumada, *Tetrahedron Lett.*, 1389 (1977); (e) K. Tamao, T. Hayashi, H. Matsumoto, H. Yamamoto, and M. Kumada, *Tetrahedron Lett.*, 2155 (1979); (f) T. Hayashi, M. Fukushima, M. Konishi, and M. Kumada, *Tetrahedron Lett.*, **21**, 79 (1980).

9. K. Tamao, A. Minato, N. Miyake, T. Matsuda, Y. Kiso, and M. Kumada, *Chem. Lett.*, 133 (1975).

10. K. Tamao, S. Kodama, T. Nakatsuka, Y. Kiso, and M. Kumada, *J. Am. Chem. Soc.*, **97**, 4405 (1975).

11. K. Tamao, M. Zembayashi, Y. Kiso, and M. Kumada, *J. Organomet. Chem.*, **55**, C91 (1973).

OXIDATION WITH BIS(SALICYLIDENE)ETHYLENEDIIMINO-COBALT(II) (SALCOMINE): 2,6-DI-*tert*-BUTYL-*p*-BENZOQUINONE

[2,5-Cyclohexadiene-1,4-dione, 2,6-bis(1,1-dimethylethyl)-]

Submitted by C. R. H. I. De Jonge,[1] H. J. Hageman,
G. Hoentjen, and W. J. Mus
Checked by K. Balasubramanian, Robert K. Boeckman, and Carl R. Johnson

1. Procedure

A 200-ml., three-necked flask equipped with a mechanical stirrer, a thermometer, and a gas-inlet tube is charged with 41.2 g. (0.200 mole) of 2,6-di-*tert*-butylphenol (Note 1) in 75 ml. of *N,N*-dimethylformamide (Note 2) and 2.5 g. (0.0075 mole) of salcomine (Note 3). With stirring, oxygen is introduced at such a rate that the temperature does not exceed 50°. This is continued for 4 hours, and at the end of the reaction the temperature drops to about 25°. The reaction mixture is then poured onto 500 g. of crushed ice and 15 ml. of 4 *N* hydrochloric acid. A yellow-brown precipitate is formed, collected by suction filtration, and washed on the filter with three 50-ml. portions of 1 *N* hydrochloric acid, with three 100-ml. portions of water, and twice with 25-ml. portions of cold ethanol. Drying under reduced pressure at 50° for 3 hours gives 43 g. of crude 2,6-di-*tert*-butyl *p*-benzoquinone as a dark-yellow crystalline solid. Recrystallization from ethanol gives 36.5 g. (83%) of pure 2,6-di-*tert*-butyl-*p*-benzoquinone, m.p. 65–66° (Notes, 4, 5).

2. Notes

1. 2,6-Di-*tert*-butylphenol purchased from Aldrich Chemical Company, Inc., was used.

2. When chloroform or methanol is used as the solvent for the oxidation of phenols, other products, originating from coupling of aryloxy radicals, *e.g.*, polyphenylene ethers and/or diphenoquinones, are also formed.[2]

3. Bis(salicylidene)ethylenediiminocobalt(II) can be prepared according to the procedure described in *Inorg. Synth.*[3]

4. 2,6-Di-*tert*-butyl-*p*-benzoquinone should be stored in a brown bottle.

5. The product has the following spectral properties; IR (CHCl$_3$) cm.$^{-1}$: 1652, 1597; ^{1}H NMR (CDCl$_3$), δ 1.33, 6.56.

3. Discussion

Various 2,6-disubstituted *p*-benzoquinones have been prepared by oxidation of the corresponding 2,6-disubstituted phenols with potassium nitrosodisulfonate[4,5] or lead dioxide in formic acid.[6] Oxidative coupling of 2,6-disubstituted phenols to poly-2,6-disubstituted phenylene ethers followed by treatment of the polymers in acetic acid with lead dioxide is reported[7] to give low yields of the corresponding 2,6-disubstituted *p*-benzoquinones.

Salcomine is a useful catalyst for the selective oxygenation of 2,6-disubstituted phenols to the corresponding *p*-benzoquinones when *N,N*-dimethylformamide is used as the solvent; laborious procedures are avoided and high yields of pure *p*-benzoquinones are obtained. Following the procedure described above, the authors have prepared 2,6-diphenyl-*p*-benzoquinone (m.p. 134−135°, yield 86%) and 2,6-dimethoxy-*p*-benzoquinone (m.p. 252°, yield 91%) from the appropriate phenols.

1. Organic and Polymer Chemistry Department, Akzo Corporate Research Laboratories, Arnhem, The Netherlands.
2. H. M. van Dort and H. J. Geursen, *Recl. Travl. Chim. Pays-Bas*, **86**, 520 (1967).
3. H. Diehl and C. C. Hack, *Inorg. Synth.*, **3**, 196 (1950).
4. H. J. Teuber and W. Rau, *Chem. Ber.*, **86**, 1036 (1953).
5. H. J. Teuber and O. Glosauer, *Chem. Ber.*, **98**, 2643 (1965).
6. C. R. H. I. de Jonge, H. M. van Dort, and L. Vollbracht, *Tetrahedron Lett.*, 1881, (1970).
7. H. Finkbeiner and A. T. Toothaker, *J. Org. Chem.*, **33**, 4347 (1968).

DIAZO TRANSFER BY MEANS OF PHASE-TRANSFER CATALYSIS: DI-*tert*-BUTYL DIAZOMALONATE

[Propanedioic acid, diazo-, bis(1,1-dimethylethyl) ester]

Submitted by HENRY J. LEDON[1]
Checked by STEVEN J. HOBBS and ROBERT M. COATES

1. Procedure

Caution! Diazomalonic esters are toxic and potentially explosive. They must be handled with care. This preparation should be carried out in a well-ventilated hood, and the distillation of di-tert-butyl diazomalonate should be conducted behind a safety shield.

A 500-ml., three-necked, round-bottomed flask equipped with a reflux condenser, a dropping funnel, an argon inlet, and a Teflon-coated magnetic stirring bar is charged with 10.8 g. (0.0500 mole) of di-*tert*-butyl malonate (Note 1), 9.9 g. (0.0502 mole) of *p*-toluenesulfonyl azide (Note 2), 0.5 g. (0.001 mole) of methyltri-*n*-octylammonium chloride (Note 3), and 200 ml. of dichloromethane (Note 4). The solution is stirred vigorously as the flask is flushed with argon for 10 minutes, then 10 ml. (0.1 mole) of aqueous 10 N sodium hydroxide is added in one portion (Note 5). The mixture is stirred for 2 hours, during which time it changes from colorless to pale yellow. A 200-ml. portion of water is added; the organic layer is separated, washed with three 500-ml. portions of water (Note 6), and dried with anhydrous magnesium sulfate. After filtration of the drying agent, the solvent is removed on a rotary evaporator using a water bath kept at 30° (Note 7). The residual yellow-orange liquid is distilled at high vacuum (Note 8). The temperature of the heating bath is gradually raised to *ca.* 70° and kept at 70–75° during the distillation. After separation of a small forerun, 7.2–7.6 g. (59–63%) of di-*tert*-butyl diazomalonate is collected, b.p. 44–45° (0.02 mm.), $n_D^{22} = 1.4568$ (Notes 9 and 10).

2. Notes

1. Di-*tert*-butyl malonate is available commercially directly from Fluka AG, Buchs, Switzerland, or from its North American representative, Tridom Chemical Inc. Alternatively this compound may be prepared from malonic acid as described in *Org. Synth., Coll. Vol. 4,* 261 (1963).

2. *p*-Toluenesulfonyl azide was prepared according to the procedure in *Org. Synth.*, **Coll. Vol. 5,** 179 (1973).

3. The submitter obtained methyltri-*n*-octylammonium chloride (Aliquat 336) from General Mills Company, Chemical Division, Kankakee, Illinois. The phase-transfer catalyst used by the checkers, which was supplied by Fluka AG through Tridom Chemical Inc., was a mixture in which the alkyl chains varied in length from *n*-octyl to *n*-decyl with the former predominating.

4. Reagent grade dichloromethane was used without further purification.

5. The sodium hydroxide solution was deoxygenated by bubbling a stream of argon through it for 10 minutes.

6. The organic layer is washed with relatively large portions of water to avoid difficulty in separating the phases. The checkers found that vigorous shaking during the extractions gave intractable emulsions. The emulsions were avoided by gentle swirling of the dichloromethane–water mixtures.

7. To avoid foaming during the distillation, the checkers removed the last traces of solvent by evacuation at 0.1 mm. and room temperature for 12–24 hours.

8. The submitter recommends that the apparatus be purged with argon prior to the distillation.

9. The checkers collected foreruns amounting to 0.3–0.7 g., b.p. 50–58° (0.003 mm.) and 40–45° (0.0006 mm.). The product was collected in two or three fractions, b.p. 53–57° (0.002–0.011 mm.), 54–58° (0.002–0.003 mm.), and 45–52° (0.0004–0.0006 mm.). Inspection and integration of the ^{1}H NMR spectra of the foreruns indicated that the fractions were mainly di-*tert*-butyl diazomalonate contaminated with 16–35% of di-*tert*-butyl malonate. The purest fractions usually crystallized on standing at room temperature to give a low-melting solid.

A GC analysis on the product by the submitter, using an 0.3 × 80 cm. column packed with 10% silicone rubber (SE-30) supported on acid-washed, 60–80 mesh Chromasorb P at 80°, exhibited a single peak. The retention times of di-*tert*-butyl malonate, di-*tert*-butyl diazomalonate, and *p*-toluenesulfonyl azide were 2, 6, and 9 minutes, respectively. The purity of the product obtained by the checkers was estimated from ^{1}H NMR spectra to be *ca.* 94%, the remainder being di-*tert*-butyl malonate.

10. The spectral properties of the product are as follows: IR (liquid film) cm.$^{-1}$: 2137 ($C{=}N_2$), 1751 ($C{=}O$), 1730 ($C{=}O$), 1686; UV (C_2H_5OH) nm. max. (log ϵ): 255 (3.68); ^{1}H NMR (CDCl$_3$), δ (multiplicity, number of protons, assignment): 1.52 [s, 18H, 2C(CH$_3$)$_3$]; ^{13}C NMR with proton decoupling (CDCl$_3$), δ (assignment): 28.5 (*C*H$_3$), 65.7 ($C{=}N_2$), 82.8 [*C*(CH$_3$)$_3$], 160.6 ($C{=}O$).

3. Discussion

The "diazo transfer reaction" between *p*-toluenesulfonyl azide and active methylene compounds is a useful synthetic method for the preparation of α-diazo carbonyl compounds.[2] However, the reaction of di-*tert*-butyl malonate and *p*-toluenesulfonyl azide to form di-*tert*-butyl diazomalonate proceeded to the extent of only 47% after 4 weeks with the usual procedure.[3] The present procedure, which utilizes a two-phase medium and methyltri-*n*-octylammonium chloride (Aliquat 336) as phase-transfer catalyst, effects this same diazo transfer in 2 hours and has the additional advantage of avoiding the use of

TABLE I
PREPARATION OF α-DIAZO CARBONYL COMPOUNDS VIA PHASE TRANSFER CATALYSIS[a]

Starting Material	Organic Phase	Aqueous Phase	Phase Transfer Catalyst[b]	Time and Temperature	Product	Yield (%)
Ethyl acetoacetate	pentane	saturated Na$_2$CO$_3$	A	15 hours, 25°	CH$_3$CO–C(N$_2$)–CO–OC$_2$H$_5$	90
Ethyl acetoacetate	pentane	3 N NaOH	A	15 hours, 25°	N$_2$=CH–CO–OC$_2$H$_5$	53
tert-Butyl acetoacetate	pentane	saturated Na$_2$CO$_3$	A	15 hours, 25°	CH$_3$CO–C(N$_2$)–CO–OC(CH$_3$)$_3$	77
	dichloromethane	3 N NaOH	B	1 hour, 0°		92
tert-Butyl acetoacetate	pentane	3 N NaOH	A	15 hours, 25°	N$_2$=CH–CO–OC(CH$_3$)$_3$	89
Phenyl acetone	benzene	10 N NaOH	A	15 hours, 0°	C$_6$H$_5$–C(N$_2$)–CO–CH$_3$	100

[a] These reactions were carried out with 0.005 mole of the carbonyl compound and 0.005 mole of p-toluenesulfonyl azide.
[b] A, tetrabutylammonium bromide; B, methyltri-n-octylammonium chloride (Aliquat 336).

anhydrous solvents.[4,5] This procedure has been employed for the preparation of diazoacetoacetates, diazoacetates, and diazomalonates (Table I).[5] Ethyl and *tert*-butyl acetoacetate are converted to the corresponding α-diazoacetoacetates with saturated sodium carbonate as the aqueous phase. When aqueous sodium hydroxide is used with the acetoacetates, the initially formed α-diazoacetoacetates undergo deacylation to the diazoacetates. Methyl esters are not suitable substrates, since they are too easily saponified under these conditions.

Although the hazardous properties of di-*tert*-butyl diazomalonate are not known with certainty, it is reasonable to assume that they are similar to those of diazoacetic esters, which are considered to be moderate explosion hazards when heated.[6] Contact with rough or metallic surfaces should be avoided. The submitter has routinely distilled 10-g. quantities of di-*tert*-butyl diazomalonate under argon with no sign of decomposition.

Diazomalonic esters serve as intermediates for the synthesis of a wide variety of compounds including cyclopropanes,[7,8] cyclopropenes,[7,9] cycloheptatrienes,[10] sulfur ylides,[11] lactones,[12] and substituted malonates.[13]

1. Laboratoire de Chimie de l'Ecole Normale Supérieure, 24, rue Lhomond, 75231 Paris Cedex 05, France. [Present address: Institut de Recherches sur la Catalyse, 79 Boulevard du 11 Novembre 1918, 69626 Villeurbanne Cedex, France.]
2. M. Regitz, J. Hocker, and A. Liedhegener, *Org. Synth.*, **Coll. Vol. 5,** 179 (1973).
3. B. W. Peace, F. Carman, and D. S. Wulfman, *Synthesis,* 658 (1971).
4. C. M. Starks, *J. Am. Chem. Soc.,* **93,** 195 (1971).
5. H. Ledon, *Synthesis,* 347 (1974).
6. N. I. Sax, "Dangerous Properties of Industrial Materials," 4th ed., Van Nostrand Reinhold, New York, 1975, p. 609.
7. For reviews see: (a) W. Kirmse, "Carbene Chemistry," 2nd ed., Academic Press, New York, 1971; (b) D. Wendisch, in E. Muller, Ed., "Methoden der Organischen Chemie" (Houben-Weyl), 4th ed., Vol. 4/3, Georg Thieme Verlag, Stuttgart, Germany, 1971; (c) V. Dave and E. Warnhoff, *Org. React.,* **18,** 217 (1970).
8. (a) B. W. Peace and D. S. Wulfman, *Synthesis,* 137 (1973); (b) E. Wenkert, M. E. Alonso, B. L. Buckwalter, and K. J. Chou, *J. Am. Chem. Soc.,* **99,** 4778 (1977).
9. M. E. Hendrick. *J. Am. Chem. Soc.,* **93,** 6337 (1971); R. Breslow, R. Winter, and M. Battiste, *J. Org. Chem.,* **24,** 415 (1959).
10. M. Jones, Jr., W. Ando, M. E. Hendrick, A. Kulczycki, Jr., P. M. Howley, K. F. Hummel, and D. S. Malament, *J. Am. Chem. Soc.,* **94,** 7469 (1972).
11. (a) For a review see W. Ando, *Acc. Chem. Res.,* **10,** 179 (1977); (b) J. Cuffe, R. J. Gillespie, and A. E. A. Porter, *J. Chem. Soc. Chem. Commun.,* 641 (1978).
12. W. Ando, I. Imai, and T. Migita, *J. Chem. Soc. Chem. Commun.,* 822 (1972).
13. H. Ledon, G. Linstrumelle, and S. Julia, *Bull. Soc. Chim. Fr.,* 2065 (1973).

DI-*tert*-BUTYL DICARBONATE

[Dicarbonic acid, bis(1,1-dimethylethyl) ester]

$$(CH_3)_3CO^-K^+ + CO_2 \xrightarrow[0°]{\text{tetrahydrofuran}} (CH_3)_3COCO_2^-K^+$$

$$(CH_3)_3COCO_2^-K^+ + Cl_2CO \xrightarrow[-5° \text{ to } -15°]{\substack{\text{tetrahydrofuran,} \\ \text{benzene}}} (CH_3)_3COCO_2CO_2CO_2C(CH_3)_3$$

$$(CH_3)_3COCO_2CO_2CO_2C(CH_3)_3 \xrightarrow[\substack{\text{carbon tetrachloride,} \\ 25°}]{\text{diazobicyclooctane}} (CH_3)_3COCO_2CO_2C(CH_3)_3 + CO_2$$

Submitted by Barry M. Pope, Yutaka Yamamoto, and D. Stanley Tarbell[1]
Checked by John C. DuBose and Herbert O. House

1. Procedure

Caution! Since the toxic gas phosgene is employed in this preparation, the reaction should be performed in an efficient hood. The glassware, which may be coated with a solution of phosgene, should be washed before it is removed from the hood.

Benzene has been identified as a carcinogen; OSHA has issued emergency standards on its use. All procedures involving benzene should be carried out in a well-ventilated hood, and glove protection is required.

A. *Di-*tert-*butyl tricarbonate.* A 1-l., three-necked flask, fitted with a mechanical stirrer, a 200-ml. pressure-equalizing dropping funnel, a calcium chloride-filled drying tube, and gas-inlet tube with a minimum internal diameter of 6 mm. (Note 1) extending nearly to the bottom of the flask, is dried either by heating with a free flame while passing anhydrous nitrogen through the apparatus, or by heating to 120° for several hours in an oven. Before use, the dropping funnel should be calibrated to indicate levels correspond-ing to 85 ml. and 105 ml. of liquid. While an atmosphere of anhydrous nitrogen (Note 2) is maintained inside the apparatus, it is allowed to cool before a mixture of 44.8 g. (0.400 mole) of alcohol-free potassium *tert*-butoxide (Note 3) and 550 ml. of anhydrous tetrahy-drofuran (Note 4) is added to the reaction flask. The mixture is stirred under an atmos-phere of anhydrous nitrogen for 5–10 minutes to obtain a solution (Notes 3, 5). The reaction flask is immersed in an ice–salt bath maintained at −5° to −20°, and all subsequent steps, including solvent removal, are performed with this cooling bath in place. A stream of anhydrous carbon dioxide (Note 2) is passed through the cold reaction solution for 30 minutes with vigorous stirring, resulting in the formation of a thick, creamy slurry in the reaction flask. While the reaction mixture is being saturated with carbon dioxide, 85 ml. of anhydrous benzene is added to the dropping funnel. A stream of phosgene is bubbled through the benzene until the total volume of the solution is 105 ml., corresponding to the addition of approximately 24 g. (0.24 mole) of phosgene (Note 6). When the addition of carbon dioxide is complete, the phosgene solution is added to the cold reaction slurry, dropwise and with vigorous stirring, over 1 hour, maintaining the temperature of the cooling bath at $-5°$ to $-15°$. During this addition the reaction mixture becomes less viscous but remains a white slurry. When the addition of the phosgene solution is complete, the cold reaction mixture is stirred for an additional 45 minutes while a stream of anhydrous nitrogen is passed through the reaction solution to sweep out most of the excess phosgene. The fittings are removed from the reaction flask, two of the three necks are stoppered, and the volume of solvents in the reaction flask is reduced from about 650 ml. to 100 ml. with a rotary evaporator. During evaporation the flask should be continuously cooled in an ice–salt bath maintained at $-5°$ to $0°$. This evaporation should be performed with either a very efficient aspirator or with a mechanical vacuum pump fitted with an efficient cold trap. Since some phosgene is still present in the reaction mixture, the exhaust from the aspirator or the vacuum pump should be discharged in the hood and any material collected in a cold trap should be emptied in the hood. The residual slurry of finely divided potassium chloride is filtered with suction in a large-diameter, fritted-glass funnel, precooled with 50 ml. of ice-cold pentane. During filtration the filter funnel may be loosely covered with an inverted large-diameter funnel through which a stream of nitrogen is passed to protect the contents of the funnel from atmospheric moisture. The residue in the reaction flask is washed into the filter funnel with 350 ml. of ice-cold pentane and washed with two additional 100-ml. portions of ice-cold pentane, leaving white potassium chloride as a residue. The combined filtrate and pentane washings are concentrated to dryness at $0°$ with a rotary evaporator, under reduced pressure supplied by a vacuum pump equipped with an efficient cold trap, yielding 33.7–39.6 g. (64–75%, Note 7) of di-*tert*-butyl tricarbonate as a colorless solid. This crude product is recrystallized by dissolving it in 1250 ml. of pentane at room temperature, and cooling the solution to $-15°$. The pentane mother liquors are concentrated with a rotary evaporator, giving two additional crops of crystalline product. The total yield is 31.2–32.8 g. (59–62%) of the pure di-*tert*-butyl tricarbonate as colorless prisms, m.p. 62–63° (dec.) (Note 8).

B.   *Di-*tert-*butyl dicarbonate*.   A solution of 20.0 g. (0.0763 mole) of di-*tert*-butyl tricarbonate in 75 ml. of carbon tetrachloride is placed in a 600-ml. beaker fitted with a magnetic stirrer, and 0.10 g. (0.89 mmole) of freshly sublimed 1,4-diazabicyclo[2.2.2]octane (Dabco) is added (Note 9), resulting in the rapid evolution of carbon dioxide. The reaction mixture is stirred at $25°$ for 45 minutes, to complete the loss of carbon dioxide (Note 10), before 35 ml. of water, containing citric acid sufficient to make the aqueous layer slightly acidic, is added. The organic layer is separated, dried over anhydrous magnesium sulfate, and concentrated at $25°$ with a rotary evaporator. The residual liquid is distilled under reduced pressure, yielding 13.3–15.1 g. (80–91%) of di-*tert*-butyl dicarbonate as a colorless liquid, b.p. 55–56° (0.15 mm.) or 62–65° (0.4 mm.) n_D^{25} 1.4071–1.4072 (Note 11).

2. Notes

1. A gas-inlet tube of smaller diameter or a tube fitted with a fritted-glass outlet tends to become clogged during this preparation and is not recommended.
2. The submitters dried this gas by passing it successively through an empty trap, through a trap containing concentrated sulfuric acid, and through another empty trap. The

checkers used this drying procedure for carbon dioxide but dried the nitrogen by passing it through a column of molecular sieves.

3. The submitters employed alcohol-free potassium *tert*-butoxide, purchased from K & K Laboratories, without further purification; the checkers employed comparable material taken from a freshly opened bottle purchased from MSA Research Corporation. The submitters report that among approximately ten different bottles of commercial potassium *tert*-butoxide used, only material from one bottle failed to form the tricarbonate. The defective material was an extremely fine powder that failed to dissolve when stirred with tetrahydrofuran. Solubility in tetrahydrofuran appears to be a good criterion for the purity of alcohol-free potassium *tert*-butoxide. The checkers have observed that a 1 : 1 complex of potassium *tert*-butoxide and *tert*-butyl alcohol is much less soluble in ethereal solvents than is alcohol-free potassium *tert*-butoxide.

4. A reagent grade of tetrahydrofuran (b.p. 65–66°) was distilled from lithium aluminum hydride before use.

5. The submitters reported that their solution had a faint blue color at this point.

6. If desired, the dropping funnel may be removed from the reaction flask and replaced with a calcium chloride drying tube during the preparation of the phosgene solution. When preparation of the phosgene solution is complete, the drying tube should be removed and quickly replaced with the dropping funnel containing the phosgene solution.

7. Although the submitters report that this crude product is suitable for use in the next step of this preparation, the checkers found that once, when using the crude product, the subsequent reaction did not go to completion unless an extra quantity of the diamine base was added. This suggests that some potentially acidic impurity such as *tert*-butyl chloroformate may be present in the crude product and could interfere with the subsequent reaction. The checkers therefore recommend that the product be purified before use in the next step of this preparation.

8. Although this tricarbonate undergoes thermal decomposition when heated above its melting point (63°), forming *tert*-butyl alcohol, isobutylene, and carbon dioxide, the product appears to be stable to storage at temperatures of 25° or less. The product exhibits IR bands (CCl$_4$) attributable to C=O stretching at 1845, 1810, and 1780 cm.$^{-1}$; the ^{1}H NMR spectrum (CCl$_4$) exhibits a singlet at δ 1.55.

9. The submitters report that both 1,4-diazabicyclo[2.2.2]octane and triethylamine have been used to catalyze this decomposition. Triethylamine was less satisfactory as a catalyst because of its relatively rapid reaction with the solvent, carbon tetrachloride, to form triethylamine hydrochloride, and because of difficulty encountered in separating triethylamine from the dicarbonate product. The 1,4-diazabicyclo[2.2.2]octane was efficiently separated from the dicarbonate product by the procedure described in which the crude product was washed with very dilute acid.

10. The progress of this reaction may be monitored either by IR, observing the disappearance of the band at 1845 cm.$^{-1}$, or by ^{1}H NMR (CCl$_4$), following the replacement of the reactant peak (CCl$_4$) at δ 1.55 by the product peak at δ 1.50.

11. The submitters report that this product solidifies when cooled and melts at 21–22° and that the product is stable when stored in a refrigerator. The product exhibits IR absorption (CCl$_4$) attributable to C=O stretching at 1810 and 1765 cm.$^{-1}$ and a ^{1}H NMR (CCl$_4$) singlet at δ 1.50. The mass spectrum of the product exhibits the following

relatively abundant fragment peaks: m/e (relative intensity), 60(10), 59(99), 57(34), 56(86), 55(47), 50(21), 44(100), 43(30), 41(91), 40(27), and 39(61).

3. Discussion

Di-*tert*-butyl tricarbonate, an example of hitherto unknown class of compounds, has been prepared only by the present procedure.[2-4] The corresponding sulfur compound, *tert*-$C_4H_9SCO_2CO_2COS$-*tert*-C_4H_9, also belongs to a new class of compounds and has been prepared by a similar procedure in comparable yields.[4,5] Both tricarbonates are smoothly converted by basic catalysts into the corresponding dicarbonates (sometimes called pyrocarbonates); kinetic studies and differences in thermal decomposition of both tricarbonates have been reported, as well as other reactions of these materials.[4,6] The amine-catalyzed decomposition of the tricarbonates to dicarbonates is believed to involve initial nucleophilic attack by the amine at the center carbonyl group of the tricarbonate.[4,6] Di-*tert*-butyl dicarbonate had been obtained previously[7] in 5% yield; no study of its properties was reported.[7] The di-*tert*-butyl dicarbonate and its sulfur analog have been shown to react with amino acids and their derivatives to form the corresponding *N-tert*-butoxycarbonyl (*t*-BOC) and *N-tert*-butylthiocarbonyl derivatives,[3] which are valuable protecting groups for amino functions. The dicarbonates described in the present synthesis are very mild reagents for the preparation of *t*-BOC and *N-tert*-butylthiocarbonyl derivatives, and may have application in selective reactions with enzymes, nucleic acids and their component nucleotides and nucleosides. Diethyl dicarbonate has been extensively studied in reactions of this type.[8,9]

Other reagents which have been found useful for the synthesis of *t*-BOC derivatives include the hazardous *tert*-butoxycarbonyl azide[10] (see warning, p. 207), *tert*-butyl phenyl carbonate,[11] and 2-*tert*-butoxycarbonyloxyimino-2-phenylacetonitrile.[12]

1. Department of Chemistry, Vanderbilt University, Nashville, Tenn. 37235. This work was supported by grants from the National Science Foundation and the National Institute of Health.
2. C. S. Dean and D. S. Tarbell, *J. Chem. Soc. D*, 728 (1969).
3. D. S. Tarbell, Y. Yamamoto, and B. M. Pope, *Proc. Natl. Acad. Sci., USA*, **69**, 730 (1972); V. F. Pozdnev, *Khim. Prir. Soedin*, **10**, 764 (1974) [*Chem. Abstr.*, **82**, 156690d (1975)]; L. Moroder, A. Hallett, E. Wünsch, O. Keller, and G. Wersin, *Hoppe Seyler's Z. Physiol. Chem.*, **357**, 1651 (1976).
4. C. S. Dean, D. S. Tarbell, and A. W. Friederang, *J. Org. Chem.*, **35**, 3393 (1970).
5. A. W. Friederang and D. S. Tarbell, *Tetrahedron Lett.*, 5535 (1968).
6. C. S. Dean and D. S. Tarbell, *J. Org. Chem.*, **36**, 1180 (1971).
7. J. H. Howe and L. R. Morris, *J. Org. Chem.*, **27**, 1901 (1962).
8. N. J. Leonard, J. J. McDonald, R. E. L. Henderson, and M. E. Reichmann, *Biochemistry*, **10**, 3335 (1971); N. J. Leonard, J. J. McDonald, and M. E. Reichmann, *Proc. Natl. Acad. Sci., USA*, **67**, 93 (1970).
9. W. B. Melchior, Jr., and D. Fahrney, *Biochemistry*, **9**, 251 (1970).
10. M. A. Insalaco and D. S. Tarbell, *Org. Synth.*, **Coll. Vol. 6**, 207 (1988); L. A. Carpino, B. A. Carpino, P. J. Crowley, C. A. Giza, and P. H. Terry, *Org. Synth.*, **Coll. Vol. 5**, 157 (1973).
11. U. Ragnarsson, S. M. Karlsson, B. E. Sandberg, and L. E. Lasson, *Org. Synth.*, **Coll. Vol. 6**, 203 (1988).
12. M. Itoh, D. Hagiwara, and T. Kamiya, *Tetrahedron Lett.*, 4393 (1975).

cis-3,4-DICHLOROCYCLOBUTENE

(Cyclobutene, 3,4-dichloro, cis-)

Submitted by R. Pettit[1] and J. Henery
Checked by J. Napierski and R. Breslow

1. Procedure

Dry chlorine gas is admitted into a solution of 104 g. (1.00 mole) of cycloöctatetraene in 150 ml. of dry carbon tetrachloride contained in a tared, 500-ml., three-necked flask equipped with a gas-inlet tube, a low-temperature thermometer, and a calcium chloride drying tube. The reaction mixture is maintained between −28° and −30° throughout the addition, which is terminated after 71 g. (1.0 mole) of chlorine has been added. After the addition, which takes approximately 1 hour, the reaction mixture is allowed to warm to 0° and 50 g. of powdered sodium carbonate is added, and the contents are shaken gently for several minutes. This treatment removes any hydrochloric acid which may have been produced during the reaction. The mixture is then filtered directly into a 1-l., round-bottomed flask containing 135 g. (0.951 mole) of dimethyl acetylenedicarboxylate. A condenser is fitted to the flask and the solution is heated at gentle reflux for 3 hours (Note 1). The solvent is removed under reduced pressure (Note 2). The crude Diels-Alder adduct, which will slowly solidify on standing, is used directly in the next step.

The crude Diels-Alder adduct is transferred to a 500-ml., pressure-equalizing dropping funnel attached to a 1-l., three-necked, round-bottomed flask; the latter is immersed in an oil bath maintained at 200° and equipped with a distillation head, condenser, and receiving flask. The pressure inside the equipment is reduced to 20 mm. A magnetic stirrer in the pyrolysis flask is started, the Diels-Alder adduct (Note 3) is added slowly to the hot flask, and the pyrolysate collected in the receiving flask. The distillation temperature during the pyrolysis varies from 135° to 152°, depending on the rate of addition of the

Diels-Alder adduct. After the addition is complete (about 1 hour), the pyrolysis is continued for a further 30 minutes or until very little material remains in the pyrolysis flask (Note 4). The crude pyrolysate is then redistilled at 12–15 mm., with all material boiling below 140° collected; this distillate consists mainly of a mixture of dichlorocyclobutene and 1,4-dichlorobutadiene, the residue being mainly dimethylphthalate.

A final distillation at 55 mm. through a 36-in., platinum spinning band column yields 49–52 g. (40–43%) of pure *cis*-3,4-dichlorocyclobutene (b.p. 70–71°, 55 mm.); the forerun (b.p. 58–62°, 55 mm.) consists mainly of 1,4-dichlorobutadiene (Note 5).

2. Notes

1. The reaction is very exothermic and usually it is necessary to remove the external source of heat for a short period as soon as reflux has started.

2. Slight warming and pressures of about 1 mm. are required to remove the last of the solvent.

3. A sun lamp situated close to the funnel may be required to prevent solidification of the Diels-Alder adduct.

4. It is necessary to keep the internal pressure close to 20 mm. (place a manometer in the line). If lower pressures are used, the Diels-Alder adduct itself will distill over; if higher pressures are maintained, the rate of removal of the dichlorocyclobutene from the hot reaction flask is reduced and extensive thermal rearrangement to 1,4-dichlorobutadiene will occur.

5. A good fractionating column is required to separate the 1,4-dichlorobutadiene from the dichlorocyclobutene. At 55 mm. the dichlorobutadiene will distill at 58–62°; after this material has been removed, the temperature will rise fairly sharply to 70°, and at this point the reflux ratio may then be reduced from 10:1 to zero and the dichlorocyclobutene collected quickly. It usually is necessary to apply heat frequently with a sun lamp to prevent solidification of the dichlorobutadiene in the exit tube of the distillation apparatus, especially if the receiving flask is kept cold.

3. Discussion

This method of preparation, due to Nenitzescu, Avram, Marica, Dinulescu, Farcasiu, Elian, and Mateescu,[2] is the only practical method available at this time for the preparation of 3,4-dichlorocyclobutene.

1. Deceased, December 10, 1981; work done at Department of Chemistry, The University of Texas at Austin, Austin, Texas 78712.
2. M. Avram, E. Marica, I. Dinulescu, M. Farcasiu, M. Elian, G. Mateescu, and C. D. Nenitzescu, *Chem. Ber.*, **97**, 372 (1964).

cis-DICHLOROALKANES FROM EPOXIDES:
cis-1,2-DICHLOROCYCLOHEXANE

(Cyclohexane, 1,2-dichloro-, cis-)

$$(C_6H_5)_3P + Cl_2 \xrightarrow[0°]{\text{benzene}} (C_6H_5)_3PCl_2 \xrightarrow[\text{benzene}]{\text{1,2-epoxycyclohexane}}$$

Submitted by JAMES E. OLIVER[1] and PHILIP E. SONNET[2]
Checked by JERROLD M. LIESCH and GEORGE BÜCHI

1. Procedure

Caution! Benzene has been identified as a carcinogen. OSHA has issued emergency standards on its use. All procedures involving benzene should be carried out in a well-ventilated hood, and glove protection is required. A hood should be employed for the chlorination.

A 1-l., three-necked flask is charged with 95 g. (0.36 mole) of triphenylphosphine (Note 1) and 500 ml. of anhydrous benzene, and fitted with a gas-inlet (Note 2), a mechanical stirrer, and a condenser with attached drying tube. The flask is cooled in an ice bath, stirring is begun, and chlorine is introduced through the gas-inlet. Dichlorotriphenylphosphorane separates as a white solid or as a milky oil; the flow of chlorine is discontinued when the mixture develops a strong lemon-yellow color (Note 3). The gas inlet is quickly replaced by an addition funnel, and a solution of 10 g. of triphenylphosphine in 60 ml. of benzene is added dropwise fairly rapidly (Note 4). A solution of 24.5 g. (0.250 mole) of 1,2-epoxycyclohexane (Note 5) in 50 ml. of benzene is then added dropwise over *ca.* 20 minutes. The ice bath is replaced by a heating mantle, and the mixture, which consists of two liquid phases, is stirred and refluxed for 4 hours. It is then cooled, and excess dichlorotriphenylphosphorane is destroyed by the slow addition of 10 ml. of methanol (Note 6). The mixture is concentrated on a rotary evaporator at *ca.* 100 mm., and the residue, which may be a white solid or a viscous oil, is triturated with 300 ml. of petroleum ether (30–60°). The solid triphenylphosphine oxide that separates is collected by suction filtration. The cake is thoroughly broken up with a spatula and washed with three 100-ml. portions of petroleum ether. The combined filtrates, from which a little more triphenylphosphine oxide precipitates, are refiltered, then washed with 250-ml. portions of aqueous 5% sodium bisulfite (Note 7) and with water. The organic phase is dried over magnesium sulfate, filtered, concentrated on a rotary evaporator at *ca.* 100 mm., and distilled through a 20-cm. Vigreux column. There is very little forerun before 27–28 g. (71–73%) of *cis*-1,2-dichlorocyclohexane is collected at 105–110° (33 mm.), n_D^{25} 1.4977 (Note 8).

2. Notes

1. Triphenylphosphine was purchased from Aldrich Chemical Company, Inc. Use of a considerable excess of triphenylphosphine ensures complete reaction and obviates the need for rigorously dried glassware and reagents. Hydrochloric acid, generated by the reaction of dichlorotriphenylphosphorane and water, can react with the epoxide to produce a *trans*-chlorohydrin, which is, however, converted to a *cis*-dichloride by dichlorotriphenylphosphorane under the conditions of the reaction.

2. A glass tube of 7-mm. diameter is recommended. If chlorine is introduced through a fritted-glass tube, the dichlorotriphenylphosphorane collects on the frit as a sticky gum.

3. A sharp endpoint is not observed. A simple test for complete chlorination is as follows: the flow of chlorine and the stirrer are stopped, and the mixture is allowed to settle. Chlorine is then admitted without stirring. If unreacted triphenylphosphine is present, a visible clouding (formation of dichlorotriphenylphosphorane) will occur at the gas–liquid interface.

4. Although a slight excess of chlorine does not appear to be deleterious, a substantial excess is avoided by adding the last portion of triphenylphosphine at this point.

5. Commercial 1,2-epoxycyclohexane, supplied by Columbia Organic Chemicals Company, Inc., was used.

6. The reaction mixture may be allowed to stand overnight before addition of methanol.

7. The distilled *cis*-dichlorocyclohexane tends to become colored if the solution is not washed with a reducing agent.

8. The checkers, using a 10-cm. Vigreux column, found that it was necessary to take a wider boiling range fraction (105–115°, 33 mm.) to obtain similar yields. The product is virtually free of *trans*-1,2-dichlorocyclohexane (the isomeric 1,2-dichlorocyclohexanes are readily separated by GC on Carbowax 20M or on diethylene glycol succinate columns).

3. Discussion

This procedure is general for the conversion of epoxides to dichlorides with inversion of configuration at each of the two carbons and, in effect, provides a method for the *cis*-addition of chlorine to a double bond.[3] *cis*-1,2-Dichlorocyclohexane has also been prepared from 1,2-epoxycyclohexane and sulfuryl chloride,[4] but the stereospecificity of the reaction appears to be extremely sensitive to reaction conditions, and the yield is lower than that obtained by the method described here. Other methods give *cis*-1,2-dichlorocyclohexane contaminated with considerable amounts of the *trans*-isomer. This method has been used to convert *cis*- and *trans*-4,5-epoxyoctanes to *meso*- and *d,l*-4,5-dichloroöctanes, respectively, and *trans*-7,8-epoxyoctadecane to *threo*-7,8-dichloroöctadecane. These conversions were carried out on smaller amounts of material, and the products were purified by column chromatography on silica gel. Yields were 51–63%.

Halogenations with dihalotriphenylphosphoranes have been reviewed briefly by Fieser and Fieser.[5] Dibromotriphenylphosphorane appears to have been studied somewhat more than the dichloro compound, but both reagents effectively convert alcohols to alkyl halides, carboxylic acids and esters to acid halides, etc. The reaction of 1,2-

epoxycyclohexane with dibromotriphenylphosphorane under conditions similar to those described here gives a mixture of *cis-* and *trans-*1,2-dibromocyclohexanes. A reagent prepared from triphenylphosphine and carbon tetrachloride has been used for similar transformations.[6]

1. Agricultural Research, Northeastern Region, Pesticide Degradation Laboratory, Agricultural Environmental Quality Institute, Beltsville Agricultural Research Center, USDA, Beltsville, Maryland 20705. Mention of a proprietary product or company does not imply endorsement by the United States Department of Agriculture (USDA).
2. Present address: Agricultural Research, Southern Region, Insect Attractants, Behavior and Basic Biology Research Laboratory, United States Department of Agriculture, 1700 SW 23rd Drive, P.O. Box 14565, Gainesville, Florida 32604.
3. P. E. Sonnet and J. E. Oliver, *J. Org. Chem.*, **41**, 3279 (1976).
4. J. R. Campbell, J. K. N. Jones, and S. Wolfe, *Can. J. Chem.*, **44**, 2339 (1966).
5. L. F. Fieser and M. Fieser, "Reagents for Organic Synthesis," Vol. I, Wiley, New York, 1968, p. 1247.
6. J. G. Calzada and J. Hooz, *Org. Synth.*, **Coll. Vol. 6**, 634 (1988).

2,3-DICYANOBUTADIENE AS A REACTIVE INTERMEDIATE BY *in situ* GENERATION FROM 1,2-DICYANOCYCLOBUTENE: 2,3-DICYANO-1,4,4a,9a-TETRAHYDROFLUORENE

(2,3-Fluorenedicarbonitrile, 1,4,4a,9a-tetrahydro-)

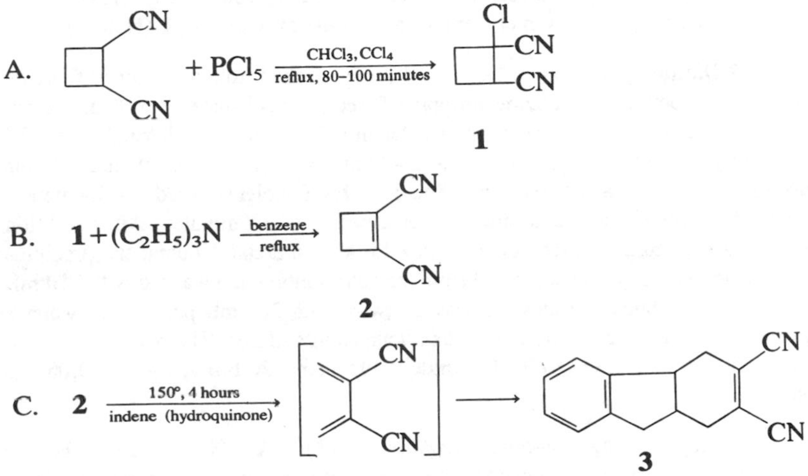

A. [structure with CN groups] $+ PCl_5 \xrightarrow[\text{reflux, 80-100 minutes}]{\text{CHCl}_3, \text{CCl}_4}$ [structure with Cl, CN, CN]

1

B. $\mathbf{1} + (C_2H_5)_3N \xrightarrow[\text{reflux}]{\text{benzene}}$ [structure with CN, CN]

2

C. $\mathbf{2} \xrightarrow[\text{indene (hydroquinone)}]{150°, 4 \text{ hours}}$ [bracketed intermediate with CN, CN] $\longrightarrow$ [fluorene structure with CN, CN]

3

Submitted by D. BELLUŠ,[1] H. SAUTER, and C. D. WEIS
Checked by A. J. ARDUENGO and WILLIAM A. SHEPPARD

1. Procedure

Caution! Benzene has been identified as a carcinogen. OSHA has issued emergency standards on its use. All procedures involving benzene should be carried out in a well-ventilated hood, and glove protection is required.

A. *1-Chloro-1,2-dicyanocyclobutane* (**1**). A 2-l., three-necked flask is equipped with a mechanical stirrer, a 500-ml. pressure-equalizing funnel, and an efficient reflux condenser provided with a gas-outlet tube connected by plastic tubing to a conical funnel inverted over a 5-l. beaker containing aqueous sodium hydroxide, for absorption of the evolved hydrogen chloride. The flask is charged with 562 g. (2.70 moles) of phosphorus pentachloride and 750 ml. of carbon tetrachloride. The rapidly stirred suspension is heated to reflux, and a solution of 159 g. (1.50 moles) of 1,2-dicyanocyclobutane (Note 1) in 120 ml. of chloroform is added dropwise over a period of 40 minutes (Note 2). After addition is complete, the solvents and the phosphorus trichloride formed during the reaction are removed by distillation at 100–150 mm. over a period of 40–60 minutes, with a bath temperature not exceeding 80° (Note 2). The residual liquid is cooled to room temperature and dissolved in 400 ml. of diethyl ether (Note 3). The etheral solution is placed in a 500-ml. dropping funnel and added over a period of 3 hours to a stirred slurry of 1.7 kg. (12 moles) of sodium hydrogen carbonate, 800 g. of crushed ice, and 500 ml. of water.

During the addition the temperature is maintained between −5° and 0° with an external ice–salt bath, and after the addition, stirring is continued for 1 hour at the same temperature. The precipitated salts are removed by suction filtration through a sintered-glass funnel of medium porosity and are thoroughly washed with 300 ml. of ether. The organic layer is separated, and the aqueous filtrate extracted with three 200-ml. portions of ether. The combined ether extracts are dried over anhydrous magnesium sulfate, filtered, and concentrated on a rotary evaporator, yielding 166–174 g. (79–83%) of a yellow oil consisting of an isomeric mixture of crude cyclobutane 1 (Note 4).

B. 1,2-*Dicyanocyclobutene* (2). A 2-l., three-necked, round-bottomed flask fitted with a 500-ml. pressure-equalizing dropping funnel, a mechanical stirrer, and a reflux condenser protected from moisture by a calcium chloride tube is charged with 131 g. (1.30 moles) of triethylamine (Note 5) and 400 ml. of benzene. The stirred solution is heated to gentle reflux, and a solution of 168.5 g. (1.199 moles) of crude cyclobutane 1 in 200 ml. of benzene is added dropwise over a period of 30 minutes. After addition is complete, the mixture is stirred under reflux for an additional 2 hours; the precipitated triethylamine hydrochloride is filtered from the cold solution and washed with 150 ml. of benzene. The combined filtrates are washed twice with 200-ml. portions of water and evaporated using a water aspirator at a bath temperature of 35°. The residue is distilled, yielding 105–110 g. (83–87%) of crude cyclobutene 2, b.p. 55–60° (0.06 mm.) (Note 6).

C. 2,3-*Dicyano*-1,4,4a,9a-*tetrahydrofluorene* (3). A 100-ml., round-bottomed flask equipped with a reflux condenser under nitrogen pressure is charged with 10.4 g. (0.100 mole) of crude cyclobutene 2, 23.3 g. (0.201 mole) of indene (Note 7), and 0.3 g. of hydroquinone. The reaction mixture is stirred and heated at 150° for 4 hours under nitrogen. The reflux condenser is replaced by a still head, and 6.3 g. (0.053 mole) of indene is distilled from the flask at a bath temperature of about 95° (11 mm.) (Note 8). The dark-colored reaction mixture is transferred to a 500-ml., round-bottomed flask, diluted with 200 ml. of benzene followed by 1 g. of decolorizing carbon, and the resulting mixture is refluxed for 2 hours. After the mixture is cooled to room temperature, the carbon is removed by filtration, and the benzene is distilled. The residual oily residue solidifies on standing and is recrystallized from 45 ml. of ethanol, yielding 15.4–15.9 g. (70–72%) of crystalline fluorene 3, m.p. 98.5–100° (Note 9).

2. Notes

1. 1,2-Dicyanocyclobutane (*cis*- and *trans*-isomer mixture) was purchased from Aldrich Chemical Company, Inc., and used without further purification.

2. The addition and distillation must be accomplished within the specified period of time; otherwise the amount of dichlorinated 1,2-dicyanocyclobutane increases considerably. The submitters found that an 80% molar excess of phosphorus pentachloride is optimum. A molar excess less than specified (under given experimental conditions) gives considerable unreacted starting material. Under forcing experimental conditions, such as longer reaction times and/or higher temperatures, the starting material can be completely consumed, even with less than 80% molar excess of phosphorus pentachloride, but a considerable amount of dichlorinated products is formed.

3. The checkers found that, because of the time required for completion of this step, a convenient modification is to cool the ether solution to $-78°$ in dry ice and store overnight at $-78°$. A solid complex of phosphorus pentachloride and cyclobutane **1**, m.p. 88–91°, precipitates in the cold ether solution. This complex may not redissolve on warming to room temperature, but the suspension in ether can be used to proceed with the second half of Step A.

4. The checkers obtained the cyclobutane **1** as a colorless crystalline solid, m.p. 45–47° (a mixture of major and minor isomers), that is relatively free of 1,2-dichloro-1,2-dicyanocyclobutane. The product had the following spectral properties: ^{1}H NMR (CDCl$_3$) δ (multiplicity): 2.35–3.15 (m), 3.40–4.20 (m); ^{13}C NMR (CDCl$_3$), δ: major isomer 21.58 (t), 38.41 (d), 37.24 (t), minor isomer 22.48 (t), 36.72 (d), 36.91 (t).

5. Reagent grade triethylamine was dried over sodium hydroxide and distilled before use (b.p. 88–89°).

6. The checkers obtained yields of 115 g. (92%) on a 1.2 mole scale and 94 g. (90%) on a 1.0 mole scale.

The product was analyzed by GC on a 1.23 m. × 0.65 cm. stainless-steel column of SE-52 on Varoport 30, which was heated to 150° and swept with helium at 60 ml. per minute. Retention times for the various components (minutes) are: cyclobutene **2**, (2.6); *trans*-1,2-dicyanocyclobutane(3.6); two isomeric 1,2-dichloro-1,2-dicyanocyclobutanes (5.1 and 5.9, respectively); *cis*-1,2-dicyanocyclobutane (9.8).

For most synthetic purposes, such as [4 + 2]- and [2 + 2]-cycloadditions,[2-7] ring-opening reactions,[2,8] and hydrolytic reactions,[2] this crude cyclobutene **2**, which contains approximately 1–3.5% of isomeric mixtures of 1,2-dichloro-1,2-dicyanocyclobutanes, can be used satisfactorily without further purification. Pure cyclobutene **2** can be prepared by treatment of the crude product with Raney cobalt, thereby removing residual quantities of isomeric 1,2-dichloro-1,2-dicyanocyclobutanes. In a typical experiment the crude product is placed in a 250-ml., round-bottomed flask and stirred with 10 g. of Raney cobalt for 4 hours at 70° under nitrogen. Distillation directly from the reaction vessel without filtering off the metal slurry yields 94–98 g. (60–63%) of cyclobutene **2** as a colorless liquid, n_D^{20} 1.4926, d_{22}^4 1.033. The Raney cobalt used by the submitters was obtained from Fluka A G, Buchs, Switzerland, as a suspension in water, and washed with tetrahydrofuran before use. Raney nickel and nickel tetracarbonyl, respectively, are also good dechlorinating reagents. The use of Raney cobalt, however, diminishes the danger of self-ignition during the preparation procedure. The spectral properties are as follows: IR (neat) cm^{-1}: 3002, 2957, 2230, 1612, 1422, 1251, 1169, 1003, and 623; UV (CH$_3$OH) nm. max. (log ε): 235 (4.06), 247 sh (3.90); ^{1}H NMR (CDCl$_3$), δ (multiplicity): 2.91 (s).

7. The indene used by the submitters was "practical grade," purchased from Fluka A G, Buchs, Switzerland. The indene used by the checkers was purchased from Aldrich Chemical Company, Inc. Both were distilled (b.p. 60–65°, 11 mm.) before use.

8. Recovered indene may be used for the next batch without further purification.

9. Fluorene **3** has the following spectral properties: IR (KBr) cm^{-1}: 2222, 1608, 1479, 1440, 773, and 742; UV (C$_2$H$_5$OH) nm. max. (log ε): 216 (4.10), 2.34 (3.99), 261 (3.16), 267 (3.11), and 274 (3.07); mass spectrum *m/e*: 220 (m^+) and 116 (base peak); ^{1}H NMR (100 MHz., CDCl$_3$): two complex multiplets for the aromatic and aliphatic protons (360 MHz., CDCl$_3$), δ (multiplicity, coupling constant J in Hz., number of protons,

assignment): 2.10 [d, J = 18; d, J = 7; and t, J = 3; one H of $CH_2(1)$ or one H of $CH_2(4)$], 2.55–2.70 [m, one H of $CH_2(1)$, one H of $CH_2(4)$, one H of $CH_2(9)$ and $CH(9a)$], 2.88 [d, J = 18; d, J = 7; d, J = 3; and d, J = 1.5; one H of $CH_2(1)$ or one H of $CH_2(4)$], 3.11 [d, J = 15; d, J = 6, one H of $CH_2(9)$], 3.42 [d, J = 5; t, J = 7, $CH(4a)$], 7.20 (m, 4, aromatic H); ^{13}C NMR ($CDCl_3$), δ (assignment): 143.8 [$C(8a)$], 141.3 [$C(4b)$], 127.6, 127.2, and 125.6 [$C(6)$, $C(7)$, and $C(8)$, not assigned individually], 126.6 and 125.9 [$C(2)$ and $C(3)$, not assigned], 123.2 [$C(5)$], 115.9 [two nitrile carbons], 40.2 [$C(4a)$], 38.4 [$C(9)$], 35.4 [$C(9a)$], 30.7 and 29.7 [$C(1)$ and $C(4)$, not assigned].

3. Discussion

Three syntheses of 1,2-dicyanocyclobutene (2) have been previously described. The first involves dehydration of cyclobutene-1,2-dicarboxamide, with no specified yield.[9] The second procedure involves a concomitant chlorination and catalytic dehydrochlorination of 1,2-dicyanocyclobutane in the gas phase, yielding 1,2-dicyanocyclobutene (2) in a mixture of several other products.[10] The third method consists of dechlorination of 1,2-dichloro-1,2-dicyanocyclobutane using metals, such as zinc copper couple,[11] Raney nickel,[11] and, especially, Raney cobalt.[2] In comparison with the third synthesis, the overall yield of the present procedure is 5–10% higher. Furthermore, the reaction is performed in less time and utilizes considerably cheaper reagents.

Pure, crystalline 2,3-dicyanobutadiene has been prepared in high yield by gas-phase thermolysis of cyclobutene (2).[2,8] Analogously, thermolysis of derivatives of cyclobutene-1,2-dicarboxylic acid appears to represent a general procedure for the synthesis of derivatives of butadiene-2,3-dicarboxylic acid, of high purity.[2,12] These butadienes take part in [4 + 2]-cycloaddition reactions either as reactive dienes[2,13,14] or as reactive dienophiles.[2,14] In the pure state, however, they tend to polymerize, and even crystalline 2,3-dicyanobutadiene slowly polymerizes, yielding a highly cross-linked polymer without losing its original crystal form. A [2 + 4]-dimer of a 2,3-dicyanobutadiene is also formed by heating the dicyanocyclobutene in solution with a polymerization inhibitor.[2] Monomeric derivatives of butadiene-2,3-dicarboxylic acid cannot be prepared in solution because of rapid dimerization.[8,14]

The present procedure, *in situ* generation and trapping of 2,3-dicyanobutadiene in the presence of olefins, overcomes these problems and affords [4 + 2]-cycloadducts in good yields, particularly in the case of olefins possessing a strained double bond.[2] Substituted 1,2-dicyanocyclohexenes, prepared by the *in situ* [4 + 2]-cycloadditions, can be dehydrogenated to new aromatic *ortho*-dinitriles. For example, 2,3-dicyanofluorene is prepared in 56% yield by heating 2,3-dicyano-1,4,4a,9a-tetrahydrofluorene (3) at 200° in dimethylmaleate in the presence of 5% palladium on charcoal. Other aromatic *ortho*-dinitriles have also been prepared by this method.[2] Because 2,3-dicyanobutadiene is an electron-deficient diene, it does not react with electron-deficient olefins, such as maleic anhydride and fumaronitrile[2,8] using this procedure. However, by generating the dicyanobutadiene in refluxing chlorobenzene in the presence of maleic anhydride and

TABLE I

[4 + 2]-Cycloaddition Reactions of 2,3-Dicyanobutadiene Formed *in situ* from
1,2-Dicyanocyclobutene[2]

Olefin	Product	Yield (%)[a]	Temperature (°C)	Time (hours)
Norbornadiene	4,5-Dicyanotricyclo-[6.2.1.0²·⁷]undeca-4,9-diene	79[b]	150	12
Acenaphthylene	8,9-Dicyano-6b,7,10,10a-tetrahydrofluoranthene	77	138	48
Cyclopentene	3,4-Dicyanobicyclo[4.3.0]-non-3-ene	65	135	16
Ethylene	1,2-Dicyanocyclohexene	58	135	16
(E)-Stilbene	1,2-Dicyano-4,5-diphenyl-cyclohexene	32	138	24
Butyl vinyl ether	1,2-Dicyano-4-butoxy-cyclohexene	28	155	16
(E)-1,2-Dichloro-ethylene	1,2-Dicyano-4,5-*trans*-dichlorocyclohexene	8	135	16
2-Vinylpyridine	1,2-Dicyano-4-(2 -pyridyl)-cyclohexene	5	138	48

[a] Yields of analytically pure products are given.
[b] A 70:24 mixture of *exo* and *endo* isomers. Some 2:1 cycloadduct was also isolated (2.4% yield).

2,5-di-*tert*-butylbenzoquinone, as an inhibitor, the [2 + 4]-cyclic dicarbonitrile adduct, m.p. 201–202.5°, was formed in a yield of 38%.[15]

1. Central Research Laboratories, Ciba-Geigy Limited, CH-4002, Basle, Switzerland.
2. D. Belluš, K. von Bredow, H. Sauter, and C. D. Weis, *Helv. Chim. Acta*, **56**, 3004 (1973).
3. D. Belluš and G. Rist, *Helv. Chim. Acta*, **57**, 194 (1974).
4. D. Belluš, H.-C. Mez, and G. Rihs, *J. Chem. Soc., Perkin Trans. II*, 884 (1974).
5. D. Belluš, H.-C. Mez, G. Rihs, and H. Sauter, *J. Am. Chem. Soc.*, **96**, 5007 (1974).
6. R. Wehrli, H. Schmid, D. Belluš, and H.-J. Hansen, *Helv. Chim. Acta*, **60**, 1325 (1977).
7. H.-D. Martin, M. Hekman, G. Rist, H. Sauter, and D. Belluš, *Angew. Chem.*, **89**, 420 (1977).
8. D. Belluš and C. D. Weis, *Tetrahedron Lett.*, 999 (1973).
9. H. Prinzbach and H.-D. Martin, *Chimia*, **23**, 37 (1969).
10. J. L. Greene and M. Godfrey, U.S. Pat. 3,336,354 (1967) [*Chem. Abstr.*, **68**, 21598v (1968)].
11. J. L. Greene, N. W. Standish, and N. R. Gray, U.S. Pat. 3,275,676 (1966) [*Chem. Abstr.*, **66**, 10637y (1967)].
12. P. Dowd and K. Kang, *Synth. Commun.*, **4**, 151 (1974).
13. E. Vogel, *Justus Liebigs Ann. Chem.*, **615**, 14 (1958).
14. U.-I. Záhorszky and H. Musso, *Justus Liebigs Ann. Chem.*, 1777 (1973).
15. W. A. Sheppard, unpublished results.

DIDEUTERIODIAZOMETHANE

(Methane-d_2, diazo-)

$$CH_2N_2 \xrightarrow[\text{tetrahydrofuran, ether, } 0°]{\text{NaOD, D}_2\text{O}} CD_2N_2$$

Submitted by P. G. GASSMAN[1] and W. J. GREENLEE
Checked by DAVID G. MELILLO and HERBERT O. HOUSE

1. Procedure

Caution! Diazomethane is toxic and explosive. The operations described in this procedure must be carried out in a good hood with an adequate shield (Note 1).

A distilled, ethereal solution (300 ml.) containing approximately 0.06 mole of diazomethane (Note 1) is prepared from 22.5 g. of a 70% dispersion (15.8 g., 0.063 mole) of bis-(N-methyl-N-nitroso)terephthalamide (Note 2), 75 ml. of aqueous 30% sodium hydroxide, 55 ml. of diethylene glycol monoethyl ether, and 375 ml. of diethyl ether by the procedure described in *Org. Synth.*, **Coll. Vol. 5,** 351 (1973). The receiving flask containing the ethereal diazomethane is capped with a rubber stopper fitted with a drying tube containing potassium hydroxide pellets to protect the solution from atmospheric moisture. The concentration of diazomethane may be determined either by titration with ethereal benzoic acid (Note 3) or spectrophotometrically (Note 4).

A dry, 250-ml. Erlenmeyer flask equipped with a Teflon-coated magnetic stirring bar is charged with 11 ml. of a solution (Note 5) containing 0.01 mole of sodium deuteroxide in 10 ml. of deuterium oxide and 1 ml. of anhydrous tetrahydrofuran. After the solution has been cooled in an ice bath, 120 ml. of the ethereal solution containing 0.039 mole of diazomethane is added, the flask is stoppered loosely with a cork, and the reaction mixture is stirred vigorously at 0° for one hour. The lower deuterium oxide layer is removed with a pipette and a fresh 11-ml. portion of the sodium deuteroxide solution is added. This mixture is then stirred for one hour at 0°, and the process is repeated until a total of four exchanges have been performed. The ethereal diazomethane solution is then decanted into a clean, dry, 250-ml. Erlenmeyer flask and dried over 10 g. of anhydrous sodium carbonate. The resulting solution (approximately 110 ml.) contains (spectrophotometric analysis, Note 4, or titration with benzoic acid, Note 3) 0.020–0.022 mole (51–56%) of dideuteriodiazomethane, which is 98–99% deuterated (Note 6).

2. Notes

1. Diazomethane is not only toxic, but also potentially explosive. Hence, one should wear heavy gloves and goggles and work behind a safety screen or a hood door with safety glass, as is recommended in the preparation of diazomethane described in *Org. Synth.*, **Coll. Vol. 4,** 250 (1963). As is also recommended, ground joints and sharp surfaces should be avoided; thus, all glass tubes should be carefully fire-polished, connections should be made with rubber stoppers, and separatory funnels should be avoided, as should

etched or scratched flasks. Explosion of diazomethane has been observed at the moment crystals (sharp edges!) suddenly separated from a supersaturated solution. Stirring with a Teflon-coated magnetic stirrer is much preferred to swirling the reaction mixture by hand (there has been at least one case of a chemist whose hand was injured by an explosion during the preparation of diazomethane in a hand-swirled reaction vessel). It is imperative that diazomethane solutions not be exposed to direct sunlight or placed near a strong artificial light because light is thought to have been responsible for some of the explosions encountered with diazomethane. Particular caution should be exercised when an organic solvent boiling higher than ether is used. Because such a solvent has a vapor pressure lower than ether, the concentration of diazomethane in the vapor above the reaction mixture is greater and an explosion is more apt to occur. Since most diazomethane explosions occur during distillation, procedures that avoid distillation offer certain advantages. An ether solution of diazomethane satisfactory for many uses can be prepared as described in *Org. Synth.*, **Coll. Vol. 2,** 165 (1943), where nitrosomethylurea is added to a mixture of ether and 50% aqueous potassium hydroxide, and the ether solution of diazomethane is subsequently decanted from the aqueous layer and dried over potassium hydroxide pellets (not sharp-edged sticks!). However, the reported potent carcinogenicity[2] of nitrosomethylurea mitigates other advantages of this procedure. Two procedures involving distillation of diazomethane, those in *Org. Synth.*, **Coll. Vol. 4,** 250 (1963) and **Coll. Vol. 5,** 351 (1973), may be recommended. In neither case is there much diazomethane present in the distilling flask. The hazards associated with diazomethane are discussed by Gutsche.[6]

2. The submitters used an undistilled ethereal solution of diazomethane, prepared from nitrosomethylurea (Note 1).[3] For use in the hydrogen–deuterium exchange reaction described, ethereal diazomethane solutions prepared by any standard preparative procedures (Note 1) appear to be equally satisfactory: [*Org. Synth.*, **Coll. Vol. 4,** 250 (1963), **Coll. Vol. 2,** 165 (1943), and **Coll. Vol. 5,** 351 (1973)].

3. The concentration of diazomethane may be determined by reaction of an aliquot of the ethereal solution with a weighed excess of benzoic acid in cold (0°) ether solution, as described in *Org. Synth.*, **Coll. Vol. 2,** 165 (1943). The unchanged benzoic acid is then determined by titration with standard aqueous 0.1 M potassium hydroxide.

4. *Caution! The following spectrophotometric analysis should be performed in a hood.* To determine the concentration of diazomethane obtained in this preparation, a 5-ml. aliquot of the distilled solution is diluted to 25 ml. with ether, and a portion of this solution is placed in a cylindrical Pyrex cell with an internal diameter of 1.0 cm. The optical density of the solution is determined at 410 nm with a suitable colorimeter such as a Bausch and Lomb Spectronic 20. From the molecular extinction coefficient, ϵ 7.2, at 410 nm for diazomethane in ether solution, the concentration of diazomethane can be calculated. In a typical preparation the optical density of the diluted solution at 410 nm was 0.46 corresponding to a diazomethane concentration of 0.064 M; thus, the concentration of the undiluted solution was 0.32 M, corresponding to a 77% yield of diazomethane.

5. It is convenient to prepare 110 ml. of this solution at a time. Since hydrogen is evolved, the solution should be prepared in a hood. A dry, 250-ml., three-necked flask is fitted with a magnetic stirrer, a rubber septum, a glass stopper, and a 125-ml. Erlenmeyer flask attached to the third neck of the reaction flask with a 10-cm. length of Gooch rubber

tubing or nylon tubing. The apparatus is flushed with nitrogen from a hypodermic needle inserted through the rubber septum. Small, freshly cut chips of metallic sodium (2.3 g., 0.10 g.-atom) are placed in the Erlenmeyer flask and 100 ml. of deuterium oxide (99.7% pure grade obtained from Columbia Organic Chemicals Company, Inc.,) is placed in the reaction flask. With a hypodermic needle inserted through the rubber septum to permit the escape of hydrogen, the sodium chips are added, slowly and with stirring, to the reaction vessel. When reaction with the sodium is complete, the solution is diluted with 10 ml. of anhydrous tetrahydrofuran and stored under a nitrogen atmosphere.

6. The deuterium content can be determined by reaction of the deuterated di-azomethane with benzoic acid- -d in anhydrous ether followed by analysis of methyl benzoate for deuterium content either by ^{1}H NMR spectroscopy or by mass spectroscopy. Benzoic acid- -d is prepared by heating a mixture of 48.6 g. (0.216 mole) of benzoic anhydride (obtained from Aldrich Chemical Company, Inc.), 0.10 g. (0.00090 mole) of anhydrous sodium carbonate, and 7.0 g. (0.35 mole) of deuterium oxide to 90° for 2 hours. The resulting mixture is distilled at atmospheric pressure in a short-path still fitted with a receiver protected from atmospheric moisture by a drying tube. After removal of a forerun, b.p. 100–101°, the benzoic acid-O-d is collected at 245–247°. During the distillation it is necessary to warm the distillation apparatus with a heat gun or an IR lamp to prevent solidification of the benzoic acid-O-d before it reaches the receiver.

Caution! The following reaction should be performed in a good hood (Note 1). A cold (0°) solution of 1.43 g. (0.0116 mole) of benzoic acid-O-d in 10 ml. of anhydrous ether is placed in a dry, 100-ml., round-bottomed flask fitted with a rubber stopper and a Teflon-coated magnetic stirring bar. The flask is cooled in an ice bath, and a sufficient amount of the ethereal dideuteriodiazomethane solution is added from a pipette, providing excess dideuteriodiazomethane in the reaction mixture. The reaction flask is stoppered loosely, and the resulting yellow solution is stirred at 0° for 10 minutes and concentrated by first warming the solution on a steam bath in the hood, then removing the last traces of solvent under reduced pressure. The residual liquid methyl benzoate (1.4–1.5 g., 90–95% yield) is analyzed for deuterium content. For a ^{1}H NMR analysis, the spectrum of the pure liquid is taken and the extent of deuteration is determined by integration of the areas under the multiplet in the region δ 7.1–8.3 (aromatic CH) and the peak at δ 3.82 (OCH_3). For mass spectroscopic analysis, the mass spectra of the deuterated sample and a sample of undeuterated methyl benzoate each are measured at an ionizing potential sufficiently low (approximately 12 eV.) to minimize the formation of an M-1 fragment at m/e 135 in the spectrum of the undeuterated sample. The relative abundances of the m/e 136 and 137 peaks in the spectrum of the undeuterated sample are then used to correct the peaks at m/e 137, 138, and 139 in the deuterated sample for contributions from the ^{13}C isotope. From the relative abundances of the m/e 136 peak and the corrected m/e 137, 138, and 139 peaks in the spectrum of the deuterated sample the relative proportions of d_0, d_1, d_2, and d_3 species in the deuterated methyl benzoate can be calculated. Both ^{1}H NMR and mass spectral analysis indicated the methyl benzoate to be 98% deuterated (6–7% d_2 species and 93–94% d_3 species). When the dideuteriodiazomethane solution was allowed to react with undeuterated benzoic acid, hydrogen–deuterium exchange occurred more rapidly than esterification. The methyl benzoate produced was 70% deuterated (^{1}H NMR analysis) and contained 4% d_0, 37% d_1, 30% d_2, and 29% d_3 species (mass spectral analysis).

3. Discussion

The exchange procedure described was developed for the preparation of dideuteriodiazomethane, for use in labeling studies. It is basically a modification of a procedure that has been used extensively;[4] however, the literature procedures give relatively little detail. This modified procedure permits the synthesis of fairly large amounts of high-purity dideuteriodiazomethane. Dideuteriodiazomethane has also been prepared from N-nitrosomethyl-d_3-urea and related trideuterated diazomethane precursors.[5] Deuterated chloroform and hydrazine hydrate have also been used to prepare dideuteriodiazomethane.[6,7]

The procedure described provides a general method for the hydrogen–deuterium exchange of simple diazoalkanes.

1. Department of Chemistry, The Ohio State University, 140 West 18th Avenue, Columbus, Ohio 43210. [Present address: Department of Chemistry, University of Minnesota, 207 Pleasant Street S. E., Minneapolis, Minnesota 55455].
2. A. Graffi and F. Hoffman, *Acta Biol. Med. Ger.*, **16**, K-1 (1966).
3. C. D. Gutsche, *Org. React.*, **8**, 391–394 (1954).
4. L. C. Leitch, P. E. Gagnon, and A. Cambron, *Can. J. Res.*, **28B**, 256 (1950); G. W. Robinson and M. McCarty, Jr., *J. Am. Chem. Soc.*, **82**, 1859 (1960); T. D. Goldfarb and G. C. Pimentel, *J. Am. Chem. Soc.*, **82**, 1865 (1960); H. Dahn, A. Donzel, A. Merbach, and H. Gold, *Helv. Chim. Acta*, **46**, 994 (1963); S. M. Hecht and J. W. Kozarich, *Tetrahedron Lett.*, 1501 (1972).
5. L. C. Leitch, P. E. Gagnon, and A. Cambron, *Can. J. Res.*, **28B**, 256 (1950); B. L. Crawford, Jr., W. H. Fletcher, and D. A. Ramsay, *J. Chem. Phys.*, **19**, 406 (1951); D. E. Milligan, and M. E. Jacox, *J. Chem. Phys.*, **36**, 2911 (1962); C. B. Moore and G. C. Pimentel, *J. Chem. Phys.*, **40**, 329 (1964); A. J. Merer, *Can. J. Phys.*, **42**, 1242 (1964).
6. S. P. McManus, J. T. Carroll, and C. L. Dodson, *J. Org. Chem.*, **33**, 4272 (1968).
7. A procedure very similar to that described in this preparation has subsequently been published: S. P. Markey and G. J. Shaw, *J. Org. Chem.*, **43**, 3414 (1978). The authors of this paper were evidently unaware of the prior publication of this *Org. Synth.* procedure [*Org. Synth.*, **53**, 38 (1973)].

DIETHYLALUMINUM CYANIDE

(Aluminum, cyanodiethyl-)

$$Al(C_2H_5)_3 + HCN \xrightarrow{C_6H_6} (C_2H_5)_2AlCN + C_2H_6$$

Submitted by W. Nagata[1] and M. Yoshioka
Checked by S. C. Welch, P. Bey, and Robert E. Ireland

1. Procedure

Caution! This preparation should be conducted in a well-ventilated hood, and neat triethylaluminum must be handled with great care.

Benzene has been identified as a carcinogen. OSHA has issued emergency standards on its use. All procedures involving benzene should be carried out in a well-ventilated hood, and glove protection is required.

A tared, 500-ml., round-bottomed flask is fitted with a vacuum take-off, and the entire assembly is connected through an adaptor containing a stopcock to an inverted cylinder of triethylaluminum, as shown in Figure 1. The assembly is connected to a nitrogen source (Note 1) through the vacuum take-off, and with the cylinder valve closed but the stopcock open, the system is alternately evacuated and filled with nitrogen four times. With the system filled with nitrogen the cylinder valve is opened and approximately 55 ml. (46 g; 0.40 mole) of triethylaluminum (Note 2) is allowed to flow into the reaction flask. The cylinder valve is then closed; the system is evacuated and filled three times with nitrogen, and the adaptor stopcock is then closed (Note 3). The reaction flask is then quickly removed, stoppered, and weighed to determine the exact amount of triethylaluminum collected. A magnetic stirring bar is added and the flask is fitted with a vacuum take-off and 250-ml., pressure-equalizing dropping funnel. The system is again placed under a nitrogen atmosphere, and the triethylaluminum is dissolved in 150 ml. of anhydrous benzene, added through the dropping funnel. The dropping funnel is charged with a solution of 11.9 g. (0.441 mole) of hydrogen cyanide (Note 4) in 100 ml. of anhydrous benzene which is added dropwise to the solution of triethylaluminum, with stirring and cooling. Preferably, the addition is carried out at a constant rate such that the hydrogen cyanide solution is added in about 2 hours. The evolution of ethane becomes slow suddenly after one molar equivalent of hydrogen cyanide is added (Note 5). After the addition is complete, the reaction mixture is allowed to stir overnight (Note 6).

After this period, the dropping funnel and the vacuum take-off are replaced by the short-path distillation assembly shown in Figure 2. The system is protected with a Drierite tube, and the benzene is distilled under reduced pressure (water aspirator). After the benzene is removed, the benzene-containing receiver is replaced with a clean, dry flask, and the system is connected to an efficient vacuum pump. The pressure in the system is reduced to 0.02 mm., and the flask is immersed deeply in an oil bath (Figure 2) heated to about 200°. After about 1 ml. of forerun is collected, diethylaluminum cyanide distils at 162° (0.02 mm.) (Note 7) and is collected in a tared, 200-ml. receiver by heating the side

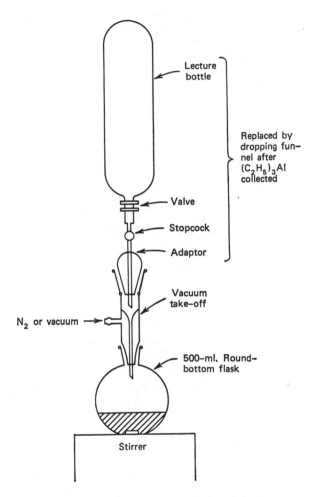

Lecture bottle

Replaced by dropping funnel after $(C_2H_5)_3Al$ collected

Valve

Stopcock

Adaptor

Vacuum take-off

N_2 or vacuum

500-ml. Round-bottom flask

Stirrer

Figure 1. Apparatus for collection of triethylaluminum.

arm and the adaptor with a stream of hot air or an IR lamp (Note 8). After all the distillate is collected in the receiver (Note 9), dry nitrogen is admitted to the evacuated apparatus, and the receiver is stoppered, giving 26.7 – 35.6 g. (60 – 80%) of diethylaluminum cyanide, usually as a pale, yellow syrup (Notes 10 and 11).

The stopper of the flask is quickly replaced with the nitrogen adaptor, and after placing the system under a nitrogen atmosphere, the diethylaluminum cyanide is treated with 130 ml. of dry benzene. The resulting mixture is allowed to stand with occasional swirling under nitrogen until the syrup goes into solution (Note 12). Sufficient dry benzene is then added to make the total volume of the solution 200 ml. After thorough mixing with a

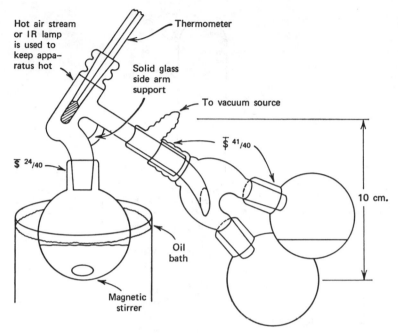

Figure 2. Short-path distillation apparatus.

magnetic stirring bar, the resulting diethylaluminum cyanide solution (13.4 – 17.8%; 1.2 – 1.6 *M*) may be divided and stored in sealed ampoules (Note 13).

2. Notes

1. The nitrogen source described in *Org. Synth.*, **Coll. Vol. 4,** 133 (1963), Figure 5, was used.

2. The exact volume of the triethylaluminum added at this point is not critical, since the exact weight is determined later. The use of a 25% solution of triethylaluminum in benzene, available from the Stauffer Chemical Company, 299 Park Avenue, New York, eliminates the tedious preparation of the triethylaluminum solution described in this procedure.

3. These precautions will ensure the removal of any adhering triethylaluminum that will flame when the apparatus is disassembled.

4. A 10% molar excess of hydrogen cyanide was employed, and the quantity added at this point was determined by the amount of triethylaluminum collected. Hydrogen cyanide can be prepared as described in *Tetrahedron Lett.*, 461 (1962).

5. The change in the rate of gas evolution is sometimes not clear, especially when the temperature of the hood is high. When the change is recognized distinctly, the addition of hydrogen cyanide solution may be stopped.

6. The reaction mixture containing about 13% (1.2 M) of diethylaluminum cyanide and a small amount of ethylaluminum dicyanide may be used for most hydrocyanation processes without further purification. Care must be taken to have no unchanged triethylaluminum, since the submitters have observed that hydrocyanation of Δ^8-11-keto steroids with diethylaluminum cyanide is greatly retarded by the presence of a small amount of unchanged triethylaluminum.[2,3]

7. The checkers collected diethylaluminum cyanide at 170 – 180° (0.35 mm.) and 167 – 175° (0.25 mm.).

8. Heating of the glassware above 150° with a hot air stream or IR lamp is needed to make the viscous product run into the receiver.

9. The pot residue contains ethylaluminum dicyanide as a nonvolatile mass, most of which may be removed with a spatula and decomposed with isopropyl alcohol and then water. The flask is then washed with running water and 20% hydrochloric acid to remove the mass completely.

10. The submitters have obtained an almost colorless syrup by reaction of purified triethylaluminum with hydrogen cyanide, followed by repeated distillation.

11. The yield depends on the efficiency of the collection of the viscous distillate in the receiver.

12. It takes considerable time (5 – 10 hours) to dissolve diethylaluminum cyanide. Magnetic stirring is not effective unless most of the material goes into solution.

13. Diethylaluminum cyanide dissolved in benzene, toluene, hexane, or isopropyl ether and stored in ampoules is stable for a long period. The cyanide is not stable in tetrahydrofuran. The anhydrous benzene used in the reaction may be replaced by diethyl or diisopropyl ether. A 1 – 2 M solution of diethylaluminum cyanide in benzene is commercially available from Alfa Products, Ventron Corporation, Danvers, Massachusetts.

3. Discussion

Formation of diethylaluminum cyanide from triethylaluminum and hydrogen cyanide was noted initially by the submitters[4] and later by Stearns,[5] but isolation and characterization of the product were first performed by the submitters.[3] An impractical process comprised of heating diethylaluminum chloride and sodium cyanide in benzene for 21 days has been reported.[6]

Diethylaluminum cyanide is a useful, potent reagent for hydrocyanation of various compounds. Features of this reagent as compared with the triethylaluminum – hydrogen cyanide reagent may be obtained from the literature.[2,3,7]

1. Shionogi Research Laboratories, Shionogi & Co., Ltd., Fukushima-ku Osaka 553, Japan.
2. W. Nagata, "Proceedings of the Symposium on Drug Research," Montreal, Canada, June 1966, p. 188.
3. W. Nagata and M. Yoshioka, *Tetrahedron Lett.*, 1913 (1966); W. Nagata, M. Yoshioka, and S. Hirai, *J. Am. Chem. Soc.*, **94**, 4635 (1972).
4. W. Nagata, M. Yoshioka, and S. Hirai, *Tetrahedron Lett.*, 461 (1962).
5. R. S. Stearns, U.S. Pat. 3,078,263 (1963).
6. R. Ehrlich and A. R. Young, *J. Inorg. Nucl. Chem.*, **28**, 674 (1966).

7. W. Nagata and M. Yoshioka, "Proceedings of the Second International Congress on Hormonal Steroids," Excerpta Medica Foundation, Amsterdam, 1967, p. 327.

DIETHYLAMINOSULFUR TRIFLUORIDE

[Sulfur, (diethylaminato)trifluoro-]

$$(C_2H_5)_2NSi(CH_3)_3 + SF_4 \xrightarrow[-60°]{\text{trichlorofluoromethane}} (C_2H_5)_2NSF_3 + FSi(CH_3)_3$$

Submitted by W. J. MIDDLETON[1] and E. M. BINGHAM[1]
Checked by RONALD F. SIELOFF, EUGENE R. KENNEDY, and CARL R. JOHNSON

1. Procedure

Caution! This procedure should be conducted in a good hood to avoid exposure to sulfur tetrafluoride. Protective gloves should be worn when handling diethylaminosulfur trifluoride since this material can cause severe HF burns.

A dry, 1-l., three-necked, round-bottomed flask is equipped with a thermometer ($-100°$ to $50°$), a magnetic stirrer, and two dry-ice condensers (one condenser is capped with a drying tube, the other with a gas-inlet tube). The apparatus is flushed with dry nitrogen, and 300 ml. of trichlorofluoromethane (Note 1) is added to the flask. As the nitrogen atmosphere is maintained, the trichlorofluoromethane is cooled to $-70°$ with an acetone – dry ice bath, and 119 g. (1.10 moles) of sulfur tetrafluoride (Note 2) is added from a cylinder through the gas-inlet tube (Note 3), which is then replaced with a 250-ml., pressure-equalizing dropping funnel charged with a solution of 145 g. (1.00 mole) of N,N-diethylaminotrimethylsilane (Note 4) in 90 ml. of trichlorofluoromethane. This solution is added dropwise, with stirring, to the sulfur tetrafluoride solution at a rate slow enough to keep the temperature of the reaction mixture below $-60°$ (about 40 minutes). The cooling bath is removed, and the reaction mixture is allowed to warm gradually to room temperature. The apparatus is prepared for distillation with a simple distillation head, and the solvent (b.p. 24°) and by-product fluorotrimethylsilane (b.p. 17°) are distilled into a well-cooled receiver by warming the reaction mixture gently to 45° with a heating mantle. The yellow to dark-brown residual liquid is transferred and distilled at reduced pressure through a spinning-band column, yielding 131 g. (81%) of di-ethylaminosulfur trifluoride as a light yellow liquid, b.p. 46 – 47° (10 mm.) (Note 5). This product can be stored for several months at room temperature in an inert plastic bottle (such as one made of polypropylene or Teflon FEP), or for short periods of time in a dry, glass bottle.

2. Notes

1. Trichlorofluoromethane (Freon 11) is available from E. I. duPont de Nemours and Company, Inc., or Matheson Gas Products.

2. Sulfur tetrafluoride is available from Air Products and Chemicals, Inc., or Matheson Gas Products.

3. If it is inconvenient to add sulfur tetrafluoride directly from a cylinder, it may first be condensed in a calibrated trap containing a boiling chip and cooled in an acetone – dry ice bath. When cooled to $-78°$, 119 g. of sulfur tetrafluoride has a volume of about 62 ml. The sulfur tetrafluoride can be added to the cooled flask by allowing it to distil slowly from the trap.

4. *N,N*-Diethylaminotrimethylsilane is available from PCR, Inc., or it can be prepared by the following procedure. A solution of 292 g. (413 ml.) of diethylamine in 1000 ml. of diethyl ether is cooled in an ice bath, and a solution of 216 g. (252 ml.) of chlorotrimethylsilane in 200 ml. of ether is added dropwise with mechanical stirring over a period of 1 hour. The precipitated solid is removed by filtration, thoroughly washed with ether, and the filtrate is fractionally distilled, yielding 175 g. (60%) of *N,N*-diethylaminotrimethylsilane as a colorless liquid, b.p. 124 – 125°.

5. The reaction of diethylaminosulfur trifluoride with water is highly exothermic; clean-up procedures should be carried out with caution.

3. Discussion

Diethylaminosulfur trifluoride is a useful and convenient reagent for replacing primary, secondary, and tertiary hydroxyl and aldehyde and ketone carbonyl oxygen[2,3] with fluorine, even in the presence of other halogens and other functional groups, such as carboxylic esters. In contrast to sulfur tetrafluoride, this reagent is a liquid easily measured and used in standard glass equipment at moderate temperatures and atmospheric pressure, and can be used on acid-sensitive compounds (such as pivaldehyde) to convert them into the fluorinated derivative. In contrast to other reagents such as SF_4, SeF_4 pyridine HF, HF·pyridine, and $(C_2H_5)_2NCF_2CFHCl$, this reagent can replace hydroxyl groups with fluorine in primary alcohols such as 2-methyl-1-propanol without causing extensive rearrangement or dehydration.

This preparation of diethylaminosulfur trifluoride is an adaptation of a procedure first described by von Halasz and Glemser.[4] The same procedure can also be used to prepare other dialkylaminosulfur trifluorides by the substitution of the diethylaminotrimethylsilane with other dialkylaminotrimethylsilanes.[2]

1. Central Research and Development Department, Experimental Station, E. I. duPont deNemours and Co., Wilmington, Del. 19898.
2. W. J. Middleton, *J. Org. Chem.*, **40**, 574 (1975); W. J. Middleton and E. M. Bingham, *Org. Synth.*, **Coll. Vol. 6**, 835 (1988).
3. L. N. Markovskij, V. E. Pashinnik, and A. V. Kirsanov, *Synthesis*, 787 (1973).
4. S. P. von Halasz and O. Glemser, *Chem. Ber.*, **104**, 1247 (1971).

DIETHYL *tert*-BUTYLMALONATE

[Propanedioic acid, (1,1-dimethylethyl)-, diethyl ester]

$$CH_2(CO_2C_2H_5)_2 + CH_3COCH_3 \xrightarrow[\substack{(CH_3CO)_2O, \\ \text{reflux}}]{ZnCl_2} (CH_3)_2C{=}C(CO_2C_2H_5)_2$$

$$(CH_3)_2C{=}C(CO_2C_2H_5)_2 + CH_3MgI \xrightarrow[\substack{(C_2H_5)_2O}]{CuCl} \xrightarrow[\substack{H_2SO_4}]{H_2O}$$

$$(CH_3)_3CCH(CO_2C_2H_5)_2$$

Submitted by E. L. Eliel,[1,2] R. O. Hutchins,[1,3] and Sr. M. Knoeber[1]
Checked by Walter J. Campbell and Herbert O. House

1. Procedure

Caution! Benzene has been identified as a carcinogen; OSHA has issued emergency standards on its use. All procedures involving benzene should be carried out in a well-ventilated hood, and glove protection is required.

Methyl iodide, in high concentrations for short periods or in low concentrations for long periods, can cause serious toxic effects in the central nervous system. Accordingly, the American Conference of Governmental Industrial Hygienists[4] has set 5 p.p.m., a level which cannot be detected by smell, as the highest average concentration in air to which workers should be exposed for long periods. The preparation and use of methyl iodide should always be performed in a well-ventilated fume hood. Since the liquid can be absorbed through the skin, care should be taken to prevent contact.

A. *Diethyl isopropylidenemalonate.* A 2-l. flask equipped with a magnetic stirrer and a reflux condenser fitted with a calcium chloride drying tube is charged with 400 g. (2.50 moles) of diethyl malonate (Note 1), 216 g. (3.73 moles) of acetone (Note 2), 320 g. (3.14 moles) of acetic anhydride, and 50 g. (0.37 mole) of anhydrous zinc chloride (assay 98.9%, Note 3). The solution is heated at reflux, with stirring, for 20 – 24 hours, then cooled and diluted with 300 – 350 ml. of benzene. The resulting dark-colored solution is washed with four 500-ml. portions of water, and the combined aqueous layers are extracted with two 100-ml. portions of benzene. The combined benzene solutions are concentrated with a rotary evaporator, and the residual liquid is fractionally distilled under reduced pressure through a 30-cm. Vigreux column or spinning band column. After separation of the unchanged diethyl malonate [119 – 135 g., b.p. 85 – 87° (9 mm.)] and intermediate fractions [b.p. 87 – 104° (9 mm.)], diethyl isopropylidenemalonate is collected as 231 – 246 g. (46 – 49%) of colorless liquid, b.p. 110 – 115° (9 – 10 mm.), n_D^{20} 1.4483 – 1.4490 (Note 4).

B. *Diethyl* tert-*butylmalonate.* A 1-l., three-necked flask fitted with a mechanical stirrer, a pressure-equalizing dropping funnel, a reflux condenser, and a nitrogen inlet tube is dried in an oven at 110 – 120°, then flushed with nitrogen and allowed to cool; an atmosphere of nitrogen is maintained in the reaction vessel throughout the subsequently described reactions. After 18.3 g. (0.753 g.-atom) of magnesium turnings have been placed in the flask, a solution of 113.5 g. (0.7993 mole) of methyl iodide (Note 5) in 200 ml. of anhydrous diethyl ether (Note 6) is added dropwise with stirring. The resulting solution of methylmagnesium iodide is cooled to 0 – 5° with an external ice – salt bath, and 1.0 g. (0.010 mole) of copper(I) chloride is introduced with stirring (Note 7). The temperature of the resulting mixture is kept at −5° to 0° while a solution of 100 g. (0.500 mole) of diethyl isopropylidenemalonate in 100 ml. of anhydrous ether (Note 6) is added, dropwise and with stirring over a period of 80 – 90 minutes. After the addition is complete, the cooling bath is removed; the reaction mixture is stirred for 30 minutes and poured onto a mixture of 500 – 1000 g. of ice and 400 ml. of 10% sulfuric acid. The ether layer is separated and the aqueous phase is extracted with three 200-ml. portions of ether. The combined ether solutions are washed with 100 ml. of saturated aqueous sodium thiosulfate (Note 8), dried over magnesium sulfate, and concentrated with a rotary evaporator. The residual liquid is distilled through a short Vigreux column, yielding 93.5 – 102 g. (87 – 94%) of diethyl *tert*-butylmalonate as a colorless liquid, b.p. 60 – 61° (0.7 mm.), n_D^{20} 1.4250 (Note 9).

2. Notes

1. A commercial grade of diethyl malonate was distilled before use, b.p. 90° (15 mm.).

2. The submitters used commercial acetone which was dried over potassium carbonate before use. The checkers dried acetone over Linde Molecular Sieves, No. 4A.

3. The checkers used a proportionate amount (52 g.) of anhydrous zinc chloride (assay, 95% minimum) purchased from the Mallinckrodt Chemical Works.

4. This procedure was described by Cope and Hancock.[5] The product has IR absorption (CCl_4) at 1730 (conjugated ester $C = O$) and 1650 cm.$^{-1}$ (conjugated $C = C$) with ^1H NMR peaks (CCl_4) at δ 1.12 (t, $J = 7$ Hz., 6H), 2.01 (s, 6H), 4.10 (q, $J = 7$ Hz., 4H), and a UV maximum (95% EtOH) at 218 nm (ϵ 10,200).

5. Reagent grade methyl iodide, obtained from J. T. Baker Chemical Company, Philadelphia, was used without further purification. The checkers used a pure grade of methyl iodide purchased from Eastman Organic Chemicals without further purification.

6. Anhydrous ether, from freshly opened containers, obtained from the Mallinckrodt Chemical Works, was used without further purification. The checkers used Baker and Adamson anhydrous ether purchased from the Industrial Chemicals Division of Allied Chemical Company.

7. The checkers found that this preparation was equally successful when a solution of 0.50 mole of diethyl isopropylidenemalonate in 100 ml. of ether was added over 25 minutes to an ethereal solution of 0.55 mole of lithium dimethylcuprate. A solution of this reagent was prepared by adding 1.10 moles of methyllithium in ether solution (purchased from Foote Mineral Co., Exton, Pennsylvania) to a cold (−10°) suspension of 0.55 mole of copper(I) iodide in 200 ml. of anhydrous ether. The reaction mixture containing a suspension of yellow methylcopper was stirred at −5° for 5 minutes, then quenched with

an aqueous mixture (pH 9) of ammonia and ammonium chloride. The product was isolated and distilled in the usual way, yielding 100 g. (93%) of the diethyl *tert*-butylmalonate, b.p. $57 - 61°$ $(0.5 - 0.7$ mm.), $n_D^{26.9}$ 1.4224.

8. The hydroiodic acid liberated during the hydrolysis step is frequently oxidized to iodine, which contaminates the final product. Washing the organic phase with aqueous sodium thiosulfate ensures the absence of this impurity in the final product.

9. The product has IR absorption (CCl_4) at 1750 cm.$^{-1}$ (ester $C=O$) with ^1H NMR peaks (CCl_4) at 1.08 (s, 9H), 1.22 (t, $J = 7$ Hz., 6H), 3.03 (s, 1H), and 4.05 (q, $J = 7$ Hz., 4H).

3. Discussion

Although diethyl *tert*-butylmalonate may be prepared in low yield by the alkylation of diethyl sodiomalonate with *tert*-butyl bromide[6] or *tert*-butyl chloride,[7] the alkylation of active methylene compounds such as malonic esters with *tert*-alkyl halides is normally a poor preparative procedure because of competing dehydrohalogenation of the alkyl halide.[8] A more satisfactory synthetic route to *tert*-alkyl derivatives of malonic esters and related compounds consists of the conjugate addition of alkyl- or arylmagnesium halides to alkylidenemalonates or alkylidenecyanoacetates.[8,9] Thus, diethyl *tert*-butylmalonate has been prepared in yields of $37 - 64\%$ by the conjugate addition of methylmagnesium iodide to isopropylidenemalonate.[10,11]

The addition of copper salts was found to favor the conjugate addition of Grignard reagents to alkylidenemalonic esters,[12] and Munch-Peterson subsequently found that the addition of 1 mole % of copper(I) chloride was generally useful for promoting the addition of Grignard reagents to α,β-unsaturated esters.[13] The preformed copper(I) reagents such as the ether-soluble lithium dialkyl- or diarylcuprates are even more effective reactants in conjugate additions to α,β-unsaturated carbonyl compounds.[14] The present preparation illustrates the use of copper(I) chloride to catalyze the addition of methylmagnesium iodide to an alkylidenemalonic ester. In this case high yields of the conjugate addition product are obtained either by this procedure or by the use of lithium dimethylcuprate (Note 7). In a related preparation[15] successive addition of three moles of methylmagnesium iodide in the presence of copper(I) bromide to $(CH_3S)_2C=C(CO_2C_2H_5)_2$ proceeds with elimination of thiomethylmagnesium halide in the first two stages.

1. Department of Chemistry, University of Notre Dame, Notre Dame, Indiana 46556.
2. Present address: Department of Chemistry, The University of North Carolina at Chapel Hill, Venable and Kenan Laboratories 045 A, Chapel Hill, North Carolina 27514.
3. Present address: Department of Chemistry, Drexel University, 23rd and Chestnut Streets, Philadelphia, Pennsylvania 19104.
4. American Conference of Governmental Industrial Hygienists (ACGIH), "Documentation of Threshold Limit Values," 3rd ed., Cincinnati, Ohio, 1971, p. 166.
5. A. C. Cope and E. M. Hancock, *J. Am. Chem. Soc.*, **60**, 2644 (1938).
6. A. W. Dox and W. G. Bywater, *J. Am. Chem. Soc.*, **58**, 731 (1936).
7. H. F. Van Woerden, *Recl. Trav. Chim. Pay-Bas*, **82**, 920 (1963).
8. A. C. Cope, H. L. Holmes, and H. O. House, *Org. React.*, **9**, 107 (1957).
9. F. S. Prout, E. P.-Y. Huang, R. J. Hartman, and C. J. Korpies, *J. Am. Chem. Soc.*, **76**, 1911 (1954).

10. S. Wideqvist, *Ark. Kemi, Mineral Geol.*, **B23**, No. 4, 1 (1946) [*Chem. Abstr.*, **41**, 1615 (1947)].

11. G. M. Lampman, K. E. Apt, E. J. Martin, and L. E. Wangen, *J. Org. Chem.*, **32**, 3950 (1967).

12. A. Brändström and I. Forsblad, *Ark. Kemi*, **6**, 561 (1954).

13. J. Munch-Peterson, *J. Org. Chem.*, **22**, 170 (1957).

14. H. O. House, W. L. Respess, and G. M. Whitesides, *J. Org. Chem.*, **31**, 3128 (1966); H. O. House and W. F. Fischer, Jr., *J. Org. Chem.*, **33**, 949 (1968).

15. Y. Ittah and I. Shahak, *Synthesis*, 320 (1976).

2-TRIMETHYLSILYLOXY-1,3-BUTADIENE AS A REACTIVE DIENE: DIETHYL *trans*-4-TRIMETHYLSILYLOXY-4-CYCLOHEXENE-1,2-DICARBOXYLATE

(4-Cyclohexene-1,2-dicarboxylic acid, 4-trimethylsilyloxy-, diethyl ester, *trans*-)

A.

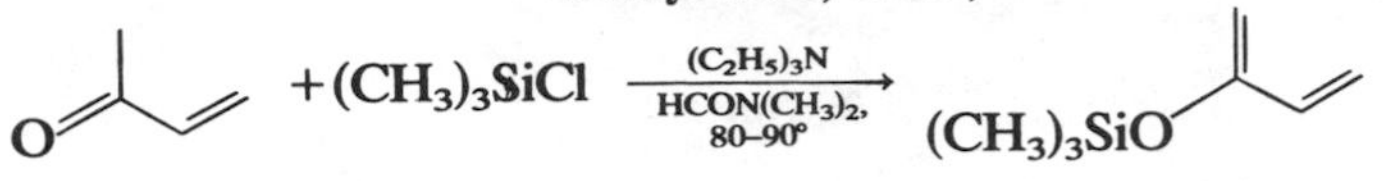

$$O= \quad +(CH_3)_3SiCl \xrightarrow[\substack{HCON(CH_3)_2, \\ 80-90°}]{(C_2H_5)_3N} \quad (CH_3)_3SiO$$

1

B. **1** + C_2H_5OOC ⟍⟍ $COOC_2H_5$ $\xrightarrow{130-150°}$ $(CH_3)_3SiO$ $COOC_2H_5$ $COOC_2H_5$

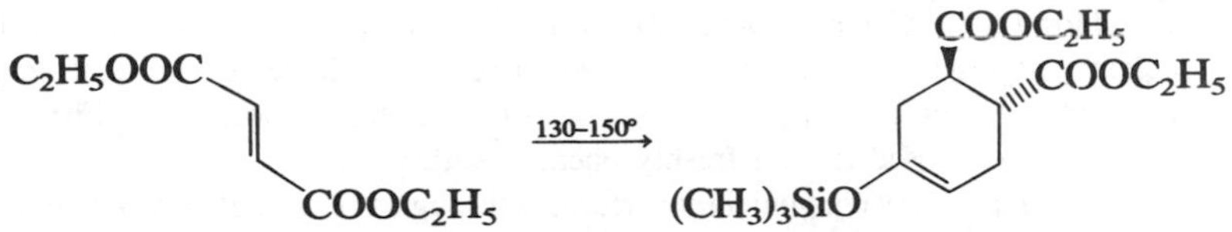

2

Submitted by Michael E. Jung and Charles A. McCombs[1]
Checked by Ping Sun Chu and George H. Büchi

1. Procedure

Caution! Part A should be carried out in a hood, since the reagents are noxious.

A. *2-Trimethylsilyloxy-1,3-butadiene* (**1**).   An oven-dried, 500-ml., three-necked, round-bottomed flask is fitted with two oven-dried addition funnels, a glass stopper, and magnetic stirrer, and placed in a 80 – 90° oil bath. Under an inert atmosphere, methyl vinyl ketone (25.0 g., 0.357 mole) in 25 ml. of N,N-dimethylformamide and chlorotrimethylsilane (43.4 g., 0.400 mole) in 25 ml. of N,N-dimethylformamide are added over 30 minutes to a magnetically stirred solution of triethylamine (40.5 g., 0.400 mole) in 200 ml. of N,N-dimethylformamide (Note 1). The reaction gradually darkens from colorless to yellow or dark brown, and supports a white precipitate of triethylamine hydrochloride. The reaction is set up to run overnight, or *ca.* 14 hours.

The reaction is cooled to room temperature, filtered (Note 2), and transferred to a 2-l. separatory funnel containing 300 ml. of pentane. To this solution is added 1 l. of *cold 5%* sodium hydrogen carbonate solution to facilitate the separation of phases and remove the N,N-dimethylformamide. The pentane layer is separated, and the aqueous layer extracted twice with 300-ml. portions of pentane. The pentane extracts are combined, washed with

200 ml. of *cold* distilled water (Note 3), dried over powdered anhydrous sodium sulfate, and filtered into a 2-1., round-bottomed flask.

The pentane and other volatiles are removed by fractional distillation, using a 5-cm., steel wool-packed column and heating the pot in a 70° oil bath (Note 4). A water aspirator vacuum is applied, and 18.2 – 22.9 g. (36 – 45%) of the diene **1** is distilled as a colorless oil, b.p. 50 – 55° (50 mm.) (Note 5). On a smaller scale, yields of up to 50% have been obtained.

B. *Diethyl* trans-4-*trimethylsilyloxy*-4-*cyclohexene*-1,2-*dicarboxylate* (**2**). A 25-ml., round-bottomed flask equipped with a reflux condenser is charged with diene **1** (7.1 g., 0.050 mole) and diethyl fumarate (5.7 g., 0.033 mole) (Note 6). The mixture is stirred under a nitrogen atmosphere in an oil bath kept at 130 – 150° for 24 hours (Note 7). Direct vacuum distillation using a short-path distillation apparatus affords a small amount of lower-boiling material, then 8.02 g. (77%) of cyclohexene **2,** b.p. 127 – 128° (0.5 mm.) (Notes 8 and 9).

2. Notes

1. The checkers obtained methyl vinyl ketone from Hoffmann-La Roche Company and chlorotrimethylsilane from Aldrich Chemical Company, Inc. Both were distilled from calcium hydride. Triethylamine, obtained from J. T. Baker Chemical Company, was distilled from calcium hydride, and *N,N*-dimethylformamide, from Fisher Scientific Company, was used from a freshly opened bottlc.

2. Filtration was conveniently performed through a plug of glass wool directly into the separatory funnel.

3. The hydrogen carbonate extractions must be performed quickly, since the product slowly hydrolyzes in the presence of water. The best yields were obtained when the phases were shaken briskly for 10 seconds and separated as soon as foaming ceased. Foaming also occurred when the pentane extracts were washed with water, but did not prevent the separation of phases.

4. The checkers used a 18-cm. Vigreux column to remove low-boiling material, and then transferred the residue to a 250-ml., round-bottomed flask for fractional distillation under reduced pressure.

5. The distilled diene **1** has been obtained with 99% purity by the submitters. The product has the following spectral properties: ^{1}H NMR: (CCl_4 with benzene as internal standard), δ (multiplicity, number of protons): 0.4 (s, 9H), 4.4 (broad s, 2H), 5.1 (d of m, 1H), 5.5 (d of d, 1H), 6.3 (d of d, 1H). The impurities were triethylamine and hexamethyldisiloxane. When stored in a serum-capped flask, in a desiccator, and removed *via* syringe, the butadiene is stable for 2 months.

6. The checkers prepared diethyl fumarate by the sulfuric acid – catalyzed esterification of fumaric acid.[7] Diethyl fumarate obtained from Aldrich Chemical Company, Inc. contained dimethyl fumarate as a contaminant.

7. The submitters allowed the reaction to proceed for 24 hours after monitoring by GC (1.83 m. × 0.64 cm., 10% S E-30/60 – 80 mesh Chromosorb W, 150°) indicated that the disappearance of diethyl fumarate was not complete after 5.5 hours. The submitters made

TABLE I
Preparation of Trimethylsilyloxycyclohexenes

$(CH_3)_3SiO$

X	Y	Z	Yield (%)
COCH$_3$	H	H	60
CO$_2$CH$_3$	H	H	46
CO$_2$CH$_3$	H	CO$_2$CH$_3$	71
CO$_2$C$_2$H$_5$	H	CO$_2$C$_2$H$_5$	77
CO$_2$C$_2$H$_5$	CO$_2$C$_2$H$_5$	H	39

no attempt to optimize the reaction time. The checkers found the reaction to be complete after 7 hours at $135 - 145°$.

8. The checkers distilled cyclohexene **2** after 7 hours reaction time using a Büchi Kügelrohr distillation oven to obtain $9.9 - 10.0$ g. $(95 - 96\%)$ of cyclohexene **2**.

9. Cyclohexene **2** has the following spectral properties: IR (liquid film) cm^{-1}: 1730, 1670; ^{1}H NMR: (CCl$_4$), δ (multiplicity, coupling constant J in Hz., number of protons, assignment): 0.16 [s, 9H, Si(CH$_3$)$_3$], 1.17 (t, $J = 7$, 6H, 2OCH$_2$CH$_3$), $2.00 - 3.16$ (m, 6H, ring CH), 4.07 (q, $J = 7$, 4H, 2OCH$_2$CH$_3$), 4.70 (broad s, 1H, olefinic CH).

3. Discussion

The first reference to 2-trimethylsilyloxy-1,3-butadiene (**1**) was a report[2] of its reaction with tetracyanoethylene by Cazeau and Frainnet without any mention of experimental details. Later, Conia[3] reported its synthesis in 50% yield with only a reference made to the usual House procedure[4] for silyl enol ethers. The diene **1** has also been prepared using lithium diisopropylamide as base and chlorotrimethylsilane in tetrahydrofuran – ether $(1:1)$ in yields up to 65%, but on a smaller scale.[5]

Butadienes substituted with alkoxy groups in the 2-position, *e.g.*, 2-ethoxy-1,3-butadiene,[6] have been prepared from methyl vinyl ketone, but required several conversions and a tedious spinning-band distillation to purify the product. This slight modification of the House procedure has been used to conveniently prepare 2-trimethylsilyloxy-1,3-butadiene from the readily available methyl vinyl ketone. This one-step procedure has provided large amounts of a new and reactive diene for Diels–Alder reactions, as illustrated in Table I.

1. Contribution No. 3759 from the Department of Chemistry, University of California, Los Angeles, California 90024.

2. P. Cazeau and E. Frainnet, *Bull. Soc. Chim. Fr.*, 1658 (1972).

3. C. Girard, P. Amice, J. P. Barnier, and J. M. Conia, *Tetrahedron Lett.*, 3329 (1974).

4. H. O. House, L. J. Czuba, M. Gall, and H. D. Olmstead, *J. Org. Chem.*, **34**, 2324 (1969).

5. M. E. Jung and C. A. McCombs, *Tetrahedron Lett.*, 2935 (1976).

6. (a) H. B. Dykstra, *J. Am. Chem. Soc.*, **57**, 2255 (1935); (b) N. A. Milas, E. Sakel, J. T. Plati, J. T. Rivers, J. K. Gladding, F. X. Grossi, Z. Weiss, M. A. Campbell, and H. F. Wright, *J. Am. Chem. Soc.*, **70**, 1597 (1948).

DIETHYL 2-(CYCLOHEXYLAMINO)VINYLPHOSPHONATE

(Phosphonic acid, [2-(cyclohexylamino)ethenyl]-, diethyl ester)

A. $BrCH_2CH(OC_2H_5)_2 + (C_2H_5O)_3P \xrightarrow{\Delta}$

$$(C_2H_5O)_2\overset{\overset{O}{\uparrow}}{P}CH_2CH(OC_2H_5)_2 + C_2H_5Br$$

B. $(C_2H_5O)_2\overset{\overset{O}{\uparrow}}{P}CH_2CH(OC_2H_5)_2 \xrightarrow{H_3O^+}$

$$(C_2H_5O)_2\overset{\overset{O}{\uparrow}}{P}CH_2CHO + 2C_2H_5OH$$

C. $(C_2H_5O)_2\overset{\overset{O}{\uparrow}}{P}CH_2CHO + \langle\text{cyclohexyl}\rangle\!-\!NH_2 \longrightarrow$

$$(C_2H_5O)_2\overset{\overset{O}{\uparrow}}{P}CH{=}CH{-}\overset{H}{N}\!-\!\langle\text{cyclohexyl}\rangle + H_2O$$

Submitted by WATARU NAGATA,[1] TOSHIO WAKABAYASHI,[1] and YOSHIO HAYASE[2]
Checked by H. A. DAVIS, K. ABE, and S. MASAMUNE

1. Procedure

A. *Diethyl 2,2-diethoxyethylphosphonate* (Note 1).   A 2-l., three-necked, round-bottomed flask fitted with a magnetic stirrer, dropping funnel, and nitrogen inlet is charged with 410 g. (2.08 moles) of bromoacetaldehyde diethyl acetal (Note 2), and a gentle stream of nitrogen is then passed continuously through the system. To the stirred solution is added dropwise 316 g. (1.90 moles) of triethyl phosphite (Note 3) over a period of 30 minutes, at 110 – 120°. The mixture is then stirred for 3 hours at 160°. The ethyl bromide evolved is trapped with a condenser and a receiver cooled in an ice bath. The low-boiling material (below 100°) is distilled under reduced pressure (22 mm.). The residual oil is fractionated under reduced pressure, and the fraction boiling at 101 – 103° (0.8 mm.) is collected, yielding 338 g. (70%) (Note 4).

B. *Diethyl formylmethylphosphonate* (Note 5). A mixture of 192 g. (0.755 mole) of diethyl 2,2-diethoxyethylphosphonate and 670 ml. of 2% hydrochloric acid is refluxed for 10 minutes under a nitrogen atmosphere. To the cooled (20 – 30°) mixture is added 240 g. of sodium chloride (Note 6). The resulting mixture is extracted with three, 500-ml. portions of dichloromethane. The combined organic extracts are washed successively with 40 ml. of 5% aqueous sodium hydrogen carbonate solution (Note 7) and 300 ml. of saturated aqueous salt solution, dried over anhydrous sodium sulfate, and distilled under reduced pressure (35 mm.) at water bath temperatures (60 – 70°). The resulting residue, weighing approximately 125 g., is fractionated under reduced pressure, and the fraction boiling at 100 – 103° (0.8 mm.) is collected, yielding 104 g. (76%) (Notes 8 and 9).

C. *Diethyl 2-(cyclohexylamino)vinylphosphonate.* A 1-l., two-necked, round-bottomed flask fitted with a magnetic stirrer, dropping funnel, and nitrogen inlet is charged with 90.0 g. (0.500 mole) of diethyl formylmethylphosphonate and 400 ml. of dry methanol. Under a nitrogen atmosphere, 49.6 g. (0.501 mole) of cyclohexylamine (Note 10) is added in portions to the stirred solution, over a period of 5 minutes. During the addition the temperature is maintained at 0 – 5° with an ice bath. The mixture is stirred for an additional 10 minutes at room temperature, and then the methanol is distilled from the mixture under reduced pressure (10 – 35 mm.) at water bath temperatures (25 – 30°). The residue is dissolved in 300 ml. of dry diethyl ether (Note 11), dried over anhydrous potassium carbonate (70 g.) (Note 12), and evaporated to dryness. The residual oil is fractionated under reduced pressure in the presence of 300 mg. of anhydrous potassium carbonate (Note 13), and the fraction boiling at 126 – 141° (0.08 mm.) is collected, giving 89 g. (68%) (Note 14). The fraction above is crystallized from dry pentane (Note 15), yielding 79 g. (60%) of diethyl 2-(cyclohexylamino)vinylphosphonate, m.p. 58 – 61° (Note 16).

2. Notes

1. This procedure is essentially the same as that described by Dawson and Burger.[3]

2. Bromoacetaldehyde diethyl acetal was distilled; b.p. 74 – 76° (25 mm.). Its preparation is described in *Org. Synth.*, **Coll. Vol. 3**, 123 (1955).

3. Triethyl phosphite was distilled; b.p. 43 – 44° (10 mm.). Its preparation is described in *Org. Synth.*, **Coll. Vol. 4**, 955 (1963).

4. The reported[3] boiling point is 128 – 130° (2 mm.). The checkers found the product to be pure by GC (UCW-98 at 150°). The ^{1}H NMR spectrum had the following characteristic peaks (CDCl$_3$): δ 1.22, 1.34 (2t, J = 7Hz., 12H, 4CH_3), 2.17 (d of d, J = 19 and 6Hz., 2H, CH_2CH), 3.6 [2q, J = 7Hz., 4H, CH(OCH_2CH$_3$)$_2$], 4.1 [2q, J = 7Hz., 4H, (CH$_3$CH_2O)$_2$PO], 4.90 [d of t, J = 6 and ~6Hz., 1H, CH$_2$CH(OC$_2$H$_5$)$_2$].

5. This procedure is essentially the same as that described by Dawson and Burger.[3]

6. It is necessary to saturate the aqueous layer with sodium chloride in order to extract the product effectively.

7. The checkers found that it was necessary to neutralize any excess acid to prevent polymerization of the product during the distillation.

8. There is no report of the boiling point of the product in the literature.[3] The product has an IR absorption (CCl$_4$ solution) at 1732 cm.$^{-1}$ (aldehyde C=O).

9. The checkers found that the product contained a 5 – 6% contamination of starting material [GC (UCW-98, 120°) and ^{1}H NMR]. The ^{1}H NMR spectrum (CDCl$_3$): δ 1.35 (t, J = 7 Hz., 6H, 2CH$_3$), 3.11 [d of d, J = 22 and 3.5 Hz., 2H, P(O)CH$_2$CH], 4.2 [2q, J = 7 Hz., 4H, (CH$_3$CH$_2$O)$_2$PO], 9.70 (d of t, J = 3 and 1 Hz., 1H, CHO).

10. Reagent grade cyclohexylamine was redistilled before use.

11. Dry ether was used to minimize the water content of the solution.

12. To avoid acid-induced dimerization[4] of diethyl 2-(cyclohexylamino)vinylphosphonate, the solution was dried thoroughly, usually overnight, with anhydrous potassium carbonate.

13. The distillation was carried out in the presence of powdered anhydrous potassium carbonate to prevent dimerization[4] of diethyl 2-(cyclohexylamino)vinylphosphonate.

14. The yields were 65 – 75% in several runs. The distilled oil was sufficiently pure for further use.

15. Reagent grade pentane passed through Merck anhydrous neutral alumina was used. Crystallization was carried out at 0° using 50 ml. of dry pentane. Filtration of the crystals was carried out under dry nitrogen. The submitters succeeded in crystallizing diethyl 2-(cyclohexylamino)vinylphosphonate only after the publication of its preparation.[4] The crystalline product is stable for several months, if stored at 0° under anhydrous conditions.

16. The checkers were not able to obtain the product in crystalline form. The ^{1}H NMR spectrum (CDCl$_3$): δ 1.3 (t, J = 7 Hz., CH$_3$CH$_2$O), 1.0 – 2.1 (m, cyclohexyl H), 4.0 (quintet, J = 7 Hz., 2 CH$_3$CH$_2$O), 4.3 – 5.0 (broad, NH), 5.9 – 7.7 (m, —CH=CH—). The number of olefinic protons was estimated to be 1.7 – 1.8 by comparison of the area in the region δ 5.9 – 7.7 with the total area in the spectrum. Although this value is slightly low, it was found that the sample was sufficiently pure to carry out further transformations. The IR spectrum has the following absorptions (neat): 3300 (N—H str.), 1620, 1210, 1058, 1035, and 955 cm.$^{-1}$.

3. Discussion

Diethyl 2-(cyclohexylamino)vinylphosphonate has proved to be an excellent reagent for conversion of aldehydes and ketones into the corresponding α,β-unsaturated aldehydes.[4,5] Formylmethylenetriphenylphosphorane[6] and diethyl 2,2-(diethoxy)ethylphosphonate[7] have been prepared and used for the conversion of aldehydes, but not ketones and hindered aldehydes, into α,β-unsaturated aldehydes. When ethyl diethylphosphonoacetate[8] or diethyl cyanomethylphosphonate[9] is used to obtain an α,β-unsaturated aldehyde, the conversion requires several stages.

1. Shionogi Research Laboratory, Shionogi & Co., Ltd., Fukushima-ku, Osaka, 553 Japan.
2. Faculty of Science, Osaka City University, Sumiyoshi-ku, Osaka, 558 Japan.
3. N. D. Dawson and A. Burger, *J. Am. Chem. Soc.*, **74**, 5312 (1952).
4. W. Nagata and Y. Hayase, *Tetrahedron Lett.*, 4359 (1968); *J. Chem. Soc., C*, 460 (1969).

5. W. Nagata, T. Wakabayashi, and Y. Hayase, *Org. Synth.*, **Coll. Vol. 6**, 448 (1988).
6. S. Trippett and D. M. Walker, *J. Chem. Soc.*, 1266 (1961).
7. H. Takahashi, K. Fujiwara, and M. Ohta, *Bull. Chem. Soc. Jpn.* **35**, 1498 (1962).
8. W. S. Wadsworth and W. D. Emmons, *J. Am. Chem. Soc.*, **83**, 1733 (1961).
9. A. K. Bose and R. T. Dahill, Jr., *J. Org. Chem.*, **30**, 505 (1965).

RADICAL ANION ARYLATION:
DIETHYL PHENYLPHOSPHONATE

(Phosphonic acid, phenyl-, diethyl ester)

$$\text{C}_6\text{H}_5\text{—I} + (\text{C}_2\text{H}_5\text{O})_2\text{PO}^-\text{Na}^+ \xrightarrow[\text{NH}_3]{h\nu} \text{C}_6\text{H}_5\text{—P}(\text{=O})(\text{OC}_2\text{H}_5)_2 + \text{NaI}$$

Submitted by JOSEPH F. BUNNETT[1] and ROBERT H. WEISS
Checked by S. C. BUSMAN and O. L. CHAPMAN

1. Procedure

Caution! Benzene has been identified as a carcinogen; OSHA has issued emergency standards on its use. All procedures involving benzene should be carried out in a well-ventilated hood, and glove protection is required.

A 2-l., three-necked, round-bottomed flask is fitted with an ammonia condenser (Note 1) with an outlet protected with a soda – lime drying tube, a dropping funnel, a nitrogen inlet, and a magnetic stirrer. As a slow stream of nitrogen is passed through the system, the condenser is charged with dry ice and 2-propanol, the funnel is briefly removed, about a liter of liquid ammonia is added straight from a commercial cylinder (Note 2), and the funnel is replaced. Bright, freshly-pared sodium metal (11.8 g., 0.513 g.-atom) is added, turning the mixture blue. Diethyl phosphonate (70.4 g., 0.510 mole) (Note 3) is cautiously added dropwise to the sodium in ammonia in the manner of a titration; the endpoint is the change from blue to colorless (Note 4). Iodobenzene (52.4 g., 0.257 mole) (Note 5) is slowly added, giving the solution a slight yellowish tint (Note 6). The dropping funnel is replaced by a stopper, the frost is wiped off the outside of the flask with a towel dampened with acetone, and the whole system is mounted in a photochemical reactor of adequate design (Note 7). *Caution! The lamps must be shielded to prevent exposure of the eyes or skin to ultraviolet radiation.* The flask is irradiated for 1 hour (Note 8), but every 20 minutes the lamps are shut off briefly while the exterior of the flask is freed of frost by spraying it, still mounted in the reactor, with 2-propanol from a wash bottle.

After irradiation, the flask is removed from the reactor, and about 50 g. (0.62 mole) of solid ammonium nitrate is added with stirring to acidify the mixture. About 200 ml. of diethyl ether is added, nitrogen flow is stopped, the condenser is removed, and the open flask is placed on a cork ring in an operating hood and allowed to stand overnight while the ammonia evaporates. The next day 300 ml. of water (Note 9) and 300 ml. of ether are added, the ether layer is separated, the water layer is extracted twice with ether, and the combined ether extracts are dried over anhydrous sodium sulfate. After evaporation of the ethyl ether, the residue is distilled under reduced pressure through a short Vigreux column. After a small forerun, 50.3 – 56.1 g. (90.4 – 92.5%) of diethyl phenylphosphonate (Note 10) is collected at 73 – 74° (0.020 mm.) (Note 11).

2. Notes

1. An ammonia condenser is an enlarged cold finger design. The interior of the "finger" is a reservoir for dry ice and 2-propanol.

2. When ammonia distilled from sodium metal is used, the yield is 3 – 5% greater, but use of ammonia straight from the tank is recommended because of the greater convenience.

3. Commercial diethyl phosphonate, formerly known as diethyl phosphite, was purchased from Aldrich Chemical Co., Inc. and used without further purification.

4. Some white foam forms as the diethyl phosphonate is added. Because water in the ammonia consumes some of the sodium, not quite all the diethyl phosphonate is required to reach the endpoint. Excess diethyl phosphonate is deleterious.

5. Commercial iodobenzene was dried over molecular sieves. Use of ratios of potassium diethyl phosphite to iodobenzene smaller than 2 : 1 gives lower yields. Bromobenzene is much less reactive than iodobenzene and gives poor yields by this procedure.

6. When distilled ammonia is used, this yellow coloration vanishes as the reaction occurs.

7. The submitters used a Rayonet Model RPR-100 Photochemical Reactor, manufactured by the Southern New England Ultraviolet Company, Hamden, Connecticut 06514, equipped with 16 Cat. No. RPR-3500A fluorescent lamps (*ca.* 24 W. each), rated to emit maximally at 350 nm. Equally good results were obtained with lamps rated to emit maximally at 300 nm. For reactions on a 0.05-mole scale excellent yields were obtained, without using the commercial photochemical reactor, by irradiating for 1 hour with two circular kitchen-type fluorescent lamps mounted on either side of the flask so as to partially encircle it.

8. The yield was not significantly improved by irradiating for 2 hours.

9. Diethyl phenylphosphonate is appreciably soluble in water. Therefore, excessive amounts of water should be avoided.

10. Spectral characterization is as follows: IR (neat) cm^{-1}: 1440 (P—C aryl), 1250 (P=O), 1020 (POC$_2$H$_5$), and 3060 (H—C aryl); ^{1}H NMR (CDCl$_3$), δ (multiplicity, coupling constant J in Hz., assignment): 1.3 (t, J = 7, CH_3), 4.13 (quintet, J = 7, CH_2), 7.33 – 8.06 (m, C$_6$$H_5$).

11. The submitters obtained a yield of 46 g. (83%), b.p. 90 – 92° (0.1 mm.).

3. Discussion

The procedure reported here is based on a reaction discovered by Bunnett and Creary,[2] and was first employed for preparative purposes by Bunnett and Traber.[3] It is attractive because of the high yield, the ease of work-up, and the cleanliness of the reaction. The reaction is believed to occur by the $S_{RN}1$ mechanism, which involves radical and radical anion intermediates.[2,4] The $S_{RN}1$ arylation of other nucleophiles, especially ketone enolate ions,[5] ester enolate ions,[6] picolyl anions,[7] and arenethiolate ions,[8] has potential application in synthesis.

This procedure has been utilized successfully with a variety of aryl iodides, but aryl bromides are much less reactive. m- and p-Diiodobenzene, m- and p-bromoiodobenzene, and p-chloroiodobenzene give the corresponding phenylenediphosphonate esters.[2,3] o-Haloiodobenzenes undergo a dark reaction leading to deiodination and other products, but the $S_{RN}1$ reaction affording phosphate esters may be made to predominate if irradiation is started quickly.[9]

The esters of arylphosphonic acids are cleaved to the acids by hydrochloric, hydrobromic, or hydroiodic acid.[10b] Arylphosphonic dichlorides ($ArPOCl_2$) are easily converted to esters by reaction with the alcohol in pyridine solution.[11]

Other methods for synthesis of arylphosphonic acids or their derivatives fall into four main categories. First, many aromatic compounds react with phosphorus trichloride under aluminum trichloride catalysis, to form aryldichlorophosphines ($ArPCl_2$).[10a,12] These add chlorine to form aryltetrachlorophosphoranes ($ArPCl_4$),[10d,12] which may be hydrolyzed to arylphosphonic dichlorides or arylphosphonic acids. This sequence may be employed for preparations on a large scale, but is subject to the orienting effects of substituents when applied to substituted benzenes.

Second, arylphosphonic acids may be prepared by the copper-catalyzed reactions of arenediazonium tetrafluoroborates with phosphorus trichloride or tribromide. This method has found wide use.[13]

Third, arylphosphonic acid derivatives have been made by organometallic reactions, such as the reaction of phenylmagnesium bromide with diethyl chlorophosphate, or of phenyllithium with phosphorodipiperididic chloride.[10c,12]

Fourth, the present procedure bears a resemblance to the photochemical reaction of aryl iodides with trialkyl phosphites, with which several dialkyl arylphosphonates have been prepared.[14] However, prolonged irradiation (>24 hours) in quartz vessels was employed.

1. Board of Studies in Chemistry, University of California, Santa Cruz, California 95064.
2. J. F. Bunnett and X. Creary, J. Org. Chem., 39, 3612 (1974).
3. J. F. Bunnett and R. P. Traber, J. Org. Chem., 43, 1867 (1978).
4. J. K. Kim and J. F. Bunnett, J. Am. Chem. Soc., 92, 7463, 7464 (1970); J. F. Bunnett, Acc. Chem. Res., 11, 413 (1978).
5. R. A. Rossi and J. F. Bunnett, J. Am. Chem. Soc., 94, 683 (1972); J. Org. Chem., 38, 1407 (1973); J. F. Bunnett and J. E. Sundberg, Chem. Pharm. Bull., 23, 2620 (1975); J. F. Bunnett and J. E. Sundberg, J. Org. Chem., 41, 1702 (1976).
6. X. Creary, unpublished work.
7. J. F. Bunnett and B. F. Gloor, J. Org. Chem., 39, 382 (1974).
8. J. F. Bunnett and X. Creary, J. Org. Chem., 39, 3173 (1974).
9. R. R. Bard, J. F. Bunnett, and R. P. Traber, J. Org. Chem., 44, 4918 (1979).
10. K. Sasse, "Methoden der Organischen Chemie" (Houben-Weyl), 4th ed., Vol. 12/1, Georg Thieme, Stuttgart, 1963, (a) p. 314; (b) p. 352; (c) p. 372; (d) p. 392.
11. T. H. Siddall, III, and C. A. Prohaska, J. Am. Chem. Soc., 84, 3467 (1962).

12. G. M. Kosolapoff, *Org. React.*, **6**, 273 (1951).
13. L. D. Freedman and G. O. Doak, *Chem. Rev.*, **57**, 479 (1957).
14. J. B. Plumb, R. Obrycki, and C. E. Griffin, *J. Org. Chem.*, **31**, 2455 (1966); C. E. Griffin, R. B. Davison, and M. Gordon, *Tetrahedron*, **22**, 561 (1966); R. Obrycki and C. E. Griffin, *J. Org. Chem.*, **33**, 632 (1968).

DIETHYL *trans*-Δ^4-TETRAHYDROPHTHALATE

(4-Cyclohexene-1,2-dicarboxylic acid, diethyl ester, *trans*-)

Submitted by THOMAS E. SAMPLE, JR.,[1] and LEWIS F. HATCH[2]
Checked by ELAINE DOTTER DONALD and RICHARD E. BENSON

1. Procedure

A 300-ml. pressure reaction vessel, capable of withstanding at least 250 p.s.i. at 125° (Note 1) is charged with 60.0 g. (0.508 mole) of 3-sulfolene (Note 2), 86.0 g. (82.0 ml., 0.500 mole) of diethyl fumarate (Note 3), and 1.0 g. of hydroquinone (Note 4). Commercial absolute ethanol (90 ml.) is added, and the mixture stirred until most of the solid is dissolved. The vessel is sealed, heated slowly to 105 – 110° (Note 5), and maintained at that temperature 8 – 10 hours (Note 6). The vessel is then allowed to cool to room temperature, opened (Note 7), and the yellowish, fluid reaction mixture is poured into a 1-l. Erlenmeyer flask.

The liquid is stirred vigorously (Note 8), and a solution of 60.0 g. (0.566 mole) of sodium carbonate in 350 ml. of water is added to the flask as rapidly as the concomitant evolution of carbon dioxide will permit. Stirring is continued for about 15 minutes after the addition is complete. The original reaction vessel is rinsed with 200 ml. of petroleum ether (b.p. 60 – 75°) (Note 9), the rinse is added to the Erlenmeyer flask containing the product, and the mixture is stirred for about 10 minutes at a rate sufficient to achieve homogenization. The resulting loose emulsion is quickly transferred to a 1-l. separatory funnel and allowed to separate; the lower (aqueous) phase is drawn off into the original Erlenmeyer flask and the upper (organic) layer containing the desired product is transferred to a 500-ml. flask. In exactly the same manner the aqueous phase is extracted by vigorous stirring with two 100-ml. portions of petroleum ether, the organic layers being added to the retention flask (Note 10). The combined organic layers are transferred from the flask to the same separatory funnel and shaken once with 100 ml. of cold, 5%, aqueous sodium carbonate and twice with 50-ml. portions of cold water. The organic layer is returned to the retention flask, dried over 10 g. of anhydrous magnesium sulfate, filtered by gravity, concentrated with a rotary evaporator (Note 11), and distilled under reduced pressure through a Vigreux column, yielding 75 – 82 g. (66 – 73%, based on diethyl fumarate) of the product, b.p. 129 – 132° (5 mm.), n_D^{25} 1.4565 – 1.4570, d_4^{25} 1.0584 – 1.0589 (Note 13).

2. Notes

1. The maximum static pressure developed at 110°, measured by the submitters, never exceeded 95 p.s.i.; however, for most of their runs they used stainless or carbon steel autoclaves with screw-on covers rated for service up to 500 p.s.i. at 250° (medium-pressure catalytic hydrogenation bombs). Gaskets or inner-disk seals for the autoclave covers were cut from soft thin copper sheet stock instead of the more usual lead or lead-alloy material. Whereas use of lead-containing seals did not seem to affect the yields, when they were used the product invariably had the offensive odor of divalent organic sulfur which was still noticeable after final distillation; substitution of copper seals eliminated this problem completely. Close-fitting, removable glass liners for the autoclave were used by the submitters for convenience in weighing and mixing the charge, and for subsequent removal of the product mixture, but are not required. Several runs were made at 100 – 105° by the submitters, without difficulty, utilizing a hydrogenation bottle of heavy borosilicate glass tightly closed with a clamped-in Neoprene stopper and placed inside an iron sleeve for safety.

2. 3-Sulfolene (butadiene cyclic sulfone), m.p. 64 – 66°, from Shell Chemical Co., or from Aldrich Chemical Co., was found satisfactory for use as received. One lot purchased from another source required solution in hot methanol and treatment with activated carbon (250 ml. of methanol and 2 g. of Norit per 100 g. of material), filtration, and crystallization to free it of color, odor, and particulate matter.

3. Diethyl fumarate from Aldrich Chemical Co., n_D^{20} 1.4406, b.p. 97.5 – 97.7° (10 mm.), was employed without treatment.

4. Eastman Kodak Co. hydroquinone, m.p. 172 – 174°, was used. 4-*tert*-Butylcatechol, m.p. 38 – 42°, from the same supplier, proved to be an equally effective polymerization inhibitor.

5. The submitters used an ordinary hot-air oven with the usual simple bimetallic temperature regulator for the majority of their experiments. In several instances, however, a large oil bath regulated with a thermostat to within ±1.0° was employed, which, in providing more exact control of the total temperature environment of the reaction, gave a greater degree of reproducibility in product yield between successive runs at any set temperature; these precise runs were performed as a result of the checkers' observations emphasizing the important effect of reaction temperature on yields.

6. Longer heating periods (12 – 18 hours) were found by the submitters to elevate the yields only slightly. On the other hand, as the checkers noted, heating for the recommended 8 – 10 hours at temperatures above 110° will raise the yields quite signifi-

cantly: the highest single product recovery obtained by the submitters was 86.8% using a strong steel autoclave (Note 1) immersed for 10 hours in a thermostat-equipped oil bath at 125°. However, the greater yield which may be realized through higher reaction temperatures should be carefully weighed against the ability of the reaction vessel to contain safely the greater pressures developed.

7. Little, if any, overpressure is released if the vessel is opened below 30°; this should be done in a hood because the contents will evolve sulfur dioxide until the sodium carbonate is added.

8. A magnetic stirrer is recommended for this operation.

9. Hexane and Skellysolve-B serve equally well; the product is miscible with aliphatic solvents, whereas 3-sulfolene is not.

10. No increase in product recovery was observed by the submitters when a 72-hour exhaustive extraction of the neutralized reaction mixture with hexane in a continuous apparatus was substituted for the procedure described.

11. The submitters observe that no more critical step in this procedure exists than the efficient removal of the rather large quantity of solvent from the product before distillation: entrainment losses during this operation usually were found chiefly responsible when the synthesis gave a low yield (recovery). The preferred apparatus for this step is a rotary evaporator equipped with a long, helical, water-cooled condenser; the distillation flask is attached to the evaporator and the product solution added portionwise as the concentration progresses; the temperature of the flask is held at 30 – 40° with a warm-water bath, or a heat lamp, while the solvent is gently removed by a water aspirator (ca. 30-mm. minimum pressure). If a suitable rotary evaporator is unavailable, a 30 – 45 cm. Vigreux column may be used without excessive product loss, providing the flask is warmed in the same manner and a gentle boiling rate is maintained through control of the aspirator suction.

12. An elaborate apparatus for the diminished-pressure distillation is not required to obtain a product of excellent purity: sensitive GC analyses could detect no more than a doubtful trace of the cis isomer in the crude mixture, and other possible contaminants boil at least 45° (760 mm.) below the product. The submitters usually employed an 8 × 300 mm., vacuum-jacketed, silvered column with a helical, Nichrome-wire heat transfer insert. They consistently obtained a product which analyzed at least 99% pure using GC conditions which proved capable of cleanly separating the product from an authentic sample of its cis-isomer [Org. Synth., Coll. Vol. 4, 304 (1963)], with the latter showing the longer retention time (terephthalic acid-terminated Carbowax 20M, 10% w/w on 60/80 mesh Chromosorb W, 6 mm. × 365 cm. aluminum column at 150°). The submitters note that distillation of the product at atmospheric pressure (b.p. 265 – 270°) is accompanied by very slight decomposition. (A b.p. of 150.5 – 151.5° at an unspecified pressure appears in the literature.[3])

13. The IR spectra of diethyl trans-Δ^4-tetrahydrophthalate and its cis-isomer [Org. Synth., Coll. Vol. 4, 304 (1963)], (cm^{-1}, neat liquids purified by GC, 0.025-mm. cells) are similar in the fundamental region; however, the following particular absorptions in the "fingerprint" region afford reliable analytical criteria for distinguishing each from (or in the presence of) the other; trans-isomer, 971 (s), 755, 743 (m, s, doublet); cis-isomer, 945 (s), 893 (w), 727 (m). The 755, 743 doublet of the trans-isomer and the 727 single absorption of the cis-isomer appear to be characteristic, and related to out-of-plane deformations of the yinylic hydrogens; bands corresponding to the 971 and 755, 743

absorptions of the *trans*-isomer are present also in the spectrum of dimethyl *trans*-Δ^4-tetrahydrophthalate.[4]

3. Discussion

A report of Backer and Blaas[5] is responsible for evolution of the present procedure; these workers were first to conduct a Diels-Alder synthesis utilizing a 3-sulfolene in place of the free diene (by heating the cyclic adduct of sulfur dioxide and 2-methyl-3-thiomethyl-1,3-butadiene with maleic anhydride). The generality of the method as a variant of the conventional diene synthesis is limited largely by the availability of the appropriate 3-sulfolene; its greatest utility, perhaps, will be realized in those diene reactions normally requiring 1,3-butadiene, since 3-sulfolene itself is now the least expensive and most widely available diene cyclic sulfone.

The submitters, in addition to their successful work with 3-sulfolene itself, have generally obtained very satisfactory results performing diene syntheses based on the present procedure using several fumaric and maleic derivatives (notably fumaric acid and fumaronitrile, and diethyl maleate) in conjunction with 3-methyl-3-sulfolene (isoprene cyclic sulfone)[6] and 2,4-dimethyl-3-sulfolene (available from Aldrich Chemical Co.), as well as 3,4-dimethyl-3-sulfolene.[7]

In preparations wherein the dienophile could undergo side reaction with the solvent alcohol, either benzene or toluene usually proved to be a satisfactory reaction medium; lower-boiling solvents, such as ether, chloroform, or acetone, were avoided by the submitters because of the high reactive pressures developed. If the dienophile is highly reactive and not appreciably volatile below about 200°, the reaction may be successfully conducted without solvent by careful fusion of an intimate mixture of the reactants in an open vessel.[5,8] *Caution! Such fusions should be performed in a good fume hood because of the copious evolution of sulfur dioxide.*

3-Sulfolene as an alternative reactant for 1,3-butadiene in diene syntheses has the following advantages: it presents no particular flammability hazard[9] and is practically nontoxic;[10] it is an odorless, crystalline, nonhygroscopic solid which may be stored without inhibitor at ordinary temperatures for years with no evidence of deterioration;[10] obviously, no low-temperature liquefaction procedures are needed for its handling, as butadiene may require; medium-pressure reaction vessels can be safely employed since autogenous pressures are relatively low owing to the thermally controlled release of the 1,3-butadiene;[11] and, because the substantial excess of free diene often employed to force the usual reaction is not required, formation of troublesome polymeric by-products is greatly diminished.

For the preparation of diethyl *trans*-Δ^4-tetrahydrophthalate, the present synthesis is superior to the Diels-Alder reaction of butadiene with diethyl fumarate[3] for the reasons given above. In addition, a higher yield is obtained at a lower reaction temperature and shorter reaction time.

Whereas *cis*-Δ^4-tetrahydrophthalic anhydride and several of its derivatives are readily available in high purity from a number of suppliers (largely because of the great ease of addition of maleic anhydride to 1,3-butadiene),[12] the *trans* analogs are much more difficult and expensive to prepare and isolate in reasonable yields and purity; hence the latter are rarely procurable at present from the usual laboratory supply sources. Thus, the procedure described for the preparation of highly pure diethyl *trans*-Δ^4-tetrahydrophthalate not only illustrates a useful variation of the diene synthesis, but provides a convenient route to a large number of *trans*-1,2-disubstituted cyclohexyl derivatives that are currently difficult to secure and are of importance for a variety of studies concerned with structural problems.

1. Texaco Inc., Bellaire Research Laboratories, Bellaire, Texas 77401 [Present address: Magcobar Division, Dresser Industries, Inc., 10205 Westheimer Road, Houston, Texas 77042.]

2. School of Science, The University of Texas, El Paso, Texas 79999. This work was supported in part by a grant from the Robert A. Welch Foundation and carried out in the Department of Chemistry at The University of Texas, Austin, Texas 78712.

3. A. A. Petrov and N. P. Sopov, *Sb. Statei Obshch. Khim.*, **2**, 853 (1953) [*Chem. Abstr.*, **49**, 5329 (1955); *Chem. Zentr.*, **127**, 2156 (1956)].

4. S. C. Sodd, Master's Thesis, The University of Texas, Austin, Texas, Synthesis and Photochemistry of (4.3.0)-Bicyclononadienes, Masters Theses in Pure and Applied Sciences, Vol. 10, Beth M. Schick, ed., Purdue University, West Lafayette, Indiana (1965); G. J. Fonken, The University of Texas, Austin, Texas, private communication.

5. H. J. Backer and T. A. H. Blaas, *Recl. Trav. Chim. Pays-Bas.*, **61**, 785 (1942). [*Chem. Abstr.*, **38**, 3646 (1944)].

6. R. L. Frank and R. P. Seven, *Org. Synth.*, **Coll. Vol. 3**, 499 (1955).

7. O. Grummitt and A. L. Endrey, *J. Am. Chem. Soc.*, **82**, 3614 (1960).

8. T. E. Sample, Jr., and L. F. Hatch, *J. Chem. Educ.*, **45**, 55 (1968); K. Alder and H. A. Dortmann, *Chem. Ber.*, **87**, 1492 (1954) [*Chem. Abstr.*, **49**, 12319 (1955)].

9. Technical Bulletin 522, "Phillips 66 Hydrocarbons and Petro-Sulfur Compounds", 6th ed., Phillips Petroleum Company, Special Products Division, Bartlesville, Oklahoma 74003, 1964, p. 184.

10. Technical Information Bulletin PD-146, "3-Sulfolene", Shell Chemical Company, Industrial Chemicals Division, New York, N.Y. 10020, 1963, pp. 2, 10.

11. L. R. Drake, S. C. Stowe, and A. M. Partansky, *J. Am. Chem. Soc.*, **68**, 2521 (1946); O. Grummitt, A. E. Ardis, and J. Fick, *J. Am. Chem. Soc.*, **72**, 5167 (1950).

12. A. C. Cope and E. C. Herrick, *Org. Synth.*, **Coll. Vol. 4**, 304 (1963).

HOMOGENEOUS CATALYTIC HYDROGENATION: DIHYDROCARVONE

[2-Cyclohexen-1-one, 2-methyl-5-(1-methylethyl)-]

Submitted by ROBERT E. IRELAND[1] and P. BEY[2]
Checked by N. HAGA and W. NAGATA

1. Procedure

Caution! Benzene has been identified as a carcinogen; OSHA has issued emergency standards on its use. All procedures involving benzene should be carried out in a well-ventilated hood, and glove protection is required.

A 500-ml., two-necked, creased flask, containing a magnetic stirring bar and connected to an atmospheric pressure hydrogenation apparatus equipped with a graduated burette to measure the uptake of hydrogen, is charged with 0.9 g. (0.9×10^{-3} mole) of freshly prepared tris(triphenylphosphine)rhodium chloride (Note 1) and 160 ml. of benzene (Note 2). One neck is stoppered with a serum cap, and the mixture is stirred magnetically (Note 3) until the solution is homogeneous. The system is then evacuated and filled with hydrogen. Using a syringe, 10 g. (0.066 mole) of carvone (Note 4) is introduced into the hydrogenation flask. The syringe is rinsed with two, 10-ml. portions of benzene, and stirring is resumed. Hydrogen uptake starts immediately (Note 5) and stops 3.5 hours later when the theoretical amount of hydrogen has been absorbed. The solution is filtered through a dry column (4 cm. diameter) of 120 g. of Florisil (60 – 100 mesh). The column is washed with 300 ml. of diethyl ether, and the combined solvent fractions are concentrated under reduced pressure. Vacuum distillation of the yellow residue through an 11-cm. Vigreux column (Note 6) affords 9.1 – 9.5 g. (90 – 94%) of dihydrocarvone; b.p. 100 – 102° (14 mm.), n_D^{24} 1.479 (Notes 7 and 8).

2. Notes

1. The tris(triphenylphosphine)rhodium chloride catalyst was prepared according to the procedure of G. Wilkinson and co-workers.[4]
2. The benzene was distilled from calcium hydride.
3. Efficient stirring is necessary to assure good surface contact during hydrogenation.
4. The carvone was distilled before use; b.p. 105 – 106° (14 mm.). The checkers used *l*-carvone obtained from Shiono Koryo K.K. (Japan).

5. With old catalyst, very erratic results with respect to the initiation time and the rate of hydrogen uptake have been observed.

6. When the hydrogenation is carried out on a smaller scale, purification can be affected by evaporative distillation in a bulb to bulb apparatus.

7. GC analysis shows contamination by less than 3% of carvone. The checkers used a 1 m. by 4 mm. glass column packed with 5% PEG 6000 on Chromosorb W (60/80 mesh). The retention times were 3.7 minutes and 2.75 minutes for carvone and dihydrocarvone, respectively, at 100° with a nitrogen flow rate of 90 ml. per minute.

8. The product shows the following spectral properties: IR (neat) 1678 cm.$^{-1}$

$$(\underset{|}{>}C{=}\underset{|}{C}{-}C{=}O)$$; UV (C$_2$H$_5$OH) λ_{max} 237 nm (ϵ = 9150); ^{1}H NMR (CDCl$_3$), δ 0.88 [d, J = 6 Hz., 6H, CH(CH$_3$)$_2$], 1.71 (d, J = 1.5 Hz., 3H, CH$_3$C$=$CH), 6.70 (m, 1H, CH$=$C).

3. Discussion[3]

This procedure is an example of the use of a soluble transition metal complex for the catalytic transfer of hydrogen to an olefin. First developed by Wilkinson and co-workers,[4] subsequent extensive investigation in those laboratories and others[5] has shown that the hydrogenation is sensitive to steric congestion; only unhindered double bonds are reduced. As a result, the rhodium complex has been found useful for the selective saturation of unhindered double bonds in polyolefinic substances, such as carvone.[6b] Unhindered double bonds may be reduced even in the presence of functions such as keto,[6a,b] nitro,[6b,7] and sulfide[6b] groups. The mechanism[4] and stereochemistry[8] of the catalysis have been investigated, and *cis*-addition of hydrogen is the general rule. The catalyst is effective for deuterium addition to unhindered olefins[9] without the extensive hydrogen – deuterium exchange observed with palladium and platinum heterogeneous catalysis. The rhodium complex causes the decarbonylation of aldehydes and acid halides; the hydrogenation of such unsaturated systems is complicated by the loss of these functional groups.[4,7] Isomerization of nonreduced olefinic bonds is also an observed side reaction.[10]

1. Division of Chemistry and Chemical Engineering, California Institute of Technology, Pasadena, California 91109.
2. Department of Chemistry, Research Laboratories, Richardson-Merrell, Inc., Strasbourg, France.
3. For further discussion and bibliography, see H. O. House, "Modern Synthetic Reactions," 2nd ed., W. A. Benjamin, New York, 1972, p. 28.
4. J. A. Osborn, F. H. Jardine, J. F. Young, and G. Wilkinson, *J. Chem. Soc.* (A), 1711 (1966).
5. For reviews, see J. Tsuji, *Adv. Org. Chem.*, **6**, 109 (1969); J. A. Osborn, *Endeavour*, **26**, 144 (1967); R. Cramer, *Acc. Chem. Res.*, **1**, 186 (1968); R. F. Heck, *Acc. Chem. Res.*, **2**, 10 (1969).
6. (a) M. Brown and L. W. Piszkiewicz, *J. Org. Chem.*, **32**, 2013 (1967); (b) A. J. Birch and K. A. M. Walker, *J. Chem. Soc.* (C), 1894 (1966).
7. R. E. Harmon, J. L. Parsons, D. W. Cooke, S. K. Gupta, and J. Schoolenberg, *J. Org. Chem.*, **34**, 3684 (1969).
8. F. H. Jardine, J. A. Osborn, and G. Wilkinson, *J. Chem. Soc.* (A), 1574 (1967).
9. J. R. Morandi and H. B. Jensen, *J. Org. Chem.*, **34**, 1889 (1969).
10. J. J. Sims, V. K. Howard, and L. H. Selman, *Tetrahedron Lett.*, 87 (1969).

trans-1,2-DIHYDROPHTHALIC ACID

(3,5-Cyclohexadiene-1,2-dicarboxylic acid, *trans*-)

Submitted by Richard N. McDonald and Charles E. Reineke[1]
Checked by Herbert A. Kirst and E. J. Corey

1. Procedure

A vigorously stirred, particle-free (Note 1) solution of 170 g. (1.02 moles) of phthalic acid and 281 g. of sodium acetate in 1.7 l. of water is cooled in an ice bath while a total of 3400 g. of 3% sodium amalgam (Note 2) is added in 50 – 100 g. portions (Note 3). With each portion of added amalgam there is also added 10 – 20 ml. (500 ml. total) of 50% acetic acid. The total addition time required is 4 – 5 hours. The solution is decanted from the mercury onto a Buchner funnel and filtered with suction through a layer of Celite 545. The cold filtrate is treated with 1.7 l. of cold 20% sulfuric acid; the acid begins to crystallize immediately. After standing for 4 hours at 20 – 22°, the acid is collected by suction filtration, washed well with ice-cold water, removing excess sulfuric acid, and dried in a vacuum desiccator over sulfuric acid, giving 124 g. (72%) of crystalline product, m.p. 210 – 213°. A purer product can be obtained by recrystallization. For minimum losses during recrystallization, the crude acid is divided into two portions, and each is added to 1.2 – 1.5 l. of rapidly stirred, boiling water to effect as rapid a solution as possible. When almost all of the solids have dissolved *ca.* 1 g. of activated charcoal (Norit) is added, the solution is filtered through a fluted filter, and the filtrate is cooled in ice to induce rapid crystallization (Note 4). The colorless crystals are collected by filtration and dried under reduced pressure, yielding 93 – 107 g. (54 – 62%) of the acid, m.p. 212 – 214°.

2. Notes

1. If any crystals of phthalic acid are present, crystallization will occur on cooling.
2. The 3% sodium amalgam was prepared by the procedure in *Org. Synth.*, **Coll. Vol. 2,** 609, (1943), Note 3, with the following explanation. The mineral oil and sodium are placed in a 1-l. heavy-wall filter flask and heated on a hot plate until the sodium is molten. To this hot, strongly swirled mixture is added mercury from a 1-l. separatory funnel. The initial amounts of mercury (about ¼ the total) are added in a fine steady stream, gradually increasing the rate of flow as the initial crackling ceases. Fire flashes can be observed if the initial addition is too rapid. Once the initial reaction subsides, the mercury should be added as rapidly as possible, and about the last third can be essentially poured in. The molten amalgam is *immediately* poured into a shallow, porcelainized, metal pan and

stirred with a wide spatula so that large chunks are not allowed to form as the amalgam cools. Once cool, the amalgam pieces are washed repeatedly with petroleum ether, removing the mineral oil.

3. Another portion of amalgam and one of acid are added as soon as the gas evolution from the previous addition subsides.

4. The submitters report that treatment with decolorizing carbon may not be necessary. Although the crude product is invariably colorless, a yellow color develops during recrystallization. The checkers found that without decolorization a yellow product of somewhat lower m.p. was obtained.

3. Discussion

The present method, based on a recent publication,[2] is a modification of that previously reported.[3] *trans*-1,2-Dihydrophthalic acid has been converted to the *cis*-anhydride by heating in acetic anhydride,[2] which on photolysis (Hanovia, type L mercury lamp) yields the photo-anhydride, bicyclo[2.2.0]hex-5-ene-2,3-dicarboxylic anhydride.[2,4] The photoanhydride has been converted to bicyclo[2.2.0]hexa-2,5-diene,[4] bicyclo[2.2.0]hex-2-ene,[2] and *exo*-bicyclo[2.2.0]hexan-2-ol[2] as well as certain derivatives of the alcohol. The present procedure also gives a more complete preparation of 3% sodium amalgam.

1. Department of Chemistry, Kansas State University, Manhattan, Kansas 66502.
2. R. N. McDonald and C. E. Reineke, *J. Org. Chem.*, **32**, 1878 (1967).
3. B. R. Landau, Ph.D. Thesis, Harvard University, 1950.
4. E. E. van Tamelen and S. P. Pappas, *J. Am. Chem. Soc.*, **85**, 3297 (1963)

5,6-DIHYDRO-2*H*-PYRAN-2-ONE AND 2*H*-PYRAN-2-ONE

Submitted by M. Nakagawa,[1] J. Saegusa,[1] M. Tonozuka,[1]
M. Obi,[1] M. Kiuchi,[1] T. Hino,[1] and Y. Ban[2]
Checked by A. Wick, D. Ehrlich, and A. Brossi

1. Procedure

A. *5,6-Dihydro-2H-pyran-2-one.* In a 500-ml., one-necked, round-bottomed flask equipped with a reflux condenser are combined 43 g. (0.50 mole) of vinylacetic acid (Note 1), 30 g. (1 mole as CH_2O) of paraformaldehyde (Note 2), 3 ml. of concentrated sulfuric acid, and 125 ml. of glacial acetic acid. This mixture is refluxed gently for 3 hours, then cooled to room temperature and swirled while 16 g. of anhydrous sodium acetate is added. Acetic acid is removed at 50 – 55° on a rotary evaporator, 100 ml. of water is added, and the flask is fitted with a two-necked adapter, a thermometer, and a magnetic stirring bar. The flask is then immersed in an ice bath, and the solution is brought to pH 8 with aqueous 20% sodium hydroxide (Note 3), which is added dropwise and with stirring at a rate such that the temperature remains below 5°. The resulting solution is transferred to a 1-l. separatory funnel and extracted with four 300-ml. portions of dichloromethane (Note 4). After being washed with one 150-ml. portion of saturated aqueous sodium chloride (Note 5), the combined organic extracts are dried over anhydrous sodium sulfate and filtered. Removal of dichloromethane with a rotary evaporator leaves a mobile yellow oil, which is distilled under reduced pressure, yielding 12.3 g. (25.1%) of 5,6-dihydro-2*H*-pyran-2-one, b.p. 114 – 117° (18 – 19 mm.) (Note 6).

B. 2H-*Pyran-2-one.*   A mixture of 9.81 g. (0.100 mole) of 5,6-dihydro-2*H*-pyran-2-one, 200 mg. of benzoyl peroxide, 18.6 g. (0.105 mole) of *N*-bromosuccinimide (Note 7), and 800 ml. of carbon tetrachloride is prepared in a 2-l., three-necked, round-bottomed flask equipped with a reflux condenser and a mechanical stirrer. The resulting suspension is stirred and heated to reflux. After 1.5 hours at reflux, most of the solid is dissolved, and the solution gives a negative test with starch – iodide paper. The reaction mixture is then allowed to cool, during which time succinimide crystallizes out. The precipitate is removed by filtration, and the filtrate is concentrated under reduced pressure, leaving crude 5-bromo-5,6-dihydro-2*H*-pyran-2-one as an oil.

This residue is stirred at room temperature while 150 ml. of triethylamine (Note 8) is added. Triethylamine hydrobromide begins to precipitate soon after the addition is started, and the resulting slurry is refluxed gently for 15 minutes. It is then cooled to room temperature, and the insoluble material is removed by filtration and washed with benzene. Concentration of the combined filtrates under reduced pressure leaves an oily residue, which is dissolved in 600 ml. of diethyl ether. The ethereal solution is transferred to a 1-l. separatory funnel, washed with two 20-ml. portions of saturated aqueous sodium chloride, dried over anhydrous sodium sulfate, and filtered. Ether is removed with a rotary evaporator, and the resulting oil is distilled at reduced pressure. A forerun of 265 mg. is collected below 103° (22 mm.) before 6.7 g. (70%) of 2*H*-pyran-2-one distils as a colorless oil, b.p. 103 – 111° (19 – 22 mm.) (Note 9).

2. Notes

1. Vinylacetic acid is available from Tokyokasei Company, Ltd., Japan or from Fluka AG, Buchs, Switzerland. Commercial material, which shows about 3% of crotonic acid in its ¹H NMR spectrum, was distilled at 90 – 92° (40 – 43 mm.) prior to use.

2. The submitters obtained paraformaldehyde from Koso Chemical Company, Inc., Japan.

3. About 180 ml. is required.

4. In these extractions, the organic layer is the lower one. If the two phases do not separate readily, fine-grained precipitates are probably at fault. These may be removed by filtration through a Büchner funnel.

5. Excess washing should be avoided, since 5,6-dihydro-2H-pyran-2-one is fairly soluble in water.

6. A forerun of approximately 180 mg. is collected below 110° (18 mm.). The IR spectrum of this material is practically identical with that of the main distillate. Reported physical constants for 5,6-dihydro-2H-pyran-2-one are: b.p. 110° (15 mm.) and n_D^{25} 1.4730.[3]

7. The submitters obtained N-bromosuccinimide from Nakarai Chemicals Ltd., Japan, and crystallized it from water prior to use (m.p. 168 – 175°).

8. Triethylamine was purified by treatment with p-toluenesulfonyl chloride and distillation.

9. The IR spectrum of this material was essentially identical to that of the redistillate, b.p. 115 – 118° (37 mm.). Reported physical constants for 2H-pyran-2-one are: b.p. 206 – 209° (atmospheric pressure), n_D^{25} 1.5272,[7] and b.p. 110° (26 mm.), n_D^{25} 1.5270.[8]

3. Discussion

5,6-Dihydro-2H-pyran-2-one has been prepared by reductive cyclization of 5-hydroxy-2-pentynoic acid, which is obtained in two steps from acetylene and ethylene oxide;[3] and by the reaction of dihydropyran with singlet oxygen.[4,5] 2H-Pyran-2-one has been prepared by pyrolysis of heavy metal salts of coumalic acid,[6] by pyrolysis of α-pyrone-6-carboxylic acid over copper,[7] and by pyrolysis of coumalic acid over copper (66 – 70% yield).[8]

The present one-step procedure for preparation of 5,6-dihydro-2H-pyran-2-one is slightly modified from that described in the original paper.[9] It is simpler and easier than the three-step method[3] used in the past and represents the most convenient synthesis currently available. The present preparation of 2H-pyran-2-one has several advantages compared to the alternatives mentioned above: simplicity of apparatus and technique, mild reaction conditions, availability of reactants, and ease of product isolation.

1. Faculty of Pharmaceutical Sciences, Chiba University, Chiba, Japan.
2. Faculty of Pharmaceutical Sciences, Hokkaido University, Sapporo, Japan.
3. L. J. Haynes and E. R. H. Jones, J. Chem. Soc., 954 (1946).
4. P. D. Bartlett, G. D. Mendenhall, and A. P. Schaap, Ann. N.Y. Acad. Sci., 171, 79 (1970).
5. E. C. Blossey, J. Am. Chem. Soc., 95, 5820 (1973).
6. H. von Pechmann, Justus Liebigs Ann. Chem., 264, 272 (1891).
7. J. Fried and R. C. Elderfield, J. Org. Chem., 6, 566 (1941).
8. H. E. Zimmerman, G. L. Grunewald, and R. M. Paudler, Org. Synth., Coll. Vol. 5, 982 (1973).
9. M. Nakagawa, M. Tonozuka, M. Obi, M. Kiuchi, T. Hino, and Y. Ban, Synthesis, 510 (1974).

2,4-DIMETHOXYBENZONITRILE

(Benzonitrile, 2,4-dimethoxy-)

$$\text{OCH}_3 \text{ (resorcinol dimethyl ether)} \xrightarrow{\text{ClSO}_2\text{NCO}} \text{CONHSO}_2\text{Cl / OCH}_3 \xrightarrow{\text{HCON(CH}_3)_2}$$

$$\text{CN / OCH}_3 + \text{SO}_3 + \text{HCl}$$

Submitted by G. Lohaus[1]
Checked by Jurgen K. Weise and Richard E. Benson

1. Procedure

Caution! Chlorosulfonyl isocyanate is a highly corrosive, irritating compound. This reaction should be carried out in an efficient hood.

A 1-l., round-bottomed flask equipped with a stirrer, a thermometer, a dropping funnel, and a reflux condenser to which is attached a drying tube containing calcium chloride, is charged with 138 g. (131 ml., 1.00 mole) of resorcinol dimethyl ether and 200 ml. of dichloromethane. The solution is stirred, and a solution of 150 g. (1.06 moles) of chlorosulfonyl isocyanate (Note 1) in 100 ml. of dichloromethane is added with stirring at 15–20° over a 25-minute period. The amide *N*-sulfonyl chloride separates as a crystalline solid, and the mixture is stirred for an hour at room temperature. The resulting mixture is cooled to 10–12° (Note 2) and 154 g. (2.1 moles) of *N,N*-dimethylformamide (Note 3) is added over a period of 5 minutes. The cooling bath is removed and the temperature gradually rises to about 30°, then falls. After 1 hour the crystals dissolve and the reaction mixture is poured onto 200 g. of ice. After the ice has melted, dichloromethane (150 ml.) is added, the mixture is shaken, and the organic layer is separated. The aqueous layer is extracted with 100 ml. of dichloromethane, and the organic phases are combined and washed with 100 ml. of water. The dichloromethane is removed by distillation, giving a white solid that is triturated with 250 ml. of cold water, recovered by filtration, and dried, yielding 155–157 g. (95–96%) of 2,4-dimethoxybenzonitrile, m.p. 91°. GC using a Chromosorb W column with 10% butadiene sulfone as the stationary phase indicates that the product has a purity of 98%. The IR spectrum shows absorption at 2220 cm.$^{-1}$ attributable to the cyano group.

2. Notes

1. Chlorosulfonyl isocyanate may be prepared as described in *Org. Synth.*, **Coll. Vol. 5**, 226 (1973); it is available from Farbwerke Hoechst AG. The checkers found it necessary to distill the product before use.

2. A less pure product is obtained if the temperature is allowed to rise at this phase of the reaction.

3. Other amides also can be used, but *N,N*-dimethylformamide generally is preferred, especially because of its low molecular weight, high solvent power, and miscibility with water. In addition, it is readily available.

TABLE I

NITRILES[2] PREPARED FROM $ClSO_2NCO$

Reactant	Product	Yield, %
		67
		20
		66
		86
		84

TABLE I (*continued*)

NITRILES[2] PREPARED FROM $ClSO_2NCO$

Reactant	Product	Yield, %
		81
		95
		89

3. Discussion

This procedure[2] is an example of a broadly applicable, simple method for introducing the cyano substituent into compounds that readily undergo electrophilic substitution. The method is characterized by mild reaction conditions, a simple workup procedure, and, in most cases, good yields. Although the method has two steps, the reaction generally can be carried out without isolation of the intermediate chlorosulfonamide. An indication of its scope is given in Table I.[2] Additional examples of the substitution reaction of chlorosulfonyl isocyanate with aromatic and heterocyclic compounds and with olefins, yielding carboxylic acid amide N-sulfonyl chlorides are reported.[2-6]

Other procedures for the preparation of 2,4-dimethoxybenzonitrile include the reaction of 2,4-dimethoxybenzamide with thionyl chloride,[3] the action of acetic anhydride on 2,4-dimethoxybenzaldoxime,[7] the reaction of diazotized 2,4-dimethoxyaniline with potassium copper cyanide,[8] and the action of cyanogen bromide on resorcinol dimethyl ether in the presence of aluminum chloride.[9]

1. Hoechst AG., Frankfurt/Main-Höchst, Germany.
2. G. Lohaus, *Chem. Ber.*, **100**, 2719 (1967).
3. F. Effenberger, R. Gleiter, L. Heider, and R. Niess, *Chem. Ber.*, **101**, 502 (1968).
4. R. Graf, *Ann. Chem.*, **661**, 111 (1963).
5. M. Seefelder, *Chem. Ber.*, **96**, 3243 (1963).
6. R. Graf, *Angew. Chem.*, **80**, 179 (1968) [*Angew. Chem., Int. Ed. Engl.*, **7**, 172 (1968)].
7. H. Baganz and I. Paproth, *Naturwissenschaften*, **40**, 341 (1953).

8. E. Späth, K. Klager, and C. Schlösser, *Ber. Dtsch. Chem. Ges.*, **64**, 2203 (1931).
9. P. Karrer, A. Rebmann, and E. Zeller, *Helv. Chim. Acta*, **3**, 261 (1920).

BIARYLS FROM SIMPLE ARENES *via* ORGANOTELLURIUM INTERMEDIATES: 4,4'-DIMETHOXY-1,1'-BIPHENYL

[1,1'-Biphenyl, 4,4'-dimethoxy-]

$$CH_3O-\!\!\!\bigcirc + TeCl_4 \xrightarrow[\text{1 hour}]{60-80°} CH_3O-\!\!\!\bigcirc-TeCl_3$$

$$CH_3O-\!\!\!\bigcirc-TeCl_3 + CH_3O-\!\!\!\bigcirc \xrightarrow[\text{6 hours}]{160°}$$

$$CH_3O-\!\!\!\bigcirc-TeCl_2-\!\!\!\bigcirc-OCH_3$$

$$CH_3O-\!\!\!\bigcirc-TeCl_2-\!\!\!\bigcirc-OCH_3 \xrightarrow[\text{200°}]{\text{Raney nickel}}$$

$$CH_3O-\!\!\!\bigcirc-\!\!\!\bigcirc-OCH_3$$

Submitted by J. BERGMAN,[1] R. CARLSSON, and B. SJÖBERG
Checked by J. DIAKUR and S. MASAMUNE

1. Procedure

Caution! Because tellurium compounds have toxic effects similar to those of arsenic compounds,[2] care should be taken not to bring tellurium tetrachloride and its reaction products into contact with the skin. Avoid breathing fumes and dust of tellurium compounds. In addition, hydrogen chloride is evolved in Step A, and pyrophoric Raney nickel is used in Step B. Therefore all manipulations described in this procedure must be carried out in an efficient fume hood.

Benzene has been identified as a carcinogen; OSHA has issued emergency standards on its use. All procedures involving benzene should be carried out in a well-ventilated hood, and glove protection is required.

A. *Bis(4-methoxyphenyl)tellurium dichloride.* A dry, 500-ml., three-necked, round-bottomed flask equipped with a thermometer and a reflux condenser fitted with a calcium chloride drying tube is charged with 27.0 g. (0.100 mole) of tellurium tetrachloride (Note 1) and 64.8 g. (0.600 mole) of dry anisole (Note 2). The mixture is heated to 160° over a period of 30 minutes and maintained at this temperature for 6 hours. The reaction mixture is allowed to cool to room temperature, and the solvent is removed with the aid of a vacuum pump. The crude solid (Note 3) is dissolved in *ca.* 250 ml. of boiling acetonitrile and filtered while hot (Note 4). Upon cooling to −25°, crystals deposit (Note 5), which weigh 35.5–38.5 g. (84–90%), m.p. 182–183° (Note 6).

B. *4,4'-Dimethoxy-1,1'-biphenyl.* A 500-ml., three-necked, round-bottomed flask is equipped with a 500-ml. dropping funnel, a stopper, and a reflux condenser fitted with a two-way stopcock, one end of which is connected to an aspirator, and the other to a cylinder of dry, oxygen-free nitrogen. To this flask are added 60 g. of Raney nickel (Note 7) and 150 ml. of benzene. The system is flushed with nitrogen, and the solvent is evaporated under reduced pressure. The Raney nickel is then degassed behind an explosion shield (Note 8) in a hood by heating to 200° (2 mm.) for 2 hours.

The catalyst is allowed to cool under nitrogen, and 400 ml. of bis(2-methoxyethyl) ether is added from the dropping funnel. The stopper is then temporarily removed to add 20.6 g. (0.0497 mole) of bis(4-methoxyphenyl)tellurium dichloride. The mixture is refluxed for 8 hours and filtered while still hot. The filtrate is evaporated under reduced pressure (10–20 mm.), giving a residue which when recrystallized from ethanol yields 8.5–9.8 g. (78–90%) of the product, m.p. 175–176° (Note 9).

2. Notes

1. The submitters used tellurium tetrachloride available from E. Merck A G. The checkers purchased the reagent from Research Organic/Inorganic Chemical Corporation.

2. The checkers distilled anisole from calcium sulfate before use. This reagent functions not only as a reactant, but also as solvent. In some similar preparations the intermediate trichloride is rather insoluble, as in the case of bis(3-methyl-4-methoxyphenyl)tellurium dichloride. The addition of co-solvents such as bis(2-methoxyethyl) ether is beneficial.[3]

3. The crude product contains the 4,4'- and the 2,4'-isomers in the ratio 99.2/0.8. See Note 5.

4. A small amount of tellurium (95 mg.) is formed during the preparation. The amount of tellurium increases slowly as the heating is prolonged. In a separate experiment pure bis(4-methoxyphenyl)tellurium dichloride was pyrolyzed at 250°. The main products were 1-chloro-4-methoxybenzene and tellurium.

5. Evaporation of the mother liquor gives a solid enriched in the 2,4'-isomer. Recrystallization of this solid from ethanol yields crystals containing 45% of the 2,4'-isomer.

6. The spectral details of the product are: ^{1}H NMR (DMSO-d_6), δ (multiplicity, number of protons, assignment): 3.80 (s, 6H, OCH_3), 7.05–8.0 (q, 8H, A_2B_2 aryl); mass spectrum m/e (relative intensity > 10% for peaks with m/e above 150): 379 (16), 377 (14), 344 (39), 342 (35), 340 (22), 272 (36), 270 (31), 237 (18), 235 (17), 233 (10), 215 (15), 214 (99), 200 (14), 199 (100), 172 (21). No peaks corresponding to the parent ion could be detected.

7. The catalyst was prepared from a nickel–aluminum (50:50) alloy using the procedure in *Org. Synth.*, **Coll. Vol. 3,** 181 (1955). The catalyst is used in large excess. Reduced amounts of catalyst resulted in decreased yields, and the product is contaminated with detectable (GC) amounts of bis(4-methoxyphenyl)telluride.

8. This operation has been performed several times without incident. However, it should be noted that the W6 and W7 forms of Raney nickel have been reported to explode [*Org. Synth.*, **Coll. Vol. 5,** 102 (1973)]. No explosions have been reported with the W2 form used in this preparation.

9. 4,4′-Dimethoxybiphenyl can also be prepared by simply refluxing bis(4-methoxyphenyl)tellurium dichloride with degassed, commercial Raney nickel. The yields are, however, lower and less reproducible,[3] and the product may contain some bis(4-methoxyphenyl)telluride.

3. Discussion

Most synthetic methods for biaryl preparation, such as the Ullmann coupling and all variants of the Grignard coupling,[4–6] require halogen-substituted aromatic compounds as starting materials. Since these components are prepared by halogenation of the appropriate precursor, either directly or indirectly, it is evident that a direct coupling method offers obvious advantages. Such reactions may be effected electrochemically,[7] and by reagents such as palladium(II) acetate,[8,9] thallium(III) trifluoroacetone,[10] and vanadium tetrachloride.[11] The applicability of these reagents and the selectivity of the reactions are often restricted, when compared with the present procedure. For a recent review, see reference 12.

Tellurium tetrachloride reacts as an electrophilic reagent with aromatic compounds bearing activating substituents, such as RO-, R_2N-, and RS- groups, providing first aryltellurium trichlorides, then diaryltellurium dichlorides, as one raises the reaction temperature. The second step should, in order to prevent formation of elemental tellurium and chlorinated aromatics, be performed at as low a temperature as possible (Note 4). This is especially important when highly reactive substrates such as 1,3-dimethoxybenzene are used. The addition of a Lewis acid to the reaction mixture brings about an acceleration of the reaction with less reactive reactants such as benzene and chlorobenzene.[3] The rate of acceleration is dramatically enhanced when the ratio of $AlCl_3/TeCl_4$ is more than 1:1. Thus, refluxing a mixture of 1 equivalent of $TeCl_3$ and 3 equivalents of $AlCl_3$ in benzene provided diphenyltellurium dichloride in 58.5% yield.[13]

The coupling reaction proceeds better when a rigorously degassed Raney nickel catalyst is used, but a nickel catalyst prepared by a much simplified procedure (Note 9) is also effective. The coupling may also be promoted by other elements, including copper and palladium.

1. Department of Organic Chemistry, Royal Institute of Technology, S-100 44 Stockholm 70, Sweden.
2. American Conference of Governmental Industrial Hygienists (ACGIH), "Documentation of Threshold Limit Values," 3rd ed. 1971, pp. 16 and 245–246; F. A. Patty in "Industrial Hygiene and Toxicology," Interscience, New York, Vol. II, 2nd ed., 1963, Chap. 24.
3. J. Bergman, *Tetrahedron*, **28,** 3323 (1972).

4. P. E. Fanta, *Synthesis,* 9 (1974).
5. A. McKillop, L. F. Elsom, and E. C. Taylor, *Tetrahedron,* **26,** 4041 (1970); L. F. Elsom, A. McKillop, and E. C. Taylor, *Org. Synth.,* **Coll. Vol. 6,** 488 (1988).
6. L. F. Elsom, J. D. Hunt, and A. McKillop, *Organomet. Chem. Rev. A,* **8,** 135 (1972).
7. A. Ronlan, K. Bechgaard, and V. D. Parker, *Acta Chem. Scand.,* **27,** 2375 (1973).
8. Y. Fujiwara, I. Moritani, K. Ikegami, R. Tanaka, and S. Teranishi, *Bull. Soc. Chem. Jpn.,* **43,** 863 (1970), and references cited therein.
9. J. M. Davidson and C. Triggs, *J. Chem. Soc. A,* 1324 (1968).
10. A. McKillop and E. C. Taylor, *Adv. Organomet. Chem.,* **11,** 147 (1973).
11. W. L. Carrick, G. L. Karapinka, and G. T. Kwiatkowski, *J. Org. Chem.,* **43,** 2388 (1969).
12. H. Sainsburg, *Tetrahedron,* **36,** 3327 (1980).
13. W. H. H. Gunther, J. Nepywoda, and J. Y. C. Chu, *J. Organomet. Chem.,* **74,** 79 (1974).

6,7-DIMETHOXY-3-ISOCHROMANONE

(3*H*-2-Benzopyran-3-one, 1,4-dihydro-6,7-dimethoxy-)

Submitted by J. FINKELSTEIN[1] and A. BROSSI[1,2]
Checked by YOSHINORI HAMADA and WATARU NAGATA

1. Procedure

A 500-ml., round-bottomed flask equipped with a mechanical stirrer, a dropping funnel, a thermometer, and a reflux condenser is charged with 49.0 g. (0.250 mole) of 3,4-dimethoxyphenylacetic acid (Note 1) and 125 ml. of acetic acid. The solution is stirred and heated at 80° on a steam bath, while 40 ml. of concentrated hydrochloric acid is added rapidly and followed immediately with 40 ml. of formalin (37% formaldehyde, by weight, in water) (Notes 2 and 3). The yellow solution is stirred and heated on a steam bath for 1 hour, during which time the reaction temperature reaches 90° (Note 4) and the solution assumes a dark-brown color. After cooling to room temperature the solution is poured, with stirring, into a mixture of 650 g. of chipped ice and 650 ml. of cold water. The mixture is transferred to a 2-l. separatory funnel, and the organic material is extracted with four 300-ml. portions of chloroform (Note 5). The combined chloroform extracts are washed with 250-ml. portions of aqueous 5% sodium hydrogen carbonate until neutral (Note 6), then with two, 250-ml. portions of water, and finally dried over anhydrous magnesium sulfate. The solvent is removed on a rotary evaporator with a water bath up to a temperature of 55° (Note 7), yielding 43.5–44.2 g. (83.7–85.1%) of crude 6,7-dimethoxy-3-isochromanone, m.p. 95–100°, as a yellow solid suitable for general synthetic purposes. A purer product is obtained by recrystallization from 55 ml. of ethanol (Notes 8–10), giving 26–27.6 g. of white crystals which, after drying at 80°, melt at 106–108° (Note 11). Upon concentration of the mother liquor to a smaller volume, an additional 1.7–3.2 g. of the isochromanone, m.p. 103–105°, is obtained. The total yield is 29.2–29.3 g. (56.2–56.4%).

2. Notes

1. The 3,4-dimethoxyphenylacetic acid was purchased from Matheson, Coleman and Bell. The checkers prepared material of m.p. 97–98° according to the procedure described in *Org. Synth.*, **Coll. Vol. 2,** 333 (1961).

2. "Baker Analyzed" reagent grade formaldehyde solution obtained from J. T. Baker Chemical Company was used. The checkers used material purchased from Wako Pure Chemical Industries, Ltd., Japan.

3. About 30 seconds each is needed for the additions of concentrated hydrochloric acid and formalin.

4. The temperature of the reaction mixture falls to 68° on addition of the reagents but rises again within 10 minutes.

5. Fisher Scientific Company Certified A.C.S. chloroform was used. The checkers used reagent grade chloroform purchased from Wako Pure Chemical Industries, Ltd., Japan.

6. The wash solution should be neutral to litmus. The checkers observed that the pH values of the first and the second wash solutions were 5–5.5 and 7.5, respectively.

7. The viscous, syrupy residue crystallizes on standing at room temperature or by addition of a small amount of methanol.

8. Anhydrous ethanol (Type 2B) was used.

9. Recrystallization can conveniently be performed with the initial syrupy residue.

10. The checkers found that a mixture of dichloromethane and ether was a more suitable solvent for crystallization of the product. The pure sample obtained from this solvent system melts at 108–109°.

11. The product has the following spectral properties; IR (CHCl$_3$) cm.$^{-1}$: 3040, 1750, 1616, 1520, 1253, and 1118; ^{1}H NMR (CDCl$_3$), δ (multiplicity, number of protons, assignment): 3.63 (s, 2H, CH$_2$CO$_2$), 3.88 (s, 6H, 2OCH$_3$), 5.25 (s, 2H, CH$_2$OCO), 6.72 (s, 1H, aromatic *H*), 6.77 (s, 1H, aromatic *H*).

3. Discussion

The reaction is essentially that described by the submitters.[3] The procedure illustrates a convenient method for the synthesis of a type of lactone which could serve as an important intermediate in the synthesis of isoquinolones, tetrahydroisoquinolines, and isoquinoline alkaloids. Several analogous and closely related lactones have been reported.

The parent, unsubstituted isochromanone reacts with a variety of aromatic amines, giving *N*-substituted 1,4-dihydro-3(2*H*)-isoquinolones,[4] and with amines, giving amides.[5] 6,7-Methylenedioxy-3-isochromanone was used as an intermediate in the synthesis of protopine and its allied alkaloids,[6] and in the synthesis of the berberine ring system.[7] The 6-methoxy analog was prepared as a potential intermediate in a camptothecin synthesis[8] and 8-methoxy-4,5,6,7-tetramethyl-3-isochromanone was used as an intermediate in the synthesis of sclerin.[9]

The compound described herein was the basis of a facile synthesis of (±)-xylopinins,[10] and its reaction with hydrazine has been reported.[11] The compound was also used in the synthesis of pseudoberberine,[12] and when reacted with 1-methyltryptamine, 1-methyl-15,16,17,18,19,20-hexahydroyohimban was obtained.[13] When isochroman-3-ones are treated with tryptamine, the synthesis of the yohimban skeleton is achieved.[14] The preparation of α-methyldopa in a rigid framework is described.[15]

The procedure may have considerable scope, as shown by the synthesis of a heterocyclic lactone, an important intermediate in the synthesis of d,l-desoxycamptothecin, which on oxidation gave camptothecin.[16] A new synthesis of benzocyclobutenes by the thermal and electron impact-induced decomposition of 3-isochromanones was described.[17]

3-Isochromanone was allowed to react with phenol ethers in polyphosphoric acid, yielding dibenzo [a,d] tropylium salts, but in formic acid homoveratric acids were obtained.[18]

The nonaromatic lactones derived from cis-[19] and trans-2-hydroxymethylcyclohexaneacetic acid[20] are important intermediates in the synthesis of indole alkaloids.

1. Chemical Research Department, Hoffmann-LaRoche, Inc., Nutley, New Jersey 07666.
2. Present address: Section on Medicinal Chemistry, U.S. Public Health Service, National Institutes of Health, Department of Health and Human Services, Bethesda, Maryland 20205.
3. J. Finkelstein and A. Brossi, J. Heterocycl. Chem., 4, 315 (1967).
4. Y. Shoo, E. C. Taylor, K. Mislow, and M. Rabian, J. Am. Chem. Soc., 89, 4910 (1967).
5. G. A. Swan, J. Chem. Soc., 1720 (1949).
6. T. S. Stevens, J. Chem. Soc., 178 (1927).
7. T. S. Stevens, J. Chem. Soc., 663 (1935).
8. T. A. Bryson, Abstr. Pap., Div. Org. Chem., Am. Chem. Soc., 164th Nat. Meet., New York, Aug. 27–Sept. 1, 1972, No. 136.
9. T. Kubota, T. Tokoroyama, T. Nishikawa, and S. Maeda, Tetrahedron Lett., 745 (1967).
10. W. Meise and F. Zymalkowski, Tetrahedron Lett., 1475 (1969).
11. G. Rosen and F. D. Popp, Can. J. Chem., 47, 864 (1969).
12. F. Shinada and T. Kono, Japan. Pat. Kokai 74 96,000 [Chem. Abstr., 82, P57993y (1975)].
13. W. Meise and H. Pfister, Arch. Pharm., 310, 495 (1977).
14. W. Meise and H. Pfister, Arch. Pharm., 310, 501 (1977).
15. S. N. Rostogi, J. S. Bindra, and N. Anand, Indian J. Chem., 9, 1175 (1971).
16. R. Volkmann, S. Danishefsky, J. Eggler, and D. M. Solomon, J. Am. Chem. Soc., 93, 5576 (1971); S. Danishefsky, S. J. Etheredge, R. Volkmann, J. Eggler, and J. Quick, J. Am. Chem. Soc., 93, 5575 (1971).
17. R. J. Spangler, B. G. Beckmann, and J. H. Kim, J. Org. Chem., 42, 2989 (1977).
18. A. V. Bicherov, G. N. Dorofeenko, and E. V. Kuznetsov, Zh. Org. Khim., 15, 588 (1979).
19. G. Stork and R. K. Hill, J. Am. Chem. Soc., 76, 949 (1954).
20. E. E. van Tamelen and M. Shamma, J. Am. Chem. Soc., 76, 950 (1954).

REGIOSELECTIVE MANNICH CONDENSATION WITH DIMETHYL(METHYLENE)AMMONIUM TRIFLUOROACETATE: 1-(DIMETHYLAMINO)-4-METHYL-3-PENTANONE

[3-Pentanone, 1-(dimethylamino)-4-methyl-]

$$(CH_3)_2NH + H-\overset{\overset{\displaystyle O}{\|}}{C}-H \xrightarrow[25^\circ]{water} (CH_3)_2NCH_2N(CH_3)_2$$

$$(CH_3)_2NCH_2N(CH_3)_2 \xrightarrow[-10^\circ \text{ to } -15^\circ]{trifluoroacetic\ acid} (CH_3)_2\overset{+}{N}{=}CH_2, \ CF_3CO_2^-$$

$$(CH_3)_2\overset{+}{N}{=}CH_2, CF_3CO_2^- + (CH_3)_2CH-\overset{\overset{\displaystyle O}{\|}}{C}-CH_3 \xrightarrow[-10^\circ \text{ to } 145^\circ]{trifluoroacetic\ acid}$$

$$(CH_3)_2CH-\overset{\overset{\displaystyle O}{\|}}{C}-CH_2CH_2N(CH_3)_2$$

Submitted by MICHEL GAUDRY,[1] YVES JASOR, and TRUNG BUI KHAC
Checked by BERNARD L. MÜLLER and GEORGE BÜCHI

1. Procedure

Caution! Trifluoroacetic acid is highly toxic; consequently Part B of this procedure must be conducted in a well-ventilated hood.

A. *Bis(dimethylamino)methane.* A 500-ml., round-bottomed flask equipped with a magnetic stirring bar and a dropping funnel is charged with 100 g. (1.0 mole) of aqueous 30% formaldehyde (Note 1). The solution is stirred and cooled in an ice bath as 225 g. (2.0 moles) of a 40% solution of dimethylamine (Note 1) in water is added dropwise. The resulting aqueous solution is allowed to stand overnight at room temperature, after which it is saturated with solid potassium hydroxide. The two layers are separated, the upper layer is dried over potassium hydroxide pellets, and the drying agent is removed. Distillation at atmospheric pressure through a Vigreux column gives 85–88 g. (83–86%) of bis(dimethylamino)methane, b.p. 81.5–83°.

B. *1-(Dimethylamino)-4-methyl-3-pentanone.* A 100-ml., two-necked, round-bottomed flask equipped with a magnetic stirring bar and a pressure-equalizing dropping funnel bearing a calcium chloride drying tube is charged with 50 ml. of anhydrous trifluoroacetic acid (Note 2). The trifluoroacetic acid is stirred and cooled in an ice–salt bath at −10° to −15° while 10.2 g. (0.100 mole) of bis(dimethylamino)methane is added over a 50-minute period (Note 3). The temperature of the resulting solution of di-methyl(methylene)ammonium trifluoroacetate is kept below −10° as 8.6 g. (0.10 mole) of 3-methyl-2-butanone (Note 4) is gradually added. The cooling bath is removed and the solution is heated in an oil bath at 65° for 1.5 hours (Note 5). The temperature of the oil bath is then raised to 145° (Note 6). After 1.5 hours the solution is cooled and the trifluoroacetic acid is neutralized by adding the contents of the flask dropwise to an ice-cold solution of 100 g. of potassium carbonate in 100 ml. of water (Note 7). The crystals are collected by filtering through a sintered-glass Büchner funnel and washed with two 50-ml. portions of dichloromethane. The aqueous filtrate is extracted with four 50-ml. portions of dichloromethane. The dichloromethane extracts are combined, washed with 50 ml. of water, dried over anhydrous sodium sulfate, and concentrated with a rotary evaporator. The concentrate, which amounts to 12.7 g. (Note 8), is distilled under reduced pressure through an 18-cm. column packed with Raschig rings (Note 9), afford-ing 7.0–8.2 g. (49–57%) of 1-(dimethylamino)-4-methyl-3-pentanone, b.p. 49° (3 mm.) (Note 10).

2. Notes

1. Formaldehyde and dimethylamine are available as aqueous 37% and 40% solu-tions, respectively, from Aldrich Chemical Company, Inc.

2. The submitters purchased trifluoroacetic acid from Prolabo, Paris, France, or E. Merck, Darmstadt, Germany, and distilled it from phosphorus pentoxide. This reagent is also available from Aldrich Chemical Company, Inc., and J. T. Baker Chemical Com-pany.

3. The reaction between bis(dimethylamino)methane and trifluoroacetic acid is very exothermic. If the temperature is carefully controlled, a colorless solution remains when the addition is complete.

4. 3-Methyl-2-butanone was purchased from Eastman Organic Chemicals and dis-tilled before use.

5. The progress of the reaction can be monitored by taking ^{1}H NMR spectra at appropriate intervals. The following absorptions for dimethyl(methylene)ammonium tri-fluoroacetate in trifluoroacetic acid disappear as the reaction progresses: δ (multiplicity, number of protons, assignment): 3.89 (broad m, 6H, 2 NCH_3), 8.07 (broad m, 2H, N=CH_2).

6. At this temperature 4-(dimethylamino)-3,3-dimethyl-2-butanone, which is formed initially, isomerizes to 1-(dimethylamino)-4-methyl-3-pentanone.

7. Removing trifluoroacetic acid by evaporation is tedious. The neutralization pro-cedure given here produces insoluble salts that are readily separated by filtration.

8. The ratio of the isomeric amino ketones in the crude product can be determined from the relative intensities of the signals for the $(CH_3)_2C$ grouping in a ^{1}H NMR spectrum taken in trifluoroacetic acid (see Note 10). In CDCl$_3$ these absorptions overlap.

9. To minimize losses of products during the distillation, the submitters used a circulating device to chill the condenser cooling water to 5–10°. In addition, the outlet to the vacuum line was located as far as possible from the drip tip, and the receivers were cooled in an ice bath.

10. The ^{1}H NMR spectrum of the product in trifluoroacetic acid shows that the

isomeric purity is greater than 90%. The ^{1}H NMR spectra for the isomeric amino ketones in both trifluoroacetic acid and CDCl$_3$, δ (multiplicity, coupling constant J in Hz., number of protons, assignment): 1-(dimethylamino)-4-methyl-3-pentanone (trifluoroacetic acid), 1.16 (d, J = 7, 6H, 2CCH_3), 2.98 (d, J = 5, 6H, 2NCH_3), 3.31 (m, 4H, CH_2CH_2); (CDCl$_3$), 1.10 (d, J = 7, 6H, 2CCH_3), 2.23 (s, 6H, 2NCH_3), 2.60 (s, 4H, CH_2CH_2); 4-(dimethylamino)-3,3-dimethyl-2-butanone (trifluoroacetic acid), 1.53 (s, 6H, 2CCH_3), 2.45 (s, 3H, COCH_3), 3.15 (d, J = 5, 6H, 2NCH_3), 3.40 (d, J = 5, 2H, CH_2N); (CDCl$_3$), 1.12 (s, 5H, 2CCH_3), 2.13 (s, 3H, COCH_3), 2.18 (s, 6H, 2NCH_3), 2.41 (s, 2H, CH_2N).

3. Discussion

The Mannich condensation has traditionally been carried out in the presence of water as a three-component condensation involving a carbonyl compound (or related carbon nucleophile), formaldehyde, and a primary or secondary amine.[2] The initial step is a condensation between the latter two reactants to form a mono- or dialkyl(methylene)ammonium ion which subsequently serves as the electrophilic partner in the reaction. With unsymmetrical ketones aminomethylation generally occurs at both positions, giving mixtures of isomeric β-amino ketones. The ratio of the isomers depends strongly on the structure of the ketone,[3] and the more highly branched β-amino ketone usually predominates.

In recent years a number of methods have been developed for the preparation of dialkyl(methylene)ammonium salts (Mannich reagents),[4-8] and their use in Mannich-type condensation reactions under anhydrous conditions has improved the scope and efficiency of this important synthetic process.[5-12] However, the orientation of the Mannich reaction may nevertheless be difficult to control. Apart from the work of the submitters, the preparation of isomerically pure Mannich bases has only been achieved by indirect methods in which specific enol derivatives are generated and allowed to react with dialkyl(methylene)ammonium salts.[9,11,13] The Mannich reaction of β-keto esters affords isomerically pure β-dimethylamino β'-keto esters which may in turn be converted to specific α-methylene ketones.[14] However, the β-amino ketones themselves are not as yet available by this method.

The submitters have found that the orientation of the reaction of Mannich reagents with unsymmetrical ketones in anhydrous solvents is highly dependent on the experimental conditions, the solvent, and the structures of the ketone and iminium ion reactants.[10] Under conditions of kinetic control, the reaction of methyl ketones with dimethyl(methylene)ammonium trifluoroacetate in trifluoroacetic acid leads to amino ketones in which the more highly substituted isomer predominates ($\geq$85% when the α'-position is tertiary and 80% when the α'-position is secondary). In contrast, reaction with diisopropyl(methylene)ammonium perchlorate in acetonitrile gives almost exclusively the less highly substituted isomer (100% when the α'-position is tertiary and 90% when it is secondary). Although the latter method directly affords the less highly substituted Mannich bases in yields greater than 80%, it cannot be utilized safely in large-scale preparative reactions owing to the hazardous nature of perchlorate salts.

The less highly substituted Mannich bases can also be prepared directly from ketones and dimethyl(methylene)ammonium trifluoroacetate by the procedure reported here, which takes advantage of the isomerization of Mannich bases in trifluoroacetic acid.[10] (In acetic acid the Mannich bases undergo elimination of dimethylamine to give α-methylene ketones.) This method is rapid and affords products, having an isomeric purity of at least 90%, without difficult separations. The 49–57% yield of 1-(dimethylamino)-4-methyl-3-pentanone obtained with this procedure compares favorably with the overall yields of amino ketones prepared by the indirect routes mentioned previously.

1-(Dimethylamino)-4-methyl-3-pentanone has been prepared by addition of isopropylmagnesium bromide to methyl 3-(dimethylamino)propionate,[15] by reduction of 1-(dimethylamino)-4-methyl-1-penten-3-one with lithium aluminum hydride,[16] and by displacement of chloride from 1-chloro-4-methyl-3-pentanone with dimethylamine.[17] Although the preparation of 1-(dimethylamino)-4-methyl-3-pentanone by Mannich condensation of 3-methyl-2-butanone with dimethylamine hydrochloride and formaldehyde has been reported,[18] the product evidently is a mixture of the two isomeric β-dimethylamino ketones.[3,17]

1. Centre d'Etudes et de Recherches de Chimie Organique Apliqueé 2-8, rue H. Dunant, 94320 Thiais, France. [Present address: Laboratoire de Chemie Organique Biologique, Universite Pierre et Marie Curie, Tour 44-45, 4, Place Jussieu, 75230 Paris Cedex 05, France.]

2. (a) M. Tramontini, *Synthesis,* 703 (1973); (b) H. O. House, "Modern Synthetic Reactions," 2nd ed., W. A. Benjamin, Menlo Park, California, 1972, pp. 654–660.

3. G. L. Buchanan, A. C. W. Curran, and R. T. Wall, *Tetrahedron,* **25,** 5503 (1969).

4. H. Böhme, E. Mundlos, and O.-E. Herboth, *Chem. Ber.,* **90,** 2003 (1957); H. Böhme and K. Hartke, *Chem. Ber.,* **93,** 1305 (1960); and other papers in this series.

5. A. Ahond, A. Cave, C. Kan-Fan, H. P. Husson, J. de Rostolan, and P. Potier, *J. Am. Chem. Soc.,* **90,** 5622 (1968); A. Ahond, A. Cave, C. Kan-Fan, and P. Potier, *Bull. Soc. Chim. Fr.,* 2707 (1970).

6. H. Volz and H. H. Kiltz, *Justus Liebigs Ann. Chem.,* **752,** 86 (1971).

7. J. Schreiber, H. Maag, N. Hashimoto, and A. Eschenmoser, *Angew. Chem. Int. Ed. Engl.,* **10,** 330 (1971).

8. For a review, see H. Böhme and M. Haake, "Methyleniminium Salts," in H. Böhme and H. G. Viehe, Eds., "Iminium Salts in Organic Chemistry," Part 1, in "Advances in Organic Chemistry," E. C. Taylor, Ed., Vol. 9, Wiley-Interscience, New York, 1976, pp. 107–223.

9. J. Hooz and J. N. Bridson, *J. Am. Chem. Soc.,* **95,** 602 (1973).

10. Y. Jasor, M. J. Luche, M. Gaudry, and A. Marquet, *J. Chem. Soc. Chem. Commun.,* 253 (1974); Y. Jasor, M. Gaudry, M. J. Luche, and A. Marquet, *Tetrahedron,* **33,** 295 (1977).

11. S. Danishefsky, T. Kitahara, R. McKee, and P. F. Schuda, *J. Am. Chem. Soc.,* **98,** 6715 (1976).

12. G. Kinast and L.-F. Tietze, *Angew Chem. Int. Ed. Engl.,* **15,** 239 (1976).

13. N. L. Holy and Y. F. Wang, *J. Am. Chem. Soc.,* **99,** 944 (1977); J. L. Roberts, P. S. Borromeo, and C. D. Poulter, *Tetrahedron Lett.,* 1621 (1977).

14. R. B. Miller and B. F. Smith, *Tetrahedron Lett.,* 5037 (1973).

15. I. N. Nazarov and R. I. Kruglikova, *Zh. Obshch. Khim.,* **27,** 346 (1957) [*J. Gen. Chem., USSR Engl. Transl.,* **27,** 387 (1957)].

16. J. C. Martin, K. R. Barton, P. G. Grott, and R. H. Meen, *J. Org. Chem.,* **31,** 943 (1966).

17. M. Brown and W. S. Johnson, *J. Org. Chem.,* **27,** 4706 (1962).

18. R. Jacquier, M. Mousseron, and S. Boyer, *Bull. Soc. Chim. Fr.,* 1653 (1956).

DIRECTED LITHIATION OF AROMATIC COMPOUNDS: (2-DIMETHYLAMINO-5-METHYLPHENYL)DIPHENYLCARBINOL

[Benzenemethanol, 2-(dimethylamino)-5-methyl-α,α-diphenyl-]

Submitted by J. V. Hay and T. M. Harris[1]
Checked by Robert A. Auerbach and Herbert O. House

1. Procedure

A dry, 500-ml., two-necked flask containing 6.75 g. (0.0500 mole) of N,N-dimethyl-p-toluidine (Note 1) and 175 ml. of anhydrous hexane (Note 2) is fitted with a Teflon-coated magnetic stirring bar, a pressure-equalizing dropping funnel capped with a rubber septum, and a nitrogen inlet tube. The reaction vessel is flushed with nitrogen, and a static nitrogen atmosphere is maintained within the apparatus for the remainder of the reaction sequence. A solution of 8.8 g. (0.076 mole of N,N,N',N'-tetramethylethylenediamine (Note 3) in 40 ml. of anhydrous hexane is added to the dropping funnel, followed by a hexane solution containing 0.076 mole of n-butyllithium (Note 4). The resulting solution, which becomes warm as the organolithium-diamine complex forms, is allowed to stand for 15 minutes, then added to the reaction mixture, dropwise and with stirring over 15–20 minutes. The resulting bright yellow, turbid reaction mixture is stirred at room temperature for 4 hours longer before a solution of 13.8 g. (0.0758 mole) of benzophenone (Note 5) in 40 ml. of anhydrous diethyl ether is added to the reaction mixture, dropwise and with stirring over 20 minutes. The resulting, deep-green solution is stirred for an additional 20 minutes, then poured into a vigorously stirred solution of 12 g. (0.20 mole) of acetic acid in 30 ml. of ether (Note 6). After the reaction solution has been successively extracted with 50 ml. of water and four 50-ml. portions of 5% hydrochloric acid, the aqueous extracts are combined (Note 7) and made basic with aqueous 10% sodium hydroxide. The alkaline mixture is heated to boiling and maintained at this temperature until the escaping vapor is no longer basic to moistened pHydrion paper (Note 8). The mixture is then cooled, and the white solid product which separates is collected on a Büchner funnel and washed with three 20-ml. portions of water. The crude product (m.p. 142–168°) is recrystallized from 250 ml. of 3:1 (v/v) hexane-ethyl acetate, giving 6.6–8.2 g. of the amino alcohol product as colorless prisms, m.p. 168–171°. Concentration of the mother liquors gives an additional 0.8–1.2 g. of product, m.p. 167–169°, for a total yield of 7.8–9.0 g. (49–57%). Although the product is sufficiently pure for most purposes, a second recrystallization from hexane-ethyl acetate raises the melting point to 169.5–172° (Note 9).

2. Notes

1. Commercial N,N-dimethyl-p-toluidine, obtained from Aldrich Chemical Company, Inc., was used without purification.

2. An A.C.S. grade of hexane, obtained from Fisher Scientific Company, was used without further purification.

3. N,N,N',N'-Tetramethylethylenediamine, purchased from Aldrich Chemical Company, Inc., was distilled from calcium hydride immediately before use; b.p. 120–122°.

4. Hexane solutions of n-butyllithium, purchased from either Alfa Inorganics, Inc., or Foote Mineral Company, were standardized by the titration procedure of Watson and Eastham.[2] A detailed procedure for this titration is provided in *Org. Synth.,* **Coll. Vol. 6,** 121 (1988).

5. Benzophenone, purchased from Fisher Scientific Company, was used without purification.

6. Reversal of this hydrolysis procedure, the addition of acetic acid to the reaction mixture, had an adverse effect on the yield of product.

7. Some of the aqueous extracts may contain small amounts of suspended particulate matter.

8. This simple steam distillation removes the unchanged N,N-dimethyl-p-toluidine present in the crude product.

9. The purified product has the following spectral properties: IR (CCl_4) 3050 cm.$^{-1}$ (associated OH); UV (95% C_2H_5OH) max (ϵ) 251 (905), 258 (823), 264 (720), and 275 nm (432); NMR ($CDCl_3$), δ 2.19 (s, 3H, C—CH_3), 2.38 [s, 6H, N(CH_3)$_2$] 6.55 (s, 1H, OH), and 7.00–7.50 (m, 13H, aromatic CH); m/e (rel. int.), 317(M^+, 100), 240(84), 225(28), 224(91), 222(43), 150(25), 134(41), 120(32), 105(41), 91(51), and 77(42).

3. Discussion

This procedure is an adaptation of one described by Hauser and co-workers.[3] The product has also been prepared from 2-bromo-N,N-dimethyl-p-toluidine by halogen-metal interchange with n-butyllithium followed by condensation with benzophenone,[3] a procedure less convenient than that presently described.

Tertiary amines such as N,N,N',N'-tetramethylethylenediamine (TMEDA) and 1,4-diazabicyclo[2.2.2]octane (DABCO) strongly catalyze metallations by alkyllithium reagents. Uncatalyzed lithiation of toluene is very poor,[4] whereas a yield of 90% has been obtained when TMEDA is employed as a catalyst.[5]

It is noteworthy that metallation of N,N-dimethyl-p-toluidine takes place at a position *ortho* to the dimethylamino group rather than on or *ortho* to the aryl methyl group.

Apparently, coordination of lithium by the amino group plays a dominant role in directing metallation even in the presence of TMEDA. Other cases have been reported in which the site of metallation is altered by addition of TMEDA.[6,7]

1. Department of Chemistry, Vanderbilt University, Nashville, Tennessee 37235.
2. S. C. Watson and J. F. Eastham, *J. Organomet. Chem.*, **9**, 165 (1967).
3. R. E. Ludt, G. P. Crowther, and C. R. Hauser, *J. Org. Chem.*, **35**, 1288 (1970).
4. H. Gilman and B. J. Gaj, *J. Org. Chem.*, **28**, 1725 (1963).
5. A. W. Langer, Jr., *Trans. N.Y. Acad. Sci.*, **27**, 741 (1965).
6. D. A. Shirley and C. F. Cheng, *J. Organomet. Chem.*, **20**, 251 (1969).
7. D. W. Slocum, G. Book, and C. A. Jennings, *Tetrahedron Lett.*, 3443 (1970).

USE OF DIPOTASSIUM NITROSODISULFONATE (FREMY'S SALT): 4,5-DIMETHYL-*o*-BENZOQUINONE

(3,5-Cyclohexadiene-1,2-dione, 4,5-dimethyl-)

Submitted by H.-J. Teuber[1]
Checked by P. A. Wehrli, F. Pigott, and A. Brossi

1. Procedure

A solution of 15 g. of sodium dihydrogen phosphate (Note 1) in 5 l. of distilled water is placed in a 6-l. separatory funnel. To this solution is added 90 g. (0.33 mole) of dipotassium nitrosodisulfonate (Fremy's salt) (Note 2). The mixture is shaken to dissolve the inorganic radical. A solution of 16 g. (0.13 mole) of 3,4-dimethylphenol (Note 3) in 350 ml. of diethyl ether is added quickly to the purple solution. As the mixture is shaken vigorously for 20 minutes (Note 4), the color of the solution changes to red-brown. The *o*-quinone thus formed is extracted in three portions with 1.2 l. of chloroform. The combined organic layers are dried over anhydrous sodium sulfate (Note 5), filtered, and evaporated under reduced pressure at 20–23° (Note 6). The residual, somewhat oily, red-brown crystals are slurried twice with 15 ml.-portions of ice-cold ether and collected on a filter. The dark-red crystals, after air drying, weigh 8.7–8.9 g. (49–50%), m.p. 105–107° (Note 7).

2. Notes

1. Monobasic sodium phosphate, $NaH_2PO_4 \cdot H_2O$, obtained from Merck & Co., Inc., was used. This buffer was found to be satisfactory for this reaction.

2. Fremy's salt may be purchased from Aldrich Chemical Company, Inc., or from Matheson, Coleman and Bell. The Fremy's salt used by the checker was prepared electrolytically.[2]

3. 3,4-Dimethylphenol was obtained from Eastman Organic Chemicals, m.p. 63–65°.

4. An efficient stirrer may be substituted for the shaking.

5. The drying was accomplished in about 5 minutes.

6. Higher temperatures may accelerate dimerization of the product.

7. The product is reported to melt at 102°.[3] This material has ^1H NMR peaks ($CDCl_3$) at δ 2.14 and 6.19 with relative intensities of 3 : 1. The IR spectrum ($CHCl_3$) shows the strongest absorption at 1670 cm^{-1} accompanied, among others, by four more bands at 1390, 1280, 1005, and 835 cm^{-1}. The product has UV maxima, nm (ϵ), ($CHCl_3$) at 260 (2600), 400 (1120), and 572 (288). It is reported that the material undergoes slow Diels-Alder dimerization.[4]

3. Discussion

o-Quinones exemplify a very important and reactive class of compounds for general organic synthesis. In the past they have been prepared from catechol derivatives by silver oxide dehydrogenation.[3] The unique oxidizing properties of Fremy's salt allow a number of readily available phenols to be converted to *o*-quinones in excellent yield.[4] The scope of this oxidation, the Teuber reaction, is the subject of numerous papers[5] which have been reviewed recently.[6]

1. H.-J. Teuber, Institut für Organische Chemie der Universität, Frankfurt/Main.
2. P. A. Wehrli and F. L. Pigott, *Inorg. Chem.*, **9**, 2614 (1970).
3. R. Willstätter and F. Müller, *Ber. Dtsch. Chem. Ges.*, **44**, 2171 (1911).
4. H.-J. Teuber and G. Staiger, *Chem. Ber.*, **88**, 802 (1955); H.-J. Teuber, U.S. Pat. 2,782,210 (1957).
5. H.-J. Teuber and S. Benz, *Chem. Ber.*, **100**, 2918 (1967) and earlier papers.
6. H. Zimmer, D. C. Lankin and S. W. Horgan, *Chem. Rev.*, **71**, 229 (1971).

1,2-DIMETHYLCYCLOBUTENES BY REDUCTIVE RING-CONTRACTION OF SULFOLANES: *cis*-7,8-DIMETHYLBICYCLO-[4.2.0]OCT-7-ENE

(Bicyclo[4.2.0]oct-7-ene, 7,8-dimethyl-, *cis*-)

A.

$$\xrightarrow[\text{reflux}]{\substack{\text{LiAlH}_4 \\ \text{tetrahydrofuran,}}}$$

B.

$$\xrightarrow[\substack{\text{pyridine,} \\ -5° \text{ to } 0°}]{\text{CH}_3\text{SO}_2\text{Cl}}$$

C.

$$\xrightarrow[\text{(CH}_3)_2\text{SO, } 120°]{\text{Na}_2\text{S}}$$

D.

$$\xrightarrow[\text{ether, } 0°]{\substack{\text{monoperphthalic} \\ \text{acid}}}$$

E.

$$\xrightarrow[\text{2. CH}_3\text{I}]{\text{1. } n\text{-butyllithium}}$$

F.

$$\xrightarrow[\substack{\text{2. LiAlH}_4, \\ \text{dioxane}}]{\text{1. } n\text{-butyllithium}}$$

Submitted by JAMES M. PHOTIS
and LEO A. PAQUETTE[1]
Checked by C.-P. MAK and G. BÜCHI

1. Procedure

Caution! Steps B and C should be performed in a hood because of the noxious odors produced.

Methyl iodide, in high concentrations for short periods or in low concentrations for long periods, can cause serious toxic effects in the central nervous system. Accordingly, the American Conference of Governmental Industrial Hygienists[2] has set 5 p.p.m., a level which cannot be detected by smell, as the highest average concentration in air to which workers should be exposed for long periods. The preparation and use of methyl iodide should always be performed in a well-ventilated fume hood. Since the liquid can be absorbed through the skin, care should be taken to prevent contact.

A. *cis*-1,2-*Cyclohexanedimethanol.* A 3-l., three-necked, round-bottomed flask fitted with a mechanical stirrer, addition funnel, and condenser is charged with 1.5 l. of anhydrous tetrahydrofuran (Note 1). With vigorous stirring, 14.8 g. (0.389 mole) of lithium aluminum hydride is added, followed by a solution of 50.0 g. (0.325 mole) of *cis*-1,2-cyclohexanedicarboxylic anhydride in 300 ml. of tetrahydrofuran introduced in a thin stream over 30 minutes. The resulting suspension is maintained at the reflux temperature for 3 hours with a heating mantle, after which heating is ceased and 100 ml. of a freshly prepared (Note 2), saturated aqueous sodium sulfate is cautiously added dropwise (Note 3). Highly granular insoluble salts, which change in appearance from gray to white, are removed by suction filtration through a Büchner funnel and washed thoroughly with diethyl ether. The combined filtrates are freed of solvent on a rotary evaporator, yielding 46.0–46.5 g. (98–100%) of the diol as a colorless, viscous oil which may slowly crystallize, m.p. 38–40° (Note 4). Pure *cis*-1,2-cyclohexanedimethanol is reported to have m.p. 42–43°.[3-5]

B. *cis*-1,2-*Cyclohexanedimethanol dimethanesulfonate.* A 5-l., three-necked, round-bottomed flask, immersed in an ice–salt bath and fitted with a mechanical stirrer and an addition funnel, is charged with a solution of 111 g. (0.969 mole) of methanesulfonyl chloride in 1.2. l of pyridine. While cooling and stirring, a solution of 46.4 g. (0.322 mole) of *cis*-1,2-cyclohexanedimethanol in 250 ml. of pyridine is added dropwise at a rate such that the temperature does not exceed 0° (Note 5). Upon completion of the addition, the mixture is stirred at −5° to 0° for an additional 2 hours. Two liters of cold, 10% hydrochloric acid is introduced at a rate which maintains the reaction mixture below 20° (Note 5). The solid which separates is isolated by suction filtration, washed sequentially with 1 l. of dilute hydrochloric acid and 2 l. of water, and air-dried, yielding 93–95 g. (96–98%) of the dimethanesulfonate, m.p. 66–67.5°. Recrystallization from methanol gives needles melting at 75–76° (Note 6).

C. *cis*-8-*Thiabicyclo*[4.3.0]*nonane.* A 3-l., three-necked, round-bottomed flask fitted with a mechanical stirrer, capillary tube, heating mantle, and 90° adapter connected to a condenser and receiving flask is charged with 240 g. (1.00 mole) of recrystallized sodium sulfide nonahydrate (Note 7) and 2 l. of dimethyl sulfoxide. As the mixture is stirred, the internal pressure is reduced to 30 mm., and heat is applied until 300–350 ml. of distillate is collected (Note 8). After cooling to 40°, the capillary and take-off adapter are replaced with a thermometer and condenser, and 95 g. (0.32 mole) of *cis*-1,2-cyclohexanedimethanol dimethanesulfonate is introduced in one portion (Note 9). The mixture is then stirred at 120° for 18 hours, cooled, and transferred to a 5-l. separatory funnel containing 1500 g. of ice. After 1 l. of hexane is added and the two-phase mixture well shaken, the aqueous phase is reextracted with hexane (500 ml.). The combined organic layers are washed with four 1-l. portions of water, dried over anhydrous magnesium sulfate, and concentrated with a rotary evaporator. The sulfide is collected by bulb-to-bulb distillation at 0.05–0.1 mm. as a colorless liquid (30.8–31.6 g., 68.0–70.5%) (Note 10).

D. cis-8-*Thiabicyclo*[4.3.0]*nonane* 8,8-*dioxide* [*benzo*[c]*thiophene* 2,2-*dioxide*, cis-*octahydro*-]. A solution of the sulfide (43.0 g., 0.303 mole) in 1 l. of ether is cooled to 0°, stirred magnetically, and treated dropwise with 1.0 l. of 0.65 N ethereal monoperphthalic acid [*Org. Synth.*, **Coll. Vol. 3**, 619 (1955)]. The mixture is kept overnight at 0°, after which time the precipitated phthalic acid is separated by filtration and the filtrate is concentrated with a rotary evaporator. Bulb-to-bulb distillation of the residual oil at 0.05–0.1 mm. affords the sulfone as a colorless liquid (48.5–50 g., 92–95%) (Note 11). This product is crystallized from ether-hexane, yielding a colorless solid, m.p. 39–41° (Note 12).

E. 7,9-*Dimethyl*-cis-8-*thiabicyclo*[4.3.0]*nonane* 8,8-*dioxide*. A 2-l., one-necked, round-bottomed flask is charged with 800 ml. of anhydrous tetrahydrofuran and 49.0 g. (0.281 mole) of *cis*-8-thiabicyclo[4.3.0]nonane 8,8-dioxide. The solution is blanketed with nitrogen and the flask is fitted with a side-arm adapter having a nitrogen inlet and a rubber septum. The contents are cooled in a 2-propanol–dry ice bath and 225 ml. of 2.5 M *n*-butyllithium in hexane (0.562 mole) is introduced by syringe (Notes 13, 14). After 5–10 minutes, 142 g. (1.00 mole) of methyl iodide is added in similar fashion and the cooling bath is removed. Upon warming to room temperature, the reaction mixture is treated slowly with 1 l. of water followed by 1 l. of ether, and the organic layer is separated, washed once with 1 l. of water, dried over anhydrous magnesium sulfate, and evaporated under reduced pressure. The resulting pale-yellow oil (49.5–51 g., 87–89.5%), which consists of a mixture of isomers, is not further purified.

F. cis-7,8-*Dimethylbicyclo*[4.2.0]*oct-7-ene*. A 1-l., one-necked, round-bottomed flask is charged with 50.5 g. (0.250 mole) of the sulfone from Part E and 200 ml. of dry dioxane (Note 15). The solution is blanketed with nitrogen, and the flask is fitted with a side-arm adapter having a nitrogen inlet and a rubber septum. With ice cooling and magnetic stirring, 150 ml. of 2.5 M *n*-butyllithium in hexane (0.375 mole) is added by syringe (Notes 13, 14). The resulting yellow-orange heterogeneous mixture is transferred under nitrogen to a 500-ml., pressure-equalizing dropping funnel and introduced over 25 minutes to a stirred refluxing mixture of lithium aluminum hydride (32.0 g., 0.843 mole) and 2 l. of dry dioxane (Notes 16, 17). Upon completion of the addition, the contents are heated at reflux temperature with a mantle for 20 hours, whereupon 100 ml. of saturated aqueous sodium sulfate is added dropwise with cooling (Note 18). The precipitated solids are separated by filtration and washed repeatedly with hexane (Note 19). The combined filtrates are diluted with an additional liter of hexane and washed with four 1-l. portions of water. The organic phase is dried over anhydrous sodium sulfate and carefully concentrated with a rotary evaporator. The residual cyclobutene is purified by distillation, yielding 10.0–12.5 g. (29.5–37%) of colorless oil, b.p. 63–65° (33 mm.) (Notes 20, 21).

2. Notes

1. The tetrahydrofuran used in these preparations was distilled from lithium aluminum hydride.

2. A sodium sulfate solution which is not freshly prepared ultimately gives a precipitate of small particle size that is exceedingly difficult and tedious to separate by vacuum filtration.

3. Because the initial reaction is extremely vigorous and exothermic, the first few milliliters must be added very cautiously. In the more advanced stages of this addition the rate of flow may be judiciously increased.

4. Recrystallization of this material from benzene–light petroleum ether gives a pure product, m.p. 42–43°.

5. This reaction is significantly exothermic. Cooling with an acetone–dry ice bath can be employed if desired to expedite the addition of diol. In any event, a temperature in excess of 20° leads to unwanted, rapid hydrolysis and formation of water-soluble by-products.

6. This dimethanesulfonate is reported to have m.p. 75–76°.[4]

7. Sodium sulfide may be conveniently recrystallized from ethanol. Unrecrystallized material may be utilized. However, significantly lower yields will result if the ensuing minor modification is not followed.

8. If unpurified sodium sulfide is employed, a significant quantity of dark insoluble material is seen to adhere to the walls of the flask. Removal of these unwanted contaminants is readily effected by decantation of the hot solution into a second 3-l., three-necked flask before crystallization begins.

9. An exotherm is witnessed and the temperature rises to 70–80°. A color change from yellow to deep purple is also seen; the extent of coloration varies with the purity of the sodium sulfide nonahydrate.

10. The submitters report a yield of 42.0–43.5 g. (93.5–96.8%). The checkers could not reproduce these results in three attempts.

This sulfide has also been prepared from the corresponding dibromide.[6]

11. The checkers performed this step on a smaller scale (*ca.* ⅔) and noted (¹H NMR spectrum) occasional contamination (up to 10%) by phthalic anhydride. This impurity causes no subsequent difficulties. Washing of the crude reaction mixture with cold aqueous sodium hydrogen carbonate resulted in serious product loss because of its appreciable solubility in this medium and, therefore, should be avoided.

12. An earlier report of this sulfone cites a melting point of 39.5–41.0°.[6]

13. *n*-Butyllithium is available from Ventron Corp.

14. A 50-ml. syringe was employed and a series of transfers was, therefore, necessary.

15. The dioxane was dried before use by distillation from calcium hydride.

16. The apparatus consisted of a 5-l., three-necked, round-bottomed flask equipped with a mechanical stirrer and reflux condenser capped with a nitrogen-inlet tube.

17. The addition rate is such that a gentle reflux is maintained without the need of external heating. Considerable evolution of gas is witnessed.

18. Hydrogen gas is vigorously evolved and the solids ultimately undergo a color change from gray to white.

TABLE I
REDUCTIVE RING CONTRACTION OF α,α'-DISUBSTITUTED SULFOLANE ANIONS WITH LITHIUM ALUMINUM HYDRIDE

Sulfolane	Product	Overall yield, %
		54
		62
		67
		22
		20
		34

19. *Caution! Because of the presence of malodorous by-products, it is recommended that the extraction and distillation be conducted in a well-ventilated hood.*

20. The ^{1}H NMR spectrum (CCl_4) consists of a broad methine signal centered at δ 2.55 and a methyl singlet at δ 1.53 superimposed upon a methylene absorption at δ 1.25–1.85. GC analysis indicated a purity of >98%.

21. The checkers performed this step on one-half scale and obtained comparable results. However, when the product was distilled at higher pressures, b.p. 94–95° (*ca.* 70 mm.), consistently lower yields were obtained, in the range of 17–18%.

3. Discussion

1-Substituted and 1,2-disubstituted cyclobutenes have previously been prepared by irradiation of 1,3-butadienes capable of photocyclization,[7–9] carbenic decomposition of acylcyclopropane tosylhydrazones,[10] photocycloaddition of α,β-unsaturated ketones to alkynes,[11] and reductive ring expansion of cyclopropane 3-carboxylates.[12] However, these and yet other less known methods[13–16] lack generality. The present procedure[17] is a versatile scheme which is widely applicable in scope.[18–20] Since a variety of five-membered ring sulfones are readily available from a number of different precursors, the method is fully applicable to a broad spectrum of structural types. Its application to the preparation of mutually stable cyclooctatetraene bond shift isomers is noteworthy.[20] The present procedure is illustrative of the general method. Other examples are given in Table I.

Such reductive ring contractions of sulfones are formally similar to two other methods capable of supplanting a sulfur atom by a carbon-carbon double bond: the Ramberg-Bäcklund[21] and Stevens rearrangements.[22] The distinguishing feature of this novel approach to cyclobutenes consists in the resulting higher level of alkyl substitution at the sp^2-hybridized centers.

1. Department of Chemistry, The Ohio State University, Columbus, Ohio 43210.
2. American Conference of Governmental Industrial Hygienists (ACGIH), "Documentation of Threshold Limit Values," 3rd ed., Cincinnati, Ohio, 1971, p. 166.
3. L. P. Kuhn, *J. Am. Chem. Soc.*, **74**, 2492 (1952).
4. G. A. Haggis and L. N. Owen, *J. Chem. Soc.*, 389 (1953).
5. L. P. Kuhn, P. von R. Schleyer, W. F. Baitinger, and L. Eberson, *J. Am. Chem. Soc.*, **86**, 650 (1964).
6. S. F. Birch, R. A. Dean, and E. V. Whitehead, *J. Org. Chem.*, **19**, 1449 (1954).
7. R. Srinivasan, *J. Am. Chem. Soc.*, **84**, 4141 (1962).
8. K. J. Crowley, *Tetrahedron*, **21**, 1001 (1965).
9. E. H. White and J. P. Anhalt, *Tetrahedron Lett.*, 3937 (1965).
10. C. D. Gutsche and D. Redmore, "Carbocyclic Ring Expansion Reactions," Academic Press, New York, 1968, p. 111.
11. P. E. Eaton, *Acc. Chem. Res.*, **1**, 50 (1968).
12. W. G. Gensler, J. J. Langone, and M. B. Floyd, *J. Am. Chem. Soc.*, **93**, 3828 (1971).
13. H. H. Freedman and A. M. Frantz, *J. Am. Chem. Soc.*, **84**, 4165 (1962).
14. R. M. Dodson and A. G. Zielske, *J. Org. Chem.*, **32**, 28 (1967).
15. Y. Hosokawa and I. Moritani, *Tetrahedron Lett.*, 3021 (1969).
16. M. S. Newman and G. Kaugars, *J. Org. Chem.*, **30**, 3295 (1965).
17. J. M. Photis and L. A. Paquette, *J. Am. Chem. Soc.*, **96**, 4715 (1974).

18. L. A. Paquette, J. S. Ward, R. A. Boggs, and W. B. Farnham, *J. Am. Chem. Soc.*, **97**, 1101 (1975).
19. L. A. Paquette, J. M. Photis, K. B. Gifkens, and J. Clardy, *J. Am. Chem. Soc.*, **97**, 3536 (1975).
20. L. A. Paquette, J. M. Photis, and G. D. Ewing, *J. Am. Chem. Soc.*, **97**, 3538 (1975).
21. L. A. Paquette, in "Mechanisms of Molecular Migrations," Vol. I, B. S. Thyagarajan, Ed., Interscience, New York, 1968, pp. 121–156.
22. R. H. Mitchell and V. Boekelheide, *J. Am. Chem. Soc.*, **96**, 1547 (1974), and references cited therein.

4,4'-DIMETHYL-1,1'-BIPHENYL

$$+ 2\,Tl + 2\,MgBr_2$$

Submitted by L. F. Elsom, Alexander McKillop,[1]
and Edward C. Taylor[2]
Checked by Ronald F. Sieloff and Carl R. Johnson

1. Procedure

*Caution! Thallium salts are very toxic. This preparation should be carried out in a well-ventilated hood. The operator should wear rubber gloves. For disposal of thallium wastes, see Note 1 in Org. Synth., **Coll. Vol. 6,** 791 (1988).*

Benzene has been identified as a carcinogen. OSHA has issued emergency standards on its use. All procedures involving benzene should be carried out in a well-ventilated hood, and glove protection is required.

A. *(4-Methylphenyl)magnesium bromide.* A 500-ml., three-necked, round-bottomed flask equipped with a reflux condenser protected by a drying tube, a mercury-sealed mechanical stirrer, and a 250-ml., pressure-equalizing dropping funnel fitted with a gas-inlet tube is thoroughly purged with dry nitrogen (Note 1) and charged with 6.25 g. (0.256 g.-atom) of magnesium turnings and 50 ml. of anhydrous tetrahydrofuran (Note 2). The dropping funnel is charged with a solution of 42.7 g. (30.5 ml., 0.250 mole) of 4-bromotoluene (Note 3) in 100 ml. of anhydrous tetrahydrofuran. Approximately 10 ml. of the 4-bromotoluene solution is added to the flask, and the contents are stirred until the Grignard reaction commences (Note 4). When the initial vigorous reaction has subsided the remainder of the 4-bromotoluene solution is added at a rate such that the mixture refluxes gently. Generally the addition is complete at the end of 1 hour, and almost all of the magnesium has dissolved. The mixture is refluxed for a further hour and cooled. The yield of (4-methylphenyl)magnesium bromide is about 95% (Note 5).

B. *4,4'-Dimethyl-1,1'-biphenyl.* A 1-l., three-necked, round-bottomed flask equipped with a reflux condenser protected by a drying tube, a mercury-sealed mechanical stirrer, and a gas-inlet tube is charged with 101 g. (0.356 mole) of thallium(I) bromide and 400 ml. of anhydrous benzene. The slurry is stirred vigorously while a stream of dry nitrogen is passed through the apparatus. The reflux condenser is temporarily removed, and the solution of (4-methylphenyl)magnesium bromide is added to the flask as rapidly as possible through a large filter funnel fitted with a *loose* plug of glass wool (Note 6). A black solid precipitates almost immediately from solution. The reflux condenser is replaced, and the contents of the flask are refluxed with stirring for 4 hours under a nitrogen atmosphere. The reaction mixture is then cooled, filtered, and the metallic thallium washed with 200 ml. of diethyl ether. The organic layer is washed once with 100 ml. of 0.1 N hydrochloric acid and once with 100 ml. of water, then dried over anhydrous sodium sulfate.

The organic solvent is removed by distillation under reduced pressure, giving 4,4'-dimethyl-1,1'-biphenyl contaminated with a small amount of bis(4-methylphenyl)thallium bromide. The crude product is dissolved in 30 ml. of benzene, and the solution is filtered through a short column of alumina (Note 7) using 250 ml. of benzene as eluent. Distillation of the benzene under reduced pressure leaves 19–21 g. (80–83%) of 4,4'-dimethyl-1,1'-biphenyl as a colorless solid, m.p. 118–120° (Note 8).

2. Notes

1. Nitrogen is dried by passage through two Drechsel bottles containing concentrated sulfuric acid and potassium hydroxide pellets, respectively.

2. Tetrahydrofuran was dried as described in *Org. Synth.,* **Coll. Vol. 4,** 259 (1963).

3. 4-Bromotoluene (purchased from Aldrich Chemical Company, Inc.) was distilled before use, b.p. 71–72° (15 mm.).

4. The Grignard reaction starts within a few minutes and should not require the use of a catalyst. If the reaction has not commenced within 5 minutes, the flask should be gently heated with a hot water bath until reaction starts.

5. The Grignard reagent may be standardized by the general procedure described for cyclohexylmagnesium chloride in *Org. Synth.,* **Coll. Vol. 1,** 187 (1932).

6. A loose plug of glass wool prevents any unreacted magnesium metal from being added to the reaction mixture. Care must be taken to ensure that the plug is loose enough to allow rapid addition of the Grignard reagent.

7. The dimensions of the alumina column are not critical. A column approximately 2.5 cm. × 12.5 cm. is recommended. The small amount of bis(4-methylphenyl)thallium bromide formed in the reaction remains on the top of the column.

8. The reaction may be conducted on two or three times the scale described with no decrease in yield.

3. Discussion

4,4'-Dimethyl-1,1'-biphenyl has been prepared by a wide variety of procedures, but few of these are of any practical synthetic utility. Classical radical biaryl syntheses such as the Gomberg reaction or the thermal decomposition of diaroyl peroxides give complex mixtures of products in which 4,4'-dimethyl-1,1'-biphenyl is a minor constituent. A radical process may also be involved in the formation of 4,4'-dimethyl-1,1'-biphenyl (13%) by treatment of 4-bromotoluene with hydrazine hydrate.[3] 4,4'-Dimethyl-1,1'-biphenyl has been obtained in moderate to good yield (68–89%) by treatment of either dichlorobis(4-methylphenyl)tellurium or 1,1'-tellurobis(4-methylbenzene) with degassed Raney nickel in 2-methoxyethyl ether.[4]

All of the useful procedures described for the preparation of 4,4'-dimethyl-1,1'-biphenyl involve coupling of either a 4-halotoluene by a metal or the corresponding Grignard reagents by a metal halide. 4-Halotoluenes can be coupled directly by treatment with lithium,[5] sodium,[6-8] magnesium,[9] or copper[10,11]; yields are, however, very low in the first three cases (5–15%) and only moderate (54–60%) when copper is employed, as in the Ullmann synthesis. Bis(1,5-cyclooctadiene)nickel(0) has also been used to couple 1-iodo-4-methylbenzene, giving the biaryl in 63% yield.[12]

The present method of preparation of 4,4'-dimethyl-1,1'-biphenyl is that described by McKillop, Elsom, and Taylor,[13] has the particular advantages of high yield and manipulative simplicity, and is, moreover, applicable to the synthesis of a variety of symmetrically substituted biaryls. 3,3'- and 4,4'-Disubstituted and 3,3',4,4'-tetrasubstituted 1,1'-biphenyls are readily prepared, but the reaction fails when applied to the synthesis of 2,2'-disubstituted-1,1'-biphenyls. The submitters have effected the following conversions with the present procedure (starting aromatic bromide, biphenyl product, % yield); bromobenzene, biphenyl, 85; 1-bromo-4-methoxybenzene, 4,4'-dimethoxy-1,1'-biphenyl, 99; 1-bromo-3-methylbenzene, 3,3'-dimethyl-1,1'-biphenyl, 85; 4-bromo-1,2-dimethylbenzene, 3,3',4,4'-tetramethyl-1,1'-biphenyl, 76; 1-bromo-4-chlorobenzene, 4,4'-dichloro-1,1'-biphenyl, 73; 1-bromo-4-fluorobenzene, 4,4'-difluoro-1,1'-biphenyl, 73.

Related procedures, in which treatment of (4-methylphenyl)magnesium halides with halides of copper(II),[14] silver(I),[15] cobalt(II),[16] or chromium(III)[17] also lead to the formation of 4,4'-dimethyl-1,1'-biphenyl, are either experimentally more difficult than, or do not give yields comparable to, the present method.

1. School of Chemical Sciences, University of East Anglia, Norwich, Norfolk NR4 7TJ, England.
2. Department of Chemistry, Princeton University, Princeton, New Jersey, 08540.
3. M. Busch and W. Schmidt, *Ber. Dtsch. Chem. Ges.*, **62**, 2612 (1929).
4. J. Bergman, *Tetrahedron*, **28**, 3323 (1972).
5. J. F. Spencer and G. M. Price, *J. Chem. Soc.*, **97**, 385 (1910).
6. T. Zincke, *Ber. Dtsch. Chem. Ges.*, **4**, 396 (1871).
7. W. Louguinine, *Ber. Dtsch. Chem. Ges.*, **4**, 514, (1871).
8. M. Weiler, *Ber. Dtsch. Chem. Ges.*, **29**, 111 (1896).
9. H. Rupe and J. Bürgin, *Ber. Dtsch. Chem. Ges.*, **44**, 1218 (1911).
10. F. Ullmann, *Justus Liebigs Ann. Chem.*, **332**, 38 (1904).
11. J. Van Alphen, *Recl. Trav. Chim. Pays-Bas*, **50**, 1111 (1931).
12. M. F. Semmelhack, P. M. Helquist, and L. D. Jones, *J. Am. Chem. Soc.*, **93**, 5908 (1971).
13. A. McKillop, L. F. Elsom, and E. C. Taylor, *Tetrahedron*, **26**, 4041 (1970).
14. E. Sakallerios and T. Kyrimis, *Ber. Dtsch. Chem. Ges.*, **57B**, 322 (1924).
15. J. H. Gardner and P. Borgstrom, *J. Am. Chem. Soc.*, **51**, 3375 (1929).
16. M. S. Kharasch and E. K. Fields, *J. Am. Chem. Soc.*, **63**, 2316 (1941).
17. G. N. Bennett and E. E. Turner, *J. Chem. Soc.*, **105**, 1057 (1914).

N,N-DIMETHYL-5β-CHOLEST-3-ENE-5-ACETAMIDE

[Cholest-3-ene-5-acetamide, *N,N*-dimethyl-, (5β)-]

Submitted by R. E. Ireland[1] and D. J. Dawson
Checked by W. Pawlak and G. Büchi

1. Procedure

A 50-ml., round-bottomed flask, equipped with a Teflon®-covered magnetic stirring bar and a reflux condenser connected to a gas-inlet tube, is charged with 970 mg. (2.51 mmoles) of cholest-4-en-3β-ol (Note 1) and 30 ml. of *o*-xylene (Note 2). The mixture is stirred to effect solution before 1.67 g. (0.0125 mole) of *N,N*-dimethylacetamide dimethyl acetal (Note 3) is added. The flask is flushed with argon, then heated (Note 4) at reflux under a positive pressure of argon with vigorous stirring for 65 hours. After cooling, the volatile materials are removed at reduced pressure (Note 5), and the yellow, oily residue (1.2 g.) is chromatographed on 60 g. of silica gel with diethyl ether (Note 6). Elution of the column with 200 ml. of ether gives a mixture of cholestadienes which is discarded; further elution with 500 ml. of ether affords 740 mg. of *N,N*-dimethyl-5β-cholest-3-ene-5-acetamide as a clear, colorless oil, which on trituration with acetone gives 740 mg. (65%) of the amide as white plates, m.p. 128–129.5°.

2. Notes

1. Cholest-4-en-3β-ol can be prepared by the procedure of Burgstahler and Nordin[2]. A melting point below 130° indicates that the material is contaminated with some of the

3α-hydroxy isomer. The material used above melted at 130.5–131° (from ethanol).

2. The Matheson, Coleman and Bell product was used without purification.

3. N,N-Dimethylacetamide dimethyl acetal was obtained from Fluka A. G. and used without purification.

4. A sand bath set into an electric heating mantle was found to be satisfactory for the long-term heating process.

5. The volatile materials were removed by rotary evaporation followed by vacuum (0.1 mm.) drying for 1 hour.

6. Merck silica gel (0.05–0.2 mm., 70–325 mesh ASTM) was used in a 2.5 × 25 cm. column. Mallinckrodt anhydrous ether was employed as the eluant.

3. Discussion

The amide–Claisen rearrangement procedure of Eschenmoser and co-workers[3] was modified for use with cholest-4-en-3β-ol.

1. Division of Chemistry and Chemical Engineering, Gates and Crellin Laboratories of Chemistry, California Institute of Technology, Pasadena, California 91109.
2. A. W. Burgstahler and I. C. Nordin, *J. Am. Chem. Soc.,* **83,** 198 (1961).
3. A. E. Wick, D. Felix, K. Steen, and A. Eschenmoser, *Helv. Chim. Acta,* **47,** 2425 (1964).

CONVERSION OF ESTERS TO AMIDES
WITH DIMETHYLALUMINUM AMIDES:
N,N-DIMETHYLCYCLOHEXANECARBOXAMIDE

$$(CH_3)_3Al \;+\; (CH_3)_2NH \xrightarrow[-10° \text{ to } 25°]{\text{benzene, hexane}} (CH_3)_2AlN(CH_3)_2 \;+\; CH_4$$

$$+ \;\; (CH_3)_2AlN(CH_3)_2 \xrightarrow[\text{reflux}]{\text{benzene, hexane}}$$

$$+ \;\; (CH_3)_2AlOCH_3$$

Submitted by Michael F. Lipton,[1] Anwer Basha,[1]
and Steven M. Weinreb[1,2]
Checked by Charles W. Hutchins and Robert M. Coates

1. Procedure

Caution! Benzene has been identified as a carcinogen; OSHA has issued emergency standards on its use. All procedures involving benzene should be carried out in a well-ventilated hood, and glove protection is required.

A dry, 300-ml., two-necked, round-bottomed flask equipped with a reflux condenser fitted with a nitrogen inlet at its top, a rubber septum, and a magnetic stirring bar is charged with 100 ml. of benzene (Note 1) and flushed briefly with nitrogen, after which 22 ml. (0.057 mole) of a 25% solution of trimethylaluminum in hexane (Note 2) is injected through the septum into the flask. The solution is stirred and cooled in an ice–salt bath at −10° to −15°, and 2.47 g. (3.64 ml., 0.0549 mole) of dimethylamine (Note 3) is added slowly with a syringe. Twenty minutes after the addition is completed, the cooling bath is removed, and the contents of the flask are allowed to stir and warm slowly to room temperature over a 45-minute period. A solution of 7.10 g. (0.0500 mole) of methyl cyclohexanecarboxylate (Note 4) in 20 ml. of benzene (Note 1) is injected through the septum. The resulting solution is heated under reflux for 22 hours, cooled to room temperature, and hydrolyzed by slow, cautious addition of 82.5 ml. (0.055 mole) of 0.67 M hydrochloric acid (Note 5). The mixture is stirred for 30 minutes to ensure complete hydrolysis. The upper organic layer is separated, and the aqueous layer is extracted with three 25-ml. portions of ethyl acetate. The organic extracts are combined, washed with sodium chloride solution, dried with anhydrous magnesium sulfate, and evaporated under reduced pressure. Distillation of the residual liquid under reduced pressure through a 10-cm. Vigreux column affords a 0.1–0.6 g. forerun of unreacted ester and 6.40–7.25 g. (83–93%) of N,N-dimethylcyclohexanecarboxamide, b.p. 100° (5.5 mm.), 57–60° (0.08 mm.) (Note 6).

2. Notes

1. The benzene was dried by distillation from calcium hydride.

2. Trimethylaluminum in hexane solution was purchased from the Alfa Division, Ventron Corporation.

3. Dimethylamine was obtained in a cylinder from the Linde Division, Union Carbide Chemical Corporation, and condensed in a dry, two-necked flask fitted with a rubber septum and cooled to −78° under nitrogen.

4. Cyclohexanecarboxylic acid is available from Aldrich Chemical Company, Inc., and conveniently esterified by the procedure of Harrison, Haynes, Arthur, and Eisenbraun.[3] A dry, 500-ml., round-bottomed flask is charged with 225 ml. of anhydrous methanol, 1.0 ml. of concentrated sulfuric acid, and 41.0 g. (0.320 mole) of cyclohexanecarboxylic acid. The flask is fitted with a Soxhlet extractor containing 53 g. of Linde type 3A molecular sieves and a condenser bearing a calcium chloride drying tube at its top. The solution is heated at reflux for 19 hours and cooled to room temperature. The sulfuric acid is neutralized by adding 3.0 g. of sodium hydrogen carbonate, the salts are filtered, and the filtrate is evaporated under reduced pressure. The remaining liquid is distilled through a 15-cm. Vigreux column at reduced pressure, affording 35.7–36.8 g. (79–81%) of methyl cyclohexanecarboxylate, b.p. 73–76° (13 mm.).

5. To avoid excessive foaming at the beginning of the hydrolysis, the checkers recommend that the hydrochloric acid solution be added 1 or 2 drops at a time. The rate of addition may be increased once the initially vigorous foaming subsides.

6. The spectral properties of the product are as follows: IR (neat) cm.$^{-1}$: 1640 (C=O); ^{1}H NMR (CDCl$_3$), δ (multiplicity, number of protons, assignment): 1.05–1.95 (m, 10H, 5CH_2), 2.50 (m, 1H, CH), 2.94 (s, 3H, NCH_3), 3.06 (s, 3H, NCH_3). A boiling

point of 107–108° (7 mm.) has been reported for N,N-dimethylcyclohexanecarbox-amide.[4]

3. Discussion

This procedure,[5] based on the work of Ishii and co-workers,[6] affords a mild and general method for converting a wide variety of esters to primary, secondary, and tertiary amides (Table I). While the preparation of the tertiary amide, N,N-dimethyl-

TABLE I
PREPARATION OF AMIDES FROM ESTERS BY AMINOLYSIS WITH DIMETHYLALUMINUM AMIDES[5]

Ester	Amine	Reaction Time (hours)[a]	Isolated Yield of Amide (%)
$\diagup\diagup\diagdown\diagdown CO_2C_2H_5$	NH_3	2[b]	70
(2,2-dimethyl-1,3-dioxolane with CO_2CH_3 and CH_3)	NH_3	16	69
$C_6H_5\diagdown\diagup CO_2CH_3$	NH_3	12	86
$Cl\diagup\diagdown\diagup CO_2C_2H_5$	NH_2 $CH_3CHCH_2CH_3$	48	76
$\diagdown\diagup\diagdown\diagup CO_2CH_3$	pyrrolidine (N–H)	45	74
$C_6H_5\diagdown\diagup CO_2C_2H_5$ $NHCOCH_3$	pyrrolidine (N–H)	40	77
$CH_3CO_2(CH_2)_3CH_3$	$C_6H_5NH_2$	40	78
$C_6H_5CO_2C_2H_5$	$C_6H_5CH_2NH_2$	25	93
$Cl\diagup\diagdown\diagup CO_2Et$	$(CH_3)_3CNH_2$	45	79

[a] The solvent was dichloromethane except as noted.
[b] Benzene was used as solvent.

cyclohexanecarboxamide, described here is carried out in benzene, aluminum amides derived from ammonia and a variety of primary amines have been prepared by reaction with trimethylaluminum in dichloromethane and utilized for aminolysis in this solvent. Although 1 equivalent of dimethylaluminum amides, prepared from amines, was generally sufficient for high conversion within 5–48 hours, best results were obtained when 2 equivalents of the aluminum reagent, prepared from ammonia, was used. Diethylaluminum amides can also effect aminolysis, but with considerably slower rates.

Although the preparation of carboxamides by direct aminolysis is a well-known and widely studied reaction,[7] the synthetic utility of this process is limited. The reactions generally require long heating periods at relatively high temperatures, and the reagents and catalysts used are usually strong bases.[8] The present procedure has the advantages of lower temperatures and moderate reaction times. The aluminum amides are conveniently prepared *in situ* and appear to be mild, nonbasic reagents, compatible with many functional groups.[5] The isolation is simple, since hydrolysis of the aluminum reagents and products affords only methane and acid-soluble aluminum salts. Another advantage is that amides of volatile amines may be prepared without the use of sealed tubes.

N,*N*-Dimethylcyclohexanecarboxamide has been prepared by acylation of dimethylamine with cyclohexanecarbonyl chloride[9] and by double alkylation of vinylidenebis(dimethylamine) with 1,5-diiodopentane to the cyclic amidinium salt followed by hydrolysis.[4]

1. This work was carried out at the Department of Chemistry, Fordham University, Bronx, New York 10458.
2. Present address: Department of Chemistry, Pennsylvania State University, University Park, Pennsylvania 16802.
3. H. R. Harrison, W. M. Haynes, P. Arthur, and E. J. Eisenbraun, *Chem. Ind. (London)*, 1568 (1968).
4. C. F. Hobbs and H. Weingarten, *J. Org. Chem.*, **39**, 918 (1974).
5. A. Basha, M. Lipton, and S. M. Weinreb, *Tetrahedron Lett.*, 4171 (1977).
6. T. Hirabayashi, K. Itoh, S. Sakai, and Y. Ishii, *J. Organomet. Chem.*, **25**, 33 (1970).
7. A. L. J. Beckwith, in J. Zabicky, Ed., "The Chemistry of Amides," Wiley-Interscience, New York, 1970, p. 96.
8. B. Singh, *Tetrahedron Lett.*, 321 (1971); K. W. Yang, J. G. Cannon, and J. G. Rose, *Tetrahedron Lett.*, 1791 (1970); H. L. Bassett and C. R. Thomas, *J. Chem. Soc.*, 1188 (1954).
9. A. C. Cope and E. Ciganek, *Org. Synth.*, **Coll. Vol. 4**, 339 (1963).

4,4-DIMETHYL-2-CYCLOHEXEN-1-ONE

(2-Cyclohexen-1-one, 4,4-dimethyl-)

Submitted by YIHLIN CHAN and WILLIAM W. EPSTEIN[1]
Checked by MICHAEL J. UMEN and HERBERT O. HOUSE

1. Procedure

A. *1-(2-Methylpropenyl)pyrrolidine*. A 200-ml., three-necked flask is equipped with a magnetic stirring bar, a heating mantle, a pressure-equalizing dropping funnel, a glass stopper, and a continuous water separator (a Dean-Stark trap, Note 1) fitted with a condenser and a nitrogen inlet tube. The reaction vessel is flushed with nitrogen, and 61.5 g. (0.853 mole) of isobutyraldehyde (Note 2) is added to the reaction flask. An additional amount of isobutyraldehyde (Note 2) is added to the continuous water separator, filling the water-collecting trap. A static nitrogen atmosphere is maintained in the reaction vessel throughout the reaction and distillation. To the reaction flask is added, dropwise and with stirring over 5 minutes, 60.6 g. (0.852 mole) of pyrrolidine (Note 3). After addition is complete, the dropping funnel is replaced with a glass stopper and the reaction mixture is refluxed with stirring for 3.5 hours during which time about 15 ml. (0.83 mole) of water collects in the water separator (Note 4). The water separator and condenser are replaced with a distillation head, and the reaction mixture is distilled under reduced pressure, yielding 99.1–100.7 g. (94–95%) of the enamine as a colorless liquid, b.p. 92–106° (115–118 mm.), n_D^{25} 1.4708–1.4738 (Note 5).

B. *4,4-Dimethyl-2-cyclohexen-1-one*. A dry, 1-l., three-necked flask is equipped with a mechanical stirrer, a pressure-equalizing dropping funnel, a nitrogen inlet tube, and an ice-water cooling bath. The apparatus is flushed with nitrogen, and a static nitrogen atmosphere is maintained in the reaction vessel throughout the reaction. 1-(2-Methylpropenyl)pyrrolidine (62.6 g., 0.501 mole) is added to the reaction flask before 42.1 g. (0.601 mole) of methyl vinyl ketone (Note 6) is added, dropwise with stirring and cooling, over 5 minutes. After the resulting mixture has been stirred with cooling for 10 minutes, the ice bath is removed and stirring at room temperature is continued for 4 hours (Note 7). The reaction mixture is again cooled with an ice-water bath, and 250 ml. of 8 *M* hydrochloric acid is added, dropwise and with stirring (Note 8). After addition is complete, the mixture is stirred with cooling for 10 minutes, then stirred at room temperature for 14 hours (Note 9). The resulting brown reaction mixture is extracted with two 300-ml. portions of diethyl ether. The residual aqueous phase is neutralized by the cautious addition of 150–155 g. of solid sodium hydrogen carbonate and extracted with two 400-ml. portions of ether. The combined ethereal extracts (Note 10) are dried over anhydrous sodium sulfate and concentrated with a rotary evaporator. The residual liquid is distilled under reduced pressure, yielding 44.2–53.0 g. (71–85%) of 4,4-dimethyl-2-cyclohexen-1-one as a colorless liquid, b.p. 73–74° (14 mm.), n_D^{25} 1.4699–1.4726 (Note 11).

2. Notes

1. An illustration of a continuous water separator is provided in *Org. Synth.*, **Coll. Vol. 3**, 502 (1955).

2. The checkers employed isobutyraldehyde from Eastman Organic Chemicals. The aldehyde, b.p. 62–63°, was freshly distilled from a few milligrams of *p*-toluenesulfonic acid.

3. The pyrrolidine, obtained from Aldrich Chemical Company, Inc., was redistilled before use; b.p. 88–89°.

4. The water should not be drained from the water separator during the course of the reaction.

5. The enamine has IR absorption (pure liquid) at 1676 cm.$^{-1}$ (enamine C=C).

6. Methyl vinyl ketone (b.p. 35–36° at 140 mm.), obtained from Aldrich Chemical Company, Inc., was distilled immediately before use.

7. The use of solvents such as anhydrous ether or benzene is not only unnecessary but also undesirable, since yields are decreased by their presence. For best results the Diels-Alder adduct, which has been characterized by Opitz and Holtmann,[2] should not be isolated for the subsequent hydrolysis and cyclization.

8. The hydrolysis product, 2,2-dimethyl-5-oxo-hexanal, can be isolated if desired by stirring the Diels-Alder adduct with either 50% acetic acid or 2 *M* hydrochloric acid followed by extraction with ether and distillation,[3] b.p. 92–94° (20 mm.).

9. Cyclization can also be accomplished by the use of an ion exchange resin.[4] On a 0.1-mole scale, 110 ml. of 1 *M* hydrochloric acid and 70 ml. of wet Amberlite 1R-120 resin (acidified with hydrochloric acid) are added to the Diels-Alder adduct, and the mixture is refluxed for 24 hours. The mixture is cooled and washed with four portions of ether. The ether extract is dried over anhydrous magnesium sulfate and distilled, separating 9.7–10.8 g. (78–87%) of 4,4-dimethyl-2-cyclohexen-1-one; b.p. 73–74° (14 mm.).

10. Washing the ethereal extract with either dilute acid, aqueous sodium hydrogen carbonate, or saturated brine only decreased the yield of product and is, therefore, omitted. An acid-catalyzed reaction of methyl vinyl ketone with isobutyraldehyde gives the product in 71% yield.[8]

11. The product exhibits a single peak (retention time 5.7 minutes) on a 4-m. GC column packed with silicone fluid QF$_1$ on Chromosorb P and heated to 191°. This material has the following spectral characteristics: IR (CCl$_4$) 1675 (conjugated C=O) and 1623 cm.$^{-1}$ (conjugated C=C); UV (95% C$_2$H$_5$OH) max 224 (10,600) and 321 nm (34); ^{1}H

NMR (CCl$_4$), δ 1.17 (s, 6H, 2CH_3), 1.7–2.6 (m, 4H, 2CH_2), 5.71 (d, J = 10 Hz., 1H, vinyl CH), and 6.65 (d of t, J = 10 and 1.5 Hz., 1H, vinyl CH); m/e (rel. int.), 124(M$^+$, 49), 96(83), 82(100), 81(56), 68(25), 67(42), 53(22), 43(21), 41(25), and 39(25).

3. Discussion

The procedure described is essentially that of Opitz and Holtmann.[2] The yield has been increased from 27% to 85% by making changes as indicated in Notes 7 and 10. 4,4-Dimethyl-2-cyclohexen-1-one has also been prepared by Michael addition of methyl vinyl ketone to isobutyraldehyde followed by ring formation in basic media with yields of 25%,[5] 35%,[6] and 43%.[7] This procedure has general utility in preparing 4-substituted or 4,4-disubstituted cyclohexen-2-ones and, as such, constitutes a useful substitute for the Robinson annelation reactions.[9,10] Unlike the latter, the alkylation step in this procedure does not require strongly basic conditions. Consequently, side reactions such as aldol condensation of the carbonyl compounds and polymerization of methyl vinyl ketone are avoided. Moreover, since the location of the double bond in the enamines is controlled by the amines used,[11] it may be possible to direct the alkylation to either side of a given carbonyl group making this procedure potentially very versatile. Similar reactions using acrolein and enamines of cyclic ketones have been utilized for the synthesis of bicyclo[n.3.1] systems.[9,12]

1. Department of Chemistry, University of Utah, Salt Lake City, Utah, 84112.
2. G. Opitz and H. Holtmann, *Justus Liebigs Ann. Chem.*, **684**, 79 (1965).
3. See also I. Flemming and M. H. Karger, *J. Chem. Soc. C*, 226 (1967).
4. R. D. Allan, B. G. Cordiner, and R. J. Wells, *Tetrahedron Lett.*, 6055 (1968).
5. E. L. Eliel and C. A. Lukach, *J. Am. Chem. Soc.*, **79**, 5986 (1957).
6. J. M. Conia and A. Le Craz, *Bull. Soc. Chim. Fr.*, 1934 (1960).
7. E. D. Bergmann and R. Corett, *J. Org. Chem.*, **23**, 1507 (1958).
8. M. E. Flaugh, T. A. Crowell, and D. S. Farlow, *J. Org. Chem.*, **45**, 5399 (1980).
9. G. Stork, A. Brizzolara, H. Landesman, J. Szmuszkovicz, and R. Terrell, *J. Am. Chem. Soc.*, **85**, 207 (1963).
10. E. D. Bergmann, D. Ginsburg, and R. Pappo, *Org. React.*, **10**, 179 (1959).
11. W. D. Gurowitz and M. A. Joseph, *J. Org. Chem.*, **32**, 3289 (1967).
12. Y. Chan, Ph.D. Dissertation, University of Utah, 1971; A. C. Cope, D. L. Nealy, P. Scheiner, and G. Wood, *J. Am. Chem. Soc.*, **87**, 3130 (1965).

REDUCTIVE AMINATION WITH SODIUM CYANOBOROHYDRIDE: *N*,*N*-DIMETHYLCYCLOHEXYL-AMINE

(Cyclohexanamine, 4,4-dimethyl-)

$$\text{O} \quad + \quad (CH_3)_2NH \cdot HCl \quad + \quad KOH \quad \xrightarrow[CH_3OH]{NaBH_3CN} \quad N(CH_3)_2$$

Submitted by Richard F. Borch[1]
Checked by K. Abe and S. Masamune

1. Procedure

A solution of 21.4 g. (0.262 mole) of dimethylamine hydrochloride in 150 ml. of methanol is prepared in a 500-ml., round-bottomed flask. Potassium hydroxide (4 g.) is added in one portion to the magnetically stirred solution (Note 1). When the pellets are completely dissolved, 19.6 g. (0.200 mole) of cyclohexanone is added in one portion. The resulting suspension is stirred at room temperature for 15 minutes before a solution of 4.75 g. (0.0754 mole) of sodium cyanoborohydride (Notes 2 and 3) in 50 ml. of methanol is added dropwise over 30 minutes to the stirred suspension. After the addition is complete, the suspension is stirred for 30 minutes. Potassium hydroxide (15 g.) is then added, and stirring is continued until the pellets are completely dissolved. The reaction mixture is filtered with suction, and the volume of the filtrate is reduced to approximately 50 ml. with a rotary evaporator while the bath temperature is kept below 45° (Notes 4 and 5). To this concentrate is added 10 ml. of water and 25 ml. of saturated aqueous sodium chloride, and the layers are separated. The aqueous layer is extracted with two 50-ml. portions of diethyl ether. The organic layer previously separated and the ethereal extracts are combined and extracted with three 20-ml. portions of 6 *M* hydrochloric acid (Note 6). The combined acid layers are saturated with sodium chloride and extracted with four 30-ml. portions of ether (Note 7). The aqueous solution is cooled to 0° in an ice bath and brought to pH >12 by addition of potassium hydroxide pellets to the stirred solution (Notes 8 and 9). The layers are separated, and the aqueous layer is extracted with two 40-ml. portions of ether. The combined organic layers are washed with 10 ml. of saturated aqueous sodium chloride, dried over anhydrous potassium carbonate, and freed of ether with a rotary evaporator (Note 4). This crude product is fractionated through a 15-cm. Vigreux column (Note 10). After 1–3 g. of a forerun, b.p. 144–155° (Note 11) is separated, the fraction boiling at 156–159° is collected, yielding 13.3–13.7 g. (52–54%) of *N*,*N*-dimethylcyclohexylamine, n_D^{25} 1.4521 (Note 12).

2. Notes

1. Precipitation of potassium chloride begins immediately; the presence of this solid does not interfere with the reaction, and removal by filtration will result in loss of dimethylamine.

2. Sodium cyanoborohydride is available as a pale brown solid from Alfa Inorganics, Inc.

3. The commercially available material can be used without further purification. Use of material purified by the published procedure[2] gives a less colored crude product, but makes no improvement in yield or purity of the final product.

4. Since the product boils at 75° (15 mm.), care should be exercised to prevent loss of material in the evaporation process.

5. It is normal for additional potassium chloride to precipitate as the evaporation continues.

6. *Caution! This addition of hydrochloric acid into a separatory funnel occurs with considerable heat evolution, causing the ether to boil. The initial addition must be carried out with gentle swirling and cooling.*

7. GC analysis shows that the ethereal extract contains solely cyclohexanol (>98%).

8. The aqueous layer in this step is saturated with ether, and the addition of potassium hydroxide must be carried out gradually to prevent the contents of the flask from boiling over.

9. Copious amounts of potassium chloride precipitate during this addition. It is not necessary to remove the salt by filtration before the ether extraction.

10. A still pot with a volume of at least 100-ml. should be used for the distillation, since foaming occurs as the distillation proceeds.

11. On a 2-m. GC column packed with 10% Apiezon L and heated to 100°, the retention times for N,N-dimethylcyclohexylamine and cyclohexanol are 15 and 4 minutes, respectively. The composition of this forerun is 80–85% of the amine and 20–15% of the alcohol.

12. GC analysis of the product shows that the product is at least 99.2% pure and is contaminated only with trace amounts of cyclohexanol. The submitter reported a 62–69% yield (15.7–17.5 g.) using the indicated scale.

3. Discussion

N,N-Dimethylcyclohexylamine has been prepared by catalytic reductive alkylation[3,4] and by the Leuckart reaction.[5] The present method is experimentally simple, requires no

TABLE I

REPRESENTATIVE REDUCTIVE AMINATIONS WITH NaBH$_3$CN[2]

Compound	Amine	Product	Yield, %
Cyclohexanone	NH$_3$	Cyclohexylamine	45
Cyclohexanone	CH$_3$NH$_2$	N-Methylcyclohexylamine	41
Cyclohexanone	CH$_3$NHCH$_3$	N,N-Dimethylcyclohexylamine	53
Acetophenone	NH$_3$	α-Phenylethylamine	77
Acetophenone	CH$_3$NH$_2$	N-Methylphenethylamine	78
Isobutyraldehyde	PhNH$_2$	N-Isobutylaniline	78
Glutaraldehyde	CH$_3$NH$_2$	N-Methylpiperidine	43

special apparatus, and is generally applicable to the synthesis of a variety of primary, secondary, and tertiary amines, as illustrated in Table I.

The submitter has found that use of sodium borohydride instead of sodium cyano-borohydride in the present procedure results in the almost exclusive formation of cyclohexanol with less than 3% of basic material.

1. Department of Chemistry, University of Minnesota, Minneapolis, Minnesota 55455.
2. R. F. Borch, M. D. Bernstein, and H. D. Durst, *J. Am. Chem. Soc.*, **93**, 2897 (1971).
3. J. D. Roberts and V. C. Chambers, *J. Am. Chem. Soc.*, **73**, 5030 (1951).
4. W. S. Emerson, *Org. React.*, **4**, 174 (1948).
5. R. D. Bach, *J. Org. Chem.*, **33**, 1647 (1968).

N,N-DIMETHYLDODECYLAMINE OXIDE

(Dodecanamine, *N,N*-dimethyl-, *N*-oxide)

$$n\text{-}C_{12}H_{25}N(CH_3)_2 + (CH_3)_3CO_2H \xrightarrow{VO(C_5H_7O_2)_2}$$

$$n\text{-}C_{12}H_{25}\overset{+}{\underset{O^-}{N}}(CH_3)_2 + (CH_3)_3COH$$

Submitted by M. N. Sheng and J. G. Zajacek[1]
Checked by William F. Fischer, Jr., and Herbert O. House

1. Procedure

Caution! Methyl iodide, in high concentrations for short periods or in low concentrations for long periods, can cause serious toxic effects in the central nervous system. Accordingly, the American Conference of Governmental Industrial Hygienists[2] has set 5 p.p.m., a level which cannot be detected by smell, as the highest average concentration in air to which workers should be exposed for long periods. The preparation and use of methyl iodide should always be performed in a well-ventilated fume hood. Since the liquid can be absorbed through the skin, care should be taken to prevent contact.

A solution of 21.3 g. (0.100 mole) of freshly distilled *N,N*-dimethyldodecylamine (Note 1), 9.6 g. (0.10 mole) of 94% *tert*-butyl hydroperoxide (Note 2), and 0.050 g. of vanadium oxyacetylacetonate (Note 3) in 27 g. (34 ml.) of *tert*-butyl alcohol is placed in a 250-ml., round-bottomed flask fitted with a thermometer, a reflux condenser, and a heating mantle. The reaction mixture is heated to approximately 65–70°, at which point an exothermic reaction begins. The heating is discontinued until the vigorous exothermic reaction subsides (about 5 minutes); the reaction mixture is then heated at reflux (the reaction mixture boils at 90°) for 25 minutes. After the resulting mixture has been cooled to room temperature, *it is analyzed* (Note 4) *to establish the absence of* tert-*butyl hydroperoxide,* then concentrated with a rotary evaporator (30–35° bath with 30–40 mm. pressure). The crude solid residue is triturated with 50 ml. of cold (0–5°), anhydrous diethyl ether and filtered under conditions which prevent exposure of the residual amine oxide to atmospheric moisture (Note 5). The solid is washed with 50 ml. of cold (0–5°) anhydrous ether and dried under reduced pressure, leaving 12.9–15.5 g. of the crystalline amine oxide, m.p. 131–131.5°. Concentration of the mother liquors and trituration of the residual paste with 25 ml. of cold (0–5°) anhydrous ether separates another 4.9–3.4 g. of the amine oxide, m.p. 130–131°. The total yield of the crystalline amine oxide (Note 6) is 17.4–18.9 g. (76–83%).

2. Notes

1. *N,N*-Dimethyldodecylamine purchased from Eastman Organic Chemicals is approximately 90% pure. This material should be fractionally distilled with a spinning band column to obtain the pure amine, b.p. 116–117° (4 mm.).

2. *tert*-Butyl hydroperoxide purchased from the Lucidol Division, Wallace and Tiernan, Inc., is approximately 92% pure. A portion of the major impurity, water, can be removed by drying the commercial material over anhydrous magnesium sulfate for 2 days, leaving material that is 94–97% pure.

3. Vanadium oxyacetylacetonate may be purchased from Alfa Inorganics, Inc., Beverly, Massachusetts.

4. Although consumption of the hydroperoxide is normally complete, the absence of this peroxide in the reaction mixture should be established by testing with moist starch-iodide paper or by iodometric titration.[3] The amine oxide content may be determined by titration with standard hydrochloric acid after any amine present has reacted with methyl iodide for 1 hour at room temperature.[4] From this volumetric analysis the submitters determined the yield of amine oxide to be 86%. The checkers found that the reaction could be followed by measuring the ^{1}H NMR spectra in *tert*-butyl alcohol solution where the *N*-methyl signals of the amine (δ 2.03) and the amine oxide (δ 2.98) are readily observed.

5. The amine oxide is exceedingly hygroscopic and must be protected from atmospheric moisture during filtration. The submitters found it convenient to use a Büchner funnel covered with a large, inverted rubber stopper. After the mixture of amine oxide and ether is added to the funnel, the mouth of the funnel is covered with the inverted rubber stor̲ ̃r, and suction is applied. The flat surface of the inverted rubber stopper forms a seal against the mouth of the funnel, preventing the entrance of moist air while the last traces of ether are removed. The checkers employed a sintered-glass funnel fitted to maintain a nitrogen atmosphere above the crystalline product.

6. The hygroscopic amine oxide should be protected from moisture during storage. The checkers found that the initial product could be recrystallized from toluene under anhydrous conditions to yield the amine oxide as white needles, m.p. 130–131°.

3. Discussion

Aqueous or alcoholic solutions of amine oxides are normally obtained by oxidizing tertiary amines with either hydrogen peroxide or a peracid.[5] For example, *N,N*-dimethyldodecylamine oxide has been prepared by treating *N,N*-dimethyldodecylamine with aqueous hydrogen peroxide.[6] The procedure illustrated in this preparation permits the oxidation of tertiary amines with *tert*-butyl hydroperoxide in organic solvents under relatively anhydrous conditions.[7] In this procedure the reaction time is short, and the method is as convenient as that with aqueous hydrogen peroxide or a peracid as the oxidant. Furthermore, isolation of the anhydrous amine oxide is often relatively simple.

1. The Research and Development Department, ARCO Chemical Company, A Division of Atlantic Richfield Company, Glenolden, Pennsylvania. [Present address: The Research and Development Department, ARCO Chemical Company, A Division of Atlantic Richfield Company, Newtown Square, Pennsylvania 19073.]
2. American Conference of Governmental Industrial Hygienists (ACGIH), "Documentation of Threshold Limit Values," 3rd ed., Cincinnati, Ohio, 1971, p. 166.
3. D. H. Wheeler, *Oil Soap (Chicago)*, **9**, 89 (1932) [*Chem. Abstr.*, **26**, 3128 (1932)].
4. L. D. Metcalfe, *Anal. Chem.*, **34**, 1849 (1962).
5. A. C. Cope and E. R. Trumbull, *Org. React.*, **11**, 378 (1960).
6. G. L. K. Hoh, D. O. Barlow, A. F. Chadwick, D. B. Lake, and S. R. Sheeran, *J. Am. Oil Chem. Soc.*, **40**, 268 (1963).
7. M. N. Sheng and J. G. Zajacek, *J. Org. Chem.*, **33**, 588 (1968).

DIMETHYL NITROSUCCINATE

[Butanedioic acid, 2-nitro-, dimethyl ester]

Submitted by S. Zen and E. Kaji[1]
Checked by M. Braun and G. Büchi

1. Procedure

Caution! Benzene has been identified as a carcinogen; OSHA has issued emergency standards on its use. All procedures involving benzene should be carried out in a well-ventilated hood, and glove protection is required.

A 1-l., three-necked, round-bottomed flask equipped with a calcium chloride drying tube, a mechanical stirrer, and a ground-glass stopper is charged with 28.2 g. (0.184 mole) of freshly distilled methyl bromoacetate, 500 ml. of anhydrous N,N-dimethylacetamide (Note 1), and 20.0 g. (0.168 mole) of methyl nitroacetate (Note 2). The solution is stirred vigorously while 146 ml. (0.168 mole) of 1.15 N sodium methoxide in methanol is added in one portion. The resulting light-yellow suspension is stirred for an additional 16 hours at room temperature during which time it changes into a clear yellow solution.

After dilution with 200 ml. of benzene, the solution is transferred to a 2-l. separatory funnel containing 800 ml. of ice water and shaken thoroughly. The aqueous layer is separated, acidified to pH 3–4 with 2–3 ml. of concentrated hydrochloric acid, and extracted with three 100-ml. portions of benzene. All the organic layers are combined and dried over anhydrous sodium sulfate. Filtration and concentration of the solution with a rotary evaporator, followed by exposure to high vacuum for 2–3 hours, affords 17.3–19.3 g. of the crude product (Note 3). Low-boiling impurities are removed by vacuum distillation (Note 4), the residual oil (14–15 g.) is transferred to a 50-ml. flask equipped with a short-path distillation apparatus, and vacuum distillation is continued. A forerun is taken until no rise in boiling point is observed before 7.2–8.5 g. (23–27%) of dimethyl nitrosuccinate is collected as a colorless oil, b.p. 85° (0.07 mm.), n_D^{20} 1.4441 (Note 5).

2. Notes

1. *N,N*-dimethylacetamide was treated with molecular sieves for 2 days, decanted, and distilled under reduced pressure, b.p. 85° (30 mm.), before use.

2. Methyl nitroacetate was prepared by the method in *Org. Synth.*, **Coll. Vol. 6,** 797 (1988). It should be distilled before use.

3. GC analysis of the crude mixture (SE-30 on Chromosorb W, 1 m., 150°) showed the presence of some low-boiling materials (including unreacted methyl nitroacetate) and a significant amount of the doubly alkylated by-product, trimethyl 2-nitro-1,2,3-propane-tricarboxylate.

4. The bath temperature should be maintained below 70–75°. Distillation was carried out using a Claisen head, and the receiving flasks were immersed in ice.

The checkers found it convenient to omit this distillation and the subsequent transfer. Instead the crude product was placed in a 25-ml. flask and carefully distilled (0.07 mm). The bath temperature was raised slowly, and a forerun was collected until the boiling point stabilized.

5. The distilled product was determined by the checkers to be 85–90% pure (GC analysis), the major impurity being the doubly alkylated by-product. Purity can be increased to 95% by redistillation. The checkers found that conducting the experiment on a ¾ scale resulted in increased yield (34%) and purity (90–93%) of once-distilled product.

For twice-distilled material: IR (liquid film) cm.$^{-1}$: 1745 strong, 1565 strong, 1430 medium strong; ^{1}H NMR (CDCl$_3$), δ (multiplicity, coupling constant J in Hz., number of protons): 3.14–3.45 (m, 2H), 3.76 (s, 3H), 3.86 (s, 3H), 5.6 (d of d, J = 6 and 8, 1H).

3. Discussion

Diethyl nitrosuccinate has been prepared by oxidation of diethyl nitrososuccinate,[2] and by the reaction of sodium nitrite with diethyl bromosuccinate, but in the latter case no experimental conditions were described.[3]

The present method is a simple, one-step procedure employing commercially available or readily accessible starting materials. Other α-nitro carboxylic esters may be prepared in this way;[4] for example, dimethyl 2-nitropentanedioate was prepared in 45–50% yield.

1. School of Pharmaceutical Sciences, Kitasato University, Tokyo, Japan.
2. J. Schmidt and K. Th. Widmann, *Ber. Dtsch. Chem. Ges.,* **42,** 497 (1909).
3. R. Gelin and S. Gelin, *C.R. Hebd. Seances Acad. Sci.,* **256,** 3705 (1963).
4. S. Zen and E. Kaji, *Bull. Chem. Soc. Jpn.,* **43,** 2277 (1970); E. Kaji and S. Zen, *Bull. Chem. Soc. Jpn.,* **46,** 337 (1973).

DIMETHYL 2,3-PENTADIENEDIOATE

[2,3-Pentadienedioic acid, dimethyl ester]

Submitted by T. A. Bryson[1] and T. M. Dolak
Checked by R. Shapiro and G. Büchi

1. Procedure

Caution! The reaction of phosphorus pentachloride with diethyl acetone-1,3-dicarboxylate should be carried out in a hood, since hydrogen chloride is evolved.

A. *Dimethyl 3-chloro-2-pentenedioate.* A dry, 500-ml., three-necked, round-bottomed flask fitted with a ground-glass stopper, a condenser provided with a gas bubbler, a gas-inlet adapter attached to a nitrogen (or argon) source, and a magnetic stirring bar, is charged with 60.0 g. (0.297 mole) of diethyl acetone-1,3-dicarboxylate (Note 1). A steady, gentle flow of nitrogen is started through the reaction vessel (Note 2), and 65.0 g. (0.313 mole) of phosphorus pentachloride (Note 3) is added in thirteen, approximately equal portions through the stoppered joint to the neat diester at 3-minute intervals with vigorous stirring (Note 4). After the addition is complete, the reaction mixture is warmed to 40° in a water bath for 30 minutes. The red solution is cooled in an ice bath and poured onto *ca.* 100 ml. of ice in a 500-ml. Erlenmeyer flask immersed in an ice bath. A 1:1 mixture of water and dichloromethane is used to rinse traces of the product from the reaction vessel into the Erlenmeyer flask, and the resulting mixture is stirred for 15 minutes (Note 5). After separating the two layers, the aqueous phase is extracted with three 100-ml. portions of dichloromethane, and the combined organic extracts are dried over anhydrous sodium sulfate. Filtration through glass wool and removal of solvents with a rotary evaporator affords *ca.* 60 g. of a red oil, which is placed in a 500-ml., round-bottomed flask containing 20 ml. of concentrated sulfuric acid in 300 ml. of anhydrous methanol (Note 6), and the solution is refluxed using a heating mantle for 18 hours. Excess methanol (200 ml.) is distilled, and the residual yellow solution is cooled to room temperature and poured into 100 ml. of water. Sodium chloride is added to saturation, and the solution is extracted with eight 100-ml. portions of diethyl ether. The combined extracts are washed successively with 150 ml. of aqueous saturated sodium hydrogen carbonate and 150 ml. of aqueous saturated sodium chloride, dried over anhydrous sodium sulfate, and filtered. Concentration of the extract with a rotary evaporator affords a yellow oil which is distilled, yielding 33.5–34.4 g. (59–60%) of dimethyl 3-chloro-2-pentenedioate[2] as a colorless liquid, b.p. 50–60° (0.02 mm.) (Note 7).

B. *Dimethyl 2,3-pentadienedioate.* A 500-ml., three-necked, round-bottomed flask, equipped with a gas-inlet adapter, a 50-ml. addition funnel, a ground-glass stopper, and a magnetic stirring bar, is charged with 27.0 g. (0.145 mole) of the diester from Part A and 100 ml. of anhydrous tetrahydrofuran (freshly distilled from sodium). The flask is flushed with nitrogen (or argon), and a positive pressure is maintained while the contents are cooled to 0° in an ice–salt bath and stirred with an efficient motor. triethylamine (22 ml., 0.16 mole, freshly distilled from calcium hydride) is added through the addition funnel over a 10-minute period, the gas-inlet adapter is replaced with a calcium chloride tube, and the mixture is stirred at 0–5° for 18 hours (Note 8). The precipitate is removed by vacuum filtration and washed with three 100-ml. portions of anhydrous diethyl ether. The combined filtrate and washings are washed successively with three 75-ml. portions of 0.1 N hydrochloric acid and 100 ml. of aqueous saturated sodium chloride, dried over anhydrous sodium sulfate, filtered, and concentrated with a rotary evaporator. The residual oil is distilled (Note 9), yielding 13.3–13.9 g. (61–64%) of dimethyl 2,3-pentadienedioate[3] (Note 10), b.p. 58° (0.02 mm.).

2. Notes

1. Diethyl acetone-1,3-dicarboxylate was purchased from the Aldrich Chemical Company, Inc., and the checkers distilled this material under reduced pressure, b.p. 135–137° (12 mm.), discarding *ca.* 10% as a forerun.

2. A continuous flow of inert gas removes hydrogen chloride and phosphoryl chloride from the reaction flask.

3. Phosphorus pentachloride was purchased by the checkers from the J. T. Baker Chemical Company, and purchased from Eastman Organic Chemicals by the submitters.

4. Warming and foaming occur during the addition, and the temperature reaches *ca.* 40–45°.

5. The checkers found that unless the aqueous workup is cooled, the dichloromethane boils vigorously.

6. The checkers used commercial "anhydrous" methanol without further drying.

7. The checkers determined the product to be a mixture of isomers (approximately 6:1) by GC analysis (15% SE-30 on Chromosorb W, 0.3 × 244 cm., 175°) and by ^{1}H NMR. The mixture was characterized as follows: IR (liquid film) cm.$^{-1}$: 1745 strong (shoulder at 1720), 1640 medium strong, 1440 medium strong; ^{1}H NMR (CDCl$_3$), δ (multiplicity, number of protons): 3.75 (s, 6H), 4.12 (s, 2H), 6.21 (s, 1H), and 6.30 (s, 1H).

8. During this time a heavy precipitate of triethylamine hydrochloride forms; the mixture first becomes yellow and eventually brown in color.

9. The allene apparently polymerizes during distillation; it yellows in the receiving flask, and becomes orange and viscous even in the refrigerator overnight. The submitters obtained higher yields by distilling the product in batches.

10. The checkers characterized the product as follows: IR (liquid film) cm.$^{-1}$: 1970 strong, 1720 strong, 1440 strong; ^{1}H NMR (CDCl$_3$), δ (multiplicity, number of protons): 3.81 (s, 6H), 6.10 (s, 2H).

3. Discussion

Dimethyl 2,3-pentadienedioate has also been prepared from the enol phosphate of diethyl acetone-1,3-dicarboxylate.[4]

1. Department of Chemistry, University of South Carolina, Columbia, S.C. 29208.
2. J. M. van der Zandon, *Recl. Trav. Chim. Pays-Bas*, **54**, 289 (1935).
3. G. Büchi and J. Carlson, *J. Am. Chem. Soc.*, **91**, 6470 (1969).
4. J. Craig and J. Moyle, *J. Chem. Soc.*, 5356 (1963).

ALLYLICALLY TRANSPOSED AMINES FROM ALLYLIC ALCOHOLS: 3,7-DIMETHYL-1,6-OCTADIEN-3-AMINE

(1,6-Octadien-3-amine, 3,7-dimethyl-)

Submitted by LANE A. CLIZBE and LARRY E. OVERMAN[1]
Checked by A. BROSSI, H. MAYER, and N. KAPPELER

1. Procedure

Caution! Part A should be carried out in a well-ventilated hood to avoid exposure to trichloroacetonitrile vapors.

A. *Geraniol trichloroacetimidate* (**1**). A dry, 250-ml., three-necked flask is equipped with a magnetic stirring bar, a pressure-equalizing dropping funnel, a thermometer, and a nitrogen inlet tube. The apparatus is flushed with nitrogen and charged with 410 mg. (0.0103 mole) of sodium hydride dispersed in mineral oil (Note 1) and 15 ml. of hexane. The suspension is stirred, and the hydride is allowed to settle. The hexane is removed with a long dropping pipette, 60 ml. of anhydrous diethyl ether is added, and a solution of 15.4 g. (0.0993 mole) of geraniol in 15 ml. of anhydrous ether is added over 5 minutes. After the evolution of hydrogen ceases (less than 5 minutes), the reaction mixture is stirred for an additional 15 minutes. The clear solution is then cooled to between −10 and 0° in an ice–salt bath. Trichloroacetonitrile (10.0 ml., 14.4 g., 0.0996 mole) is added dropwise to the stirred solution, while the reaction temperature is maintained below 0° (Note 2). Addition is completed within 15 minutes, and the reaction mixture is allowed to warm to room temperature. The light amber mixture is poured into a 250-ml., round-bottomed flask, and the ether is removed with a rotary evaporator. Pentane [150 ml., containing 0.4 ml. (0.01 mole) of methanol] is added, the mixture is shaken vigorously for 1 minute, and a small amount of dark, insoluble material is removed by gravity filtration. The residue is washed two times with pentane (50 ml. total), and the combined filtrate is concentrated with a rotary evaporator, yielding 27–29 g. (90–97%) of nearly pure (Note 3) imidate **1**.

B. *3,7-Dimethyl-3-trichloroacetamido-1,6-octadiene* (**2**). A 500-ml., round-bottomed flask equipped with a condenser, a magnetic stirring bar, and a calcium chloride drying tube is charged with imidate **1** and 300 ml. of xylene. The solution is refluxed for 8 hours (Note 4). After cooling to room temperature the dark xylene solution is filtered through a short column (4.5 cm. in diameter) packed with silica gel (70 g.) and toluene. The column is eluted with an additional 250 ml. of toluene, and the combined light yellow eluant is concentrated with a rotary evaporator. Vacuum distillation through a 15-cm. Vigreux column yields 20–22 g. (67–74% for the two steps) of octadiene **2** as a colorless liquid, b.p. 94–97° (0.03 mm.) (Note 5).

C. *3,7-Dimethyl-1,6-octadien-3-amine* (**3**). A 500-ml., round-bottomed flask equipped with a magnetic stirring bar, a condenser, and a nitrogen inlet tube is charged with 9.0 g. (0.030 mole) of octadiene **2,** 160 ml. of 95% ethanol, and 150 ml. of aqueous 6 N sodium hydroxide. The air is replaced with nitrogen (Note 6), and the solution is stirred at room temperature for 40 hours (Note 7). Ether (300 ml.) is added, the organic layer is separated, and the aqueous layer is washed twice with 50 ml. of ether. After drying over anhydrous potassium carbonate and filtration, the organic extracts are concentrated with a rotary evaporator, affording a white, semisolid residue, which is extracted four times with 50 ml. of boiling hexane. The extract is concentrated with a rotary evaporator, and the residual yellow liquid is distilled under reduced pressure using a short-path apparatus, yielding 2.99–3.45 g. (65–75%) of amine **3**, b.p. 58–61° (2.6 mm.) (Notes 8 and 9).

2. Notes

1. The reagents used in this procedure were obtained from the following sources: geraniol (99+%), Aldrich Chemical Company, Inc.; sodium hydride (58% dispersion in

mineral oil), Alfa Products, Division of the Ventron Corporation; trichloroacetonitrile, Aldrich Chemical Company, Inc.; xylene (a mixture of isomers, b.p. 137–144°), Mallinckrodt Chemical Works; anhydrous ether, Mallinckrodt Chemical Works; Silica Gel (Grade 62), Grace Davidson Chemical. The reagents used by the checkers were obtained from the following sources: geraniol (98.7%), sodium hydride (58% dispersion in mineral oil), trichloroacetonitrile, and xylene (a mixture of isomers, b.p. 137–143°), Fluka A G, Chemische Fabrik, Buchs, Switzerland; Silica gel (70–230 mesh ASTM), E. Merck A G, Darmstadt, Germany.

2. For secondary or tertiary alcohols, the yields are improved by adding the alcohol–alkoxide solution dropwise to a solution of trichloroacetonitrile and ether at 0°.

3. The crude imidate 1 is sufficiently pure (checked by ^{1}H NMR) for most purposes and can be used in the next step without further purification. IR (neat) cm^{-1}: 3340 weak (N—H), 1660 strong (C=N); ^{1}H NMR (CCl$_4$), δ (multiplicity, coupling constant J in Hz., number of protons, assignment): 1.58 (s, 3H, CH$_3$), 1.66 (s, 3H, CH$_3$), 1.73 (s, 3H, CH$_3$), 4.77 (d, $J = 7$, 2H, H_1), 5.1 (broad s, 1H, H_6), 5.5 (approx. t, $J = 7$, 1H, H_2), 8.3 (broad s, 1H, NH). The crude imidate 1 may be distilled rapidly through a short Vigreux column to give 24–28 g. (80–93%) of distilled product, b.p. 109–111° (0.1 mm.). However, the checkers found by ^{1}H NMR analysis that this product already contains substantial amounts of octadiene 2 formed by thermal rearrangement during distillation.

4. The reaction can be conveniently monitored in the IR by observing the decrease in the C=N stretching absorption at 1660 cm.$^{-1}$ or by TLC [silica, developed with hexane–ethyl acetate (9:1) or (4:1)].

5. The octadiene 2 appears pure by ^{1}H NMR and elemental analysis. A TLC [silica, developed with hexane–ethyl acetate (4:1), visualized with 5% ceric ammonium nitrate in 20% sulfuric acid and heating] shows a major spot at $R_f = 0.4$ and a very small impurity at 0.7. The trace impurity may be removed by crystallization from hexane at −78°. The checkers found that a GC (2% Silicone OV-17 on Gaschrom Q., 80–100 mesh, 200 × 0.22 cm., 120°) showed one major peak (97.0–97.2%) at a relative retention value of 1.00 (17.3 minutes, nitrogen, 30 ml. per minute), and a minor peak at 0.94 (<3%). The octadiene 2 has the following spectral properties: IR (neat) cm.$^{-1}$: 3423 and 3355 weak (N—H), 1722 strong (C=O), 1504 strong (CO—NH—, amide II band), 983 weak and 915 medium (CH=CH$_2$); ^{1}H NMR (CCl$_4$), δ (multiplicity, coupling constant J in Hz., number of protons, assignment): 1.49 (s, 3H, CH$_3$), 1.60 (s, 3H, CH$_3$), 1.67 (s, 3H, CH$_3$), 4.9–5.3 (m, 3H, H_1 and H_6), 5.96 (approx. d of d, $J = 9.5$ and 18.5, 1H, H_2), 6.6 (broad s, 1H, NH); ^{13}C NMR (acetone-d_6), δ (assignment): 159.8 (C=O), 141.1 (C_2), 132.4 (C_7), 123.4 (C_6), 113.2 (C_1), 93.3 (CCl$_3$), 58.7 (C_3), 38.8 (C_4), 25.7 (CH$_3$), 23.9 (CH$_3$), 22.5 (C_5), 17.5 (CH$_3$).

6. The apparatus described in Org. Synth., Coll. Vol. 4, 132 (1963) was used to maintain a nitrogen atmosphere.

7. The reaction time may be cut to less than 1 hour by running the reaction at reflux, but the yield is 5–10% lower. The checkers found that under these conditions the weight yield of the distilled product was 67–70%. However, GC (3% Silicone GE-SE-30 on Gaschrom Q, 80–100 mesh, 200 × 0.22 cm., 60°) showed that the product contained substantial amounts (19–23%) of an unidentified by-product having a relative retention value of 1.22 (nitrogen, 30 ml. per minute) and a minor peak at 0.96 (<2%). The checkers also found that the reaction can be conveniently monitored by TLC [silica, hexane–ethyl acetate (4:1)].

8. The checkers found that the crude amine **3** (linalylamine) is more conveniently distilled at a pressure of 11–12 mm.

9. A GC (10% Carbowax 20 M–2% KOH on Chromosorb W, AW, 80–100 mesh, 180 × 0.3 cm., 90°) showed one major peak (99%) at a relative retention value of 1.00 (3.5 minutes, nitrogen, 50 ml. per minute), and a minor peak at 1.1 (<1%). Temperature programming to 200° detected the presence of several higher-boiling trace impurities (<1%). The checkers found that a GC (3% Silicone GE-SE-30 on Gaschrom Q, 80–100 mesh, 200 × 0.22 cm., 60°) showed one major peak (97.9–98.4%) at a relative retention value of 1.00 (11.8 minutes, nitrogen, 30 ml. per minute), and a minor peak at 0.96 (<2.1%). IR (neat) cm.$^{-1}$: 3345 and 3301 weak (N—H), 996 and 912 medium (CH=CH$_2$); ^{1}H NMR (CCl$_4$), δ (multiplicity, coupling constant J in Hz., number of protons, assignment): 1.08 (s, 3H, CH_3), 1.55 (s, 3H, CH_3), 1.63 (s, 3H, CH_3), 4.7–5.2 (m, 3H, H_1 and H_6), 5.84 (approx. d of d, J = 10.1 and 17.8, 1H, H_2).

3. Discussion

This procedure illustrates a general method for the preparation of rearranged allylic amines from allylic alcohols.[2,3] The method is experimentally simple and has been used to prepare a variety of allylic *n*-, *sec*-, and *tert*-carbonyl amines, as illustrated in Table I. The only limitation encountered so far is a competing ionic elimination reaction which becomes important for trichloroacetimidic esters of 3-substituted-2-cyclohexenols.[3,4] The rearrangement is formulated as a concerted [3,3]-sigmatropic rearrangement on the basis of its stereo- and regiospecificity[3,5] which are similar to those observed in related [3,3]-sigmatropic processes.[6] In certain cases the allylic imidate rearrangement may be accomplished at or below room temperature by the addition of catalytic amounts of mercury(II) salts.[2,3]

TABLE I

CONVERSION OF ALLYLIC ALCOHOLS INTO REARRANGED TRICHLOROACETAMIDES[3]

Alcohol	Trichloroacetamide Product	Overall Isolated Yield (%)
(*E*)-2-Hexen-1-ol	3-Trichloroacetamido-1-hexene	72
Cinnamyl alcohol	3-Phenyl-3-trichloroacetamido-1-propene	76
1-Hepten-3-ol	(*E*)-1-Trichloroacetamido-2-heptene	74
2-Cyclohexen-1-ol	3-Trichloroacetamido-1-cyclohexene	61
Linalool	3,7-Dimethyl-1-trichloroacetamido-2,6-octadiene[a]	83
3,5,5-Trimethyl-2-cyclohexen-1-ol	3,5,5-Trimethyl-3-trichloroacetamido-1-cyclohexene	10–43
3-(4,4-Ethylenedioxybutyl)-2-cyclohexen-1-ol	3-(4,4-Ethylenedioxybutyl)-3-trichloroacetamido-1-cyclohexene	20

[a] A mixture of *E* and *Z* isomers.

The experimental procedure for the addition of an alcohol to trichloroacetonitrile, a modification of the procedure of Cramer,[7] appears to be totally general. For hindered secondary or tertiary alcohols it is more convenient to form the catalytic alkoxide with potassium hydride, and it is essential that the alcohol–alkoxide solution be added to (inverse addition) a solution of trichloroacetonitrile in ethyl ether at 0°.[3] The thermal rearrangement of allylic imidates was first reported by Mumm and Möller[8] in 1937, and has been noted in scattered reports since that time.[9] In these cases the imidates were not available by a general route (as is the case for trichloroacetimidates), and as a result this rearrangement had not become a generally useful synthetic method. Alternative methods for the allylic transposition of oxygen and nitrogen functions include the thermolysis of allylic sulfamate esters,[10] phenylurethanes,[11] oxime O-allyl ethers,[12] and N-chlorosulfonylurethanes,[13] as well as the S_N2' reaction of certain allylic alcohol esters with amines.[14] These procedures are much less attractive than the method reported here, usually affording mixtures of allylic isomers.[10–14] Other routes for the preparation of 3,7-dimethyl-1,6-octadien-3-amine have not, to our knowledge, been reported.

1. Department of Chemistry, University of California, Irvine, California, 92717.
2. L. E. Overman, *J. Am. Chem. Soc.*, **96**, 597 (1974).
3. L. E. Overman, *J. Am. Chem. Soc.*, **98**, 2901 (1976).
4. L. E. Overman, *Tetrahedron Lett.*, 1149 (1975).
5. Y. Yamamoto, H. Shimoda, J. Oda, and Y. Inouye, *Bull. Chem. Soc. Jpn.*, **49**, 3247 (1976).
6. S. J. Rhoads and N. R. Raulens, *Org. React.*, **22**, 1 (1975).
7. F. Cramer, K. Pawelzik, and H. J. Baldauf, *Chem. Ber.*, **91**, 1049 (1958).
8. O. Mumm and F. Möller, *Ber. Dtsch. Chem. Ges.*, **70**, 2214 (1937).
9. W. M. Lauer and R. G. Lockwood, *J. Am. Chem. Soc.*, **76**, 3974 (1954); W. M. Lauer and C. S. Benton, *J. Org. Chem.*, **24**, 804 (1959); R. M. Roberts and F. A. Hussein, *J. Am. Chem. Soc.*, **82**, 1950 (1960); D. St. C. Black, F. W. Eastwood, R. Okraglik, A. J. Poynton, A. M. Wade, and C. H. Welker, *Aust. J. Chem.*, **25**, 1483 (1972).
10. E. H. White and C. A. Elliger, *J. Am. Chem. Soc.*, **87**, 5261 (1965).
11. M. E. Synerholm, N. W. Gilman, J. W. Morgan, and R. K. Hill, *J. Org. Chem.*, **33**, 1111 (1968).
12. A. Eckersley and N. A. J. Rogers, *Tetrahedron Lett.*, 1661 (1974).
13. J. B. Hendrickson and I. Joffee, *J. Am. Chem. Soc.*, **95**, 4083 (1973).
14. G. Stork and W. N. White, *J. Am. Chem. Soc.*, **78**, 4609 (1965); G. Stork and A. F. Kreft III, *J. Am. Chem. Soc.*, **99**, 3850 (1977).

BICYCLIC KETONES FOR THE SYNTHESIS OF TROPINOIDS:
2α,4α-DIMETHYL-8-OXABICYCLO[3.2.1]OCT-6-EN-3-ONE

(8-Oxabicyclo[3.2.1]oct-6-en-3-one, 2α,4α-dimethyl-)

A. $CH_3CH_2\overset{\displaystyle O}{\overset{\|}{C}}CH_2CH_3 + 2Br_2 \xrightarrow[-10-0°]{PBr_3 \text{ (catalyst)}} CH_3CHBr\overset{\displaystyle O}{\overset{\|}{C}}CHBrCH_3 + 2HBr$

1

B. $\mathbf{1}$ + (furan) $\xrightarrow[CH_3CN, 50°]{2NaI, 2Cu}$ (bicyclic structure)

2

Submitted by M. R. Ashcroft and H. M. R. Hoffmann[1]
Checked by D. M. Lokensgard and O. L. Chapman

1. Procedure

Caution! This reaction should be carried out in an efficient hood. 2,4-Dibromo-3-pentanone is a potent lachrymator and a readily absorbed skin irritant. Contact with the skin produces a sensation of sunburn and should be treated immediately by washing with a soap solution, followed by washing with sodium hydrogen carbonate solution.

A.   *2,4-Dibromo-3-pentanone* (**1**).   A three-necked, 250-ml. flask is fitted with a stirrer, a dropping funnel, and a condenser protected by a calcium chloride drying tube. Bromine (160 g., 1.00 mole) is added rapidly to a stirred solution of 45 g. (0.52 mole) of 3-pentanone (Note 1) and 1 ml. of phosphorus tribromide maintained between −10° and 0° with an acetone–dry ice bath in an efficient hood. Toward the end of the reaction very large amounts of hydrogen bromide are evolved, and the rate of addition must be controlled to allow the hood to exhaust the gas. Alternatively, a gas trap may be used. Depending on the efficiency of the hood, the addition should take 20–40 minutes. The flask is then evacuated with a water pump, removing dissolved hydrogen bromide, and the reaction mixture is immediately fractionally distilled through a 40-cm. column packed with glass helices (or, more quickly, with a Dufton column)[2] under reduced pressure. The dibromoketone **1**, a mixture of *dl-* and *meso*-isomers,[3a] distills at 67–82° (10 mm.), and 91 g. (72%) of product is collected as a colorless liquid (Note 2).

B.   *2α,4α-Dimethyl-8-oxabicyclo[3.2.1]oct-6-en-3-one* (**2**).   A 1-l., three-necked, round-bottomed flask is fitted with a 100-ml. dropping funnel having a nitrogen inlet tube, a magnetic stirrer, a thermometer, and an efficient double-surface condenser carrying a nitrogen outlet tube connected to a bubbler, and placed on a combined hotplate–magnetic stirring unit in a heat-resistant glass dish, acting as a water bath. Dry acetonitrile (200 ml.) (Note 3) is introduced into the flask, followed by 90 g. (0.60 mole) of dried powdered sodium iodide (Note 4), with vigorous stirring under a slow stream of nitrogen. When the stirring bar rotates steadily, 20 g. (0.31 g.-atom) of powdered copper bronze (Note 5) is added, followed by 28 g. (30 ml., 0.41 mole) of freshly-distilled furan (Note 6). The dropping funnel is then charged with a solution of 24.4 g. (0.100 mole) of dibromoketone **1** in 50 ml. of dry acetonitrile, which is rapidly added to the stirred reaction mixture (Note 7). On addition of dibromoketone **1**, the temperature rises to 45–50°, and a characteristic oatmeal-colored precipitate forms. After about 2 hours the temperature begins to drop, and the reaction is maintained at 50–60° with the water bath for a total reaction time of 4 hours (Note 8).

The flask is cooled to 0° with crushed ice (Note 9), and 150 ml. of dichloromethane is added with stirring. The reaction mixture is poured into a 2-l. beaker containing 500 ml. of water and 500 ml. of crushed ice; material remaining in the flask is rinsed into the beaker with 10 ml. of dichloromethane. The mixture is stirred thoroughly, further salts being precipitated, until the ice just melts, and filtered into a cooled filter flask under reduced pressure through a sintered or Büchner funnel and a kieselguhr filter-aid cake (Note 10). The beaker and filter cake are washed with 50 ml. of dichloromethane, and the clear combined filtrate is transferred to a 2-l. separatory funnel while still cold (Note 11).

The mixture is shaken vigorously, the lower layer is separated and stored in ice, and the aqueous layer is extracted with two, 50-ml. portions of dichloromethane. The combined, organic extracts are shaken with 100 ml. of ice-cold, concentrated aqueous ammonia (35% w/w) filtered through a filter-aid cake, and separated (Note 12). The extraction and filtration are repeated with fresh ammonia solution using the same filter (Note 13). The filter is washed with 50 ml. of dichloromethane, and the organic layer is separated and dried over anhydrous magnesium sulfate. The dried solution is filtered, the filter is washed with 50 ml. of dichloromethane, and the solvent is removed on a rotary evaporator at 30°. The flask containing the residual oil is cooled to 0° before exposure to air (Note 14).

The light-yellow oil is dissolved in 60 ml. of 30% anhydrous diethyl ether in pentane and treated with 2 g. of anhydrous sodium sulfate and 0.5 g. of decolorizing carbon. The mixture is swirled for a few minutes, allowed to settle, and filtered by gravity through three sheets of fine filter paper into a 100-ml., round-bottomed flask with a 14/20 joint. The filter is washed with 10 ml. of pentane, and the flask is sealed by wiring on a 14-mm. serum cap. The flask is placed on a cork ring, lowered into an insulated container (large Dewar bottle, styrofoam box, etc.) half filled with dry ice, and cooled slowly to −78°. When recrystallization is complete, a nitrogen supply is connected to the flask *via* a syringe needle; the supernatant liquid is then withdrawn by syringe and replaced with 50 ml. of pentane, previously cooled to −78°.

The flask is swirled, washing the crystals (Note 15), and the pentane is withdrawn. The flask is connected to a vacuum (water pump) *via* the nitrogen inlet and warmed to room temperature. The crude cycloadduct **2** (6.1–7.3 g., 40–48%) is isolated as colorless needles, m.p. 43.5–45°, from the first recrystallization (Note 16). Pure 2α,4α-dimethyl-8-oxabicyclo[3.2.1]oct-6-en-3-one (**2**) can be obtained by recrystallization from pentane at −78° with minimal loss, m.p. 45–46° (Note 17).

2. Notes

1. 3-Pentanone is available from Aldrich Chemical Company, Inc.

2. A slight coloration has no effect on the yield of the subsequent reactions. The dibromoketone **1** should be stored cold in a well-stoppered bottle (dark) under nitrogen and is best handled cold to minimize spread of lachrymator vapors.

3. Commercial acetonitrile from Hopkins and Williams (GPR grade, given analysis 0.1% water, 0.02% acid) was used. Further purification of the solvent had no effect on the yield. The checkers used MCB reagent grade acetonitrile, refluxed over and distilled from calcium hydride.

4. The sodium iodide was dried at 150° for at least 3 hours, cooled in a desiccator, and finely powdered in a mortar. Although the use of less sodium iodide (*e.g.*, 33 g., 0.22 mole) gives similar yields, the separation of the aqueous phase on extraction with dichloromethane and the following work-up are easier under the given conditions.

5. The copper bronze was supplied by BDH Chemicals Ltd. (Poole, England) as an extremely fine powder. The use of more granular electrolytic copper had no effect on the yield, but made magnetic stirring more difficult.

6. Commercial furan was refluxed over and distilled from calcium hydride and anhydrous potassium carbonate prior to use. Furan, b.p. 31°, is more volatile than diethyl ether, and precautions must be taken to minimize losses through evaporation.

7. Addition of dibromoketone **1** has been carried out all at once, slowly over a period of half an hour, and over 1 hour with no apparent change in yield.

8. Although a filtered sample of the reaction mixture analyzed by ^{1}H NMR (CDCl$_3$) shows no more dihaloketone after a reaction time of 2 hours (excess of furan was discernible), the reaction mixture must be heated for an additional 2 hours to destroy traces of dihaloketone which make the subsequent work-up and analytical TLC of the product mixture difficult. When the reaction was carried out with less sodium iodide (33 g., 0.22 mole), the presence of diiodoketone in the final product was noted by formation of an iodine color and rapid decomposition of the cycloadduct to a black solid; also, the pentane washings developed an iodine color on exposure to light.

9. Since the dihaloketones may induce the decomposition of the product, it is essential that the solution be cooled in ice before allowing entry of air. Otherwise, the oily cycloadduct becomes brown, and polymeric material has to be removed, before crystallization, by passage down a 2 × 5 cm. column of silica gel (impregnated with 12% silver nitrate solution and redried).

10. Filtration through Hopkins and Williams kieselguhr filter-aid cake speedily removes even colloidal copper halides and breaks up any emulsion.

11. If the temperature is allowed to rise above 0°, the cycloadduct decomposes and the yellow copper compound present liberates blue-green copper(II) salts which make the work-up difficult.

12. The checkers, using 15 ml. of acetonitrile instead of 10 ml. of dichloromethane for rinsing the reaction vessel, found that no solid formed and filtration was not necessary. In this case three extractions with ammonia solution are required before addition of more ammonia fails to produce a blue color.

13. Using the same filter ensures that the ammonia solution is saturated with halide salts, which aid final separation.

14. The oil is essentially pure cycloadduct 2, but owing to traces of impurity, crystallization may be difficult at 0°. Analytical TLC on alumina, using low-polarity solvents such as pentane or carbon tetrachloride, was not successful in the presence of traces of dihaloketone, although it gave high resolution with related compounds. When dehalogenation is complete, however, the resolution is restored (see also Note 9).

15. If product 2 appears to be colored at this point, the pentane solution can simply be warmed to dissolve the crystals and cooled to recrystallize.

16. Approximately 0.4 g. of product 2 remaining in the supernatant solution can be recovered by repeated recrystallizations from the ether–pentane solution; however, the presence of 10 impurities (TLC) makes this rather impractical.

17. IR (CCl_4) cm.$^{-1}$: 1721 (very strong); ^1H NMR ($CDCl_3$), δ (multiplicity, coupling constant J in Hz., number of protons, assignment): 0.98 (d, $J = 7$, 6H, 2CH_3), 2.8 (d of q, $J = 5$ and 7, 2H, $CHCOCH$), 4.86 (d, $J = 5$, 2H, $CHOCH$), 6.35 (broad s, 2H, 2 olefinic H).[3b]

3. Discussion

The reaction of 1,3-dibromo-1,3-diphenyl-2-propanone with sodium iodide in the presence of furan and cyclopentadiene, giving bridged seven-membered rings has been reported by Cookson,[4a] who followed up earlier work by Fort.[4b] Chidgey[5a,b] demonstrated that the reaction can be extended to simple dibromoketones such as 1 and improved by using metallic copper to remove molecular iodine liberated during the reaction. Mechanistically, the reaction seems to involve two very fast S_N2 displacements of bromide by iodide, as seen by the precipitation of sodium bromide. A subsequent slower nucleophilic attack of excess iodide ion on the positively polarized iodine of the diiodoketone is envisioned, yielding an allylic iodide which forms an allyl cation in a fairly facile S_N1 reaction. The allyl cation is trapped by furan, giving the oxygen-bridged seven-membered ring.[5b] The secondary–tertiary dibromoketone, $CH_3CHBrCOCBr(CH_3)_2$, can also be used as an allyl cation precursor in this reaction, but the yields with the primary–tertiary dibromoketone, $CH_2BrCOCBr(CH_3)_2$, are less satisfactory. Zinc-induced cycloaddition works well,[6] in this instance, and also in the case of the ditertiary dibromoketone, $(CH_3)_2CBrCOCBr(CH_3)_2$, which fails to undergo the initial S_N2 displacement with sodium iodide. Hence, although the sodium iodide–copper procedure is probably less general than the zinc[3b] and silver ion–promoted[5c] cycloaddition, it is experimentally convenient and yields preferentially the thermodynamically more stable adduct *via* an allyl cation in a W-configuration and compact transition state.[5b] The reaction involves inexpensive starting materials, proceeds under homogeneous conditions, and can be scaled-up readily.

The procedure described here has been used for the preparation of sensitive 6,7-dehydrotropinones in modest to good yields.[6] If in the present experiment furan is replaced by cyclopentadiene, an epimeric mixture of *cis*-diequatorial and *cis*-diaxial 2,4-dimethylbicyclo[3.2.1]oct-6-en-3-one is formed in almost 90% yield.[7] Instead of zinc or sodium iodide/copper, diiron nonacarbonyl may also be used as a reducing agent for dibromoketones.[8] Recently, $2\alpha,4\alpha$-dimethyl-8-oxabicyclo[3.2.1]oct-6-en-3-one (2) has been used as a precursor for the synthesis of ($\pm$)-nonactic acid, the building block of the macrotetrolide antibiotic nonactin.[9]

$$CH_3CHBrCOCHBrCH_3 \xrightarrow[S_N2 \text{ (very fast)}]{2NaI} CH_3CHICOCHICH_3 + 2NaBr$$

1. Chemistry Department, University College, London WC1H 0AJ, England. [Present address: Institut für Organische Chemie der Technischen Universität, Schneiderberg 1B, D-3000 Hannover 1, Germany.] This work was supported by the Science Research Council and by the Petroleum Research Fund, administered by the American Chemical Society.

2. See A. I. Vogel, "A Text-Book of Practical Organic Chemistry," 3rd ed., Longmans, London, 1959, p. 91. The all-glass Dufton column is a plain tube into which a solid glass spiral, wound around a central tube or rod, is placed. The spiral should fit tightly inside the tube to prevent leakage of vapor between the walls of the column and the spiral.

3. (a) H. M. R. Hoffmann and J. G. Vinter, *J. Org. Chem.*, **39**, 3921 (1974); (b) H. M. R. Hoffmann, K. E. Clemens, and R. H. Smithers, *J. Am. Chem. Soc.*, **94**, 3940 (1972).

4. (a) R. C. Cookson, M. J. Nye, and G. Subrahmanyam, *J. Chem. Soc. C*, 473 (1967); (b) A. W. Fort, *J. Am. Chem. Soc.*, **84**, 2620, 4979 (1962).

5. (a) R. Chidgey, Ph.D. Thesis, University of London, 1975; (b) H. M. R. Hoffmann, *Angew. Chem. Int. Ed. Engl.*, **12**, 819 (1973); (c) H. M. R. Hoffmann, D. R. Joy, and A. K. Suter, *J. Chem. Soc. B*, **57** (1968); R. Schmid and H. Schmid, *Helv. Chim. Acta*, **57**, 1883 (1974); H. Mayr and B. Grubmüller, *Angew. Chem.*, **90**, 129 (1978).

6. G. Fierz, R. Chidgey, and H. M. R. Hoffmann, *Angew. Chem. Int. Ed. Engl.*, **13**, 410 (1974).

7. D. I. Rawson, unpublished work; A. Busch and H. M. R. Hoffmann, *Tetrahedron Lett.*, 2379 (1976); D. I. Rawson, B. K. Carpenter, and H. M. R. Hoffmann, *J. Am. Chem. Soc.*, **101**, 1786 (1979). See also H. M. R. Hoffmann and H. Vathke, *Chem. Ber.*, **113**, 3416 (1980); H. M. R. Hoffman and J. Matthei, *Chem. Ber.*, **113**, 3837 (1980).

8. R. Noyori, S. Makino, T. Okita, and Y. Hayakawa, *J. Org. Chem.*, **40**, 806 (1975); R. Noyori, Y. Baba, S. Makino, and H. Takaya, *Tetrahedron Lett.*, 1741 (1973); R. Noyori, *Acc. Chem. Res.*, **12**, 61 (1979).

9. M. J. Arco, M. H. Trammell, and J. D. White, *J. Org. Chem.*, **41**, 2075 (1976).

2,2-DIMETHYL-4-PHENYLBUTYRIC ACID

(Benzenebutanoic acid, α,α-dimethyl-)

$$(CH_3)_2CHCO_2Na + LiN[CH(CH_3)_2]_2 \rightarrow [(CH_3)_2CCO_2]^{2-} Li^+Na^+$$

$$[(CH_3)_2CCO_2]^{2-} Li^+Na^+ + C_6H_5CH_2CH_2Br \rightarrow$$
$$C_6H_5CH_2CH_2C(CH_3)_2CO_2H$$

Submitted by P. L. Creger[1]
Checked by Paul Kalicky and Ronald Breslow

1. Procedure

A 500-ml., three-necked flask is equipped with a mechanical stirrer and a two-necked adapter carrying a Friedrich condenser, and a thermometer which contacts the flask contents. The third neck of the flask carries a pressure-equalizing dropping funnel which is exchanged for a serum stopper as required. The condenser is attached to a suitable source of nitrogen. The reaction flask is placed in a heating mantle and charged with 7.75 g. (0.0767 mole) of diisopropylamine (Note 1), 3.68 g. (0.082 mole) of 54% sodium hydride in mineral oil (Note 2), and 75 ml. of tetrahydrofuran (Note 3). From the dropping funnel, 6.6 g. (0.075 mole) of isobutyric acid (Note 4) is added to the stirred mixture over 5 minutes. The internal temperature rises to 50–60° and hydrogen evolution is completed by heating the mixture to reflux for 15 minutes. After cooling to 0° with an externally applied ice–salt bath (Note 5), 52 ml. of a standard solution of n-butyllithium in heptane (1.45 mmole/ml.; 0.075 mole) (Note 6) is added through the stopper by injection (Note 7) at a temperature below 10°. The ice bath is retained for 15 minutes before the mixture is heated to 30–35° for 30 minutes to complete the metalation. The resulting turbid solution is cooled, and 13.9 g. (0.0751 mole) of (2-bromoethyl)benzene (Note 8) is added from the dropping funnel over 20 minutes at 0°. A colorless precipitate of sodium bromide begins to separate almost immediately. The ice bath is retained for 30 minutes, after which the mixture is heated to 30–35° for 1 hour.

At the conclusion of the reaction period, 100 ml. of water is added at a temperature below 15° (Note 9). The aqueous layer is separated and the reaction flask and organic layer are washed with a mixture of 50 ml. of water and 75 ml. of diethyl ether. The aqueous layers are combined, back-extracted with 50 ml. of ether, and acidified to Congo red with 6 N hydrochloric acid. The product is extracted with two 75-ml. portions of ether, washed with 50 ml. of saturated sodium chloride, and dried over anhydrous magnesium sulfate, and the solvent is evaporated. The remaining traces of solvent may be removed in a rotary evaporator, yielding 12–12.8 g. (83–90%) of crude 2,2-dimethyl-4-phenylbutyric acid, m.p. 89–94°, which is suitable for many purposes. Recrystallization from 75 ml. of hexane at room temperature followed by refrigeration yields 8.4–9.7 g. (58–67%) of product as colorless needles, m.p. 98–99.5° (Note 10). Recrystallization of the filtrate residue from 20 ml. of hexane yields a small second crop, 1.2–1.6 g. (8–11%), m.p. 95–97°. The combined yield totals 10–11 g. (70–76%) (Note 11).

2. Notes

1. Diisopropylamine supplied by Matheson, Coleman and Bell was distilled from calcium hydride, b.p. 83–85°.

2. Sodium hydride supplied by Ventron Corp., is satisfactory. It is unnecessary to remove the mineral oil.

3. Tetrahydrofuran was obtained in drum quantities from E. I. duPont de Nemours Co., and transferred under nitrogen pressure to 1-gallon containers for stock. This material could be used without special treatment. The quality of the tetrahydrofuran should be determined [*Org. Synth.*, **Coll. Vol. 5,** 976 (1973)] if there is no assurance of the absence of gross contamination. A small excess of sodium hydride was used in the reaction to remove traces of moisture which may have been introduced during measurement.

4. Isobutyric acid supplied by Matheson, Coleman and Bell is satisfactory.

5. A brisk nitrogen flow is required to exclude air when cooling.

6. *n*-Butyllithium in 1-mole serum cap bottles from Foote Mineral Co., was used. The less volatile hexane or heptane solutions were preferred.

7. The *n*-butyllithium solution was forced into a 100-ml. syringe through a 1½-in., 19 gauge hypodermic needle by slightly pressurizing (½ to 1½ p.s.i.) the storage bottle with nitrogen. Nitrogen was admitted through a second, 2-in., 20 gauge needle inserted at an upward angle through the stopper into the void space of the inclined bottle. The pressure in the bottle should be released and the bottle should be returned to an upright position before the syringe is withdrawn. This operation should be conducted behind a safety shield.

8. (2-Bromoethyl)benzene was used as supplied by Matheson, Coleman and Bell.

9. At the beginning, water should be added cautiously since a small quantity of unreacted sodium hydride is present.

10. The literature[2-7] reports m.p. 97–98°.

11. NMR analysis of the filtrate residue indicates that it is *ca.* 90% product. The submitters carried out the procedure on four times this scale, with mechanical stirring.

3. Discussion

2,2-Dimethyl-4-phenylbutyric acid has been prepared by Clemmensen reduction[2-6] or by hydriodic acid-phosphorus reduction[7] of 3-benzoyl-2,2-dimethylpropionic acid, and by catalytic reduction[6] of 2,2-dimethyl-4-phenyl-3-butenoic acid.

The preparation of 2,2-dimethyl-4-phenylbutyric acid is a specific example of a generally applicable procedure for alkylating dialkylacetic acids. The present procedure represents an improvement on one described earlier[8] in that one equivalent of *n*-butyllithium is required. Sufficient experience has been accumulated by the submitter over several years to recommend the present method as a possible alternative to the multistep Haller-Bauer sequence.[9] The procedure offers the obvious advantages of being short, avoiding use of blocking groups for the carboxyl group which must be removed later, and affording ease of workup while still providing preparative yields of product. The scale of the reaction can be increased to 1–2 moles with only a moderate increase in the size of the equipment.

The same procedure has been used to monoalkylate alkylacetic acids[10] and may be used to selectively monoalkylate methylated benzoic acids.[11] The metalated intermediates generated from alkylacetic acids are generally less soluble in the reaction medium specified, and heterogeneous mixtures result. The physical state of the reaction mixture has no apparent effect on the success of the subsequent alkylation[10] so long as metalation is complete. Homogeneous solutions may be obtained at the expense of operational convenience by suitable changes in the cation used,[11,12] or by use of hexamethylphosphoric triamide as co-solvent.[12,13] For warning concerning hexamethylphosphoric triamide, see *Science*, **190**, 422 (1975).

Metalated carboxylic acids have been employed as intermediates in a broad range of synthetic applications. The early literature has been summarized,[14] contrasting prior experience with the present method for generating carboxylic acid dianions. Likewise, more recent applications have been reviewed.[15,16]

1. Department of Chemistry, Division of Medical and Scientific Affairs, Parke, Davis and Company, Ann Arbor, Michigan 48106. [Present address: Pharmaceutical Research Division, Warner-Lambert Company, Ann Arbor, Michigan 48105.]
2. G. R. Clemo and H. G. Dickenson, *J. Chem. Soc.*, 255 (1937).
3. R. D. Desai and M. A. Wali, *Proc. Indian Acad. Sci.*, **6A**, 135 (1937) [*Chem. Abstr.*, **32**, 509⁵ (1938)].
4. S. C. Sengupta, *J. Prakt. Chem.*, **151**, 82 (1938) [*Chem. Abstr.*, **32**, 8402¹ (1938)].
5. E. Rothstein and R. W. Saville, *J. Chem. Soc.*, 1946 (1949).
6. E. N. Marvell and A. O. Geiszler, *J. Am. Chem. Soc.*, **74**, 1259 (1952).
7. E. Rothstein and M. A. Saboor, *J. Chem. Soc.*, 425 (1943).
8. P. L. Creger, *J. Am. Chem. Soc.*, **89**, 2500 (1967).
9. K. E. Hamlin and A. W. Weston, *Org. React.*, **9**, 1 (1957).
10. P. L. Creger, *J. Am. Chem. Soc.*, **92**, 1397 (1970).
11. P. L. Creger, *J. Am. Chem. Soc.*, **92**, 1396 (1970).
12. P. L. Creger, U.S. Pat. 3,413,288 (1968).
13. P. E. Pfeffer and L. S. Silbert, *J. Org. Chem.*, **35**, 262 (1970).
14. P. L. Creger, *J. Org. Chem.*, **37**, 1907 (1972).
15. P. L. Creger, *Ann. Rep. Med. Chem.*, **12**, 278 (1977).
16. A. P. Krapcho and E. A. Dundulis, *J. Org. Chem.*, **45**, 3236 (1980).

CYCLOPENTENONES FROM α,α'-DIBROMOKETONES AND ENAMINES: 2,5-DIMETHYL-3-PHENYL-2-CYCLOPENTEN-1-ONE

(2-Cyclopenten-1-one, 2,5-dimethyl-3-phenyl-)

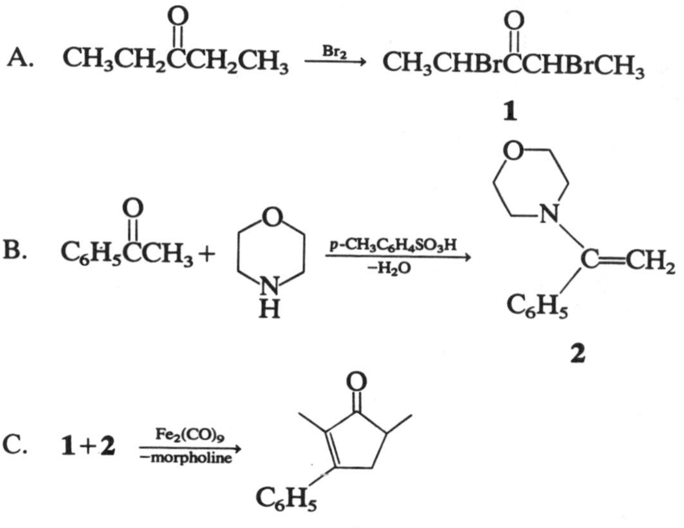

A. $CH_3CH_2\overset{O}{\overset{\|}{C}}CH_2CH_3 \xrightarrow{\text{Br}_2} CH_3CHBr\overset{O}{\overset{\|}{C}}CHBrCH_3$

1

B. $C_6H_5\overset{O}{\overset{\|}{C}}CH_3 +$ [morpholine] $\xrightarrow[-H_2O]{p\text{-CH}_3\text{C}_6\text{H}_4\text{SO}_3\text{H}}$ [morpholino-C(C_6H_5)=CH_2]

2

C. **1+2** $\xrightarrow[-\text{morpholine}]{\text{Fe}_2(\text{CO})_9}$ [structure **3**]

3

Submitted by R. Noyori,[1] K. Yokoyama, and Y. Hayakawa
Checked by Michael J. Haire and William A. Sheppard

1. Procedure

Caution! Benzene has been identified as a carcinogen; OSHA has issued emergency standards on its use. All procedures involving benzene should be carried out in a well-ventilated hood, and glove protection is required. The reaction in Part C should be carried out in a well-ventilated hood because iron carbonyls are highly toxic.

A. *2,4-Dibromo-3-pentanone* (**1**). A 300-ml., three-necked, round-bottomed flask equipped with a magnetic stirrer, a thermometer, and a 125-ml., pressure-equalizing dropping funnel, connected to a trap for absorbing hydrogen bromide evolved during the reaction (Note 1), is charged with 43.0 g. (0.500 mole) of diethyl ketone and 100 ml. of 47% hydrobromic acid. The dropping funnel is charged with 160 g. (1.00 mole) of bromine, which is added with stirring over a 1-hour period, causing the temperature of the reaction mixture to increase to 50–60°. After addition is complete, stirring is continued for an additional 10 minutes, before 100 ml. of water is added to the reaction mixture. The separated heavy organic layer is washed with 30 ml. of saturated aqueous sodium bisulfite. The brownish organic solution is dried over calcium chloride and distilled under reduced pressure through a 15-cm. vacuum-jacketed Vigreux column, yielding 85.2–92.5 g. (70–76%) of 2,4-dibromo-3-pentanone (1) (Note 2), b.p. 51–57° (3 mm.), as a slightly yellow liquid.

B. *α-Morpholinostyrene* (2).   A mixture of 75.0 g. (0.625 mole) of acetophenone, 81.0 g. (0.930 mole) of morpholine (Note 3), 200 mg. of *p*-toluenesulfonic acid, and 250 ml. of benzene is placed in a 500-ml., round-bottomed flask equipped with a water separator (Note 4), under a reflux condenser protected by a calcium chloride drying tube. Separation of water begins with reflux and is complete after 180 hours. After the mixture is cooled to room temperature, 200 mg. of sodium ethoxide is added to remove *p*-toluenesulfonic acid, and the mixture is concentrated with a vacuum rotary evaporator (50°, 80–100 mm.). The crude oily product is distilled under reduced pressure through a 15-cm. vacuum-jacketed Vigreux column. After 40–50 ml. of a mixture of morpholine and acetophenone is recovered as a forerun at 40–90° (20 mm.), 67.5–75.4 g. (57–64%) of α-morpholinostyrene is collected as a pale yellow liquid, b.p. 85–90° (0.03 mm.) (Note 5).

C. *2,5-Dimethyl-3-phenyl-2-cyclopenten-1-one* (3).   A 1-l., three-necked, round-bottomed flask equipped with a sealed mechanical stirrer, a rubber septum, and a bubbler filled with liquid paraffin is charged with 40.0 g. (0.110 mole) of diiron nonacarbonyl (Note 6) and 250 ml. of dry benzene (Note 7). The system is flushed with nitrogen, and 56.8 g. (0.300 mole) of α-morpholinostyrene (Note 8) and 24.4 g. (0.100 mole) of 2,4-dibromo-3-pentanone (1) are injected by syringe through the rubber septum. The flask is immersed in a 32° bath, and the reaction mixture is stirred under a nitrogen atmosphere (Note 9). After 20 hours 230 g. of silica gel (Note 10) and 100 ml. of benzene are added. The resulting slurry is stirred at 32° for an additional 2.5 hours (Note 11). The whole mixture is poured onto a silica gel pad (Notes 10 and 12) with the aid of 200 ml. of diethyl ether, and the pad is washed with 1 l. of ether (Notes 13 and 14). The combined organic solutions are concentrated on a vacuum rotary evaporator (Note 13), giving 35–45 g. of the desired cyclopentenone 3, a brown oil contaminated by acetophenone formed by decomposition of the excess enamine. The oil is distilled under reduced pressure with a short-path distillation apparatus (Note 15). The forerun of 20–25 g., b.p. 35–50° (0.1 mm.), is recovered acetophenone. At 100–105° (0.02 mm.), 12.0–12.4 g. [64–67% yield (Note 14)] of cyclopentenone 3 is obtained as a pale yellow oil, which crystallizes on cooling with ice water. Recrystallization from hexane gives an analytical sample as colorless needles, m.p. 57–59° (Note 16).

2. Notes

1. See Figure 7 in *Org. Synth.*, **Coll. Vol. 1,** 95 (1941).
2. The submitter reported a yield of 116 g. (95%). Care should be taken to prevent the dibromoketone from coming into contact with the skin; allergic reactions have been observed in several cases. Also, the checkers found the crude product to have lachrymatory properties. Immediate use after distillation is recommended if high yield is to be obtained in the next step.

3. An excess of morpholine is required because a considerable amount is lost with the water that separates during the reaction.

4. See Figure 12 in *Org. Synth.*, **Coll. Vol. 3,** 381 (1955).

5. The distilled product is 97% pure and contaminated with 3% acetophenone (NMR analysis). Since the enamine is easily hydrolyzed and deteriorates on long standing, use of a freshly-distilled material is recommended. The checkers found that α-morpholinostyrene contaminated with 20% acetophenone could be used for the next step without any significant reduction in yield.

6. Diiron nonacarbonyl is available from Alpha Inorganics, Inc., or Strem Chemicals, Inc. The submitters made the complex through photolysis of iron pentacarbonyl by the method of King.[2] Procedures for preparation are also given by Braye and Hübel,[3] who use the name diiron enneacarbonyl.

7. The submitters used benzene distilled from lithium aluminum hydride, but the checkers used ACS-grade benzene as well as benzene distilled from lithium aluminum hydride, with no significant change in yield.

8. The submitters obtained a markedly lower yield of product when an excess of the enamine was not used.

9. Evolution of carbon monoxide begins a few minutes after mixing the starting materials, continuing for *ca.* 3 hours. Cessation of gas evolution does not necessarily mean completion of the cyclocoupling reaction.

10. The submitters used Merck Kieselgel 60 (70–230 mesh ASTM).

11. This procedure is for elimination of morpholine from the labile primary product, 2,5-dimethyl-3-morpholino-3-phenylcyclopentanone.

12. The 150 g. of silica gel is packed in a 13 (diameter) × 12-cm. (length) glass filter.

13. The filter cake and the distillate must be treated with nitric acid to decompose the contaminates of iron carbonyl complexes. This treatment should be done very carefully in a well-ventilated hood, because carbon monoxide is evolved vigorously.

14. The submitters reported a yield of 14.8–15.6 g. (80–84%) based on the starting dibromide. The submitters report that the cyclopentenone product seems to absorb on silica gel, and 1 l. of ether is required to attain complete extraction. A smaller quantity of ether wash was used by the checkers. ^{1}H NMR analysis of the crude mixture before distillation indicated the formation of the cyclopentenone **3** in 83–87% yield.

15. See Figure 2 in *Org. Synth.*, **Coll. Vol. 6,** 436 (1988).

16. The spectral properties of pure product are as follows: IR (CCl_4) cm^{-1}: 1696 (conjugated C=O) and 1626 (conjugated C=C); UV (C_2H_5OH) nm. max. (ϵ): 220 (5220) and 279 (11,200); ^{1}H NMR (CCl_4), δ (multiplicity, coupling constant J in Hz., number of protons, assignment): 1.22 (d, $J = 7.0$, 3H, CHCH_3), 1.91 (t, $J = 2.0$, 3H, vinyl CH_3), 2.1–2.7 (m, 2H, CHCH$_3$ and a methylene H *cis* to CH$_3$), 3.14 (d of d of q, $J = 18$, 7.5, and 2.0, 1H, methylene H *trans* to CH$_3$), and 7.38 (m, 5H, C$_6H_5$).

3. Discussion

The starting materials, 2,4-dibromo-3-pentanone[4] and α-morpholinostyrene,[5] have been prepared in satisfactory yields by modifying known procedures. The procedure for the 3 + 2 → 5 cyclocoupling reaction is essentially that described originally by the submitters.[6] The main advantages of this procedure are the directness, the availability of

TABLE I
Iron Carbonyl-Promoted Cyclopentenone Synthesis[a]

Dibromide	Enamine	Product[b]	Yield (%)[c]
$CH_3CHBrCCHBrCH_3$ (C=O)	morpholine enamine	trimethyl-cyclopentenone	79
$(CH_3)_2CHCHBrCCHBrCH(CH_3)_2$ (C=O)	C_6H_5 morpholine enamine	diisopropyl C_6H_5 cyclopentenone	72
$CH_3CHBrCCHBrCH_3$ (C=O)	cyclopentenyl morpholine enamine	bicyclic dimethyl cyclopentenone	74
$CH_3CHBrCCHBrCH_3$ (C=O)	cyclohexenyl morpholine enamine	bicyclic dimethyl cyclopentenone	100

TABLE I (continued)
Iron Carbonyl-Promoted Cyclopentenone Synthesis[a]

Dibromide	Enamine	Product[b]	Yield (%)[c]
$(CH_3)_2CHCHCHBrCCHBrCH(CH_3)_2$ (with C=O)			73
$CH_3CHBrCCHBrCH_3$ (with C=O)			100
$CH_3CHBrCCHBrCH_3$ (with C=O)			90

[a]Reference 6.
[b]A mixture of epimers, when possible, is obtained.
[c]Isolated yield based on starting dibromide.

524

starting materials, and the wide generality for preparation of 2,5-dialkyl-2-cyclopenten-1-ones, as shown in Table I. Enamines derived from aldehydes, open-chain ketones, and cyclic ketones can be employed. The method has been extended to the synthesis of spiro[4.n]alkenones and certain azulene derivatives.[6] A reaction mechanism for the cyclocoupling reaction has been advanced.[7] The reactive oxyallyl intermediates generated from dibromoketones and iron carbonyls can be trapped efficiently by enamines,[6] aromatic olefins,[8] 1,3-dienes,[9,10] furans,[10,11] carboxamides,[12] and alkyl 1H-pyrrole-1-carboxylates.[13] Intramolecular trapping of the reactive species by olefin or furan has also been achieved.[14] At present, dibromides of methyl ketones cannot be used as starting materials except in intramolecular cyclocoupling reactions. However, polybromoketones, including $\alpha,\alpha,\alpha',\alpha'$-tetrabromoacetone, serve as a precursors of the reactive species in certain cases, and the coupling reactions have been applied to various naturally occurring products.[14-19]

1. Department of Chemistry, Nagoya University, Chikusa, Nagoya 464, Japan.
2. R. B. King, *Organomet. Synth.*, **1**, 93 (1965).
3. E. H. Braye and W. Hubel, *Inorg. Synth.*, **8**, 178 (1966).
4. C. Rappe, *Acta Chem. Scand.*, **16**, 2467 (1962).
5. S. Hünig, K. Hübner, and E. Benzing, *Chem. Ber.*, **95**, 926 (1962).
6. R. Noyori, K. Yokoyama, S. Makino, and Y. Hayakawa, *J. Am. Chem. Soc.*, **94**, 1772 (1972); Y. Hayakawa, K. Yokoyama, and R. Noyori, *J. Am. Chem. Soc.*, **100**, 1799 (1978).
7. R. Noyori, Y. Hayakawa, M. Funakura, H. Takaya, S. Murai, R. Kobayashi, and S. Tsutsumi, *J. Am. Chem. Soc.*, **94**, 7202 (1972); R. Noyori, Y. Hayakawa, H. Takaya, S. Murai, R. Kobayashi, and N. Sonoda, *J. Am. Chem. Soc.*, **100**, 1759 (1978).
8. R. Noyori, K. Yokoyama, and Y. Hayakawa, *J. Am. Chem. Soc.*, **95**, 2722 (1973); Y. Hayakawa, K. Yokoyama, and R. Noyori, *J. Am. Chem. Soc.*, **100**, 1791 (1978).
9. R. Noyori, S. Makino, and H. Takaya, *J. Am. Chem. Soc.*, **93**, 1272 (1971).
10. H. Takaya, S. Makino, Y. Hayakawa, and R. Noyori, *J. Am. Chem. Soc.*, **100**, 1765 (1978).
11. R. Noyori, Y. Baba, S. Makino, and H. Takaya, *Tetrahedron Lett.*, 1741 (1973).
12. R. Noyori, Y. Hayakawa, S. Makino, N. Hayakawa, and H. Takaya, *J. Am. Chem. Soc.*, **95**, 4103 (1973); Y. Hayakawa, H. Takaya, S. Makino, N. Hayakawa, and R. Noyori, *Bull. Chem. Soc. Jpn.*, **50**, 1990 (1977).
13. R. Noyori, Y. Baba, and Y. Hayakawa, *Tetrahedron Lett.*, 1049 (1974).
14. R. Noyori, M. Nishizawa, F. Shimizu, Y. Hayakawa, K. Maruoka, S. Hashimoto, H. Yamamoto, and H. Nozaki, *J. Am. Chem. Soc.*, **101**, 220 (1979).
15. R. Noyori, Y. Baba, and Y. Hayakawa, *J. Am. Chem. Soc.*, **96**, 3336 (1974); Y. Hayakawa, Y. Baba, S. Makino, and R. Noyori, *J. Am. Chem. Soc.*, **100**, 1786 (1978).
16. R. Noyori, S. Makino, T. Okita, and Y. Hayakawa, *J. Org. Chem.*, **40**, 806 (1975); Y. Hayakawa, M. Sakai, and R. Noyori, *Chem. Lett.*, 509 (1975); H. Takaya, S. Makino, Y. Hayakawa, and R. Noyori, *J. Am. Chem. Soc.*, **100**, 1778 (1978).
17. R. Noyori, T. Souchi, and Y. Hayakawa, *J. Org. Chem.*, **40**, 2681 (1975).
18. Y. Hayakawa, F. Shimizu, and R. Noyori, *Tetrahedron Lett.*, 993 (1978).
19. R. Noyori, T. Sato, and Y. Hayakawa, *J. Am. Chem. Soc.*, **100**, 2561 (1978); T. Sato, R. Ito, Y. Hayakawa, and R. Noyori, *Tetrahedron Lett.*, 1829 (1978).

ALKYLATIONS OF ALDEHYDES *via* REACTION OF THE MAGNESIOENAMINE SALT OF AN ALDEHYDE: 2,2-DIMETHYL-3-PHENYLPROPIONALDEHYDE

(Benzenepropanal, α,α-dimethyl-)

$$(CH_3)_2CHCHO + (CH_3)_3CNH_2 \xrightarrow[\text{1 hour}]{K_2CO_3} (CH_3)_2CHCH{=}NC(CH_3)_3$$

$$(CH_3)_2CHCH{=}NC(CH_3)_3 + C_2H_5MgBr \xrightarrow[\text{12–14 hours}]{\text{tetrahydrofuran}}$$

$$\left[\begin{array}{c} (CH_3)_2C{=}CH{-}N{-}C(CH_3)_3 \\ | \\ MgBr \end{array} \right] + C_2H_6$$

$$\left[\begin{array}{c} (CH_3)_2C{=}CHN{-}C(CH_3)_3 \\ | \\ MgBr \end{array} \right] + C_6H_5CH_2Cl \xrightarrow[\text{2. 10\% aq. HCl}]{\text{1. reflux, 20 hours}}$$

$$C_6H_5CH_2C(CH_3)_2CHO$$

Submitted by G. Stork[1] and S. R. Dowd[2]
Checked by D. R. Williams and G. Büchi

1. Procedure

A. N-(2-*Methylpropylidene*)-tert-*butylamine*. A 100-ml., three-necked, round-bottom flask equipped with a condenser, a nitrogen inlet tube, a 50-ml. dropping funnel, and a magnetic stirring bar is evacuated through a mercury bubbler, flamed dry, and flushed with nitrogen three times. The flask is charged with 36.0 g. (0.500 mole) of *tert*-butylamine (Note 1), and 36.5 g. (0.501 mole) of isobutyraldehyde (Note 1) is placed in the dropping funnel. Half of the isobutyraldehyde is added slowly through the dropping funnel before the remaining half-volume is added rapidly. The milky solution is allowed to stand at room temperature for 1 hour; the water layer is then pipeted out, and excess anhydrous potassium carbonate is added. Filtration and distillation of this reaction mixture gives 32.0 g. (50%) of N-(2-methylpropylidene)-*tert*-butylamine, b.p. 50° (75 mm.) (Note 2).

B. 2,2-*Dimethyl-3-phenylpropionaldehyde*. A 100-ml., three-necked, round-bottom flask equipped with an ether condenser, a nitrogen inlet tube, a 50-ml. Herschberg dropping funnel, and a magnetic stirring bar is evacuated through a mercury bubbler, flamed dry, and flushed with nitrogen three times. The system is left under a slight positive pressure of nitrogen, and all the reactants are added under a stream of nitrogen. A solution of 0.05 mole of ethylmagnesium bromide in 37 ml. of tetrahydrofuran (Note 3) is placed in the flask. A solution of 6.35 g. (0.0567 mole) of N-(2-methylpropylidene)-*tert*-butylamine (Note 4) in 5 ml. of tetrahydrofuran (Note 5) is then added from the dropping funnel. The resulting mixture is refluxed for 12–14 hours until 1 mole-equivalent of gas has evolved (Note 6). The reaction mixture is cooled to room temperature; 6.30 g.

(0.0498 mole) of benzyl chloride (Note 1) is added from the dropping funnel; and the solution is refluxed for 20 hours, at which time the pH is 9–10 (pHydrion paper). To the cooled solution, which contains a large amount of solid, is added 20–30 ml. of 10% hydrochloric acid. The clear, yellow-brown solution is then refluxed for 2 hours, cooled, saturated with solid sodium chloride, and extracted five times with diethyl ether. The organic extracts are washed once with 25 ml. of 5% hydrochloric acid, then repeatedly with brine until the washings are neutral. The organic layer is dried over anhydrous magnesium sulfate and filtered, and the solvent is removed at atmospheric pressure through a 12-in. Vigreux column fitted with a partial take-off head. Distillation of the residue (Note 7) through a 20-in. vacuum-jacketed fractionating column affords 5.1– 5.4 g. (63–66%) of 2,2-dimethyl-3-phenylpropionaldehyde, b.p. 70–73° (1.5 mm.) (Note 8).

2. Notes

1. These reagents were obtained from Eastman Organic Chemicals, but not distilled prior to use.

2. The checkers found that the yield could be improved (37–38 g., 58–60%) if the reaction mixture was allowed to remain over anhydrous potassium carbonate for 8–12 hours.

3. Ethylmagnesium bromide was prepared in dry tetrahydrofuran and stored no longer than 1 week in a 250-ml. tube fitted with a 3-way vacuum stopcock and a dropping buret. The solution is decanted into the buret, and the correct volume is transferred to the reaction flask with positive nitrogen pressure. The tetrahydrofuran was purified by distillation from lithium aluminum hydride. See *Org. Synth.*, **Coll. Vol. 5,** 976 (1973), for warning regarding the purification of tetrahydrofuran.

4. The aldimine is freshly distilled [b.p. 50° (75 mm.)] prior to use.

5. A vigorous reaction may result. At this stage of the reaction, control is maintained with an ice–water bath.

6. The volume of gas evolved is estimated with an inverted cylinder filled with water attached by rubber tubing to the outlet of the mercury bubbler.

7. GC analysis at 155° on a 5 ft. 5% SE-30 column shows only 2,2-dimethyl-3-phenylpropionaldehyde and solvent.

8. The IR spectrum (CHCl$_3$) showed absorption at 2705, 1725, and 1605 cm^{-1}. The 2,4-dinitrophenylhydrazone, recrystallized from ethanol–ethyl acetate as long, yellow-orange needles, melted at 150–152° (reported[3] 154–155°).

3. Discussion

This procedure illustrates the mono-alkylation of α-substituted aldehydes by the metalloenamine method.[4] The preparation of the aldimine has been adapted from the procedure of Tiollais[5] and is useful in the preparation of aldimines from low-boiling components. The readily prepared aldimine is treated with an alkyl Grignard reagent generating the magnesioenamine halide salt, which can be alkylated with a variety of alkylating agents at the α-position. The yields are high, monoalkylation is the exclusive reaction, and there is no rearrangement when using an allylic halide. The general method

TABLE I
Reaction of Various Aldimine and Ketimine Magnesium Bromide Salts with Alkylating Agents in Tetrahydrofuran

Imine	Halide	Yield, %
N-(2-methylpropylidene)-tert-butylamine	n-butyl iodide	65
N-(heptylidene)-tert-butylamine	n-butyl iodide	47
N-(propylidene)-tert-butylamine	n-butyl bromide	60
N-(cyclohexylidene)cyclohexylamine	n-butyl iodide	78
	2-iodopropane	61
	benzyl chloride	60
N-(cyclopentylidene)cyclohexylamine	n-butyl iodide	72
N-(cycloheptylidene)cyclohexylamine	n-butyl iodide	75

is applicable to the alkylation of ketones *via* the magnesium bromide salts of the corresponding ketimines (Table I).

In a variation of this method, metalloenamines have been generated from the aldimine and lithium diisopropylamide in ether and alkylated in a limited number of cases.[6] A further variation, using lithium dialkylamides in hexamethylphosphoric triamide, has been shown by Cuvigny and Normant[7] to give good yields of alkylated aldehydes. However, secondary alkyl halides fail to react, giving the dehydrohalogenation product instead. Another approach, that of Meyers and co-workers,[8] involves the alkylation of the lithio salt of 2-methyldihydro-1,3-oxazines, but suffers from the necessity of carrying out several low-temperature steps and a pH-controlled borohydride reduction.

1. Department of Chemistry, Columbia University, New York, New York 10027.
2. Present address: Department of Biological Sciences, Carnegie-Mellon University, Pittsburgh, Pennsylvania 15213.
3. G. Opitz, H. Heilmann, H. Mildenberger, and H. Suhr, *Justus Liebigs Ann. Chem.*, **649**, 36 (1961).
4. G. Stork and S. R. Dowd, *J. Am. Chem. Soc.*, **85**, 2178 (1963).
5. R. Tiollais, *Bull. Soc. Chim. Fr.*, **14**, 708 (1947).
6. G. Wittig, H.-D. Frommeld, and P. Suchanek, *Angew. Chem. Int. Ed. Engl.*, **2**, 683 (1963); G. Wittig and H.-D. Frommeld, *Chem. Ber.*, **97**, 3548 (1964).
7. Th. Cuvigny and H. Normant, *Bull. Soc. Chim. Fr.*, 3976 (1970).
8. A. I. Meyers, A. Nabeya, H. W. Adickes, and I. R. Politzer, *J. Am. Chem. Soc.*, **91**, 763 (1969).

3,5-DINITROBENZALDEHYDE

(Benzaldehyde, 3,5-dinitro-)

Submitted by J. E. SIGGINS,[1] A. A. LARSEN,[2]
J. H. ACKERMAN,[1] and C. D. CARABATEAS[1]
Checked by D. J. BICHAN and PETER YATES

1. Procedure

Caution! Benzene has been identified as a carcinogen; OSHA has issued emergency standards on its use. All procedures involving benzene should be carried out in a well-ventilated hood, and glove protection is required.

A 3-l., three-necked, round-bottomed flask is equipped with an efficient stirrer, a pressure-equalizing dropping funnel with a nitrogen inlet, and a Y-tube fitted with a low temperature thermometer and a nitrogen outlet. The outlet is vented through a bubbler tube, maintaining a slight positive pressure. The flask and dropping funnel are flamed in a stream of dry nitrogen (Note 1). To the flask is added 115.0 g. (0.4989 mole) of 3,5-dinitrobenzoyl chloride (Note 2) followed by 500 ml. of dry diglyme (Note 3). The solution is stirred vigorously, and the flask is immersed in a cooling bath at −78° (Note 4). A diglyme solution of lithium aluminum tri-*tert*-butoxyhydride (Note 5) is prepared in the following manner. Dry diglyme (450 ml.) is added with vigorous stirring to an Erlenmeyer flask containing 140.0 g. (0.5512 mole) of lithium aluminum tri-*tert*-butoxyhydride. After standing overnight, the resulting suspension is filtered under a blanket of dry nitrogen through a thick layer of Celite packed tightly on a Büchner funnel (Note 6). The flask containing the filtrate is kept stoppered until the reducing agent is transferred to the dropping funnel. Dropwise addition of this solution is started when the contents of the reaction flask reach −72°. There is a color change and a temperature rise of a few degrees. The rate of addition is adjusted, maintaining the temperature of the mixture between −78° and −68° (Note 7). After addition is complete the mixture is stirred at −78° for 30 minutes longer.

The cold reaction mixture is poured slowly with stirring into a 3-l. beaker containing 150 ml. of concentrated hydrochloric acid, 300 ml. of saturated aqueous sodium chloride, and 150 g. of ice. A white precipitate starts to separate (Note 8). An additional 150 ml. of saturated aqueous sodium chloride is added to the beaker and, after a minute, an upper layer begins to appear. The contents are transferred to a 2-l. separatory funnel and allowed to stand for 15 to 30 minutes while an upper brown layer separates. The upper layer is reserved while the lower layer is extracted with several portions of benzene, totalling 900 ml. The upper layer and the benzene extracts are combined and washed with seven 1-l. portions of water containing 10 ml. of concentrated hydrochloric acid. The benzene layer is washed successively with 100-ml. portions of aqueous 2% sodium hydrogen carbonate until the washings are basic, dried over 100 g. of anhydrous sodium sulfate, treated with 1 g. of charcoal, and filtered. The filtrate is concentrated at reduced pressure, yielding 59–62 g. (60–63%) of crude 3,5-dinitrobenzaldehyde, as a tan solid, m.p. 76–80°. Trituration in an ice bath with cold dry diethyl ether (ca. 0.3 ml./g.) gives a spongy solid, m.p. 85–87° (lit.[3] 85°), with losses of 5–10%, sufficiently pure for most uses. Further purification may be effected by recrystallization from toluene-hexane.

2. Notes

1. These operations are best done the day before the experiment is performed.

2. Since commercial 3,5-dinitrobenzoyl chloride is contaminated with 3,5-dinitrobenzoic acid, it was treated with thionyl chloride in boiling benzene under dry nitrogen. The product obtained after evaporation under vacuum melted at 68–70° and was stored over phosphorus pentoxide and potassium hydroxide in a vacuum desiccator.

3. "Diglyme" is the dimethyl ether of diethylene glycol. Commercial diglyme (Ansul Ether 141, Ansul Chemical Company, Marinette, Wisconsin) was used after drying over lithium aluminum hydride followed by distillation under reduced pressure; b.p. 59–61° (15 mm.).

4. An insulated bucket such as the "Nicer" available from B.F. Goodrich Company contains the mixture of dry ice and 2-propanol. Acetone foams excessively and has a high vapor pressure.

5. This hydride is obtained from Ventron, Inc., Beverly, Massachusetts.

6. Suspended particles will plug the dropping funnel in the subsequent operation. Two funnels may be used if the filtration is too slow.

7. At elevated temperatures reduction to the alcohol takes place. The addition time varies from 75 to 100 minutes.

8. The supernatant liquid is a brilliant yellow. A troublesome blue color may appear in an occasional run.

3. Discussion

3,5-Dinitrobenzaldehyde has been made previously by reducing 4-bromo-3,5-dinitrobenzaldehyde with copper(I) hydride.[3]

The method described is that of Brown.[4] Lithium aluminum tri-tert-butoxyhydride reduction of acid chlorides to aldehydes is a synthetic method of wide utility, and the yields of 20 aromatic and 10 aliphatic aldehydes so prepared have been tabulated.[5] This reagent is also used widely to reduce steroid aldehydes and ketones to alcohols, frequently at 0°.[6] Having only one active hydrogen, it has been used for the partial reduction of diketones to hydroxy ketones.[7] With proper temperature control, it does not affect oxido,[8] ester,[6,9] acetal,[10] nitrile, or nitro groups, or lactone rings,[8] all of which react with lithium aluminum hydride. In contrast to some other complex hydrides, this reagent may reduce a ketone stereoselectively, resulting in a high relative yield of the more stable epimeric alcohol[11] and an improved absolute yield.[7,12]

1. Sterling-Winthrop Research Institute, Rensselaer, New York.
2. Mead Johnson Research Center, Evansville, Indiana 47721.
3. H. H. Hodgson and E. W. Smith, *J. Chem. Soc.*, 315 (1933).
4. H. C. Brown and R. F. McFarlin, *J. Am. Chem. Soc.*, **80**, 5372 (1958).
5. H. C. Brown and B. C. Subba Rao, *J. Am. Chem. Soc.*, **80**, 5377 (1958).
6. K. Heusler, P. Wieland, and C. Meystre, *Org. Synth.*, **Coll. Vol. 5**, 692 (1973).
7. R. E. Ireland and J. A. Marshall, *J. Org. Chem.*, **27**, 1620 (1962).
8. C. Tamm, *Helv. Chim. Acta*, **43**, 338 (1960).
9. A Bowers, E. Denot, L. C. Ibáñez, M. E. Cabezas, and H. J. Ringold, *J. Org. Chem.*, **27**, 1862 (1962).
10. J. A. Zderic and J. Iriarte, *J. Org. Chem.*, **27**, 1756 (1962).
11. J. Fajkoš, *Collect. Czech. Chem. Commun.*, **24**, 2284 (1959).
12. O. H. Wheeler and J. L. Mateos, *Chem. Ind. (London)*, 395 (1957).

2,3-DIPHENYL-1,3-BUTADIENE

(1,3-Butadiene, 2,3-diphenyl)

$$2\ CH_3SOCH_3 + 2\ NaH \longrightarrow 2\ CH_3SOCH_2^{\ominus}Na^{\oplus} + 2\ H_2$$

$$C_6H_5C{\equiv}CC_6H_5 \xrightarrow{CH_3SOCH_2^{\ominus}Na^{\oplus}} C_6H_5C{-}C{-}C_6H_5$$

$$\underset{CH_2}{\overset{\|}{}}\ \underset{CH_2}{\overset{\|}{}}$$

Submitted by Issei Iwai and Junya Ide[1]
Checked by James B. Sieja and Richard E. Benson

1. Procedure

A 300-ml., three-necked, round-bottomed flask is fitted with a sealed mechanical stirrer, a thermometer, and a reflux condenser to which is attached a T-tube connected to a source of pure nitrogen. The remaining joint of the T-tube is connected to a bubbling device so that the rate of nitrogen flow can be observed throughout the course of the reaction. The flask is flushed with nitrogen and charged with 30 ml. of anhydrous dimethyl sulfoxide (Note 1) and 2.4 g. of about 50% sodium hydride in oil (*ca*. 0.05 mole) (Note 2). Stirring is begun and the contents of the flask are heated to 75° for 30 minutes (Note 3) under a slight pressure of nitrogen. The flask is cooled in a water bath to 30°, before a solution of 4.45 g. (0.0250 mole) of diphenylacetylene in 20 ml. of anhydrous dimethyl sulfoxide is added dropwise with stirring to the dark-gray solution. During the addition, the temperature of the reaction mixture gradually rises until it approaches 40°. After the addition is completed, the reaction mixture is heated to 65° and held at this temperature for 2.5 hours. The resulting red-brown reaction mixture is cooled to room temperature and poured into 500 ml. of an ice and water mixture with stirring. After the ice has melted, the mixture is extracted with five 150 ml. portions of diethyl ether. The ether extracts are combined, washed with three 100-ml. portions of water, then dried over sodium sulfate. The ether is removed by distillation at reduced pressure, and the product (about 6.0 g.) is chromatographed on 180 g. of alumina (Note 5). Elution with 1:7 (v/v) benzene:n-hexane gives ten 50-ml. fractions, of which the first two contain nearly all the mineral oil from the sodium hydride reagent. Fractions 3 through 8 are combined, and the ether is removed by distillation, yielding 1.2–1.4 g. (22–25%) of slightly impure 2,3-diphenyl-1,3-butadiene (Note 6). Recrystallization from methanol gives 0.55–0.70 g. (10.7–13.6%) of pure 2,3-diphenyl-1,3-butadiene, m.p. 47–48° (Note 7).

The ^{1}H NMR spectrum (60 MHz., CCl$_4$) shows a complex multiplet centered at δ 7.2 attributable to the aromatic protons, and two doublets centered at δ 5.4 and 5.2, respectively, attributable to the olefinic protons.

2. Notes

1. Commercially available dimethyl sulfoxide was freshly distilled in the presence of calcium hydride, b.p. 56–57° (5 mm.).

2. Sodium hydride in oil (about 50%), available from Metal Hydrides Inc., Beverly, Massachusetts, was used.

3. The formation of dimsyl(methylsulfinyl) anion is essentially complete at this time.

4. The preparation of diphenylacetylene is described in *Org. Synth.*, **Coll. Vol. 3,** 350 (1955) and **Coll. Vol. 4,** 377 (1963). The checkers purchased it from Eastman Organic Chemicals.

5. Aluminum Oxide Woelm neutral, activity grade 1, available from M. Woelm, Eschwege, Germany, was used. The column dimensions were 2.9 cm. × 29 cm., and the alumina was packed with n-hexane.

6. This grade of 2,3-diphenyl-1,3-butadiene is satisfactory for most purposes.

7. Crude 2,3-diphenyl-1,3-butadiene is unstable. The pure product should be stored in the dark in a refrigerator. The submitters have found it to be stable for at least one year under these conditions.

3. Discussion

This method provides a simple, one-step synthesis of 2,3-diphenyl-1,3-butadiene from the readily available diphenylacetylene and illustrates an unusual, relatively uninvestigated reaction. The scope of the reaction is unknown, but it appears that the procedure could be applied to disubstituted acetylenes having aryl substituents unaffected by the strong basic conditions of the reaction.

A conventional preparation of 2,3-diphenyl-1,3-butadiene involves dehydration of *meso*-2,3-diphenyl-2,3-butanediol by acidic reagents such as acetic anhydride,[2–4] acetyl bromide,[4] sulfanilic acid,[5] and potassium hydrogen sulfate.[6] Other procedures have been summarized[7] previously.

1. Sankyo Company Limited, Central Research Laboratories, Shinagawa-ku, Tokyo, 140 Japan.
2. W. Thörner and T. Zincke, *Ber. Dtsch. Chem. Ges.*, **13,** 641 (1880).

3. J. M. Johlin, *J. Am. Chem. Soc.*, **39**, 291 (1917).
4. C. F. H. Allen, C. G. Eliot, and A. Bell, *Can. J. Res.*, **B17**, 75 (1939).
5. Y. S. Zal'kind and P. Mosunov, *Zh. Obshch. Khim.*, **10**, 517 (1940) [*Chem. Abstr.*, **34**, 7887 (1940); *Chem. Zentr.*, **111, II**, 1863 (1940).]
6. K. Alder and J. Haydn, *Justus Liebigs Ann. Chem.*, **570**, 201 (1950).
7. L. I. Smith and M. M. Falkof, *Org. Synth.*, **Coll. Vol. 3**, 350 (1955); A. C. Cope, D. S. Smith, and R. J. Cotter, *Org. Synth.*, **Coll. Vol. 4**, 377 (1963).

REAGENTS FOR SYNTHESIS OF ORGANOSELENIUM COMPOUNDS: DIPHENYL DISELENIDE AND BENZENESELENENYL CHLORIDE

Submitted by Hans J. Reich,[1] Martin L. Cohen, and Peter S. Clark
Checked by Anna Vinogradoff, Albert W. M. Lee, and Robert V. Stevens

1. Procedure

Caution! Most selenium compounds are toxic; care should be exercised to avoid contact with skin. All operations in this procedure should be conducted in a well-ventilated hood.

A. *Diphenyl diselenide.* A 2-l., three-necked, round-bottomed flask is equipped with a mechanical stirrer, a pressure-equalizing dropping funnel, and an efficient reflux condenser mounted with a combined inlet-outlet assembly connected to a nitrogen source and a bubbler. The apparatus is flamed dry while a slow stream of nitrogen is passed through the system. In the cooled flask a solution of phenylmagnesium bromide is prepared from 160 g. (1.02 mole) of bromobenzene, 24.0 g. (0.988 g.-atom) of magnesium, and 550 ml. of anhydrous diethyl ether. The dropping funnel is removed, and an Erlenmeyer flask containing 70 g. (0.89 g.-atom) of selenium (Note 1) is attached to the neck of the flask with a section of nylon tubing (Note 2). The selenium is added in portions at a rate sufficient to maintain a vigorous reflux (Note 3). The addition requires 15–30 minutes, after which the mixture is stirred and heated at reflux for another 30 minutes. The Erlenmeyer flask and nylon tubing are removed, and 3 g. (0.2 mole) of water is added to hydrolyze any excess Grignard reagent. The mixture is stirred and cooled in an ice bath while 74.3 g. (23.8 ml., 0.465 mole) of bromine is added dropwise at a rate such that the ether does not reflux (Note 4). Cooling and stirring are continued as a solution of 53.5 g. (1.00 mole) of ammonium chloride in 140 ml. of water is added slowly. The mixture is filtered by gravity into a 1-l., round-bottomed flask, and the granular precipitate is washed thoroughly with three 100-ml. portions of ether. The combined filtrates are evaporated, the remaining solid is dissolved insofar as possible in 500 ml. of hot hexane, and a small amount of insoluble material is separated by gravity filtration. The filtrate is allowed to crystallize at room temperature and then at 6°. The yellow microcrystalline solid is collected, washed with 30 ml. of pentane, and dried in the air, 89–97 g. (64–70%) of diphenyl diselenide, m.p. 60–62° (Notes 5 and 6).

B. *Benzeneselenenyl chloride.* A 1-l., three-necked, round-bottomed flask equipped with a magnetic stirring bar, a thermometer, a gas-inlet tube, and a reflux condenser is charged with 50 g. (0.16 mole) of diphenyl diselenide and 350 ml. of hexane (Note 7). The mixture is warmed to 40–50°, dissolving the solid. The resulting solution is stirred while chlorine gas is passed through the gas-inlet tube into the flask 1 cm. above the surface of the liquid, at a rate sufficient to maintain the temperature between 40° and 50°. Chlorination is continued until 11.3 g. (0.159 mole) of the gas is absorbed (Note 8). The solution is heated to reflux, filtered by gravity, and allowed to cool slowly at room temperature, then at 6° (Note 9). The mother liquor is decanted, the large, deep-orange crystals are washed with 25 ml. of pentane, and the residual solvent is evaporated by passing a slow stream of nitrogen over the crystals for 30 minutes, giving 51–54 g. (84–88%) of benzeneselenenyl chloride, m.p. 60–62° (Note 10).

2. Notes

1. The gray powdered form of selenium should be used. The submitters purchased this material from Research Organic/Inorganic Chemical Corporation, 11686 Sheldon Street, Sun Valley, California 91352, and from Var-lac-oid Chemical Company, 666 South Front Street, Elizabeth, New Jersey 07202.

2. Alternatively the flask may be equipped with a Y-shaped adapter bearing a straight condenser on its vertical branch and fitted with a nitrogen inlet in its lower branch. A slow stream of nitrogen is passed through the nitrogen inlet while the portions of selenium are added through the top of the condenser. The top of the condenser is stoppered between additions.

3. Careful exclusion of oxygen during this operation is important.

4. The addition of bromine requires about 30 minutes, after which the maintenance of a nitrogen atmosphere is no longer necessary.

5. The submitters have carried out this procedure on a 3-mole scale and obtained similar yields. A melting point of 63.5° is reported[2] for diphenyl diselenide.

6. The appearance of a red coloration in the product indicates the presence of excess selenium. Free selenium begins to separate when the product is contaminated by more than *ca*. 1% of diphenyl triselenide. Even material that crystallizes as a brick-red solid may contain only 5% excess selenium. The procedure described here gives diphenyl diselenide containing less than 0.5% free selenium. More of this contaminant will be present, however, if the formation of the Grignard reagent is incomplete, or if oxygen is allowed to enter the flask during the addition of selenium.

7. Technical grade hexane is adequate.

8. The amount of chlorine absorbed can be measured by the increased weight of the flask. The progress of the reaction can also be monitored in the following manner. A white ring of phenylselenium trichloride forms on the wall of the flask just above the surface of the liquid during the chlorination. If diphenyl diselenide remains in the solution, the solid dissolves when the flask is tipped slightly to immerse the phenylselenium trichloride below the surface of the solution. The solid no longer dissolves after the reaction is complete. The remaining ring of phenylselenium trichloride can be removed by adding another small portion of diphenyl diselenide.

9. The solution should not be cooled below 0°, since impurities, including diphenyl selenide dichloride, may also crystallize.

10. The submitters obtained 54–57 g. (88–93%) of product, m.p. 62–64° (lit.,[3] m.p. 64–65°).

3. Discussion

Diphenyl diselenide has been prepared by disproportionation of phenyl selenocyanate in the presence of potassium hydroxide[4,5] or ammonia,[4] and by air oxidation of benzeneselenol.[6,7] The preparation of benzeneselenol is described in an earlier volume in this series.[8] In the present procedure phenylselenomagnesium bromide, formed from phenylmagnesium bromide and selenium,[8] is oxidized directly to diphenyl diselenide with bromine.[9] Thus, the liberation of the malodorous and toxic hydrogen selenide and benzeneselenol is avoided. Benzeneselenenyl chloride has been prepared by thermal elimination of ethyl chloride from ethyl phenyl selenide dichloride,[3,10] by thermal elimination of chlorine from phenylselenium trichloride,[11] and by chlorinolysis of diphenyl diselenide with either sulfuryl chloride[12,13] or chlorine.[9,13]

Diphenyl diselenide and benzeneselenenyl chloride have been utilized as intermediates for the preparation of several phenyl-substituted organoselenium reagents (Table I).[14] Benzeneselenenyl bromide is available by direct brominolysis of diphenyl diselenide.[9,15] The reaction of benzeneselenenyl halides with silver acetate and silver trifluoroacetate has been employed to generate benzeneselenenyl acetate[13] and trifluoroacetate[16] *in situ*. *N,N*-Dialkyl benzeneselenenamides have been isolated from the reaction of secondary amines with benzeneselenenyl chloride or bromide.[17] Oxidation of benzeneselenenyl chloride and diphenyl diselenide with ozone affords benzeneseleninyl chloride and benzeneseleninic anhydride, respectively.[9,18] The highly nucleophilic selenenylating reagent, selenophenoxide, is liberated in solution readily by reduction of diphenyl diselenide with sodium borohydride in ethanol[19] or with other reducing agents.[14] Solutions of selenophenol are conveniently prepared by reduction with hypophosphorous acid.[20]

Diphenyl diselenide, benzeneselenenyl chloride, and organoselenium compounds derived from them have served as convenient reagents for introducing the phenylseleno group. The reaction of organolithium and Grignard reagents with diphenyl diselenide affords phenyl selenides.[21] The phenylseleno group has been introduced into the α-position of aldehydes, ketones, esters, nitriles, sulfones, and related compounds by reaction of enol derivatives, enolate anions, or carbanions with diphenyl diselenide or benzeneselenenyl chloride.[9,14,15a] The addition of benzeneselenenyl halides,[13,22] acetate,[13] and trifluoroacetate[16] to olefins affords alkyl phenyl selenides substituted in the β-position with halo, acetoxy, and trifluoroacetoxy groups.

TABLE I
Organoselenium Reagents Prepared from Diphenyl Diselenide or Benzeneselenenyl Chloride

Organoselenium Reagent	M.p. or B.p. (·)	Reference(s)
C_6H_5SeBr	62	4, 9, 15
$C_6H_5SeOCCH_3$ (with $\overset{O}{\|}$ above C)	a	13
$C_6H_5SeOCCF_3$ (with $\overset{O}{\|}$ above C)	a	16
$C_6H_5SeN\overset{CH_3}{\underset{CH_3}{}}$	39–40 (0.1 mm.)	17
$C_6H_5-\overset{O}{\underset{\|}{Se}}-Cl$	75	9, 18
$C_6H_5-\overset{O}{\underset{\|}{Se}}-O-\overset{O}{\underset{\|}{Se}}-C_6H_5$	120–122	18
C_6H_5SeNa	a	14, 19, 20
C_6H_5SeH	a	20

^{a} This reagent was generated in solution and used without isolation.

1. Department of Chemistry, University of Wisconsin, Madison, Wisconsin 53706.
2. F. Krafft and R. E. Lyons, *Ber. Dtsch. Chem. Ges.*, **27**, 1761 (1894).
3. D. G. Foster, *Recl. Trav. Chim. Pays-Bas*, **53**, 405 (1934).
4. O. Behaghel and H. Seibert, *Ber. Dtsch. Chem. Ges.*, **65**, 812 (1932).
5. H. Reinboldt and E. Giesbrecht, unpublished results cited by H. Reinboldt in E. Müller, Ed., "Methoden der Organischen Chemie" (Houben-Weyl), 4th ed., Vol. 9, Georg Thieme Verlag, Stuttgart, Germany 1955, p. 1096.
6. M. T. Rogers and T. W. Campbell, *J. Am. Chem. Soc.*, **69**, 2039 (1947).
7. K. B. Sharpless and M. W. Young, *J. Org. Chem.*, **40**, 947 (1975).
8. D. G. Foster, *Org. Synth.*, **Coll. Vol. 3**, 771 (1955).
9. H. J. Reich, J. M. Renga, and I. L. Reich, *J. Am. Chem. Soc.*, **97**, 5434 (1975).
10. D. G. Foster, *J. Am. Chem. Soc.*, **55**, 822 (1933).
11. O. Behaghel and K. Hofmann, *Ber. Dtsch. Chem. Ges.*, **72**, 582 (1939).
12. O. Behaghel and H. Seibert, *Ber. Dtsch. Chem. Ges.*, **66**, 708 (1933).

13. K. B. Sharpless and R. F. Lauer, *J. Org. Chem.*, **39**, 429 (1974).
14. For recent reviews see: H. J. Reich, *Acc. Chem. Res.*, **12**, 22 (1979); H. J. Reich, "Organoselenium Oxidations," in W. Trahanovsky, Ed., "Oxidation in Organic Chemistry," Part C, Academic Press, New York, 1978, pp. 1–130; D. L. J. Clive, *Tetrahedron*, **34**, 1049 (1978).
15. (a) K. B. Sharpless, R. F. Lauer, and A. Y. Teranishi, *J. Am. Chem. Soc.*, **95**, 6137 (1973); (b) G. Bergson and S. Wold, *Ark. Kemi*, **19**, 215 (1962).
16. H. J. Reich, *J. Org. Chem.*, **39**, 428 (1974); D. L. J. Clive, *J. Chem. Soc. Chem. Commun.*, 695 (1973); D. L. J. Clive, *J. Chem. Soc. Chem. Commun.*, 100 (1974).
17. H. J. Reich and J. M. Renga, *J. Org. Chem.*, **40**, 3313 (1975).
18. G. Ayrey, D. Barnard, and D. T. Woodbridge, *J. Chem. Soc.*, 2089 (1962).
19. K. B. Sharpless and R. F. Lauer, *J. Am. Chem. Soc.*, **95**, 2697 (1973); H. J. Reich, S. Wollowitz, J. E. Trend, F. Chow, and D. F. Wendelborn, *J. Org. Chem.*, **43**, 1697 (1978).
20. W. H. H. Gunther, *J. Org. Chem.*, **31**, 1202 (1966); W. G. Salmond, M. A. Barta, A. M. Cain, and M. C. Sobala, *Tetrahedron Lett.*, 1683 (1977).
21. H. Reinboldt, in E. Müller, Ed., "Methoden der Organischen Chemie" (Houben-Weyl), 4th ed., Vol. 9, Georg Thieme Verlag, Stuttgart, Germany 1955, pp. 997–999, 1101.
22. S. R. Raucher, *J. Org. Chem.*, **42**, 2950 (1977).

AROMATIC HYDROCARBONS FROM AROMATIC KETONES AND ALDEHYDES: 1,1-DIPHENYLETHANE

(Benzene, phenylethyl-)

$$C_6H_5COC_6H_5 \xrightarrow[\text{diethyl ether, 25°}]{CH_3Li} \left[(C_6H_5)_2\overset{\overset{O\ Li}{|}}{C}CH_3 \right] \xrightarrow[\text{NH}_4Cl,\ -33°]{Li,\ NH_3\ \text{(liq.)}} (C_6H_5)_2CHCH_3$$

Submitted by SHARON D. LIPSKY and STAN S. HALL[1]
Checked by ROBERT E. IRELAND, PAULA J. CLENDENING, KATHRYN D. CROSSLAND, and ALVIN K. WILLARD

1. Procedure

A 500-ml., three-necked, round-bottomed flask, equipped with a 5-cm., glass-coated, magnetic stirring bar, a Dewar condenser connected to a static argon line (Note 1), and a pressure-equalizing dropping funnel, is sealed with a rubber septum (Note 2). After flushing with argon, 30 ml. of anhydrous diethyl ether and 19.5 ml. of 1.89 M (0.0368 mole) methyllithium solution (Note 3) are injected through the septum into the flask (Note 4). A solution of 4.54 g. (0.0249 mole) of benzophenone (Note 5) in 35 ml. of anhydrous ether is placed in the dropping funnel and added to the reaction mixture over a 20-minute period. The mixture is allowed to stir for one half hour (Note 6). The septum is removed, and the side arm is quickly adapted with Tygon® tubing leading through a tower of solid potassium hydroxide to a tank of anhydrous ammonia. After approximately 75 ml. of ammonia is slowly distilled into the flask (Note 7), the tubing is removed, and 0.525 g. (0.0761 g.-atom, *ca.* 15 cm. added as 0.5 cm. pieces) of lithium wire (Note 8) is quickly added, and the flask is stoppered. After 15 minutes the dark blue color is discharged by the continuous addition of excess ammonium chloride (*ca.* 5 g. over a 15-minute period) (Note 9). The argon-inlet tube is disconnected, and the ammonia is allowed to evaporate. The residue is then partitioned between 100 ml. of aqueous saturated sodium chloride and 100 ml. of ether. The aqueous layer is separated and extracted with two 50-ml. portions of ether. The combined ether extracts are dried over anhydrous magnesium sulfate. Removal of the ether on a rotary evaporator yields 4.36–4.49 g. of crude product (Note 10). Filtration of the product through 60 g. of Woelm alumina (Grade III) with 150 ml. of petroleum ether affords 4.16–4.31 g. (92–95%) of 1,1-diphenylethane (Note 11), which on evaporative distillation in a Kügelrohr oven gives 4.12–4.21 g. (91–93%) of 1,1-diphenylethane, b.p. 100° (0.25 mm.), n_D^{28} 1.5691 (Note 12).

2. Notes

1. The entire reaction sequence is performed under an argon atmosphere. A T-tube and an oil bubbler are utilized, and the system is operated at a moderate flow-rate throughout the synthesis.

2. All of the glassware is oven dried and cooled to room temperature in a large box desiccator, or the assembled glassware can be flamed dry under an argon atmosphere and allowed to cool.

3. Methyllithium in ether solution is available from Foote Mineral Company and the Aldrich Chemical Company.

4. If methyllithium of a different molarity is used, the total volume should be adjusted to 50 ml. by varying the amount of ether used.

5. Benzophenone is available from Matheson, Coleman and Bell.

6. Toward the end of this sequence 2-propanol and dry ice are added to the condenser in preparation for the reduction step.

7. To prevent splattering, the apparatus is tilted slightly to allow the condensing ammonia to run down the walls of the flask.

8. Evidently surface area is important, since when the 15 cm. of lithium wire was added as 1-cm. pieces, the reduction was incomplete. Lithium wire (0.32 cm., 0.02% sodium) available from Alpha Inorganics, Inc., was wiped free of oil and rinsed with petroleum ether immediately prior to use.

9. Ammonium chloride is most conveniently introduced by attaching a glass tube filled with the salt to a side arm with Tygon® tubing. When the ammonium chloride is to be added, the tube is raised and tapped gently to introduce the quenching agent smoothly. Often a very vigorous reaction occurs after a considerable (10-minute) induction period. Should this step start to become violent, the addition and the vigorous stirring should be momentarily stopped to avoid eruption.

10. GC on a 200 cm. by 0.6 cm. column packed with 10% Apiezon L on Chromosorb W (AW, DMCS) using a flame-detector instrument, at a 40 ml./minute helium carrier gas flow rate, gives a trace peak at 9.9 minutes (diphenylmethane), a major peak at 11.7 minutes (1,1-diphenylethane), and a trace peak at 15.4 minutes (1,1-diphenylethanol) when the oven is held at 190° for 10 minutes and then programmed at 10°/minute to 290°.

11. The sample is sufficiently pure at this point to use for most purposes. The chromatography step is an efficient means to remove any 1,1-diphenylethanol that was not reduced.

TABLE I

Aromatic Hydrocarbons from Aromatic Ketones and Aldehydes

Aromatic Carbonyl Compound	Organolithium Reagent	Product	Yield (%)
	CH_3Li		95
	CH_3Li		94
	CH_3Li		95
	CH_3Li		78
	C_4H_9Li	C_5H_{11}	$86^{a,e}$
	C_4H_9Li	C_4H_9	$89^{a,e}$
	C_4H_9Li	C_4H_9	$76^{b,e}$
	C_4H_9Li	C_4H_9	$70^{a,e}$

539

TABLE I (*continued*)

AROMATIC HYDROCARBONS FROM AROMATIC KETONES AND ALDEHYDES

Aromatic Carbonyl Compound	Organolithium Reagent	Product	Yield(%)
	C_6H_5Li		93[c,e]
	C_6H_5Li		89[d,e]
	C_6H_5Li		97[c,e]
	C_6H_5Li		97[d,e]

[a]The organolithium reagent was generated *in situ* in ether from 1-bromobutane.
[b]Commercial butyllithium (Foote Mineral Company, *ca.* 15% in hexane) and six equivalents of lithium were used.
[c]The organolithium reagent was generated *in situ* in ether from bromobenzene.
[d]Commercial phenyllithium (the Aldrich Chemical Co., 2 *M* in cyclohexane-ether) and six equivalents of lithium were used.
[e]Reaction conducted on a 0.005 mole scale using as solvent 20 ml. of ether and 20 ml. of ammonia. Yield after column chromatography.

12. The spectral properties of the product are as follows; ^{1}H NMR (CCl₄), δ (multiplicity, coupling constant *J* in Hz., number of protons, assignment): 1.54 (d, *J* = 7, 3H, *CH₃*), 4.03 (q, *J* = 7, 1H, *CH*), 7.12 (s, 10H, 2*C₆H₅*); mass spectrum *m/e* (relative intensity): 182 (M, 32), 167 (100).

3. Discussion

This procedure illustrates a general method for preparing aromatic hydrocarbons by the tandem alkylation-reduction of aromatic ketones and aldehydes.[2] Additional examples are given in Table I. The advantages of the method are that the entire sequence is carried out in the same reaction vessel without isolation or purification of intermediates, the procedure consumes only a few hours, and in most cases the isolated yield of the aromatic hydrocarbon is excellent.

The method may be modified so that the organolithium reagent is generated *in situ* in ether from the corresponding bromide. Best results were obtained by having from the outset all of the lithium wire necessary to generate the organolithium reagent and reducing the intermediate benzyl alkoxide present.[3] Commercial organolithium reagents such as *n*-butyllithium in hexane or phenyllithium in cyclohexane–ether were satisfactory when twice as much lithium is used for the reduction step. In some cases, by running the alkylation step at −78° to minimize competing side reactions,[4] higher yields than those listed in Table I can be realized. In addition to the present method, other procedures have been reported for the synthesis of 1,1-diphenylethane.[5-7]

This alkylation-reduction sequence has now been extended to benzylidene carbonyls,[8] complex carbonyls,[9] and esters and lactones;[10] and has been used to synthesize α-cyclopropyl aromatic hydrocarbons,[11] β,γ-unsaturated aromatic hydrocarbons,[12] and 1,4-dienes from α,β,γ,δ-unsaturated ketones.[13]

1. Department of Chemistry, Rutgers University, Newark, New Jersey 07102.
2. S. S. Hall and S. D. Lipsky, *J. Chem. Soc. D,* 1242 (1971); *J. Org. Chem.,* **38,** 1735 (1973).
3. The lithium wire is cut into 0.5-cm. pieces and hammered to a foil immediately prior to use.
4. J. D. Buhler, *J. Org. Chem.,* **38,** 904 (1973).
5. J. S. Reichert and J. A. Nieuwland, *J. Am. Chem. Soc.,* **45,** 3090 (1923).
6. E. Späth, *Monatsh. Chem.,* **34,** 1965 (1913).
7. J. Böeseken and M. C. Bastet, *Recl. Trav. Chim. Pays-Bas,* **32,** 184 (1913).
8. S. S. Hall, *J. Org. Chem.,* **38,** 1738 (1973).
9. S. S. Hall and F. J. McEnroe, *J. Org. Chem.,* **40,** 271 (1975).
10. S. T. Srisethnil and S. S. Hall, *J. Org. Chem.,* **42,** 4266 (1977).
11. S. S. Hall, C.-K. Sha, and F. Jordan, *J. Org. Chem.,* **41,** 1494 (1976).
12. F. J. McEnroe, C.-K. Sha, and S. S. Hall, *J. Org. Chem.,* **41,** 3465 (1976).
13. J. S. R. Zilenovski and S. S. Hall, *J. Org. Chem.,* **44,** 1159 (1979).

NUCLEOPHILIC α-sec-AMINOALKYLATION:
2-(DIPHENYLHYDROXYMETHYL)PYRROLIDINE

[2-Pyrrolidinemethanol, α,α-diphenyl-, (±)-]

Submitted by D. ENDERS, R. PIETER, B. RENGER, and D. SEEBACH[1]
Checked by C. HUTCHINS and M. F. SEMMELHACK

1. Procedure

Caution! Since N-*nitrosopyrrolidine is a potent carcinogen and is produced as an intermediate, this entire "one-pot" procedure should be performed in a well-ventilated hood. Wearing of disposable polyethylene gloves is recommended.*

Benzene has been identified as a carcinogen; OSHA has issued emergency standards on its use. All procedures involving benzene should be carried out in a well-ventilated hood, and glove protection is required.

A dry, 250-ml., one-necked, round-bottomed flask equipped with a magnetic stirrer and a three-way stopcock is charged with 4 g. (0.05 mole) of ethyl nitrite (Note 1), 4 g. of dry tetrahydrofuran (Note 2), and 2.35 g. (0.0331 mole) of pyrrolidine (Note 3). The stopcock is closed (Note 4), and the mixture is stirred at room temperature for 2 days. Excess ethyl nitrite, tetrahydrofuran, and the ethanol formed are removed from the N-nitrosopyrrolidine (Note 5) by stirring at 25° under reduced pressure (10 mm., water aspirator, Note 6) for 2 hours. The stopcock is fitted with a rubber septum, the air in the system is replaced with dry argon (Notes 4 and 7), and 50 ml. of tetrahydrofuran is injected by syringe. A solution of lithium diisopropylamide is prepared in a separate, dry, 100-ml. flask by adding 21.1 ml. (0.0340 mole) of a 1.61 M solution of n-butyllithium in hexane (Note 8) to a solution of 3.44 g. (4.76 ml., 0.0341 mole) of diisopropylamine (Note 9) in 25 ml. of tetrahydrofuran at −78° (methanol–dry ice bath) with stirring under argon. The solution is warmed to 0° in 15 minutes, then added dropwise with a syringe within 4 minutes to the nitrosamine solution, stirred at −78°. Stirring of the yellow to orange solution is continued at this temperature for 25 minutes. A solution of 5.46 g. (0.0300 mole) of benzophenone in 12 ml. of tetrahydrofuran is added dropwise by syringe (Note 10), and the mixture is stirred for 12 hours at −78°, then warmed to 0° within 2 hours. After addition of 0.6 ml. (0.03 mole) of water, the flask is transferred from the

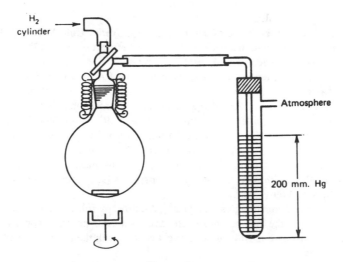

Figure 1.

argon line to a rotary evaporator (within the hood). Solvents and diisopropylamine are removed under reduced pressure in a 40° bath (Note 11). The remaining solid is dissolved with slight warming in 120 ml. of dry methanol (Note 12), before 3.9 g. (66 equivalents) of Raney nickel (Note 13) is rinsed into the solution with 30 ml. of dry methanol. The reaction vessel is equipped again with the three-way stopcock, and the air in the flask is replaced with hydrogen (Note 7). The flask is filled five times with hydrogen from a balloon; during this operation vigorous stirring of the Raney nickel–methanol suspension is necessary. The flask is attached to a mercury bubbler to maintain a positive hydrogen pressure (200 mm.) supplied from a cylinder, as shown in Figure 1. The reaction mixture is stirred for 3 hours at room temperature while a slow stream of hydrogen is passed through the system. The major part of the solution is decanted and filtered, and the remaining Raney-nickel suspension is extracted by refluxing three times for 10 minutes each with 20 ml. of methanol (Note 14). The combined methanol solutions are concentrated under reduced pressure. The residue is dissolved in 150 ml. of diethyl ether and 100 ml. of water, the layers are separated (Note 15), and the aqueous layer is extracted three times with 50-ml. portions of ether. The combined extracts are dried over sodium carbonate and concentrated in a rotary evaporator to a total volume of 150 ml. Dry hydrogen chloride gas is bubbled into the solution with stirring until the mixture is acidic. The almost colorless precipitate of the hydrochloride is filtered, washed two times with 30-ml. portions of dry ether, and dried in a desiccator under reduced pressure for 3 hours, giving 5.99–6.11 g. (58–60%, based on benzophenone) of the product, m.p. 244–249° (dec.). Recrystallization from methanol–acetone gives 5.06–5.20 g. (58–60%) of analytically pure product, m.p. 267–269° (dec.) (Note 16). The free base is obtained by treatment of the hydrochloride with 10% aqueous sodium hydroxide and extraction with ether, m.p. 82–83° (Note 17).

2. Notes

1. Ethyl nitrite was prepared as described in *Org. Synth.*, **Coll. Vol. 2,** 204 (1943), or purchased from Merck-Schuchardt and distilled before use, b.p. 17°. The volatile nitrite can be easily handled as a 50% tetrahydrofuran solution and stored in a refrigerator.

2. Technical grade tetrahydrofuran, available from BASF-A G or Fisher Scientific Company, was dried by distillation, first from potassium hydroxide then from lithium aluminum hydride and used for all operations in this procedure. For a warning note regarding the purification of tetrahydrofuran, see *Org. Synth.*, **Coll. Vol. 5,** 976 (1973).

3. Pyrrolidine, b.p. 87–88°, obtained from Aldrich Chemical Company, Inc., or BASF-A G was distilled from potassium hydroxide before use.

4. The three-way stopcock with standard-tapered joint must be securely fastened to the neck of the flask with wire, rubber bands, or springs [see Figure 1 and *Org. Synth.*, **Coll. Vol. 6,** 316, 869 (1988).]

5. Nitrosamines are strong carcinogens[2,3]; *N*-nitrosopyrrolidine causes liver tumors in rats.[2,4] Although the one-pot procedure described here prevents contact with the nitrosamine, utmost care must be used to avoid contact with the reaction mixture during all manipulations.

6. At the beginning of the evacuation the pressure should be lowered slowly to prevent bumping.

7. This was done by alternately evacuating and filling with dry argon three times; during the reaction a pressure of about 50 mm. above atmospheric was maintained using a mercury bubbler.

8. Purchased from Metallgesellschaft, Frankfurt, or Alfa-Products, Division of the Ventron Corporation. The content of the solution was determined prior to use by acidimetric titration.

9. The diisopropylamine, b.p. 83–84°, available from Fluka A G, BASF-A G, or Aldrich Chemical Company, Inc., was purified by refluxing over potassium hydroxide and subsequent distillation. It was stored over calcium hydride.

10. Benzophenone, m.p. 47–49°, was purchased from Riedel-de-Haen-A G or from Fisher Scientific Company. The reaction mixture turns green and then blue during the addition, because of the formation of ketyl radicals.

11. The checkers found it more convenient to remove the volatile material at this stage by warming the stirred mixture at 40° and using a water aspirator vacuum (10 mm.). About 3 hours were required.

12. Methanol was dried by heating at reflux for 3 hours over magnesium, then distilling.

13. The Raney nickel reagent was prepared by addition of 9.5 g. of sodium hydroxide pellets over 8–10 minutes to a stirred suspension of 7.8 g of nickel–aluminum alloy (50% Ni, 50% Al powder, purchased from Merck-Schuchardt) in 120 ml. of distilled water, contained in a 250-ml. beaker. Fifteen minutes after the addition was completed, the beaker was immersed into a 70° water bath for 20 minutes. The water was decanted, and the catalyst was washed sequentially with two 20-ml. portions of distilled water and two 20-ml. portions of methanol.

14. *Caution! The dry Raney nickel catalyst is pyrophoric. The residues can be destroyed by allowing them to ignite and burn on filter paper in a safe place.*

TABLE I

α-Substituted Secondary Amines via Electrophilic Substitution[8]

Starting Materials	Nitrosamine	Metallation Time (minutes)	Product[a]	Yield (%)
Benzaldehyde, dimethylamine	$CH_3-\underset{NO}{N}-CH_3$	10	$CH_3-\underset{H}{N}-CH_2-\underset{OH}{CH}-C_6H_5$	80
Benzylbromide, methylisopropylamine	$(CH_3)_2CH-\underset{NO}{N}-CH_3$	10	$(CH_3)_2CH-\underset{H}{N}-CH_2-CH_2-C_6H_5$	80
Benzophenone, methylisopropylamine	$(CH_3)_2CH-\underset{NO}{N}-CH_3$	10	$(CH_3)_2CH-\underset{H}{N}-CH_2-\underset{OH}{C}(C_6H_5)_2$	75
Piperonal, methyl tert-butylamine	$(CH_3)_3C-\underset{NO}{N}-CH_3$	10	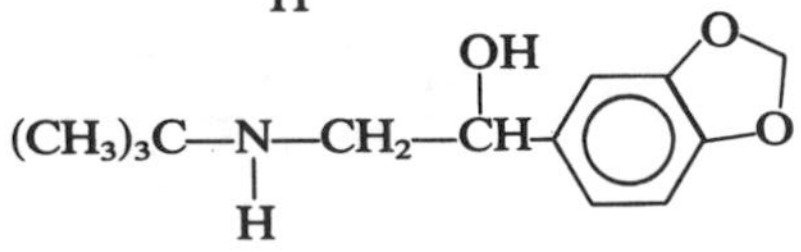 $(CH_3)_3C-\underset{H}{N}-CH_2-\underset{OH}{CH}-$	80
Carbon dioxide, methyl tert-butylamine	$(CH_3)_3C-\underset{NO}{N}-CH_3$	10	$(CH_3)_3C-\underset{H}{N}-CH_2-COOH$	75[b]

545

TABLE I (continued)
α-Substituted Secondary Amines via Electrophilic Substitution[8]

Starting Materials	Nitrosamine	Metallation Time (minutes)	Product[a]	Yield (%)
Benzaldehyde, diethylamine	$C_2H_5-N-C_2H_5$ $\|$ NO	10	$C_2H_5-N-CH(CH_3)-CH(C_6H_5)OH$ $\|$ H	75[b]
Benzaldehyde, dihexylamine	$C_6H_{13}-N-C_6H_{13}$ $\|$ NO	10	$C_6H_{13}-N-CH(C_5H_{11})-CH(C_6H_5)OH$ $\|$ H	80[b]
Benzophenone, azetidine	azetidine $N-NO$	7	azetidine, $C(C_6H_5)_2OH$, $N-H$	75[b]
Benzophenone, piperidine	piperidine $N-NO$	180	piperidine, $C(C_6H_5)_2OH$, $N-H$	50

Cyclohexanone, piperidine	180		55
Iodopropane, 3-hydroxy-piperidine	240		40
Benzophenone, perhydroazepine	20		80[b]

[a] Isolated and characterized as hydrochlorides.

[b] Overall yield of stepwise procedure; the denitrosation was performed by bubbling gaseous hydrogen chloride into a benzene solution of the nitrosamine. This cleavage of nitrosamines is usually not as clean as the one with Raney nickel.

15. Upon dissolving the residue in 200 ml. of ether and 100 ml. of water, the checkers obtained an emulsion that cleared slowly on standing for 2–3 hours.

16. The literature reports m.p. >240°,[5] >250°,[6] and 262–263°.[7] The yield given includes a small second crop obtained by recrystallization of the filtrate residue. The reported yield and m.p. data were obtained by the checkers. The submitters report 6.50–6.95 g. (75–80%, m.p. 260–265°) before recrystallization and 5.20–5.62 g. (60–65%, m.p. 267–269°) for analytically pure product.

17. The m.p. is reported to be 81–82°[6] and 83°.[5] The spectral properties are: IR spectrum (KI) cm^{-1}: strong absorptions at 3360, 3080, 3060, 3020, 2980–2800, 1595, 1490, 1450, 1400, 1190, 1100, 1060, 1030, 990, 900, 750, 700, 660, and 635; ^{1}H NMR (CDCl$_3$), δ (multiplicity, number of protons): 1.60 (m, 4H), 2.95 (m, 2H, with overlapping broad peak for O*H* and N*H*), 4.18 (m, 1H), 7.00–7.65 (m, 10H). The compound has psychostimulating activity.[7]

3. Discussion

The procedure described here is an example of the "nitrosamine method" for the electrophilic substitution of **1** to **3**, *via* the intermediate anion **2**, as outlined in detail in a recent review article.[8]

$$X = NO$$

Currently, this is the only method that allows reversible enhancement of the acidity of α-nitrogen C-protons in a large variety of secondary amines, using nitroso-substituted nitrogen (X = NO, see examples in Table I). Nucleophile **2** is synthetically equivalent to α-aminocarbanion **4**, while the inherent reactivity of a carbon adjacent to an amino nitrogen is electrophilic (see the immonium ion **5**).[8,9] Lithiated nitrosamines are also

useful because their formation and reactions with electrophiles occur in high yield, as well as with a high degree of regio- and stereoselectivity (see Table I). For further information see the review article[8] and other recent publications.[10] This method has the disadvantage of carcinogenic nitrosamine intermediates, but this one-pot procedure reduces the danger of contact.

1. Institüt für Organische Chemie der Justus Liebig-Universität, 6300 Giessen, Germany. Present address: Eidg. Techische Hochschule, Laboratorium für Organische Chemie, Universitätsstrabe 16, CH-8092 Zürich, Switzerland.
2. H. Druckrey, R. Preussmann, S. Ivankovic, and D. Schmaehl, *Z. Krebsforsch,* **69,** 103 (1967).

3. P. N. Magee and J. M. Barnes, *Adv. Cancer Res.*, **10**, 163 (1967).
4. M. Greenblatt and W. Lijinsky, *J. Nat. Cancer Inst.*, **48**, 1687 (1972).
5. J. Kapfhammer and A. Matthes, *Hoppe-Seylers Z. Physiol. Chem.*, **223**, 43 (1933).
6. S. O. Winthrop and L. G. Humber, *J. Org. Chem.*, **26**, 2834 (1961).
7. A. M. Likhosherstov, K. S. Raevskii, A. S. Lebedeva, A. M. Kritsyn, and A. P. Skoldinov, *Khim. Farm. Zh.*, **1**, 30 (1967) [*Chem. Abstr.*, **67**, 90,642w (1967)].
8. Review: D. Seebach and D. Enders, *Angew. Chem.*, **87**, 1 (1975); *Angew. Chem. Int. Ed. Engl.*, **14**, 15 (1975). Preliminary communications: D. Seebach and D. Enders, *Angew. Chem.*, **84**, 350, 1186, 1187 (1972) [*Angew. Chem. Int. Ed. Engl.*, **11**, 301, 1101, 1102 (1972)]; D. Seebach, D. Enders, B. Renger, and W. Bruegel, *Angew. Chem.*, **85**, 504 (1973) [*Angew. Chem. Int. Ed. Engl.*, **12**, 495 (1973)]. Full papers with experimental detail: D. Seebach and D. Enders, *J. Med. Chem.*, **17**, 1225 (1974); *Chem. Ber.*, **108**, 1293 (1975); D. Seebach, D. Enders, and B. Renger, *Chem. Ber.*, **110**, 1852 (1977); B. Renger, H.-O. Kalinowski, and D. Seebach, *Chem. Ber.*, **110**, 1866 (1977); D. Seebach, D. Enders, R. Dach, and R. Pieter, *Chem. Ber.*, **110**, 1879 (1977); B. Renger and D. Seebach, *Chem. Ber.*, **110**, 2334 (1977).
9. D. Seebach and M. Kolb, *Chem. Ind. (London)*, 687 (1974), 910 (1975).
10. R. R. Fraser and T. B. Grindley, *Can. J. Chem.*, **53**, 2465 (1975); R. R. Fraser, T. B. Grindley, and S. Passannanti, *Can. J. Chem.*, **53**, 2473 (1975); D. H. R. Barton, R. D. Bracho, A. A. L. Gunatilaka, and D. A. Widdowson, *J. Chem. Soc., Perkin Trans. I*, 579 (1975); J. E. Baldwin, S. E. Branz, R. F. Gomez, P. L. Kraft, A. J. Sinskey, and S. R. Tannenbaum, *Tetrahedron Lett.*, 333 (1976). R. R. Fraser and S. Passananti, *Synthesis*, 540 (1976).

DIPHENYLKETENE

(Ethenone, diphenyl-)

$$(C_6H_5)_2CHCO_2H \xrightarrow[C_6H_6, \text{ reflux}]{SOCl_2} (C_6H_5)_2CHCOCl + SO_2 + HCl$$

$$(C_6H_5)_2CHCOCl \xrightarrow[(C_2H_5)_2),0°]{(C_2H_5)_3N} (C_6H_5)_2C{=}C{=}O + (C_2H_5)_3N{\cdot}HCl$$

Submitted by EDWARD C. TAYLOR,[1] ALEXANDER McKILLOP,[2] and GEORGE H. HAWKS
Checked by C. J. MICHEJDA, D. D. VON RIESEN, R. W. COMNICK,
and HENRY E. BAUMGARTEN

1. Procedure

Caution! Benzene has been identified as a carcinogen; OSHA has issued emergency standards on its use. All procedures involving benzene should be carried out in a well-ventilated hood, and glove protection is required.

A. *Diphenylacetyl chloride.* A 500-ml., three-necked flask equipped with a dropping funnel and a reflux condenser carrying a calcium chloride drying tube is charged with 50.0 g. (0.236 mole) of diphenylacetic acid (Note 1) and 150 ml. of thiophene-free, anhydrous benzene. The mixture is heated under reflux, and 132 g. (80.1 ml., 1.11 mole) of thionyl chloride is added dropwise over 30 minutes. Refluxing is continued for 7 additional hours before the benzene and excess thionyl chloride are removed by distillation under reduced pressure. The pale yellow oil which remains contains a little thionyl chloride, best removed by adding 100 ml. of anhydrous benzene and again distilling under reduced pressure. The residue is dissolved in 150 ml. of refluxing, anhydrous hexane (Note 2). The hot solution is treated with charcoal and filtered, and the filtrate is chilled to 0° in a sealed flask. The product, which crystallizes as colorless plates (Note 3), is filtered, washed with a little cold hexane, dried at 25° under vacuum for 2 hours, and stored in a tightly stoppered bottle, giving 42–45 g. (77–84%) of product, m.p. 51–53°. Concentration of the hexane mother liquors to about 50 ml. followed by chilling to 0° and addition of a seed crystal gives an additional 2.5–4.0 g. (5–8%) of product of equal purity, for a total yield of 44.5–49 g. (82–94%) of diphenylacetyl chloride, m.p. 51–53° (Note 4).

B. *Diphenylketene.* A 500-ml., three-necked flask equipped with a magnetic stirring bar, a gas-inlet tube, a calcium chloride drying tube, and a dropping funnel is charged with a solution of 23.0 g. (0.0997 mole) of diphenylacetyl chloride in 200 ml. of anhydrous diethyl ether. The flask is cooled in an ice bath, dry nitrogen is passed through the system, and 10.1 g. (0.100 mole) of triethylamine is added dropwise over 30 minutes to the stirred solution; triethylamine hydrochloride precipitates as a colorless solid, and the ether becomes bright yellow in color. When addition of the triethylamine is complete, the flask is tightly stoppered and stored overnight at 0°. The hydrochloride is collected on a 9-cm., sintered glass funnel and washed with anhydrous ether until the washings are colorless. The ether is removed under reduced pressure; the residual red oil is transferred to a 50-ml., distilling flask fitted with a 10-cm. Vigreux column and distilled (Note 5), giving 10.2–10.8 g. (53–57%) of diphenylketene as an orange oil, b.p. 118–120° (1 mm.) (Note 6). It can be stored at 0° in a tightly stoppered bottle for several weeks without decomposition.

2. Notes

1. Superior grade diphenylacetic acid, m.p. 147–148° (Matheson, Coleman and Bell) was used without further purification.

2. Commercial hexane, A.C.S. grade (Matheson, Coleman and Bell) was dried by distillation from potassium hydroxide.

3. Diphenylacetyl chloride crystallizes best when a seed crystal is added to the cold hexane solution. If, after several hours at 0°, crystallization has not commenced, scratching with a glass rod is sufficient to induce crystallization.

4. The checkers found it necessary to recrystallize the diphenylacetyl chloride twice to obtain the reported melting point.

5. Most of the diphenylketene distils cleanly at 118–119° (1 mm.) but strong heating is necessary for distillation of the final portion from the polymeric pot residue.

The checkers used a variety of different distillation setups. The best results were obtained when the oil being distilled filled the flask to about two-thirds its capacity, the Vigreux column was no longer than 10 cm. in length, the whole apparatus was kept as small as possible, and the distillation was conducted as rapidly as possible.

6. The submitters obtained yields of 73–84% on the scale described and yields of up to 70% on a scale twice that described. From IR analysis of the crude (undistilled) product

the checkers concluded that this material represented a yield in excess of 80%. Thus, the critical step appears to be the distillation. The checkers have used the crude (undistilled) product for some applications, but this procedure has not been uniformly successful and is not recommended.

3. Discussion

Diphenylacetyl chloride has been obtained from the reaction of diphenylacetic acid with phosphorus pentachloride,[3] phosphorus oxychloride and phosphorus pentachloride,[4] or thionyl chloride.[5] It has also been prepared by treatment of diphenylketene with hydrogen chloride.[6] The methods of preparation of diphenylketene have been reviewed earlier in this series.[7] To those cited should be added the debromination of α-bromodiphenylacetyl bromide with triphenylphosphine.[8] The procedure above is a modification of that described by Staudinger.[9]

The present preparation consists of two very simple steps, uses relatively inexpensive starting materials, and does not involve hazardous or toxic chemicals or special apparatus. An important advantage is that the diphenylketene, until it is finally distilled, is never exposed to temperatures greater than 30–35°; hence polymerization is minimized (cf. ref. 6).

1. Department of Chemistry, Princeton University, Princeton, New Jersey 08540.
2. Present address: School of Chemical Sciences, University of East Anglia, Norwich NR4 7TJ, United Kingdom.
3. F. Klingemann, *Justus Liebigs Ann. Chem.*, **275**, 50 (1893).
4. A. Bistrzycki and A. Landtwing, *Ber. Dtsch. Chem. Ges.*, **41**, 686 (1908).
5. W. A. Bonner and C. J. Collins, *J. Am. Chem. Soc.*, **75**, 5372 (1953).
6. H. Staudinger, *Ber. Dtsch. Chem. Ges.*, **38**, 1735 (1905); *Justus Liebigs Ann. Chem.*, **356**, 51 (1907).
7. L. I. Smith and H. H. Hoehn, *Org. Synth.*, **Coll. Vol. 3**, 356 (1955).
8. S. D. Darling and R. L. Kidwell, *J. Org. Chem.*, **33**, 3974 (1968).
9. H. Staudinger, *Ber. Dtsch. Chem. Ges.*, **44**, 1619 (1911).

ALKENES *via* HOFMANN ELIMINATION: USE OF ION-EXCHANGE RESIN FOR PREPARATION OF QUATERNARY AMMONIUM HYDROXIDES: DIPHENYLMETHYL VINYL ETHER

(Benzene, 1,1'-[(ethenyloxy)methylene]bis-)

$$(C_6H_5)_2CHOCH_2CH_2N(CH_3)_2 + CH_3I \xrightarrow{25°} (C_6H_5)_2CHOCH_2CH_2\overset{+}{N}(CH_3)_3\overset{-}{I}$$

$$\xrightarrow[\text{resin,methanol}]{\text{Anion-exchange}} [(C_6H_5)_2CHOCH_2CH_2\overset{+}{N}(CH_3)_3\overset{-}{O}H]$$

$$\xrightarrow{\Delta} (C_6H_5)_2CHOCH=CH_2$$

Submitted by CARL KAISER and JOSEPH WEINSTOCK[1]
Checked by P. MÜLLER and G. BÜCHI

1. Procedure

Caution! Methyl iodide, in high concentrations for short periods or in low concentrations for long periods, can cause serious toxic effects in the central nervous system. Accordingly, the American Conference of Governmental Industrial Hygienists[2] has set 5 p.p.m., a level which cannot be detected by smell, as the highest average concentration in air to which workers should be exposed for long periods. The preparation and use of methyl iodide should always be performed in a well-ventilated fume hood. Since the liquid can be absorbed through the skin, care should be taken to prevent contact.

A 250-ml., three-necked, round-bottomed flask equipped with a sealed mechanical stirrer, a dropping funnel, and a reflux condenser is charged with 13.3 g. (0.0522 mole) of 2-(diphenylmethoxy)-*N*,*N*-dimethylethylamine (Note 1) and 50 ml. of acetone. The solution is stirred, and 8.1 g. (0.057 mole) of methyl iodide in 15 ml. of acetone is added dropwise over 5 minutes (Note 2). After the addition is complete, the mixture is stirred for 30 minutes, then cooled to 0–10° with an ice bath. The crystalline product is filtered and washed with 15 ml. of acetone and 30 ml. of diethyl ether, yielding 20.0–20.2 g. (97–98%) of colorless, crystalline methiodide, m.p. 194–196°.

An excess (60 g., *ca.* 0.26 equivalent) of anion exchange resin (OH⁻ form, Note 3) in a 500-ml. Erlenmeyer flask is stirred with 200 ml. of methanol (Note 4) for 5 minutes. The methanolic slurry of resin is transferred to a 6.5 cm. × 25 cm. chromatography column, using 50–100 ml. of methanol to aid in the transfer. The resin column is washed with 750 ml. of methanol, added gradually so as to maintain about a 1–2.5 cm. solvent head above the upper resin level (Note 5). About two-thirds of the resin slurry is poured from the column (using about 100 ml. of methanol to aid the transfer) into a suspension of 19.9 g. (0.0499 mole) of the methiodide in 50 ml. of methanol (Note 6). The mixture is stirred and heated gently on a water bath, dissolving the crystalline methiodide. The resulting resin suspension is poured onto the column containing the remaining one-third of the resin. Additional methanol (*ca.* 50 ml.) is required to facilitate transferral. The column is eluted with about 500 ml. of methanol until the eluent no longer affords an alkaline reaction to pH paper (Note 7). The methanolic eluent is concentrated under reduced pressure (10–25 mm.), and the residual liquid (Note 8) is gradually heated to 100° under the water-aspirator vacuum. Following completion of thermal decomposition, as evidenced by the end of gas evolution (*ca.* 5–10 minutes), the residue is dissolved in 250 ml. of ether (Note 9). The ether solution is washed with 100 ml. of 0.2 N sulfuric acid and 100 ml. of water, dried over anhydrous magnesium sulfate, and filtered. The filtrate is concentrated, and distillation of the residue gives 8.5–9.0 g. (81–86%) of diphenylmethyl vinyl ether as a colorless liquid, b.p. 163–167° (18 mm.), n_D^{25} 1.5716 (Notes 10 and 11).

2. Notes

1. The submitters used 2-(diphenylmethoxy)-N,N-dimethylethylamine, b.p. 150–165° (2 mm.),[3] obtained from Searle Chemicals, Inc., or from the hydrochloride, m.p. 161–162°, which is available commercially from Gane's Chemical Works, Inc., New York, New York, under the generic name, diphenhydramine.

2. The reaction exotherm is just sufficient to cause moderate reflux.

3. A strongly basic, polystyrene, alkyl quaternary amine (hydroxide form) of medium porosity was employed. Research grade Rexyn 201 (OH) (purchased from Fisher Scientific Company) and Amberlite IRA-400 (purchased from Mallinckrodt Chemical Works) were found to be satisfactory. Chloride-form resins must be converted to the hydroxide form before use, as described below (Note 7).

4. It is necessary to wash the resin with methanol prior to packing of the column. If this is not done, swelling of the resin on treatment with the solvent may cause explosion of the column.

5. If the resin was not washed exhaustively with methanol, significant amounts of benzhydrol (α-phenylbenzenemethanol) and diphenylmethyl methyl ether were obtained in the final product.

6. Stirring of the methiodide with the anion exchange resin prior to introduction into the column is necessary, because of the insolubility of this quaternary salt in methanol. For methanol-soluble methiodides, a solution of the salt may be added directly to the methanol-washed resin column.

7. The recovered resin can be reconverted to the hydroxide form by eluting a column of the material with aqueous 10% sodium hydroxide until it is free of halide ion (silver nitrate–nitric acid test), then with water until the eluent is no longer alkaline to pH paper.

8. Heating should be carried out in a 1-l. (oversized) flask because decomposition is accompanied by considerable foaming.

9. A small amount of insoluble material, which is mainly unreacted 2-(diphenylmethoxy)-N,N-dimethylethylamine methiodide, can be removed at this point.

10. The product has the following spectral properties; IR (neat) cm.$^{-1}$: 1670, 1200, 770, 710; ^{1}H NMR (CDCl$_3$), δ (multiplicity, approx. coupling constant J in Hz., number of protons): 3.97 (d of d, $J_{XY} = 2$, $J_{AX} = 7$, 1H), 4.27 (d of d, $J_{XY} = 2$, $J_{AY} = 14$, 1H),

5.7 (s, 1H), 6.36 (q, $J_{AX} = 7$, $J_{AY} = 14$, 1H), 7.2 (s, 10H). The distilled product was about 98% pure by GC analysis on a 60 cm. × 0.6 cm. aluminum column packed with 10% SE-30 silicon rubber on Gas Chrom Z, 100–200 mesh, operated at 180°. The retention time was about 2.0 minutes. Minor amounts of benzhydrol and diphenylmethyl methyl ether (retention times 2.5 minutes and 1.8 minutes, respectively) accounted for the remainder of the distillate. The checkers found that GC analysis on a 1.8-m. column packed with 15% SE-30 on GAW, 60–80 mesh at 180°, at an injector temperature of 250° resulted in extensive decomposition. A satisfactory analysis, however, could be performed by lowering the injector temperature to 180°.

11. Reppe[4] reports b.p. 120° (15 mm.) for diphenylmethyl vinyl ether.

3. Discussion

Diphenylmethyl vinyl ether has also been prepared from benzhydrol and acetylene under high-pressure conditions.[4] In the described method, an adaptation of the procedure of Weinstock and Boekelheide,[5] improved yields of the alkene are obtained by using more convenient experimental conditions.

The described method for converting a quaternary halide to the corresponding hydroxide, utilizing an anion-exchange resin, has general application in the Hofmann elimination reaction.[6] It has been used extensively in the submitters' laboratories for the synthesis of a variety of alkenes[7] and for the preparation of a number of ethyl 1-benzylcyclopropanecarboxylates *via* abnormal Hofmann elimination of diethyl [2-(*N,N*-dimethylamino)ethyl]benzylmalonates.[8] It offers several notable advantages over more conventional methods for preparing quaternary hydroxides. Formation of quaternary hydroxides from iodides with bases (*e.g.*, silver oxide) that form insoluble iodides has disadvantages due to the expense of the reagent and, in some instances, the oxidizing power of silver salts in basic solution. Thallous ethoxide has been used to avoid the oxidation effect; however, it is expensive[9–11] and toxic. Quaternary methosulfates may be hydrolyzed to sulfates and converted to the hydroxide with barium hydroxide,[12] but this method has not found general application. The described procedure of exchange of hydroxide ion for halide is suitable for even very sensitive compounds and obviates most of the objectionable features of the precipitation methods.[5,6] In the event that methanol is undesirable, the conversion may be carried out in water.[5]

1. Smith Kline and French Laboratories, Philadelphia, Pennsylvania 19101.
2. American Conference of Governmental Industrial Hygienists (ACGIH), "Documentation of Threshold Limit Values," 3rd ed., Cincinnati, Ohio, 1971, p. 166.
3. G. Rieveschl, Jr., (to Parke Davis and Company), U.S. Pat. 2,421,714 (1947) [*Chem. Abstr.*, **41**, 5550h (1947)].
4. W. Reppe, *Justus Liebigs Ann. Chem.*, **601**, 81 (1956) [*Chem. Abstr.*, **51**, 9579b (1957)].
5. J. Weinstock and V. Boekelheide, *J. Am. Chem. Soc.*, **75**, 2546 (1953).
6. A. C. Cope, *Org. React.*, **11**, 317 (1960).
7. C. Kaiser and C. L. Zirkle (to Smith Kline and French Laboratories). U.S. Pat. 3,462,491, August 19, 1969.
8. C. Kaiser, C. A. Leonard, G. C. Heil, B. M. Lester, D. H. Tedeschi, and C. L. Zirkle, *J. Med. Chem.*, **13**, 820 (1970).
9. H. Wieland, C. Schöpf, and W. Hermsen, *Justus Liebigs Ann. Chem.*, **444**, 40 (1925).

10. B. Witkop, *J. Am. Chem. Soc.*, **71**, 2559 (1949).
11. F. v. Bruchhausen, H. Oberembt, and A. Feldhaus, *Justus Liebigs Ann. Chem.*, **507**, 144 (1933).
12. J. v. Braun and E. Anton, *Ber. Dtsch. Chem. Ges.*, **64**, 2865 (1931).

2,3-DIPHENYLVINYLENE SULFONE

(Thiirene, diphenyl-, 1,1-dioxide)

$$\underset{\text{Br}}{\underset{|}{\text{C}_6\text{H}_5\text{CH}}}\text{SO}_2\underset{\text{Br}}{\underset{|}{\text{CHC}_6\text{H}_5}} \xrightarrow{\text{N(C}_2\text{H}_5)_3} \text{C}_6\text{H}_5\underset{\text{SO}_2}{\overset{}{\triangle}}\text{C}_6\text{H}_5$$

Submitted by Louis A. Carpino[1] and Louis V. McAdams III
Checked by Timothy P. Higgs and Ronald Breslow

1. Procedure

Caution! Benzene has been identified as a carcinogen; OSHA has issued emergency standards on its use. All procedures involving benzene should be carried out in a well-ventilated hood, and glove protection is required.

To a magnetically stirred solution of 6.36 g. (0.0157 mole) of crude α,α'-dibromodibenzyl sulfone (m.p. 135–150°; *Org. Synth.*, **Coll. Vol. 6**, 403 (1988)) in 40 ml. of dichloromethane contained in a 100-ml., round-bottomed flask fitted with a reflux condenser there is added in one portion 5.05 g. (0.0500 mole) of triethylamine. The solution is heated at reflux with stirring for 3 hours (Note 1), filtered, and the precipitate washed with 5 ml. of cold (0°) dichloromethane. The combined dichloromethane solution is washed with two 20-ml. portions of 3 N hydrochloric acid followed by one 10-ml. portion of water. Removal of the solvent by distillation from a water bath at 30° with the aid of a water aspirator gives 3.0 g. (80%) of the vinylene sulfone as a tan solid, m.p. 116–126° (dec.). The crude solid is washed with 5 ml. of cold (0°) ethanol and recrystallized from about 10 ml. of benzene, yielding 2.4–2.5 g. (63–67%) of the pure sulfone as tiny, snow-white needles, m.p. 123–126° (dec.) (Note 2).

2. Notes

1. After about 45 minutes triethylamine hydrobromide began to precipitate.
2. The submitters obtained a similar percent yield on ten times the scale.

3. Discussion

2,3-Diphenylvinylene sulfone has been prepared by dehydrohalogenation of either α,α'-dibromodibenzyl sulfone[2] or α,α'-dichlorodibenzyl sulfone,[3] as well as by oxidation of 2,3-diphenylvinylene sulfoxide.[4] The procedure described here illustrates a general technique for the synthesis of diarylvinylene sulfones and has been extended to a number of substituted derivatives.[4,5] Aliphatic derivatives have been synthesized in other ways.[2b,6,7] The vinylene sulfones are a group of compounds of considerable theoretical interest.[2,8]

1. Department of Chemistry, University of Massachusetts, Amherst, Massachusetts 01003.
2. (a) L. A. Carpino and L. V. McAdams, III, *J. Am. Chem. Soc.*, **87**, 5804 (1965); (b) L. A. Carpino, L. V. McAdams, III, R. H. Rynbrandt, and J. W. Spiewak, *J. Am. Chem. Soc.*, **93**, 476 (1971).
3. J. C. Philips, J. V. Swisher, D. Haidukewych, and D. Morales, *J. Chem. Soc. D*, 22 (1971).
4. L. A. Carpino and H.-W. Chen, *J. Am. Chem. Soc.*, **93**, 785 (1971) and **101**, 390 (1979).
5. M. H. Rosen and G. Bonet, *J. Org. Chem.*, **39**, 3805 (1974).
6. L. A. Carpino and R. H. Rynbrandt, *J. Am. Chem. Soc.*, **88**, 5682 (1966).
7. L. A. Carpino and J. R. Williams, *J. Org. Chem.*, **39**, 2320 (1974).
8. (a) C. Müller, A. Schweig, and H. Vermeer, *J. Am. Chem. Soc.*, **100**, 8056 (1978) and **97**, 982 (1975); (b) H. L. Hase, C. Müller, and A. Schweig, *Tetrahedron*, **34**, 2983 (1978); (c) F. de Jong, A. J. Noorduin, T. Bouwman, and M. J. Janssen, *Tetrahedron Lett.*, 1209 (1974); (d) D. T. Clark, *Int. J. Sulfur Chem.*, (C) **7**, 11 (1972); (e) H. L. Ammon, L. Fallon, and L. A. Plastas, *Acta Crystallogr.*, Sect. B, **32**, 2171 (1976).

1,3-DITHIANE

$$\text{HS(CH}_2)_3\text{SH} + \text{H}_2\text{C(OCH}_3)_2 \xrightarrow{\text{BF}_3 \cdot (\text{C}_2\text{H}_5)_2\text{O}} \underset{S \quad \quad S}{\bigcirc} + 2\,\text{CH}_3\text{OH}$$

Submitted by E. J. Corey[1] and D. Seebach[2]
Checked by J. C. Reilly and W. D. Emmons

1. Procedure

A 1-l., three-necked, round-bottomed flask with ground-glass fittings is charged with a mixture of 36 ml. of boron trifluoride diethyl etherate, 72 ml. of glacial acetic acid, and 120 ml. of chloroform (Note 1). The flask is equipped with a spiral reflux condenser, an efficient mechanical stirrer, and a dropping funnel (Note 2). The chloroform solution is heated and maintained at reflux with vigorous stirring, and a solution of 32 g. (30 ml., 0.30 mole) of 1,3-propanedithiol and 25 g. (29 ml., 0.33 mole) of methylal (dimethoxymethane) in 450 ml. of chloroform (Note 1) is added at a constant rate over 8 hours. The mixture is allowed to cool to room temperature, washed successively with four 80-ml. portions of water, twice with 120 ml. of 10% aqueous potassium hydroxide, and twice with 80-ml. portions of water. The chloroform solution obtained is dried over potassium carbonate and concentrated in a 500-ml., round-bottomed flask under reduced pressure with a rotating flask evaporator (Note 3). The residue, which crystallizes on cooling to room temperature, is dissolved in 60 ml. of methanol by heating to the boiling point. The hot solution is filtered rapidly through a prewarmed funnel, allowed to cool slowly to room temperature, then kept overnight at −20°. The colorless crystals are collected by filtration through a prechilled Büchner funnel, washed with cold methanol (−20°), and dried under reduced pressure (Notes 3 and 4), yielding 28–29 g. of product, m.p. 53–54°. Solvent is removed from the mother liquor, and the residue is recrystallized as described, giving an additional 1.5–2.0 g.; the total yield of recrystallized 1,3-dithiane is 29.5–31.0 g. (82–86% based on propane-1,3-dithiol). A purer sample (Note 5) can be prepared by subsequent sublimation of the recrystallized product at 0.1–0.5 mm. (45–48° bath temp.) (Note 6). The yield of pure product is 28–30 g. (78–84%), m.p. 53–54°. The residue from the sublimation is a brown syrup weighing less than 1 g.

2. Notes

1. The following components were used as supplied: boron trifluoride etherate, Eastman white label; chloroform and methylal, Fischer reagent; glacial acetic acid, du Pont reagent; 1,3-propanedithiol, Aldrich Chemical Co. The submitters have scaled up this preparation by a factor of 5 without difficulty.

2. Neither the boron trifluoride etherate nor the acetic acid should be added through the dropping funnel, since the presence of even small amounts of acid in the funnel would catalyze undesired condensation of the reagents to be added later. Furthermore the reagents added from the dropping funnel should fall directly into the boiling liquid.

3. Owing to the volatility of 1,3-dithiane, the pressure should be above 30 mm., and the operation should not be prolonged.

4. The crude product can also be distilled; b.p. 95° (20 mm.).

5. This additional purification is recommended if the 1,3-dithiane is to be used in organometallic reactions.[3]

6. Using a sublimator large enough to contain all the recrystallized product obtained, the sublimation can be completed in *ca.* 2 hours at 0.1 mm.

3. Discussion

Although many workers have reported studies with 1,3-dithiane,[4–8] no satisfactory description of its preparation has been published. Generally, 1,3-propanedithiol and formaldehyde[4,6,8] have been used as components; in one instance, 1,3-dibromopropane was treated with sodium thiosulfate to form a precursor of the dithiol which was used as such with formalin.[5] The formation of linear condensation products is a serious side reaction under such conditions.

The procedure given here is a simple and efficient method for producing 1,3-dithiane, a valuable intermediate in the synthesis *via* lithio derivatives of a wide variety of compounds, including aldehydes, ketones, α-hydroxyketones, 1,2-diketones, and α-keto acid derivatives.[3]

1. Department of Chemistry, Harvard University, Cambridge, Massachusetts 02138.
2. Present address: Laboratorium fur Organische Chemie, ETH-Zentrum, CH-8092, Zurich, Switzerland.
3. E. J. Corey and D. Seebach, *Angew. Chem.*, **77**, 1134, 1135 (1965) [*Angew. Chem., Int. Ed. Engl.*, **4**, 1075, 1077 (1965)].
4. W. Autenrieth and K. Wolff, *Ber. Dtsch. Chem. Ges.*, **32**, 1375 (1899).

5. D. T. Gibson, *J. Chem. Soc.*, 12 (1930).
6. E. E. Campaigne and G. F. Schaefer, *Bol. Col. Quim. P. R.*, **9**, 25 (1952) [*Chem. Abstr.*, **46**, 1088d (1952)].
7. H. Friebolin, S. Kabuss, W. Maier, and A. Lüttringhaus, *Tetrahedron Lett.*, 683 (1962).
8. S. Oae, W. Togaki, and A. Ohno, *Tetrahedron*, **20**, 427 (1964).

2,2'-DITHIENYL SULFIDE

(Thiophene, 2,2'-thiobis-)

Submitted by E. Jones and I. M. Moodie[1]
Checked by Jerry G. Kohlhoff, Wayland E. Noland, and William E. Parham

1. Procedure

Caution! Benzene has been identified as a carcinogen; OSHA has issued emergency standards on its use. All procedures involving benzene should be carried out in a well-ventilated hood, and glove protection is required.

A three necked, 250-ml. flask fitted with a reflux condenser stirrer and a thermometer pocket with a nitrogen inlet is charged with 100 ml. of *N,N*-dimethylformamide, 5.6 g. (0.10 mole) of potassium hydroxide (Note 1), and 7.15 g. (0.0500 mole) of freshly precipitated copper(I) oxide[2] (Note 2). 2-Bromothiophene (16.4 g., 0.101 mole) (Note 3) is added, and the apparatus is flushed with nitrogen (Note 4). 2-Thiophenethiol[3] [11.6 g., 0.100 mole; *Org. Synth.*, **Coll. Vol. 6,** 979 (1988)] is then added slowly through the condenser; an exothermic reaction begins, and the temperature may rise to 50−60°. The flask is heated in an oil bath at 130−140° for 16 hours (Note 5). The mixture is cooled to room temperature before the contents of the flask are poured into 100 ml. of 6 *N* hydrochloric acid in ice. The mixture is stirred vigorously for 2 hours (Note 6); the oily black paste obtained is removed by filtration and thoroughly extracted in a Soxhlet extractor with benzene until a colorless extract is obtained. The filtrate is also extracted with two 100-ml. portions of benzene. These extracts are combined, washed with water until neutral, and dried over anhydrous sodium sulfate. Removal of the solvent gives a yellow-colored oil which, on vacuum distillation, yields 11.5−12.5 g. (58−63%) of 2,2'-dithienyl sulfide as a pale yellow oil, b.p. 75−78° (0.06 mm.) n_D^{25} 1.6643.

2. Notes

1. The presence of an equivalent of potassium or sodium hydroxide is necessary to promote the substitution reaction of the 2-bromothiophene with copper(I) oxide.

2. Freshly precipitated copper(I) oxide was dried at 110° before use. Commercial grades were found to give lower yields (*ca.* 35–40%) of the desired sulfide. The checkers dried copper(I) oxide in a vacuum oven at 100° for 6 hours.

3. 2-Bromothiophene was supplied by Columbia Organic Chemicals, Inc. 2-Chloro- or 2-iodothiophenes may be used; however, the former gives poorer yields of the sulfide.

4. Nitrogen is passed through the system to provide an inert atmosphere, preventing possible oxidation of the 2-thiophenethiol to the corresponding disulfide.

5. Reaction is sluggish at temperatures below 130°.

6. Vigorous stirring helps to break up the thick sludge formed on addition of the reaction mixture to the acid and also removes soluble inorganic salts.

3. Discussion

The submitters have also prepared 2,2'-dithienyl sulfide in 34% yield by condensation of 2-thiophenethiol with 2-bromothiophene in the presence of copper(I) oxide in a quinoline–pyridine mixture.[4] Challenger and Harrison[5] have obtained 2,2'-dithienyl sulfide in 50–55% yield by treatment of 2-thienylmagnesium bromide with excess sulfur. This sulfide may also be obtained in 20% yield by condensation of thiophene with sulfur monochloride, followed by pyrolysis of the resultant disulfide.[6]

This synthetic process offers a route to the preparation of the isomeric dithienyl sulfides[7] (2,3- and 3,3-) which cannot be prepared readily by any of the standard methods. Thus, condensation of 2-thiophenethiol with 3-bromothiophene or 3-thiophenethiol with 2-bromothiophene gives 2,3'-dithienyl sulfide in 63.0 and 73.5% yields, respectively. Similarly, 3,3'-dithienyl sulfide is obtained in 48% yield. The method has also been extended to the synthesis of the isomeric *bis*-(thienylthio)thiophenes in 40–50% yield.[4]

This method may also be used for the preparation in high yields of other aromatic sulfides.[8]

1. Work done at the former Arthur D. Little Research Institute, Inveresk, Midlothian, Scotland.
2. A. King, "Inorganic Preparations," rev. ed., Geo. Allen and Unwin, London, 1950, p. 40.
3. W. H. Houff and R. D. Schuetz, *J. Am. Chem. Soc.,* **75,** 6316 (1953).
4. E. Jones and I. M. Moodie, unpublished observations.
5. F. Challenger and J. B. Harrison, *J. Inst. Pet.,* **21,** 135 (1935).
6. E. Koft, U.S. Pat. 2,571,370 (1951).
7. E. Jones and I. M. Moodie, *Tetrahedron,* **21,** 2413 (1965).
8. R. G. R. Bacon and H. A. O. Hill, *J. Chem. Soc.,* 1108 (1964).

SELECTIVE EPOXIDATION OF TERMINAL DOUBLE BONDS: 10,11-EPOXYFARNESYL ACETATE

[2,6-Nonadien-1-ol, 9-(3,3-dimethyloxiranyl)-3,7-dimethyl-, acetate, (*E*,*E*)-]

Submitted by R. P. HANZLIK[1]
Checked by A. GRIEDER and G. BÜCHI

1. Procedure

A.   *Farnesyl acetate*.   A solution of 25 g. (0.11 mole) of farnesol (Note 1) in 40 ml. of dry pyridine (Note 2) is prepared in a stoppered, 250-ml. Erlenmeyer flask, and 40 ml. of acetic anhydride is added in four portions over a 15-minute period. The mixture is stirred well, allowed to stand for 6 hours then poured onto 250 g. of ice. Water is added (400 ml.), and the mixture is extracted with five 100-ml. portions of petroleum ether (b.p.

60–68°). The organic extracts are combined and washed in succession with two 50-ml. portions each of water, 5% sulfuric acid, and saturated aqueous sodium hydrogen carbonate, dried over anhydrous magnesium sulfate (*ca*. 50 g.), and concentrated on a rotary evaporator, yielding 28–29 g. (94–98%) of farnesyl acetate as a colorless oil (Note 3).

B. 10-*Bromo*-11-*hydroxy*-10,11-*dihydrofarnesyl acetate*. Farnesyl acetate (29 g., 0.11 mole) is dissolved in 1 l. of *tert*-butyl alcohol (Note 4) contained in a 3-l. Erlenmeyer flask. Water is added (500 ml.), and the solution is cooled to about 12° with an external ice water bath. Maintaining this temperature, rapid magnetic stirring is begun, and more water is added until a saturated solution is obtained. The second addition of water may be rapid initially, but the saturation point must be approached carefully, like the end point of a titration. A total of about 1200 ml. of water is required for the specified quantities of farnesyl acetate and *tert*-butyl alcohol. The solution must remain clear and homogeneous at about 12°, and if the saturation point is accidentally passed by adding too much water, *tert*-butyl alcohol should be added to remove the turbidity.

External cooling is discontinued, and 21.4 g. (0.120 mole) of *N*-bromosuccinimide (Note 5) is added. Stirring is continued until all of the solid is dissolved (*ca*. 1 hour). The resulting solution, which may be pale yellow, is concentrated with a rotary evaporator (bath temperature 40–45°) to a volume of about 300 ml. and extracted with five 120-ml. portions of diethyl ether. The combined ether extracts are dried over anhydrous magnesium sulfate (20–50 g.), and removal of solvent at reduced pressure provides an oil, which is purified by column chromatography on silica gel (Note 6). The pure bromohydrin acetate is obtained as a colorless oil in amounts of up to 26 g., a 65% yield based on farnesyl acetate (Notes 7 and 8).

C. 10,11-*Epoxyfarnesyl acetate*. The bromohydrin acetate prepared in Part B is dissolved in 300 ml. of methanol, the solution is placed in a 500-ml. flask, and excess solid potassium carbonate (three times the molar amount of bromohydrin acetate) is added (Note 9). The mixture is stirred for 12 hours and then concentrated to *ca*. 100 ml. on a rotary evaporator (bath at 40–45°). Water is added (200 ml.), and the mixture is extracted with four 100-ml. portions of petroleum ether (b.p. 60–68°). The combined extracts are dried over anhydrous magnesium sulfate (*ca*. 40 g.) and evaporated, leaving 10,11-epoxyfarnesol as a colorless oil (Note 10).

This material is acetylated with 35 ml. of pyridine and 35 ml. of acetic anhydride for 6 hours at room temperature. The mixture is poured onto 250 g. of ice and extracted with five 75-ml. portions of ether. The combined ether extracts are washed successively with three 10-ml. portions each of saturated aqueous sodium hydrogen carbonate and water then dried over magnesium sulfate. It is important not to use strong acids such as hydrochloric or sulfuric to remove pyridine, as was done in Part A, since they can destroy the acid-sensitive product.

Concentration of the ethereal solution at reduced pressure gives the epoxyacetate as a colorless oil more viscous than water. The overall yield based on farnesyl acetate is near 60% (Note 11). This material is reasonably pure if the preparation has been executed carefully, but it can be further purified by column chromatography (Note 12) or distillation (Note 13).

2. Notes

1. Farnesol was obtained from Fluka AG (Buchs, CH9470, Switzerland) as a mixture of 65% $(E),(E)$- and 35% $(Z),(E)$-isomers. It is also available from the Aldrich Chemical Company, Inc. This procedure works equally well with pure $(E),(E)$-farnesol, which may be obtained from the above mixture by careful distillation, at reduced pressure, through a Nester-Faust Teflon spinning-band column.

2. Pyridine was distilled from sodium hydroxide.

3. IR (neat) cm.$^{-1}$: 1740, 1240; ^{1}H NMR (CCl$_4$), δ (multiplicity, coupling constant J in Hz., number of protons, assignment): 2.00 (s, 3H, O$_2$CCH$_3$), 4.46 (d, $J = 7$, 2H, CH$_2$O), 4.49–5.4 (m, 3H, olefinic H).

4. Tetrahydrofuran or glyme work equally well in place of tert-butyl alcohol as a co-solvent, but should be distilled under dry nitrogen from lithium aluminum hydride[2] or sodium and benzophenone [*Org. Synth.*, **Coll. Vol. 5,** 201 (1973)] prior to use, to destroy peroxides which may be present.

5. The N-bromosuccinimide should be nearly white and uncontaminated by free bromine. If necessary, it may be recrystallized from hot water and stored in a refrigerator.

6. The submitters mixed active anhydrous silica gel with water (12% w/w) and stored it in a sealed container for at least 24 hours prior to use. A ratio of 60–80 g. of silica gel per gram of crude product was used for column chromatographic separations, and a column was chosen that would give a 10:1 height:diameter ratio of adsorbent. Columns were wet-packed with distilled petroleum ether (b.p. 60–68°), and after the crude product had been applied a step-gradient was run through rapidly: 2% ether, 5% ether, and 10% ether (v/v) in petroleum ether. The column was then eluted with 20% v/v ether in petroleum ether until the bromohydrin acetate was obtained.

The checkers obtained roughly 30 g. of crude product in each run. Freshly opened Woelm silica gel (obtained from ICN Pharmaceuticals, 26201 Miles Ave., Cleveland, Ohio 44128) was deactivated as above, and 1800 g. was wet-packed with petroleum ether in a 65-mm.-internal-diameter column. In the first run the column was eluted as above, but a considerable amount of solvent was required to collect the product. Therefore, in the second run the crude product was applied to the column as a solution in petroleum ether, and 1-l. portions of 20% v/v ether:petroleum ether, 30% ether, 40% ether, 50% ether, 60% ether, and 70% ether were run through. None of these six fractions contained a significant weight of material. Elution with 2 l. of 80% v/v ether:petroleum ether provided the bromohydrin acetate.

Fractions may be monitored by TLC on silica gel, developing with 10% v/v ethyl acetate in hexane and visualizing with iodine vapor. The following R_f values were observed: farnesol, 0.07; farnesyl acetate, 0.35; bromohydrin acetate, 0.20.

7. In each of two runs, the checkers obtained 25 g. (63% yield).

8. IR (neat) cm.$^{-1}$: 3520, 3450, 1740, 1235; ^{1}H NMR (CCl$_4$), δ (multiplicity, coupling constant J in Hz., number of protons, assignment): 1.32 [s, 6H, OC(CH$_3$)$_2$], 2.00 (s, 3H, O$_2$CCH$_3$), 3.7–4.0 (m, 1H, CHBr), 4.50 (d, $J = 7$, 2H, CH$_2$O), 4.9–5.5 (m, 2H olefinic H).

9. Using a smaller amount of K$_2$CO$_3$ made the reaction much slower and did not avoid or reduce the accompanying loss of the acetate group. Pyridine at room temperature is not sufficiently basic to form the epoxide. Other bases were not tested.

10. IR (neat) cm.$^{-1}$: 3450, 2940, 1450, 1375; ^{1}H NMR (CCl$_4$), δ (multiplicity, coupling constant J in Hz., number of protons, assignment): δ 1.25 and 1.29 [2s, 6H, OC(CH$_3$)$_2$], 2.55 (t, $J = 6$, 1H, OCH), 3.97 (d, $J = 7$, 2H, OCH$_2$), 5.0–5.5 (m, 2H, olefinic CH).

11. In each of two runs, the checkers obtained 19 g. (61% yield). IR (neat) cm.$^{-1}$: 2950, 1740, 1380, 1235; ^{1}H NMR (CCl$_4$), δ (multiplicity, coupling constant J in Hz., number of protons, assignment): 1.21 and 1.23 [2s, 6H, OC(CH$_3$)$_2$], 1.97 (s, 3H, O$_2$CCH$_3$), 2.50 (t, $J = 6$, 1H, OCH), 4.48 (d, $J = 7$, 2H, OCH$_2$), 5.0–5.5 (m, 2H, olefinic CH).

12. Deactivation of silica gel and preparation of the column is carried out as in Note 6, except that the checkers consider 20 g. of silica gel per gram of crude product to be adequate in this case. Running through a gradient of petroleum ether containing increasing amounts of ether, the submitters found that the product was eluted with 15% (v/v) ether, but the checkers found that 25% (v/v) ether was required.

13. The submitters recommend distillation at 100° at less than 0.005 mm. The checkers distilled 1–2 g.-samples at 0.05 mm. (b.p. 113°) and at 0.15 mm. (b.p. 117°), and in both cases a clean product was obtained with high recovery.

3. Discussion

10,11-Epoxyfarnesol was first prepared by van Tamelen, Storni, Hessler, and Schwartz[3] using essentially this procedure, which is based on the findings of van Tamelen and Curphey[4] that N-bromosuccinimide in a polar solvent is a considerably more selective oxidant than others tried. This method has been applied to produce terminally epoxidized mono-, sesqui-, di-, and triterpene systems for biosynthetic studies and bioörganic synthesis.[5] It has also been applied successfully in a simple synthesis of tritium-labeled squalene and squalene-2,3-oxide[6] and in the synthesis of *Cecropia* juvenile hormone.[7]

The oxidation procedure described above is intended to illustrate the selectivity that may be achieved using this system for functionalizing only one of several superficially similar double bonds in a molecule. In the case of acyclic terpenes, the yield of the desired terminal monobromohydrin decreases from 75–80% with geraniol (C$_{10}$H$_{18}$O) to 30–35% with squalene (C$_{30}$H$_{50}$), due to competing formation of polybromohydrins, allylic bromides, and bromocyclized material. The significant point, however, is that in all cases more than 95% of the monobromohydrin produced results from attack at the terminal double bond. A mechanistic investigation[8] showed that the N-bromosuccinimide was not merely providing a source of bromine or hypobromous acid, and that the reaction was promoted by acid and inhibited by base.

1. Department of Medicinal Chemistry, School of Pharmacy, University of Kansas, Lawrence, Kansas 66045.
2. L. F. Fieser and M. Fieser, "Reagents for Organic Synthesis," Vol. 1, Wiley, New York, 1967, p. 1140.
3. E. E. van Tamelen, A. Storni, E. J. Hessler, and M. Schwartz, *J. Am. Chem. Soc.*, **85**, 3295 (1963).
4. E. E. van Tamelen and T. J. Curphey, *Tetrahedron Lett.*, 121 (1962).
5. For leading references see E. E. van Tamelen and R. J. Anderson, *J. Am. Chem. Soc.*, **94**, 8225 (1972); E. E. van Tamelen, *Acc. Chem. Res.*, **1**, 111 (1968).

6. R. Nadeau and R. Hanzlik, "Synthesis of Labeled Squalene and Squalene-2,3-Oxide," in *Methods in Enzymology*, Vol. 15, R. B. Clayton, Ed., Academic Press, New York, 1969, p. 346.
7. E. J. Corey, J. A. Katzenellenbogen, N. W. Gilman, S. A. Roman, and B. W. Erickson, *J. Am. Chem. Soc.*, **90**, 5618 (1968); E. E. van Tamelen and J. P. McCormick, *J. Am. Chem. Soc.*, **92**, 737 (1970).
8. E. E. van Tamelen and K. B. Sharpless, *Tetrahedron Lett.*, 2655 (1967).

1-ETHOXY-1-BUTYNE

(1-Butyne, 1-ethoxy-)

$$ClCH_2CH(OCH_2CH_3)_2 \xrightarrow[\text{NH}_3]{\text{NaNH}_2} NaC\equiv COCH_2CH_3$$

$$\xrightarrow{C_2H_5Br} CH_3CH_2C\equiv COCH_2CH_3$$

Submitted by MELVIN S. NEWMAN[1] and W. M. STALICK[2]
Checked by P. J. KOCIENSKI and G. BÜCHI

1. Procedure

A 1-l., three-necked, round-bottomed flask equipped with an efficient dry-ice condenser (Note 1), a mechanical stirrer, and a gas-inlet tube is immersed in an acetone–dry ice bath, and 600 ml. of anhydrous ammonia is introduced. After replacing the inlet tube with a stopper, the cooling bath is lowered but kept beneath the flask. To the slowly stirred ammonia is added 0.5 g. of hydrated iron(III) nitrate. Air, dried by passing through calcium chloride, is bubbled through the solution for about 10 seconds (Note 2) after a small piece of freshly cut sodium is added (Note 3). Once hydrogen evolution has ceased, the blue color is discharged, leaving a finely divided, black precipitate. Small pieces of freshly cut sodium are then added over a 20 minute period until 36.0 g. (1.56 g.-atom) has been added. After the formation of sodium amide is complete (Note 4), the stopper is replaced with a pressure-equalizing dropping funnel containing 76.1 g. (0.499 mole) of chloroacetaldehyde diethyl acetal (Note 5), and the addition is made over a period of 20 minutes (Note 6). After 30–60 minutes the mixture becomes light gray, and 120 g. (1.10 mole) of freshly distilled ethyl bromide is added rapidly through the addition funnel (Note 7). The mixture is stirred vigorously for 2.5 hours, after which 30 ml. of cooled, saturated ammonium chloride solution is *cautiously* added through the addition funnel, followed by 120 ml. of pentane and an additional 370 ml. of the cooled, saturated ammonium chloride solution. The contents of the flask are transferred to a 2-l. separatory funnel (in the hood); the lower aqueous layer is removed and extracted with two 75-ml. portions of pentane (Note 8). The combined organic layers are filtered through glass wool to dissipate any emulsions, dried over magnesium sulfate, and filtered through a coarse-fritted funnel with gentle suction into a 500-ml., round-bottomed flask. A magnetic stirring bar is added, and the pentane is removed by distillation at atmospheric pressure through a 20 × 2 cm. column packed with glass beads (Note 9) and fitted with a well-cooled fractionating head. With an acetone–dry ice trap between the receiver and the vacuum source, the yellow residue is distilled under reduced pressure (Note 10), with rapid magnetic stirring, into dry-ice-cooled receivers, giving 1-ethoxy-1-butyne as a clear, colorless liquid at 43–45° (50 mm.). A lower-boiling fraction collection at 20–42° (50 mm.) is combined with any material removed from the dry-ice trap and redistilled, yielding additional product, for a total of 30.2–32.3 g. (62–66%) (Note 11).

2. Notes

1. A gas bubbling device is attached to the dry-ice condenser. A simple apparatus consists of two, 500-ml. filtering flasks equipped with one-hole neoprene stoppers and glass tubing extending to the bottom of the flasks. The two flasks are connected through the glass tubing by a short piece of Tygon tubing. About 150 ml. of mineral oil is then placed in the flask distant from the condenser.

2. The checkers found that air was not necessary to initiate the formation of sodium amide. See Ref.[3]

3. The sodium is cut under dry pentane just before introducing each sample into the flask.

4. Complete conversion into sodium amide is indicated by cessation of gas evolution and disappearance of the blue color of the solution. This generally requires 20–30 minutes and results in a gray suspension of sodium amide in a dark-gray reaction medium.

5. Chloroacetaldehyde diethyl acetal was used as obtained from Aldrich Chemical Company, Inc.

6. During any of the additions in this preparation excessive foaming may occur. This may be effectively diminished by interrupting the addition of the reagent or by brief immersion of the reaction flask in the acetone–dry-ice bath. If foaming has reached the condenser, be certain that the condenser is not plugged before proceeding.

7. The addition of ethyl bromide is accompanied by vigorous reflux of the ammonia and should be carefully monitored.

8. Since 1-ethoxy-1-butyne is very volatile, extreme care should be taken during the work-up to minimize loss of product due to evaporation. Extractions should be accompanied by careful and frequent venting of the separatory funnel to prevent excessive pressure.

9. The checkers used a 20-cm. Widmer column for the distillation.

10. The pot temperature should be kept below 80°. Distillation must be conducted at temperatures below 90° to preclude dimerization.[4]

11. IR (CCl_4) cm.$^{-1}$, strong peaks: 1245, 1230, 1015, 855. ^{1}H NMR (CCl_4), δ (multiplicity, number of protons, assignment): 1.23 (overlapped t, 6H, 2CH_3), 2.13 (q, 2H, C≡CCH_2), 3.99 (q, 2H, OCH_2).

3. Discussion

The synthesis of 1-ethoxy-1-butyne has been reported previously, but the preparations have required multistep sequences. Two of the procedures use 1,2-dibromo-1-ethoxy butane which is dehydrohalogenated in two successive steps, first by an amine base and then by either powdered potassium hydroxide[5] or sodium amide;[6] no yields are given. The other procedure starts with 1,2-dibromoethyl ethyl ether which, upon treatment with

N,N-diethylaniline, yields 2-bromovinyl ethyl ether. When 2-bromovinyl ethyl ether is allowed to react with lithium amide in ammonia, followed by alkylation with diethyl sulfate, 1-ethoxy-1-butyne is isolated in about 55% yield.[7]

Some studies seeking preferred conditions for this reaction have been reported.[8] Optimum yields of 1-ethoxy-1-propyne and 1-ethoxy-1-butyne are found when the product is worked up before allowing the ammonia solvent to evaporate, as the product evidently volatilizes with the ammonia. An experiment with 1-ethoxy-1-propyne showed a marked increase in yield when ammonia predried over calcium hydride was used instead of ammonia directly obtained from a cylinder. A twofold excess of ethyl bromide is required to obtain a good yield of 1-ethoxy-1-butyne, since elimination apparently competes with alkylation in this case.

1. Department of Chemistry, The Ohio State University, Columbus, Ohio 43210. This research was conducted at The Ohio State University and supported in part by National Science Foundation Grant No. 12445.
2. Department of Chemistry, George Mason University, Fairfax, Virginia 22030.
3. L. F. Fieser and M. F. Fieser, "Reagents for Organic Synthesis," Vol. I., Wiley-Interscience, New York, 1967, p. 1034.
4. J. Nieuwenhuis and J. F. Arens, *Recl. Trav. Chim. Pays-Bas,* **77,** 761 (1958).
5. J. Ficini, *Bull. Soc. Chim. Fr.,* 1367 (1954).
6. H. Olsman, *Proc. K. Ned. Akad. Wet., Ser. B, Phys. Sci.,* **69,** 645 (1966).
7. L. Brandsma, "Preparative Acetylenic Chemistry," Elsevier, New York, 1971, pp. 41 and 191.
8. M. S. Newman, J. R. Geib, and W. M. Stalick, *Org. Prep. Proced. Int.,* **4,** 89 (1972).

DEMETHYLATION OF METHYL ARYL ETHERS: 4-ETHOXY-3-HYDROXYBENZALDEHYDE

(Benzaldehyde, 4-ethoxy-3-hydroxy-)

Submitted by ROBERT E. IRELAND[1] and DAVID M. WALBA[2]
Checked by R. ANDERSEN and G. BÜCHI

1. Procedure

Caution! Benzene has been identified as a carcinogen; OSHA has issued emergency standards on its use. All procedures involving benzene should be carried out in a well-ventilated hood, and glove protection is required.

A. *4-Ethoxy-3-methoxybenzaldehyde ethylene acetal.* A 500-ml., three-necked flask with vertical necks is fitted with a magnetic stirring bar and a 30-ml. water separator, to which is attached a condenser topped with an argon inlet. The flask is charged with 300 mg. (0.00158 mole) of *p*-toluenesulfonic acid monohydrate, 270 ml. of benzene (Note 1), 33 g. (30 ml., 0.53 mole) of ethylene glycol, and 5.0 g. (0.028 mole) of 4-ethoxy-3-methoxybenzaldehyde (Note 2), then placed under a positive pressure of argon, which is maintained throughout the reaction. Vigorous stirring is begun, and the solution is brought to reflux with an oil bath at 110°.

After 20 hours at reflux, the mixture is cooled to room temperature with a water bath and poured with vigorous stirring into 500 ml. of 10% aqueous potassium carbonate contained in a 1-l. separatory funnel. The benzene layer is washed successively with two 250-ml. portions of 10% aqueous potassium carbonate and 250 ml. of brine containing potassium carbonate (Note 3), dried over sodium sulfate for 10 minutes, and filtered. Removal of the solvent on a rotary evaporator with a 25–35° water bath provides a solid yellow residue, which is dried under vacuum.

To obtain a crystalline product, a solution of the residue in 30 ml. of benzene containing a few drops of triethylamine (Note 4) is placed in a 250-ml. Erlenmeyer flask, heated gently on a steam bath, and diluted with 150 ml. of hexane. Heating is continued for about 5 minutes (Note 5), after which the solution is allowed to cool to room temperature, seeded, and put in a freezer at −15° for at least 5 hours. The resulting solid is collected by suction filtration, washed with cold hexane, and vacuum dried, giving 5.8 g. (94%) of light cream-colored crystals, m.p. 75–77°.

B.   *4-Ethoxy-3-hydroxybenzaldehyde.*   A 100-ml., three-necked flask containing a magnetic stirring bar and fitted with an argon inlet, a rubber septum, and a ground-glass stopper in the center neck is evacuated, flame-dried, and allowed to cool under a positive pressure of argon, which is maintained throughout the following sequence. Using a syringe, 30 ml. of dry tetrahydrofuran (Note 6) and 5.0 ml. (0.029 mole) of diphenylphosphine (Note 7) are added through the septum. The resulting solution is stirred and cooled with an ice bath, and 15 ml. (0.032 mole) of cold 2.1 *M* n-butyllithium–hexane solution (Note 8) is added by syringe over *ca.* 3 minutes. Stirring is continued as the red solution is allowed to warm to room temperature over about 30 minutes before 5.0 g. (0.022 mole) of 4-ethoxy-3-methoxybenzaldehyde ethylene acetal is added through the center neck. The flask is stoppered, and the mixture is stirred at room temperature for 2 hours.

The reaction mixture is then poured into a 500-ml. Erlenmeyer flask containing 200 ml. of vigorously stirred water, 10 ml. of 10% aqueous sodium hydroxide is added, and the mixture is transferred to a 500-ml. separatory funnel. The reaction vessel and Erlenmeyer flask are rinsed with water, and the rinsings are also poured into the funnel. Alkali-insoluble impurities are removed by washing the basic aqueous phase with four 100-ml. portions of diethyl ether, which are combined and backextracted with two 50-ml. portions of 10% aqueous sodium hydroxide. The combined aqueous layers are then put into a 1-l. Erlenmeyer flask, cooled in an ice bath, and acidified with concentrated hydrochloric acid to a Congo red end point. During acidification the clear yellow basic solution becomes cloudy white. This milky suspension is stirred without cooling for 3 minutes then extracted with 200 ml. of ether and two 100-ml. portions of ether. The combined ether layers are washed successively with 100 ml. of water and 100 ml. of saturated aqueous sodium chloride, dried over magnesium sulfate, and filtered. Removal of solvent with a rotary evaporator provides a residue that is vacuum dried, yielding 3.58–3.60 g. (97–98%) of a slightly yellow solid, m.p. 121.5–126°. One recrystallization from 20 ml. of benzene gives almost white crystals which are vacuum dried, affording 3.21–3.29 g. (87–88%) of 4-hydroxy-3-ethoxybenzaldehyde, m.p. 125.5–127° (Notes 9 and 10).

2. Notes

1. Reagent grade benzene was used without further purification.

2. Practical grade 4-ethoxy-3-methoxybenzaldehyde was obtained by the submitters from MC and B Manufacturing Chemists and by the checkers from Aldrich Chemical Company, Inc. This material was purified by distillation (b.p. 125–135°/0.1 mm.), followed by one recrystallization from cyclohexane (100 ml./10 g. crude solid). Colorless crystals, m.p. 60–62°, were obtained after filtration and vacuum drying. Purification of 20 g. of the commercial material gave about 15 g. of recrystallized product.

3. This solution is prepared by dilution of 25 ml. of 10% aqueous potassium carbonate to 250 ml. with saturated aqueous sodium chloride.

4. Triethylamine was distilled from calcium hydride prior to use and added to the benzene to protect the sensitive acetal from hydrolysis.

5. It is not necessary to boil the solution. This heating merely prevents crystals from coming out of solution too fast on addition of the hexane.

6. Tetrahydrofuran was purified and dried according to the procedure described in *Org. Synth.*, **Coll. Vol. 5,** 976 (1973).

7. Commercial diphenylphosphine obtained from Orgmet, Inc., may be used without further purification. Alternatively, the material may be prepared from triphenylphosphine as follows. A 2-l., three-necked flask containing a magnetic stirring bar and fitted with an argon inlet is charged with 120 g. (0.458 mole) of triphenylphosphine and 1 l. of dry tetrahydrofuran (Note 6). To the stirred solution is added 18.42 g. (2.670 g.-atoms, 542 cm. of 0.32 cm.-diameter wire) of lithium wire, which has been washed with hexane and dried carefully with a paper towel. *(Caution! If the towel is rubbed against the lithium too fast, a fire will result.)* Lithium is added by cutting 3–5-mm. segments directly into the center neck of the flask with scissors. A slow argon flow is maintained throughout the addition, which requires about 20 minutes. The flask is stoppered, and the red solution is stirred for 2.5 hours under argon.

The solution is then filtered through a piece of glass wool (fitted loosely in a funnel) into a 2-l. beaker containing 600 g. of crushed ice. A glass rod is used to stir the mixture, and the reaction vessel and filter are rinsed with ether. The resulting two clear phases are transferred to a 2-l. separatory funnel and extracted with four 200-ml. portions of ether. The combined ether layers are washed with 250 ml. of 5% hydrochloric acid, 250 ml. of water, and two 250-ml. portions of saturated aqueous sodium chloride, then dried over magnesium sulfate for ½ hour. Since diphenylphosphine is susceptible to air oxidation, especially in dilute ether solution, the extractions should be carried out as quickly as possible. After gravity filtration, the ether solution is concentrated on a rotary evaporator, and the residue is vacuum dried, yielding 83.7 g. (99%) of crude product. Pure material is then obtained by distillation. The submitters used a small (14/20 joints, 15 cm. long) Vigreux column and observed b.p. 90–103° (0.06 mm.). Using a 20-cm. Vigreux column, the checkers observed b.p. 95–115° (0.06 mm.). In either case, the yield of clear liquid was 62 g. (74%). If the product is stored under argon in a bottle sealed with a rubber serum cap, it is stable for months at room temperature. *Caution! Care should be taken not to get any diphenylphosphine on a paper towel, as it may ignite spontaneously.*

8. *n*-Butyllithium in hexane was obtained from Ventron Corporation and stored in a refrigerator under argon. The solution was titrated with 2-butanol in xylene, using 1,10-phenanthroline as indicator.

9. The literature m.p. for the colorless crystals is 127–128°.[3]

10. IR (CHCl$_3$) cm.$^{-1}$: 3550 (OH), 1680 (C=O), 1610, 1580, 1510, 1470; ^{1}H NMR (CDCl$_3$), δ (multiplicity, coupling constant J in Hz., number of protons): 1.47 (t, J = 7,

3. Discussion

4-Ethoxy-3-hydroxybenzaldehyde (isobourbonal) has been prepared in good yield

The present procedure illustrates the facile demethylation of methyl aryl ethers with lithium diphenylphosphide.[4] This reaction is specific for methyl ethers and may be carried out in the presence of ethyl ethers in high yield.[5] Use of excess reagent allows cleavage in the presence of enolizable ketones.[6] In the present case, the cleavage may be performed without protection of the aldehyde, but two equivalents of reagent are required, and the yield is reduced to *ca.* 60%.

The exact time and temperature required for complete reaction must be determined for each individual compound. It has been observed that nucleophilic demethylation of methyl *o*-alkoxyaryl ethers is accelerated relative to anisole,[7] and this reaction is no exception. Lithium diphenylphosphide cleavage of anisole is complete in about 4 hours in refluxing tetrahydrofuran, whereas the present reaction is complete within 2 hours at 25°.

1. Department of Chemistry, California Institute of Technology, Pasadena, California 91109.
2. Present address: Department of Chemistry, University of Colorado, Campus Box 215, Boulder, Colorado 80309.
3. T. Kametani, H. Iida, and C. Kobayashi, *J. Heterocycl. Chem.*, **7**, 339 (1970).
4. F. G. Mann and M. J. Pragnell, *J. Chem. Soc.*, 4120 (1965).
5. R. E. Ireland and S. Welch, *J. Am. Chem. Soc.*, **92**, 7232 (1970).
6. Gloria Pfister, unpublished results, this laboratory.
7. G. I. Feutrill and R. N. Mirrington, *Aust. J. Chem.*, **25**, 1719, 1731 (1972).

CARBENE GENERATION BY α-ELIMINATION WITH LITHIUM 2,2,6,6-TETRAMETHYLPIPERIDIDE: 1-ETHOXY-2-p-TOLYL-CYCLOPROPANE

[Benzene, 1-(2-ethoxycyclopropyl)-4-methyl]

Submitted by Charles M. Dougherty[1] and Roy A. Olofson[2]
Checked by Mark W. Johnson and Robert M. Coates

1. Procedure

Caution! Benzene has been identified as a carcinogen; OSHA has issued emergency standards on its use. All procedures involving benzene should be carried out in a well-ventilated hood, and glove protection is required.

A 250-ml., three-necked, round-bottomed flask equipped with a 50-ml. pressure-equalizing dropping funnel capped by a rubber septum, an efficient reflux condenser connected to a nitrogen inlet, and a magnetic stirrer (Note 1) is charged with 7.02 g. (0.0500 mole) of α-chloro-p-xylene (Note 2) and 45.6 g. (0.633 mole) of ethyl vinyl ether (Note 3). A solution of 7.06 g. (0.0501 mole) of 2,2,6,6-tetramethylpiperidine (Note 4) in 15 ml. of dry diethyl ether is injected through the septum into the dropping funnel. Lithium 2,2,6,6-tetramethylpiperidide is generated *in situ* by injecting 46.5 ml. (0.0502 mole) of a 1.08 M solution of methyllithium in ether (Note 5) through the septum over a 5–10-minute period (Notes 6 and 7). After another 10 minutes, the contents are added dropwise to the vigorously stirred solution in the flask at a rate that maintains a gentle reflux. When the *ca.* 2-hour addition period is complete, the white slurry is stirred overnight at room temperature (Note 8). Water (10 ml.) is added dropwise to the stirred suspension, and the contents of the flask are poured into a separatory funnel containing 100 ml. of ether and 100 ml. of water. The aqueous layer is separated and extracted with two 100-ml. portions of ether. The combined ether solutions are washed successively with 10% aqueous citric acid (Note 9), 5% aqueous sodium hydrogen carbonate, and water, dried with anhydrous calcium chloride, filtered and evaporated with a rotary evaporator.

The residual liquid is distilled at reduced pressure, affording 6.6–7.0 g. (75–80%) of 1-ethoxy-2-*p*-tolylcyclopropane, b.p. 116–118° (10 mm.), 95–96° (3.2 mm.) (Note 10).

2. Notes

1. The glassware is dried in an oven at approximately 125° and assembled while still warm. The nitrogen inlet, which consists of a T-tube assembly connected to an oil bubbler, is attached, and the apparatus is allowed to cool while being swept with a stream of dry nitrogen. The septum is placed on top of the dropping funnel, and the nitrogen flow adjusted to maintain a slight positive pressure of nitrogen within the apparatus during the reaction.

2. *α*-Chloro-*p*-xylene was obtained from Aldrich Chemical Company, Inc., and purified by distillation under reduced pressure.

3. Ethyl vinyl ether was supplied by Aldrich Chemical Company, Inc., and distilled from sodium. If simple alkenes are used in place of ethyl vinyl ether, the submitters find that the yields of cyclopropanes are improved by dilution of the olefin with one or two volumes of ethyl ether.

4. 2,2,6,6-Tetramethylpiperidine, furnished by Aldrich Chemical Company, Inc., Fluka A G, and ICN Life Sciences Group, is sometimes contaminated with traces of water, hydrazine, and/or 2,2,6,6-tetramethyl-4-piperidone. These impurities may be removed by drying with sodium hydroxide or potassium hydroxide pellets, filtering, and distilling at atmospheric pressure, b.p. 153–154°. The purified amine can be stored indefinitely under a nitrogen atmosphere.

5. Methyllithium in ethyl ether from Ventron Corporation was used. Directions for the preparation of ethereal methyllithium from methyl bromide are also available [see *Org. Synth.*, **Coll. Vol. 6**, 901 (1988).] The checkers standardized the solution immediately before use by diluting a 2.5-ml. aliquot with 10 ml. of benzene and titrating with a 1 *M* solution of 2-butanol in xylene according to the procedure of Watson and Eastham[3] [see *Org. Synth.*, **Coll. Vol. 6**, 121 (1988)], with 1,10-phenanthroline as indicator. The submitters report that the yield of arylcyclopropane is lower if a commercially available solution of *n*-butyllithium in hydrocarbon solvents is used.

6. The methane generated is vented by passage through the oil bubbler.

7. Since the reaction between methyllithium and 2,2,6,6-tetramethylpiperidine is relatively slow at lower temperatures, lithium 2,2,6,6-tetramethylpiperidide is best prepared at room temperature. The reagent may, however, be used over a wide range of temperatures.

8. Approximately the same yields are obtained if the product is isolated after 2–3 hours.

9. The use of aqueous citric acid avoids the formation of insoluble gelatinous precipitates, which result when hydrochloric acid is employed. Sulfuric acid is a suitable alternative to citric acid but must be used in substantial excess to prevent precipitation. 2,2,6,6-Tetramethylpiperidine may be recovered from the citric acid extract by making the aqueous solution basic and extracting with ether.

10. The product, a mixture of *cis*- and *trans*-isomers in the ratio of about 2:1, has the following spectral properties: IR (liquid film) cm^{-1} (strong): 1510, 1440, 1370, 1340, 1120, 1080; ^{1}H NMR (CCl$_4$), δ (multiplicity, number of protons, assignment): 0.63–1.3

(m, 5H, cyclopropyl CH_2 and OCH_2CH_3), 1.4–2.0 (m, 1H, cyclopropyl CH), 2.23 (s, 1H, *trans*-aromatic CH_3), 2.28 (s, *ca.* 2H, *cis*-aromatic CH_3), 2.8–3.7 (m, 3H, $CHOCH_2CH_3$), 6.7–7.2 (m, 4H, C_6H_4). The following specific absorptions in the ¹H NMR spectrum may be used to estimate the ratio of the two isomers, δ (multiplicity, coupling constant J in Hz., number of protons, assignment); *cis*-isomer: 0.92 (t, $J = 7$, 3H, OCH_2CH_3), 7.02 (center of $AA'BB'$ m, 4H, C_6H_4); *trans*-isomer: 1.14 (t, $J = 7$, 3H, OCH_2CH_3), 6.88 (center of $AA'BB'$ m, 4H, C_6H_4).

3. Discussion

This procedure describes the generation of the strong, nonnucleophilic amide base, lithium 2,2,6,6-tetramethylpiperidide, which is used in the regioselective abstraction of a proton from a very weak carbon acid containing other sites reactive toward nucleophilic attack.[4,5] In contrast, most other strong bases undergo preferential alkylation with benzyl halides. 1-Ethoxy-2-*p*-tolylcyclopropane is one of over a dozen aryl cyclopropanes, cyclopropenes, and cyclopropanone ketals that have been prepared by this method[5] (Table I). An analog, 1-methoxy-2-phenylcyclopropane, has been obtained in 8% yield from the reaction of methyllithium with dichloromethyl methyl ether in styrene.[6] The alkene is present in large excess, as is commonly the case for reactions involving short-lived carbene intermediates. In the present procedure ethyl vinyl ether serves as both solvent and reactant. For best results with alkenes lacking alkoxy substituents, approximately two volumes of ether or tetrahydrofuran should be used as diluent. Alkoxy,[7,8] acyloxy,[9] alkenyl,[5,10,11] trialkylsilyl,[10] and trialkylstannyl[10] carbenes have been generated and trapped *in situ* with alkenes and alkynes by this method, affording a variety of substituted cyclopropanes.

Lithium 2,2,6,6-tetramethylpiperidide has also been used to advantage in a number of other types of reactions. This base reacts with aryl halides[5,10,12,13] (and, less cleanly, with aryl sulfonates[14]), giving benzynes, which have been trapped with thiolates,[5] acetylides,[5,14] enolates,[10,13,14] and conjugated dienes.[10,12,14] Replacement of halogen by

TABLE I
PREPARATION OF CYCLOPROPANES FROM ALKYL HALIDES, ALKENES, AND
LITHIUM 2,2,6,6-TETRAMETHYLPIPERIDIDE

Product	Yield (%)	Product	Yield (%)
C_2H_5O⊲ —CH_3, CH_3	66[7]	$(CH_3)_2N\overset{O}{\overset{\|}{C}}O$~⊲—$CH=CH_2$	21[9]
C_2H_5O⊲ —CH_3, —CH_3	66[7]	$ClCH_2CH_2O$~⊲▷	64[8]

ORGANIC SYNTHESES

TABLE I (*continued*)
PREPARATION OF CYCLOPROPANES FROM ALKYL HALIDES, ALKENES, AND
LITHIUM 2,2,6,6-TETRAMETHYLPIPERIDIDE

Product	Yield (%)	Product	Yield (%)
ClCH$_2$CH$_2$O—△—CH=CH$_2$	74[8]	(CH$_3$)$_3$CCO—△—O	39[9]
(CH$_3$)$_2$CHO—△—CH=CH$_2$, OSi(CH$_3$)$_3$	46[7]	ClCH$_2$CH$_2$O—△	62[8]
(CH$_3$)$_3$CCO—△—CH$_3$, CH$_3$, CH$_3$	35[9]	(CH$_3$)$_2$NCO—△	20 (24)[9]

hydrogen, a major reaction observed between other dialkylamide bases and aryl halides, does not occur with lithium 2,2,6,6-tetramethylpiperidide.[5] While alkyl benzoates undergo selective deprotonation at the *ortho*-position upon treatment with this amide base,[15] methyl thiobenzoate and *N,N*-dimethylbenzamide are metallated at the methyl group, forming dipole-stabilized carbanions.[16] The organolithium intermediates produced condense with the remaining ester or amide, affording various aryl ketones. Lithiation of dibromomethane[17] and at the α-position of isocyanides[18] with lithium 2,2,6,6-tetramethylpiperidide produces an organolithium intermediate reactive toward carbonyl compounds. In the synthesis of enol carbonates from ketone enolates and chloroformates, this base is the only one to accomplish the reaction in high yield.[19] Lithium 2,2,6,6-tetramethylpiperidide has been shown to be the base of choice for irreversible ketone enolate formation[20] and has been used to discriminate sterically between two potential enolate sites to yield alkylation products with extremely high regioselectivity.[21] The enolate anions of esters[5,22] and dianions of β-keto esters[23] and propiolic acid[24] have been formed by reaction with lithium 2,2,6,6-tetramethylpiperidide. It is a superior base for metallation of selenides,[25] selenoacetals, and selenoketals.[26] Other reactions in which this hindered base has proved effective include the conversion of an epoxide to an enolate anion,[27] the generation of certain α-lithioörganoboranes,[28] the preparation of the highly strained tetracyclo(4.2.0.0^{2,4}.0^{3,5})oct-7-ene from the appropriate tosylhydrazone,[29] and the insertion of magnesium into bacteriopheophytin α.[30] In many of these reactions, other bases, including less hindered amide bases such as lithium diisopropylamide, gave either lower yields or different products entirely.

1. Department of Chemistry, Herbert H. Lehman College, City University of New York, Bronx, New York 10468.
2. Department of Chemistry, The Pennsylvania State University, University Park, Pennsylvania 16802.
3. S. C. Watson and J. F. Eastham, *J. Organomet. Chem.*, **9**, 165 (1967).
4. R. A. Olofson and C. M. Dougherty, *J. Am. Chem. Soc.*, **95**, 581 (1973).
5. R. A. Olofson and C. M. Dougherty, *J. Am. Chem. Soc.*, **95**, 582 (1973).
6. W. K. Kirmse and H. Schütte, *Chem. Ber.*, **101**, 1674 (1968).
7. R. A. Olofson, K. D. Lotts, and G. N. Barber, *Tetrahedron Lett.*, 3779 (1976).
8. G. N. Barber and R. A. Olofson, *Tetrahedron Lett.*, 3783 (1976).
9. R. A. Olofson, K. D. Lotts, and G. N. Barber, *Tetrahedron Lett.*, 3381 (1976).
10. R. A. Olofson, D. H. Hoskin, and K. D. Lotts, *Tetrahedron Lett.*, 1677 (1978).
11. R. A. Moss and R. C. Munjal, *Synthesis*, **11**, 425 (1979).
12. K. L. Shepard, *Tetrahedron Lett.*, 3371 (1975).
13. I. Fleming and T. Mah, *J. Chem. Soc. Perkin Trans. I*, 964 (1975).
14. I. Fleming and T. Mah, *J. Chem. Soc. Perkin Trans. I*, 1577 (1976).
15. C. J. Upton and P. Beak, *J. Org. Chem.*, **40**, 1094 (1975).
16. P. Beak and R. Farney, *J. Am. Chem. Soc.*, **95**, 4771 (1973); P. Beak, B. G. McKinnie, and D. B. Reitz, *Tetrahedron Lett.*, 1839 (1977).
17. H. Taguchi, H. Yamamoto, and H. Nozaki, *Tetrahedron Lett.*, 2617 (1976).
18. U. Schöllkopf, F. Gerhart, I. Hoppe, R. Harms, K. Hantke, K.-H. Scheunemann, E. Eilers, and E. Blume, *Justus Liebigs Ann. Chem.*, 183 (1976).
19. R. A. Olofson, J. Cuomo, and B. A. Bauman, *J. Org. Chem.*, **43**, 2073 (1978).
20. X. Creary, *J. Org. Chem.*, **45**, 2419 (1980); X. Creary and A. J. Rollin, *J. Org. Chem.*, **44**, 1798 (1979).
21. J. D. White and Y. Fukuyama, *J. Am. Chem. Soc.*, **101**, 226 (1979).
22. S. R. Wilson and R. S. Meyers, *J. Org. Chem.*, **40**, 3309 (1975).
23. See also: S. N. Huckin and L. Weiler, *J. Am. Chem. Soc.*, **96**, 1082 (1974).
24. H. H. Meyer, *Justus Liebigs Ann. Chem.*, 337 (1978); B. S. Pitzele, J. S. Baran, and D. H. Steinman, *J. Org. Chem.*, **40**, 269 (1975).
25. H. Reich, *Acc. Chem. Res.*, **12**, 22 (1979).
26. D. van Ende, A. Cravador, and A. Krief, *J. Organomet. Chem.*, **177**, 1 (1979).
27. L. S. Trzupek, T. L. Newirth, E. G. Kelley, N. E. Sbarbati, and G. M. Whitesides, *J. Am. Chem. Soc.*, **95**, 8118 (1973).
28. M. W. Rathke and R. Kow, *J. Am. Chem. Soc.*, **94**, 6854 (1972); R. Kow and M. W. Rathke, *J. Am. Chem. Soc.*, **95**, 2715 (1973); D. S. Matteson and R. J. Moody, *J. Am. Chem. Soc.*, **99**, 3196 (1977).
29. G. E. Gream, L. R. Smith, and J. Meinwald, *J. Org. Chem.*, **39**, 3461 (1974).
30. M. R. Wasielewski, *Tetrahedron Lett.*, 1373 (1977).

ESTERIFICATION OF CARBOXYLIC ACIDS
WITH TRIALKYLOXONIUM SALTS:
ETHYL AND METHYL 4-ACETOXYBENZOATES

[Benzoic acid, 4-(acetyloxy)-, ethyl and methyl esters]

$$R = CH_3 \text{ or } C_2H_5$$

Submitted by Douglas J. Raber,[1] Patrick Gariano, Jr., Albert O. Brod,
Anne L. Gariano, and Wayne C. Guida
Checked by H. Fliri and G. Büchi

1. Procedure

A. *Ethyl 4-acetoxybenzoate.* A 100-ml., one-necked, round-bottomed flask is charged with 2.09 g. (0.0110 mole) of triethyloxonium tetrafluoroborate (Notes 1 and 2), 75 ml. of dichloromethane (Note 3), and 1.80 g. (0.0100 mole) of 4-acetoxybenzoic acid (Note 4). A magnetic stirring bar is added, and the solution is stirred while 1.4 g. (1.9 ml., 0.011 mole) of diisopropylethylamine (Note 5) is introduced with a syringe (Note 6). The flask is then stoppered and allowed to stand at room temperature for 16–24 hours.

Work-up is initiated by extracting the reaction mixture with three 50-ml. portions of 1 N hydrochloric acid, three 50-ml. portions of aqueous 1 N potassium hydrogen carbonate (Notes 7 and 8), and 50 ml. of saturated aqueous sodium chloride. The organic solution is dried over sodium sulfate (Note 9), filtered, and concentrated on a rotary evaporator. Purification of the residue by bulb-to-bulb distillation (Note 10) at about 140° (5 mm.) provides 1.77–1.98 g. (85–95%) of ethyl 4-acetoxybenzoate as a colorless, viscous liquid (Note 11).

B. *Methyl 4-acetoxybenzoate.* A 100-ml., one-necked, round-bottomed flask is charged with 1.63 g. (0.0110 mole) of trimethyloxonium tetrafluoborate (Notes 2 and 12), 75 ml. of dichloromethane (Notes 3 and 13), and 1.80 g. (0.0100 mole) of 4-acetoxybenzoic acid (Note 4). A magnetic stirring bar is added, and the suspension is stirred while 1.4 g. (1.9 ml., 0.011 mole) of diisopropylethylamine (Note 5) is introduced with a syringe (Note 6). The flask is then stoppered, and stirring is continued at room temperature for 16–24 hours, during which time the oxonium salt dissolves. Work-up exactly as that described in Part A is followed by bulb-to-bulb distillation (Note 10) at about 140° (5 mm.), yielding 1.65–1.84 g. (85–95%) of methyl 4-acetoxybenzoate, m.p. 78–80° (Note 14).

2. Notes

1. Triethyloxonium tetrafluoroborate was prepared according to *Org. Synth.*, **Coll. Vol. 5,** 1080 (1973). The submitters stored this material at $-20°$ under ether in a tightly-stoppered jar and found that the use of a dry box or an inert atmosphere was not required in its handling. A sample of the oxonium salt–ether slurry was transferred to the tared reaction flask, ether was removed on a rotary evaporator, and the resulting solid was weighed and used without further purification. The checkers stored the dry oxonium salt under argon at $-15°$, maintained an argon atmosphere in all operations involving this reagent, and used dry solvent (Note 3). They obtained 98% yields in two runs of Part A and two runs of Part B.

2. A 10% molar excess of the oxonium salt with regard to the carboxylic acid gives slightly higher yields than does an equimolar quantity.

3. The submitters used reagent grade dichloromethane without purification. The checkers dried dichloromethane by distillation from phosphorus pentoxide.

4. This is prepared from 4-hydroxybenzoic acid and acetic anhydride following a procedure for 2-acetoxybenzoic acid.[2] The crude product is conveniently purified by stirring with chloroform (about 15 ml. per gram of acid) and removing any insoluble residue by filtration. Evaporation of the chloroform gives the desired material, m.p. $186-188°$ (lit.,[3] m.p. $189-190°$).

5. This product was obtained from Aldrich Chemical Company, Inc. In many cases, the use of other amines may be satisfactory. Nevertheless, the use of a hindered base minimizes destruction of the oxonium salt by side reaction with the amine. Substitution of triethylamine for diisopropylamine in the procedure gave lower yields of the ester.

6. The use of a syringe affords a convenient method both for measuring the desired quantity of amine and adding it to the reaction mixture. In general, a mildly exothermic reaction takes place during addition of the amine. The submitters, working in an open vessel, suggest that if the reaction is scaled up or if the solution is more concentrated, care should be taken to add the amine gradually so that the reaction mixture does not boil over. For large-scale reactions they recommend the use of a dropping funnel. The checkers, working under argon and introducing the amine through a rubber septum, noted a considerable increase in pressure during the addition. Thus, with this experimental setup a suitable pressure vent is required.

7. Any unreacted carboxylic acid may be recovered by neutralization and extraction of the hydrogen carbonate solution.

8. In preparing ethers of phenols, aqueous 1 N sodium hydroxide should be substituted for the sodium hydrogen carbonate solution.

9. In many instances, the dichloromethane solution can be dried adequately by a simple filtration through coarse filter paper.

10. The "Kügelrohr" apparatus sold by Rinco Instrument Company, Inc., or any comparable bulb-to-bulb distillation apparatus is satisfactory. Fractional distillation is unnecessary.

11. The product crystallizes on standing overnight at $-20°$ and melts at $30-32°$. GC analysis showed it to be at least 99% pure. Ethyl 4-acetoxybenzoate has been reported to melt at $34°$.[4]

12. This compound was prepared according to *Org. Synth.*, **Coll. Vol. 6,** 1019

(1988). Both submitters and checkers stored and handled this material using the techniques outlined for triethyloxonium tetrafluoborate in Note 1.

13. Trimethyloxonium tetrafluoroborate is only slightly soluble in dichloromethane. However, the use of a two-phase mixture presents no difficulties in either experimental procedure or yield.

14. GC analysis showed this material to be at least 99% pure. The melting point of methyl 4-acetoxybenzoate has been reported[5] as 81–81.6°. In both their runs, the checkers obtained a distilled product which melted from 60° to 74°, resolidified at 74°, and then remelted at 78–79°. A sample recrystallized from hexane showed the same behavior.

3. Discussion

This procedure provides a convenient method for the esterification of a wide variety of carboxylic acids.[6] The reaction proceeds smoothly with sterically hindered acids[6a] and with those which contain various functional groups.[6b] Esters are obtained in high purity using Kugelrohr distillation as the sole purification technique. In cases where traces of dichloromethane present no problems, the crude product is usually pure enough to be used directly in subsequent reactions. Methyl and ethyl ethers of phenols may also be prepared by this procedure (see Note 8).

Examples of polyfunctional carboxylic acids esterified by this method are shown in Table I. Yields are uniformly high, with the exception of those cases (maleic and fumaric acids) where some of the product appears to be lost during work-up as a result of water solubility. Even with carboxylic acids containing a second functional group (e.g., amide, nitrile) which can readily react with the oxonium salt, the more nucleophilic carboxylate anion is preferentially alkylated. The examples described in detail above illustrate the esterification of an acid containing a labile acetoxy group, which would not survive other procedures such as the traditional Fischer esterification.

The great utility of the trialkyloxonium salts is illustrated by the fact that high yields of esters are obtained using reagent which has been stored for up to 6 months under the submitters' conditions (Notes 1 and 12). Thus, either trimethyl- or triethyloxonium tetrafluoroborate can be prepared in quantity, stored, and used for esterification as required.

Other examples of esterification with trialkyloxonium salts have been reported.[7,8] The present procedure offers the advantages that the reactive carboxylate ion is generated *in situ* and that a low-boiling, nonaqueous solvent is employed, simplifying the experimental procedure considerably. A related method which utilizes a hindered amine with dimethyl sulfate as the alkylating agent has been reported. The present procedure is carried out under somewhat milder conditions and avoids the use of highly toxic reagents.

The only other esterification method which rivals the present procedure in convenience, mildness of conditions, selectivity, and yield is the preparation of methyl esters with diazomethane.[10] Esterification with trialkyloxonium salts, however, allows preparation of both methyl and ethyl esters and avoids the toxicity and explosion hazard[11] of diazomethane.

Furthermore, recent studies indicate that esterifications involving triethyloxonium tetrafluoroborate are often very rapid. For example, subsequent to the checking of this

TABLE I

ESTERIFICATION OF CARBOXYLIC ACIDS WITH TRIALKYLOXONIUM FLUOBORATES

Acid	Triethyloxonium Fluoborate		Trimethyloxonium Fluoborate	
	Yield (%)[a]	Purity (%)[b]	Yield (%)[a]	Purity (%)[b]
2,4,6-trimethylbenzoic acid (CH_3, CH_3, CH_3, CO_2H)	90	> 99	90	> 99
benzoic acid (CO_2H)	91	> 99	92	> 99
maleic/fumaric acid $HO_2C\,C{=}C\,CO_2H$ (H,H)	77[c]	> 99	80[c]	> 99
$HO_2C\,C{=}C\,CO_2H$ (H,H)	70[c]	> 99	74[c]	> 99
$N{\equiv}C-CH_2-CO_2H$	91	> 99	82	95
CO_2H, $C-N(C_2H_5)_2$ ($\|$ O)	88	> 99	86	> 99
CO_2H, $O-C-CH_3$ ($\|$ O)	89	95	85	> 99
CO_2H, $C-C_6H_5$ ($\|$ O)	95	> 99	95	> 99

[a]Yield of distilled or crystallized ester.
[b]By GC.
[c]In the esterification of dibasic acids a corresponding increase in equivalents of oxonium salt is employed; the product is the diester.

procedure the submitters have found that the reaction time of Part A may be shortened from 16–24 hours to 0.5 hour with no decrease in yield. The longer reaction time is still recommended for esterifications involving the trimethyl salt, such as that of Part B, because of the heterogeneous nature of the reaction mixture in these cases.

1. Department of Chemistry, University of South Florida, Tampa, Florida 33620.
2. A. I. Vogel, "A Textbook of Practical Organic Chemistry," 3rd ed., Longman, London, 1956, p. 996.
3. F. D. Chattaway, *J. Chem. Soc.*, 2495 (1931).
4. D. Vorländer and W. Selke, *Z. Phys. Chem. Abt. A*, **129**, 434 (1927) [*Chem. Abstr.*, **24**, 4198 (1930)].
5. C. G. Mitton, R. L. Schowen, M. Gresser, and J. Shapley, *J. Am. Chem. Soc.*, **91**, 2036 (1969).
6. (a) D. J. Raber and P. Gariano, *Tetrahedron Lett.*, 4741 (1971); (b) D. J. Raber, P. Gariano, Jr., A. O. Brod, A. Gariano, W. C. Guida, A. R. Guida, and M. D. Herbst, *J. Org. Chem.*, **44**, 1149 (1979).
7. H. Meerwein, G. Hinz, P. Hofmann, E. Kroning, and E. Pfeil, *J. Prakt. Chem.*, **147**, 257 (1937).
8. T. Hamada and O. Yonemitsu, *Chem. Pharm. Bull.*, **19**, 1444 (1971).
9. F. H. Stodola, *J. Org. Chem.*, **29**, 2490 (1964).
10. L. F. Fieser and M. Fieser, "Reagents for Organic Synthesis," Vol. 1, Wiley, New York, 1967, p. 192.
11. T. J. DeBoer and H. J. Backer, *Org. Synth.*, **Coll. Vol. 4**, 250 (1963).

ortho-ALKYLATION OF ANILINES:
ETHYL 4-AMINO-3-METHYLBENZOATE

[Benzoic acid, 4-amino-3-methyl-, ethyl ester]

Submitted by P. G. Gassman[1,2] and G. Gruetzmacher
Checked by M. Savitsky, R. R. Schmidt, III,
and G. Büchi

1. Procedure

*Caution! Part A must be conducted in a hood due to the noxious odor of dimethyl
sulfide. In Part B, the usual precautions associated with the pyrophoric reagent Raney
nickel must be observed (see Note 8).*

A. *Ethyl 4-amino-3-(methylthiomethyl)benzoate.* A 1-l., three-necked, round-
bottomed flask is fitted with a mechanical stirrer, a condenser topped with a gas-inlet
tube, a two-necked adapter holding a low-temperature thermometer, and a 100-ml.,
pressure-equalizing dropping funnel. The flask is charged with 16.50 g. (0.1000 mole) of
ethyl 4-aminobenzoate (Benzocaine, Note 1), 300 ml. of acetonitrile, and 100 ml. of
dichloromethane, flushed with nitrogen, and immersed in a 40% aqueous methanol–dry-
ice bath maintained between −40° and −50° (Note 2). When the reaction mixture has
come to −40°, a solution of 10.85 g. (0.1000 mole) of *tert*-butyl hypochlorite [*Org.
Synth., Coll. Vol. 5,* 184 (1973)] in 25 ml. of dichloromethane is added dropwise over a
15-minute period. The addition funnel is rinsed with 25 ml. of dichloromethane, the
reaction solution is stirred for another 5 minutes, and 23 g. (25 ml., 0.34 mole) of

dimethyl sulfide (Note 3) is added at a rate that allows the vigorously stirred reaction mixture to be maintained below −30°.

Shortly after the addition is complete, a voluminous white precipitate appears, and the resulting slurry is stirred and maintained at −50° to −40° for 4 hours. Cooling and stirring are continued as 18 g. (25 ml., 0.18 mole) of triethylamine (Note 4) is added dropwise, during which time the reaction mixture first goes to a clear solution, then becomes cloudy. The resulting mixture is stirred at −50° to −40° for 1 hour.

The cooling bath is then replaced with a steam bath, and the reaction mixture is refluxed for 16 hours. It is then cooled, transferred to a one-necked, 1-l., round-bottomed flask, and concentrated to dryness on a rotary evaporator. The dark residue is dissolved in a mixture of 200 ml. of water, 200 ml. of dichloromethane, and 20 ml. of triethylamine, and the aqueous phase is separated and washed with two 200-ml. portions of dichloromethane. The organic phases are combined and washed with 300 ml. of saturated aqueous sodium chloride, dried over anhydrous magnesium sulfate, and filtered. Removal of the solvent on a rotary evaporator gives a red oil, which solidifies on storage at 0−5° (Note 5). Recrystallization of this solid from 40 ml. of absolute ethanol gives 7.6−8.4 g. (34−37%) of ethyl 4-amino-3-(methylthiomethyl)benzoate, m.p. 83−85°. A second crop of 1.1−2.5 g. of crystalline material, m.p. 78−83°, may be obtained by concentration of the mother liquors (Note 6).

B. *Ethyl 4-Amino-3-methylbenzoate.* A 1-l., three-necked, round-bottomed flask equipped with a mechanical stirrer, a condenser, and a nitrogen-inlet tube is charged with 11.25 g. (0.05000 mole) of ethyl 4-amino-3-(methylthiomethyl)benzoate, 300 ml. of absolute ethanol, and 17 teaspoons (*ca.* 50 g.) of W-2 Raney nickel (Note 7). The reaction mixture is stirred at 25° for one hour before stirring is discontinued, and the ethanolic solution is decanted from the catalyst (Note 8). The catalyst is then washed with one 300-ml. portion of absolute ethanol and one 500-ml. portion of dichloromethane, the solvent being removed by decanting in each case. The organic solutions are combined and concentrated on a rotary evaporator, giving a solid which is dissolved in 200 ml. of dichloromethane, dried over anhydrous magnesium sulfate, filtered, and taken to dryness on a rotary evaporator, yielding 7.0−7.6 g. (77−84%) of ethyl 4-amino-3-methylbenzoate as a white solid, m.p. 75−78°.

2. Notes

1. Benzocaine was purchased from Mallinckrodt Chemical Works and recrystallized from absolute ethanol prior to use to give material of m.p. 91−92°.

2. Fairly precise temperature control is required, since the reaction mixture begins to solidify at about −50°.

3. Dimethyl sulfide was purchased from MC and B Manufacturing Chemists and used without further purification.

4. Triethylamine was purchased from the J. T. Baker Chemical Company and used without further purification.

5. Solidification is facilitated by dissolving the oil in a minimum amount of diethyl

ether (*ca.* 100 ml.) and removing the ether on a rotary evaporator. Should the oil still refuse to solidify, this process is repeated several times with pentane.

6. The second crop of crystalline material is contaminated with a small amount of ethyl *p*-aminobenzoate and usually red-orange in color. It is of sufficient purity to be used in Part B.

7. W-2 Raney nickel was purchased from W. R. Grace & Company. Prior to use it was washed with distilled water until the washings were neutral, then washed three times with absolute ethanol and stored under ethanol until needed [*Org. Synth.*, **Coll. Vol. 3**, 181 (1955)].

8. Activated Raney nickel is pyrophoric and should never be allowed to become dry. Thus, decanting is preferred to filtration, and when decanting, a small amount of solvent must always be left behind to cover the catalyst powder. For safe (but environmentally unsound) disposal, the spent catalyst should be slurried in water and flushed down the drain under running water.

3. Discussion

This procedure illustrates a general method for the *ortho*-alkylation of anilines.[3] It can be utilized for both anilines and mono-*N*-substituted anilines, with a variety of functional groups on the aromatic ring. By substituting α-thioketones and α-thioesters for dialkyl sulfides, the method has been extended to produce indoles[4] and oxindoles,[5] respectively. An example of the indole synthesis appears elsewhere in this volume. Ethyl 4-amino-3-methylbenzoate has been reported previously.[6]

1. Department of Chemistry, The Ohio State University, Columbus, Ohio 43210.
2. Present address: Department of Chemistry, University of Minnesota, Minneapolis, Minnesota 55455.
3. P. G. Gassman and G. Gruetzmacher, *J. Am. Chem. Soc.*, **95**, 588 (1973); P. G. Gassman and C. T. Huang, *J. Am. Chem. Soc.*, **95**, 4453 (1973); P. G. Gassman, T. J. Van Bergen, and G. D. Gruetzmacher, *J. Am. Chem. Soc.*, **95**, 5608 (1973); P. G. Gassman and D. R. Amick, *Tetrahedron Lett.*, 889 (1974); P. G. Gassman and G. D. Gruetzmacher, *J. Am. Chem. Soc.*, **96**, 5487 (1974); P. G. Gassman, G. D. Gruetzmacher, and T. J. Van Bergen, *J. Am. Chem. Soc.*, **96**, 5512 (1974); P. G. Gassman and R. L. Parton, *Tetrahedron Lett.*, 2055 (1977); P. G. Gassman and R. J. Balchunis, *Tetrahedron Lett.*, 2235 (1977); P. G. Gassman and W. M. Schenk, *J. Org. Chem.*, **42**, 3240 (1977); P. G. Gassman and R. L. Parton, *J. Chem. Soc. Chem. Commun.*, 694 (1977); P. G. Gassman and H. R. Drewes, *J. Am. Chem. Soc.*, **100**, 7600 (1978); P. G. Gassman and D. R. Amick, *J. Am. Chem. Soc.*, **100**, 7611 (1978).
4. P. G. Gassman and T. J. van Bergen, *J. Am. Chem. Soc.*, **95**, 590 (1973).
5. P. G. Gassman and T. J. van Bergen, *J. Am. Chem. Soc.*, **95**, 2718 (1973).
6. F. J. Viliani, U.S. Patent 2,764,519 (1956) [*Chem. Abstr.*, **51**, 4443e (1957)].

ETHYL 5β-CHOLEST-3-ENE-5-ACETATE

[Cholest-3-ene-5-acetic acid, ethyl ester, (5β)-]

$$+ \ CH_3C(OC_2H_5)_3 \xrightarrow{\text{heat}}$$

Submitted by R. E. IRELAND[1] and D. J. DAWSON
Checked by W. PAWLAK and G. BÜCHI

1. Procedure

A 100-ml., Claisen distillation flask with two 14/20 standard taper joints and a thermometer-inlet is equipped with a gas-inlet adapter, a receiver, a thermometer, and a magnetic stirring bar. A 40-ml. calibration mark is made on the flask, and 970 mg. (2.50 mmoles) of cholest-4-en-3β-ol (Note 1) is introduced. Triethyl orthoacetate is then distilled under argon into the flask to the 40-ml. mark (Note 2). The mixture is stirred, effecting solution while the flask is purged with argon, then the top joint is sealed with a thermometer (Figure 1). The stirred solution is heated under a positive pressure of argon so that the vapor reflux level is just below the side arm of the flask; the temperature on the lower thermometer is 142–147°; the upper thermometer temperature is kept between 25 and 70° (Note 3). After 8 days of reflux, during which time a small amount of the volatile material distills into the receiver, the reaction flask is cooled, and all the volatile materials are removed at reduced pressure (Note 4). The residue (1.3 g. of a pale yellow oil) is chromatographed on 120 g. of silica gel with 10% diethyl ether in petroleum ether as the eluant (Note 5). The side products eluted with the first 240 ml. of the solvent are discarded; further elution with 120 ml. of the solvent affords 690 mg. of ethyl 5β-cholest-3-ene-5-acetate as a clear, colorless oil. Trituration of this product with acetone produces 560–690 mg. (49–60%) of the ester as white plates, m.p. 89–92.5°.

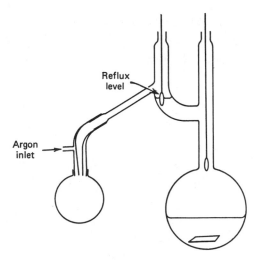

Figure 1.

2. Notes

1. Cholest-4-en-3β-ol can be prepared by the procedure of Burgstahler and Nordin.[2] A melting point below 130° indicates that the material is contaminated with some of the 3α-hydroxy isomer. The material used above melted at 130.5–131° (from ethanol).

2. The Matheson, Coleman and Bell product was used without purification. After a 10-ml. forerun, the triethyl orthoacetate was distilled (b.p. 142–147°) directly into the reaction flask.

3. A sand bath in an electric heating mantle was found to be satisfactory for the long-term heating process.

4. The volatile materials were removed by rotary evaporation followed by vacuum (0.1 mm.) drying for 1 hour.

5. Merck silica gel (0.05–0.2 mm., 70–325 mesh ASTM) was used in a 3.5 × 26 cm. column. Mallinckrodt anhydrous ether and Baker petroleum ether (b.p. 30–60°) were employed as eluants.

3. Discussion

The ester–Claisen rearrangement procedure of Johnson and co-workers[3] was modified for use with cholest-4-en-3β-ol.

1. Division of Chemistry and Chemical Engineering, Gates and Crellin Laboratories of Chemistry, California Institute of Technology, Pasadena, California 91109.
2. A. W. Burgstahler and I. C. Nordin, *J. Am. Chem. Soc.*, **83**, 198 (1961).
3. W. S. Johnson, L. Werthemann, W. R. Bartlett, T. J. Brocksom, T.-t. Li, D. J. Faulkner, and M. R. Peterson, *J. Am. Chem. Soc.*, **92**, 741 (1970).

FREE-RADICAL CYCLIZATION:
ETHYL 1-CYANO-2-METHYLCYCLOHEXANECARBOXYLATE

(Cyclohexanecarboxylic acid, 1-cyano-2-methyl-, ethyl ester)

Submitted by MARC JULIA and MICHEL MAUMY[1]
Checked by EDWARD J. ZAIKO and HERBERT O. HOUSE

1. Procedure

Caution! Since hydrogen is evolved in this procedure, it should be performed in an efficient hood.

A.   *Ethyl (E)-2-cyano-6-octenoate.*   A solution of 38.4 g. (0.384 mole) of (E)-4-hexen-1-ol (Note 1) in 160 ml. of anhydrous pyridine is placed in a 1-l. Erlenmeyer flask equipped with a magnetic stirring bar, and the solution is cooled in an ice bath. Over 1 hour, 91.5 g. (0.478 mole) of p-toluenesulfonyl chloride (Note 2) is added portionwise and with stirring to the reaction mixture while maintaining the temperature at 0–5°. The resulting slurry is allowed to stand overnight (Note 3) in a refrigerator, then poured into 500 g. of an ice-water mixture. The reaction mixture is extracted with four 150-ml. portions of diethyl ether, and the combined ethereal extracts are washed successively with four 150-ml. portions of 2 N sulfuric acid, 50 ml. of aqueous saturated sodium hydrogen carbonate, and two 50-ml. portions of water. The ethereal solution is dried over anhydrous sodium sulfate, and the solvent removed on a rotary evaporator at 25°, yielding 76–88 g. (78–91%) of crude (E)-4-hexen-1-yl p-toluenesulfonate as a pale yellow oil (Note 4).

A dry, 2-l., three-necked, round-bottomed flask is equipped with a sealed mechanical stirrer, a thermometer, and a pressure-equalizing dropping funnel fitted with a calcium chloride drying tube. Sodium hydride dispersion (Note 5), [19.2 g. of a 50% w/w suspension in mineral oil, 9.6 g. (0.40 mole) of sodium hydride] is placed in the flask, and 100 ml. of anhydrous pentane is added. After the dispersion has been stirred, the sodium hydride is allowed to settle, and the supernatant liquid is removed with a pipet or a siphon. This washing operation is repeated with two additional 100-ml. portions of pentane, and

400 ml. of anhydrous *N,N*-dimethylformamide (Note 6) is then added to the reaction flask. To the vigorously stirred suspension of sodium hydride in *N,N*-dimethylformamide is added, dropwise over 30 minutes, 68 g. (0.60 mole) of ethyl cyanoacetate (Note 7). The mixture is stirred until a clear solution is obtained before all of the previously prepared crude (*E*)-4-hexen-1-yl *p*-toluenesulfonate, dissolved in 100 ml. of anhydrous *N,N*-dimethylformamide (Note 6), is added in one portion. The resulting solution is slowly heated to 100°, with continuous stirring, over a period of 3 hours. During this time the reaction becomes dark red, and crystalline sodium *p*-toluenesulfonate separates. The mixture is allowed to stand overnight at room temperature, then transferred to a 2-l., one-necked, round-bottomed flask, and most of the solvent is removed on a rotary evaporator. The residual semisolid is mixed with 700 ml. of water and extracted with three 250-ml. portions of ether. The combined ethereal extracts are dried over anhydrous sodium sulfate and concentrated on a rotary evaporator. The residual liquid is fractionally distilled under reduced pressure through a 12-cm. Vigreux column. After removal of a low-boiling forerun, b.p. 54–57° (0.2 mm.), containing mainly ethyl cyanoacetate, 34.5–38.4 g. (46–51%) of colorless ethyl (*E*)-2-cyano-6-octenoate is collected, b.p. 84–86° (0.2 mm.), n_D^{25} 1.4458 (Notes 8 and 9).

B. *Ethyl 1-cyano-2-methylcyclohexanecarboxylate.* A 2-l., three-necked, round-bottomed flask is equipped with a heating mantle, a Teflon®-coated magnetic stirring bar, a 1-l. pressure-equalizing dropping funnel fitted with a Teflon® stopcock (Note 10), a stopper, and a reflux condenser fitted with a nitrogen-inlet tube. The apparatus is flushed with nitrogen, and 200 ml. of freshly distilled cyclohexane and 0.30 g. (0.0012 mole) of benzoyl peroxide are added to the flask. While a static nitrogen atmosphere is maintained in the flask (Note 11), a solution of 5.00 g. (0.0270 mole) of ethyl (*E*)-2-cyano-6-octenoate in 800 ml. of freshly distilled cyclohexane is added from the dropping funnel to the stirred, refluxing benzoyl peroxide solution over a period of 40 hours (Note 10). Three additional 0.30-g. (0.0012-mole) portions of benzoyl peroxide (total: 1.2 g., 0.0050 mole, Note 11), are added at 12-hour intervals during the addition of the unsaturated ester. After the addition of the unsaturated ester and the benzoyl peroxide is complete, the reaction mixture is refluxed with stirring for an additional 20 hours (Note 12). The resulting colorless to pale-yellow solution is concentrated on a rotary evaporator to approximately 200 ml., diluted with 250 ml. of ether, and washed with 100 ml. of aqueous saturated iron(II) sulfate to destroy any unchanged benzoyl peroxide. The organic layer is then washed successively with two 200-ml. portions of aqueous saturated sodium hydrogen carbonate and two 200-ml. portions of water and dried over anhydrous calcium chloride. The solvent is removed on a rotary evaporator, and the residual liquid is fractionally distilled under reduced pressure through a 5-cm. Vigreux column. After separation of a 0.16–0.25-g. (3–5%) forerun, b.p. 63–64° (0.2 mm.), n_D^{25} 1.4550–1.4560, containing (Note 13) primarily the cyclized product, 3.68–3.82 g. (74–76%) of colorless ethyl 1-cyano-2-methylcyclohexanecarboxylate is collected, b.p. 64–75° (0.2 mm.), n_D^{25} 1.4532–1.4539 (Notes 14 and 15).

2. Notes

1. The preparation of this unsaturated alcohol is described in *Org. Synth.*, **Coll. Vol. 6**, 675 (1988).

2. The checkers employed a commercial sample of p-toluenesulfonyl chloride (obtained from Eastman Organic Chemicals) without further purification.

3. The checkers found that a longer period of standing before isolation lowered the yield of the sulfonate ester.

4. IR (CCl$_4$) cm.$^{-1}$: 1375, 1195, 1185, 975 [(E)CH=CH], 940.

5. The submitters employed a dispersion of sodium hydride in mineral oil obtained from Prolabo, Paris. The checkers employed 17 g. of mineral oil dispersion containing 57% sodium hydride obtained from Alfa Inorganics, Inc.

6. The submitters dried commercial N,N-dimethylformamide over anhydrous barium oxide for 5 days, then distilled the solvent at atmospheric pressure, b.p. 155°. The checkers allowed commercial N,N-dimethylformamide to stand over activated Linde 4A Molecular Sieves for several hours, then decanted the solvent and distilled it under reduced pressure, b.p. 43° (6 mm.).

7. The checkers employed commercial ethyl cyanoacetate (purchased from Eastman Organic Chemicals) without purification.

8. The checkers found this fractional distillation to be simplified if the initial crude product was first subjected to a rapid short-path distillation under reduced pressure to remove the bulk of the dialkylated material and other high molecular weight components.

9. The spectral properties of the product are as follows; IR (CCl$_4$) cm.$^{-1}$: 2250 (C≡N), 1745 (C=O), 970 [(E) CH=CH]; ^{1}H NMR (CCl$_4$), δ (multiplicity, coupling constant J in Hz., number of protons, assignment): 1.0–2.4 (m, 9H, 3CH$_2$ and allylic CH$_3$), 1.31 (t, J = 7, 3H, OCH$_2$CH$_3$), 3.44 [t, J = 6.5, 1H, CH(CN)], 4.25 (q, J = 7, 2H, OCH$_2$CH$_3$), 5.1–5.8 (m, 2H, CH=CH). GC analysis of the checkers' product on a 5-m. column packed with ethylene glycol isophthalate on Chromosorb P operated at 188° shows 2 products in the ratio of approximately 95:5, and retention times of 33.8 minutes and 37.0 minutes, respectively.

10. Because of difficulties in adjusting an ordinary glass stopcock to avoid leakage and to maintain a drop rate that would add the unsaturated cyanoester solution over a 40-hour period, the checkers recommend the use of a funnel equipped with a Teflon® stopcock.

11. The submitters report this reaction to be a radical chain process that requires less than 0.2 mole of benzoyl peroxide per mole of starting material. The checkers can offer additional evidence of the radical chain nature of the reaction from their finding that the cyclization reaction is almost completely inhibited if the refluxing solution is not protected from atmospheric oxygen.

12. The reaction solution can be analyzed to establish complete consumption of the starting unsaturated cyanoester by injecting an aliquot of the reaction solution onto a 5-m. GC column packed with ethylene glycol isophthalate suspended on Chromosorb P operated at 190°. Under these conditions the retention times of the starting unsaturated cyanoester and the cyclized cyanoester are 31.2 minutes and 28.0 minutes, respectively.

13. GC analysis (Note 12) of the forerun indicated the presence of ethyl 1-cyano-2-methylcyclohexanecarboxylate and minor amounts of five or more lower boiling impurities.

14. The product, which exhibits a single GC peak (Note 12), is presumably a mixture of stereoisomers. The spectral properties of the product are as follows; IR (CCl$_4$) cm.$^{-1}$: 2250 (C≡N), 1745 (C=O); ^{1}H NMR (CCl$_4$), δ (multiplicity, coupling constant J in

Hz., number of protons, assignment): 0.97 (d, J = 6.2, 3H, CH_3), 1.0–2.5 [m, 9H, $CH(CH_2)_4$], 1.32 (t, J = 7, 3H, OCH_2CH_3), 4.25 (q, J = 7, 2H, OCH_2CH_3). The methyl doublet at δ 0.97 is accompanied by a second weak doublet (J = 6.8 Hz.) at δ 1.02 that is presumably attributable to the second stereoisomer of ethyl 1-cyano-2-methylcyclohexanecarboxylate; mass spectrum m/e (relative intensity): 195 (M, 5), 141 (27), 136 (40), 126 (34), 123 (62), 122 (80), 108 (100), 98 (85), 95 (62), 94 (44), 82 (27), 81 (41), 70 (51), 67 (56), 55 (50), 53 (32), 42 (27), 41 (58), 39 (31).

15. The submitters report that this free radical cyclization was also effected by heating a solution of 5.00 g. (0.256 mole) of ethyl (E)-2-cyano-6-octenoate and 1.25 g. (0.00856 mole) of di-*tert*-butyl peroxide in 500 ml. of freshly distilled cyclohexane at 140° in an autoclave for 30 hours. The solution was concentrated and the residue was distilled to yield 3.4 g. (68%) of ethyl 1-cyano-2-methylcyclohexanecarboxylate.

3. Discussion

Ethyl 1-cyano-2-methylcyclohexanecarboxylate has been prepared by catalytically hydrogenating the Diels-Alder adduct of butadiene and ethyl 2-cyano-2-butenoate[2] and by the procedure described in this preparation.[3,4] This procedure illustrates a general method for the preparation of alicyclic compounds by the cyclization of δ-ethylenic carbon radicals **1**.[5] Whereas the primary 5-hexen-1-yl radical **1** (R=R'=R''=H) cyclizes to

form methylcyclopentane (*via* **3**),[6] the 1-cyano-1-carboethoxy disubstituted radicals **1** (R'=CN, R''=$CO_2C_2H_5$) lead predominantly, and sometimes exclusively, to six-membered rings **2**. Monosubstituted radicals **1** (R''=H) often give mixtures of both isomers **2** and **3**.[4,5]

1. École Normale Superieure, Laboratoire de Chimie, 24, rue Lhomond, 75231 Paris Cedex 05, France. [Present address: Laboratoire de Recherches Organiques de l'E.S.P.C.I., Université Pierre et Marie Curie, 10, Rue Vauquelin, 75231 Paris Cedex 05, France].
2. C. J. Morell and W. G. Stoll, *Helv. Chim. Acta*, **35**, 2556 (1952).
3. M. Julia, J. M. Surzur, and L. Katz, *Bull. Soc. Chim. Fr.*, 1109 (1964).
4. M. Julia and M. Maumy, *Bull. Soc. Chim. Fr.*, 2415, 2427 (1969).
5. For a recent review, see M. Julia, *Acc. Chem. Res.*, **4**, 386 (1971).
6. R. C. Lamb, P. W. Ayers, and M. K. Toney, *J. Am. Chem. Soc.*, **85**, 3483 (1963).

2,2-(ETHYLENEDITHIO)CYCLOHEXANONE

(1,4-Dithiaspiro[4,5]decan-6-one)

$$\text{(structure)} + CH_3C_6H_4SO_2S(CH_2)_2SSO_2C_6H_4CH_3 \xrightarrow[CH_3OH\ 65°]{KOCOCH_3} \text{(structure)}$$

Submitted by R. B. WOODWARD,[1] I. J. PACHTER,[2] and M. L. SCHEINBAUM[3]
Checked by J. G. GREEN and S. MASAMUNE

1. Procedure

Caution! Benzene has been identified as a carcinogen; OSHA has issued emergency standards on its use. All procedures involving benzene should be carried out in a well-ventilated hood, and glove protection is required.

A 300-ml., one-necked flask equipped with a reflux condenser, the top of which is attached to a nitrogen inlet tube, is charged with 3.85 g. (0.0338 mole) of 2-hydroxymethylenecyclohexanone (Note 1), 10 g. (0.025 mole) of ethylene dithiotosylate (Note 2), and 10 g. of potassium acetate in 150 ml. of methanol. The mixture is refluxed under nitrogen for 3 hours with stirring, the solvent is removed from the reaction mixture on a rotary evaporator, and the residue is extracted with three 50-ml. portions of diethyl ether. The combined ethereal extracts are washed with cold, aqueous 2 *N* sodium hydroxide (Note 3) until the aqueous layer is basic to litmus, then with 50 ml. of saturated aqueous sodium chloride. The ethereal layer is dried over anhydrous magnesium sulfate, filtered, and concentrated on a rotary evaporator. The oily residue is diluted with 1 ml. of benzene and 3 ml. of cyclohexane and transferred to a chromatographic column (14 × 2 cm.) prepared with 50 g. of alumina (Note 4) and a 3:1 mixture of cyclohexane and benzene. With this solvent system the desired product moves with the solvent front, and the first 100 ml. of eluent contains 85% of the total product. Further elution with approximately 100 ml. of the same solvent mixture removes the rest of the material before a second component begins to come off. Evaporation of the solvent from the combined 200 ml. of eluent leaves an oily residue which crystallizes on standing, yielding 2.76–3.04 g. (57–64%) of crude 2,2-(ethylenedithio)cyclohexanone. Recrystallization from approximately 50 ml. of pentane affords 2.1–2.6 g. (45–55%) of needles, m.p. 56–57° (Note 5).

2. Notes

1. 2-Hydroxymethylenecyclohexanone was prepared by both the submitters and checkers by a procedure similar to, but slightly modified from, that described in *Org. Synth.*, **Coll. Vol. 4,** 536 (1963). To a cooled (ice bath), stirred suspension of 10.2 g. (0.189 mole) of commercial sodium methoxide in 75 ml. of anhydrous benzene under nitrogen was added dropwise, but rapidly (*ca.* 2 minutes), a mixture of 9.8 g. (0.10 mole)

of distilled cyclohexanone and 14.8 g. (0.200 mole) of distilled ethyl formate. After addition, the reaction was allowed to warm to room temperature and left overnight. Ice water (100 ml.) was added to the resulting suspension. The aqueous layer was separated, and the benzene layer was washed three times with 50 ml. of cold, aqueous 0.1 N sodium hydroxide. The aqueous layers were combined, and the product was isolated according to the procedure referenced above. This modified version provided slightly higher yields of the product than that recorded in *Org. Synth.*, and the ease of handling sodium methoxide, compared with sodium metal, is advantageous.

2. Ethylene dithiotosylate, m.p. 73–73.5°, as described in *Org. Synth.*, **Coll. Vol. 6,** 1016 (1988), is employed.

3. Treatment with alkali removes the various acidic by-products and their salts (acetate, sulfinate, and formate) and also serves to hydrolyze and remove unreacted starting materials.

4. The checkers used "Aluminum Oxide" purchased from J. T. Baker Chemical Company.

5. The ^{1}H NMR spectrum of the product (CDCl$_3$): δ 1.83 (m, 4H), 2.42 (m, 2H), 2.73 (m, 2H), 3.30 (s, 4H).

3. Discussion

The procedure for the preparation of a dithiolane from a hydroxymethylene derivative of a ketone and ethylene dithiotosylate (ethane-1,2-dithiol di-*p*-toluenesulfonate) can be varied to produce dithianes, when the latter reagent is replaced with trimethylene dithiotosylate.[4,5] Dithiotosylates also react with enamine derivatives, producing dithiaspiro compounds.[5,6]

1. 1965 Nobel Laureate in Chemistry; deceased July 8, 1979; formerly at the Department of Chemistry, Harvard University, Cambridge, Massachusetts 02138.
2. Present address: Bristol Laboratories, Division of Bristol-Myers Company, Syracuse, New York 13201.
3. Present address: Sterling-Winthrop Research Institute, Rensselaer, New York 12144.
4. R. B. Woodward, I. J. Pachter, and M. L. Scheinbaum, *Org. Synth.*, **Coll. Vol. 6,** 1016 (1988).
5. R. B. Woodward, I. J. Pachter, and M. L. Scheinbaum, *J. Org. Chem.*, **36,** 1137 (1971).
6. R. B. Woodward, I. J. Pachter, and M. L. Scheinbaum, *Org. Synth.*, **Coll. Vol. 6,** 1014 (1988).

A GENERAL SYNTHESIS OF 4-ISOXAZOLECARBOXYLIC ESTERS: ETHYL 3-ETHYL-5-METHYL-4-ISOXAZOLECARBOXYLATE

(4-Isoxazolecarboxylic acid, 3-ethyl-5-methyl-, ethyl ester)

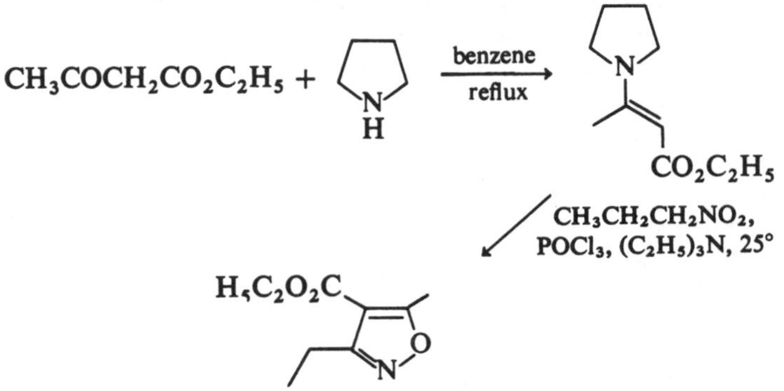

Submitted by JOHN E. McMURRY[1]
Checked by U. P. HOCHSTRASSER and G. BÜCHI

1. Procedure

Caution! The following reactions should be performed in an efficient hood to protect the experimentalist from noxious vapors (pyrrolidine, phosphorus oxychloride, and triethylamine).

Benzene has been identified as a carcinogen; OSHA has issued emergency standards on its use. All procedures involving benzene should be carried out in a well-ventilated hood, and glove protection is required.

A. *Ethyl β-pyrrolidinocrotonate.*   Ethyl acetoacetate (130 g., 1.00 mole) (Note 1) and pyrrolidine (71 g., 1.0 mole) are dissolved in 400 ml. of benzene and placed in a 1-1., one-necked flask fitted with a Dean-Stark water separator on top of which is a condenser fitted with a nitrogen inlet tube. The reaction mixture is placed under a nitrogen atmosphere (Note 2), then brought to and maintained at a vigorous reflux for 45 minutes, at which time the theoretical amount of water (18 ml.) has been collected. The benzene is removed with a rotary evaporator, yielding 180 g. (98%) of highly pure ethyl β-pyrrolidinocrotonate, which may be used without distillation (Note 3).

B. *Ethyl 3-ethyl-5-methyl-4-isoxazolecarboxylate.*   Ethyl β-pyrrolidinocrotonate (183 g., 1.00 mole), 1-nitropropane (115 g., 116 ml., 1.29 mole), and triethylamine (400 ml.) are dissolved in 1 l. of chloroform and placed in a 5-l., three-necked flask fitted with a 500 ml., pressure-equalizing dropping funnel and a gas-inlet tube. The flask is cooled in an ice bath, and its contents are placed under a nitrogen atmosphere. While the contents of the flask are stirred magnetically, a solution of 170 g. (1.11 mole) of phosphorus oxychloride in 200 ml. of chloroform is added slowly from the dropping funnel. After 3 hours, addition is complete and the ice bath is removed. The reaction mixture is allowed to warm to room temperature and stirring is continued for an additional 15 hours.

The reaction mixture is poured into a 4-l. separatory funnel and washed with 1 l. of cold water. The chloroform layer is washed with 6 N hydrochloric acid until the amine bases are removed and the wash remains acidic (Note 4). The chloroform extracts are washed successively with 5% aqueous sodium hydroxide (Note 5) and saturated brine, dried over anhydrous magnesium sulfate, and filtered. The solvent is removed with a rotary evaporator, and the product is distilled under vacuum, yielding 122–130 g. (68–71%) of ethyl 3-ethyl-5-methyl-4-isoxazolecarboxylate, b.p. 72° (0.5 mm.), n_D^{23} 1.4615 (Note 6).

2. Notes

1. The following reagents were used as supplied: pyrrolidine and triethylamine, Aldrich Chemical Company, Inc.; ethyl acetoacetate, Eastman Organic Chemicals (white label); nitropane, Matheson, Coleman and Bell (practical); phosphorus oxychloride, Matheson, Coleman and Bell (reagent).

2. Ethyl β-pyrrolidinocrotonate is typical of most enamines in that it discolors rapidly when exposed to air and therefore must be handled under an inert atmosphere.

3. Distillation is unnecessary and inadvisable, since discoloration usually occurs and there are product losses.

4. If this acid wash is not done thoroughly, the triethylamine hydrochloride remaining will sublime during distillation of the product and coat the still with a fluffy white powder. This impurity can be removed from the distillate, however, by a simple water wash.

5. This alkaline wash removes traces of ethyl acetoacetate which might form by hydrolysis of unreacted starting material during the preceding acid wash.

6. The product had the following spectral data: IR: 1725, 1605, 1300 cm.$^{-1}$; ^{1}H NMR (CCl$_4$): δ 1.3 (m, 6H, 2CH_3), 2.6 (s, 3H, C=CCH_3), 2.7 (q, 2H, CH_2CH$_3$), 4.2 (q, 2H, OCH_2CH$_3$).

3. Discussion

This procedure is illustrative of a general method for preparing a wide range of pure 3,5-disubstituted-4-isoxazolecarboxylic esters and (by hydrolysis) their acids,[2] free from positional isomers. A wide range of primary nitro compounds and enamino esters can be used,[2,3] and the esters thus obtained are useful as reagents in the isoxazole annelation reaction.[3,4] The only other general synthesis of these compounds involves chloromethylation and oxidation of a suitable 4-unsubstituted isoxazole,[5] but suffers from two difficulties: low yields and the unavailability of starting isoxazole. Most methods of isoxazole formation yield a mixture of positional isomers,[6] but the present method is quite selective. It has been shown that the reaction of a primary nitro compound with a dehydrating agent such as phosphorus oxychloride produces an intermediate nitrile oxide[7,8] which then undergoes a 1,3 dipolar cycloaddition to the enamine. This addition is remarkably selective with respect to orientation and no isomer formation is detected.[9] The intermediate isoxazoline then loses pyrrolidine enroute to the final product.

$$CH_3CH_2CH_2NO_2 + POCl_3 \longrightarrow CH_3CH_2C{\equiv}N{\rightarrow}O$$

1. Division of Natural Science, University of California, Santa Cruz, Santa Cruz, California 95060. [Present address: Department of Chemistry, Cornell University, Ithaca, New York 14853.]
2. G. Stork and J. E. McMurry, *J. Am. Chem. Soc.*, **89**, 5461 (1967).
3. G. Stork and J. E. McMurry, *J. Am. Chem. Soc.*, **89**, 5464 (1967).
4. G. Stork, S. Danishevsky, and M. Ohashi, *J. Am. Chem. Soc.*, **89**, 5459 (1967).
5. N. K. Kochetkov, E. D. Khomutova, and M. V. Bazilevskii, *J. Gen. Chem. U.S.S.R., Engl. Transl.*, **28**, 2762 (1958).
6. For a review of isoxazole chemistry, see; N. K. Kochetkov and S. D. Sokolov, "Advances in Heterocyclic Chemistry," Vol. 2, Academic Press, New York, 1963, pp. 365–421.
7. T. Mukaiyama and T. Hoshino, *J. Am. Chem. Soc.*, **82**, 5339 (1960).
8. G. B. Bachman and L. E. Strom, *J. Org. Chem.*, **28**, 1150 (1963).
9. G. Bianchi, and P. Grunanger, *Tetrahedron*, **21**, 817 (1965).

3-ALKYL-1-ALKYNES SYNTHESIS:
3-ETHYL-1-HEXYNE

(1-Hexyne, 3-ethyl-)

$$C_4H_9C\equiv CH \quad \xrightarrow[\text{2. } C_2H_5Br, \text{ 0--25}°]{\text{1. } C_4H_9Li, \text{ pentane}} \quad \diagup\!\!\diagdown\!\!\diagup\!\!\diagup C\equiv CH$$

Submitted by A. J. QUILLINAN and F. SCHEINMANN[1]
Checked by Y. KITA and G. BÜCHI

1. Procedure

A dry, 2-l., three-necked, round-bottomed flask is fitted with a nitrogen inlet, a reflux condenser provided with a gas-outlet connected to a gas-bubbler, a rubber septum, and a magnetic stirrer. After being charged with 400 ml. of pure, dry pentane (Note 1) and 41 g. (0.50 mole) of 1-hexyne (Note 1), the flask is flushed with nitrogen and immersed in a cold bath (Note 2), and the contents are stirred. The nitrogen atmosphere is maintained, and a solution of n-butyllithium in hexane or pentane (500 ml. of a 2.5 N solution, or 1.25 mole) is transferred to the flask with a 100-ml. syringe or a cannula (Note 3). The mixture is allowed to warm to 10° and stirred for 30 minutes, until the initially formed precipitate has completely dissolved.

The clear yellow solution is recooled to 0° in an ice bath before a solution of 88 g. (0.80 mole) of freshly-distilled ethyl bromide (Note 4) in 100 ml. of pure pentane is added dropwise with stirring over a 30-minute period while the solution warms to room temperature. Formation of a precipitate begins after an hour and is virtually complete after 6 hours (Note 5). After the mixture has been stirred for 2 days, 400 ml. of 4 N hydrochloric acid is carefully added with cooling (ice bath) and stirring. The layers are separated, and the organic phase (Note 6) is washed with 15 ml. of water, dried over anhydrous potassium carbonate, and filtered. Low-boiling materials are removed by distillation through an efficient Vigreux column (Note 7). The residue is distilled using a spinning-band apparatus (Note 8), yielding 35.2–35.8 g. (64–65%) of 3-ethyl-1-hexyne as a colorless, pungent oil (Note 9), b.p. 113–114°, n_D^{20} 1.4101 (Note 10).

2. Notes

1. Solvents must be pure since higher-boiling impurities will accumulate in the reaction product, making the final distillation much more difficult and reducing the purity of the product. Commercial pentane was purified by redistillation through an efficient Vigreux column (Note 7) and collected at 33–34°. 1-Hexyne is available from Fluka A G, Tridom Chemical, Inc., and Koch-Light Laboratories, Ltd., or by synthesis from sodium acetylide and 1-bromobutane (L. Brandsma, "Preparative Acetylene Chemistry," Elsevier, Amsterdam, 1971, p. 45.).

2. The temperature is not critical; the submitters used a bath temperature of -35 to $-20°$, but claimed that an acetone–dry-ice bath is satisfactory or that an ice-water bath with slow addition of n-butyllithium can also be used.

3. All operations involving alkyllithium reagents should be carried out under an inert gas, since they tend to ignite spontaneously in air.

4. Commercial ethyl bromide was dried over anhydrous potassium carbonate, filtered, and redistilled prior to use.

5. The reaction may be conveniently terminated here, but reaction for two days gives a slightly higher yield at the expense of some hydrocarbon impurity formed from the slow reaction of excess alkyllithium with the alkyl bromide. If this impurity can contaminate the final product, the side reaction may be suppressed by conducting the reaction in the minimum amount of solvent necessary to dissolve the dilithium complex.

6. Dark coloration is usually a result of insufficient acidity; the aqueous phase at the separation stage should have a pH between 2 and 4.

7. A vacuum-jacketed Vigreux column of 1.5-cm. internal diameter and $ca.$ 90-cm. length is satisfactory. With shorter or less efficient columns, redistillation of the distillate may be necessary to reduce losses in yield.

8. The submitters used a Büchi spinning-band distillation apparatus by Abegg ($ca.$ 30 theoretical plates resolution), operated at a reflux ratio of $10-15:1$ and having a capacity of $50-100$ ml. The checkers found that this distillation required four days to obtain the pure product.

9. 3-Ethyl-1-hexyne is an irritant to the membranes of the nose and throat and should only be handled in a well-ventilated hood. Other 3-substituted 1-alkynes prepared in this series[2] have pleasant or neutral odors, but direct contact should be avoided, since toxicity data are not available.

10. The previously unknown 3-ethyl-1-hexyne was further characterized by the submitters as follows: IR cm.$^{-1}$: 3338, 2975, 2940, 2881, 2103, 1460, 1380, 1240; ^{13}C NMR (CDCl$_3$), δ 88.0, 69.23, 37.12, 33.29, 28.23, 20.69, 13.97, 11.71; ^{1}H NMR (CDCl$_3$), δ (multiplicity, coupling constant J in Hz., number of protons, assignment): 0.99 (t, $J = 6$, 6H, 2CH_3), 1.2–1.7 (m, 6H, 3CH_2), 2.00 (d, $J = 2$, 1H, C≡CH), 2.3 (m, 1H, tertiary CH).

Hydrogenation at 20° over Adams catalyst, platinum oxide, at atmospheric pressure afforded the known 3-ethylhexane,[3] b.p. 118–119° (lit.,[2] b.p. 119°).

In an analogous synthesis,[2a] 3-butyl-1-heptyne was also characterized by mercuric oxide–sulfuric acid hydration[4] to the known 3-butyl-2-heptanone,[5,6] which also formed the known semicarbazone.[6]

3. Discussion

This procedure has been shown[2] to be extremely general and applicable to the reaction of a wide variety of straight-chain 1-acetylenes, 4-substituted 1-acetylenes, and α,ω-diacetylenes with primary halides, sterically hindered primary halides, secondary halides, and α,ω-dihalides.

Most of the compounds formed are new and were formerly inaccessible,[7] available only by dehydrohalogenation of geminal or 1,2-dibromides which are often unavailable themselves.[7] Alcoholic potassium hydroxide[8,9] or sodamide in liquid paraffin[7,10] under forceful conditions has been used for this elimination, but yields are generally not good.[7-10]

The procedure described here is characterized by good yields and mild conditions, and affords an easy route to a pure compound from readily available starting materials. Since tertiary aliphatic acetylenes do not form readily under these conditions, the excess of alkyllithium used is not particularly critical. The small amount of by-products that also forms is readily removed at the distillation stage.

1. Department of Chemistry and Applied Chemistry, University of Salford, Salford M5 4WT, England.
2. (a) A. J. Quillinan, E. A. Khan, and F. Scheinmann, *J. Chem. Soc. Chem. Commun.*, 1030 (1974); (b) J. Klein and S. Brenner, *J. Org. Chem.*, **36**, 1319 (1971); (c) J. Klein and J. Y. Becker, *Tetrahedron*, **28**, 5385 (1972); (d) J. Klein and J. Y. Becker, *J. Chem. Soc., Perkin Trans. II*, 599 (1973).
3. L. Clarke and E. R. Riegel, *J. Am. Chem. Soc.*, **34**, 674 (1912).
4. R. M. Roberts, J. C. Gilbert, L. B. Rodewald, and A. S. Wingrove, "An Introduction to Modern Experimental Organic Chemistry", Holt, Rinehart and Winston, New York, 1969.
5. K. Hess and R. Bappert, *Justus Liebigs Ann. Chem.*, **441**, 151 (1925).
6. W. B. Renfrow, Jr., *J. Am. Chem. Soc.*, **66**, 144 (1944).
7. T. L. Jacobs, *Org. React.*, **5**, 1 (1949).
8. V. Sawitsch, *C.R. Hebd. Seances Acad. Sci.*, **52**, 399 (1861).
9. (M.) W. Morkownikoff, *Bull. Soc. Chim. Fr.*, 90 (1861).
10. B. Gredy, *Bull. Soc. Chim. Fr.*, [5], **2**, 1951 (1935).

β-HYDROXY ESTERS FROM ETHYL ACETATE AND ALDEHYDES OR KETONES: ETHYL 1-HYDROXYCYCLOHEXYLACETATE

(Cyclohexaneacetic acid, 1-hydroxy, ethyl ester)

$$HN[Si(CH_3)_3]_2 + n\text{-}C_4H_9Li \longrightarrow LiN[Si(CH_3)_3]_2 + n\text{-}C_4H_{10}$$

$$LiN[Si(CH_3)_3]_2 + CH_3CO_2C_2H_5 \longrightarrow$$
$$LiCH_2CO_2C_2H_5 + HN[Si(CH_3)_3]_2$$

Submitted by MICHAEL W. RATHKE[1]
Checked by Y. HOYANO and S. MASAMUNE

1. Procedure

Caution! The first step of the reaction should be conducted in a well-ventilated hood since butane is liberated.

A. *Lithium bis(trimethylsilyl)amide* (Note 1).   A dry, 500-ml., three-necked flask, fitted with a pressure-equalizing dropping funnel in the center neck and a stopcock in each side neck, is equipped for magnetic stirring and maintained under a static nitrogen pressure by attaching a nitrogen source to one stopcock and a mercury bubbler to the other. In the flask is placed 153 ml. of a hexane solution containing 0.250 mole of *n*-butyllithium (Note 2), and stirring is started. The flask is immersed in an ice-water bath, and 42.2 g. (0.263 mole) of hexamethyldisilazane (Note 3) is added dropwise over a period of 10 minutes. The ice-bath is removed, and the solution is stirred for 15 minutes longer. The hexane is removed under reduced pressure by replacing the mercury bubbler with heavy rubber tubing connected to a dry-ice condenser and an oil pump. During this step, the flask is immersed in a water bath at 40–50°, and stirring is continued as long as possible. After complete evaporation of the hexane, white crystals of lithium bis-(trimethylsilyl)amide appear. The flask is again subjected to a static pressure of nitrogen (Note 4), and 225 ml. of tetrahydrofuran (Note 5) is added to dissolve the crystals.

B. *Ethyl lithioacetate.*   The reaction flask is immersed in an acetone–dry-ice bath, and the solution is stirred for 15 minutes to achieve thermal equilibration. After this time, 22.1 g. (0.250 mole) of ethyl acetate is added dropwise over a 10-minute period. Stirring is continued for an additional 15 minutes to complete formation of ethyl lithioacetate (Note 6).

C. *Ethyl 1-hydroxycyclohexylacetate.* A solution of 24.6 g. (0.250 mole) of cyclohexanone (Note 7) in 25 ml. of tetrahydrofuran is added dropwise to the reaction mixture over a 10-minute period. After an additional 5 minutes, the reaction mixture is hydrolyzed by adding 75 ml. of 20% hydrochloric acid in one portion (Note 8). The cooling bath is removed, and the stirred solution is allowed to reach room temperature.

The organic layer is separated, the aqueous layer is extracted with two 50-ml. portions of diethyl ether, and the combined extracts are dried over anhydrous sodium sulfate. The solvent is removed with a rotary evaporator (Note 9), and the almost-colorless residue is distilled under reduced pressure through a 10-cm. Vigreux column, yielding 37–42 g. (79–90%) of the β-hydroxy ester as a colorless liquid, b.p. 77–80° (1 mm.); n_D^{24} 1.4555–1.4557 (Note 10).

2. Notes

1. The preparation of lithium bis(trimethylsilyl)amide is adapted from an earlier procedure.[2] The original procedure specifies addition of *n*-butyllithium to hexamethyldisilazane in diethyl ether followed by a reflux period. It is generally more convenient to add the hexamethyldisilazane to the *n*-butyllithium; satisfactory results are obtained without using ether or refluxing.

2. A 1.63 *M* solution of *n*-butyllithium in hexane was purchased from Foote Mineral Company.

3. Hexamethyldisilazane was obtained from Pierce Chemical Company and used without further purification.

4. Lithium bis(trimethylsilyl)amide is hydrolyzed rapidly by moist air. It is therefore essential to break the vacuum by admitting nitrogen rather than air.

5. The submitters used reagent grade tetrahydrofuran (available from Fisher Scientific Company) from a freshly opened bottle. The checkers used tetrahydrofuran purified by distillation from lithium aluminum hydride. See *Org. Synth., Coll. Vol. 5,* 976 (1973) for warning regarding purification of this solvent.

6. Solutions of ethyl lithioacetate prepared by this method are stable indefinitely at −78°, but decompose rapidly if allowed to reach room temperature.

7. White label cyclohexanone (Eastman Organic Chemicals) was used without further purification.

8. The yield of the β-hydroxy ester is somewhat lower if the reaction mixture is allowed to reach room temperature prior to hydrolysis.

9. Continuing the evaporation process for some time after removal of the solvent is helpful in removing any residual hexamethyldisilazane (b.p. 125°) together with its hydrolysis product, hexamethyldisiloxane (b.p. 100°).

10. ^{1}H NMR (CDCl$_3$), δ 1.26 (t, $J = 7$ Hz., 3H, OCH$_2$C*H*$_3$), 1.52 (broad s, 10H, 5C*H*$_2$), 2.45 (s, 2H, C*H*$_2$CO$_2$), 3.33 (s, 1H, O*H*), 4.19 (q, $J = 7$ Hz., 2H, OC*H*$_2$CH$_3$).

3. Discussion

Ethyl 1-hydroxycyclohexylacetate has been prepared by the Reformatsky reaction of cyclohexanone with zinc and ethyl bromoacetate $(56-71\%)$[3] and by the condensation of ethyl acetate with cyclohexanone in liquid ammonia, using two equivalents of lithium amide (69%).[4]

This preparation illustrates a general method for the preparation of β-hydroxy esters from ethyl acetate and aldehydes or ketones.[5] The procedure is simpler and less time-consuming than other methods and the yields are usually higher. In addition, the β-hydroxy esters are obtained in a high state of purity.

The procedure is especially suited to small-scale preparations (25 mmoles or less) where the necessity of evaporating hexane from the lithium bis(trimethylsilyl)amide is much less of a handicap. In such cases, it is convenient to equip the reaction vessel with a septum-inlet and transfer all reagents with a syringe.

1. Department of Chemistry, Michigan State University, East Lansing, Michigan 48823.
2. E. H. Amonoo-Neizer, R. A. Shaw, D. O. Skovlin, and B. C. Smith, *J. Chem. Soc.*, 2997 (1965).
3. R. L. Shriner, *Org. React.*, **1,** 17 (1942).
4. W. R. Dunnavant and C. R. Hauser, *J. Org. Chem.*, **25,** 503 (1960).
5. M. W. Rathke, *J. Am. Chem. Soc.*, **92,** 3222 (1970).

INDOLES FROM ANILINES:
ETHYL 2-METHYLINDOLE-5-CARBOXYLATE

[1H-Indole-5-carboxylic acid, 2-methyl, ethyl ester]

Submitted by P. G. Gassman[1] and T. J. van Bergen
Checked by J. L. Belletire and G. Büchi

1. Procedure

Caution! Part A must be conducted in an efficient hood to avoid exposure to methanethiol and chloroacetone, both of which are highly irritating. In Part C, the usual precautions associated with the pyrophoric reagent Raney nickel (Note 12) should be observed.

A. *Methylthio-2-propanone.*[2] A 2-l., three-necked, round-bottomed flask is equipped with a sealed mechanical stirrer and a two-necked adapter holding a thermometer and a condenser topped with a silica gel drying tube. After 700 ml. of anhydrous methanol has been added, the third neck is stoppered, and the flask is immersed in an ice water bath. Stirring is begun, and 108 g. (2.00 moles) of sodium methoxide (Note 1) is added in small portions at a rate sufficiently gradual to prevent a large exotherm (Note 2). When all of the methoxide has dissolved and the temperature has returned to 5°, the stopper is replaced with a 200-ml., pressure-equalizing, jacketed addition funnel (Note 3). This funnel, charged previously with 130 ml. (2.20 moles) of methanethiol (Note 4), contains a 2-propanol-dry ice slurry in the cooling jacket, and is topped with a silica gel drying tube. Stirring and cooling are continued while the methanethiol is run into the flask over a 20-minute period, and for 15 minutes thereafter. The jacketed addition funnel is then replaced by a standard, 200-ml., pressure-equalizing addition funnel, which is used to add 185 g. (2.00 moles) of chloroacetone (Note 5) to the reaction mixture over 1 hour. When this addition is complete, the ice bath is removed, and the suspension is stirred overnight at room temperature. The insoluble material, largely inorganic salts, is removed by filtration through Celite, and the filter cake is washed with two 150-ml. portions of absolute methanol. After methanol has been removed from the combined filtrates by distillation, the residue is distilled through a 300-mm. Vigreux column, yielding 155–158 g. (74–76%) of methylthio-2-propanone, b.p. 153–154°, n_D^{23} 1.4728.

B. *Ethyl 2-methyl-3-methylthioindole-5-carboxylate.* A 1-l., three-necked, round-bottomed flask is equipped with a sealed mechanical stirrer, a 100-ml., pressure-equalizing addition funnel, and a two-necked adapter holding a low-temperature thermometer and a gas-inlet tube. The flask is charged with 16.5 g. (0.100 mole) of ethyl 4-aminobenzoate (Benzocaine) (Note 6) and 500 ml. of dichloromethane (Note 7), and a positive pressure of dry nitrogen is established while the solution is stirred and cooled to −70° with a 2-propanol–dry-ice bath. The resulting suspension is stirred vigorously, and a solution of 10.8 g. (0.0995 mole) of *tert*-butyl hypochlorite [*Org. Synth.*, **Coll. Vol. 5,** 184 (1973)] in 50 ml. of dichloromethane (Note 7) is added dropwise over a 10-minute period. The reaction mixture is stirred for 1 hour at −70°, followed by dropwise addition of a solution of 10.4 g. (0.100 mole) of methylthio-2-propanone in 50 ml. of dichloromethane (Note 7) over 10 minutes. A slight exotherm (*ca.* 5°) is noted during the addition, resulting in a clear yellow solution. After stirring for another hour at −70°, during which time a suspension of precipitated salts forms, a solution of 10.1 g. (0.100 mole) of triethylamine (Note 8) in 30 ml. of dichloromethane (Note 7) is added dropwise over 10 minutes. After an additional 15 minutes at −70°, the cooling bath is removed.

When the reaction mixture has warmed to room temperature, stirring is made more vigorous, and 100 ml. of water is added. The layers are separated, and the organic phase is dried over anhydrous magnesium sulfate and filtered. Removal of dichloromethane with a rotary evaporator leaves an oily residue, which is cooled in an ice bath, inducing crystallization. The resulting solid is stirred with 50 ml. of diethyl ether for 30 minutes at 0° (Note 9), collected by filtration, and washed with 25 ml. of ether at 0°. Concentration of the filtrate to 15 ml. and chilling at *ca.* 5° overnight yields a second crop of crystalline product, bringing the total crude yield to 14.7–18.0 g. (59–73%), m.p. 120.5–124° (Note 10). Recrystallization from absolute ethanol (5 ml. per 2 g.) yields 12.8–17.5 g. (51–70%) of ethyl 2-methyl-3-methylthioindole-5-carboxylate, m.p. 125.5–127°.

C. *Ethyl 2-methylindole-5-carboxylate.* A solution of 10.0 g. (0.0402 mole) of ethyl 2-methyl-3-methylthioindole-5-carboxylate in 300 ml. of absolute ethanol is placed in a 500-ml., three-necked, round-bottomed flask fitted with a mechanical stirrer. An excess (15 teaspoons), (Note 11) of freshly washed W-2 Raney nickel (Note 12) is added, and the mixture is stirred for one hour. Stirring is then stopped, the liquid phase is decanted, and the catalyst is washed twice by stirring for 15 minutes with 100-ml. portions of absolute ethanol and decanting the solvent. The combined ethanolic solutions are concentrated on a rotary evaporator. The residual solid is dissolved in 150 ml. of warm dichloromethane, dried over anhydrous magnesium sulfate, and filtered; the drying agent is washed with 40 ml. of dichloromethane. Concentration of the combined filtrates with a rotary evaporator gives 7.5–8.1 g. (93–99%) of ethyl 2-methylindole-5-carboxylate, m.p. 140.5–142.0° (Note 13).

2. Notes

1. Sodium methoxide was purchased from MC and B Manufacturing Chemists and used without further purification. Accurate weighing is important, since the checkers noted that an excess of methoxide relative to chloroacetone, however slight, led to little or no product.

2. A convenient technique is to add the methoxide from an Erlenmeyer flask, connected to the previously stoppered neck of the reaction flask by a piece of flexible rubber tubing (Gooch tubing).

3. The addition funnel used in this preparation is depicted in Figure 1.

4. Methanethiol was purchased from Matheson Gas Products. *Caution! This compound has a powerful, vile odor, even at extremely low concentrations.* The gas, which liquefies at 6°, may be condensed by calibrating the addition funnel at 130 ml. with a grease pencil, inserting a cold-finger condenser topped with a gas-inlet tube in the female joint, and protecting the male joint from moisture by a drying tube. A 2-propanol–dry-ice slurry is placed in the condenser and the cooling jacket, and methanethiol is introduced through the gas-inlet until the condensed liquid reaches the calibration mark.

5. Chloroacetone was purchased from Distillation Products (Eastman Organic Chemicals) and distilled prior to use. *Caution! This compound is an intensely powerful lachrymator.*

6. Ethyl *p*-aminobenzoate was purchased from the Aldrich Chemical Company, Inc., and used without further purification.

7. Commercial dichloromethane was distilled prior to use.

8. Triethylamine was purchased from the J. T. Baker Chemical Company and used without further purification.

9. If the ether is added prior to crystallization, a slightly reduced yield results.

10. Occasionally the product is contaminated with an impurity (1–2%), which appears as tiny red needles. This material has been tentatively identified as diethyl azobenzene-4,4′-dicarboxylate.

11. A level teaspoonful contains about 3 g. of nickel.

12. W-2 Raney nickel was purchased from W. R. Grace and Co. Prior to use it was washed with distilled water until neutral, then three times with absolute ethanol [*Org. Synth.*, **Coll. Vol. 3,** 181 (1955)]. This material may ignite spontaneously if allowed to become dry. Thus, in decanting, a small amount of solvent must be left behind to cover

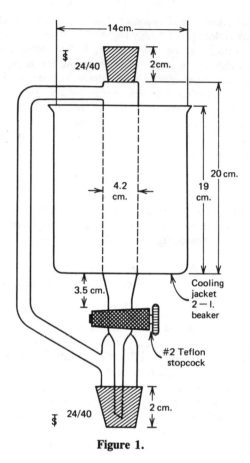

Figure 1.

the catalyst. Although environmentally unsound, spent catalyst is discarded by slurrying in water and flushing the slurry down the drain with running water.

13. Little change in melting point results when the product is recrystallized from benzene.

3. Discussion

This procedure is an example of a new indole synthesis, which can be utilized to prepare 1-, 2-, 4-, 5-, 6-, or 7-substituted indoles.[3] Indoles substituted on the phenyl ring with nitro, ethoxycarbonyl, chloro, methyl, and acetoxy groups have been prepared; hydrogen, methyl, and phenyl groups have been placed in the 2-position; and the method has been used to prepare 1-methylindoles. A similar procedure substituting α-thioesters for α-thioketones yields oxindoles in good, overall yields.[4] The major advantages of this sequence are the availability of starting materials and the high overall yields of indoles and oxindoles realized.[5]

1. Department of Chemistry, The Ohio State University, Columbus, Ohio 43210. [Present address: Department of Chemistry, University of Minnesota, Minneapolis, Minnesota 55455.]
2. C. K. Bradsher, F. C. Brown, and R. J. Grantham, *J. Am. Chem. Soc.*, **76**, 114 (1954).
3. P. G. Gassman and T. J. van Bergen, *J. Am. Chem. Soc.*, **95**, 590 (1973); P. G. Gassman, T. J. van Bergen, and G. D. Gruetzmacher, *J. Am. Chem. Soc.*, **95**, 5608 (1973); P. G. Gassman, D. P. Gilbert, and T. J. van Bergen, *J. Chem. Soc. Chem. Commun.*, 201 (1974); P. G. Gassman, T. J. van Bergen, D. P. Gilbert, and B. W. Cue, Jr., *J. Am. Chem. Soc.*, **96**, 5495 (1974); P. G. Gassman, G. D. Gruetzmacher, and T. J. van Bergen, *J. Am. Chem. Soc.*, **96**, 5512 (1974); P. G. Gassman, D. P. Gilbert, and T.-Y. Luh, *J. Org. Chem.*, **42**, 1340 (1977).
4. P. G. Gassman and T. J. van Bergen, *J. Am. Chem. Soc.*, **95**, 2718 (1973); P. G. Gassman, T. J. van Bergen, and G. D. Gruetzmacher, *J. Am. Chem. Soc.*, **95**, 5608 (1973); P. G. Gassman and T. J. van Bergen, *J. Am. Chem. Soc.*, **96**, 5508 (1974); P. G. Gassman, G. D. Gruetzmacher, and T. J. van Bergen, *J. Am. Chem. Soc.*, **96**, 5512 (1974); P. G. Gassman, D. P. Gilbert, and T.-Y. Luh, *J. Org. Chem.*, **42**, 1340 (1977); P. G. Gassman, B. W. Cue, Jr., and T.-Y. Luh, *J. Org. Chem.*, **42**, 1344 (1977); P. G. Gassman and R. L. Parton, *J. Chem. Soc. Chem. Commun.*, 694 (1977); P. G. Gassman and K. M. Halweg, *J. Org. Chem.*, **44**, 628 (1979).
5. For a summary of other indole syntheses, see R. J. Sundberg, "The Chemistry of Indoles," Academic Press, New York, 1970.

STEREOSELECTIVE SYNTHESIS OF TRISUBSTITUTED OLEFINS: ETHYL 4-METHYL-(*E*)-4,8-NONADIENOATE

(4,8-Nonadienoic acid, 4-methyl, ethyl ester, *trans-*)

$$CH_2{=}CHCH_2CH_2Cl \xrightarrow[\text{ether}]{\text{Mg}} CH_2{=}CHCH_2CH_2MgCl$$

$$CH_2{=}C(CH_3)CHO \downarrow \text{ether}$$

$$\overset{\displaystyle OH}{\underset{\displaystyle |}{}} \quad \overset{\displaystyle CH_3}{\underset{\displaystyle |}{}}$$

$$CH_2{=}CHCH_2CH_2CH{-}C{=}CH_2$$

$$CH_3C(OC_2H_5)_3 \diagup \begin{array}{l} CH_3CH_2CO_2H, \\ 138\text{--}140° \end{array}$$

$$\left[CH_2{=}CHCH_2CH_2{-}\overset{H}{\underset{}{C}}\underset{O}{\diagdown}\overset{OC_2H_5}{\underset{\underset{CH_3}{|}}{C}}{=}CH_2 \right.$$
$$\left. \overset{}{\underset{}{C}}{=}CH_2 \right]$$

$$\searrow$$

$$\underset{CH_2{=}CHCH_2CH_2}{\overset{H}{\diagdown}}C{=}C\overset{CH_2CH_2CO_2C_2H_5}{\underset{CH_3}{\diagup}}$$

Submitted by Ronald I. Trust and Robert E. Ireland[1]
Checked by David G. Melillo and Herbert O. House

1. Procedure

A. *2-Methyl-1,6-heptadien-3-ol.* A dry, three-necked, 1-l., round-bottomed flask fitted with a mechanical stirrer, a reflux condenser with a nitrogen inlet tube, and a 125-ml., pressure-equalizing dropping funnel capped with a rubber septum is charged with 15.7 g. (0.646 g.-atom) of magnesium turnings. The flask is dried by heating with a flame while a stream of dry nitrogen is passed through the reaction vessel from the condenser and allowed to exit from a hypodermic needle inserted in the rubber septum. After drying, the hypodermic needle is removed and the flask is allowed to cool; a static nitrogen atmosphere is maintained in the reaction vessel for the remainder of the reaction. A small crystal of iodine and 450 ml. of anhydrous diethyl ether are added (Note 1). A solution of 49.4 g. (0.546 mole) of 4-chloro-1-butene (Note 2) in 50 ml. of anhydrous ether is then added from the dropping funnel, dropwise and with stirring. Sufficient external heat is applied to the reaction flask to keep the temperature of the reaction mixture at about 30°. After approximately 10–50% of the chloride solution has been added, a spontaneous reaction ensues as evidenced by the disappearance of the yellow iodine color, the appearance of a gray color in the reaction solution, and the commencement of gentle refluxing. The external heat is removed, and the remainder of the chloride solution is added at a rate that maintains gentle refluxing. After the addition is complete, the reaction mixture is refluxed for 30 minutes before a solution of 40.1 g. (0.572 mole) of methacrolein (Note 3) in 50 ml. of anhydrous ether is added, dropwise with stirring and refluxing, over 45 minutes. Since the reaction with methacrolein is exothermic, the application of external heat may not be necessary to maintain refluxing during this addition. During the addition the reaction mixture usually becomes cloudy. When the addition is complete, the reaction mixture is refluxed with stirring for 1.5 hours before it is cooled in an ice water bath and 250 ml. of 5% hydrochloric acid is added slowly and with stirring (Note 4). The organic layer is separated and the aqueous layer is extracted with four 200-ml. portions of ether. The combined organic solutions are washed successively with 200 ml. of saturated aqueous sodium hydrogen carbonate and 200 ml. of saturated aqueous sodium chloride and dried over anhydrous sodium sulfate. The resulting ether solution is concentrated, and the residual liquid is distilled under reduced pressure, yielding 37.2–47.3 g. (54–69%) of 2-methyl-1,6-heptadien-3-ol as a colorless liquid, b.p. 85–88° (33 mm.), n_D^{25} 1.4531–1.4535 (Note 5).

B. *Ethyl 4-methyl-(E)-4,8-nonadienoate.* A 500-ml., one-necked, round-bottomed flask containing a magnetic stirring bar is fitted with a Claisen adapter, two thermometers, and a receiving flask as illustrated in Figure 1. To the flask is added 186 g. (1.15 moles) of ethyl orthoacetate (Note 6), 25.2 g. (0.200 mole) of 2-methyl-1,6-heptadien-3-ol, and 0.70 g. (0.0094 mole) of propionic acid (Note 7). The mixture is heated with stirring to keep the temperature above the liquid at 138–142°. Heating is continued until ethanol no longer distils from the reaction flask (approximately one hour is required). The reaction mixture is allowed to cool to room temperature, and the excess ortho ester and propionic acid are removed by distillation under reduced pressure (approximately 50–60° at 20 mm.). The colorless to yellow residual liquid is then distilled under reduced pressure (0.25 mm., Note 8), giving 32.6–34.6 g. (83–88%) of ethyl 4-methyl-(*E*)-4,8-nonadienoate as a colorless liquid, b.p. 54–55° (0.25 mm.), n_D^{25} 1.4504 (Note 9).

2. Notes

1. After drying, exposure of the reaction vessel and its contents to the atmosphere should be minimized. The iodine crystal should be added by lifting the dropping funnel, then replacing it quickly. The ether (anhydrous grade from Mallinckrodt Chemical Works) should be distilled from lithium aluminum hydride immediately before use and transferred to the reaction vessel with a stainless-steel cannula or a large hypodermic syringe inserted through the rubber septum.

2. 4-Chloro-1-butene is commercially available from Chemical Samples Company. The checkers employed this material without further purification. The submitters used

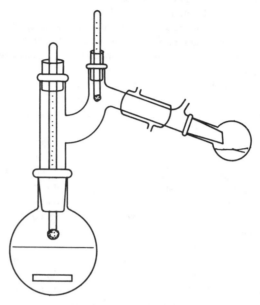

Figure 1. Apparatus for the preparation of ethyl 4-methyl-(E)-4,8-nonadienoate.

material prepared from 3-buten-1-ol by a modified procedure of Roberts and Mazur.[2] Since material prepared according to the literature is invariably contaminated with thionyl chloride, which will interfere with formation of the Grignard reagent, the following modification is recommended. A two-necked, 200-ml., round-bottomed flask is equipped with a magnetic stirring bar, a 60-ml., pressure-equalizing dropping funnel, and a reflux condenser fitted with a calcium chloride drying tube. The flask is charged with 49.8 g. (0.691 mole) of 3-buten-1-ol and 1.57 ml. of anhydrous pyridine (distilled from calcium hydride). With stirring and external cooling (ice water bath), 82 g. (49 ml., 0.69 mole) of thionyl chloride (Matheson, Coleman and Bell commercial grade was used without further purification) is added dropwise over 3.5 hours. On completion of the addition, the mixture is heated under reflux for one hour. The external heating is then momentarily discontinued, and the condenser and dropping funnel are replaced by a stopper and short-path distilling head with receiver. Distillation of the mixture gives an opaque, colorless liquid (b.p. 68°). The crude product is washed with two 20-ml. portions of saturated aqueous sodium hydrogen carbonate solution (frothing) and 20 ml. of saturated brine, then dried over magnesium sulfate and filtered. The filtrate is distilled, giving 43.4 g. (67–69%) of 4-chloro-1-butene as a colorless liquid, b.p. 68–70°.

   3-Buten-1-ol, although commercially available from Aldrich Chemical Company, Inc., can be prepared economically and in large quantities by the addition of paraformaldehyde to allylmagnesium bromide [*Org. Synth.*, **Coll. Vol. 4,** 748 (1963)] in ether according to procedures outlined for a similar synthesis [*Org. Synth.*, **Coll. Vol. 1,** 188

(1944)]. In the present case, the submitters found it convenient to add the paraformaldehyde (Matheson, Coleman and Bell commercial grade was dried overnight under reduced pressure and in the presence of phosphorus pentoxide) directly to the allylmagnesium bromide solution. After a reaction period of 6 hours at reflux, the previously described [*Org. Synth.*, **Coll. Vol. 1**, 188 (1944)] isolation procedure gave 3-buten-1-ol in 56% yield.

3. Technical grade (90%) methacrolein (Aldrich Chemical Company, Inc.) was distilled (b.p. 67–69°) immediately before use.

4. Since the methacrolein is used in excess, frothing is no problem as there is no Grignard reagent remaining after the reaction is completed. Addition of 5% hydrochloric acid causes some coagulation of magnesium salts in the aqueous layer, which can be redissolved by addition of more 5% hydrochloric acid.

5. The product has the following spectral characteristics: IR (CCl_4), 3620 (free OH), 3480 (associated OH), 1645 (C=C), and 910 cm.$^{-1}$ (CH=CH$_2$); UV (95% C_2H_5OH) end absorption 210 nm (ϵ 208); ^{1}H NMR (CCl_4), δ 1.2–2.3 (m, 4H, 2CH_2), 1.70 (s, 3H, CH_3), 2.93 (broad, 1H, OH), 4.00 (t, J = 6 Hz., 1H, OCH), 4.6–5.2 (m, 4H, vinyl CH), and 5.5–6.1 (m, 1H, vinyl CH); m/e (rel. int.), 111(28), 84(29), 83(21), 71(51), 71(100), 69(23), 67(28), 57(30), 55(51), 43(71), 41(49), and 39(37).

6. Ethyl orthoacetate, available from Aldrich Chemical Company, Inc., was distilled before use. A large forerun was collected, consisting of hydrolysis products of the ortho ester. Material boiling at 135–142° is suitable for use in the reaction. It is convenient to transfer the material to the reaction flask with a stainless-steel cannula to avoid its exposure to atmospheric moisture. A fivefold excess of the ortho ester is needed, since the first step in the reaction is probably the reversible acid-catalyzed exchange of 2-methyl-1,6-heptadien-3-ol with ethanol.[3]

7. Practical grade propionic acid (Matheson, Coleman and Bell) was distilled before use (b.p. 141°).

8. On two occasions, the submitters noticed a sublimable solid crystallizing in the distilling head just before the product began to distill. The distilling head was rinsed with ether, dried, and replaced, and the distillation was continued. The checkers observed the same phenomenon.

9. The product has the following spectral properties: IR (CCl_4) 1735 (ester C=O), 1645 (C=C), and 920 cm.$^{-1}$ (CH=CH$_2$); UV (95% C_2H_5OH) end absorption 210 nm (ϵ 1960); ^{1}H NMR ($CDCl_3$), δ 1.24 (t, J = 7 Hz., 3H, OCH$_2$CH_3), 1.63 (broad, 3H, C=CCH_3), 1.9–2.2 (m, 4H, 2CH_2), 2.37 (broad, 4H, 2CH_2), 4.14 (q, J = 7 Hz., 2H, OCH_2CH$_3$), 4.8–5.4 (m, 3H, vinyl CH), 5.5–6.2 (m, 1H, vinyl CH); m/e (rel. int.), 196 (M$^+$, 4), 155(67), 151(30), 113(34), 109(100), 108(33), 85(47), 81(80), 67(74), 55(41), 53(31), 43(32), and 41(30). In C_6D_6 the allylic CH$_3$ signal of the major component present, the *trans*-isomer, is found at δ 1.50 and is accompanied by a minor peak at δ 1.61 attributable[4] to 3–4% of the *cis*-olefin in the product.

3. Discussion

The use of the Claisen rearrangement and several other methods for the stereoselective synthesis of trisubstituted olefins has been reviewed.[4] In allyl vinyl ethers of type A, the stereochemistry of the rearrangement is determined largely by the steric requirements of

R_1, which can be either axial or equatorial in the transition state.[5] When R_3 = H, the *trans/cis* ratio is approximately equal to the equatorial/axial equilibrium ratio of R_1-cyclohexane at the reaction temperature. When R_3 is larger than hydrogen, the steric effect is even greater, due to a potential 1,3-interaction which would develop in the transition state if R_1 were axial. No significant effect of R_2 on the *trans/cis* ratio has been observed.

The use of ethyl orthoacetate in the formation of vinyl ethers where R_3 = OC_2H_5 has been described.[3,6] The method described herein appears to be quite general, in that a variety of esters of type B (R_3 = OC_2H_5; R_2 = CH_3) may be prepared by merely varying the Grignard reagent used in preparing the starting allyl alcohol. The only limitation is the use of alcohols that are unsymmetrically bis-allylic, from which mixtures of structural isomers may be obtained.

Stereoselectivity in the synthesis of trisubstituted olefins is necessary for the study of biosynthetic routes to polyisoprenoids, the nonenzymatic cyclization of polyolefinic substrates, and the study of insect hormones.

1. Department of Chemistry, California Institute of Technology, Pasadena, California 91109.
2. J. D. Roberts and R. H. Mazur, *J. Am. Chem. Soc.*, **73**, 2509 (1951).
3. W. S. Johnson, L. Werthemann, W. R. Bartlett, T. J. Brocksom, T. Li, D. J. Faulkner, and M. R. Petersen, *J. Am. Chem. Soc.*, **92**, 741 (1970).
4. D. J. Faulkner, *Synthesis*, 175 (1971).
5. D. J. Faulkner and M. R. Petersen, *Tetrahedron Lett.*, 3243 (1969).
6. W. S. Johnson, M. B. Gravestock, and B. E. McCarry, *J. Am. Chem. Soc.*, **93**, 4332 (1971).

METALATION OF 2-METHYLPYRIDINE DERIVATIVES: ETHYL 6-METHYLPYRIDINE-2-ACETATE

(Pyridineacetic acid, 6-methyl-, ethyl ester)

Submitted by WILLIAM G. KOFRON[1] and LEONA M. BACLAWSKI
Checked by R. E. IRELAND and R. A. FARR

1. Procedure

Caution! This preparation should be carried out in a good hood to avoid exposure to ammonia.

A 1-l., three-necked flask is fitted with an acetone-dry ice condenser, a glass stirrer, and a glass stopper (Note 1). Potassium amide is prepared in 400 ml. of liquid ammonia with 8.0 g. (0.20 g.-atom) of potassium metal (Note 2). The glass stopper is replaced with an addition funnel containing 32.1 g. (0.300 mole) (Note 3) of 2,6-lutidine dissolved in *ca.* 20 ml. of anhydrous diethyl ether. The lutidine solution is added to the amide solution, and the funnel is rinsed with a little ether, which is also added. The resulting orange potassiolutidine solution is stirred for 30 minutes, then cooled in an acetone–dry-ice bath (Note 3). As rapidly as possible, 11.8 g. (0.100 mole) of freshly distilled diethyl carbonate is added, the cooling bath is removed, and the solution changes to a green color. After 5 minutes the reaction mixture is neutralized by the addition of 10.7 g. (0.200 mole) of ammonium chloride. The green color is discharged; the final mixture is gray. The condenser is removed, and the ammonia is allowed to evaporate (Note 4). The residue is stirred with 500 ml. of ether, filtered, and extracted with an additional 100 ml. of ether. The combined ethereal extracts are concentrated with a rotary evaporator, and the residual oil is distilled using a modified Claisen flask. 2,6-Lutidine (22 g., 69%) is collected at $46-56°$ (10 mm.), and $10.7-13.4$ g. ($59-75\%$) of the ester is collected at 87° (0.7 mm.) as a bright yellow liquid, n_D^{25} 1.4995, d_4^{20} 1.0608 (Note 5).

2. Notes

1. A Teflon stirrer should not be used since Teflon is attacked by alkali metals, metal amides, and carbanions.

2. The preparation of potassium amide is described in *Org. Synth.*, **Coll. Vol. 5,** 187 (1973). *Caution! Potassium may form an explosive red or orange peroxide coating. Potassium is a silver-gray metal with a blue-violet cast, and any potassium showing an orange or red color, or with an appreciable oxide coating, should be considered extremely hazardous.*[3]

3. The use of an extra mole of 2,6-lutidine and rapid addition of diethyl carbonate both decrease formation of urethane. The use of a dry-ice bath to cool the reaction mixture permits rapid addition of diethyl carbonate without excessive foaming. If urethane is formed from diethyl carbonate and ammonia, the yield of product is decreased and the distillation is difficult.

4. A steam bath or hot air gun may be used with care to speed up evaporation of the ammonia.

5. The ester has been reported[4] to boil at 132° (18 mm.). It was reported as a colorless oil, giving a hydrochloride melting at $112-115°$. Ethyl pyridine-2-acetate was reported in *Org. Synth.*, **Coll. Vol. 3,** 413 (1955) as a light-yellow liquid, b.p. $135-137°$ (28 mm.). The ^{1}H NMR spectrum of the present product is in accord with the structure assigned, and the hydrochloride melts at 115°.

3. Discussion

Ethyl pyridine-2-acetate[5] and ethyl 6-methylpyridine-2-acetate have previously been prepared by carboxylation of the lithio derivatives of α-picoline and 2,6-lutidine, respectively. Use of diethyl carbonate to acylate the organometallic derivative avoids the intermediacy of the (unstable) carboxylic acid, and the yield is better. In the present procedure potassium amide is used as the metalating agent; the submitters report that the same esters may be formed by metalation with sodium amide (43% yield) or *n*-butyllithium (39% yield). The latter conditions also yield an appreciable amount of the acid (which decarboxylates).

1. Department of Chemistry, University of Akron, Akron, Ohio 44325.
2. See also L. F. Fieser and M. Fieser, "Reagents for Organic Synthesis," Vol. 1, Wiley, New York, 1967, p. 907.
3. See J. F. Short, *Chem. Ind. (London)*, 2132 (1964) and references therein; see also D. P. Mellor, *Chem. Ind. (London)*, 723 (1965); and M. S. Bil, *Chem. Ind. (London)*, 812 (1965).
4. V. Boekelheide and W. G. Gall, *J. Org. Chem.*, **19,** 499 (1954).
5. R. B. Woodward and E. C. Kornfeld, *Org. Synth.*, **Coll. Vol. 3,** 413 (1955).

ETHYL 1-NAPHTHYLACETATE

(1-Naphthaleneacetic acid, ethyl ester)

$$1\text{-}C_{10}H_7COCl + CH_2N_2 + (C_2H_5)_3N \rightarrow$$
$$1\text{-}C_{10}H_7COCHN_2 + (C_2H_5)_3\overset{+}{N}H\ Cl^-$$

$$1\text{-}C_{10}H_7COCHN_2 + C_2H_5OH \xrightarrow[(C_2H_5)_3N]{C_6H_5CO_2Ag} 1\text{-}C_{10}H_7CH_2CO_2C_2H_5 + N_2$$

Submitted by Ving Lee and Melvin S. Newman[1]
Checked by Gordon F. Hambly and Peter Yates

1. Procedure

Caution! Diazomethane is hazardous. Follow the directions for its safe handling given in Org. Synth., **Coll. Vol. 4,** *250 (1963) and* **Coll. Vol. 5,** *351 (1973). The intermediate, 1-(diazoacetyl)naphthalene, is a very strong skin irritant.*

A.   1-*(Diazoacetyl)naphthalene.*   A solution of 30.5 g. (0.160 mole) of 1-naphthoyl chloride (Notes 1 and 2) in 50 ml. of dry diethyl ether (Note 3) is added over 30 minutes to a magnetically stirred, ice-cooled solution of 6.72 g. (0.160 mole) of diazomethane [*Org. Synth.,* **Coll. Vol. 2,** 165 (1943); (Note 4)] and 16.1 g. (0.160 mole) of dry triethylamine (Note 5) in 900 ml. of dry ether. The mixture is stirred for 3 hours in the cold, and the triethylamine hydrochloride, removed by filtration, is washed twice with 30–50 ml. portions of dry ether (Note 6). The ether is removed from the combined filtrate and washings on a rotary evaporator. The yellow solid residue is dissolved in 75 ml. of dry ether, and the solution is cooled with an acetone–dry-ice mixture. The solid deposited is collected by filtration on a glass fritted-disk funnel, and the adhering ether is removed under reduced pressure as the temperature is allowed to reach room temperature, giving 26.6–28.8 g. (85–92%) of yellow 1-(diazoacetyl)naphthalene, m.p. 52–53° (Notes 7 and 8).

B.   *Ethyl 1-naphthylacetate.*   A solution of 15.7 g. (0.0801 mole) of 1-(diazoacetyl)naphthalene in 50 ml. of absolute ethanol is placed in a 100-ml., two-necked flask equipped with a Teflon-coated magnetic stirring bar, a serum stopper cap, and a reflux condenser connected to a gas-collecting device. The solution is heated to reflux, and 1 ml. of a freshly prepared catalyst solution made by dissolving 1 g. of silver benzoate (Note 9) in 10 ml. of triethylamine is added by injection through the serum cap. Evolution of nitrogen occurs and the mixture turns black. Addition of a second milliliter of catalyst solution is made when the evolution of nitrogen almost stops. This procedure is continued until further additions cause no further evolution of nitrogen (Note 10). The reaction mixture is refluxed for 1 hour, cooled, and filtered. The solvents are removed from the filtrate on a rotary evaporator. The residue is taken up in 75 ml. of ether, and the solution is washed twice in turn with aqueous 10% sodium carbonate, water, and saturated brine. Each aqueous extract is extracted with ether, and the combined ethereal extracts and solution are dried by filtration through anhydrous magnesium sulfate. After removal of the ether, distillation affords 14.4–15.8 g. (84–92%) of colorless ethyl 1-naphthylacetate, b.p. 100–105° (0.1–0.2 mm.) (Note 11).

2. Notes

1. The submitters prepared pure 1-naphthoyl chloride, b.p. 95–96° (0.2 mm.), from pure 1-naphthoic acid in 95% yield by treatment with thionyl chloride or phosphorus pentachloride. The 1-naphthoic acid used was prepared by carbonation of 1-naphthylmagnesium bromide. As commercial 1-bromonaphthalene is impure, fractionation through a 17 × 600 mm. column is needed to obtain pure 1-bromonaphthalene, b.p. 105–108° (0.1–0.2 mm.), as indicated by GC.

2. The checkers prepared 1-naphthoyl chloride by treatment of 1-naphthoic acid, m.p. 157–161°, obtained from Aldrich Chemical Co., with phosphorus pentachloride.

3. All dry ether used was freshly distilled from ethylmagnesium bromide.

4. Solutions of diazomethane in ether were titrated with benzoic acid.

5. Triethylamine was dried by storage over anhydrous barium oxide.

6. About 90% of the theoretical yield of triethylamine hydrochloride is obtained.

7. This compound is a severe skin irritant; hence, great care should be exercised to avoid any contact. For best yields this crystallization is recommended, since the yield of ethyl 1-naphthylacetate is reduced by about 20% if the crude product is used in the rearrangement step. A sample of crystallized 1-(diazoacetyl)naphthalene, m.p. 52–53°, that had been stored in a screw-top bottle in a refrigerator for about 2 weeks afforded the same yield of ethyl 1-naphthylacetate as a freshly prepared sample.

8. The checkers obtained 26.1–26.5 g. (83–84.5%) of 1-(diazoacetyl)naphthalene, m.p. 47–49.5°, when 1-naphthoyl chloride prepared from commercial 1-naphthoic acid was used (cf. Notes 1 and 2). Recrystallization from hexane gave 24.6 g. (78%) of 1-(diazoacetyl)naphthalene, m.p. 49.5–52°, that was used in Part B.

9. The silver benzoate was made by reaction of silver nitrate with sodium benzoate in water. The submitters dried the silver benzoate and recrystallized it from N-methylpyrrolidone or N,N-dimethylformamide. The checkers dried it in an oven at 130° for 1 hour immediately before use, but did not recrystallize it. Any precipitate present after dissolving the silver benzoate in triethylamine is removed by filtration or centrifugation.

10. Usually 3–4 additions are required. The total time of reaction should not be more than 45 minutes. The checkers added the catalyst solution in 0.5-ml. portions; nitrogen evolution was initially very vigorous, and only four such additions were required.

11. When propanol is used instead of ethanol, comparable results are obtained: propyl 1-naphthylacetate, b.p. 115–118° (0.1–0.2 mm.).

3. Discussion

Ethyl 1-naphthylacetate has been prepared by ethanolysis of 1-naphthylacetonitrile under acidic conditions[2] and by the Arndt-Eistert reaction of 1-(diazoacetyl)naphthalene with ethanol and silver oxide.[3]

The method described here represents a modified Arndt-Eistert reaction as developed by Newman and Beal,[4] gives results that are more reproducible than those of the original

Arndt-Eistert reaction and, in general, allows the rearrangement to be carried out successfully on larger-scale runs. The use of triethylamine in the formation of diazo ketones makes possible the use of only one equivalent of diazomethane.[5]

1. Evans Chemistry Laboratory, Ohio State University, Columbus, Ohio 43210.
2. M. Julia and M. Baillargé, *Bull. Soc. Chim. Fr.*, 928 (1957).
3. F. Arndt and B. Eistert, *Ber. Dtsch. Chem. Ges.*, **68**, 200 (1935).
4. M. S. Newman and P. F. Beal III, *J. Am. Chem. Soc.*, **72**, 5163 (1950).
5. M. S. Newman and P. F. Beal III, *J. Am. Chem. Soc.*, **71**, 1506 (1949).

ETHYL (*E*)-3-NITROACRYLATE

[2-Propenoic acid, 3-nitro-, ethyl ester, (*E*)-]

This procedure [*Org. Synth.*, **56**, 65 (1977)] has been replaced by an improved procedure for Methyl (*E*)-3-Nitroacrylate.

γ-KETOESTERS FROM ALDEHYDES *VIA* DIETHYL ACYLSUCCINATES: ETHYL 4-OXOHEXANOATE

[Hexanoic acid, 4-oxo-, ethyl ester]

Submitted by Pius A. Wehrli[1] and Vera Chu
Checked by William B. Farnham and William A. Sheppard

1. Procedure

A. *Diethyl propionylsuccinate* (**1**). A solution of 412 g. (2.40 mole) of diethyl maleate (Note 1), 278 g. (4.79 mole) of freshly distilled propionaldehyde (Note 2), and 1.2 g. (0.0048 mole) of benzoyl peroxide in a normal, 2-l. Pyrex flask is heated under reflux while undergoing irradiation with a UV lamp (Note 3). The initial reflux temperature is 60°. After 2 hours another 1.2 g. of benzoyl peroxide is added. Strong reflux and irradiation are maintained throughout the entire reaction period. After 18 hours of reflux, the internal pot temperature reaches 68°, at which point the last 1.2 g. of benzoyl peroxide is added. The reaction is continued for a total of 30 hours, at which time the pot temperature reaches 74.5°. The reflux condenser is then replaced with a distillation head, and the excess propionaldehyde (119 g.) is distilled under atmospheric pressure, b.p. 48–49°. Succinate 1 is distilled under reduced pressure. The main fraction, b.p. 145–151.5° (15–16 mm), provides 417–449 g. (75–81%) of product having sufficient purity for use in the next step (Note 4).

B. *Ethyl 4-oxohexanoate* (2). A 1-l., three-necked, round-bottomed flask equipped with a mechanical stirrer, thermometer, and Claisen condenser connected to a gas-measuring device (Note 5) is charged with 276 g. (1.20 mole) of succinate 1 and 74.1 g. (1.20 mole) of boric acid (Note 6). The initially heterogeneous mixture is stirred and immersed in a 150° oil bath. Within 1 hour 36 g. of distillate (mainly ethanol) and approximately 2.3 l. of gas collect. As the temperature is raised to 170°, the rate of carbon dioxide evolution increases, a total of 24.9 l. of gas being collected after 1.5 hours. At this time gas evolution has almost ceased, and the reaction mixture has a clear, light-yellow appearance. The contents of the flask are cooled to room temperature, poured onto 1.5 l. of ice, and extracted with three 500-ml. portions of toluene. The combined organic layers are dried over anhydrous magnesium sulfate, and the solvent is removed under reduced pressure. The product is distilled through a 10-cm. Vigreux column, yielding 156–162 g. (82–85%) of ethyl 4-oxohexanoate, b.p. 109–112° (18 mm.). GC analysis indicates the material to be 99.2% pure (Note 7).

2. Notes

1. Diethyl maleate, practical grade, available from Eastman Organic Chemicals, was used without further purification.
2. Propionaldehyde was obtained from Aldrich Chemical Company, Inc., and must be distilled before use.
3. The checkers used a 275-W. General Electric sunlamp. The submitters used a 140-W. Hanovia Ultraviolet Quartz lamp of a type no longer available.
4. The succinate 1 has the following ^{1}H NMR spectra (CDCl$_3$), δ (multiplicity, number of protons, assignment): 1.0–1.45 (m, 9H, 3CH_3), 2.6–3.5 (m, 4H, 2CH_2), 3.9–4.4 (m, 5H, 2OCH_2CH$_3$ and CH).
5. As a gas-measuring device, the submitters used an inverted, calibrated, 10-l. bottle, filled with saturated sodium chloride, resting in an enamel bucket big enough to hold the volume to be displaced. The checkers used a gas meter. However, the rate of gas evolution can be estimated by using a simple gas bubbler.
6. Reagent grade boric acid, available from Aldrich Chemical Company, Inc., was used.
7. GC analysis was performed on a Hewlett-Packard Model 5720 with dual flame detector; column 1.85 m. × 0.313 cm. outer diameter, stainless steel; 10% UCW-98 on Diatoport 5, programmed at 30° per minute from 50–250°. The purity was calculated by an area comparison. Ester 2 has the following ^{1}H NMR spectrum (CDCl$_3$), δ (multipli-

city, coupling constant J in Hz., number of protons, assignment): 1.05 (t, $J = 7$, 3H, CH_3), 1.2 (t, $J = 7$, 3H, CH_3), 2.4–2.9 (m, 6H, 3CH_2), 4.2 (q, $J = 7$, 2H, OCH_2CH$_3$).

3. Discussion

γ-Ketoesters, notably 5-substituted ethyl levulinates, have been prepared *via* radical addition of aldehydes to diethyl maleate to give acylated diethyl succinates.[2] These intermediates in turn had to be saponified,[2] decarboxylated,[2] and reesterified to give the corresponding 4-oxocarboxylates. A more direct method[3] utilizes the free radical addition of butyraldehyde to methyl acrylate, but the reported yield is low (11%).

The present method[4] is simple, versatile, and efficient in contrast to earlier methods, which were multistep or preparatively unsatisfactory. Various 5-substituted 4-oxocarboxylates can be prepared by this procedure.[4]

γ-Ketoesters, in general, and levulinic acid or esters, in particular, have extensive utility.[5] For example, they can serve as central intermediates for γ-butyrolactones,[6] 1,4-diols,[7] thiophenes,[8] pyrrolidones,[9] and 2-alkyl-1,3-cyclopentanediones.[10]

1. Chemical Research Department, Hoffmann–La Roche Inc., Nutley, New Jersey 07110.
2. T. M. Patrick, Jr., *J. Org. Chem.*, **17**, 1009 (1952).
3. E. C. Ladd, U.S. Pat. 2,577,133 (1951) [*Chem. Abstr.*, **46**, 6147h (1952)].
4. P. A. Wehrli and V. Chu, *J. Org. Chem.*, **38**, 3436 (1973).
5. R. H. Leonard, *Ind. Eng. Chem.*, **48**, 1330 (1956).
6. H. A. Schuette and P. Sah, *J. Am. Chem. Soc.*, **48**, 3163 (1926).
7. A. Müller and H. Wachs, *Monatsh. Chem.*, **53**, 420 (1929).
8. N. R. Chakrabarty and S. K. Mitra, *J. Chem. Soc.*, 1385 (1940).
9. R. L. Frank, W. R. Schmitz, and B. Zeidman, *Org. Synth.*, **Coll. Vol. 3**, 328 (1955).
10. U. Hengartner and V. Chu, *Org. Synth.*, **Coll. Vol. 6**, 774 (1988).

ETHYL PYRROLE-2-CARBOXYLATE

(1*H*-Pyrrole-2-carboxylic acid, ethyl ester)

Submitted by Denis M. Bailey, Robert E. Johnson,
and Noel F. Albertson[1]
Checked by A. Brossi and P. Wehrli

1. Procedure

A. *2-Pyrrolyl trichloromethyl ketone.* A 3-l., three-necked, round-bottomed flask equipped with a sealed mechanical stirrer, a dropping funnel, and an efficient reflux condenser with a calcium chloride drying tube is charged with 225 g. (1.23 moles) of trichloroacetyl chloride and 200 ml. of anhydrous diethyl ether. The solution is stirred while 77 g. (1.2 moles) of freshly distilled pyrrole in 640 ml. of anhydrous ether is added over 3 hours (Note 1); the heat of reaction causes the mixture to reflux. Following the addition, the mixture is stirred for 1 hour before 100 g. (0.724 mole) of potassium carbonate in 300 ml. of water is slowly added through the dropping funnel (Note 2). The layers are separated, and the organic phase is dried with magnesium sulfate, treated with 6 g. of Norit, and filtered. The solvent is removed by distillation on a steam bath, and the residue is dissolved in 225 ml. of hexane. The dark solution is cooled on ice to induce crystallation. The tan solid is collected and washed with 100 ml. of cold hexane, giving 189–196 g. (77–80%) of the ketone, m.p. 73–75° (Note 3).

B. *Ethyl pyrrole-2-carboxylate.* A 1-l., three-necked, round-bottomed flask equipped with a sealed mechanical stirrer and powder funnel is charged with 1.0 g. (0.44 g.-atom) of sodium and 300 ml. of anhydrous ethanol. When the sodium is dissolved, 75 g. (0.35 mole) of 2-pyrrolyl trichloromethyl ketone is added portionwise over a 10-minute period (Note 4). Once the addition is complete, the solution is stirred for 30 minutes, then concentrated to dryness using a rotary evaporator. The oily residue is partitioned between 200 ml. of ether and 25 ml. of 3 *N* hydrochloric acid. The ether layer is separated, and the aqueous layer is washed once with 100 ml. of ether. The combined ether solutions are washed once with 25 ml. of saturated sodium hydrogen carbonate solution, dried with magnesium sulfate, and concentrated by distillation. The residue is fractionated at reduced pressure, giving 44.0–44.5 g. (91–92%) of ethyl pyrrole-2-

carboxylate as a pale yellow oil, b.p. 125–128° (25 mm.) (Note 5). The yield based on pyrrole is 70–74%. Upon standing at room temperature the product crystallizes, m.p. 40–42°.

2. Notes

1. If the addition time is shortened to 1 hour, the yield is decreased by about 5%.

2. Excessive frothing will occur if the potassium carbonate solution is added too fast.

3. A similar run on a scale 3.3 times as large with a 3-hour addition time gave the ketone in 74% yield.

4. The solution becomes warm during the addition, and the final color of the solution is reddish-brown.

5. A similar run on a scale three times as large gave the ester in 96% yield.

3. Discussion

Pyrrole-2-carboxylic acid esters have been prepared from ethyl chloroformate and pyrrolylmagnesium bromide[2] or pyrrolyllithium,[3] by hydrolysis and decarboxylation of dimethyl pyrrole-1,2-dicarboxylate followed by re-esterification of the 2-acid[4] and by oxidation of pyrrole-2-carboxaldehyde followed by esterification with diazomethane.[4]

Methods of acylating pyrrole similar to the present, using oxalyl chloride,[5] trifluoroacetic anhydride,[6] carbamic acid chloride,[7] and trichloroacetyl chloride,[8] have been reported. In the last preparation, it was necessary to separate the product from highly colored by-products with alumina chromatography. 2-Pyrrolyl trichloromethyl ketone has also been prepared by the reaction of pyrrolylmagnesium halide with trichloroacetyl chloride.[9]

The present procedure provides a facile and versatile synthesis, on large scale, of a variety of pyrrole-2-carboxylic acid derivatives without necessitating the use of moisture-sensitive organometallic reagents. The use of alcohols other than ethanol in the alcoholysis reaction provides virtually any desired ester. Ammonia or aliphatic amines readily give amides in high yields, and aqueous base can be used to give the free acid.

1. Sterling-Winthrop Research Institute, Rensselaer, New York 12144.
2. F. K. Signaigo and H. Adkins, *J. Am. Chem. Soc.*, **58,** 1122 (1936).
3. A. Treibs and A. Dietl, *Justus Liebigs Ann. Chem.*, **619,** 80 (1958).
4. P. Hodge and R. W. Rickards, *J. Chem. Soc.*, 2543 (1963).
5. J. L. Archibald and M. E. Freed, *J. Heterocycl. Chem.*, **4,** 335 (1967).
6. W. D. Cooper, *J. Org. Chem.*, **23,** 1382 (1958).
7. A. Treibs and R. Derra, *Justus Liebigs Ann. Chem.*, **589,** 174 (1954).
8. A. Treibs and F.-H. Kreuzer, *Justus Liebigs Ann. Chem.*, **721,** 105 (1969).
9. G. Sanna, *Gazz. Chim. Ital.*, **63,** 479 (1933) [*Chem. Abstr.*, **28,** 763 (1934)].

THIAZOLES FROM ETHYL ISOCYANOACETATE AND THIONO ESTERS: ETHYL THIAZOLE-4-CARBOXYLATE

(4-Thiazolecarboxylic acid, ethyl ester)

$$\bar{Cl}\ \overset{+}{N}H_3CH_2C\overset{O}{\underset{OC_2H_5}{\big\langle}} + H-C\overset{O}{\underset{OCH_3}{\big\langle}} \xrightarrow[\text{reflux}]{(C_2H_5)_3N} \overset{O}{\underset{H}{\big\rangle}}C-NHCH_2C\overset{O}{\underset{OC_2H_5}{\big\langle}}$$

$$\overset{O}{\underset{H}{\big\rangle}}C-NHCH_2C\overset{O}{\underset{OC_2H_5}{\big\langle}} \xrightarrow[\substack{\text{dichloromethane,}\\0°}]{POCl_3,\ (C_2H_5)_3N} C\equiv N-CH_2-C\overset{O}{\underset{OC_2H_5}{\big\langle}}$$

$$HC\overset{OC_2H_5}{\underset{OC_2H_5}{\overset{|}{-}}}OC_2H_5 \xrightarrow[\substack{\text{hydroquinone,}\\\text{acetic acid}}]{H_2S,\ H_2SO_4} H-C\overset{S}{\underset{OC_2H_5}{\big\langle}}$$

$$C\equiv N-CH_2-C\overset{O}{\underset{OC_2H_5}{\big\langle}} + H-C\overset{S}{\underset{OC_2H_5}{\big\langle}} \xrightarrow[\substack{\text{ethanol,}\\45-50°}]{NaCN} \text{(thiazole ring)}\overset{O}{\underset{OC_2H_5}{\overset{\|}{C}}}$$

Submitted by G. D. Hartman and L. M. Weinstock[1]
Checked by Louis E. Benjamin, Sr., Norman W. Gilman, and Gabriel Saucy

1. Procedure

A. N-*Formylglycine ethyl ester*. A 1-l., three-necked, round-bottomed flask fitted with a mechanical stirrer, a pressure-equalizing dropping funnel, and a reflux condenser bearing a calcium chloride drying tube is charged with 69.5 g. (0.495 mole) of glycine ethyl ester hydrochloride and 250 ml. of methyl formate (Note 1). The suspension is stirred and heated at reflux while 55.0 g. (0.544 mole) of triethylamine (Note 1) is added. The resulting mixture is stirred and heated under reflux for 20 hours, cooled to room temperature, and filtered through a Büchner funnel, removing triethylamine hydrochloride. The filtrate is concentrated on a rotary evaporator, and the remaining clear oil is distilled under reduced pressure, yielding 51.7–61.5 g. (79–94%) of N-formylglycine ethyl ester, b.p. 94–97° (0.05 mm., Note 2).

B. *Ethyl isocyanoacetate*. A 3-l., three-necked, round-bottomed flask equipped with a thermometer, a mechanical stirrer, and a pressure-equalizing dropping funnel bearing a nitrogen inlet is charged with 65.5 g. (0.500 mole) of N-formylglycine ethyl ester, 125.0 g. (1.234 moles) of triethylamine, and 500 ml. of dichloromethane, and the apparatus is flushed with nitrogen. The resulting solution is stirred and cooled to 0° to −2° in an ice–salt bath, and 76.5 g. (0.498 mole) of phosphorus oxychloride (Note 3) is added dropwise over 15–20 minutes while the temperature is kept at 0°. The mixture becomes reddish brown as it is stirred and cooled at 0° for an additional 1 hour. The ice–salt bath is removed and replaced by an ice-water bath. Stirring is continued as a solution of 100 g. of anhydrous sodium carbonate in 400 ml. of water is added dropwise at a rate such that the temperature of the mixture is maintained at 25–30° (Note 4). The two-phase mixture is stirred for another 30 minutes, after which water is added until the volume of the aqueous layer is brought to 1 l. The aqueous layer is separated and extracted with two 250-ml. portions of dichloromethane. The dichloromethane solutions are combined, washed with saturated sodium chloride solution, and dried over anhydrous potassium carbonate. Evaporation of the solvent under reduced pressure and distillation of the remaining brown oil afford 43–44 g. (76–78%) of ethyl isocyanoacetate, b.p. 89–91° (11 mm.) (Note 5).

C. *O-Ethyl thioformate*. *Caution! Hydrogen sulfide gas is highly toxic; this procedure should be conducted in a well-ventilated hood.* A 1-l., three-necked, round-bottomed flask is equipped with a mechanical stirrer, a gas-inlet tube with a fritted-glass tip extending near to the bottom of the flask, and a gas-outlet connected to a scrubber flask containing 0.5–1.0 l. of aqueous 20% sodium hydroxide (Note 6). The flask is charged with 333 g. (2.25 moles) of triethyl orthoformate (Note 7), 330 ml. of glacial acetic acid, 3.2 g. of hydroquinone, and 0.4 ml. of concentrated sulfuric acid. The resulting solution is stirred and cooled in an ice bath as hydrogen sulfide gas is passed through the gas-inlet tube into the solution (Note 8). After the solution becomes saturated with hydrogen sulfide, the contents of the flask are poured into a 4-l. beaker containing a mechanically stirred mixture of 2.3 l. of ice and 340 ml. of diethyl ether. The mixture is poured into a separatory funnel, and the layers are separated. The organic layer is washed successively with two 100-ml. portions of aqueous saturated sodium hydrogen carbonate solution, two 100-ml. portions of water, three 80-ml. portions of aqueous saturated sodium hydrogen carbonate, and two 80-ml. portions of aqueous saturated sodium chloride. The solution is dried over anhydrous sodium sulfate and distilled at atmospheric pressure through a 45-cm. Vigreux column, affording 60–76.7 g. (30–38%) of O-ethyl thioformate as a yellow liquid, b.p. 87–89° (Notes 9 and 10).

D. *Ethyl thiazole-4-carboxylate*. A 250-ml., three-necked, round-bottomed flask fitted with a thermometer, a mechanical stirrer, and a pressure-equalizing dropping funnel bearing a calcium chloride drying tube is charged with 0.25 g. (0.0051 mole) of sodium cyanide and 10 ml. of absolute ethanol. The suspension is stirred vigorously at room temperature as a solution of 4.52 g. (0.0439 mole) of ethyl isocyanoacetate and 3.60 g. (0.0400 mole) of O-ethyl thioformate in 15 ml. of absolute ethanol is added slowly. Because the reaction is exothermic the temperature of the mixture should be kept below 45° by adjusting the addition rate and, if necessary, cooling the flask in an ice bath (Note

11). When the addition is completed, the contents of the flask are stirred and heated at 50° for another 30 minutes. The solvent is removed by rotary evaporation, and the resulting dark oil is extracted with three 60-ml. portions of hot hexane (Note 12). The combined hexane extracts are concentrated with a rotary evaporator until the product begins to separate, and the concentrate is cooled in an ice bath, yielding 5.1–5.5 g. (81–87%) of ethyl thiazole-4-carboxylate as off-white needles, m.p. 52–53° (Note 13).

2. Notes

1. Glycine ethyl ester hydrochloride, methyl formate, and triethylamine were purchased from Aldrich Chemical Company, Inc., and were used without purification.

2. The submitters obtained 63 g. (96%), b.p. 104–106° (0.1 mm).

3. The submitters recommend that phosphorus oxychloride either be taken from a previously unopened bottle or distilled before use.

4. Foaming generally occurs during the addition.

5. Essentially no forerun need be taken prior to collection of the product. The last few milliliters of distillate were slightly yellow, and 2–3 g. of intractable material remained in the distillation flask. A boiling point of 76–78° (4 mm.) has been reported[2] for ethyl isocyanoacetate. The submitters have found the distilled product to be stable for up to 6 months when stored under nitrogen at −20°.

6. The checkers used a gas-inlet tube without a fritted-glass tip and aqueous 30% sodium hydroxide as scrubber solution.

7. Triethyl orthoformate is available from Aldrich Chemical Company, Inc.

8. Hydrogen sulfide is admitted into the flask slowly at first, with bubble formation kept to a minimum. Initially the gas dissolves readily. As the solution becomes saturated, more bubbling action is apparent in the flask and the scrubber.

9. The submitters reported a yield of 56.5 g. (28%), b.p. 90–92°. The product is often contaminated with ca. 5% of ethyl formate; however, this impurity does not interfere with Part D. The submitters state that the product is stable when stored under nitrogen at −20°.

10. The checkers found that glassware in which the product was stored acquired a disagreeable odor that was difficult to remove with aqueous sodium hydroxide.

11. Controlling the temperature in this manner prevents discoloration of the final product and improves its yield.

12. The product was obtained as tan needles when boiling hexane was used by the checkers. Off-white needles were isolated when the temperature of the hexane was 50–55°.

13. The submitters reported a yield of 5.8 g. (92%) of product as white needles, m.p. 53–54°. However, the checkers obtained colorless needles only after recrystallizing the product from hexane. The melting points of the discolored needles obtained initially by the checkers and the recrystallized material were identical (lit.,[3,4] m.p. 57° and 52–54°). The spectral properties of the product are as follows: IR (CHCl$_3$) cm.$^{-1}$: 3130, 3030, 1724 (C=O), 1500, 1270; ^{1}H NMR (CDCl$_3$), δ (multiplicity, coupling constant J in Hz., number of protons, assignment): 1.39 (t, J = 7, 3H, OCH$_2$CH_3), 4.43 (q, J = 7, 2H, OCH_2CH$_3$), 8.33 (d, J = 2.5, 1H, H at C-5), 8.98 (d, J = 2.5, 1H, H at C-2).

3. Discussion

Ethyl thiazole-4-carboxylate has been prepared by hydrogenolysis of ethyl 2-bromothiazole-4-carboxylate with Raney nickel in ethanol,[3] by desulfurization of ethyl 2-mercaptothiazole-4-carboxylate with hydrogen peroxide in concentrated hydrochloric acid,[4] and by condensation of ethyl bromopyruvate with thioformamide in ether.[5,6]

The present procedure is illustrative of a mild and general method for preparing thiazoles substituted at the 4-position with electron-withdrawing substitutents such as ethoxycarbonyl,[7] cyano,[7] and p-toluenesulfonyl.[8] Thus, condensation of ethyl isocyanoacetate with various thiono esters affords the parent ethyl thiazole-4-carboxylate as well as a series of analogs bearing substituents in the 5-position (Table I).[7] A similar reaction of α-isocyanoacetonitrile with O-ethyl thioformate gave the cyano analog in 23% yield. However, the instability of the latter isocyanide hampers the utility of this reaction.

The mechanism of the condensation in Part D probably involves thioformylation of the

TABLE I

THIAZOLES FROM CONDENSATION OF ETHYL ISOCYANOACETATE AND
α-ISOCYANOACETAMIDE WITH THIONO ESTERS

Thiazole	M.p. or B.p. (°)	Yield (%)
CO$_2$C$_2$H$_5$ / CH$_3$ thiazole	89–90	82
CO$_2$C$_2$H$_5$ / CH$_2$CH$_2$CH$_3$ thiazole	85–87 (0.15 mm.)	68
CO$_2$C$_2$H$_5$ / C$_6$H$_5$ thiazole	183–185	22
CO$_2$C$_2$H$_5$ / CH$_2$C$_6$H$_5$ thiazole	49–50	75
CN thiazole	55–56	23

metallated isocyanoacetate followed by intramolecular 1,1-addition of the tautomeric enethiol to the isonitrile. This thiazole synthesis is analogous to the formation of oxazoles from acylation of metallated isonitriles with acid chlorides or anhydrides.[9,10] Interestingly, ethyl formate does not react with isocyanoacetate under the conditions of this procedure. Ethyl and methyl isocyanoacetate have been prepared in a similar manner by dehydration of the corresponding N-formylglycine esters with phosgene[2] and trichloromethyl chloroformate,[11] respectively. The phosphorus oxychloride method described here was provided to the submitters by Professor U. Schöllkopf[12] and is based on the procedure of Böhme and Fuchs.[13] The preparation of O-ethyl thioformate was developed from a report by Ohno, Koizuma, and Tsuchihaski.[14]

4-Substituted thiazoles are important intermediates, as in the synthesis of thiabendazole [2-(4-thiazolyl)benzimidazole], a leading anthelminthic utilized for control of gastrointestinal nematodes in ruminants.[15] Other thiazoles have displayed significant pharmacologic activity as antiinflammatory[16] and antibacterial agents.[17] Recently, 4-substituted thiazoles were implicated as intermediates in the energy transfer mechanism of firefly bioluminescence.[18]

1. Merck, Sharp, and Dohme Research Laboratories, Division of Merck & Co., Inc., West Point, Pennsylvania 19486.
2. I. Ugi, U. Fetzer, U. Eholzer, H. Knupfer, and K. Offermann, *Angew. Chem. Int. Ed. Engl.,* **4,** 472 (1965).
3. H. Erlenmeyer and C. J. Morel, *Helv. Chim. Acta,* **28,** 362 (1948).
4. J. J. D'Amico and T. W. Bartram, *J. Org. Chem.,* **25,** 1336 (1960).
5. M. Erne, F. Ramirez, and A. Burger, *Helv. Chim. Acta,* **34,** 143 (1951).
6. For a review, see J. M. Sprague and A. H. Land, "Thiazoles and Benzothiazoles," in R. C. Elderfield, Ed., "Heterocyclic Compounds," Vol. 5, Wiley New York, 1957, pp. 484–722.
7. G. D. Hartman and L. M. Weinstock, *Synthesis,* 681 (1976); G. D. Hartman, U.S. Pat. 4,021,438 (1977) [*Chem. Abstr.,* **87,** 53264s (1977)].
8. O. H. Oldenziel and A. M. Van Leusen, *Tetrahedron Lett.,* 2777 (1972).
9. U. Schöllkopf and R. Schröder, *Angew Chem. Int. Ed. Engl.,* **10,** 333 (1971); R. Schröder, U. Schöllkopf, E. Blume, and I. Hoppe, *Justus Liebigs Ann. Chem.,* 533 (1975); M. Suzuki, T. Iwasaki, M. Miyoshi, K. Okumura, and K. Matsumoto, *J. Org. Chem.,* **38,** 3571 (1973); A. M. van Leusen, B. E. Hoogenboom, and H. Siderius, *Tetrahedron Lett.,* 2369 (1972).
10. For a review on α-metallated isocyanides, see U. Schöllkopf, *Angew. Chem. Int. Ed. Engl.,* **16,** 339 (1977).
11. G. Skorna and I. Ugi, *Angew. Chem. Int. Ed. Engl.,* **16,** 259 (1977).
12. U. Schöllkopf, private communication.
13. H. Böhme and G. Fuchs, *Chem. Ber.,* **103,** 2775 (1970).
14. A. Ohno, T. Koizumi, and G. Tsuchihashi, *Tetrahedron Lett.,* 2083 (1968).
15. H. D. Brown, A. R. Matzuk, I. R. Ilves, L. H. Peterson, S. A. Harris, L. H. Sarett, J. R. Egerton, J. J. Yakstis, W. C. Campbell, and A. C. Cuckler, *J. Am. Chem. Soc.,* **83,** 1764 (1961).
16. D. A. Evans, S. African Pat. 69 04,621 (1970) [*Chem. Abstr.,* **74,** 100028 m (1971)].
17. E. G. Curphey, *Ind. Chem.,* **34,** 85 (1958) [*Chem. Abstr.,* **52,** 9078b (1958)].
18. K. Okada, H. Iio, I. Kubota, and T. Goto, *Tetrahedron Lett.,* 2771 (1974).

CARBOXYLATION OF AROMATIC COMPOUNDS: FERROCENECARBOXYLIC ACID

(Ferrocene, carboxy-)

Submitted by PERRY C. REEVES[1]
Checked by J. J. MROWCA, M. M. BORECKI,
and WILLIAM A. SHEPPARD

1. Procedure

A. *(2-Chlorobenzoyl)ferrocene.* A thoroughly dried, 1-l., three-necked, round-bottomed flask is equipped with a mechanical stirrer, a funnel for the addition of air-sensitive solids (Note 1), and a two-necked adapter holding a thermometer and a gas-inlet tube. Throughout the ensuing reaction the system is maintained under positive pressure of dry nitrogen. The flask is charged with 18.6 g. (0.100 mole) of ferrocene (Note 2), 17.5 g. (0.100 mole) of 2-chlorobenzoyl chloride (Note 3), and 200 ml. of dichloromethane, and the addition funnel contains 14.0 g. (0.105 mole) of anhydrous aluminum chloride. Stirring is begun, and the flask is immersed in an ice bath. When the solution has been chilled thoroughly, the aluminum chloride is added in small portions at such a rate that the reaction mixture remains below 5°. The appearance of a deep blue color indicates that the reaction is occurring. This addition requires about 20 minutes, and after its completion stirring is continued for 30 minutes with ice cooling and for 2 hours at room temperature.

The reaction mixture is cooled again in ice, 200 ml. of water is added cautiously, and the resulting two-phase mixture is stirred vigorously for 30 minutes. After transferring the mixture to a separatory funnel, the layers are separated, and the aqueous layer is extracted with two 50-ml. portions of dichloromethane. The combined dichloromethane solutions are washed once with 50 ml. of water, twice with 50-ml. portions of 10% aqueous sodium hydroxide, and dried over magnesium sulfate, filtered and evaporated to dryness at reduced pressure, yielding 30.4–30.9 g. (94–96%) of (2-chlorobenzoyl)ferrocene as a viscous, red liquid, which gradually solidifies (Note 4).

B. *Ferrocenecarboxylic acid.* A dry, 500-ml., three-necked, round-bottomed flask equipped with a mechanical stirrer and a reflux condenser topped with a nitrogen-inlet tube is charged with 250 ml. of 1,2-dimethoxyethane (Note 5) and 46.0 g. (0.411

mole) of potassium *tert*-butoxide (Note 6). A nitrogen atmosphere is established in the system, and 2.2 ml. (0.12 mole) of water (Note 6) is added with stirring, producing a slurry, to which the crude (2-chlorobenzoyl)ferrocene is added. The red solution is stirred and refluxed under nitrogen. As the reaction proceeds the color fades to tan, and after 1 hour at reflux the reaction mixture is cooled and poured into 1 l. of water. The resulting solution is washed with three 150-ml. portions of diethyl ether, which are combined and back-extracted with two 50-ml. portions of 10% aqueous sodium hydroxide. The aqueous phases are then combined and acidified with concentrated hydrochloric acid. The precipitate is collected by filtration and air dried, yielding 17.0–19.2 g. (74–83% from ferrocene) of ferrocenecarboxylic acid as an air-stable yellow powder, m.p. 214–216° (dec., Note 7).

2. Notes

1. If such a funnel is not available, an Erlenmeyer flask connected to the reaction flask by a length of thin-walled rubber tubing (Gooch tubing) may be substituted. In this case, the reaction mixture must not be stirred so vigorously that liquid is splashed up into the neck of the flask, which would cause aluminum chloride to cake there and prevent it from falling into the flask.

2. Ferrocene was purchased from Strem Chemicals Incorporated, Danvers, Massachusetts.

3. 2-Chlorobenzoyl chloride was purchased from Aldrich Chemical Company, Inc., and used as received with a stated purity of 95%.

4. The crude material contains approximately 5% of unreacted ferrocene. Recrystallization from heptane affords pure (2-chlorobenzoyl)ferrocene as scarlet needles, m.p. 99–100°; however, the crude material may be used without purification for Part B.

5. 1,2-Dimethoxyethane was distilled from calcium hydride immediately prior to use.

6. Potassium *tert*-butoxide was purchased from Columbia Organic Chemicals Company, Columbia, South Carolina. The molar ratio of potassium *tert*-butoxide to water is critical, and the amounts specified represent optimum quantities for cleavage of 0.10 mole of ketone.[2,3]

7. The decomposition point was obtained in a sealed capillary tube and not corrected. As the solid is heated, it first changes from yellow to brownish red and then decomposes to a dark red liquid. The decomposition temperature of this compound has been reported to be 208.5°,[4] 224–225°,[5] and 225–230°.[6]

The crude product is suitable for most purposes. It may be recrystallized from toluene (1 g. in 15–20 ml. of solvent, 80% recovery in the first crop), giving material melting at 220–222° (dec.).

3. Discussion

The carboxylic acids of organometallic systems, important synthetic intermediates, have been prepared by many different synthetic methods. Ferrocenecarboxylic acid has been studied the most extensively,[7] and the best laboratory syntheses previously reported involve hydrolyses of cyanoferrocene[8] or *S*-methylferrocenethiocarbonate.[9]

The present synthesis[10] consists of two simple steps, uses readily available and inexpensive starting materials, and produces pure material in high overall yield. It is based on two observations: that nonenolizable ketones may be cleaved to carboxylic acids by potassium *tert*-butoxide – water,[2] and that aryl 2-chlorophenyl ketones may be cleaved with loss of the 2-chlorophenyl group, giving only one of the two possible acids.[11] Other compounds prepared by this route include carboxycyclopentadienyltricarbonylmanganese (79%)[10] and several substituted benzoic acids:[11] biphenyl-4-carboxylic acid (64%), 3,4-dimethylbenzoic acid (57%), 2,4,6-trimethylbenzoic acid (59%), 3,4-dimethoxybenzoic acid (73%), and 2,4-dimethoxybenzoic acid (60%). In cases where the cleavage reaction proceeds in low yield, substitution of 2,6-dichlorobenzoyl chloride for 2-chlorobenzoyl chloride may be helpful. With thiophene, for example, the yield of carboxylic acid was increased from 10% to 72% by this modification.[11]

1. Department of Chemistry, Southern Methodist University, Dallas, Texas 75275. [Present address: Abilene Christian University, Abilene, Texas 79601.] This work was supported by the Robert A. Welch Foundation.
2. G. A. Swan, *J. Chem. Soc.*, 1408 (1948).
3. P. G. Gassman, J. T. Lumb, and F. V. Zalar, *J. Am. Chem. Soc.*, **89**, 946 (1967).
4. K. L. Rinehart, K. L. Motz, and S. Moon, *J. Am. Chem. Soc.*, **79**, 2749 (1957).
5. R. L. Schaaf, *J. Org. Chem.*, **27**, 107 (1962).
6. V. Weinmayr, *J. Am. Chem. Soc.*, **77**, 3009 (1955).
7. D. E. Bublitz and K. L. Rinehart, *Org. React.*, **17**, 1 (1969).
8. A. N. Nesmeyanov, E. G. Perevalova, and L. P. Jaryeva, *Chem. Ber.*, **93**, 2729 (1960).
9. D. E. Bublitz and G. H. Harris, *J. Organomet. Chem.*, **4**, 404 (1965).
10. E. R. Biehl and P. C. Reeves, *Synthesis*, 360 (1973).
11. M. Derenberg and P. Hodge, *Tetrahedron Lett.*, 3825 (1971).

FLUORINATIONS WITH PYRIDINIUM POLYHYDROGEN FLUORIDE REAGENT: 1-FLUOROADAMANTANE

(Tricyclo[3.3.1.1³,⁷]decane, 1-fluoro-)

A. ⬡ +(HF)$_x$ $\longrightarrow$ C$_5$H$_5$N·HF(HF)$_x$
                                              1

B. (OH adamantane structure) +**1** $\longrightarrow$ (F adamantane structure)

                                              2

Submitted by GEORGE A. OLAH[1] and MICHAEL WATKINS
Checked by PAUL G. WILLIARD and G. BÜCHI

1. Procedure

Caution! Proper precautions must be used when handling anhydrous hydrogen fluoride and pyridinium polyhydrogen fluoride. Hydrogen fluoride is extremely corrosive to human tissue, contact resulting in painful, slow-healing burns. Laboratory work with HF should be conducted only in an efficient hood, with the operator wearing a full-face shield and protective clothing (Note 1).

A. *Pyridinium polyhydrogen fluoride* (**1**). A tared, 250-ml., polyolefin bottle is equipped with a polyolefin gas-inlet and drying tube inserted through holes in the cap and sealed with Teflon tape. The bottle is charged with 37.5 g. (0.475 mole) of pyridine (Note 2) and cooled in an acetone–dry ice bath. After the pyridine has solidified, 87.5 g. (4.37 moles) of anhydrous hydrogen fluoride (Note 3) is condensed from a cylinder into the bottle through the inlet tube. The amount of hydrogen fluoride is determined by weighing the bottle. After the hydrogen fluoride has cooled, the bottle is cautiously swirled with cooling until the solid dissolves (Note 4). The solution can now be safely allowed to warm to room temperature.

B. *1-Fluoroadamantane* (**2**). A 250-ml., polyolefin bottle is equipped with a Teflon-coated magnetic stirring bar and a polyolefin drying tube inserted through a hole in the cap and sealed with Teflon tape. The bottle is charged with 5.0 g. (0.033 mole) of 1-adamantanol (Note 5) and 50 ml. of pyridinium polyhydrogen fluoride (**1**). The solution is allowed to stir for 3 hours at ambient temperature, after which 150 ml. of petroleum ether is added, and stirring is continued for another 15 minutes. The resulting two-phase solution is transferred to a 250-ml., polyolefin separatory funnel, and the bottom layer is discarded (Note 6). The organic layer is washed successively with 50 ml. of water, 50 ml. of a saturated sodium hydrogen carbonate solution and 50 ml. of water, then dried over magnesium sulfate. After the organic layer is filtered, the solvent is removed under reduced pressure (Note 7), yielding 4.5–4.6 g. (88–90%) of adamantane 2 as a white powder, m.p. 225–227° (sublimes in a sealed capillary) (Note 8), which can be purified by vacuum sublimation or by recrystallization from methanol-carbon tetrachloride.

2. Notes

1. The recommended first-aid for a hydrogen fluoride burn is to flood with water, pack with ice, and get medical attention as quickly as possible. Local medical personnel should be alerted and prepared when work with hydrogen fluoride is planned. Directions for proper medical treatment[2] are given in *Org. Synth.*, **Coll. Vol. 5,** 66 (1973).

2. The submitters used A.C.S. certified reagent grade pyridine from Fisher Scientific Company which was distilled from potassium hydroxide prior to use.

3. The submitters obtained anhydrous hydrogen fluoride from Harshaw Chemical Company, and the checkers purchased this reagent from Matheson Gas Products.

4. The dissolution is an extremely exothermic process that can be violent if the bath temperature is not carefully controlled at −78°. A preferred procedure developed by Dr. A. E. Feiring, Central Research and Development Department, Du Pont Experimental Station, involves keeping the pyridine as cold as possible without freezing (*ca.* −40°), then slowly condensing the hydrogen fluoride into the vessel so that the entire mixture remains liquid during the preparation. Stirring is also helpful.

5. This was obtained from Aldrich Chemical Company, Inc., and used without further purification.

6. The inorganic layer can be safely disposed by slow addition to large amounts of ice-cold water.

7. Since 1-fluoroadamantane sublimes easily, the water bath should be controlled at about 32°, and the vacuum evaporation of the solvent limited to as short a time as possible.

8. GC and IR analysis indicate no detectable amount of starting alcohol. ^{1}H NMR of adamantane 2 (CDCl$_3$) yields a series of multiplets centered at δ 1.62, 1.86, and 2.18.

3. Discussion

1-Fluoroadamantane can also be prepared by halogen exchange of the bromide with silver fluoride[3] or zinc fluoride.[4] The procedure described is more convenient and economical than the halogen exchanges, and has found successful application in the preparation of a wide variety of secondary and tertiary fluorides from the corresponding alcohols, with yields generally within the range of 70–90%[5] (Table I).

The hydrogen fluoride–pyridine reagent is an effective complement to the dimethylaminosulfur trifluoride (DAST) reagent[6] in the preparation of alkyl fluorides from alcohols. DAST is also useful for the conversion of carbonyl groups to difluoromethylene functions. The hydrogen fluoride–pyridine reagent, however, can also be used for the

TABLE I
PREPARATION OF TERTIARY- AND SECONDARY-ALKYL FLUORIDES FROM
ALCOHOLS WITH HYDROGEN FLUORIDE–PYRIDINE REAGENT

Alcohol	Temp- erature	Reaction Time (hours)	Alkyl Fluoride	b.p.(m.p.)	Yield (%)
Isopropyl	50	3.0	Isopropyl	−9 to −7	30
sec-Butyl	20	3.0	sec-Butyl	25−26	70
tert-Butyl	0	1.0	tert-Butyl	12	50
3-Ethyl-3-pentyl	0	0.5	3-Ethyl-3-fluoro- pentane	30−33 (60 mm.)	95
3-Methyl-3- heptyl	−70	0.5	3-Fluoro-3- methylheptane	35 (40 mm.)	85
3-Methyl-4- heptyl	0	2.0	4-Fluoro-3- methylheptane		35
Cyclohexyl	20	2.0	Cyclohexyl fluoride	100−102	99
2-Norbornyl	20	1.0	2-Fluoronor- bornane	(56−59)	95
2-Adamantyl	20	0.5	2-Fluoroada- mantane	(254−255)	98
α-Phenylethyl	20	0.5	1-Fluoro-1- phenylethane	46 (15 mm.)	65

TABLE II
HYDROFLUORINATION OF ALKENES WITH HYDROGEN FLUORIDE-PYRIDINE
REAGENT

Alkene	Reaction Temp- erature	Product	b.p.(m.p.)	Yield (%)
Propene	20	Isopropyl fluoride	−11 to −9	35
Cyclopropane	20	n-Propyl fluoride	−3 to −1	75
2-Butene	0	sec-Butyl fluoride	24−25	40
2-Methylpropene	0	tert-Butyl fluoride	11−13	60
Cyclopentene	0	Cyclopentyl fluoride	51−52 (200 mm.)	65
Cyclohexene	0	Cyclohexyl fluoride	102−104	80
Cycloheptene	0	Cycloheptyl fluoride	70−71 (200 mm.)	90
Norbornene	0	2-Fluoronorbornane	(56−59)	65
1-Hexyne	0	2,2-Difluorohexane	85−87	70
3-Hexyne	0	3,3-Difluorohexane	84−86	75

hydrofluorination of alkenes,[7] alkynes,[7] cyclopropanes,[7] and diazo compounds,[8] the halofluorination of alkenes,[9] the preparation of fluoroformates from carbamates,[10] the preparation of α-fluorocarboxylic acids from α-amino acids,[11] and as a deprotecting reagent in peptide chemistry.[12] Examples of the hydrofluorination of alkenes with hydrogen fluoride–pyridine are given in Table II.

1. Hydrocarbon Research Institute, Department of Chemistry, University of Southern California, Los Angeles, California 90007.
2. C. M. Sharts and W. A. Sheppard, *Org. React.*, **21**, 192, 220–223 (1974).
3. R. C. Fort and P. v. R. Schleyer, *J. Org. Chem.*, **30**, 789 (1965).
4. K. S. Bhandari and R. E. Pincock, *Synthesis*, 655 (1977).
5. G. A. Olah, M. Nojima, and I. Kerekes, *Synthesis*, 786 (1973).
6. W. J. Middleton and E. M. Bingham, *Org. Synth.*, **Coll. Vol. 6**, 440, 835 (1988).
7. G. A. Olah, M. Nojima, and I. Kerekes, *Synthesis*, 779 (1973).
8. G. A. Olah and J. Welch, *Synthesis*, 896 (1974).
9. G. A. Olah, M. Nojima, and I. Kerekes, *Synthesis*, 780 (1973).
10. G. A. Olah and J. Welch, *Synthesis*, 654 (1974).
11. G. A. Olah and J. Welch, *Synthesis*, 652 (1974).
12. S. Matsuura, C. H. Niu, and J. S. Cohen, *J. Chem. Soc. Chem. Commun.*, 451 (1976).

ALDEHYDES FROM AROMATIC NITRILES:
4-FORMYLBENZENESULFONAMIDE

(Benzenesulfonamide, 4-formyl-)

$$\text{Ni}-\text{Al}/\text{HCO}_2\text{H}$$

Submitted by T. van Es and B. Staskun[1]
Checked by A. Brossi, L. A. Dolan, and A. Laurenzano

1. Procedure

A 2-l., two-necked, round-bottomed flask fitted with a mechanical stirrer and a reflux condenser is charged with 40.0 g. (0.232 mole) of 4-cyanobenzenesulfonamide (Note 1), 600 ml. of 75% (v/v) formic acid, and 40 g. of Raney nickel alloy (Note 2). The stirred mixture is heated under reflux for 1 hour (Note 3). The mixture is filtered with suction through a Büchner funnel coated with a filter aid (Note 4), and the residue is washed with two 160-ml. portions of 95% ethanol. The combined filtrates are evaporated with a rotary evaporator (Note 5). The solid residue (Note 6) is dissolved in 400 ml. of boiling water and freed from a small amount of insoluble material by decantation through a plug of glass wool placed in a filter funnel. The filtrate is chilled in an ice bath, and the precipitate is collected by filtration with suction, washed with a small amount of cold water, and dried at 50° under vacuum, yielding about 32 g. of crude product, m.p. 112–114°.

The product is dissolved in 800 ml. of hot 95% ethanol, 15.5 g. of activated carbon (Note 7) is added, and the mixture is swirled periodically while it is allowed to cool for 1 hour. The activated carbon is removed by filtration with suction through a bed of filter aid (Note 4), the filter cake is washed with 50 ml. of 95% ethanol, and the combined filtrates are evaporated with a rotary evaporator. The residue is dissolved in 225 ml. of boiling water, and the hot solution is decanted through glass wool placed in a filter funnel. The filtrate is cooled to 0°, the product is collected by filtration with suction, washed with a small amount of cold water, and dried in a vacuum oven at 50°, yielding 25.6–28.0 g. (62.9–68.8%) of 4-formylbenzenesulfonamide, m.p. 117–118° (Note 8).

2. Notes

1. The checkers used 4-cyanobenzenesulfonamide purchased from Aldrich Chemical Company, Inc., m.p. 167–169°. The submitters prepared this material by diazotization of the corresponding amine,[2] followed by cyanation.[3] The product was crystallized from water, m.p. 152–154°.

2. The checkers used material purchased from Harshaw Chemical Company. The submitters used nickel–aluminum alloy (50% Ni, 50% Al) supplied by British Drug Houses Ltd.

3. The reaction proceeds with some frothing; this is more appreciable and vigorous in mixtures containing a higher proportion of water, as when the reduction is conducted in 50% formic acid.[4]

4. Hyflo Supercel, a filter aid purchased from Johns-Manville Corporation, was used by the checkers.

5. This procedure is to be avoided with steam-volatile aldehydes (e.g., 4-chlorobenzaldehyde), in which case the reduction product is isolated by solvent extraction.[4]

6. The product is contaminated with nickel salts; its IR spectrum showed little, if any, unchanged nitrile.

7. Norite A, purchased from Matheson, Coleman and Bell, was used.

8. The melting point of 4-formylbenzenesulfonamide has been reported as 118–120°,[4,5] 122°,[6] and 123–124°.[7]

3. Discussion

4-Formylbenzenesulfonamide has been prepared by chromic acid oxidation of p-toluenesulfonamide,[5] the Sommelet reaction on 4-chloromethylbenzenesulfonamide,[8] and by the Stephen reduction of 4-cyanobenzenesulfonamide.[5] The present method provides a general procedure for the synthesis of substituted aromatic aldehydes as illustrated in Table I.

Some studies seeking preferred conditions for this reaction have been made. Optimum yields are obtained when the amount of water present is appreciable, and it was noted that the rate of hydrogen evolution increases with increasing water content. A 75% formic acid

TABLE I
ALDEHYDES FROM AROMATIC NITRILES[4]

Nitrile	Aldehyde	Yield, %[a]
C_6H_5CN	C_6H_5CHO	97
$4\text{-}ClC_6H_4CN$	$4\text{-}ClC_6H_4CHO$	100
$4\text{-}CH_3OC_6H_4CN$	$4\text{-}CH_3OC_6H_4CHO$	93
$2\text{-}C_{10}H_7CN$	$2\text{-}C_{10}H_7CHO$	95

[a]Determined as the 2,4-dinitrophenylhydrazone derivative.

system appears to be generally preferred. Under the reaction conditions examined by the submitters, olefins, ketones, esters, amides, and acids are inert, but nitro compounds are reduced to the formamide derivative.

A related method for the synthesis of aldehydes from nitriles has also been studied.[9] This method, which has been found to be extremely effective for the reduction of hindered nitriles to aldehydes, uses moist, preformed Raney nickel catalyst in formic acid. Compounds synthesized by this method are illustrated in Table II.

TABLE II
ALDEHYDES FROM NITRILES WITH RANEY NICKEL CATALYST IN FORMIC ACID[9]

Nitrile	Aldehyde	Yield, %[a]
C_6H_5CN	C_6H_5CHO	72
$2\text{-}CH_3C_6H_4CN$	$2\text{-}CH_3C_6H_4CHO$	65–75
$2\text{-}ClC_6H_4CN$	$2\text{-}ClC_6H_4CHO$	70–83
$2\text{-}CH_3OC_6H_4CN$	$2\text{-}CH_3OC_6H_4CHO$	80
$2,6\text{-}(CH_3O)_2C_6H_3CN$	$2,6\text{-}(CH_3O)_2C_6H_3CHO$	60–65
$C_6H_5CH{=}CHCN$	$C_6H_5CH{=}CHCHO$	64

[a]Determined as the 2,4-dinitrophenylhydrazone derivative.

1. Department of Chemistry, University of the Witwatersrand, Johannesburg, South Africa.
2. C. H. Andrewes, H. King, and J. Walker, *Proc. R. Soc. London, Ser. B*, **133**, 20 (1946).
3. R. C. Iris, R. D. Leyva, and C. Ramirez, *Rev. Inst. Salubr. Enferm. Trop. Mexico City*, **7**, 95 (1946) [*Chem. Abstr.* **41**, 4117g (1947)].
4. T. van Es and B. Staskun, *J. Chem. Soc.*, 5775 (1965).
5. H. Burtod P. F. Hu, *J. Chem. Soc.*, 601 (1948).
6. S. Koizuka and K. Ichiriki, Japan. Pat. 180,234 (1949) [*Chem. Abstr.*, **46**, 5085e (1952)].
7. T. P. Sycheva and M. N. Shchukina, *Sb. Statei Obshch. Khim.*, **1**, 527 (1953) [*Chem. Abstr.*, **49**, 932c (1955)].
8. S. J. Angyal, P. J. Morris, J. R. Tetaz, and J. G. Wilson, *J. Chem. Soc.* 2141 (1950).
9. B. Staskun and O. G. Backeberg, *J. Chem. Soc.*, 5880 (1964).

GERANYL CHLORIDE

[2,6-Octadiene, 1-chloro-3,7-dimethyl-, *(E)*-]

Submitted by Jose G. Calzada and John Hooz[1]
Checked by K.-K. Chan, A. Specian, and A. Brossi

1. Procedure

A dry, 300-ml., three-necked flask is equipped with a magnetic stirring bar and reflux condenser (to which is attached a Drierite-filled drying tube) and charged with 90 ml. of carbon tetrachloride (Note 1) and 15.42 g. of geraniol (0.1001 mole) (Note 2). To this solution is added 34.09 g. (0.1301 mole) of triphenylphosphine (Note 3), and the stirred reaction mixture is heated under reflux for 1 hour. This mixture is allowed to cool to room temperature; dry pentane is added (100 ml.), and stirring is continued for an additional 5 minutes.

The triphenylphosphine oxide precipitate is filtered and washed with 50 ml. of pentane. The solvent is removed from the combined filtrate with a rotary evaporator under water aspirator pressure at room temperature. Distillation of the residue through a 2-cm. Vigreux column attached to a short-path distillation apparatus (Note 4) provides 13.0–14.0 g. (75–81%) of geranyl chloride, b.p. 47–49° (0.4 mm.), $n_D^{23} = 1.4794$ (Note 5).

2. Notes

1. Carbon tetrachloride was dried over magnesium sulfate and distilled from phosphorus pentoxide through a 25-cm. Vigreux column. Lower yields were obtained when either the glassware or reagents were not dried.

2. Geraniol was purchased from Koch-Light Laboratories (>98% pure), dried over potassium carbonate and distilled through an 8-cm. Vigreux column, b.p. 108–109° (8 mm.). The checkers used geraniol purchased from Aldrich Chemical Co., Inc., and distilled it prior to use.

3. Triphenylphosphine, m.p. 80–81°, was obtained from Eastman Organic Chemicals and kept in a drying pistol held at approximately 65° (1 mm.) for 18 hours prior to use. When only 10–20% excess triphenylphosphine was employed the yield of geranyl chloride was approximately 75%, but the small amount of unreacted geraniol which remained rendered product isolation more difficult.

4. A "Bantam-ware" distilling head and condenser assembly from Kontes Glass Co. (K-287100) was used. Foaming may occur due to incomplete removal of solvent. This can be avoided by cooling the distillation flask to approximately −50° and gradually lowering

the pressure to 10 mm. The pot temperature is then allowed to increase gradually to room temperature, and the distillation then proceeds without difficulty.

5. The pure geranyl chloride has characteristic IR absorption (liquid film) at 845, 1255, and 1665 cm.$^{-1}$. The use of these absorptions to assay mixtures of geranyl and linalyl chloride has been discussed in detail.[7] The ^{1}H NMR spectrum (100 MHz., CCl$_4$) shows absorption at δ 1.61 and 1.67 [2s, 6H, C=C(CH$_3$)$_2$], 1.71 [d, J = 1.4 Hz., 3H, C=C(CH$_3$)CH$_2$], 2.05 (m, 4H, 2CH$_2$), 3.98 (d, J = 8 Hz., 2H, CH$_2$Cl), 5.02 [m, 1H, CH=C(CH$_3$)$_2$], and 5.39 (t of partially resolved m, 1H, C=CHCH$_2$Cl).

3. Discussion

Geranyl chloride has been prepared by allylic rearrangement of linalool using hydrogen chloride in toluene solution at 100° or phosphorus trichloride in the presence of potassium carbonate at 0°.[2] The conversion of geraniol to geranyl chloride has been reported using hydrogen chloride in toluene,[2] phosphorus trichloride or phosphorus pentachloride in petroleum ether,[3] and thionyl chloride and pyridine.[4-6] These methods give mixtures[7,8] of geranyl and linalyl chloride, which are difficult to separate; tedious fractionation[5] is required to isolate the geranyl chloride. Procedures which give a pure product involve treatment of geraniol (a) in ether–hexamethylphosphoric triamide (HMPA) with methyllithium, followed by p-toluenesulfonyl chloride and lithium chloride in ether–HMPA;[9] (b) in 2,4,6-collidine with lithium chloride in N,N-dimethylformamide, followed by methanesulfonyl chloride;[10] and (c) in pentane with methanesulfonyl chloride at −5°, followed by the addition of pyridine.[8]

The present procedure is representative of a fairly general method of converting alcohols to chlorides using carbon tetrachloride and a tertiary phosphine. The reaction occurs under mild, essentially neutral conditions and, as illustrated by the present synthesis, may be employed to convert allylic alcohols to the corresponding halide without allylic rearrangement.

Carbon tetrachloride serves as both solvent and halogen source. Several trivalent phosphorus reagents may be employed, including triphenylphosphine,[11,12] tri-n-octylphosphine,[12] tri-n-butylphosphine,[13] and trisdimethylaminophosphine.[13] The latter "more nucleophilic" phosphines react more rapidly and under milder conditions than triphenylphosphine. When triphenylphosphine is employed, the by-product, triphenylphosphine oxide, usually precipitates completely and is easily removed by filtration. After evaporation of the solvent, the product is isolated in high purity by distillation. On occasion, difficulties may be encountered in separating the alkyl halide from the accompanying oxide. The presence of residual soluble organophosphine oxide may pose a serious problem when attempting to isolate sensitive (e.g., allylic or optically active) halides and can lead to loss of product, racemization, etc. This difficulty is usually resolved, as illustrated in the present synthesis, by adding a diluent such as pentane to ensure precipitation. Although precise conditions will undoubtedly depend on the specific substrate at hand, it is usually desirable to employ a modest excess of the organophosphine. This is especially helpful for the preparation of sensitive halides since, by ensuring complete consumption of alcohol, it simplifies the isolation procedure (Note 3) by avoiding the possible necessity of a careful fractionation step.

Triphenylphosphine reacts with carbon tetrachloride[14] or carbon tetrabromide[15] *in the absence of alcohols,* forming the corresponding triphenylphosphine dihalomethylene ylide and triphenylphosphine dihalide.

$$2(C_6H_5)_3P + CX_4 \rightarrow (C_6H_5)_3P{=}CX_2 + (C_6H_5)_3PX_2$$
$$X = Cl,Br$$

The mechanism has been viewed as involving either the formation of a pentacovalent phosphorus intermediate,[14,15] or alternatively, by initial nucleophilic attack on halogen,

$$(C_6H_5)_3P + CX_4 \rightarrow (C_6H_5)_3P\begin{array}{c} X \\ \diagup \\ \diagdown \\ CX_3 \end{array}$$

$$(C_6H_5)_3P + (C_6H_5)_3P\begin{array}{c} X \\ \diagup \\ \diagdown \\ CX_3 \end{array} \rightarrow (C_6H_5)_3P{=}CX_2 + (C_6H_5)_3PX_2$$

forming an intermediate phosphonium species.[16] These mixtures react with carbonyl compounds, providing a useful route to

$$(C_6H_5)_3P + CX_4 \rightarrow (C_6H_5)_3\overset{+}{P}X\overset{-}{C}X_3$$
$$(C_6H_5)_3\overset{+}{P}X\overset{-}{C}X_3 \rightarrow (C_6H_5)_3\overset{+}{P}CX_3\overset{-}{X}$$
$$(C_6H_5)_3P + (C_6H_5)_3\overset{+}{P}CX_3\overset{-}{X} \rightarrow (C_6H_5)_3P{=}CX_2 + (C_6H_5)_3PX_2$$

1,1-dihaloalkenes.[14,15,17]

The success of the present method depends critically on the *initial* presence of an alcohol to trap the intermediate phosphonium species.[12] If the alcohol is added last, the $R_3P{-}CX_4$ reaction described above (an exothermic process for the more nucleophilic phosphines) may go to completion, in which case little or no alkyl halide is formed.[13] Since the reaction displays several characteristics of an S_N2 process, it is thought to proceed by the pathway illustrated:

$$R_3P + CX_4 \rightarrow R_3\overset{+}{P}X\overset{-}{C}X_3$$
$$R_3\overset{+}{P}X\overset{-}{C}X_3 + R'OH \rightarrow R_3\overset{+}{P}O\overset{-}{R}' + HCX_3$$
$$R_3\overset{+}{P}X\overset{-}{O}R' \rightarrow R_3\overset{+}{P}OR'\overset{-}{X}$$
$$R_3\overset{+}{P}OR'\overset{-}{X} \rightarrow R'X + R_3PO$$

Yields of chlorides are good to excellent for primary and secondary alcohols, but a competing olefin-forming elimination process renders the method of limited value for preparing tertiary chlorides.[12] An adaptation of the procedure using carbon tetrabromide allows the synthesis of alkyl bromides. Some examples are the preparation of n-pentyl bromide (97%) and benzyl bromide (96%).[12] Farnesyl bromide has been prepared in 90% yield from farnesol.[22]

The advantages of the carbon tetrahalide – organophosphine – alcohol preparation of halides are simplicity of experimental procedure, good yields, relatively mild, essentially neutral reaction conditions, and absence of allylic rearrangements. The reaction proceeds with inversion of configuration and is a useful simple device for converting optically active alcohols to chiral halides in high optical purity.[12,23]

1. Department of Chemistry, University of Alberta, Edmonton, Alberta, Canada. T6G 2G2.
2. J. L. Simonsen and L. N. Owen, "The Terpenes," Vol. I, 2nd ed., Cambridge University Press, 1953, pp. 50, 63.
3. L. Ruzicka, *Helv. Chim. Acta*, **6**, 492 (1923).
4. M. O. Forster and D. Cardwell, *J. Chem. Soc.*, 1338 (1913).
5. D. Barnard and L. Bateman, *J. Chem. Soc.*, 926 (1950).
6. R. M. Carman and N. Dennis, *Aust. J. Chem.*, **20**, 783 (1967).
7. D. Barnard, L. Bateman, A. J. Harding, H. P. Koch, N. Sheppard, and G. B. B. M. Sutherland, *J. Chem. Soc.*, 915 (1950).
8. C. A. Bunton, D. L. Hachey, and J.-P. Leresche, *J. Org. Chem.*, **37**, 4036 (1972).
9. G. Stork, P. A. Grieco, and M. Gregson, *Tetrahedron Lett.*, 1393 (1969); *Org. Synth.*, **Coll. Vol. 6**, 638 (1988).
10. E. W. Collington and A. I. Meyers, *J. Org. Chem.*, **36**, 3044 (1971).
11. I. M. Downie, J. B. Holmes, and J. B. Lee, *Chem. Ind. (London)*, 900 (1966); J. B. Lee and T. J. Nolan, *Can. J. Chem.*, **44**, 1331 (1966).
12. J. Hooz and S. S. H. Gilani, *Can. J. Chem.*, **46**, 86 (1968).
13. I. M. Downie, J. B. Lee, and M. F. S. Matough, *Chem. Commun.*, 1350 (1968).
14. R. Rabinowitz and R. Marcus, *J. Am. Chem. Soc.*, **84**, 1312 (1962).
15. F. Ramirez, N. B. Desai, and N. McKelvie, *J. Am. Chem. Soc.*, **84**, 1745 (1962).
16. B. Miller, in M. Grayson and E. J. Griffith, Eds. "Topics in Phosphorus Chemistry," Vol. 2, Wiley-Interscience, New York, 1965, p. 133.
17. Reagents of the type R_3PX_2 (R = C_4H_5, n-C_4H_9, OC_6H_5) have independently been employed to convert alcohols to halides and phenols to aryl halides.[18–20] The reaction of $(C_6H_5O)_3P$ with methyl iodide and an alcohol gives good yields of iodides. Replacing the methyl iodide by benzyl bromide or chloride permits the synthesis of alkyl bromides and chlorides.[21]
18. L. Horner, H. Oediger, and H. Hoffmann, *Justus Liebigs Ann. Chem.*, **626**, 26 (1959).
19. H. Hoffmann, L. Horner, H. G. Wippel, and D. Michael, *Chem. Ber.*, **95**, 523 (1962).
20. G. A. Wiley, R. L. Hershkowitz, B. M. Rein, and B. C. Chung, *J. Am. Chem. Soc.*, **86**, 964 (1964).
21. H. N. Rydon, *Org. Synth.*, **Coll. Vol. 6**, 830 (1988).
22. E. H. Axelrod, G. M. Milne, and E. E. van Tamelen, *J. Am. Chem. Soc.*, **92**, 2139 (1970).
23. D. Brett, I. M. Downie, J. B. Lee, and M. F. S. Matough, *Chem. Ind. (London)*, 1017 (1969).

ALLYLIC CHLORIDES FROM ALLYLIC ALCOHOLS:
GERANYL CHLORIDE

[2,6-Octadiene, 1-chloro-3,7-dimethyl-, (E)-]

Submitted by GILBERT STORK,[1] PAUL A. GRIECO,[2] and MICHAEL GREGSON
Checked by P. A. ARISTOFF and R. E. IRELAND

1. Procedure

Caution! Hexamethylphosphoric triamide (HMPA) vapors have been reported to cause cancer in rats.[3] All operations with hexamethylphosphoric triamide should be performed in a good hood, and care should be taken to keep the liquid off the skin.

A dry, 1-l., three-necked, round-bottomed flask is equipped with an overhead mechanical stirrer, a 125-ml. pressure equalizing dropping funnel fitted with a rubber septum, and a nitrogen inlet tube. The system is flushed with nitrogen, and 15.4 g. (0.100 mole) of geraniol (Note 1), 35 ml. of dry hexamethylphosphoric triamide (Note 2), 100 ml. of anhydrous diethyl ether (Note 3), and 50 mg. of triphenylmethane (Note 4) are placed in the flask. The stirred solution is cooled to 0° with an ice bath, and 63 ml. (0.1 mole) of 1.6 M methyllithium in ether (Note 5) is injected into the addition funnel. The methyllithium solution is added dropwise over a period of 30 minutes. After the addition is complete, the funnel is rinsed by injecting 5 ml. of dry ether.

A solution of 20.0 g. (0.105 mole) (Note 6) of p-toluenesulfonyl chloride in 100 ml. of anhydrous ether is injected into the addition funnel and added over a period of 30 minutes to the stirred, red, 0° reaction mixture. The red color immediately disappears upon addition. After addition is complete, 4.2 g. (0.0990 mole) of anhydrous lithium chloride (Note 7) is added. The reaction mixture is warmed to room temperature and stirred overnight (18–20 hours), during which time lithium p-toluenesulfonate precipitates.

After a total of 20–22 hours, 100 ml. of ether is added, followed by 100 ml. of water. The layers are separated, and the organic phase is washed four times with 100-ml. portions of water, and finally with 100 ml. of saturated sodium chloride. After drying the organic phase over anhydrous magnesium sulfate, the solvent is removed on a rotary evaporator. The crude product is transferred to a 50-ml. flask and distilled through a 20-cm. Vigreux column, yielding 14.1–14.6 g. (82–85%) of geranyl chloride as a colorless liquid, b.p. 78–79° (3.0 mm.) (Notes 8 and 9).

2. Notes

1. Geraniol (+99%) can be purchased from the Aldrich Chemical Company, Inc.
2. Hexamethylphosphoric triamide was purchased from the Fisher Scientific Com-

pany and the Aldrich Chemical Company, Inc., and distilled from calcium hydride prior to use.

3. Anhydrous ether, available from J. T. Baker Chemical Company, can be used without further drying.

4. Triphenylmethane is available from Eastman Organic Chemicals. Although not necessary, it was used as an indicator to check the molarity of the methyllithium used.

5. Methyllithium (prepared from methyl chloride), available from Foote Mineral Company, can be used without further purification. Attention should be drawn to the following: methyllithium purchased from Alfa Inorganics is prepared from methyl bromide and, thus, produces a mixture of geranyl bromide and chloride.

6. p-Toluenesulfonyl chloride available from either the Aldrich Chemical Company, Inc., or Matheson, Coleman and Bell, Inc., was used without further purification.

7. Available from Alfa Inorganics. If necessary, finely powdered lithium chloride can be dried by heating under vacuum (0.1 mm.) at 100° for several hours.

8. Our sample of geranyl chloride was identical (IR, ^{1}H NMR, and mass spectrum) to a sample prepared by an alternate route (Professor John Hooz, Department of Chemistry, University of Alberta).

9. The IR spectrum (neat) shows major absorptions at 2970, 2920, 2855, 1660, 1450, 1375, 1380, 1255, 835, and 660 cm.$^{-1}$ The ^{1}H NMR spectrum (CCl$_4$) has a four-line multiplet at δ 1.55–1.85, characteristic of the olefinic methyl protons, two peaks at δ 2.0–2.2, due to the four allylic methylene protons, a d at δ 4.02 ($J = 7.0$ Hz.), due to the allylic methylene protons adjacent to the chlorine, a very broad t at δ 5.09, and a broad t at δ 5.45. ($J = 7.0$ Hz.), both due to the vinyl protons.

3. Discussion

The reaction described here illustrates a general procedure for the preparation of allylic chlorides from allylic alcohols without rearrangement and under conditions allowing the retention of sensitive groups.[4] For example, the sensitive acetal alcohol I with geraniol geometry was similarly treated with ether-hexamethylphosphoric triamide, with methyllithium in ether, and then with p-toluenesulfonyl chloride and lithium chloride. Workup afforded the corresponding chloride II in 80% yield with no detectable rearrangement. The method was equally successful with the cis-isomer of I.

In addition, 85–90% yields of neryl chloride can be obtained from nerol, the geometrical isomer of geraniol. A modification[5] of the above method has appeared which employs methanesulfonyl chloride and a mixture of lithium chloride, N,N-dimethylformamide, and 2,4,6-collidine at 0°; however, its applicability to compounds possessing sensitive groups was not demonstrated.

Initial attempts at preparing γ,γ-disubstituted allyl chlorides employing thionyl chloride in the presence of tri-n-butylamine[6] led to appreciable amounts of rearranged (tertiary) halides.

1. Department of Chemistry, Columbia University, New York, New York 10027.
2. Present address: Department of Chemistry, Indiana University, Bloomington, Indiana 47405.
3. J. A. Zapp, Jr., *Science*, **190**, 422 (1975).
4. G. Stork, P. A. Grieco, and M. Gregson, *Tetrahedron Lett.*, 1393 (1969).
5. E. W. Collington and A. I. Meyers, *J. Org. Chem.*, **36**, 3044 (1971).
6. See W. G. Young, F. F. Caserio, Jr., and D. D. Brandon, Jr., *J. Am. Chem. Soc.*, **82**, 6163 (1960).

GLUTACONALDEHYDE SODIUM SALT
FROM HYDROLYSIS OF
PYRIDINIUM-1-SULFONATE

[2-Pentenedial, ion (1⁻), sodium]

Submitted by JAN BECHER[1]
Checked by J. P. O'BRIEN, S. TEITEL, and G. SAUCY

1. Procedure

A 500-ml., three-necked, round-bottomed flask fitted with a mechanical stirrer and a thermometer is charged with 42 g. (1.1 moles) of sodium hydroxide dissolved in 168 ml. of water. The contents of the flask are cooled to −20° and stirred vigorously as 48 g. (0.30 mole) of pyridinium-1-sulfonate (Note 1), which has been previously chilled to −20°, is added in one portion. The mixture is stirred for 20 minutes while the temperature is kept below −5° (Note 2). The cooling bath is removed, and the stirred mixture is warmed gradually to 20° over 20 minutes. The temperature of the dark orange mixture is then raised to 55–60°, but after 1 hour lowered again to −5°. The brown crystals that separate are filtered by suction, pressed into a compact filter cake, and washed with three 100-ml. portions of acetone (Note 3), yielding 46–52 g. of crude product after drying on filter paper overnight or at 50° (1 mm.) for 1 hour (Note 4).

If further purification is desired, the crude product is added to 1 1. of methanol in a 2-l., three-necked, round-bottomed flask equipped with a reflux condenser and a mechanical stirrer. The mixture is stirred and heated under reflux for 30 minutes. A 10-g. portion of activated carbon is added, and after 5 minutes the hot mixture is filtered. The light yellow-red filtrate is concentrated to a volume of 50 ml. under reduced pressure and cooled to 0°. The resulting orange crystals are filtered, washed with two 25-ml. portions of acetone, and dried for 1 hour at 50° (1 mm.), affording 24–27 g. (50–58%) of glutaconaldehyde sodium salt dihydrate (Notes 5 and 6).

2. Notes

1. Pyridinium-1-sulfonate was prepared according to the procedure of Sisler and Audrieth.[2] The submitter reports that this procedure may be conveniently carried out at 5 times the specified scale. The reagent should be dry and used soon after its preparation. The checkers found that a technical grade of pyridinium-1-sulfonate (sulfur trioxide pyridine complex) purchased from Aldrich Chemical Company, Inc., gave substantially lower yields of product.

2. The initial exothermic reaction that occurs at this point produces the intermediate glutaconaldehyde iminesulfonate disodium salt shown in the scheme. It separates as a yellow, unstable precipitate that may be isolated by filtering, washing with ice-cold isopropyl alcohol, and drying. The yield of the disodium salt is 64 g. (96%).

3. The acetone washes serve to remove colored by-products.

4. The crude product is relatively stable and sufficiently pure for most purposes.

5. The submitter advises that the product be dried at room temperature for 17 hours prior to analysis. An analysis including a Karl Fischer titration for water content was reported by the checkers. Analysis calculated for $C_5H_5O_2Na \cdot (H_2O)_2$: C, 38.46; H, 5.82; H_2O, 23.08. Found: C, 38.67; H, 5.91; H_2O, 23.40. The m.p. of the product is higher than 350° and its spectral characteristics are as follows: IR (KBr) cm.$^{-1}$: 3320 (H_2O), 1723, 1715 (C=O), 1530 (C—O); ^1H NMR (DMSO-d_6), δ (multiplicity, coupling constant J in Hz., number of protons, assignment): 5.07 (d of d, J = 9 and 13, 2H, H_2 and H_4), 7.03 (t, J = 13, 1H, H_3), 8.58 (d, J = 9, 2H, H_1 and H_5); UV (aqueous 0.1 M sodium hydroxide) nm. max. (log ϵ): 363 (4.75).

6. The water of hydration that accompanies the glutaconaldehyde sodium salt described in this procedure may interfere with applications requiring anhydrous conditions. Consequently the submitter has provided the following alternative procedure for preparing the anhydrous potassium salt. Pyridinium-1-sulfonate (108 g., 0.679 mole) is added to a solution of 155 g. (3.88 moles) of potassium hydroxide in 378 ml. of water in a 1-l. flask; the solution is stirred and cooled to −20°. After 1 hour, the temperature is slowly raised to 20° over 4 hours. The mixture is heated at 30–40° for 30 minutes and cooled to 5°. The crude product that precipitates is filtered, washed with two 100-ml. portions of acetone, and dried in the air, giving 120 g. of yellow-brown crystals. This material is heated at reflux in 2.5 l. of methanol, 5 g. of activated carbon is added, the carbon is filtered, and the filtrate is concentrated under reduced pressure to a volume of 100 ml. The pale yellow crystals of glutaconaldehyde potassium salt are collected, washed with acetone, and dried, yielding 53–57 g. (57–62%). Analysis of the potassium salt indicates the empirical

formula $C_5H_5O_2K$, and the salt melts above 350°. The ^1H NMR spectrum is identical to that of the sodium salt, and the UV spectrum in aqueous 0.1 M potassium hydroxide solution exhibits a maximum at 362 nm. (log ϵ, 4.84).

The sodium and potassium salts of glutaconaldehyde are soluble only in polar solvents such as water, dimethyl sulfoxide, N,N-dimethylformamide, pyridine, and methanol. However, the stable tetrabutylammonium salt is soluble in relatively nonpolar solvents such as chloroform and ethyl acetate. It may be prepared from the potassium salt in the following manner. A 1-l. Erlenmeyer flask equipped with a magnetic stirring bar is charged with a solution of 13.6 g. (0.100 mole) of crude glutaconaldehyde potassium salt in 200 ml. of water and a solution of 33.9 g. (0.100 mole) of tetrabutylammonium hydrogen sulfate in 200 ml. of ice-cold water, the pH of which was adjusted to 10 by adding aqueous 2 M sodium hydroxide. The resulting mixture is stirred for 5 minutes in an ice bath and extracted with three 400-ml. portions of dichloromethane, previously dried by filtration through anhydrous potassium carbonate. The combined dichloromethane extracts are dried over 20 g. of anhydrous potassium carbonate and evaporated under reduced pressure. A 100-ml. portion of toluene is added, and the mixture is again evaporated under reduced pressure, removing residual water. The yield of dry, nearly colorless crystals of glutaconaldehyde tetrabutylammonium salt monohydrate is 23–25.1 g. (64–70%), m.p. 105–108°. Analysis corresponds to the empirical formula $C_{21}H_{41}NO_2 \cdot H_2O$, and the salt may be recrystallized from ethyl acetate.

3. Discussion

The sodium salt of glutaconaldehyde, first described by Baumgarten[3] in 1924, has been mentioned several times in the literature subsequently, but without full details of its preparation. The present procedure involves base-catalyzed hydrolysis of pyridinium-1-sulfonate at low temperature to glutaconaldehyde iminesulfonate dianion, followed by a second hydrolysis of the iminesulfonate at 55–60°, affording glutaconaldehyde sodium salt as a dihydrate.[4] The anhydrous potassium salt[5] and the monohydrated tetrabutylammonium salt may be prepared by similar procedures (see Note 6). In addition to being anhydrous, the potassium salt is more stable than the sodium salt; however, the sodium salt has the advantage of being more soluble in dimethyl sulfoxide and N,N-dimethylformamide. Analogous glutaconaldehyde iminesulfonate dianions with methyl and methoxy substituents at the 4-position are obtained by regiospecific ring opening of 3-methyl and 3-methoxy pyridinium-1-sulfonates.[6]

The reaction of glutaconaldehyde anion with benzoyl chloride and acetic anhydride gives the corresponding enol esters.[3,7] 4-Methyl- and 4-methoxyglutaconaldehyde enol benzoates are available by benzoylation of the corresponding iminesulfonate dianions and subsequent hydrolysis (Table I).[6] Halogenation of glutaconaldehyde anion or its enol benzoate gives a series of 2-halo and 2,4-dihalo derivatives (Table I).[6,8]

Glutaconaldehyde anion serves as an interesting intermediate for the synthesis of heterocyclic compounds. Pyrylium perchlorate has been prepared from glutaconaldehyde and 70% perchloric acid in ether at −55°.[9] The reactions of glutaconaldehyde anion with alkyl and aryl isothiocyanates and isoselenocyanates evidently occur initially at the 2-position of the former, leading to a variety of N-substituted 3-formyl-2(1H)-pyridinethiones and the corresponding selenones (Table I).[10] A five-membered heterocy-

TABLE I

GLUTACONALDEHYDE ENOL BENZOATES[6-8] AND 1-SUBSTITUTED 3-FORMYL-2(1H)-PYRIDINETHIONES[10] PREPARED FROM GLUTACONALDEHYDE ANION AND ITS DERIVATIVES

R	R'	M.p. (°)	Yield (%)	R	M.p. (°)	Yield (%)^d
H	H	119–121	87	CH$_3$	126–128	58
CH$_3$	H	138–139	61^a	C$_2$H$_5$	109–110	61
CH$_3$O	H	123–124	27^a	C$_3$H$_7$	88–90	65
H	Br	128–129	72^b	i-C$_3$H$_7$	113–115	75
H	Cl	126–128	55^c	C$_6$H$_5$	180–182	95
H	I	131–141	58^c	3-FC$_6$H$_4$	171–173	97
Cl	Cl	114–116	66^c	4-FC$_6$H$_4$	191–193	82
Br	Br	98–100	65^c			

^{a}This ester was prepared by benzoylation of the corresponding glutaconaldehyde iminesulfonate dianion and subsequent hydrolysis.
^{b}This ester was prepared by bromination of glutaconaldehyde enol benzoate.
^{c}This ester was prepared by halogenation of glutaconaldehyde anion followed by benzoylation.
^{d}The 3-formyl-2(1H)-pyridinethiones were prepared by reaction of glutaconaldehyde anion with the corresponding isothiocyanates (RN=C=S).

cle, 2-isoxazolin-5-yl acetaldehyde oxime, is formed from reaction with hydroxylamine.[11] The chemistry of glutaconaldehyde is closely related to the chemistry of 5-amino-2,4-pentadienal, derivatives of which are interesting sources for a variety of polyenes. A review on glutaconaldehyde and 5-amino-2,4-pentadienal has recently been published.[12]

1. Department of Chemistry, Odense University, DK 5230, Odense M. Denmark.
2. H. H. Sisler and J. F. Audrieth, *Inorg. Synth.*, **2**, 173 (1946).
3. P. Baumgarten, *Ber. Dtsch. Chem. Ges.*, **57**, 1622 (1924).
4. P. Baumgarten, *Ber. Dtsch. Chem. Ges.*, **59**, 1166 (1926).
5. S. S. Malhotra and M. C. Whiting, *J. Chem. Soc.*, 3812 (1960).
6. J. Becher, N. Haunsø, and T. Pedersen, *Acta Chem. Scand., Ser. B.*, **29**, 124 (1975).
7. J. Becher, *Acta Chem. Scand.*, **26**, 3627 (1972).
8. J. Becher and M. C. Christensen, *Tetrahedron*, **35**, 1523 (1979).
9. F. Klages and H. Träger, *Chem. Ber.*, **86**, 1327 (1953).
10. J. Becher and E. G. Frandsen, *Acta Chem. Scand., Ser. B*, **30**, 863 (1976); J. Becher and E. G. Frandsen, *Tetrahedron*, **33**, 341 (1977); J. Becher, E. G. Frandsen, C. Dreier, and L. Henriksen, *Acta Chem. Scand., Ser. B*, **31**, 843 (1977).

11. J. Becher and J. P. Jacobsen, *Acta Chem. Scand.*, Ser. B, **31**, 184 (1977).
12. J. Becher, *Synthesis*, 589 (1980).

ALDEHYDES FROM PRIMARY ALCOHOLS BY OXIDATION WITH CHROMIUM TRIOXIDE: HEPTANAL

$$CrO_3 + 2 \; \underset{N}{\bigcirc} \longrightarrow CrO_3(C_5H_5N)_2$$

$$CH_3(CH_2)_5CH_2OH \xrightarrow[\text{CH}_2\text{Cl}_2,\ 25°]{\text{CrO}_3(\text{C}_5\text{H}_5\text{N})_2} CH_3(CH_2)_5CHO$$

Submitted by J. C. Collins[1] and W. W. Hess[2]
Checked by R. T. Uyeda and R. E. Benson

1. Procedure

Caution! The reaction of chromium trioxide with pyridine is extremely exothermic; the preparation should be conducted in a hood, observing the precautions noted.

A. *Dipyridine chromium(VI) oxide (Note 1).* A dry, 1-l., three-necked flask fitted with a sealed mechanical stirrer, a thermometer, and a drying tube, is charged with 500 ml. of anhydrous pyridine (Note 2), which is stirred and cooled to approximately 15° (Note 3) with an ice bath. The drying tube is periodically removed and 68 g. (0.68 mole) of anhydrous chromium(VI) oxide (Note 4) is added in portions through the neck of the flask over a 30-minute period. The chromium trioxide should be added at such a rate that the temperature does not exceed 20° and in such a manner that the oxide mixes rapidly with the pyridine and does not adhere to the side of the flask (Note 5). As the chromium trioxide is added, an intensely yellow, flocculent precipitate separates from the pyridine and the viscosity of the mixture increases. When the addition is complete, the mixture is allowed to warm slowly to room temperature with stirring. Within one hour the viscosity of the mixture decreases and the initially yellow product changes to a deep red, macro-crystalline form that settles to the bottom of the flask when stirring is discontinued. The supernatant pyridine is decanted from the complex and the crystals are washed several times by decantation with 250-ml. portions of anhydrous petroleum ether. The product is collected by filtration on a sintered glass funnel and washed with anhydrous petroleum ether, avoiding contact with the atmosphere as much as possible. The complex is dried at 10 mm. until it is free-flowing, leaving 150–160 g. (85–91%) of dipyridine chromium-(VI) oxide[3] as red crystals. The product is extremely hygroscopic; contact with moisture converts it rapidly to the yellow dipyridinium dichromate.[4] It is stored at 0° in a brown bottle (Note 6).

B. *General oxidation procedure for alcohols.* A sufficient quantity of a 5% solution of dipyridine chromium(VI) oxide (Note 1) in anhydrous dichloromethane (Note 7) is prepared to provide a sixfold molar ratio of complex to alcohol, an excess usually required for complete oxidation to the aldehyde. The freshly prepared, pure complex dissolves completely in dichloromethane at 25° at 5% concentration, giving a deep red solution, but solutions usually contain small amounts of brown, insoluble material when prepared from crude complex (Note 8). The alcohol, either pure or as a solution in anhydrous dichloromethane, is added to the red solution in one portion with stirring at room temperature or lower. The oxidation of unhindered primary (and secondary) alcohols proceeds to completion within 5 to 15 minutes at 25° with deposition of brownish-black, polymeric, reduced chromium–pyridine products (Note 9). When deposition of reduced chromium compounds is complete (monitoring the reaction by GC or TLC is helpful), the supernatant liquid is decanted from the (usually tarry) precipitate, which is rinsed thoroughly with dichloromethane (Note 10).

The combined dichloromethane solutions may be washed with dilute hydrochloric acid, sodium hydrogen carbonate solution, and water, or filtered directly through a filter aid, or passed through a chromatographic column to remove traces of pyridine and chromium salts. The product is obtained by removal of dichloromethane; any pyridine that remains can often be removed under reduced pressure.

C. *Heptanal.* A dry, 1-l. three-necked round-bottomed flask is equipped with a mechanical stirrer, and 650 ml. of anhydrous dichloromethane (Note 7) is added. Stirring is begun and 77.5 g. (0.300 mole) of dipyridine chromium(VI) oxide (Note 1) is added at room temperature, followed by 5.8 g. (0.050 mole) of 1-heptanol (Note 11) in one portion. After stirring for 20 minutes, the supernatant solution is decanted from the insoluble brown gum, which is washed with three 100-ml. portions of ether. The ether and dichloromethane solutions are combined and washed successively with 300 ml. of aqueous 5% sodium hydroxide, 100 ml. of 5% hydrochloric acid (Note 12), two 100-ml. portions of saturated aqueous sodium hydrogen carbonate, and, finally, with 100 ml. of saturated aqueous sodium chloride. The organic layer is dried over anhydrous magnesium sulfate, and the solvent is removed by distillation. Distillation of the residual oil at reduced pressure through a small Claisen head separates 4.0–4.8 g. (70–84%) of heptanal, b.p. 80–84° (65 mm.), n_D^{25} 1.4094 (Note 13).

2. Notes

1. Dipyridine chromium(VI) oxide is available from Eastman Organic Chemicals. To be an effective reagent, it must be anhydrous. It should form a red solution on dissolution in anhydrous dichloromethane.

2. Commercial reagent grade pyridine was used. The checkers used material available from Allied Chemical Corporation, B and A grade.

3. To avoid the accumulation of excess, unchanged chromium trioxide and rapid temperature rise when it does react, *the initial temperature of the pyridine should never be below 10°.*

4. Reagent grade chromium(VI) oxide was dried over phosphorus pentoxide. The checkers used material available from Allied Chemical Corporation, B and A grade.

5. A glassine paper cone or glass funnel inserted in the drying tube neck of the flask during additions proved satisfactory, provided the cone or funnel was replaced frequently. The paper must be discarded carefully, since it may inflame. Adding the chromium trioxide from a flask through rubber tubing proved dangerous because it caused local excesses of the oxide below and in the neck of the flask. Pyridine added to chromium trioxide spontaneously ignites causing spot fires that extinguish themselves rapidly if the pyridine temperature is below 20° and stirring is efficient. Such fires should and can be avoided.

6. Since the complex itself loses pyridine under reduced pressure and darkens with surface decomposition, it should not be stored under vacuum or over acidic drying agents. Minimal exposure to the atmosphere is required to prevent hydration of the complex. The checkers found that a free-flowing product was obtained on drying for one hour.

7. Commercial dichloromethane was dried by distillation from phosphorus pentoxide. The dichloromethane may also be decanted from phosphorus pentoxide prior to use. Small amounts of suspended phosphorus pentoxide do not seem to interfere with the oxidation.

8. If the complex does not dissolve in dichloromethane, forming a red solution, either the complex has been hydrated in handling, or the dichloromethane is not anhydrous.

9. After the alcohol and complex are thoroughly mixed, the mixture may be stirred near its surface to avoid fouling of the stirrer by the thick, chromium-containing reduction product. Alternatively, the mixture may be swirled periodically to collect the reduction product on the side of the flask.

10. The reduced chromium precipitate is soluble in saturated sodium hydrogen carbonate, but no additional aldehyde was obtained on extracting this hydrogen carbonate solution with ether.

11. 1-Heptanol, obtained from Aldrich Chemical Company, Inc., was distilled before use, b.p. 176°.

12. A second washing with 100 ml. of 5% hydrochloric acid will reduce the amount of pyridine present in the final product without significantly decreasing the yield.

13. The product shows a strong band at 1720 cm.$^{-1}$ (C=O) in the IR. GC analysis indicated a purity of about 94–98%, with pyridine as the major impurity.

3. Discussion

Chromic acid, in a variety of acidic media, has been used extensively for the oxidation of primary alcohols to aldehydes but rarely has provided aldehydes in greater than 50% yield.[5] Chromium trioxide in pyridine was introduced as a unique, nonacidic reagent for alcohol oxidations and has been used extensively to prepare ketones,[6] but has been applied with only limited success to the preparation of aldehydes. While 2-methoxybenzaldehyde was obtained in 89% yield, 4-nitrobenzaldehyde and heptanal were obtained in 28% and 10% yields, respectively.[7]

Using the preformed dipyridine chromium(VI) oxide in dichloromethane, the rate of chromate ester formation and decay to the aldehyde[8] is enhanced at least twentyfold over the rate observed in pyridine solution.[4] Isolation of products is facile, and aldehydes appear to be relatively stable to excess reagent. The reagent has been used extensively to prepare acid-sensitive aldehydes, particularly intermediates in the total synthesis of prostaglandins[9] and steroids.[10] An 85% yield was reported for the conversion of 2-vinylcyclopropylcarbinol to the aldehyde.[11] Although excess reagent is required for the oxidations (usually sixfold), the reaction conditions are so mild and isolation of products so easy that the complex will undoubtedly find broad use as a specialty reagent. Isolation of the complex can be avoided by *in situ* preparation of the chromium oxide/pyridine complex.[12]

Other general syntheses of aldehydes from primary alcohols involve the use of dimethyl sulfoxide[13] with a dehydrating agent such as dicyclohexylcarbodiimide and phosphoric acid (or pyridinium trifluoroacetate),[14] diethylcarbodiimide,[15] or sulfur trioxide.[16] Alternatively, dimethyl sulfoxide has been used with derivatives of the alcohol such as the chloroformate,[17] the iodide,[18] and the tosylate.[19] Tertiary butyl chromate[20] and lead tetraacetate in pyridine[21] have been employed to oxidize aliphatic primary alcohols to aldehydes, while manganese dioxide[22] has been used to prepare aromatic and α,β-unsaturated aldehydes. More recently, pyridinium chlorochromate, pyridinium dichromate and chromium trioxide-3,5-dimethylpyrazole complex have been reported[23] to be effective reagents for the oxidation of primary alcohols to aldehydes in aprotic solvents.

1. Vice President for Chemistry, Sterling-Winthrop Research Institute, Rensselaer, New York 12144.
2. Department of Chemistry, Illinois Wesleyan University, Bloomington, Illinois 61701.
3. H. H. Sisler, J. D. Bush, and O. E. Accountius, *J. Am. Chem. Soc.*, **70**, 3827 (1948).
4. J. C. Collins, W. W. Hess, and F. J. Frank, *Tetrahedron Lett.*, 3363 (1968).
5. K. B. Wiberg, "Oxidation in Organic Chemistry," Part A, Academic Press, New York, 1965, p. 142; J. Carnduff, *Q. Rev. (London)*, **20**, 169 (1966).
6. G. I. Poos, G. E. Arth, R. E. Beyler, and L. H. Sarett, *J. Am. Chem. Soc.*, **75**, 422 (1953).
7. J. R. Holum, *J. Org. Chem.*, **26**, 4814 (1961).
8. W. Nagata, T. Wakabayashi, Y. Hayase, M. Narisada, and S. Kamata, *J. Am. Chem. Soc.*, **92**, 3202 (1970); **93**, 5740 (1971).
9. E. J. Corey, N. M. Weinshenker, T. K. Schaaf, and W. Huber, *J. Am. Chem. Soc.*, **91**, 5675 (1969).
10. W. S. Johnson, C. A. Harbert, and R. D. Stipanovic, *J. Am. Chem. Soc.*, **90**, 5279 (1968).
11. A. W. Burgstahler and C. M. Groginsky, *Trans. Kans. Acad. Sci.*, **72**, 486 (1969).
12. R. Ratcliffe and R. Rodehorst, *J. Org. Chem.*, **35**, 4000 (1970).
13. W. W. Epstein and F. W. Sweat, *Chem. Rev.*, **67**, 247 (1967).
14. K. E. Pfitzner and J. G. Moffatt, *J. Am. Chem. Soc.*, **85**, 3027 (1963); *J. Am. Chem. Soc.*, **87**, 5670 (1965).
15. A. F. Cook and J. G. Moffatt, *J. Am. Chem. Soc.*, **89**, 2697 (1967).
16. J. R. Parikh and W. von E. Doering, *J. Am. Chem. Soc.*, **89**, 5505 (1967).
17. D. H. R. Barton, B. J. Garner, and R. H. Wightman, *J. Chem. Soc.*, 1855 (1964).
18. A. P. Johnson and A. Pelter, *J. Chem. Soc.*, 520 (1964).
19. N. Kornblum, W. J. Jones, and G. J. Anderson, *J. Am. Chem. Soc.*, **81**, 4113 (1959).
20. R. V. Oppenauer and H. Oberrauch, *An. Asoc. Quim. Argent.*, **37**, 246 (1949) [*Chem. Abstr.*, **44**, 3871 (1950)]; H.-W. Bersch and A. v. Mletzko, *Arch. Pharm. (Weinheim Ger)*, **291**, 91 (1958); T. Suga, K. Kihara, and T. Matsuura, *Bull. Chem. Soc. Jpn.*, **38**, 893 (1965); T. Suga and T. Matsuura, *Bull. Chem. Soc. Jpn.*, **38**, 1503 (1965).
21. R. E. Partch, *Tetrahedron Lett.*, 3071 (1964).

22. R. M. Evans, *Quart. R. (London)*, **13**, 61 (1959).

23. E. J. Corey and D. L. Boger, *Tetrahedron Lett.*, 2461 (1978); E. J. Corey and G. Schmidt, *Tetrahedron Lett.*, 399 (1979); E. J. Corey and G. W. J. Fleet, *Tetrahedron Lett.*, 4499 (1973).

CONVERSION OF NITRO TO CARBONYL
BY OZONOLYSIS OF NITRONATES:
2,5-HEPTANEDIONE

Submitted by JOHN E. McMURRY[1] and JACK MELTON
Checked by ROBERT M. COATES and ROBERT W. MASON

1. Procedure

A. *5-Nitroheptan-2-one.* A 500-ml., three-necked flask equipped with a magnetic stirring bar, a 50-ml. addition funnel, and a condenser fitted with a nitrogen-inlet tube is flushed with nitrogen and charged with 36.0 g. (35.7 ml., 0.414 mole) of 1-nitropropane, 14.4 g. (20.0 ml., 0.142 mole) of diisopropylamine (Note 1), and 200 ml. of chloroform. The resulting solution is stirred and heated to 60°, and 28 g. (0.40 mole) of methyl vinyl ketone is added dropwise over a 2-hour period. The reaction mixture is then stirred for another 16 hours at 60°, allowed to cool to room temperature, and transferred to a 500-ml. separatory funnel, where it is washed with two 30-ml. portions of water and 30 ml. of 5% hydrochloric acid. After drying over anhydrous sodium sulfate, the chloroform solution is concentrated with a rotary evaporator. Distillation of the residue under reduced pressure gives 39.1 g. (61%) of 5-nitroheptan-2-one, b.p. 65–70° (0.2 mm.), n_D^{20} 1.4403 (Note 2).

B. *2,5-Heptanedione.* Methanolic sodium methoxide is prepared by cautiously adding small pieces of freshly cut sodium (5.67 g., 0.247 g.-atom) to 200 ml. of cold methanol (Note 3) in a 500-ml., three-necked flask equipped with a mechanical stirrer, a 50-ml. addition funnel, and a condenser fitted with a nitrogen-inlet tube. The resulting solution is stirred and cooled in an ice bath while 38.6 g. (0.243 mole) of 5-nitroheptan-2-one is added over 15 minutes, after which stirring is continued for another 15 minutes at 0°. The ice bath is then replaced with an acetone–dry ice bath, the nitrogen-inlet tube and addition funnel are removed, and a fritted-glass dispersion tube is inserted into the solution. With continued cooling and vigorous stirring, an ozone–oxygen mixture is bubbled through the solution for 5 hours (Notes 4 and 5).

After ozone generation has been stopped, pure oxygen is passed through the reaction mixture, removing excess ozone. Dry ice cooling is continued while 21 g. (0.34 mole) of dimethyl sulfide is added in one portion (Note 6), and the mixture is then allowed to come to ambient temperature overnight (18 hours). Methanol is removed with a rotary evaporator, and the residual liquid is dissolved in 250 ml. of diethyl ether. This solution is percolated through a short mat of silica gel (50 g.), removing polar impurities, then concentrated with a rotary evaporator, leaving a residue which is stirred with 30 ml. of 5% hydrochloric acid for 45 minutes (Note 7). Chloroform (40 ml.) is added, the organic layer is separated, and the aqueous phase is further extracted with two 30-ml. portions of chloroform. The organic extracts are combined, washed with saturated aqueous sodium hydrogen carbonate, dried over anhydrous sodium sulfate, filtered, and concentrated with a rotary evaporator. Vacuum distillation of the residue with a Kugelrohr apparatus gives 22.7 g. (73%) of 2,5-heptanedione, b.p. 90° (20 mm.), n_D^{20} 1.4313 (Note 8).

2. Notes

1. Nitropropane from MC and B Manufacturing Chemists and methyl vinyl ketone and diisopropylamine from Aldrich Chemical Company, Inc., were used as supplied.

2. IR (neat) cm.$^{-1}$: 1715, 1545; ^{1}H NMR (CCl$_4$), δ (multiplicity, coupling constant J in Hz., number of protons): 0.97 (t, $J = 7$, 3H), 2.13 (s, 3H), 4.38 (m, 1H).

3. By using 300 ml. of methanol at this point, the checkers were able to avoid crust formation on the gas-dispersion tube during ozonolysis (see Note 5).

4. Ozone was generated using a Welsbach ozonator, with a total gas flow of 1.0 l. per minute at 115 volts. This corresponds to an ozone flow of 0.104 mole per hour; thus, the time theoretically required to generate one equivalent of ozone in this reaction is 2.3 hours. The use of excess ozone is permissible only for secondary nitronates (see Discussion).

5. The precipitate that forms during ozonolysis sometimes impedes stirring, and in some cases it may be necessary to dilute the slurry with another 100 ml. of methanol after the first hour. Solid can also clog the gas-dispersion tube. The submitters scraped the fritted-glass tip occasionally to maintain a constant flow rate, whereas the checkers prepared a more dilute solution of nitronate anion (Note 3).

6. Dimethyl sulfide is added as a safety precaution, to reduce any highly oxidized and potentially dangerous by-products that might have formed during ozonolysis.

7. Acid treatment hydrolyzes dimethyl ketal by-products, which form to the extent of 5–10% during the reaction.

8. IR (neat) cm.$^{-1}$: 1710; ^{1}H NMR (CCl$_4$), δ (multiplicity, coupling constant J in Hz., number of protons): 1.01 (t, $J = 7$, 3H), 2.11 (s, 3H), 2.58 (s, 4H), 3.41 (q, $J = 7$, 2H). GC analysis by the checkers (6.3 mm. by 3 m. column of 20% SE-30 on Chromosorb W, 185°, 60 ml. of helium per minute) showed the presence of two minor impurities with retention times of 1.0 and 2.7 minutes. The major product, 2,5-heptanedione, had a retention time of 2.4 minutes.

3. Discussion

There are various methods available for transforming a nitro group into a carbonyl group, including the Nef reaction (strongly acidic),[2] permanganate oxidation of nitronate anions (basic, oxidative),[3] persulfate oxidation of nitronates (basic, oxidative),[4] treatment with a mixture of organic and inorganic nitrite (neutral, oxidative),[5] and treatment of either a free nitro compound or a nitronate anion with aqueous titanium(III) (neutral, reductive).[6] Each method is limited, however, by poor yield, inconvenience, or lack of generality.

With the proviso that the substrate not contain a reactive carbon-carbon double bond, the present ozonolysis procedure[7] appears to provide a convenient and efficient method for carrying out the desired transformation. As can be seen in Table I, both primary and secondary nitronates undergo the reaction. If a primary nitronate is to be used, however, one equivalent of ozone must be slowly metered in since use of an excess leads to overoxidation.

TABLE I

R_2CHNO_2	$\xrightarrow{CH_3O^-} \xrightarrow{O_3}$	$R_2C{=}O$	Yield (%)

			73
			68
			65
			88

1. Thimann Laboratories, University of California, Santa Cruz, California 95064. [Present address: Department of Chemistry, Cornell University, Ithaca, New York 14853.]
2. W. E. Noland, *Chem. Rev.*, **55**, 137 (1955).
3. H. Shechter and F. T. Williams, *J. Org. Chem.*, **27**, 3699 (1962).
4. A. H. Pagano and H. Shechter, *J. Org. Chem.*, **35**, 295 (1970).
5. N. Kornblum and P. A. Wade, *J. Org. Chem.*, **38**, 1418 (1973).

6. J. E. McMurry and J. Melton, *J. Org. Chem.*, **38**, 4367 (1973).
7. J. E. McMurry, J. Melton, and H. Padgett, *J. Org. Chem.*, **39**, 259 (1974). We have recently been informed that ozonolysis of nitronate anions was first reported in a paper by F. Asinger, *Ber. Dtsch. Chem. Ges.*, **77**, 73 (1944).

MACROCYCLIC POLYAMINES:
1,4,7,10,13,16-HEXAÄZACYCLOÖCTADECANE

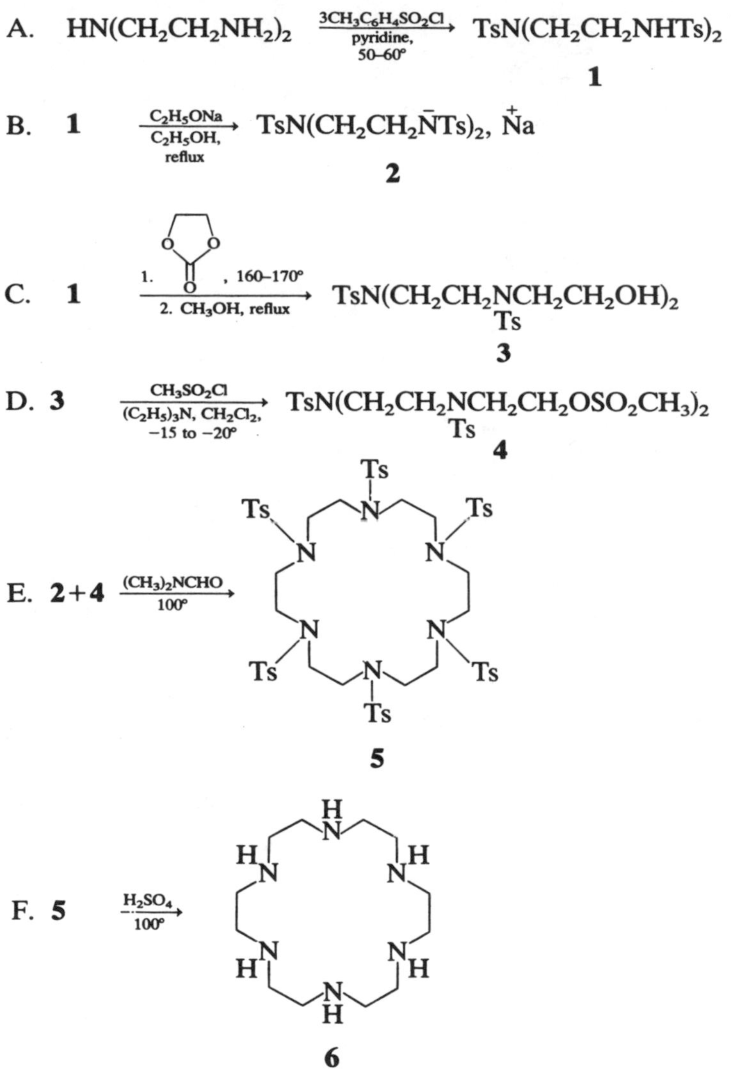

A. $HN(CH_2CH_2NH_2)_2 \xrightarrow[\substack{pyridine, \\ 50-60°}]{3CH_3C_6H_4SO_2Cl} TsN(CH_2CH_2NHTs)_2$

1

B. **1** $\xrightarrow[\substack{C_2H_5OH, \\ reflux}]{C_2H_5ONa} TsN(CH_2CH_2\bar{N}Ts)_2, \overset{+}{Na}$

2

C. **1** $\xrightarrow[\text{2. }CH_3OH, \text{ reflux}]{\substack{1. \text{ (ethylene carbonate)}, 160-170°}} TsN(CH_2CH_2NCH_2CH_2OH)_2$
                     Ts

3

D. **3** $\xrightarrow[\substack{(C_2H_5)_3N, CH_2Cl_2, \\ -15 \text{ to } -20°}]{CH_3SO_2Cl} TsN(CH_2CH_2NCH_2CH_2OSO_2CH_3)_2$
                    Ts **4**

E. **2 + 4** $\xrightarrow[100°]{(CH_3)_2NCHO}$

5

F. **5** $\xrightarrow[100°]{H_2SO_4}$

6

$$Ts = p\text{-}CH_3C_6H_4SO_2-$$

Submitted by T. J. Atkins,[1] J. E. Richman, and W. F. Oettle
Checked by K. Bernauer, F. Schneider, and A. Brossi

1. Procedure

A. *N,N',N''-Tris(p-tolylsulfonyl)diethylenetriamine* (1). A 5-l., three-necked, round-bottomed flask equipped with a mechanical stirrer, reflux condenser, thermometer, and addition funnel is charged with 1150 g. (6.037 moles) of p-toluenesulfonyl chloride (Note 1) and 3 l. of pyridine. The mixture is stirred and warmed to 50°, dissolving the solid; the flask is immersed in a 30° water bath; and a solution of 206 g. (2.00 moles) of diethylenetriamine (Note 1) in 300 ml. of pyridine is added through the addition funnel at a rate that maintains a reaction temperature of 50–60° (1 hour). The reaction mixture is kept at 50–60° for 30 minutes longer, cooled, and divided into two equal portions in 4-l. Erlenmeyer flasks. The pyridine solutions are mechanically stirred as 1000 ml. of water is slowly poured into each. After stirring overnight and cooling in an ice bath for 2 hours, the white solid is collected by filtration, thoroughly washed with ice-cold 95% ethanol, and dried in a vacuum oven at 100°, yielding 950–1015 g. (84–90%) of triamine 1, m.p. 173–175°.

B. *N,N',N''-Tris(p-tolylsulfonyl)diethylenetriamine-N,N''-disodium salt* (2). A 3-l., three-necked, round-bottomed flask equipped with a mechanical stirrer, reflux condenser, and addition funnel is charged with 1 l. of absolute ethanol and 425 g. (0.752 mole) of triamine 1 under nitrogen. The stirred slurry is heated to reflux, the heat source is removed, and 1000 ml. of 1.5 N sodium ethoxide solution (Note 2) is added through the addition funnel as rapidly as possible. The solution is decanted from any undissolved residue into an Erlenmeyer flask. The disodium salt 2, which crystallizes on standing overnight, is filtered under nitrogen, washed with absolute ethanol, and dried in a vacuum oven at 100°, yielding 400–440 g. (87–96%).

C. *3,6,9-Tris(p-tolylsulfonyl)-3,6,9-triazaundecane-1,11-diol* (3). A 2-l., three-necked, round-bottomed flask equipped with a mechanical stirrer, thermometer, reflux condenser, and heating mantle is charged with 226 g. (0.400 mole) of triamine 1, 77.5 g. (0.881 mole) of ethylene carbonate (Note 1), and 0.7 g. of powdered potassium hydroxide. The stirred mixture is heated at 160–170° for 4 hours (Note 3). The reaction mixture is then allowed to cool to 90°, and 500 ml. of methanol is added through the condenser as rapidly as possible. The solution is refluxed for 30 minutes, treated with 5 g. of activated carbon, and filtered through Celite. Water (120–140 ml.) is added dropwise to the stirred filtrate until the cloud point is reached. After crystallization is complete, diol 3 is collected and washed with 3:1 water–ethanol and dried in a vacuum oven at 50°, yielding 225–240 g. (86–92%) of colorless product, m.p. 108–112° (Note 4).

D. *3,6,9-Tris(p-tolylsulfonyl)-3,6,9-triazaundecane-1,11-dimethanesulfonate* (4). A 3-l., three-necked, round-bottomed flask equipped with a mechanical stirrer, addition funnel, nitrogen inlet, and low-temperature thermometer is charged with a dried solution of 200 g. (0.306 mole) of diol 3 and 100 ml. of triethylamine in 1500 ml. of dichloromethane (Note 5). The stirred solution is held at −15 to −20° in an acetone-dry ice bath as 74 g. (50 ml., 0.65 mole) of methanesulfonyl chloride (Note 1) is added over 10 minutes. The dry-ice bath is replaced with an ice bath, and the solution is stirred for 30 minutes, poured into a mixture of 1 l. of crushed ice and 500 ml. of 10% hydrochloric acid, and shaken. The layers are separated, and the organic layer is washed with two 500-ml. portions of water and 500 ml. of saturated salt solution, then dried over anhydrous magnesium sulfate. The solution is filtered and evaporated to dryness under reduced pressure giving a white solid, which is dissolved in 250 ml. of dichloromethane and crystallized by addition of 500 ml. of ethyl acetate and cooling in an ice bath, yielding 215–235 g. (87–95%) of methanesulfonate **4**, m.p. 146–148.

E. 1,4,7,10,13,16-*Hexakis*(p-*tolylsulfonyl*)-1,4,7,10,13,16-*hexaäzacyclooctadecane* (**5**). A 5-l., three-necked, round-bottomed flask equipped with a mechanical stirrer, thermometer, and an addition funnel is charged with 151 g. (0.248 mole) of sodium salt **2** and 2000 ml. of *N*,*N*-dimethylformamide. The stirred solution is held at 100° as a solution of 200 g. (0.247 mole) of sulfonate **4** in 800 ml. of *N*,*N*-dimethylformamide is added dropwise over 3 hours. After 30 minutes the heat source is removed, and 500 ml. of water is added through the addition funnel. After cooling to room temperature and stirring overnight, the cyclic hexamine **5** is collected by filtration, washed with 95% ethanol, and dried in a vacuum oven at 100°, yielding 206–225 g. (70–77%), m.p. 260–290° (Note 6).

F. 1,4,7,10,13,16-*Hexaäzacyclooctadecane* (**6**). A 3-l., three-necked, round-bottomed flask equipped with mechanical stirrer, nitrogen inlet, and addition funnel is charged with 200 g. (0.169 mole) of cyclic hexamine **5** and 500 ml. of concentrated (97%) sulfuric acid. The stirred mixture is held at 100° for 70 hours, then cooled in ice as 1300 ml. of anhydrous diethyl ether is slowly added. The precipitated polyhydrosulfate salt is filtered under nitrogen and washed with anhydrous ether (Note 7). The salt is then stirred in 200 ml. of water and cooled in ice as 71 ml. of aqueous 50% sodium hydroxide is added to neutralize the solution. Activated carbon (3 g.) is added, and the solution is heated to 80° and filtered through Celite. The filtrate is cooled in ice and reacidified to pH 1 by adding 42 ml. of concentrated sulfuric acid. The white, nonhygroscopic tris(sulfuric acid) salt of **6** that precipitates is collected and washed with 95% ethanol.

To the salt and 200 ml. of water in a 1-l., round-bottomed flask equipped with an efficient magnetic stirrer and cooled in ice is added 400 ml. of 50% sodium hydroxide solution. The resulting mixture is continuously extracted with tetrahydrofuran for 4 days (Note 8). The extract is concentrated to dryness at reduced pressure, and 1,4,7,10,13,16-hexaäzacyclooctadecane (**6**) is recrystallized from acetonitrile (30 ml. per g.), giving 19–22 g. (49–50%) of long, white needles, m.p. 147–150° (Note 9).

2. Notes

1. This reagent was purchased from Aldrich Chemical Company, Inc.

2. Sodium ethoxide was prepared under nitrogen just prior to use by dissolving 34.5 g. of sodium metal in 1000 ml. of absolute ethanol.

3. At 100–120° the solid begins to dissolve and carbon dioxide evolution commences.

4. Pure diol **3**, m.p. 110–112°, may be obtained by recrystallization from toluene (10 ml. per g.), but further purification is unnecessary for use in Step D.

5. This solution should be dried over 4-Å molecular sieves overnight.

6. The submitters obtained a yield of 190–210 g. (65 – 71%), m.p. 290–315° (dec.).

7. At this stage the grayish salt is quite hygroscopic and should be carefully kept from air to prevent difficulty in filtering.

8. A large excess of base is needed to reduce the water solubility of the amine. The precipitated solids contain product; any lumps should be broken up and the aqueous slurry efficiently stirred during the extraction.

9. The submitters found m.p. 154–156°.

3. Discussion

Macrocyclic polyamines and amine-ethers can be readily prepared without high-dilution techniques by this improved, general procedure.[2] Previous methods employed high-dilution techniques[3-5] or transition metal templates.[6] By adapting the present procedures, macrocycles of up to 24 members may be designed and directly synthesized in high yields from readily available starting materials.[2,7-9]

The critical cyclization step gives 50–85% yields when hydrocarbon segments between heteroatoms are short (see Table I) and when relatively equal segments of the target macrocycle are condensed. Methane- and di-p-toluenesulfonate esters[10,11] give markedly better yields than dihalides (see Table II). Cyclizations in N,N-dimethylformamide solvent are generally more convenient, although comparable yields are obtained in dimethyl sulfoxide and hexamethylphosphoric triamide [see Science, 190, 422 (1975) for a toxicity warning concerning the latter compound]. Ether linkages or selectively substituted nitrogens may replace N-tosyl groups along the chain without seriously affecting cyclization yields. Other macrocyclic amines and amine-ethers[2,7-9,12] prepared by these methods are listed in Tables III and IV.

TABLE I

CONDENSATIONS OF TERMINAL ALKANE DITOSYLATES, TsO—$(CH_2)_n$—OTs

n	Yield (%)
2	71
3	84
4	81
5	55
6	40–50

TABLE II
Yields of Cyclization for Various Leaving Groups

X	Yield (%)
–OTs	80
–OMs	66
–Cl	42
–Br	40
–I	25

The hydroxyethylation of *sec*-sulfonamides has been adapted from Niederprüm, Voss, and Wechsberg.[13] Many new terminal diols for cyclization can be readily prepared by this method.

1. Central Research and Development Department, Experimental Station, E. I. du Pont de Nemours and Company, Wilmington, Delaware 19898.
2. J. E. Richman and T. J. Atkins, *J. Am. Chem. Soc.*, **96**, 2268 (1974).
3. (a) H. Stetter and E.-E. Roos, *Chem. Ber.*, **87**, 566 (1954); (b) H. Stetter and J. Marx, *Justus Liebigs Ann. Chem.*, **607**, 59 (1957); (c) H. Stetter and K.-H. Mayer, *Chem. Ber.*, **94**, 1410 (1961).
4. H. E. Simmons and C. H. Park, *J. Am. Chem. Soc.*, **90**, 2428 (1968).
5. B. Dietrich, J. M. Lehn, J. P. Sauvage, and J. Blanzat, *Tetrahedron*, 1629 (1973).
6. N. F. Curtis, *Coord. Chem. Rev.*, **3**, 3 (1968).
7. W. Rasshofer, W. Wehner, and F. Vögtle, *Justus Liebigs Ann. Chem.*, 916 (1976).
8. W. Rasshofer and F. Vögtle, *Justus Liebigs Ann. Chem.*, 1340 (1977).
9. E. Buhleier, W. Rasshofer, W. Wehner, F. Luppertz, and F. Vögtle, *Justus Liebigs Ann. Chem.*, 1344 (1977).
10. The preparation of the dimesylate in the procedure is essentially that of R. K. Crossland and K. L. Servis, *J. Org. Chem.*, **35**, 3195 (1970).

TABLE III
Yields of Macrocyclic Amine–Ethers

Amine–Ether	Yield (%)	Amine–Ether	Yield (%)	Amine–Ether	Yield (%)
	32[b] 79[c]		63[b]		66[b] 47[c]
	38[b]		52[b] 54[c]		80[a]
	90[a] 63[b]		28[b] 69[c]		58[c]

TABLE III (continued)
YIELDS OF MACROCYCLIC AMINE–ETHERS

Amine-Ether	Yield (%)	Amine-Ether	Yield (%)	Amine-Ether	Yield (%)
	29[b]		69[b]		23[b] 43[c]
	74[c]		71[b] 88[c]		52[b] 65[c]
	7[b]		85[b]		44[b]

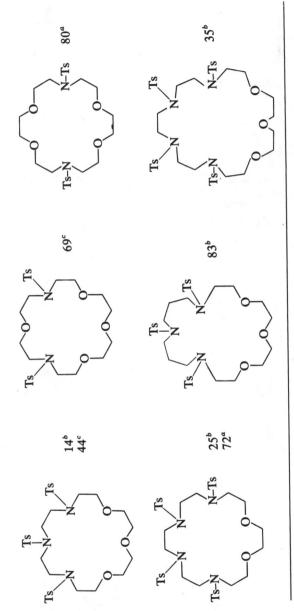

14^b
44^c

69^c

80^a

25^b
72^a

83^b

35^b

[a] Reference 2.
[b] Reference 7.
[c] T. J. Atkins, unpublished results.

TABLE IV
Yields of Macrocyclic Polyamines

Polyamine	Yield (%)	Polyamine	Yield (%)	Polyamine	Yield (%)
	58		67		24
			77		70

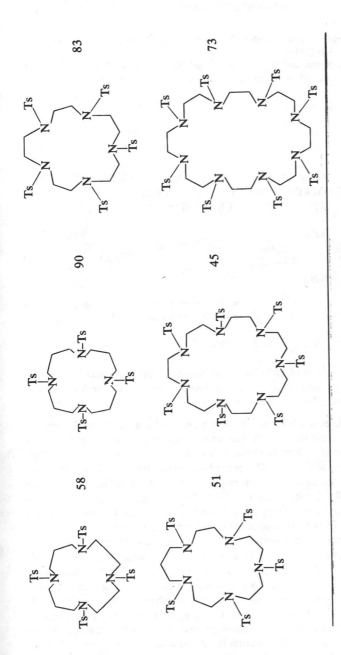

83

73

90

45

58

51

11. Ditosylates can be prepared by the procedure described in C. S. Marvel and V. C. Sekera, *Org. Synth.*, **Coll. Vol. 3,** 366 (1955).

12. T. J. Atkins, J. E. Richman, and W. F. Oettle, unpublished results.

13. H. Niederprüm, P. Voss, and M. Wechsberg, *Justus Liebigs Ann. Chem.*, 11 (1973).

CUPROUS ION-CATALYZED OXIDATIVE CLEAVAGE OF AROMATIC *o*-DIAMINES BY OXYGEN: (Z,Z)-2,4-HEXADIENEDINITRILE

Submitted by JIRO TSUJI[1] and HIROSHI TAKAYANAGI
Checked by KYO OKADA and WATARU NAGATA

1. Procedure

Caution! Benzene has been identified as a carcinogen. OSHA has issued emergency standards on its use. All procedures involving benzene should be carried out in a well-ventilated hood, and glove protection is required.

A 1-l. three-necked, round-bottomed flask equipped with a mechanical stirrer, a gas-inlet tube, and a dropping funnel is charged with 200 ml. of pyridine (Note 1) and 9.9 g. (0.10 mole) of copper(I) chloride (Note 2) which partially dissolves in pyridine, forming a yellow suspension (Note 3). Oxygen is bubbled into the suspension rapidly for 10 minutes; the suspension changes into a deep-green turbid solution (Note 4). A solution of 27 g. (0.25 mole) of 1,2-benzenediamine (Note 5) in 300 ml. of pyridine is added slowly from the dropping funnel during 2 hours, while vigorous stirring and bubbling of oxygen are continued (Notes 6, 7). The reaction mixture is transferred to a 1-l., round-bottomed flask, and pyridine is removed under reduced pressure (20 mm.) using a rotary evaporator until a deep-green solid residue is obtained, to which 400 ml. of 6 N hydrochloric acid and 400 ml. of dichloromethane are added. The mixture is shaken until the solid is dissolved, and the lower layer is separated (Note 8). The upper layer is extracted with three 100-ml. portions of dichloromethane. The combined dichloromethane solution is washed with 100 ml. of 5% aqueous sodium hydrogen carbonate, dried over anhydrous sodium sulfate, and evaporated. The brown residue is dissolved in 600 ml. of warm benzene, and the solution is filtered through filter paper. When the benzene is evaporated, 23–24 g. (88–93%) of crude product is obtained as a brownish solid, which is recrystallized twice from methanol (7 ml. for 1 g. of the crude material), yielding 19–20 g. (73–77%) of (Z,Z)-2,4-hexadienedinitrile as colorless needles, m.p. 128–129° (Note 9).

2. Notes

1. Commercial pyridine dried over potassium hydroxide pellets is satisfactory.

2. Reagent grade copper(I) chloride was obtained from Wako Pure Chemical Co., Osaka, Japan.

3. Powdered cuprous chloride should be added in small portions with efficient stirring in order to prevent coagulation.

4. When oxygen is bubbled too long, the solution becomes viscous and separation of solid mass occurs, but the mass dissolves with addition of 1,2-benzenediamine.

5. Reagent grade 1,2-benzenediamine was obtained from Wako Pure Chemical Co., Osaka, Japan.

6. The moment a drop of the diamine solution hits the reaction mixture, a spot becomes violet then turns to deep-green again. The addition should be slow, so that the violet color does not persist. The yeild of (Z,Z)-2,4-hexadienedinitrile decreases drastically if the addition is too fast.

7. The reaction is slightly exothermic, but no precaution is necessary for a small-scale experiment. It is advisable to cool the flask in a water bath when a large-scale synthesis is carried out.

8. As both layers are black and the interface is not easy to distinguish, careful separation is necessary. In addition, black amorphous material forms at the interface, making the separation difficult. It can be coagulated by standing or removed by filtration.

9. The IR spectrum (CHCl$_3$) shows bands at 2230 (medium strong), 1348, and 940 (medium) cm.$^{-1}$. The ^{1}H NMR spectrum (CDCl$_3$) shows absorption at δ 5.73 and 7.33 (AA'XX' pattern).

3. Discussion

A practical method of synthesizing (Z,Z)-2,4-hexadienedinitrile is the oxidative cleavage of 1,2-benzenediamine. Various oxidizing agents such as nickel peroxide,[2] lead tetraäcetate,[3] and silver oxide[4] are used in more than stoichiometric amounts, but the yields are below 50%. In comparison, the present method described by Takahashi, Kajimoto, and Tsuji[5] gives a very high yield and requires less than a stoichiometric quantity of copper(I) chloride. This procedure can also be applied satisfactorily to the preparation of mucononitrile derivatives from 1,2-benzenediamines substituted with an electron-donating group, but no reaction takes place with the derivatives substituted with an electron-withdrawing group.

1. Department of Chemical Engineering, Tokyo Institute of Technology, Ōokayama, Meguro-ku, Tokyo 152, Japan.
2. K. Nakagawa and H. Onoue, *Tetrahedron Lett.*, 1433 (1965).
3. K. Nakagawa and H. Onoue, *Chem. Commun.*, 396 (1965).
4. B. Ortiz, P. Villanueva, and F. Walls, *J. Org. Chem.*, **37**, 2748 (1972).
5. H. Takahashi, T. Kajimoto, and J. Tsuji, *Synth. Commun.*, 181 (1972); T. Kajimoto, H. Takahashi, and J. Tsuji, *J. Org. Chem.*, **41**, 1389 (1976).

HEXAFLUOROACETONE IMINE

(2-Propanimine, 1,1,1,3,3,3-hexafluoro-)

$$(CF_3)_2CO \xrightarrow{NH_3} (CF_3)_2\overset{\displaystyle OH}{\underset{\displaystyle |}{C}}-NH_2 \xrightarrow{POCl_3} (CF_3)_2C{=}NH$$

Submitted by W. J. MIDDLETON[1] and H. D. CARLSON
Checked by L. SCERBO and W. D. EMMONS

1. Procedure

Caution! This procedure should be conducted in a good hood to avoid exposure to ammonia and hexafluoroacetone.

A 3-l., four-necked, round-bottomed flask is equipped with a thermometer ($-50°$ to $150°$), a dry ice-cooled reflux condenser (protected from the atmosphere through a T-tube that is also connected to a nitrogen source and a Nujol bubbler), and a gas-inlet tube above the liquid level. The flask and condenser are heated in an air oven at $125°$ for several hours and flamed under nitrogen with a Bunsen burner. Pyridine (1.2 l.), previously dried over potassium hydroxide pellets, is added to the flask. A nitrogen atmosphere is maintained in the system, the pyridine is cooled to $-40°$, and 462 g. (2.78 moles) of hexafluoroacetone (b.p. $-28°$) is added from a cylinder through the gas-inlet tube over 30 minutes (Notes 1, 2, and 3). Liquid ammonia (58.3 ml., 47.6 g., 2.80 moles), previously distilled into a cold trap and measured at $-78°$, is distilled into the pyridine solution over a period of 1 hour (Note 4). During this addition the bath is held at $-45°$ to $-40°$, keeping the solution at $-25°$ to $-30°$.

As soon as the ammonia has been added, the gas-inlet tube is replaced with a 250-ml. pressure-equalized dropping funnel and the reaction mixture is heated with a heating mantle to $40°$ over 30 minutes or as quickly as possible (Note 5). The dry ice–cooled condenser is then replaced with a 24-in. water-cooled bulb condenser with Tygon tubing joining the top of the condenser to a 300-ml. cold trap protected from the atmosphere with a calcium chloride drying tube and cooled in a bath maintained at $-30°$. The condenser is cooled with $18–20°$ water. The heating mantle is turned off, and the dropping funnel is charged with 394 g. (235 ml., 2.57 moles) of phosphorus oxychloride, which is added dropwise at a rate to maintain a gentle reflux. The imine (b.p. $16°$) collects in the cold trap. When addition is complete, the reaction mixture is heated to $100°$ over 20 minutes and maintained at this temperature for 30 minutes. The cold trap collects $320–360$ g. of crude liquid product (Note 6), which is distilled through a Podbielniak still with a reflux head temperature of about $0°$ (Note 7), yielding $254–291$ g. ($55–65\%$) of the purified imine, b.p. $15.5–17°$ (Note 8). It can be stored indefinitely in a stainless-steel cylinder.

2. Notes

1. Hexafluoroacetone may be obtained from E. I. du Pont de Nemours and Company, Inc., or Allied Chemical Corporation.

2. If it is inconvenient to add the hexafluoroacetone directly from a cylinder, it may first be condensed in a calibrated trap containing a boiling chip and cooled in an acetone–dry ice bath. When cooled to $-78°$, 462 g. of liquid hexafluoroacetone has a volume of *ca.* 280 ml. The hexafluoroacetone can be added to the reaction mixture by allowing it to boil slowly from the trap.

3. The bath is maintained at $-40°$, and the pyridine solution is maintained below $-20°$. Acetone, to which pieces of dry ice are added periodically to adjust the temperature, is a convenient cooling bath.

4. The ammonia can be prevented from bumping with a magnetic stirrer in the trap. Heat for the distillation can be obtained from the air of the laboratory and by occasionally flushing the outside wall of the trap with room-temperature acetone.

5. Prolonged heating decreases the yield.

6. For storage until Podbielniak distillation can be carried out, the product can be drained into an evacuated, 500-ml. stainless-steel cylinder.

7. The checkers used a 60-cm., vacuum-jacketed column packed with glass helices, with satisfactory results.

8. This preparation has been run in the submitters' laboratory at four times this scale with yields as high as 67%.

3. Discussion

Hexafluoroacetone imine has been prepared by the reaction of hexafluoroacetone with triphenylphosphine imine,[2] by the pyrolysis of *N*-phenyl-2,2-diaminohexafluoropropane,[2,3] by the reaction of hexafluorothioacetone with hydrazoic acid,[4] and by the reaction of ammonia and phosphorus oxychloride with hexafluoroacetone.[4,5] The latter method, described here, is the most convenient, as it does not require preparation of several intermediates or use of pressure equipment. This method has also been used to prepare the imines of other fluoroketones, including those of chloropentafluoroacetone, dichlorotetrafluoroacetone, and perfluorodiethyl ketone.[5] Substitution of methylamine for ammonia in this procedure gives the *N*-methyl imine.[5]

1. Central Research Department, Experimental Station, E. I. du Pont de Nemours and Company, Inc., Wilmington, Delaware 19898.
2. Yu. V. Zeifman, N. P. Gambaryan, and I. L. Knunyants, *Dokl. Akad. Nauk SSSR*, **153**, 1334 (1963).
3. Yu. V. Zeifman, N. P. Gambaryan, and I. L. Knunyants, *Izv. Akad. Nauk SSSR Ser. Khim.*, 450 (1965).
4. W. J. Middleton, U.S. Pat. 3,226,439 (1965).
5. W. J. Middleton and C. G. Krespan, *J. Org. Chem.*, **30**, 1398 (1965).

3-TRIMETHYLSILYL-3-BUTEN-2-ONE AS MICHAEL ACCEPTOR FOR CONJUGATE ADDITION-ANNELATION: cis-4,4a,5,6,7,8-HEXAHYDRO-4a,5-DIMETHYL-2(3H)-NAPHTHALENONE

[2(3H)-Napthalenone, 4,4a,5,6,7,8-hexahydro-, 4a,5-dimethyl-, cis-, (±)-]

Submitted by Robert K. Boeckman, Jr.,[1] David M. Blum, and Bruce Ganem[2]

Checked by Seiichi Inoue and Robert M. Coates

1. Procedure

A 100-ml., three-necked flask fitted with an argon inlet, a rubber septum, a magnetic stirrer, and a 25-ml. pressure-equalizing dropping funnel (Note 1) is charged with 1.9 g. (0.010 mole) of copper(I) iodide (Note 2) and 40 ml. of anhydrous diethyl ether. The mixture is stirred and cooled in an ice bath while 10 ml. (0.02 mole) of a 2 M solution of methyllithium in ether (Note 3) is injected through the septum into the flask. The resulting straw-yellow solution of lithium dimethylcuprate is cooled to −78° in an acetone−dry ice bath, and a solution of 1.10 g. (0.0100 mole) of 2-methyl-2-cyclohexenone (Note 4) in 10 ml. of ether is injected into the flask, with stirring, over a 2–3 minute period. The cooling bath is allowed to warm slowly to about −20° over ca. 1 hour (Note 5) before a solution of 2.13 g. (0.0150 mole) of 3-trimethylsilyl-3-buten-2-one (Note 6) in 10 ml. of ether is added dropwise over 5 minutes. The stirred mixture is cooled between −20° and −30° for another hour by occasional addition of dry ice to the cooling bath. The contents of the flask are poured into 100 ml. of an ammonium chloride−ammonium hydroxide buffer solution (Note 7) that has been cooled to 0° (Note 8). The ether layer is extracted with two or three additional 100-ml. portions of buffer solution (Note 9), and the combined aqueous solutions are extracted with two 75-ml. portions of ether. The combined ether extracts are washed once with saturated sodium chloride, dried with anhydrous magnesium sulfate, and evaporated at reduced pressure.

The residual yellow liquid (2.7–3.4 g.) is dissolved in a mixture of 40 ml. of methanol and 5 ml. of 4% aqueous potassium hydroxide, then heated at reflux under an argon atmosphere for 4 hours. The methanol is evaporated from the cooled solution under reduced pressure, and the residue is dissolved in 50 ml. of ether. The ether solution is washed once with water, dried with anhydrous magnesium sulfate, and evaporated. Distillation of the residual liquid with a Kugelrohr apparatus (Note 10) at 0.5 mm. affords, after separation of a 0.1–0.2 g. forerun collected at an oven temperature of 50°, 0.76–1.02 g. (43–57%) of the octalone at an oven temperature of 85–90° (Notes 11 and 12).

2. Notes

1. The apparatus is flamed dry and maintained under an atmosphere of argon during the reactions.
2. The submitters purified copper(I) iodide by precipitation from a concentrated aqueous solution of potassium iodide as described by Kauffman and Teter,[3] and dried it at 100° over phosphorus pentoxide at high vacuum. Copper(I) iodide from Fisher Scientific Company was used by the checkers after drying at high vacuum.
3. A solution of methyllithium in ether may be purchased from Ventron Corporation. Directions for the preparation of ethereal methyllithium from methyl bromide are also available [*Org. Synth.*, **Coll. Vol. 6**, 901 (1988)]. The solution should be standardized before use by a titration procedure such as that of Watson and Eastham[4] [*Org. Synth.*, **Coll. Vol. 6**, 121 (1988)].
4. 2-Methyl-2-cyclohexenone is prepared by the method in *Org. Synth.*, **Coll. Vol. 4**, 162 (1963).
5. This warming operation was effected by the checkers by removing the original cooling bath and replacing it with another one that had been cooled to −30°. The bath temperature was then allowed to rise to −20° over *ca.* 1 hour.
6. 3-Trimethylsilyl-3-buten-2-one was prepared by the method in *Org. Synth.*, **Coll. Vol. 6**, 1033 (1988).
7. The buffer solution is prepared by adding enough concentrated ammonium hydroxide to 10% ammonium chloride to raise the pH to 8.
8. The checkers recommend that the mixture be stirred for 30 minutes at 0° to dissolve the pasty precipitate which forms and facilitate the following extractions.
9. The extractions with buffer solution should be continued until the characteristic blue color of copper(II) is no longer visible in the aqueous layer.
10. Kugelrohr distillation ovens produced by Büchi Glasapparatefabrik are available from Brinkmann Instruments, Inc.
11. The checkers collected the 0.1–0.2 g. forerun at an oven temperature of about 70–85° and the main fraction over a 5–10° range between about 105 and 120°. Combination of the main fraction from two runs gave 2.03 g., which, on redistillation with the Kugelrohr apparatus at 0.4 mm., provided a 0.2 g. forerun collected with an oven temperature of 45–53° and a main fraction totalling 1.5 g., collected at 88–93°.
12. The product has the following spectral properties: IR (thin film) cm^{-1}: 2960 (C—H), 1685 (C=O), 1620 (C=C); ^{1}H NMR (CCl$_4$), δ (multiplicity, number of protons, assignment): 0.92 (m, 3H, CH$_3$), 1.11 (s, 3H, CH$_3$), 5.64 (broad s, 1H, vinyl-*H*), minor absorption of an unidentified impurity at 5.75 (m, 0.1–0.2H, vinyl-*H*).

A GC analysis of the product by the submitters using a 1.8-m. column packed with

TABLE I
Synthesis of Polycyclic Ketones[8-11]

Reactant	Product	Yield (%)
		57
		54
		53
		57
		67
		>60

20% Carbowax 20 M suspended on Chromosorb P and operated at 150° with a flow rate of 30 ml. per minute showed a peak for the major component having a retention time of 16 minutes and two minor peaks having retention times of 4 and 7 minutes, with relative areas amounting to 6% and 2% of the major peak, respectively. The stereochemical purity of the product was shown to be >95% *cis* by the submitters by GC analysis using a 50-ft. capillary column coated with Carbowax 20 M and heated to 140°; these conditions give separate peaks for a 3:2 mixture of the *cis* and *trans* isomers of the two octalones.[5,6]

3. Discussion

4,4a,5,6,7,8-Hexahydro-4a,5-dimethyl-2(3H)-naphthalenone, an important intermediate in the total syntheses of the sesquiterpenes, aristolone,[5,6] and fukinone,[7] has been prepared in 15% yield as a 3:2 mixture of *cis* and *trans* isomers by the Robinson annelation reaction between 2,3-dimethylcyclohexanone and methyl vinyl ketone.[5,6] This *cis*-octalone has also been synthesized stereoselectively *via* the crystalline enol lactone, *cis*-3,4,4a,5,6,7-hexahydro-4a,5-dimethyl-2H-1-benzopyran-2-one.[6] In the procedure reported here the *cis*-octalone is prepared from 2-methyl-2-cyclohexen-1-one in a simple, three-step procedure consisting of conjugate addition with lithium dimethylcuprate, Michael addition of the resulting enolate anion to 3-trimethylsilyl-3-buten-2-one, and cyclization of the intermediate diketone.[8] Reasonable structures for the intermediates are proposed in the lead equation but have not been experimentally established. This procedure also illustrates the general utility of 1-trimethylsilylvinyl ketones in regio- and stereoselective enolate trapping reactions[9] and the application of the method in the synthesis of polycyclic ketones (Table I).[8-11]

1. Department of Chemistry, Wayne State University, Detroit, Michigan, 48202 [Present address: Department of Chemistry, University of Rochester, Rochester, New York 14627].
2. Department of Chemistry, Cornell University, Ithaca, New York 14853.
3. G. B. Kauffman and L. A. Teter, *Inorg. Synth.*, **7**, 9 (1963).
4. S. C. Watson and J. F. Eastham, *J. Organomet. Chem.*, **9**, 165 (1967).
5. C. Berger, M. Franck-Neumann, and G. Ourisson, *Tetrahedron Lett.*, 3451 (1968).
6. E. Piers, R. W. Britton, and W. de Waal, *Can. J. Chem.*, **47**, 4307 (1969).
7. E. Piers and R. D. Smillie, *J. Org. Chem.*, **35**, 3997 (1970).
8. R. K. Boeckman, Jr., *J. Am. Chem. Soc.*, **95**, 6867 (1973).
9. G. Stork and B. Ganem, *J. Am. Chem. Soc.*, **95**, 6152 (1973).
10. R. K. Boeckman, Jr., *J. Am. Chem. Soc.*, **96**, 6179 (1974).
11. G. Stork and J. Singh, *J. Am. Chem. Soc.*, **96**, 6181 (1974).

NITRONES FOR INTRAMOLECULAR 1,3-DIPOLAR CYCLOADDITIONS: HEXAHYDRO-1,3,3,6-TETRAMETHYL-2,1-BENZISOXAZOLINE

(2,1-Benzisoxazole, 1,3,3a,4,5,6,7,7a-octahydro-1,3,3,6-tetramethyl-)

$$(CH_3)_2C\!=\!CHCH_2CH_2CH(CH_3)CH_2CHO + CH_3NHOH\cdot HCl \xrightarrow[\text{reflux}]{\text{NaOCH}_3,\ \text{toluene,}}$$

Submitted by NORMAN A. LeBel and DOROTHY HWANG[1]
Checked by CHRISTOPHER K. VANCANTFORT and ROBERT M. COATES

1. Procedure

A 1-l., three-necked, round-bottomed flask is fitted with a mechanical stirrer, a reflux condenser attached to a Dean-Stark water separator, and a 250-ml. dropping funnel. The flask is charged with 25.0 g. (0.16 mole) of 3,7-dimethyl-6-octenal (Note 1) and 500 ml. of toluene. The solution is heated to reflux with stirring, and a solution of N-methylhydroxylamine, methanol, and toluene is added (see below).

To a cooled and magnetically stirred solution of 23.4 g. (0.280 mole) of N-methylhydroxylamine hydrochloride (Notes 2 and 3) in 40 ml. of methanol is added 15.3 g. (0.282 mole) of sodium methoxide. The cooling bath is removed, and the mixture is stirred at room temperature for 15 minutes. The mixture is filtered rapidly through a 35-mm., coarse, sintered-glass funnel, and the filter cake is washed with 10 ml. of methanol. The filtrates are combined, refiltered, and mixed with 150 ml. of toluene.

The two-phase mixture containing N-methylhydroxylamine is added dropwise to the refluxing toluene solution of the aldehyde over 3 hours. During this time the distillate is collected and discarded in 25-ml. portions until the last such portion collected and discarded is clear (Note 4). Reflux and stirring are continued for an additional 3 hours, and then the clear reaction mixture is allowed to cool. The product is extracted with three 80-ml. portions of 10% hydrochloric acid. The extracts are combined, and the pH of the solution is adjusted to >12 by the slow addition of 30% aqueous potassium hydroxide. The basic mixture is extracted with two 120-ml. portions of pentane (Note 5), and the combined extracts are washed once with 100 ml. of water and dried over anhydrous potassium carbonate. The pentane is removed on a rotary evaporator, and the residue is distilled through a short Vigreux column under reduced pressure, yielding 19.1–19.6 g. (64–67%) (Note 6) of hexahydro-1,3,3,6-tetramethyl-2,1-benzisoxazoline, b.p. 90–92° (9 mm.) (Note 7).

TABLE I

INTRAMOLECULAR 1,3-DIPOLAR CYCLOADDITIONS OF NITRONES TO ALKENES

Carbonyl Compound	Hydroxylamine	Product	Yield (%)
$CH_2=CH(CH_2)_3CHO$	CH_3NHOH		41[4]
	C_2H_5NHOH		77[4]
	CH_3NHOH		87[7]
	CH_3NHOH		78[7]

671

TABLE I (continued)

Intramolecular 1,3-Dipolar Cycloadditions of Nitrones to Alkenes

Carbonyl Compound	Hydroxylamine	Product	Yield (%)
CH_2=$CH(CH_2)_3CCH_3$ (with C=O)	CH_3NHOH		80[8]
	$(CH_3)_2CHNHOH$		70–80[9]
	C_6H_5NHOH		88[10]

672

cyclohex-3-enyl-CH₂CHO	CH₃NHOH		55.4[11]
aromatic CHO / COOCH₃ structure	CH₃NHOH		80[12]
aromatic CH=CH₂ / CH₂CHO structure	CH₃NHOH		10[13]

673

2. Notes

1. Matheson, Coleman and Bell technical grade 3,7-dimethyl-6-octenal (citronellal), b.p. 87–90° (10 mm.), is used after a simple distillation.

2. This quantity is a 0.72 molar excess. A molar excess of at least 0.5 is needed to maximize the yield.

3. N-Methylhydroxylamine hydrochloride, m.p. 83–84°, purchased from Aldrich Chemical Company, Inc., was used directly. Alternatively, the hydrochloride can be prepared by the reduction of nitromethane with zinc dust and ammonium chloride.[2]

4. Water and methanol are removed by this procedure, so that a higher reaction temperature can be achieved.

5. One to four additional extractions improve the yield slightly.

6. The yield is lowered by the presence of 3,7-dimethyl-7-octenal in the technical grade 3,7-dimethyl-6-octenal (citronellal) used.

7. A GC analysis carried out by the submitters using a 1.85 m. × 0.32 cm. stainless-steel column packed with 10% Polyglycol E-20M on Chromosorb W at 150° indicated that the product is a mixture of trans,trans and cis,trans stereoisomers in a ratio of 89:11. The spectral properties of the product are: IR (thin film) cm^{-1}: 1462, 1387, 1348, 1287, 1277, 1193, 1179, 1136, 1124, 935, 915, 877, 810; ^{1}H NMR (CDCl$_3$), δ (multiplicity, coupling constant J in Hz., number of protons, assignment): 1.00 (d, $J = 7$, 3H, CH$_3$), 1.08 (s, 3H, CH$_3$), 1.28 (s, 3H, CH$_3$), 2.60 (s, 3H, NCH$_3$).

3. Discussion

This procedure is an adaptation[3] of the original method,[4] and avoids the isolation and purification of N-methylhydroxylamine. Nitrones undergo 1,3-dipolar cycloadditions with a wide variety of dipolarophiles (see recent review[5]). The intramolecular variation represents a useful synthetic approach, as carbocyclic or heterocyclic rings are generated together with the five-membered isoxazolidines.[6] The intermediate nitrone is usually not preformed (present example), although intermolecular cycloaddition is rarely a problem. N-alkyl-, N-alkenyl-, and N-arylhydroxylamines have been used with aldehydes and ketones to generate the nitrones in situ, and some typical examples are listed in Table I. Cyclic azomethine imine oxides are also important substrates for intramolecular cycloadditions.[6]

Although isoxazolidines are less basic than the analogous amines, N-alkylisoxazolidines can form quaternary ammonium salts. Reductive cleavage of isoxazolidines and the methiodides can be effected with various reagents (zinc–acetic acid, hydrogen–palladium, lithium aluminum hydride), and the yields of 1,3-aminoalcohols are generally excellent.[4] Other reagents that result in modifications of the isoxazolidine ring include peroxyacids,[14] strong bases,[15] triplet photosensitizers,[15] and cyanogen bromide.[7]

1. Department of Chemistry, Wayne State University, Detroit, Michigan 48202.
2. E. Beckmann, Justus Liebigs Ann. Chem., 365, 201 (1909).
3. N. A. LeBel and E. G. Banucci, J. Org. Chem., 36, 2440 (1971).
4. N. A. LeBel, M. E. Post, and J. J. Whang, J. Am. Chem. Soc., 86, 3759 (1964).
5. D. St. C. Black, R. F. Crozier, and V. C. Davis, Synthesis, 205, (1975).

6. For a survey of intramolecular 1,3-dipolar cycloadditions, see A. Padwa, *Angew. Chem. Int. Ed. Engl.*, **15,** 123 (1976).

7. R. J. Newland, *Diss. Abstr. Int. B.*, **35,** 3250 (1975).

8. M. Raban, F. B. Jones, Jr., E. H. Carlson, E. Banucci, and N. A. LeBel, *J. Org. Chem.*, **35,** 1496 (1970).

9. N. A. LeBel, G. M. J. Slusarczuk, and L. A. Spurlock, *J. Am. Chem. Soc.*, **84,** 4360 (1962).

10. N. A. LeBel, G. H. Greene, and P. R. Peterson, unpublished work.

11. N. A. LeBel, N. D. Ojha, J. R. Menke, and R. J. Newland, *J. Org. Chem.*, **37,** 2896 (1972).

12. W. Oppolzer and H. P. Weber, *Tetrahedron Lett.*, 1121 (1970).

13. W. Oppolzer and K. Keller, *Tetrahedron Lett.*, 4313 (1970).

14. N. A. LeBel, *Trans. N. Y. Acad. Sci.*, **27,** 858 (1965).

15. N. A. LeBel, T. A. Lajiness, and D. B. Ledlie, *J. Am. Chem. Soc.*, **89,** 3076 (1967).

(E)-4-HEXEN-1-OL

Submitted by RAYMOND PAUL, OLIVIER RIOBÉ, and MICHEL MAUMY[1]
Checked by EDWARD J. ZAIKO and HERBERT O. HOUSE

1. Procedure

Caution! All operations described in this procedure should be performed in an efficient hood, because toxic chlorine and methyl bromide are used in Steps A and B, respectively, and hydrogen is evolved in Step C.

A. *2,3-Dichlorotetrahydropyran.* A 1-l., three-necked, round-bottomed flask fitted with a glass-inlet tube extending nearly to the bottom of the flask, a low-temperature thermometer, an exit tube attached to a calcium chloride drying tube, and a Teflon®-coated magnetic stirring bar is charged with a solution of 118 g. (1.40 moles) of dihydropyran (Note 1) in 400 ml. of anhydrous diethyl ether. While the solution is stirred continuously it is cooled to −30° with an acetone–dry ice bath. Anhydrous chlorine (Note 2) is passed through the solution and introduced at such a rate that the temperature of the reaction solution does not rise above −10° (Note 3). Completion of the addition process (*ca*. 1 hour) is indicated by a rapid development of a yellow color (excess chlorine) in the reaction solution and a distinct decrease in the temperature of the reaction mixture. When the addition is complete, several drops of dihydropyran are added to discharge the yellow color, and the colorless solution is stored at −30° (Note 4) until it is used in the next step.

B. *3-Chloro-2-methyltetrahydropyran*. A dry, 4-l., three-necked, round-bottomed flask fitted with a powerful mechanical stirrer, a reflux condenser protected by a calcium chloride drying tube, and a gas-inlet tube extending nearly to the bottom of the flask is charged with 51 g. (2.11 g.-atoms) of magnesium turnings and 1.2 l. of anhydrous ether. Methyl bromide (200 g., 2.15 moles) is allowed to distill (Note 5) into the continuously stirred reaction mixture at such a rate as to maintain gentle refluxing. The formation of an ethereal solution of methylmagnesium bromide requires approximately 2 hours. The gas-inlet tube is then replaced with a dry, 1-l. dropping funnel, protected with a calcium chloride drying tube. The reaction mixture is cooled with stirring in an ice–salt bath. The cold, ethereal 2,3-dichlorotetrahydropyran solution is placed in the dropping funnel and added dropwise, with continuous stirring and cooling, to the solution of methylmagnesium bromide at such a rate that the reaction solution does not reflux too vigorously. When this addition is complete, the resulting slurry is refluxed with stirring for 3 hours, then cooled in an ice bath. To the resulting cold (0°), vigorously stirred suspension is slowly added 900 ml. of cold 15% hydrochloric acid. The organic layer is separated, and the aqueous phase is extracted with two 200-ml. portions of ether. The combined ethereal solution is dried over anhydrous potassium carbonate, then concentrated by distillation at atmospheric pressure. The residual liquid is distilled under reduced pressure through a 12-cm. Vigreux column, separating 122−136 g. (65−72%) of a mixture of *cis*- and *trans*-3-chloro-2-methyltetrahydropyran as a colorless liquid, boiling over the range 48−95° (17−18 mm.) (Note 6), sufficiently pure for use in the next step (Note 7).

C. *(E)-4-Hexen-1-ol*. A dry, 3-l., three-necked, round-bottomed flask fitted with a powerful mechanical stirrer, a 500-ml. dropping funnel, and a reflux condenser protected by a calcium chloride drying tube is charged with 53 g. (2.3 g.-atoms) of finely divided sodium (Note 8) and 1.2 l. of anhydrous ether. 3-Chloro-2-methyltetrahydropyran (136 g., 1.01 moles) is added dropwise to the rapidly stirred, ethereal suspension of sodium. When the reaction commences (Note 9), the reaction mixture turns blue. After the reaction has started, the remaining chloro-ether is added, dropwise and with stirring over approximately 90 minutes, at such a rate that a brisk reflux is maintained. The resulting mixture, which remains dark blue throughout the addition of the chloro-ether, is refluxed with stirring for an additional hour (Note 10), then cooled in an ice–salt bath. The cold reaction mixture is stirred vigorously as 30 ml. of absolute ethanol is added dropwise and with caution to the reaction mixture, after which, 700 ml. of water is added dropwise, with stirring and cooling. After the organic layer has been separated, the aqueous phase is extracted with two 200-ml. portions of ether, and the combined ether extracts are dried over anhydrous potassium carbonate and concentrated by distillation at atmospheric pressure. The residual liquid is distilled under reduced pressure through a 12-cm. Vigreux column, separating 89–94 g. (88–93%) of *(E)*-4-hexen-1-ol as a colorless liquid, b.p. 70–74° (19 mm.), n_D^{25} 1.4389 (Note 11).

2. Notes

1. Dihydropyran (purchased from Eastman Organic Chemicals) was distilled before use; b.p. 84–86°.

2. The chlorine, obtained from a compressed gas cylinder, was passed through a wash bottle containing concentrated sulfuric acid before being passed into the reaction solution.

3. This addition of chlorine has been carried out at −5–0°, but the yield is slightly decreased and the time required for the addition is greatly extended.

4. 2,3-Dichlorotetrahydropyran may be stored for a few hours at 0°, but the yield in the subsequent step is decreased and partial decomposition may have occurred.

5. The checkers employed a sealed ampoule of methyl bromide (b.p. 5°, obtained from Eastman Organic Chemicals), which was cooled to 0° and opened. After a boiling chip had been added to the ampoule, it was connected to the gas-inlet tube of the reaction apparatus with rubber tubing and warmed in a water bath to distill the methyl bromide into the reaction vessel.

6. The submitters report that the boiling point of this mixture is 50–70° (18 mm.) or 60–80° (45 mm.), $n_D^{19.5}$ 1.4596. However, the checkers found that the product obtained from the initial distillation should be collected over a wider range [45–90° (17–18 mm.)], because the boiling point of the final portion of the product is raised by the higher molecular weight residue that remains in the stillpot. The isomers have been isolated by fractional distillation through a 90-cm. Crismer column.[2] The physical constants for the lower boiling *trans*-isomer: b.p. 56° (23 mm.), n_D^{21} 1.4543; for the higher-boiling *cis*-isomer: b.p. 72° (23 mm.), n_D^{21} 1.4646.

7. The checkers found that a fraction, b.p. 45–71° (18 mm.), had the following spectral properties; IR (CCl$_4$): no absorption in the 3300–1600 cm.$^{-1}$ region attributable to OH, C=O, or C=C vibrations; ^{1}H NMR (CDCl$_3$), δ (multiplicity, number of protons, assignment): 1.0–2.5 (m, 7H, CH$_3$ and 2CH$_2$), 3.1–4.2 (m, 4H, CHCl and CH$_2$OCH). TLC analysis of this fraction on silica gel plates using chloroform as eluent indicated the presence of a major component (the *cis*- and *trans*-isomers), R_f = 0.60, and a minor unidentified component, R_f = 0.14.

8. The finely divided sodium was prepared under boiling toluene or boiling xylene by agitation of the molten sodium with a Vibromixer. After the dispersion had cooled and the sodium had settled, the toluene or xylene was decanted, and the finely divided sodium was washed with two portions of anhydrous ether.

9. The reaction generally commences with the addition of approximately 5% of the chloro-ether; if not, the mixture should be heated to boiling to initiate the reaction. If a much larger proportion of the chloro-ether has been added before the reaction commences, initiation of the reaction may be too violent to control. The *cis*-isomer was found to be more reactive toward sodium than the *trans*-isomer.

10. The submitters reported that the blue color of the mixture fades during the reflux period, leaving a cream-white colored reaction mixture. In the checkers' runs (performed under a nitrogen atmosphere) the blue color was not discharged until ethanol was added to the reaction mixture.

11. The product exhibits the following spectral properties; IR (CCl_4) cm.$^{-1}$: 3620 (OH), 3330 broad (associated OH) 970 [(E) CH=CH]; ^{1}H NMR ($CDCl_3$) δ (multiplicity, coupling constant J in Hz., number of protons, assignment): 1.3–2.4 (m, 4H, 2CH_2), 1.64 (d of d, $J = 1$ and $J = 5$, 3H, CH_3), 3.05 (broad, 1H, OH), 3.62 (t, $J = 6.5$, 2H, CH_2O), 5.1–5.9 (m, 2H, CH=CH); mass spectrum m/e (relative intensity): 100 (M, 3), 82 (38), 67 (100), 55 (47), 54 (22), 41 (95), 39 (36), and 31 (28). The submitters report that their product exhibited a single GC peak on several columns. GC analysis of the sample obtained by the checkers using a 6-m. column packed with 1,2,3-tris(β-cyanoethoxy)propane on Chromosorb P revealed one major peak having a retention time of 20.2 minutes and a second minor peak (6% of the total peak area) having a retention time of 22.6 minutes. The mass spectrum of this minor peak exhibited the following abundant peaks: m/e (relative intensity), 100 (M, 3) 82 (38), 67 (100), 55 (42), 54 (25), 41 (97), 40 (100), 39 (35), and 31 (30). Hence, this minor component appears to be an isomer of the major product, (E)-4-hexen-1-ol.

3. Discussion

This procedure illustrates a general method for the stereoselective synthesis of (E)-disubstituted alkenyl alcohols. The reductive elimination of cyclic β-halo-ethers with metals was first introduced by Paul,[3] and one example, the conversion of tetrahydrofurfuryl chloride to 4-penten-1-ol, is described in an earlier volume of this series.[4] In 1947 Paul and Riobé[5] prepared 4-nonen-1-ol by this method; subsequently, the general method has been used to obtain alkenyl alcohols with other substitution patterns.[2,6-8] (E)-4-Hexen-1-ol has been prepared by this method[9] and in lower yield by an analogous reaction with 3-bromo-2-methyltetrahydropyran.[10]

The coupling reaction of 2,3-dichlorotetrahydropyran and Grignard reagents, RMgBr, has been effected with a number of reagents including those where R = CH_3,[2] C_2H_5,[2] $CH_3CH_2CH_2$,[2] $CH_3(CH_2)_2CH_2$,[11] C_6H_5,[11] $CH_3(CH_2)_4CH_2$,[12] β-cyclohexenylethyl,[6] β-phenylethyl,[7,13,14] vinyl,[15] and ethynyl.[16]

1. Laboratoire de Synthèse et Électrochimie Organiques, Université Catholique de l'Quest, Angers, and Laboratoire de Chimie Organique, E.S.P.C.I., 10, Rue Vauquelin, 75005, Paris, France. [Present address: Laboratoire de Recherches Organiques de l'E.S.P.C.I., Université Pierre et Marie Curie, 10, Rue Vauquelin, 75231 Paris Cedex 05, France.]
2. O. Riobé, Ann. Chim. (Paris), 4, 593 (1949).
3. R. Paul, Bull. Soc. Chim. Fr., 53, (4), 424 (1933).
4. L. A. Brooks and H. R. Snyder, Org. Synth., Coll. Vol. 3, 698 (1955).
5. R. Paul and O. Riobé, C.R. Hebd. Seances Acad. Sci., 224, 474 (1947).
6. M. Julia and F. LeGoffic, Bull. Soc. Chim. Fr., 1129 (1964).
7. J. C. Chottard and M. Julia, Bull. Soc. Chim. Fr., 3700 (1968).
8. W. E. Parham and H. E. Holmquist, J. Am. Chem. Soc., 76, 1173 (1954).
9. L. Crombie and S. H. Harper, J. Chem. Soc., 1707 (1950).

10. R. C. Brandon, J. M. Derfer, and C. E. Boord, *J. Am. Chem. Soc.*, **72,** 2120 (1950); C. L. Stevens, B. Cross, and T. Toda, *J. Org. Chem.*, **28,** 1283 (1963).
11. R. Paul, *C.R. Hebd. Seances Acad. Sci.*, **218,** 122 (1944).
12. R. Paul and S. Tchelitcheff, *Bull. Soc. Chim. Fr.*, **15,** 1199 (1948).
13. M. F. Ansell and M. E. Selleck, *J. Chem. Soc.*, 1238 (1956).
14. O. Riobé, *Bull. Soc. Chim. Fr.*, 1138 (1963).
15. H. Normant, *Bull. Soc. Chim. Fr.*, 1769 (1959).
16. L. Gouin, *Ann. Chim. (Paris)*, **5,** 529 (1960).

FRAGMENTATION OF α,β-EPOXYKETONES TO ACETYLENIC ALDEHYDES AND KETONES: PREPARATION OF 2,3-EPOXY-CYCLOHEXANONE AND ITS FRAGMENTATION TO 5-HEXYNAL

(5-Hexynal)

Submitted by Dorothee Felix, Claude Wintner, and A. Eschenmoser[1]
Checked by Robert E. Ireland and David M. Walba

1. Procedure

A. *2,3-Epoxycyclohexanone.* A 300-ml., three-necked, round-bottomed flask equipped with a magnetic stirring bar and a thermometer is charged with a solution of 9.60 g. (0.100 mole) of 2-cyclohexen-1-one (Note 1) in 100 ml. of methanol. After the solution is cooled to 1–3° in an ice bath, 34 g. (30 ml., 0.30 mole) of 30% aqueous hydrogen peroxide (Note 2) is added. The mixture is stirred vigorously and kept well cooled in the ice bath, and 0.15 ml. (0.75 mmole) of 20% aqueous sodium hydroxide is added in one portion. The temperature of the reaction mixture rises to approximately 30° within a few minutes then falls again to 3 – 5°. Fifteen minutes after addition of the sodium hydroxide solution, the cold reaction mixture is poured into a 1-1. separatory funnel containing 150 g. of ice and 200 ml. of saturated aqueous sodium chloride. The resulting suspension is extracted with 200 ml. of dichloromethane. After two further extractions of the aqueous layer with 150-ml. portions of dichloromethane, the combined organic extracts are dried over anhydrous magnesium sulfate, and the solvent is removed by distillation through a 30-cm. Vigreux column (Note 3). Distillation of the residue through a 15-cm. Vigreux column under reduced pressure affords a forerun fraction of approximately 0.7 g., b.p. 60 – 75° (11 mm.) (Note 4), then 8.4 – 8.6 g. (75 – 77% of pure 2,3-epoxycyclohexanone (Notes 5 and 6), b.p. 75 – 77° (11 mm.), n_D^{20} 1.4748, d_4^{20} 1.129.

B. *5-Hexynal*.   To a solution of 5.60 g. (0.0500 mole) of 2,3-epoxycyclohexanone in 120 ml. of benzene in a 500-ml., round-bottomed flask is added 10.82 g. (0.5152 mole) of *trans*-1-amino-2,3-diphenylaziridine [*Org. Synth.*, **Coll. Vol. 6**, 56 (1988)]. Initially, after brief swirling at room temperature, the reaction mixture is a colorless, homogeneous solution; however it rapidly turns yellow and cloudy due to separation of water. After 2 hours the benzene and water are removed as an azeotrope on a rotary evaporator with the bath maintained at approximately 30°. The resulting crude mixture of diastereomeric hydrazones weighs 15.4 g. (Note 7) and is subjected directly to fragmentation (Note 8).

The fragmentation is conveniently performed in a 100-ml., wide-necked (Note 9), round-bottomed flask with a side arm. The side arm is equipped with a capillary fitted with a balloon filled with argon or nitrogen. The capillary should be set so that it does not dip into the reaction mixture. The reaction flask is equipped with a magnetic stirring bar and fitted with a short assembly for distillation under reduced pressure. A dropping funnel extends through the stillhead to a level about that of the neck of the flask, and the receiver is cooled in an ice bath. A trap is positioned between the receiver and the vacuum source and cooled in a 2-propanol – dry ice mixture. The reaction flask is immersed in an oil bath at 150 – 155°, and the apparatus is evacuated to a pressure of 11 mm. The crude hydrazone mixture is dissolved in 20 ml. of diethyl phthalate (Note 10) and carefully added in small portions, from the funnel, into the heated flask, with the stirrer in operation. There is a rapid evolution of nitrogen, and 5-hexynal begins to distil. Addition of the entire hydrazone solution requires approximately 2 hours, after which the funnel is washed with 5 ml. of diethyl phthalate, and the temperature of the reaction flask is raised to 160 – 165°. The reaction is complete when there is no further evolution of nitrogen. The last traces of product can be driven into the receiver by warming the stillhead with a heat gun.

The contents of the receiver and trap are combined with the aid of a few drops of dichloromethane and distilled through a 15-cm. Vigreux column under reduced pressure. After only a few drops of forerun, 2.87 – 3.17 g. (60 – 66%) of 5-hexynal, b.p. 61 – 62° (30 mm.), n_D^{20} 1.4447, d_4^{20} 0.875 (Notes 11 – 13), is collected. GC analysis (Note 14) shows this material to contain 3 – 5% of unidentified impurities with longer retention times (Note 15).

2. Notes

1. 2-Cyclohexen-1-one is easily prepared by a two-step procedure from cyclohexane-1,3-dione as described in *Org. Synth.*, **Coll. Vol. 5**, 294, 539, (1973). It is available from Fluka AG CH-9470 Buchs.

2. A brand name of this reagent is Merck Perhydrol.

3. To avoid loss of the volatile epoxide, removal of the dichloromethane on a rotary evaporator is not recommended.

4. GC analysis at 120° on a 2.2-m. column packed with 10% diethylene glycol succinate showed the forerun fraction to contain approximately 50% product. The other fraction is pure 2,3-epoxycyclohexanone.

5. The IR [(CHCl$_3$) 1710 cm.$^{-1}$ strong (C=O)] and UV [(C$_2$H$_5$OH) max. 298 nm. (ϵ 15)] spectra demonstrate the absence of the enone system. ^{1}H NMR spectrum, δ (multiplicity, coupling constant J in Hz., number of protons): 1.5–3.0 (m, 6H), 3.32 (d, $J = 4$, 1H), 3.60 (m, 1H).

6. A further 1.06 g. (9.4%) of product may be obtained by Kugelrohr distillation of the residue.

7. This crude product does not show carbonyl absorption at 1710 cm.$^{-1}$ in the IR spectrum due to unreacted 2,3-epoxycyclohexanone.

8. If the crude hydrazone mixture is not to be used immediately, it must be stored in a refrigerator.

9. (E)-Stilbene sublimes during the pyrolysis and may block a narrow aperture.

10. "Pract"-grade diethyl phthalate (obtainable from Fluka AG) should be redistilled at 126–129° (1 mm.) before use. It is stable under the reaction conditions and does not co-distil with 5-hexynal.

11. 5-Hexynal is very susceptible to air oxidation.

12. The product has the following IR spectrum cm.$^{-1}$: 3310 (C≡CH), 2725 (CH=O), 2115 (C≡C), 1720 (C=O).

13. Use of a Kugelrohr (70–80°, 30 mm.) is also satisfactory.

14. GC analysis was performed on a 2.2-m. 10% diethylene glycol succinate column at 80°.

15. The checkers found that GC analysis of one sample using a 3055 cm. by 0.3 cm. column packed with 10% SF-96 on Chromosorb P operated at 70° with a 60 ml./minute helium carrier gas flow rate gave five minor impurity peaks, two at shorter retention times, and three at longer retention times. None of these impurities was present in greater than 1.1%; total impurities were 3%.

3. Discussion

2,3-Epoxycyclohexanone has been prepared in 30% yield[2] by epoxidation of 2-cyclohexen-1-one with alkaline hydrogen peroxide, using a procedure described for isophorone oxide (4,4,6-trimethyl-7-oxabicyclo[4.1.0]heptan-2-one).[3] A better yield (66%) was obtained using tert-butyl hydroperoxide and Triton B in benzene.[4] The procedure described here is simple and rapid.

The N-aminoaziridine version[5] of the α,β-epoxyketone→alkynone fragmentation is a possible alternative in situations where the simple tosylhydrazone version[6,7] fails. The tosylhydrazone method often gives good yields at low reaction temperatures, but it tends to be unsuccessful with the epoxides of acyclic enones or those not fully substituted at the β-carbon atom. For example, it has been reported[7] that 2,3-epoxycyclohexanone does not produce 5-hexynal by the tosylhydrazone route. The N-aminoaziridine method can also be recommended for the preparation of acetylenic aldehydes as well as ketones.

Both trans-1-amino-2,3-diphenylaziridine and 1-amino-2-phenylaziridine give α,β-

epoxyhydrazones that fragment in the desired manner between 100° and 200°, the choice of reagent being dictated by the ease of separation of the alkynone from the by-products, (E)-stilbene and styrene, respectively. The diphenylaziridine is especially useful when the alkynone is relatively volatile and easily separable by distillation from (E)-stilbene, as is the case in the present example. The phenylaziridine is less bulky and more stable to acid than the diphenyl derivative, and may be tried with sterically hindered epoxyketones. The fragmentation is often run in an inert, high-boiling solvent to reduce resinification, but in many cases it can be achieved by pyrolysis of the neat, crude hydrazone with concomitant distillation of the product.

The limitations of the reaction have not been systematically investigated, but the inherent lability of the aziridines can be expected to become troublesome in the case of epoxyketones which do not readily undergo hydrazone formation. The use of acid catalysis is curtailed by the instability of the aziridines, particularly the diphenylaziridine, in acidic media. Because of their solvolytic lability, the hydrazones are best formed in inert solvents: a procedure proved helpful in some cases is to mix the aziridine and the epoxyketone in anhydrous benzene, then to remove the benzene on a rotary evaporator at room temperature; water formed in the reaction is thus removed as the azeotrope. This process is repeated, if necessary, until no carbonyl band remains in the IR spectrum of the residue.

Additional examples and a mechanistic discussion are available.[5] Borrevang[8] has reported a closely related fragmentation involving diazirine derivatives of cyclic α,β-epoxyketones.

1. Laboratorium für Organische Chemie, Eidgenössische Technische Hochschule, CH-8006 Zürich, Switzerland.
2. H. O. House and R. L. Wasson, *J. Am. Chem. Soc.*, **79**, 1488 (1957).
3. R. L. Wasson and H. O. House, *Org. Synth.*, **Coll. Vol. 4**, 552 (1963).
4. N. C. Yang and R. A. Finnegang, *J. Am. Chem. Soc.*, **80**, 5845 (1958).
5. D. Felix, R. K. Müller, U. Horn, R. Joos, J. Schreiber, and A. Eschenmoser, *Helv. Chim. Acta*, **55**, 1276 (1972).
6. A. Eschenmoser, D. Felix, and G. Ohloff, *Helv. Chim. Acta*, **50**, 708 (1967); D. Felix, J. Schreiber, G. Ohloff, and A. Eschenmoser, *Helv. Chim. Acta*, **54**, 2896 (1971).
7. M. Tanabe, D. F. Crowe, R. L. Dehn, and G. Detre, *Tetrahedron Lett.*, 3739 (1967); M. Tanabe, D. F. Crowe, and R. L. Dehn, *Tetrahedron Lett.*, 3943 (1967).
8. P. Borrevang, J. Hjort, R. T. Rapala, and R. Edie, *Tetrahedron Lett.*, 4905 (1968).

γ-HYDROXY-α,β-UNSATURATED ALDEHYDES
VIA 1,3-BIS(METHYLTHIO)ALLYLLITHIUM:
trans-4-HYDROXY-2-HEXENAL

(2-Hexenal, 4-hydroxy-)

$$\underset{\text{CH}_2\text{—CH—CH}_2\text{Cl}}{\overset{\text{O}}{\triangle}} + 2\,\text{CH}_3\text{SH} + 2\,\text{NaOH} \xrightarrow[\text{below 50}°]{\text{methanol}} \underset{\text{CH}_2\text{—CH—CH}_2}{\overset{\text{SCH}_3 \quad \text{OH} \quad \text{SCH}_3}{}}$$

$$\underset{\text{CH}_2\text{—CH—CH}_2}{\overset{\text{SCH}_3 \quad \text{OH} \quad \text{SCH}_3}{}} + \text{NaH} + \text{CH}_3\text{I} \xrightarrow[\text{20–25}°]{\text{tetrahydrofuran}} \underset{\text{CH}_2\text{—CH—CH}_2}{\overset{\text{SCH}_3 \quad \text{OCH}_3 \quad \text{SCH}_3}{}}$$

$$\underset{\text{CH}_2\text{—CH—CH}_2}{\overset{\text{SCH}_3 \quad \text{OCH}_3 \quad \text{SCH}_3}{}} + 2[(\text{CH}_3)_2\text{CH}]_2\text{NLi} \xrightarrow[0°]{\text{tetrahydrofuran–}n\text{-hexane}}$$

$$\text{Li}_{\oplus}\left[\underset{\text{CH} \ominus \text{CH}}{\overset{\text{SCH}_3 \quad \text{CH} \quad \text{SCH}_3}{}}\right] \xrightarrow[-75\text{–}25°]{\text{CH}_3\text{CH}_2\text{CHO}} \underset{\overset{|}{\text{OH}}}{\text{CH}_3\text{CH}_2\text{CH—CH—CH}{=}\text{CH}} \overset{\text{SCH}_3 \quad \text{SCH}_3}{}$$

$$\underset{\overset{|}{\text{OH}}}{\text{CH}_3\text{CH}_2\text{CH—CH—CH}{=}\text{CH}} \overset{\text{SCH}_3 \quad \text{SCH}_3}{} + \text{HgCl}_2 + \text{CaCO}_3 \xrightarrow[55°]{\text{tetrahydrofuran–water}}$$

$$\underset{\overset{|}{\text{OH}} \quad \overset{|}{\text{H}}}{\text{CH}_3\text{CH}_2\text{CH—C}{=}\text{C—CHO}} \overset{\text{H}}{}$$

Submitted by Bruce W. Erickson[1]
Checked by Susumu Kamata and Wataru Nagata

1. Procedure

Caution! This preparation requires the use of a good hood.

Methyl iodide, in high concentrations for short periods or in low concentrations for long periods, can cause serious toxic effects in the central nervous system. Accordingly, the American Conference of Governmental Industrial Hygienists[2] has set 5 p.p.m., a level which cannot be detected by smell, as the highest average concentration in air to which workers should be exposed for long periods. The preparation and use of methyl iodide should always be performed in a well-ventilated fume hood. Since the liquid can be absorbed through the skin, care should be taken to prevent contact.

A. 1,3-*Bis(methylthio)-2-propanol*. A solution of 44 g. (1.1 moles) of sodium hydroxide in 300 ml. of methanol is placed in a 1-l., four-necked flask equipped with a dry ice reflux condenser, a mechanical stirrer, a thermometer, a gas-inlet, and an ice bath. While the solution is stirred and cooled, 50 g. (56 ml., 1.0 mole) of methanethiol (Note 1) is distilled from a lecture bottle into the solution at such a rate that the temperature is maintained below 20°. The gas-inlet is then replaced with a 60-ml., glass-stoppered, pressure-equalizing dropping funnel, and, while the reaction mixture is stirred and cooled, 44.4 g. (37.6 ml., 0.480 mole) of epichlorohydrin (Note 2) is added dropwise at such a rate that the temperature is maintained below 50° (Note 3).

The reaction mixture is stirred at 25° for 1 hour, diluted with 500 ml. of water, and extracted with two 200-ml. portions of diethyl ether. The combined extract is washed with five 100-ml. portions of water and 100 ml. of saturated aqueous sodium chloride, dried over anhydrous magnesium sulfate (Note 4), filtered, and evaporated at 50° (30 mm.). The residual liquid is distilled under reduced pressure without a column (Note 5), and the fraction boiling at 110–111° (7 mm.) or 101–102° (4 mm.) affords 61.0–63.6 g. (84–87%) of 1,3-bis(methylthio)-2-propanol (Notes 6, 7).

B. 1,3-*Bis(methylthio)-2-methoxypropane*. A dry, 1-l., three-necked flask equipped with a mechanical stirrer, a 60-ml. pressure-equalizing dropping funnel, and a thermometer is charged with 13.8 g. (0.339 mole) of sodium hydride dispersed in mineral oil (Note 8). The mineral oil is removed by washing the dispersion with five 100-ml. portions of hexane (Note 9), which is removed with a pipet after the sodium hydride has been allowed to settle (Note 10).

When most of the hexane has been removed, 500 ml. of dry tetrahydrofuran (Notes 11 and 12) is added. While the reaction mixture is stirred, 33.4 g. (0.220 mole) of 1,3-bis(methylthio)-2-propanol is added dropwise over 15 minutes. After the evolution of hydrogen ceases, the stirred mixture is cooled with a water bath. When the mixture reaches 20°, 34.2 g. (15.0 ml., 0.241 mole) of methyl iodide (Note 2) is added dropwise over 5 minutes. The dropping funnel is replaced with a glass stopper, and the reaction mixture is stirred at 20–25° for 24 hours.

The mixture is concentrated to about 200 ml. at 50° (30 mm.), diluted with 200 ml. of ether, and washed with 100 ml. of saturated aqueous sodium chloride, two 100-ml. portions of 0.5 *M* aqueous ammonium chloride, and 100 ml. of saturated aqueous sodium chloride. Each of the aqueous washes is extracted with the same 100-ml. portion of ether. The combined ethereal solution is dried over anhydrous magnesium sulfate (Note 4), filtered, and evaporated at 25° (10 mm). The residual material is distilled under reduced pressure without a column (Note 5), and the fraction boiling at 96–97° (9 mm.) or 89–90° (7 mm.) affords 31.4–35.7 g. (86–97%) of 1,3-bis(methylthio)-2-methoxypropane (Notes 13, 14).

C. 1,3-*Bis(methylthio)-1-hexen-4-ol*. A dry, 500-ml., three-necked flask containing a Teflon®-coated magnetic stirring bar and equipped with a 100-ml., pressure-equalizing dropping funnel, a thermometer, and a side arm connected to a nitrogen bubbler system is charged with 11.65 g. (0.1153 mole) of diisopropylamine (Note 15) and 175 ml. of dry tetrahydrofuran (Note 11). The flask is flushed with nitrogen, and a slight positive pressure is maintained with a slow stream of nitrogen during the following operation. The solution is magnetically stirred and cooled to about −75° with an acetone-dry ice bath. After 15 minutes, 77.0 ml. (0.112 mole) of 1.45 *M* solution of *n*-

butyllithium in hexane (Note 16) is added dropwise, using the pressure-equalizing dropping funnel. While the resulting solution of lithium diisopropylamide is stirred at −75°, 9.15 g. (0.0551 mole) of 1,3-bis(methylthio)-2-methoxypropane is added through a dropping funnel. The acetone–dry ice bath is replaced with an ice-water bath, and the solution is stirred at 0° for 2.5 hours. The solution slowly becomes deep purple in color as 1,3-bis(methylthio)allyllithium is generated (Note 17).

The solution is stirred and cooled to −75° with an acetone–dry ice bath, and 2.90 g. (0.0500 mole) of propionaldehyde (Note 18) is added through a dropping funnel. The solution is stirred at −75° for 5 minutes. The acetone–dry ice bath is replaced with a water bath, and the solution is stirred at 20° for 30 minutes. The solution is diluted with 250 ml. of ether and washed with two 100-ml. portions of 5 *M* aqueous ammonium chloride, two 100-ml. portions of water, and 100 ml. of saturated aqueous sodium chloride. Each of the aqueous washes is extracted with the same 250-ml. portion of ether. The combined ethereal solution is dried over anhydrous magnesium sulfate (Note 4), filtered, and evaporated at 40° (25 mm.). The residual liquid is distilled under reduced pressure through a 10 × 0.7 cm., unpacked, vacuum-jacketed column, and the material boiling at 93–98° (2 mm.) furnishes 7.90–7.99 g. (82–83%) of 1,3-bis(methylthio)-1-hexen-4-ol (Notes 19, 20).

D.   trans-4-*Hydroxy-2-hexenal*.   A 500-ml., one-necked flask containing a Teflon®-coated magnetic stirring bar is charged with 3.85 g. (0.0200 mole) of 1,3-bis(methylthio)-1-hexen-4-ol, 80 ml. of tetrahydrofuran (Note 11), and 6.00 g. (0.0600 mole) of powdered calcium carbonate. The mixture is stirred, and a solution of 16.4 g. (0.0603 mole) of mercury(II) chloride in 140 ml. of tetrahydrofuran and 40 ml. of water is added. The mixture is stirred and heated at 50–55° with a water bath for 15 hours. The reaction mixture is filtered (Note 21) with suction through a pad of diatomaceous earth in a sintered-glass funnel (Note 22). The filter cake is washed with a mixture of 200 ml. of pentane and 200 ml. of dichloromethane. The combined filtrate is washed with three 100-ml. portions of saturated aqueous sodium chloride. Each of the aqueous washes is extracted with a mixture of 100 ml. of pentane and 100 ml. of dichloromethane. The combined organic solution is dried over anhydrous magnesium sulfate (Note 4), filtered, and evaporated at 25° (25 mm.). When almost all the solvent is removed, the residue is slurried with a mixture of 4 ml. of chloroform and 1 ml. of ether as soon as possible. Purification of the product is effected by chromatography on a 3 × 20 cm. column of 50 g. of silicic acid (Note 23) by elution with 4 : 1 chloroform–ether. Each fraction (50 ml.) is monitored by TLC, and the fractions containing *trans*-4-hydroxy-2-hexenal uncontaminated with any by-product are collected. The solvent is evaporated at 25° (25 mm. and 1 mm.), yielding 1.37–1.41 g. (60–62%) of product, pure by GC (Note 24) and ¹H NMR spectroscopy (Note 25). Distillation without a column yields 1.32 g. (58%) of *trans*-4-hydroxy-2-hexenal, b.p. 48–51° (0.01–0.03 mm.), 2,4-dinitrophenylhydrazone, m.p. 198–199° (Note 26).

2. Notes

1.   The submitter used methanethiol (b.p. 8°) obtained from Matheson Gas Products. The weight of the methanethiol distilled from the lecture bottle into the reaction flask was

measured by difference. The pungent odor of this mercaptan is contained by slow distillation of the reagent into the flask and the use of a well-ventilated hood. The checkers used methanethiol (b.p. 8°, d_4^0 0.8948) obtained from Toyo Chemical Industries Co. (Japan) and changed the above procedure as follows: methanethiol distilled from the lecture bottle was first trapped in an ice-cooled graduated flask equipped with a gas-inlet and a dry ice reflux condenser; 56 ml. of the liquid was then redistilled into the reaction mixture.

2. The submitter used reagent grade material obtained from Eastman Kodak Co., and the checkers used reagent grade material from Kanto Chemical Co., Inc. (Japan).

3. A large quantity of sodium chloride precipitates during addition of the epichlorohydrin.

4. The checkers used anhydrous sodium sulfate as a drying agent instead of anhydrous magnesium sulfate.

5. The checkers used a 15 × 1 cm., unpacked, vacuum-jacketed column for the distillation.

6. For GC analysis of the product, the submitter used a 3 ft. × 0.125 in. stainless-steel column of 5% LAC-446 (cross-linked diethylene glycol-adipic ester) on Diatoport S (60–80 mesh), swept with prepurified nitrogen at 30 ml. per minute. The retention time was 1.75 minutes at 140°. The checkers used a 1 m. × 4 mm. glass column of 5% OV-17 on Chromosorb W (60–80 mesh), swept with prepurified nitrogen at 90 ml. per minute. The retention time was 2.65 minutes at 150°.

7. The ^{1}H NMR spectrum (CCl$_4$) of the product [(CH$_3^a$SCH$_2^b$)$_2$CH^{c}OH^d] shows peaks at δ 2.12 (s, 6H, H^a), 2.63 (d, $J_{b,c}$ = 6 Hz., 4H, H^b), 3.86 (pentet, 1H, H^c), and 3.78 (s, 1H, H^d).

8. Sodium hydride was obtained as a 59% dispersion in mineral oil from Metal Hydrides, Inc.

9. The submitter used a reagent grade mixture of isomeric hexanes (b.p. 68–70°) obtained from Fisher Scientific Co., and the checkers used the same grade mixture of isomeric hexanes obtained from Wako Pure Chemical Industries Ltd. (Japan).

10. About 90% of the mineral oil was removed by this procedure. Because some sodium hydride is lost in the pipet, an excess is initially employed.

11. The submitter used tetrahydrofuran obtained from Fisher Chemical Co. and distilled from lithium aluminium hydride under nitrogen shortly before use; see *Org. Synth.*, **Coll. Vol. 5**, 926 (1973) for warning concerning the purification of tetrahydrofuran. The checkers used tetrahydrofuran obtained from Wako Pure Chemical Industries Ltd. (Japan) and distilled from sodium hydride under nitrogen shortly before use.

12. When dry ether was used in place of dry tetrahydrofuran, only about 15% of the alcohol was methylated in 36 hours.

13. Using the columns described in Note 6, the retention times were 0.67 minute at 140° and 1.00 minute at 120° (submitter) and 2.30 minutes at 150° and 7.05 minutes at 120° (checkers), respectively.

14. The ^{1}H NMR spectrum (CCl$_4$) of the product [(CH$_3^a$SCH$_2^b$)$_2$CH^{c}OCH$_3^d$] exhibits absorption at δ 2.13 (s, 6H, H^a), 2.68 (d, $J_{b,c}$ = 5.4 Hz., 4H, H^b), 3.41 (s, 3H, H^d), and 3.50 (pentet, 1H, H^c).

15. The submitter used diisopropylamine from Aldrich Chemical Co., and the check-

ers used diisopropylamine from Wako Pure Chemical Industries Ltd. (Japan). It was distilled from calcium hydride before use.

16. The submitter used a 1.24 M solution of n-butyllithium in pentane obtained from Foot Mineral Co., and the checkers used a 1.45 M solution of n-butyllithium in hexane obtained from Wako Pure Chemical Industries Ltd. (Japan). The nominal titer of active alkyl on the bottle agreed well with the value found by titration[3] with 0.80 M (submitter) or 0.50 M (checkers) solution of 2-butanol in xylene using 1,10-phenanthroline as indicator.

17. Nearly quantitative generation of 1,3-bis(methylthio)allyllithium was proved, as this solution yielded 1,3-bis(methylthio)propene (88–89%) and 1,3-bis(methylthio)-1-butene (89%) by reaction with methanol and methyl iodide, respectively. The checkers found that lithium diisopropylamide can be replaced by n-butyllithium without any trouble for the generation of 1,3-bis(methylthio)allyllithium, simplifying the procedure considerably at least in this particular case. Subsequent reaction with propionaldehyde gave 1,3-bis(methylthio)-1-hexen-4-ol in 85% yield, and no appreciable amount of by-product, such as the addition product of n-butyllithium with propionaldehyde or with the intermediate 1,3-bis(methylthio)propene, was formed.

18. The submitter used propionaldehyde from Aldrich Chemical Co., and the checkers used material obtained from Tokyo Chemical Industry Co. Ltd. (Japan). It was distilled from calcium hydride shortly before use.

19. For GC analysis of the products, the submitter used a 3 ft. × 0.125 in. stainless-steel column of 5% LAC-446 (cross-linked diethylene glycol-adipic ester) on Diatoport S (60–80 mesh), heated at 140° and swept with prepurified nitrogen at 30 ml. per minute. The four isomers were observed as three peaks at retention times of 3.08, 3.69, and 4.07 minutes. The checkers used a 1 m. × 4 mm. glass column of 10% DEGS on Gas Chrome Q (60–80 mesh), heated at 200° and swept with prepurified nitrogen at 75 ml. per minute. The four isomers were observed at retention times of 3.30, 3.85, 4.10, and 5.20 minutes.

20. The ^{1}H NMR spectrum (CCl$_4$) of the product [CH^a(SCH$_3^b$)=CH^{c}CH^d(SCH$_3^e$)—CH^f(OH)CH$_2^g$CH$_3^h$] exhibits absorption at δ 0.94 (t, $J_{g,h}$ = 7 Hz., 3H, H^h), 1.4 (m, 2H, H^g), 2.00, 2.03, and 2.05 (s, 3H, H^e), 2.22 and 2.26 (s, 3H, H^b), 2.6–3.8 (m, 2H, H^d and H^f), 5.0–5.8 (m, 1H, H^c), and 5.9–6.3 (m, 1H, H^a). The olefinic multiplets indicate[4] that two *trans* diastereomers and two *cis* diastereomers are present in the ratios 25:23:41:11, respectively.

21. At this stage, the submitter added 2 g. of sodium hydrogen carbonate to buffer the liquid phase near neutrality before filtration. The checkers found that this operation can be omitted without any trouble.

22. Diatomaceous earth is used to avoid clogging the sintered glass filter with the insoluble material in the reaction mixture.

23. To remove the remaining mercury(II) chloride and some by-products the submitter filtered a carbon tetrachloride solution of the product through a 1-cm. column of Merck silicic acid in an experiment of one-twentieth scale. For the larger-scale preparation the checkers found it necessary to carry out the chromatographic purification using silicic acid (M. Woelm Eshwege, Grade II) as described.

24. Using the same columns described in Note 19, the product shows a retention time of 1.23 minutes at 140° (submitter) and 1.45 minutes at 200° (checkers).

25. The ^1H NMR spectrum (CCl_4) of the γ-hydroxy-α,β-unsaturated aldehyde [$O{=}CH^aCH^b{=}CH^cCH^d(OH)CH_2^eCH_3^f$] shows absorption at δ 0.96 (t, 3H, H^f), 1.2–1.9 (m, 2H, H^e), 2.22 (s, 1H, OH), 4.30 (m, 1H, H^d), 6.23 (d of d, 1H, H^b), 6.85 (d of d, 1H, H^c), and 9.49 (d, 1H, H^a), with coupling constants J in Hz.: $J_{a,b} = 7.6$, $J_{b,c} = 15.5$, $J_{b,d} = 1.0$, $J_{c,d} = 4.3$, and $J_{e,f} = 7.0$. IR (CCl_4) cm^{-1}: 3615 medium, 3470 medium and broad (OH), 2805 weak, 2720 weak (CH of aldehyde), 1693 strong (C=O), 1640 weak (C=C).

26. *trans*-4-Hydroxy-2-hexenal is unstable and decomposed on standing at room temperature, giving polymeric and dehydrated compounds. The checkers prepared the 2,4-dinitrophenylhydrazone, m.p. 198–199°,[5] for its identification in the usual way.

3. Discussion

1,3-Bis(methylthio)-2-methoxypropane is important as the precursor of 1,3-bis(methylthio)allyllithium,[6] a symmetrical nucleophile synthetically equivalent to the currently unknown (and probably intrinsically unstable) 3-lithio derivative of acrolein (Li—CH=CH—CH=O).

Reaction of 1,3-bis(methylthio)-2-methoxypropane with 2 moles of lithium diisopropylamide[6] (or *n*-butyllithium) effects (a) the elimination of methanol, forming 1,3-bis(methylthio)propene, and (b) the lithiation of this propene, generating 1,3-bis(methylthio)allyllithium in solution. Its conjugate acid, 1,3-bis(methylthio)propene, can be regenerated by protonation with methanol, and has also been prepared (a) in 31% yield by reaction of methylthioacetaldehyde with the lithio derivative of diethyl methylthiomethylphosphonate,[6] (b) in a low yield by acid-catalyzed pyrolysis of 1,1-bis(methylthio)-3-methoxypropane,[7] and (c) in low yield by acid-catalyzed coupling of vinyl chloride with chloromethyl methyl sulfide.[8]

A variety of α,β-unsaturated aldehydes are available by alkylation of 1,3-bis(methylthio)allyllithium and hydrolysis of the product. For example, *trans*-2-octenal is obtained in 75% yield overall on alkylation with 1-bromopentane and hydrolysis with mercuric chloride.[6]

Addition of 1,3-bis(methylthio)allyllithium to aldehydes, ketones, and epoxides followed by mercury(II) ion-promoted hydrolysis furnishes hydroxyalkyl derivatives of acrolein[6] that are otherwise available in lower yield by multistep procedures. For example, addition of 1,3-bis(methylthio)allyllithium to acetone proceeds in 97% yield, giving a tertiary alcohol that is hydrolyzed with mercury(II) chloride and calcium carbonate to *trans*-4-hydroxy-4-methyl-2-pentenal in 41% yield.[6] Addition to an epoxide and

TABLE I

CONVERSION OF EPOXIDES INTO δ-ACETOXY-*trans*-α,β-UNSATURATED
ALDEHYDES

Epoxide	Coupling	Acetylation	Hydrolysis	Overall
	Yield, %			
Propylene oxide	97	100	81	78
Cyclopentene oxide	99	92	59	54
Cyclohexene oxide	96	100	85	82

hydrolysis affords a δ-hydroxy-α,β-unsaturated aldehyde.[9] Similarly, addition of 1,3-bis(methylthio)allyllithium to an epoxide, acetylation of the hydroxyl group, and hydrolysis with mercury(II) chloride and calcium carbonate provides a δ-acetoxy-*trans*-α,β-unsaturated aldehyde,[6] as indicated in Table I. Cyclic *cis*-epoxides give aldehydes in which the acetoxy group is *trans* to the 3-oxopropenyl group.

1. The Rockefeller University, New York, New York 10021.
2. American Conference of Governmental Industrial Hygienists (ACGIH), "Documentation of Threshold Limit Values," 3rd ed., Cincinnati, Ohio, 1971, p. 166.
3. S. C. Watson and J. F. Eastham, *J. Organomet. Chem.*, **9**, 165 (1967).
4. B. W. Erickson, Ph.D. Thesis, Harvard University, Cambridge, Mass., 1970; [*Diss. Abst. Int. B*, **31**, 6500 (1971)].
5. H. Esterbauer and W. Weger, *Monatsh. Chem.*, **98**, 1884 and 1994 (1967).
6. E. J. Corey, B. W. Erickson, and R. Noyori, *J. Am. Chem. Soc.*, **93**, 1724 (1971).
7. J. Hine, L. G. Mahone, and C. L. Liotta, *J. Org. Chem.*, **32**, 2600 (1967).
8. T. Ichikawa, H. Owatari, and T. Kato, *J. Org. Chem.*, **35**, 344 (1970).
9. E. J. Corey and R. Noyori, *Tetrahedron Lett.*, 311 (1970).

17β-HYDROXY-5-OXO-3,5-*seco*-4-NORANDROSTANE-3-CARBOXYLIC ACID

(1*H*-Benz[*e*]indene-6-propanoic acid, dodecahydro-3-hydroxy-3*a*,6-dimethyl-7-oxo-)

Submitted by L. Milewich[1] and L. R. Axelrod
Checked by A. P. King and R. E. Benson

1. Procedure

A 3-1., three-necked, round-bottomed flask fitted with an efficient mechanical stirrer, a 100-ml. dropping funnel, and a 500-ml. dropping funnel is charged with a solution of 10.0 g. (0.0303 mole) of testosterone acetate [17β-(acetyloxy)-androst-4-en-3-one] (Note 1) in 600 ml. of *tert*-butyl alcohol and a solution of 5.6 g. (0.041 mole) of anhydrous potassium carbonate in 150 ml. of water. To the flask are added 100 ml. of a solution prepared from 40.0 g. (0.187 mole) of sodium metaperiodate (Note 2) in 500 ml. of water, and 10 ml. of a solution prepared from 0.8 g. (0.005 mole) of potassium permanganate in 100 ml. of water. Stirring is begun, and the remaining portions of these solutions are transferred to the appropriate dropping funnels. The metaperiodate solution is added over a period of about 30 minutes, and the permanganate solution is added as needed to maintain a pink color (Note 3). Stirring is continued for an additional 1.5 hours (Note 4) before a solution of 20.0 g. of sodium bisulfite in 40 ml. of water is slowly added (Note 5). After stirring for 20 minutes the suspension is filtered through a 10-g. pad of Celite® filter aid (Note 6) on a coarse sintered-glass filter, and the residual cake is washed with two 50-ml. portions of *tert*-butyl alcohol. The filtrates are combined and concentrated by distillation (Note 7) to a volume of about 200 ml. After cooling, 25 ml. of 10% sulfuric acid is added. The resulting mixture is extracted with three 300-ml. portions of diethyl ether. The combined ethereal extracts are washed, first with two 50-ml. portions of water, twice with solutions prepared from 5.0 g. of sodium bisulfite in 20 ml. of water, and finally twice with 50 ml. of water. Crushed ice is added to the ethereal layer, and the mixture is extracted with three 70-ml. portions of 10% aqueous sodium hydroxide. The aqueous layers are combined and washed with 50 ml. of ether. The aqueous layer is transferred to a 2-1. separatory funnel, crushed ice is added, and 300 ml. of 10% sulfuric acid is introduced carefully. The separatory funnel is shaken gently, and the product separates as a gum or stringy mass (Note 8). The mixture is extracted with four 200-ml.

portions of dichloromethane, and the extracts are combined and washed with three 15-ml. portions of water. The organic layer is dried over anhydrous sodium sulfate, filtered, and evaporated under reduced pressure. The resulting material is triturated with 50 ml. of acetone, and after a brief heating period crystals begin to separate. The mixture is cooled and filtered, yielding 6.0–6.5 g. (63–70%) of 17β-hydroxy-5-oxo-3,5-*seco*-4-norandrostane-3-carboxylic acid, m.p. 204–205° (Note 9). Additional product can be obtained from the filtrate by concentration and cooling.

2. Notes

1. Testosterone acetate was obtained from Steraloids, Inc.
2. Sodium metaperiodate was purchased from J. T. Baker Chemical Company.
3. During these additions the temperature of the reaction mixture rises to about 35°.
4. At this point the reaction mixture has a pH of about 5.2.
5. When the bisulfite solution is added, the solution acquires a deep brown color and iodine fumes develop.
6. The product was purchased from Johns-Manville Corporation.
7. Evaporation is carried out under reduced pressure with a bath temperature of 75°. A deep iodine-colored fraction distills first, followed by a colorless distillate.
8. The checkers found it convenient to separate the aqueous phase from the sticky amorphous mass and later dissolve the product in the dichloromethane solution that was used for the extractions.
9. The reported melting points are 206.5–207°,[2] 204–205.5°,[3] 200–202°,[4] 204–206°,[5] and 192°.[6] The product has the following spectral properties; IR (KBr) cm.$^{-1}$: 3390, 1717, 1700, and 1056; ^{1}H NMR (CDCl$_3$), δ: 0.82 (C$_{18}$–CH$_3$), 1.14 (C$_{19}$–CH$_3$), and additional broad absorptions; $[\alpha]_D^{28}$ +36°.

3. Discussion

17β-Hydroxy-5-oxo-3,5-*seco*-4-norandrostane-3-carboxylic acid has been prepared by ozonolysis of testosterone[2–4] or testosterone acetate, followed by alkaline hydrolysis,[5] and by the oxidation of testosterone acetate with ruthenium tetroxide.[6]

The present procedure corresponds to the method[7] described earlier for the synthesis of 5-oxo-3,5-*seco*-4-norcholestane-3-carboxylic acid and is useful for preparing large quantities of the title compound.

1. Southwest Foundation for Research and Education, Division of Biological Growth and Development, Department of Biochemistry, San Antonio, Texas 78284. [Present address: Department of Obstetrics and Gynecology, The University of Texas, Health Science Center at Dallas, Dallas, Texas 75235.]
2. C. C. Bolt, *Recl. Trav. Chim. Pays-Bas*, **57**, 905 (1938).
3. F. L. Weisenborn, D. C. Remy, and T. L. Jacobs, *J. Am. Chem. Soc.*, **76**, 552 (1954).
4. H. J. Ringold and G. Rosenkranz, *J. Org. Chem.*, **22**, 602 (1957).
5. P. N. Rao and L. R. Axelrod, *J. Chem. Soc.*, 1356 (1965).
6. D. M. Piatak, H. B. Bhat, and E. Caspi, *J. Org. Chem.*, **34**, 112 (1969).
7. J. T. Edwards, D. Holder, W. H. Lunn, and I. Puskas, *Can. J. Chem.*, **39**, 599 (1961).

DIRECTED ALDOL CONDENSATIONS: *threo*-4-HYDROXY-3-PHENYL-2-HEPTANONE

[2-Heptanone, 4-hydroxy-3-phenyl-, (R*,R*)-]

$$C_6H_5CH_2COCH_3 \xrightarrow[\text{2. } (CH_3CO)_2O, \, 5-30°]{\text{1. NaH, } CH_3OCH_2CH_2OCH_3, \, 20°}$$

2-Acetoxy-trans-1-phenylpropene / threo product reaction:

$$\xrightarrow[\substack{\text{2) ZnCl}_2, \, (C_2H_5)_2O \\ \text{3) n-C}_3H_7CHO, \, 0\text{-}10°; \\ \overline{N}H_4Cl, \, H_2O, \, 0°}]{\substack{\text{1) CH}_3Li \, (2 \text{ equiv.}) \\ CH_3OCH_2CH_2OCH_3}}$$

Submitted by ROBERT A. AUERBACH, DAVID S. CRUMRINE, DAVID L. ELLISON, and HERBERT O. HOUSE[1]

Checked by WATARU NAGATA and NORBUHIRO HAGA

1. Procedure

Caution! Since hydrogen is liberated, this preparation should be performed in a hood.

A.   2-*Acetoxy*-trans-1-*phenylpropene*.   A dry, 500-ml., three-necked flask is fitted with a mechanical stirrer, a pressure-equalizing dropping funnel, and a rubber septum, and the apparatus is arranged so that the flask may be cooled intermittently with an ice bath. After the reaction vessel has been flushed with nitrogen (admitted through a hypodermic needle in the rubber septum) a static nitrogen atmosphere is maintained in the reaction vessel for the remainder of the reaction. The flask is charged with 35 g. of a 57% dispersion of sodium hydride (20 g., 0.83 mole) in mineral oil (Note 1). The mineral oil is washed from the hydride with 200 ml. of anhydrous pentane. The supernatant pentane layer is removed with a stainless-steel cannula inserted through the rubber septum (Note 2). The residual sodium hydride is mixed with 250 ml. of anhydrous 1,2-dimethoxyethane (Note 3) before 65 g. (0.48 mole) of phenylacetone (Note 4) is added dropwise and with stirring over 50–60 minutes. During this addition an open hypodermic needle should be inserted in the rubber septum to permit the escape of hydrogen, and intermittent cooling with an ice bath may be necessary to keep the reaction solution from boiling. The resulting mixture is stirred for 3 hours while it is allowed to cool, then the mixture is allowed to stand for approximately 2 hours, permitting the excess sodium hydride to settle. The supernatant liquid is transferred under positive nitrogen pressure through a stainless-steel cannula (Note 2) into a 1-l., three-necked flask containing 108 g. (100 ml., 1.00 mole) of cold (0°), freshly distilled acetic anhydride (b.p. 140°) and fitted with a mechanical stirrer, a thermometer, an ice bath, and a rubber septum into which are inserted a hypodermic needle to admit nitrogen and a cannula to transfer the enolate solution.

The enolate solution is added slowly with cooling and vigorous stirring so that the temperature of the reaction mixture remains below 30°. After all the supernatant enolate solution has been transferred, the residual slurry of sodium hydride is washed with an additional 50-ml. portion of anhydrous 1,2-dimethoxyethane (Note 3); these washings are also added to the acetic anhydride solution. The resulting viscous mixture is stirred at room temperature for an additional 30 minutes and poured cautiously into a mixture of 500 ml. of pentane, 500 ml. of water, and 130 g. (1.54 moles) of sodium hydrogen carbonate. When hydrolysis of the excess acetic anhydride and neutralization of the acetic acid are complete, the pentane layer is separated, and the aqueous phase is extracted with 100 ml. of pentane. The combined pentane solutions are dried over anhydrous magnesium sulfate and concentrated with a rotary evaporator. Distillation of the residual orange liquid through a 20–30-cm. Vigreux column (Note 5) provides 61.7–80.6 g. (73–95%) of 2-acetoxy-*trans*-1-phenylpropene, b.p. 82–89° (1 mm.), n_D^{25} 1.5320–1.5327 (Note 6).

B.   threo-4-*Hydroxy-3-phenyl-2-heptanone*. A dry, 500-ml., three-necked flask is fitted with a Teflon®-coated magnetic stirring bar, a gas-inlet tube equipped with a stopcock, a low-temperature thermometer, and a rubber septum and mounted to permit the use of an external cooling bath. The apparatus is flushed with nitrogen, and a static nitrogen atmosphere is maintained in the reaction vessel throughout the reaction. After 10–20 mg. of 2,2'-bipyridyl has been added to the reaction flask as an indicator, an ethereal solution containing 0.412 mole of halide-free methyllithium (Note 7) is added to the reaction flask with a hypodermic syringe or stainless-steel cannula inserted through the rubber septum. The diethyl ether is removed under reduced pressure (Note 8) while the flask is warmed to 40° with a water bath, the reaction vessel is refilled with nitrogen, and 120 ml. of anhydrous 1,2-dimethoxyethane is added (Note 3). The resulting purple solution is cooled to −10 to −20° with a 2-propanol–dry ice bath before 35.2 g. (0.200 mole) (Note 9) of 2-acetoxy-*trans*-1-phenylpropene is added from a hypodermic syringe dropwise and with stirring over 15 minutes while the temperature of the reaction mixture is kept in the range −20 to +10°. The resulting red-brown solution is stirred for an additional 10 minutes at −10 to 0° before 285 ml. of an ethereal solution containing 0.202 mole of anhydrous zinc chloride (Note 10) is added to the cold (−10 to +10°) reaction mixture from a hypodermic syringe dropwise and with stirring over 10 minutes. The reddish-yellow cloudy reaction mixture (Note 11) is stirred at 0° for 10 minutes before 14.50 g. (0.2014 mole) of freshly distilled butyraldehyde (Note 12) is added rapidly (30 seconds) and with stirring to the cold (−5 to +10°) reaction mixture. After the mixture has been stirred at 0–5° for 4 minutes, it is poured with vigorous stirring into a cold (0–5°) mixture of 500 ml. of 4 *M* ammonium chloride and 200 ml. of ether. The ether layer is separated, and the aqueous phase is extracted with two 200-ml. portions of ether. The combined organic solutions are washed successively with two 100-ml. portions of 1 *M* ammonium chloride and with two 50-ml. portions of saturated aqueous sodium chloride, and the combined aqueous washings are extracted with an additional 100-ml. portion of ether. The combined ether solutions are dried over anhydrous magnesium sulfate and concentrated under reduced pressure (water aspirator) with a rotary evaporator, removing the solvents and residual 1,2-dimethoxyethane. The residual liquid, which may crystallize on standing (Note 13), is triturated with 50 ml. of pentane, and the crystalline solid that separates is collected on a filter. The filtrate is concentrated under reduced pressure and again triturated with pentane, yielding an additional crop of the crude product. The combined crops of the crude *threo*-aldol product total 26.2–28.4 g. (64–69%), m.p. 57–62°. The crude product is recrystallized from 125–150 ml. of hexane. After the solution has been cooled to 0°, 21.3–24.1 g. of the *threo*-aldol product is collected as white needles, m.p. 71.5–72.5° (Note 14). The mother liquors are concentrated and cooled, separating additional fractions of the product (0.5–0.8 g.), m.p. 71–72°. The total yield of the *threo*-aldol product is 22.1–24.6 g. (53–60%).

2. Notes

1. The submitters used a 57% dispersion of sodium hydride in mineral oil obtained from Alfa Inorganics, Inc., and the checkers used a 50% dispersion of sodium hydride in mineral oil obtained from Metal Hydrides, Inc.

2. As a stainless-steel cannula was not available, the checkers made a minor modification in the operation without any trouble. They transferred the supernatant pentane and the solution of the sodium enolate using a Luer-lock hypodermic syringe with a stainless-steel needle preflushed with nitrogen, sweeping the apparatus with nitrogen during this operation.

3. The submitters distilled 1,2-dimethoxyethane (b.p. 83°) from lithium aluminum hydride immediately before use. The checkers distilled from sodium hydride immediately before use.

4. The submitters used a commercial sample of phenylacetone obtained from Aldrich Chemical Company, Inc.; the checkers used material of the same grade obtained from Maruwaka Chemical Industries Ltd. (Japan) without further purification.

5. The checkers used a 15 × 1 cm., unpacked, vacuum-jacketed column instead of Vigreux column for the distillation.

6. The results of GC analysis of the products made by the submitters are as follows: On a 3-m. GC column, packed with silicone fluid QF_1 supported on Chromosorb P, and heated to 190°, the product exhibits peaks at 5.8 minutes corresponding to 2–3% phenylacetone, at 7.5 minutes corresponding to 97–98% of the enol acetate (*cis* and *trans* isomers not resolved), and at 8.0 minutes corresponding to a trace (<1%) of 3-phenyl-2,4-pentanedione. On a second, 7-m. GC column, packed with silicone fluid DC-710 on Chromosorb P and heated to 190°, the product exhibits peaks at 21.0 minutes corresponding to phenylacetone, at 39.0 minutes corresponding to the *trans*-enol acetate (97–98% of the product), and at 42.2 minutes corresponding to the *cis*-enol acetate (2–3% of the product). The checkers used a 45 m. × 0.25 mm. stainless-steel column (Golay type) coated with Apiezon L, heated to 150° and swept with helium at 1.5 kg./cm.[2] The product exhibits peaks at 5.5 minutes corresponding to phenylacetone (2–3% of product), at 14.2 minutes corresponding to the *trans*-enol acetate (91–92% of the product), and at 15.8 minutes corresponding to the *cis*-enol acetate (5–6% of the product). The product has IR absorption (CCl_4) at 1765 (enol ester C=O) and 1685 cm^{-1} (enol ester C=C) with UV maxima (95% C_2H_5OH) at 248.5 nm (ϵ 18,000) and 325 nm (ϵ 415) and ^1H NMR peaks (CCl_4) at δ 2.01 (partially resolved m, 6H, CH_3CO and vinyl CH_3), 5.82 (partially resolved m, 1H, vinyl CH), and 7.0–7.4 (m, 5H, C_6H_5). The mass spectrum of the product has a parent ion at m/e 176 with abundant fragment peaks at m/e 134, 91, 45, 43, and 39.

7. The submitters used an ether solution of halide-free methyllithium, purchased from Foote Mineral Company, while the checkers prepared the compound from methyl chloride and lithium metal in ether according to the literature.[2] The solution was standardized before use by the titration procedure described in *Org. Synth., Coll. Vol. 6,* 121 (1988). The checkers observed that use of a halide-containing ether solution of methyllithium resulted in a considerable decrease in yield of the product, principally due to difficulty in following the subsequent procedure described in the text.

8. A convenient apparatus for evacuating the reaction vessel and refilling it with nitrogen is described in *Org. Synth., Coll. Vol. 6,* 121 (1988).

9. If the violet color of the reaction solution is completely discharged, indicating that all the methyllithium has been consumed, addition of the enol acetate should be stopped at that point. The actual concentration of enolate anion in the solution can be calculated from the amount of enol acetate added.

10. To prepare an ethereal solution of anhydrous zinc chloride (m.p. 283°), the submitters placed 50.0 g. (0.369 mole) of pulverized zinc chloride, obtained from either Mallinckrodt Chemical Works or Fisher Scientific Company, in a 1-l., round-bottomed flask, and the vessel was evacuated to about 1 mm. pressure. The flask was heated strongly with a burner with swirling until as much of the solid as practical had been melted. The evacuated flask was cooled and shaken *(Caution! Perform this operation behind a safety shield in a hood and with heavy gloves to protect the operator's hands in case the flask should implode)* to break up the large lumps of zinc chloride. This fusion under reduced pressure should be repeated three times. To the resulting anhydrous zinc chloride was added 500 ml. of anhydrous diethyl ether, freshly distilled from lithium aluminum hydride. The mixture was refluxed for 3 hours under a static nitrogen atmosphere and allowed to stand until the undissolved solid had settled. The resulting supernatant solution was transferred with a stainless-steel cannula under positive nitrogen pressure (Note 2) into a second dry flask or Schlenk tube capped with a rubber septum. Aliquots of this solution, diluted with aqueous ammonia, can be titrated with standard EDTA solution to a Erichrome Black T endpoint to determine the zinc content.[3] Alternatively, the chloride ion concentration of aliquots can be determined by a Volhard titration. Typical values found for these ether solutions are 0.73 M in zinc ion and 1.38 M in chloride ion, or 0.69–0.73 M in zinc chloride. If the final solution is significantly less concentrated than 0.7 M in zinc chloride, it is probable that the dehydration of the solid zinc chloride was not complete. In this event, the submitters recommend that a fresh solution of zinc chloride be prepared with more attention to the initial dehydration of the solid zinc chloride. The checkers used pulverized zinc chloride, obtained from Wako Pure Chemical Industries Ltd. (Japan).

11. The white precipitate that separates is a part of the lithium chloride formed in the reaction mixture. Separation of the material is not necessary.

12. The submitters used a commercial grade of butyraldehyde from Eastman Organic Chemicals; the checkers used butyraldehyde of the same grade from Wako Pure Chemical Industries Ltd. (Japan) and distilled it before use, b.p. 72–74°.

13. The ^{1}H NMR spectrum (C_6D_6) of the crude product exhibits benzylic CH doublets at δ 3.42 (J = 5.3 Hz., attributable to 4–10% of the *erythro* aldol isomer) and 3.58 (J = 9.4 Hz., attributable to 90–96% of the *threo*-aldol isomer). This mixture may be

separated by chromatography on acid-washed silicic acid, permitting the isolation of both the *threo* and the *erythro* diastereoisomers.[4]

14. The *threo*-hydroxy ketone exhibits IR absorption (CCl_4) at 3540 (associated OH) and 1705 cm.$^{-1}$ (C=O) with a series of weak (ϵ 300 or less) UV maxima (95% C_2H_5OH) in the region 240–270 nm as well as a maximum at 286 nm (ϵ 345). The ^1H NMR spectrum (CCl_4) of the product shows resonance at δ 0.6–1.9 [m, 7H, $(CH_2)_2CH_3$], 2.03 (s, 3H, CH_3CO), 3.35 (s, 1H, O*H*), 3.65 (d, J = 9.5 Hz., 1H, benzylic C*H*), 4.0–4.4 (m, 1H, C*H*O), and 7.1–7.5 (m, 5H, C_6H_5). The mass spectrum of the product exhibits the following relatively abundant peaks: m/e (relative intensity), 206 (M$^+$, 0.1), 188 (8), 146 (20), 135 (26), 134 (100), 117 (52), 92 (48), 91 (76), 65 (31), 44 (36), and 43 (60).

3. Discussion

The present procedures illustrate general methods for the use of preformed lithium enolates[5] as reactants in the aldol condensation[4] and for quenching alkali metal enolates in acetic anhydride, forming enol acetates with the same structure and stereochemistry as the starting metal enolate.[6] The aldol product, *threo*-4-hydroxy-3-phenyl-2-heptanone, has been prepared only by this procedure.

The methods previously used to obtain single aldol products (or their dehydrated derivatives) from reactants where several aldol products are possible[7] include the reaction of bromozinc enolates, from α-bromoketones, with aldehydes;[8] the reaction of bromo-magnesium enolates, from either α-bromoketones, ketones and bromomagnesium amides or sterically hindered Grignard reagents, with aldehydes;[9,10] and the reaction of α-lithio derivatives of imines with aldehydes or ketones.[11] Like the present procedure, each of these methods relies upon trapping the intermediate β-keto alkoxide derivative as a metal chelate in an aprotic reaction solvent. The present procedure increases the versatility of the aldol condensation by utilizing the variety of specific lithium enolates that can be generated from unsymmetrical ketones.[5] In this procedure the lithium enolate is treated successively with anhydrous zinc chloride and an aldehyde, forming the zinc(II) chelate of a β-keto alkoxide. The optimum quantity of zinc chloride is that amount required to form zinc(II) salts of all strong bases in the reaction mixture. Thus, 1 mole of zinc chloride should be added for each mole of lithium enolate (and accompanying lithium *tert*-butoxide) formed from an enol acetate as in the present example. If the lithium enolate is formed from the ketone and lithium diisopropylamide or from a trimethylsilyl enol ether and methyllithium, then 0.5 mole of zinc chloride should be used for each mole of lithium enolate. The optimum reaction solvent is either ether or ether–1,2-dimethoxyethane mixtures, with a reaction temperature of -10 to $+10°$ and a reaction time of 2–5 minutes. Longer reaction times and higher reaction temperatures may lead to a variety of by-products resulting from polycondensation and dehydration. The aldol products are efficiently isolated by *adding the reaction mixtures* to a cold (0–5°), aqueous solution of ammonium chloride followed by rapid separation of the aldol products. Since many of the aldol products are especially prone to epimerization, dehydration, or reversal of the aldol condensation, they should not be exposed to strong acids or strong bases. Mixtures of stereoisomeric aldol products with similar physical properties can usually be separated by chromatography on acid-washed silicic acid.[4,12]

In several cases (including the present example) where diastereoisomeric aldol pro-

TABLE I[4]

DIRECTED ALDOL CONDENSATIONS WITH PREFORMED LITHIUM ENOLATES IN
THE PRESENCE OF ZINC CHLORIDE

Enolate	Aldol Product	Yield, %	Stereoisomer ratio, *threo/erythro*
$(CH_3)_3C-\overset{O^-\ Li^+}{\underset{\|}{C}}=CH_2$	$(CH_3)_3CCOCH_2\overset{OH}{\underset{\|}{C}}HC(CH_3)_3$	82	—
		76	4/1
$C_6H_5CH=\overset{O^-\ Li^+}{\underset{\|}{C}}-CH_3$	$C_6H_5\overset{COCH_3}{\underset{\|}{C}}H-\underset{\underset{OH}{\|}}{C}H-C_3H_7\text{-}n$	80	9/1
	$(CH_3)_3C-$	84	2/1[a]
$n\text{-}C_4H_9CH=\overset{O^-\ Li^+}{\underset{\|}{C}}-CH_3$	$n\text{-}C_4H_9\overset{COCH_3}{\underset{\|}{C}}H-\underset{\underset{OH}{\|}}{C}H-C_6H_5$	80	1/1

[a] The aldol product contained 70% of isomers with an axial α-hydroxybenzyl substituent.

ducts are possible, there is a preference for the formation of the *threo*-diastereoisomer. This stereochemical preference presumably arises because the six-membered cyclic zinc chelate of the *threo*-isomer can exist in a chair conformation with both substituents in equatorial positions. Table I summarizes the results obtained from several aldol condensations performed by the present procedure.

1. School of Chemistry, Georgia Institute of Technology, Atlanta, Georgia 30332.
2. H. J. Berthold, G. Groh, and K. Stoll, *Z. Anorg. Allg. Chem.*, **37**, 53 (1969).

3. W. Biederman and G. Schwarzenbach, *Chimia*, **2**, 56 (1948); G. Schwarzenbach and H. Flaschka, "Complexometric Titrations," 2nd. ed., Methuen, London, 1969, p. 260.

4. H. O. House, D. S. Crumrine, A. Y. Teransishi, and H. D. Olmstead, *J. Am. Chem. Soc.*, **95**, 3310 (1973).

5. For examples and leading references, see H. O. House, M. Gall, and H. D. Olmstead, *J. Org. Chem.*, **36**, 2361 (1971).

6. H. O. House, R. A. Auerbach, M. Gall, and H. D. Olmstead, *J. Org. Chem.*, **38**, 514 (1973).

7. For a general review of the aldol condensation, see A. T. Nielsen and W. J. Houlihan, *Org. React.*, **16**, 1 (1968).

8. T. A. Spencer, R. W. Britton, and D. S. Watt, *J. Am. Chem. Soc.*, **89**, 5727 (1967).

9. A. T. Nielsen, C. Gibbons, and C. A. Zimmerman, *J. Am. Chem. Soc.*, **73**, 4696 (1951).

10. J. E. Dubois and P. Fellmann, *C. R. Hebd. Seances Acad. Sci. Ser. C*, **274**, 1307 (1972).

11. G. Wittig and A. Hesse, *Org. Synth.*, **Coll. Vol. 6**, 901 (1988); G. Wittig and H. Rieff, *Angew. Chem. Int. Ed. Engl.* **7**, 7 (1968).

12. H. Brockmann and H. Muxfeldt, *Chem. Ber.*, **89**, 1379 (1956).

MACROLIDES FROM CYCLIZATION OF ω-BROMOCARBOXYLIC ACIDS: 11-HYDROXYUNDECANOIC LACTONE

(Oxacyclododecan-2-one)

$$Br(CH_2)_{10}CO_2H \xrightarrow[\substack{\text{dimethyl sulfoxide,} \\ 100°·}]{K_2CO_3} (CH_2)_{10} \begin{array}{c} O \\ | \\ C=O \end{array}$$

Submitted by C. GALLI and L. MANDOLINI[1]
Checked by KAORU MORI and CARL R. JOHNSON

1. Procedure

A 1-l., three-necked, round-bottomed flask equipped with an internal thermometer, mechanical stirrer, dropping funnel, and calcium chloride drying tube is charged with 500 ml. of dimethyl sulfoxide and 15 g. (0.11 mole) of potassium carbonate (Note 1). The mixture is heated to 100°, and a solution of 10.0 g. (0.0377 mole) of 11-bromoundecanoic acid (Note 2) in 200 ml. of dimethyl sulfoxide is added dropwise with vigorous stirring over 1 hour. After cooling at room temperature, the mixture is decanted and filtered free of any suspended solid material (Note 3) through a Büchner funnel with occasional suction. The solid residue is rinsed with 50 ml. of dimethyl sulfoxide, and the washings are added to the original filtrate. The resulting clear solution is diluted with 250 ml. of water and extracted with three 250-ml. portions of petroleum ether. The combined organic layers are washed with 200 ml. of water, dried over anhydrous sodium sulfate, and concentrated, leaving *ca.* 7 g. of crude material. Simple distillation at reduced pressure from a small Claisen flask yields 5.5–5.8 g. (79–83%) of pure 11-hydroxyundecanoic lactone as a colorless, musk-smelling liquid, b.p. 124–126° (13 mm.), n_D^{19} 1.4721 (Notes 4 and 5). The residue is ground with 5 ml. of hexane and filtered, affording 0.4–0.7 g.

(6–10%) of the 24-membered dilactone 1,13-dioxacyclotetracosane-2,14-dione as white crystals, m.p. 71.5–72° (from hexane) (Note 6).

2. Notes

1. Reagent grade dimethyl sulfoxide and anhydrous potassium carbonate were used.

2. 11-Bromoundecanoic acid, available from Aldrich Chemical Company, Inc., was used without further purification.

3. Filtration is optional; however, it does reduce the extent of emulsion formation during the subsequent extractions.

4. The submitters report that the pure lactone and dilactone can also be obtained by chromatography of the crude product on silica gel with chloroform as the eluant.

5. The product is pure by GC and TLC; IR (CCl$_4$) cm^{-1}: 1740; ^{1}H NMR (CCl$_4$), δ (multiplicity, number of protons, assignment): 1.2–1.8 (m, 16H), 2.30 (broad t, 2H, CH_2CO), 4.14 (broad t, 2H, CH_2O).

6. Stoll and Rouvè[2] report m.p. 71.5–72°.

3. Discussion

Available methods for the synthesis of macrolides include the cyclization of long-chain bifunctional precursors,[3] depolymerization processes,[4] ring-enlargement reactions,[5] and special methods such as the thermal decomposition of tricycloalkylidene peroxides.[6] The method reported here is essentially that of the submitters.[7] Its improvements result from a quantitative approach to the cyclization of a series of ω-bromo fatty acids under conditions well defined from the kinetic point of view. A unique feature of this procedure in comparison with other methods involving cyclization of α,ω-bifunctional precursors, which are generally run under Ziegler's high-dilution conditions, is that high rates of feed can be used, so that the special devices usually employed for the slow addition of the reagent into the reaction medium are not required. The synthesis is characterized by relatively mild reaction conditions and simple work-up. Moreover, it is suited for relatively large-scale preparations. Up to 50 g. of 11-bromoundecanoic acid can be cyclized in more than 70% yield in a single run, employing no more than 1 l. of solvent and an addition time of 3–4 hours.

The reaction illustrates a typical preparation of a macrolide. Lactones with more than 12 members can be obtained in even better yields. For example, 15-hydroxypentadecanoic lactone (m.p. 35–37°) and 17-hydroxyheptadecanoic lactone (m.p. 40–41°) were prepared by the submitters in about 95% yield, practically pure, with no trace of the corresponding dilactones.

Recent progress in chemistry and biochemistry of macrolides was recently reviewed.[8]

1. Centro di Studio sui Meccanismi di Reazione del Consiglio Nazionale delle Ricerche, c/o Istituto di Chimica Organica, Università di Roma, Piazzale Aldo Moro, 5, 00185, Rome, Italy.
2. M. Stoll and R. Rouvè, *Helv. Chim. Acta,* **18,** 1119 (1935).
3. E. J. Corey and K. C. Nicolau, *J. Am. Chem. Soc.,* **96,** 5614 (1974), and references therein.
4. E. W. Spanagel and W. H. Carothers, *J. Am. Chem. Soc.,* **58,** 654 (1936).
5. I. J. Borowitz, V. Bandurco, M. Heyman, R. D. G. Rigby, and S. Ueng, *J. Org. Chem.,* **38,** 1234 (1973); C. H. Hassall, *Org. React.,* **9,** 73 (1957).

6. P. R. Story and P. Busch, *Adv. Org. Chem.*, **8**, 67 (1972).

7. C. Galli and L. Mandolini, *Gazz. Chim. Ital.*, **105**, 367 (1975).

8. S. Masamune, G. S. Bates, and J. W. Corcoran, *Angew. Chem. Int. Ed. Engl.*, **16**, 585 (1977); K. C. Nicolau, *Tetrahedron*, **33**, 683 (1977); T. G. Back, *Tetrahedron*, **33**, 3041 (1977).

DIRECT IODINATION OF POLYALKYLBENZENES: IODODURENE

(Benzene, 3-iodo-1,2,4,5-tetramethyl-)

$$7 \quad \begin{array}{c} H_3C \\ H_3C \end{array} \bigcirc \begin{array}{c} CH_3 \\ CH_3 \end{array} + 3I_2 + HIO_4 \cdot 2H_2O$$

$$\downarrow$$

$$7 \quad \begin{array}{c} I \\ H_3C \\ H_3C \end{array} \bigcirc \begin{array}{c} CH_3 \\ CH_3 \end{array} + 6H_2O$$

Submitted by H. Suzuki[1]

Checked by Robert E. Ireland and Robert Czarny

1. Procedure

A 200-ml., three-necked flask equipped with a reflux condenser, a thermometer, a glass stopper, and a magnetic stirring bar is charged with 13.4 g. (0.101 mole) of durene (Note 1), 4.56 g. (0.0215 mole) of periodic acid dihydrate, and 10.2 g. (0.0402 mole) of iodine. A solution of 3 ml. of concentrated sulfuric acid and 20 ml. of water in 100 ml. of glacial acetic acid is added to this mixture. The resulting purple solution is heated at 65–70° with stirring for approximately 1 hour until the color of iodine disappears. The reaction mixture is diluted with approximately 250 ml. of water, and the white-yellow solid that separates (Note 2) is collected on a Büchner funnel and washed three times with 100-ml. portions of water. The product is dissolved in a minimum amount of boiling acetone (about 125 ml. is required); the solution is cooled to room temperature and subsequently stored overnight in a refrigerator. The product is collected by rapid filtration through a Büchner funnel, yielding 20.8–22.6 g. (80–87%) of iododurene as colorless, fine needles, m.p. 78–80°.

TABLE I
IODOARENES PREPARED FROM THE CORRESPONDING ARENES

Iodoarenes	Yield, %	Iodoarenes	Yield, %
	10 81		14 94
	10 89		14 84
	10 85[a]		14 69
	10 85		14 86

TABLE I (*continued*)

Iodoarenes Prepared from the Corresponding Arenes

Iodoarenes	Yield, %	Iodoarenes	Yield, %
10	84	**13**	72
11	86	**14**	85[b]
12	85		

[a] Based on unrecovered hydrocarbon. The reaction proceeds quite slowly.
[b] A solution of periodic acid in acetic acid is added dropwise, with stirring, to a mixture of carbazole, iodine, and 80% (v/v) acetic acid. It is necessary to separate the product from colored substance by chromatography over alumina, using benzene as the eluant.

2. Notes

1. Durene (m.p. 79–80°), prepared according to the procedure in *Org. Synth.*, **Coll. Vol. 2**, 248 (1943), was used by the submitter. Commercially available durene, which melted at 79 – 80° after purification by the *Org. Synth.* procedure above, was used by the checkers.

2. Some crystals of iododurene that have formed during the heating period tend to take on a purple coloration because of occluded iodine. This impurity is readily removed by the recrystallization procedure.

3. Discussion

The present procedure is the most convenient method for the direct, high-yield preparation of mono-, di-, or triiodo derivatives from various polyalkylbenzenes and their derivatives. It is also applicable to some moderately activated heteroaromatic systems. However, the reaction fails with compounds bearing strongly deactivating substituent groups. Shorter reaction times and higher degree of product purity are assured by the use of periodic acid as an oxidizing agent. A feature of the reagent is that iodine is oxidized by periodic acid and periodic acid is reduced by iodine, both forming an active iodinating species which reacts with an aromatic compound, eventually leading to the formation of only the desired iodination product and water. A brief review of the iodine/periodic acid reagent has recently appeared.[2] The preparation of tetraiodo and more highly iodinated derivatives of alkylbenzenes by this procedure is difficult, and the Jacobsen reaction for the disproportionation of diiodo compounds by the action of sulfuric acid is preferred.[3] Polyiodo derivatives are useful for the characterization of polyalkylbenzenes and their derivatives, liquids available only in a small quantities, since the introduction of iodine atoms increases the molecular weight and converts a liquid hydrocarbon into a highly crystalline solid with a moderate melting range.[4] Table I illustrates the iodoarenes prepared from the corresponding arenes by conditions similar to that described herein.

Iododurene has been prepared by treatment of durene with iodine and mercury(II) oxide,[5] sulfur iodide and nitric acid,[6] iodine and zinc chloride[7] or copper(II) chloride,[8] or iodoanisole and sulfuric acid.[9]

1. Department of Chemistry, Kyoto University, Kyoto 606, Japan. [Present address: Department of Chemistry, Ehime University, Matsuyama 790, Japan.]
2. A. J. Fatiadi, in J. S. Pizey, Ed., "Synthetic Reagents," Vol. 4, Halsted Press, Wiley, New York, 1981, pp. 184–187.
3. H. Suzuki and R. Goto, *Bull. Chem. Soc. Jpn.*, **36**, 389 (1963).
4. H. Suzuki and Y. Haruta, *Bull. Chem. Soc. Jpn.*, **46**, 589 (1973).
5. A. Töhl, *Ber. Dtsch. Chem. Ges.*, **25**, 1521 (1892).
6. A. Edinger and P. Goldberg, *Ber. Dtsch. Chem. Ges.*, **33**, 2875 (1900).
7. R. M. Keefer and L. J. Andrews, *J. Am. Chem. Soc.*, **78**, 5623 (1956).
8. W. C. Baird and J. H. Surridge, *J. Org. Chem.*, **35**, 3436 (1970).
9. H. Suzuki, T. Sugiyama, and R. Goto, *Bull. Chem. Soc. Jpn.*, **37**, 1858 (1964).
10. H. Suzuki, K. Nakamura, and R. Goto, *Bull. Chem. Soc. Jpn.*, **39**, 128 (1966).
11. H. Suzuki, *Bull. Chem. Soc. Jpn.*, **44**, 2871 (1971).
12. H. Suzuki, K. Ishizaki, and T. Hanafusa, *Bull. Chem. Soc. Jpn.*, **48**, 2609 (1975).

13. H. Suzuki, T. Iwao, and T. Sugiyama, *Bull. Inst. Chem. Res., Kyoto Univ.*, **52**, 561 (1974).
14. H. Suzuki and Y. Tamura, *Nippon Kagaku Zasshi*, **92**, 1021 (1971).

trans-IODOPROPENYLATION OF ALKYL HALIDES:
(*E*)-1-IODO-4-PHENYL-2-BUTENE

[2-Butene, 1-iodo-4-phenyl, (*E*)-]

Submitted by KOICHI HIRAI[1] and YUKICHI KISHIDA
Checked by S. F. MARTIN and G. BÜCHI

1. Procedure

Caution! Benzene has been identified as a carcinogen; OSHA has issued emergency standards on its use. All procedures involving benzene should be carried out in a well-ventilated hood, and glove protection is required.

Hexamethylphosphoric triamide (HMPA) vapors have been reported to cause cancer in rats.[2] All operations with hexamethylphosphoric triamide should be performed in a good hood, and care should be taken to keep the liquid off the skin.

Methyl iodide, in high concentrations for short periods or in low concentrations for long periods, can cause serious toxic effects in the central nervous system. Accordingly, the American Conference of Governmental Industrial Hygienists[3] has set 5 p.p.m., a level which cannot be detected by smell, as the highest average concentration in air to which workers should be exposed for long periods. The preparation and use of methyl iodide should always be performed in a well-ventilated fume hood. Since the liquid can be absorbed through the skin, care should be taken to prevent contact.

A. *4,5-Dihydro-2-(2-propenylthio)thiazole.* A solution of 11.9 g. (0.100 mole) of 2-mercapto-2-thiazoline (Note 1) in 60 ml. of tetrahydrofuran is prepared in a 200-ml., one-necked, round-bottomed flask fitted with a 25-ml., pressure-equalizing dropping funnel. Allyl bromide (12.1 g., 8.74 ml., 0.100 mole) is added in one portion at room temperature, and 5 minutes later 10.1 g. (13.9 ml., 0.100 mole) of triethylamine is added dropwise over a 10-minute period. After the dropping funnel has been replaced with a condenser, the mixture is refluxed gently for 4 hours. The resulting slurry is cooled and filtered, removing triethylamine hydrobromide, which is washed with 20 ml. of fresh tetrahydrofuran, and the combined organic solutions are concentrated with a rotary evaporator. The residue is dissolved in 100 ml. of diethyl ether, and this solution is washed with three 20-ml. portions of 5% aqueous potassium hydroxide and two 20-ml. portions of water. After drying over anhydrous magnesium sulfate, the ethereal solution is filtered and concentrated with a rotary evaporator. Vacuum distillation of the residue gives 10.9–11.1 g. (69–70%) of 4,5-dihydro-2-(2-propenylthio)thiazole, b.p. 51–54° (0.02 mm., bath temperature 75–80°, Note 2); n_D^{20} 1.5864 (Note 3).

B. *4,5-Dihydro-2-[(1-phenylmethyl-2-propenyl)thio]thiazole.* A dry, 100-ml., four-necked, round-bottomed flask is fitted with a thermometer, a rubber septum, a pressure-equalizing dropping funnel, and a mechanical stirrer (Note 4). A positive pressure of nitrogen, applied either through an adapter inserted in the top of the funnel or through a needle inserted in the septum, is maintained throughout the reaction. A solution of 2.0 g. (0.012 mole) of the thiazole prepared in part A in 24 ml. of dry tetrahydrofuran (Note 5) is placed in the flask, stirred vigorously, and cooled in an acetone–dry ice bath. When the internal temperature reaches −55°, 6.0 ml. (0.013 mole) of a 2.1 *M* solution of *n*-butyllithium in *n*-hexane (Note 6) is added by syringe at a rate such that the temperature remains below −55°; approximately 5 minutes is required. After stirring for an additional 20 minutes at −55° to −60°, a solution of 2.15 g. (0.0125 mole) of benzyl bromide (Note 7) in 2 ml. of dry tetrahydrofuran is added at a rate such that the temperature remains below −55°, which again requires approximately 5 minutes. The solution is stirred for another 50 minutes at −55° to −60°, allowed to warm to 0° over 30 minutes, and poured onto 70 ml. of ice water. The resulting mixture is extracted with three 25-ml. portions of ethyl acetate; the extracts are combined, washed with two 15-ml. portions of saturated aqueous sodium chloride, and dried over anhydrous magnesium sulfate.

Removal of solvent from the extracts leaves a residue that is purified by dry-column chromatography.[4] The residue is dissolved in 40 ml. of acetone in a 300-ml., round-bottomed flask, 30 g. of silica gel (Note 8) is added, and the acetone is removed with a rotary evaporator. The resulting solid mixture is placed on top of 360 g. of dry silica gel (Note 8) packed in flexible nylon tubing (Note 9), and the column is developed with 420 ml. of 10:1 (v/v) benzene–acetone. Approximately 150 ml. of solvent drips from the bottom of the column toward the end of development; this eluent is collected in 25-ml. fractions and checked for product by TLC (Note 10). The column itself is then cut into 2-cm. sections, the silica gel in each section is eluted with three 25-ml. portions of ethyl acetate, and the eluent from each section is analyzed by TLC (Note 10). Combination of all the product-containing fractions yields 1.2–1.5 g. (40–47%) of the benzylated compound as an oil, n_D^{20} 1.6083 (Notes 11 and 12).

C. (E)-1-*Iodo-4-phenyl-2-butene*. A 20-ml., round-bottomed flask is charged with 2.0 g. (0.008 mole) of the thiazole prepared in Part B, 5 ml. of methyl iodide, and 2 ml. of N,N-dimethylformamide. The resulting solution is heated at 75–80° for 2.5 hours under a nitrogen atmosphere (Note 13), cooled, and poured into 10 ml. of water. Extraction with three 12-ml. portions of ether separates the product from water-soluble by-products. The extracts are combined, washed with 8 ml. of 1% aqueous sodium thiosulfate and two 8-ml. portions of water, dried over anhydrous magnesium sulfate, and filtered. Removal of ether by distillation at 30° (100 mm.) leaves 1.5–1.7 g. (74–82%) of (E)-1-iodo-4-phenyl-2-butene (Notes 14 and 15).

2. Notes

1. This product was purchased from the Aldrich Chemical Company, Inc., and used without further purification.

2. Rapid distillation is required to avoid a [3,3]-sigmatropic rearrangement, which gives N-allylthiazolidine-2-thione.

3. IR (neat) cm.$^{-1}$: 1570, 995, 965, 920; ^{1}H NMR (CDCl$_3$), (multiplicity, coupling constant J in Hz., number of protons, assignment): 3.39 (t, $J = 7$, 2H, ring CH$_2$), 3.75 (d, $J = 6$, 2H, allylic CH$_2$), 4.20 (t, $J = 7$, 2H, ring CH$_2$), 5.0–5.4 (m, 2H, CH=CH$_2$), 5.6–6.3 (m, 1H, CH=CH$_2$).

4. A three-necked flask may be used if magnetic stirring is substituted for mechanical stirring or if benzyl bromide is added with a syringe instead of a dropping funnel.

5. Tetrahydrofuran was distilled from lithium aluminum hydride immediately prior to use. See *Org. Synth., Coll. Vol. 5*, 976 (1973) for precautions.

6. This product was purchased from Sankyo Kasei, Inc., Tokyo (submitters) and Ventron Corporation (checkers).

7. This product was purchased from the Aldrich Chemical Company, Inc., and distilled prior to use, b.p. 126–128° (80 mm.).

8. Woelm silica gel for dry-column chromatography, activity III/30 mm. (according to Brockmann and Schodder,[5]) was supplied by M. Woelm, Eschwege, Germany.

9. Woelm nylon column DCC-5 was used, giving a packed column 66–67 cm. high and 32 mm. in diameter.

10. Merck precoated silica gel F$_{254}$ plates, layer thickness 0.25 mm., were used. Developed with 10:1 (v/v) benzene–acetone and visualized with UV light, the product appears at R_f 0.58–0.67. Normally the product is found in the lower third of the column, and occasionally some is found in the last fractions of eluent collected during development. However, since the exact position of this material on the column depends critically on the way in which the column is packed, a thorough check of all fractions is advisable.

11. IR (neat) cm.$^{-1}$: 1570, 995, 965, 920, 740, 700; ^{1}H NMR (CDCl$_3$), δ (multiplicity, coupling constant J in Hz., number of protons, assignment): 2.9–3.4 (m, 4H, ring CH$_2$ and C$_6$H$_5$CH$_2$), 4.14 (t, $J = 8$, 2H, ring CH$_2$), 4.3–4.7 (m, 1H, allylic CH), 4.9–5.3 (m, 2H, CH$_2$=CH), 5.5–6.2 (m, 1H, CH$_2$=CH), 7.23 (broad s, 5H, C$_6$H$_5$).

12. The submitters also tried to purify the crude product by distillation (130–145° at 0.005 mm.), but under these conditions decomposition occurred. It is possible to substitute thick-layer chromatography for the dry column. Using Merck silica gel F_{254} precoated plates, layer thickness 2 mm., and developing with 10 : 1 (v/v) benzene–acetone, the submitters report a 73% yield of pure product.

13. *N*-Methyl-2-methylthiothiazolium iodide (m.p. 132°) usually precipitates as the reaction proceeds.

14. (*E*)-1-Iodo-4-phenyl-2-butene is reported to decompose on attempted distillation at 4 mm.[6] The crude product, which is suitable for subsequent reactions (see Note 15), may be purified by thick-layer chromatography. Using the plates described in Note 10 and developing with hexane, the product is found at R_f 0.5 as an oil, n_D^{20} 1.5940; IR (neat) cm.$^{-1}$: 1655 weak, 1600 medium, 1460 strong, 1150 strong, 960 strong (*trans*-CH=CH—), 740 strong, 690 strong; ^{1}H NMR (CDCl$_3$), δ (multiplicity, number of protons, assignment): 3.2–3.5 (m, 2H, C$_6$H$_5$C*H*$_2$), 3.7–4.0 (m, 2H, C*H*$_2$I), 5.6–5.9 (m, 2H, vinylic C*H*), 7.23 (broad s, 5H, C$_6$*H*$_5$).

15. This material may be converted directly to a phosphonium salt: 1.40 g. (0.00545 mole) of the crude iodide is dissolved in 20 ml. of benzene, and 1.42 g. (0.00542 mole) of triphenylphosphine is added. On standing, 2.5 g. (77%) of the triphenylphosphonium salt precipitates as a colorless 1 : 1 complex with benzene, m.p. 135–137°. Recrystallization from methanol–benzene raises the melting point to 140–142°. Analysis calculated for C$_{28}$H$_{29}$PI·C$_6$H$_6$: C, 68.23; H, 5.39. Found: C, 68.15; H, 5.28.

3. Discussion

(*E*)-1-Iodo-4-phenyl-2-butene has been prepared previously by addition of phenyl and chloro units [generated by decomposition of phenyldiazonium chloride in the presence of a catalytic amount of copper(II) chloride] across the conjugated system of butadiene, followed by treatment with ethanolic potassium iodide.[6]

The present preparation illustrates a general and convenient method for the *trans*-iodopropenylation of an alkyl halide.[7] The iodopropenylated material is not usually stable but is a useful synthetic intermediate. For example, it forms a stable crystalline triphenylphosphonium salt for use in the Wittig reaction, and under Kornblum reaction conditions it gives an (*E*)-α,β-unsaturated aldehyde. In addition to the phosphonium salt described in Note 15, the following have been prepared: (4-methoxyphenyl-2-butenyl)triphenylphosphonium iodide, m.p. 123–127°; (2-octenyl)triphenylphosphonium iodide, m.p. 98°; and (2-octadecenyl)triphenylphosphonium iodide, m.p. 50°.

The alkylation of 4,5-dihydro-2-(2-propenylthio)thiazole is noteworthy, since in general the coupling of an alkyllithium and an alkyl halide gives many by-products due to halogen–metal interconversion.[8] In the present case, alkylation α to sulfur occurred cleanly and may be attributed to a five-membered chelating effect.[9] In some cases, addition of 1/10–1/20 volume of hexamethylphosphoric triamide to the tetrahydrofuran solution facilitates the alkylation. Representative alkyl halides examined and the yields of products isolated are as follows: (a) amyl bromide, (63%); (b) decyl bromide, (70%); (c) anisyl chloride, (57%); (d) phenethyl bromide, (59%); and (e) cyclohexyl bromide, (52%).[7,9a] The same type of alkylation occurred successfully with anions of the 4,5-

dihydro derivatives of 2-(methylthio)thiazole, 2-[(3-phenyl-2-propenyl)thio]thiazole, and 2-(phenylmethylthio)thiazole, but attempts to alkylate 4,5-dihydro-2-(ethylthio)thiazole, 2-(methylthio)- or 2-(2-propenylthio)benzo[d]thiazole were unsuccessful.

The final step, C—S bond cleavage with allylic rearrangement, incorporates two useful features. First, it is stereospecific, producing only the (E)-iodopropenylated product. Second, the sulfur-containing moiety is converted to a water-soluble product; thus, the desired material may be isolated in reasonable purity by a simple water–ether partitioning of the crude reaction mixture. The present procedure represents an improvement over a previous iodomethylation sequence (lithiation of thioanisole, alkylation, and cleavage with methyl iodide and sodium iodide in N,N-dimethyl formamide at 75°), in which the product must be separated from thioanisole.[10]

1. Central Research Laboratories, Sankyo Company Ltd., 1-2-58 Hiromachi, Shinagawa-ku, Tokyo 140, Japan.
2. J. A. Zapp, Jr., *Science*, **190**, 422 (1975).
3. American Conference of Governmental Industrial Hygienists (ACGIH), "Documentation of Threshold Limit Values," 3rd ed., Cincinnati, Ohio, 1971, p. 166.
4. B. Loev and M. M. Goodman, *Chem. Ind. (London)*, 2026 (1965).
5. H. Brockman and H. Schodder, *Ber. Dtsch. Chem. Ges.*, **74B**, 73 (1941).
6. A. V. Dombrovskĭi and A. P. Terent'ev, *Zh. Obshch. Khim.*, **26**, 2776 (1956) [*Chem. Abstr.* **51**, 7337d (1957)].
7. K. Hirai and Y. Kishida, *Tetrahedron Lett.*, 2743 (1972).
8. R. G. Jones and H. Gilman, *Org. React.*, **6**, 339 (1951).
9. (a) K. Hirai, H. Matsuda, and Y. Kishida, *Tetrahedron Lett.*, 4359 (1971); (b) T. Mukaiyama, K. Narasaka, K. Maekawa, and M. Furusato, *Bull. Chem. Soc.*, **44**, 2285 (1971).
10. E. J. Corey and M. Jautelat, *Tetrahedron Lett.*, 5787 (1968).

2-IODO-*p*-XYLENE

(Benzene, 2-iodo-1,4-dimethyl)

$$\text{CH}_3\text{-benzene-CH}_3 + \text{Tl(OCOCF}_3)_3 \xrightarrow{25^\circ} \text{CH}_3\text{-benzene-Tl(OCOCF}_3)_2\text{-CH}_3 + \text{CF}_3\text{COOH}$$

$$\text{CH}_3\text{-benzene-Tl(OCOCF}_3)_2\text{-CH}_3 + 2\text{ KI} \xrightarrow{0^\circ} \text{CH}_3\text{-benzene-I-CH}_3 + \text{TlI} + 2\text{CF}_3\text{COOK}$$

Submitted by Edward C. Taylor,[1] Frank Kienzle,[1] and Alexander McKillop[2]
Checked by Gordon S. Bates and S. Masamune

1. Procedure

Caution! Thallium salts are very toxic. This preparation should be carried out in a well-ventilated hood. The operator should wear rubber gloves. For disposal of thallium wastes, see Note 1 in Org. Synth., **Coll. Vol. 6,** 791 (1988).

A 500-ml., round-bottomed flask equipped with a magnetic stirring bar and a glass stopper is charged with 110 ml. of trifluoroacetic acid (Note 1) and 54.34 g. (0.1008 mole) of solid thallium(III) trifluoroacetate (Note 2). A clear solution is obtained after 30 minutes of vigorous stirring. Upon addition of 10.6 g. (0.100 mole) of *p*-xylene (Note 3), the reaction mixture turns brown (Note 4). After vigorous stirring for 20 minutes, the trifluoroacetic acid is removed on a rotary evaporator with a bath temperature of 35°, and the residue is dissolved in 100 ml. of diethyl ether. The solvent is again evaporated, and the solid residue is dissolved in 100 ml. of ether. With ice cooling (Note 5), a solution of 33.2 g. (0.200 mole) of potassium iodide in 100 ml. of water is added in one portion. After the resulting dark suspension is stirred vigorously for 10 minutes, a solution of 3 g. of sodium bisulfite in 30 ml. of water is added (Note 6). Yellow thallium(I) iodide is removed by filtration after another 10 minutes of vigorous stirring and washed thoroughly with 150 ml. of ether. The aqueous layer is separated and extracted with two 60-ml. portions of ether. The combined ether extracts are washed once with 10% aqueous sodium hydroxide (Note 7) and twice with 20 ml. of water. After being dried over anhydrous magnesium sulfate for 1 hour, the ether is removed on a rotary evaporator. Distillation under reduced pressure yields 18.5–19.6 g. (80–84%) of pure 2-iodo-*p*-xylene, b.p. 110–113° (19 mm.) (Notes 8–10).

2. Notes

1. This chemical is available from Aldrich Chemical Company, Inc., Halocarbon Products Corporation, Allied Chemical Corporation, or Eastman Organic Chemicals.

2. Both the submitters and the checkers used thallium(III) trifluoroacetate prepared from thallium(III) oxide and trifluoroacetic acid.[3] Although this material may be purchased from Aldrich Chemical Company, Inc., and Eastman Organic Chemicals, the submitters recommend that the reagent be prepared immediately prior to use.

3. This reagent is obtainable from major chemical suppliers.

4. The submitters report that *p*-xylylthallium bis(trifluoroacetate) precipitates after 5 minutes. The checkers did not obtain this precipitate until the bulk of the solvent had been evaporated.

5. The reaction of aqueous potassium iodide and *p*-xylylthallium bis(trifluoroacetate) is exothermic and the ether boils off unless the reaction mixture is cooled.

6. The sodium bisulfite is added to reduce any free iodine formed in this reaction. Due to the presence of trifluoroacetic acid in the reaction mixture, sulfur dioxide evolves upon addition of the bisulfite. If not added in small portions, this operation may cause overflow of the reaction mixture.

7. The sodium hydroxide solution should be added slowly, since the reaction with the acidic ether extract is exothermic and may cause the ether to boil. The ether extract should be washed with aqueous sodium hydroxide until the aqueous layer remains basic to litmus. This extraction is self-indicating; the ether turns from a bright yellow to a light brown and color appears in the aqueous phase.

8. There is usually a lower boiling fraction of 0.1–0.3 g. consisting mainly of unreacted *p*-xylene, along with 1.0–1.6 g. of a dark brown residue.

9. The purity of the product may be checked by GC. The submitters used a 10-m. column with 30% QF-1 on 45/60 Chrom W. The checkers used a 2-m. column of 10% UCW-98 on WAW DMCS operated at 150°.

10. The overall time needed for this preparation is less than 5 hours. The product decomposes slowly and should be refrigerated in the dark.

3. Discussion

This procedure for the synthesis of 2-iodo-*p*-xylene is slightly modified from that of Taylor and McKillop.[3] The reaction is generally applicable to a wide range of aromatic substrates,[3,4] and, with some modifications, to thiophenes. A critical feature of this synthesis is that the entering iodine substituent always replaces thallium at the same position on the aromatic ring. The great preference of the thallium electrophile for the *para*-position in activated aromatic substrates leads, therefore, to iodo-compounds of high isomeric purity. With substituents capable of chelating with the thallium(III) electrophile, thallation may occur by an intramolecular delivery route, resulting in exclusive *ortho*-substitution in optimum cases. Furthermore, aromatic electrophilic thallation is reversible, and under conditions of thermodynamic rather than kinetic control, *meta*-

substitution often predominates. The preparation of aromatic iodo-compounds *via* aryl-thallium bis(trifluoroacetate) intermediates thus possesses the additional advantage of potential orientation control.[4]

2-Iodo-*p*-xylene has been prepared by the action of potassium iodide on diazotized *p*-xylidine (2,5-dimethylaniline) (21% yield),[5] from the reaction of *p*-xylene with molecular iodine in concentrated nitric acid (50% yield)[6] or in ethanol–sulfuric acid in the presence of hydrogen peroxide (64% yield),[7] and with molecular iodine in glacial acetic acid–sulfuric acid in the presence of iodic acid as a catalyst (85% yield).[8]

1. Department of Chemistry, Princeton University, Princeton, New Jersey 08540.
2. School of Chemical Sciences, University of East Anglia, Norwich, Norfolk NR4 7TJ, England.
3. A. McKillop, J. D. Hunt, M. J. Zelesko, J. S. Fowler, E. C. Taylor, G. McGillivray, and F. Kienzle, *J. Am. Chem. Soc.*, **93**, 4841 (1971).
4. E. C. Taylor, F. Kienzle, R. L. Robey, A. McKillop, and J. D. Hunt, *J. Am. Chem. Soc.*, **93**, 4845 (1971).
5. G. T. Morgan and E. A. Coulson, *J. Chem. Soc.*, 2203 (1929).
6. R. L. Datta and N. R. Chatterjee, *J. Am. Chem. Soc.*, **39**, 435 (1917).
7. L. Jurd, *Aust. J. Sci. Res., Ser. A*, **3**, 587 (1950); [*Chem. Abstr.*, **45**, 6592i. (1951)].
8. H. O. Wirth, O. Königstein, and W. Kern, *Justus Liebigs Ann. Chem.*, **634**, 84 (1960).

cis-α,β-UNSATURATED ACIDS: ISOCROTONIC ACID

[2-Butenoic acid, (Z)-]

$$CH_3CH_2COCH_3 + 2\ Br_2 \longrightarrow CH_3CHBrCOCH_2Br + 2\ HBr$$

$$CH_3CHBrCOCH_2Br \xrightarrow[\text{2. HCl}]{\text{1. KHCO}_3} \underset{H}{\overset{CH_3}{\diagdown}}C=C\underset{H}{\overset{COOH}{\diagup}}$$

Submitted by C. Rappe[1]
Checked by A. F. Kluge and J. Meinwald

1. Procedure

Caution! 1,3-Dibromo-2-butanone is a powerful lachrymator and a vesicant. This preparation should be carried out in a hood and contact of this compound with the skin should be avoided.

A. 1,3-*Dibromo-2-butanone.* A mixture of 72.1 g. (90.0 ml., 1.00 mole) of 2-butanone and 100 ml. of precooled (5°) 48% hydrobromic acid is prepared in a 1-l., three-necked, round-bottomed flask equipped with a dropping funnel, a condenser (Note 1), and a Teflon stirrer. The flask is immersed in ice water, and when the temperature of the mixture reaches 5°, 319.6 g. (102.5 ml., 1.998 moles) of bromine is added dropwise at a rate such that the temperature does not rise above 10° and unreacted bromine does not accumulate (Note 2). After addition of the bromine is complete, 400 ml. of water is added, and the heavier organic layer is separated and immediately (Note 3) fractionated under reduced pressure (Note 4) through a 25-cm. Widmer column, giving 115–134 g. (50–58%) of pure 1,3-dibromo-2-butanone, b.p. 91–94° (13 mm.), n_D^{25} 1.5252 (Notes 5 and 6).

B. *Isocrotonic acid.* A solution of 100 g. (1.00 mole) of potassium hydrogen carbonate (Note 7) in 1 l. of water is placed in a 2-l., three-necked, round-bottomed flask equipped with a condenser, a dropping funnel, and a Teflon stirrer. 1,3-Dibromo-2-butanone (46.0 g., 0.200 mole) is added over a 5-minute period (Note 8). The mixture is stirred thoroughly, and after 2–3 hours (Note 9) when constant titration values against methyl orange are obtained, the solution is extracted with two 100-ml. portions of diethyl ether (Note 10) and acidified to pH 1–2 by dropwise addition of dilute hydrochloric acid (Note 11). The aqueous solution is re-extracted with six 100-ml. portions of ether, and the ether phase is dried overnight in a refrigerator over magnesium sulfate.

The ethereal solution is filtered with suction, and the ether is removed with a rotary evaporator connected to a 500-ml., acetone–dry ice trap. A water bath maintained at 5–10° is used to facilitate the removal of the ether (Notes 12 and 13).

The yield of crude isocrotonic acid is 11.8–13.2 g. (69–77%). It is sufficiently pure for most purposes although NMR analysis (Note 14) shows that the crude acid contains a small amount of the stable *trans*-isomer (Note 15). The crude product cannot be stored without isomerization. For purification, 13.0 g. of the product is dissolved in 25 ml. of petroleum ether (b.p. 40–65°) at 5°. When left at −15° for days, crystals separate and are filtered at 5°, yielding 9.3 g. of product, m.p. 12.5–14°, n_D^{25} 1.4453, which can be stored in the dark at 30° for 3 weeks or at 5° for years with no detectable isomerization (Note 16).

2. Notes

1. The hydrogen bromide evolved from the condenser should be absorbed in a gas trap.

2. Accumulation of bromine results in an uncontrolled reaction and a decrease in the yield. This step requires 6–8 hours.

3. The crude product soon begins to decompose if it is not distilled immediately.

4. Since corrosive vapor is evolved, a water pump should be used.

5. The distilled product might be highly colored (violet, green, and blue), but this has no effect on its further use.

6. The dihaloketone purified in this way is stable for years when stored at 5°.

7. The yield is slightly lower when other bases such as sodium hydrogen carbonate, sodium carbonate, or potassium carbonate are used.

8. The reaction is slightly exothermic.

9. Less time is required when stronger bases are used.

10. The nonacidic by-products are discarded.

11. Because of vigorous foaming, the addition must be made slowly and with care. The end-point can also be detected by a fading color of the reaction mixture. The checkers performed this acidification in the reaction flask with mechanical stirring which minimized the foaming.

TABLE I

PREPARATION OF BROMOKETONES AND cis-α,β-UNSATURATED ACIDS FROM METHYL KETONES

$$RCH_2COCH_3 \xrightarrow{2\ Br_2} RCHBrCOCH_2Br \xrightarrow[\text{2. HCl}]{\text{1. } KHCO_3} \begin{array}{c} R \\ \diagdown \\ C=C \\ \diagup \quad \diagdown \\ H \quad\quad H \end{array} \!\!\! \begin{array}{c} COOH \end{array}$$

	Bromoketone		cis-α,β-Unsaturated Acid				
R	Yield, %	n_D^{25}	Time h	Yield, %	b.p., ° mm.	n_D^{25}	m.p., °
C₂H₅	57	1.5176	3	68	39–41(0.4)	1.4473	–43
n-C₃H₇	51	1.5080	24	61	71–73(0.2)	1.4495	0–1
2-C₃H₇	58	1.5099	24	85	59.5–60(0.2)	1.4420	15.5–17.5
n-C₄H₉	65	1.5043	20	64	69–70(0.4)	1.4515	–19
tert-C₄H₉	49	1.5071	48	75	60–61(0.8)	1.4432	11–12
n-C₅H₁₁	40	1.5001	48	50	75–76(0.3)	1.4530	—
(CH₃)₂CH(CH₂)₂	35	1.4997	48	58	93–94(0.2)	1.4518	—
n-C₆H₁₃	41	1.4983	96	28	91–92(0.8)	1.4549	2–3

713

12. The submitter used the following procedure for removal of ether. A 250-ml., two-necked, round-bottomed flask, fitted with a dropping funnel, is equipped for distillation under reduced pressure (water pump). The ethereal solution is added dropwise (Note 13) and, when all the solution is added and the pressure has dropped to 10 mm., the last traces of ether are removed with an oil pump (0.4 mm.) for a period of 30 minutes.

13. This is to avoid isomerization which is easily initiated at elevated temperature.

14. NMR spectroscopy is an excellent tool for distinguishing between the isomers.[2]

15. The checkers detected the presence of approximately 10% of the *trans*-acid by NMR analysis.

16. The crude acid could be distilled in 5–10 ml. portions at 1 mm. without isomerization (b.p. 36°), but these samples were found to be more sensitive to isomerization.

3. Discussion

Isocrotonic acid can be prepared by the stereospecific *cis*-hydrogenation of tetrolic acid[3] or, mixed with the *trans*-isomer, by reduction of 3-chloro-*cis*-crotonic acid with sodium amalgam.[4] The *cis*-acid can also be prepared in small amounts by isomerization of the *trans*-acid.[5] The method described herein is much less laborious than the older procedures.[2]

The reaction is an example of a stereospecific Favorskii rearrangement,[6] and seems to have general applicability for the preparation of *cis*-α,β-unsaturated acids.[7] Only a limited number of the higher homologues have previously been prepared by the more laborious stereospecific *cis*-hydrogenation of the corresponding acetylenic acid,[1,8,9] and, moreover, in some cases, they seem to have been mixtures of the two geometric isomers. The rearrangements, starting with commercially available methyl ketones, yield the *cis*-isomer exclusively as determined by NMR spectroscopy.[7] The higher homologues can be purified by distillation with minimal losses. Purified in this manner, the samples can be stored at 5° for years without detectable isomerization. The yields and physical constants of the bromoketones and the *cis*-α,β-unsaturated acids are given in Table I.

1. Department of Organic Chemistry, University of UMEÅ, S-901 87 UMEÅ, Sweden.
2. C. Rappe, *Acta. Chem. Scand.*, **17**, 2766 (1963).
3. M. Bourguel, *Bull. Soc. Chim. Fr.*, [4] **45**, 1067 (1929).
4. A. Michael and O. Schulthess, *J. Prakt. Chem.*, [2] **46**, 236 (1892).
5. R. Stoenmer and E. Robert, *Ber. Dtsch. Chem. Ges.*, **55**, 1030 (1922)
6. A. S. Kende, *Org. React.*, **11**, 261, (1960).
7. C. Rappe and R. Adeström, *Acta. Chem. Scand.*, **19**, 383 (1965).
8. H. Silwa and P. Maitte, *Bull Soc. Chim. Fr.*, 369 (1962).
9. J. A. Knight and J. H. Diamond, *J. Org. Chem.*, **24**, 400 (1959).

TRICHLOROMETHYL CHLOROFORMATE AS A PHOSGENE EQUIVALENT: 3-ISOCYANATOPROPANOYL CHLORIDE

(Propanoyl chloride, 3-isocyanato-)

Submitted by KEISUKE KURITA[1] and YOSHIO IWAKURA
Checked by WILLIAM F. BURGOYNE, CHRISTOPHER
VANCANTFORT, and ROBERT M. COATES

1. Procedure

Caution! Trichloromethyl chloroformate is toxic. These reactions should both be carried out in a well-ventilated hood (Note 1).

A. *Trichloromethyl chloroformate.* A 100-ml., three-necked, round-bottomed Pyrex flask is equipped with a thermometer, a reflux condenser protected at the top with a calcium chloride tube, and a gas-inlet tube with a coarse fritted-glass tip extending almost to the bottom of the flask. In the flask are placed 37.8 g. (0.400 mole) of freshly distilled methyl chloroformate (Note 2) and a Teflon-coated magnetic stirring bar. The flask is illuminated with a 100-W., high-pressure, mercury-vapor lamp (Note 3) placed beside it (Notes 4 and 5). The methyl chloroformate is stirred and irradiated as a slow stream of chlorine (Note 6) is passed into the flask through the gas-inlet tube (Note 7). When the temperature reaches 30° due to the exothermic reaction, the flask is immersed in a water bath (Note 8). The chlorine is then passed into the solution more rapidly so as to maintain the temperature at 30–35° (Note 9). After *ca.* 6.5–7 hours the colorless solution assumes the pale yellow-green color of chlorine, which indicates that the end point of the reaction has been reached (Note 10). Distillation under reduced pressure affords 65–72 g. (82–91%) of trichloromethyl chloroformate as a colorless liquid, b.p. 53–55° (53 mm.) (Note 11).

B. *3-Isocyanatopropanoyl chloride.* A 500-ml., two-necked flask is equipped with a thermometer and a reflux condenser protected at its top by a calcium chloride tube. A Teflon-coated magnetic stirring bar, 250 ml. of anhydrous dioxane (Note 12), 12.6 g. (0.100 mole) of finely pulverized 3-aminopropanoic acid hydrochloride (Note 13), and 23.8 g. (14.4 ml., 0.120 mole) of trichloromethyl chloroformate (Note 14) are placed in the flask in the order specified. The mixture is stirred and heated at 55–60°. After *ca.* 5

hours, the solid has completely dissolved, leaving a clear solution. The heating is discontinued after a total of 7 hours (Note 15), and the solvent is removed under reduced pressure. The residual oil is distilled rapidly under reduced pressure, yielding 11.2–12.4 g. (84–93%) of distillate at 75–85° (20 mm.) (Note 16). Redistillation affords 10.5–11.8 g. (79–88%) of 3-isocyanatopropanoyl chloride as a colorless liquid, b.p. 92–94° (25 mm.) (Note 17).

2. Notes

1. Vapors of all of the polychlorinated methyl chloroformates are toxic.[2-4] Trichloromethyl chloroformate has a phosgene-like odor and is known to decompose to phosgene at elevated temperature[5] or on contact with iron(III) oxide or charcoal.[6]

2. Methyl chloroformate is available from Aldrich Chemical Company, Inc.

3. The checkers used a 200-W., high-pressure, mercury-vapor lamp and the corresponding transformer which are available from the Hanovia Lamp Division, Canrad-Hanovia, Inc., 100 Chestnut St., Newark 5, New Jersey 07105. The lamp was suspended vertically in a cylindrical, double-walled, Pyrex jacket cooled by flowing water. The inside diameter, outside diameter, and length of the cooling jacket were 3, 4, and 22 cm., respectively. The cooling jacket was clamped in place ca. 5 cm. from the reaction vessel to allow the cooling bath to be raised into position. The use of the 200-W. lamp did not alter the reaction time.

4. It is advisable to wrap the entire apparatus with aluminum foil to avoid exposure to UV light. The reaction solution can be observed through a small hole in the aluminum foil which is is shielded from the direct radiation of the lamp.

5. The irradiation should be started before the flow of chlorine gas is begun to avoid the risk of explosion.

6. The checkers used chlorine from a lecture bottle supplied by the Linde Division, Union Carbide Chemical Corp.

7. Chlorine should be introduced slowly at first to prevent an accumulation of unreacted chlorine in the solution and avoid the risk of a rapid, exothermic reaction. The accumulation of chlorine is indicated by the appearance of its characteristic yellow-green color.

8. The checkers used a 15 × 7.5 cm. Pyrex crystallizing dish as a transparent water bath.

9. Once the chlorine gas flow is properly adjusted, the solution remains colorless and the temperature stays in the 30–35° range without further adjustment until the end point is approached. However, the color and temperature should be observed frequently during the 6.5- to 7-hour reaction time.

Although this reaction can be carried out at higher temperatures, the yields are reduced, probably owing to loss of the volatile methyl chloroformate, b.p. 71°, and/or decomposition of the product. For example, the yields of trichloromethyl chloroformate are ca. 75% and 55% when the chlorination is carried out at 50–55° and 85–90°, respectively.

10. The checkers judged that the end point had been reached when the yellow-green color persisted for 2–3 minutes after the chlorine flow had been stopped. At the first appearance of the yellow-green color, the gas stream was shut off, and the color faded

within *ca.* 15 sec. The chlorine flow was resumed at a slow rate until the end point was reached.

11. The product has the following spectral properties: IR (neat) cm.$^{-1}$: 1815 (C=O), 1054, 968, 912, 814, 764; ^{13}C NMR (CDCl$_3$), δ (assignment): 108.37 (CCl$_3$), 143.93 (C=O). Trichloromethyl chloroformate is stable at room temperature, but decomposes to phosgene when heated above 300°[5] or on contact with iron(III) oxide or charcoal.[6] Decomposition to carbon tetrachloride and carbon dioxide occurs on exposure to alumina, aluminum chloride, or iron(III) chloride.[5-8]

12. The checkers dried and purified dioxane by distillation from the sodium–benzophenone ketyl.

13. 3-Aminopropanoic acid (β-alanine) is available from Aldrich Chemical Company, Inc., and from the Nutritional Biochemical Division, ICN Products. The hydrochloride salt is prepared in the following manner. A solution of 89 g. (1.0 mole) of 3-aminopropanoic acid in 200 ml. of water is acidified by addition of 100 ml. (1.20 mole) of concentrated hydrochloric acid and concentrated to a white solid with a rotary evaporator. The solid is pulverized to a fine powder and dried at 60° under reduced pressure, yielding 113–125 g. (90–100%) of 3-aminopropanoic acid hydrochloride.

14. Although the reaction can be carried out with an equimolar amount of trichloromethyl chloroformate, a longer time (15–20 hours) is required to reach completion, and the yield is reduced somewhat. If a 1.5–2.0-fold excess of trichloromethyl chloroformate is used, the reaction time is decreased to *ca.* 5 hours and the yield is increased to 90–95%.

15. The checkers heated the suspension for a total of 10 hours, 7–8 hours having been required to dissolve the solid completely. The reaction time may depend on the particle size of the hydrochloride salt and the rate of stirring.

16. The submitters advise that the distillation be carried out rapidly to avoid the formation of a tarry residue.

17. The spectral properties of 3-isocyanatopropanoyl chloride are as follows: IR (liquid film) cm.$^{-1}$: 2278 (N=C=O), 1795 (O=C—Cl); ^{1}H NMR (CDCl$_3$), δ (multiplicity, coupling constant J in Hz., number of protons, assignment): 3.17 (t, J = 6, 2H, CH_2CH$_2$N=C=O), 3.67 (t, J = 6, 2H, CH$_2$CH_2N=C=O); ^{13}C NMR (CDCl$_3$), δ (assignment): 38.3 (CH$_2$CH$_2$N=C=O), 47.6 (CH$_2$CH$_2$N=C=O), 123.0 (N=C=O), 171.6 (O=C—Cl). A small peak at δ 67.1 in the ^{13}C NMR spectrum of the product obtained by the checkers was attributed to a small amount of dioxane.

3. Discussion

The chlorination of methyl chloroformate in sunlight was first reported by Hentschel, but without a detailed description of either the procedure or the results.[5] The first step of the present procedure for the preparation of trichloromethyl chloroformate utilizes a UV light source and affords a simple, reproducible way to obtain this reagent. Although trichloromethyl chloroformate may also be synthesized by photochemical chlorination of methyl formate,[9-11] the volatility of methyl formate causes losses during the reaction and increases the hazard of forming an explosive mixture of its vapor and chlorine gas. The preparation of trichloromethyl chloroformate by chlorination of methyl chloroformate in the dark with diacetyl peroxide as initiator has been reported;[12] however, the procedure consists of several steps, and the overall yield is rather low.

Trichloromethyl chloroformate is synthetically useful as a substitute for phosgene, which, owing to its high volatility and toxicity, presents a severe hazard in the laboratory. Although trichloromethyl chloroformate is toxic, it is a dense and less volatile liquid (b.p. 128°, d_{15}^{15} 1.65), having a vapor pressure of only 10 mm. at 20°. Consequently it is more easily handled in a safe manner than phosgene.

Trichloromethyl chloroformate has proven effective in the preparation of N-carboxy-α-amino acid anhydrides from amino acids,[13] and various compounds having isocyanate, acid chloride, and chloroformate groups.[14,15] For example, trichloromethyl chloroformate may be used instead of phosgene in the preparation of 2-tert-butoxycarbonyloxyimino-2-phenylacetonitrile.[15] The use of this reagent is illustrated here by the synthesis of 3-isocyanatopropanoyl chloride from 3-aminopropanoic acid hydrochloride.

3-Isocyanatopropanoyl chloride has also been prepared by the reaction of 3-aminopropanoic acid hydrochloride with phosgene;[16] however, the yield is only 36%, and hydrogen chloride must be introduced to increase the yield to 92%. The present procedure effects this reaction without additional hydrogen chloride and avoids the hazards of handling phosgene. This procedure has been successful in the synthesis of isocyanato acid chlorides and isocyanato chloroformates from amino acids and amino alcohols, respectively.[14] For example, 6-isocyanatohexanoyl chloride can be prepared in good yield with trichloromethyl chloroformate, although it is obtained in only trace amounts with phosgene unless additional hydrogen chloride is used. Isocyanato acid chlorides such as 3-isocyanatopropanoyl chloride, having two different, highly reactive electrophilic groups, are novel reagents for introducing amino acid residues into organic compounds[14] and polymers.[17]

1. Department of Industrial Chemistry, Faculty of Engineering, Seikei University, Musashino-shi, Tokyo, Japan.
2. A. Mayer, H. Magne, and L. Plantefol, *C.R. Hebd. Seances Acad. Sci.*, **172**, 136 (1921).
3. A. Desgrez, H. Guillemard, and Savès, *C.R. Hebd. Seances Acad. Sci.*, **171**, 1177 (1920).
4. A. Desgrez, H. Guillemard, and A. Labat, *C.R. Hebd. Seances Acad. Sci.*, **172**, 342 (1921).
5. W. Hentschel. *J. Prakt. Chem.*, [2] **36**, 99 (1887).
6. H. P. Hood and H. R. Murdock, *J. Phys. Chem.*, **23**, 498 (1919).
7. W. Hentschel, *J. Prakt. Chem.*, [2] **36**, 305 (1887).
8. A. Kling, D. Florentin, A. Lassieur, and E. Schmutz, *C.R. Hebd. Seances Acad. Sci.*, **169**, 1166 (1919).
9. W. Hentschel, *J. Prakt. Chem.*, [2] **36**, 209 (1887).
10. V. Grignard, G. Rivat, and E. Urbain, *C.R. Hebd. Seances Acad. Sci.*, **169**, 1074 (1919).
11. H. C. Ramsperger and G. Waddington, *J. Am. Chem. Soc.*, **55**, 214 (1933).
12. S. Yura, *Kogyo Kagaku Zasshi*, **51**, 157 (1948) [*Chem. Abstr.*, **45**, 547b (1951)].
13. M. Oya, R. Katakai, H. Nakai, and Y. Iwakura, *Chem. Lett.*, 1143 (1973).
14. K. Kurita, T. Matsumura, and Y. Iwakura, *J. Org. Chem.*, **41**, 2070 (1976).
15. See *Org. Synth., Coll. Vol. 6*, 199 (1988).
16. Y. Iwakura, K. Uno, and S. Kang, *J. Org. Chem.*, **31**, 142 (1966).
17. K. Hayashi, S. Kang, and Y. Iwakura, *Makromol. Chem.*, **86**, 64 (1965).

HYDROBORATION OF OLEFINS: (+)-ISOPINOCAMPHEOL

[(+)-3-Pinanol]

This procedure [*Org. Synth.*, **52**, 59 (1972)] has been replaced by an improved procedure for (−)-Isopinocampheol.

(−)-ISOPINOCAMPHEOL

(Bicyclo[3.1.1]heptan-3-ol, 2,6,6-trimethyl-, [1R-(1α,2β,3α,5α)]-)

Submitted by C. F. LANE and J. J. DANIELS[1]
Checked by D. M. RYCKMAN and R. V. STEVENS

1. Procedure

A 300-ml., three-necked flask (Note 1), equipped with a magnetic stirring bar, thermometer, pressure-equalizing dropping funnel fitted with septum inlet adapter (Note 2), and reflux condenser fitted with a hose adapter leading to a mineral oil bubbler (Note 3), is charged with 10.0 ml. (0.100 mole) of borane–methyl sulfide complex (Note 4) and 30 ml. of tetrahydrofuran (Notes 5 and 6). The flask is immersed in an ice-water bath as 27.2 g. (31.7 ml., 0.200 mole) of (+)-α-pinene (Note 7) is added dropwise at 0–3° to the well-stirred reaction mixture over a period of 15 minutes. The (−)-diisopinocampheylborane [(−)-di-3-pinanylborane] precipitates as a white solid as the reaction proceeds. Following addition, the reaction mixture is stirred for 3.5 hours at 0°. Under a slow stream of nitrogen, the outlet hose adapter on the reflux condenser is connected with rubber vacuum hose to a vacuum trap which is then cooled in an acetone–dry ice bath. The dimethyl sulfide and tetrahydrofuran are bulb-to-bulb vacuum-distilled (0.1 mm.) with the reaction flask in a room temperature water bath. When only a dry, white solid residue remains, the vacuum is released with nitrogen. The flask is again placed under a slight positive pressure of nitrogen. The solid is slurried in 36 ml. of tetrahydrofuran (Note 5) at room temperature. An additional 4.08 g. (4.76 ml., 0.030 mole) of (+)-α-pinene (Note 7) is added. The resulting slurry is stirred at room tempera-ture for 5 minutes and then stored under nitrogen in a closed system in a cold room at 4° for 3 days (Note 8). The flask is then removed from the cold room and immersed in an ice-water bath. Under a slow stream of nitrogen, the outlet adapter on the reflux condenser is again connected to the mineral oil bubbler. The excess hydride is destroyed by the slow, dropwise addition of 8 ml. of methanol (Note 9), followed by the addition in one portion of 36.6 ml. of 3 M aqueous sodium hydroxide. The borinic acid intermediate is now oxidized by the dropwise addition of 24 ml. of 30% aqueous hydrogen peroxide (Note 10) to the well-stirred reaction mixture at 35° ± 3° (Note 11). After the hydrogen peroxide addition is complete, the ice-water bath is replaced with a warm-water bath and the reaction mixture is stirred for one hour at 50–55° (Note 12) and then cooled to room temperature. The aqueous layer is saturated with sodium chloride and 50 ml. of diethyl ether is added. The upper organic layer is removed, and the aqueous layer is extracted with two 100-ml. portions of ether. The organic layer and extracts are combined, dried over anhydrous potassium carbonate, filtered, and concentrated to an oil on a rotary evaporator at 60° (15 mm.) (Note 13). The crude product is fractionally distilled using a 30-cm. column packed with glass helices, giving 24.7 g. (80%) of (−)-isopinocampheol, b.p. 60–65° (0.1 mm.) (Note 14). The distillate crystallizes completely in the receiver, m.p. 49–55°, 97.5% purity by GC, $[\alpha]_D^{19}$ −34.3° (C, 20 in ethanol) (Note 15). Slurrying 4.7 g. in 2.3 ml. of pentane at room temperature, cooling to −78°, collecting on a filter, and air drying gives 3.8 g. of crystalline (−)-isopinocampheol, m.p. 52–55°, purity 99.2% by GC, $[\alpha]_D^{19}$ −34.9° (C, 20 in ethanol).

2. Notes

1. The apparatus is dried in an oven and assembled hot while being flushed with nitrogen. A slow stream of nitrogen is continued until the apparatus is cool. Alternatively, the apparatus can be assembled and then flame-dried while flushing with nitrogen.

2. A suitable septum inlet adapter is available from Aldrich Chemical Company (product number Z10,130-3).

3. A suitable bubbler is available from Aldrich (product number Z10,121-4) and used to maintain a slight positive pressure of nitrogen in the reaction vessel.

4. Borane–methyl sulfide complex was obtained from Aldrich and used as received.

5. An anhydrous grade of tetrahydrofuran was obtained from Aldrich and used as received.

6. Borane–methyl sulfide and anhydrous tetrahydrofuran are extremely moisture-sensitive. All transfers must be done under a nitrogen atmosphere, with syringe tech-niques being the most convenient.[2]

7. (+)-α-Pinene ($[\alpha]_D^{21}$ +47.1°) was obtained from Aldrich and was short-path vacuum-distilled under nitrogen from a small amount of lithium aluminum hydride.

8. The equilibrated (−)-diisopinocampheylborane obtained at this point can be utilized directly for asymmetric hydroboration.[3]

9. Since the pinene is hydroborated only to the dialkylborane stage (R_2BH), methanolysis liberates a large amount of hydrogen. The rate of evolution is controlled by the slow addition of methanol, and some foaming is observed. *The hydrogen must be vented to an efficient hood.* Water can be used to destroy the hydride, but methanol addition is easier to control and the final mixture is homogenous once methanolysis is complete.

10. Thirty percent aqueous hydrogen peroxide was obtained from Aldrich and used as received.

11. The oxidation is exothermic and can be quite vigorous. It should be controlled by the slow, dropwise addition of hydrogen peroxide. Cooling in an ice-water bath is necessary. However, a reaction temperature of around 35° must be maintained.

12. The additional one hour of heating is necessary to destroy excess hydrogen peroxide. Oxygen is evolved and some foaming occurs.

13. The oil solidifies upon cooling but can be easily remelted in a warm-water bath for transfer to a small distillation flask.

14. An air-cooled condenser is used for the distillation of isopinocampheol.

15. The enantiomeric excess of (−)-isopinocampheol was determined to be greater than 95% by 200 MHz ^{1}H NMR using the chiral shift reagent, tris[3-heptafluoropropylhydroxymethylene)-d-camphorato], europium(III). Addition of the lanthanide reagent (40 mg., 0.034 mmol.) to the chiral alcohol (30 mg., 0.21 mmol.) produced a shift of 7.5 p.p.m. in the peak centered around δ 4.05. Only one broad peak was observed.

Treatment of (±)-isopinocampheol in a similar manner gave two broad singlets of equal intensity centered around δ 10.75 (separated by 0.15 p.p.m.), attributable to the diastereomeric protons.

3. Discussion

Isopinocampheol has been prepared by hydroboration of α-pinene using *in situ* generated diborane with diglyme as the solvent.[4]

The present procedure employs a recently developed method which provides a product of greatly improved enantiomeric purity (>99%).[5] Also, this preparation utilizes commercially available borane−methyl sulfide and α-pinene of 92% enantiomeric purity. An equilibration is used to improve the optical purity of the intermediate dialkylborane.

Isopinocampheol is only of limited interest. More importantly this procedure provides optically pure (−)-diisopinocampheylborane, a very versatile reagent which has been used widely for the synthesis of many chiral products.[3]

1. Aldrich-Boranes, Inc., Sheboygan Falls, Wisconsin 53085.
2. C. F. Lane, *Aldrichimica Acta,* **10,** 11 (1977).
3. For a review of asymmetric syntheses using chiral organoboranes, see H. C. Brown, P. K. Jadhav, and A. K. Mandal, *Tetrahedron,* **37,** 3547 (1981).
4. G. Zweifel and H. C. Brown, *Org. Synth.,* **52,** 59 (1972).
5. H. C. Brown, M. C. Desai, and P. K. Jadhav, *J. Org. Chem.,* **47,** 5065 (1982).

REACTION OF ARYL HALIDES WITH
π-ALLYLNICKEL HALIDES: METHALLYLBENZENE

[Benzene, (2-methyl-2-propenyl)-]

Submitted by Martin F. Semmelhack[1] and Paul M. Helquist[2]
Checked by Bradley E. Morris and Richard E. Benson

1. Procedure

Caution! Nickel carbonyl is a flammable, volatile (b.p. 43°), highly toxic reagent. Safety glasses, gloves, and an apron should be worn when handling this reagent and the first step of this preparation should be conducted in an efficient hood (Note 1).

Benzene has been identified as a carcinogen; OSHA has issued emergency standards on its use. All procedures involving benzene should be carried out in a well-ventilated hood, and glove protection is required.

A 1-l., three-necked flask is equipped with a reflux condenser, a pressure-equalizing dropping funnel, a three-way stopcock, and a large magnetic stirring bar. The system is flushed with argon (Note 2), and 380 ml. of benzene (Note 3) is placed in the flask. From an inverted lecture cylinder 50.8 g. (38.5 ml., 0.298 mole) of nickel carbonyl (Note 4) is introduced into the addition funnel. The nickel carbonyl is added to the benzene while an atmosphere of argon is maintained, and the flask is immersed in an oil bath at 50°. With a syringe, 10.04 g. (0.07437 mole) of methallyl bromide (Note 5) is added over 10 minutes. After a short induction period, evolution of carbon monoxide becomes rapid and a deep red color appears. The exit gas is led from the top of the condenser through a gas bubbler tube to monitor the rate of gas evolution. As the gas evolution becomes vigorous, the bath temperature is raised to 70° and maintained at this temperature for 30 minutes after gas evolution ceases (the total time after addition of methallyl bromide is 1.5 hours). The resulting solution is allowed to cool to 25°, and the benzene and excess nickel carbonyl are removed under reduced pressure (water aspirator), applying an oil bath at 30° as needed to maintain a rapid rate of evaporation (Note 6). The residual, red, solid π-methallylnickel bromide (>85%) is used directly in the next step (Note 7).

A solution of 9.95 g. (0.0634 mole) of bromobenzene (Note 8) in 100 ml. of oxygen-free N,N-dimethylformamide is added, under an argon atmosphere at 25°, to a solution of the crude nickel complex (an 85% yield is assumed) in 65 ml. of oxygen-free N,N-dimethylformamide, over a 15-minute period. After the addition is complete, the reaction mixture is stirred at 25° for 12 hours, then warmed to 60° for one hour. Complete reaction of the nickel complex is indicated by a red to emerald green color change, characteristic of a solution of nickel dibromide in N,N-dimethylformamide. After being cooled to 25°, the solution is poured into a mixture of 250 ml. of water and 250 ml. of petroleum ether (b.p. 30–60°); 2 ml. of 12 M hydrochloric acid is added (Note 10), and the mixture is filtered through Celite filter aid, facilitating separation of the layers. The organic layer is separated, washed with two 100-ml. portions of water, dried over anhydrous magnesium sulfate, and concentrated with a rotary evaporator at water aspirator pressure, affording 8.0–9.6 g. of a clear, colorless liquid. Distillation through a short Vigreux column gives 5.58–6.02 g. (67–72% yield based on bromobenzene) of methallylbenzene as a colorless liquid, b.p. 67–68° (19 mm.), n_D^{25} 1.5064 (Note 11).

2. Notes

1. The treatment for nickel carbonyl poisoning involves intramuscular injection of BAL (2,3-dimercapto-1-propanol).[3]

2. Argon is preferred by the submitters for its high density which allows opening of the reaction vessel without significant displacement of the inert atmosphere by air. A nitrogen atmosphere was used by the checkers and was equally effective in preventing oxidation of the π-allylnickel complex.

3. Anhydrous, air-free benzene was prepared by distillation under argon, discarding a 20% forerun. The checkers used benzene from a freshly opened bottle (Fisher Scientific Company).

4. Nickel carbonyl, available from Matheson Gas Products, was used by the checkers.

5. Methallyl bromide is prepared from methallyl chloride (Eastman Organic Chemicals) with a halide exchange reaction. A solution of 148.3 g. (1.639 moles) of methallyl chloride and 213.8 g. (2.460 moles) of lithium bromide in 1 l. of dry acetone is refluxed for 5 hours. The mixture is filtered, and the filtrate is distilled through a 30-cm. Vigreux column, affording 69.2 g. (31.3%) of methallyl bromide, b.p. 88–93°, n_D^{25} 1.4672. The purity was 98 ± 2% by GC analysis on a 20% Carbowax column at 65°.

6. Nickel carbonyl is drawn into the aspirator flow during this operation. In many laboratories the hood plumbing is connected with the general plumbing line, and vapors of highly toxic nickel carbonyl may diffuse back to sinks at the laboratory bench. If such an arrangement is suspected, the solvent and excess nickel carbonyl can be collected by employing a cold trap (−78° or −196°) between the reaction mixture and the aspirator. Care should be used in the disposal of this mixture.

7. Pure π-methallylnickel bromide can be obtained by dissolving the residue in 150

ml. of oxygen-free anhydrous diethyl ether, filtering under argon, concentrating the filtrate until crystals begin to form, and cooling at $-78°$ for 12 hours. The crystals are isolated by removing the liquid *via* suction through a syringe needle under a positive pressure of argon, yielding 12.1 g. (85%) of dark-red crystals. The ^{1}H NMR spectrum can be obtained only by rigorous exclusion of oxygen from the sample and filtration of the sample as the last stage of sample preparation. The ^{1}H NMR spectrum (C_6D_6) shows three singlets at δ 2.07 (3H), 2.82 (2H), and 2.83 (2H).

8. Bromobenzene was used as supplied by Aldrich Chemical Company, Inc. Purification by distillation under argon did not change the yield of methallylbenzene. The checkers used the product available from Eastman Organic Chemicals (white label).

9. *N,N*-Dimethylformamide was distilled from calcium hydride at 71° (32 mm.) and stored under argon. The checkers used a freshly opened bottle of the product (white label) grade) available from Eastman Organic Chemicals.

10. The hydrochloric acid solution is added to speed solution of the nickel salts, which would otherwise lead to emulsions during separation. If no emulsion is encountered after mixing the petroleum ether and water solutions, no hydrochloric acid need be added. Similarly, the filtration through Celite filter aid is intended to remove finely divided nickel metal and other insoluble particles which complicate the washing procedure. If no particles are present, the filtration step should be omitted.

11. The product consists of 99% methallylbenzene and 1% 2,5-dimethyl-1,5-hexadiene, by ^{1}H NMR analysis. The ^{1}H NMR spectrum of methallylbenzene (CCl_4) shows peaks at δ 1.63 (broad s, 3H, allylic CH_3), 3.25 (broad s, 2H, allylic CH_2), 4.75 (m, 2H, C—CH_2), and 7.15 (s, 5H, C_6H_5). The ^{1}H NMR spectrum of 2,5-dimethyl-1,5-hexadiene (CCl_4) shows peaks at δ 1.70 (broad s, 6H, 2CH_3), 2.12 (broad s, 4H, 2 allylic CH_2), and 4.75 (m, 4H, 2CH_2=C). A small forerun contained 0.30 g. (3% yield) of methallylbenzene and a larger quantity of 2,5-dimethyl-1,5-hexadiene. The distillation residue is composed of 0.24–0.34 g. (3–4% yield) of methallylbenzene and 0.38–0.52 g. (8–10% yield) of biphenyl. The distillation fractions may be analyzed by GC with a 180 cm. by 6.4 mm. column packed with 10% SE-30 on Chromosorb G. The retention times for methallylbenzene and 2,5-dimethyl-1,5-hexadiene are 4.0 minutes and 2.0 minutes, respectively, at 125°.

3. Discussion

The simple example outlined above, replacement of halogen by a methallyl group, could be carried out in an equally direct way using phenylmagnesium bromide and methallyl halide. However, the Grignard reaction is complicated by formation of the conjugated, isomeric β,β-dimethylstyrene,[4] or by a rearrangement to *trans*-2-butenylbenzene.[5] In no case has this approach afforded methallylbenzene in greater than 50% yield. Dehydration of (2-hydroxy-2-methylpropyl)benzene also produces a mixture of methallylbenzene (68%) and β,β-dimethylstyrene (32%).[6] Elimination of benzoic acid from the benzoate ester of (2-hydroxy-2-methylpropyl)benzene gives the same ratio of products, although the combined yield (86%) is lower.[7] The Wittig reaction of methylenetriphenylphosphorane with 1-phenyl-2-propanone produces methallylbenzene in only 2% yield.[8]

The preparation illustrates the procedure for formation of π-allylnickel halides and their reaction with aryl halides.[9] The complexes can be obtained from allylic chlorides, bromides, and iodides[10-12] even when the allylic halides bear alkyl, carboalkoxyl, or alkenyl side chains.[11] The coupling step is generally applicable to aryl, alkyl, and vinyl bromides or iodides;[9] organic chlorides are usually unreactive with π-allylnickel halides. Other polar aprotic solvents (hexamethylphosphoric triamide, dimethyl sulfoxide, N-methylpyrrolidone) have been used.[13] Protic solvents lead to the destruction of the π-allylnickel complex by slow protonation of the allyl ligand.[13] No reaction occurs between aryl, alkyl, or vinyl halides and π-allylnickel bromide in less polar solvents such as tetrahydrofuran, 1,2-dimethoxyethane, ether, or hydrocarbons. The π-allylnickel bromide complexes are very reactive with allyl halides, but halogen–metal exchange precedes coupling and a mixture of products is obtained as illustrated in the following example.[14]

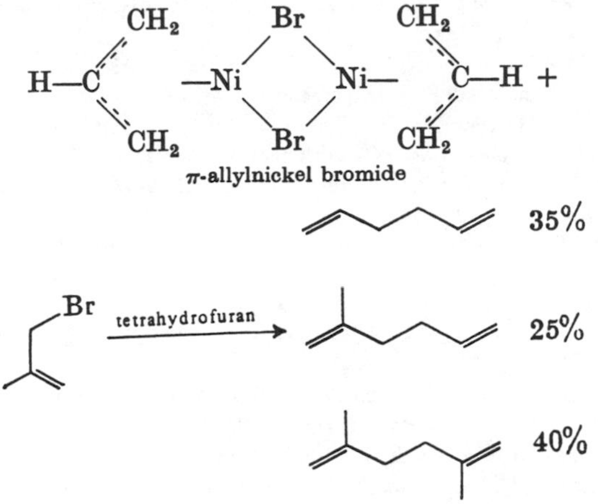

π-allylnickel bromide

35%

25%

40%

The π-allylnickel complex from *trans*-geranyl bromide reacts with alkyl halides, giving a mixture of *cis* and *trans* products,[9] the double bond that participates in the π-allyl group is isomerized during the sequence of reactions:

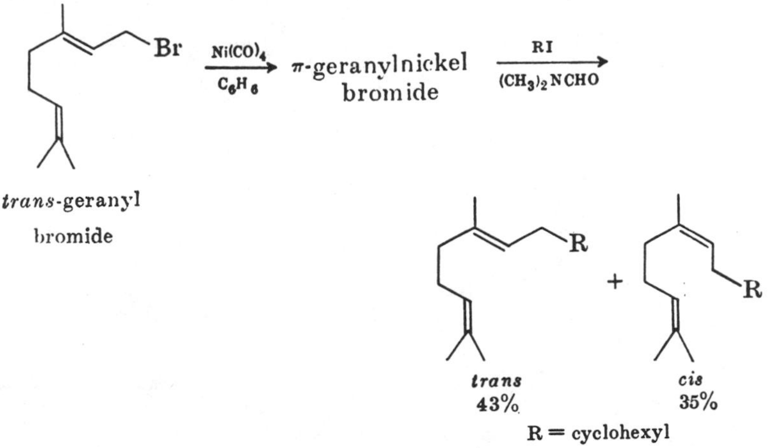

trans-geranyl bromide

π-geranylnickel bromide

trans 43%

cis 35%

R = cyclohexyl

Similarly, *trans*-4-iodocyclohexanol reacts with π-methallylnickel bromide, producing a mixture of epimeric 4-methallylcyclohexanols.[9] Note that the hydroxyl group has no significant effect on this reaction.

The advantages of π-allylnickel halides reside in their nonnucleophilic and nonbasic character which allow especially selective reactions with organic halides in the presence of a large number of other functional groups. Carbonyl and hydroxyl groups react much more slowly than organic halides (especially iodides) with π-allylnickel halides, while olefins, nitriles, alkyl chlorides, and aromatic hydrocarbons are inert to these reagents.[9,13]

1. Department of Chemistry, Cornell University, Ithaca, New York 14850 [Present address: Department of Chemistry, Princeton University, Princeton, New Jersey 08544].
2. Present address: Department of Chemistry, State University of New York at Stony Brook, Stony Brook, New York 11794.
3. P. G. Stecher, "The Merck Index," 8th ed., Merck and Co., Rahway, N.J., 1968, p. 372.
4. C. M. Buess, J. V. Karavinos, P. V. Kunz, and L. C. Gibbons, *Natl. Advis. Comm. Aeronaut. Tech. Notes,* No. 1021 (1946).
5. K. W. Wilson, J. D. Roberts, and W. G. Young, *J. Am. Chem. Soc.,* **71,** 2019 (1949).
6. L. Beránek, M. Kraus, K. Kochloefl, and V. Bažant, *Collect. Czech. Chem. Commun.,* **25,** 2513 (1960) [*Chem Abstr.,* **55,** 3464 (1961)].
7. R. Onesta and G. Castelfranchi, *Chim. Ind. (Milan),* **42,** 735 (1960) *Chem Abstr.,* [**55,** 24646 (1961)].
8. C. Rüchardt, *Chem. Ber.,* **94,** 2599 (1961).
9. E. J. Corey and M. F. Semmelhack, *J. Am. Chem. Soc.,* **89,** 2755 (1967); M. F. Semmelhack, *Org. React.,* **19,** 119 (1972).
10. E. O. Fischer and G. Bürger, *Z. Naturforsch. Teil B,* **16,** 77 (1961).
11. M. F. Semmelhack, Ph.D. Dissertation, Harvard University, 1967.
12. G. Wilke, B. Bogdanović, P. Hardt, P. Heimbach, W. Keim, M. Kröner, W. Oberkirch, K. Tanaka, E. Steinrücke, D. Walter, and H. Zimmermann, *Angew. Chem., Int. Ed. Engl.,* **5,** 151 (1966).
13. See reference 11 and unpublished results of M. F. Semmelhack and E. J. Corey.
14. E. J. Corey, M. F. Semmelhack, and L. S. Hegedus, *J. Am. Chem. Soc.,* **90,** 2416 (1968).

SULFONYL CYANIDES: METHANESULFONYL CYANIDE

$$CH_3SO_2Cl \xrightarrow[\text{water, } 25°]{Na_2SO_3, \ NaHCO_3} CH_3SO_2Na \xrightarrow[\text{water, } 10-15°]{ClCN} CH_3SO_2CN$$

Submitted by M. S. A. VRIJLAND[1]
Checked by Y. SUGIMURA and G. BÜCHI

1. Procedure

Caution! Since cyanogen chloride is highly toxic, the preparation and isolation of the sulfonyl cyanide should be conducted in a well-ventilated hood.

Benzene has been identified as a carcinogen; OSHA has issued emergency standards on its use. All procedures involving benzene should be carried out in a well-ventilated hood, and glove protection is required.

A 2-l., three-necked, round-bottomed flask equipped with a sealed mechanical stirrer, a pressure-equalizing dropping funnel capped with a gas-outlet, and a thermometer is charged with 126 g. (0.500 mole) of sodium sulfite heptahydrate, 84.0 g. (1.00 mole) of sodium hydrogen carbonate, and 1 l. of water (Note 1). Stirring is begun, and 57.3 g. (0.500 mole) of freshly distilled methanesulfonyl chloride (Note 2) is added dropwise over 30 minutes. The slightly exothermic reaction is accompanied by the evolution of carbon dioxide. After stirring for 2 hours, gas evolution ceases, and a clear, colorless solution of sodium methanesulfinate (Note 3) is obtained.

The dropping funnel is removed, the solution is cooled to 10° by the addition of ice, and 50 ml. (1.0 mole) of liquid cyanogen chloride (Note 4) is added in one portion with vigorous stirring. Addition of ice keeps the mixture at or below 15°. Within 1 minute the reaction mixture becomes turbid, and methanesulfonyl cyanide separates as a heavy, colorless oil. The mixture is stirred for an additional 15 minutes before 200 ml. of benzene is added (Note 5). After 3 minutes of stirring, the layers are separated in a 2-l. separatory funnel, and the aqueous layer is extracted with two 100-ml. portions of benzene. The combined extracts are washed with water and dried overnight over anhydrous calcium chloride. Filtration and removal of solvent with a rotary evaporator *in a hood* affords an almost pure product (Note 6) which is distilled, yielding 35.4–37.8 g. (67–72%) of methanesulfonyl cyanide, b.p. 68–69° (15 mm.),n_D^{20} 1.4301 (Note 7), which may be stored in a well-stoppered bottle, kept at or below 0°, for prolonged times without loss in purity (Note 8).

2. Notes

1. Excess sodium sulfite or sodium hydrogen carbonate should be avoided, since either would react with the sulfonyl cyanide once formed.

2. The procedure given is applicable to many other sulfonyl chlorides as well (see Table I). Solid sulfonyl chlorides are added as such. When heavy frothing occurs in the

reduction (*e.g.*, with 4-nitrobenzenesulfonyl chloride), addition of 50 ml. of chloroform to the reaction mixture will eliminate the foam without reducing the final yield. When the sulfonyl chlorides were prepared according to Meerwein and co-workers,[2] it was found advantageous to use the crude, damp sulfonyl chlorides, since these are more easily reduced than the dried or recrystallized materials.

TABLE I
PREPARATION OF SULFONYL CYANIDES
FROM SULFONYL CHLORIDES[a]

R =	RSO$_2$Cl from	m.p.	b.p.	Yield, %
Methyl	Commerce	—	68–69° (15 mm.)	72
Ethyl	RSCN + Cl$_2$[3]	—	80–80.5° (18 mm.)	84
Propyl	R$_2$S$_2$ + Cl$_2$[4]	—	81–81.5° (18 mm.)	76
Benzyl	RSC(NH)NH$_2$·HCl[5]	89.5–91°		91
Cyclohexyl	RH + SO$_2$Cl$_2$[6]		72–73° (0.4 mm.)	85
4-Methoxyphenyl	RH + SO$_2$Cl$_2$[7]	66–68°		88
p-Tolyl	Commerce	49.5–51°		89
Phenyl	Commerce	19–20°	118–119° (15 mm.)	92
4-Chlorophenyl	RN$\frac{1}{2}$Cl$^-$ + SO$_2$[2]	57.5–59°		65[b]
4-Cyanophenyl	RN$^+_2$Cl$^-$ + SO$_2$[2]	123–125°		79[b]
4-Nitrophenyl	RN$^+_2$Cl$^-$ + SO$_2$[2]	122–123.5°		66[b]

[a]The preparations were performed on a 0.25 to 1-mole scale.
[b]Overall yield from the corresponding aniline as starting material.

3. When crude sulfonyl chlorides were used as starting materials, on completion of the reduction, and before the addition of cyanogen chloride, the reaction mixture was washed with a suitable solvent (benzene or dichloromethane, or, in some cases, chloroform) to remove organic impurities. In the case of higher-melting, crystalline sulfonyl chlorides, heating to 50° may be necessary to complete their reduction. The solution of the sulfinate salt may be kept overnight, if desired, with no decrease in the yield of sulfonyl cyanide.

4. Cyanogen chloride is commercially available in gas cylinders. It is liquefied by passing the gas through a condenser cooled with ice water. Where difficult to obtain, it may be prepared by passing chlorine gas through a stirred suspension of sodium tetrakis-(cyano-C)zincate prepared *in situ* from sodium cyanide and zinc sulfate.[8]

5. If the benzenesulfinates were substituted with electron-withdrawing groups, *e.g.*, 4-nitro- and 4-cyanobenzenesulfinate, the yields were slightly improved when the reaction time with cyanogen chloride was lengthened to 1 hour.

The higher-melting sulfonyl cyanides which separate as solids should be dried when dissolved in a suitable solvent, *e.g.*, benzene. 4-Nitrobenzenesulfonyl cyanide is not readily extracted from the reaction mixture; it is collected on a Büchner funnel, pressed as dry as possible, dissolved in benzene, washed with water, and dried over anhydrous calcium chloride.

6. Solid sulfonyl cyanides show a melting point not more than 1–2° below that of recrystallized material. They may be used without further purification. Analytically pure samples are obtained by recrystallization from dry benzene, dry petroleum ether, or a mixture of the two.

7. The product was further characterized as follows: IR (liquid film) cm.$^{-1}$: 2195 strong, 1370 strong, 1170 strong; ^{1}H NMR resonance (CDCl$_3$), δ 3.43 (s).

8. Contrary to the findings of Cox and Ghosh,[9] methanesulfonyl cyanide may be distilled without decomposition. Samples of benzene-, 4-methoxybenzene-, and 4-chlorobenzenesulfonyl cyanides were kept for over a year without loss in purity.

3. Discussion

Whereas sulfonyl halides have been known for a long time and, especially the chlorides, have become of great synthetic value, sulfonyl cyanides were unknown until 1968. They were first prepared by van Leusen and co-workers from the reaction of sulfonylmethylenephosphoranes with nitrosyl chloride.[10] The same group also investigated part of their chemistry.[11a–e] Since then, two more, completely different methods of synthesis have been published: one, involving the reactions of sulfinates with cyanogen chloride,[9] and another, the oxidation of thiocyanates.[12]

The procedure given above for the preparation of methanesulfonyl cyanide is essentially a combination of the sulfite reduction of a sulfonyl chloride, as originally described by Bere and Smiles,[13] and the sulfinate–cyanogen chloride reaction, first published by Cox and Ghosh.[9]

Some sulfinates are commercially available and may be used as starting materials for the preparation of sulfonyl cyanides. Yields, however, are not significantly better than when the much cheaper and more readily available sulfonyl chlorides are used as starting materials. Good to excellent results are obtained, even when starting from rather impure sulfonyl chlorides.[14] Illustrative examples are given in Table I.

Sulfonyl cyanides have an activated cyano group and show many interesting reactions. With a range of N-, O-, S-, and C-nucleophiles, transfer of the cyano group to these nucleophiles is observed.[11a,15,16] Hydroxylamine, hydrazine, and phenylhydrazine (α-effect nucleophiles) add to the cyano group of sulfonyl cyanides, yielding products that could be converted into substituted 1,2,4-oxadiazoles[17] and 1,2,4-triazoles,[11e,15] respectively. Dienes show Diels-Alder cycloadditions with sulfonyl cyanides.[11b–d,15,18] 1,3-Dipolar cycloadditions to the cyano group give rise to substituted tetrazoles (from azides), to substituted 1,2,3-triazoles (from diazo compounds), or to substituted 1,2,4-oxadiazoles (from nitrile N-oxides).[11b,15] Sulfonyl cyanides undergo free-radical additions to alkenes.[15,19] Chlorine and sulfenyl chlorides add to the cyano group of sulfonyl cyanides.[20a,b,21]

1. Twente University of Technology, Enschede, The Netherlands.
2. H. Meerwein, G. Dittmar, P. Göllner, K. Hafner, F. Mensch, and O. Steinfort, *Chem. Ber.*, **90**, 841 (1957).
3. T. B. Johnson and I. B. Douglass, *J. Am. Chem. Soc.*, **61**, 2548 (1939).

4. T. Zincke and A. Dahm, *Ber. Dtsch. Chem. Ges.*, **45**, 3457 (1912).
5. J. M. Sprague and T. B. Johnson, *J. Am. Chem. Soc.*, **59**, 1837 (1937); *Caution!* See K. Folkers, A. Russell, and R. W. Bost, *J. Am. Chem. Soc.*, **63**, 3530 (1941).
6. M. S. Kharasch and A. T. Read, *J. Am. Chem. Soc.*, **61**, 3089 (1939).
7. M. S. Morgan and L. H. Cretcher, *J. Am. Chem. Soc.*, **70**, 375 (1948).
8. H. Schröder, *Z. Anorg. Allg. Chem.*, **297**, 296 (1958); *cf.* G. H. Coleman, R. W. Leeper, and C. C. Schulze, *Inorg. Synth*, **2**, 90 (1946).
9. J. M. Cox and R. Ghosh, *Tetrahedron Lett.*, 3351 (1969).
10. A. M. van Leusen, A. J. W. Iedema, and J. Strating, *Chem. Commun.*, 440 (1968).
11. (a) A. M. van Leusen and J. C. Jagt, *Tetrahedron Lett.*, 967 (1970); (b) A. M. van Leusen and J. C. Jagt, *Tetrahedron Lett.*, 971 (1970); (c) J. C. Jagt and A. M. van Leusen, *Recl. Trav. Chim. Pays-Bas*, **92**, 1343 (1973); (d) J. C. Jagt and A. M. van Leusen, *J. Org. Chem.*, **39**, 564 (1974); (e) J. C. Jagt and A. M. van Leusen, *Recl. Trav. Chim. Pays-Bas*, **94**, 12 (1975).
12. R. G. Pews and F. P. Corson, *J. Chem. Soc. D.*, 1187 (1969).
13. C. M. Bere and S. Smiles, *J. Chem. Soc.*, **125**, 2359 (1924); *cf.* S. Smiles and C. M. Bere, *Org. Synth.*, **Coll. Vol. 1,** 7 (1932).
14. Attempts to prepare sulfonyl cyanides from the corresponding sulfonyl chlorides according to the procedure described were unsuccessful when applied to mono-, di-, and trichloromethanesulfonyl chloride, to dimethylsulfamoyl chloride, and to ethylene- and 2,4-dinitrophenylsulfonyl chloride.
15. J. C. Jagt, Ph.D. Thesis, Groningen University, The Netherlands, 1973.
16. F. P. Corson and R. G. Pews, *J. Org. Chem.*, **36**, 1654 (1971).
17. U. Treuner, *Synthesis*, 559 (1972).
18. R. G. Pews, E. B. Nyquist, and F. P. Corson, *J. Org. Chem.*, **35**, 4096 (1970).
19. R. G. Pews and T. E. Evans, *J. Chem. Soc. D.*, 1397 (1971).
20. (a) M. S. A. Vrijland and J. Th. Hackmann, *Tetrahedron Lett.*, 3763 (1970); (b) M. S. A. Vrijland, *Tetrahedron Lett.*, 837 (1974).
21. H. Kristinsson, *Tetrahedron Lett.*, 4489 (1973).

1,6-METHANO[10]ANNULENE

(Bicyclo[4.4.1]undeca-1,3,5,7,9-pentaene)

A. $\xrightarrow[\text{C}_2\text{H}_5\text{OH}]{\text{Na/NH}_3}$

B. $\xrightarrow[\text{(CH}_3)_3\text{COK}]{\text{CHCl}_3}$

C. $\xrightarrow{\text{Na/NH}_3}$

D. $\xrightarrow[\text{Dioxane}]{\text{DDQ}}$

Submitted by E. Vogel,[1] W. Klug, and A. Breuer
Checked by R. E. Ireland, R. A. Farr, H. A. Kirst, T. C. McKenzie, R. H. Mueller, R. R. Schmidt, III, D. M. Walba, A. K. Willard, and S. R. Wilson

1. Procedure

Caution! This reaction should be carried out in an efficient hood.

A. 1,4,5,8-*Tetrahydronaphthalene* (*isotetralin*). A 12-l. (Note 1),. three-necked, round-bottomed flask is immersed in an acetone–dry ice bath and fitted with a dry ice condenser (Note 2), a tube-sealed stirrer (Note 3), a drying tube (potassium hydroxide), and a gas delivery-tube running to the bottom of the flask. Ammonia (3 l.) is condensed (Note 4) into the flask. The gas delivery-tube is removed, and 192.3 g. (8.361 g.-atoms) of sodium is added in small portions (Note 5), with vigorous stirring, over a period of 1 hour. The flask is then fitted with a dropping funnel, through which a solution of 192.3 g.

(1.502 mole) of naphthalene in a mixture of 750 ml. of diethyl ether and 600 ml. of ethanol is added dropwise to the blue solution over 3 hours. After the addition is complete, the reaction mixture is stirred at −78° (Note 6) for another 6 hours. The cooling bath is removed, and the ammonia is allowed to evaporate overnight. The remaining white, solid residue is processed, with ice cooling and stirring under a nitrogen atmosphere, by slow addition of 120 ml. of methanol to destroy unreacted sodium, then 4−5 l. of ice water to dissolve the salts (Note 7). The reaction mixture is extracted with 1 l. of ether. Evaporation of the ether phase at room temperature under reduced pressure gives a coarse, white solid which is collected on a sintered-glass funnel and washed with water. Recrystallization from methanol (about 1.6 l.) using a heated funnel followed by drying of the crystals under reduced pressure (Note 8) gives 148−158 g. (75−80%) of isotetralin, m.p. 52−53° (purity ~98%) (Note 9). Pure 1,4,5,8-tetrahydronaphthalene is reported to have m.p. 58°.[2]

B. 11,11-*Dichlorotricyclo*[4.4.1.0^{1,6}]*undeca*-3,8-*diene*. A 3-l., three-necked, round-bottomed flask fitted with a tube-sealed stirrer, a pressure-equalizing dropping funnel, and a Claisen-adapter bearing an inlet tube for argon and a low temperature thermometer is charged with a solution of 132.2 g. (1.000 mole) of 1,4,5,8-tetrahydronaphthalene (isotetralin) in 1.3 l. of anhydrous ether (Note 10). To this solution is added 150 g. (1.33 mole) of potassium *tert*-butoxide (Note 11) under an argon atmosphere, and the resulting suspension is cooled to −30° with an acetone–dry ice bath and stirred efficiently. While these conditions are maintained, a solution of 119.5 g. (1.000 mole) of chloroform in 150 ml. of ether is added dropwise over 90 minutes (Note 12). The mixture is stirred for another 30 minutes at −30° before the temperature is allowed to rise above 0°. Following this, 300−350 ml. of ice water is added to dissolve the salts; the two layers formed are separated (Note 13). The organic layer is washed with two 300-ml. portions of water, while the aqueous layer is extracted with two 200-ml. portions of ether. The ether phases are combined, dried over magnesium sulfate, and filtered. The ether is removed with a rotary evaporator, and the residual liquid (or solid) is distilled under reduced pressure.

The distillation is expediently carried out using a 500-ml., round-bottomed flask, an electrically heated 1.5 × 30 cm. column packed with V4A wire spirals (4 mm.) (Note 14), a short, air-cooled condenser (Note 15), and an ice-cooled, three-necked, 250-ml. receiver flask. During the distillation, the liquid is stirred magnetically and heated with an oil bath. The first fraction, b.p. 55−58° (1 mm.), yields *ca.* 50 g. of isotetralin (Note 16), more of which is collected when the column is heated to about 100°. The temperature in the head of the column thereby rises to 90−95°, and it is necessary to change the receiver flask. The second fraction, collected at 95−102° (1 mm.), yields *ca.* 108 g. of 1:1-adducts, which consists of 92% of the desired tricyclo product and 8% of the side-addition product. The residue mainly contains the 2:1-adducts. The fraction containing the 1:1-adducts is recrystallized from methanol (about 500 ml.), giving 87−97 g. (40−45%, based on isotetralin) of 11,11-dichlorotricyclo[4.4.1.0^{1,6}]undeca-3,8-diene as long, colorless needles, m.p. 88−89° (Note 17).

C. *Tricyclo*[4.4.1.0^{1,6}]*undeca*-3,8-*diene*. Ammonia (800 ml.) is condensed into a 2-l., three-necked round-bottomed flask, immersed in an acetone–dry ice bath and fitted with a dry ice condenser, a tube-sealed stirrer, a drying tube, and a gas delivery tube running to the bottom of the flask. The gas delivery tube is removed, and with vigorous stirring 56 g. (2.4 g.-atoms) of sodium is added in small portions with vigorous stirring over a period of 30 minutes (Note 18). The flask is then fitted with a dropping funnel, through which a solution of 81.4 g. (0.378 mole) of 11,11-dichlorotricyclo[4.4.1.0^{1,6}]-undeca-3,8-diene in 500 ml. of anhydrous ether is added over 1 hour, while cooling and stirring are maintained. After addition is complete, the acetone–dry ice bath is removed, and the ammonia is allowed to evaporate overnight. The flask is placed in the acetone–dry ice bath again, and a gentle stream of argon is passed continuously through the system. With stirring, a mixture of 90 ml. of methanol and 90 ml. of ether is added dropwise. The bath temperature is then allowed to rise to 0° and, with continued stirring, 500 ml. of ice water is added slowly. The reaction mixture is transferred to a 2-l. separatory funnel, and the two layers are separated. The organic layer is washed with 200 ml. of water, the aqueous layer is extracted with three 150-ml. portions of pentane (Note 19), and the combined ether–pentane phases are dried over magnesium sulfate. After filtration of the drying agent (Note 20) the solvent is removed by distillation through a 30-cm. Vigreux column. The remaining liquid is transferred to a 250-ml., round-bottomed flask and distilled under reduced pressure through a packed column (Note 21), yielding 46.9–49.7 g. (85–90%) of tricyclo[4.4.1.0^{1,6}]undeca-3,8-diene, collected as a colorless liquid at 80–81° (11 mm.), n_D^{20} 1.5180 (Note 22).

D. *1,6-Methano[10]annulene.* A 2-l., three-necked, round-bottomed flask fitted with a tube-sealed stirrer, a reflux condenser protected with a calcium chloride drying tube, and an inlet tube for argon is charged with 900 ml. of anhydrous dioxane (Note 23). To this solvent is added, with stirring, 149 g. (0.656 mole) of 2,3-dichloro-5,6-dicyano-1,4-benzoquinone (DDQ) (Note 24). When the DDQ has dissolved, 43.8 g. (0.300 mole) of tricyclo[4.4.1.0^{1,6}]undeca-3,8-diene and 10 ml. of glacial acetic acid are added. The system is then flushed with argon, and the stirred mixture is heated under reflux for 5 hours. The reaction starts within a few minutes, as evidenced by effervescing of the solution and massive precipitation of the hydroquinone. At the same time the originally red-brown color of the mixture turns almost black. Following the reflux period, the bulk of the dioxane (600–650 ml.) is removed by distillation through a 15-cm. Vigreux column while stirring is maintained. The resulting pasty mixture is cooled, and 150 ml. of *n*-hexane is added. The solid is suction filtered, washed with 500 ml. of warm *n*-hexane, and dried at 100°, giving *ca.* 144 g. (95%) of pure 2,3-dichloro-5,6-dicyanohydroquinone (Note 25). The filtrate and washings are combined and passed through a 5 × 30 cm. column of neutral alumina (Note 26), which is eluted with *n*-hexane (Note 27). The solvent is removed by distillation through a 30-cm. Vigreux column, and the residual liquid is distilled from a 250-ml., round-bottomed flask through a packed column (Note 28), yielding 36.2–37.0 g. (85–87%) of faintly yellow 1,6-methano[10]annulene, b.p. 68–72° (1 mm.), which may crystallize in the receiver-flask, m.p. 27–28° (Note 29).

2. Notes

1. It is advisable to use a 10- or 12-l. flask for runs on this scale because the reaction mixture may effervesce if the naphthalene solution is added too quickly.
2. It is necessary to use a dry ice condenser to shorten the time required to condense

the ammonia (4 hours compared with 6 hours without the condenser). The ammonia tank was warmed with an air gun during the distillation. The condenser was removed after the ammonia was collected.

3. It is necessary to use a strong stirring motor since the reaction mixture becomes, temporarily, rather viscous.

4. One should not pour the liquified ammonia directly out of the cylinder since particles of iron compounds might be carried along, catalyzing the formation of sodium amide. For the exclusion of moisture it is also necessary to use a drying tower (potassium hydroxide) between the cylinder and the flask.

5. The sodium should be cut into small particles to increase the speed of dissolution and diminish the danger of stirrer blockage.

6. During this period the reaction mixture might turn white. In this case, another portion of sodium must be added until the solution becomes blue again.

7. The white residue should be worked up as soon as possible. On standing the residue gradually turns brown-red due to the formation of decomposition products; isolation of isotetralin then becomes difficult, and the yield may drop sharply. The submitters evaporated any remaining ether from the reaction flask at reduced pressure and filtered the water slurry of isotetralin, obtaining the same yield after crystallization.

8. Isotetralin should not be kept under vacuum longer than necessary since the compound has a relatively high vapor pressure.

9. A second extraction of the aqueous phase with ether yields an additional 1.5 g. of material. A second crop, (29.4 g., m.p. 49–52°) of isotetralin can be obtained from the mother liquors of the recrystallization.

10. All solvents used should be anhydrous.

11. The yield of 11,11-dichlorotricyclo[4.4.1.0^{1,6}]undeca-3,8-diene strongly depends on the quality of the potassium *tert*-butoxide used. Commercially available, sublimed potassium *tert*-butoxide was employed. When freshly sublimed potassium *tert*-butoxide is utilized, yields of up to 45% of 11,11-dichlorotricyclo[4.4.1.0^{1,6}]undeca-3,8-diene can be obtained. Potassium *tert*-butoxide, prepared by the method of Doering,[3] gave yields comparable to those achieved with the commercial product.

12. The stated reaction temperature should be maintained carefully. Raising the temperature above −30° noticeably reduces the regio-selectivity of the addition of di-chlorocarbene, whereas lowering the temperature causes the yield of 1:1-adducts to drop due to partial crystallization of isotetralin.

13. The formation of emulsions may render it difficult to discern the two rather dark layers. In this case it is helpful to acidify with dilute sulfuric acid.

14. The checkers used an electrically heated, 1.5 × 30 cm. Vigreux column and obtained the same results.

15. To prevent isotetralin and the 1:1-adducts from solidifying in the condenser external heating with an IR lamp was applied.

16. The recovered isotetralin can be reused.

17. The product is approximately 99% pure by GC (SE-30 on kieselguhr, 150°). After two or three recrystallizations from methanol, 11,11-dichlorotricyclo[4.4.1.0^{1,6}]undeca-3,8-diene shows m.p. 90–91°.

18. For the preparation of the solution of sodium in liquid ammonia, compare part A.

19. If emulsions occur, it is advisable to acidify with dilute sulfuric acid to attain separation of the two layers.

20. The drying agent should be washed well with pentane.

21. The column used for this distillation is described in part B.

22. Tricyclo[4.4.1.0^{1,6}]undeca-3,8-diene was shown to be approximately 99% pure by GC (SE-30 on kieselguhr, 150°).

23. The use of anhydrous solvents is necessary to avoid hydrolytic decomposition of 2,3-dichloro-5,6-dicyano-1,4-benzoquinone.

24. Commercially available 2,3-dichloro-5,6-dicyano-1.4-benzoquinone was employed. 1,6-Methano[10]annulene was obtained in equally good yields, when 2,3-dichloro-5,6-dicyano-1,4-benzoquinone, prepared by the method of Walker and Waugh,[4] was utilized.

25. 2,3-Dichloro-5,6-dicyano-1,4-benzoquinone is readily regenerated in good yield from the hydroquinone by oxidation with nitric acid.[4]

26. Brockmann alumina, activity grade II–III, M. Woelm, 344 Eschwege, West Germany.

27. By contrast to the filtrate and washings, which are rather dark, the eluent is yellow due to the color of 1,6-methano[10]annulene.

28. The column used for this distillation is described in part B.

29. The purity of the 1,6-methano[10]annulene was shown by GC (SE-30 on kieselguhr, 150°) to be higher than 99%. Recrystallization of the hydrocarbon from methanol raises its melting point to 28–29°.

3. Discussion

The procedure described for the Birch reduction of naphthalene is a modification of the methods previously developed by Birch,[5] Hückel,[2] and Grob.[6] Apart from this reduction, no other practical approaches to isotetralin are known. The scale employed in the present procedure is not mandatory to achieve optimum yields. Equally good yields were realized when the runs were halved or enlarged up to fourfold. In the latter case, however, the apparatus reaches pilot plant dimensions.

Tricyclo[4.4.1.0^{1,6}]undeca-3,8-diene, the strategic intermediate in the synthesis of 1,6-methano[10]annulene from naphthalene, can alternatively be obtained in one step by the reaction of isotetralin with the Simmons–Smith reagent.[7] The two-step preparation of tricyclo[4.4.1.0^{1,6}]undeca-3,8-diene utilized here has the following merits: (1) dichlorocarbene adds to the central double bond of isotetralin with exceptionally high regioselectivity (as compared to that of methylene transfer reagents), giving 11,11-dichlorotricyclo[4.4.1.0^{1,6}]undeca-3,8-diene as a readily isolable, crystalline compound; and (2) the transformation of the dichloro compound into tricyclo[4.4.1.0^{1,6}]undeca-3,8-diene with sodium in liquid ammonia[8] is a simple operation and affords the product in high yield and purity. The dichlorocarbene employed in the two-step cyclopropanation of isotetralin was generated by the original method of Doering and Hoffmann.[3] Other sources of dichlorocarbene, notably the methods of Parham and Schweizer[9] and of Makosza and Wawrzyniewicz,[10] have also been tried, but did not lead to improved yields of adduct.

The rapid conversion of tricyclo[4.4.1.0^{1,6}]undeca-3,8-diene to 1,6-methano[10]annulene by the high potential quinone, DDQ, is yet another illustration of the usefulness of this agent as a means of dehydrogenation of hydroaromatic compounds.[11] If DDQ is not available, it is recommended that tricyclo[4.4.1.0^{1,6}]undeca-3,8-diene be aromatized by a bromination-dehydrobromination sequence similar to that described in the synthesis of 1,6-oxido[10]annulene;[12] both aromatization methods give essentially the same yield of 1,6-methano[10]annulene.

The synthesis of 1,6-methano[10]annulene outlined above is an improved version of the method first suggested by Vogel and Roth.[13] 1,6-Methano[10]annulene represents a Hückel-type aromatic $(4n + 2)\pi$-system and is similar to benzene or naphthalene in both its physical and chemical properties.[14] The aromatic nature of the hydrocarbon is born out most impressively by its ^{1}H NMR spectrum which exhibits an AA'BB'-system for the vinylic protons at relatively low field (δ 6.8–7.5) and a singlet for the bridge protons at relatively high field (δ 0.5). 1,6-Methano[10]annulene may serve as a starting material for the preparation of other molecules of current interest, such as the bicyclo[5.4.1]dodeca-pentaenylium ion[14] and benzocyclopropene.[15]

1. Institut für Organische Chemie der Universität Köln, West Germany.
2. W. Hückel and H. Schlee, *Chem. Ber.*, **88**, 346 (1955).
3. W. v. E. Doering and A. K. Hoffmann, *J. Am. Chem. Soc.*, **76**, 6162 (1954).
4. D. Walker and T. D. Waugh, *J. Org. Chem.*, **30**, 3240 (1965).
5. A. J. Birch and G. Subba Rao, "Advances in Organic Chemistry," Vol. 8, Wiley-Interscience, New York, 1972, p. 1.
6. C. A. Grob and P. W. Schiess, *Helv. Chim. Acta*, **43**, 1546 (1960).
7. P. H. Nelson and K. G. Untch, *Tetrahedron Lett.*, 4475 (1969).
8. E. E. Schweizer and W. E. Parham, *J. Am. Chem. Soc.*, **82**, 4085 (1960).
9. W. E. Parham and E. E. Schweizer, *J. Org. Chem.*, **24**, 1733 (1959).
10. M. Makosza and M. Wawrzyniewicz, *Tetrahedron Lett.*, 4659 (1969).
11. D. Walker and J. D. Hiebert, *Chem. Rev.*, **67**, 153 (1967).
12. E. Vogel, W. Klug, and A. Breuer, *Org. Synth.*, **Coll. Vol. 6**, 862 (1988).
13. E. Vogel and H. D. Roth, *Angew. Chem.*, **76**, 145 (1964).
14. E. Vogel, in "Aromaticity," *Chem. Soc. Spec. Publ.*, W. D. Ollis, ed., **21**, p. 113 (1967).
15. E. Vogel, W. Grimme, and S. Korte, *Tetrahedron Lett.*, 3625 (1965); the preparation of benzocyclopropene on a large scale is best effected by the recent procedure of W. E. Billups, A. J. Blakeney, and W. Y. Chow, *Org. Synth.*, **Coll. Vol. 6**, 87 (1988).

VINYL SULFIDES FROM THIOACETALS WITH COPPER(I) TRIFLUOROMETHANESULFONATE: (Z)-2-METHOXY-1-PHENYLTHIO-1,3-BUTADIENE

(Benzene, [(2-methoxy-1,3-butadienyl)thio]-, (Z)-)

Submitted by Theodore Cohen,[1] Robert J. Ruffner,
David W. Shull, Elaine R. Fogel, and J. R. Falck
Checked by J. Bisaha and M. F. Semmelhack

1. Procedure

Caution! Part A should be carried out in an efficient hood to minimize exposure to the foul-smelling thiophenol.

Benzene has been identified as a carcinogen; OSHA has issued emergency standards on its use. All procedures involving benzene should be carried out in a well-ventilated hood, and glove protection is required.

A.   *Bis(phenylthio)methane.*   A dry, 2-1., one-necked flask equipped with a magnetic stirring bar and a 250-ml., pressure-equalizing dropping funnel mounted with a combined inlet-outlet assembly for introducing argon (Note 1) is charged with 1.5 1. of distilled dichloromethane and 79.9 g. (110 ml., 0.791 mole) of triethylamine (Note 2), and purged with argon. The solution is stirred and cooled in an ice bath as 87 g. (81 ml., 0.79 mole) of thiophenol (Note 3) is added over 20–30 minutes. The mixture is allowed to warm to 20° and stirred at this temperature for another 3 hours. Triethylamine hydrochloride is removed by filtration through a fritted-glass Büchner funnel, and the filtrate is washed with two 200-ml. portions of 10% aqueous sodium hydroxide, two 200-ml. portions of 2 *N* hydrochloric acid, and one 300-ml. portion of water. The dichloromethane solution is dried over anhydrous magnesium sulfate and evaporated under reduced pressure. Recrystallization of the residue from petroleum ether gives 51.2–59.6 g. (56–65%) of bis(phenylthio)methane as white crystals, m.p. 35–37° (Note 4).

B.   *4,4-Bis(phenylthio)-3-methoxy-1-butene.*   A 250-ml., two-necked, round-bottomed flask equipped with a magnetic stirring bar, a rubber septum, and a reflux condenser bearing an inlet-outlet assembly for argon (Note 1) is dried in an oven at 110° and cooled under a stream of argon passed through the septum with a syringe needle and vented through the argon bubbler. A solution of 5.00 g. (0.0215 mole) of bis(phenylthio)-methane in 150 ml. of tetrahydrofuran (Note 5) is placed in the flask and cooled to −20° with a carbon tetrachloride–dry ice bath. A 15.8-ml. (0.0217 mole) aliquot from a 1.37 *M* solution of *n*-butyllithium in hexane (Note 6) is injected through the septum with a syringe. The resulting deep-yellow solution is stirred and cooled at −20° for 1 hour and then cooled to −78° with an acetone–dry ice bath. The color of the solution is discharged immediately when 1.4 g. (1.7 ml., 0.025 mole) of acrolein (Note 7) is added by syringe at −78°. Stirring and cooling are continued for 15 minutes, after which 2.90 g. (2.19 ml., 0.0230 mole) of dimethyl sulfate (Note 8) is added, and the cooling bath is removed. The solution is stirred at room temperature for 16 hours and heated at reflux for 2 hours. A 3-ml. portion of water is added to the cooled mixture, most of the tetrahydrofuran is removed by rotary evaporation, and the concentrate is partitioned between 30 ml. of water and 30 ml. of diethyl ether. The organic layer is washed three times with 10-ml. portions of concentrated ammonia and once with water. The ethereal solution is dried over anhydrous magnesium sulfate and evaporated under reduced pressure. Crystallization of the viscous yellow residue from 30 ml. of 95% ethanol at −20° gives 3.11–3.36 g. (48–52%) of 4,4-bis(phenylthio)-3-methoxy-1-butene as a light-yellow solid, m.p. 45–48°, which is of adequate purity for use in the next step (Note 9).

C.   *(Z)-2-Methoxy-1-phenylthio-1,3-butadiene.*   A 250-ml., two-necked, round-bottomed flask equipped with a magnetic stirring bar, a rubber septum, and a reflux condenser bearing a combined inlet-outlet assembly for argon (Note 1) is flushed with argon and charged with 3.27 g. [0.00650 mole, 0.0126 g.-atom of copper(I)] of bis[copper(I) trifluoromethanesulfonate] benzene complex (Note 10), 0.036 g. (0.18 mmole) of 3-*tert*-butyl-4-hydroxy-5-methylphenyl sulfide (Note 11), and 70 ml. of benzene (Note 12). With syringes 1.20 g. (0.00397 mole) of 4,4-bis(phenylthio)-3-methoxy-1-butene, 1.33 g. (1.80 ml., 0.0103 mole) of *N,N*-diisopropylethylamine (Note 13), and 14 ml. of tetrahydrofuran (Note 5) are injected through the septum into the flask. The suspension is stirred and heated under reflux for 4.75 hours (Note 14), after which the starting thioacetal has completely reacted, as judged by TLC (Note 15). Water (2 ml.) is added to the cooled mixture, the insoluble material is removed by filtration through Celite, and the flask is rinsed with several portions of ether. A 40-ml. portion of water is added to the filtrate, the layers are separated, and the aqueous layer is extracted with three 25-ml. portions of ether. The organic solutions are combined, dried over anhydrous magnesium sulfate, and evaporated under reduced pressure. Bulb-to-bulb distillation of the residual brown oil (0.782 g.) in a Kugelrohr apparatus (Note 16) with an oven temperature of 85–95° (0.005 mm.) provides 0.421–0.486 g. (55–64%) of (Z)-2-methoxy-1-phenylthio-1,3-butadiene as a light-yellow oil (Notes 17 and 18).

2. Notes

1. The inlet-outlet assembly is connected to both a source of argon and a bubbler which serves as exit for the inert gas. Argon is passed through the apparatus for 30 minutes, and the system is then kept under a slight positive pressure of inert gas by maintaining a slow flow of argon through the bubbler.

2. Triethylamine was purchased from Eastman Organic Chemicals and distilled from calcium hydride before use.

3. Thiophenol was purchased from Aldrich Chemical Company, Inc., and distilled, b.p. 75–77° (30 mm.), 168–169° (760 mm.).

4. The submitters obtained 59.6–64.1 g. (65–70%) of product, m.p. 36–37°, after recrystallization from ethanol. Reported melting points for bis(phenylthio)methane are 34–35°,[2] 38–40°,[3] and 39.5–40.5°.[4] The ^{1}H NMR spectrum (CCl$_4$) exhibits a two-proton singlet at δ 4.30 and a 10-proton multiplet at δ 7.10–7.56.

5. Tetrahydrofuran was distilled from lithium aluminum hydride by the submitters and collected in a flask containing molecular sieves. For a warning regarding this method of purifying tetrahydrofuran, see *Org. Synth.*, **Coll. Vol. 5**, 976 (1973).

6. n-Butyllithium in hexane was purchased from Alfa Division, Ventron Corporation.

7. Acrolein was purchased by the submitters from Cationics, Division of Columbia Organic Chemicals Company, Inc. (Columbia, South Carolina) and distilled immediately before use, b.p. 51–53°.

8. Dimethyl sulfate was used as supplied by Eastman Organic Chemicals. The submitters obtained lower yields when methyl iodide was substituted for dimethyl sulfate.

9. The yield and melting point data given are those of the checkers. The purity of the product was estimated to be at least 98% from analysis of its ^{1}H NMR spectrum. The submitters report that the crude product crystallized on standing in a freezer and that one recrystallization from absolute ethanol afforded 3.6–4.1 g. (55–63%) of product, m.p. 49.5–51°. The ^{1}H NMR spectrum (CCl$_4$) exhibits the following absorptions, δ (multiplicity, coupling constant J in Hz., number of protons, assignment): 3.23 (s, 3H, OCH_3), 3.48–3.97 (m, 1H, CHOCH$_3$), 4.37 [d, J = 4.0, 1H, CH(SC$_6$H$_5$)$_2$], 5.03–5.43 (m, 2H, CH=CH_2), 5.67–6.13 (m, 1H, CH=CH$_2$), 6.90–7.57 (m, 10H, 2SC$_6H_5$).

10. Bis[copper(I) trifluoromethanesulfonate] benzene complex was prepared by a modification of a procedure reported by Salomon and Kochi[5] as described in the following

paragraph. The copper(I) oxide used was purchased from J. T. Baker Chemical Company, and trifluoromethanesulfonic anhydride was prepared by the procedure in *Org. Synth., Coll. Vol.* **6,** 324 (1988). The submitters have found once-distilled anhydride to be satisfactory provided that the weight of phosphorus pentoxide used was approximately equal to the weight of sulfonic acid, and the reaction mixture was stirred vigorously. When twice-distilled anhydride was used in the procedure below, reaction times as long as 13 hours were required before decolorization occurred. The submitters suggest that the enhanced rates observed with once-distilled anhydride may be attributed to the presence of the sulfonic acid. Although trifluoromethanesulfonic anhydride is available from Aldrich Chemical Company, Inc., the commercial reagent has not been used in this procedure.

A dry, 1-l., two-necked, round-bottomed flask equipped with a magnetic stirring bar, a rubber septum, and a reflux condenser bearing a combined inlet-outlet assembly for argon (Note 1) is purged with argon and charged with 18.0 g. (0.126 mole) of copper(I) oxide and 600 ml. of degassed benzene (Note 12). With a syringe, 42.7 g. (25.5 ml., 0.151 mole) of trifluoromethanesulfonic anhydride is injected through the septum into the flask. The suspension is stirred and heated under reflux until nearly all of the red copper(I) oxide has dissolved. Although 3–5 hours is normally sufficient, reaction times as long as 19 hours were required on some occasions (see preceding paragraph). The hot suspension is filtered through a Büchner funnel in an argon-filled glove bag kept dry with a dish of phosphorus pentoxide. The filtrate is allowed to cool in the glove bag for 1 hour, after which the crop of fine white crystals is collected on a fritted-glass Büchner funnel. The funnel is tightly covered with aluminum foil and then placed in a vacuum desiccator containing anhydrous calcium sulfate and phosphorus pentoxide. The desiccator is removed from the glove bag, evacuated overnight, refilled with argon, and returned to the glove bag. The dry bis[copper(I) trifluoromethanesulfonate] benzene complex, 37.6–49.5 g. (60–79%), is transferred to vials in the glove bag. The product maintains its activity indefinitely when protected from moisture and air.

The procedure described above provides a sufficient quantity of bis[copper(I) trifluoromethanesulfonate] benzene complex for several reactions at the scale used in Part C. If bis[copper(I) trifluoromethanesulfonate] benzene complex for a single reaction is desired, the same procedure can be followed at the appropriate scale without the use of the glove bag. In this case, the decolorized solution is not filtered but instead cooled, and the product is crystallized in the reaction vessel. The supernatant benzene is decanted, and the crystals are washed in the flask with fresh benzene. The bis[copper(I) trifluoromethanesulfonate] benzene complex is then used without drying in the same flask.

11. 3-*tert*-Butyl-4-hydroxy-5-methylphenyl sulfide was purchased from Aldrich Chemical Company, Inc., and serves as a radical inhibitor, preventing the polymerization of the product. Lower yields of product were obtained when hydroquinone was used as the inhibitor.

12. Benzene was freshly distilled from calcium hydride and collected in a flask containing molecular sieves.

13. *N,N*-Diisopropylethylamine supplied by Aldrich Chemical Company, Inc., was distilled prior to use. The amine is added to prevent polymerization of the diene by acid generated during the reaction. If the product is not sensitive to acid, the amine may be omitted.

14. The temperature at which elimination of thiophenol occurs depends on the substituents on the sulfur-bearing carbon.[6] Thioketals react rapidly at 25°. In some cases the elimination of thiophenol from the less reactive thioacetals may also be performed at 25°. However, in the present case the combined inductive effects of the vinyl and methoxy groups evidently destabilize the incipient carbonium ion and necessitate a higher temperature for the reaction.

15. TLC was carried out on plates coated with silica gel, using 1:10 (v/v) ether–hexane as developing solvent.

16. Kugelrohr distillation ovens manufactured by Büchi Glasaparatefabrik are available from Brinkmann Instruments, Inc., Westbury, New York.

17. The yields and boiling point range given are those reported by the checkers. When the checkers used starting thioacetal that had been purified by both column chromatography and recrystallization (m.p. 50–52°), the yield was 0.486 g. (64%). Using recrystallized thioacetal, m.p. 49.5–51°, the submitters obtained 0.532–0.646 g. (70–85%) of product.

The ^{1}H NMR spectrum (CCl$_4$) of the diene exhibits the following absorptions, δ (multiplicity, number of protons, assignment): 3.70 (s, 3H, OCH$_3$), 4.87–5.52 (m, 2H, CH=CH$_2$), 5.64 (s, 1H, CHSC$_6$H$_5$), 5.83–6.31 (m, 1H, CH=CH$_2$), 6.98–7.37 (m, 5H, SC$_6$H$_5$). TLC on silica gel with 3:2 (v/v) benzene–hexane as developing solvent showed a single spot. The single sharp peak for the methoxy group in the ^{1}H NMR spectrum and the absence of a spot at a high R_f value from TLC established that the product was not contaminated by its E-isomer. A mixture of the two isomers in which the E-isomer predominates can be prepared by heating a solution of the product in dichloromethane at reflux for 4 hours. The stereochemistry of the original product is assigned as Z on the basis of its high reactivity in Diels-Alder reactions and on the exclusive formation of an adduct with cis stereochemistry from reaction with methyl vinyl ketone.[7] The E-isomer undergoes Diels-Alder reactions much more slowly.[8]

18. The product is stable for months when mixed with a small amount of the radical inhibitor, 3-tert-butyl-4-hydroxy-5-methylphenyl sulfide, and stored in a freezer. In the absence of the inhibitor, it isomerizes to a mixture of E- and Z-isomers over a period of some months.

3. Discussion

Part C of the present procedure illustrates a mild method for effecting the elimination of thiophenol from thioacetals and thioketals under essentially neutral conditions. The reaction of simple thioacetals and thioketals with bis[copper(I) trifluoromethanesulfonate] benzene complex in benzene-tetrahydrofuran at room temperature affords vinyl sulfides in high yield (Table I).[6,7] The reaction presumably occurs by coordination of the thiophilic copper(I) reagent with sulfur, heterolysis to a phenylthio-stabilized carbonium ion with formation of the insoluble copper(I) thiophenoxide, and finally proton loss, giving the vinyl sulfide. Since trifluoromethanesulfonic acid is generated in stoichiometric quantity during the reaction, the medium becomes highly acidic. If the reactant or product is unstable to acid, as is the case in the present procedure, the pH can be kept neutral by adding N,N-diisopropylethylamine.

TABLE I

Thioacetal or Thioketal	Vinyl Sulfide	Yield (%)
CH_3CH_2CH with SC_6H_5 and SC_6H_5	CH_3—CH=CH—SC_6H_5	91
$(CH_3)_2CHCH$ with SC_6H_5 and SC_6H_5	CH_3, CH_3 —C=CH—SC_6H_5	85
C_6H_5S, SC_6H_5 on CH_3—C—C_6H_5	CH_2=C with SC_6H_5 and C_6H_5	85
cyclohexane with SC_6H_5 and SC_6H_5	cyclohexene—SC_6H_5	92
cyclohexyl—CH with SC_6H_5 and SC_6H_5	cyclohexylidene=CH—SC_6H_5	92
C_6H_5S, SC_6H_5 on $(CH_3)_2CH$—C—CH_3	CH_3, CH_3 —C=C with SC_6H_5 and CH_3	94
SC_6H_5 on CH_3CHCH_2CH with SC_6H_5 and SC_6H_5	CH_2=CH—CH=CH—SC_6H_5	84
C_6H_5S, SC_6H_5 on $C_6H_5SCH_2CH_2$—C—CH_3	CH_2=C(SC_6H_5)—CH=CH$_2$	76

The reaction of crotonaldehyde and methyl vinyl ketone with thiophenol in the presence of anhydrous hydrogen chloride effects conjugate addition of thiophenol as well as acetal formation. The resulting β-phenylthio thioacetals are converted to 1-phenylthio- and 2-phenylthio-1,3-butadiene, respectively, upon reaction with 2 equivalents of copper(I) trifluoromethanesulfonate (Table I).[7] The copper(I)-induced heterolysis of carbon–sulfur bonds has also been used to effect pinacol-type rearrangements of bis(phenylthio)methyl carbinols.[9] Thus, the addition of bis(phenylthio)methyllithium to ketones and aldehydes followed by copper(I)-induced rearrangement results in a one-carbon ring expansion or chain-insertion transformation, giving α-phenylthio ketones. Monothioketals of 1,4-diketones are cyclized to 2,5-disubstituted furans by the action of copper(I) trifluoromethanesulfonate.[6,10]

The most common procedure previously employed to effect the elimination of thiols from thioacetals has been heating in the presence of a protic acid.[11] For example, propionaldehyde diethyl thioacetal is converted to 1-ethylthio-1-propene on heating at 175° in the presence of phosphoric acid.[11a] The relatively high temperature and acidic conditions of such procedures are, however, distinct disadvantages of this method. Another approach consists of oxidation of a thioacetal to the mono S-oxide and thermal elimination of a sulfenic acid at 140–150°.[12]

Vinyl sulfides have found numerous synthetic applications. Vinyl sulfides unsubstituted in the 1-position are metallated readily, and the resulting 1-phenythio- or 1-alkylthiovinyllithium reagents have been utilized for nucleophilic acylation and other applications.[13,14] The phenylthio-substituted 1,3-butadienes serve as interesting functionalized dienes in Diels-Alder reactions.[7,15,16] For example, (Z)-2-methoxy-1-phenylthio-1,3-butadiene and methyl vinyl ketone afford an adduct with the normally inaccessible "meta" relationship between the methoxy and acetyl substituents.[7] The isomeric 2-methoxy-3-phenylthio- and 1-methoxy-4-phenylthio-1,3-butadienes have been prepared recently by thermal ring opening of the appropriate cyclobutenes.[16]

When methacrolein is substituted for acrolein in the present procedure, (Z)-2-methoxy-3-methyl-1-phenylthio-1,3-butadiene is produced in good yield.[17] However, when a β-alkyl group is present on the enal, a rearranged product, a 4-alkyl-1-methoxy-2-phenylthio-1,3-butadiene, is produced as a pair of geometric isomers.[17]

1. Department of Chemistry, University of Pittsburgh, Pittsburgh, Pennsylvania 15260.
2. L. Field and C. H. Banks, *J. Org. Chem.*, **40**, 2774 (1975).
3. A. W. Harriott and D. Picker, *Synthesis*, 447 (1975).
4. E. J. Corey and D. Seebach, *J. Org. Chem.*, **31**, 4097 (1966).
5. R. G. Salomon and J. K. Kochi, *J. Am. Chem. Soc.*, **95**, 1889, 3300 (1973).
6. T. Cohen, G. Herman, J. R. Falck, and A. J. Mura, Jr., *J. Org. Chem.*, **40**, 812 (1975).
7. T. Cohen, A. J. Mura, Jr., D. W. Shull, E. R. Fogel, R. J. Ruffner, and J. R. Falck, *J. Org. Chem.*, **41**, 3218 (1976).
8. T. Cohen and D. W. Shull, unpublished results.
9. T. Cohen, D. Kuhn, and J. R. Falck, *J. Am. Chem. Soc.*, **97**, 4749 (1975).
10. A. J. Mura, Jr., Ph.D. Thesis, University of Pittsburgh, 1976.
11. (a) H. J. Boonstra, L. Brandsma, A. W. Weigman, and J. F. Arens, *Recl. Trav. Chim. Pays-Bas*, **78**, 252 (1959); (b) L. Brandsma, P. Vermeer, J. G. A. Kooijman, H. Boelens, and J. T. M. Maessen, *Recl. Trav. Chim. Pays-Bas*, **91**, 729 (1972).
12. A. Deljac, Z. Stefanac, and K. Balenovic, *Tetrahedron*, Suppl. **8**, 33 (1966).

13. For a review see B.-T. Gröbel and D. Seebach, *Synthesis*, 357 (1977).
14. K. Oshima, K. Shimoji, H. Takahashi, H. Yamamoto, and H. Nozaki, *J. Am. Chem. Soc.*, **95**, 2694 (1973); I. Vlattas, L. D. Vecchia, and A. O. Lee, *J. Am. Chem. Soc.*, **98**, 2008 (1976); B. Harirchian and P. Magnus, *J. Chem. Soc. Chem. Commun.*, 522 (1977); R. R. Schmidt and B. Schmid, *Tetrahedron Lett.*, 3583 (1977).
15. D. A. Evans, C. A. Bryan, and C. L. Sims, *J. Am. Chem. Soc.*, **94**, 2891 (1972).
16. B. M. Trost and A. J. Bridges, *J. Am. Chem. Soc.*, **98**, 5017 (1976); B. M. Trost, S. A. Godleski, and J. Ippen, *J. Org. Chem.*, **43**, 4559 (1978).
17. T. Cohen and Z. Kosarych, *Tetrahedron Lett.*, **21**, 3955 (1980).

6-METHOXY-β-TETRALONE

[2(1H)-Naphthalenone, 3,4-dihydro-6-methoxy-]

Submitted by JAMES J. SIMS,[1] L. H. SELMAN and M. CADOGAN
Checked by ROBERT E. IRELAND and RONALD I. TRUST

1. Procedure

A 2-l., three-necked flask equipped with a mechanical stirrer, a reflux condenser fitted with a calcium chloride drying tube, and a pressure-equalizing dropping funnel is charged with 53.4 g. (0.400 mole) of anhydrous aluminum chloride (Note 1) and 800 ml. of dichloromethane (Note 2). The flask is placed in an acetone–dry ice bath, and the mixture is stirred for a few minutes before slowly adding a solution of 36.9 g. (0.200 mole) of (4-methoxyphenyl)acetyl chloride (Note 3) in 200 ml. of dichloromethane over 45 minutes. When the addition is complete, the funnel is replaced with a gas-inlet tube (Note 4), and ethylene is bubbled vigorously into the flask for about 10 minutes. The gas-inlet tube is replaced with a stopper, the cooling bath is removed, and the reaction mixture is allowed to warm to room temperature, then stirred for 3–3.5 hours (Note 5). The reaction mixture is cooled in an ice bath while 250 ml. of ice water is added *carefully* (Note 6). The mixture is stirred until all of the solid material is dissolved. The yellow organic layer is separated, washed two times with 150-ml. portions of 5% hydrochloric acid and two times with 150-ml. portions of saturated sodium hydrogen carbonate. The organic layer is dried over magnesium sulfate and filtered. The solvent is distilled with a rotary evaporator, keeping the bath temperature under 60°. Distillation (Note 7) of the yellow residue through a 15-cm. Vigreux column gives 21–24 g. (60–68%) of 6-methoxy-β-tetralone, b.p. 114–116° (0.2 mm.), which solidifies to a white solid on standing in the refrigerator, m.p. 33.5–35° (Note 8).

2. Notes

1. A 100% molar excess of aluminum chloride is necessary to obtain an acceptable yield in a short time. The reaction of phenylacetyl chloride with ethylene requires only 1 mole of aluminum chloride per mole of acid chloride.

2. Matheson, Coleman and Bell dichloromethane, b.p. 39.5–40.5°, was used without purification. The use of this solvent rather than carbon disulfide is the major improvement of this procedure over the published one.[2]

3. (4-Methoxyphenyl)acetyl chloride is prepared from (4-methoxyphenyl)acetic acid (Aldrich Chemical Company, Inc., m.p. 85–86.5°) by the procedure of Buckles and Cooper.[3] Thionyl chloride (50 ml.) is added to 100 g. (0.603 mole) of (4-methoxyphenyl)acetic acid in a 500-ml. round-bottomed flask containing a few boiling stones and a magnetic stirring bar, and fitted with a calcium chloride drying tube. The contents of the flask are stirred slowly for 24 hours at room temperature. Dry benzene (80 ml.) is added and removed by distillation on a rotary evaporator; the process is then repeated. The yellow-green liquid is transferred to a 200-ml. flask with a small volume of benzene and, after removal of the solvent under vacuum, distillation of the liquid through a 15-cm. Vigreux column affords 102–108 g. (92–97%) of (4-methoxyphenyl)acetyl chloride, b.p. 80–88° (0.5 mm.).

4. A glass tube (6 mm. o.d.), flanged at one end and bent to direct gas bubbles in the direction of stirring is used. A fritted disk will become clogged during the addition and should not be used.

5. This reaction time was found to give the best yield of pure product. The progress of the reaction should be checked by either IR spectroscopy or GC. A small aliquot (1–2 ml.) is worked up in a test tube by quenching with water, separating the organic phase, and drying over magnesium sulfate. The IR spectrum will show a disappearance of the acid chloride carbonyl peak at 1786 cm.$^{-1}$ and appearance of the 6-methoxy-β-tetralone carbonyl peak at 1701 cm.$^{-1}$ GC was carried out in a 185 cm. × 3.2 mm. column packed with 5% by weight SE-30 on Diatoport S (60–80 mesh). The reaction may also be followed visually. The initial yellow suspension changes to green and finally to red-brown with a green cast. The reaction is essentially complete with the precipitation of dark green aluminum salts.

6. Much heat is generated on addition of water to the dark red-brown mixture. The addition should be dropwise until the heat is dissipated.

7. If the distillation is not carried out promptly, the crude product should be placed under nitrogen in a freezer. The tetralone seems to keep well under nitrogen at low temperature in glassware that has been rinsed with ammonium hydroxide and dried in an oven; the distillation flask and column were also routinely treated this way.

8. This material contains a small amount of impurity (2–5%). A higher grade material may be obtained by discarding a 1–2 g. forerun.

3. Discussion

This procedure, an improvement over the method of Burckhalter and Campbell,[2] represents the most convenient method of preparing 6-methoxy-β-tetralone, a valuable intermediate for the synthesis of natural products, and provides a general method for the synthesis of substituted β-tetralones.[2,4]

6-Methoxy-β-tetralone has been synthesized from 6-methoxy-3,4-dihydronaphthalenecarboxylic acid by the Curtius reaction.[5] Other preparations include the Birch reduction of 6-methoxy-2-naphthol,[6,7] oxidation of 6-methoxytetralin,[6,8] and synthesis from 6-methoxy-α-tetralone.[9] The other practical approaches depend upon tedious preparations of naphthalene derivatives.

1. Department of Plant Pathology, University of California, Riverside, California 92502.
2. J. H. Burckhalter and J. R. Campbell, *J. Org. Chem.*, **26**, 4232 (1961). See also, J. Colonge and J. Chambion, *C. R. Hebd. Seances Acad. Sci.*, **224**, 128 (1947).
3. L. F. Fieser and M. Fieser, "Reagents for Organic Synthesis," Vol. 1, Wiley, New York, 1967, p. 1159.
4. A. Rosowsky, J. Battaglia, K. K. N. Chen, and E. J. Modest, *J. Org. Chem.*, **33**, 4288 (1968).
5. G. P. Crowley and R. Robinson, *J. Chem. Soc.*, 2001 (1938).
6. N. A. Nelson, R. S. P. Hsi, J. M. Schuck, and L. D. Kahn, *J. Am. Chem. Soc.*, **82**, 2573 (1960).
7. R. L. Kidwell and S. D. Darling, *Tetrahedron Lett.*, 531 (1966).
8. W. Salzer, *Hoppe-Seyler's Z. Physiol. Chem.*, **274**, 46 (1942).
9. W. Nagata and T. Terasawa, *Chem. Pharm. Bull.*, **9**, 267 (1961) [*Chem. Abstr.*, **55**, 27227b (1961)].

METHYL GROUPS BY REDUCTION OF AROMATIC CARBOXYLIC ACIDS WITH TRICHLOROSILANE- TRI-*n*-PROPYLAMINE: 2-METHYLBIPHENYL

[1,1′-Biphenyl, 2-methyl]

Submitted by GEORGE S. LI, DAVID F. EHLER, and R. A. BENKESER[1]

Checked by KYO OKADA and WATARU NAGATA

1. Procedure

Caution! This procedure should be conducted in a well-ventilated hood to avoid inhalation of trichlorosilane and hydrogen chloride.

A 300-ml., three-necked, round-bottomed flask is equipped with a magnetic stirrer, a thermometer, a glass stopper, and an efficient condenser attached to a nitrogen line with a gas bubbler (Note 1). The system is flushed with dry nitrogen, then charged with 19.8 g. (0.100 mole) of biphenyl-2-carboxylic acid (Note 2), 80 g. (60 ml., 0.59 mole) of trichlorosilane (Notes 3 and 4), and 80 ml. of acetonitrile (Note 5). A low nitrogen flow is maintained as the mixture is stirred and heated to reflux (40–45°) for 1 hour or until gas evolution ceases and the carboxylic acid has dissolved. The solution is then cooled in an acetone–dry ice bath to at least 0°, and the glass stopper is replaced with a 100-ml., pressure-equalizing dropping funnel charged with 37.8 g. (0.264 mole) of tri-*n*-propylamine (Notes 6 and 7), which is emptied rapidly into the stirred solution. The cooling bath is removed, and the flask contents are allowed to stir until the reaction ceases to be exothermic. A heating mantle is then used to maintain reflux for 16 hours (Note 8), during which time the temperature rises to 70–75°.

As soon as reflux is terminated, the solution is poured rapidly into a 1-l. Erlenmeyer flask, allowed to cool, and diluted with enough anhydrous diethyl ether (Note 9) to make the total volume about 850 ml. (Note 10). After the flask has been sealed and refrigerated for one hour, the precipitate is removed by rapid filtration (water aspirator) through a 150-ml. Büchner funnel and washed with three 50-ml. portions of anhydrous ether. The clear yellow filtrate is concentrated as follows. A 300-ml., one-necked, round-bottomed flask is fitted with a magnetic stirring bar and a 100-mm. Vigreux column topped with a distillation head. The filtrate is placed in a 1-l., pressure-equalizing dropping funnel inserted into the top of the distillation head. Approximately 100 ml. is run down through the column into the flask, which is then heated. As ether is removed by distillation, the remainder of the filtrate is dripped into the flask at a constant rate, and in this way the solution is concentrated into a small flask in a continuous operation. Distillation is continued until most of the ether has been removed, and the resulting murky solution is heated at 40° (80 mm.), removing the remaining volatiles.

The Vigreux column, dropping funnel, and distilling head are then replaced by a 100-ml., pressure-equalizing dropping funnel charged with 100 ml. of methanol. With the top of the funnel left open to the atmosphere, the methanol is added slowly to the oily flask contents (Note 11). After vigorous boiling has ceased, the solution is heated under reflux for one hour, cooled in an ice bath, then treated slowly with a solution of 56 g. (1.0 mole) of potassium hydroxide in 25 ml. of water and 50 ml. of methanol.[2] The resulting mixture is heated under reflux for 19 hours (Note 8), dissolved in 600 ml. of water, and extracted three times with 100 ml. of ether. The extracts are combined, washed once with 50 ml. of 5 N hydrochloric acid, and dried over anhydrous magnesium sulfate. Ether is removed by distillation as described above, using a 300-ml., pressure-equalizing dropping funnel, a 50-ml., round-bottomed distilling flask, and a 100-mm. Vigreux column. Vacuum distillation of the remaining liquid gives 12.5–13.4 g. (74–80%) of 2-methylbiphenyl, b.p. 76–78° (0.5 mm.), n_D^{20} 1.5920 (Notes 12 and 13).

2. Notes

1. All glassware is thoroughly dried by flame or in an oven prior to use.

2. Biphenyl-2-carboxylic acid was purchased from Aldrich Chemical Company, Inc., and used without further purification.

3. The submitters used trichlorosilane supplied by Union Carbide Corporation. The checkers obtained trichlorosilane from Tokyo Chemical Industries Company, Ltd., Japan.

4. Good results may sometimes be achieved with a 4:1 or 5:1 mole ratio of trichlorosilane to carboxylic acid. Excess trichlorosilane is needed to compensate for losses of this volatile reactant over extended reflux periods.

5. The submitters used reagent grade acetonitrile (Mallinckrodt Chemical Works) dried prior to use by storage over Matheson Linde type 4A molecular sieves. The checkers used reagent grade acetonitrile obtained from Ishizu Pharmaceutical Company, Japan, dried prior to use by storage over 4A molecular sieves obtained from Nakarai Chemicals, Ltd., Japan.

6. The submitters used tri-n-propylamine obtained from Aldrich Chemical Company, Inc., and the checkers used tri-n-propylamine obtained from Wako Pure Chemical Industries, Ltd., Japan. Both groups stored the reagent over Linde type 4A molecular sieves prior to use.

7. To ensure a homogeneous reducing medium, the tri-n-propylamine:trichlorosilane ratio should be about 1:2.

8. Overnight reflux was chosen partly for convenience. Similar results are possible with somewhat shorter reaction times.

9. The submitters employed anhydrous ether obtained from Mallinckrodt Chemical Works; the checkers used anhydrous ether obtained from Wako Pure Chemical Industries, Ltd., Japan and distilled it from sodium hydride under nitrogen shortly prior to use.

10. The volume of ether added should be sufficient to precipitate most of the tri-n-propylamine hydrochloride in solution. The checkers diluted to a total volume of about 1 l. to precipitate the salt more efficiently.

11. Vigorous evolution of hydrogen chloride is observed as the methanol is added.

12. The literature[3] value for a carefully purified sample of 2-methylbiphenyl is n_D^{20} 1.5914.

13. ^{1}H NMR (CDCl$_3$), δ (multiplicity, number of protons, assignment): 2.55 (s, 3H, CH_3), 7.27 (s, 4H, C_6H_4), 7.33 (s, 5H, C_6H_5). IR (CHCl$_3$) cm.$^{-1}$: 1600 medium, 1480 medium strong (aromatic), 1380 medium (CH$_3$). GC analysis of the product (1.5 m. by 0.5 cm. glass column, KF-54 on Chromosorb W, 60–80 mesh) showed a single peak with a retention time of 2.60 minutes at 170°.

3. Discussion

2-Methylbiphenyl has been prepared by diazotization of o-toluidine and coupling with benzene (8%);[4] by reaction of o-tolylmagnesium bromide with cyclohexanone, followed by dehydration of the resulting alcohol and dehydrogenation (30–50%);[5] and by coupling o-tolyllithium with chlorobenzene in the presence of piperidine (51%).[6] The present procedure gives 2-methylbiphenyl in much improved yield.

In a more general sense, this reduction method provides a convenient pathway for converting an aromatic carboxyl group to a methyl group (see Table I).[7] Previously, this transformation has been achieved by reduction of the acid to the alcohol with lithium aluminum hydride, conversion of the alcohol to the tosylate, and a second reduction either with lithium aluminum hydride, or Raney nickel and hydrogen.[8] Alcohols of the benzylic type have also been reduced directly with hydrogen under pressure in the presence of various catalysts,[9] and benzoic acids have been reduced to toluenes with rhenium-type catalysts and hydrogen at high temperatures and pressures.[10]

TABLE I

REDUCTION OF AROMATIC ACIDS TO SUBSTITUTED BENZENES[7]

Starting Acid	Product (% Yield)
Benzoic	Toluene (78)
m-Toluic	m-Xylene (82)
p-Toluic	p-Xylene (74)
3,5-Dimethylbenzoic	Mesitylene (82)
4-Chlorobenzoic	p-Chlorotoluene (94)
4-Bromobenzoic	p-Bromotoluene (94)
Phthalic	o-Xylene (64)

1. Chemistry Department, Purdue University, West Lafayette, Indiana 47907.
2. C. Eaborn and S. H. Parker, *J. Chem. Soc.*, 126 (1955). These authors note that similar

cleavages have been effected in 4–8 hours with an excess of refluxing potassium hydroxide–methanol–water.

3. I. A. Goodman and P. H. Wise, *J. Am. Chem. Soc.*, **72**, 3076 (1950).
4. M. Gomberg and J. C. Pernert, *J. Am. Chem. Soc.*, **48**, 1372 (1926).
5. I. R. Sherwood, W. F. Short, and R. Stansfield, *J. Chem. Soc.*, 1832 (1932). See also M. Orchin, *J. Am. Chem. Soc.*, **67**, 499 (1945).
6. R. Huisgen, J. Sauer, and A. Hauser, *Chem. Ber.*, **91**, 2366 (1958).
7. R. A. Benkeser, K. M. Foley, J. M. Gaul, and G. S. Li, *J. Am. Chem. Soc.*, **92**, 3232 (1970).
8. N. G. Gaylord, "Reduction with Complex Metal Hydrides," Interscience, New York, 1956, p. 855.
9. R. L. Shriner and R. Adams, *J. Am. Chem. Soc.*, **46**, 1683 (1924).
10. H. S. Broadbent, G. C. Campbell, W. J. Bartley, and J. H. Johnson, *J. Org. Chem.*, **24**, 1847 (1959); H. S. Broadbent and D. W. Seegmiller, *J. Org. Chem.*, **28**, 2347 (1963).

1-*d*-ALDEHYDES FROM ORGANOMETALLIC REAGENTS: 2-METHYLBUTANAL-1-*d*

(Butanal-1-*d*, 2-methyl-)

Submitted by G. E. Niznik, W. H. Morrison, III, and H. M. Walborsky[1]
Checked by Frank E. Herkes and Richard E. Benson

1. Procedure

A. 1,1,3,3-*Tetramethylbutyl isonitrile.* A 3-l., three-necked, round-bottomed flask fitted with a Hershberg stirrer, a 500-ml. pressure-equalizing addition funnel, and a nitrogen-inlet tube is flamed dry under a nitrogen atmosphere and allowed to cool. The nitrogen-inlet tube is replaced with a low-temperature thermometer, and the nitrogen line is attached to a Y-tube placed on the addition funnel. To the flask are added 118 g. (0.752 mole) of N-(1,1,3,3-tetramethylbutyl)formamide (Note 1) and 1500 ml. of N,N-dimethylformamide (Note 2). The addition funnel is charged with a premixed (Note 3) solution of 89 g. (55 ml., 0.75 mole) of thionyl chloride and 250 ml. of N,N-dimethylformamide. The flask is immersed in an acetone–dry ice bath, and moderately fast stirring is started. When the temperature of the flask reaches $-50°$, the solution in the funnel is added at a rate such that the temperature ranges between $-55°$ and $-50°$ (about 10 minutes are required for the addition). After the addition is complete, the bath is removed momentarily, allowing the reaction temperature to rise to $-35°$. The bath is then replaced, and 159 g. (1.50 mole) of dry sodium carbonate (Note 4) is added directly to the mixture (Note 5). After the addition the bath is removed, and the flask contents are stirred for an additional 6 hours at room temperature (Note 6). The reaction mixture (Note 7) is poured into a 6-l. Erlenmeyer flask containing 3 l. of ice water (Note 8). The reaction flask is rinsed with 300 ml. of pentane and sufficient water to dissolve the inorganic material that may be present (Note 9). The washings are added to the Erlenmeyer flask. The mixture is stirred vigorously for 5 minutes, and the layers are separated. The upper layer is washed twice with 100-ml. portions of water and dried over anhydrous sodium sulfate. The solution is filtered, and the pentane is removed by distillation. The crude product is distilled through a 1.5 × 15 cm. Vigreux column; the fraction collected at 55.5–56.6° (11 cm.), yields 86–90 g. (82–87%) of 1,1,3,3-tetramethylbutyl isonitrile, n_D^{30} 1.4178, d^{25} 0.7944. The compound shows strong absorption in the IR (CCl$_4$) at 2130 cm.$^{-1}$ attributable to the isonitrile function. The ^{1}H NMR spectrum (neat, external tetramethylsilane reference) shows peaks at δ 1.08 (s, 9H, C(CH$_3$)$_3$), 1.43 (t, $J_{^{14}N-H}$ = 2 Hz., 6H, C(CH$_3$)$_2$), and 1.58 (t, $J_{^{14}N-H}$ = 2.3 Hz., 2H, CH$_2$).

B. N-(2-*Methylbutylidene-1-d*)-1,1,3,3-*tetramethylbutylamine* (Note 10). A 1-1., three-necked, round-bottomed flask fitted with a Teflon paddle stirrer, a 500-ml. pressure-equalizing addition funnel, and a nitrogen-inlet tube is flamed dry under a nitrogen atmosphere and allowed to cool. The nitrogen-inlet tube is replaced with a thermometer, and the nitrogen line is attached to a Y-tube placed on the addition funnel. A solution of 27.8 g. (35.1 ml., 0.200 mole) of 1,1,3,3-tetramethylbutyl isonitrile in 300 ml. of anhydrous diethyl ether is added to the flask. The flask is cooled to 0° with an ice–salt bath, and 0.2 mole of 2-butyllithium in hexane (Note 11) is transferred to the addition funnel with a syringe. The alkyllithium solution is added to the stirred (Note 12) solution at such a rate that the temperature never exceeds 5°. After the addition is complete, the mixture is stirred for 15 minutes as the temperature slowly drops to $-5°$, and 8 ml. (0.4 mole) of deuterium oxide (Note 13) is injected rapidly into the reaction mixture (Note 14). The ice bath is removed; the mixture is stirred for 30 minutes and filtered through a Büchner funnel into a 1-l., round-bottomed flask. The reaction flask is rinsed with pentane, and the rinse is added to the flask. After evaporation of the solvent, 33.7–34.9 g. (85–88%) of the aldimine is collected by distillation through a 1.5 × 15 cm. Vigreux column, b.p. 52.5–54° (1.5 mm.), $n_D^{24.5}$ 1.4321. The IR spectrum (neat) shows strong absorption at 1663 cm.$^{-1}$, attributable to the isonitrile function.

C. *2-Methylbutanal-1-d*. A 1-l., three-necked, round-bottomed flask is equipped with a dropping funnel, a gas-inlet tube for steam, and a Dean-Stark trap attached to a condenser through which acetone cooled to $-15°$ is circulated (Note 15). A solution of 50.4 g. (0.400 mole) of oxalic acid dihydrate in 200 ml. of water is added to the flask and heated at reflux. Steam is introduced into the flask, and when it begins to condense in the Dean-Stark trap, the aldimine from part B is added dropwise from the funnel. The aldehyde and water collect in the trap, and the water layer is periodically removed. After the distillation of the aldehyde is complete, the product is drained from the trap; the water layer is separated and discarded. The oil is washed with three 25-ml. portions of saturated sodium chloride and dried over anhydrous calcium sulfate. The crude aldehyde is decanted from the drying agent and distilled through a short Vigreux column, giving 13.0–13.3 g. (87–88%) of high-quality 2-methylbutanal-1-d, b.p. 92–93°, n_D^{30} 1.3896 (Note 16). The IR spectrum (neat) shows strong adsorption at 1721 cm.$^{-1}$ attributable to the carbonyl function.

2. Notes

1. *N*-(1,1,3,3-tetramethylbutyl)formamide is prepared in 86–90% yield by refluxing 194 g. (1.50 moles) of 1,1,3,3-tetramethylbutylamine with 138 g. (3.00 moles) of formic acid in 400 ml. of toluene. Azeotropic distillation using a Dean-Stark trap gradually removes all water and excess formic acid. The toluene is removed by distillation at atmospheric pressure, and the product is distilled at reduced pressure, b.p. 76–77° (1 mm.), n_D^{25} 1.4521, yielding 203–214 g. (86–90%).

2. Industrial grade *N,N*-dimethylformamide is purified by distillation, first at atmospheric pressure to remove most of the water in the initial small fraction, and then by distillation at reduced pressure from barium oxide, b.p. 63° (30 mm.).

3. A temperature rise of about 30° is observed.

4. Commercial anhydrous sodium carbonate is dried in a vacuum oven at 130° for 1 hour.

5. The checkers used a solid addition funnel and added the sodium carbonate under nitrogen over a 10-minute period.

6. The mixture can be left stirring overnight since the isonitrile is stable to the reaction conditions. Alternatively, a hot-water bath can be used, with very fast stirring, to heat the mixture quickly to 35°. The bath is then removed, and after 1 hour of additional stirring, the mixture is ready for the workup procedure.

7. *Isonitriles are presumed to be toxic*, and it is recommended that the workup procedure be performed in a hood. Unlike most isonitriles, however, 1,1,3,3-tetramethylbutyl isonitrile (TMBI) is not malodorous. It has a sweetish pine odor, which becomes unpleasant only after continued inhalation.

8. The addition of the reaction mixture to water is exothermic.

9. The inorganic salts are not always soluble in the amount of water specified.

10. This procedure uses the commercially available 2-butyllithium reagent. The submitters state that the corresponding Grignard reagent may also be used. A 300-ml., three-necked, round-bottomed flask is fitted with an addition funnel, a reflux condenser, a magnetic stirring bar, and a nitrogen-inlet tube. Magnesium turnings (3.65 g., 0.150

mole) and 80 ml. of anhydrous tetrahydrofuran (Note 17) are added to the flask, and a nitrogen atmosphere is maintained. 2-Bromobutane (20.6 g., 0.150 mole), 0.25 ml. of 1,2-dibromoethane, and 70 ml. of tetrahydrofuran are placed in the addition funnel. Stirring is begun, and the solution is added dropwise at a rate which sustains refluxing. After the addition is complete, the solution is stirred until room temperature is reached. The amount of Grignard reagent prepared is determined (Note 18). To this Grignard solution is added 14.2 g. (0.102 mole) of 1,1,3,3-tetramethylbutyl isonitrile (Note 19). After stirring for 4–6 hours (Note 20), the solution is cooled in an ice bath to 0°. To the rapidly stirred solution (Note 21) is injected 6.0 ml. (0.30 mole) of deuterium oxide (Note 13). The ice bath is removed, and the mixture is stirred for 10 minutes before 50 ml. of water is added (Note 22). The contents of the reaction flask are decanted into a 1-l. separatory funnel containing 200 ml. of ether. The aqueous layer is separated, and the ether layer is washed with 100 ml. of saturated sodium chloride. The magnesium salts remaining in the reaction vessel are washed twice with 100-ml. portions of ether. The ether extracts are washed with 100 ml. of saturated sodium chloride. The combined ether solutions are dried over anhydrous sodium sulfate and evaporated with a rotary evaporator, giving the crude aldimine. Distillation (see Part B) yields 13.7 g. (67%) of the pure aldimine. Hydrolysis and steam distillation (as described in Note 15) yield 5.85 g. (65%, overall) of the aldehyde (Note 23).

11. The commercially available organolithium reagent is titrated with benzoic acid, using triphenylmethane as an indicator according to the procedure of Eppley and Dixon.[2] The checkers used product available from Alfa Inorganics, Inc.

12. During the addition of the alkyllithium the mixture becomes gelatinous. As this happens, the stirring rate is increased to ensure thorough mixing.

13. Deuterium oxide having an isotopic purity of >99% was used. The product available from Columbia Organic Chemicals Company, Inc., was used by the checkers.

14. The stirring should be very rapid at this point or frothing will occur. The flask temperature will reach 30° during deuteriolysis. It is important that the temperature of the ice–salt bath remains at −10° to −15°.

15. The submitters recommend the following procedure for steam distillation of low-boiling aldehydes. A 500-ml., three-necked flask is fitted with two addition funnels, one of which has a double-bore stopcock for external drainage. This addition funnel is fitted with a cold finger (−5°) and an inlet tube leading to a bubbler and a nitrogen source. The aldimine is placed in the other addition funnel. While the aldimine is added dropwise to the refluxing oxalic acid solution, the distillate passes up the equalizing pressure tube of the collecting funnel and is condensed by the cold finger. The water layer is periodically drained back into the flask. After distillation, the aldehyde is washed with saturated sodium chloride, then drained from the funnel through the external tube.

16. The submitters found that GC analysis of the undistilled aldehyde, conducted on a column packed with 16% LS-40 on Chromosorb P/AW at 100°, indicated a purity of 98.6%. ^{1}H NMR (CCl$_4$) analysis indicated an isotopic purity of 97.9% (trace of impurity at δ 9.60 due to CHO. If the crude aldimine is hydrolyzed, the aldehyde is obtained in 96% overall yield; however, the purity is only 94% by GC analysis. The checkers found no detectable impurity by GC in the distilled aldehyde, and the NMR spectrum indicated very high isotopic purity.

17. Alkyl magnesium halides form intermediates with tetramethylbutyl isonitrile which are not very soluble in ether. If it is necessary to prepare the Grignard reagent in ether, the ether should either be diluted with tetrahydrofuran or replaced by tetrahydrofuran. See the warning note in *Org. Synth.*, **Coll. Vol. 5,** 976 (1973), for purification of tetrahydrofuran.

18. The molarity is determined[3] by adding an excess of standardized acid and back-titrating with base. The moles of Grignard reagent present are determined based on a volume of 150 ml. When ether is used to prepare the Grignard reagent, an actual measurement of the volume is necessary. This can be done conveniently by transferring the solution back into the addition funnel with a large graduated syringe. The Grignard reagent content averages 0.102 mole.

19. The molar amount of 1,1,3,3-tetramethylbutyl isonitrile used is equivalent to the Grignard reagent content.

20. After 4 hours, periodic aliquots are taken and worked up. The disappearance of the isonitrile absorption at 2130 cm.$^{-1}$ and the appearance of the imine absorption at 1665 cm.$^{-1}$ are used for analysis. Usually within 6 hours the isonitrile peak vanishes or remains at a very low constant intensity indicative of completion of the reaction.

21. The sudden quenching with deuterium oxide minimizes the exchange between the 1-metalloaldimine and the active hydrogen at the C-2 position of the already-deuteriated aldimine. Performing the reaction in refluxing tetrahydrofuran produces an exchange (approximately 10%) with incorporation of deuterium in the C-2 position. If only the 1-H-aldehyde is desired, 50 ml. of water is added dropwise.

22. Saturated ammonium chloride solution slowly hydrolyzes the aldimine to the aldehyde, which, in this case, is undesirable.

23. NMR analysis shows that deuterium incorporation at the C-1 position is 95.3% and at C-2, 5%.

TABLE I

ALDEHYDES FROM 1,1,3,3-TETRAMETHYLBUTYL ISONITRILE
AND ORGANOMETALLIC REAGENTS

Organometallic Reagent	Aldehyde, % Yield
n-Butyllithium	93[a]
Phenyllithium	55[a]
2-Butylmagnesium bromide	67 (96[b,c])
tert-Butylmagnesium bromide	48[b]
n-Hexylmagnesium bromide	62[b]
2-Phenylethylmagnesium bromide	63 (80[b,c])
Cyclopentylmagnesium bromide	66 (89[b,c])

[a]Reference 6.
[b]Reference 5.
[c]Percent deuterium at C-1 as determined by NMR.

3. Discussion

This reaction illustrates a general procedure for the preparation of 1-d-aldehydes from aliphatic and alicyclic[4] Grignard[5] and lithium[6] reagents. The use of the lithium reagent is normally preferred because of higher yields and greater isotopic purity if the 1-d-aldehyde is desired. For the synthesis of aromatic aldehydes, the use of the lithium reagent is specifically preferred since aromatic Grignard reagents react poorly with 1,1,3,3-tetramethylbutyl isonitrile. The aldehydes prepared by this method are illustrated in Table I.

The intermediate aldimines can also be alkylated[7] or used as condensing agents[8] by removal of the α-hydrogen atom. The metalloaldimine is a useful intermediate for the preparation of α-keto acids, acyloins, β-hydroxy ketones, and silyl ketones.[5,6]

1. Chemistry Department, Florida State University, Tallahassee, Florida 32306.
2. R. L. Eppley and J. A. Dixon, *J. Organomet. Chem.*, **8**, 176 (1967).
3. L. F. Fieser and M. Fieser, "Reagents for Organic Synthesis," Vol. 1, Wiley, New York, 1967, p. 417.
4. Benzyl and vinyllithium reagents do not add to TMBI. For a discussion of this reaction see, G. E. Niznik, W. H. Morrison and H. M. Walborsky, *J. Org. Chem.*, **39**, 600 (1974); and M. P. Periasamy and H. M. Walborsky, *Org. Prep. Proced. Int.*, **11**, 293 (1979).
5. H. M. Walborsky, W. H. Morrison, III, and G. E. Niznik, *J. Am. Chem. Soc.*, **92**, 6675 (1970).
6. H. M. Walborsky and G. E. Niznik, *J. Am. Chem. Soc.*, **91**, 7778 (1969).
7. G. Stork and S. R. Dowd, *J. Am. Chem. Soc.*, **85**, 2178 (1963).
8. G. Wittig, H. D. Frommeld, and P. Suchanek, *Angew. Chem.*, **75**, 978 (1963) [*Angew. Chem., Int. Ed. Engl.*, **2**, 683 (1963)].

PREPARATION OF VINYL TRIFLUOROMETHANESULFONATES: 3-METHYL-2-BUTEN-2-YL TRIFLATE

(Methanesulfonic acid, trifluoro-, 1,2-dimethyl-1-propenyl ester)

$$CF_3SO_3H \xrightarrow[-H_2O]{P_2O_5} (CF_3SO_2)_2O$$

$$(CH_3)_2CHCOCH_3 + (CF_3SO_2)_2O + \text{(pyridine)} \xrightarrow[-78 \text{ to } 25°]{\text{pentane}}$$

$$(CH_3)_2C{=}CCH_3 + (CH_3)_2CHC{=}CH_2 + \text{(pyridinium)} NH^{\oplus} \ ^{\ominus}OSO_2CF_3$$
$$\underset{OSO_2CF_3}{|} \qquad \underset{OSO_2CF_3}{|}$$

Submitted by Peter J. Stang[1] and Thomas E. Dueber
Checked by Wayne Jaeger and Herbert O. House

1. Procedure

A. *Trifluoromethanesulfonic anhydride.* A dry, 100-ml., round-bottomed flask is charged with 36.3 g. (0.242 mole) of trifluoromethanesulfonic acid (Note 1) and 27.3 g. (0.192 mole) of phosphorus pentoxide (Note 2). The flask is stoppered and allowed to stand at room temperature for at least 3 hours. During this period the reaction mixture changes from a slurry to a solid mass. The flask is fitted with a short-path distilling head and heated first with a stream of hot air from a heat gun and then with the flame from a small burner. The flask is heated until no more trifluoromethanesulfonic anhydride distills, b.p. 82–115°, yielding 28.4–31.2 g. (83–91%) of the anhydride, a colorless liquid. Although this product is sufficiently pure for use in the next step, the remaining acid may be removed from the anhydride by the following procedure. A slurry of 3.2 g. of phosphorus pentoxide in 31.2 g. of the crude anhydride is stirred at room temperature in a stoppered flask for 18 hours. After the reaction flask has been fitted with a short-path distilling head, it is heated with an oil bath, yielding 0.7 g. of forerun, b.p. 74–81°, followed by 27.9 g. of the pure trifluoromethanesulfonic acid anhydride, b.p. 81–84° (Note 3).

B. *3-Methyl-2-buten-2-yl triflate.* A solution of 2.58 g. (0.0300 mole) of 3-methyl-2-butanone (Note 4) and 2.78 g. (0.0352 mole) of anhydrous pyridine (Note 5) in 10 ml. of anhydrous n-pentane (Note 6) is placed in a dry, 50-ml. Erlenmeyer flask, and the flask is stoppered with a rubber septum. After the solution has been cooled in an acetone–dry ice cooling bath, 9.72 g. (5.80 ml., 0.0395 mole) of trifluoromethanesulfonic anhydride is added with a hypodermic syringe, dropwise and with swirling over 2–3 minutes. The resulting mixture, from which a white solid separates initially, is allowed to warm to room temperature over 22–24 hours. During this period the reaction mixture becomes red in color (Note 7), and a viscous, red semisolid separates. The supernatant pentane solution is decanted, and the residual viscous semisolid is washed with two 10-ml. portions of pentane. While the combined pentane solutions are stored over anhydrous potassium carbonate, the remaining red semisolid is dissolved in 5 ml. of saturated aqueous sodium bicarbonate and extracted with three 5-ml. portions of pentane. The combined pentane solutions (including the solid potassium carbonate) are washed rapidly (Note 8) with 5 ml. of cold saturated aqueous sodium hydrogen carbonate and dried over anhydrous potassium carbonate. After the orange pentane solution has been concentrated to a volume of approximately 10 ml. with a rotary evaporator, it is transferred to a small distilling apparatus. The remaining pentane is removed by distillation at atmospheric pressure (Note 9). The residual liquid is fractionally distilled under reduced pressure, yielding 2.94–2.97 g. (45%) of a colorless, liquid fraction containing 3-methyl-2-buten-2-yl triflate and a lesser amount of 3-methyl-1-buten-2-yl triflate, b.p. 58–66° (22 mm.), n_D^{25} 1.3832–1.3898 (Note 10). Fractional distillation of this mixture with a 40-cm. spinning band column separated higher boiling fractions, b.p. 45–47° (12 mm.), that contained (Note 10) 98% of the 3-methyl-2-buten-2-yl triflate as a colorless liquid, n_D^{25} 1.3838 (Note 11).

2. Notes

1. Trifluoromethanesulfonic acid, purchased from Eastman Organic Chemicals, was used without purification. Trifluoromethanesulfonic acid in large quantities is available from 3M Company as Fluorocarbon Acid FC-24.

2. The phosphorus pentoxide should be protected from atmospheric moisture by weighing this reagent in a dry, stoppered flask. On a larger scale the yields are better and the reaction easier if the phosphorus pentoxide is premixed with an equal volume of Celite (filter aid).

3. This product exhibits one major GC peak (retention time 2.3 minutes, silicone fluid QF_1 on Chromosorb P) as well as one minor, unidentified, more rapidly eluted impurity. The product has strong IR absorption (CCl_4) at 1470, 1240, and 1130 cm.$^{-1}$.

4. 3-Methyl-2-butanone, purchased from Eastman Organic Chemicals, was used without purification. In general, commercially available ketones may be used without further purification.

5. A reagent grade of pyridine, purchased from Fisher Scientific Company, was dried over anhydrous potassium carbonate, distilled, and collected at 112–113°.

6. A commercial sample of n-pentane was distilled from calcium hydride to separate the pure solvent, b.p. 35–36°.

7. The intensity of color developed in the reaction mixture is an approximate indication of the extent of reaction. With reactive triflates a dark, almost tarry-looking mass develops in a few hours, while with the slower forming triflates several days at room temperature may be required for adequate color development. In no case was any product isolated when fairly dark color had not developed in the reaction mixture.

8. Only relatively stable vinyl triflates should be washed with water. More reactive triflates such as α-styryltriflate do not survive washing with water.

9. For the more volatile triflates, removal of the solvent should be accomplished by

distillation to minimize loss of the volatile product. Also, care should be taken not to overheat the residual product since overheating can result in decomposition.

10. The fractions from this distillation may be analyzed by GC, employing a column packed with Carbowax 20 M suspended on Chromosorb P. The retention times for the various components (minutes) are: pentane (1.6), 3-methyl-2-butanone (4.2), 3-methyl-1-buten-2-yl triflate (6.7), and 3-methyl-2-buten-2-yl triflate (9.5).

11. The pure, more highly substituted olefin, n_D^{25} 1.3840, could also be separated by preparative GC, and the unchanged ketone could be separated from the two triflate isomers by chromatography on a silica gel column with pentane as the eluant. The pure product has IR absorption (CCl$_4$) at 1700 (enol C=C), 1210 and 1140 cm.$^{-1}$ (SO$_2$) with end absorption in the UV (heptane, ϵ 700 at 210 nm) and the following broad singlets in the ^{1}H NMR spectrum (C$_6$H$_6$): δ 1.42 (3H, CH$_3$), 1.63 (3H, CH$_3$), and 1.82 (3H, CH$_3$). The mass spectrum of the product exhibits the following relatively abundant peaks, m/e (relative intensity): 218 (M$^+$, 58), 69 (75), 57 (92), 43 (100), 41 (44), and 39 (24).

3. Discussion

Vinyl trifluoromethanesulfonates (triflates) are a new class of compounds, unknown before 1969, that have been used most extensively in solvolytic studies to generate vinyl cations[2-9] and unsaturated carbenes.[10] Three methods have been used to prepare these sulfonic esters. The first, involving the preparation and decomposition of acyltriazines,[11] requires several steps to prepare the acyltriazines and is limited to the preparation of fully substituted vinyl triflates. The second method involves the electrophilic addition of trifluoromethanesulfonic acid to acetylenes[12-14] and, consequently, is not applicable to the preparation of trisubstituted vinyl triflates and certain cyclic vinyl triflates. However, this second procedure is relatively simple and often gives purer products in higher yield than the discussed reaction with ketones. Table I lists vinyl triflates prepared by this procedure.

The third procedure, illustrated by this preparation, involves the reaction of ketones with trifluoromethanesulfonic anhydride in a solvent such as pentane, dichloromethane, or carbon tetrachloride and in the presence of a base such as pyridine, lutidine, or anhydrous sodium carbonate[4-7,14,15] and, most recently, the sterically hindered,

TABLE I

VINYL TRIFLATES PREPARED BY THE REACTION OF ACETYLENES WITH
TRIFLUOROMETHANESULFONIC ACID

Substrate	Product	b.p.	Yield, %
CH$_3$C≡CH	CH$_2$=C(CH$_3$)OSO$_2$CF$_3$	25–27° (12 mm.)	60–80
C$_2$H$_5$C≡CC$_2$H$_5$	cis and trans	68.5–69.5° (25 mm.)	40–60
	C$_2$H$_5$CH=C(C$_2$H$_5$)OSO$_2$CF$_3$		
(CH$_3$)$_3$CC≡CH	CH$_2$=C(OSO$_2$CF$_3$)C(CH$_3$)$_3$	45–50° (15 mm.)	40–60
C$_6$H$_5$C≡CH	CH$_2$=C(C$_6$H$_5$)OSO$_2$CF$_3$	44–45° (0.3 mm.)	20–60
CH$_3$(CH$_2$)$_3$C≡CH	CH$_2$=C(OSO$_2$CF$_3$)CH$_2$(CH$_2$)$_2$CH$_3$	67–69° (15 mm.)	70
(CH$_3$)$_2$CHC≡CH	CH$_2$=C(OSO$_2$CF$_3$)CH(CH$_3$)$_2$	37–40° (15 mm.)	62

nonnucleophilic 2,6-di-*tert*-butyl-4-methylpyridine.[16] This procedure, which presumably involves either acid- or base-catalyzed enolization of the ketone followed by acylation of the enol with the acid anhydride, has also been used to prepare other vinyl sulfonate esters such as tosylates[8] or methanesulfonates.[9] Vinyl tosylates (but not vinyl triflates) may also be prepared from the ditosylate derivatives of 1,2-diols.[17]

Examples of the use of this procedure to prepare vinyl triflates from ketones are provided in Table II. Often mixtures of *cis* and *trans* isomers as well as the various double bond isomers of vinyl triflates are obtained by this procedure, and the amounts of these isomers produced may vary with the base and solvent used.[18] Also, small amounts of unchanged ketone may contaminate the initial crude product. Consequently, separation procedures such as preparative GC or efficient fractional distillation may be required to obtain a single vinyl triflate isomer.

TABLE II
VINYL TRIFLATES PREPARED FROM KETONES

Ketone	Reaction Conditions	Time, Days	Product	Yield, %
$XC_6H_4COCH_3$ (X = H, 4-Cl, 3-Cl, 4-CF_3, 4-NO_2)	CH_2Cl_2, Na_2CO_3	3–21	$XC_6H_4\overset{\underset{\textstyle OSO_2CF_3}{\vert}}{C}=CH_2$	15–45
	CH_2Cl_2, Na_2CO_3	14		28
	CCl_4, pyridine	5		54
$C_6H_5CH(CH_3)COCH_3$	CCl_4, pyridine	2–4	$C_6H_5C(CH_3)=C(CH_3)OSO_2CF_3$ (*cis*, 18%; *trans*, 51%) + $C_6H_5CH(CH_3)C(OSO_2CF_3)=CH_2$ (31%)	53
	CCl_4, P_2O_5	1	$C_6H_5C(CH_3)=C(CH_3)OSO_2CF_3$ (*trans*, 80%; *cis*, 20%)	32
$(CH_3)_2CH\overset{\underset{\textstyle}{O}}{\overset{\Vert}{C}}CH_3$	CH_2Cl_2, Na_2CO_3	10	$(CH_3)_2C=C(CH_3)OSO_2CF_3$ (**1**, 70%) + $(CH_3)_2CHC(OSO_2CF_3)=CH_2$ (**2**, 30%)	33
	CCl_4, pyridine	1	**1** (90%) + **2** (10%)	58

1. Department of Chemistry, the University of Utah, Salt Lake City, Utah 84112.
2. P. J. Stang, *Acc. Chem. Res.*, **11**, 107 (1978).

3. P. J. Stang, Z. Rappoport, M. Hanack, L. R. Subramanian, "Vinyl Cations," Academic Press, New York, 1979.
4. W. D. Pfeifer, C. A. Bahn, P. v. R. Schleyer, S. Bocher, C. E. Harding, K. Hummel, M. Hanack, and P. J. Stang, *J. Am. Chem. Soc.*, **93**, 1513 (1971).
5. M. A. Imhoff, R. H. Summerville, P. v. R. Schleyer, A. G. Martinez, M. Hanack, T. E. Dueber, and P. J. Stang, *J. Am. Chem. Soc.*, **92**, 3802 (1970).
6. T. C. Clarke, D. R. Kelsey, and R. G. Bergman, *J. Am. Chem. Soc.*, **94**, 3626 (1972).
7. P. J. Stang and T. E. Dueber, *J. Am. Chem. Soc.*, **95**, 2683 (1973).
8. N. Frydman, R. Bixon, M. Sprecher, and Y. Mazur, *J. Chem. Soc. D.*, 1044 (1969).
9. W. E. Truce and L. K. Liu, *Tetrahedron Lett.*, 517 (1970).
10. P. J. Stang, *Chem. Rev.*, **78**, 383 (1978).
11. W. M. Jones and D. D. Maness, *J. Am. Chem. Soc.*, **91**, 4314 (1969).
12. P. J. Stang and R. H. Summerville, *J. Am. Chem. Soc.*, **91**, 4600 (1969).
13. A. G. Martinez, M. Hanack, R. H. Summerville, P. v. R. Schleyer, and P. J. Stang, *Angew. Chem., Int. Ed. Engl.*, **9**, 302 (1970).
14. R. H. Summerville, C. A. Senkler, P. v. R. Schleyer, T. E. Dueber, and P. J. Stang, *J. Am. Chem. Soc.*, **96**, 1100 (1974).
15. T. E. Dueber, P. J. Stang, W. D. Pfeifer, R. H. Summerville, M. A. Imhoff, P. v. R. Schleyer, K. Hummel, S. Böcher, C. E. Harding, and M. Hanack, *Angew. Chem., Int. Ed. Engl.*, **9**, 521 (1970).
16. P. J. Stang and W. Treptow, *Synthesis*, 283 (1980).
17. P. E. Peterson and J. M. Indelicato, *J. Am. Chem. Soc.*, **90**, 6515 (1968).
18. P. J. Stang, M. Hanack, and L. R. Subramanian, *Synthesis*, 85 (1982).

PREPARATION AND REDUCTIVE CLEAVAGE OF ENOL PHOSPHATES: 5-METHYLCOPROST-3-ENE

[Cholest-3-ene, 5-methyl-, (5β)-]

Submitted by D. C. MUCHMORE[1]
Checked by DAVID G. MELILLO and HERBERT O. HOUSE

1. Procedure

A. *Diethyl 5-methylcoprost-3-en-3-yl phosphate.* A dry, 100-ml., three-necked flask equipped with a magnetic stirring bar, a pressure-equalizing dropping funnel, a nitrogen inlet tube, and a rubber septum is charged with 384 mg. (0.00201 mole) of copper(I) iodide (Note 1) and 20 ml. of anhydrous diethyl ether (Note 2). After the reaction vessel has been flushed with nitrogen, a static oxygen-free nitrogen atmosphere is maintained in the reaction vessel throughout the remainder of the reaction. The reaction mixture is cooled in an ice bath and an ether solution, containing 0.0040 mole of methyllithium (Note 3), is added with a hypodermic syringe, dropwise and with stirring. As the methyllithium is added, the initial yellow precipitate of polymeric methylcopper(I) redissolves, forming a colorless to pale-yellow solution of lithium dimethylcuprate (Note 4). To the resulting cold solution is added, dropwise and with stirring over 20 minutes, a solution of 576 mg. (0.00150 mole) of cholest-4-en-3-one (Note 5) in 20 ml. of ether (Note 2). During the addition of the enone, a yellow precipitate of polymeric methylcopper(I) separates from the reaction solution. After the addition is complete, the cooling bath is removed, and the reaction mixture is stirred for 2 hours at room temperature. The dropping funnel is replaced with a second dry dropping funnel which contains a loose plug of glass wool above the stopcock. The reaction mixture is again cooled in an ice bath and a mixture of 4.0 ml. of triethylamine (Note 6) and 2.00 g. (0.0115 mole) of diethyl phosphorochloridate (Note 7) is added from the dropping funnel to the reaction mixture, rapidly and with stirring. After this addition, the cooling bath is removed, and stirring is continued for one hour. Saturated aqueous sodium hydrogen carbonate is added to hydrolyze any remaining organometallic reagents before the reaction mixture is transferred to a separatory funnel and washed successively with two 50-ml. portions of cold (0°) 1 M ammonium hydroxide and a 50-ml. portion of water. The aqueous washes are extracted in turn with a 30-ml. portion of ether. The combined ether solutions are dried over anhydrous sodium sulfate and concentrated with a rotary evaporator. A solution of the residual liquid in 3 ml. of ether is applied to a 2.5 cm. by 15 cm. chromatographic column packed with a slurry of 50 g. of silica gel (Note 8) in ether. The column is eluted with ether. After the first 70 ml. of eluent has been collected and discarded, the next 120 ml. of ether eluent is collected and concentrated with a rotary evaporator, yielding 420–480 mg. of crude phosphate ester (Note 9), a colorless liquid, sufficiently pure for use in the following procedure.

B. *5-Methylcoprost-3-ene.* A dry, 100-ml., three-necked flask equipped with a polyethylene-coated magnetic stirring bar, two gas-inlet tubes, and a pressure-equalizing dropping funnel is immersed in a 2-propanol–dry ice cooling bath maintained at −15° to −20°. The reaction vessel is flushed with either helium or argon, and a static atmosphere of one of these gases is maintained in the reaction vessel throughout the reaction. Ethylamine (Note 10) is distilled through a tower of sodium hydroxide pellets into the cold reaction flask until 50 ml. of the liquid amine has been collected. A 70-mg. (0.010 g.-atom) piece of lithium wire is cleaned by dipping it successively into methanol and pentane and added to the reaction flask. The resulting cold (−15°) mixture is stirred for 10 minutes to dissolve the lithium before a solution of the diethyl 5-methylcoprost-3-en-3-yl phosphate and 0.50 ml. (0.39 g., 0.0053 mole) of *tert*-butyl alcohol (Note 11) in 15 ml. of tetrahydrofuran (Note 2) is added, dropwise and with stirring over 15 minutes, to the cold, blue lithium–amine solution. The blue solution is stirred for an additional 15 minutes before 1 ml. of saturated ammonium chloride is added to consume excess lithium. The resulting colorless mixture is warmed, evaporating ethylamine, and the residue is diluted with 90 ml. of 10% aqueous sodium hydroxide and extracted with two 30-ml. portions of pentane. The combined organic solutions are washed with 50 ml. of aqueous sodium chloride, dried over anhydrous sodium sulfate, and concentrated with a rotary evaporator. The residual, viscous liquid is subjected to evaporative distillation, from a 25-ml. flask into a male 14–20 standard-taper glass joint as shown in Figure 1. The air bath is heated to 150–180° while the pressure in the system is maintained at 0.05 mm. to 0.4 mm. Distillation of 5-methylcoprost-3-ene yields 260–295 mg. (45–51%) of colorless liquid, n_D^{25} 1.5115–1.5123 (Note 12).

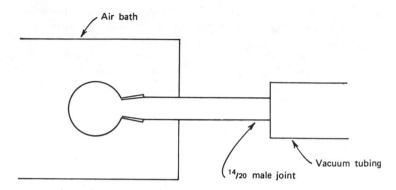

Figure 1.   Apparatus for evaporative distillation.

2. Notes

1. A purified grade of copper(I) iodide, purchased from Fisher Scientific Company, was used without purification.

2. Reagent grade ether and tetrahydrofuran were distilled from lithium aluminum hydride immediately prior to use.

3. Ethereal solutions of methyllithium are available from either Foote Mineral Company or Alpha Inorganics, Inc. These solutions should be titrated immediately before use with 2-butanol and 2,2-bipyridyl as an indicator [*Org. Synth.,* **Coll. Vol. 5,** 211 (1973); *Org. Synth.,* **Coll. Vol. 6,** 121 (1988)].[2] In a typical run, 2.44 ml. of ethereal 1.64 *M* methyllithium was employed.

4. The appearance of a brown to black precipitate indicates either oxidative or thermal decomposition of the cuprate. If such decomposition has occurred, it is best to prepare the reagent again with greater care, avoiding molecular oxygen and/or excessive reaction temperatures.

5. A commercial sample of cholest-4-en-3-one from Eastman Organic Chemicals was used without further purification. The preparation of this ketone has also been described in *Org. Synth.,* **Coll. Vol. 4,** 192 (1963).

6. The glass wool filter collects any of the insoluble triethylammonium chloride which might be formed. A reagent grade of triethylamine (b.p. 88°) was distilled from calcium hydride prior to use. *N,N,N′,N′*-Tetramethylethylenediamine has also been found to be a satisfactory Lewis base for less reactive enolates[3,4].

7. Commercial diethyl phosphorochloridate, b.p. 60–62° (1.5 mm.), purchased from Eastman Organic Chemicals, was distilled prior to use in this reaction.

8. A good grade of silica gel, such as that available from E. Merck and Company, Darmstadt, is appropriate for this chromatography.

9. The ^{1}H NMR spectrum (CDCl$_3$) of the crude product has absorption at δ 0.6–2.3 (m, *ca.* 52 H, aliphatic C*H*), 3.9–4.4 (m, 4H, 2C*H*$_2$O), and 5.1 (m, 1H, vinyl *H*).

10. Ethylamine (b.p. 17°) is available from Eastman Organic Chemicals.

11. A commercial grade of *tert*-butyl alcohol (b.p. 83°) should be distilled from calcium hydride before use.

12. The product exhibits end absorption in the UV (95% C_2H_5OH) with ϵ 330 at 210 nm and a series of ^{1}H NMR peaks (CDCl₃) at δ 0.67, 0.82, 0.85, 0.88, and 0.92 (18H, 6CH_3) with a multiplet at δ 1.0–2.2 and a partially resolved multiplet attributable to two vinyl protons. This latter absorption corresponds approximately to signals at δ 5.28 (d of t, J = 8 and 1 Hz., 1H) and 5.60 (d of t, J = 8 and 2.8 Hz., 1H). The mass spectrum of the product has the following abundant peaks: m/e (rel. int.), 384 (100, M$^+$), 369 (69), 355 (70), 229 (27), 122 (28), 109 (28), 107 (60), 95 (33), 93 (34), 81 (72), 55 (30), and 43 (31).

The submitters report that ozonolysis of the product at −10° in a mixture of ethyl acetate and acetic acid, followed by reaction with hydrogen peroxide, formed 3,4-*seco*-5-methylcoprostan-3,4-dioic acid as crystals from ethyl acetate, m.p. 168–172° with prior softening at 130°.

3. Discussion

The conjugate addition of lithium dimethylcuprate and other organocopper reagents to α,β-unsaturated ketones is a reaction which has had wide application and has been fairly well studied.[5] In order that the positional specificity which has been conferred upon the enolate anions generated by such additions might be maintained, these intermediates have been intercepted with acetic anhydride,[5a] chlorotrimethylsilane,[6] diethyl phosphorochloridate,[4,7] tetramethyldiamidophosphorochloridate,[3,4] alkyl halides,[8] aldehydes,[9] ketones,[10] acid chlorides,[11] and Michael acceptors.[12]

The reductive fission of enol phosphates to olefins is a modification of the procedure used by Kenner and Williams[13] to deoxygenate phenols. The enol phosphates, which have been reduced by the action of lithium in ammonia or alkylamines and by the action of titanium metal,[14] have been prepared by treatment of α-bromoketones with triethyl phosphite,[4,15] by interception of enolates generated by the addition of lithium dimethylcuprate to α,β-unsaturated ketones,[4,7] by interception of enolates resulting from treatment of unsaturated ketones with lithium in ammonia,[7] and by phosphorylation of enolates of ketones.[16]

1. Division of Chemistry and Chemical Engineering, Gates & Chellin Laboratories, California Institute of Technology, Pasadena, California 91109. [Present address: Institute of Molecular Biology, University of Oregon, Eugene, Oregon 97403.]
2. S. C. Watson and J. F. Eastham, *J. Organomet. Chem.*, **9**, 165 (1967).
3. R. E. Ireland, D. Muchmore, and U. Hengartner, *J. Am. Chem. Soc.*, **94**, 5098 (1972).
4. D. Muchmore, Ph.D. dissertation, California Institute of Technology, Pasadena, 1971.
5. (a) H. O. House, W. L. Respess, and G. M. Whitesides, *J. Org. Chem.*, **31**, 3128 (1966); (b) H. O. House and W. F. Fischer, Jr., *J. Org. Chem.*, **33**, 949 (1968); (c) G. Posner, *Org. React.*, **19**, 1 (1972).
6. (a) G. Stork and P. F. Hudrlik, *J. Am. Chem. Soc.*, **90**, 4462 (1968); (b) H. O. House, L. J. Czuba, M. Gall, and H. D. Olmstead, *J. Org. Chem.*, **34**, 2324 (1969).
7. R. E. Ireland and G. Pfister, *Tetrahedron Lett.*, 2145 (1969).
8. (a) P. A. Grieco and R. Finkelhor, *J. Org. Chem.*, **38**, 2100 (1973); (b) R. K. Boeckman, Jr., *J. Org. Chem.*, **38**, 4450 (1973).
9. K. Heng and R. Smith, *Tetrahedron Lett.*, 589 (1975).
10. F. Näf, R. Decorzant, and W. Thommen, *Helv. Chim. Acta,* **58**, 1808 (1975).

11. T. Tanaka, S. Kurozumi, T. Toru, M. Kobayashi, S. Miura, and S. Ishimoto, *Tetrahedron Lett.*, 1535 (1975).
12. R. K. Boeckman, Jr., *J. Am. Chem. Soc.*, **95**, 6867 (1973).
13. G. W. Kenner and N. R. Williams, *J. Chem. Soc.*, 522 (1955).
14. S. Welch, *J. Org. Chem.*, **43**, 2715 (1978).
15. M. Fetizon, M. Jurion, and N. T. Anh, *J. Chem. Soc. D*, 112 (1969).
16. I. Borowitz, S. Firstenberg, F. Caspar, and R. Crouch, *J. Org. Chem.*, **36**, 3282 (1971).

OXYMERCURATION-REDUCTION: ALCOHOLS FROM OLEFINS: 1-METHYLCYCLOHEXANOL

(Cyclohexanol, 1-methyl-)

Submitted by J. M. JERKUNICA and T. G. TRAYLOR[1]
Checked by A. K. WILLARD and R. E. IRELAND

1. Procedure

A 3-l, three-necked flask fitted with a thermometer and a mechanical stirrer is charged with 95.7 g. (0.300 mole) of mercury(II) acetate (Note 1) and 300 ml. of water. After the acetate dissolves, 300 ml. of diethyl ether is added. While this suspension is stirred vigorously 28.8 g. (0.300 mole) of 1-methylcyclohexene (Notes 2 and 3) is added, and stirring is continued for 30 minutes at room temperature (Note 4). A solution of 150 ml. of 6 N sodium hydroxide is added followed by 300 ml. of 0.5 M sodium borohydride in 3 N sodium hydroxide. The borohydride solution is added at a rate such that the reaction mixture can be maintained at or below 25° with an ice bath.

The reaction mixture is stirred at room temperature for 2 hours, after which time the mercury is found as a shiny liquid. The supernatant liquid is separated from the mercury (Note 5), the ether layer is separated, and the aqueous solution is extracted with two 100-ml. portions of ether. The combined ether solutions are dried over magnesium sulfate and distilled, giving 24.1–25.8 g. (70.5–75.4%) of 1-methylcyclohexanol, b.p. 154.5–156°; n_D^{21} 1.4596 (Note 6).

2. Notes

1. The mercury(II) acetate was purchased from Mallinckrodt Chemical Works.

2. 1-Methylcyclohexene was purchased from K & K Laboratories and used without further purification.

3. Sometimes a yellow color [mercury(II) oxide] appears at this point and disappears as the reaction proceeds. If the yellow color does not disappear in about 10 minutes, 1.5 ml. of 70% perchloric acid per mole of mercury(II) acetate may be added to accelerate the reaction. Under these conditions even unreactive olefins are completely oxymercurated in about an hour.

4. The checkers found that extending the time of oxymercuration to 2 hours did not improve the yield.

5. The checkers found that the reaction mixture could be decanted from the mercury only if the mixture was allowed to stand for at least one hour after stirring was stopped. An alternate procedure, which proved quite satisfactory, was filtration of the entire reaction mixture through a Celite pad immediately after the stirring was stopped.

6. The distilled product slowly deposits mercury. In an effort to determine whether extent of this deposition is reduced by extending the time of reduction, the checkers found that stirring the crude alcohol with Celite for 15 hours, followed by filtration and distillation, did not diminish the amount of mercury deposited. However, in a typical run where the yield of distilled alcohol was 24.6 g. (72.0%), after standing for 24 hours, the distillate was decanted from the deposited mercury and redistilled to give 21.4 g. of 1-methylcyclohexanol, which did not deposit mercury upon standing for one week at room temperature. The yield of twice-distilled alcohol was 62.6%.

3. Discussion

This method of preparing alcohols is an adaptation of an oxymercuration procedure of Sand and Genssler[2] and the reduction methods of Henbest and Nicholls.[3] Other methods for preparing 1-methylcyclohexanol are oxymercuration followed by reduction in tetrahydrofuran-water;[4] reaction of cyclohexanone with methylmagnesium halides;[5] and reduction of 1-methylcyclohexene epoxide or methylenecyclohexane epoxide with lithium aluminum hydride.[6]

Although the reaction proceeds faster in tetrahydrofuran–water[4] or in acetone–water, ether was used as solvent in this reaction for convenience of product separation and purification. However, the oxymercuration is acid catalyzed;[7] oxymercuration of unreactive olefins such as cis-cyclooctene can be accelerated by adding acid (Note 3). Rapid stirring also accelerates the reaction. An additional advantage of using ether is that less olefin is produced during the reduction using this solvent than when tetrahydrofuran is used. Elimination can be a serious side reaction during the reduction, amounting to 30% of total demercurated product when the oxymercurial from cis-cyclooctene is reduced in tetrahydrofuran–water. In ether–water, however, less than 10% olefin is produced.

As a general procedure, if the olefin is impure, the oxymercuration-reduction process may include an olefin purification step. Alternatively, this process may be used to purify the olefin for other purposes.[2,4c] In such cases, acetone is substituted for ether, and after oxymercuration for the same length of time as suggested above, the solution is poured with stirring into two volumes of water containing one equivalent each of sodium hydrogen carbonate and sodium chloride. The mercury derivative is filtered, recrystallized from ethanol–water, ether, dioxane, or ethyl acetate–heptane,[8] then either reduced, as described above (in 70–80% yield), producing pure alcohol, or deoxymercurated with cold 6 N hydrochloric acid,[2] ethereal lithium aluminum hydride[9] (added cautiously), or high concentrations of alkali halides,[4c,9,10] producing the pure olefin.

Oxymercuration may also be used to prepare ethers, acetates, amines, or amides (Markownikoff adducts). Several excellent procedures for these syntheses have been published by Brown and co-workers.[4b]

1. Department of Chemistry, University of California, San Diego, LaJolla, California 92093.
2. J. Sand and O. Genssler, *Ber. Dtsch. Chem. Ges.*, **36**, 3705 (1903).
3. H. B. Henbest and B. Nicholls, *J. Chem. Soc.* 227 (1959).
4. (a) H. C. Brown and P. Geoghegan, Jr., *J. Am. Chem. Soc.*, **89**, 1522 (1967); (b) H. C. Brown, J. H. Kawakami, and S. Misumi, *J. Org. Chem.*, **35**, 1360 (1970) (Other reduction studies are summarized here); (c) H. C. Brown and P. Geoghegan, Jr., *J. Org. Chem.*, **35**, 1844 (1970).
5. M. Barbier and M. F. Hügel, *Bull. Soc. Chim. Fr.*, 951 (1961).
6. M. Mousseron, R. Jacquier, M. Mousseron-Canet, and R. Zagdoun, *C. R. Hebd. Seances Acad. Sci.*, **235**, 177 (1952).
7. W. Kitching, *Organomet. Chem. Rev.*, **3**, 61 (1968).
8. T. G. Traylor and A. W. Baker, *J. Am. Chem. Soc.*, **85**, 2746 (1963).
9. T. G. Traylor, Thesis, University of California, Los Angeles, 1952.
10. T. G. Traylor and S. Winstein, *Abstr. Pap., Am. Chem. Soc., Div. Org. Chem. 135th Natl. Meet.*, Boston, April 1959, Pap. 82–0.

REDUCTIVE CLEAVAGE OF ALLYLIC ALCOHOLS, ETHERS, OR ACETATES TO OLEFINS: 3-METHYLCYCLOHEXENE

(Cyclohexene, 3-methyl-)

Submitted by I. Elphimoff-Felkin[1] and P. Sarda
Checked by M. L. Lee and G. Büchi

1. Procedure

A. *3-Methyl-2-cyclohexen-1-ol.* A solution of 33.6 g. (0.305 mole) of 3-methyl-2-cyclohexen-1-one (Note 1) in 600 ml. of anhydrous diethyl ether is placed in a 2-l., three-necked, round-bottomed flask fitted with a mechanical stirrer, a reflux condenser attached to a source of dry nitrogen, and a pressure-equalizing dropping funnel. The solution is stirred and cooled in an ice bath while 471 ml. (0.0825 mole) of a 0.175 M solution of lithium aluminum hydride in ether (Note 2) is added dropwise. When the addition is complete the reaction mixture is stirred at 0° for another 15 minutes. Cooling and gentle stirring are continued while moist ether is added through the dropping funnel until gas is no longer evolved. The resulting slurry is filtered, and the filtrate is washed with saturated aqueous sodium chloride and dried over magnesium sulfate. Removal of ether on a water bath and distillation of the residue under reduced pressure provide 33.7 g. (98%) of 3-methyl-2-cyclohexen-1-ol, b.p. 94–95° (31 mm.).

B. *Amalgamated zinc.* Zinc powder (206 g., 3.15 moles) is placed in a 1-l. beaker, covered with 250 ml. of 10% hydrochloric acid, and stirred for 2 minutes. The acid is then decanted and replaced by distilled water, the mixture is stirred, and the supernatant is decanted. Washing is continued in this way until the water is neutral to litmus. A warm solution of 40 g. (0.15 mole) of mercury(II) chloride in 250 ml. of distilled water is then poured onto the zinc, and the mixture is stirred gently for 10 minutes. After filtration, the powder is washed with 250 ml. of distilled water, five 250-ml. portions of 95% ethanol, and five 250-ml. portions of anhydrous ether. Drying under vacuum gives 196 g. of zinc amalgam.

C. *3-Methylcyclohexene*. A 1-l., round-bottomed, three-necked flask equipped with a mechanical stirrer, a reflux condenser connected to a source of dry nitrogen, and a pressure-equalizing addition funnel is charged with 196 g. (3 moles) of dry amalgamated zinc powder (Note 3), 22.4 g. (0.200 mole) of 3-methyl-2-cyclohexen-1-ol, and 280 ml. of anhydrous ether. The flask is placed in an ethanol–water–dry ice bath maintained at −15° throughout the reaction. The reaction mixture is stirred gently for 5 minutes, then stirred vigorously while 153 ml. (0.40 mole) of 2.6 M hydrogen chloride in anhydrous ether (Note 4) is added dropwise over 1.5 hours. When the addition is complete, stirring is continued for an additional 15 minutes, after which the reaction medium is neutral to moist litmus (Note 5).

Decanting the reaction mixture separates residual zinc, which is washed thoroughly with two 200-ml. portions of ether. The ethereal solutions are combined, washed sequentially with two 50-ml. portions of water, 50 ml. of 10% aqueous sodium hydrogen carbonate, and two 50-ml. portions of saturated aqueous sodium chloride (Note 6), and dried over magnesium sulfate. After filtration, ether is removed by careful distillation through a Dufton column at atmospheric pressure (Note 7). When the residual solution is approximately 100 ml. in volume, it is transferred to a smaller apparatus and distilled slowly at atmospheric pressure. After a forerun of ether, 13.2–14.4 g. (68–75%) of 3-methylcyclohexene distils at 103–104° (Note 8).

2. Notes

1. The unsaturated ketone[2] can be purchased from Ega Chemie K.G. or from Frinton Laboratories, P.O. Box 301, South Vineland, New Jersey.

2. The concentration of this solution was established by decomposing the lithium aluminum hydride with excess iodine according to the following equation:

$$\text{LiAlH}_4 + 2\text{I}_2 \rightarrow 2\text{H}_2 + \text{LiI} + \text{AlI}_3$$

The amount of unreacted iodine was determined by titration with sodium thiosulfate; the amount of iodine initially present was determined by a separate blank titration.[3]

Working solutions were prepared by dissolving 40 g. of iodine in 1 l. of anhydrous benzene and by dissolving 248 g. of sodium thiosulfate pentahydrate in 1 l. of water (*ca.* 1 M); an accurately prepared 0.100 N aqueous sodium thiosulfate solution is also required. For titration, a 25.0-ml. portion of the iodine solution was stirred vigorously in a 500-ml. Erlenmeyer flask, and 1.00 ml. of the lithium aluminum hydride solution was added rapidly, followed by 20 ml. of water, 2 drops of acetic acid, and 5.00 ml. of the 1 M sodium thiosulfate solution. With continued stirring, the solution was titrated to the colorless end point by adding V_a ml. of the 0.100 N sodium thiosulfate solution.

Separately, another 25.0-ml. portion of the iodine solution was stirred in a 500-ml. Erlenmeyer flask and treated with 5.00 ml. of the 1 M aqueous sodium thiosulfate, 20 ml. of water, and 2 drops of acetic acid. Titration was continued to the end point by adding, with continuous stirring, V_b ml. of 0.100 N aqueous sodium thiosulfate. The molarity of the lithium aluminum hydride solution was then calculated from the following equation:

$$M = 0.025(V_b - V_a)$$

3. A large excess of zinc is used to make the reaction faster and to reduce the formation of polymeric by-products.

4. The solution of dry hydrogen chloride in ether was prepared as follows. Commercial hydrogen chloride gas was dried by passing it through an empty safety trap, a wash bottle of concentrated sulfuric acid, a calcium chloride tube, and another empty safety trap. Anhydrous ether was cooled in an ice bath, and the hydrogen chloride was bubbled through rapidly. Gas uptake was followed by weighing the ethereal solution occasionally, and the concentration of the final solution was determined by alcalimetric titration. The optimum concentration of hydrogen chloride is 2.5–3 M. Use of excess acid led to overreduction in the case of 2-phenyl-2-cyclohexen-1-ol, the desired 1-phenylcyclohexene being contaminated by phenylcyclohexane.

5. If the total reaction time is less than approximately 1.75 hours, starting material remains. Therefore, if the ethereal hydrogen chloride is added in less than 1.5 hours, the subsequent stirring must be lengthened accordingly.

6. If zinc chloride is not removed by washing, it causes polymerization of the olefin during distillation.

7. Slow distillation of the ether is essential in order to prevent the low-boiling olefin from codistilling with the solvent. For higher-boiling olefins the ether can be removed on a water bath prior to distillation of the olefin under reduced pressure.

8. GC analysis, using an HMDS-treated Chromosorb W column with 7% Craig polyester as the stationary phase, indicated the product to have a purity of 97%. The 3% impurity is most probably the isomeric 1-methylcyclohexene.

3. Discussion

The model procedure described above is applicable to allylic alcohols, ethers, and acetates. The submitters' results for the conversion of several such compounds to the corresponding olefins, performed on a smaller scale, are summarized in Table I. Reductive cleavage of allylic alcohols, ethers, and acetates has often been reported in the literature. Typical reagents used are sodium in liquid ammonia,[4,5] zinc and acetic acid,[6] chloroaluminum hydride,[7] and propylmagnesium bromide in the presence of dichlorobis(triphenylphosphine)nickel.[8] In all of these procedures, however, when two or more isomeric olefins can be formed, the thermodynamically more stable olefin generally predominates. The advantage of the present procedure[9] is that it leads, depending on the structure of the starting material, either exclusively or predominantly to the less stable isomer (see Table I). These results have been interpreted[9] by assuming that reduction takes place through an intermediate that behaves like an allylic metal halide. Studies of allylic metal halides such as crotyl zinc halides[10] and the crotyl Grignard reagent[11] suggest that such an intermediate would be protonated predominantly at the more substituted end of the allylic system. The method has been applied recently to the synthesis of some natural products.[12]

TABLE I
OLEFINS FROM ALLYLIC ALCOHOLS, ETHERS, AND ACETATES[9]

Starting Material	Total yield of Olefins (%)[a]	Major Olefin (relative %)[b]
	85–90	70
	85–90	70
	85–90	70
	85–90	70
	70	80
	80	80
	not determined	80
(cis + trans)[c]	70	(racemic)

TABLE I (continued)

OLEFINS FROM ALLYLIC ALCOHOLS, ETHERS, AND ACETATES[9]

Starting Material	Total yield of Olefins (%)[a]	Major Olefin (relative %)[b]
	70	(racemic)
	80	

[a]Distilled product.
[b]Determined by GC.
[c]From lithium aluminum hydride reduction of (−)carvone.

1. Institut de Chimie des Substances Naturelles, C.N.R.S., 91190 Gif-sur-Yvette, France.
2. M. W. Cronyn and G. H. Riesser, *J. Am. Chem. Soc.*, **75**, 1664 (1953).
3. H. Felkin, *Bull. Soc. Chim. Fr.*, **347**. (1951).
4. A. J. Birch, *Q. Rev. Chem. Soc.*, **84**, 69 (1950).
5. A. S. Hallsworth, H. B. Henbest, and T. I. Wrigley, *J. Chem. Soc.*, 1969 (1957).
6. L. F. Fieser and M. Fieser, "Steroids," Reinhold, New York, 1959, p. 204.
7. J. H. Brewster and H. O. Bayer, *J. Org. Chem.*, **29**, 116 (1964).
8. H. Felkin and G. Swierczewski, *C.R. Hebd. Seances Acad. Sci. Ser. C*, **266**, 1611 (1968).
9. I. Elphimoff-Felkin and P. Sarda, *Tetrahedron Lett.*, 725 (1972).
10. R. Gaudemar, *Bull. Soc. Chim. Fr.*, 1475 (1958); *Bull. Soc. Chim. Fr.*, 974 (1962); C. Agami, M. Andrac-Taussig, and C. Prevost, *Bull. Soc. Chim. Fr.*, 1915, 2596 (1966).
11. R. A. Benkeser, *Synthesis*, 347 (1971).
12. G. Büchi, A. Hauser, and J. Limacher, *J. Org. Chem.*, **42**, 3323 (1977); E. J. Corey, D. A. Clark, G. Goto, A. Marfat, and C. Mioskowski, *J. Am. Chem. Soc.*, **102**, 1436 (1980).

γ-KETOESTERS TO PREPARE CYCLIC DIKETONES: 2-METHYL-1,3-CYCLOPENTANEDIONE

(1,3-Cyclopentanedione, 2-methyl-)

$$\text{NaOCH}_3, \text{CH}_3\text{OH} \xrightarrow[\text{(CH}_3)_2\text{SO}]{\text{xylene, reflux,}}$$

Submitted by U. Hengartner[1] and Vera Chu
Checked by William B. Farnham and William A. Sheppard

1. Procedure

A 3-1., three-necked, round-bottomed flask fitted with a dropping funnel (Note 1), a mechanical stirrer, and a distillation head with a thermometer and efficient Liebig condenser is charged with 1.4 l. of xylene (Note 2). The xylene is stirred and heated to boiling with a heating mantle, while 179 g. of a solution containing 43 g. (0.80 mole) of sodium methoxide in methyl alcohol (Note 3) is added over 20 minutes. During this period 450 ml. of solvent is distilled.

After addition is complete, 300 ml. of xylene is added and the distillation continued until the vapor temperature rises again to 138°. During that time an additional 250 ml. of distillate is collected, leaving a white suspension to which 18 ml. of dimethylsulfoxide is added. A solution of 100 g. (0.633 mole) of ethyl 4-oxohexanoate (Note 4) in 200 ml. of xylene is then added (Note 1) to the vigorously stirred sodium methoxide suspension over 25 minutes, while 900 ml. of distillate is collected continuously, maintaining the vapor temperature at 134–137°. The orange-colored mixture is stirred and heated for an additional 5 minutes, then cooled to room temperature. Addition of 165 ml. of water with vigorous stirring over a 5-minute period (Note 5) gives two clear phases, which are cooled in an ice bath and acidified by adding 82 ml. (0.98 mole) of 12 N hydrochloric acid with vigorous stirring. After the mixture is stirred at 0° for another 1.5 hours, the crystalline product is collected by suction filtration and carefully washed successively with 100-ml. and 50-ml. portions of ice-cooled diethyl ether (Note 6).

The crude product is dissolved in 1 l. of boiling water, and the solution is filtered quickly through a preheated fritted-disk funnel (Note 7). The filtrate is concentrated on a hot plate at atmospheric pressure to a volume of 550–600 ml. and allowed to stand at 0° overnight. The crystals are collected by filtration and dried at 85°, yielding 50.0–50.6 g. (70–71%) of 2-methyl-1,3-cyclopentanedione, m.p. 210–211° (Note 8).

2. Notes

1. A 500-ml. dropping funnel, provided with a calcium sulfate drying tube, was used.
2. Reagent grade xylene, b.p. 138–141°, from Fisher Scientific Company, was used.

3. The sodium methoxide solution was prepared as follows: 203 g. of methyl alcohol, available from Fisher Scientific Company, is placed in a 500-ml., two-necked flask under an inert atmosphere. The flask is equipped with a magnetic stirring bar and a reflux condenser provided with a calcium sulfate drying tube. Freshly cut, clean sodium (23 g., 1.0 mole) is added in small pieces at such a rate that reflux is maintained. The mixture is stirred until all the sodium has reacted.

4. Ethyl 4-oxohexanoate was prepared by the method in *Org. Synth.*, **Coll. Vol. 6,** 615 (1988).

5. The temperature of the mixture is kept at 25–35°.

6. The ether washings contain 11 g. of a brown viscous oil containing various condensation products.

7. A small amount of insoluble, tarry material is removed by this filtration. Preheating the funnel is necessary, since the product crystallizes easily on cooling. Dark-colored impurities in crude 2-methyl-1,3-cyclopentanedione can be removed by recrystallization from methanol.

8. 2-Methyl-1,3-cyclopentanedione, [*Org. Synth.*, **Coll. Vol. 5,** 747 (1973)] m.p. 210–212°, exists in the enol form in solution. It has the following spectral properties: UV (0.1 N HCl) nm. max. (ϵ): 252 (19,000); ^{1}H NMR (dimethyl sulfoxide-d_6) δ (multiplicity, number of protons, assignment): 1.51 (s, 3H, CH_3), 2.39 (s, 4H, 2CH_2).

3. Discussion

2-Methyl-1,3-cyclopentanedione is a key intermediate in the total synthesis of steroids.[2] A number of methods have been described for its preparation, among them the condensation of succinic acid with propionyl chloride,[3] and that of succinic anhydride with 2-buten-2-ol acetate,[4] both in the presence of aluminum chloride. It has also been obtained from 3-methylcyclopentane-1,2,4-trione by catalytic hydrogenation[5] and Wolff–Kishner reduction.[6] The base-promoted cyclization of ethyl 4-oxohexanoate and diethyl propionylsuccinate with tertiary alkoxides was first reported by Bucourt.[7] The present cyclization process provides an experimentally simple route to 2-methyl-1,3-cyclopentanedione. Using the same procedure, ethyl 4-oxohexanoate has been cyclized to give 2-ethyl-1,3-cyclopentanedione in 46% yield.

1. Chemical Research Department, Hoffmann-La Roche Inc., Nutley, New Jersey 07110.
2. For leading references, see R. T. Blickenstaff, A. C. Ghosh, and G. C. Wolf, "Total Synthesis of Steroids," Academic Press, New York, 1974.
3. H. Schick, G. Lehmann, and G. Hilgetag, *Chem. Ber.*, **102,** 3238 (1969).
4. V. J. Grenda, G. W. Lindberg, N. L. Wendler, and S. H. Pines, *J. Org. Chem.*, **32,** 1236 (1967).
5. M. Orchin and L. W. Butz, *J. Am. Chem. Soc.*, **65,** 2296 (1943).
6. J. P. John, S. Swaminathan, and P. S. Venkataramani, *Org. Synth.*, **Coll. Vol. 5,** 747 (1973).
7. R. Bucourt, A. Pierdet, G. Costerousse, and E. Toromanoff, *Bull Soc. Chim. Fr.*, 645 (1965).

SULFIDE CONTRACTION *via* ALKYLATIVE COUPLING:
3-METHYL-2,4-HEPTANEDIONE

(2,4-Heptanedione, 3-methyl-)

$$2(CH_3)_2NCH_2CH_2CH_2MgCl + C_6H_5PCl_2 \xrightarrow{\text{tetrahydrofuran}}$$
$$C_6H_5P[CH_2CH_2CH_2N(CH_3)_2]_2$$

$$CH_3CH_2CH_2COSH + CH_3\overset{\overset{\text{O}}{\|}}{C}CHCH_3 \xrightarrow{(C_2H_5)_3N} CH_3CH_2CH_2CO-S-CHCOCH_3$$
$$\underset{Br}{} \qquad\qquad \underset{CH_3}{}$$

$$CH_3CH_2CH_2CO-S-CHCOCH_3 + C_6H_5P[CH_2CH_2CH_2N(CH_3)_2]_2 \xrightarrow[\substack{\text{acetonitrile,}\\70°}]{\text{LiBr}}$$
$$\underset{CH_3}{}$$

$$CH_3CH_2CH_2\overset{\overset{\text{S}}{\|}}{C}OCHCOCH_3 + C_6H_5P[CH_2CH_2CH_2N(CH_3)_2]_2$$
$$\underset{CH_3}{}$$

Submitted by P. LOELIGER[1] and E. FLÜCKIGER[2]
Checked by K. MATSUO and G. BÜCHI

1. Procedure

Caution! To avoid exposure to toxic dichlorophenylphosphine vapors, the Grignard reaction should be conducted in a hood.

Benzene has been identified as a carcinogen; OSHA has issued emergency standards on its use. All procedures involving benzene should be carried out in a well-ventilated hood, and glove protection is required.

A. *Bis(3-dimethylaminopropyl)phenylphosphine.* A 3-l. separatory funnel is charged with 395 g. (2.50 moles) of 3-chloro-*N,N*-dimethyl-1-propylamine hydrochloride (Note 1), and a cold solution of 179 g. of potassium hydroxide in 540 ml. of water is added. The mixture is extracted three times with 300-ml portions of 5:1 diethyl ether-dichloromethane. The organic extracts are washed with 300 ml. of aqueous 2 N potassium hydroxide, combined, and dried over anhydrous sodium sulfate. The solvent is removed by distillation through a 25-cm. Vigreux column at atmospheric pressure, and the residual liquid is distilled under reduced pressure through a 13-cm. Vigreux column, giving 263–276 g. (87–91%) of 3-chloro-*N,N*-dimethyl-1-propylamine as a colorless liquid, b.p. 72–73° (100 mm.) (Notes 2 and 3), which is used immediately in the Grignard reaction (Note 4).

A 3-l., four-necked, round-bottomed flask equipped with a sealed mechanical stirrer, a pressure-equalizing dropping funnel, a thermometer, and a condenser fitted with a nitrogen-inlet tube is charged with 48.6 g. (2.00 g.-atoms) of magnesium turnings (Note 5). The flask is flushed with dry nitrogen and thoroughly dried with a heat gun, and 300 ml. of anhydrous tetrahydrofuran (Note 6) is added. The Grignard reaction is initiated by adding about 10% of a solution of 243.0 g. (1.98 moles) of 3-chloro-N,N-dimethyl-1-propylamine in 300 ml. of anhydrous tetrahydrofuran (Note 6), and 4 ml. of ethyl bromide while gently heating the flask with the drier (Note 7). The remainder of the 3-chloro-N,N-dimethyl-1-propylamine solution is added over a period of approximately 1 hour so as to maintain gentle reflux. The reaction mixture is heated at reflux for 3 hours, after which time most of the magnesium has reacted. The dark gray solution is cooled to 0° before a solution of 107.3 g. (81.29 ml., 0.5994 mole) of dichlorophenylphosphine (Note 8) in 200 ml. of anhydrous tetrahydrofuran (Note 6) is added dropwise, with efficient stirring, over a 1 hour period so that the temperature does not exceed 5° (Note 9). A greenish precipitate is formed locally where the phosphine is added. After the addition is complete, the reaction mixture is stirred and heated at reflux for 2 hours, during which time a heavy, greenish precipitate is formed. After cooling to room temperature, 600 ml. of ether (Note 10) is added, and the reaction mixture is left standing overnight, during which time the precipitate separates to the bottom of the flask. The solution is decanted into a 3-l. separatory funnel containing 300 ml. of 40% aqueous potassium hydroxide and 1 kg. of ice. The remainder of the reaction product is suction filtered with the aid of 1200 ml. of 5:1 ether–dichloromethane through a 3-cm. layer of Celite® (Note 11). The filtrate is added to the separatory funnel, and the organic layer is separated and washed twice with 600-ml. portions of saturated aqueous sodium chloride. The aqueous layer is extracted four times with 700-ml. portions of 5:1 ether–dichloromethane. The combined organic extracts are dried over anhydrous sodium sulfate, and the solvent is removed with a rotary evaporator. The crude yellow oil is distilled at high vacuum through a 14-cm. Vigreux column, yielding 109–116 g. (65–69%, based on phenylphosphonous dichloride) of bis(3-dimethylaminopropyl)phenylphosphine as a colorless liquid, b.p. 100–108° (0.005 mm.) (Note 12). Redistillation furnishes 94–97 g. (56–58%) of product, b.p. 102–105° (0.005 mm.) (Note 13), n_D^{24} 1.5265.

B. *S*-(2-*Oxobut*-3-*yl*) *Butanethioate*. A 750-ml., four-necked, round-bottomed flask equipped with a sealed mechanical stirrer, a pressure-equalizing dropping funnel, a thermometer, and a condenser fitted with a nitrogen-inlet tube is charged with 10.4 g. (0.100 mole) of thiobutyric acid (Note 14) in 300 ml. of anhydrous ether (Note 10). With stirring, 10.1 g. (0.100 mole) of triethylamine (Note 15) is added in one portion. Over a 15-minute period (Note 16), 15.1 g. (0.100 mole) of 3-bromo-2-butanone (Note 17) is added dropwise from the dropping funnel. The solution is heated at reflux with stirring for 1.5 hours and filtered through Celite®; the precipitate is washed with 60 ml. of ether. The ether solution is concentrated on a rotary evaporator. The residual orange-yellow oil is dissolved in 20 ml. of 5:1 benzene–ether and filtered through 70 g. of silica gel (Note 18), using 500 ml. of this solvent mixture as eluent. The solvent is removed on a rotary evaporator, yielding 17.0–17.4 g. (98–100%) of the thiol ester as a pale yellow oil which can be used without further purification in the next step. (Notes 19 and 20).

C. 3-*Methyl*-2,4-*heptanedione*. A dry, 500-ml., three-necked, round-bottomed flask equipped with a magnetic stirring bar, a pressure-equalizing dropping funnel, a thermometer, and a condenser fitted with a nitrogen-inlet tube is charged with 17.8 g. (0.204 mole) of anhydrous lithium bromide (Note 21). Under a nitrogen atmosphere, 34.8 g. (0.218 mole) of S-(2-oxobut-3-yl)butanethioate dissolved in 120 ml. of anhydrous acetonitrile (Note 22) is added to the flask. With stirring, the mixture is heated with a drier until a homogeneous solution is obtained. From the dropping funnel, 67 g. (69 ml., 0.24 mole) of redistilled bis(3-dimethylaminopropyl)phenylphosphine is added in one portion to the warm ($ca.$ 60°) solution. The temperature rises to about 70°, and after 1–2 minutes a thick, white precipitate appears. The reaction mixture is stirred at 70° for 15 hours (Note 23). After cooling to room temperature, the reaction mixture is transferred with 600 ml. of 5:1 ether–dichloromethane into a separatory funnel containing 900 ml. of cold 1 N hydrochloric acid. The organic layer is separated and washed three times with 500-ml. portions of saturated aqueous sodium chloride. The aqueous phase is washed twice with 600-ml. portions of 5:1 ether–dichloromethane. The combined organic layer is dried over anhydrous sodium sulfate, and the solvent removed on a rotary evaporator, the temperature of the bath not exceeding 30°. The crude yellow oil is distilled through a 10-cm. Vigreux column under reduced pressure, yielding 23.5–24.7 g. (83–87%) of 3-methyl-2,4-heptanedione as a colorless liquid, b.p. 74–76° (9 mm.), n_D^{23} 1.4455 (Note 24).

2. Notes

1. 3-Chloro-N,N-dimethyl-1-propylamine hydrochloride was purchased from Fluka AG CH-9470 Buchs or Aldrich Chemical Company, Inc.

2. The fractions may be analyzed by GC for absence of solvent; a 300 cm. by 0.3 cm. glass column packed with XE-60 (1.5% w/w) coated on Chromosorb G AW DCMS (80/100 mesh) was employed.

3. The spectral properties of the product are as follows; IR (neat) cm.$^{-1}$: 1470, 1465, 1265, 1040; ^{1}H NMR (CDCl$_3$), δ (multiplicity, coupling constant J in Hz., number of protons): 1.8–2.6 (m, 4H), 2.2 (s, 6H), 3.6 (t, J = 6, 2H).

4. It is advisable to distill the solvent one day, store the residue overnight under nitrogen at 0°, and distill the product the next morning, allowing ample time for the following Grignard reaction. On standing at room temperature a white solid precipitates.

5. Magnesium turnings were purchased from E. Merck & Company, Inc., Darmstadt, Germany or J. T. Baker Chemical Company.

6. Tetrahydrofuran (purchased from Fluka AG or J. T. Baker Chemical Company) was distilled from sodium hydride prior to use. For warnings regarding the purification of tetrahydrofuran, see $Org. Synth.$, **Coll. Vol. 5**, 976 (1973).

7. It is advisable to have available an ice bath for cooling, should the reaction become violent.

8. Dichlorophenylphosphine was purchased from Fluka AG or Aldrich Chemical Company, Inc.

9. The reaction is very exothermic and cooling with an ice–sodium chloride bath is necessary.

10. Anhydrous ether was purchased from Fluka AG or J. T. Baker Chemical Company.

11. During this operation, the reaction vessel is washed with 5:1 ether–dichloromethane several times.

12. The colorless forerun weighs 12–20 g.; the dark brown residue weighs 13–22 g.

13. The reported b.p. is 102–105° (0.005 mm.); a full spectroscopic characterization is given in the original paper.[3]

14. The thiobutyric acid was prepared[4] as follows: A rapid stream of hydrogen sulfide is passed, with vigorous stirring at −30°, through 200 ml. of anhydrous pyridine, contained in a four-necked, round-bottomed flask equipped with a sealed mechanical stirrer, a gas-inlet tube, a pressure-equalizing dropping funnel, and a thermometer. Over approximately 1 hour, 50 g. of 1-butyryl chloride is added dropwise to this solution. Approximately 400 ml. of 5 N sulfuric acid is added slowly until the pH is ~5. The organic acid, which separates as a yellow oil, is taken up in ether and dried over anhydrous sodium sulfate. After removal of the ether on a rotary evaporator, the product is distilled through a 14-cm. Vigreux column under a nitrogen atmosphere, yielding 23.1–32.1 g. (43–65%, not optimized by the submitters) of thiobutyric acid as a colorless liquid, b.p. 119–121°.

15. Triethylamine was purchased from Fluka AG or J. T. Baker Chemical Company.

16. A colorless precipitate of triethylamine hydrobromide is formed immediately. The temperature rises to about 35°.

17. 3-Bromo-2-butanone was purchased from Fisher Scientific Company. The submitters prepared it according to the literature[5] and checked its purity (>95%) by GC (Note 2).

18. Silica gel (70–230 mesh ASTM) purchased from E. Merck & Company, Inc., Darmstadt, Germany was used in a 2.5-cm. diameter column.

19. GC analysis (Note 2) indicated <2% impurities. IR (neat) cm.$^{-1}$: 1720, 1695.

20. The submitters obtained a similar yield on ten times the scale.

21. The absence of water in the lithium bromide is of great importance. Traces of water lower the yield of product by 10–20%. Lithium bromide dihydride (purchased from E. Merck & Company, Inc., Darmstadt or City Chemical Corporation) was dissolved three times in anhydrous 1:1 acetonitrile–benzene, and the solvents were removed each time with a rotary evaporator. The lithium bromide was dried under high vacuum at 100° for 1 hour, ground to a fine powder with a mortar and pestle while still warm, and again dried at 100°, as above, for 3 hours.

22. Acetonitrile was purchased from Fluka AG or J. T. Baker Chemical Company and distilled from potassium carbonate immediately prior to use.

23. The reaction is followed best by GC analysis (Note 2). Traces of water seem to slow down the rate of the reaction.

24. By GC analysis (Note 2) the product is >98% pure. In the literature,[3] a full spectroscopic characterization is given. IR (neat) cm.$^{-1}$: 1725, 1700, 1600, 1360.

3. Discussion

This procedure illustrates a broadly applicable method which is essentially that found in the literature[3] for the synthesis of enolizable β-dicarbonyl compounds.[3] Although there are various methods for the preparation of β-dicarbonyl systems,[6] sulfide contraction widens the spectrum of available methods. The procedure can also be utilized in the synthesis of aza and diaza analogs of β-dicarbonyl systems. Eschenmoser[3] has utilized the method to produce vinylogous amides and amidines in connection with the total synthesis of corrins and vitamin B_{12}.[7]

S-Alkylation of a thiocarboxylic acid with an α-halogenated carbonyl compound gives a thiol ester in which the two carbons to be connected are linked *via* a sulfur bridge (see the scheme below). Enolization and formation of the episulfide creates the desired carbon–carbon bond. Removal of atomic sulfur by a thiophile, either a phosphine or a phosphite, liberates the β-dicarbonyl compound. The addition of base is necessary in most cases; however, in the vinylogous amidine systems[7] electrophilic catalysis was employed. Normally a teritary alkoxide is utilized in the contraction. The addition of anhydrous lithium bromide or lithium perchlorate allows the reaction to proceed with the use of a tertiary amine as the base. Presumably, the lithium salts complex with the carbonyl groups, enhancing the enolization and/or contraction step.

This procedure also incorporates the use of bis(3-dimethylaminopropyl)phenylphosphine as a combined amine–phosphine reagent. The merit of using this basic phosphine as opposed to a tertiary amine and a phosphine lies in the ease of workup. Excess phosphine and phosphine sulfide can be removed by extraction with dilute acid.

Since the new carbon–carbon bond is formed intramolecularly in the sulfide extrusion method, its main potential lies in cases where intermolecular condensations fail.[8,9]

1. Present address: F. Hoffmann-La Roche & Co., CH-4002 Basle, Switzerland.
2. SOCAR AG, Dübendorf, Switzerland.
3. M. Roth, P. Dubs, E. Götschi, and A. Eschenmoser, *Helv. Chim. Acta,* **54,** 710 (1971).
4. A. Fredga and H. Bauer, *Ark. Kemi,* **2,** 115 (1951).
5. J. R. Catch, D. F. Elliott, D. H. Hey, and E. R. H. Jones, *J. Chem. Soc.,* 272, 278 (1948).
6. For a review see H. O. House, "Modern Synthetic Reactions," W. A. Benjamin, Menlo Park, California, 1972, pp. 734–785.
7. (a) Y. Yamada, D. Miljkovic, P. Wehrli, B. Golding, P. Loeliger, R. Keese, K. Müller, and A. Eschenmoser, *Angew. Chem.,* **81,** 301 (1969) [*Angew. Chem. Int. Ed. Engl.,* **8,** 343 (1969)]; A. Eschenmoser, *Q. Rev. Chem. Soc.,* **24,** 366 (1970); A. Eschenmoser, *Spec. Lect.,* 23rd *Int. Congr.* IUPPC, Vol. II, Butterworth and Company, London, 1971, pp. 69–106. (b) R. B. Woodward, *Pure Appl. Chem.,* **17,** 519 (1968); *Pure Appl. Chem.,* **25,** 283 (1971); *Pure Appl. Chem.,* **33,** 145 (1973).
8. I. Felner and K. Schenker, *Helv. Chim. Acta,* **53,** 754 (1970).
9. A. Gossauer and W. Hirsch, *Tetrahedron Lett.,* 1451 (1973).

ISOXAZOLE ANNELATION REACTION:
1-METHYL-4,4a,5,6,7,8-HEXAHYDRONAPHTHALEN-2(3*H*)-ONE

[2(3*H*)-Naphthalenone, 4,4a,5,6,7,8-hexahydro-1-methyl-]

Submitted by JOHN E. MCMURRY[1]
Checked by U. P. HOCHSTRASSER and G. BÜCHI

1. Procedure

Caution! Lithium aluminum hydride can react with explosive violence on contact with water or when overheated, and great care must be taken in its handling.

Benzene has been identified as a carcinogen; OSHA has issued emergency standards on its use. All procedures involving benzene should be carried out in a well-ventilated hood, and glove protection is required.

A. *3-Ethyl-4-hydroxymethyl-5-methylisoxazole.* A slurry of lithium aluminum hydride (21.0 g., 0.553 mole) in anhydrous diethyl ether is prepared by cautiously adding the powdered reagent (Note 1) to 2.5 l. of freshly prepared anhydrous ether in a 5-l., three-necked flask fitted with a reflux condenser, a 500-ml. pressure-equalizing addition funnel, and a strong mechanical stirrer. The contents of the flask are then placed under a nitrogen atmosphere *via* a gas-inlet tube attached to the top of the condenser. Ethyl 3-ethyl-5-methyl-4-isoxazolecarboxylate (124 g., 0.678 mole) (Note 2), dissolved in 300 ml. of dry ether, is placed in the addition funnel and added dropwise over 4 hours to the lithium aluminum hydride slurry (Note 3). The reaction is refluxed gently for 4 hours, then placed in an ice bath. Quenching of excess reagent and hydrolysis of aluminate salts is effected by *cautious, slow* addition of 20 ml. of water, followed by 30 ml. of aqueous 10% sodium hydroxide and another 30 ml. of water (Note 4). The ether layer is filtered from granular aluminum salts, poured into a 2-l. separatory funnel, and washed with 250 ml. of saturated brine. The organic extract is dried over anhydrous magnesium sulfate and filtered, and the solvent is removed with a rotary evaporator. The residual oil is distilled, yielding 76–82 g. (80–86%) of 3-ethyl-4-hydroxymethyl-5-methylisoxazole, b.p. 99–101° (0.15 mm.); n_D^{23} 1.4835; IR: 3450, 1640 cm.$^{-1}$; ^{1}H NMR (CCl$_4$): δ 1.2 (t, 3H, CH$_2$CH$_3$), 2.2 (s, 3H, C=CCH$_3$), 2.6 (q, 2H, CH$_2$CH$_3$), 4.2 (s, 2H, CH$_2$O).

B. *4-Chloromethyl-3-ethyl-5-methylisoxazole.* *Caution! The following reaction should be carried out in a fume hood to avoid thionyl chloride vapors.*

3-Ethyl-4-hydroxymethyl-5-methylisoxazole (54 g., 0.38 mole) is dissolved in 70 ml. of dichloromethane and placed in a 500-ml., one-necked flask fitted with a 100-ml., pressure-equalizing addition funnel and a magnetic stirrer. The flask is placed in an ice bath, and its contents are stirred while a solution of 53 g. of thionyl chloride (32 ml., 0.45 mole) in 50 ml. of dichloromethane is added dropwise. Addition is complete in 1 hour, and the reaction is allowed to warm to room temperature and stirred for an additional hour. The solvent is removed with a rotary evaporator and the dark residual liquid is distilled, yielding 47–49 g. (78–81%) of 4-chloromethyl-3-ethyl-5-methylisoxazole, b.p. 77–78° (1.5 mm.); n_D^{23} 1.4845; IR: 1620, 680 cm.$^{-1}$; ^{1}H NMR (CCl$_4$): δ 1.3 (t, 3H, CH$_2$CH$_3$), 2.3 (s, 3H, C=CCH$_3$), 2.6 (q, 2H, CH$_2$CH$_3$), 4.3 (s, 2H, CH$_2$Cl).

C. *2-(3-Ethyl-5-methyl-4-isoxazolylmethyl)cyclohexanone.* Sodium hydride (10.0 g. of a 60% slurry in mineral oil, 0.25 mole) is degreased in a flame-dried, 1-l., three-necked flask fitted with a 250-ml., pressure-equalizing addition funnel and a condenser through which a stream of nitrogen is blown. The sodium hydride is washed by adding 20 ml. of dry benzene, stirring magnetically, allowing it to settle, and drawing off the supernatant benzene wash with a syringe. The washing process is repeated four more times before 100 ml. of dry benzene is added, followed with 100 ml. of dry *N,N*-dimethylformamide (Note 5). The contents of the flask are covered with a nitrogen atmosphere and a solution of ethyl 2-cyclohexanonecarboxylate (41.0 g., 0.241 mole) (Note 6) in 100 ml. of 1:1 benzene–dimethylformamide is added slowly over 45 minutes, with cooling, keeping the reaction mixture near room temperature (Note 7). A solution of 4-chloromethyl-3-ethyl-5-methylisoxazole (32 g., 0.20 mole) in 100 ml. of 1:1 benzene–dimethylformamide is then added over 30 minutes, and the reaction is stirred for 2 days at room temperature. The reaction is diluted with 300 ml. of ether, poured into a 1-l. separatory funnel, washed three times with 100-ml. portions of water and once with brine, dried over anhydrous magnesium sulfate and filtered. The solvents are removed with a rotary evaporator, and the residual oil is dissolved in 150 ml. of glacial acetic acid and placed in a 500-ml., one-necked flask fitted with a magnetic stirrer and a reflux condenser. Hydrochloric acid (150 ml., 6 *N*) is added. The mixture is refluxed for

36 hours (Note 8), then concentrated with a rotary evaporator. The residue is taken up in 500 ml. of ether, poured into a 1-l. separatory funnel, and washed twice with 100-ml. portions of water, once with 5% aqueous sodium hydroxide, and once with brine. After drying over anhydrous magnesium sulfate and filtration, the organic extracts are concentrated with a rotary evaporator and distilled (Note 9), yielding 33–35 g. (75–80%) of 2-(3-ethyl-5-methyl-4-isoxazolylmethyl)cyclohexanone, b.p. 130° (0.001 mm.); n_D^{23} 1.4970; IR: 1710, 1630 cm.$^{-1}$; ^{1}H NMR (CCl$_4$): δ 1.2 (t, 3H, CH$_2$CH$_3$), 1.5–2.2 (m, 11H, 5CH$_2$ and CH, 2.3 (s, 3H, C=CCH$_3$), 2.5 (q, 2H, CH$_2$CH$_3$).

D. *1-Methyl-4,4a,5,6,7,8-hexahydronaphthalen-2(3H)-one.* *Caution! Sodium ethoxide formation should be carried out in a hood since a large volume of hydrogen gas is evolved.*

2-(3-Ethyl-5-methyl-4-isoxazolylmethyl)cyclohexanone (27.6 g., 0.125 mole) is dissolved in 250 ml. of ethanol in a Parr hydrogenation bottle, and 20 g. of freshly prepared W-4 Raney nickel catalyst (Note 10) is added. Hydrogenation is started at an initial hydrogen pressure of 25 p.s.i. Cleavage of the isoxazole ring is complete after 6 hours, after which time the reaction is stopped and the solution is filtered free of catalyst (Note 11). The catalyst is washed with ether and absolute ethanol, and the combined organic filtrates are concentrated with a rotary evaporator (Note 12).

The viscous, residual liquid is dissolved in 25 ml. of absolute ethanol and a stream of nitrogen is bubbled through the solution for 15 minutes, removing dissolved oxygen (Note 13). A solution of sodium ethoxide is then prepared by cautiously dissolving freshly cut sodium (11.5 g., 0.500 mole) in 150 ml. of absolute ethanol, under a nitrogen atmosphere, in a 500-ml., three-necked flask fitted with a reflux condenser topped with a gas-inlet, a magnetic stirrer, and a rubber serum cap on one of the sidearms. When ethoxide formation is complete, the deoxygenated solution of the hydrogenated isoxazole is injected into the stirred reaction mixture through the rubber serum cap with a syringe. The solution is refluxed until the UV spectrum of a small aliquot withdrawn with a syringe through the serum cap shows the absence of absorption at 345 nm. (Note 14). This requires about 30 hours.

A solution of 15 ml. of glacial acetic acid and 30 ml. of water is deoxygenated as described above and slowly injected with a syringe into the reaction. Refluxing is continued for 6 hours, the flask is cooled, and its contents are poured into a 1-l. separatory funnel along with 200 ml. of water. The solution is extracted four times with 100-ml. portions of ether, and the combined ether extracts are washed successively with 100 ml. of 6 N hydrochloric acid, 100 ml. of water, and 100 ml. of brine. The organic extracts are dried over anhydrous magnesium sulfate, filtered, concentrated with a rotary evaporator, and distilled, yielding 13.2–13.8 g. (65–67%) of 1-methyl-4,4a,5,6,7,8-hexahydronaphthalen-2(3H)-one, b.p. 85–90° (0.5 mm.); n_D^{23} 1.5120; IR: 1670, 1605 cm.$^{-1}$; ^{1}H NMR (CCl$_4$): δ 1.0–2.5 (m, 13H, 6CH$_2$ and CH), 1.7 (s, 3H, C=CCH$_3$).

2. Notes

1. The reagents used in this procedure were obtained from the following sources: lithium aluminum hydride, Alfa Inorganics, Inc.; thionyl chloride, Matheson, Coleman and Bell; sodium hydride, Metal Hydrides, Inc. The nitrogen was prepurified.

2. See *Org. Synth.*, **Coll. Vol. 6,** 592 (1988).

3. The addition must be done cautiously and the reaction watched constantly to see that efficient stirring is maintained. When the addition is approximately half-complete, doughy lumps, which tend to form on top of the solution, impede the stirring.

4. This quenching procedure is mentioned in the literature.[2]

5. *N,N*-Dimethylformamide was dried and purified by distillation from anhydrous copper sulfate.

6. Ethyl cyclohexanonecarboxylate was purchased from Aldrich Chemical Company, Inc., and contains approximately 40% methyl ester. The amount used takes this fact into account.

7. *N,N*-Dimethylformamide begins to decompose if the temperature rises too much.

8. The submitter stated that 24 hours of refluxing was sufficient for complete decarboxylation; however, the checkers found that after 24 hours at reflux, approximately 30% of the ester remained. Analysis was performed by GC (6-ft. column, 10% silicon rubber, 210°).

9. The distillation is most conveniently done in a short-path distillation apparatus with a mercury diffusion pump.

10. The W-4 Raney nickel is prepared according to the literature.[3]

11. *Caution! Since the catalyst is highly pyrophoric when dry, do not dry it completely.*

12. The hydrogenated isoxazole is quite sensitive to air and heat and should be used as soon as possible to prevent decomposition.

13. Oxygen must be rigorously avoided, particularly in smaller scale reactions, to prevent oxidation of the dihydropyridine intermediate to the corresponding pyridine.

14. The absorption maximum at 345 nm. corresponds to an acetyldihydropyridine intermediate (see Discussion) and disappears when the acetyl group is cleaved by ethoxide. Thus, the reaction can be readily followed spectroscopically.

3. Discussion

The isoxazole annelation reaction[4,5] is a general method for fusing a new cyclohexanone ring onto an existing system and complementary to the well-known Robinson annelation.[6] It has several major advantages:

1. The isoxazole ring serves as a "masked" or protected 3-oxobutyl side chain which can be positioned *alpha* to the existing ketone at an early stage in a complex synthesis. The isoxazole ring is stable to acids, bases, and hydride reducing agents[7] but can be cleanly and selectively cleaved by hydrogenolysis. Thus, at an appropriate time, the 3-oxobutyl side chain can be unmasked and annelation completed.

2. Although the present procedure attaches the isoxazole *via* alkylation of a β-keto ester, there are several different methods by which attachment could have been effected. Both alkylation of a cyclohexanone enamine[8] and direct alkylation of an enone anion followed by hydrogenation of the enone double bond have been used successfully.[4,5]

3. Since a wide range of 3-substituted-4-chloromethylisoxazoles can be easily prepared, the isoxazole annelation sequence allows one to construct a variety of substituted cyclohexenone systems.

The mechanism of the annelation sequence is of some interest and has been shown to proceed through the following path:[9]

The anhydrous, deoxygenated sodium ethoxide solution readily dehydrates carbinol-amide 2 to the acyldihydropyridine 3, but prevents hydrolysis or oxidation of 3. Base-catalyzed double bond migrations can lead to 4, the imine of a β-diketone, and the acetyl fragment can then be cleaved. Addition of water to the reaction causes hydrolysis of the cross-conjugated dienamine 5 to diketone 6, which then cyclizes.

The present procedure is illustrative of the general method which finds its utility largely in the construction of more complex polycyclic systems. The specific compound synthesized herein can be made more conveniently by standard Robinson annelation techniques.[10]

1. Division of Natural Science, University of California, Santa Cruz, Santa Cruz, California 95060 [Present address: Department of Chemistry, Cornell University, Ithaca, New York 14853].

2. L. Fieser and M. Fieser, "Reagents for Organic Synthesis," Vol. 1, New York, 1967, p. 584.
3. A. A. Pavlic and H. Adkins, *J. Am. Chem. Soc.*, **68**, 1471 (1946).
4. G. Stork, S. Danishevsky, and M. Ohashi, *J. Am. Chem. Soc.*, **89**, 5459 (1967).
5. G. Stork and J. E. McMurry, *J. Am. Chem. Soc.*, **89**, 5464 (1967).
6. E. C. du Feu, J. McQuillin, and R. Robinson, *J. Chem. Soc.*, 53 (1937).
7. For a review of isoxazole chemistry, see N. K. Kochetkov and D. Sokolov, "Advances in Heterocyclic Chemistry," Vol. 2, Academic Press, New York, 1963, pp. 365–421.
8. G. Stork and M. Ohashi, personal communication.
9. G. Stork and J. E. McMurry, *J. Am. Chem. Soc.*, **89**, 5463 (1967).
10. G. Stork, A. Brizzolara, R. Landesman, J. Szmuskovicz, and R. Terrell, *J. Am. Chem. Soc.*, **85**, 207 (1963).

ADDITION OF ORGANOLITHIUM REAGENTS TO ALLYL ALCOHOL: 2-METHYL-1-HEXANOL

(1-Hexanol, 2-methyl-)

Submitted by J. K. Crandall[1] and A. C. Rojas
Checked by D. E. Berthet and G. Büchi

1. Procedure

A 500-ml., three-necked, round-bottomed flask is fitted with a gas-inlet tube, a rubber septum, a reflux condenser connected to a mineral oil bubbler, and a sealed mechanical stirrer. The system is flamed with a Bunsen burner as it is flushed with dry nitrogen. The reaction vessel is cooled under nitrogen in an ice bath, and 7.25 g. (0.125 mole) of 2-propen-1-ol (Note 1), 70 ml. of pentane (Note 2), and 1.16 g. (0.0100 mole) of *N*,*N*,*N'*,*N'*-tetramethyl-1,2-ethanediamine (Note 3) are added successively through the rubber septum with a syringe. While maintaining a positive nitrogen pressure, 180 ml. of 1.5 *M* (0.270 mole) *n*-butyllithium in pentane (Note 4) is added from a syringe over a 20-minute period (Note 5). The ice bath is removed, and the reaction mixture is stirred for an additional hour (Note 6). The ice bath is then restored, the gas-inlet tube replaced with a pressure-equalizing dropping funnel, and 70 ml. of water is added, cautiously at first, and then more rapidly after the exothermic reaction ceases. The resulting mixture is transferred to a separatory funnel, the aqueous layer is separated and discarded, and the pentane layer is washed with a 10-ml. portion of 3 N hydrochloric acid, then two 10-ml. portions of water. The organic layer is dried over anhydrous magnesium sulfate and filtered. The solvent is removed by distillation through a 20-cm. Vigreux column. Distillation of the residual oil through a short-path distillation apparatus yields 9.3–9.6 g. (64–66%) (Note 7) of 2-methyl-1-hexanol, b.p. 166–167° (Notes 8 and 9).

2. Notes

1. Commercial 2-propen-1-ol was purchased from Aldrich Chemical Company, Inc., and was distilled prior to use (b.p. 94.5–95°).
2. Technical grade pentane was distilled from concentrated sulfuric acid.
3. Commercial N,N,N',N'-tetramethyl-1,2-ethanediamine was obtained from Aldrich Chemical Company, Inc., and distilled prior to use (b.p. 119.5°).
4. Commercial solutions of n-butyllithium were obtained from Foote Mineral Company.
5. During the addition of n-butyllithium, a gel forms. If the solution is not well agitated during this period, the yield is somewhat lower.
6. Extending the reaction time did not increase the yield.
7. The checker's yield was 10.5–10.7 g. (72–74%) (see Note 5).
8. The purity of the product is greater than 99% as determined by GC analysis using a 6-m. column of 30% Carbowax 20M on 60–80 Chromosorb W. The major impurity (<1%) was shown to be 3-heptanol by comparison of GC retention times and mass spectral fragmentation patterns with those of an authentic sample.
9. The spectral properties of the product are as follows; IR (neat) cm.$^{-1}$: 3268, 1377, 1037; ^{1}H NMR (CCl$_4$), δ (multiplicity, number of protons): 0.88 (m, 6H), 1.38 (m, 7H), 3.33 (unresolved d, 2H), 5.14 (broad s, 1H).

3. Discussion

This procedure illustrates a convenient method of converting allyl alcohol to 2-substituted-1-propanols by the addition of an organolithium reagent.[2] A variety of organolithiums have given moderate to high yields of the corresponding alcohols. The indicated organolithium species is a demonstrated intermediate which can, in principle, be employed in a host of further synthetic conversions.[2] Substituted allylic alcohols, however, do not undergo analogous conversions efficiently, except when the substituent is at the carbinol carbon.[2]

2-Methyl-1-hexanol has also been prepared by the reaction of 2-hexylmagnesium halides with formaldehyde,[3] the reduction of 2-methylhexanoic acid or its ester,[4,5] and by hydroformylation of 1-hexene[6–8] among others.

1. Department of Chemistry, Indiana University, Bloomington, Indiana 47401.
2. J. K. Crandall and A. C. Clark, *J. Org. Chem.*, **37**, 4236 (1972).
3. N. D. Zelinsky and E. S. Przewalsky, *Zh. Russ. Fiz.-Khim. O-va.*, **40**, 1105 (1908) [*Chem. Abstr.*, **3**, 307 (1909)].
4. P. A. Levene and L. A. Mikeska, *J. Biol. Chem.*, **84**, 571 (1929).
5. M. Leclercq, J. Billard, and J. Jacques, *Mol. Cryst. Liq. Cryst.*, **8**, 367 (1969).

6. I. Wender, R. Levine, and M. Orchin, *J. Am. Chem. Soc.*, **72**, 4375 (1950).
7. T. Asahara, H. Sekiguchi, and C. Kimura, *Yuki Gosei Kagaku Kyokai Shi*, **10**, 538 (1952) [*Chem. Abstr.*, **47**, 11123i (1953)].
8. British Petroleum Company Ltd., Fr. Pat. 1,549,414 (1968) [*Chem. Abstr.*, **72**, 2995p (1970)].

CONVERSION OF PRIMARY ALCOHOLS TO URETHANES *via* THE INNER SALT OF METHYL (CARBOXYSULFAMOYL)TRIETHYLAMMONIUM HYDROXIDE: METHYL *n*-HEXYLCARBAMATE

[Carbamic acid, hexyl-, methyl ester]

Submitted by EDWARD M. BURGESS,[1] HAROLD R. PENTON, JR.,
E. ALAN TAYLOR, and W. MICHAEL WILLIAMS
Checked by JAMES E. NOTTKE and RICHARD E. BENSON

1. Procedure

Caution! Chlorosulfonyl isocyanate is highly corrosive. This preparation should be carried out in an efficient hood, and rubber gloves should be worn during the first step. Benzene has been identified as a carcinogen; OSHA has issued emergency standards on its use. All procedures involving benzene should be carried out in a well-ventilated hood, and glove protection is required.

A. *Methyl (chlorosulfonyl)carbamate.* A dry, two-necked, 500-ml., round-bottomed flask fitted with a magnetic stirring bar, a 125-ml., pressure-equalizing dropping funnel, and a reflux condenser to which is attached a calcium chloride drying tube is charged with a solution of 70.8 g. (43.6 ml., 0.580 mole) of chlorosulfonyl isocyanate (Note 1) in 150 ml. of anhydrous benzene (Note 2). A solution of 16.0 g. (20.2 ml., 0.500 mole) of anhydrous methanol (Note 3) in 25 ml. of anhydrous benzene (Note 2) is placed in the dropping funnel. The flask is immersed in a water bath (Note 4), stirring is begun, and the methanol–benzene solution is added dropwise over 0.5 hour. Cold water is added to the bath as required to maintain a temperature of 25–30°. The reaction mixture is stirred for an additional 0.5 hour before 125 ml. of olefin-free hexane (Note 5) is added from the addition funnel over a 5-minute period while cooling the flask to 0–5° with an ice bath. The moisture-sensitive product is removed by filtration, washed twice with 40 ml. of hexane, and dried under reduced pressure, giving 76–80 g. (88–92%) of methyl (chlorosulfonyl)carbamate as white crystals, m.p. 72–74° (Note 6). This material should be stored in a brown bottle protected from light (Note 7).

B. *Inner salt of methyl (carboxysulfamoyl)triethylammonium hydroxide.* A two-necked, 500-ml., round-bottomed flask is fitted with a magnetic stirring bar, a 500-ml., pressure-equalizing dropping funnel, and a condenser to which a calcium chloride drying tube is attached. A solution of 23.0 g. (31.8 ml., 0.225 mole) of anhydrous triethylamine (Note 8) in 50 ml. of anhydrous benzene (Note 2) is placed in the flask, stirring is begun, and a solution of 17.4 g. (0.100 mole) of methyl (chlorosulfonyl)carbamate in 225 ml. of dried benzene (Note 9) is added dropwise over 1 hour. During addition the flask is cooled with a water bath maintained at 10–15°. The resulting mixture is stirred at 25–30° for an additional 0.5 hour then filtered to remove triethylamine hydrochloride (13.8 g.). Evaporation of the filtrate under reduced pressure leaves 22–23 g. of light tan needles, m.p. 70–72° (dec.), which is dissolved in 160 ml. of anhydrous tetrahydrofuran (Note 10) at 30°. On cooling, 20.0–20.6 g. (84–86%) of the inner salt of methyl (carboxysulfamoyl)triethylammonium hydroxide precipitates as colorless needles, m.p. 70–72° (dec.) (Note 11).

C. *Methyl n-hexylcarbamate.* In a dry, 100-ml., round-bottomed flask fitted with a reflux condenser, to which a calcium chloride drying tube is attached, are placed a boiling chip, 14.8 g. (0.0622 mole) of the inner salt of methyl (carboxysulfamoyl)triethylammonium hydroxide, and 6.0 g. (0.058 mole) of freshly distilled 1-hexanol (Note 12). After a mildly exothermic reaction (occasionally there is a 5-minute induction period), the viscous, yellow reaction mixture is heated with an oil bath at 95° for 1 hour. The mixture is cooled to 30°, diluted with 50 ml. of water, and extracted with three 50-ml. portions of dichloromethane. The organic extracts are combined, washed successively with 100 ml. of 5% hydrochloric acid and 50 ml. of water, and dried over anhydrous magnesium sulfate. After filtration, dichloromethane is removed with a rotary evaporator. The residue is triturated with 50 ml. of anhydrous diethyl ether and filtered; the recovered solid is triturated with two further 50-ml. portions of ether. The three ethereal filtrates are combined and concentrated with a rotary evaporator, affording 8.0 g. of crude product. Fractionation of this oil through a short-path distillation apparatus gives 4.8–4.9 g. (51–52%) of methyl n-hexylcarbamate as a colorless oil, b.p. 59–60° (0.08 mm.); n_D^{20} 1.4361 (Note 13).

2. Notes

1. The preparation of chlorosulfonyl isocyanate is described in *Org. Synth.*, **Coll. Vol. 5,** 226 (1973). This compound is highly corrosive, reacts explosively with water, and may be contaminated with cyanogen chloride.

2. Throughout this preparation the submitters used reagent grade materials distilled prior to use. The checkers used ACS reagent grade benzene available from Fisher Scientific Company.

3. The checkers used ACS reagent grade methanol available from Fisher Scientific Company.

4. The water bath should not be positioned around the flask until after the solution of chlorosulfonyl isocyanate has been added.

5. The checkers used spectro-grade reagent available from Phillips Petroleum Company.

6. ^1H NMR (CD$_3$CN) δ: 3.64 (s).

7. The checkers observed a violent decomposition when product stored in a clear glass container was inadvertently exposed to sunlight.

8. Triethylamine was dried by distillation from phosphorus pentoxide at atmospheric pressure. The checkers used reagent grade material available from Eastman Organic Chemicals.

9. The compound dissolves readily in benzene on warming to 40°.

10. The checkers used ACS reagent grade material available from Fisher Scientific Company, taken from a freshly opened bottle.

11. IR (CHCl$_3$) cm.$^{-1}$: 1690 (C=O), 1345, 1110 (SO$_2$), 1260 (C—O); ^1H NMR (CDCl$_3$), δ (multiplicity, coupling constant J in Hz., number of protons): 1.15 (t, $J = 7$, 9H), 3.29 (q, $J = 7$, 6H), 3.66 (s, 3H).

12. The checkers used practical grade material (available from Eastman Organic Chemicals) distilled immediately prior to use.

13. IR (neat) cm.$^{-1}$: 3400, 2950, 1700, 1520, 1255, 1190, 774; ^1H NMR (neat), δ (multiplicity, number of protons): 0.8–1.5 (m, 13H), 3.6 (s, 3H), 6.14 (broad t, 1H).

3. Discussion

The above procedure describes the only known preparation of the inner salt of methyl (carboxysulfamoyl)triethylammonium hydroxide and illustrates the use of this reagent to convert a primary alcohol to the corresponding urethane.[2] Hydrolysis of the urethane would provide the primary amine. The method is limited to primary alcohols; secondary and tertiary alcohols are dehydrated to olefins under these conditions, often in synthetically useful yields.[2]

Other sequences that transform primary alcohols to primary amines include: (a) conversion of the alcohol to a cyanate, rearrangement to an isocyanate, and hydrolysis,[3] and (b) conversion of the alcohol to an *N*-alkylformamide *via* the Ritter reaction, followed by hydrolysis.[4]

1. School of Chemistry, Georgia Institute of Technology, Atlanta, Georgia 30332. This work was supported by a grant (GM-12672) from the National Institutes of Health.
2. E. M. Burgess, H. R. Penton, Jr., and E. A. Taylor, *J. Org. Chem.*, **38**, 26 (1973).
3. J. W. Timberlake and J. C. Martin, *J. Org. Chem.*, **33**, 4054 (1968).
4. L. I. Krimen and D. J. Cota, *Org. React.*, **17**, 213 (1969).

METHYL 2-ALKYNOATES FROM 3-ALKYL-2-PYRAZOLIN-5-ONES: METHYL 2-HEXYNOATE

(2-Hexynoic acid, methyl ester)

$$CH_3CH_2CH_2\overset{\overset{\displaystyle O}{\|}}{C}CH_2COOC_2H_5 + H_2NNH_2 \longrightarrow$$

$$+ \ C_2H_5OH$$

$+ \ 2Tl(NO_3)_3 + CH_3OH \longrightarrow$

$$CH_3CH_2CH_2C{\equiv}CCOOCH_3 + 2TlNO_3$$

$$+ \ 4\ HNO_3 + N_2$$

Submitted by Edward C. Taylor,[1] Roger L. Robey,[1] David K. Johnson,[1] and Alexander McKillop[2]

Checked by F. Kienzle and A. Brossi

1. Procedure

Caution! Thallium compounds are highly toxic.[3] However, they may be safely handled if prudent laboratory procedures are practiced. Rubber gloves and laboratory coats should be worn and reactions should be carried out in an efficient hood. In addition, thallium wastes should be collected and disposed of separately (Note 1).

A. 3-(1-*Propyl*)-2-*pyrazolin-5-one*. A 500-ml., round-bottomed flask equipped with a magnetic stirring bar and a reflux condenser is charged with 23.7 g. (0.150 mole) of ethyl 3-oxohexanoate (Note 2), 250 ml. of ethanol, and 9.8 g. (0.17 mole) of 85% aqueous hydrazine hydrate (Note 3). The mixture is stirred for 2 hours at 0° and 2 hours at reflux, then reduced in volume to 50–100 ml. on a rotary evaporator. The resulting suspension is cooled to 0–5° and suction filtered, giving 14–16 g. (77–83%) of 3-(1-propyl)-2-pyrazolin-5-one as colorless crystals, m.p. 204–206°, which are dried for 1–2 hours over anhydrous calcium chloride and used without further purification (Note 4).

B. *Methyl 2-hexynoate.* A 1-l., round-bottomed flask equipped with a magnetic stirring bar and a reflux condenser is charged with 12.62 g. (0.1002 mole) of 3-(1-propyl)-2-pyrazolin-5-one and 500 ml. of methanol (Note 5). To this solution, 93.20 g. (0.2097 mole) of thallium(III) nitrate trihydrate (Note 6) is slowly added so as to avoid foaming. The reaction mixture is stirred for 20 minutes at room temperature and 20 minutes at reflux (Notes 7 and 8), then reduced to approximately half its volume by evaporation on a rotary evaporator. It is then cooled to 0–5° and filtered through fluted filter paper, removing precipitated thallium(I) nitrate. The filter cake is washed with 150 ml. of chloroform, and 250 ml. of water is added to the filtrate. The chloroform layer is separated, and two additional extractions with 100 ml. of chloroform are carried out. The combined chloroform layers are washed once with 100 ml. of 5% aqueous sodium hydrogen carbonate, twice with 100 ml. of water, and dried over anhydrous magnesium sulfate. The chloroform is removed on a rotary evaporator, and the residue is filtered through a 2 cm. by 12 cm. column of 100–200 mesh Florisil (Note 9) using approximately 250 ml. of chloroform as eluent. The chloroform is removed on a rotary evaporator, and the resulting pale yellow liquid is vacuum distilled through a 19-cm., unpacked column (Note 10), yielding 8.63–9.24 g. (68–73%) of methyl 2-hexynoate, b.p. 47–50° (5 mm.), as a colorless to slightly yellow liquid (Note 11).

2. Notes

1. The submitters recommend collection of solid wastes in an appropriate solid waste container, and liquid wastes (filtrates containing thallium residues, etc.) in suitably labeled bottles or cans. For the disposal of thallium wastes, a commercial organization specializing in the disposal of toxic materials was employed.

2. Ethyl 3-oxohexanoate is available under the name of ethyl butyrylacetate from Aldrich Chemical Company, Inc.

3. This product is available from Matheson, Coleman and Bell.

4. The pyrazolinone should be colorless. If it is not, it may be washed with a minimum of ice-cold ethanol. This procedure is convenient and yields material of adequate purity for the subsequent reaction. Additional pyrazolinone may be obtained by evaporating the filtrate and recrystallizing the residue from ethanol.

5. Commercially available anhydrous methanol was used without further treatment.

6. Thallium(III) nitrate trihydrate is best prepared fresh by dissolving, with stirring, 200 g. (0.439 mole) of thallium(III) oxide (available from American Smelting and Refining, Denver, Colorado) in 400 ml. of concentrated nitric acid. The submitters have found the proportion of 1 g. of thallium(III) oxide to 2 ml. of nitric acid to be best. Any suspended matter is removed by suction filtration through a medium fritted-glass funnel. The filtrate is cooled in an ice bath with mechanical stirring, yielding thallium(III) nitrate trihydrate as a fine white powder. The precipitate is separated by suction filtration through a medium fritted-glass funnel, pressed as dry as possible, and dried for approximately 6 hours in a vacuum desiccator over phosphorus pentoxide and potassium hydroxide.

Longer drying times result in thallium(III) nitrate trihydrate of poorer quality. These crystals of thallium(III) nitrate trihydrate often occlude a considerable amount of nitric acid, with a consequent decrease in reactivity. To assure removal of occluded nitric acid, the submitters recommend grinding the initially dried material to a fine powder with a mortar and pestle and redrying in a vacuum desiccator, again over phosphorus pentoxide and potassium hydroxide, for an additional 6 hours. The resulting extremely reactive thallium(III) nitrate trihydrate should be stored in a desiccator, since it rapidly turns brown upon contact with moist air. Thallium residues may conveniently be removed with aqueous 1 N hydrochloric acid.

7. The reaction mixture first turns muddy brown, due to the hydrolysis of thallium(III) nitrate to thallium(III) hydroxide and thallium(III) oxide, and then yellow with the separation of colorless thallium(I) nitrate.

8. The reduction of thallium(III) to thallium(I) may be followed with potassium iodide–starch paper. A drop of solution is placed on the paper and allowed to dry. Thallium(III) gives a purple color when the paper is moistened with water, due to the oxidation of iodide to iodine by thallium(III). Thallium(I) gives a lemon-yellow color due to the formation of thallium(I) iodide.

9. This product is available from Floridin Company, Berkley Springs, West Virginia 25411. The checkers found that this filtration was not necessary.

10. Best results were obtained with an oil bath maintained at 80–85°. The bath temperature should never exceed 100°.

11. The spectral properties of the product are as follows; IR (film) cm.$^{-1}$: 2230 strong, 1718 strong, 1428 strong, 1261 strong, 1075 strong; ^{1}H NMR (neat), δ (multiplicity, coupling constant J in Hz., number of protons, assignment): 1.01 (t, J = 7.2, 3H, CH_3), 1.63 (m, 2H, CH_2), 2.34 (t, J = 6.8, 2H, CH_2C≡C), 3.68 (s, 3H, OCH_3). GC analysis may be conveniently carried out using 10% Carbowax 20M on 60/80 Diatoport S.

3. Discussion

Methyl 2-hexynoate has been prepared by the esterification of 2-hexynoic acid, which was prepared by the carboxylation of sodium hexynylide.[4] α,β-Alkynoic acids have

TABLE I^a
METHYL 2-ALKYNOATES FROM 2-PYRAZOLIN-5-ONES
SUBSTITUTED AT POSITION 3

Substituent	Yield of Ester (%)
CH$_3$—	53
CH$_3$CH$_2$—	70
(CH$_3$)$_2$CHCH$_2$—	79
CH$_3$(CH$_2$)$_3$CH$_2$—	79
CH$_3$(CH$_2$)$_4$CH$_2$—	78
C$_6$H$_5$—	67
4–ClC$_6$H$_4$—	43

^{a} Yields are for 0.01 mole reactions.

generally been obtained by either carboxylation of metal alkynylides or by elimination reactions.[5] In particular, they have been prepared by the elimination of enol brosylates and tosylates,[6] an intramolecular Wittig reaction involving triphenylphosphinecarbomethoxymethylene and carboxylic acid chlorides,[7] and the base-promoted elimination reaction of 3-substituted-4,4-dichloro-2-pyrazolin-5-ones.[8]

The present method[9] affords the methyl ester directly in high yields from 2-pyrazolin-5-ones, which are readily prepared in nearly quantitative yields from readily accessible β-keto-esters. In addition, the reaction is simple to carry out, conditions are mild, and the product is easily isolated in a high state of purity. A limitation of the reaction is that only the methyl ester can be made, as other alcohols have been found to give poor yields and undesirable mixtures of products. Table I illustrates other examples of the reaction.[10]

1. Department of Chemistry, Princeton University, Princeton, New Jersey 08540.
2. School of Chemical Sciences, University of East Anglia, Norwich, Norfolk NR4 7TJ, England.
3. E. C. Taylor and A. McKillop, *Acc. Chem. Res., 3,* 338, (1970).
4. A. O. Zoss and G. F. Hennion, *J. Am. Chem. Soc., 63,* 1151 (1941).
5. T. F. Rutlege, "Acetylenic Compounds," Reinhold, New York, 1968, p. 32.
6. J. C. Craig, M. D. Bergenthal, I. Fleming, and J. Harley-Mason, *Angew. Chem. Int. Ed. Engl., 8,* 429 (1969).
7. G. Märkl, *Chem. Ber., 94,* 3005 (1961).
8. L. A. Carpino, P. H. Terry, and S. D. Thatte, *J. Org. Chem., 31,* 2867 (1966).
9. E. C. Taylor, R. L. Robey, and A. McKillop, *Angew. Chem. Int. Ed. Engl., 11,* 48 (1972)
10. R. L. Robey, Ph.D. Thesis, Princeton University, Princeton, New Jersey, 1972, p. 98.

METHYL (*trans*-2-IODO-1-TETRALIN)CARBAMATE

[Carbamic acid, (1,2,3,4-tetrahydro-2-iodo-1-naphthalenyl)-, methyl ester, *trans*-]

$$AgNCO + I_2 \longrightarrow AgI + INCO$$

Submitted by C. H. Heathcock[1] and A. Hassner[2]
Checked by William G. Kenyon and Richard E. Benson

1. Procedure

A 1-l., three-necked, round-bottomed flask fitted with a mechanical stirrer, a thermometer, and a calcium chloride drying tube is immersed in an ice-salt bath and charged with 38 g. (0.25 mole) of silver cyanate (Note 1), 34.7 g. of 75% 1,2-dihydronaphthalene (0.20 mole, Note 2), and 400 ml. of anhydrous diethyl ether. Stirring is begun, and when the temperature of the contents of the flask has reached 0°, 50.8 g. (0.200 mole) of iodine is added in one portion. The brown mixture is stirred vigorously for 2 hours at 0–5°, then for 6 hours at room temperature. The resulting mixture, which still retains the color of iodine, is filtered through a layer of filter aid. The filtrate is transferred to a 1-l. separatory funnel and washed with 75-ml. portions of 15% sodium bisulfite solution until the ether layer is nearly colorless. The resulting ether solution is concentrated to 200 ml. using a rotary evaporator at room temperature (20 mm.).

A solution of lithium methoxide, prepared from 0.015 g. of lithium in 200 ml. of methanol, is added to the ether solution, and the resulting mixture is allowed to stand at room temperature for 1 hour. The volume is reduced to 200 ml. by distillation using a rotary evaporator at room temperature (20 mm.) before the solution is added to 600 ml. of an ice-water mixture containing 3 g. of sodium bisulfite. The solid product that separates is collected by filtration, washed with water, and air-dried. The crude carbamate (57–64 g.) is dissolved in 180 ml. of hot methanol. The resulting mixture, which is slightly cloudy, is filtered rapidly through a coarse, fluted filter paper. The filtrate is warmed to redissolve the product, and 30 ml. of water is added slowly while the solution is heated. The flask is allowed to stand overnight, then cooled to 0°. The resulting solid crystalline is collected by filtration, washed with 10 ml. of ice-cold, 4:1 (v/v) methanol–water, and air-dried, yielding 39.6–41.0 g. (60–62%) of methyl (*trans*-2-iodo-1-tetralin)carbamate, m.p. 125.5–126.5°. The IR spectrum (KBr) shows strong absorption at 3220 and 1685 cm.$^{-1}$.

2. Notes

1. The purity of the silver cyanate used seems to be critical. Best results are obtained using product prepared in the following manner. A solution of 100 g. (0.588 mole) of silver nitrate in 3 l. of distilled water is added to a solution of 49.5 g. (0.611 mole) of potassium cyanate in 700 ml. of distilled water. The white precipitate is recovered by filtration and washed successively with distilled water, methanol, and ether. The product is protected from light and air-dried overnight on a filter funnel attached to an aspirator. The product is then dried over phosphorus pentoxide under vacuum for at least 24 hours.

2. The submitters used technical grade dihydronaphthalene of 83% purity (Columbia Organic Chemicals Company, Inc.) or of 75% purity (Aldrich Chemical Company, Inc.) as indicated by GC. The checkers used the Aldrich product.

3. Discussion

The addition of iodine isocyanate to olefins is a general reaction leading stereospecifically to *trans*-β-iodoisocyanates, convertible to *trans*-β-iodocarbamates or ureas.[8] The procedure described here is essentially that of Hassner and Heathcock.[3] The method is applicable to unsaturated alcohols, esters, ketones, and dienes, but not to conjugated unsaturated esters or ketones. The effect of solvent on the rate of reaction for the addition of iodine isocyanate to cyclohexene has been studied;[4] the rate of reaction in dichloromethane was found to be much greater than that in ether. Stereochemical and regiochemical effects as well as possible rearrangements during the addition have been evaluated.[5] β-Iodocarbamates serve as useful intermediates in the synthesis of aziridines,[3,6,7] azepines,[8] 1,2-diamines,[9] carbamates,[5] oxazolidones,[10] and amino alcohols.[10]

The conversion of methyl (*trans*-2-iodo-1-tetralin)carbamate to 1,2,3,4-tetrahydronaphthalene(1,2)imine is described in *Organic Syntheses*.[7]

1. Department of Chemistry, University of California, Berkeley, California 94720.
2. Department of Chemistry, State University of New York Binghamton, Binghamton, New York 13901.
3. A. Hassner and C. Heathcock, *Tetrahedron*, **20**, 1037 (1964).
4. C. G. Gebelein, *Chem. Ind. (London)*, 57 (1970).
5. A. Hassner, R. P. Hoblitt, C. Heathcock, J. E. Kropp, and M. Lorber, *J. Am. Chem. Soc.*, **92**, 1326 (1970).
6. G. Drefahl, K. Ponsold, and G. Köllner, *J. Prakt. Chem.*, [4] **23**, 136 (1964).
7. C. H. Heathcock and A. Hassner, *Org. Synth.*, **Coll. Vol. 6**, 967 (1988).
8. L. A. Paquette, and D. E. Kuhla, *Tetrahedron Lett.*, 4517 (1967).
9. G. Swift and D. Swern, *J. Org. Chem.*, **32**, 511 (1967).
10. A. Hassner, M. E. Lorber, and C. Heathcock, *J. Org. Chem.*, **32**, 540 (1967).

METHYL NITROACETATE

(Acetic acid, nitro-, methyl ester)

$$2\ CH_3NO_2 + 2\ KOH \xrightarrow{160°} KO_2N\!=\!CHCOOK + NH_3 + 2\ H_2O$$

$$KO_2N\!=\!CHCOOK + H_2SO_4 + CH_3OH \xrightarrow{-15°}$$
$$O_2NCH_2COOCH_3 + K_2SO_4 + H_2O$$

Submitted by S. ZEN, M. KOYAMA, and S. KOTO[1]
Checked by M. ANDO and G. BÜCHI

1. Procedure

Caution! Benzene has been identified as a carcinogen; OSHA has issued emergency standards on its use. All procedures involving benzene should be carried out in a well-ventilated hood, and glove protection is required.

A. *Dipotassium salt of nitroacetic acid.* A 3-l., three-necked, round-bottomed flask equipped with a sealed mechanical stirrer, a condenser fitted with a calcium chloride drying tube, and a pressure-equalizing dropping funnel is charged with a fresh solution of 224 g. of potassium hydroxide in 112 g. of water. From the dropping funnel is added, over 30 minutes (Note 1), 61 g. (1.0 mole) of nitromethane. The reaction mixture is heated to reflux for 1 hour in an oil bath maintained at approximately 160° (Note 2). After cooling to room temperature, the precipitated crystalline product is filtered, washed several times with methanol, and dried in a vacuum desiccator under reduced pressure, yielding 71.5–80.0 g. (79–88%) of the dipotassium salt of nitroacetic acid, m.p. 262° (dec.).

B. *Methyl nitroacetate.* A 2-l., three-necked, round-bottomed flask equipped with a sealed mechanical stirrer, a pressure-equalizing dropping funnel fitted with a calcium chloride drying tube, and a thermometer is charged with 70 g. (0.39 mole) of finely powdered dipotassium salt of nitroacetic acid (Note 4) and 465 ml. (11.6 moles) of methanol.

The reaction mixture is cooled to −15° ± 3° and 116 g. (1.16 moles) of concentrated sulfuric acid is added with vigorous stirring over approximately 1 hour at such a rate that the reaction temperature is maintained at −15°. The reaction mixture is allowed to warm to room temperature over a 4-hour period and stirred for another 4 hours at room temperature. The precipitate is removed by suction filtration, and the filtrate is concentrated on a rotary evaporator at 30–40°. The residual oil is dissolved in benzene and washed with water. The organic layer is dried over anhydrous sodium sulfate, and the benzene is removed by distillation. Further distillation under reduced pressure yields 30–32 g. (66–70%) of methyl nitroacetate, b.p. 80–82° (8 mm.), 111–113° (25 mm.) (Note 5).

2. Notes

1. The reaction mixture heats to 60–80° during the addition of nitromethane. The mixture may require external heating to maintain this temperature. The initial, yellowish color begins to turn red-brown and gradually deepens as ammonia gas is liberated.

2. The reaction mixture should not be stirred mechanically during this period in order to avoid decomposition of the product.

3. This crude product is rather pure. It can and should be employed for the esterification step without further purification. Elemental analyses for $C_2HO_4NK_2$ were as follows; *calculated:* C, 13.26; H, 0.56; N, 7.73; K, 43.16%, *found:* C, 13.27; H, 0.57; N, 7.80; K, 42.68%. This is a hygroscopic crystalline powder and should be used immediately after drying. There is a report[2] regarding an explosion of the dry dipotassium salt prepared by another method. There is no evidence that this procedure produces the same unstable impurities.

4. This must be ground into a fine powder with a mortar and pestle immediately prior to use.

5. The spectral properties of the product are as follows; IR (neat) cm.$^{-1}$: 1776, 1760; ^1H NMR ($CDCl_3$), δ (multiplicity, number of protons, assignment): 3.83 (s, 3H, OCH_3), 5.20 (s, 2H, CH_2); n_D^{20} 1.4260.

3. Discussion

Methyl nitroacetate has been prepared from nitromethane *via* the dipotassium salt of nitroacetic acid by the classical Steinkopf method,[3] but in lower yield. The dipotassium salt was obtained in 45% yield. The method has been improved by Matthews and Kubler,[4] but the salt must be recrystallized prior to esterification.

This procedure[5] is an improvement in that the reaction time is reduced and the yield is improved by increasing the concentration of alkali.

The acid-catalyzed esterification has been accomplished with either hydrochloric acid[3] or sulfuric acid;[6] an improvement on the Steinkopf method has been reported,[7] but the procedure lacks the simplicity of the present method.

Application of sulfuric acid as the catalyst is considered more practical for esterification because of its higher boiling point, its incompatibility with benzene, and the stability of nitroacetic acid in the reaction mixture, which allows omission of the final neutralization step.

The ethyl ester can also be prepared from ethyl acetoacetate (ethyl 3-oxobutanoate) by the method of Rodionov[8] as well as *via* Steinkopf's method.[3] Ethyl nitroacetate can be prepared in >70% yield[5] from the dipotassium salt, ethanol, and sulfuric acid, using anhydrous magnesium sulfate to avoid the Nef reaction.[9] The propyl and 2-propyl esters can also be obtained by this method.

1. School of Pharmaceutical Sciences, Kitasato University, Tokyo, Japan.
2. D. A. Little, *Chem. Eng. News.* **27**, 1473 (1949).
3. W. Steinkopf, *Ber. Dtsch. Chem. Ges.*, **42**, 2026, 3925 (1909); *Justus Liebigs Ann. Chem.*, **434**, 21 (1923).
4. V. E. Matthews and D. G. Kubler, *J. Org. Chem.*, **25**, 266 (1960).

5. S. Zen, M. Koyama, and S. Koto, *Kogyo Kagaku Zasshi*, **74**, 70 (1971).
6. H. Feuer, H. B. Hass, and K. S. Warren, *J. Am. Chem. Soc.*, **71**, 3078 (1949).
7. S. Umezawa and S. Zen, *Bull. Chem. Soc. Jpn.*, **36**, 1143 (1963).
8. V. M. Rodionov, E. V. Machinskaya, and V. M. Belikov, *Zh. Obshch. Khim.*, **18**, 917 (1948).
9. W. E. Noland, *Chem. Rev.*, **55**, 137 (1955).

METHYL (*E*)-3-NITROACRYLATE

[2-Propenoic acid, 3-nitro-, methyl ester (*E*)-]

$$\overset{I_2, N_2O_4}{\underset{ether, 0°}{\longrightarrow}} \quad O_2NCH_2\underset{\underset{I}{|}}{C}HCO_2CH_3$$

$$O_2NCH_2\underset{\underset{I}{|}}{C}HCO_2CH_3 \quad \overset{CH_3CO_2Na}{\underset{ether. 0°}{\longrightarrow}}$$

Submitted by John E. McMurry,[1] John H. Musser,
Ian Fleming,[2] J. Fortunak, and C. Nübling
Checked by Wayland E. Noland and
David D. McSherry

1. Procedure

Caution! Part A should be carried out in an efficient fume hood to protect the operator from poisonous nitrogen dioxide vapors.

A. *Methyl 2-iodo-3-nitropropionate.* A dry, 5-l., three-necked, round-bottomed flask is fitted with a mineral oil-containing liquid-sealed mechanical stirrer, a gas-inlet tube connected to a drying tower containing 1 : 1 sodium hydroxide–Drierite, and an open neck that will later be connected to a rubber septum. The flask is flushed gently by passing dry nitrogen through the drying tower and gas-inlet tube as methyl acrylate (298.4 g., 331 ml., 3.47 moles) (Note 1) and anhydrous diethyl ether (2500 ml.) (Note 1) are added through the open neck. The flask is fitted with the rubber septum and then cooled to 0° by covering three-fourths of the flask with ice in a large ice bath made from a picnic cooler insulated on top with glass wool. The gas-inlet tube is removed just long enough to permit the addition of iodine (250 g., 0.98 mole) (Note 1). The mixture is stirred for about 15 minutes and then dinitrogen tetroxide (76.6 ml., 1.24 moles) (Notes 1 and 2) is introduced rapidly with a syringe (Note 3) through the rubber septum. The reaction solution is stirred at 0° for 30 hours. The solution is divided into three batches of about 1-l., which are cooled in a freezer to −4°. One batch at a time is removed and added to a 2-l. separatory funnel where it is washed with 80 ml. of aqueous 70% saturated sodium sulfite solution that had been precooled to −4° (Note 4). The ethereal solutions are washed alternately so that each batch is exposed to room temperature for the minimum time. The sodium sulfite washes are repeated two or three times on each batch until the color of the ethereal solution remains light yellow when kept at −4° (Note 5). The cooled ethereal solutions are then dried by washing each with precooled (to −4°) aqueous solutions containing 80 ml. of saturated sodium chloride and 15 ml. of saturated sodium sulfite. The ethereal solutions are then combined in a 4-l. Erlenmeyer flask and dried over anhydrous sodium sulfate at −4° overnight in a freezer. The dried solution is then filtered to remove the sodium sulfate, and concentrated by evaporating it in portions in a 1-l. round-bottomed flask on a rotating evaporator fitted with a cold-finger condenser containing a dry-ice bath (Note 6). The solution should be concentrated until it contains approximately equal volumes of product and ether. The solution is then cooled to −78° for 30 minutes (Notes 6 and 7), causing the separation of 403.5–409 g. of ether–wet crystals. The crystals are filtered and stored in a freezer, and the filtrate is eluted through a chromatographic column containing 700 g. of silica gel (Baker Analyzed, 60–200 mesh, 7-cm. column diameter), using 30% diethyl ether/petroleum ether (b.p. 30–60°) as eluent. The solution containing the first band to be eluted, which is yellow, is evaporated on a rotating evaporator fitted with a cold finger condenser cooled with dry ice as described above. This gives a residue of 35–54.5 g. of yellowish crystals, which are a 7:3 mixture of methyl 2-iodo-3-nitropropionate and methyl (E)-3-nitroacrylate. The ether–wet product from the original filtration and the eluted product are combined in a 1-l. round-bottomed flask and dried at 0° under a vacuum (0.3–0.5 mm.) for 3–4 hours, giving 396–401 g. (78–79%), m.p. 32–35°, of methyl 2-iodo-3-nitropropionate containing a small amount of methyl (E)-3-nitroacrylate. This mixture may be used directly in the next step. The product can be recrystallized from 30% ether/petroleum ether (b.p. 30–60°) to obtain thermally unstable white plates, m.p. 33–35° (Note 8).

B. *Methyl (E)-3-nitroacrylate*. Anhydrous diethyl ether (2800 ml.) is placed in a dry 5-l. three-necked, round-bottomed flask fitted with a mineral-oil, liquid-sealed, mechanical stirrer and a drying tower. The contents are cooled to 0° by covering three-fourths of the flask with ice in a large ice bath made from a picnic cooler insulated on top with glass wool. Powdered anhydrous sodium acetate (95 g., 1.16 moles) (Note 9) is added to the flask, followed by methyl 2-iodo-3-nitropropionate (150 g., 0.579 mole). The flask that contained the ester is rinsed with cold anhydrous diethyl ether (200 ml.), which is then added to the 5-l. flask. The flask is stoppered and its contents are stirred vigorously at 0° for 40 hours (Note 10). The flask is then removed from the ice bath and may be kept, until needed, in a freezer for 5–7 days. The ethereal solution is then decanted and filtered to remove the sodium acetate. The residue in the flask is rinsed with three 150-ml. portions of ether; the rinse solutions are also filtered and added to the original ethereal solution. The ethereal solution is then divided into three batches (about 1-l. each), each of which is added successively to a 2-l. separatory funnel. Each batch is washed with three 80-ml. portions of aqueous 70% saturated sodium hydrogen carbonate solution, followed by one wash with 80 ml. of saturated brine. The ether solutions are then combined in a 4-l. Erlenmeyer flask and dried overnight over anhydrous sodium sulfate in a freezer at −4°. The solution is filtered to remove the drying agent and then evaporated in portions in a 1-l. round-bottomed flask on a rotating evaporator fitted with a cold-finger condenser cooled with dry ice as described in Part A. The yellow, solid residue is dissolved in a minimum of 50% diethyl ether/petroleum ether (b.p. 30–60°) and cooled to

−20° for 30 minutes (Note 7). The yellow crystals are collected by filtration and washed with cold diethyl ether. The filtrate and ether wash are concentrated and cooled to −70° for 30 minutes, giving a small second crop of crystals. The combined yield of methyl (*E*)-3-nitroacrylate is 67–69 g. (88–91% from dried methyl 2-iodo-3-nitropropionate) as yellow plates, m.p. 33–35° (Note 11). *Caution! Methyl (E)-3-nitroacrylate is a potent lachrymator, like nitroolefins in general. Contact with the eyes, or accidental transfer from hands to eyes, should be avoided. If lachrymation occurs, the eyes should be washed thoroughly with water until the irritation stops.*

2. Notes

1. The submitters used the following reagents as supplied: iodine, ether, and sodium acetate from the Mallinckrodt Chemical Works; dinitrogen tetroxide from Matheson Gas Products, Inc.; and methyl acrylate from the Aldrich Chemical Company, Inc. The checkers obtained their reagents from the same sources, except that the iodine (resublimed) was obtained from Aldrich. The dinitrogen tetroxide was purchased in a lecture bottle.

2. Dinitrogen tetroxide is often contaminated with dinitrogen trioxide, giving the condensed liquid a blue-green color. This can be oxidized by passing dry air through the condensed liquid at ⁻10° for 1–2 hours until the characteristic amber-brown color of dinitrogen tetroxide persists.

3. Slower introduction of dinitrogen tetroxide by either dropwise addition of an ethereal solution or entrainment in a stream of nitrogen gas gives similar results. The direct injection method was found to be easiest. The syringe should be precooled in a dry-ice bath to avoid back pressure during the dinitrogen tetroxide additions.

4. The sodium sulfite solution is made by preparing a saturated solution at room temperature, cooling it to −4° and decanting it from the resulting precipitate, and diluting with cold water to 70% saturation.

5. If kept at higher temperatures, the solution will darken to red and eventually brown due to thermal decomposition with release of iodine.

6. Diethyl ether was used as the liquid phase of the slurry in the dry-ice baths.

7. The low crystallization temperatures require the use of a drying tower connected to the crystallization flask to prevent condensation of atmospheric moisture.

8. ^{1}H NMR (60 mHz, CDCl$_3$): δ 3.81 (s, 3H, OCH$_3$), 4.56–5.25 (m, 3H, CH$_2$CH).

9. The anhydrous sodium acetate (purchased from the Aldrich Chemical Company, Inc.) should be dried by heating at 135° overnight (12–16 hours), cooled in a vacuum desiccator, and then ground to a fine powder which is again dried in a vacuum desiccator for several hours.

10. The progress of the reaction can be monitored by ^{1}H NMR analysis of concentrated aliquots obtained by evaporation of the solvent and following the appearance of the vinyl signals of the product at δ 6.95 and 7.55 (Note 12) and the disappearance of the A$_2$B multiplet of the starting material at δ 4.56–5.25 (Note 8).

11. The progress of the reaction can be monitored by ^{1}H NMR analysis of concentrated aliquots obtained by evaporation of the solvent and following the appearance of the vinyl signals of the product at δ 6.95 and 7.55 (Note 12) and the disappearance of the A$_2$B multiplet of the starting material at δ 4.56–5.25 (Note 8).

12. ^{1}H NMR (60 MHz, CDCl$_3$): δ 3.8 (s, 3H, OCH$_3$), 6.95 (d, $J = 14$ hz, 1H, =CHNO$_2$—), 7.55 (d, $J = 14$ Hz, 1H, =CHNO$_2$): IR (CCl$_4$): cm.$^{-1}$ 3100 s, 3000 m, 2950 s, 2880 m (all CH), 1720 s (C=O), 1640 ms (C=C), 1530 s and 1350 s (NO$_2$).

3. Discussion

This procedure for the synthesis of methyl (E)-3-nitroacrylate replaces the procedure for the corresponding ethyl ester given in Volume 56 of this series.[3] Both procedures were adapted from those for methyl (E)-3-nitroacrylate of Stevens and Emmons[4] for the addition and of Schechter, Conrad, Daulton, and Kaplan[5] for the elimination. Four major changes have been made in the procedures: (1) rapid introduction of dinitrogen tetroxide, (2) limited purification of the methyl 2-iodo-3-nitropropionate, (3) use of finely powdered anhydrous sodium acetate, and (4) carrying out the elimination at 0° for 40 hours followed by 5–7 days in a freezer. With these modifications, the preparation is reproducible and proceeds in 69–72% overall yield from iodine.

The compound has also been prepared by the reaction of methyl acrylate with dinitrogen tetroxide and oxygen in ether at 0°, followed by hydrolysis (of the 2-nitrito and 2-nitrato esters), neutralization, and distillation, in 13% overall conversion.[6] The methyl 2-hydroxy-3-nitropropionate obtained as a coproduct in 27% conversion was dehydrated by refluxing with acetyl chloride to give additional methyl (E)-3-nitroacrylate in 43% conversion.[6] The compound has also been prepared by the reaction of methyl acrylate with nitrosyl chloride in ether in a sealed tube at room temperature to give methyl 2-chloro-3-nitropropionate in 32% yield, which was dehydrochlorinated with anhydrous sodium acetate in ether at −5° and +20° to methyl (E)-3-nitroacrylate in 37% yield.[7]

The homolog, ethyl (E)-3-nitroacrylate, has been shown to be an extremely reactive receptor in the Michael reaction. It has been used in the synthesis of the α methylenebutyrolactone moiety[8] characteristic of many sesquiterpenes, as shown below.

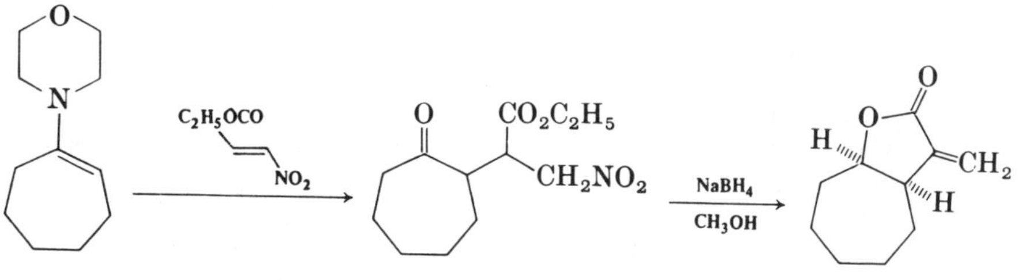

Methyl (E)-3-nitroacrylate is a reactive dienophile in the Diels-Alder reaction. With cyclopentadiene it gives the corresponding adduct, methyl 6-nitro-2-norbornene-5-carboxylate, in 43% yield in refluxing benzene[9] or in quantitive yield in ether at 0°,[10] as shown below. The adduct was shown by NMR analysis to consist of an 86:14 mixture of endo-:exo-nitro stereoisomers, which could not be separated by crystallization.[10]

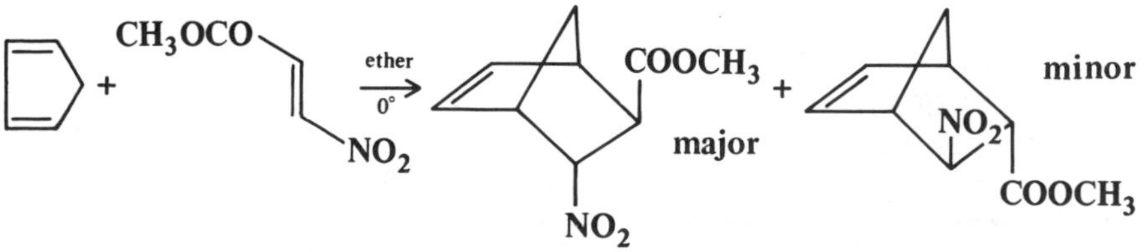

1. Natural Sciences I, University of California, Santa Cruz, California 95064. [Present address: Department of Chemistry, Baker Laboratory, Cornell University, Ithaca, New York 14853.]
2. University Chemical Laboratory, Cambridge, England CB2 1EW.
3. J. E. McMurry and J. H. Musser, *Org. Synth.*, **56**, 65 (1977).
4. T. E. Stevens and W. D. Emmons, *J. Am. Chem. Soc.*, **80**, 338 (1958).
5. H. Shechter, F. Conrad, A. L. Daulton, and R. B. Kaplan, *J. Am. Chem. Soc.*, **74**, 3052 (1952).
6. H. Shechter and F. Conrad, *J. Am. Chem. Soc.*, **75**, 5610 (1953).
7. K. A. Ogloblin and V. P. Semenov, *J. Org. Chem. USSR (Engl. Transl.)*, **1**, 1378 (1965).
8. J. W. Patterson and J. E. McMurry, *J. Chem. Soc. D*, 488 (1971).
9. S. S. Novikov, G. A. Schwecheimer, and A. A. Dudinskaya, *Bull. Acad. Sci. USSR, Div. Chem. Sci. (Engl. Transl.)*, 640 (1961).
10. N. F. Blom, D. M. F. Edwards, J. S. Field, and J. P. Michael, *J. Chem. Soc. Chem. Commun.*, 1240 (1980).

2-METHYL-2-NITROSOPROPANE AND ITS DIMER

(Propane, 2-methyl 2-nitroso-)

$$t\text{-}C_4H_9\text{—}NH_2 \xrightarrow[\text{H}_2\text{O},55°]{\text{KMnO}_4} t\text{-}C_4H_9\text{—}NO_2$$

$$\xrightarrow[\text{H}_2\text{O},(\text{C}_2\text{H}_5)_2\text{O}]{\text{Al-Hg}} t\text{-}C_4H_9\text{—}NHOH \xrightarrow[\text{H}_2\text{O},-20\text{ to }25°]{\text{NaOBr}}$$

$$t\text{-}C_4H_9\text{—}N\text{=}O \rightleftharpoons t\text{-}C_4H_9$$

Submitted by A. CALDER, A. R. FORRESTER,[1] and S. P. HEPBURN
Checked by DAVID S. CRUMRINE and HERBERT O. HOUSE

1. Procedure

A. *2-Methyl-2-nitropropane.* To a well-stirred suspension of 650 g. (4.11 moles) of potassium permanganate in 3 l. of water, contained in a 5-l., three-necked flask fitted with a reflux condenser, a mechanical stirrer, a thermometer, and a 250-ml. dropping funnel, is added, dropwise and with stirring over a 10- minute period, 100 g. (1.37 moles) of *tert*-butylamine (Note 1). When the addition is complete, the reaction mixture is heated to 55° over a period of approximately 2 hours, and maintained at 55° with continuous stirring for 3 hours. The dropping funnel and reflux condenser are replaced with a stopper and a still head fitted for steam distillation, and the product is steam distilled from the reaction mixture (Note 2). The liquid product is separated from the denser water layer, diluted with 250 ml. of diethyl ether, and washed successively with two 50-ml. portions of 2 *M* hydrochloric acid and 50 ml. of water. After the ethereal solution has been dried over anhydrous magnesium sulfate, the solution is fractionally distilled at atmospheric pressure, removing the ether. The residual crude product (Note 3) totals 106–128 g. and is sufficiently pure for use in the next step. In a typical run, distillation of 124 g. of the crude product affords 110 g. (78%) of the pure 2-methyl-2-nitropropane as a colorless liquid, b.p. 127–128°, n_D^{25} 1.3992, which slowly solidifies on standing to a waxy solid, m.p. 25–26° (Note 4).

B. N-tert-*Butylhydroxylamine*. *Caution! Since hydrogen may be liberated during the reduction with aluminum amalgam, the reaction should be conducted in a hood. Also, the aluminum amalgam may be pyrophoric. Consequently, it should be used immediately and not allowed to become dry.*

Aluminum foil (30 g. or 1.1 g.-atoms, thickness 0.002–0.003 cm.) is cut into strips 5 × 25 cm., and each strip is rolled into a cylinder about 1 cm. in diameter. Each of the aluminum foil cylinders is amalgamated by immersing it in a solution of 8.0 g. (0.030 mole) of mercury(II) chloride in 400 ml. of water for 15 seconds. Each amalgamated cylinder is then rinsed successively in ethanol and ether, and added to a mixture of 1.5 l. of ether and 15 ml. of water (Note 5) contained in a 3-l., three-necked flask fitted with a dropping funnel, a mechanical stirrer, and two efficient reflux condensers in series. The reaction mixture is stirred vigorously, and 60 g. (0.58 mole) of 2-methyl-2-nitropropane is added dropwise at a rate such that the ether refluxes briskly. The reaction usually exhibits a 5- to 7-minute induction period, after which a vigorous reaction occurs and cooling with an ice bath is necessary. After addition of the nitro compound is complete, the reaction mixture is stirred for an additional 30 minutes. The stirrer is then stopped and the gelatinous precipitate is allowed to settle. The colorless reaction solution is decanted through a glass wool plug into a 2-l. separatory funnel and washed with two 250-ml. portions of 2 *M* aqueous sodium hydroxide (Note 6). The precipitate in the reaction flask is washed with two 500-ml. portions of ether, and these washings are combined and washed with the aqueous sodium hydroxide solution (Note 6). The combined ethereal solutions are dried over anhydrous sodium sulfate and concentrated with a rotary evaporator. The residual crystalline solid is dried under reduced pressure (10–15 mm.) at room temperature, leaving 33.7 – 38.7 g. (65–75%) of the crude hydroxylamine product, m.p. 59–60°, which is sufficiently pure for use in the next step. The crude product may be recrystallized from pentane, yielding the pure N-*tert*-butylhydroxylamine as white plates, m.p. 64–65° (Note 7).

C. 2-*Methyl-2-nitrosopropane*. A solution of sodium hypobromite is prepared by adding, dropwise and with stirring over a 5-minute period, 57.5 g. (18.5 ml., 0.360 mole) of bromine to a solution of 36.0 g. (0.900 mole) of sodium hydroxide in 225 ml. of water. The resulting yellow solution, contained in a 1-l., three-necked flask fitted with a mechanical stirrer, a thermometer, and an acetone–dry ice cooling bath, is cooled to −20°. A suspension of 26.7 g. (0.300 mole) of N-*tert*-butylhydroxylamine in 50 ml. of water is added to the reaction flask, with continuous stirring, as rapidly as possible without allowing the temperature of the reaction mixture to exceed 0°. The reaction solution is again cooled to −20° before the cooling bath is removed and the mixture is stirred for 4 hours while the reaction mixture warms to room temperature. The solid product, the nitroso dimer which has separated, is collected on a sintered glass funnel, pulverized, and washed with 1 l. of water (Note 8). The residual solid is dried at room temperature under reduced pressure (10–15 mm.), leaving 19.6–22.2 g. (75–85%) of the 2-methyl-2-nitrosopropane dimer, m.p. 80–81° (Note 9). The product is sufficiently pure to be stored (Note 10) for use as a free radical trapping reagent.

2. Notes

1. *tert*-Butylamine, purchased from Aldrich Chemical Company, Inc., may be used without purification.

2. Approximately 1 l. of distillate needs to be collected to remove the product from the reaction mixture.

3. The principal contaminant is residual ether.

4. The purified product exhibits IR bands (CCl$_4$) at 1545 cm.$^{-1}$ (broad) and 1355 cm.$^{-1}$ (NO$_2$) with a UV maximum (95% C$_2$H$_5$OH) at 279 nm (ϵ 24) and a ^{1}H NMR (CCl$_4$) singlet at δ 1.58 [(CH$_3$)$_3$C]. The mass spectrum has the following abundant fragment peaks: *m/e* (rel. int.), 57 (100), 41 (74), 39 (45), and 29 (57).

5. Since water is one of the reactants in this reduction, it is necessary that at least a stoichiometric quantity of water is present.

6. Since the hydroxylamine product is readily oxidized by air to the blue nitroso compound, these manipulations should be performed rapidly to minimize exposure of the product to atmospheric oxygen. Any nitroso compound formed at this stage will co-distil with the ether and is difficult to recover.

7. The product has IR absorption (CCl$_4$) at 3600 cm.$^{-1}$ and 3250 (broad) cm.$^{-1}$ (OH and NH) with ^{1}H NMR (CCl$_4$) singlets at δ 1.09 [9H, C(CH$_3$)] and 5.86 (2H, N*H* and O*H*). The mass spectrum has the following abundant peaks: *m/e* (rel. int.), 89 (M$^+$, 11), 74 (96), 58 (41), 57 (100), 56 (52), 42 (41), 41 (74), 39 (34), 29 (54), and 28 (39).

8. Thorough washing to remove the last traces of alkali is essential, or the nitroso dimer will decompose to volatile products on standing.

9. When the colorless nitroso dimer is dissolved in various solvents, it partially dissociates, forming a blue solution which contains an equilibrium mixture of monomer and dimer. In C$_6$D$_6$ and CCl$_4$, the ^{1}H NMR spectrum of the initial solutions of dimer changes rapidly and equilibrium is established within 20–30 minutes. From ^{1}H NMR measurements at about 40° the equilibrium mixtures in CCl$_4$ and C$_6$D$_6$ contain 80–81% of the monomer;[2] the ^{1}H NMR singlets attributable to *tert*-butyl groups are observed at δ 1.24 (monomer) and 1.57 (dimer) in CCl$_4$ and at δ 0.97 (monomer) and 1.49 (dimer) in C$_6$D$_6$. The IR spectrum of the equilibrated mixture (CCl$_4$) exhibits absorption at 1565 cm.$^{-1}$ attributable to the N=O group of the monomer; this peak is not observed in the IR spectrum (KBr) of the dimer. The mass spectrum of the product exhibits the following abundant fragment peaks: *m/e* (rel. int.), 72 (10), 57 (100), 56 (23), 55 (21), 42 (22), 41 (97), 39 (55), 30 (49), 29 (74), and 28 (53). A water solution of the dimer initially is colorless and exhibits a UV maximum at 287 nm (ϵ 8000). On standing, the solution slowly turns blue. A solution of the dimer in C$_2$H$_5$OH, after standing for 20–30 minutes, exhibits maxima at 292 nm (ϵ 682 dimer) and 686 nm (ϵ 14.5 monomer).

10. The submitters report that if the product is stored at 0° in the dark, it may be kept indefinitely.

3. Discussion

The oxidation of *tert*-butylamine to 2-methyl-2-nitropropane is an example of a procedure previously illustrated in *Org. Synth.*[3] N-*tert*-Butylhydroxylamine has previously been prepared by acid-catalyzed hydrolysis of 2-*tert*-butyl-3-phenyloxazirane[4] and by oxidation of *tert*-butylamine.[5] The procedure described here is based on a method mentioned briefly by Smith and co-workers.[6] 2-Methyl-2-nitrosopropane has been prepared directly by oxidation of *tert*-butylamine,[2,5] but is usually obtained by oxidation of the hydroxylamine.[7] 2-Methyl-2-nitrosopropane has also been prepared by electrolytic reduction[8] or by zinc dust reduction[9] of 2-methyl-2-nitropropane.

2-Methyl-2-nitrosopropane is an excellent scavenger of free radicals and is now widely used in "spin trapping" experiments[10,11] (although it has certain disadvantages).[12] In this technique, a reactive radical is trapped by the nitroso compound and identified by analysis of the e.s.r. spectrum of the so-formed stable nitroxide radical. The perdeuterated derivative of 2-methyl-2-nitrosopropane has also been recommended for this purpose.[13] *tert*-Butylhydroxylamine, an intermediate in the present procedure, may also be used to synthesize *tert*-butylphenylnitrone which has been used as a "spin-trapping" reagent.[11] The reaction of 2-methyl-2-nitrosopropane with aryl Grignard reagents has been used to prepare N-aryl-N-*tert*-butylhydroxylamines.[14]

1. Department of Chemistry, University of Aberdeen, AB9 2UE, Old Aberdeen, Scotland.
2. J. C. Stowell, *J. Org. Chem.*, **36**, 3055 (1971).
3. N. Kornblum and W. J. Jones, *Org. Synth.*, **Coll. Vol. 5**, 845 (1973).
4. W. D. Emmons, *J. Am. Chem. Soc.*, **79**, 5739 (1957).
5. E. Bamberger and R. Seligman, *Ber. Dtsch. Chem. Ges.*, **36**, 685 (1903); R. J. Holman and M. J. Perkins, *J. Chem. Soc. C*, 2195 (1970).
6. P. A. S. Smith, H. R. Alul, and R. L. Baumgarten, *J. Amer. Chem. Soc.*, **86**, 1139 (1964).
7. W. D. Emmons, *J. Am. Chem. Soc.*, **79**, 6522 (1957).
8. P. E. Iversen and H. Lund, *Tetrahedron Lett.*, 4027 (1967).
9. F. D. Greene and J. F. Pazos, *J. Org. Chem.*, **34**, 2269 (1969).
10. M. J. Perkins, P. Ward, and A. Horsfield, *J. Chem. Soc. B*, 395 (1970); C. Lagercrantz and S. Forshult, *Acta Chem. Scand.*, **23**, 708 (1969).
11. O. H. Griffith and A. S. Waggoner, *Acct. Chem. Res.*, **2**, 17 (1969); E. G. Janzen, *Acct. Chem. Res.*, **2**, 279 (1969); **4**, 31 (1971); M. J. Perkins, *Chem. Soc., Spec. Publ.*, No. **24**, 97 (1970).
12. A. R. Forrester and S. P. Hepburn, *J. Chem. Soc. C*, 701 (1971).
13. R. J. Holman and M. J. Perkins, *J. Chem. Soc. C*, 2324 (1971).
14. A. R. Forrester and S. P. Hepburn, *J. Chem. Soc. C*, 1277 (1970) and other papers in this series.

NUCLEOPHILIC ACYLATION WITH DISODIUM TETRA-CARBONYLFERRATE: METHYL 7-OXOHEPTANOATE AND METHYL 7-OXOÖCTANOATE

(Octanoic acid, 7-oxo-, methyl ester)

$$Fe(CO)_5 \xrightarrow[\substack{1,4\text{-dioxane,} \\ \text{reflux}}]{\substack{Na, \\ \text{benzophenone}}} Na_2Fe(CO)_4 \cdot (\text{dioxane})_{1.5}$$

Submitted by RICHARD G. FINKE[1,2] and THOMAS N. SORRELL[1,3]
Checked by RICHARD T. TAYLOR and MARTIN F. SEMMELHACK

1. Procedure

Caution! Iron pentacarbonyl and carbon monoxide are highly toxic; consequently all parts of this procedure should be carried out in a well-ventilated hood. Iron pentacarbonyl is easily recognized by its musty odor. Since disodium tetracarbonylferrate is very pyrophoric, the reagent must be kept under a dry inert atmosphere at all times.

Methyl iodide, in high concentrations for short periods and in low concentrations for long periods, can cause serious toxic effects in the central nervous system. Accordingly, the American Conference of Governmental Industrial Hygienists[4] has set 5 p.p.m., a level which cannot be detected by smell, as the highest average concentration in air to which workers should be exposed for long periods. The preparation and use of methyl iodide should always be performed in a well-ventilated fume hood. Since the liquid can be absorbed through the skin, care should be taken to prevent contact.

A. *Disodium tetracarbonylferrate sesquidioxanate* (Note 1).   A dry, 2-l., three-necked, round-bottomed flask is equipped with a mechanical stirrer (Note 2), a three-way stopcock with one branch connected to a nitrogen source, and a Y-shaped adapter fitted with a reflux condenser vented through an oil bubbler and a pressure-equalizing dropping funnel capped by a rubber septum. The apparatus is flushed with nitrogen (Note 3) for 15 minutes and charged with 600 ml. of dry, deoxygenated dioxane (Notes 4 and 5), 10.6 g. (0.461 g.-atoms) of sodium, and 9.1 g. (0.050 mole) of benzophenone. The solution is stirred vigorously and heated under reflux with a heating mantle until the deep blue color of the benzophenone ketyl appears. With a gas-tight syringe, 45.3 g. (29.8 ml., 0.231 mole) of iron pentacarbonyl (Note 6) is injected into the dropping funnel. The blue solution is then titrated to a white or slightly yellow end point by adding iron pentacarbonyl to the refluxing solution over 2.5 hours (Note 7). The suspension is heated at reflux for another 45 minutes, then cooled to room temperature. Precipitation of the disodium tetracarbonylferrate sesquidioxanate as a white powder is completed by adding 600 ml. of dry, deoxygenated hexane (Notes 5 and 8). The Y-shaped adapter and the mechanical stirrer are removed under a rapid stream of nitrogen and quickly replaced with a tight-fitting rubber septum and a gas-dispersion tube with a fritted-glass tip (see Figure 1). The solvent is forced up the gas-dispersion tube and out of the flask with nitrogen pressure (Note 9). The product is washed in the flask with two 400-ml. portions of hexane added *via* a cannula (Note 5), and the supernatant solvent is removed with nitrogen pressure in the same manner. The disodium tetracarbonylferrate sesquidioxanate is used directly in Parts B and C without drying or weighing (Note 10).

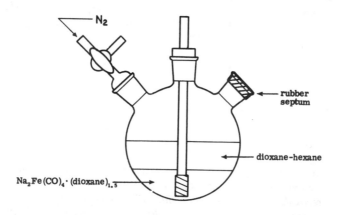

Figure 1.

B. *Methyl 7-oxoheptanoate*.   In the same three-necked, round-bottomed flask containing 72–78 g. (0.21–0.23 mole) of disodium tetracarbonylferrate sesquidioxanate, the gas-dispersion tube is replaced with a mechanical stirrer, and 1.5 l. of dry, deoxygenated tetrahydrofuran (Note 11) is added. The light tan suspension is stirred vigorously as 41.8 g. (0.200 mole) of methyl 6-bromohexanoate (Note 12) is added in one portion by syringe. The nitrogen is flushed from the flask with carbon monoxide (Notes 13 and 14) admitted through the other branch of the three-way stopcock, and the suspension is stirred under 10 p.s.i. of carbon monoxide for at least 14 hours, during which time the solid dissolves. A rapid flow of nitrogen is swept through the flask while the septum is removed and replaced quickly with a pressure-equalizing dropping funnel. The dropping funnel is flushed with nitrogen and charged with 50 ml. of glacial acetic acid, added dropwise to the orange solution (Note 15). Stirring is continued for 20 minutes, after which the deep red solution is concentrated to a volume of *ca.* 400 ml. with a rotary evaporator in the hood *(Caution! Some iron pentacarbonyl is present)* and poured into 2 l. of water. The mixture is extracted with four 400-ml. portions of diethyl ether, and the combined organic solutions are washed with 400 ml. of water. The ethereal solution is mixed with 400 ml. of 2 *M* hydrochloric acid, and 68 g. of iron(III) chloride is added in small portions until carbon monoxide evolution subsides and the organic layer becomes green from triiron dodecacarbonyl. The organic layer is washed with successive 400-ml. portions of 2 *M* hydrochloric acid, water, saturated sodium hydrogen carbonate, and saturated sodium chloride. After being dried over anhydrous sodium sulfate, the ethereal solution is concentrated to a green oil with a rotary evaporator. Iron-containing by-products such as triiron dodecacarbonyl and iron pentacarbonyl are removed by rapid chromatography on a 7 × 40 cm. column prepared in hexane with 400 g. of silica gel (Note 16). The green triiron dodecacarbonyl is first eluted with *ca.* 3 l. of hexane before the product is eluted with 3 l. of 2:1 (v/v) ether–hexane. The ether–hexane eluate is dried over anhydrous magnesium sulfate and evaporated. Distillation of the residual oil through a 15-cm. Vigreux column at reduced pressure affords a small forerun of 1 ml. or less and 17.9–20.0 g. (57–63%) of methyl 7-oxoheptanoate, b.p. 65–80° (0.1 mm.), n_D^{20} 1.4388 (Note 17).

C. *Methyl 7-oxoöctanoate*.   The 2-l., three-necked, round-bottomed flask containing 72–78 g. (0.21–0.23 mole) of disodium tetracarbonylferrate from Part A is flushed rapidly with nitrogen while the gas-dispersion tube is removed, a magnetic stirring bar is placed inside, and a pressure-equalizing dropping funnel is attached. The other branch of the three-way stopcock is connected to a bubbler, the dropping funnel is capped with a rubber septum, and 600 ml. of dry, deoxygenated *N*-methylpyrrolidinone (Notes 5 and 18) is added. The suspension is stirred and maintained under a static nitrogen atmosphere as 41.8 g. (0.200 mole) of methyl 6-bromohexanoate (Note 12) is injected into the dropping funnel with a gas-tight syringe and added dropwise into the flask. The resulting solution (Note 19) is stirred for 30 minutes at room temperature and cooled in an ice bath before 64 g. (28 ml., 0.45 mole) of methyl iodide is added over 20 minutes. The ice bath is removed, and stirring is continued for 20–40 hours (Note 14). The dark-red mixture is poured into 3 l. of saturated aqueous sodium chloride and extracted with three 400-ml. portions of ether and one 400-ml. portion of hexane. The combined organic solutions are washed with one 400-ml. portion of water and one 400-ml. portion of saturated sodium chloride. After 400 ml. of 2 M hydrochloric acid is added, the iron by-products are oxidized with iron(III) chloride exactly as described in Part B. The ethereal solution is evaporated under reduced pressure, the residual oil is applied to a column of silica gel packed in hexane, and the organic product is separated from the iron by-products by chromatography as described in Part B. The ether–hexane eluate is dried over anhydrous magnesium sulfate, the solvent is evaporated, and the remaining liquid is distilled at reduced pressure, giving 24.0–24.8 g. (70–72%) of methyl 7-oxoöctanoate, b.p. 112–127° (10 mm.), n_D^{20} 1.4360 (Note 20).

2. Notes

1. Approximately 4.5 hours are required to complete Part A.

2. A mechanical stirrer is necessary on any scale owing to the formation of a thick slurry toward the end of the reaction.

3. The submitters passed the nitrogen through *ca.* 100 g. of BASF catalyst R3-11 contained in a metal tube and heated at 160°, to remove oxygen, and through a column of Linde type 3A molecular sieves, to remove water. The catalyst (catalog number 18-3000-00) and pertinent literature were obtained from Chemical Dynamics Corporation, P.O. Box 395, South Plainfield, New Jersey 07080. The checkers used "prepurified" nitrogen and argon in separate runs without additional drying or oxygen scavenging.

4. Dioxane was heated at reflux with sodium overnight under nitrogen, benzophenone was added, and the solvent was distilled after appearance of the deep-blue color of the benzophenone ketyl.

5. The submitters recommend that the solvent be distilled under nitrogen into a two-necked receiving flask fitted with a three-way stopcock. The receiving flask is separated from the distillation apparatus under a rapid nitrogen flow and fitted quickly with a rubber septum. The solvent is then transferred to the reaction vessel by needlestock techniques[5] as follows: A stainless-steel cannula with a 2-mm. inside diameter and both ends sharpened is inserted through the septum into the receiving flask above the surface of the liquid, and a stream of nitrogen is passed briefly through the stopcock and out the cannula, removing air. The other end of the cannula is then inserted through the septum on the reaction vessel, the end of the cannula in the receiver is pushed below the surface of the liquid, and the solvent is forced into the reaction vessel with nitrogen pressure.

6. Iron pentacarbonyl was purchased from PCR, Inc., Gainesville, Florida, and stored under nitrogen.

7. Approximately 95% of the iron pentacarbonyl is added within 2 hours, and the remaining 5% is then added dropwise over the next 30 minutes. The blue color should never be completely discharged prior to the end point, particularly toward the end of the reaction, since the remaining solution may be deactivated. Avoiding premature discharge of the blue color is especially important in small-scale preparations. At the end point 1 ml. or less of the iron pentacarbonyl remains in the dropping funnel.

The checkers carried out the reaction on a smaller scale in two runs, adding 13.1 g. (8.62 ml., 0.0669 mole) of iron pentacarbonyl as a solution in 50 ml. of dry dioxane.

8. Reagent grade hexane was dried and deoxygenated by distillation from calcium hydride under nitrogen.

9. Alternatively, 300–400 ml. of the dioxane may be transferred by this method into

another flask before the hexane is added. The recovered dioxane may then be used in another preparation without purification.

10. The yield is typically 72–78 g. (90–100%). An excess of the reagent is not detrimental to the procedures in Parts B and C. The submitters have doubled the scale of this procedure with no change in the yield. For smaller-scale reactions the submitters recommend that the reagent be purchased from Alfa Division, Ventron Corporation. The checkers used the commercially available reagent successfully in one run, the material having been transferred to the reaction vessel in a dry box.

11. Tetrahydrofuran was dried and deoxygenated by distillation from calcium hydride under nitrogen.

12. 6-Bromohexanoic acid was purchased from Aldrich Chemical Company, Inc., and esterified with sulfuric acid and methanol. Methyl 6-bromohexanoate was obtained as a colorless liquid, b.p. 92–94° (5 mm.), n_D^{20} 1.4510, judged to be greater than 99% pure according to GC analysis on dimethylsilicone (OV-101) as liquid phase.

13. A cylinder of carbon monoxide equipped with a suitable regulator calibrated in pounds per square inch (p.s.i.) is connected to the three-way stopcock. All joints and the septum must be secured with clamps or wire. A vertical tube containing mercury was connected to an exit tube from the reaction flask by the checkers. A pressure of carbon monoxide was maintained against a 500-mm. column of mercury.

14. Alkyl tetracarbonyl iron(0) reagents in solution decompose more rapidly with increasing concentration and temperature, especially above 0°. Carbon monoxide must be added without delay to convert this intermediate to the more stable acyl iron compound.

15. Subsequent operations may be conducted in air. However, the procedure should not be interrupted until the ethereal solution is drying over sodium sulfate.

16. Silica gel of mesh 60–200 was supplied by Davison Chemical Division, W. R. Grace and Company, Baltimore, Maryland, and dried at 70° before use. The flow rate of hexane during the chromatography was 4 l. per hour.

17. The checkers obtained 13.6 g. (43%) of product, b.p. 65–75° (0.1 mm.). The submitters determined the purity of the product to be greater than 90% by GC using 10% dimethylsilicone (OV-101) as liquid phase at 200°. The ^1H NMR spectrum (CDCl$_3$) shows absorptions at δ (multiplicity, coupling constant J in Hz., number of protons, assignment): 1.48 (m, 6H, 3CH_2), 2.32 (t, 2H, CH_2CO$_2$CH$_3$), 2.40 (t of d, J = 2 and 8, 2H, CH_2CH=O), 3.65 (s, 3H, CO$_2$CH_3), 9.62 (t, J = 2, 1H, CH=O).

18. N-Methylpyrrolidinone was heated at reflux over calcium hydride under reduced pressure for at least 24 hours by the submitters, then distilled from the calcium hydride under reduced pressure. The checkers stirred the solvent in the presence of calcium hydride at 100° for 48 hours prior to distillation at 150 mm.

19. The homogeneous solution, if free of impurities, is bright yellow. Usually, however, the color is dark red or orange, evidently owing to the presence of trace amounts of impurities.

20. Starting with 13.1 g. (8.62 ml., 0.0669 mole) of iron pentacarbonyl, the checkers obtained 7.3–7.4 g. (71–75%) of product, b.p. 92–95° (2 mm.) A GC analysis by the submitters using dimethylsilicone (OV-101) as liquid phase at 200° showed the purity of the product to be greater than 95%. The spectral properties of the product are as follows: IR (thin film) cm.$^{-1}$: 1735 (ester C=O), 1715 (ketone C=O); ^1H NMR (CDCl$_3$), δ (multiplicity, coupling constant J in Hz., number of protons, assignment): 1.46 (m, 6H,

3CH$_2$), 2.12 (s, 3H, CH$_2$COCH$_3$), 2.32 (t, J = 8, 2H, CH$_2$CO$_2$CH$_3$), 2.44 (t, J = 8, 2H, CH$_2$COCH$_3$), 3.67 (s, 3H, CO$_2$CH$_3$).

3. Discussion

This procedure describes the preparation of disodium tetracarbonylferrate sesquidioxanate by reduction of iron pentacarbonyl with sodium and the use of the reagent for nucleophilic formylation and acylation of the primary bromide, methyl 6-bromohexanoate. Disodium tetracarbonylferrate serves as a synthetic equivalent of carbon

monoxide dianion, $\overset{O}{\underset{\ominus C \ominus}{\|}}$, in most preparations. Thus, the reagent reacts with alkyl

halides and p-toluenesulfonates, forming anionic alkyltetracarbonyl iron(0) complexes which combine in a second step with various electrophiles, giving carbonyl compounds.[6] The reagent donates the new carbonyl carbon, which becomes bonded to both the alkyl group and the electrophile in the final product. If the electrophile is a proton from acetic acid or an alkyl group from a second alkyl halide, the overall transformations amount to nucleophilic formylation[6] and acylation,[7] respectively. The preparation of carboxylic acids, esters, and amides by formal nucleophilic carboxylation is accomplished by use of oxygen and water (or sodium hypochlorite and water), iodine and alcohols, and iodine and amines, respectively, as the electrophiles.[8] The general reactions are summarized by the following equations and several specific examples[7-9] are presented in Table I.

The initial reaction between the alkyl halide (or p-toluenesulfonate) and disodium tetracarbonylferrate behaves as a typical S$_N$2-type substitution.[6,10] Thus, this step proceeds smoothly with primary and secondary reactants, but the tertiary analogs fail owing to elimination. Allylic substrates are also incompatible, since these undergo elimination, forming stable iron tricarbonyl-1,3-diene complexes. The initial substitution with secondary p-toluenesulfonates occurs more efficiently than with the corresponding halides, and the stereochemistry results from clean inversion. The solvents used are generally either

TABLE I

NUCLEOPHILIC ACYLATION AND CARBOXYLATION OF ALKYL HALIDES AND
p-TOLUENESULFONATES WITH DISODIUM TETRACARBONYLFERRATE[7-9]

Alkyl Halide or p-Toluenesulfonate[a]	Electrophile	Product	Yield (%)
$C_6H_5(CH_2)_2Br$	CH_3CO_2H	$C_6H_5(CH_2)_2-\overset{\overset{\displaystyle O}{\|\|}}{C}-H$	86[b]
$(CH_3)_2C=CH(CH_2)_2Br$	CH_3CO_2H	$(CH_3)_2C=CH(CH_2)_2-\overset{\overset{\displaystyle O}{\|\|}}{C}-H$	81[b]
$\underset{\underset{\displaystyle CH_3(CH_2)_5CHCH_3}{\|}}{Br}$	CH_3CO_2H	$CH_3(CH_2)_5\underset{\underset{\displaystyle CHCH_3}{\|}}{\overset{\overset{\displaystyle O}{\diagdown}}{C}}{\overset{\displaystyle H}{\diagup}}$	50[b,c]
$CH_3(CH_2)_7Br$	CH_3CH_2I	$CH_3(CH_2)_7-\overset{\overset{\displaystyle O}{\|\|}}{C}-CH_2CH_3$	80
$\underset{\underset{\displaystyle CH_3(CH_2)_5CHCH_3}{\|}}{OTs}$	CH_3I	$CH_3(CH_2)_5\underset{\underset{\displaystyle CHCH_3}{\|}}{\overset{\overset{\displaystyle O}{\diagdown}}{C}}{\overset{\displaystyle CH_3}{\diagup}}$	79
$CH_3(CH_2)_5Br$	O_2, H_2O	$CH_3(CH_2)_5-\overset{\overset{\displaystyle O}{\|\|}}{C}-OH$	82
$Cl(CH_2)_6Br$	O_2, H_2O	$Cl(CH_2)_6-\overset{\overset{\displaystyle O}{\|\|}}{C}-OH$	84
$CH_3(CH_2)_7Br$	I_2, C_2H_5OH	$CH_3(CH_2)_7-\overset{\overset{\displaystyle O}{\|\|}}{C}-OC_2H_5$	84
$CH_3(CH_2)_4Br$	$I_2, (C_2H_5)_2NH$	$CH_3(CH_2)_4-\overset{\overset{\displaystyle O}{\|\|}}{C}-N(C_2H_5)_2$	80[b]

[a]The structural abbreviation Ts is used for p-toluenesulfonate.
[b]Yield determined by GC.
[c]Isomeric octenes identified as by-products.

tetrahydrofuran or N-methylpyrrolidinone, the reactions being as much as 10^4 times faster in the latter. The initially formed alkyl iron intermediate rearranges to an acyl iron complex either prior to or during the subsequent reaction with the electrophile.[6-9,11] The presence of carbon monoxide or triphenylphosphine enhances the rate of rearrangement to the more stable acyl iron intermediate. The reactions show high selectivity and are compatible with functional groups such as chloro, cyano, and esters. The reagent reacts with acid chlorides, forming the acyl iron complexes directly, which may then be hydrolyzed to aldehydes or utilized for nucleophilic acylation.

Methyl 7-oxoheptanoate has previously been synthesized from cycloheptanone in two steps with a 42% overall yield.[12] 7-Oxoöctanoic acid, the methyl ester of which is the product of Part C, has been prepared by base-induced ring cleavage of 2-acetylcyclohexanone.[13]

1. Department of Chemistry, Stanford University, Stanford, California 94305.
2. Present address: Department of Chemistry, University of Oregon, Eugene, Oregon 97403.
3. Present address: Department of Chemistry, University of North Carolina, Chapel Hill, North Carolina 27514.
4. American Conference of Governmental Industrial Hygienists (ACGIH), "Documentation of Threshold Limit Value," 3rd ed., Cincinnati, Ohio, 1971, p. 166.
5. D. F. Shriver, "The Manipulation of Air Sensitive Compounds," McGraw-Hill, New York, 1969, p. 157.
6. J. P. Collman, Acc. Chem. Res., 8, 342 (1975).
7. M. P. Cooke, Jr., J. Am. Chem. Soc., 92, 6080 (1970).
8. J. P. Collman, S. R. Winter, and D. R. Clark, J. Am. Chem. Soc., 94, 1788 (1972).
9. J. P. Collman, S. R. Winter, and R. G. Komoto, J. Am. Chem. Soc., 95, 249 (1973).
10. J. P. Collman, R. G. Finke, J. N. Cawse, and J. I. Brauman, J. Am. Chem. Soc., 99, 2515 (1977).
11. J. P. Collman, J. N. Cawse, and J. I. Brauman, J. Am. Chem. Soc., 94, 5905 (1972); J. P. Collman, R. G. Finke, J. N. Cawse, and J. I. Brauman, J. Am. Chem. Soc., 100, 4766 (1978).
12. D. Taub, R. D. Hoffsommer, C. H. Kuo, H. L. Slates, Z. S. Zelawski, and N. L. Wendler, Tetrahedron, 29, 1447 (1973).
13. C. R. Hauser, F. W. Swamer, and B. I. Ringler, J. Am. Chem. Soc., 70, 4023 (1948).

ALDEHYDES FROM ALLYLIC ALCOHOLS AND PHENYLPALLADIUM ACETATE: 2-METHYL-3-PHENYLPROPIONALDEHYDE

(Benzenepropanal, α-methyl-)

$$C_6H_5HgOCOCH_3 + Pd(OCOCH_3)_2 \longrightarrow [C_6H_5PdOCOCH_3] + Hg(OCOCH_3)_2$$

$$[C_6H_5PdOCOCH_3] + \underset{\overset{\displaystyle CH_3}{\displaystyle |}}{CH_2{=}CCH_2OH} \longrightarrow$$

$$\left[\underset{\overset{\displaystyle |}{\displaystyle PdOCOCH_3}}{\overset{\overset{\displaystyle CH_3}{\displaystyle |}}{C_6H_5CH_2CCH_2OH}} \right] \longrightarrow$$

$$\underset{\overset{\displaystyle CH_3}{\displaystyle |}}{C_6H_5CH_2CHCHO} + CH_3COOH + Pd$$

Submitted by R. F. Heck[1]
Checked by Robert A. Clement and Richard E. Benson

1. Procedure

A slurry comprised of 33.6 g. (0.0998 mole) of commercial phenylmercury(II) acetate, 200 ml. of acetonitrile, and 14.4 g. (16.8 ml., 0.200 mole) of methallyl alcohol (Note 1) is prepared in a 500-ml., three-necked flask fitted with a mechanical stirrer, a condenser, and a thermometer. The slurry is stirred and cooled in an ice bath, and 22.4 g. (0.0998 mole) of powdered palladium(II) acetate (Note 2) is added over 1 minute. Stirring is continued with cooling for 1 hour, then at room temperature for 3 more hours (Note 3). The temperature of the reaction mixture reaches a maximum of 27° after removal of the ice bath.

The black reaction mixture is diluted with about 100 ml. of diethyl ether and poured onto 200 g. of ether-wet alumina (Woelm, Activity Grade 1) in a 45 × 2.5 cm. glass chromatographic column. The product is washed through the alumina with about 1 l. of ether. The brown eluate is concentrated by distilling the ether through a 45-cm. Vigreux column on a steam bath at atmospheric pressure. When the ether has been distilled, a slight vacuum is applied, removing most of the acetonitrile. After the volume reaches about 50 ml., the mixture is filtered into a 100-ml. distillation flask, removing some precipitated palladium metal. The flask is rinsed with 10-ml. of ether, and the rinse is combined with the product. The flask is equipped with a 10 cm. Vigreux column for distillation at reduced pressure. After removal of the solvent, 8.1–8.5 g. (55–58%) of 2-methyl-3-phenylpropionaldehyde is collected, b.p. 75–85° (3 mm.) (Note 4), n_D^{25} 1.5113 (Note 5).

2. Notes

1. Methallyl alcohol was obtained from Eastman Organic Chemicals.
2. Palladium(II) acetate was purchased from Engelhard Industries.
3. The yield improves slightly with stirring overnight; the checkers obtained the aldehyde in 69% yield in this manner.
4. The bulk of the product has b.p. 77–80° (3 mm.).
5. The product is 90–95% pure by GC and NMR analyses. The checkers estimated the purity to be at least 95% by these criteria. The ^{1}H NMR spectrum (CDCl$_3$) shows peaks at δ 0.95 (d, J = 6.5 Hz., 3H), ~2.7 (complex m, 3H), 7.20 (s, 5H) and 9.65 (d, J = 1.5 Hz., 1H).

3 Discussion

The formation of 3-aryl-substituted aldehydes and 3-aryl-substituted ketones by the reaction of "arylpalladium salts" with allylic alcohols is general.[2] Illustrations of the preparation of two aldehydes and two ketones are given in Table I.

The presence of nitro, carboalkoxy, carboxyl, chloro, formyl, alkyl, and acyl groups does not interfere with the reaction. A single alkoxy group also does not interfere, but if two or more are present the yields are markedly decreased. The reaction is inhibited by the presence of unhindered, basic nitrogen substituents, by the phenolic group, and probably by the thiol group.

A variation of this procedure involves the use of a catalytic amount of palladium(II) chloride with copper(II) chloride as a reoxidant.[2] This method, however, generally gives lower yields and less pure products. Another related preparation uses palladium(II) acetate with two equivalents of triphenylphosphine, catalyzing the reaction of iodo-[3] or bromo-[4] benzene with methallyl alcohol in the presence of weak bases. Phenyldiazonium salts also may be used to react with methallyl alcohol and a palladium(0) catalyst to form the propionaldehyde.

Other preparations of 2-methyl-3-phenylpropionaldehyde include the pyrolysis of a mixture of the calcium salts of 2-methyl-3-phenylpropionic acid and formic acid,[5] the

TABLE I

3-ARYLCARBONYL COMPOUNDS FROM ALLYLIC ALCOHOLS AND
"PHENYLPALLADIUM ACETATE"[2]

Allylic Alcohol	Product	Yield, %	Boiling Point, °C.
CH$_2$=CHCH$_2$OH	C$_6$H$_5$CH$_2$CH$_2$CHO	35	220–225°a
trans-CH$_3$CH=CHCH$_2$OH	C$_6$H$_5$CH(CH$_3$)CH$_2$CHO	36	67–75° (1 mm.)
trans-CH$_3$CH=CHCH(OH)CH$_3$	C$_6$H$_5$CH(CH$_3$)CH$_2$COCH$_3$	51	70–75° (3 mm.)
(CH$_3$)$_2$C=CHCH(OH)CH$_3$	C$_6$H$_5$C(CH$_3$)$_2$CH$_2$COCH$_3$	29	83–87° (2 mm.)

a Purification by careful distillation is necessary in this example to remove cinnamaldehyde which is also formed in the reaction (b.p. 252°).

pyrolysis of the glycidic ester obtained from 2-phenyl-2-propanone and ethyl chloroacetate,[7] the hydroformylation of allylbenzene,[8] the benzylation of 2-ethylthiazoline followed by reduction with aluminum amalgam and cleavage with mercury(II) chloride,[9] and the reaction of phenylmagnesium bromide with 2-vinyl-5,6-dihydro-1,3-oxazine followed by methylation and hydrolysis.[10]

1. Contribution No. 1504 from the Research Center, Hercules Incorporated, Wilmington, Delaware 19899. [Present address: Department of Chemistry, University of Delaware, Newark, Delaware 19711.]
2. R. F. Heck, *J. Am. Chem. Soc.*, **90,** 5526 (1968).
3. R. F. Heck and J. B. Melpolder, *J. Org. Chem.*, **41,** 265 (1976).
4. A. J. Chalk and S. A. Magennis, *J. Org. Chem.*, **41,** 273 (1976).
5. K. Kikukawa and T. Matsuda, *Chem. Lett.*, 159 (1977).
6. W. V. Miller and G. Rohde, *Ber. Dtsch. Chem. Ges.*, **23,** 1079 (1890).
7. G. Darzens, *C. R. Hebd. Seances Acad. Sci.*, **139,** 1214 (1904).
8. R. Lai and E. Ucciani, *C. R. Hebd. Seances Acad. Sci., Ser. C*, **275,** 1033 (1972).
9. J. C. Durandetta and A. I. Meyers, *J. Org. Chem.*, **40,** 201 (1975).
10. H. W. Adickes, A. C. Kovelesky, G. R. Malone, A. I. Meyers, A. Nebeya, R. L. Nolen, I. R. Politzer, and R. C. Portnoy, *J. Org. Chem.*, **38,** 36 (1973).

ENDOCYCLIC ENAMINE SYNTHESIS: *N*-METHYL-2-PHENYL-Δ²-TETRAHYDROPYRIDINE

(Pyridine, 1,2,3,4-tetrahydro-1-methyl-5-phenyl-)

$$C_6H_5COCH_3 + CH_3NH_2 \xrightarrow[(C_2H_5)_2O,\ C_6H_{14},\ -30°]{TiCl_4} C_6H_5\overset{\overset{\displaystyle NCH_3}{\|}}{C}CH_3$$

$$C_6H_5\overset{\overset{\displaystyle NCH_3}{\|}}{C}CH_3 + BrCH_2CH_2CH_2Cl \xrightarrow[\text{tetrahydrofuran},\ -30°]{LiN(i\text{-}C_3H_7)_2} C_6H_5\overset{\overset{\displaystyle NCH_3}{\|}}{C}(CH_2)_3CH_2Cl$$

$$C_6H_5\overset{\overset{\displaystyle NCH_3}{\|}}{C}(CH_2)_3CH_2Cl \xrightarrow{\text{reflux}}$$

Submitted by D. A. Evans[1] and L. A. Domeier
Checked by R. Decorzant and G. Büchi

1. Procedure[2]

A. N-(α-*Methylbenzylidene*)*methylamine*. Approximately 70 ml. of methylamine, passed through a potassium hydroxide trap, is condensed into a dry, premarked, nitrogen-purged, 1-l. flask (Note 1) equipped with a mechanical stirrer, an acetone–dry ice condenser with drying tube, and a 250-ml., pressure-equalizing addition funnel topped by a gas-inlet connection. The flask is cooled in a methanol–ice bath, and a solution of 48 g. (0.40 mole) of acetophenone (Note 2) in 200 ml. of dry diethyl ether (Note 3) is added through the addition funnel. The addition funnel is rinsed with 25 ml. of dry ether, purged with nitrogen, and charged with 220 ml. of 1 *M* titanium tetrachloride in hexane (Note 4), which is added to the cooled flask over a 1.5-hour period (Note 5). After stirring an additional 30 minutes in the methanol–ice bath and 30 minutes at room temperature, the mixture is filtered through a Büchner funnel into a 1-l., round-bottomed flask (Note 6), and the solid material is rinsed with an additional 100 ml. of ether. The solvents are removed on a rotary evaporator, and the yellow residue is transferred to a 100 ml. round-bottomed flask. Distillation through a short, vacuum-jacketed, Vigreux column yields 37–47 g. (70–88%) of the colorless imine, b.p. 93–95° (11 mm.) (Notes 7 and 8).

B. N-*Methyl-2-phenyl-Δ²-tetrahydropyridine*. A 500-ml., nitrogen-purged flask equipped with serum cap, reflux condenser with nitrogen inlet connection, thermometer, and stirring bar, is charged with 100 ml. of dry tetrahydrofuran (Note 9) and 21.0 ml. (0.155 mole) of diisopropylamine (Note 10). The solution is cooled to −30° with an acetone bath to which dry ice was added as needed, and 72.4 ml. (0.155 mole) of 2.14 *M* n-butyllithium in hexane (Note 11) is added while keeping the temperature below 0°. After cooling the mixture to −40°, 20.6 g. (0.155 mole) of the imine (from part A) is added *via* syringe over a period of about 2 minutes. The resulting yellow solution is maintained at −40° to −30° for 15 minutes then cooled to −60°. To the cold solution is added 25.2 g. (16.5 ml., 0.160 mole) of 1-bromo-3-chloropropane (Note 12) in one portion *via* syringe while the temperature is maintained below −40°. The reaction mixture is maintained between −60° and −50° for 5 minutes, the bath is removed, and the mixture is allowed to warm to room temperature. The reaction mixture is refluxed 3 hours to effect ring closure (Note 13). After the addition of 150 ml. of 10% aqueous potassium carbonate to the cooled solution, the reaction mixture is stirred several minutes and transferred to a nitrogen-purged separatory funnel. The reaction flask is rinsed with 100 ml. of 1:1 benzene–ether which is added to the separatory funnel, and the entire mixture is diluted with 150 ml. of water. After shaking, the aqueous layer is removed, the organic layer is washed with 100 ml. of brine, shaken with anhydrous granular sodium sulfate, and filtered into a 1-l., round-bottomed flask. The solvents are removed on a rotary evaporator, and the residue is transferred to a 100-ml., round-bottomed flask. Short-path distillation under high vacuum yields 18.8–21.7 g. (70–81%) of pale yellow enamine, b.p. 87–88° (4 mm.) (Notes 14 and 15).

2. Notes

1. All three necks of the flask should be vertical and not set at an angle, preventing the accumulation of large amounts of the methylamine complex of titanium tetrachloride on the sides of the reaction flask.

2. Acetophenone was purchased from Matheson, Coleman and Bell and used without further purification.

3. Anhydrous ether available from Mallinckrodt Chemical Co. can be used without further drying.

4. Titanium tetrachloride (purified grade) was purchased from J. T. Baker. A 1 *M* titanium tetrachloride solution was prepared by diluting 55 ml. of titanium tetrachloride (1.73 g./ml.) to a volume of 500 ml. with hexane which had been passed through 40–50 g. of basic alumina (Activity I).

5. The addition funnel should be thoroughly, but gently, flushed with nitrogen before being charged with the titanium tetrachloride solution. A slight flow of nitrogen should be maintained throughout the addition, preventing the diffusion of methylamine into the funnel where it will form a red insoluble titanium tetrachloride–amine complex.

6. If done quickly, the filtration need not be done under nitrogen, with no effect on the yield in the case of this particular imine.

7. ^{1}H NMR (CCl$_4$): 2.1 (s, 3H, NCH_3), 3.2 (s, 3H, vinylic CH_3), 7.2 (m, 3H, aryl CH), 7.7 (m, 2H, aryl CH). IR (CCl$_4$) cm.$^{-1}$: 1645, 1450, 1370, 1290. Purity was confirmed by GC on a 4-ft., 10% Carbowax 20 *M* column at 165°.

8. The imine should be stored under nitrogen and exposed to the air as little as possible during handling.

9. Tetrahydrofuran was freshly distilled from lithium aluminum hydride. See *Org. Synth.*, **Coll. Vol. 5,** 976 (1973) for a note concerning the hazards involved in purifying tetrahydrofuran.

10. Diisopropylamine was purchased from Aldrich Chemical Co. and distilled from calcium hydride prior to use.

11. *n*-Butyllithium in hexane was purchased from Ventron Corp.

12. 1-Bromo-3-chloropropane was purchased from Matheson, Coleman and Bell and distilled from phosphorus pentoxide prior to use.

13. If the mixture is not refluxed, the intermediate imine may be isolated.

14. ^{1}H NMR (CCl$_4$): 1.5–2.1 (m, 4H, 2CH_2), 2.4 (s, 3H, NCH_3), 3.0 (m, 2H, NCH_2), 4.8 (t, J = 4 Hz., 1H, vinylic CH), 7.3 (m, 5H, C$_6H_5$). IR (CCl$_4$) cm^{-1}: 1640, 1605, 1500, 1460, 1375, 1360, 1130, 1040. Purity was confirmed by GC on a 4-ft., 10% Carbowax 20 M column at 165°.

15. The enamine should be refrigerated under nitrogen and used within a few days.

3. Discussion

N-Methyl-2-phenyl-Δ^2-tetrahydropyridine and similar compounds have previously been prepared by the hydrolysis and decarboxylation of α-benzoyl-*N*-methyl-2-piperidone[3] and by the addition of phenyl Grignard reagents to *N*-methyl-2-piperidone, followed by dehydration.[4] Both of these methods require that a heterocyclic ring already be present in the system. In contrast, this procedure offers a new, flexible route to the construction of five- or six-membered heterocyclic rings which may easily be incorporated into larger polycyclic products. Several examples[5] of this process demonstrate this utility:

A wide variety of more complex endocyclic enamines are thus made available as synthetic intermediates.

1. Department of Chemistry, University of California, Los Angeles, California 90024. [Present address: Division of Chemistry and Chemical Engineering, The Chemical Laboratories, California Institute of Technology, Pasadena, California 91125].

2. This procedure is based on the work of Weingarten: H. Weingarten, J. P. Chupp, and W. A. White, *J. Org. Chem.*, **32**, 3246 (1967).

3. K. H. Büchel, H. J. Schulze-Steinem, and F. Korte, U.S. Pat. 3,247,213 (1966); K. H. Büchel and F. Korte, *Chem. Ber.*, **95**, 2438 (1962).

4. R. Lakes and O. Grossmann, *Coll. Czech. Chem. Commun.*, **8**, 533 (1936).

5. D. A. Evans, *J. Am. Chem. Soc.*, **92**, 7593 (1970).

HYDROGENOLYSIS OF CARBON-HALOGEN BONDS WITH CHROMIUM(II)-EN PERCHLORATE: NAPHTHALENE FROM 1-BROMONAPHTHALENE

Submitted by Ruth S. Wade and C. E. Castro[1]
Checked by Norton P. Peet and Herbert O. House

1. Procedure

A 250-ml., three-necked flask equipped with a magnetic stirring bar, nitrogen inlet, and outlet stopcocks is charged with 60 ml. of N,N-dimethylformamide (Note 1) and 6.01 g. (0.100 mole) of ethylenediamine (Note 2). The outlet stopcock is connected to a trap containing mercury or Nujol, and the third neck of the flask is fitted with a rubber septum. While the solution in the flask is stirred, the system is flushed with nitrogen for 30 minutes; a static nitrogen atmosphere is maintained in the reaction vessel during the remainder of the reaction. An aqueous solution containing 0.03 mole of chromium(II) perchlorate (Note 3) is added to the reaction vessel with a hypodermic syringe, forming a purple solution of the chromium(II)-en complex. To this solution is added, with a hypodermic syringe, a solution of 1.66 g. (0.00802 mole) of 1-bromonaphthalene (Note 4) in 20 ml. of oxygen-free (Note 5) N,N-dimethylformamide (Note 1). The reaction solution is stirred for 70 minutes (Note 6), during which time the color changes from purple to deep red, and poured into a solution of 40 g. of ammonium sulfate in 400 ml. of 0.4 M hydrochloric acid. The resulting emulsion is extracted with five 60-ml. portions of diethyl ether. The combined ethereal extracts are washed with two 25-ml. portions of water, dried over potassium carbonate, and concentrated. The residue crystallizes, yielding 0.96–1.00 g. (93–98%) of naphthalene, m.p. 77–80°. Recrystallization from ethanol affords pure naphthalene as white plates, m.p. 80–81°.

2. Notes

1. Baker reagent grade N,N-dimethylformamide was used without purification.

2. The submitters employed, without purification, 98% ethylenediamine obtained from Mallinckrodt Chemical Works; the checkers employed material from Eastman Organic Chemicals, which was redistilled (b.p. 117–118°) before use. The amount of ethylenediamine employed is sufficient to provide three equivalents of diamine for each mole of chromium(II) and to neutralize any acid remaining in the chromium(II) perchlorate solution.

3. The submitters employed a 1.64 M solution of chromium(II) perchlorate, prepared by stirring a mixture of 5.7 g. of pure chromium metal pellets (United Mineral and Chemical Corporation, 129 Hudson St., New York, 10013) with 60 ml. of 20% perchloric acid under a nitrogen atmosphere at 30° for 12 hours.[2,3] The rate of dissolution of the chromium metal is increased if the metal is washed successively with concentrated hydrochloric acid and water just before it is added to the perchloric acid. The checkers employed a 0.519 M solution of chromium(II) perchlorate, prepared in a comparable manner with chromium metal obtained from the Mining and Metals Division, Union Carbide Corporation. The deep-blue solution of chromium(II) perchlorate is transferred to a storage vessel with a siphon or a hypodermic syringe, and the solution is stored under a nitrogen atmosphere in a vessel fitted with a rubber septum. Provided this solution is protected from oxygen, it is stable for long periods of time; aliquots for standardization or reaction are conveniently removed with a hypodermic syringe. The solution is standardized by adding 5.00-ml. aliquots to excess 1 M iron(III) chloride followed by titration of the iron(II) ion produced with standard cerium(IV) sulfate solution, using phenanthroline as an indicator. [*Org. Synth., Coll. Vol. 5,* 993 (1973)].

The submitters had recommended use of only slightly more [2.3 moles of chromium(II) complex per mole of halide] than the stoichiometric amount of chromium(II) complex in this reduction. However, because these concentrations of reagents lead to a very slow reaction rate in the last 5–10% of the reduction (Note 6), the checkers found it more convenient to employ excess reducing agent [3.8 moles of chromium(II) complex per mole of halide].

4. 1-Bromonaphthalene, m.p. 2–4°, obtained either from the Aldrich Chemical Company, Inc., or from Matheson, Coleman and Bell was used without further purification.

5. A slow stream of nitrogen was passed through the N,N-dimethylformamide for 30 minutes to remove any dissolved oxygen.

6. The progress of this reaction may be followed by quenching aliquots of the reaction solution in acidic aqueous ammonium sulfate followed by extraction with ether and analysis of the ethereal extract by GC. With a 1.2-m. GC column packed with silicone fluid, No. 710, on Chromosorb P and heated to 215°, the retention times of naphthalene and 1-bromonaphthalene were 1.9 minutes and 6.7 minutes, respectively. The submitters employed a 30-cm. GC column packed with Porpak P for this analysis.

Since the presence of even 5–10% of unchanged 1-bromonaphthalene makes purification of the naphthalene difficult, it is important that the reduction be complete before the product is isolated. With reaction conditions described in this preparation [0.100 mole of ethylenediamine, 0.0080 mole of 1-bromonaphthalene, 60 ml. (0.031 mole) of 0.519 M chromium(II) perchlorate, and 80 ml. of N,N-dimethylformamide], the checkers found·

that reduction was usually complete in less than 15 minutes. Under the conditions [0.032 mole of ethylenediamine, 0.0040 mole of 1-bromonaphthalene, 5.5 ml. (0.0090 mole) of 1.64 M chromium(II) perchlorate, and 40 ml. of N,N-dimethylformamide] originally suggested by the submitters, a reduction time of approximately 3 hours was required for complete reduction.

3. Discussion

1-Bromonaphthalene has been reduced to naphthalene in good yield by hydrogenation over Raney nickel in methanolic potassium hydroxide,[4] by triphenyltin hydride in benzene,[5] by magnesium in 2-propanol,[6] by sodium hydrazide and hydrazine in ether,[7] and by copper(I) acetate in pyridine.[8]

The present procedure illustrates the ease of reduction of aryl, vinyl, and primary alkyl halides to the corresponding hydrocarbons with the chromium(II)-en reagent.[9] This reagent will also convert epoxides and aliphatic halides, with good leaving groups in the β-position, to olefins.[9] Although the reduction of alkyl halides with this en complex is chemically similar to reductions with solutions of other chromium(II) salts in aqueous N,N-dimethylformamide,[10,11] the en complexes of the chromium(II) ion are more reactive than the aquated chromium(II) ion.[9] The checkers have found that the potential measured between platinum and calomel electrodes in a solution of chromium(II) perchlorate in aqueous N,N-dimethylformamide is increased by the addition of ethylenediamine until three equivalents of the diamine have been added. However, presumably at least one of the six coordination sites on the chromium(II) ion must be vacant at the time reduction of a halide occurs, permitting transfer of the halogen atom from the substrate to the chromium ion.[9]

Chromium(II) perchlorate is the salt of choice for preparing the en complex in N,N-dimethylformamide. At comparable concentrations chromium(II) sulfate[10] is insoluble, and chromium(II) chloride[2,11] is only partially soluble in the reaction solution.

1. Department of Nematology, University of California, Riverside, California 92502.
2. H. Lux and G. Illmann, *Chem. Ber.*, **91**, 2143 (1958).
3. D. G. Holah and J. P. Fackler, *Inorg. Synth.*, **10**, 29 (1967).
4. H. Kämmerer, L. Horner, and H. Beck, *Chem. Ber.*, **91**, 1376 (1958).
5. L. A. Rothman and E. I. Becker, *J. Org. Chem.*, **25**, 2203 (1960).
6. D. Bryce-Smith, B. J. Wakefield, and E. T. Blues, *Proc. Chem. Soc. London*, 219 (1963).
7. T. Kaufmann, H. Henkler, and H. Zengel, *Angew. Chem.*, **74**, 248 (1962).
8. R. G. R. Bacon and H. A. O. Hill, *J. Chem. Soc.*, 1112 (1964).
9. J. K. Kochi, D. M. Singleton, and L. J. Andrews, *Tetrahedron*, **24**, 3503 (1968).
10. A. Zurqiyah and C. E. Castro, *Org. Synth.*, **Coll. Vol. 5**, 993 (1973).
11. J. R. Hanson and E. Premuzic, *Angew. Chem., Int. Ed. Engl.*, **7**, 247 (1968).

THIOPHENOLS FROM PHENOLS: 2-NAPHTHALENETHIOL

Submitted by MELVIN S. NEWMAN[1] and FREDERICK W. HETZEL[2]
Checked by W. SCHILLING, R. KEESE, and A. ESCHENMOSER

1. Procedure

Caution! Benzene has been identified as a carcinogen; OSHA has issued emergency standards on its use. All procedures involving benzene should be carried out in a well-ventilated hood, and glove protection is required.

A. *O-2-Naphthyl dimethylthiocarbamate.* A solution of 21.6 g. (0.150 mole) of 2-naphthol (Note 1) in 100 ml. of water containing 8.4 g. (0.15 mole) of potassium hydroxide is cooled below 10° in a 500-ml., three-necked flask equipped with a stirrer, a thermometer, and a 125-ml. addition funnel. A solution of 24.8 g. (0.201 mole) of *N,N*-dimethylthiocarbamyl chloride (Note 2) in 40 ml. of dry tetrahydrofuran (Note 3) is added over 20–30 minutes to the stirred solution at such a rate that the temperature never exceeds 12°. After the addition is complete, the cooling bath is removed and stirring is continued for 10 minutes. The reaction mixture is made alkaline with 50 ml. of 10% potassium hydroxide and shaken three times with 100-ml. portions of benzene. The organic layers are combined, washed with saturated sodium chloride, and dried by filtration through anhydrous magnesium sulfate. The solvent is removed by distillation, giving the crude product. Crystallization from 75 ml. of absolute methanol yields 23.5–

25.2 g. (68–73%) of O-2-naphthyl dimethylthiocarbamate, as colorless crystals, m.p. 90–90.5°.

B. *2-Naphthalenethiol*. A 250-ml. flask, fitted with a diffusion tube[3] and swept with nitrogen, is charged with 23.1 g. (0.100 mole) of O-2-naphthyl dimethylthiocarbamate (Note 4). The flask is heated at 270–275° for 45 minutes in a salt bath (Note 5). After cooling, a solution of 8.4 g. (0.15 mole) of potassium hydroxide in 10 ml. of water and 75 ml. of ethylene glycol is added to the flask. The diffusion tube is replaced with a condenser, and the mixture is heated at reflux for 1 hour (Note 6). The cooled reaction mixture is poured onto 150 g. of ice. After the ice has melted, the mixture is shaken two times with 150-ml. portions of chloroform. The chloroform layers are discarded, and the aqueous layer is cautiously acidified with concentrated hydrochloric acid (Note 7) and shaken three times with 75-ml. portions of chloroform. The organic layers are combined and dried by filtration through anhydrous magnesium sulfate. The solvent is removed by distillation, yielding 13–15 g. of crude product. Distillation yields 10.3–12.8 g. (71–80%) of pure 2-naphthalenethiol, b.p. 92–94° (0.4 mm.), m.p. 80–81° (Note 8).

2. Notes

1. Practical 2-naphthol, obtained from Matheson, Coleman and Bell, was recrystallized twice from benzene, m.p. 123–124°.

2. N,N-Dimethylthiocarbamyl chloride can be prepared as described in *Org. Synth.*, **Coll. Vol. 4,** 310 (1963), or by rapidly adding 740 g. (10.5 moles) of chlorine dissolved in 3 l. of carbon tetrachloride to a stirred, refluxing suspension of 2400 g. (9.982 moles) of tetramethylthiram disulfide (Note 9) in 5 l. of carbon tetrachloride. After the addition is complete, approximately one-half of the solvent is removed by distillation. The reaction mixture is cooled, filtered, removing the precipitated sulfur, and further concentrated. The residue is distilled, yielding 1980 g. (80%) of N,N-dimethylthiocarbamyl chloride, b.p. 65–68° (0.2 mm.).

3. See *Org. Synth.*, **Coll. Vol. 5,** 976 (1973) for a warning note regarding the purification of tetrahydrofuran.

4. Crude O-2-naphthyl dimethylthiocarbamate should not be used in this step as the yield is markedly decreased.

5. The salt bath is described in *Org. Synth.*, **Coll. Vol. 4,** 498 (1963).

6. The hydrolysis should be performed in a hood because of the vigorous evolution of dimethylamine.

7. The dilute acid solution should be added slowly since foaming results from the evolution of carbon dioxide.

8. Purification can be accomplished by recrystallization from methanol, but the overall yield of pure material is 65–70%.

9. Tetramethylthiram disulfide was obtained from the Pennwalt Corporation.

3. Discussion

The procedure described is a good, general method[4] for obtaining a thiophenol from the respective phenol. It employs three steps: conversion of a phenol to the O-aryl dialkylthiocarbamate by treatment with N,N-dialkylthiocarbonyl chloride; pyrolysis of the O-aryl dialkylthiocarbamate to the S-aryl dialkylthiocarbamate; and hydrolysis of the latter to the aryl mercaptan.

Previous preparations of 2-naphthalenethiol have included reduction of 2 naphthylsulfonyl chloride with zinc and acid[5,6] or phosphorus and iodine.[7,8] Alternatively, 2-naphthyldiazonium chloride has been converted to the thiol using potassium ethyl xanthate and sodium carbonate.[9]

1. Department of Chemistry, Ohio State University, Columbus, Ohio 43210.
2. B. F. Goodrich Research Center, Brecksville, Ohio 44141.
3. L. F. Fieser and M. Fieser, "Reagents for Organic Synthesis," Vol. 1, Wiley, New York, 1967, p. 105.
4. M. S. Newman and H. A. Karnes, *J. Org. Chem.*, **31**, 3980 (1966).
5. Y. Schaafsma, A. F. Bickel, and E. C. Kooyman, *Recl. Trav. Chim. Pays-Bas*, **76**, 180 (1957).
6. J. Jacques, *Bull. Soc. Chim. Fr.*, 231 (1955).
7. J. Kiss and E. Vinkler, *Acta Phys. Chem.*, **3**, 75 (1950) [*Chem. Abstr.*, **47**, 111a (1953)].
8. H. Pitt, U.S. Pat. 2,947,788 (1960) [*Chem. Abstr.*, **55**, 462a (1961)].
9. O. Dann and M. Kokorudz, *Chem. Ber.*, **91**, 172 (1958).

(S)-(−)-α-(1-NAPHTHYL)ETHYLAMINE

[(S)-1-Naphthalenemethanamine, α-methyl-]

$$\alpha\text{-}C_{10}H_7\underset{\underset{NH_2}{|}}{CH}\text{---}CH_3 \xrightarrow[\substack{\text{with} \\ \text{(-)-DAG.} \\ \text{acetone, reflux}}]{\text{Resolution}} \alpha\text{-}C_{10}H_7\underset{H_3C}{\overset{}{\diagup}}\overset{NH_2}{\underset{H}{\diagdown}}C$$

racemate   [S](−)-isomer

(−)-DAG: (−)-2,3:4,6-di-O-isopropylidene-2-keto-L-gulonic acid hydrate [L-$xylo$-2-hexulosonic acid, bis-O-(1-methylethylidene)-]

Submitted by E. Mohacsi and W. Leimgruber[1]
Checked by P. E. Georghiou, J. D. Lock, Jr., and S. Masamune

1. Procedure

Caution! Benzene has been identified as a carcinogen; OSHA has issued emergency standards on its use. All procedures involving benzene should be carried out in a well-ventilated hood, and glove protection is required.

A. *(S)-(−)-α-(1-Naphthyl)ethylamine.* A mixture of 58.44 g. (0.1849 mole) of (−)-2,3:4,6-di-O-isopropylidene-2-keto-L-gulonic acid hydrate [(−)-DAG) (Note 1) and 1.7 l. of acetone (Note 2) is placed in a 3-l. Erlenmeyer flask. A boiling chip is added, and the mixture is heated to a gentle boil. To the resulting hot solution is added cautiously but rapidly, over a 1-minute period, 34.24 g. (0.2002 mole) of racemic α-(1-

naphthyl)ethylamine (Note 3) in 100 ml. of acetone. The mixture is allowed to stand at room temperature for approximately 4 hours. The (-)-amine (-)-DAG salt is filtered with suction, washed with 100 ml. of acetone, and dried in a vacuum oven at 60° to constant weight, yielding 73–76 g. of the crude (-)-amine (-)-DAG salt, m.p. 205–207° (dec.), (Note 4), $[\alpha]_D^{25}$ −14.2° (c 1.01%, methanol). The crude salt and 4.2 l. of ethanol (Note 5) are placed in a 5-l., round-bottomed flask fitted with a reflux condenser and a mechanical stirrer. The mixture is stirred and heated at reflux for about 4 hours, during which time a clear solution is obtained. The condenser is then placed in a descending position and approximately 1.4 l. of the ethanol is distilled at atmospheric pressure (Note 6), and stirring is continued (Note 7) at room temperature for about 16 hours. The purified salt is collected on a filter and dried to constant weight, yielding 36–37 g. of white needles, m.p. 216–218° (dec., Note 4), $[\alpha]_D^{25}$ −17.5° (c 1.02%, methanol). For recrystallization, the crude salt and 4.0 l. of ethanol are used. Removal of 3.0 l. of the solvent yields 33.5–33.9 g. (75–76%) of pure (-)-amine (-)-DAG salt as white needles, m.p. 219–221° (dec., Note 4), $[\alpha]_D^{25}$ −18.5° (c 0.90%, methanol).

To a slurry of 33.5–33.9 g. of the pure (-)-amine (-)-DAG salt in 130 ml. of water is added 56 ml. of 2 N aqueous sodium hydroxide, and the resulting oily suspension is extracted with four 80-ml. portions of diethyl ether. The combined ether extracts are washed with 50 ml. of water and dried over anhydrous magnesium sulfate. After filtration and removal of the ether on a rotary evaporator, the crude base is distilled under reduced pressure through a 20-cm. Vigreux column (Note 8), affording 10.9–11.7 g. (85–90% yield, based on the amount of the salt used) of the pure (-)-amine as a colorless liquid, b.p. 156–157° (11 mm.), n_D^{24} 1.6211–1.6212, d_4^{24} 1.056, $[\alpha]_D^{25}$ −80.1° (neat), $[\alpha]_D^{25}$ −60.4° (c 10.0%, methanol), $[\alpha]_D^{25}$ −59.3° (c 0.65%, methanol) (Notes 9 and 10).

B. *Recovery of* (-)-DAG. The basic aqueous solution (about 186 ml.), obtained after removal of the (-)-amine by ether extraction, is placed in a 600-ml. beaker and cooled to 0–5° with an ice bath. The solution is magnetically stirred and carefully acidified with 2 N hydrochloric acid at 0–5° to approximately pH 2 (Note 11). The precipitated (-)-DAG is filtered without delay (Note 12), washed with 20 ml. of ice water, and air dried to constant weight, yielding 20.0–20.9 g. (91–94%, yield based on the amount of the salt used) of (-)-DAG, m.p. 103° (dec.), $[\alpha]_D^{25}$ −21.6° (c 2.28%, methanol) (Notes 13 and 14).

2. Notes

1. (-)-2,3 : 4,6-Di-O-isopropylidene-2-keto-L-gulonic acid hydrate [(-)-DAG] is available from the Commercial Development Dept., Hoffmann-La Roche Inc., Nutley, New Jersey 07110.

2. Mallinckrodt A.R. grade acetone was used. The checkers found that the use of 1.5 l. of acetone, as originally recommended by the submitters, resulted in an immediate precipitation of the (-)-DAG salt upon addition of the amine, inducing incomplete mixing of the two reagents.

3. Practical grade racemic α-(1-naphthyl)ethylamine (purchased from Norse Laboratories, Inc., Santa Barbara, California 93103) was distilled before use, b.p. 117–118° (2 mm.).

4. The melting point was measured in an evacuated, sealed capillary and found to deviate slightly from this value on occasion.

5. The checkers used ethanol containing a maximum of 0.1% (w/w) of water and a maximum of 0.001% (w/w) of benzene.

6. The salt begins to crystallize toward the end of this operation.

7. It is advisable to maintain stirring, avoiding the formation of lumps, thus assuring a uniform product.

8. The distillation apparatus was first flushed with nitrogen, as the amine formed a white crystalline solid on contact with atmospheric carbon dioxide.

9. Reported physical constants[2,3] of the amine are: b.p. 153° (11 mm.), d_4^{25} 1.055, $[\alpha]_D^{25}$ −80.8° (neat).

10. A 100 MHz. ^{1}H NMR spectrum (CDCl$_3$) of the amine in the presence of an equal amount of the chiral shift reagent, tris[3-(trifluoromethylhydroxymethylene)-d-camphorato]europium(III)[4] (submitters), or in the presence of an equal amount of tris[3-(heptafluoropropylhydroxymethylene)-d-camphorato]europium(III) (checkers), revealed that the product contained no detectable enantiomeric isomer.

11. The pH was measured with a Beckman Zeromatic pH meter.

12. (-)-DAG is unstable in aqueous acid.

13. TLC of (-)-DAG on Silica Gel G using the solvent system, benzene:methanol:acetone:acetic acid (70:20:5:5), shows one spot, $R_f \sim 0.7$.

14. The reported[5] m.p. is 98–99°.

3. Discussion

(S)-(-)-α-(1-Naphthyl)ethylamine has been prepared by resolution of the racemic amine with camphoric acid in unspecified yield.[2,3]

The procedure presented herein allows the preparation of the same optically active amine in approximately 70% yield by the use of a new resolving agent, (-)-2,3:4,6-di-O-isopropylidene-2-keto-L-gulonic acid hydrate, [(-)-DAG], whose utility for the resolution of a variety of amines has been thoroughly demonstrated.[6]

(-)-DAG is an ascorbic acid derivative with the following structure:

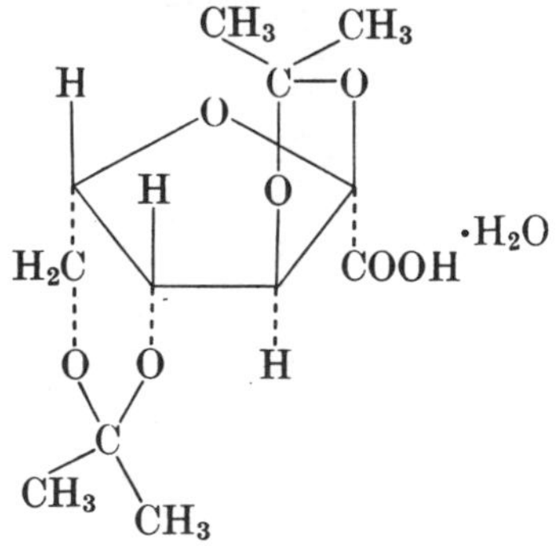

(-)-DAG

It is an attractive resolving agent, because it is relatively inexpensive and commercially available on a ton scale for industrial applications. One of the remarkable properties of (-)-DAG, lacking in other acidic resolving agents, is its water-insolubility, which permits the recovery of the resolving agent in a simple and efficient manner.

Among the amines that have been resolved with (-)-DAG: α-phenylethylamine,[6] [(R-(R*, R*)]-2-amino-1-(4-nitrophenyl)-1,3-propanediol,[6] 1,2,3,4,5,6,7,8-octahydro-1-(4-methoxyphenylmethyl)isoquinoline,[6] 3-methoxymorphinan,[6] 1,2,3,4-tetrahydro-7-methoxy-4-phenylisoquinoline,[6] 3-hydroxy-N-methylmorphinan,[6] 1,2,3,4-tetrahydro-6,7-dimethoxy-1-[3,4-(methylenedioxy)phenyl]isoquinoline,[7] 1,2,3,4-tetrahydro-6,7-dimethoxy-1-(3,4,5-trimethoxyphenyl)isoquinoline,[7] 2-[4-(phenylmethoxy)phenyl]-2-(3,4-dimethoxyphenyl)ethanamine,[8] N-norlaudanosine,[9] and 1,2,3,4-tetrahydro-6,7-dimethoxy-1-[3-methoxy-4-(phenylmethoxy)phenyl]isoquinoline.[10]

1. Deceased July 8, 1981; Work done at Chemical Research Department, Hoffmann-La Roche Inc., Nutley, New Jersey 07110.
2. E. Samuelsson, Sven. Kem. Tidskr. 34, 7 (1922) [Chem. Abstr., 16, 2140 (1922)].
3. E. Samuelsson, Thesis, University of Lund, Lund, Sweden, 1923 [Chem. Abstr., 18, 1833 (1924)].
4. H. L. Goering, J. N. Eikenberry, and G. S. Koermer, J. Amer. Chem. Soc., 93, 5913 (1971).
5. T. Reichstein and A. Grüssner, Helv. Chim. Acta, 17, 311 (1934).
6. C. W. Den Hollander, W. Leimgruber, and E. Mohacsi, U.S. Pat. 3,682,925 (1972).
7. A. Brossi and S. Teitel, Helv. Chim. Acta, 54, 1564 (1971).
8. A. Brossi and S. Teitel, J. Org. Chem., 35, 3559 (1970).
9. S. Teitel, J. O'Brien, and A. Brossi, J. Med. Chem., 15, 845 (1972).
10. J. G. Blount, V. Toome, S. Teitel, and A. Brossi, Tetrahedron, 29, 31 (1973).

ALKYL IODIDES: NEOPENTYL IODIDE AND IODOCYCLOHEXANE

(Propane, 1-iodo-2,2-dimethyl- and Cyclohexane, iodo-)

$$
\text{A.} \quad \underset{\underset{CH_3}{|}}{\overset{\overset{CH_3}{|}}{CH_3CCH_2OH}} + CH_3I + (C_6H_5O)_3P \longrightarrow
$$

$$
\underset{\underset{CH_3}{|}}{\overset{\overset{CH_3}{|}}{CH_3CCH_2I}} + (C_6H_5O)_2POCH_3 + C_6H_5OH
$$

$$
\text{B.} \quad (C_6H_5O)_3P + CH_3I \longrightarrow [(C_6H_5O)_3PCH_3]I
$$

$$
[C_6H_5O)_3PCH_3]I + C_6H_{11}OH \longrightarrow
$$

$$
C_6H_{11}I + (C_6H_5O)_2POCH_3 + C_6H_5OH
$$

Submitted by H. N. RYDON[1]
Checked by W. FUHRER, R. KEESE, and A. ESCHENMOSER

Note. Two procedures are given. Procedure A is the simplest to perform, but B is preferred for sensitive alcohols and in cases where elimination to olefins is expected, for example, with all tertiary and many secondary alcohols. Procedure A is best for sterically hindered alcohols, for example, neopentyl alcohol.

1. Procedures

Caution! Methyl iodide, in high concentrations for short periods or in low concentrations for long periods, can cause serious toxic effects in the central nervous system. Accordingly, the American Conference of Governmental Industrial Hygienists[2] has set five p.p.m., a level which cannot be detected by smell, as the highest average concentration in air to which workers should be exposed for long periods. The preparation and use of methyl iodide should always be performed in a well-ventilated fume hood. Since the liquid can be absorbed through the skin, care should be taken to prevent contact.

A. *Neopentyl iodide.* A 500-ml., two-necked, round-bottomed flask fitted with a reflux condenser equipped with a calcium chloride drying tube is charged with 136 g. (115 ml., 0.439 mole) of triphenyl phosphite, 35.2 g. (0.400 mole) of neopentyl alcohol, and 85 g. (37 ml., 0.60 mole) of methyl iodide (Note 1). A thermometer of sufficient length extends into the liquid contents of the flask. The mixture is heated under gentle reflux with an electric heating mantle until the temperature of the refluxing liquid rises from its initial value of $75-80°$ to about $130°$, and the mixture darkens and begins to fume. The time required is about 24 hours. It is necessary to adjust the heat input from the mantle from time to time as the reaction proceeds and the reflux rate diminishes (Note 2).

The reaction mixture is distilled under reduced pressure through a 13-cm. Vigreux column. The fraction boiling below 65° (50 mm.) is collected and washed with 50 ml. of water, then with 50-ml. portions of cold 1 N sodium hydroxide until the washings no longer contain phenol (Note 3). The product is washed again with 50 ml. of water, dried over calcium chloride and redistilled, yielding 51–60 g. (64–75%) of neopentyl iodide, b.p. 54–55° (55 mm.), n_D^{21} 1.4882 (Note 4).

B. *Iodocyclohexane*. A 500-ml., two-necked, round-bottomed flask fitted with a reflux condenser equipped with a calcium chloride drying tube is charged with 124 g. (107 ml., 0.400 mole) of triphenyl phosphite and 85 g. (37 ml., 0.60 mole) of methyl iodide (Note 1). A thermometer of sufficient length extends into the liquid contents of the flask. The mixture is heated under gentle reflux with a heating mantle until the internal temperature has risen to about 120°. At this point the mixture is dark and viscous (Note 5). The flask is cooled, and 40 g. (0.40 mole) of cyclohexanol is added to the oily methyl-triphenoxyphosphonium iodide. The mixture is shaken gently until homogeneous (Note 6) and allowed to stand overnight at room temperature (Note 7). The mixture is distilled through a 13-cm. Vigreux column (Note 8), yielding 62.5–63 g. (74–75%) of iodocyclohexane, b.p. 66–68° (12 mm.), n_D^{22} 1.5475.

2. Notes

1. The use of 1.2 instead of 1.5 equivalents of methyl iodide proved beneficial in some runs carried out by the checkers; some difficulty was experienced in reaching the recommended final temperature.

2. To obtain the yields cited, it is essential that the reaction temperature reaches the indicated, final value, but heating should not be unnecessarily prolonged. The reaction is conveniently monitored by IR spectroscopy. As the reaction proceeds, a broad, strong band at 865 cm.$^{-1}$ with a shoulder at 880 cm.$^{-1}$ disappears, and another broad, strong band at 945 cm.$^{-1}$ and a sharp, medium band at 1310 cm.$^{-1}$ appear.

3. Testing with iron(III) chloride is recommended.

4. The product contains about 5% of *tert*-amyl iodide (^{1}H NMR); this is in agreement with the finding of Kornblum and Iffland,[3] who describe a simple way of removing this impurity.

5. As in procedure A, the heat input should be increased from time to time, and it is essential to attain the recommended final temperature; unnecessarily prolonged heating after this is reached should be avoided as the phosphonium iodide decomposes at high temperatures.

6. With some alcohols (*e.g.*, cyclohexanol) there is no appreciable rise in temperature, with others (*e.g.*, *tert*-amyl alcohol) it may be considerable, in which case the mixture should be cooled with water. In addition, there may be an induction period of up to 1 hour.

7. The reaction may be followed by IR spectroscopy: a strong, broad band at 1040 cm.$^{-1}$ disappears, and a similar band appears at 945 cm.$^{-1}$. The reaction appears to be complete after 6 hours.

8. It is not necessary to remove phenol from the reaction mixture if the alkyl iodide has a boiling point well below that of phenol. For isolation of higher boiling alkyl iodides phenol should be removed by dissolving the reaction mixture in ether (400 ml.) and washing as in procedure A.

3. Discussion

The methods described above are applicable to almost any alcohol. Procedure A is best for sterically hindered alcohols, while procedure B is especially useful for alcohols sensitive to rearrangement or alkene formation. The submitter's results for the conversion of a number of alcohols to the iodides on a preparative scale are summarized in Table I.

TABLE I

IODIDES FROM ALCOHOLS

Iodide	Procedure	Yield, %	Boiling Point, °C.
n-Butyl iodide	A	80	126–128° (760 mm.)
n-Hexyl iodide	A	75	64–66° (15 mm.)
2-Phenylethyl iodide	A	95	94–95° (1.5 mm.)
1,3-Diiodo-2,2-dimethylpropane[a]	A	75	70–71° (0.1 mm.)
tert-Amyl iodide	B	80	50–52° (50 mm.)
Ethyl 3-iodopropionate	B	90	65–66° (8 mm.)

[a]Reference 4.

Landauer and Rydon[5] describe the application of the method, on a smaller scale, to numerous other iodides; they also prepared alkyl bromides and alkyl chlorides by substituting benzyl bromide and benzyl chloride, respectively, for methyl iodide.

The preparation of alkyl iodides by the phosphorus and iodine method is described in an earlier volume of *Organic Syntheses*.[6]

1. Department of Chemistry, The University, Exeter EX4 4QD, England.
2. American Conference of Governmental Industrial Hygienists (ACGIH), "Documentation of Threshold Limit Values," 3rd ed., Cincinnati, Ohio, 1971, p. 166.
3. N. Kornblum and D. C. Iffland, *J. Am. Chem. Soc.*, **77**, 6653 (1955).
4. A. Campbell and H. N. Rydon, *J. Chem. Soc.*, 3002 (1953).
5. S. R. Landauer and H. N. Rydon, *J. Chem. Soc.*, 2224 (1953).
6. W. W. Hartman, J. R. Byers, and J. B. Dickey, *Org. Synth.*, **Coll. Vol. 2,** 322 (1943); H. S. King, *Org. Synth.*, **Coll. Vol. 2,** 399 (1943).

SULFIDE SYNTHESIS IN PREPARATION OF DIALKYL AND ALKYL ARYL SULFIDES: NEOPENTYL PHENYL SULFIDE

(Benzene, [(2,2-dimethylpropyl)thio]-)

$$(CH_3)_3CCH_2Br + C_6H_5S\overset{-}{N}\overset{+}{a} \xrightarrow[H_2O]{(C_4H_9)_3C_{16}H_{33}\overset{+}{P}\overset{-}{Br}} C_6H_5SCH_2C(CH_3)_3 + NaBr$$

Submitted by D. Landini[1] and F. Rolla
Checked by Ronald L. Sobczak and S. Masamune

1. Procedure

A 100-ml., two-necked flask fitted with a reflux condenser, a gas-inlet, and a magnetic stirrer is charged with 15.1 g. (12.0 ml., 0.100 mole) of 1-bromo-2,2-dimethylpropane (Note 1), aqueous sodium benzenethiolate (0.1 mole) (Note 2), and 1.67 g. (0.00329 mole) of tributylhexadecylphosphonium bromide (Notes 3 and 4). This mixture is heated at 70° with vigorous stirring under nitrogen (Note 5) for 3.5 hours (Note 6). After the mixture has cooled to room temperature, the organic layer is separated, and the aqueous phase is extracted with two 20-ml. portions of diethyl ether. The combined organic phases are washed with 20 ml. of 10% aqueous sodium chloride and dried over calcium chloride. After removal of the solvent, the resulting, residual oil is distilled through a 10-cm. Vigreux column, giving 14.1–15.3 g. (78–85%) of colorless neopentyl phenyl sulfide (Note 7), b.p. 85–87° (5 mm.), 96–98° (8 mm.); n_D^{24} 1.5365 (Note 8).

2. Notes

1. 1-Bromo-2,2-dimethylpropane (neopentyl bromide) was obtained from Fluka A G or Tridom Chemical Inc.
2. Aqueous sodium benzenethiolate was prepared by adding 11.0 g. (10.2 ml., 0.100 mole) of commercial benzenethiol (listed as thiophenol by Aldrich Chemical Company, Inc., and Tridom Chemical Inc.) to an ice-cold solution of 4.0 g. of sodium hydroxide in 25 ml. of water.
3. The tributylhexadecylphosphonium bromide was prepared by heating 0.1 mole of 1-bromohexadecane and 0.1 mole of tributylphosphine at 60–70° for three days, according to Starks' procedure.[2] The product, while hot, was poured into 300 ml. of hexane and the mixture was stirred for 15 minutes. After cooling of the mixture to 0°, a solid product crystallized, was filtered on a Büchner funnel, and dried under reduced pressure, m.p. 54–56° (84%).
4. When the reaction was carried out using 0.033 mole equivalent of tricaprylylmethylammonium chloride (aliquat 336), obtained from General Mills Company, Chemical Division, Kankakee, Illinois, as catalyst, the reaction required about 10 hours for completion.
5. The nitrogen flow must be as slow as possible to avoid loss of 1-bromo-2,2-dimethylpropane.
6. The reaction time depends on the concentration of the catalyst; *e.g.,* with 0.1 and

0.01 mole equivalents of phosphonium salt, the reaction required 1 and 10 hours, respectively.

7. The catalyst could be recovered (80–90%) from the distillation residue, which also contained some neopentyl phenyl sulfide and diphenyl disulfide. These products were eliminated from the residue by column chromatography on silica (8 g. for 1 g. of phosphonium salt; eluent, ether). Extraction of the silica with two 25-ml. portions of boiling ethanol and evaporation of the solvent afforded the phosphonium salt, m.p. 48–51°. This material could be reused without further purification.

8. The product showed the following ^{1}H NMR spectrum (CDCl$_3$) δ (multiplicity, number of protons, assignment): 1.03 [s, 9H, (CH_3)$_3$C], 2.88 (s, 2H, CH_2), 7.02–7.52 (m, 5H, C$_6H_5$).

3. Discussion

This procedure[3] illustrates a simple and general method for the preparation of primary and secondary dialkyl and alkyl aryl thioethers via alkylation of sodium sulfide, sodium alkyl- or arylthiolates with alkyl chlorides or bromides. The method is an example of phase-transfer catalysis, characterized by mild reaction conditions, high yields, and simple work-up procedure.

Dineopentyl and neopentyl phenyl sulfides are obtained from 1-bromo-2,2-dimethylpropane. Some other examples are given in Table I.

TABLE I

PREPARATION OF DIALKYL AND ALKYL PHENYL SULFIDES

Alkyl Halide	Nucleophile	Catalyst (mole equivalent)	Temperature (°C)	Time (minutes)	Yield of Sulfide[a](%)
1-Chloroöctane	Na$_2$S[b]	0.1	70	40	91
2-Chloroöctane	Na$_2$S[b]	0.1	70	300	90
1-Bromoöctane	Na$_2$S[b]	0.1	70	20	91
2-Bromoöctane	Na$_2$S[b]	0.1	70	80	91
Neopentylbromide	Na$_2$S[b]	0.1	70	500	81[c]
1-Chloroöctane	C$_2$H$_5$SNa[d]	0.033	40	40	90
2-Chloroöctane	C$_2$H$_5$SNa[d]	0.033	70	250	88
1-Bromoöctane	C$_2$H$_5$SNa[d]	0.033	40	15	91
2-Bromoöctane	C$_2$H$_5$SNa[d]	0.033	70	120	89
1-Chloroöctane	C$_6$H$_5$SNa[d]	0.033	40	30	92
2-Chloroöctane	C$_6$H$_5$SNa[d]	0.033	70	180	90
1-Bromoöctane	C$_6$H$_5$SNa[d]	0.033	40	10	91
2-Bromoöctane	C$_6$H$_5$SNa[d]	0.033	70	60	90

[a]Isolated products.
[b]Mole ratio of Na$_2$S to alkyl halide is 0.6.
[c]Reaction carried out under nitrogen.
[d]Mole ratio of sodium salt to alkyl halide is 1.

Neopentyl sulfides have been prepared by alkylation of sodium sulfide with neopentyl tosylate in high-boiling polar solvents,[4,5] or in low yields by reduction of alkyl 2,2-dimethylpropanethioate with lithium aluminum hydride in a large excess of boron trifluoride-diethyl etherate.[6]

1. Centro C.N.R. e Istituto di Chimica Industriale dell'Universita', Via C. Golgi 19, Milano 20133, Italy.
2. C. M. Starks, *J. Am. Chem. Soc.,* **93,** 195 (1971).
3. D. Landini and F. Rolla, *Synthesis,* 565 (1974).
4. F. G. Bordwell, B. M. Pitt, and M. Knell, *J. Am. Chem. Soc.,* **73,** 5004 (1951).
5. W. E. Parham and L. D. Edwards, *J. Org. Chem.,* **33,** 4150 (1968).
6. E. L. Eliel and R. A. Daignault, *J. Org. Chem.,* **29,** 1630 (1964).

4-NITROBENZYL FLUORIDE

[Benzene, 1-(fluoromethyl)-4-nitro-]

$$O_2N\text{—}\langle\bigcirc\rangle\text{—}CH_2OH + (C_2H_5)_2NSF_3 \xrightarrow[10°]{\text{dichloromethane}}$$

$$O_2N\text{—}\langle\bigcirc\rangle\text{—}CH_2F + (C_2H_5)_2NSOF + HF$$

Submitted by W. J. MIDDLETON[1] and E. M. BINGHAM
Checked by EUGENE R. KENNEDY, RONALD F. SIELOFF,
and CARL R. JOHNSON

1. Procedure

Caution! Protective gloves should be worn when handling diethylaminosulfur trifluoride because this material can cause severe HF burns.

A dry, 1-l., three-necked, round-bottomed flask is fitted with a 500-ml. dropping funnel, thermometer, a magnetic stirrer, and a reflux condenser protected from the atmosphere with a drying tube. The apparatus is flushed with dry nitrogen, and 150 ml. of dry dichloromethane and 21 ml. (0.16 mole) of diethylaminosulfur trifluoride [*Org. Synth.,* **Coll. Vol. 6,** 440 (1988)] are added to the flask. The contents of the flask are cooled to 10°, and a solution of 23.0 g. (0.150 mole) of 4-nitrobenzyl alcohol (Note 1) in 450 ml. of dichloromethane is added dropwise at a fast rate (45 minutes). The reaction mixture is allowed to come to room temperature and poured into a beaker containing 300 g. of ice, decomposing any unreacted diethylaminosulfur trifluoride. The organic layer is separated, and the water layer is extracted twice with 45-ml. portions of dichloromethane. The organic layer and extracts are combined, washed with 150 ml. of water, and dried over anhydrous magnesium sulfate. Evaporation to dryness under reduced pressure gives 20.9–22.1 g. (90–95%) of crude product. Recrystallization from 500 ml. of pentane yields 15.5 g. (67%) of 4-nitrobenzyl fluoride as colorless needle-shaped crystals, m.p. 36–37° (Note 2).

2. Notes

1. 4-Nitrobenzyl alcohol is available from Eastman Organic Chemicals or Aldrich Chemical Company, Inc.

2. An additional quantity of product of lesser purity can be obtained as a second crop by evaporation of the pentane.

3. Discussion

This procedure is an example of a broadly applicable, simple method for replacing the hydroxyl group of functionally substituted and unsubstituted primary, secondary, and tertiary alcohols with fluorine. Diethylaminosulfur trifluoride,[2] the fluorinating reagent used in this procedure, is less likely to cause rearrangements or dehydration than other reagents sometimes used for this purpose (SF_4, HF, HF·pyridine, SeF_4·pyridine, and $(C_2H_5)_2NCF_2CHClF$).[3] Furthermore, diethylaminosulfur trifluoride is a liquid that can be measured easily and used in standard glass equipment at moderate temperatures and atmospheric pressure.

Some alcohols that have been converted into the corresponding fluorides by reactions with diethylaminosulfur trifluoride include 1-octanol, 2-methyl-2-butanol, 2-butanol, cyclooctanol, ethylene glycol, crotyl alcohol, 2-phenylethanol, 2-bromoethanol, ethyl lactate, and ethyl α-hydroxynaphthaleneacetate.[3]

4-Nitrobenzyl fluoride has also been prepared in 40–60% yield by the reaction of 4-nitrobenzyl bromide with mercuric fluoride[4] and in mixture with the *ortho* and *meta* isomers by the nitration of benzyl fluoride.[5]

1. Central Research and Development Department, Experimental Station, E. I. duPont deNemours and Co., Wilmington, Del. 19898.
2. W. J. Middleton and E. M. Bingham, *Org. Synth.*, **Coll. Vol. 6,** 440 (1988).
3. W. J. Middleton, *J. Org. Chem.*, **40,** 574 (1975).
4. J. Bernstein, J. S. Roth, and W. T. Miller, *J. Am. Chem. Soc.*, **70,** 2310 (1948).
5. C. K. Ingold and C. H. Ingold, *J. Chem. Soc.*, 2249 (1928).

1-NITROCYCLOÖCTENE

(Cycloöctene, 1-nitro-)

Submitted by WOLFGANG K. SEIFERT[1]
Checked by E. LEWARS, P. H. McCABE, and PETER YATES

1. Procedure

Caution! Dinitrogen tetroxide is very toxic (Note 1), and nitro nitrites are unstable. The reaction must be carried out in a well-ventilated hood with an adequate shield.

Benzene has been identified as a carcinogen; OSHA has issued emergency standards on its use. All procedures involving benzene should be carried out in a well-ventilated hood, and glove protection is required.

Sodium-dried diethyl ether (150 ml.) is placed in a 1-l., four-necked flask equipped with a fritted gas-inlet extending to its bottom, a sealed mechanical stirrer (Note 2), a 100-ml., pressure-equalizing dropping funnel, a thermometer, and a dry ice condenser protected with a phosphorus pentoxide drying tube. Cycloöctene (44.4 g., 53.5 ml., 0.404 mole) (Note 3) is placed in the dropping funnel, and the system is swept with dry oxygen. Dinitrogen tetroxide (39.3 g., 27.1 ml. at $-9°$, 0.427 mole) (Note 4) is condensed (Note 5) in a graduated, calibrated trap that is protected with a phosphorus pentoxide drying tube and has been swept with dry oxygen.

The flask is cooled to $-10°$, and the dinitrogen tetroxide is distilled with a warm water bath from the trap into the ether, with slow stirring; the transfer is aided by a minimal flow of dry oxygen. The solution is allowed to warm to 0–5°, and the oxygen flow rate is increased to 10 ml. per minute (Note 6). The cycloöctene is dropped into the dinitrogen tetroxide solution, with vigorous stirring, over a 30-minute period. The reaction is exothermic, and the temperature is kept at 9–12° by cooling with a methanol–dry ice bath at $-20°$. The dropping funnel is rinsed with 25 ml. of ether, and the yellow solution (Note

7) is stirred for an additional 30 minutes at 10° with continued oxygen flow. Triethylamine (121 g., 1.20 moles) (Note 8) is added, with stirring, over a 12-minute period; the temperature of the reaction mixture is kept at 4–12° by maintaining the bath at −4° (Note 9). The mixture is kept at room temperature for an additional 30 minutes, diluted with 150 ml. of ether, and cooled to 0–5°. The excess triethylamine is neutralized with an ice-cold solution of 72 g. of acetic acid in 200 ml. of water, with stirring. The reaction mixture is transferred to a 2-l. separatory funnel and extracted with three 400-ml. portions of ether. The combined ethereal extracts are washed with two 200-ml. portions of water, three 150-ml. portions of saturated aqueous sodium hydrogen carbonate, and again with water (Note 10).

Most of the ether is removed at room temperature with a rotary evaporator. The water that separates is removed with the aid of a small separatory funnel, and the remaining ether, traces of water, and cyclooctane (Note 11) are distilled at room temperature (10 mm.) over 3 hours, yielding 59–61 g. of crude 1-nitrocyclooctene as a yellow oil (Notes 12 and 13). Chromatography on silica gel (Note 14) with successive elution with n-hexane and benzene gives 39–40 g. (63–64%) of 1-nitrocyclooctene (Note 15). Distillation (Note 16) gave an analytically pure sample, b.p. 60° (0.2 mm.), n_D^{20} 1.5116 (Note 17).

2. Notes

1. Concentrations of dinitrogen tetroxide of 100–150 p.p.m. are dangerous for exposures of 30–60 minutes, and concentrations of 200–700 p.p.m. may be fatal after even very short exposures.

2. The checkers found that magnetic stirring could be used in place of mechanical stirring.

3. Cyclooctene (95% pure) from Columbia Carbon Co., a division of Cities Service, was used without further purification.

4. A slight excess of dinitrogen tetroxide over olefin is necessary for maximum yields.

5. For good yields all reagents must be absolutely dry. For condensation of dry dinitrogen tetroxide free of dinitrogen trioxide, streams of dry oxygen (run through a flow meter, a calcium chloride tube, and concentrated sulfuric acid) and dinitrogen tetroxide (99.5% pure from the Matheson Company), are combined and run slowly through a phosphorus pentoxide tube before condensation. The freezing point of dinitrogen tetroxide is −9.3°, and a convenient cooling bath for condensation is methanol–dry ice at −8° to −10°.

6. The use of oxygen in this reaction prevents formation of undesirable by-products, e.g., nitro nitroso compounds.[2] For the preparation of 1-nitro-1-octadecene the optimum mole ratio of olefin to oxygen was found[3] to be 1/50 to 1/150, compared with 1/30 in this procedure.

7. Nitro nitrites are unstable, and it is safe practice to keep them in solution until they are converted to nitro alcohols[2] or nitro olefins.[3]

8. Commercial triethylamine (Eastman Kodak Co.) was used without further purification. Stoichiometric amounts of triethylamine based on olefin produced poor yields of nitro olefins, owing to the slow rate of elimination of the nitro nitrite compounds.[3] A twofold molar ratio of triethylamine to olefin was sufficient to produce

1-nitro-1-octadecene from 1-nitro-2-octadecyl nitrite in 92% yield.[3] The same excess applied to the crude reaction product from cycloöctene and dinitrogen tetroxide resulted in only an 80% yield of 1-nitrocycloöctene.

9. At the end of the exothermic elimination reaction the color turns brown with simultaneous precipitation of the triethylammonium salts.

10. The checkers washed the ethereal extracts with saturated brine and dried them over anhydrous magnesium sulfate before removal of ether.

11. Cycloöctane is the major impurity in the starting material.

12. The submitter estimated the crude product to be 95% pure by IR spectroscopy: $\epsilon_{6.59\mu}/\epsilon_{3.40\mu} = 3.00$ (CCl$_4$: matched 0.1-mm. cells; analytical absorbances, 0.2–0.7).

13. For further reactions involving reduction,[4] the crude product can be used.

14. The silica gel column was 14 × 2.5 in. I.D.; a shorter column may suffice.

15. The checkers found for this product: n_D^{26} 1.5106; ^{1}H NMR (CDCl$_3$), δ 1.6 (m, 8 H), 2.3 (m, 2 H), 2.7 (m, 2 H), and 7.24 (t, $J \equiv 9$ Hz., 1 H).

16. It is safe practice to remove the peroxide that may be formed in this free radical reaction by chromatography before distillation. The submitter distilled an aliquot (12.7 g.) of the hexane eluate, giving 9.9 g. of product, $\epsilon_{6.59\mu}/\epsilon_{3.40\mu} = 3.12$ (Calcd. for C$_8$H$_{13}$NO$_2$: C, 61.91; H, 8.44; N, 9.03. Found: C, 61.84; H, 8.27; N, 8.80).

17. Slow decomposition with simultaneous precipitation of a solid occurred on standing for several weeks at 23°. Immediate analysis and use of the product are advised.

3. Discussion

The major advantage of the present method, the only method reported[3] for the preparation of 1-nitrocycloöctene, is the convenience of converting an olefin to a 1-nitro olefin in good yield without isolation of any intermediate. The submitter has also used this method[3] successfully for the preparation of 1-nitro-1-octadecene from 1-octadecene.

In the past the products from the addition of dinitrogen tetroxide to olefins have been hydrolyzed and converted to 1-nitro olefins by various methods, e.g., acetylation of the isolated nitro alcohol and elimination of acetic acid with potassium carbonate,[5,6] dehydration of the nitro alcohol with phthalic anhydride[7] or potassium hydrogen sulfate,[8] and base-catalyzed elimination of nitrous and nitric acid from dinitro compounds and nitro nitrates, respectively.[2] Besides representing longer syntheses, these routes require separation of the nitro alcohol from the dinitro compound and, since these substances occur in approximately equal amounts, 50% of the yield is lost in the first step. Furthermore, in the case of the higher 1-olefins, this separation is difficult[9,10] and tedious.[3] 1-Nitro 1-olefins have been employed in the preparation of saturated nitro compounds and oximes.[4]

1. Chevron Research Company, Richmond, California 94802.
2. H. Baldock, N. Levy, and C. W. Scaife, *J. Chem. Soc.*, 2627 (1949), and previous papers.
3. W. K. Seifert, *J. Org. Chem.*, **28**, 125 (1963); U.S. Pat. 3,035,101 (1962) [*Chem. Abstr.*, **57**, 13609 (1962)].
4. W. K. Seifert and P. C. Condit, *J. Org. Chem.*, **28**, 265 (1963); W. K. Seifert, U.S. Pat. 3,156,723 (1964) [*Chem. Abstr.*, **62**, 3954 (1965)].
5. E. Schmidt and G. Rutz, *Ber. Dtsch. Chem. Ges.*, **61**, 2142 (1928).
6. H. Schwartz and G. Nelles, U.S. Pat. 2,257,980 (1941) [*Chem. Abstr.*, **36**, 494 (1942)].

7. G. D. Buckley and C. W. Scaife, *J. Chem. Soc.*, 1471 (1947).
8. H. Wieland and E. Sakellarios, *Ber. Dtsch. Chem. Ges.*, **52**, 898 (1919).
9. C. R. Porter and B. Wood, *J. Inst. Pet. London*, **38**, 877 (1952).
10. C. R. Porter and B. Wood, *J. Inst. Pet. London*, **37**, 388 (1951).

FORMATION AND PHOTOCHEMICAL WOLFF REARRANGEMENT OF CYCLIC α-DIAZO KETONES: D-NORANDROST-5-EN-3β-OL-16-CARBOXYLIC ACIDS

[D-Norandrost-5-ene-16-carboxylic acids, 3β-hydroxy-, (3β,16α) and (3β,16β)-]

Submitted by THOMAS N. WHEELER and J. MEINWALD[1]
Checked by R. A. BLATTEL, D. G. B. BOOCOCK, and PETER YATES

1. Procedure

Caution! Benzene has been identified as a carcinogen; OSHA has issued emergency standards on its use. All procedures involving benzene should be carried out in a well-ventilated hood, and glove protection is required.

A. 16-*Oximinoandrost-5-en-3β-ol-17-one*. A 2-l., three-necked, round-bottomed flask fitted with a reflux condenser, a mechanical stirrer, and a pressure-equalizing dropping funnel is charged with 750 ml. of anhydrous *tert*-butyl alcohol (Note 1). As the *tert*-butyl alcohol is slowly stirred, a stream of dry nitrogen is passed through the flask and 12.2 g. (0.312 g.-atom) of potassium metal is added cautiously. The flask is surrounded by a water bath maintained at 70° to assist in dissolving the potassium metal. After 1.5 hours the stirred mixture is homogeneous, the water bath is removed, and the reaction mixture is allowed to cool to room temperature. To the potassium *tert*-butoxide solution is slowly added 45.0 g. (0.156 mole) of dehydroisoandrosterone (Note 2), and stirring is continued for one hour until the gold-colored mixture is again homogeneous. To the reaction mixture is now added, dropwise, 36.5 g. (42.0 ml., 0.312 mole) of isoamyl nitrite (Note 3), and stirring is continued overnight at room temperature.

The deep-orange reaction mixture is diluted with an equal volume of water, poured into a 2-l. separatory funnel, and acidified with 3 *M* hydrochloric acid. The addition of 400 ml. of diethyl ether assists in effecting the separation of the clear, yellow, aqueous, lower layer from the fluffy-white ethereal suspension that forms the upper layer. This suspension is filtered through a 250-ml. coarse sintered glass funnel, and the precipitate of oximino ketone is washed with ether several times. After drying overnight in a vacuum desiccator at −5°, 48.0–48.5 g. (79%, Notes 5 and 6) of a white product, m.p. 245–247° (dec.), is obtained; its ^{1}H NMR spectrum (pyridine) shows it to be a 1:1 solvate of the oximino ketone with *tert*-butyl alcohol (Note 4). This product is used without further purification in the synthesis of the α-diazo ketone (Note 7).

B. 16-*Diazoandrost-5-en-3β-ol-17-one*. A 1-l., three-necked, round-bottomed flask is fitted with a mechanical stirrer, a 50-ml., pressure-equalizing dropping funnel, and a thermometer. As stirring is initiated, 375 ml. of methanol and 72 ml. of 5.0 *M* aqueous sodium hydroxide (0.36 mole) is added to the flask, followed by 18.0 g. (0.0460 mole) of the 1:1 solvate of 16-oximinoandrost-5-en-3β-ol-17-one with *tert*-butyl alcohol. The oximino ketone readily dissolves, giving a yellow solution. To the reaction mixture is added 28.3 ml. of concentrated aqueous ammonia (0.425 mole), and the flask is surrounded by an ice bath, maintaining the reaction temperature at 20°. Through the dropping funnel 133 ml. of 3.0 *M* aqueous sodium hypochlorite (0.40 mole) is added dropwise. The sodium hypochlorite solution should be kept near 0°; 25-ml. portions should be added to the addition funnel and the remaining solution should be kept in an ice bath (Note 8). It is important that the rate of addition of the sodium hypochlorite, and the position of the ice bath be adjusted so as to maintain the temperature of the reaction mixture at 20° ± 1° (Note 9). As soon as all of the sodium hypochlorite has been added, the ice bath is removed, and the reaction mixture is allowed to warm to room temperature and stirred for 6 hours.

The reaction mixture is diluted with an equal volume of water and extracted with 400-ml. and 200-ml. portions of dichloromethane. The combined dichloromethane extracts are washed with three 250-ml. portions of 20% aqueous sodium chloride, dried over anhydrous magnesium sulfate, and concentrated, leaving a yellow solid. Recrystallization from acetone gives 8.0–9.3 g. (55–64%) of crystalline α-diazo ketone, m.p. 200–202° (dec.) (Note 10).

C. D-*Norandrost-5-en-3β-ol-16α- and 16β-carboxylic acids*. In a solution of 500 ml. of 1,4-dioxane, 1250 ml. of ether, and 250 ml. of water contained in a 3-l.

three-necked, round-bottomed flask is dissolved 7.50 g. (0.0239 mole) of 16-diazoandrost-5-en-3β-ol-17-one. The flask is fitted with a reflux condenser, a quartz immersion well, and a nitrogen inlet. After the reaction vessel has been flushed with nitrogen, the diazo ketone solution is irradiated for 48 hours with a 450-watt Hanovia lamp with a Corex filter (Note 11). The photolysis mixture is decanted in portions into a 2-l. separatory funnel, washed three times with 500-ml. portions of water, removing the dioxane, and dried over magnesium sulfate. The ether is evaporated, leaving a pale-yellow residue. The residue is digested with 125 ml. of boiling dichloromethane under reflux for 30 minutes. The dichloromethane solution is allowed to cool to room tempera-ture and filtered, separating about 1.4 g. of the crude α-isomer as a white powder. This solid is recrystallized by dissolving it in a large volume of methanol (125 ml.) and concentrating the solution to a small volume (25 ml.), yielding 1.2 g. (17%) of D-norandrost-5-en-3β-ol-16α-carboxylic acid as a white solid, m.p. 271–274° (Note 12). The β-isomer is most readily obtained by concentrating the dichloromethane mother liquor and dissolving the residue in a mixture of 75 ml. of methanol and 25 ml. of ether. This solution is treated with an excess of diazomethane in ether at room temperature. After one hour at room temperature, the excess diazomethane is removed with a stream of nitrogen, the solvent is evaporated, and the solid residue is chromatographed on 175 g. of Woelm neutral alumina Activity Grade II. Elution with a 3:1 (v/v) benzene–ether mixture gives 3.9 g. of a white solid, which is recrystallized from ether-heptane, giving 3.0–3.1 g. (39–41%) of white, crystalline methyl D-norandrost-5-en-3β-ol-16β-carboxylate, m.p. 161–163° (Notes 13 and 14).

2. Notes

1. Anhydrous *tert*-butyl alcohol may be conveniently prepared by distilling it from calcium hydride into a receiver containing Type 4A molecular sieves.

2. Dehydroisoandrosterone is available from Searle Chemicals, Inc.

3. Isoamyl nitrite of sufficient purity may be prepared by the method in *Org. Synth.*, **Coll. Vol. 2,** 108 (1943). The isoamyl nitrite is stored over anhydrous magnesium sulfate until used.

4. The ^{1}H NMR spectrum (pyridine) included signals at δ 0.97 (s, 3H), 1.03 (s, 3H), 1.42 (s, 9H) and 5.37 (m, 1H).

5. The submitters, working on twice the scale described, obtained 90.0 g. (74%) of the solvate, m.p. 245–247°.

6. This oximino ketone has been previously prepared by a somewhat different procedure[2] and recrystallized from 2-propanol, m.p. 247–248°.

7. Other oximino ketones may be too soluble in ether to permit utilization of this isolation procedure. In this case, the submitters, working on twice the scale described, utilized the following procedure. After acidification of the reaction mixture, the oximino ketone is extracted into 1000 ml. of ether, and the ethereal extract is washed with four 100-ml. portions of saturated aqueous sodium hydrogen carbonate and exhaustively extracted with 0.5 *M* aqueous potassium hydroxide in 250-ml. portions until acidification of the basic extract gives no oximino ketone. A stream of nitrogen is bubbled through the combined basic extracts, removing any dissolved ether before the solution is cooled in an

ice bath and acidified with 3 *M* hydrochloric acid. The precipitate is collected by suction filtration and dried in a vacuum desiccator.

8. A procedure for preparing concentrated sodium hypochlorite solution is given by Coleman and Johnson.[3] Common bleach solution, such as Clorox, may also be used, although the volume of the solution is considerably increased.

9. If the temperature of the reaction mixture is maintained below 20°, an appreciable amount of colorless α-dichloro ketone is obtained.[4] If the temperature rises above 20°, the chloramine decomposes before it has time to react with the oximino ketone. The generation of chloramine *in situ* is quite exothermic, and care must be taken to maintain the temperature at 20°.

10. The submitters, working on twice the scale described, obtained 21.6 g. (74%) of product, m.p. 201–202° (dec.).

11. A Pyrex filter has been used; however, the photolysis appears to proceed more cleanly through Corex. The photolysis is considered complete when the IR spectrum of a sample shows no diazo absorption at 2065 cm^{-1}.

12. Utilizing a somewhat different procedure, Mateos and Pozas have also obtained the α-carboxylic acid, m.p. 272–275°.[5]

13. The reported melting point of the β-methyl ester is 163–164°.[5]

14. The submitters, working on twice the scale described, obtained 2.8 g. (19%) of the α-acid, m.p. 272–275°, and 9.2 g. (61%) of the methyl ester of the β-acid, m.p. 161–163°.

3. Discussion

The earliest methods for preparing cyclic α-diazo ketones involved the oxidation of the monohydrazones prepared from α-diketones, generally using mercury(II) oxide.[6,7] Recent modifications of this procedure include the use of calcium hypochlorite in aqueous sodium hydroxide or "activated" manganese dioxide as oxidants.[8] The latter reagent, especially, seems preferable to mercury(II) oxide. The base-catalyzed decomposition of the monotosylhydrazones of α-diketones has been used to prepare α-diazo ketones. Such reactions have been performed in aqueous sodium hydroxide,[9,10] with basic aluminum oxide in dichloromethane,[11] and with a variety of other bases. A promising and novel approach to cyclic α-diazo ketones involves the reaction of α-hydroxymethylene ketones with diethylamine and tosyl azide, giving high yields of the α-diazo ketone.[12]

The present procedure for the synthesis of an α-diazo ketone is a modification of the Forster reaction,[13] which has been recently exploited by numerous workers.[10,14–18] The synthesis is convenient, generally applicable to cyclic ketones, and offers moderate yields (60–70%) of pure α-diazo ketones.

The photochemical Wolff rearrangement represents a generally useful ring contraction technique.[19,20]

1. Department of Chemistry, Cornell University, Ithaca, New York 14850.
2. F. Stodola, E. C. Kendall, and B. F. McKenzie, *J. Org. Chem.*, **6**, 841 (1941).
3. G. H. Coleman and H. L. Johnson, *Inorg. Synth.*, **1**, 59 (1939).
4. T. N. Wheeler, Ph.D. thesis, Cornell University, 1969.

5. J. L. Mateos and R. Pozas, *Steroids,* **2,** 527 (1963).
6. T. Curtius and K. Thun, *J. Prakt. Chem.,* [2] **44,** 161 (1891).
7. C. D. Nenitzescu and E. Solomonica, *Org. Synth.,* **Coll. Vol. 2,** 496 (1943).
8. H. Morrison, S. Danishefsky, and P. Yates, *J. Org. Chem.,* **26,** 2617 (1961).
9. M. P. Cava and R. L. Little, *Chem. Ind. (London),* 367 (1957).
10. M. P. Cava, R. L. Little, and D. R. Napier, *J. Am. Chem. Soc.,* **80,** 2257 (1958).
11. J. M. Muchowski, *Tetrahedron Lett.,* 1773 (1966); J. K. Crandall, Ph.D. thesis, Cornell University, 1963.
12. M. Rosenberger, P. Yates, J. B. Hendrickson, and W. Wolf, *Tetrahedron Lett.,* 2285 (1964); M. Regitz, *Angew Chem. Int. Ed. Engl.,* **6,** 733 (1967); J. B. Hendrickson and W. A. Wolf, *J. Org. Chem.,* **33,** 3610 (1968).
13. M. O. Forster, *J. Chem. Soc.,* **107,** 260 (1915).
14. J. Meinwald, G. G. Curtis, and P. G. Gassman, *J. Am. Chem. Soc.,* **84,** 116 (1962).
15. M. P. Cava and E. Moroz, *J. Am. Chem. Soc.,* **84,** 115 (1962).
16. G. Muller, C. Huynh, and J. Mathieu, *Bull. Soc. Chim. Fr.,* 296 (1962).
17. J. L. Mateos and O. Chao, *Bol. Inst. Quim. Univ. Nac. Auton. Mex.,* **13,** 3 (1961) [*Chem. Abstr.,* **57,** 9914 (1962)].
18. P. M. Weintraub, Ph.D. thesis, Ohio State University, Columbus, Ohio, 1964 [*Diss. Abstr.,* **25,** 6246 (1965)].
19. F. Weygand and H. J. Bestmann, *Angew. Chem.,* **72,** 535 (1960).
20. W. Reid and H. Mengler, *Fortschr. Chem. Forsch.,* **5,** 1 (1965).

HIGHLY REACTIVE MAGNESIUM
FOR THE PREPARATION
OF GRIGNARD REAGENTS:
1-NORBORNANECARBOXYLIC ACID

(Bicyclo[2.2.1]heptane-1-carboxylic acid)

Submitted by Reuben D. Rieke,[1] Stephen E. Bales,
Phillip M. Hudnall, Timothy P. Burns, and Graham S. Poindexter
Checked by Dennis P. Stack and Robert M. Coates

1. Procedure

Caution! Potassium is highly reactive. Although it may be handled safely in air if it is covered with a hydrocarbon solvent such as heptane or mineral oil, it will spark and ignite flammable organic vapors on contact with water. The magnesium formed in this reaction is highly reactive and pyrophoric (Note 1). Accordingly, Parts C and D of this procedure should be carried out behind a safety shield.

A. *2,2-Dichloronorbornane.* A 1-l., round-bottomed flask equipped with a mechanical stirrer and a calcium sulfate drying tube is charged with 91.3 g. (0.665 mole) of phosphorus trichloride and 90.0 g. (0.817 mole) of norcamphor (Note 2). The solution is stirred and cooled to 0° in an ice–salt bath, then 193 g. (0.927 mole) of phosphorus pentachloride is added in portions over a 1-hour period. The mixture is allowed to warm to room temperature and stand overnight. The contents of the flask are poured carefully onto 1000 g. of crushed ice. The mixture is thoroughly dispersed and extracted with four 500-ml. portions of pentane. The combined pentane layers are washed with two 600-ml.

portions of water, and the aqueous layers are extracted with one 500-ml. portion of pentane. The pentane extracts are combined, dried over anhydrous sodium sulfate, and evaporated under reduced pressure. Distillation of the residual brown liquid affords 111–114 g. (82–85%) of 2,2-dichloronorbornane as a clear liquid, b.p. 70–74° (14 mm.), which solidifies on standing (Note 3).

B. 1-*Chloronorbornane*. A 5-l., three-necked, round-bottomed flask equipped with a mechanical stirrer, a condenser fitted with a drying tube, and a stopper is charged with 230 g. (1.39 moles) of 2,2-dichloronorbornane and 3 l. of pentane (Note 4). The solution is stirred as 87.0 g. (0.652 mole) of aluminum chloride is added over 4.5 hours. The mixture is stirred for 40 hours, during which time hydrogen chloride gas evolves and a brown sludge accumulates on the walls of the flask. The supernatant pentane solution is decanted, and the brown sludge remaining in the flask is thoroughly extracted with four 200-ml. portions of pentane. The combined pentane extracts are washed with three 600-ml. portions of water and one 600-ml. portion of saturated sodium chloride. The combined aqueous washes are extracted with two additional 500-ml. portions of pentane, which are combined with the preceding pentane solution and dried over anhydrous sodium sulfate. The pentane solution is concentrated by distillation through a 3 × 30 cm. Vigreux column, and the residual liquid is distilled, affording 110–114 g. (60–63%) of 1-chloronorbornane as a colorless liquid, b.p. 70–74° (55 mm.), which solidifies on standing at room temperature (Note 5).

C. *Active magnesium*. A 200-ml., three-necked, round-bottomed flask equipped with a Teflon-coated magnetic stirring bar, stopper, rubber septum, and condenser connected to an argon inlet (Note 6) is charged with 1.5 g. (0.038 g.-atom) of freshly cut potassium (Notes 7, 8, and 19), 2.01 g. (0.0211 mole) of anhydrous magnesium chloride (Note 9), 3.55 g. (0.0214 mole) of anhydrous potassium iodide (Note 10), and 50 ml. of tetrahydrofuran (Note 11). The mixture is stirred vigorously (Note 12) and heated to reflux with an electric heating mantle (Note 13). A black precipitate starts to form within a few minutes. After 3 hours at reflux temperature, the reduction should be complete (Note 14), producing active magnesium as a black powder that settles very slowly when the stirring is stopped (Note 15).

D. 1-*Norbornanecarboxylic acid*. The mixture of active magnesium metal and potassium salts is allowed to cool to room temperature, after which 1.25 g. (0.00958 mole) of 1-chloronorbornane is injected with a syringe through the septum into the flask (Note 16). The reaction mixture is heated under reflux for 6 hours and cooled to room temperature. A large excess of freshly sublimed dry ice chunks is added quickly to the Grignard reagent through the extra neck of the flask. The mixture is stirred vigorously, warmed to room temperature, acidified with 50-ml. of 20% hydrochloric acid, and extracted with three 100-ml. portions of diethyl ether. The combined ether layers are extracted with 100 ml. of 10% aqueous sodium hydroxide. The alkaline solution is acidified with concentrated hydrochloric acid, and the acidic solution is extracted with two 100-ml. portions of ether. The ether extracts are combined, washed with two 50-ml. portions of water, dried over anhydrous sodium sulfate, and evaporated, giving 0.80–

0.94 g. (60–70%) of 1-norbornanecarboxylic acid as a slightly yellow crystalline solid, m.p. 106–109° (Notes 17 and 18).

2.Notes

1. Although the submitters have never had a fire or explosion caused by active magnesium or other activated metals, they suggest extreme caution in working with these reactive materials, especially while the worker familiarizes him- or herself with the characteristics of each step in the procedure. If the active magnesium is wet with solvent when removed from the reaction vessel, it does not ignite spontaneously. If, however, the magnesium is allowed to dry first, it begins to glow when exposed to air. The submitters advise that the magnesium powder be kept under an argon atmosphere at all times.

2. The submitters purchased norcamphor from Aldrich Chemical Company, Inc. The checkers prepared this compound according to *Org. Synth., Coll. Vol. 5,* 852 (1973).

3. The spectral characteristics of 2,2-dichloronorbornane are as follows: IR (CCl₄) cm.⁻¹: 1449, 1307, 1072, 966, 933, 713; ¹H NMR (CCl₄), δ (number of protons): 1.1–2.8 (10H).

4. Pentane was dried by distillation from aluminum chloride.

5. 1-Chloronorbornane has the following spectral properties: IR (CCl₄) cm.⁻¹: 1451, 1310, 1298, 1037, 992, 947, 905, 838; ¹H NMR (CCl₄), δ (multiplicity, number of protons, assignment): 1.2–1.9 (broad m, 10H, 5C*H*₂) and 2.2 (broad s, 1H, C*H*); ¹³C NMR (CDCl₃): δ (off-resonance multiplicity, assignment): 31.0 (m, C-2 and C-6 or C-3 and C-5), 34.8 (d, C-4), 38.4 (m, C-2 and C-6 or C-3 and C-5), 46.8 (t, C-7), 70.0 (s, C-1).

6. The apparatus is dried in an oven and maintained under an argon atmosphere during the reaction. The submitters recommend against the use of nitrogen, since there are indications that nitrogen reacts with active magnesium. Argon, used as supplied by Matheson Gas Products or the Linde Division of Union Carbide Corporation, was delivered to the gas-inlet through a combination of glass and Tygon tubing. A minimum of Tygon tubing is advised to avoid the diffusion of air into the argon stream.

7. Purified grade potassium from J. T. Baker Chemical Company has been found by the submitters to give the most consistent results. The checkers used potassium metal from Allied Chemical Corporation. Very impure potassium or sodium generally gives magnesium powder with much reduced reactivity. Sodium may be used in place of potassium provided that the boiling point of the solvent chosen (Note 11) is higher than the melting point of the metal.

8. The potassium is usually cut into two or three pieces under hexane or heptane and placed wet in a tared flask that has been purged with argon. The flask is evacuated, removing the hydrocarbon, filled again with argon, and weighed to determine the exact amount of potassium used by the checkers varied from 1.4 to 1.6 g., the weights of the other reagents being adjusted proportionately. With this procedure the pieces of potassium are shiny and relatively free from oxide coating. Alternatively, the potassium cuttings may be wiped free of solvent, quickly weighed in air, and placed in the flask. The submitters recommend that the first procedure be used.

9. Anhydrous magnesium chloride from Alfa Division, Ventron Corporation, was

used as supplied by both the submitters and checkers. The submitters have subsequently had success with anhydrous magnesium chloride and bromide purchased from Cerac, Inc., P.O. Box 1178, Milwaukee, Wisconsin 53201. The checkers were unsuccessful in several attempts to prepare suitably active magnesium from analytical grade anhydrous magnesium chloride, purchased from Research Organic/Inorganic Chemical Corporation. The submitters stress that the reagent must be anhydrous. It may be stored in a desiccator containing anhydrous calcium sulfate and, if required, dried overnight in an oven at 120°. Anhydrous magnesium chloride cannot, however, be prepared by heating the hexahydrate under vacuum, since hydrogen chloride is released before dehydration is complete. The submitters have prepared active magnesium from anhydrous magnesium bromide and iodide; however, highly insoluble magnesium salts such as the fluoride or sulfate are not reduced. A small excess of magnesium chloride is used in this procedure to ensure that the potassium is completely consumed.

The submitters have also provided the following unchecked procedure, which is suitable for preparing both anhydrous magnesium chloride and bromide. The magnesium turnings and 1,2-dibromoethane used were purchased from J. T. Baker Chemical Company and Aldrich Chemical Company, Inc., respectively. A 200-ml., three-necked, round-bottomed flask equipped with a magnetic stirring bar, two stoppers, and a condenser connected to an argon inlet (Note 6) is charged with 0.35 g. (0.014 g.-atom) of magnesium turnings, 50 ml. of tetrahydrofuran (Note 11), and 3.0 g. (0.016 mole) of 1,2-dibromoethane. The suspension is warmed gently, initiating the reaction. After the initially exothermic reaction subsides, the mixture is heated at reflux for 50 minutes. The solvent is evaporated under a reduced pressure of argon or nitrogen, leaving a white solid. The flask is then evacuated and heated in an oil bath at 150° for 1 hour. The dry magnesium bromide is ready for preparing active magnesium in the same flask.

10. Potassium iodide (>99% purity) from Allied Chemical Corporation or Mallinckrodt Chemical Works is finely ground with a mortar and pestle, dried overnight in an oven at 120°, and stored in a desiccator. The molar ratio of potassium iodide to magnesium chloride is not highly critical and may vary from 0.05 to 2.0. However, the optimum ratio is 1:1, as specified in the procedure. If the potassium iodide is omitted, the black magnesium powder produced reacts with bromobenzene at −78°. However, since the magnesium prepared in this way does not react with fluorobenzene in refluxing tetrahydrofuran, it is evidently less reactive than that produced in the presence of potassium iodide.

11. The submitters purified the tetrahydrofuran prior to use by distillation from lithium aluminum hydride. For a warning concerning potential hazards of this procedure, see *Org. Synth., Coll. Vol. 5,* 976 (1973). The checkers distilled the solvent from the sodium ketyl of benzophenone.

The submitters have found that diglyme and 1,2-dimethoxyethane are also effective solvents. The reactivity of the magnesium obtained with 1,2-dimethoxyethane as solvent is slightly reduced. Hydrocarbons, amines, and dioxane proved to be ineffective solvents, owing to the insolubility of the magnesium salts and consequent incomplete reduction.

12. Efficient stirring is essential for the generation of highly reactive magnesium. If the stirring is not effective, the reduction may not be complete after the 3-hour reaction time. The remaining unreacted potassium is a fire hazard during the isolation of the

product. If the scale of the reaction is increased, measures should be taken to ensure that effective stirring can be maintained throughout the reaction period. The submitters recommend that, as a precaution, the scale be increased gradually.

13. The mildly exothermic reduction may result in excessive foaming which carries potassium particles up into the condenser. This problem is avoided by using a relatively large flask (in this case, 200 ml. instead of 100 ml.) and by carefully controlling the temperature at the beginning of the reduction.

14. The reduction appears to be essentially complete in 30–45 minutes. However, a reaction time of 3 hours is recommended to ensure complete consumption of the potassium (Note 12).

15. Although the submitters have found that the active magnesium may be stored under argon for several days, they advise that the preparation be used within a few hours to obtain the maximum reactivity. Most of the reactions carried out by the submitters with the active magnesium were performed in the same flask and solvent used for the reduction. Attempts to evaporate the tetrahydrofuran and replace it with different solvents resulted in magnesium suspensions of reduced reactivity. The active magnesium may be conveniently transferred to another reaction vessel, if desired, as a slurry under an atmosphere of argon.

16. The solid chloride was melted by warming on a steam bath and drawn into a syringe that had been warmed briefly in an oven.

17. The submitters reported a melting point of 114–116°. The checkers obtained analytically pure material with a recovery of 80% after decolorization with activated carbon and recrystallization from 2–3 ml. of hexane at 0°. The product was also purified with comparable efficiency by sublimation at 85–90° (10 mm.). A small amount of a yellow, volatile impurity was removed from the cold finger before the product began to sublime. The melting point of the product after purification by the checkers was 110–112°. (lit.,[2] m.p. 114–116°).

18. The spectral properties of the product are as follows: IR (KBr) cm.$^{-1}$: 2960 (OH), 1693 (C=O), 1422, 1312, 1262, 952, 734; ^{1}H NMR (CCl$_4$), δ (multiplicity, number of protons, assignment): 1.1–1.9 (m, 10H, 5 CH_2), 2.2 (broad s, 1H, CH), 12.5 (s, CO$_2$$H$); ^{13}C NMR (CDCl$_3$), δ (off-resonance multiplicity, assignment): 30.0 and 33.0 (m, C-2, C-3, C-5, C-6), 37.8 (d, C-4), 42.4 (t, C-7), 52.2 (s, C-1), 183.8 (s, carboxyl C).

19. The submitters have recently developed the following unchecked procedure, which is suitable for preparing highly reactive magnesium powder, using lithium as a reducing agent.[3] This procedure avoids potassium and produces a magnesium powder equal in reactivity to that obtained using potassium as the reducing agent. A 50-ml., two-necked, round-bottomed flask equipped with a Teflon-coated magnetic stirring bar, rubber septum, and condenser connected to an argon inlet (Note 6) is charged with 0.224 g. (0.0325 g.-atom) of freshly cut lithium (Note 20), 1.57 g. (0.0165 mole) of anhydrous magnesium chloride (Note 9), 0.436 g. (0.00341 mole) of naphthalene and 10 ml. of tetrahydrofuran (Note 11). The mixture is stirred vigorously at room temperature for 24 hours (Note 21). After complete reduction, the highly reactive magnesium appears as a dark gray to black powder which slowly settles after stirring is stopped. In some cases, the tetrahydrofuran has a slight olive green color due to a small amount of lithium naphtha-

lide. This can be ignored when the highly reactive magnesium is reacted. If desired this can be removed by withdrawing the tetrahydrofuran with a syringe and adding fresh, dry tetrahydrofuran or other solvent.

20. Lithium (99.9%, rod, 1.27 cm. dia.) from Alfa has been used extensively in our studies. The lithium is cut under oil, rinsed in hexane, and transferred to a tared 24/40 adapter with a stopcock and rubber septum which has been filled with argon. The adapter is evacuated, removing the hexane, filled with argon and weighed. The lithium is then transferred to the reaction vessel under an argon stream.

21. It is important that the reaction be stirred vigorously and that the lithium make frequent contact with the stirring bar, as the lithium has a tendency to become coated with magnesium and stop the reduction from continuing. If reduction does stop, it can be initiated again by gently rubbing the piece of lithium against the wall of the flask with a metal spatula, the rubber septum can be temporarily removed under a stream of argon to carry out this procedure.

3. Discussion

The procedures for the preparation of 2,2-dichloronorbornane and 1-chloronorbornane are based on those of Bixler and Nieman.[2] The active magnesium generated in Part C of this procedure[4-6] is useful for the formation of Grignard reagents from alkyl and aryl halides that do not react, or react only slowly, with magnesium turnings or magnesium activated by previously known methods. Prior to the development of this procedure, four basic modifications of the usual methods for preparing Grignard reagents were utilized for relatively unreactive halides: the use of (1) higher reaction temperatures by variation of the solvent, (2) more strongly coordinating solvents such as tetrahydrofuran,[7-9] (3) various procedures to activate the surface of the magnesium,[10-14] and (4) magnesium slurries prepared by co-condensation of magnesium vapor and solvent.[15]

Activation of magnesium in the third method has been effected by reduction of the size of the metal particles[13] and chemical reactions. The Gilman procedure,[10] which consists of adding iodine to activate the magnesium surface, is representative of the latter technique. Ethyl bromide and 1,2-dibromoethane have been employed in catalytic amounts to activate the metal surface, and in stoichiometric proportions for entrainment.[11] Certain transition metal halides have also proved to be effective catalysts.[12] The magnesium preparations obtained by the co-condensation method are quite active, though considerably less active than those generated by the reduction process.[16] An alternative procedure has recently been published for the reduction of magnesium halides to activated magnesium with sodium naphthalene radical anion.[17]

Some results from an investigation into the reactions of activated magnesium with various halides and dihalides, some of which react with difficulty under the conditions of normal Grignard preparations, are given in Table I.[6] A number of important features can be noted from the table, including the facile formation of di-Grignard reagents and allyl- and vinylmagnesium halides. Alkyl and aryl fluorides are easily converted to the corresponding magnesium fluorides. The formation of Grignard reagents may be effected at temperatures of $-78°$ or below with active magnesium, thus allowing Grignard reactions to be carried out with unstable compounds.

The Grignard reagents prepared from activated magnesium appear to react normally

TABLE I

REACTION OF VARIOUS HALIDES WITH ACTIVATED MAGNESIUM

Halide	Magnesium/ Halide[a] Ratio	Temper- ature (°)	Time (minutes)	Yield of Product (%)		
				Mono- Grignard[b]	Di- Grignard[b]	Carbox- ylic Acid[c]
1,4-Dibromo- benzene	4	25	15		100	
1,4-Bromo- chloroben- zene	4	25	15	100	10	
	4	25	120	100	100	
1,4-Dichloro- benzene	2	25	180	90	0	80
	4	25	120	100	30	
Fluorobenzene	4	66	60			69
tert-Butyl chloride	2	25	10	100		52
1-Chloronorbor- nane	1.7	66	360	74		63
Methallyl chloride	2	25	60			82
2-Bromo- propene	2	25	5	100		71

[a]With the aryl halides and 1-chloronorbornane the activated magnesium was formed in the presence of potassium iodide.
[b]Yield of hydrocarbon determined by GC after hydrolysis.
[c]Isolated yield based on halide after carbonation.

with electrophiles. Thus, reactions with proton donors, ketones, and carbon dioxide afford hydrocarbons, alcohols, and carboxylic acids, respectively. The reductive coupling of ketones to pinacols had also been accomplished with activated magnesium.[16]

1-Norbornanecarboxylic acid has been prepared by concurrent rearrangement and hydrogenolysis of endo-2-bromo-2-norbornanecarboxylic acid,[18] by sequential reduction and hydrolysis of exo-2-bromo-1-norbornanecarboxamide,[19] by ozonolysis of 1-(4-methoxyphenyl)norbornane,[20] and by carbonation of 1-norbornyllithium.[2]

1. Department of Chemistry, University of Nebraska—Lincoln, Lincoln, Nebraska 68588. This work was supported in part by grants from the Research Corporation, National Science Foundation, North Carolina Board of Science and Technology, the Alfred P. Sloan Foundation, the Army Research Office, and the Division of Sciences, Department of Energy (Contract No. DE-AC02-80ER-1063).
2. R. L. Bixler and C. Niemann, J. Org. Chem., 23, 742 (1958).
3. R. D. Rieke, P. T.-Z. Li, T. P. Burns, and S. T. Uhm, J. Org. Chem., 46, 4323 (1981).
4. R. D. Rieke and P. M. Hudnall, J. Am. Chem. Soc., 94, 7178 (1972).
5. R. D. Rieke and S. E. Bales, J. Chem. Soc. Chem. Commun., 879 (1973).
6. R. D. Rieke and S. E. Bales, J. Am. Chem. Soc., 96, 1775 (1974).
7. H. Normant, C. R. Hebd. Seances Acad. Sci., 239, 1510 (1954); H. Normant, Bull. Soc. Chim. Fr., 1444 (1957).

8. H. E. Ramsden, A. E. Balint, W. R. Whitford, J. J. Walburn, and R. Cserr, *J. Org. Chem.*, **22**, 1202 (1957); H. E. Ramsden, J. R. Leebrick, S. D. Rosenberg, E. H. Miller, J. J. Walburn, A. E. Balient, and R. Cserr, *J. Org. Chem.*, **22**, 1602 (1957).
9. C. S. Marvel and R. G. Woolford, *J. Org. Chem.*, **23**, 1658 (1958).
10. H. Gilman and N. B. St. John, *Recl. Trav. Chim. Pays-Bas*, **49**, 717 (1930); H. Gilman and R. H. Kirby, *Recl. Trav. Chim. Pays-Bas*, **54**, 577 (1935).
11. D. E. Pearson, D. Cowan, and J. D. Beckler, *J. Org. Chem.*, **24**, 504 (1959).
12. W. L. Respess and C. Tamborski, *J. Organomet. Chem.*, **18**, 263 (1969).
13. R. C. Fuson, W. C. Hammann, and P. R. Jones, *J. Am. Chem. Soc.*, **79**, 928 (1957).
14. E. C. Ashby, S. H. Yu, and R. G. Beach, *J. Am. Chem. Soc.*, **92**, 433 (1970); S. H. Yu and E. C. Ashby, *J. Org. Chem.*, **36**, 2123 (1971).
15. K. J. Klabunde, H. F. Efner, L. Satek, and W. Donley, *J. Organomet. Chem.*, **71**, 309 (1974).
16. R. D. Rieke, *Top. Curr. Chem.*, **59**, 1 (1975); R. D. Rieke, *Acc. Chem. Res.*, **10**, 301 (1977).
17. R. T. Arnold and S. T. Kulenovic, *Synth. Commun.*, **7**, 223 (1977).
18. H. Kwart and G. Hull, *J. Am. Chem. Soc.*, **80**, 248 (1958).
19. W. R. Boehme, *J. Am. Chem. Soc.*, **81**, 2762 (1959).
20. D. C. Kleinfelter and P. von R. Schleyer, *J. Org. Chem.*, **26**, 3740 (1961).

$\Delta^{9,10}$-OCTALIN

(Naphthalene, 1,2,3,4,5,6,7,8-octahydro-)

Submitted by Edwin M. Kaiser[1] and Robert A. Benkeser[2]
Checked by Frederick J. Sauter and Herbert O. House

1. Procedure

Caution! This preparation should be performed in a hood to avoid exposure to amine vapors.

A. *$\Delta^{9,10}$- and $\Delta^{1,9}$-Octalin.* Naphthalene (25.6 g. or 0.200 mole, Note 1) is dissolved in a mixture of 250 ml. of anhydrous ethylamine and 250 ml. of anhydrous dimethylamine (Note 2) in a 1-l., three-necked flask fitted with a mechanical stirrer (Note 3) and a dry ice condenser and cooled in an ice bath. To this solution is added rapidly (Note 4), piecewise with stirring, 11.55 g. (1.674 g.-atoms) of lithium metal wire cut into 0.5-cm. lengths (Note 5). After the addition is complete, the cooling bath is removed and the blue solution is stirred for 14 hours, additional dry ice being added to the condenser as required. The dry ice condenser is replaced with a water condenser and the mixture is allowed to stand overnight, during which time the volatile amine solvents evaporate (Note 6). The top of the condenser is protected with an anhydrous calcium sulfate drying tube, maintaining anhydrous conditions in the reaction vessel during the evaporation of solvent (Note 7). The reaction flask is placed in an ice bath and the grayish-white residue is hydrolyzed by the *cautious,* dropwise addition of 500 ml. of water, with occasional stirring. After the resulting suspension has been filtered with suction, the residual solid is washed with four 25-ml. portions of diethyl ether. The ether layer is separated, and the aqueous phase of the filtrate is extracted with four additional 25-ml. portions of ether. The combined ethereal solutions are dried over calcium sulfate and concentrated with a rotary evaporator. The residual liquid is distilled under reduced pressure, separating 19–20 g. (70–74%) of the product, b.p. 72–77° (14 mm.), n_D^{23} 1.4978. Analysis of the product by GC (Note 8) shows the presence of 80–83% of $\Delta^{9,10}$-octalin and 17–20% of $\Delta^{1,9}$-octalin. This mixture should be either stored under a nitrogen atmosphere or immediately subjected to the described purification procedure (Note 9).

B. *Purification of $\Delta^{9,10}$-octalin.* A 1-l., three-necked flask is equipped with a magnetic stirrer, a pressure-equalizing dropping funnel, and a reflux condenser fitted with a nitrogen inlet tube, to maintain a nitrogen atmosphere in the reaction vessel throughout the portions of this preparation where anhydrous conditions are employed. A solution of 2.35 g. (0.0618 mole) of sodium borohydride and 11.55 g. (0.1650 mole) of 2-methyl-2-butene (Note 10) in 100 ml. of anhydrous tetrahydrofuran is added to the reaction flask. A solution of 11.75 g. (0.08275 mole) of boron trifluoride diethyl etherate in 22 ml. of anhydrous tetrahydrofuran is then added, dropwise and with stirring, over a 45-minute period. Although the initial rate of addition must be slow to control the exothermic reaction, the rate can be increased during the latter part of the reaction. To the resulting tetrahydrofuran solution of bis(3-methyl-2-butyl)borane is added, dropwise and with stirring over 10 minutes, the mixture of $\Delta^{9,10}$- and $\Delta^{1,9}$-octalins (19–20 g. containing *ca.* 0.03 mole of $\Delta^{1,9}$-octalin) obtained in Part A. The reaction mixture is stirred at room temperature for 3.5 hours and treated with 50 ml. of water. After 35 ml. of 3 M aqueous sodium hydroxide has been added, dropwise and with stirring over 10 minutes, 35 ml. of 30% aqueous hydrogen peroxide is added dropwise over a 45-minute period, with continuous stirring. The resulting mixture is heated to 45° (Note 11) with stirring for 5 hours, then allowed to cool to room temperature. The organic layer is separated, washed with four 30-ml. portions of water, and dried over calcium sulfate. After the organic solution has been tested with moist starch-iodide paper to ensure the absence of peroxides, the volatile solvents are removed with a rotary evaporator, and the residual liquid is distilled under reduced pressure in an apparatus fitted with a capillary tube to admit nitrogen as an ebullator (Note 12). After 1–2 g. of forerun has been separated, 9–13 g. (33–48% based on the starting naphthalene) of $\Delta^{9,10}$-octalin is collected at 75–77° (14 mm.), n_D^{20} 1.4990. The product is at least 99% $\Delta^{9,10}$-octalin, by GC analysis (Note 8), and is stored under a nitrogen atmosphere (Note 9).

2. Notes

1. Naphthalene purchased from Eastman Organic Chemicals was used without purification.

2. Ethylamine (b.p. 16.6°) and dimethylamine (b.p. 7.4°) may be distilled into the reaction flask from cylinders. The checkers employed amines, sealed in ampules, purchased from Eastman Organic Chemicals. After the ampules had been cooled to 0° in an ice bath, they were opened and the contents were added to flasks cooled in ice baths. Small portions (0.5–1.0 g.) of sodium metal were added to each of the cold amines. The cooling baths were then removed, and the amines were allowed to distill from the sodium into the reaction flask.

3. Stirrers with Teflon paddles should not be used. Although the reaction proceeds satisfactorily, the Teflon parts are blackened (and presumably partially degraded) by the reaction solution.

4. The reaction is very exothermic during the initial addition of lithium metal. The use of an external ice bath reduces the number of times additional dry ice must be added to the acetone–dry ice condenser.

5. The lithium wire is coated with a hydrocarbon grease. As 0.5-cm. pieces are cut with scissors, they should be dropped into a beaker of anhydrous hexane to dissolve the grease. The pieces of bare lithium wire are then removed with forceps, drained briefly by touching them to a dry towel, and weighed in a second beaker of anhydrous hexane. The pieces of metal are then removed from the second beaker with forceps and added to the reaction mixture.

6. If evaporation of the solvent is not complete, the remainder can be removed by immersing the reaction flask in a bath of warm water.

7. If anhydrous conditions are not maintained, the product becomes contaminated with a tarry material.

8. A GC column packed with Apiezon L suspended on Chromosorb P was employed for this analysis. In a typical analysis at 140° the retention times were: $\Delta^{1,9}$-octalin, 46.2 minutes; $\Delta^{9,10}$-octalin, 49.2 minutes. The ^{1}H NMR spectrum (CCl$_4$ at 60 mHz.) of this mixture exhibits a multiplet at δ 0.8–2.6 (aliphatic CH) with a weak, partially resolved multiplet at δ 5.25 (vinyl CH of $\Delta^{1,9}$-octalin).

9. The octalins usually react rapidly with oxygen, forming unidentified, oxygenated products. The checkers found that, if the crude product is not protected from air oxidation, the yield of $\Delta^{9,10}$-octalin isolated from the purification procedure is lowered substantially.

10. 2-Methyl-2-butene, purchased either from Phillips Petroleum Company or Aldrich Chemical Co., was used without purification.

11. The checkers found a large oil bath to be a convenient method for heating the reaction mixture to 45°.

12. The mixture tends to foam badly during this distillation.

3. Discussion

Among the preparative routes to the octalin mixtures, the acid-catalyzed dehydration of 2-decalol[3] and the metal-amine reduction of naphthalene[4] appear most satisfactory. Apart from the purification method described in this preparation, pure $\Delta^{9,10}$-octalin has also been obtained by reaction of the octalin mixture with nitrosyl chloride. After separation of the adducts by fractional crystallization, the pure $\Delta^{9,10}$-octalin has been regenerated from its nitrosyl chloride adduct.[3,5]

Lithium dissolved in amines of low molecular weight constitutes a useful and convenient reagent for reducing aromatic hydrocarbons to monoölefins.[6] Although mixtures of isomeric olefins are usually obtained with primary amine solvents, the use of secondary amines as co-solvents dramatically increases the selectivity of these reductions so that the more thermodynamically stable olefin usually becomes the predominant product. Thus, in the reduction of naphthalene, the yield of $\Delta^{9,10}$-octalin increases from 52% in pure ethylamine to 80–82% in an ethylamine–dimethylamine mixture. As another example, the reduction of *tert*-butylbenzene with lithium in pure ethylenediamine yields a product mixture 70% in 1-*tert*-butylcyclohexene.[7] When a mixture of ethylenediamine and morpholine is used as the reaction solvent, the product mixture is 84% 1-*tert*-butylcyclohexene.[8]

Aromatic hydrocarbons may also be reduced to monoölefins by calcium in low molecular weight amines.[9] For example, this metal in methylamine–ethylenediamine converts naphthalene to $\Delta^{9,10}$- and $\Delta^{1,9}$-octalin in yields identical to those obtained from the lithium–amine system used in this preparation. Calcium may even be employed to effect such reductions in ether provided small amounts of hexamethylphosphoric triamide (HMPA) are present; for a warning concerning the use of HMPA, see J. A. Zapp, Jr., *Science,* **190,** 422 (1975).

The separation procedure[4] described in this preparation illustrates the *in situ* generation of a tetrahydrofuran solution of bis(3-methyl-2-butyl)borane and use of this sterically hindered borane to react with a trisubstituted olefin ($\Delta^{1,9}$-octalin) in preference to a more hindered tetrasubstituted olefin ($\Delta^{9,10}$-octalin).[10] The alkylborane adduct produced from $\Delta^{1,9}$-octalin is oxidized with alkaline hydrogen peroxide to 1-decalol, which is easily separated from the olefinic product.

1. Department of Chemistry, University of Missouri, Columbia, Missouri 65211.
2. Department of Chemistry, Purdue University, W. Lafayette, Indiana 47907.
3. W. G. Dauben, E. C. Martin, and G. J. Fonken, *J. Org. Chem.,* **23,** 1205 (1958).
4. R. A. Benkeser and E. M. Kaiser, *J. Org. Chem.,* **29,** 955 (1964).
5. A. S. Hussey, J. F. Sauvage, and R. H. Baker, *J. Org. Chem.,* **26,** 256 (1961).
6. R. A. Benkeser, R. E. Robinson, D. M. Sauve, and O. H. Thomas, *J. Am. Chem. Soc.,* **77,** 3230 (1954).
7. R. A. Benkeser, R. K. Agnihotri, and M. L. Burrous, *Tetrahedron Lett.,* 1 (1960); R. A. Benkeser, R. K. Agnihotri, M. L. Burrous, E. M. Kaiser, J. M. Mallan, and P. W. Ryan, *J. Org. Chem.,* **29,** 1313 (1964).
8. T. J. Hoogeboom, M. S. thesis, Purdue University, 1965.
9. R. A. Benkeser and J. Kang, *J. Org. Chem.,* **44,** 3737 (1979).
10. H. C. Brown and G. Zweifel, *J. Am. Chem. Soc.,* **83,** 1241 (1961).

2,2,7,7,12,12,17,17-OCTAMETHYL-21,22,23,24-TETRAOXAPERHYDROQUATERENE

(21,22,23,24-Tetraoxapentacyclo-[16.2.1.1^{3,6}.1^{8,11}.1^{13,16}]tetracosane, 2,2,7,7,12,12,17,17-octamethyl-)

Submitted by Maurice Chastrette,[1] Francine Chastrette,
and Jean Sabadie
Checked by William V. Phillips, Herbert O. House, James C. Kauer,
and William A. Sheppard

1. Procedure

A.   2,2,7,7,12,12,17,17-*Octamethyl*-21,22,23,24-*tetraoxaquaterene*.   A   1-liter, three-necked flask fitted with a mechanical stirrer, a dropping funnel, and a condenser is charged with 27.2 g. (0.400 mole) of freshly distilled furan (Note 1), 24 ml. of absolute ethanol, 57.3 g. (0.200 mole) of the lithium perchlorate–1,2-dimethoxyethane complex (Note 2), and 16 ml. of reagent grade concentrated hydrochloric acid. The resulting pale-yellow to pale-tan solution is stirred continuously and heated to 63° with an oil bath before 58.1 g. (0.80 mole) of acetone (Note 3) is added dropwise over 1 hour. During this addition the heat of reaction causes the solution to reflux, and some white solid begins to separate. After the addition is complete, stirring and heating are continued until refluxing ceases (about 30 minutes). The temperature of the bath is then raised to 78°, and the mixture is refluxed with stirring for an additional 30 minutes. The reaction mixture, a dark red-brown solution containing a pasty white precipitate, is cooled to room temperature with continuous stirring, and 25 ml. of water and 300 ml. of diethyl ether are added. The resulting mixture is stirred at room temperature for 1 hour, converting the precipitate into a finely divided white solid, and the mixture is filtered with suction, employing a medium-porosity, sintered-glass funnel. The residual crude product is washed thoroughly (Note 4) with portions of ether (total volume 100 ml.) followed by portions of 95% ethanol (total volume 50 ml.) and allowed to air-dry. The crude product (14.2–14.8 g.), m.p. 231–240°, is recrystallized from 140–150 ml. of reagent grade chloroform, yielding 7.3–7.4 g. of product as white needles, m.p. 241.5–243°. The mother liquors are concentrated to one-third their initial volume, separating 2.9–3.1 g. of a second crop of crystalline product, m.p. 240–242°, for a total yield of the tetraoxaquaterene of 10.2–10.5 g. (24–25%) (Note 5).

B.   2,2,7,7,12,12,17,17-*Octamethyl*-21,22,23,24-*tetraoxaperhydroquaterene*.   A 400-ml., stainless-steel, shaking autoclave is charged with 4.0 g. (0.0092 mole) of the tetraoxaquaterene from Part A, 200 ml. of ethanol, and 400 mg. of 5% palladium on charcoal (Note 6). The autoclave is filled with hydrogen at an initial pressure of 170 atm. and heated with shaking for 4 hours at 105°. The catalyst and a white solid are removed by filtration (Note 7), the solid is dissolved in 100 ml. of warm chloroform, the solution is filtered, the chloroform is evaporated, and the white solid obtained is dried under reduced pressure at 60° (Note 8), leaving the tetraoxaperhydroquaterene as a white solid, m.p. 204–209° (Note 9), in a yield of 2.85–2.97 g. (69–72%).

2. Notes

1. Furan, purchased from Aldrich Chemical Company, Inc., was distilled before use; b.p. 31–32°.
2. Anhydrous lithium perchlorate (60 g.), obtained from Ventron Corporation, was placed in a 500-ml. flask under dry nitrogen, and 85 ml. of anhydrous 1,2-dimethoxyethane was added. The resulting warm mixture was swirled and heated on the steam bath under dry nitrogen until all of the solid dissolved. The resulting solution was cooled and gradually crystallized to a solid mass. Additional 1,2-dimethoxyethane (155 ml.) was added, the mixture was stirred, and supernatant liquid was removed with a sintered-glass filter stick. The resulting solid was vacuum dried briefly in a vacuum dessicator (0.3 mm.) for 20 minutes then sealed under vacuum overnight. The resulting, solvated lithium perchlorate, $LiClO_4(C_4H_{10}O_2)_2$, 108 g, may be stored in a brown bottle under dry nitrogen until needed.

When lithium perchlorate was initially dissolved in the entire 240 ml. of 1,2-dimethoxyethane and processed as above, the yield was considerably lower. The submitters report that if less lithium perchlorate is used in the preparation of the octamethyltetraoxaquaterene the yield of the product is lowered.

3. Reagent grade acetone was used. The submitters state that an excess of acetone is necessary. When a 1 : 1 mole ratio of acetone to furan was used, they obtained a 21% yield of crude product; when a 1 : 2 ratio of acetone to furan was employed, the yield of product was less then 5%.

4. The submitters report that if this material is not washed thoroughly to remove the soluble, low molecular weight, linear polymers present the crude product will melt and/or darken at much lower temperatures than 231–240°.

5. The product has the following spectral properties; ^1H NMR (CDCl$_3$), δ (multiplic-

ity, number of protons, assignment): 1.48 (s, 24H, 8CH_3), 5.90 (s, 8H, 8CH); mass spectrum m/e (relative intensity): 432 (M$^+$, 40) 418 (34), 417 (100), 201 (28), 186 (31), 149 (55), 85 (46), 83 (67), 75 (21), 60 (21), 47 (20), 45 (24), 43 (53), and 41 (26).

6. Catalyst obtained from Engelhard Industries was used. The submitters used 200 mg. of Fluka 10% palladium on charcoal catalyst with 5 g. of starting material in 250 ml. of ethanol and obtained a total yield of 2.3 g. (46%), m.p. 208–211°.

7. The ethanolic filtrate can be concentrated to 10–15 ml. under reduced pressure to obtain 0.3 g. (7%) of crude product, m.p. 187–202°. Unchanged starting material, if present, is concentrated in this second fraction and may be detected by the furan resonance at δ 5.85 in the ^{1}H NMR spectrum or by a sharp IR absorption at 772 cm.$^{-1}$, not present in the product. Elemental analyses of these second crops suggested that other impurities were also present.

8. The solid tenaciously holds a small amount of chloroform which can be detected by ^{1}H NMR (δ 7.25). Vacuum drying overnight at 60° removes this impurity.

9. In one isolated case the checkers found that no hydrogen uptake occurred, and unreacted starting material was recovered. This erratic result may have resulted from accidental poisoning of the catalyst by contaminants present in the autoclave or associated valves and lines. If multiple runs are carried out, the products of each should be checked by IR or ^{1}H NMR spectroscopy before being combined. The ^{1}H NMR spectrum (CDCl$_3$) shows singlets at δ 0.74, 0.82, 0.93, and 1.04 (24H, 8CH_3) and multiplets at 1.2–1.9 (16H, 8CH_2), 1.9–2.8 (4H, 2CH_2), and 3.0–4.2 (8H, 4CH_2); IR (Nujol) cm.$^{-1}$ strong absorptions: 1078, 1039, medium: 999, 991, 552, 528, 520, and weaker: 1285, 1248, 1204, 977, 884, 840, 719, 659, 598, 560. The checkers concluded from examination of the spectra that a variable mixture of isomers is obtained from the hydrogenation.

3. Discussion

The unsaturated tetraoxaquaterene (accompanied by linear condensation products) was first synthesized in 18.5% yield by the acid-catalyzed condensation of furan with acetone in the absence of added lithium salts.[2] Other ketones also condensed with furan, giving analogous products in 6–12% yield.[2–4] A corresponding macrocycle was also prepared in 9% yield from pyrrole and cyclohexanone.[4] The macrocyclic ether products have also been obtained by condensation of short, linear condensation products of 2, 3, or 4 furan rings with a carbonyl compound.[5]

The method described here gives higher yields of the macrocyclic tetraethers and allows the product from furan and cyclohexanone to be formed directly in 5–10% yield, whereas this product was previously obtained only by an indirect route. Lithium perchlorate undoubtedly accelerates the reaction; after short reaction times, the product was isolated in 20% yield when the salt was present and in only 5% yield when the salt was absent. The lithium cation presumably acts as a template which coordinates with the oxygen atoms of the furan units, favoring cyclization instead of linear polymerization.[6] The hydrogenated macrocycle has been shown to form complexes with lithium salts.[6,7]

1. Laboratoire de Chimie Organique Physique, Laboratoire de Chimie Organique II, Université Claude-Bernard Lyon I, 43, Boulevard du 11 Novembre 1918, F 69622 Villeurbanne Cedex, France.

2. R. G. Ackman, W. H. Brown, and G. F. Wright, *J. Org. Chem.*, **20**, 1147 (1955).
3. R. E. Beals and W. H. Brown, *J. Org. Chem.*, **21**, 447 (1956).
4. W. H. Brown, B. J. Hutchinson, and M. H. MacKinnon, *Can. J. Chem.*, **49**, 4017 (1971).
5. W. H. Brown and W. N. French, *Can. J. Chem.*, **36**, 537 (1958).
6. M. Chastrette and F. Chastrette, *J. Soc. Chem. Commun.*, 534 (1973).
7. Y. Kobuke, H. Hanje, K. Horiguchi, M. Asada, Y. Makayama, and J. Furukawa, *J. Am. Chem. Soc.*, **98**, 7414 (1976).

ORCINOL MONOMETHYL ETHER

(Phenol, 3-methoxy-5-methyl-)

Submitted by R. N. Mirrington and G. I. Feutrill[1]
Checked by H. Gurien, G. Kaplan, and A. Brossi

1. Procedure

Caution! Benzene has been identified as a carcinogen; OSHA has issued emergency standards on its use. All procedures involving benzene should be carried out in a well-ventilated hood, and glove protection is required.

A. *Orcinol dimethyl ether.* A 1-l., three-necked flask fitted with a mechanical stirrer, a condenser, and a 100-ml. dropping funnel is charged with 124 g. (0.984 mole) of anhydrous potassium carbonate, 410 ml. of acetone (Note 1), and 42.6 g. (0.344 mole) of orcinol monohydrate (Note 2). The stirrer is started, and 94.5 g. (70.9 ml., 0.750 mole) of dimethyl sulfate is added from the dropping funnel to the pink mixture over a period of 2 minutes. The mixture warms appreciably and begins to reflux after an additional 5 minutes. When the spontaneous boiling has subsided (15–20 minutes after addition of the dimethyl sulfate), the stirred mixture is heated gently under reflux for 4 hours longer. The condenser is then arranged for distillation and 200 ml. of acetone is distilled. A 50-ml.

portion of concentrated aqueous ammonia is added to the reaction mixture; stirring and heating are continued for 10 minutes. The mixture is diluted with water to a total volume of approximately 750 ml., the layers are separated, and the organic layer is combined with two 150-ml. ethereal extracts of the aqueous layer. The organic phase is washed with 50 ml. of water, twice with 50-ml. portions of 3 N sodium hydroxide solution (Note 3), once with 50 ml. of saturated aqueous sodium chloride, and dried over magnesium sulfate. After evaporation of the ether at atmospheric pressure, the residual liquid is distilled under reduced pressure, yielding 42.9–43.7 g. (94–96%) of orcinol dimethyl ether, b.p. 133–135° (40 mm.) (Notes 4 and 5).

Caution! Because hydrogen is evolved and large volumes of foul-smelling ethyl methyl sulfide are liberated, this step should be conducted in a well-ventilated hood.

B. *Orcinol monomethyl ether.* A 1-1., three-necked flask equipped with a magnetic stirrer, a condenser, a dropping funnel, and a nitrogen inlet is charged with 250 ml. of dry N,N-dimethylformamide (Note 6) and 22 g. (0.55 mole) of sodium hydride (60% oil dispersion). The suspension is stirred under an atmosphere of dry nitrogen and cooled with an ice bath while a solution of 31 g. (37 ml., 0.50 mole) of ethanethiol (Note 7) in 150 ml. of dry N,N-dimethylformamide (Note 6) is added slowly from the dropping funnel over a period of 20 minutes. The ice bath is removed and stirring is continued for an additional 10 minutes. A solution of 38.0 g. (36.5 ml., 0.250 mole) of orcinol dimethyl ether in 100 ml. of dry N,N-dimethylformamide (Note 6) is added in one lot, and the mixture is refluxed under an atmosphere of dry nitrogen for 3 hours (Notes 8 and 9). The mixture is cooled, poured into 1.8 1. of cold water, and extracted with two 250-ml. portions of petroleum ether (b.p. 50–70°), which are discarded. The aqueous layer is acidified with 330 ml. of ice-cold 4 N hydrochloric acid and extracted with three 250-ml. portions of ether. The combined ethereal extracts are washed with 100 ml. of saturated aqueous sodium chloride and dried over magnesium sulfate. After the ether is distilled at atmospheric pressure, the residual liquid is distilled under reduced pressure, yielding 28–30.5 g. (81–88%) of orcinol monomethyl ether, b.p. 89–90° (0.2 mm.) or 156–158° (25 mm.) (Notes 10 and 11.)

2. Notes

1. Technical acetone containing about 1% water is quite satisfactory.
2. British Drug Houses Ltd. reagent grade orcinol monohydrate was used without further purification.
3. If the first washing is colorless, as is usually the case, the second washing is unnecessary. Washing with sodium hydroxide solution should be continued until the washings are colorless.
4. A similar run using 100 g. of orcinol monohydrate afforded 102 g. (95%) of orcinol dimethyl ether, b.p. 67.5–68.5° (0.2 mm.).
5. GC analysis of the product on two columns (silicone gum rubber SE-30 and OV-1) indicated the presence of traces of two other compounds with retention times longer than that of orcinol dimethyl ether. These impurities, which were most likely C-methylated materials,[2] totaled less than 0.5% of the product.

6. N,N-Dimethylformamide, b.p. 58° (25 mm.), was distilled from calcium hydride under a reduced pressure of nitrogen immediately before use.

7. British Drug Houses Ltd. reagent grade ethanethiol was distilled from calcium hydride before use (b.p. 36°).

8. The mixture may become gelatinous during this time, but stirring is not necessary.

9. A polythene tube leading from the top of the condenser to the back of the hood is advisable, preventing any diffusion of the by-product, ethyl methyl sulfide, into the laboratory. Alternatively, this by-product may be collected, if desired, by passing the vapors through a cold trap (dry ice in acetone).

10. This distillate, which is sufficiently pure for most reactions, solidifies after standing for 4–6 hours. A sample crystallizes from benzene-petroleum ether as off-white prisms, m.p. 61–62°, and is relatively free of sulfurous odor.

11. ^{1}H NMR (CCl$_4$): δ 2.19 (s, 3H, CH_3), 3.63 (s, 3H, OCH_3), 6.17 (m, 3H, C$_6H_3$), 6.38 (broad s, 1H, OH).

3. Discussion

Previous preparations of orcinol monomethyl ether have been effected by partial methylation of oricinol with methyl iodide and potassium hydroxide[3] or sodium ethoxide,[4] or with dimethyl sulfate and sodium hydroxide.[5] These procedures required tedious purification steps and the pure monomethyl ether was obtained in 37% yield at best.[5]

This procedure is characterized by the easy isolation of a high-purity product in excellent yield. The reaction illustrates a general method[6] for the conversion of aryl methyl ethers to the corresponding phenols, and has proved to be of special advantage with acid-sensitive substrates.[6,7]

A unique feature of this procedure is the selective monodemethylation of the dimethyl ether. The scope of this reaction is illustrated[6] in part by the preparation in high yield of 4-methoxyphenol, guaiacol, and phloroglucinol dimethyl ether from the respective fully O-methylated compounds. An exception is pyrogallol trimethyl ether which affords pyrogallol 1-monomethyl ether in high yield.[6]

1. Department of Organic Chemistry, University of Western Australia, Nedlands, W.A., 6009. [Present address: Department of Organic Chemistry, University of Melbourne, Parkville, Victoria 3052, Australia]

2. D. D. Ridley, E. Ritchie, and W. C. Taylor, *Aust. J. Chem.*, **21**, 2979 (1968).

3. F. Tiemann and F. Streng, *Ber. Dtsch. Chem. Ges.*, **14**, 1999 (1881).

4. F. Henrich, *Monatsh. Chem.*, **22**, 232 (1901).

5. F. Henrich and G. Nachtigall, *Ber. Dtsch. Chem. Ges.*, **36**, 889 (1903).

6. G. I. Feutrill and R. N. Mirrington, *Tetrahedron Lett.*, 1327 (1970); *Aust. J. Chem.*, **25**, 1719 (1972).

7. G. I. Feutrill, R. N. Mirrington, and R. J. Nichols, *Aust. J. Chem.*, **26**, 345 (1973); G. I. Feutrill and R. N. Mirrington, *Aust. J. Chem.*, **26**, 357 (1973).

1,6-OXIDO[10]ANNULENE

(11-Oxabicyclo[4.4.1]undeca-1,3,5,7,9-pentaene)

Submitted by E. VOGEL, W. KLUG,[1] and A. BREUER
Checked by D. KNOPP, U. SCHWIETER, and A. BROSSI

1. Procedure

A. 11-*Oxatricyclo[4.4.1.0¹·⁶]undeca-3,8-diene*. A 1-l., three-necked, round-bottomed flask equipped with a sealed mechanical stirrer, a pressure-equalizing dropping funnel, and a thermometer is charged with 66.1 g. (0.501 mole) of 1,4,5,8-tetrahydronaphthalene [*Org. Synth.*, **Coll. Vol. 6,** 731 (1988)] and 200 ml. of anhydrous dichloromethane. To the resulting solution is added 75 g. of anhydrous sodium acetate. After the suspension is cooled with an ice bath, 104.5 g. (0.55 mole) of commercial 40% peracetic acid (Notes 1 and 2) is added dropwise over a period of 20–30 minutes, while maintaining the temperature at approximately 15° (Note 3) and stirring vigorously. To the reaction mixture is added, without delay, 500 ml. of water, to dissolve the sodium acetate and extract the bulk of the acetic acid. The organic layer is washed successively with two 100-ml. portions of 5% aqueous sodium hydroxide and two 100-ml. portions of water and dried over anhydrous potassium carbonate. The solvent is removed on a rotary evaporator, leaving a solid residue. Two recrystallizations from approximately 50 ml. of petroleum ether (b.p. 40–60°), cooling the solution to −40°, yields 58.5–62.0 g. (79–84%) of 11-oxatricyclo[4.4.1.0^{1,6}]undeca-3,8-diene as white needles, m.p. 58–61° (Note 4).

B. 3,4,8,9-*Tetrabromo*-11-*oxatricyclo*[4.4.1.0^{1,6}]*undecane*. A 1-l., three-necked, round-bottomed flask fitted with a sealed mechanical stirrer, a pressure-equalizing dropping funnel, and a calcium chloride drying tube is charged with 59.2 g. (0.400 mole) of 11-oxatricyclo[4.4.1.0^{1,6}]undeca-3,8-diene and 500 ml. of anhydrous diethyl ether (Note 5). The resulting solution is cooled in an acetone–dry ice bath, and 120 g. (0.75 mole) of bromine (Note 6) is added with stirring over a period of 1.5 hours (Note 7). After the addition is complete, the ether is removed on a rotary evaporator. The solid residue is dissolved in 800 ml. of hot chloroform. To this solution is added, with gentle stirring, 150 ml. of hot petroleum ether (b.p. 60–90°). The resulting, clear mixture, from which the product begins to crystallize, is allowed to cool to room temperature and then stand in a refrigerator at −40° overnight, completing the crystallization. The yield of white crystalline tetrabromide obtained is 115–125 g. (61–67%), m.p. 151–153° (Note 8). An additional 18–25 g. of product, m.p. 149–152°, is recovered from the mother liquor by concentration to about one quarter of the volume, giving a total yield of 136–144 g. (73–77%) of material sufficiently pure to be used in the following step.

C. 1,6-*Oxido*[10]*annulene*. A 2-l., three-necked, round-bottomed flask equipped with a sealed mechanical stirrer and a reflux condenser protected by a calcium chloride drying tube is charged with 81 g. (1.5 moles) of sodium methoxide (Note 9) and 600 ml. of anhydrous ether. To this slurry is added, with stirring, 117 g. (0.250 mole) of finely powdered 3,4,8,9-tetrabromo-11-oxatricyclo[4.4.1.0^{1,6}]undecane. The reaction mixture is refluxed with stirring for 10 hours and allowed to stand overnight. Following this, 500 ml. of water is added slowly, dissolving the solids. The ether layer is separated, and the aqueous layer is extracted with two 100-ml. portions of ether. The combined ethereal solution is washed with 250 ml. of water and dried over anhydrous potassium carbonate. Removal of the ether on a rotary evaporator affords a brown oil, which when distilled gives 34.4–35.1 g. of yellow 1,6-oxido[10]annulene, b.p. 77° (0.02 mm.) (Note 10). This material readily solidifies at room temperature, and two recrystallizations at −40° from 225 ml. of (5:1) pentane–ether yields 18.3–18.5 g. (51%) of 1,6-oxido[10]annulene as pale yellow needles, m.p. 51–52° (Notes 11 and 12).

2. Notes

1. Satisfactory 40% peracetic acid is obtainable from Buffalo Electrochemical Corporation, Food Machinery and Chemical Corporation, Buffalo, New York. The specifications given by the manufacturer for its composition are: peracetic acid, 40%; hydrogen peroxide, 5%; acetic acid, 39%; sulfuric acid, 1%; water, 15%. Its density is 1.15 g./ml. The peracetic acid concentration should be determined by titration. A method for the analysis of peracid solutions is based on the use of ceric sulfate as a titrant for the hydrogen peroxide present, followed by an iodometric determination of the peracid present.[2] The checkers found that peracetic acid of a lower concentration (27.5%) may

also be used without a decrease in yield. The product was found to be sufficiently pure, after only one recrystallization from 60 ml. of petroleum ether (b.p. 40–60°) and cooling overnight to −18°, to be used in the next step.

2. Alternatively *m*-chloroperbenzoic acid may be used.[3]

3. The yields of the desired product decrease substantially if the temperature exceeds 20°.

4. The reported m.p. is 64°.[4] GC analysis using a 1-m. column containing 20% Reoplex 400 on Diatoport S 6080, operated at 160°, indicates the purity of the product to be ~98%.

5. The rate of bromination of 11-oxatricyclo[4.4.1.0^{1,6}]undeca-3,8-diene is markedly higher in ether than in dichloromethane or chloroform. The former solvent thus permits the reaction to be carried out at relatively low temperatures.

6. Bromine was freshly distilled from phosphorus pentoxide.

7. The addition of bromine should not be started before the solution has cooled to approximately −70°.

8. The tetrabromide apparently consists of a mixture of stereoisomers. After several recrystallizations from chloroform–petroleum ether (60–90°) the major isomer, m.p. 160–162°, is obtained.

9. As reported by Shani and Sondheimer,[3,5] the dehydrohalogenation of the tetrabromide with potassium hydroxide in ethanol at 50–55° affords a mixture, readily separated by chromatography on alumina, of 1,6-oxido[10]annulene and the isomeric 1-benzoxepin. The latter compound is also formed during chromatography of 1,6-oxido[10]annulene on silica gel.[6]

10. To avoid acid-catalyzed rearrangements of 1,6-oxido[10]annulene it is recommended that the distillation flask be treated with a base before use.

11. The reported m.p. is 52–53°.[6] The purity of the product is greater than 99% as established by TLC, using plates prepared with Silica Gel Si F, obtained from Riedel-De Haen AG, 3016 Seelze, West Germany. If 1,6-oxido[10]annulene is to be used for spectroscopic investigations, care should be taken that its potential contaminants, such as naphthalene, 1-bromonaphthalene, α-naphthol, and 1-benzoxepin, are absent, as checked by TLC. The checkers could not obtain the reported yield of 24.5–25.6 g. (68–71%). Likewise, in experiments where the ether was replaced with tetrahydrofuran or dioxane, the yield given by the submitters could not be obtained.

12. The spectral properties of the product are as follows; ^{1}H NMR (CDCl$_3$), δ (multiplicity, assignment): 7.25–7.75 (m, AA′BB′, aromatic *H*); UV (95% C$_2$H$_5$OH) nm. max. (ε): 255 (74,000), 299 (6900), 393 (240) complex band; mass spectrum (250°, 70 eV) *m/e* (relative intensity > 10%): 144 (M, 43), 116 (40), 115 (100), 89 (15), 63 (15), 51 (10), 39 (11).

3. Discussion

The preparation of 1,6-oxido[10]annulene, described simultaneously by Sondheimer and Shani[3,5] and by Vogel, Biskup, Pretzer, and Böll,[6] is illustrative of the rather general synthesis of aromatic 1,6-bridged [10]annulenes from 1,4,5,8-tetrahydronaphthalene. In addition to the present compound, the following bridged [10]annulenes have thus far been obtained by this approach: 1,6-methano[10]annulene,[7,8,9] the 11,11-dihalo-1,6-methano[10]annulenes,[9,10] and 1,6-imino[10]annulene.[11]

The epoxidation of 1,4,5,8-tetrahydronaphthalene exemplifies the well-known selectivity of peracids in their reaction with alkenes possessing double bonds that differ in the degree of alkyl substitution.[12] Regarding the method of aromatization employed in the conversion of 11-oxatricyclo[4.4.1.0^{1,6}]undeca-3,8-diene to 1,6-oxido[10]annulene, the two-step bromination–dehydrobromination sequence is given preference to the one-step DDQ-dehydrogenation, which was advantageously applied in the synthesis of 1,6-methano[10]annulene,[7,9] since it affords the product in higher yield and purity.

1,6-Oxido[10]annulene closely resembles 1,6-methano[10]annulene in many of its spectral properties, particularly in its ^{1}H NMR, UV, IR, and ESR spectra,[13] but is chemically less versatile than the hydrocarbon analog due to its relatively high sensitivity toward proton and Lewis acids.

1. Institut für Organische Chemie der Universität Köln, 5 Köln, Greinstrasse 4, West Germany.
2. F. P. Greenspan and D. G. MacKellar, *Anal. Chem.*, **20**, 1061 (1948).
3. A. Shani and F. Sondheimer, *J. Am. Chem. Soc.*, **89**, 6310 (1967).
4. W. Hückel and H. Schlee, *Chem. Ber.*, **88**, 346 (1955).
5. F. Sondheimer and A. Shani, *J. Am. Chem. Soc.*, **86**, 3168 (1964).
6. E. Vogel, M. Biskup, W. Pretzer, and W. A. Böll, *Angew. Chem.*, **76**, 785 (1964) [*Angew. Chem. Int. Ed. Engl.*, **3**, 642 (1964)].
7. E. Vogel, W. Klug, and A. Breuer, *Org. Synth.*, **Coll. Vol. 6**, 731 (1987).
8. E. Vogel and H. D. Roth, *Angew. Chem.*, **76**, 145 (1964) [*Angew. Chem. Int. Ed. Engl.*, **3**, 228 (1964)].
9. P. H. Nelson and K. G. Untch, *Tetrahedron Lett.*, 4475 (1969).
10. V. Rautenstrauch, H.-J. Scholl, and E. Vogel, *Angew. Chem.*, **80**, 278 (1968) [*Angew. Chem. Int. Ed. Engl.*, **7**, 288 (1968)].
11. E. Vogel, W. Pretzer, and W. A. Böll, *Tetrahedron Lett.*, 3613 (1965).
12. D. Swern, "Organic Peroxides," Vol. 2, Wiley-Interscience, New York, 1971, p. 452.
13. F. Gerson, E. Heilbronner, W. A. Böll, and E. Vogel, *Helv. Chim. Acta*, **48**, 1494 (1965).

CYANIDE-CATALYZED CONJUGATE ADDITION OF ARYL ALDEHYDES: 4-OXO-4-(3-PYRIDYL)BUTYRONITRILE

(3-Pyridinebutanenitrile, γ-oxo)

Submitted by H. Stetter,[1] H. Kuhlmann, and G. Lorenz
Checked by Benjamin G. Padilla and George Büchi

1. Procedure

Caution! Sodium cyanide is highly toxic. Care should be taken to avoid direct contact of the chemical or its solutions with the skin, and impervious gloves should be worn to handle the reagent.

A 1-l., three-necked, round-bottomed flask equipped with a mechanical stirrer, a reflux condenser fitted with a potassium hydroxide drying tube, and a pressure-equalizing dropping funnel mounted with a nitrogen-inlet tube is charged with 4.9 g. (0.18 mole) of finely ground sodium cyanide (Note 1) and 500 ml. of dry *N,N*-dimethylformamide (Note 2). The flask is immersed in a water bath kept at 35°, stirring is begun, and the apparatus is purged thoroughly with dry nitrogen (Note 3). After 15 minutes 107.1 g. (1.001 mole) of 3-pyridinecarboxaldehyde (Note 4) is added dropwise over a period of 30 minutes. The dark brown solution is stirred for another 30 minutes (Note 5), after which 39.8 g. (0.751 mole) of freshly distilled acrylonitrile is added over 1 hour. The solution, now red-orange in color and quite viscous (Note 6), is stirred for 3 hours, 6.6 g. (0.11 mole) of acetic acid is added, and stirring is continued for 5 additional minutes. The solvent is removed with a rotary evaporator, the residue is dissolved in 500 ml. of water, and the solution is extracted continuously (Note 7) with 500 ml. of chloroform for 12 hours (Note 8). The solvent is evaporated under reduced pressure, and the residual liquid is distilled under reduced pressure through a short-path distillation apparatus. An initial fraction, consisting mainly of 3-pyridinecarboxaldehyde, collects in the cold trap. The product, b.p. 150–152° (0.1 mm.), solidifies in the condenser and is freed by heating the water in the cooling jacket to nearly 100°, yielding 94–101 g. (78–84%) of the light yellow, solid distillate. Recrystallization from 400 ml. of 2-propanol gives 77–82 g. (64–68% based on acrylonitrile) of 4-oxo-4-(3-pyridyl)butyronitrile as yellow-tinged, white crystals, m.p. 70–72° (Note 9).

2. Notes

1. Analytical grade *(pro analysi)* sodium cyanide, purchased by the submitters from Merck, Darmstadt, Germany, was dried for 24 hours in a vacuum desiccator containing potassium hydroxide pellets. The checkers obtained sodium cyanide from Fisher Scientific Company and dried the reagent in the same manner.

2. The submitters purified technical grade N,N-dimethylformamide by distillation from powdered calcium hydride. The checkers used N,N-dimethylformamide that had been dried over Linde type 4A molecular sieves. A small amount of dimethylamine in the solvent does not interfere with the reaction.

3. The drying tube was connected to a Nujol bubbler. A nitrogen atmosphere was maintained during the reaction by passing nitrogen through the apparatus at a rate of $ca.$ one bubble per second.

4. 3-Pyridinecarboxyaldehyde (nicotinaldehyde) was supplied by Aldrich-Europe, Beerse, Belgium. The checkers purified this reagent by fractional distillation, b.p. 95–97° (15 mm.). The submitters stress that 3-pyridinecarboxaldehyde should be completely free from contamination by the acid. They stirred 150 g. of the aldehyde with 100 g. of potassium carbonate and 300 ml. of ethanol for 12 hours, filtered the suspended solid, and fractionally distilled the filtrate through a 30-cm. Vigreux column, using a water aspirator. However, the checkers found that the recovery of aldehyde from this procedure was very low, and recommend vacuum distillation instead. 3-Pyridinecarboxaldehyde is a powerful skin irritant and should be handled with protective gloves.

5. The solution, in which some sodium cyanide remains suspended, becomes quite thick at this stage owing to formation of the benzoin-type dimer of 3-pyridinecarboxaldehyde. An adequate amount of N,N-dimethylformamide should be used as solvent to ensure that the dimer does not crystallize.

6. Although the solution becomes very viscous at this point, stirring is still possible and should be continued.

7. Continuous extraction is only necessary if the product is appreciably soluble in water. Products such as those shown in Table I may be isolated by extraction in a separatory funnel.

TABLE I

γ-KETONITRILES PREPARED BY CYANIDE-CATALYZED CONJUGATE ADDITION OF ARYL ALDEHYDES TO α,β-UNSATURATED NITRILES

$\underset{\underset{R_2 \quad R_3}{\mid \quad \mid}}{R_1-\overset{\overset{\displaystyle O}{\parallel}}{C}-CH-CH-CN}$	Distilled Yield (%)[a]	Recrystallized Yield (%)[b]	B.p. (°) (pressure, mm.)	M.p. (°)
$R_1 = C_6H_5$, $R_2 = R_3 = H$	71	50[c]	131–134 (0.2)	72–73
$R_1 = C_6H_5$, $R_2 = H$, $R_3 = CH_3$	73–76	34–37[d]	113–115 (0.1)	42–43
$R_1 = C_6H_5$, $R_2 = CH_3$, $R_3 = H$	62–64	45–47[d]	111–113 (0.15)	58–59
$R_1 = R_2 = C_6H_5$, $R_3 = H$	83	56[e]	157–159 (0.05)	83–84
$R_1 = 4\text{-}ClC_6H_4$, $R_2 = R_3 = H$	88–89	71–72[c]	152–154 (0.1)	70–71

[a] The distilled products were almost pure.
[b] The recrystallized products were pure, but considerable losses were entailed.
[c] Recrystallized from aqueous ethanol (decolorized with activated carbon).
[d] Recrystallized from ethyl acetate–petroleum ether at −30°.
[e] Recrystallized from 2-propanol.

8. The submitters state that the solution need not be dried, since water is removed by azeotropic distillation as the chloroform is evaporated. However, the checkers dried the chloroform solution with anhydrous magnesium sulfate prior to evaporation.

9. The checkers dried the product in a vacuum desiccator for 24 hours to remove all the 2-propanol and obtained 77–79 g. (64–66%), m.p. 70–72°. The yield reported by the submitters was 82–89 g. (68–74%), m.p. 73–74° (lit.,[2] m.p. 66–67°). The product's ^{1}H NMR spectrum (90MHz., CDCl$_3$), δ (multiplicity, coupling constant J in Hz., number of protons, assignment): 2.84 (t, J = 6.6, 2H, CH$_2$CH$_2$CN), 3.40 (t, J = 6.6, 2H, CH$_2$CH$_2$CN), 7.49 (d of d, J = 4.4 and 7.3, 1H, H_5), 8.24 (d of t, J = 2.0 and 7.3, 1H, H_4), 8.82 (d of d, J = 2.0 and 4.5, 1H, H_6), 9.16 (d, J = 2.0, 1H, H_2); mass spectrum m/e (relative intensity): 160 (M$^+$, 8), 106 (74), 78 (100), 51 (100).

3. Discussion

4-Oxo-4-(3-pyridyl)butyronitrile has been prepared in three steps by Leete, Chedekel, and Bodem,[2] and in one step by Stetter and Schreckenberg[3] using a method closely related to the present procedure. This compound serves as precursor in syntheses of myosmine[2] and various nicotine analogs. Other general methods for the preparation of γ-ketonitriles include the addition of hydrogen cyanide to α,β-unsaturated ketones,[4] the reaction of potassium cyanide with the hydrochlorides of Mannich bases from ketones,[5] and a variety of new methods for nucleophilic acylation.[6]

The addition of 3-pyridinecarboxyaldehyde to acrylonitrile is only one example of a wide range of reactions involving the conjugate addition of aldehydes to electron-deficient olefins. The reaction is not limited to α,β-unsaturated nitriles;[3,7] For example, γ-diketones[7,8] and γ-keto esters[7,9] may be similarly prepared by addition of aldehydes to α,β-unsaturated ketones and esters. Important advantages of this method are the simplicity of the procedure, the catalytic nature of the reaction, and low-cost reagents.

The γ-ketonitriles shown in Table I were prepared by the cyanide-catalyzed procedure described here. While this procedure is generally applicable to the synthesis of γ-diketones, γ-keto esters, and other γ-ketonitriles, the addition of 2-furancarboxaldehyde is more difficult, and a somewhat modified procedure should be employed.[10] Although the cyanide-catalyzed reaction is generally limited to aromatic and heterocyclic aldehydes, the addition of aliphatic aldehydes to various Michael acceptors may be accomplished in the presence of thioazolium ions,[7,11] which are also effective catalysts for the additions.[7,12]

The mechanism of the cyanide- and thioazolium ion-catalyzed conjugate addition reactions[7] is considered to be analogous to the Lapworth mechanism for the cyanide-catalyzed benzoin condensation. Thus, the cyano-stabilized carbanion, resulting from deprotonation of the cyanohydrin of the aldehyde, is presumed to be the actual Michael donor. After conjugate addition to the activated olefin, cyanide is eliminated, forming the product and regenerating the catalyst.

1. Institut für Organische Chemie der Rheinisch-Westfälischen Technischen Hochschule Aachen, West Germany.
2. E. Leete, M. R. Chedekel, and G. B. Bodem, *J. Org. Chem.*, **37**, 4465 (1972).
3. H. Stetter and M. Schreckenberg, *Chem. Ber.*, **107**, 210 (1974); H. Stetter and M. Schreckenberg, Ger. Pat. 2,262,343 (1972) [*Chem. Abstr.*, **81**, 105015j (1974)].

4. C. F. H. Allen, M. R. Gilbert, and D. M. Young, *J. Org. Chem.*, **2**, 227 (1937).
5. E. B. Knott, *J. Chem. Soc.*, 1190 (1947).
6. For a review, see O. W. Lever, Jr., *Tetrahedron*, **32**, 1943 (1976).
7. H. Stetter, *Angew. Chem.*, **88**, 695 (1976) [*Angew. Chem. Int. Ed. Engl.*, **15**, 639 (1976)].
8. H. Stetter and M. Schreckenberg, *Chem. Ber.*, **107**, 2453 (1974).
9. H. Stetter, M. Schreckenberg, and K. Wiemann, *Chem. Ber.*, **109**, 541 (1976).
10. H. Stetter and H. Kuhlmann, *Tetrahedron*, **33**, 353 (1977).
11. H. Stetter and H. Kuhlmann, *Chem. Ber.*, **109**, 2890 (1976).
12. H. Stetter and H. Kuhlmann, *Angew. Chem.*, **86**, 589 (1974) [*Angew. Chem. Int. Ed. Engl.*, **13**, 539 (1974)]; H. Stetter and H. Kuhlmann, Ger. Pat. 2,437,219 (1974) [*Chem. Abstr.*, **84**, 164172t (1976)].

ALDEHYDES FROM *sym*-TRITHIANE: *n*-PENTADECANAL

(Pentadecanal)

Submitted by D. Seebach[1] and A. K. Beck
Checked by A. Brossi, N. W. Gilman, and G. Walsh

1. Procedure

A. *2-Tetradecyl*-sym-*trithiane*. A 1-l., round-bottomed, side-armed flask containing a magnetic stirring bar is charged with 25.0 g. (0.180 mole) of finely ground, pure, *sym*-trithiane (Note 1). The flask is equipped with a three-way stopcock and a rubber septum on the side arm. The air in the flask is replaced with dry nitrogen (Note 2). Tetrahydrofuran (Note 3) (350 ml.) is added by syringe, and the resulting slurry is stirred vigorously in a cooling bath at −30° (Note 4). After the addition of 0.190 mole of *n*-butyllithium (1.5–2.5 molar in *n*-hexane) (Note 5), the mixture is stirred for 1.5–2.5 hours, keeping the bath temperature between −25° and −15°. After this period of time the trithiane is dissolved (Note 6), and dry ice is added (no excess!) to the bath until the temperature is about −70°. To this cooled solution is rapidly added 50.0 g. (49.5 ml., 0.180 mole) of 1-bromotetradecane (myristyl bromide) (Note 7) by syringe, and the resulting mixture is stirred overnight, during which time the bath temperature rises to 0–25° and a heavy, colorless precipitate separates. Stirring is continued for 1 hour at room temperature before the mixture is poured into a 2-l. separatory funnel containing 800 ml. of water and 500 ml. of carbon tetrachloride. After shaking, the layers are separated and the aqueous layer is shaken with two additional 500-ml. portions of carbon tetrachloride. Some undissolved trithiane is filtered from the combined organic layers, which are washed with water and dried over anhydrous potassium carbonate. The solvent is removed by evaporation, yielding 54–59 g. of crude, solid 2-tetradecyl-*sym*-trithiane, after drying under reduced pressure (Note 8).

B. *Pentadecanal dimethyl acetal.* The crude material obtained from Part A is placed in a 2-l., three-necked flask fitted with an overhead stirrer, a reflux condenser with drying tube, and a stopper. Methanol (1 l., reagent grade) is added, the stirrer is started, and 40 g. (0.18 mole) of mercury(II) oxide and 100 g. (0.368 mole) of mercury(II) chloride are introduced. The mixture is heated under reflux for 4.5 hours and filtered through a Büchner funnel after cooling. The residue is washed with 300 ml. of pentane (Note 9), and the combined organic solutions are poured into 1 l. of water. The layers are separated, and the lower aqueous layer is shaken with two 500-ml. portions of pentane. The combined organic layers are quickly washed with 10% ammonium acetate solution (Note 10) and water and dried over sodium sulfate. The pentane is evaporated under reduced pressure, giving 30.0–32.5 g. of the crude acetal as a mobile, slightly yellow oil.

C. n-*Pentadecanal.* The crude acetal from Part B is dissolved in 600 ml. of tetrahydrofuran, and 150 ml. of water containing 2 g. of *p*-toluenesulfonic acid monohydrate is added. The resulting pale mixture is heated at reflux for 1 hour and cooled. The hydrolysate is poured into 600 ml. of water and extracted with three 300-ml. portions of pentane (Note 9). The colorless pentane extracts are combined, washed three times with saturated sodium hydrogen carbonate solution and once with water, and dried over sodium sulfate. Evaporation of the solvent furnishes an oil which upon distillation under reduced pressure (Note 11) yields 18.7–22.5 g. of *n*-pentadecanal, b.p. 103–106° (0.2 mm.). The overall yield from 1-bromotetradecane is 47–55%. The product solidifies eventually and should be kept under an inert atmosphere in the refrigerator.

2. Notes

1. It is essential that the *sym*-trithiane be of good purity. Commercial *sym*-trithiane can be purified by extraction from a thimble in a hot extractor using 300 ml. of toluene for 30 g. of trithiane. After cooling the extract to 0°, *sym*-trithiane is recovered by filtration and recrystallized from toluene. In one run the checkers used *sym*-trithiane as obtained from Eastman Organic Chemicals and observed a 10% decrease in yield of *n*-pentadecanal.

2. This is done by evaporating and filling with dry nitrogen three times; during the reaction a pressure of about 50 mm. is maintained against the atmosphere using a mercury bubbler.

3. Tetrahydrofuran is distilled from a blue solution of benzophenone ketyl, obtained by refluxing tetrahydrofuran in the presence of sodium wire, some potassium, and

TABLE I

ALDEHYDES FROM 2-LITHIO-1,3,5-TRITHIANE AND ALKYL HALIDES

Halide	2-Alkyl-1,3,5-Trithiane Yield, %[a,b,c]	Aldehyde Dimethyl Acetal Yield, %[a,b]	Aldehyde Product	Aldehyde Yield, %[a,e,f]
1-Bromopentane	96	66[d]	Hexanal	43
(S)-(+)-1-Iodo-2-methylbutane	98	67[d,g]	(S)-(−)-3-Methylpentanal	41
1-Bromoheptane	100	60[d]	Octanal	45
1-Bromodecane	100	99[c]	Undecanal	65
1-Bromohexadecane	100	60[c]	Heptadecanal	32
Benzyl Bromide	100	32[d]	Phenylacetaldehyde	20

[a] Based on halide.
[b] Reaction conducted on 100 mmoles scale.
[c] Crude product.
[d] Distilled product.
[e] Reaction conducted on 5–10 mmoles scale.
[f] Yield of distilled or recrystallized product.
[g] (+)-Iodide with optical purity of 89% gave acetal with $\alpha_D + 7.6°$ (neat, $l = 100$ mm.).

benzophenone. See *Org. Synth.*, **Coll. Vol. 5,** 976 (1973) for a warning note regarding the purification of tetrahydrofuran.

4. A 2-l. Dewar cylinder was used.

5. The checkers used 120 ml. of 1.6 *M* *n*-butyllithium in hexane, obtained from Foote Mineral Company.

6. If the trithiane, apart from a few crystals, does not dissolve entirely, the workup procedure is complicated. The crude tetradecyltrithiane must then be purified (by dissolving in 500 ml. of carbon tetrachloride at 30°, filtering, and precipitating with 1.5 l. of methanol) before conversion to the acetal.

7. A commercial product (Aldrich Chemical Company, Inc., or Matheson, Coleman and Bell) proved satisfactory without further purification. The purity should be checked by refractive index and/or GC.

8. The crude tetradecyltrithiane contains 4–6% of *sym*-trithiane, m.p. 69–71°; recrystallization (Note 6) gives pure product, m.p. 76.3–76.6°.

9. Low-boiling ligroin can be used as well.

10. A white precipitate is formed during the first and second washing.

11. A short-path distillation apparatus with a cold finger, but no condenser, should be used since the product may crystallize. The distillation is carried out under nitrogen or argon (balloon at capillary).

3. Discussion

The procedure described here provides a convenient route to aldehydes, with trithiane serving as an inexpensive, "masked" carbonyl group.[2-4] The reaction is limited, however, to the use of primary alkyl halides, aldehydes, and ketones for elaboration of the carbon chain through attack on the metallated trithiane. Examples of aldehydes synthesized by this method are given in Table I.

The S-acetal is converted to the O-acetal in anhydrous methanol because hydrolysis of monosubstituted trithianes in aqueous methanol furnishes a mixture of the free aldehyde and its O-acetal derivative. It is advantageous to store aldehydes as the O-acetal derivatives since free aldehydes are susceptible to polymerization and oxidation.

This method has several common features with the dithiane method[5] that is useful for the synthesis of aldehydes and ketones.[4] This latter method is illustrated by the synthesis of cyclobutanone [*Org. Synth.*, **Coll. Vol. 6,** 316 (1988)].

n-Pentadecanal has also been prepared by pyrolysis of α-hydroxy-[6] and α-methoxypalmitic acid,[7] from α-bromopalmitic acid chloride and sodium azide,[8] and from α-hydroxypalmitic acid and lead tetraäcetate.[9]

1. Institut für Organische Chemie der Universität (T.H.) 7500 Karlsruhe, West Germany [Present address: Laboratorium Für Organische Chemie, Eidgenössische Technische Hochschule, Universitatstrasse 16, CH-8006 Zurich, Switzerland].
2. D. Seebach and E. J. Corey, unpublished work, 1965.
3. D. Seebach and D. Steinmüller, *Angew. Chem.*, **80,** 617 (1968) [*Angew. Chem. Int. Ed. Engl.*, **7,** 619 (1968)].
4. D. Seebach, *Synthesis*, 17 (1969); cf. D. Seebach, *Angew. Chem.*, **81,** 690 (1969) [*Angew. Chem. Int. Ed. Engl.*, **8,** 639 (1969)].
5. E. J. Corey and D. Seebach, *Angew. Chem.*, **77,** 1134, 1135 (1965) [*Angew. Chem. Int. Ed. Engl.*, **4,** 1075, 1077 (1965)].
6. See literature cited in *Beilstein*, **1,** p. 716, 2nd Suppl., **1,** p. 770.
7. M. Prostenik, N. Z. Stanacev, and M. Munk-Weinert, *Croat. Chem. Acta*, **34,** 1 (1962) [*Chem. Abstr.*, **57,** 7910c (1962)].
8. M. S. Newman, *J. Am. Chem. Soc.*, **57,** 732 (1935).
9. W. M. Lauer, W. J. Gensler, and E. Miller, *J. Am. Chem. Soc.*, **63,** 1153 (1941).

(PENTAFLUOROPHENYL)ACETONITRILE

[Benzeneacetonitrile, 2,3,4,5,6-pentafluoro-]

$$C_6F_6 + NCCH_2CO_2C_2H_5 \xrightarrow[\text{N,N-dimethylformamide, } 110-120°]{K_2CO_3}$$

$$C_6F_5CH(CN)CO_2C_2H_5 + KF + H_2O + CO_2$$

$$C_6F_5CH(CN)CO_2C_2H_5 \xrightarrow[\text{water, reflux}]{H_2SO_4, CH_3COOH} C_6F_5CH_2CN + C_2H_5OH + CO_2$$

Submitted by ROBERT FILLER[1] and SARAH M. WOODS[2]
Checked by ANDREW E. FEIRING and WILLIAM A. SHEPPARD

1. Procedure

A. *Ethyl cyano(pentafluorophenyl)acetate.* A 2-l., four-necked flask equipped with mechanical stirrer, addition funnel, thermometer, and condenser is charged with 650 ml. of *N*,*N*-dimethylformamide (Note 1) and 140 g. (1.01 mole) of anhydrous potassium carbonate. The rapidly stirred mixture is heated to 152–154° and 113 g. (1.00 mole) of ethyl cyanoacetate is added dropwise over 10–15 minutes without further heating. The temperature of the mixture is allowed to drop to 110–120° and maintained within this range while 186 g. (1.00 mole) of hexafluorobenzene (Note 2) is added dropwise over 1 hour. The dark mixture is stirred for 3 hours after the addition is complete, poured into 3 l. of ice water contained in a 5-l. Erlenmeyer flask, and acidified *(Caution! Foaming)* with 20% sulfuric acid. After being cooled overnight in the refrigerator, the top, aqueous layer is decanted from a lower viscous organic layer. The organic layer is dissolved in 600 ml. of diethyl ether, washed with water, and aqueous 10% sodium hydrogen carbonate, and dried over anhydrous magnesium sulfate. The ether is removed on a rotary evaporator, affording 217 g. (78%) of dark oil which crystallizes on standing (Note 3). An analytical sample is prepared by dissolving 2 g. of the crude material in 5 ml. of boiling 95% ethanol. Hexane is added until mixture becomes turbid. Crystallization occurs when the mixture is cooled with vigorous stirring in an acetone–dry ice bath. The solid is quickly collected on a Büchner funnel and transferred to a sublimator. Sublimation at 30° (0.5–1.0 mm.) affords white crystals, m.p. 38–38.5°, of analytically pure ethyl cyano(pentafluorophenyl)acetate (Note 4).

B. *(Pentafluorophenyl)acetonitrile.* A 1-l., one-necked flask equipped with magnetic stirrer and a reflux condenser is charged with 139.5 g. (0.500 mole) of crude ethyl cyano(pentafluorophenyl)acetate, 350 ml. of aqueous 50% acetic acid, and 12.5 ml. of concentrated sulfuric acid. The mixture is heated at reflux for 15 hours. After cooling to room temperature, the mixture is diluted with an equal volume of water and cooled in an ice bath for 1 hour. The top layer is decanted from a dark organic layer which settles to the bottom of the flask. The organic phase is dissolved in 200 ml. of ether and washed with water and aqueous 10% sodium hydrogen carbonate. After being dried over anhydrous magnesium sulfate, the ether is removed on a rotary evaporator. The residue is distilled through a 25-cm. jacketed Vigreux column, affording 74–78 g. (71–75%) of (penta-fluorophenyl)acetonitrile as a colorless liquid, b.p. 105° (8 mm.), n_D^{25} 1.4370 (Note 5).

2. Notes

1. Technical grade *N,N*-dimethylformamide was stirred over anhydrous cupric sulfate, filtered, and distilled under reduced pressure. The submitters used reagent grade *N,N*-dimethylformamide without purification.

2. Hexafluorobenzene was purchased from PCR, Inc., Gainesville, Florida, and distilled (b.p. 80–81°) before use.

3. In one run the checkers obtained only 135 g. of crude product by this procedure. The aqueous solution which was decanted from the crude product was divided into three portions and each portion was extracted with one 250-ml. portion of ether. The combined ether extracts were washed with water and aqueous 10% sodium hydrogen carbonate, dried over anhydrous magnesium sulfate, and concentrated on the rotary evaporator, affording an additional 83 g. of crude product, for a total of 218 g.

4. ^{1}H NMR (CCl$_4$), δ (multiplicity, number of protons): 1.38 (t, 3H) 4.35 (q, 2H), 5.05 (s, 1H); IR (CHCl$_3$) cm.$^{-1}$: 3003, 2933, 2257, 1760, 1661, 1527, 1513; ^{19}F NMR (CCl$_4$, CFCl$_3$ internal standard): δ 141.2 (sym. m, 2 *ortho* F), 151.8 (t of t, $J_{1,2}$ = 20.3 Hz., $J_{1,3}$ = 2.5 Hz., *para* F), 161.1 (m, 2 *meta* F). The pKa in dimethyl sulfoxide is 5.06 ± 0.02.[3]

5. ^{1}H NMR (CCl$_4$), δ: 3.75 (s, with fine structure); IR (neat) cm.$^{-1}$: 2985, 2273, 1667, 1527, 1515; ^{19}F NMR (CCl$_4$, CFCl$_3$ internal standard), δ 142.4 (sym. m, 2 *ortho* F), 153.8 (t, with fine structure, J = 20 Hz., *para* F), 161.7 (m, 2 *meta* F). The pKa in dimethyl sulfoxide is 15.8 ± 0.3.[3]

3. Discussion

The formation of ethyl cyano(pentafluorophenyl)acetate illustrates the *intermolecular* nucleophilic displacement of fluoride ion from an aromatic ring by a stabilized carbanion. The reaction proceeds readily as a result of the activation imparted by the electron-withdrawing fluorine atoms.[4] The selective hydrolysis of a cyano ester to a nitrile has been described.[5] (Pentafluorophenyl)acetonitrile[6] has also been prepared by cyanide displacement on (pentafluorophenyl)methyl halides. However, this direct displacement is always accompanied by an undesirable side reaction, yielding 15–20% of 2,3-bis(pentafluorophenyl)propionitrile. The reaction of one equivalent of hexafluorobenzene with one equivalent of lithioacetonitrile (prepared from acetonitrile and *n*-butyllithium) provides a low yield (7–10%) of (pentafluorophenyl)acetonitrile and about a 22% yield of bis(pentafluorophenyl)acetonitrile (m.p. 65°; the pKa in dimethyl sulfoxide is 7.95 ± 0.04[3]). The yield of the latter compound can be increased by use of excess lithioacetonitrile.[7]

(Pentafluorophenyl)acetonitrile is a useful intermediate to 4,5,6,7-tetrafluoroindole.[8] The nitrile is readily converted to 2-(pentafluorophenyl)ethylamine hydrochloride in 80% yield by catalytic hydrogenation in dilute hydrochloric acid. Although the salt is stable, the amine undergoes a facile intermolecular nucleophilic aromatic substitution reaction, even at room temperature. However, freshly distilled 2-(pentafluorophenyl)ethylamine is converted by heating in the presence of anhydrous potassium fluoride in N,N-dimethylformamide to 4,5,6,7-tetrafluoroindoline (62% yield) by intramolecular nucleophilic displacement of fluoride ion.[9] The indoline is aromatized by treatment with activated manganese dioxide,[10] giving 4,5,6,7-tetrafluoroindole (82% yield). The anion of (pentafluorophenyl)acetonitrile is converted to bis(pentafluorophenyl)acetonitrile on treatment with hexafluorobenzene *(vide supra)*.[11]

1. Department of Chemistry, Illinois Institute of Technology, Chicago, Illinois 60616.
2. Department of Chemistry, Roosevelt University, Chicago, Illinois 60605.
3. F. G. Bordwell and R. Filler, to be published.
4. W. P. Norris, *J. Org. Chem.*, **37**, 147 (1972), reported an analogous reaction with octafluorotoluene.
5. A. Kalir and Z. Pelah, *Israel J. Chem.*, **4**, 155 (1966) [*Chem. Abstr.*, **66**, 55328 (1967)].
6. A. K. Barbour, M. W. Buxton, P. L. Coe, R. Stephens, and J. C. Tatlow, *J. Chem. Soc.*, 808 (1961).
7. G. L. Cantrell and R. Filler, unpublished observations.
8. R. Filler, S. M. Woods, and A. F. Freudenthal, *J. Org. Chem.*, **38**, 811 (1973).
9. V. P. Petrov, V. A. Barkhash, G. S. Shchegoleva, T. D. Petrova, T. I. Savchenko, and G. G. Yakobson, *Dokl. Chem. (Engl. Transl.)*, **178**, 113 (1968).
10. A. B. A. Jansen, J. M. Johnson, and J. R. Surtees, *J. Chem. Soc.*, Suppl. 1, 5573 (1964).
11. R. Filler, A. E. Fiebig, and M. Y. Pelister, *J. Org. Chem.*, **45**, 1290 (1980).

PENTAFLUOROPHENYLCOPPER TETRAMER, A REAGENT FOR SYNTHESIS OF FLUORINATED AROMATIC COMPOUNDS

[Copper, tetrakis(pentafluorophenyl)tetra]

$$C_6F_5Br \xrightarrow[\text{ether}]{\text{Mg}} C_6F_5MgBr \xrightarrow[\text{2. dioxane}]{\text{1. CuBr}} (C_6F_5Cu)_2 \cdot \text{dioxane}$$

$$(C_6F_5Cu)_2 \cdot \text{dioxane} \xrightarrow[10^{-6}\text{ mm.}]{100-128°} (C_6F_5Cu)_4$$

Submitted by ALLAN CAIRNCROSS,[1] WILLIAM A. SHEPPARD,[2] and EDWARD WONCHOBA
Checked by WILLIAM J. GUILFORD, CYNTHIA B. HOUSE, and ROBERT M. COATES

1. Procedure

Caution! Benzene has been identified as a carcinogen; OSHA has issued emergency standards on its use. All procedures involving benzene should be carried out in a well-ventilated hood, and glove protection is required.

A 1-l., four-necked, round-bottomed flask fitted with a condenser bearing a nitrogen inlet, a pressure-equalizing dropping funnel, a thermometer, and a mechanical stirrer is charged with 5.40 g. (0.222 g.-atom) of magnesium turnings (Note 1). The flask is flame dried while being flushed with nitrogen and kept dry and oxygen-free with a static nitrogen atmosphere throughout the preparation (Note 2). After 150 ml. of diethyl ether (Note 3) is introduced into the flask with a syringe, 54.9 g. (28.2 ml., 0.222 mole) of bromopentafluorobenzene (Note 4) is added dropwise over *ca.* 45 minutes at a rate that maintains a gentle reflux. The reflux is maintained for another 15 minutes by heating at 35° (Note 1). The resulting black solution is cooled to room temperature, and 63.1 g. (0.440 mole) of powdered, anhydrous copper(I) bromide (Note 5) is added in three 21-g. portions at 1-minute intervals. An exothermic reaction occurs after each addition (Note 6).

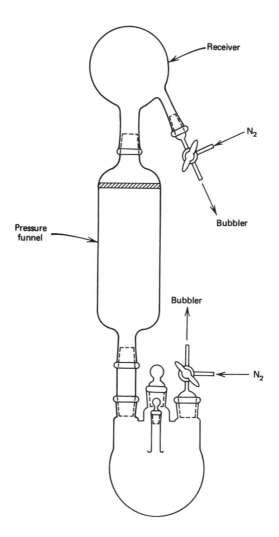

Figure 1.

The brown mixture is stirred for 30 minutes, 100 ml. of ether is added, and the mixture heated at reflux for another 30 minutes. The brown suspension is diluted with 400 ml. of ether, after which 100 ml. of 1,4-dioxane (Note 7) is added carefully over 15 minutes, moderating the mildly exothermic reaction. The light-gray suspension is stirred for 30 minutes.

The dropping funnel is replaced with a three-way stopcock with an attached nitrogen source and a bubbler open to the system (see Figure 1). The thermometer and mechanical stirrer are replaced with ground-glass plugs (Note 8), and the condenser is replaced with a 1-l., 90-mm., medium-porosity fritted-disk nitrogen pressure funnel attached to a 1-l., round-bottomed, two-necked flask fitted with a three-way stopcock on the side arm (Figure 1). All joints are either clamped or taped together, and the apparatus is carefully inverted, pouring the slurry into the funnel. The mixture is filtered with nitrogen pressure into the 1-l. receiving flask (Note 9), which is maintained under a nitrogen atmosphere. The solid filter cake is rinsed with three 50-ml. portions of 4:1 (v/v) ether–dioxane injected with a long needle syringe through the three-way stopcock. The four-necked flask is rinsed by manipulation of the needle, and the apparatus and the solids are dispersed in each rinse. The pale-yellow filtrate is evaporated to dryness under reduced pressure with a warm water bath at 40°, and the powdery white solid is dried at 10-mm. pressure for 4 hours at 25°, yielding 38.3–48.8 g. (63–80%) of bis(pentafluorophenylcopper)dioxane complex (Note 10). The color of this dioxane complex varies from tan to white. The complex is transferred to a 200-ml., round-bottomed flask under a nitrogen atmosphere. The flask is evacuated to a pressure of 0.001 mm., immersed in an oil bath, and slowly heated to 100° over 1 hour (Note 11). The temperature is slowly and constantly increased to 128° during a second hour and maintained at 128° for 4 hours (Note 12). Pure pentafluorophenylcopper tetramer is obtained as a gray to tan powder weighing 29.8–34.7 g. (58–68%) (Note 13).

2. Notes

1. Magnesium metal turnings, from Fisher Scientific Company or Mallinckrodt Chemical Works, were used. Excess magnesium or incomplete formation of the Grignard must be avoided, since any free magnesium reacts when the copper salts are added, producing a dark product in lower yield and of questionable purity.

2. All glassware must be flame dried, and an absolute nitrogen atmosphere must be maintained during each step, since pentafluorophenylcopper hydrolyzes easily and undergoes oxidative coupling in air. The checkers used dry grade nitrogen from a cylinder, supplied by the Linde Division, Union Carbide Corporation.

3. Anhydrous ether, purchased from Fisher Scientific Company or Mallinckrodt Chemical Works, was dried over Linde type 3A molecular sieves. The checkers degassed the solvent immediately before use by evacuating and filling the container with nitrogen three times.

4. Bromopentafluorobenzene, obtained from either PCR, Inc., or Columbia Organic Chemicals Company, Inc., was used without purification.

5. Anhydrous copper(I) bromide, from Fisher Scientific Company, was powdered and used without drying by the submitters. The checkers dried the copper(I) bromide at 140° for 2 hours under reduced pressure before use.

6. The copper(I) bromide can also be added gradually from a solid addition apparatus such as a 50-ml. Erlenmeyer flask connected to a ground-glass adapter with a short piece of gooch tubing.

7. Spectral grade 1,4-dioxane, from MC and B Manufacturing Chemists or Mallinckrodt Chemical Works, was dried over Linde type 4A molecular sieves. The checkers degassed the solvent prior to use as described in Note 3.

8. The checkers used rubber septa, secured with wire bands.

9. The filtering operation can also be conveniently done in a high-quality nitrogen atmosphere dry box by vacuum filtration through a 90-mm., 60-ml., medium-porosity, fritted-disk Büchner funnel.

10. Pentafluorophenylcopper is first isolated as a 1:1 complex with dioxane and is usually white. Half of the complexed dioxane is very labile and is usually lost during vacuum drying, giving the 2:1 complex. Excessive heating can cause loss of additional dioxane (lower apparent yield) and eventual decomposition of the product.

If the product is isolated in a nitrogen dry box, the filtrate is evaporated without heat until a small amount of solvent remains. The precipitate is collected cold and rinsed with cold ether. The nearly white 1:1 pentafluorophenylcopper·dioxane complex is obtained with no significant loss of yield.

11. The rate of heating is critical. The temperature must be increased very gradually to remove the dioxane without causing decomposition of the product. A vacuum of 0.1 to 1.0 mm is sufficient to remove the last of the dioxane although 0.001 mm was usually used by the submitters. Occasionally, cold traps become plugged with frozen dioxane (m.p. 12°), resulting in loss of vacuum, overheating, and decomposition of the pentafluorophenyl dioxane complexes.

12. If the product is heated to 130° or higher, decomposition of product occurs with the formation of a copper mirror.

13. The spectral properties of pentafluorophenylcopper tetramer are as follows: IR (Nujol) cm.$^{-1}$: 1630 medium; 1391 medium; 1353 medium; 1275 medium; 1090, 1081, and 1071 strong triplet; 978 strong; 785 medium; ^{19}F NMR (tetrahydrofuran with CFCl$_3$ as internal reference): δ (multiplicity, coupling constant J in Hz., number of fluorines, assignment): 107.2 (20-line m, 2F, *ortho* F), 153.4 (t of t, $J = \sim$1.3 and 20, 1F, *para* F), 162.3 (17-line m, 2F, *meta* F). Absorptions at 820–900, 1100–1125, and 1290 cm.$^{-1}$ in the IR spectrum and at δ 3.05 in the ^{1}H NMR spectrum indicate that dioxane is still present.

The pentafluorophenylcopper tetramer is usually analytically pure as isolated and melts at 200° with decomposition. If any significant decomposition occurs during the final drying, the product can be purified by dissolution in ether, filtration to remove copper metal, and precipitation by addition of hexane. It can also be recrystallized from benzene. When kept in a sealed container under nitrogen at room temperature, pentafluorophenyl copper tetramer appears to be stable for reasonable periods. It can be stored indefinitely at −78° under an atmosphere of carbon dioxide.

3. Discussion

Pentafluorophenylcopper is representative of a series of fluorinated organocopper compounds that are highly soluble in organic solvents, more thermally stable than their hydrocarbon analogs, and useful as synthetic intermediates.[3-5] Pentafluorophenylcopper has been used to introduce the pentafluorophenyl group[6-7] and as a reagent for an improved, Ullman diphenyl ether synthesis.[8] It is also an effective catalyst for decarboxylation of aromatic acids,[9,10] rearrangement of bicyclic hydrocarbons,[11] and decomposition of alkyldiazo compounds.[3,4] It also is an excellent reagent for the preparation of anhydrous copper salts of carboxylic acids[9] and can be used for coating substrates with copper by thermal decomposition.[12]

Pentafluorophenylcopper exists as a tetramer.[13] It forms complexes with a variety of reagents and solvents as well as "ate" complexes; a representative list is given in Table I. For many syntheses the crude reaction mixtures of copper(I) halide with either pentafluorophenylmagnesium bromide, or pentafluorophenyllithium,[7] or the pentafluorophenylcopper·dioxane complex[6] react as well as the solvent-free tetramer.

TABLE I
COMPLEXES OF PENTAFLUOROPHENYLCOPPER[3,4]

Complex	Properties
C_6F_5Cu·benzonitrile	m.p. 101°
C_6F_5Cu·tributylamine	liquid at room temperature
$(C_6F_5Cu)_2$·dioxane	m.p. 200–220° (dec.)
C_6F_5Cu·quinoline	m.p. 170–176°
$(C_6F_5Cu)_2$·1,5-cycloöctadiene	dec. 160°
C_6F_5Cu·dimethylacetylene	dec. 145°
$(C_6F_5Cu)_2$·butadiene	dec. 215°
C_6F_5Cu·tetraethylammonium cyanide	dec. 160°

A selection of synthetic uses of pentafluorophenylcopper is given in Table II. Two unchecked experimental procedures illustrating the use of pentafluorophenylcopper tetramer and the dioxane complex to introduce the pentafluorophenyl group are given below. In coupling reactions hexane is usually the preferred solvent, particularily with alkyl halides that can readily form carbonium ion intermediates. Aromatic solvents are often alkylated during coupling, giving undesired by-products.

A. *(Pentafluorophenyl)benzene.* A 100-ml., round-bottomed flask equipped with a magnetic stirring bar and a reflux condenser bearing a nitrogen inlet is flushed with nitrogen and charged with a solution of 2.40 g. (0.00259 mole) of pentafluorophenylcopper tetramer in 25 ml. of benzene; 2.12 g. (0.0104 mole) of iodobenzene is then added. A static nitrogen atmosphere is maintained in the flask as the solution is heated to reflux. Copper(I) iodide starts to precipitate almost immediately. After 2 hours at reflux, the mixture is cooled and filtered, separating 1.67 g. of copper(I) iodide. The filtrate is evaporated, and the remaining pale-brown residue is sublimed at 100° (0.1 mm.), affording 2.01 g. (79%) of (pentafluorophenyl)benzene as a colorless solid, m.p. 110.0–112.4°. The presence of about 3% of decafluorobiphenyl in the product is revealed by GC analysis. (Pentafluorophenyl)benzene may be further purified by column chromatography

TABLE II

SUMMARY OF REACTIONS OF PENTAFLUOROPHENYLCOPPER

Reactant	Form of $C_6F_5Cu^a$	Conditions	Product (% yield)	Reference
H_2O	B	ether, 25°, 1 hour	C_6F_5H (87)	4
None	B	200°	C_6F_5—C_6F_5 (60)	4
Br_2	B	hexane, 0°	C_6F_5Br(4) + C_6F_5—C_6F_5 (47)	3a
$CuBr_2$	B	hexane, reflux	C_6F_5—C_6F_5 (68)	3a
CO_2	A or B	neat or aprotic solvent, −78° to 25°	no reaction	3a
CH_3I	A	hexane, 25°, 5 days	CH_3—C_6F_5 (39)	3a
$C_6H_5CH_2Br$	B	ether, 25°, 4 hours	$C_6H_5CH_2C_6F_5$ (40)	3a
1-Bromobicyclo-[2.2.2]octane	A	hexane, reflux, 20 hours	1-pentafluorophenylbicyclo[2.2.2]octane (83)	3
$C_2H_5O_2CCHN_2$	A	tetrahydrofuran, 0°; hydrolysis	$(C_2H_5O_2)CCH_2C_6F_5$ (43)	3a
C_6H_5I	A	benzene, reflux, 2 hours	C_6F_5—C_6H_5 (87)	3a
$3\text{-}FC_6H_4I$	B	benzene, reflux, 2 hours	$3\text{-}FC_6H_4$—C_6F_5 (73)	6
$O_2NC_6H_4I$	B	meta: benzene, reflux, 2 hours	$O_2NC_6H_4C_6F_5$ meta (85)	6
		para: benzene, reflux, 2 hours	para (85)	6
		ortho: ether, reflux (exothermic)	ortho (73)	4
$4\text{-}(CH_3)_2NC_6H_4I$	B	benzene, reflux, 2 hours	$4\text{-}(CH_3)_2NC_6H_4C_6F_5$ (26)	6
$4\text{-}C_2H_5O_2CC_6H_4I$	B	benzene, reflux, 2 hours	$4\text{-}C_2H_5O_2CC_6H_4C_6F_5$ (97)	6

$CF_2{=}CFI$	C	tetrahydrofuran, 25–55°, 5 hours	$C_6F_5CF{=}CF_2$ (55)	14
$CBr_2{=}CHBr$	C	tetrahydrofuran, −5°, 3 hours	$C_6F_5C{\equiv}CC_6F_5$ (43)	15
$CH_3\overset{O}{C}Cl$	D	tetrahydrofuran–hexane, 0°, several hours	$C_6F_5\overset{O}{C}CH_3$ (84)	7b
$Cl\overset{O\ O}{C}CCl$	D	tetrahydrofuran, 0° (exothermic)	$(C_6F_5\overset{O}{C})_2$ (71)	16
$-(CH_2\overset{O}{C}Cl)_2$	C	tetrahydrofuran, −5°, 3 hours	$(C_6F_5\overset{O}{C}CH_2)_2$ (71)	15
C_6H_5I	C	tetrahydrofuran, 66°, 10 hours	$C_6H_5{-}C_6F_5$ (74)	14
$C_6H_5\overset{O}{C}Cl$	C	tetrahydrofuran, −5°, 3 hours	$C_6F_5\overset{O}{C}C_6H_5$ (77)	15
$(C_2H_5)_3SiC{\equiv}CBr$	C	tetrahydrofuran, 25°, 1 hour; reflux, 10 hours	$(C_2H_5)_3SiC{\equiv}CC_6F_5$ (85)	17

[a] A, $(C_6F_5Cu)_4$; B, $(C_6F_5Cu)_2{\cdot}$dioxane; C, reagent prepared *in situ* from C_6F_5MgX and CuX; D, reagent prepared *in situ* from C_6F_5Li and CuX.

on acid-washed alumina (Woelm, activity grade I) with hexane as an eluent, the decafluorobiphenyl impurity being eluted first. The purified product melts at 111.3–112.0°.

B.   1-*(Pentafluorophenyl)adamantane*.   A 500 ml., three-necked flask equipped with a thermometer, a magnetic stirring bar, and a reflux condenser bearing a nitrogen inlet is maintained under nitrogen, and charged with 32.5 g. (0.00590 mole) of bis(pentafluorophenylcopper)·dioxane complex, 24.9 g. (0.116 mole) of 1-bromoadamantane, and 175 ml. of spectral grade hexane. The mixture is stirred and slowly warmed until the onset of an exothermic reaction which causes the mixture to reflux for approximately 15 minutes. After the exothermic reaction subsides, the mixture is heated at reflux overnight, stirred briefly with 3 ml. of water, and filtered, separating 17.0 g. of copper(I) bromide. The filtrate is concentrated, and the residue is recrystallized from ethanol, affording 32.6 g. (93%) of colorless crystals of 1-pentafluorophenyladamantane. After sublimation at 100° (0.1 mm.) the product melts at 109.9–111.0°.

1. Central Research & Development Department, E. I. du Pont de Nemours and Company, Experimental Station, Wilmington, Delaware 19898.
2. Deceased Nov. 2, 1978.
3. (a) A. Cairncross and W. A. Sheppard, *J. Am. Chem. Soc.*, **90**, 2186 (1968); (b) A. Cairncross and W. A. Sheppard, U.S. Pat. 3,700,693 (1972) [*Chem. Abstr.*, **78**, 30001d (1973)];
4. A Cairncross and W. A. Sheppard, Abstract P152, Division of Organic Chemistry, 155th National American Chemical Society Meeting, San Francisco, California, April 1968.
5. For recent reviews of organocopper reagents, see A. E. Jukes, *Adv. Organomet. Chem.*, **12**, 215 (1974); G. H. Posner, *Org. React.*, **22**, 253 (1975).
6. W. A. Sheppard, *J. Am. Chem. Soc.*, **92**, 5419 (1970).
7. (a) C. Tamborski, E. J. Soloski, and R. J. DePasquale, *J. Organomet. Chem.*, **15**, 494 (1968); (b) A. E. Jukes, S. S. Dua, and H. Gilman, *J. Organomet. Chem.*, **21**, 241 (1970); A. E. Jukes, S. S. Dua, and H. Gilman, *J. Organomet. Chem.*, **24**, 791 (1970).
8. M. P. Cava and A. Afzali, *J. Org. Chem.*, **40**, 1553 (1975).
9. A. Cairncross, J. R. Roland, R. M. Henderson, and W. A. Sheppard, *J. Am. Chem. Soc.*, **92**, 3187 (1970).
10. B. M. Trost and P. L. Kineson, *J. Org. Chem.*, **37**, 1273 (1972).
11. P. G. Gassman, G. R. Meyer, and F. J. Williams, *J. Am. Chem. Soc.*, **94**, 7741 (1972).
12. A. Cairncross and W. A. Sheppard, U.S. Pat. 3,817,784 (1974) [*Chem. Abstr.*, **81**, 128814h (1975)].
13. A. Cairncross, H. Omura, and W. A. Sheppard, *J. Am. Chem. Soc.*, **93**, 248 (1971).
14. R. J. DePasquale and C. Tamborski, *J. Org. Chem.*, **34**, 1736 (1969).
15. A. F. Webb and H. Gilman, *J. Organomet. Chem.*, **20**, 281 (1969).
16. S. S. Dua, A. E. Jukes, and H. Gilman, *Organomet. Chem. Synth.* **1**, 87 (1970/1971).
17. F. Waugh and D. R. M. Walton, *J. Organomet. Chem.*, **39**, 275 (1972).

trans-3-PENTEN-2-ONE

(3-Penten-2-one, *trans*-)

$$\text{CH}_3\text{CH}{=}\text{CH}_2 + \text{CH}_3\text{COCl} \xrightarrow[\Delta]{\text{AlCl}_3} \text{CH}_3\text{COCH}_2\overset{\overset{\displaystyle\text{Cl}}{\displaystyle|}}{\text{CH}}\text{CH}_3$$

$$\text{CH}_3\text{COCH}_2\overset{\overset{\displaystyle\text{Cl}}{\displaystyle|}}{\text{CH}}\text{CH}_3 \ +$$

$$\downarrow \Delta$$

$$\text{CH}_3\text{COCH}{=}\text{CHCH}_3 \ +$$

Submitted by H. C. ODOM and A. R. PINDER[1]
Checked by WALTER J. CAMPBELL and HERBERT O. HOUSE

1. Procedure

Caution: Since both hydrogen chloride and propene may escape from the reaction vessel during this preparation, the reaction should be performed in a hood.

A 2-l., three-necked flask is equipped with an efficient mechanical stirrer, a gas-inlet tube extending almost to the bottom of the flask, and an efficient reflux condenser fitted with a calcium chloride drying tube. After the apparatus has been dried in an oven, 800 ml. of dichloromethane (Note 1) and 157 g. (142 ml., 2.00 moles) of acetyl chloride (Note 2) are added to the flask. This solution is stirred while 320 g. (2.40 moles) of powdered, anhydrous aluminum chloride (Note 3) is added in portions over a 15-minute period. As soon as this addition is complete, a stream of propene gas (Note 4) is passed through the continuously stirred reaction solution at a rate sufficient to maintain a gentle reflux. The gas flow is continued until no more heat is evolved and refluxing ceases (10–30 hours, Note 4), at which time the flask is nearly full and the contents separate into two layers after stirring is stopped. The contents of the flask are poured cautiously onto about 1.5 kg. of ice (Note 5), and the upper organic layer is separated. The aqueous phase is shaken with three 100-ml. portions of dichloromethane (Note 6), and the combined organic solutions are washed with 50 ml. of water and dried over anhydrous magnesium sulfate.

The resulting dark brown solution is placed in a 2-l., round-bottomed flask equipped with a thermometer, a magnetic stirring bar, a heating mantle, and an assembly consisting of a distilling head, a condenser, and a receiver which permits distillation under reduced pressure. The bulk of the dichloromethane and volatile hydrocarbons are distilled from the mixture at water-aspirator pressure while sufficient heat is supplied with the heating mantle, maintaining the temperature of the mixture at about 0° (Note 7). When the bulk of the solvent has been removed, a 1-l., round-bottomed flask, cooled in a 2-propanol–dry ice bath, is attached to the apparatus as a receiver, and the pressure is reduced to 1 mm. or less with a vacuum pump. With the heating mantle, the temperature of the viscous liquid in the distillation flask is raised slowly from about 0° to 45° over a period of 90 minutes, distilling the volatile products (dichloromethane, low molecular-weight hydrocarbons, 4-chloropentan-2-one, and 3-penten-2-one) (Note 7). The resulting distillate (400–500 g. of pale green liquid) is mixed with 256 g. (1.98 moles) of quinoline (Note 8) and heated to boiling. To remove the remaining dichloromethane and other low-boiling materials (Note 9), liquid is allowed to distil from the mixture until the temperature of the distilling liquid reaches 110–120°. The remaining solution is refluxed for 30 minutes, then cooled and diluted with the previously removed distillate and 200 ml. of pentane. The resulting solution is washed with successive 250-ml. portions of 10% hydrochloric acid until the aqueous washings are acidic. The aqueous washings are combined, acidified, and shaken with three 100-ml. portions of pentane. The combined organic solutions are washed with 50 ml. of saturated aqueous sodium hydrogen carbonate and dried over anhydrous magnesium sulfate. The resulting organic solution is fractionally distilled through a 30 cm. Vigreux column, and 42–63 g. (25–37%) of crude product is collected in the fraction boiling at 119–124°. This contains 3-penten-2-one of 86–92% purity (Note 10). If greater purity is desired, the crude product may be distilled through a 60-cm., spinning band column. Since this distillation may be accompanied by partial isomerization of the α,β-unsaturated ketone to the lower boiling β,γ-isomer (Note 10), the product from the fractional distillation should be subjected to an acid-catalyzed equilibration. In a typical purification 79.4 g. of a mixture of penten-2-one isomers, b.p. 117–119°, from a fractional distillation, is mixed with 400 mg. of p-toluenesulfonic acid and refluxed for 30 minutes. The resulting mixture is diluted with 100 ml. of diethyl ether and washed with 50 ml. of saturated aqueous sodium hydrogen carbonate and dried over anhydrous magnesium sulfate. The resulting ether solution is fractionally distilled through a 16-cm. Vigreux column, giving 60.4 g. of trans-3-penten-2-one of 97% purity, b.p. 121.5–124°, n_D^{25} 1.4329 (Note 11).

2. Notes

1. Dichloromethane was dried over calcium chloride before use.
2. The submitters used a practical grade of acetyl chloride obtained from Eastman Organic Chemicals. The checkers used reagent grade acetyl chloride obtained from the Industrial Chemicals Division, Allied Chemical Corporation.
3. The reagent grade of powdered, anhydrous aluminum chloride employed was obtained from the Specialty Chemicals Division, Allied Chemical Corporation.
4. A chemically pure grade of propene, obtained from Matheson Gas Products, was employed. A large excess of propene is used since much of the olefin is converted to

polymeric products. The submitters report obtaining markedly lower yields of product when an excess of propene was not used.

5. At this point there should be only a relatively mild exothermic reaction as the anhydrous aluminum salts are hydrolyzed and solvated.

6. In these extractions the organic layer is the lower one.

7. The submitters had originally distilled the volatile products from this mixture, containing mainly polymeric material, by a more conventional procedure. However, the checkers found the problem of foaming during distillation so severe that the alternative, low-temperature distillation procedure was adopted.

8. The practical grade of quinoline employed was obtained from Aldrich Chemical Company, Inc.

9. The reaction solution must reach a temperature of approximately 90° for rapid dehydrochlorination to occur. If low-boiling impurities prevent the reaction mixture from reaching this temperature, the final product may be contaminated with the intermediate β-chloroketone. On a GC column packed with Carbowax 20M suspended on Chromosorb P, the retention times for the α,β-unsaturated ketone and the intermediate β-chloroketone are 4.9 minutes and 12.0 minutes, respectively. A sample of 4-chloro-2-pentanone, collected from this GC column, has IR absorption (CHCl$_3$) at 1720 cm.$^{-1}$ (C=O) with ^{1}H NMR absorption (CDCl$_3$) at δ 1.55 (d, J = 7 Hz., 3H, CH_3), 2.20 (s, 3H, CH_3CO), 2.5–3.3 (m, 2H, CH_2), and 4.45 (sextuplet, J =7 Hz., 1H, CHCl).

10. The product was analyzed with a 2-m. GC column packed with Carbowax 20M suspended on Chromosorb P. In chromatograms obtained from this column at 100°, the retention times of 4-penten-2-one and 3-penten-2-one are 2.6 and 3.9 minutes, respectively. The crude product contains several additional low-boiling components with GC retention times in the range 1.6–2.8 minutes. Any 4-penten-2-one present as an impurity exhibits IR absorption (CCl$_4$) at 1720 cm.$^{-1}$ (nonconjugated C=O).

11. The pure *trans*-3-penten-2-one has IR absorption (CCl$_4$) at 1680 and 1705 cm.$^{-1}$ (cisoid and transoid conformers[2] of the conjugated C=O), 1635 cm.$^{-1}$ (conjugated C=C), and 970 cm.$^{-1}$ (*trans*-CH=CH) with a UV maximum (95% C$_2$H$_5$OH) at 220 nm (ϵ 11,000) and ^{1}H NMR absorption (CDCl$_3$) at δ 1.88 (d of d, J = 1.5 and 7 Hz., 3H, CH_3), 2.22 (s, 3H, CH_3CO), 6.10 (d of partially resolved m, J = 16 Hz., 1H, α-vinyl CH), and 6.85 (d of q, J = 7 and 16 Hz., 1H, β-vinyl CH). The mass spectrum of the product has the following relatively abundant peaks: m/e (relative intensity), 84(M$^+$, 36), 69(100), 43(57), 41(78), and 39(33).

3. Discussion

trans-3-Penten-2-one has been prepared by the dehydration of 4-hydroxy-2-pentanone with heat,[3] acetic anhydride,[3] sulfuric acid,[4] or iodine.[5] It has also been obtained by fractional distillation of a commercial product from the aldol condensation of acetaldehyde and acetone.[6] Preparations are described involving reactions of acetyl bromide with propene, in the presence of anhydrous aluminum bromide,[7] and of acetic anhydride or acetyl chloride with propene, in the presence of anhydrous aluminum chloride.[8] Other preparative methods include the oxidation of *trans*-3-penten-2-ol with chromic acid[6] and the Wittig reaction between acetylmethylenetriphenylphosphorane and acetaldehyde.[9]

The present procedure, an adaption of one described previously,[8] illustrates the acylation of an olefin in the presence of a Lewis-acid catalyst. Although this method may lead to complex mixtures of acylated products when higher molecular-weight olefins are acylated in the presence of excess aluminum chloride, the application of this procedure to propene gives a single, monomeric, acetylated product, accompanied by a complex mixture of low molecular-weight hydrocarbons and unidentified, higher molecular-weight materials. The relatively low boiling point of the monoacetylated product permits its ready separation from most of the components of this mixture, after which it is de-hydrochlorinated to the desired product. The product of this reaction is of sufficient purity to serve as a synthetic intermediate in annelation reactions with cycloalkanones.[10]

1. Department of Chemistry, Clemson University, Clemson, South Carolina 29631.
2. R. L. Erskine and E. S. Waight, *J. Chem. Soc.*, 3425 (1960).
3. L. Claisen, *Ber. Dtsch. Chem. Ges.*, **25**, 3164 (1892); *Justus Liebigs Ann. Chem.*, **306**, 322 (1899).
4. L. P. Kyriakides, *J. Am. Chem. Soc.*, **36**, 530 (1914); A. L. Wilds and C. Djerassi, *J. Am. Chem. Soc.*, **68**, 1715 (1946).
5. W. S. Rapson, *J. Chem. Soc.*, 1626 (1936).
6. H. O. House, D. D. Traficante, and R. A. Evans, *J. Org. Chem.*, **28**, 348 (1963). See also S. T. Young, J. R. Turner, and D. S. Tarbell, *J. Org. Chem.*, **28**, 928 (1963); J. E. Baldwin, *J. Org. Chem.*, **30**, 2423 (1965).
7. S. Krapiwin, *Bull. Soc. Imp. Nat. Moscou*, 1 (1908) [*Chem. Zentralbl.*, **81**, **I**, 1335 (1910); *Chem. Abstr.*, **5**, 1281 (1911)].
8. N. Jones and H. T. Taylor, *J. Chem. Soc.*, 1345 (1961). In this paper the boiling point of the product is given erroneously as 103°.
9. H. O. House, W. L. Respess, and G. M. Whitesides, *J. Org. Chem.*, **31**, 3128 (1966). These authors noted that *trans*-3-penten-2-one prepared by oxidation of *trans*-3-penten-2-ol is often contaminated with unchanged alcohol. The submitters concur with this observation.
10. See, for example, J. A. Marshall, H. Faubl, and T. M. Warne, Jr., *Chem. Commun.*, 753 (1967); R. M. Coates and J. E. Shaw, *Chem. Commun.*, 47 (1968); H. C. Odom and A. R. Pinder, *J. Chem. Soc. Chem. Commun.*, 26 (1969); J. A. Marshall and R. A. Ruden, *Tetrahedron Lett.*, 1239 (1970).

ARENE OXIDE SYNTHESIS: PHENANTHRENE-9,10-OXIDE

(Phenanthro[9,10-*b*]oxirene, 1*a*,9*b*-dihydro-)

A.

B.   **1**

Submitted by CECILIA CORTEZ and RONALD G. HARVEY[1]
Checked by JAMES JACKSON and ORVILLE L. CHAPMAN

1. Procedure

Caution! Benzene has been identified as a carcinogen; OSHA has issued emergency standards on its use. All procedures involving benzene should be carried out in a well-ventilated hood, and glove protection is required.

A.   trans-9,10-*Dihydro*-9,10-*phenanthrenediol* (**1**).   Phenanthrenequinone (6 g., 0.03 mole) (Note 1) is placed in a fritted-glass (coarse porosity) extraction thimble of a Soxhlet apparatus over a 1-l. flask containing a suspension of 3 g. of lithium aluminum hydride in 500 ml. of anhydrous diethyl ether (Note 2). Extraction of the quinone over a period of 16 hours affords a green solution (Note 3). The reaction is quenched by the cautious addition of water (Note 4) and neutralized with glacial acetic acid. The ether layer is separated, and the aqueous layer is extracted with two 200-ml. portions of ether. The combined ether extracts are washed consecutively with aqueous sodium hydrogen carbonate and water, then dried over magnesium sulfate. Evaporation of the solvent under reduced pressure gives the crude product (Note 5), which is recrystallized from benzene, giving 3.8–4.1 g. (62–68%) of pure diol **1** as fluffy, white needles, m.p. 185–190°, (Note 6).

B.   *Phenanthrene-9,10-oxide* (**2**).   A solution of 10.6 g. (0.0500 mole) of **1** and 13 g. of N,N-dimethylformamide dimethyl acetal (Note 7) in 40 ml. of N,N-dimethylformamide (Note 8) and 100 ml. of dry tetrahydrofuran (Note 9) is heated at reflux for 16 hours. The solution is then allowed to cool to room temperature, and 200 ml. of water and 100 ml. of ether are added. The organic layer is separated, the aqueous layer is washed with two 200-ml. portions of ether, and the combined ether phases are dried over magnesium sulfate. Evaporation of the solvent under reduced pressure gives 9.6 g. of a yellow solid. Trituration with 25 ml. of hexane removes colored impurities, and recrystallization from benzene–cyclohexane (Note 10) gives 5.6–6.2 g. (58–64%) of oxide **2** as off-white plates, m.p. 125° (dec.) (Note 11). A second crop of 1.0 g can be obtained, for an overall yield of 68–74% (Note 12).

2. Notes

1. Phenanthrenequinone, free of anthraquinone, is available from Aldrich Chemical Company, Inc., or from J. T. Baker Chemical Company and should be recrystallized from benzene before use.

2. Use of more efficient solvents (tetrahydrofuran, isopropyl ether, dimethoxyethane) or more soluble metal hydride reagents (sodium borohydride, lithium tributoxy aluminum hydride, sodium bis(2-methoxyethyl) aluminum hydride) favors the alternative reduction pathway to the hydroquinone.

3. The checkers noted that use of a paper thimble resulted in increased time for extraction. The submitters recommend use of a glass thimble, since prolonged heating can lead to lower yields. It is easier to determine when extraction is complete with a transparent thimble. Other quinones may require longer extraction periods.

4. Care must be taken to add water cautiously and slowly, since the reaction between water and lithium aluminum hydride is vigorous. The reaction is quenched when the solution stops refluxing.

5. The crude product may darken on drying because of the presence of minor amounts of the air-sensitive hydroquinone by-product.

6. Large-scale reactions usually result in lower yields. The checkers obtained product, m.p. 189–191°, in runs with slightly lower yields.

7. N,N-Dimethylformamide dimethyl acetal, obtained from Aldrich Chemical Company, Inc., was redistilled before use.

8. N,N-Dimethylformamide was distilled under reduced pressure and stored over molecular sieves, type 4Å.

9. Tetrahydrofuran was distilled from lithium aluminum hydride. For a warning concerning potential hazards of this procedure, see *Org. Synth.,* **Coll. Vol. 5,** 976 (1973).

10. Excessive heating during recrystallization should be avoided because it can lead to thermal decomposition of the product.

11. Because of the relative facility of thermal rearrangement to phenols, melting points of arene oxides are not an entirely reliable index of purity. The checkers found variation from 119 to 135° (dec.). Purification by chromatography on activity IV alumina is also possible, but residence time on the column should be held to a minimum.

12. The ^{1}H NMR spectrum (CDCl$_3$) of pure **2** showed a characteristic oxiranyl singlet peak at δ 4.67 (s, 2H) and an aromatic signal at 7.2–7.8 (m, 8H).

3. Discussion

The method employed here is essentially that reported earlier,[2] modified by subsequent experience.[3] In the second step, N,N-dimethylformamide dimethyl acetal acts as a dehydrating agent, giving the epoxide, and is converted to N,N-dimethylformamide and methanol. Phenanthrene-9,10-oxide has also been prepared by cyclization of 2,2'-biphenyldicarboxaldehyde with hexamethylphosphorus triamide[4] and by dehydrohalogenation of 10-chloro-9,10-dihydro-9-phenanthrenyl acetate, obtained through reaction of the corresponding 2-alkoxy-1,3-dioxolane with trimethylsilyl chloride.[5] The present procedure is simpler, requiring fewer steps from readily available starting materials; both alternative procedures start with phenanthrene. The product is relatively easy to purify, since the only by-products are N,N-dimethylformamide and methanol (an important consideration with molecules sensitive to decomposition), and appears to be more stable on storage than the compound obtained via the dialdehyde route.

The cyclization method utilized in this synthesis appears quite general in its applicability, having been applied successfully in our laboratory[3] to the preparation of the K-region arene oxides[6] of benz[a]anthracene, chrysene, dibenz[a,h]anthracene, benzo[c]phenanthrene, pyrene, 1-methylphenanthrene, benzo[a]pyrene, and 7,12-dimethylbenz[a]anthracene, among others. The latter two are potent carcinogens; the K-region oxides of these have been shown to be formed metabolically and exhibit significant biological activity.[7]

The K-region quinones required as starting materials in this synthesis are in certain cases (e.g., phenanthrene, chrysene, benzo[c]phenanthrene) available from direct oxidation of the parent hydrocarbons with chromic acid. When oxidation occurs preferentially elsewhere in the molecule, the K-region dihydroaromatic derivatives can often be converted to the corresponding quinone through oxidation with dichromate in acetic acid–acetic anhydride;[8] yields, however, are only in the 20–30% range. Alternatively, the K-region quinones may be obtained from the hydrocarbons through oxidation with osmium tetroxide to the corresponding cis-diols, followed by a second oxidation with pyridine–sulfur trioxide and dimethyl sulfoxide,[2,3] generally the most useful procedure. A significant advantage is that all possible K-region oxidized derivatives (cis-diols, quinones, trans-diols, phenols,[3] and hydroquinones[9]) with intact ring systems can be obtained directly or by appropriate modification of the general sequence. The disadvantage of this method, and of any alternative procedure[4,5] involving the cis-diol, is the hazardous and expensive osmium tetroxide employed.

1. Ben May Laboratory for Cancer Research, University of Chicago, Chicago, Illinois 60637.
2. S. H. Goh and R. G. Harvey, J. Am. Chem. Soc., **95**, 242 (1973).
3. R. G. Harvey, S. H. Goh, and C. Cortez, J. Am. Chem. Soc., **97**, 3468 (1975).
4. M. S. Newman and S. Blum, J. Am. Chem. Soc., **86**, 5598 (1964).
5. P. Dansette and D. M. Jerina, J. Am. Chem. Soc., **96**, 1224 (1974).
6. The K-region of a polycyclic aromatic hydrocarbon is typified by the 9,10-bond of phenanthrene. According to the Schmidt–Pullman electronic theory, an unsubstituted K-region is a requirement for carcinogenic activity; see A. Pullman and B. Pullman, "La Cancerisation par les Substances Chimiques et la Structure Moleculaire," Masson, Paris, 1955.

7. For leading references see K. W. Jennette, A. M. Jeffrey, S. H. Blobstein, F. A. Beland, R. G. Harvey, and I. B. Weinstein, *Biochemistry,* **16,** 932 (1977).
8. H. Cho and R. G. Harvey, *Tetrahedron Lett.,* 1491 (1974).
9. H. Cho and R. G. Harvey, *J. Chem. Soc., Perkin Trans. I,* 836 (1976).

FREE-RADICAL ALKYLATION OF QUINONES: 2-PHENOXYMETHYL-1,4-BENZOQUINONE

[2,5-Cyclohexadiene-1,4-dione, 2-(phenoxymethyl)-]

Submitted by NIELS JACOBSEN[1]
Checked by R. J. DeFRANCO and R. E. BENSON

1. Procedure

A 250-ml., three-necked flask fitted with a mechanical stirrer, a thermometer, and a 25-ml., graduated, pressure-equalizing dropping funnel is charged with 7.60 g. (0.0500 mole) of phenoxyacetic acid (Note 1), 5.40 g. (0.0500 mole) of 1,4-benzoquinone (Note 2), 1 g. (0.006 mole) of silver nitrate (Note 3), and 125 ml. of water (Note 4). The mixture is stirred and heated to 60–65° with a heating mantle until dissolution is complete. The resulting solution is stirred vigorously while a solution of 13.7 g. (0.0601 mole) of ammonium peroxydisulfate (Note 5) in 25 ml. of water is added at a rate of 0.5 ml. per minute for the first 40 minutes, then at a rate of 0.25 ml. per minute for the last 20 minutes. Throughout the addition, the reaction mixture is maintained at 60–65° (Notes 6 and 7).

After the addition is complete the mixture is stirred for 5 minutes at 65° then cooled to 5–10° in an ice bath. The precipitated solid is collected by suction filtration (Note 8), washed with 50 ml. of cold water, and pressed, removing most of the liquid. Inorganic contaminants, usually present in small amounts, are removed by dissolving the solid in 350 ml. of boiling acetone and filtering the hot solution through fluted filter paper. Concentration of the filtrate on a rotary evaporator gives a dark red crude product (10.5–11.4 g.), which is dissolved in 220–240 ml. of boiling 95% ethanol. On cooling the solution to 5°, the alkylated quinone crystallizes in brownish-yellow needles, which are collected by filtration and air-dried, yielding 6.7–8.0 g., m.p. 135–137°. Recrystallization from 30 ml. of ethanol per gram of product gives 6.6–7.4 g. (61–69%) of 2-phenoxymethyl-1,4-benzoquinone, m.p. 137–138° (Note 9).

TABLE I

SUBSTITUTED QUINONES DERIVED BY ALKYLATION

Parent Quinone	Acid	Derived Substituted Quinone	Yield (%)[a]	Reference
1,4-Benzoquinone	Pivalic	2-tert-Butyl	67[b]	2
1,4-Benzoquinone	Phenylacetic	2-Benzyl	87	2
1,4-Benzoquinone	α-Chloropropionic	2-(α-Chloroethyl)	45[b]	2
1,4-Naphthoquinone	Methoxyacetic	2-Methoxymethyl	50	2
1,4-Naphthoquinone	Adipic	2-(ω-Carboxybutyl)	51	2
2-Methyl-1,4-naphthoquinone	Cyclopropanecarboxylic	2-Cyclopropyl-3-methyl	37	3
2-Acetoxy-1,4-naphthoquinone	4-Methyl-3-pentenoic	2-Acetoxy-3-(γ,γ-dimethylallyl)	73	4

[a]Yields are based on the parent quinone.
[b]Modified procedure (see Discussion).

2. Notes

1. The submitter used Fluka *puriss* grade phenoxyacetic acid. The checkers used material available from Eastman Organic Chemicals.

2. The submitter used Fluka *purum* grade benzoquinone, recrystallized once from petroleum ether (b.p. 60°), m.p. 111–113°. The checkers used Fisher purified grade material without recrystallization.

3. The submitter used reagent grade silver nitrate available from Merck & Company, Inc.

4. In the case of a water-insoluble quinone or carboxylic acid, acetonitrile can be used as a co-solvent.[2]

5. Fluka *purum* grade ammonium peroxydisulfate was used by the submitter. The checkers used ACS reagent grade material available from Fisher Scientific Company.

6. The reaction is slightly exothermic, but it is necessary to heat the mixture occasionally in order to maintain it at 60–65°.

7. The checkers found that increasing the addition rate of persulfate solution to 1.5 ml. per minute, while giving a somewhat lower initial yield (62% after one recrystallization), resulted in a product of sufficient purity (m.p. 137–138°) as to require no further recrystallization.

8. This work-up procedure applies only when the crude product can be crystallized from the reaction mixture. If the product is partly soluble in the reaction medium or if it separates as a gum, an extraction procedure is employed.

9. IR (CHCl$_3$) cm.$^{-1}$: 1660 strong, 1600 medium, 1590 medium; UV (95% C$_2$H$_5$OH) nm. max. (ϵ): 220 (12,700), 248 (18,600), 269 shoulder, 276 shoulder; ^{1}H NMR (CDCl$_3$), δ (multiplicity, coupling constant J in Hz., number of protons): 4.9 (d, $J = 2$, 2H, CH_2O), 6.7–7.5 (m, 8H).

3. Discussion

The procedure described above has been used to prepare various, alkylated 1,4-benzoquinones and 1,4-napthoquinones,[2,3] including some naturally occurring quinones.[4] A few examples are listed in Table I, showing the scope of the method.

The reaction is a free-radical alkylation in which radicals are derived from a carboxylic acid by decarboxylation with silver peroxydisulfate. It has the advantage that the reaction medium can be adjusted so that the monoalkylated product precipitates as it is formed, thereby suppressing di- or polyalkylation.[5]

The reaction fails if the decarboxylation produces a radical that is easily oxidized, such as an α-hydroxyalkyl radical.[2] In intermediate cases, such as *tert*-alkyl or α-alkoxyalkyl radicals,[2] the yield based on the parent quinone is usually improved by using an excess of the peroxydisulfate and carboxylic acid to compensate for the loss of radicals due to oxidation (footnote *b*, Table I).

1. Department of Organic Chemistry, University of Aarhus, 8000 Aarhus C, Denmark. [Present address: A/S Cheminova, P. O. Box 9, DK-7620 Lemvig, Denmark.]
2. N. Jacobsen and K. Torssell, *Justus Liebigs Ann. Chem.*, **763**, 135 (1972).
3. J. Goldman, N. Jacobsen, and K. Torssell, *Acta Chem. Scand.*, **28**, 492 (1974).

4. N. Jacobsen and K. Torssell, *Acta Chem. Scand.*, **27**, 3211 (1973).
5. B. M. Bertilsson, B. Gustafsson, I. Kühn, and K. Torssell, *Acta Chem. Scand.*, **24**, 3590 (1970); N. Jacobsen, *J. Chem. Soc., Perkin Trans. II*, 569 (1979).

3-PHENYL-2*H*-AZIRINE-2-CARBOXALDEHYDE

(2*H*-Azirine-2-carboxaldehyde, 3-phenyl-)

Submitted by Albert Padwa,[1] Thomas Blacklock, and Alan Tremper
Checked by W. F. Oettle, E. R. Holler, and William A. Sheppard

1. Procedure

Caution! Although the organic azide intermediates used in this procedure have not shown any explosive hazard under the experimental conditions, they should always be handled with adequate shielding and normal protective equipment such as face shield and leather gloves.

A. (*1-Azido-2-iodo-3,3-dimethoxypropyl*)*benzene*. A dry, 1-l., three-necked, round-bottomed flask fitted with an efficient magnetic stirrer and two 250-ml. pressure-equalizing dropping funnels is charged with 75 g. (1.1 moles) of sodium azide and 450 ml. of dry acetonitrile (Note 1). The mixture is stirred and cooled in an ice–salt bath ($-5°$ to $0°$), and 83 g. (0.51 mole) of iodine monochloride (Note 2) is added dropwise from one of the addition funnels over 10–20 minutes. The solution is stirred for an additional 5–10 minutes before 81 g. (0.45 mole) of cinnamaldehyde dimethyl acetal (Note 3) is added from the other dropping funnel over a 15–20 minute period, while the cooling bath temperature is maintained at 0–5°. The resulting red-brown mixture is stirred for 12 hours at room temperature, poured into 500 ml. of water, and extracted with three 500-ml. portions of diethyl ether. The combined organic extracts are washed successively with 700 ml. of 5% aqueous sodium thiosulfate (Note 4) and 1 l. of water. The ether solution in dried over magnesium sulfate. The solvent is removed with a rotary evaporator, giving the azide product as a orange oil (Note 5), 150–156 g. (97–98%), of sufficient purity to be used for the next step.

B. (1-*Azido*-3,3-*dimethoxy*-1-*propenyl*)*benzene*. A 2-l., one-necked, round-bottomed flask equipped with a magnetic stirrer and powder funnel is charged with 156 g. (0.450 mole) of the iodoazide from Part A and 1500 ml. of anhydrous ether. The solution is stirred and cooled in an ice–salt bath ($-5°$ to $0°$), and 62 g. (0.55 mole) of potassium *tert*-butoxide (Note 6) is added. The powder funnel is replaced with a calcium chloride drying tube and the mixture is stirred for 4–5 hours at 0°, at which time 350 ml. of water is added while the mixture is still cold. The ethereal layer is separated, washed with three 350-ml. portions of water, and dried over magnesium sulfate. The solvent is removed with a rotary evaporator without heating, leaving 67–75 g. (68–76%) of (1-azido-3,3-dimethoxy-1-propenyl)-benzene as a dark oily liquid (Note 7), which can be used without further purification for Part C (Note 8).

C. 2-(*Dimethoxymethyl*)-3-*phenyl*-2H-*azirine*. The crude product (71–75 g., 0.32–0.34 mole) obtained from Part B is heated at reflux in 1 l. of chloroform in a 2-l., round-bottomed flask for 12 hours (Note 9). The solvent is removed with a rotary evaporator and the crude residue is distilled, giving 48–61 g. (78–93%) of 2-(dimethoxymethyl)-3-phenyl-2*H*-azirine, b.p. 103–105° (0.27 mm.) as a colorless oil (Note 10).

D. 3-*Phenyl*-2H-*azirine*-2-*carboxaldehyde*. The product from Part C (59.0 g., 0.31 mole) is placed in a 3-l., three-necked, round-bottomed flask fitted with a mechanical stirrer, a reflux condenser, and a thermometer of sufficient length to extend into the liquid contents of the flask. After addition of 600 ml. of 1,4-dioxane (Note 11) and 800 ml. of 20% acetic acid, the mixture is stirred and heated sufficiently to bring the temperature of the reaction mixture up to 90° over a period of one hour (Note 12). The temperature of the reaction mixture is held at 90° for an additional 5 minutes, then the flask is rapidly cooled in an ice–salt bath ($-5°$ to $0°$). The product is extracted with four 1-l. portions of ether, and the combined organic extracts are washed successively with 1 l. of 5% aqueous sodium hydrogen carbonate and 1-l. of saturated aqueous sodium chloride. After the ether layer has been dried over anhydrous magnesium sulfate, the solvent is removed with a rotary evaporator, and a mixture of 5 ml. of ether and 10 ml. of pentane is added. The residual oil is allowed to stand in a refrigerator (0–3°) for 12 hours, completing the crystallization of the crude product. The crystalline solid is collected on a cold filter and sublimed at 35° (0.01 mm.), giving 13.3 g. (30%) (Note 13) of 3-phenyl-2*H*-azirine-2-carboxaldehyde, m.p. 49–51° (Note 14).

2. Notes

1. Reagent grade acetonitrile (J. T. Baker Chemical Company) was used without further purification.

2. Iodine monochloride, purchased from J. T. Baker Chemical Company, was used without further purification.

3. Cinnamaldehyde dimethyl acetal was prepared by the method used to prepare the corresponding diethyl acetal.[2] A mixture of 66.0 g. (0.50 mole) of *trans*-cinnamaldehyde (Aldrich Chemical Company, Inc.), 100 g. (1.06 mole) of trimethyl orthoformate (Eastman Organic Chemicals), 450 ml. of anhydrous methanol (J. T. Baker Chemical Company), and 0.5 g. of *p*-toluenesulfonic acid monohydrate (Fisher Scientific Company) is stirred at room temperature for 24 hours. At the end of this time, the alcohol is removed with a rotary evaporator, and the residue is distilled, giving 81–83 g. (91–93%) of cinnamaldehyde dimethyl acetal, b.p. 93–96° (0.2 mm.).

4. The orange color of the ethereal solution is completely discharged after washing with 5% aqueous sodium thiosulfate.

5. The product has the following spectral properties: IR (neat) cm.$^{-1}$: 2120 (strong N_3 absorption); ^{1}H NMR (CDCl$_3$), δ (multiplicity, coupling constant J in Hz., number of protons, assignment): 3.38 (s, 3H, OCH_3), 3.46 (s, 3H, OCH_3), 3.93 (d, $J = 4$, 1H, 1- or 3-CH), 4.38 (d of d, $J = 9$ and 4, 1H, CHI), 4.78 (d, $J = 9$, 1H, 1- or 3-CH), 7.33 (s, 5H, C$_6H_5$).

6. Potassium *tert*-butoxide, purchased from Columbia Organic Chemicals Company, Inc., was sublimed at 150° (0.02 mm.) before use and was added in one portion.

7. The submitters reported a yield of 94–96 g. (97–98%). The spectral properties of the product are: IR (neat) cm.$^{-1}$: 2151 and 1642; ^{1}H NMR (CDCl$_3$), δ (multiplicity, coupling constant J in Hz., number of protons, assignment): 3.26 (s, 6H, 2 OCH_3), 4.78 [d, $J = 8$, 1H, CH(OCH$_3$)$_2$], 5.60 (d, $J = 8$, 1H, CH), 7.45 (s, 5H, C$_6H_5$).

8. The intermediate vinyl azide should either be used immediately or stored cold in a vented container, since it slowly evolves nitrogen on standing at room temperature.

9. The reaction can be conveniently monitored by IR spectroscopy by observing the intensity of the band at 2150 cm.$^{-1}$ (N$_3$).

10. The spectral properties are: IR (neat) cm.$^{-1}$: 1754 (azirine); ^{1}H NMR (CDCl$_3$), δ (multiplicity, coupling constant J in Hz., number of protons, assignment): 2.38 (d, $J = 3$, 1H, CH), 3.35 (s, 3H, OCH_3), 3.47 (s, 3H, OCH_3), 4.39 [d, $J = 3$, 1H, CH(OCH$_3$)$_2$], 7.3–8.0 (m, 5H, C$_6H_5$).

11. 1,4-Dioxane available from Fisher Scientific Company was used without further purification.

12. The mixture is brought to 90° by heating at a rate of 1° per minute. The mixture *must not be overheated*, or else the final product will be very difficult to crystallize.

13. Starting with 45.3 g. (0.237 mole) of the dimethyl acetal from Part C, the checkers obtained 10.2 g. (30%) of the product.

14. The submitters reported a yield of 35–38 g. (55–60%) based on 78–84 g. of starting material and using appropriate proportions of reagents. Their product had m.p.

45–47°. The spectral properties of the azirine product are: IR (KBr) cm.$^{-1}$: 1786 and 1709; ^{1}H NMR (CDCl$_3$), δ (multiplicity, coupling constant J in Hz., number of protons, assignment): 2.89 (d, $J = 7$, 1H, CH), 7.5–8.0 (m, 5H, C$_6$H$_5$), 9.04 (d, $J = 7$, 1H, CHO).

3. Discussion

The formation of substituted azirines by the thermal decomposition of vinyl azides is a general reaction.[3] Iodine azide offers an excellent route to vinyl azides;[4,5] it adds to many olefinic compounds, giving α iodoazides which can easily eliminate hydrogen iodide upon treatment with base. The direction of iodine azide addition is consistent with electrophilic attack of I$\oplus$, giving a cyclic iodonium ion which is opened by azide ion. The presence of the dimethyl acetal moiety in the system above does not not interfere with the iodine azide reaction. This procedure does not work with *trans*-cinnamaldehyde, owing to a competing aldol condensation in the elimination step.

The aldehyde functionality present in 3-phenyl-2H-azirine-2-carboxaldehyde reacts selectively with amines, and Grignard and Wittig reagents, yielding a variety of substituted azirines,[6] which have been used, in turn, to prepare a wide assortment of heterocyclic rings such as oxazoles, imidazoles, pyrazoles, pyrroles, and benzazepins.[6,7]

In addition to the present method, 2H-azirines can be prepared by a modified Neber reaction,[8–10] or by heating 4,5-dihydro-1,2,5-oxazaphospholes.[11–14]

1. Chemistry Department, State University of New York at Buffalo, Buffalo, N.Y. 14214. [Present address: Department of Chemistry, Emory University, Atlanta, Georgia 30322].
2. J. Klein and E. D. Bergmann, *J. Am. Chem. Soc.*, **79**, 3452 (1957).
3. F. W. Fowler, *Adv. Heterocycl. Chem.*, **13**, 45 (1971).
4. A. Hassner and L. A. Levy, *J. Am. Chem. Soc.*, **87**, 4203 (1965).
5. F. W. Fowler, A. Hassner, and L. A. Levy, *J. Am. Chem. Soc.*, **89**, 2077 (1967).
6. A. Padwa, J. Smolanoff, and A. Tremper, *Tetrahedron Lett.*, 29 (1974).
7. A. Padwa and J. Smolanoff, *Tetrahedron Lett.*, 33 (1974).
8. R. F. Parcell, *Chem. Ind. (London)*, 1396 (1963).
9. S. Sato, *Bull. Chem. Soc. Jpn.*, **41**, 1440 (1968).
10. G. Alvernhe, S. Arsenyiadis, R. Chaabouni, and A. Laurent, *Tetrahedron Lett.*, 355 (1975).
11. H. J. Bestmann and R. Kunstmann, *Angew. Chem. Int. Ed. Engl.*, **5**, 1039 (1966).
12. H. J. Bestmann and R. Kunstmann, *Chem. Ber.*, **102**, 1816 (1969).
13. R. Huisgen and J. Wulff, *Tetrahedron Lett.*, 917 (1967).
14. R. Huisgen and J. Wulff, *Chem. Ber.*, **102**, 1833 (1969).

PHASE-TRANSFER ALKYLATION OF NITRILES: 2-PHENYLBUTYRONITRILE

(Benzeneacetonitrile, α-ethyl-)

$$C_6H_5CH_2CN + C_2H_5Br \xrightarrow[\text{[(C}_2\text{H}_5)_3\text{NCH}_2\text{C}_6\text{H}_5]^+\text{Cl}^-]{\text{NaOH}} C_6H_5\underset{\underset{C_2H_5}{|}}{C}HCN$$

Submitted by M. MAKOSZA[1] and A. JOŃCZYK
Checked by HAROLD W. WAGNER and RICHARD E. BENSON

1. Procedure

Caution! Benzene has been identified as a carcinogen; OSHA has issued emergency standards on its use. All procedures involving benzene should be carried out in a well-ventilated hood, and glove protection is required.

A 3-l., four-necked, round-bottomed flask equipped with a mechanical stirrer, a dropping funnel, a thermometer, and an efficient reflux condenser is charged with 540 ml. of 50% aqueous sodium hydroxide, 257 g. (253 ml., 2.20 moles) of phenylacetonitrile (Note 1), and 5.0 g. (0.022 mole) of benzyltriethylammonium chloride (Note 2). Stirring is begun, and 218 g. (150 ml., 2.00 moles) of ethyl bromide (Note 3) is added dropwise over a period of approximately 100 minutes at 28–35°. If necessary, the flask may be cooled with a cold-water bath to keep the temperature of the mixture at 28–35°. After the addition of ethyl bromide is complete, stirring is continued for 2 hours, then the temperature is increased to 40° for an additional 30 minutes. The reaction mixture is cooled to 25°, 21.2 g. (20.3 ml., 0.200 mole) of benzaldehyde (Note 4) is added, and stirring is continued for 1 hour. The flask is immersed in a cold-water bath, and 750 ml. of water and 100 ml. of benzene are added. The layers are separated, and the aqueous phase is extracted with 200 ml. of benzene. The organic layers are combined and washed successively with 200 ml. of water, 200 ml. of dilute hydrochloric acid (Note 5), and 200 ml. of water. The organic layer is dried over anhydrous magnesium sulfate, and the solvent is removed by distillation under reduced pressure. The product is distilled through a Vigreux column, giving 225–242 g. (78–84%) of 2-phenylbutyronitrile, b.p. 102–104° (7 mm.), n_D^{25} 1.5065–1.5066 (Notes 6–8).

2. Notes

1. The checkers used phenylacetonitrile obtained from Aldrich Chemical Company, Inc., and distilled it before use. It may also be purified according to the directions given in *Org. Synth.*, **Coll. Vol. 1**, 108 (1948).

2. Benzyltriethylammonium chloride is available from Fisher Scientific Company. The preparation of this reagent is described in *Org. Synth.*, **Coll. Vol. 6**, 232 (1988).

3. Ethyl bromide (available from Fisher Scientific Company) was distilled before use.

4. Benzaldehyde (available from Fisher Scientific Company) was distilled before use. It is added at this point to convert any unreacted phenylacetonitrile to the high-boiling α-phenylcinnamonitrile (Note 7).

5. The acid solution was prepared by adding 1 volume of acid to 5 volumes of water.

6. The checkers obtained a forerun of 7–12 g. of product having n_D^{25} 1.5065–1.5066.

7. The α-phenylcinnamonitrile (Note 4) present in the distillation flask can be recovered. The residue is broken up with 75 ml. of methanol, the mixture stirred and cooled, and the product recovered by filtration. Recrystallization from methanol gives 17–20 g. of crystalline material, m.p. 86–88°. The ^{1}H NMR spectrum (CDCl$_3$) shows complex multiplets at δ 7.20–8.00.

8. GC analysis on a column packed with silicone gum nitrile on acid-washed Gas Chrome Red, 80–100 mesh and heated at 150°, shows that the product is about 97% pure. The material has the following spectral properties; IR (neat) cm.$^{-1}$: 2250 (C≡N), 1610, 1590 shoulder and 1500 (aromatic C≡C), 1385 (C—CH$_3$), and 760 and 697 (monosubstituted aromatic); ^{1}H NMR (CCl$_4$), δ (multiplicity, coupling constant J in Hz., number of protons, assignment): 0.99 (t, J = 7, 3H, CH$_3$), 1.82 (m, 2H, CH$_2$), 3.70 (t, J = 7, 1H, CH), 7.32 (s, 5H, C$_6$H$_5$).

3. Discussion

This reaction is illustrative of a general procedure for the tetraalkylammonium salt-catalyzed alkylation of active methylene functions in the presence of concentrated aqueous alkali. This catalytic method has been used to alkylate arylacetonitriles with monohaloalkanes,[2] dihaloalkanes,[3] α-chloroethers,[4] chloronitriles,[5] haloacetic acid esters,[6] and halonitro aromatic compounds.[7] It has also been used to alkylate ketones,[8] 1H-indene,[9] 9H-fluorene,[10] and the Reissert compound.[11] The reaction is inhibited by alcohols and iodide ion.[2]

Methods for the alkylation of nitriles have been reviewed.[12] These procedures, as well as those applied to other active methylenes, generally involve the use of dangerous and expensive condensing agents (sodium amide, metal hydrides, triphenylmethide, potassium $tert$-butoxide, etc.) and strictly anhydrous organic solvents (ether, benzene, N,N-dimethylformamide, dimethyl sulfoxide, etc.) or liquid ammonia. The catalytic method is much simpler and generally gives good yields of purer products. Because of its high selectivity[13] it is particularly adapted to the synthesis of pure monoalkyl derivatives of phenylacetonitrile which have also been obtained by alkylation of ethyl cyanophenylacetate[14] or cyanophenylacetic acid,[15] followed by elimination of the ethoxycarbonyl or carboxyl groups.

The catalytic conditions (aqueous concentrated sodium hydroxide and tetraälkylammonium catalyst) are very useful in generating dihalocarbenes from the corresponding haloforms. Dichlorocarbene thus generated reacts with alkenes, giving high yields of dichlorocyclopropane derivatives,[16] even in cases where other methods have failed,[17] and with some hydrocarbons, yielding dichloromethyl derivatives.[18] Similar conditions are suited for the formation and reactions of dibromocarbene,[19] bromofluoro- and chlorofluorocarbene,[20] and chlorothiophenoxy carbene,[21] as well as the Michael addition of trichloromethyl carbanion to unsaturated nitriles, esters, and sulfones.[22]

This method exemplifies a broad class of processes that proceed via transfer of reacting species between two liquid phases. Such processes may require a catalyst that can combine with species present in one phase and effect their transfer in this form to the second phase where the main reaction occurs. Starks[23] has termed such a process

TABLE I
ALKYLATIONS IN AQUEOUS MEDIUM

Compound	Alkylation Agent	Product	(%) Yield	Reference
$C_6H_5CH_2CN$	$(C_6H_5)_2CHCl$	$(C_6H_5)_2CHCH(C_6H_5)CN$	94	2
$C_6H_5CH_2CN$	$Br(CH_2)_4Br$	(cyclopentane ring bearing CN and C_6H_5)	88	3
$C_6H_5CH(C_2H_5)CN$	$C_6H_5CH_2Cl$	$C_6H_5CH_2C_2H_2C(C_6H_5)CN$ (C_2H_5)	94	2
$(C_6H_5)_2CHCN$	$BrCH_2CH_2Br$	$(C_6H_5)_2C(CH_2CH_2Br)CN$	91	3
$C_6H_5CH(CH_3)CN$	$ClCH_2OCH_3$	$C_6H_5C(CH_2OCH_3)CN$ (CH_3)	68	4
$C_6H_5CH(CH_3)CN$	$4\text{-}ClC_6H_4NO_2$	$C_6H_5C(4\text{-}NO_2C_6H_4)CN$ (CH_3)	82	11
$C_6H_5CH(C_2H_5)CN$	$ClCH_2COOC_4H_9\text{-}tert$	$C_6H_5C((CH_2COOC_4H_9\text{-}tert)CN$ (C_2H_5)	77	6
$C_6H_5CH_2COCH_3$	$BrCH_2CH_2Br$	(cyclopropane ring bearing C_6H_5 and $COCH_3$)	54	8
(indane structure)	$Br(CH_2)_4Br$	(spiro cyclopentane–indene structure)	64	9

899

"phase-transfer catalysis" and has demonstrated its utility in reactions involving inorganic anions. For example, he has shown that the rates of some displacement, oxidation, and hydrolysis reactions conducted in two-phase systems are dramatically enhanced by the presence of ammonium and phosphonium salts. However, in reactions involving weakly active methylenes, the catalyst seems to be more than a simple transfer agent; it is necessary for carbanion formation.

The versatility of this method for the alkylation of compounds containing active methylene groups is illustrated by Table I. Review articles have recently appeared,[24] and the application to the Hofmann carbylamine reaction is described in *Org. Synth.*, **Coll. Vol. 6,** 232 (1988).

1. Institute of Organic Chemistry and Technology, Polish Academy of Sciences, Warsaw, Poland.
2. M. Makosza and B. Serafin, *Rocz. Chem.*, **39,** 1223, 1401, 1595, 1805 (1965) [*Chem. Abstr.*, **64,** 12595h, 17474g, 17475c, 17475g (1966)].
3. M. Makosza and B. Serafin, *Rocz. Chem.*, **40,** 1647, 1839 (1966)[*Chem. Abstr.* **66,** 94792x, 11435a (1967)].
4. M. Makosza, B. Serafinowa, and M. Jawdosiuk, *Rocz. Chem.*, **41,** 1037 (1967) [*Chem. Abstr.*, **68,** 39313h (1968)].
5. J. Lange and M. Makosza, *Rocz. Chem.*, **41,** 1303 (1967) [*Chem. Abstr.*, **68,** 29374q (1968)].
6. M. Makosza, *Rocz. Chem.*, **43,** 79 (1969) [*Chem. Abstr.*, **70,** 114776h (1969)].
7. M. Makosza, *Tetrahedron Lett.*, 673 (1969); M. Makosza and M. Ludwikow, *Bull. Acad. Pol. Sci.*, *Ser. Sci. Chim.*, **19,** 231 (1971) [*Chem. Abstr.*, **75,** 48646r (1971)].
8. A. Jończyk, B. Serafin, and M. Makosza, *Tetrahedron Lett.*, 1351 (1971).
9. M. Makosza, *Tetrahedron Lett.*, 4621 (1966).
10. M. Makosza, *Bull. Acad. Pol. Sci.*, *Ser. Sci. Chim.*, **15,** 165 (1967) [*Chem. Abstr.*, **67,** 64085x (1967)].
11. M. Makosza, *Tetrahedron Lett.*, 677 (1969).
12. A. C. Cope, H. L. Holmes, and H. O. House, *Org. React.*, **9,** 107 (1957); M. Makosza, *Wiad. Chem.*, **21,** 1 (1967) [*Chem. Abstr.*, **67,** 53161t (1967)]; M. Makosza, *Wiad. Chem.*, **23,** 35, 759 (1969)[*Chem. Abstr.*, **70,** 96065u (1969)]; [*Chem. Abstr.*, **72,** 110907v (1970)].
13. M. Makosza, *Tetrahedron*, **24,** 175 (1968).
14. R. Delaby, P. Reynaud, and F. Lily, *Bull. Soc. Chim. Fr.*, 864 (1960).
15. E. M. Kaiser and C. R. Hauser, *J. Org. Chem.*, **31,** 3873 (1966).
16. M. Makosza and M. Wawrzyniewicz, *Tetrahedron Lett.*, 4659 (1969).
17. E. V. Dehmlow and J. Schönefeld, *Justus Liebigs Ann. Chem.*, **744,** 42 (1971).
18. I. Tabushi, Z. Yoshida, and N. Takahashi, *J. Am. Chem. Soc.*, **92,** 6670 (1970); E. V. Dehmlow, *Tetrahedron*, **27,** 4071 (1971).
19. M. Makosza and M. Fedorynski, *Bull. Acad. Pol. Sci.*, *Ser. Sci. Chim.*, **19,** 105 (1971) [*Chem. Abstr.*, **75,** 19745s (1971)].
20. P. Weyerstahl, G. Blume, and C. Müller, *Tetrahedron Lett.*, 3869 (1971).
21. M. Makosza and E. Bialecka, *Tetrahedron Lett.*, 4517 (1971).
22. M. Makosza and I. Gajos, *Bull. Acad. Pol. Sci.*, *Ser. Sci. Chim.*, **20,** 33 (1972) [*Chem. Abstr.*, **76,** 153179j (1972)].
23. C. M. Starks, *J. Am. Chem. Soc.*, **93,** 195 (1971).
24. J. Dock, *Synthesis*, 441 (1973); E. V. Dehmlow, *Angew. Chem. Int. Ed. Engl.*, **13,** 170 (1974 .

DIRECTED ALDOL CONDENSATIONS: β-PHENYLCINNAMALDEHYDE

(2-Propenal, 3,3-diphenyl-)

$$C_6H_{11}NH_2 + CH_3CHO \xrightarrow[-H_2O]{Na_2SO_4} C_6H_{11}N{=}CHCH_3$$

$$\xrightarrow[-(i\text{-}C_3H_7)_2NH]{(i\text{-}C_3H_7)_2NLi,\ (C_2H_5)_2O} C_6H_{11}N{=}CHCH_2Li \xrightarrow[(C_2H_5)_2O,\ 25°]{(C_6H_5)_2CO}$$

$$\xrightarrow[-C_6H_{11}NH_2]{(CO_2H)_2} (C_6H_5)_2C{=}CHCHO$$

Submitted by G. Wittig[1] and A. Hesse
Checked by Allan Y. Teranishi and Herbert O. House

1. Procedure

A. N-*Ethylidenecyclohexylamine.* A dry, 500-ml., round-bottomed flask fitted with a magnetic stirrer is flushed with nitrogen and stoppered with a rubber septum. With a hypodermic syringe 99.2 g. (1.00 mole) of freshly distilled cyclohexylamine (b.p. 133–134°) is added to the flask. After the amine has been cooled to approximately −20° (Note 1) with an acetone–dry ice bath, 44.1 g. (1.00 mole) of freshly distilled acetaldehyde (b.p. 21°) is added from a hypodermic syringe dropwise and with stirring over a 15-minute period. During the initial phase of this addition a white solid separates but redissolves as the addition is continued. The resulting cold solution is stirred at −20° for approximately 45 minutes, at which time a large amount of white solid separates and further stirring is impractical. The resulting mixture is allowed to stand at −20° for 15 minutes before 15 g. of anhydrous sodium sulfate is added and the mixture is allowed to melt and warm to room temperature. The resulting mixture is gravity filtered, and the residue is washed with approximately 15 ml. of diethyl ether. The combined filtrates are dried over 5 g. of anhydrous magnesium sulfate and filtered. The filtrate is distilled under reduced pressure, separating 95–99 g. (76–79%) of N-ethylidenecyclohexylamine as a colorless liquid, b.p. 47–48° (12 mm.) or 54–55° (16 mm.), n_D^{20} 1.4579; n_D^{25} 1.4560. This product should either be used immediately in the next step or stored in a refrigerator (5–10°) under a nitrogen atmosphere.

B. N-(3-*Hydroxy*-3,3-*diphenylpropylidene)cyclohexylamine.* A dry, 250-ml., round-bottomed flask or dry 250-ml. Schlenk tube fitted with a magnetic stirrer is flushed with oxygen-free nitrogen (Note 2), stoppered with a rubber septum, and cooled in an ice bath. A slight positive pressure of oxygen-free nitrogen (Note 2) is maintained in the vessel throughout the reaction with a nitrogen line connected both to a pressure relief valve and a hypodermic needle which is inserted through the rubber septum. With a hypodermic syringe a solution of 2.53 g. (3.60 ml. or 0.0250 mole) of pure diisopropylamine (Note 3) in 25 ml. of absolute ether (Note 4) is added to the cold reaction vessel. An ethereal solution containing 0.025 mole of methyllithium (Note 5) is added from a hypodermic syringe dropwise and with stirring. During this addition a vigorous evolution of methane is observed. After the solution of lithium diisopropylamide has been stirred at 0° for 5–10 minutes a negative Gilman color test[2] for methyllithium is obtained. A solution of 3.13 g. (0.0250 mole) of N-ethylidenecyclohexylamine in 20 ml. of absolute ether (Note 4) is added from a hypodermic syringe, dropwise and with stirring, to the cold (0°) solution of lithium diisopropylamide, and the resulting solution is stirred for 10 minutes (Note 6). This solution is then cooled to −70° with a methanol–dry ice bath, and a solution of 4.55 g. (0.0250 mole) of benzophenone in 25 ml. of absolute ether is added to the cold (−70°) reaction vessel with a hypodermic syringe. The resulting solution is allowed to warm to room temperature and stand for 24 hours, during which time a white solid separates. The reaction mixture is cooled to 0° in an ice bath, treated with approximately 50 ml. of water, and stirred at 0° for 30 minutes. The cold mixture is filtered with suction, removing the white crystalline product, and the organic phase from the filtrate is separated, dried over anhydrous sodium sulfate, and concentrated under reduced pressure. The combined residues from the filtration and concentration of the organic phase of the filtrate are recrystallized from hexane, separating 6.80–7.06 g. (89–92%) of N-(3-hydroxy-3,3-diphenylpropylidene)cyclohexylamine as white needles, m.p. 127–128° (Note 7).

C. *β-Phenylcinnamaldehyde.* A mixture of 1.54 g. (0.00520 mole) of N-(3-hydroxy-3,3-diphenylpropylidene)cyclohexylamine and 10 g. (0.11 mole) of oxalic acid is subjected to steam distillation, which is continued until a clear distillate is obtained; this requires about 2 hours. The steam distillate is extracted with two 25-ml. portions of ether, and the combined ethereal extracts are dried over anhydrous magnesium sulfate and concentrated under reduced pressure. The residual crude product (approximately 1.0 g., m.p. 42–44°) is recrystallized from pentane, yielding 0.80–0.88 g. (78–85%) of β-phenylcinnamaldehyde as pale yellow needles, m.p. 46–47° (Note 8).

2. Notes

1. The amine (m.p. −21°) should be cooled with stirring until it just begins to freeze. At this point the temperature of the external cooling bath should be maintained at −20° by the periodic addition of pieces of dry ice.

2. One suitable arrangement for the purification of nitrogen is described by H. Metzger and E. Müller.[3] The checkers used a prepurified grade of nitrogen without further purification.

3. Diisopropylamine (b.p. 83–84°, available from Fluka AG or from Eastman Organic Chemicals) was purified by refluxing it over either sodium wire or sodium hydride for approximately 30 minutes, then distilling the amine into a dry receiver under a nitrogen atmosphere. Because of the relatively low boiling point of the amine, a dispersion of

sodium hydride in mineral oil, available from Metal Hydrides, Inc., Beverly, Massachusetts, can be used directly in this purification without prior removal of the mineral oil.

4. The submitters purified the ether by refluxing it over sodium wire until the blue color of benzophenone ketyl persisted when benzophenone was added, and distilling the ether into a dry receiver under a nitrogen atmosphere. The checkers further purified an absolute grade of ether obtained from Mallinckrodt Chemical Works by distilling it from lithium aluminum hydride under a nitrogen atmosphere.

5. An ethereal solution of methyllithium may be prepared in the following manner. A dry, 1-l., three-necked flask is fitted with a magnetic stirrer, a gas-inlet tube, and a dry ice reflux condenser. In the flask are placed 800 ml. of absolute ether (Note 4) and 16 g. (2.3 g.-atoms) of pieces of lithium wire. Over a period of 4–5 hours, 100 g. (1.05 moles) of methyl bromide is distilled into the reaction flask with continuous stirring. The resulting mixture is stirred for an additional hour and allowed to stand overnight under a nitrogen atmosphere, permitting the insoluble particles to settle. The supernatant liquid is transferred under a nitrogen atmosphere to a dry storage buret or some other dry vessel capped with a rubber septum. Alternatively, an ethereal solution of methyllithium may be purchased from Foote Mineral Co., Exton, Pennsylvania.

Aliquots of the methyllithium solution should be removed from the storage buret or storage vessel for standardization. The checkers employed the titration procedure of Watson and Eastham [*Org. Synth.*, **Coll. Vol. 5,** 211 (1973)][4] with either 2,2'-dipyridyl or o-phenanthroline as an indicator for standardizing the methyllithium solution.

6. When an aliquot of this solution is subjected to a Gilman color test,[2] a wine-red color is obtained.

7. The product has IR absorption (CHCl$_3$) at 3250 (broad, associated OH) and 1665 cm.$^{-1}$ (C=N) with ^{1}H NMR peaks (CDCl$_3$) at δ 1.0–3.0 (m, 11H, C$_6$H$_{11}$), 3.12 (d, $J = 4$ Hz., 2H, CH$_2$), 7.0–7.7 (m, 10H, 2C$_6$H$_5$), and 7.78 (t, $J = 4$ Hz., 1H, CH=N).

8. The product has IR absorption (CCl$_4$) at 2720, 2745, and 2840 cm.$^{-1}$ (aldehyde CH) and at 1675 cm^{-1} (conjugated C=O) with UV maxima (95% C$_2$H$_5$OH) at 224 nm (ϵ 13,500) and 300 nm (ϵ 16,500) and ^{1}H NMR peaks (CCl$_4$) at δ 6.22 (d, $J = 8$ Hz., 1H, CH), 7.1–7.5 (m, 10H, 2C$_6$H$_5$), and 9.40 (d, $J = 8$ Hz., 1H, CHO).

3. Discussion

Until recently it has not been possible to control the aldol condensation;[5] the enolate anion derived from an aldehyde could not be condensed with a carbonyl group of a ketone, because of the rapidity of the self-condensation of the aldehyde. This problem can be circumvented if the aldehyde is first converted to the corresponding azomethine derivative. The anion derived from this "protected" aldehyde can be added to another carbonyl group, giving an easily crystallized β-hydroxy imine adduct. Subsequent dehydration and concurrent removal of the imino protecting group yields an α,β-unsaturated aldehyde. The overall procedure, utilizing an organometallic intermediate, constitutes a new method for effecting an aldol condensation. With reaction conditions illustrated in this preparation other carbonyl compounds with acidic α-hydrogens, such as acetone or β-ionone, can also be used, because the deprotonation of the carbonyl compound by the metalated Schiff base is largely suppressed. This directed aldol condensation is useful for the preparation of naturally occurring α,β-unsaturated carbonyl compounds or intermediates useful in the syntheses of these substances.[9] The method can also be applied to the synthesis of α,β-unsaturated ketones if a ketimine is used as the azomethine component in the condensation.[9] Although the condensation is also successful with acetaldimine derivatives which contain one α-alkyl substituent, only very poor yields of condensation products are obtained when two α-alkyl substituents are present.[9] This limitation is possibly the result of a retarded rate of proton abstraction from the imine, due to the steric hindrance offered by the α-alkyl substituents.[10]

Although the reaction of aldehydes with β-carbonylmethylene phosphoranes constitutes a good synthetic route to α,β-unsaturated carbonyl compounds,[11-14] this procedure is normally not applicable to ketones. This limitation has recently been overcome by the reaction of ketones with the cyclohexylimine derivative of β-carbonylmethylene-phosphonates.[15]

With the aid of the directed aldol condensation procedure, N-(3-hydroxy-3,3-diphenylpropylidene)cyclohexylamine has been prepared for the first time. Previous methods employed for the synthesis of β-phenylcinnamaldehyde include the application of the Sommelet reaction to 3,3-diphenylallyl bromide with isolation of the aldehyde as its semicarbazone,[16] the reaction of β,β-diphenylvinylmagnesium bromide with N-methylformanilide followed by hydrolysis,[17] and the reaction of this same Grignard reagent with ethyl orthoformate followed by hydrolysis of the resulting acetal.[18] This unsaturated aldehyde has also been prepared by the formylation of 1,1-diphenylethylene with N-methylformanilide and phosphorus oxychloride,[19] by the oxidation of β-phenylcinnamyl alcohol with manganese dioxide,[20] and by the rearrangement of 1,1-diphenyl-2-propyn-1-ol in an ethylene glycol solution containing boron trifluoride and mercury(II) oxide, followed by hydrolysis of the intermediate acetal.[21]

1. Institut für Organische Chemie, Universität Heidelberg, Heidelberg, Germany.
2. H. Gilman and F. Schulze, *J. Am. Chem. Soc.*, **47**, 2002 (1925).
3. H. Metzger and E. Müller, in "Methoden der organischen Chemie" (Houben-Weyl), 4th ed., E. Müller, Ed., Vol 1/2, George Thieme Verlag, Stuttgart, 1959, p. 327.
4. S. C. Watson and J. F. Eastham, *J. Organomet. Chem.*, **9**, 165 (1967).
5. A. T. Nielsen and W. J. Houlihan, *Org. React.*, **16**, 1 (1968).
6. G. Wittig, H. Pommer, and W. Stilz, Ger. Pat. 1,199,252 [*Chem. Abstr.*, **63**, 1739 (1965)].
7. G. Wittig and H. D. Frommeld, *Chem. Ber.*, **97**, 3548 (1964).
8. (a) G. Wittig, H. D. Frommeld, and P. Suchanek, *Angew. Chem. Int. Ed. Engl.*, **2**, 683 (1963); (b) G. Wittig and H. Reiff, *Angew, Chem., Int. Ed. Engl.*, **7**, 7 (1968); (c) G. Wittig, *Rec. Chem. Prog.*, **28**, 45 (1967).
9. G. Wittig and P. Suchanek, *Tetrahedron*, **22**, 347 (1966).
10. H. Reiff, Dissertation, Universität Heidelberg, 1966.
11. S. Trippett and D. M. Walker, *J. Chem. Soc.*, **1961**, 1266.
12. H. Takahashi, K. Fujiwara, and M. Ohta, *Bull. Chem. Soc. Jpn.*, **35**, 1498 (1962).
13. A. K. Bose and R. T. Dahill, Jr., *J. Org. Chem.*, **30**, 505 (1965).
14. A. Maercker, *Org. React.*, **14**, 270 (1965).
15. W. Nagata and Y. Hayase, *Tetrahedron Letters*, **1968**, 4359.
16. K. Ziegler and P. Tiemann, *Ber. Dtsch. Chem. Ges.*, **55**, 3406 (1922).
17. G. Wittig and R. Kethur, *Ber. Dtsch. Chem. Ges.*, **69**, 2078 (1936).
18. E. P. Kohler and R. G. Larsen, *J. Am. Chem. Soc.*, **57**, 1448 (1935).
19. H. Lorenz and R. Wizinger, *Helv. Chim. Acta*, **28**, 607 (1945).

20. R. Heilmann and R. Glenat, *Bull. Soc. Chim. Fr.*, [5] **22**, 1586 (1955).
21. H. Siemer and K. Stack, Ger. Pat. 1,001,674 [*Chem. Abstr.*, **54**, 1454 (1960)].

ALDEHYDES FROM 2-BENZYL-4,4,6-TRIMETHYL-5,6-DIHYDRO-1,3(4*H*)-OXAZINE: 1-PHENYLCYCLOPENTANECARBOXALDEHYDE

(Cyclopentanecarboxaldehyde, 1-phenyl-)

Submitted by IEVA R. POLITZER and A. I. MEYERS[1]
Checked by DENNIS R. RAYNER and RICHARD E. BENSON

1. Procedure

A. 2-(1-*Phenylcyclopentyl*)-4,4,6-*trimethyl*-5,6-*dihydro*-1,3(4H)-*oxazine*. A 1-l., three-necked flask is equipped with a magnetic stirring bar, a 125-ml. pressure-equalizing funnel fitted with a rubber septum, and a nitrogen inlet tube. The system is flushed with nitrogen, and 500 ml. of dry tetrahydrofuran (Note 1) and 21.7 g. (0.100 mole) of

2-benzyl-4,4,6-trimethyl-5,6-dihydro-1,3(4H)-oxazine (Note 2) are added to the flask. The stirred solution is cooled to −78° with an acetone–dry ice bath, and 49 ml. (0.11 mole) of a 2.25 M solution of n-butyllithium in n-hexane (Note 3) is injected into the addition funnel. The n-butyllithium solution is added over a period of 15 minutes, and the funnel is rinsed by injecting 5 ml. of dry tetrahydrofuran. The yellow to orange solution is allowed to stir at −78° for 30 minutes (Note 4).

1,4-Dibromobutane (23.8 g., 0.110 mole) (Note 5) is injected into the addition funnel and added to the solution with stirring over a period of about 15 minutes. The funnel is rinsed by injection of 5 ml. of dry tetrahydrofuran, and the reaction is stirred at −78° for 45 minutes. n-Butyllithium (55 ml., 0.12 mole) in n-hexane is injected into the addition funnel and added to the solution over a period of 15 minutes. The reaction is stirred at −78° for 1 hour and stored at −20° overnight (Note 6). The mixture is poured into about 300 ml. of ice water and acidified to pH 2–3 with 9 N hydrochloric acid. The acidic solution is shaken with three 200-ml. portions of diethyl ether, and the ether extracts are discarded. The aqueous layer is made basic by careful addition of 40% sodium hydroxide (Note 7). The resulting mixture is shaken with four 200-ml. portions of ether, and the ether extracts are dried over anhydrous potassium carbonate. The ether is removed with a rotary evaporator, giving 24.4–25.8 g. (90–95%) of crude 2-(1-phenylcyclopentyl)-4,4,6-trimethyl-5,6-dihydro-1,3(4H)-oxazine, which is sufficiently pure for use in the following step.

B.   2-(1-*Phenylcyclopentyl*)-4,4,6-*trimethyltetrahydro*-1,3-*oxazine*.   A   600-ml. beaker containing a magnetic stirring bar is charged with 200 ml. of tetrahydrofuran, 200 ml. of 95% ethanol, and 25.0 g. (0.0922 mole) of the oxazine obtained in Part A. The mixture is stirred and cooled to −35 to −40° with an acetone bath to which dry ice is added as needed. A 9 N hydrochloric acid solution is added dropwise to the stirred solution until an approximate pH of 7 is obtained as determined by pH paper. A solution of sodium borohydride is prepared by dissolving 5.0 g. (0.13 mole) in a minimum amount of water (5–8 ml.) to which 1 drop of 40% sodium hydroxide is added (Note 8). The sodium borohydride solution and 9 N hydrochloric acid solution are alternately added dropwise to the stirred solution so that a pH 6–8 is maintained (Note 9). During the addition care is taken to maintain a temperature between −35 and −45°. After addition of the borohydride solution is complete, the reaction mixture is stirred at −35° for 1 hour. A pH of 7 is maintained by occasional addition of 9 N hydrochloric acid (Note 10). The reaction mixture is then stored at −20° overnight.

The reaction mixture is poured into 300 ml. of water, and the resulting mixture is made basic with 40% sodium hydroxide. The mixture is shaken three times with 200-ml. portions of ether, and the combined ether extracts are washed with 10 ml. of saturated sodium chloride. After drying over potassium carbonate, the ether is removed with a rotary evaporator, giving 22.9–25.0 g. (91–99%) of product, which is used without purification in the next step (Note 11).

C.   1-*Phenylcyclopentanecarboxyaldehyde*.   The crude tetrahydroöxazine (25.0 g., 0.0916 mole) from Part B is heated at reflux with 300 ml. of water containing 37.8 g. (0.300 mole) of oxalic acid dihydrate for 3 hours. The solution is cooled, and the aldehyde is extracted with four 150-ml. portions of petroleum ether (b.p. 40–60°). The organic extracts are combined, washed with 10 ml. of saturated sodium hydrogen carbonate, and dried with anhydrous powdered magnesium sulfate. The petroleum ether is removed with a rotary evaporator, and the product is distilled through a Vigreux column, giving 7.8–8.7 g. (50–55%) of 1-phenylcyclopentanecarboxaldehyde, b.p. 70–73° (0.1 mm.) $n_D^{26.5}$ 1.5350, IR spectrum (neat) 1720 cm.$^{-1}$ (C=O) (Note 12).

2. Notes

1. Tetrahydrofuran is dried by distillation from lithium aluminum hydride. [See *Org. Synth.*, **Coll. Vol. 5,** 976 (1973) for warning regarding the purification of tetrahydrofuran.]

2. 2-Benzyl-4,4,6-trimethyl-5,6-dihydro-1,3(4*H*)-oxazine is available from Columbia Organic Chemicals Company, Inc. The product was distilled, b.p. 80–82° (0.25 mm.). It may be prepared according to a modification of the method of Ritter and Tillmanns.[2] A 2-l. flask equipped with a thermometer, a stirrer, and a 250-ml. addition funnel is charged with 200 ml. of concentrated sulfuric acid (95–98%). Stirring is begun, the acid is cooled to 0–5° with an ice bath, and 128.7 g. (1.100 moles) of phenylacetonitrile is added from the funnel at such a rate that the temperature is maintained at 0–5°. After the addition is complete, 118 g. (1.00 mole) of 2-methyl-2,4-pentanediol is added at a rate to maintain a reaction temperature of 0–5°. The mixture is stirred for an additional hour and poured onto 700 g. of crushed ice. The aqueous solution is washed with two 75-ml. portions of chloroform, and the extracts discarded. The aqueous solution is made alkaline with 40% sodium hydroxide, with ice being added periodically during the neutralization to keep the solution temperature below 35°. The yellow oil is separated, and the aqueous solution is washed three times with 75-ml. portions of ether. The oil and ether extracts are combined, and the solution is dried over anhydrous potassium carbonate. The ether is removed with a rotary evaporator, and the product is distilled through a 25-cm. Vigreux column, yielding 107–115 g. (49–53%) of 2-benzyl-4,4,6-trimethyl-5,6-dihydro-1,3(4*H*)-oxazine as a straw-yellow liquid, b.p. 78–80° (0.25 mm.); n_D^{27} 1.5085; IR spectrum (neat) 1660 and 1600 cm.$^{-1}$, ^{1}H NMR (CDCl$_3$) δ 1.08 (s, 6H), 3.32 (s, 2H), 4.04 (m, 1H) and 7.08 (m, 5H).

3. *n*-Butyllithium in *n*-hexane is available from Lithium Corporation of America, Inc.

4. If less solvent is used, the anion may appear as a yellow precipitate, but it is also usable in this form.

5. 1,4-Dibromobutane was distilled, b.p. 40–45° (0.5 mm.).

6. Alternatively, the reaction mixture may be allowed to reach room temperature over a period of 2–3 hours.

7. Ice may be added to keep the mixture cool during the neutralization.

8. The aqueous borohydride suspension is warmed, and the borohydride lumps crushed to achieve a homogeneous solution. The product available from Alfa Inorganics, Inc., was used by the checkers.

9. It is convenient to introduce the acid and hydride solutions from two burets or dropping funnels placed above the beaker. A total volume of about 15 ml. of 9 *N* hydrochloric acid is required.

10. The reaction can be conveniently monitored by IR spectroscopy, observing the intensity of the band at 1650 cm.$^{-1}$ (C=N). The submitters observed almost complete disappearance of this band, whereas the checkers found it still present in medium intensity in their product. In those instances where the α-carbon bears three alkyl substituents, steric effects retard the rate of addition, and in some cases (i.e., α,α,α-triethyl or larger groups) the C=N bond is resistant to reduction.

11. The intensity of the IR band at 1650 cm.$^{-1}$ should be weak.

12. The ^{1}H NMR spectrum (CCl$_4$) shows peaks at δ 1.5–2.7 (m, 8H, 4CH_2), 7.2 (s, 5H, C$_6H_5$), and 9.3 (s, 1H, CHO). GC on a 240 cm. × 6 mm. column packed with 10% SE-30 on Chromosorb P at 120° gives a single peak with a retention time of 1.3 minutes.

TABLE I

ALDEHYDES FROM 2-BENZYL-4,4,6-TRIMETHYL-5,6-DIHYDRO-1,3(4H)-OXAZINE [4-7]

Alkylating Agent	Aldehyde	Yield, %
CH$_3$I	C$_6$H$_5$CHCHO $\mid$ CH$_3$	65
CH$_3$I (2.0 equiv.)	CH$_3$ $\mid$ C$_6$H$_5$C—CHO $\mid$ CH$_3$	48
CH$_3$CH$_2$CH$_2$Br	C$_6$H$_5$CHCHO $\mid$ CH$_2$CH$_2$CH$_3$	64
CH$_2$=CHCH$_2$Br	C$_6$H$_5$CHCDO $\mid$ CH$_2$CH=CH$_2$	70[a]
BrCH$_2$CH$_2$Br		57
Br(CH$_2$)$_3$Br		49
	C$_6$H$_5$—C—CHO 	54
	C$_6$H$_5$—C—CHO $\parallel$ C$_6$H$_5$—CH	63

[a] Reduction was performed with NaBD$_4$.

3. Discussion

This procedure illustrates a general method for preparing α-phenyl aldehydes.[3] Additional examples are given in Table I.

A limitation to this reaction sequence is that the reduction of the dihydroöxazine fails if bulky substituents are present. Thus, the alkylated oxazines A and B were not reduced under the reaction conditions specified in this procedure.

This technique may also be modified to prepare acetaldehyde derivatives, using 2,4,4,6-tetramethyl-5,6-dihydro-1,3(4H)-oxazine[3-7] and 2-carboethoxy acetaldehydes, using 2-(carboethoxymethyl)-4,4,6-trimethyl-5,6-dihydro-1,3(4H)-oxazine.[3] Functionalized aldehydes and dialdehydes may also be obtained by suitable modification.[8] Generally, the intermediates can be used without purification, and the overall yields of the aldehydes range from 50 to 70%.

In addition to the present method, 1-phenylcyclopentanecarboxaldehyde has been prepared by the reduction of N-acylaziridines obtained from 1-phenylcyclopentanecarboxylic acid.[9]

1. Department of Chemistry, Wayne State University, Detroit, Michigan 48202. [Present address: Department of Chemistry, Colorado State University, Fort Collins, Colorado 80523.]
2. E.-J. Tillmanns and J. J. Ritter, *J. Org. Chem.,* **22,** 839 (1957).
3. A. I. Meyers, H. W. Adickes, I. R. Politzer, and W. N. Beverung, *J. Am. Chem. Soc.,* **91,** 765 (1969); A. I. Meyers, A. Nabeya, H. W. Adickes, I. R. Politzer, G. R. Malone, A. C. Kovelesky, R. L. Nolen, and R. C. Portnoy, *J. Org. Chem.,* **38,** 36 (1973).
4. J. M. Fitzpatrick, G. R. Malone, I. R. Politzer, H. W. Adickes, and A. I. Meyers, *Org. Prep. Proced.,* **1,** 193 (1969).
5. A. I. Meyers, A. Nabeya, H. W. Adickes, and I. R. Politzer, *J. Am. Chem. Soc.,* **91,** 763 (1969).
6. A. I. Meyers, A. Nabeya, H. W. Adickes, J. M. Fitzpatrick, G. R. Malone, and I. R. Politzer, *J. Am. Chem. Soc.,* **91,** 764 (1969).
7. H. W. Adickes, I. R. Politzer, and A. I. Meyers, *J. Am. Chem. Soc.,* **91,** 2155 (1969).
8. A. I. Meyers, G. R. Malone, and H. W. Adickes, *Tetrahedron Lett.,* 3715 (1970).
9. J. W. Wilt, J. M. Kosturik, and R. C. Orlowski, *J. Org. Chem.,* **30,** 1052 (1965).

AMINES FROM MIXED CARBOXYLIC-CARBONIC ANHYDRIDES: 1-PHENYLCYCLOPENTYLAMINE

(Cyclopentanamine, 1-phenyl-)

Submitted by CARL KAISER and JOSEPH WEINSTOCK[1]
Checked by A. BROSSI, R. A. MICHELI, and L. PORTLAND

1. Procedure

A 1-l., three-necked, round-bottomed flask equipped with an air stirrer, dropping funnel, and low-temperature thermometer is charged with 38.0 g. (0.200 mole) (Note 1) of 1-phenylcyclopentanecarboxylic acid and 150 ml. of acetone. The mixture is stirred, and 22.3 g. (30.6 ml., 0.221 mole) of triethylamine is added over 5 minutes (a 2° rise in temperature is observed). The solution is chilled to −5 to 0° in an ice–salt bath, and 24.0 g. (21.1 ml., 0.221 mole) of ethyl chlorocarbonate (Note 2) in 50 ml. of acetone is added slowly (25 minutes), maintaining the temperature between −5 to 0°. After the addition is complete, the cold mixture is stirred for an additional 15 minutes. A solution of 26.0 g.

(0.400 mole) of sodium azide in 75 ml. of water is added over a 25-minute period while the temperature is kept at −5 to 0°. The mixture is stirred for 30 minutes longer at this temperature, poured into 750 ml. of ice water, and shaken with four 250-ml. portions of toluene (Note 3). The combined toluene extracts are dried over anhydrous magnesium sulfate and transferred to a 2-l., three-necked, round-bottomed flask equipped with a two-necked, Claisen-type adapter, stirrer, and reflux condenser. The stirred solution is heated cautiously (Note 4) under reflux for 1 hour on a steam bath (nitrogen evolution is observed initially). The toluene is then removed at 50° with a rotary evaporator (aspirator pressure). The flask containing the residual, oily isocyanate is again fitted with the Claisen-type adapter, stirrer, and reflux condenser. The oil is stirred, cooled in an ice bath, and 300 ml. of 8 N hydrochloric acid (Note 5) is added. The cooling bath is removed, the stirred mixture is gradually heated on a steam bath (Note 6) until carbon dioxide evolution has subsided (30 minutes), and the solution is heated under reflux for 10 minutes. The flask is evacuated with a water aspirator and warmed in a bath at 50° for about 10 minutes. About 10–20 ml. of distillate is collected before a crystalline product separates (Note 7). Ice water (200 ml.) is added to the flask with cooling in an ice bath, and 1 l. of 2.5 N sodium hydroxide is added slowly to pH 12. The mixture is shaken with three 200-ml. portions of diethyl ether, the combined extracts are dried over anhydrous magnesium sulfate, and the ether is removed by distillation at 50° with a water aspirator, yielding 29.7 g. of an oil. Distillation of the crude product gives 24.5–26.1 g. (76–81%) of 1-phenylcyclopentylamine, b.p. 112–114° (9 mm.), $n_D^{20.5}$ 1.5439 (Note 8).

2. Notes

1. 1-Phenylcyclopentanecarboxylic acid,[2,3] m.p. 157–159°, was obtained from Aldrich Chemical Company, Inc.

2. A commercial grade of ethyl chlorocarbonate, stabilized with calcium carbonate, was employed without purification.

3. Approximately 3.6–3.9 g. of 1-phenylcyclopentanecarboxylic acid (m.p. 158–159.5°) can be recovered by acidification of the aqueous solution which remains after washing with toluene.

4. Rearrangement of the azide must be carried out carefully as the reaction is exothermic, and a large volume of nitrogen is evolved. The submitters have encountered no difficulties if the described dilution (0.2 mole of azide in 1 l. of toluene) is employed. The steam bath should be replaced by a cooling bath if the solution refluxes vigorously.

5. Approximately 8 N hydrochloric acid was prepared by addition of 100 ml. of water to 200 ml. of 37.5% hydrochloric acid.

6. Heating should be gradual and in a large reaction vessel as hydrolysis of isocyanate is accompanied by evolution of carbon dioxide and considerable foaming may occur. If excessive foaming occurs, the steam bath should be removed.

7. The checkers carried the synthesis through to this point without interruption. Since this required about 8 hours on the scale described, the equipment and chemicals were made ready the preceding day.

8. ^{1}H NMR spectrum (CDCl$_3$) δ 1.30 (s, 2H, NH_2), 1.83 (broad s, 8H, 4CH_2), and 7.33 (m, 5H, C$_6$$H_5$). The distilled product was of high purity, as determined by GC

analysis. A 122 cm. × 6.4 mm. O.D. stainless-steel column containing 3% SE-30 on Diatoport S (80–100 mesh) and programmed from 100–200° at 10°/minute was used. The retention time is about 5 minutes.

3. Discussion

1-Phenylcyclopentylamine has also been prepared from 1-phenylcyclopentanecarboxylic acid by the Hofmann degradation of the intermediate amide[4,5] and from the intermediate carboxylic acid chloride by the Curtius reaction.[6] In the method described, using the mixed carboxylic–carbonic anhydride,[7] improved yields of the amine are obtained.

The usual procedure of preparing acid azides, which involves treating an acid chloride with sodium azide,[8,9] suffers from the disadvantage that it is often difficult to obtain pure acid chlorides in good yields from acids which either decompose or undergo isomerization in the presence of mineral acids.[7] Synthesis of the azide by way of the ester and hydrazide[10] has been used to circumvent this difficulty, but is much less convenient. The present procedure permits ready formation of acid azides in excellent yields from mixed carboxylic–carbonic anhydrides and sodium azide under very mild conditions.

A possible limitation to this procedure, however, is that it is dependent upon the relative reactivity of the two carbonyl groups of the mixed carboxylic–carbonic anhydride toward the azide anion. Although either carbonyl group may be attacked by the azide ion, an attack on the more electrophilic carbonyl group is usually strongly favored and high yields of the acid azides generally result. Steric considerations may be important, but preference for azide attack on even a somewhat hindered carboxylic carbonyl is illustrated by the present example in which this group is proximal to phenyl and cyclopentyl groups.

This modification of the Curtius reaction has been used extensively in many laboratories and has been found to be generally applicable. Some examples from the literature include the stereoselective synthesis of a wide variety of cyclopropylamine derivatives from the corresponding acids,[11–13] the stereoselective preparation of some substituted norbornylamines from easily isomerized acids,[14] the preparation of some 1-aminocyclobutanecarboxylic acids from the corresponding acid esters,[15] the preparation of a substituted cyclobutanone from the corresponding cyclobutane-1,1-dicarboxylic acid *via* the 1,1-diamine,[16] and the preparation of a variety of heterocyclic amines from the corresponding acids.[17–19]

1. Smith Kline and French Laboratories, Philadelphia, Pennsylvania 19101.
2. F. H. Case, *J. Am. Chem. Soc.*, **56**, 715 (1934).
3. J. W. Wilt and Brother H. Philip, *J. Org. Chem.*, **24**, 616 (1959).
4. H. Yoshikawa, Japan. Pat. 473 (1964) [*Chem. Abstr.*, **60**, 10595e (1964)].
5. A. Kalir and Z. Pelah, *Isr. J. Chem.*, **5**, 223 (1967) [*Chem. Abstr.*, **68**, 39170j (1968)].
6. P. M. G. Bavin, *J. Med. Chem.*, **9**, 52 (1966).
7. J. Weinstock, *J. Org. Chem.*, **26**, 3511 (1961).
8. C. Naegeli and G. Stefanovitsch, *Helv. Chim. Acta*, **11**, 609 (1928).
9. H. Lindemann, *Helv. Chim. Acta*, **11**, 1027 (1928).
10. P. A. S. Smith, *Org. React.*, **3**, 337 (1946).
11. J. Finkelstein, E. Chiang, F. M. Vane, and J. Lee, *J. Med. Chem.*, **9**, 319 (1966).

12. C. Kaiser, B. M. Lester, C. L. Zirkle, A. Burger, C. S. Davis, T. J. Delia, and L. Zirngibl, *J. Med. Pharm. Chem.*, **5**, 1243 (1962).
13. A. R. Patel, *Acta Chem. Scand.*, **20**, 1424 (1966).
14. G. I. Poos, J. Kleis, R. R. Wittekind, and J. D. Rosenau, *J. Org. Chem.*, **26**, 4898 (1961).
15. A. Burger and S. E. Zimmerman, *Arzneim.-Forsch.*, **16**, 1571 (1966). [*Chem. Abstr.*, **67**, 43491m (1967)].
16. C. Beard and A. Burger, *J. Org. Chem.*, **26**, 2335 (1961).
17. P. F. Juby, R. B. Babel, G. E. Bocian, N. M. Cladel, J. C. Godfrey, B. A. Hall, T. W. Hudyma, G. M. Luke, J. D. Matiskella, W. F. Minor, T. A. Montzka, R. A. Partyka, R. T. Standridge, and L. C. Cheney, *J. Med. Chem.*, **10**, 491 (1967).
18. G. E. Hall and J. Walker, *J. Chem. Soc. C*, 1357 (1966).
19. A. Burger, R. T. Standridge, and E. J. Ariens, *J. Med. Chem.*, **6**, 221 (1963).

cis-2-PHENYLCYCLOPROPANECARBOXYLIC ACID

(Cyclopropanecarboxylic acid, 2-phenyl-, *cis*-)

$$C_6H_5CH=CH_2 + N_2CHCO_2C_2H_5 \longrightarrow C_6H_5\text{—}\triangle\text{—}CO_2C_2H_5$$

Submitted by CARL KAISER, JOSEPH WEINSTOCK,[1]
and M. P. OLMSTEAD
Checked by WILLIAM E. PARHAM, WAYLAND E. NOLAND,
PAUL CAHILL, and THOMAS S. STRAUB

1. Procedure

Caution! Benzene has been identified as a carcinogen; OSHA has issued emergency standards on its use. All procedures involving benzene should be carried out in a well-ventilated hood, and glove protection is required.

A. *Ethyl* cis- *and* trans-*2-phenylcyclopropanecarboxylate.* Xylene (500 ml., Note 1) is heated to reflux in a 2-l. flask equipped with a mechanical stirrer, dropping funnel, and reflux condenser. A solution of 179 g. (1.57 moles) of ethyl diazoacetate [*Org.*

Synth., **Coll. Vol. 4,** 424 (1963)] (Note 2) and 163 g. (1.57 moles) of styrene (Note 3) is placed in the dropping funnel and added dropwise to the refluxing, stirred xylene over a period of 90 minutes. After addition is complete, the solution is stirred and heated at the reflux temperature for an additional 90 minutes (Note 4). Xylene is removed under reduced pressure, and the residual red oil is distilled through a short Vigreux column. The fraction boiling at 85–93° (0.5 mm.) is collected, yielding 155 g. (52%) of colorless product, n_D^{26} 1.5150, n_D^{20} 1.5166 (Notes 5 and 6).

B.   cis-2-*Phenylcyclopropanecarboxylic acid.*   A 1-l., three-necked flask equipped with a dropping funnel, stirrer, and a short Vigreux column (Note 7), to which is attached a partial take-off distilling head with reflux condenser, is charged with 155 g. (0.816 mole) of ethyl *cis*- and *trans*-2-phenylcyclopropanecarboxylate, 200 ml. of ethanol, 65 ml. of water, and 24.5 g. (0.612 mole) of sodium hydroxide pellets (Note 8). The mixture is heated at the reflux temperature for 5 hours during which time 200 ml. of ethanol is slowly distilled and replaced by an equal volume of water added through the dropping funnel. Heating is discontinued, 250 ml. of water and 150 ml. of benzene are added, and the mixture is stirred for 2–3 minutes. The layers are separated, and the aqueous layer is washed with two 50-ml. portions of benzene (Note 9).

The benzene extracts are placed in the apparatus used above, and 130 ml. of water and 13 g. (0.32 mole) of sodium hydroxide pellets are added. Benzene (200 ml.) is distilled and then the 200 ml. of ethanol obtained in the initial hydrolysis is added. The reflux-distillation process is continued for 5 hours, during which time 250 ml. of distillate is obtained. The mixture is cooled, and 65 ml. of benzene and 30 ml. of concentrated hydrochloric acid are added. The layers are separated, and the aqueous solution is washed twice with 35-ml. portions of benzene. The combined benzene extracts are concentrated and dried by distillation of 90 ml. of benzene. The hot concentrate is decanted from a trace of salt, 70 ml. of petroleum ether is added, and the resulting solution is stored overnight at 0°. Filtration of the solid from the cold petroleum ether solution and washing with a small volume of cold 50:50 benzene–petroleum ether yields 19.5–23.8 g. (14.6–20.7% based on the mixed ester used, or 38–46.4% based on the *cis* ester, Note 10) of *cis*-2-phenylcyclopropanecarboxylic acid, m.p. 106–109° (Note 11). An additional 2–3 g. of the *cis* acid may be obtained by concentrating the mother liquors to low volume, adding petroleum ether, and chilling the resulting solution for several days.

2. Notes

1. The checkers used xylene distilled from sodium.
2. Ethyl diazoacetate is available from Aldrich Chemical Co. *Diazoacetic esters are potentially explosive and, therefore, must be handled with caution.* [See *Org. Synth.,* **Coll. Vol. 4,** 424 (1963).]
3. Redistilled styrene, b.p. 52–3° (28 mm.) was used by the submitters. The checkers used reagent grade styrene obtained from Eastman Organic Chemicals without further purification.
4. Refluxing was discontinued after nitrogen evolution ceased.
5. GC of several samples of the mixed ester revealed a composition of 55–65% *trans*-, 30–40% *cis*-ester, and 5% impurities.

6. The checker obtained 204 g. of product (68% yield), n_D^{26} 1.5160, with the approximate composition 39% *cis*–60% *trans*-ester.

7. The checkers used a 50-cm. Vigreux column.

8. The amount of base used is 0.75 mole per mole of mixed esters. This is slightly more than necessary to saponify all the *trans*-ester present. Since the *trans*-ester is saponified more rapidly than the *cis*-ester, this affords an effective separation of the isomer. This procedure is a modification of that of Walborsky and Plonsker.[2]

9. *trans*-2-Phenylcyclopropanecarboxylic acid may be obtained from the aqueous solution in the following way. The aqueous solution is treated with 65 ml. of concentrated hydrochloric acid, and the mixture is extracted with one 130-ml. portion of benzene and two 20-ml. portions of benzene. The carefully separated benzene layers are combined and dried by distilling 100 ml. of benzene. The resulting solution is decanted from the small amount of salt present and diluted with 200 ml. of petroleum ether. The resulting solution is cooled at 0° overnight, and the precipitate is collected and washed with a small amount of a cold, 50:50 mixture of petroleum ether–benzene, yielding 63–65 g. (80% based on *trans*-ester originally present) of needles, m.p. 87–93°. Recrystallization of this product several times from carbon tetrachloride–petroleum ether gives the pure *trans*-acid, m.p. 93°.

10. The submitters obtained 25–28 g. of *cis*-acid (19–21% based on the mixed ester used).

11. Burger and Yost[3] reported m.p. 106–107° for pure *cis*-2-phenylcyclopropanecarboxylic acid and m.p. 93° for pure *trans*-2-phenylcyclopropanecarboxylic acid.

3. Discussion

The method described is a modification of that described by Walborsky and Plonsker.[2] It is based on the more rapid hydrolysis, due to less steric hindrance, of the *trans*- over the *cis*-ester. The *cis*-acid has also been obtained by fractional crystallization of the mixed acids.[3]

The present method offers a convenient preparation of *cis*-2-phenylcyclopropanecarboxylic acid that is amenable to large-scale work. The process above has been carried out by the submitters on a twentyfold scale with essentially the same results. The method also provides an example of the separation of two isomers based on differences in reaction rate.

1. Smith Kline and French Laboratories, Philadelphia, Pennsylvania 19101.
2. H. M. Walborsky and L. Plonsker, *J. Am. Chem. Soc.*, **83**, 2138 (1961).
3. A. Burger and W. L. Yost, *J. Am. Chem. Soc.*, **70**, 2198 (1948).

SUBSTITUTION OF ARYL HALIDES WITH COPPER(I) ACETYLIDES: 2-PHENYLFURO[3,2-*b*]PYRIDINE

(Furo[3,2-*b*]pyridine, 2-phenyl-)

$$Cu^{II}(NH_3)_4{}^{+2} \xrightarrow{\text{HONH}_3{}^+ \text{Cl}^-} Cu^I(NH_3)_2{}^+$$

$$Cu^I(NH_3)_2{}^+ + C_6H_5-C{\equiv}CH \longrightarrow C_6H_5-C{\equiv}C-Cu + NH_4{}^+ + NH_3$$

Submitted by D. C. Owsley and C. E. Castro[1]
Checked by Michael J. Umen and Herbert O. House

1. Procedure

A. *Copper(I) phenylacetylide.* A 2-l. Erlenmeyer flask fitted with a large magnetic stirring bar (Note 1) and an ice-water cooling bath is charged with a solution of 25.0 g. (0.100 mole) of copper(II) sulfate pentahydrate (Note 2) in 100 ml. of concentrated aqueous ammonia. The solution is stirred with cooling for 5 minutes while a stream of nitrogen is passed over the solution (Note 3). Water (400 ml.) is added, and stirring and cooling under a nitrogen atmosphere (Note 3) are continued for 5 minutes. Solid hydroxylamine hydrochloride (13.9 g., 0.200 mole, Note 4) is added to the reaction solution, with continuous stirring and cooling under nitrogen, over 10 minutes (Note 5). A solution of 10.25 g. (0.1005 mole) of phenylacetylene (Note 6) in 500 ml. of 95% ethanol is then added rapidly to the pale blue solution. The reaction flask is swirled by hand, copper(I) phenylacetylide separates as a copious yellow precipitate, and an additional 500 ml. of water is added. After the mixture has been allowed to stand for 5 minutes, the precipitate is collected on a sintered glass filter (Note 7) and washed successively with five 100-ml. portions of water, five 100-ml. portions of absolute ethanol, and five 100-ml. portions of anhydrous diethyl ether. The copper(I) acetylide is dried in a 250-ml., round-bottom flask heated to 65° for 4 hours under reduced pressure on a rotary evaporator, yielding 14.8–16.4 g. (90–99%) of a bright yellow solid. The dry acetylide may be stored under nitrogen in a brown bottle (Note 8).

B. *2-Phenylfuro[3,2-b]pyridine.* A 300-ml., three-necked flask fitted with a nitrogen inlet stopcock, a magnetic stirring bar, and a condenser attached to a nitrogen outlet stopcock and a mercury trap is charged with 2.47 g. (0.0150 mole) of copper(I) phenylacetylide. The system is purged with nitrogen for 20 minutes before 80 ml. of pyridine (Note 9) is added. The resulting mixture is stirred for 20 minutes under a nitrogen atmosphere (Note 10), and 3.30 g. (0.0149 mole) of 3-hydroxy-2-iodopyridine (Note 11) is added. The mixture, which changes in color from yellow to dark green as the acetylide dissolves (Note 12), is warmed in an oil bath at 110–120° for 9 hours with continuous stirring under a nitrogen atmosphere (Note 10). The reaction solution is transferred to a 500-ml., round-bottom flask and concentrated to a volume of 20 ml. at 60–70° (20–80 mm.) with a rotary evaporator. The pyridine solution is treated with 100 ml. of concentrated aqueous ammonia, and the resulting deep-blue mixture is stirred for 10 minutes and extracted with five 100-ml. portions of ether. The combined ethereal extracts are washed with three 250-ml. portions of water, dried over anhydrous magnesium sulfate, and concentrated with a rotary evaporator. The crude product, 2.6–2.76 g. of orange semisolid, is dissolved in 100 ml. of boiling cyclohexane. The solution is filtered, concentrated to a volume of about 30 ml., and cooled in an ice bath. The partially purified product crystallizes as 2.3–2.7 g. of orange solid, m.p. 83–89°. Further purification is effected by sublimation at 110–120° (0.01–0.2 mm.), yielding 2.2–2.4 g. (75–82%) of a yellow solid, m.p. 90–91° (Note 13).

2. Notes

1. An 8-cm., Teflon-coated stirring bar is convenient.

2. A reagent grade copper(II) sulfate pentahydrate, purchased from either Mallinckrodt Chemical Works or J. T. Baker Chemical Company, may be employed.

3. A nitrogen atmosphere is maintained above the reaction solution throughout the preparation of the copper(I) acetylide.

4. Material of satisfactory purity was obtained either from J. T. Baker Chemical Company or from Matheson, Coleman and Bell.

5. Too rapid an addition of the hydroxylamine salt results in precipitation of a dark solid that dissolves slowly. If solids do separate, they should be pulverized to hasten solution.

6. Phenylacetylene, purchased either from K & K Laboratories or from Aldrich Chemical Company, Inc., was used without purification.

7. A 600-ml., coarse porosity, sintered glass filter is recommended to shorten the filtration time. The filtration may also be hastened by periodically scraping the bottom of the funnel with a spatula.

8. The submitters report that the acetylide is stable for years under these conditions.

9. A reagent grade of pyridine, purchased from either J. T. Baker Chemical Company or Fisher Scientific Company, was employed.

10. Oxygen will convert the acetylide to 1,4-diphenylbutadiyne.[4a]

11. This material, obtained from Aldrich Chemical Company, Inc., was used without purification.

12. Although the reaction mixture becomes homogeneous in this example, the submitters report that only partial solution occurs in other successful substitution reactions. The solubilities of the acetylides and the heterogeneous character of the cyclization have been described.[2]

13. The product exhibits UV maxima (95% C_2H_5OH solution) at 312 nm (ϵ 32,900) and 326 nm (ϵ 27,100) with ^1H NMR peaks (acetone-d_6) at δ 7.1–8.1 (m, 8H) and 8.49 (d of d, J = 1.4 and 4.7 Hz., 1H). The mass spectrum has the following relatively abundant peaks: m/e (rel. int.), 196 (25), 195 (100, M$^+$), 166 (13), 139 (8), 102 (5), and 39 (6).

3. Discussion

Copper(I) acetylides can be prepared from ammoniacal copper(I) iodide and acetylenes.[3,4b] The generation of fresh solutions of the copper(I) salts results in a higher purity acetylide.

The substitution of aryl halides by copper(I) acetylides provides a convenient, high-yield route to aromatic acetylenes.[4] Aliphatic acetylenes can also be obtained under forcing conditions.[5] The procedure is also useful for the preparation of conjugated acetylenic ketones and alkynyl sulfides.[2] Moreover, the reaction provides the basis for the facile synthesis of an exceedingly broad scope of heterocycles. Thus, a halide bearing an adjacent nucleophilic substituent can be cyclized by the copper(I) salt. The example described is illustrative of the preparation of indoles,[4a] benzo[b]thiophenes,[6] phthalides,[4a] benzofurans,[4a] 3(H)-isobenzofurans,[2] furans,[5] 1(H)-2-benzopyrans,[2] 1(H)-thieno[3,4-b]-2-pyranones,[7] furo[3,2-b]pyridines,[7] furo[3,2-c]pyridines,[7] pyrrolo[3,2-b]pyridines,[2] and 4,5-dihydro-4-keto[3] benzoxepins. The furo[3,2-b]pyridine system has only been prepared by this route.[7]

1. Department of Nematology, University of California, Riverside, California 92502.
2. C. E. Castro, R. H. Havlin, V. K. Honwad, A. M. Malte, and S. W. Mojé, *J. Am. Chem. Soc.*, **91**, 6464 (1969).
3. V. A. Sazonova, and N. Ya. Kronrod, *J. Gen. Chem. U.S.S.R.* (Engl. Transl.), **26**, 2093 (1956).
4. (a) C. E. Castro, E. J. Gaughan, and D. C. Owsley, *J. Org. Chem.,* **31**, 4071 (1966); (b) C. E. Castro and R. D. Stephens, *J. Org. Chem.,* **28**, 2163, 3313 (1963); (c) S. A. Kandil and R. E. Dessy, *J. Am. Chem. Soc.,* **88**, 3027 (1966); (d) M. D. Rausch, A. Siegel, and L. P. Klemann, *J. Org. Chem.,* **31**, 2703 (1966); (e) R. E. Atkinson, R. F. Curtis, D. M. Jones, and J. A. Taylor, *Chem. Commun.,* 718. (1967); (f) R. E. Atkinson, R. F. Curtis, and J. A. Taylor, *J. Chem. Soc.,* C, 578 (1967).
5. K. Gump, S. W. Mojé, and C. E. Castro, *J. Am. Chem. Soc.,* **89**, 6770 (1967).
6. A. M. Malte and C. E. Castro, *J. Am. Chem. Soc.,* **89**, 6770 (1967).
7. S. A. Mladenovič and C. E. Castro, *J. Heterocycl. Chem.,* **5**, 227 (1968).

KETONES AND ALCOHOLS FROM ORGANOBORANES:
PHENYL HEPTYL KETONE, 1-HEXANOL, AND
1-OCTANOL

(1-Octanone, 1-phenyl-)

A. $n\text{-}C_4H_9CH{=}CH_2 + BH_3 \longrightarrow (C_6H_{13})_3B$

B. $(C_6H_{13})_3B + N_2CHCOC_6H_5 \xrightarrow{H_2O} n\text{-}C_7H_{15}COC_6H_5$

C. $(C_6H_{13})_3B \xrightarrow[H_2O_2]{NaOH} n\text{-}C_6H_{13}OH + n\text{-}C_4H_9CH(OH)CH_3$

D. $(CH_3)_2C{=}CHCH_3 + BH_3 \longrightarrow [(CH_3)_2CHCH(CH_3)]_2BH$

E. $[(CH_3)_2CHCH(CH_3)]_2BH + n\text{-}C_6H_{13}CH{=}CH_2 \longrightarrow$

$$[(CH_3)_2CHCH(CH_3)]_2B{-}n\text{-}C_8H_{17}$$

F. $[(CH_3)_2CHCH(CH_3)]_2B{-}n\text{-}C_8H_{17} \xrightarrow[H_2O_2]{NaOH}$

$$(CH_3)_2CHCH(OH)CH_3 + n\text{-}C_8H_{17}OH$$

Submitted by HIROMICHI KONO and JOHN HOOZ[1]
Checked by DENNIS R. MURAYAMA, JACK EMERT,
J. M. PECORARO, and RONALD BRESLOW

1. Procedure

A. *Trihexylborane.* A dry, 1-1., three-necked flask is equipped with a magnetic stirring bar, a reflux condenser fitted with a drying tube, a pressure-equalizing dropping funnel to which is attached a rubber septum cap, and a three-way, parallel sidearm connecting tube fitted with a thermometer and an inlet tube (containing a stopcock), permitting introduction of a dry nitrogen atmosphere. The apparatus is flushed with nitrogen and charged with 27.8 g. (0.331 mole) of 1-hexene (Note 1) and 150 ml. of anhydrous tetrahydrofuran (Note 2) with a hypodermic syringe, and 103 ml. (0.110 mole) of a 1.07 M solution of borane in tetrahydrofuran (Note 3) is added dropwise over a 20-minute period to the stirred solution, while the reaction temperature is maintained below *ca.* 20° with an ice bath. After the addition, the reaction mixture is stirred for an additional one hour at room temperature. The resulting solution of trihexylborane (Note 4) is ready for use in the next step.

B. *Phenyl heptyl ketone.* To the solution prepared in Section A. is added 18 ml. (1.0 mole) of water. The nitrogen flow is ceased, and the drying tube is quickly replaced with a stopcock attached with Tygon tubing, to a gas-measuring tube. A solution of 14.6 g. (0.100 mole) of diazoacetophenone (Note 5) in 125 ml. of tetrahydrofuran is added to the stirred solution over a period of one hour. After the addition is complete, the mixture is stirred vigorously for one hour at room temperature, then heated to reflux for one hour. The resulting mixture is cooled to approximately 25° with an ice bath (Notes 6, 7, and 8). A solution of 73 ml. (0.2 mole) of 3 N sodium acetate solution is added, followed by the dropwise addition of 23 ml. (0.22 mole) of 30% hydrogen peroxide, maintaining the reaction temperature below *ca.* 20° (ice-cooling). The cooling bath is then removed, and the mixture is stirred at room temperature for one hour.

The resulting mixture is saturated with sodium chloride. The organic phase is separated, washed with three 50-ml. portions of saturated brine solution, dried over sodium sulfate, and concentrated on a rotary evaporator. Distillation of the residue through a 7-cm. Vigreux column separates 15.34–15.99 g. (75–80%) of phenyl heptyl ketone, b.p. 118–120° (0.60 mm.) n_D^{26} 1.5034 (Note 9).

C. 1-*Hexanol.* To a solution of trihexylborane in a 500-ml. three-necked flask [prepared from 25.3 g. (0.301 mole) of 1-hexene in 150 ml. of tetrahydrofuran and 84 ml. of a 1.20 *M* solution of borane in tetrahydrofuran, as described in Section A] is added 34 ml. (0.1 mole) of a 3 *N* solution of sodium hydroxide. This is followed by the dropwise addition of 36 ml. (0.35 mole) of 30% hydrogen peroxide at a rate such that the reaction temperature is maintained at approximately 35° (water bath). After being stirred at room temperature for one hour, the mixture is poured into 100 ml. of water. The organic phase is separated, and the aqueous phase is extracted with 50 ml. of diethyl ether. The combined organic extracts are washed with three 50-ml. portions of saturated brine solution and dried over Drierite. After the bulk of the solvent is removed by distillation, the residue is fractionated with a 24-in., Teflon spinning band column (Note 10), yielding 7.7–8.2 g. (25.1–26.7%) of 1-hexanol of 95% purity, b.p. 145–153° and 10.1–15.4 g. (33.3–50.3%) of pure 1-hexanol, b.p. 153–155° (Note 11). The total yield of material with >95% purity is 58.4–77%.

D. *Bis(3-methyl-2-butyl)borane (disiamylborane).*[2] A dry, 500-ml., three-necked flask is equipped as described in Section A. The apparatus is flushed with nitrogen, and the flask is charged with 92 ml. (0.11 mole) of a 1.2 *M* solution of borane in tetrahydrofuran. The flask is cooled with an ice bath before a solution of 15.4 g. (0.220 mole) of 2-methyl-2-butene (Note 12) in 40 ml. of anhydrous tetrahydrofuran is added to the stirred solution over a 30-minute period. After the addition is complete, the reaction mixture is kept below *ca.* 10° for 2 hours. The resulting solution is used directly in the next step.

E. *Addition of disiamylborane to 1-octene.* To the solution prepared in Section D is added a solution of 11.2 g. (0.100 mole) of 1-octene (Note 13) in 20 ml. of anhydrous tetrahydrofuran over a 30-minute period, while the reaction temperature is maintained below *ca.* 20°. The ice bath is removed and stirring is continued for one hour at room temperature.

F. 1-*Octanol.* The stirred solution prepared in Section E is cooled below *ca.* 10° with an ice bath, and a solution of 34 ml. (0.1 mole) of 3 *N* sodium hydroxide is introduced. This is followed by the dropwise addition of 36 ml. (0.35 mole) of 30% hydrogen peroxide, added at a rate such that the reaction temperature is maintained between 30–35°. After the addition is complete, the mixture is stirred at room temperature for 1.5 hours. The reaction mixture is then extracted with 100 ml. of ether. The ether extract is washed with four 100-ml. portions of water and dried over Drierite. After removal of solvent on a rotary evaporator, the residue is distilled through a 3-cm. Vigreux column, giving, as a forerun, 9.6–10.5 g. of 3-methyl-2-butanol, b.p. 110–115°, followed by 8.5–9.1 g. (65–70%) of 1-octanol, b.p. 182–186° (Note 14).

2. Notes

1. 1-Hexene (99%), purchased from Aldrich Chemical Company, Inc., was stored over molecular sieves and distilled prior to use.

2. Commercial tetrahydrofuran, purchased from British Drug House (Canada) Ltd. or Fisher Scientific Company, was refluxed over sodium metal, distilled from sodium metal, then redistilled from lithium aluminum hydride under a nitrogen atmosphere. [See *Org. Synth.*, **Coll. Vol. 5,** 976 (1973) for warning regarding the purification of tetrahydrofuran.]

3. A commercial one molar solution of borane in tetrahydrofuran, obtained from Alfa Inorganics, Inc., was standardized by measuring the amount of hydrogen evolved on titration with 40% aqueous ethylene glycol.

4. The trihexylborane solution contains approximately 94% primary and 6% secondary boron-bound alkyl groups.[3,4]

5. Diazoacetophenone was prepared as described in *Org. Synth.*, **Coll. Vol. 6,** 386 (1988).

6. Approximately 95% of the theoretical amount of nitrogen is evolved.

7. GC indicates a 92% yield of product. Using a 10 ft. by 0.25 in. column packed with 20% NPGSE (Neopentyl Glycol Sebacate Ester) suspended on Chromosorb W heated to 235° and a helium flow rate of 60 ml. per minute, the submitters found a retention time of 21 minutes for phenyl heptyl ketone.

8. Distillation of the product from the crude reaction mixture at this stage gives somewhat lower yields. Therefore, residual organoboranes are oxidized prior to isolation of product.

9. 1-Hexanol, 18.5–20.3 g. (78–86%), boiling at approximately 35° (0.75 mm.), is obtained as a forerun. The checkers found that two distillations were required to give a product of >95% purity.

10. A forerun of approximately 1.0 g., b.p. 135–145°, comprised largely of 2-hexanol (75%), is discarded. The submitters used a stainless-steel spinning band column with equivalent results.

11. The product may be analyzed by use of a GC column packed with 20% SF-96 suspended on Chromosorb WAW, 5 ft. by 0.25 in., operated at 105°. Using a helium flow rate of 60 ml. per minute, the submitters found a retention time of 4 minutes.

12. Commercial 2-methyl-2-butene (99%), purchased from Chemical Samples Company, 4692 Kenny Road, Columbus, Ohio 43220, was used as received.

13. 1-Octene (97%), b.p. 122–123°, purchased from the Aldrich Chemical Company, Inc., was stored over molecular sieves and distilled prior to use.

14. The product may be analyzed by using a GC column packed with 20% SF-96 suspended on Chromosorb WAW, 5 ft. by 0.25 in., operated at 140°. Using a helium flow rate of 60 ml. per minute the submitters found a retention time of 3.5 minutes.

3. Discussion

Phenyl heptyl ketone has been prepared by the Friedel-Crafts acylation of benzene with octanoyl chloride.[5] It is also a product of the thermal decomposition of the mixed iron(II) salts of benzoic and octanoic acids.[6]

The present preparation of phenyl heptyl ketone illustrates the formation of a homologated ketone from the reaction of a trialkylborane with an α-diazoketone. It is representative of a fairly general reaction between an organoborane and a stabilized diazo compound, as illustrated in the accompanying equation, yielding the corresponding ketone,[7] diketone,[8] nitrile,[9] ester,[9] or aldehyde.[10]

$$R_3B + N_2CHA \xrightarrow[H_2O]{-N_2} RCH_2A$$

$$A = COCH_3, COC_6H_5, CO(CH_2)_nCOCHN_2, CN, CO_2C_2H_5, CHO$$

The extent of reaction is conveniently monitored by measuring the quantity of nitrogen evolved. Organoboranes derived from terminal olefins react readily (>90% gas evolution) at room temperature or below, whereas more highly hindered organoboranes react more sluggishly (ca. 3–6 hours of reflux) to complete the liberation of nitrogen.

The enol borinate intermediates are rapidly hydrolyzed to product in the presence of water.[10,11] Since neither the organoborane nor diazo compound reacts with water appreciably under the experimental conditions, hydrolysis is conveniently accomplished in situ by adding water to the organoborane solution prior to the addition of diazo substrate. Although the product may be isolated from the crude mixture by extraction and distillation, an oxidation step (to convert residual organic boron-containing material to boric acid) is employed, since it gives somewhat higher isolated yields.

An adaptation of the procedure, employing deuterium oxide as the hydrolytic medium, permits the synthesis of α-deuterio ketones and esters in high isotopic purity. α,α-

$$R_3B + N_2CHA \xrightarrow{D_2O} R-\underset{\underset{D}{|}}{C}H-A$$

$$A = COCH_3, COC_6H_5, CO_2C_2H_5$$

Dideuterio ketones and esters are also produced in high purity using the appropriate α-deuteriodiazocarbonyl precursor.[12]

A useful extension of the facile, in situ hydrolysis is the alkylation of cyclic α-diazo

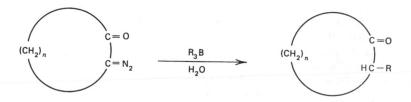

ketones[13] ($n = 3, 4, 5, 6$). This adaptation obviates the necessity of the several separate (yield-lowering) steps required for the removal of activating or blocking groups by other alkylation methods.[14]

The principal disadvantage of this procedure is that only one alkyl group of the trialkylborane is constructively utilized. The reaction is also sensitive to steric factors. Although yields are excellent for terminal olefins, the reaction becomes more sluggish and yields of ketone decrease progressively as steric effects in the trialkylborane are increased. The method is of limited utility for rare olefins. However, the overall simplicity, mild reaction conditions, and absence of any isomeric contaminants recommend the method for reactions involving rarer diazocarbonyl substrates.

Apart from the oxidation of trihexylborane,[15] 1-hexanol has been prepared by a previous *Organic Syntheses*[16] procedure involving the reaction of ethylene oxide with *n*-butylmagnesium bromide; alternate methods of synthesis are reviewed therein.

The present preparation illustrates the hydroboration[17] of a terminal olefin and the oxidation of the resultant trialkylborane.

The hydroboration of an olefin involves a *cis* addition of a boron–hydrogen bond to an alkene linkage, and for unsymmetric olefins occurs in an anti-Markownikoff fashion. 1-Alkenes and simple 1,2-disubstituted olefins undergo rapid conversion to the corresponding trialkylborane, whereas addition of diborane to tri- and tetrasubstituted olefins may be conveniently terminated at the respective di- and monoalkylborane stage. 1-Alkenes yield trialkylboranes in which there is a preponderant (approximately 94%) addition of the boron atom to the terminal carbon.[3,4]

The oxidation of a trialkylborane may be effected by perbenzoic acid or by aqueous hydrogen peroxide in the presence of alkali.[19] A detailed systematic study of the reaction parameters (oxidation temperature, base concentration, hydrogen peroxide concentration) of the latter method has led to the development[3] of a standard and common procedure for oxidizing organoboranes, and is illustrated in the present procedure.

The oxidation step occurs with retention of configuration of the carbon atom undergoing migration. The mechanism is believed to proceed as illustrated in the following equation.[3] As a result, the sequence involving the hydroboration of an olefin followed by

$$H_2O_2 + OH^- \rightleftharpoons HOO^- + H_2O$$

treatment with sodium hydroxide–hydrogen peroxide constitutes a useful device for effecting the overall anti-Markownikoff *cis* hydration of an olefin.

The principal disadvantage of this procedure resides in its application to terminal olefins. Since the hydroboration step produces *ca.* 94% primary boron-bound alkyl groups, the maximum purity of primary carbinol is obviously limited to *ca.* 94%. Isolation of primary alcohol free of the contaminant secondary alcohol requires a tedious, yield-lowering fractionation procedure. This difficulty may be circumvented by employing a more selective hydroborating reagent, disiamylborane, as illustrated in the synthesis of 1-octanol.

1-Octanol has previously been prepared from ethyl caprylate by catalytic hydrogenolysis,[20] and by the Bouveault-Blanc method using sodium and alcohol in toluene.[21] Other preparative methods include the reaction between *n*-hexylmagnesium bromide and ethylene oxide,[22] and the oxidation of trioctylborane.[23]

The hydroboration of a trisubstituted olefin, exemplified by the reaction of 2-methyl-2-butene with diborane, is conveniently stopped at the dialkylborane stage, producing disiamylborane. As a result of its rather large steric requirements this reagent selectively hydroborates terminal olefins, placing *ca.* 99% of the boron atom on the terminal carbon.

$$(CH_3)_2C{=}CHCH_3 + BH_3 \longrightarrow \left[(CH_3)_2CH{-}\overset{\overset{\displaystyle CH_3}{|}}{CH} \right]_2 BH$$

Consequently, oxidation produces essentially homogeneous 1-alkanol. This procedure is the method of choice for converting terminal olefins to primary alcohols without the accompanying formation of isomers.

$$R_2BH + R'CH{=}CH_2 \longrightarrow R'{-}CH_2CH_2{-}BR_2$$
$$R = (CH_3)_2CH{-}\overset{\displaystyle |}{CH}(CH_3)$$

The advantages of a hydroboration-oxidation synthesis of alcohols are simplicity of procedure, relatively mild reaction conditions, high overall yields, absence of skeletal rearrangements, and production of a carbinol in which there is an overall *cis* addition of water to a double bond in an anti-Markownikoff sense.

1. Department of Chemistry, University of Alberta, Edmonton, Alberta, Canada T6G ZGZ.
2. This trivial term, which now finds common usage, was coined as a contraction of the only *sec*-isoamyl structure possible, $(CH_3)_2CHCH(CH_3)$.[3]
3. H. C. Brown, "Hydroboration," W. A. Benjamin, New York, 1962.
4. G. Zweifel and H. C. Brown, *Org. React.*, **13**, 1 (1963).
5. F. L. Breusch and M. Oguzer, *Chem. Ber.*, **87**, 1225 (1954).
6. C. Granito and H. P. Schultz, *J. Org. Chem.*, **28**, 879 (1963).
7. J. Hooz and S. Linke, *J. Am. Chem. Soc.*, **90**, 5936 (1968).
8. J. Hooz and D. M. Gunn, *J. Chem. Soc. D*, 139 (1969).
9. J. Hooz and S. Linke, *J. Am. Chem. Soc.*, **90**, 6891 (1968).
10. J. Hooz and G. F. Morrison, *Can. J. Chem.*, **48**, 868 (1970).
11. D. J. Pasto and P. W. Wojtkowski, *Tetrahedron Lett.*, 215 (1970).
12. J. Hooz and D. M. Gunn, *J. Am. Chem. Soc.*, **91**, 6195 (1969).

13. J. Hooz, D. M. Gunn and H. Kono, *Can. J. Chem.*, **49**, 2371 (1971).
14. H. O. House, "Modern Synthetic Reactions." W. A. Benjamin, New York, 1965, p. 195.
15. H. C. Brown and G. Zweifel, *J. Am. Chem. Soc.*, **82**, 4708 (1960).
16. E. E. Dreger, *Org. Synth.*, **Coll. Vol. 2**, 306 (1941).
17. Although the term hydroboration is most commonly employed[3,4,14] to denote the addition of a boron–hydrogen linkage to carbon–carbon multiple bonds, it has also been used "for the two-step oxidative process to distinguish it from the process of reduction involving H-B addition and protonolysis."[18]
18. L. F. Fieser and M. Fieser, "Reagents for Organic Synthesis," Vol. 1, Wiley, New York, 1967, p. 203.
19. J. R. Johnson and M. G. Van Campen, Jr., *J. Am. Chem. Soc.*, **60**, 121 (1938).
20. K. Folkers and H. Adkins, *J. Am. Chem. Soc.*, **54**, 1145 (1932).
21. C. S. Marvel and A. L. Tanenbaum, *J. Am. Chem. Soc.*, **44**, 2645 (1922).
22. R. C. Huston and A. H. Agett, *J. Org. Chem.*, **6**, 123 (1944).
23. H. C. Brown and B. C. Subba Rao, *J. Am. Chem. Soc.*, **81**, 6423 (1959).

1-PHENYL-1,4-PENTADIYNE AND 1-PHENYL-1,3-PENTADIYNE

(Benzene, 1,4-pentadiynyl- and Benzene, 1,3-pentadiynyl-)

$$C_6H_5C\equiv CMgBr + BrCH_2C\equiv CH \xrightarrow[\text{THF}]{\text{CuCl}}$$

$$C_6H_5C\equiv CCH_2C\equiv CH \xrightarrow[\text{C}_2\text{H}_5\text{OH}]{\text{NaOH}} C_6H_5C\equiv CC\equiv CCH_3$$

Submitted by H. TANIGUCHI, I. M. MATHAI, and SIDNEY I. MILLER[1]
Checked by MARION F. HABIBI and RICHARD E. BENSON

1. Procedure

A. *Phenylethynylmagnesium bromide.* A 1-l., four-necked flask fitted with a sealed mechanical stirrer, a reflux condenser carrying calcium chloride and soda lime tubes, a nitrogen gas inlet, and a dropping funnel is charged with 19 g. (0.81 g.-atom) of magnesium. The flask is flushed with prepurified nitrogen, the stirrer is started, and 109 g. of ethyl bromide (1.00 mole) in 350 ml. of anhydrous tetrahydrofuran (Note 1) is added. After the magnesium has dissolved (Note 2), 102 g. (1.00 mole) of phenyl-acetylene (Note 3) in 150 ml. of tetrahydrofuran is added over a period of *ca.* 30 minutes at a rate that maintains a gentle reflux (Note 4). The reaction mixture is then heated at reflux for *ca.* 1.5 hours (Note 5).

B. *1-Phenyl-1,4-pentadiyne.* Anhydrous copper(I) chloride (2 g.) is added to the flask, and heating under reflux is continued for 20 minutes. At this point 96 g. (0.81 mole) of propargyl bromide in 120 ml. of tetrahydrofuran (Notes 1 and 6) is added over 30–40 minutes at a rate that maintains a gentle reflux. The mixture, containing a yellow solid, is heated another 30–40 minutes, allowed to cool to ambient temperature, and poured slowly into 2 1. of ice-water slush containing 50 ml. of concentrated sulfuric acid. The whole mixture, which should be acidic at this point, is stirred thoroughly and extracted five times with 200-ml. portions of diethyl ether (Note 7). The ether extracts are combined and washed with 100-ml. portions of water until the water layer is no longer acidic to litmus paper (3–4 washings are usually required). The ether layer is dried over magnesium sulfate overnight, separated from the desiccant, and concentrated by distillation.

The product is distilled under nitrogen using a 25-cm. Vigreux column (Notes 8 and 9). After removal of unreacted propargyl bromide and phenylacetylene, 51–64 g. (45–57%) of 1-phenyl-1,4-pentadiyne is collected as the main fraction, b.p. 64–66° (0.45 mm.), n_D^{25} 1.5713 (Notes 9 and 10). The colorless product is best stored under nitrogen at ca. −78°. The ^{1}H NMR spectrum (60 mHz., neat, external tetramethylsilane reference) shows peaks at δ 1.76 (t, 1H), 2.92 (d, 2H), and 6.87 (m, 5H) (Note 11).

C. 1-*Phenyl*-1,3-*pentadiyne*. A flask containing a magnetic stirrer and a solution of 2 g. of sodium hydroxide in 50 ml. of ethanol is flushed with nitrogen, and 10 g. (0.071 mole) of 1-phenyl-1,4-pentadiyne is added. The flask is stoppered, and the contents are stirred for ca. 2 hours. The brown solution is then poured into 200 ml. of water, and the mixture extracted four times with 50-ml. portions of ether. The extracts are combined and washed with two 50-ml. portions of water. The ether layer is dried over magnesium sulfate, filtered, and concentrated by distillation. The product is distilled as described in Part B, yielding 5.4–7.5 g. (54–75%) of 1-phenyl-1,3-pentadiyne as a colorless oil, b.p. 62° (0.15 mm.), n_D^{25} 1.6324. The ^{1}H NMR spectrum (60 mHz., neat, external tetramethylsilane reference) shows peaks at δ 1.47 (s, 3H) and 6.87 (m, 5H) (Note 11).

2. Notes

1. Reagent grade tetrahydrofuran available from Fisher Scientific Company was used by the checkers. For warning regarding the purification of tetrahydrofuran see *Org. Synth.*, **Coll. Vol. 5,** 976 (1973).

2. The flask may be warmed to hasten the dissolution of magnesium.

3. Phenylacetylene was distilled under nitrogen, b.p. 91–92° (17.5 mm.).

4. An ice-water bath should be kept at hand to cool the flask should the refluxing become too vigorous.

5. If the reaction is slow, the solution is kept at reflux overnight under an atmosphere of nitrogen. More tetrahydrofuran (50 ml.) may be added if stirring becomes difficult. It is desirable that procedure B follow A as soon as possible.

6. Propargyl bromide was redistilled, b.p. 82–83°.

7. The diyne discolors at room temperature and on exposure to air. Therefore, it is desirable to proceed without delay in the workup steps.

8. The checkers found it advantageous to do a crude preliminary distillation using a vapor-bath still. The resulting distillate can be redistilled using a spinning band column. This procedure appeared to avoid an exothermic reaction that occurred when the bath temperature rose above 110° (Note 9).

9. After the 1,4-diyne has distilled, overheating the pot contents is to be avoided since an exothermic reaction can occur.

10. The checkers observed yields of 44–50%, conducting the reaction on a scale one-half that described here.

11. Where they differ from those reported previously,[2] the values of the physical properties are those obtained by the checkers.

3. Discussion

The synthetic route described here has been used for various "skipped" diynes and related 1,3-diynes, among which are precursors or analogs of naturally occurring polyynes.[2] The copper(I) chloride-promoted coupling reaction in tetrahydrofuran provides the best and often the sole route to 1,4-diynes.[2] From these it is simple to proceed to the 1,3-diyne and, possibly, the isomeric allene.[2,3] Because all of these compounds appear to be sensitive to heat and oxygen the relatively mild reaction conditions are noteworthy.

1-Phenyl-1,4-pentadiyne has been prepared by coupling in tetrahydrofuran without a copper chloride catalyst in 22% yield,[4] and in ether with a copper chloride catalyst in 28% yield.[5] In general, the coupling of propargyl halides with various metallic acetylides, *e.g.*, sodium, silver, gives 1,4-pentadiynes in low yields at best.[2] For example, 1,4-pentadiyne was prepared from propargyl bromide and acetylenemagnesium bromide in tetrahydrofuran in 20% yield.[6] Strongly basic reactants such as alkali acetylides or basic conditions for workup and purification, *e.g.*, column chromatography over alumina, promote the isomerization of the 1,4-diyne.[2,3] The evolution of the present method, which emphasizes the use of tetrahydrofuran as solvent, a copper(I) salt, a short coupling time, and neutral reaction and workup conditions, has been described in detail.[2]

1-Phenyl-1,3-pentadiyne has been prepared by the dehydrobromination of the corresponding butadiene tetrabromide.[7] Other unsymmetrical 1,3-diynes have been prepared,

$$R-C{\equiv}C-H \\ + \qquad \xrightarrow[\text{base}]{Cu^+} R-C{\equiv}C-C{\equiv}C-R' \qquad (1) \\ R'-C{\equiv}C-Br$$

as described in scheme 1.[8] It is, of course, typical of crossed-coupling reactions that some should proceed as desired in (2),[2] and that others should go astray, as in (3).[2] The

$$C_6H_5(C{\equiv}C)_2MgBr + 4\text{-}CH_3C_6H_4SO_2CH_2C_6H_5 \rightarrow \\ C_6H_5(C{\equiv}C)_2CH_2C_6H_5 \qquad (2)$$

conversion of 1,4- to 1,3-diynes by strong base in ethanol or methanol at reflux has been described.[4,5] As judged by the isomerization rates in sodium ethoxide–ethanol, heating

$$C_6H_5(C{\equiv}C)_2MgBr + C_6H_5CH_2Br \xrightarrow{Cu_2Cl_2} (C_6H_5CH_2)_2 \qquad (3)$$

appears to be unnecessary. Typical rate constants are given in (4).[3]

$$C_6H_5C{\equiv}CCH_2C{\equiv}CH + C_2H_5O^- \xrightarrow[26°]{1.43\ M^{-1}\ sec^{-1}}$$

$$C_6H_5C{\equiv}C-CH{=}C{=}CH_2 \xrightarrow[26°]{0.47\ M^{-1}\ sec^{-1}}$$

$$C_6H_5C{\equiv}C-C{\equiv}C-CH_3 \qquad (4)$$

The present route to conjugated diynes is the method of choice *if* the corresponding skipped diyne or allene is available.[2,3] With naturally occurring polyynes this is often the case. Depending on the 1,4-diyne, the base-catalyzed isomerization may be successfully stopped (by acidification), yielding the intermediate allenylacetylene. Since the isolation of the allene may or may not be desired, it seems prudent to monitor the progress of isomerization of new 1,4-diynes, *e.g.*, by ^{1}H NMR, and establish the approximate lifetimes of transient species. In this way allenes as well as conjugated diynes may be obtained.[3]

1. Department of Chemistry, Illinois Institute of Technology, Chicago, Illinois 60616.
2. H. Taniguchi, I. M. Mathai, and S. I. Miller, *Tetrahedron*, **22**, 867 (1966).
3. I. M. Mathai, H. Taniguchi, and S. I. Miller, *J. Am. Chem. Soc.*, **89**, 115 (1967).
4. L. Groizeleau-Miginiac, *C.R. Hebd. Seances Acad. Sci.*, **248**, 1190 (1959).
5. A. A. Petrov and K. A. Molodova, *Zh. Obshch. Khim.*, **32**, 3510 (1962). [*J. Gen. Chem. USSR (Eng. Transl.)*, **32**, 3445 (1962)]
6. A. J. Ashe, III, and P. Shu, *J. Am. Chem. Soc.*, **93**, 1804 (1971).
7. C. Prévost, *Justus Liebigs Am. Chem.*, [10] **10**, 356 (1928).
8. B. Eglington and W. McCrae, *Adv. Org. Chem.*, **4**, 225 (1963).

PHENYLATION WITH DIPHENYLIODONIUM CHLORIDE: 1-PHENYL-2,4-PENTANEDIONE

(2,4-Pentanedione, 1-phenyl-)

$$CH_3COCH_2COCH_3 \xrightarrow[NH_3]{NaNH_2} NaCH_2COCH\overset{Na}{C}OCH_3$$

$$\downarrow (C_6H_5)_2ICl$$

$$C_6H_5CH_2COCH_2COCH_3 \xleftarrow{H_3O^+} C_6H_5CH_2COCH\overset{Na}{C}OCH_3$$

Submitted by K. Gerald Hampton,[1] Thomas M. Harris,[2] and Charles R. Hauser[3]
Checked by William N. Washburn and Ronald Breslow

1. Procedure

Caution: This preparation should be carried out in an efficient hood to avoid exposure to ammonia.

A 1-l., three-necked flask is equipped with an air condenser (Note 1), a ball-sealed mechanical stirrer, and a glass stopper. The stopper is removed, and 800 ml. of anhydrous liquid ammonia is introduced from a cylinder through an inlet tube. The tube is removed and replaced by the stopper. A small piece of sodium is added to the stirred ammonia. After the appearance of a blue color a few crystals of iron(III) nitrate hydrate (about 0.25 g.) are added, followed by small pieces of freshly cut sodium until 18.4 g. (0.800 g.-atom) has been added. After the formation of sodium amide is complete (Note 2), the glass stopper is replaced with a pressure-equalizing dropping funnel containing 40.0 g.

(0.400 mole) of 2,4-pentanedione (Note 3) in 30 ml. of anhydrous diethyl ether. The top of the addition funnel is fitted with a nitrogen inlet tube. The reaction flask is immersed at least 3 inches into an acetone–dry ice bath (Note 4), and the slow introduction of dry nitrogen through the inlet tube is begun simultaneously. After the reaction mixture is cooled thoroughly (about 20 minutes), the 2,4-pentanedione solution is added in small portions (Note 4) over 10 minutes. The cooling bath is removed, and the nitrogen flow is stopped. After 30 minutes the addition funnel is removed, and 63.3 g. (0.200 mole) of diphenyliodonium chloride (Note 5) is added through Gooch tubing from an Erlenmeyer flask over 15–25 minutes (Note 6). The reaction mixture is stirred for 6 hours, during which time the ammonia gradually evaporates. The Gooch tubing is replaced with an addition funnel, and 400 ml. of anhydrous ether is added. The remaining ammonia is removed by cautious heating on a warm-water bath. After the ether has distilled gently for 15 minutes, the flask is cooled in an ice-water bath, and 200 g. of crushed ice is added. A mixture of 60 ml. of concentrated hydrochloric acid and 10 g. of crushed ice is then added. The reaction mixture is stirred until the solid material has dissolved (Note 7) and transferred to a separatory funnel, the flask being washed with a little ether and dilute hydrochloric acid. The ethereal layer is separated, and the aqueous layer (Notes 8 and 9) is shaken three times with 50-ml. portions of ether. The combined ethereal extracts are dried over anhydrous magnesium sulfate. After filtration and removal of the solvent, the residual oil is purified by vacuum distillation, giving 21.0–22.5 g. (60–64%) (Note 10) of 1-phenyl-2,4-pentanedione, b.p. 133–136° (10 mm.), as a colorless to light yellow liquid (Note 11).

2. Notes

1. The flask is insulated with cloth towels to reduce the rate of ammonia evaporation. In addition the towels exclude light, thus reducing the photolytic production of iodine from diphenyliodonium chloride and the reaction products.

2. Conversion to sodium amide is indicated by the disappearance of the blue color. This generally requires about 20 minutes.

3. 2,4-Pentanedione, obtained from Aldrich Chemical Company, Inc., was dried over potassium carbonate and distilled before use, the fraction b.p. 133–135° being used.

4. The addition of 2,4-pentanedione to liquid ammonia is highly exothermic. Also, ammonia vapor reacts with the β-diketone, producing an insoluble ammonium salt which tends to clog the tip of the addition funnel. Cooling the reaction mixture in an acetone–dry ice bath reduces the vigor of the reaction and minimizes the clogging of the addition funnel. The 2,4-pentanedione should be added in spurts which fall on the surface of the reaction mixture rather than on the wall of the flask.

5. Diphenyliodonium chloride prepared by the method of Beringer and co-workers was used,[4] but it can also be prepared by the method described in *Org. Synth.*, **Coll. Vol. 3**, 355 (1955), or may be purchased from Aldrich Chemical Company, Inc.

6. If the addition is too fast, the reaction mixture will foam out of the flask.

7. A little diphenyliodonium salt may remain which will not dissolve. It will settle between the layers.

8. The aqueous layer should be acidic to litmus paper. If it is basic, more hydrochloric acid should be added until an acidic test is obtained.

9. The ethereal solution is usually dark, but should not have a purple color. A purple color indicates the presence of iodine. Iodine can arise by a light-catalyzed reaction in the latter stages of the reaction and during isolation of the product. For this reason the reaction should be shielded from strong light. In addition it is advisable for the ether employed in the reaction mixture to be peroxide-free. If iodine is present in the reaction product, it must be removed by extraction with aqueous sodium thiosulfate solution, since an adequate separation is not obtained by distillation.

10. The submitters have obtained the product in yields as high as 92% by a similar procedure and 85–91% by this procedure on this scale. The yield is calculated with the assumption that only one of the phenyl groups of diphenyliodonium ion is available for phenylation. This is not rigorously true; however, the magnitude of error is not great.[5]

11. A forerun of 2,4-pentanedione and iodobenzene, b.p. 32–92° (35 mm.), is obtained before the pressure is reduced to 10 mm. The purity of the product may be demonstrated by GC at 130° using a 180-cm. column packed with silicone gum rubber (Hewlett-Packard Co.). The chromatogram obtained showed only traces of iodobenzene and 3-phenyl-2,4-pentanedione.

3. Discussion

The method described is that of Hampton, Harris, and Hauser[5] and is an improvement over the benzyne method, which gives poor yields.[5,6] This β-diketone has been prepared by Claisen condensation of ethyl phenylacetate with acetone,[7] but the yield is poorer and the product has been shown by GC to be impure.[5] The β-diketone has also been prepared by hydrolysis of 4-methoxy-5-phenyl-3-penten-2-one[8] and by hydrolysis and decarboxylation of ethyl 1-acetyl-2-oxo-3-phenylbutyrate[9] but these compounds are more difficult to obtain than the starting materials used in the present synthesis.

This procedure represents a novel, convenient, and fairly general method of preparing γ-aryl-β-diketones. By this method the submitters have phenylated the dianion of 1-phenyl-1,3-butanedione (61%), 2,4-heptanedione (98%), 2,4-nonanedione (78%), 2,4-tridecanedione (53%), and 3,5-heptanedione (50%).[5] Substituted diaryliodonium salts have also been used to produce 1-(4-chlorophenyl)-2,4-pentanedione (44%), 4-(4-methylphenyl)-1-phenyl-1,3-butanedione (44%), and 1-(4-methylphenyl)-2,4-nonanedione (21%).[5] Under these conditions no more than a trace, if any, of arylation at the α-position of the β-diketones was observed by GC analysis.

Although the phenylation of monoanions of β-diketones does not proceed at a significant rate under the present conditions, phenylation of monoanions using diphenyliodonium salts under somewhat more vigorous conditions has been observed. The monoanions of 5,5-dimethyl-1,3-cyclohexanedione,[10,11] dibenzoylmethane,[10] tribenzoylmethane,[10] 1,3-indandione,[12] 2-mesityl-1,3-indandione,[13] and 2-phenyl-1,3-indandione[12] have been phenylated to give the mono- or diphenylated products. Keto esters such as ethyl 1,3-indandione-2-carboxylate,[12] ethyl cyclohexanone-2-carboxylate,[14] and other esters such as ethyl phenylacetate,[14] ethyl diphenylacetate,[14] diethyl acetamidomalonate,[14] diethyl ethylmalonate,[14] diethyl phenylmalonate,[14] and diethyl malonate[15] have been arylated in fair to good yields. Kornblum and Taylor have also found that nitroalkanes can be phenylated in 54–69% yield. These include 1-

nitropropane, 2-nitropropane, 2-nitrobutane, 2-nitroöctane, nitrocyclohexane, and ethyl 1-nitrocaproate.[16]

1. Chemistry Department, Texas A & M University, College Station, Texas 77843. This research was conducted at Texas A & M University and supported in part by the Petroleum Research Fund of the American Chemical Society.
2. Chemistry Department, Vanderbilt University, Nashville, Tennessee 37203.
3. Chemistry Department, Duke University, Durham, North Carolina 27706 (deceased January 6, 1970).
4. F. M. Beringer, E. J. Geering, I. Kuntz, and M. Mausner, *J. Phys. Chem.*, **60**, 141 (1956).
5. K. G. Hampton, T. M. Harris, and C. R. Hauser, *J. Org. Chem.*, **29**, 3511 (1964).
6. C. R. Hauser and T. M. Harris, *J. Am. Chem. Soc.*, **80**, 6360 (1958).
7. C. R. Hauser and R. M. Manyik, *J. Org. Chem.*, **18**, 588 (1953), G. T. Morgan and C. R. Porter, *J. Chem. Soc.*, **125**, 1269 (1924).
8. L. I. Smith and J. S. Showell, *J. Org. Chem.*, **17**, 836 (1952).
9. E. Fischer and C. Bülow, *Ber. Dtsch. Chem. Ges.*, **18**, 2131 (1885).
10. F. M. Beringer, P. S. Forgione, and M. D. Yudis, *Tetrahedron*, **8**, 49 (1960).
11. O. Ia. Neiland, G. Ia. Vanag, and E. Iu. Gudrinietse, *J. Gen. Chem. USSR. (Engl. Transl.)* **28**, 1256 (1958).
12. F. M. Beringer, S. A. Galton, and S. J. Huang, *J. Am. Chem. Soc.*, **84**, 2819 (1962).
13. F. M. Beringer and S. A. Galton, *J. Org. Chem.*, **28**, 3417 (1963).
14. F. M. Beringer and P. S. Forgione, *J. Org. Chem.*, **28**, 714 (1963).
15. F. M. Beringer and P. S. Forgione, *Tetrahedron*, **19**, 739 (1963).
16. N. Kornblum and H. J. Taylor, *J. Org. Chem.*, **28**, 1424 (1963).

1-PHENYL-4-PHOSPHORINANONE

(4-Phosphorinanone, 1-phenyl-)

Submitted by THEODORE E. SNIDER, DON L. MORRIS,
K. C. SRIVASTAVA, and K. D. BERLIN[1]
Checked by JOHN R. BERRY and RICHARD E. BENSON

1. Procedure

A. *Bis(2-cyanoethyl)phenylphosphine.* A 250-ml., three-necked flask is equipped with a magnetic stirrer, a thermometer, a pressure-equalizing dropping funnel, and a reflux condenser, with the entire system flushed with nitrogen. To the flask is added under an atmosphere of nitrogen 50.0 g. (0.454 mole) of phenylphosphine (Note 1), 50 ml. of acetonitrile, and 10 ml. of 10 N potassium hydroxide (Note 2). An ice-water bath is prepared for immediate cooling of the reaction flask. To the reaction mixture is added dropwise 50.0 g. (0.943 mole) of acrylonitrile (Note 3) with stirring and cooling over a period of 45–60 minutes. The rate of addition is controlled so that the temperature of the solution never exceeds 35° (Note 4). After the addition is complete, the solution is stirred at room temperature for an additional 2.5 hours. The reaction mixture is diluted with 100 ml. of ethanol and chilled to 0°. The product starts to crystallize, and the mixture is allowed to stand until crystallization is complete. The heavy slurry is filtered, and the crystalline product is washed with 200 ml. of cold ethanol and dried at 60° (2 mm.), yielding 74–84 g. (76–86%) of bis(2-cyanoethyl)phenylphosphine, m.p. 71–74° (Note 5). An additional 5–9 g. of product may be recovered from the combined washings and filtrate by concentration of the solution with subsequent chilling, bringing the total yield to 79–91 g. (80–93%).

B.   4-*Amino*-1,2,5,6-*tetrahydro*-1-*phenylphosphorin*-3-*carbonitrile*.   A   nitrogen-flushed, 1-l., three-necked flask equipped with a mechanical stirrer, a pressure-equalizing addition funnel, and a reflux condenser is charged with 25 g. (0.22 mole) of potassium *tert*-butoxide (Note 6) and 200 ml. of toluene (Note 7). The mixture is heated to reflux, and a solution of 43.2 g. (0.200 mole) of bis(2-cyanoethyl)phenylphosphine in 400 ml. of toluene is added dropwise with stirring over a period of 40–50 minutes (Note 8). After the addition is complete, the mixture is stirred and heated at reflux for an additional 3 hours. The mixture is cooled to room temperature, 250 ml. of water is added, and the resulting mixture is stirred for 30 minutes while the product is washed with two 50-ml. portions of cold ethanol and dried at 78° (1 mm.), yielding 36–38 g. (84–88%) of 4-amino-1,2,5,6-tetrahydro-1-phenylphosphorin-3-carbonitrile, m.p. 134.5–137° (Note 9). A small amount of product can be recovered from the filtrate (Note 10).

C.   1-*Phenyl*-4-*phosphorinanone*.   A solution of 35 g. (0.16 mole) of 4-amino-1,2,5,6-tetrahydro-1-phenylphosphorin-3-carbonitrile in 400 ml. of 6 N hydrochloric acid is heated at a vigorous reflux under nitrogen for 30 hours in a 1-l., three-necked flask equipped with a mechanical stirrer and a reflux condenser (Note 11). The mixture is then cooled with an ice bath, and 300 ml. of cold 10 N potassium hydroxide is added with stirring over a period of 10 minutes (Note 12). The resulting solution is stirred an additional 10 minutes (Note 13) and extracted with 300 ml. of diethyl ether. The ether layer is separated and washed two times with 100-ml. portions of water, then dried. The solvent is removed by distillation using a rotary evaporator, and the resulting oil crystallizes on standing, giving 23–26 g. of crude product, m.p. 38–43°. Distillation through a short-path column yields 21.5–21.7 g. (68–69%) of pure 1-phenyl-4-phosphorinanone, m.p. 43.5–44°, b.p. 120–122° (0.02 mm.) (Note 14).

2. Notes

1.   Phenylphosphine is available from Pressure Chemical Company and Strem Chemicals Inc., and best stored in a dry box under nitrogen. The compound is extremely air sensitive and malodorous. The container should be handled in the hood while wearing rubber gloves. Satisfactory preparations of phenylphosphine have been described.[2]

2.   Potassium hydroxide solution is prepared by adding 5.6 g. of potassium hydroxide to sufficient water, giving a final volume of 10 ml.

3.   Practical grade acetonitrile, available from Eastman Organic Chemicals, is satisfactory.

4.   The optimum reaction temperature is approximately 30°. A yellow product results at higher reaction temperatures, while lower reaction temperatures lead to an uncontrollable reaction resulting from the base-initiated polymerization of acrylonitrile.

5.   This product is of satisfactory purity for the next step. If a purer product is desired, bis(2-cyanoethyl)phenylphosphine may be recrystallized from hot ethanol or

distilled, b.p. 215–223° (0.2 mm.).[3] The IR adsorption maxima (KBr), cm.$^{-1}$, occur at 3086, 2242, 1481, 1429, 1333, 750, 716, and 694. ^{1}H NMR spectrum (CDCl$_3$) shows complex multiplets centered at δ 7.2 (8H) and 7.6 (5H). The ^{31}P NMR spectrum (40.5 MHz, C$_2$H$_5$OH) has a signal at −21.4 p.p.m. relative to 85% phosphoric acid.

6. Potassium *tert*-butoxide is available from MSA Research Corp.

7. Reagent grade toluene was dried by standing over sodium ribbon.

8. The reaction mixture becomes quite viscous when the addition is about two-thirds complete.

9. The product, as isolated, is pure enough for conversion to 1-phenyl-4-phosphorinanone. If a higher degree of purity is desired, the product may be recrystallized from ethanol–water or chromatographed on alumina. IR absorption maxima (KBr), cm^{-1} occur at 3401, 3344, 3236, 2874, 2169, 1642, 1603, 1399, 1323, 1188, 830, 781, 737, and 690. The ^{1}H NMR spectrum (CDCl$_3$, containing a small amount of dimethylsulfoxide-d_6) shows peaks at δ 1.8–3.0 (m, 2H), 5.15 (broad s, 2H), and 7.4–7.8 (m, 5H).

10. A small amount of product can be recovered from the filtrate by extracting the water layer with chloroform.

11. A white precipitate forms in the reaction medium after approximately 6 hours of reaction time. This precipitate may be the hydrochloride salt of 1-phenyl-4-phosphorinanone, m.p. >200°.

12. Rapid addition of base seems to result in higher yields than a more cautious addition, even though the temperature of the solution increases to about 40°. The solution must be strongly basic for efficient extraction.

13. Impure 1-phenyl-4-phosphorinanone may crystallize at this point. If crystallization occurs, the solid is recovered by filtration and washed thoroughly with two 20-ml. portions of water. The material is dried in a desiccator over phosphorus pentoxide, giving a product of m.p. 42.5–44°. To obtain a product satisfactory for distillation, the checkers found it necessary to dissolve the material in ether and wash it with water before distillation.

14. GC analysis of the product on a column containing 10% SE-30 on acid-washed Chromosorb U indicated one component, injected as a 20% solution in ethanol at 230° and helium flow of 15 ml. per minute with a retention time of 760 seconds. The ^{1}H NMR spectrum (CDCl$_3$) shows multiplets centered at δ 2.4 (8H) and 7.4 (5H). IR absorption maxima (KBr), cm^{-1}, occur at 3077 (=CH), 2976 and 2924 (CH), 1704 (C=O), 1595 and 1486 (aromatic C=C), 1433 (P-phenyl), 754 and 701 (monosubstituted phenyl). The ^{31}P NMR spectrum (40.5 MHz., CHCl$_3$) shows a signal at −39.3 p.p.m. relative to 85% phosphoric acid.

3. Discussion

This reaction sequence illustrates a broadly applicable synthetic route to a functionalized phosphorus heterocycle and has been utilized for the synthesis of 1-phenyl-,[4,5] 1-ethyl-,[4] and 1-methyl-4-phosphorinanone.[6]

The cyanoethylation of phenylphosphine has been carried out in the presence of a basic catalyst,[3] and at high temperature.[7,8] (2-Cyanoethyl)phenylphosphine has been reported as a contaminant but this difficulty has not been observed in the procedure reported herein.

The intermediates, bis(2-cyanoethyl)phenylphosphine and 4-amino-1,2,5,6-tetrahydro-1-phenylphosphorin-3-carbonitrile, are easily isolated and characterized, and show little or no oxidation when exposed to the air. 1-Phenyl-4-phosphorinanone is a highly-crystalline material, more sensitive to air oxidation than its two precursors. However, 1-phenyl-4-phosphorinanone may be stored in a well-capped bottle for several months without appreciable oxidation. It is best stored in a dark bottle in a dry box under nitrogen.

An extension of this type of synthetic sequence is illustrated by the cyclization of 2-cyanoethyl(2-cyanophenyl)phenylphosphine to the corresponding 2-enaminenitrile followed by hydrolysis with acid, yielding 2,3-dihydro-1-phenyl-4(1H)phosphinolinone.[9] The only other reported synthesis of this class of compounds involves the addition of phenylphosphine to substituted divinyl ketones.[10]

Phosphorinanones have been utilized as substrates for the preparation of alkenes,[11] amines,[12] indoles,[5,13] and in the synthesis of a series of secondary and tertiary alcohols via reduction,[10a] and by reaction with Grignard[6,11] and Reformatsky[11,14] reagents. Phosphorinanones have also been used as precursors to a series of 1,4-disubstituted phosphorins.[15] The use of 4-amino-1,2,5,6-tetrahydro-1-phenylphosphorin-3-carbonitrile for the direct formation of phosphorino[4,3-d] pyrimidines has been reported.[16] The ^{13}C NMR spectra of 1-phenylphosphorinanone has been reported.[17]

1. Department of Chemistry, Oklahoma State University, Stillwater, Oklahoma 74074. We gratefully acknowledge partial support of this work by a grant from the National Cancer Institute, CA 11967-08A1.

2. R. J. Horvat and A. Furst, *J. Am. Chem. Soc.*, **74**, 562 (1952); L. D. Freedman and G. O. Doak, *J. Am. Chem. Soc.*, **74**, 3414 (1952); F. G. Mann and I. T. Millar, *J. Chem. Soc.*, 3039 (1952).

3. M. M. Rauhut, I. Hechenbleikner, H. A. Currier, F. C. Schaefer, and V. P. Wystrach, *J. Am. Chem. Soc.*, **81**, 1103 (1959).

4. R. P. Welcher, G. A. Johnson, and V. P. Wystrach, *J. Am. Chem. Soc.*, **82**, 4437 (1960).

5. M. J. Gallagher and F. G. Mann, *J. Chem. Soc.*, 5110 (1962).

6. L. D. Quin and H. E. Shook, Jr., *Tetrahedron Lett.*, 2193 (1965).

7. B. A. Arbuzov, G. M. Vinokurova, and I. A. Perfil'eva, *Dokl. Chem. (Engl. Transl.)*, **127**, 657 (1959).

8. F. G. Mann and I. T. Millar, *J. Chem. Soc.*, 4453 (1952).

9. M. J. Gallagher, E. C. Kirby, and F. G. Mann, *J. Chem. Soc.*, 4846 (1963).

10. (a) R. P. Welcher and N. E. Day, *J. Org. Chem.*, **27**, 1824 (1962); (b) R. P. Welcher, U.S. Pat. 3,105,096 (1963); [*Chem. Abstr.*, **60**, 5553 (1964)].

11. H. E. Shook, Jr., and L. D. Quin, *J. Am. Chem. Soc.*, **89**, 1841 (1967).

12. Don L. Morris and K. D. Berlin, *Phosphorus*, **2**, 305 (1972).

13. M. J. Gallagher and F. G. Mann, *J. Chem. Soc.*, 4855 (1963).

14. L. D. Quin and D. A. Mathewes, *Chem. Ind. (London)*, 210 (1963).

15. G. Markl and H. Olbrich, *Angew. Chem.*, **78**, 598 (1966) [*Angew. Chem. Int. Ed. Engl.*, **5**, 589 (1966).]

16. Theodore E. Snider and K. D. Berlin, *Phosphorus*, **2**, 43 (1972).

17. S. D. Venkataramu, K. D. Berlin, S. E. Ealick, J. R. Baker, S. Nichols, and D. van der Helm, *Phosphorus Sulfur*, **7**, 133 (1979).

4-PHENYL-1,2,4-TRIAZOLINE-3,5-DIONE

[3H-1,2,4-Triazole-3,5(4H)-dione, 4-phenyl-]

$$H_2NNH_2 + CO(OC_2H_5)_2 \longrightarrow H_2NNHCO_2C_2H_5$$

$$H_2NNHCO_2C_2H_5 + C_6H_5NCO \longrightarrow C_6H_5NHCONHNHCO_2C_2H_5$$

Submitted by R. C. Cookson,[1] S. S. Gupte,
I. D. R. Stevens, and C. T. Watts
Checked by Y. Chao and R. Breslow

1. Procedure

Caution! Benzene has been identified as a carcinogen; OSHA has issued emergency standards on its use. All procedures involving benzene should be carried out in a well-ventilated hood, and glove protection is required.

A. *Ethyl hydrazinecarboxylate.* To 100 g. (96.9 ml., 2.00 moles) of 100% hydrazine hydrate, contained in a 1-l., round-bottomed flask, is added 236 g. (243 ml., 2.00 moles) of diethyl carbonate (Note 1). The flask is fitted with a calcium chloride-containing drying tube and shaken vigorously, mixing the two liquids. After about 5 minutes, the milky emulsion becomes warm, and shaking is continued until a clear solution is obtained (approximately 20 minutes). The flask is equipped with a reflux condenser fitted with a calcium chloride-containing drying tube and heated on a steam bath for 3.5 hours. The reaction mixture is transferred to a 500-ml., round-bottomed flask and is distilled through a 15-cm. Vigreux column under reduced pressure, yielding 161–176 g. (77–85%) of a colorless liquid collected at 102–103° (18 mm.) or 117–118° (40 mm.) (Note 2), n_D^{22} 1.4495; IR cm^{-1} 1640, 1725, 3350. The product, which may crystallize on standing, m.p. 45–47°, need not be purified for the next step.

B. *4-Phenyl-1-carbethoxysemicarbazide*. A 1-l., three-necked, round-bottomed flask equipped with a liquid-sealed mechanical stirrer (Note 3), a constant-pressure dropping funnel, and a reflux condenser fitted with a drying tube containing silica gel is charged with a solution of 52 g. (0.50 mole) of ethyl hydrazinecarboxylate in 550 ml. of dry benzene (Note 4). After the solution is cooled with an ice bath, stirring is begun and 59.7 g. (54.5 ml., 0.501 mole) of phenyl isocyanate is added dropwise to the solution over a 45-minute period. After about one-half of the isocyanate has been added, a white precipitate of the product appears, and the reaction mixture becomes progressively thicker. After addition is complete the ice bath is removed; the mixture is stirred at room temperature for 2 hours, then heated under reflux for 2 hours. The suspension is allowed to cool to room temperature, and 4-phenyl-1-carbethoxysemicarbazide is isolated by suction filtration, washed with 500 ml. of benzene, and dried in a vacuum desiccator, yielding 108 g. (97%) of 4-phenyl-1-carbethoxysemicarbazide, m.p. 151–152°. The product is not further purified for use in the next step, but may be recrystallized from ethyl acetate to yield white crystals, m.p. 154–155°; IR cm.$^{-1}$ 1645, 1687, 1797, and 3300 (Note 6).

C. *4-Phenylurazole*. A 250-ml. Erlenmeyer flask is charged with 100 ml. of aqueous 4 *M* potassium hydroxide and 44.6 g. (0.200 mole) of 4-phenyl-1-carbethoxy-semicarbazide. The suspension is warmed on a steam bath, the flask being swirled occasionally to wash the solid off the sides. After 1.5 hours most of the solid has dissolved, and the hot solution is filtered. After cooling to room temperature, the solution is acidified with concentrated hydrochloric acid (about 33 ml. is required). The mixture is again cooled to room temperature and the precipitated 4-phenylurazole is isolated by suction filtration. The mother liquor is evaporated to dryness on a rotary evaporator, and the residue is extracted twice with 100-ml. portions of boiling absolute ethanol (Note 7). The ethanol solutions are combined, filtered, and evaporated to dryness on a rotary evaporator, and the additional 4-phenylurazole recovered is combined with that obtained above. The product is crystallized from 95% ethanol (about 80 ml.), yielding 30.0–33.5 g. (85–95%) of 4-phenylurazole, m.p. 209–210°; IR cm.$^{-1}$ 1685 and 3120 (Notes, 8, 9, and 10).

D. *4-Phenyl-1,2,4-triazoline-3,5-dione*. A 100-ml., three-necked, round-bottomed flask equipped with a dropping funnel, a gas-inlet tube, a calcium chloride-containing drying tube, and a magnetic stirrer is flushed with oxygen-free nitrogen (Note 11) and charged with 12 ml. of ethyl acetate (Note 12) and 4.4 g. (0.025 mole) of 4-phenylurazole (Note 13). The stirrer is started, and 2.5 g. (2.8 ml., 0.023 mole) of *tert*-butyl hypochlorite (Notes 14 and 15) is added to the flask over a period of approximately 20 minutes, the reaction mixture being maintained close to room temperature with a cold-water bath (Notes 16 and 17). After the addition is complete, the resulting suspension is stirred for 40 minutes at room temperature. The reaction mixture is transferred to a 100-ml., round-bottomed flask, and the solvent is removed on a rotary evaporator, keeping the temperature below 40°. The last traces of solvent are removed with a high-vacuum pump (about 0.1 mm.). The product is sublimed (Note 18) onto an ice-cooled cold finger under vacuum (100° at 0.1 mm.), yielding 2.7–2.8 g. (62–64%) of the triazoline as carmine-red crystals which decompose (165–175°) before melting; IR cm.$^{-1}$ 1760 and 1780; UV (dioxane) nm (ϵ) 247 (2300), 310 (1020), and 532 (171) (Notes 19 and 20).

2. Notes

1. Both the hydrazine hydrate and diethyl carbonate were British Drug Houses Ltd. or Matheson Laboratory reagent grade and used without further purification.

2. A forerun of approximately 100 ml., boiling below 80° (18 mm.), containing ethanol, water, and unreacted starting materials, is also collected.

3. An efficient stirrer should be employed, since the reaction mixture becomes quite viscous. If efficient mixing is not maintained a violent reaction can occur. This is especially important when using aliphatic isocyanates.

4. British Drug Houses Ltd. or Amend Drug & Chemical Co., Inc., reagent grade benzene, dried over sodium wire, is adequate.

5. British Drug Houses Ltd. or Matheson Laboratory reagent grade phenyl isocyanate was used without further purification. When using other isocyanates, care should be taken to ensure their purity as the yield is greatly dependent upon this, commercially available 4-nitrophenyl isocyanate being a case in point.

6. The submitters report a similar yield on a scale three times that illustrated here. This method has been employed for the preparation of the 4-methyl- (100%, m.p. 143° from ethyl acetate), 4-tert-butyl- (100%, m.p. 147° from ethyl acetate), and 4-(4-nitrophenyl)-1-carbethoxysemicarbazide (90%, m.p. 219° from methanol). Because of the impure nature of commercial 4-nitrophenyl isocyanate, the product from that reaction may be contaminated with 4-nitrophenylurea. It can be used in the impure form for preparing the corresponding urazole, as the contaminant is alkali insoluble.

7. The extraction procedure increases the yield of 4-phenylurazole by about 6%. This step is unnecessary when preparing 4-(4-nitrophenyl)urazole, as it is insoluble in water.

8. This method has been used to prepare 4-methyl- (90%, m.p. 240° from methanol) and 4-(4-nitrophenyl)urazole (80%, m.p. 264° from ethanol).

9. 4-tert-Butyl-1-carbethoxysemicarbazide can be cyclized by refluxing with 4% sodium ethoxide in ethanol for 4 hours, followed by acidification with an ethanolic solution of hydrogen chloride. Filtration, evaporation of the filtrate, and crystallization from ethyl acetate yields 4-tert-butylurazole (89%, m.p. 168°).

10. 4-Benzalaminourazole (m.p. 255°) can be prepared from 4-aminourazole[2] by condensation with benzaldehyde.

11. A gentle stream of nitrogen is maintained through the apparatus during the entire reaction. Hydrogen chloride is evolved and adequate precautions should be taken to prevent exposure to the gas.

12. Ethyl acetate was purified by Fieser's method.[3]

13. The 4-phenylurazole should be ground with a pestle and mortar before use.

14. tert-Butyl hypochlorite was prepared by the method described in Org. Synth., **Coll. Vol. 4,** 125 (1963).

15. An excess of tert-butyl hypochlorite should not be used, as it cannot be removed and interferes with the sublimation of the product.

16. When preparing the 4-(4-nitrophenyl)- and 4-benzalamino- analogs, the reaction mixture should be maintained at 0–5°.

17. As soon as the first drop of hypochlorite is added, the reaction mixture becomes red in color, with the color deepening as the addition proceeds.

18. The impure material has a limited stability and should be sublimed as quickly as possible. The scale of the reaction should not be greatly increased unless an efficient large subliming apparatus is available. The submitters report similar yields on experiments four times this scale.

19. The product has a shelf life of several months if stored in the dark in a refrigerator.

20. This method has been used to prepare 4-methyl- [sublimed at 50° (0.1 mm.), 85%, m.p. 104°], 4-tert-butyl- [50° (0.1 mm.), 80%, m.p. 119°], 4-(4-nitrophenyl)- [100° (0.1 mm.), 25%, m.p. 130°], and 4-benzalamino-1,2,4-triazoline-3,5-dione [100° (0.1 mm.), 75%].

3. Discussion

Ethyl hydrazinecarboxylate has been prepared from hydrazine hydrate and ethyl N-tricarboxylate in good yield.[4] The method described here is comparable in efficiency, but has the added advantage that both starting materials are commercially available.

Methods for preparing 4-phenyl-1-carbethoxysemicarbazide and 4-phenylurazole have been described in principle by Zinner and Deucker.[5] 4-Phenylurazole has also been prepared from biurea and aniline hydrochloride;[6,7] however, the method is unreliable, with yields varying from 0 to 20%. 4-Substituted urazoles have also been made by heating the corresponding N,N'-disubstituted diamides of hydrazodicarboxylic acid,[8] but the results are difficult to reproduce.

4-Phenyl-1,2,4-triazoline-3,5-dione has been prepared by oxidizing 4-phenylurazole with lead dioxide,[6] and with ammoniacal silver nitrate followed by treatment with an ethereal solution of iodine.[7] The yields are low for both methods. 4-Substituted triazolinediones can also be made by oxidation of the corresponding urazole with fuming nitric acid[8] or dinitrogen tetroxide.[9] Oxidation with tert-butyl hypochlorite in acetone has also been described;[10,11] however, it yields an unstable product, even after sublimation. Dioxane[11] and ethyl acetate are preferred as solvents for the reaction, since the product is obtained in a stable form. The latter solvent is superior since 4-phenylurazole has a greater solubility in it.

In common with other azodicarboxylic acid derivatives, 4-phenyl-1,2,4-triazoline-3,5-dione has many uses. It undergoes Diels-Alder reactions with most dienes[10-13] and is, in fact, the most reactive dienophile so far reported.[14,15] As with the formation of all Diels-Alder adducts the reaction is reversible, and in the case of the adduct with 3β-acetoxy-17-cyano-5,14,16-androstatriene, the reverse reaction can be made to proceed under especially mild conditions.[13] An instance has also been reported of the dione photochemically catalyzing other retro Diels-Alder reactions.[16] Along with the proved use of azodicarboxylic ester,[17,18] the dione should be potentially important in the preparation of strained ring compounds.

4-Phenyl-1,2,4-triazoline-3,5-dione also undergoes "addition-abstraction" reactions (e.g., with acetone[16]). As would be expected for such a species, it will oxidize alcohols to the corresponding aldehydes or ketones.[19] This oxidation is especially mild (room temperature in benzene, chlorobenzene or ethyl acetate) and is, as such, a valuable method of oxidizing or preparing compounds sensitive to acid, base, or heat.

1. Department of Chemistry, University of Southampton, Southampton, SO9 5NH, England.
2. L. F. Audrieth and E. B. Mohr, *Inorg. Synth.*, **4**, 29 (1953).
3. L. F. Fieser, "Experiments in Organic Chemistry," 3rd ed., Heath, Boston, Mass., 1955, p. 287.
4. C. F. H. Allen and A. Bell, *Org. Synth.*, **Coll. Vol. 3**, 404 (1955).
5. G. Zinner and W. Deucker, *Arch. Pharm. Weinheim, Ger.*, **294**, 370 (1961) [*Chem. Abstr.*, **55**, 22298h (1961)].
6. J. Thiele and O. Stange, *Justus Liebigs Ann. Chem.*, **283**, 1 (1894).
7. F. Arndt, L. Lowe, and A. Tarlan-Akön, *Istanbul Univ. Fen Fak. Mecm.*, *Seri A*, **13**, 127 (1948) [*Chem. Abstr.*, **42**, 8190d (1948)].
8. M. Furdik, S. Mikulasek, M. Livar, and S. Priehradny, *Chem. Zvesti*, **21**, 427 (1967) [*Chem. Abstr.*, **67**, 116858y (1967)].
9. J. C. Stickler and W. H. Pirkle, *J. Org. Chem.*, **31**, 3444 (1966).
10. R. C. Cookson, S. S. H. Gilani, and I. D. R. Stevens, *Tetrahedron Lett.*, 615 (1962).
11. R. C. Cookson, S. S. H. Gilani, and I. D. R. Stevens, *J. Chem. Soc. C*, 1905 (1967).
12. S. S. H. Gilani and D. J. Triggle, *J. Org. Chem.*, **31**, 2397 (1966).
13. A. J. Solo, H. Sachdev, and S. S. H. Gilani, *J. Org. Chem.*, **30**, 769 (1965).
14. J. Sauer, *Angew. Chem.*, **79**, 76 (1967) [*Angew. Chem. Int. Ed. Engl.*, **6**, 16 (1967).]
15. 4-(4-Nitrophenyl)-1,2,4-triazoline-3,5-dione is even more reactive (M. Burrage, R. C. Cookson, S. S. Gupte, and I. D. R. Stevens, *J. Chem. Soc., Perkin Trans. 2*, 1375 (1975)).
16. S. S. H. Gilani, Ph.D. thesis, University of Southampton, England, 1963.
17. O. Diels, J. H. Blom, and W. Koll, *Justus Liebigs Ann. Chem.*, **443**, 242 (1925).
18. R. Criegee and A. Rimmelin, *Chem. Ber.*, **90**, 414 (1957).
19. R. C. Cookson, I. D. R. Stevens, and C. T. Watts, *Chem. Commun.*, 744 (1966).

2-PHENYL-2-VINYLBUTYRONITRILE

(Benzeneacetonitrile, α-ethenyl-α-ethyl-)

$$C_6H_5CHCN + HC\equiv CH \xrightarrow[\text{dimethyl sulfoxide, 60° to 70°}]{[(C_2H_5)_3NCH_2C_6H_5]^+Cl^-, KOH} C_6H_5CCN$$

with C_2H_5 substituent on the left nitrile and $CH{=}CH_2$ and C_2H_5 substituents on the product.

Submitted by M. Makosza,[1] J. Czyzewski, and M. Jawdosiuk
Checked by John C. Sauer and Richard E. Benson

1. Procedure

Caution! Benzene has been identified as a carcinogen; OSHA has issued emergency standards on its use. All procedures involving benzene should be carried out in a well-ventilated hood, and glove protection is required.

A 1-l., four-necked, round-bottomed flask equipped with a sealed mechanical stirrer, a thermometer, and a gas-inlet tube is charged with 145 g. (1.00 mole) of 2-phenylbutyronitrile (Note 1), 2.3 g. (0.010 mole) of benzyltriethylammonium chloride (Note 2), and 50 ml. of dimethyl sulfoxide (Note 3). The gas-inlet tube is adjusted to extend below the surface of the liquid, and a gas-exit tube is attached to the flask. A slow stream of acetylene (Note 4) is passed through the gas-inlet tube into the flask, removing the air. After 5 minutes, 56 g. of finely powdered potassium hydroxide is added and stirring is begun. Acetylene is introduced at the rate of 15–20 l./hour. An exothermic reaction occurs; the temperature rises to 70–80° and is held in this range with a cold water bath (Note 5). After 40–60 minutes, a warm bath is required to maintain the temperature of the reaction mixture at 60–70°, and stirring is continued for an additional 20–30 minutes (Note 6). The mixture is cooled to room temperature, the inlet tube is replaced with a pressure-equalizing dropping funnel, and 500 ml. of water is added slowly (Note 7). The resulting dark-brown mixture is transferred to a separatory funnel and washed twice with 200-ml. portions of benzene. The benzene layers are combined and washed successively with 200 ml. of water, 100 ml. of 10% hydrochloric acid, and 200 ml. of water. The organic layer is dried over anhydrous magnesium sulfate, and the benzene is removed by distillation at reduced pressure. The residual oil is distilled through a short Vigreux column, giving 125–135 g. of crude product, b.p. 115–125° (13 mm.). Redistillation of this product through a Vigreux column gives 101–107 g. (59–63%) of colorless 2-phenyl-2-vinylbutyronitrile, b.p. 110° (8 mm.), n_D^{25} 1.5157 (Note 8).

2. Notes

1. The preparation of 2-phenylbutyronitrile is described in *Org. Synth.*, **Coll. Vol. 6,** 897 (1988).

2. The checkers used the product available from Aldrich Chemical Company, Inc. The preparation of this reagent is described in *Org. Synth.*, **Coll. Vol. 6,** 232 (1988).

3. The checkers used the product available from Fisher Scientific Company.

4. The checkers used acetylene available from Matheson Gas Products. The gas was purified by passing it through concentrated sulfuric acid, through a tower filled with potassium hydroxide pellets, then into a 1-l. safety flask which was connected to the gas-inlet tube with rubber tubing. The checkers used a rotameter, calibrated with air, to determine the flow rate of acetylene.

5. The checkers attempted to keep the temperature at 65–70°.

6. The reaction may be monitored by GC. The submitters used a 2-m. column containing silicone oil on diatomite support (190°).

7. The water should be added slowly, since the mixture is saturated with acetylene and the gas may be evolved vigorously.

8. The checkers found the product to be at least 95% pure on a GC column containing 10% silicone 200 on nonacid washed Chromosorb W operated at 125°. The spectral properties of the product are as follows; IR (neat) cm.$^{-1}$: 1639, 1000, 930 (CH=CH$_2$); ^{1}H NMR (neat), δ (multiplicity, coupling constant J in Hz., number of protons, assignment): 0.92 (t, $J = 7$, 3H, CH$_3$), 1.87 (q, $J = 7$, 2H, CH$_2$), 5.00–6.20 (m, 3H, CH=CH$_2$), 7.17–7.57 (m, 5H, C$_6$H$_5$).

3. Discussion

This procedure, which involves the addition of an anion derived from a nitrile to an unactivated acetylenic bond under rather mild conditions, is a convenient, general method for the synthesis of α-vinylnitriles (see Table I). The reaction proceeds smoothly in either

TABLE I

α-Vinylnitriles Derived from Acetylenes

Nitrile	Acetylene	Product	b.p.	Yield (%)
$C_6H_5CH(C_5H_{11})CN$	$HC{\equiv}CH$	$C_6H_5C(C_5H_{11})CN$ $\quad\ \ \|\|$ $\quad CH{=}CH_2$	139° (8 mm.)	88
$C_6H_5CH(2\text{-}C_3H_7)CN$	$HC{\equiv}CC_6H_5$	$C_6H_5C(2\text{-}C_3H_7)CN^a$ $\quad\ \ \|\|$ $\quad CH{=}CHC_6H_5$	145° (0.8 mm.)	83
$(C_6H_5)_2CHCH(C_6H_5)CN$	$HC{\equiv}CSC_4H_9$	$(C_6H_5)_2CHC(C_6H_5)CN^b$ $\qquad\qquad \|\|$ $\qquad CH{=}CHSC_4H_9$	113°^{c}	96
$(C_6H_5)_2CHCN$	$HC{\equiv}COC_2H_5$	$(C_6H_5)_2C{-}C{=}CH_2$ $\qquad\quad\ \| \qquad \|$ $\qquad\ CN\ \ OC_2H_5$	146° (0.6 mm.)	77

^{a}Product is a 2:1 mixture of (Z) and (E) isomers; (Z) isomer, m.p. 62°.
^{b}Only the (Z) isomer is obtained.
^{c}Melting point.

dimethyl sulfoxide or hexamethylphosphoric triamide [*see* J. A. Zapp, Jr., *Science*, **190,** 422 (1975) *for a toxicity warning concerning this compound*] with a tetraalkylammonium salt as catalyst. The products thus prepared are obtained in yields higher[2] than those obtained under conventional conditions, which generally require higher temperatures and elevated pressures.[3,4]

1. Institute of Organic Chemistry, Polish Academy of Sciences, Warsaw, Poland.
2. M. Makosza, Pol. Pat. **55113** (1968) [*Chem. Abstr.*, **70,** 106006s (1969)]; M. Makosza, *Tetrahedron Lett.*, 5489 (1966); M. Makosza and M. Jawdosiuk, *Bull. Acad. Pol. Sci., Ser. Sci. Chim.*, **16,** 589 (1968) [*Chem. Abstr.*, **71,** 30193y (1969)].
3. P. P. Karpukhin and A. I. Levchenko, *Zh. Prikl. Khim. Leningrad*, **32,** 1354 (1959) [*Chem. Abstr.*, **54,** 450c (1960)]; P. P. Karpukhin, A. I. Levchenko, and E. V. Dudko, *Zh. Prikl. Khim. Leningrad*, **34,** 1117 (1961) [*Chem. Abstr.*, **55,** 22259f (1961)].
4. M. Seefelder, *Justus Liebigs Ann. Chem.*, **652,** 107 (1962).

BORANES IN FUNCTIONALIZATION OF OLEFINS TO AMINES: 3-PINANAMINE

(Bicyclo[3.1.1]heptan-3-amine, 2,6,6-trimethyl-)

A. $(NH_2OH)_2 \cdot H_2SO_4 + 2ClSO_3H \longrightarrow \overset{+}{N}H_3OSO_3^- + 2HCl + H_2SO_4$

1

B.

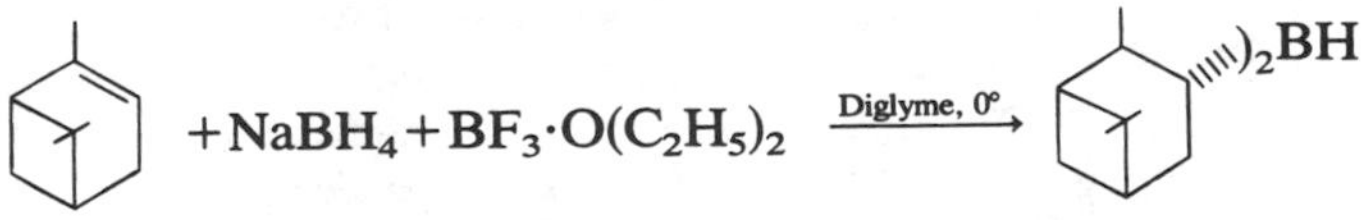

2

1 + 2

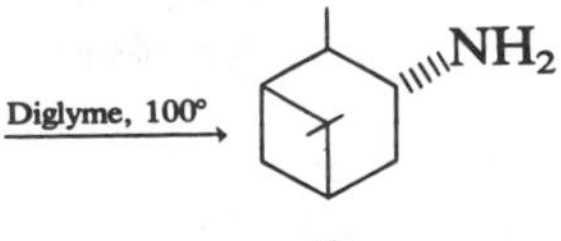

3

Submitted by MICHAEL W. RATHKE[1] and ALAN A. MILLARD
Checked by ARNOLD BROSSI and JUN-ICHI MINAMIKAWA

1. Procedure

A. *Hydroxylamine-O-sulfonic acid* (**1**). A 500-ml., three-necked, round-bottomed flask is fitted with a mechanical stirrer, dropping funnel, and calcium chloride drying tube. Finely powdered hydroxylamine sulfate (26.0 g., 0.158 mole) (Note 1) is placed in the flask, and 60 ml. (107 g., 0.92 mole) of chlorosulfonic acid (Note 1) is added dropwise over 20 minutes with vigorous stirring (Note 2). After the addition is complete, the flask, with stirring, is placed in a 100° oil bath for 5 minutes. The pasty mixture is cooled to room temperature, and the flask is placed in an ice bath. To the stirred mixture 200 ml. of diethyl ether is slowly added over 20–30 minutes (Note 3). During the ether addition, the pasty contents change to a colorless powder which is collected by suction on a Büchner funnel. The powder is washed with 300 ml. of tetrahydrofuran, then with 200 ml. of ether. The product **1**, after drying, weighs 34–35 g. (95–97%). Iodometric titration shows the product is 96–99% pure (Note 4) and adequate for the following reaction.

    B.   3-*Pinanamine* (3).   A 1000-ml., three-necked, round-bottomed flask is fitted with a gas-inlet tube, a reflux condenser connected to a mineral oil bubbler, and a sealed mechanical stirrer. The system is flamed with a Bunsen burner while being flushed with dry nitrogen. The reaction vessel is then cooled under a nitrogen stream in an ice bath while a slight positive pressure of nitrogen is maintained. A solution of 3.12 g. (0.0824 mole) of sodium borohydride (Note 5) in 100 ml. of diglyme (Note 6) is added to the flask, followed by 27.25 g. (0.2004 mole) of (±)-α-pinene (Note 7). Hydroboration is achieved by dropwise addition of 15.6 g. (0.110 mole) of boron trifluoride diethyl etherate (Note 8) over a 15-minute period. Di-3-pinanylborane (**2**) precipitates as a white solid. The ice bath is removed, and the reaction mixture is stirred at room temperature for 1 hour. Hydroxylamine-*O*-sulfonic acid (**1**) (24.9 g., 0.220 mole) in 100 ml. of diglyme is added dropwise to the mixture over a 5-minute period (Note 9). The mixture is then heated in a 100° oil bath for 3 hours. The mixture is cooled to room temperature, and 80 ml. of concentrated hydrochloric acid is added over a 5-minute period. The mixture is poured into 800 ml. of water and extracted with two 100-ml. portions of ether. The ether layers are discarded, and the aqueous layer is made alkaline with sodium hydroxide pellets (60–65 g. is needed). The aqueous layer is extracted with two 100-ml. portions of ether, the combined ether extracts are dried over anhydrous sodium sulfate, and the drying agent is removed by filtration. The filtrate is transferred to a 500-ml., ice-cooled flask fitted with a magnetic stirring bar. A solution of 85–88% phosphoric acid (12 g., 0.10 mole) in 100 ml. of ethanol is added to the flask over 10 minutes with stirring. The precipitated colorless crystals are collected with suction on a Büchner funnel, and the salt is suspended in 300 ml. of hot water contained in a 1-l. flask. The mixture is heated and magnetically stirred in a 120–130° oil bath until all the salt has dissolved (*ca.* 20–30 minutes) then quickly filtered with suction. Pure phosphate salt immediately precipitates as colorless plates, which are collected on a Büchner funnel and dried in a desiccator. The yield is 16.6 g. (33.1%). A second crop of 4.4 g. can be obtained by concentrating the mother liquor to about half its original volume. The total yield of pure phosphate salt is 21.0 g. (41.8%), m.p. 275–280° (dec.) (Note 10). The salt is easily converted to free amine **3** by the following procedure: 10 g. (0.040 mole) of the salt is dissolved in 40 ml. of aqueous 3 *M* sodium hydroxide and extracted with two 50-ml. portions of ether. The combined extracts are dried over anhydrous sodium sulfate, the drying agent is removed by filtration, and the solvent is removed under reduced pressure with a rotary evaporator. The residual oil is distilled, giving 5.9 g. (93% from phosphate salt) of amine **3** as a colorless liquid, b.p. 83° (13 mm.) (Note 11).

2. Notes

1. Commercial hydroxylamine sulfate and chlorosulfonic acid, obtained from East-man Kodak Company, were used directly. The checkers found that commercially available hydroxylamine-O-sulfonic acid is sometimes of low purity; therefore, the use of freshly prepared reagent is recommended.

2. Hydrogen chloride gas is evolved during the addition. The reaction should be carried out in a hood, and an aqueous base scrubber is recommended.

3. Rapid addition of the ether must be avoided because of its high reactivity with chlorosulfonic acid.

4. Iodometric titration was carried out as follows: About 100 mg. of hydroxylamine-O-sulfonic acid was exactly weighed and dissolved in 20 ml. of distilled water. Sulfuric acid (10 ml. of 10% solution) and 1 ml. of saturated potassium iodide solution were then added. After the solution was allowed to stand for 1 hour, liberated iodine was titrated with 0.1 N sodium thiosulfate solution until the iodine color disappeared. The following stoichiometric relation was used: 0.1 N $Na_2S_2O_3$ (1 ml.) = 5.66 mg. $H_3\overset{+}{N}OSO_3^-$. Hydroxylamine-O-sulfonic acid should be stored in tightly sealed bottles in a refrigerator.

5. Commercial sodium borohydride was obtained from Ventron Corporation and used directly.

6. Commercial diglyme (dimethyl ether of diethylene glycol) was obtained from Ansul Chemical Company, Marinette, Wisconsin, and purified by distillation from lithium aluminum hydride at 62–63° (15 mm.) [*Org. Synth.*, **Coll. Vol. 6**, 719 (1988)].

7. (±)-α-Pinene, b.p. 54° (22 mm.), was obtained from Aldrich Chemical Company, Inc., and distilled before use.

8. Commercial boron trifluoride etherate, b.p. 46° (10 mm.), available from Matheson, Coleman and Bell, was distilled from calcium hydride before use.

9. *Caution: Since* (±)-α-*pinene is hydroborated to the dialkylborane state* (R_2BH), *a large amount of hydrogen is evolved on addition of hydroxylamine-O-sulfonic acid. Consequently, the addition should be carried out dropwise and adequate ventilation should be provided.*

10. The phosphate salt has the empirical formula $C_{10}H_{19}N \cdot H_3PO_4$.

11. The product showed one peak on GC (3% SE-30, 70°C).

3. Discussion

Hydroxylamine-O-sulfonic acid can also be prepared from hydroxylamine sulfate and 30% fuming sulfuric acid (oleum).[2] The present procedure is essentially that of F. Sommer et al.[3]

The hydroboration–amination sequence in diglyme is a general procedure for the conversion of olefins to primary amines without rearrangement and with predictable stereochemistry.[4] An alternative procedure, using tetrahydrofuran as solvent and either hydroxylamine-O-sulfonic acid or chloramine, can be applied to terminal olefins and relatively unhindered internal and alicyclic olefins.[5] O-Mesitylenesulfonylhydroxylamine also gave desired amines in comparable yield.[6] Alternative procedures for the hydroboration of olefins use commercially available solutions of diborane in tetrahydrofuran[7] or dimethylsulfide.[8]

Olefins may be converted to primary amines by the Ritter reaction[9] or by reaction with mercury(II) nitrate in acetonitrile.[10] In both cases regiospecificity for the formal addition of ammonia across the double bond is opposite to that observed in the hydroboration–amination sequence.

1. Department of Chemistry, Michigan State University, East Lansing, Michigan 44824.
2. H. J. Matsuguma and L. F. Audrieth, *Inorg. Synth.*, **5**, 122 (1957).
3. F. Sommer, O. F. Schulz, and M. Nassau, *Z. Anorg. Allg. Chem.*, **147**, 142 (1925).
4. M. W. Rathke, N. Inoue, K. R. Varma, and H. C. Brown, *J. Am. Chem. Soc.*, **88**, 2870 (1966).
5. H. C. Brown, W. R. Heydkemp, E. Breuer, and W. S. Murphy, *J. Am. Chem. Soc.*, **86**, 3565 (1964).
6. Y. Tamura, J. Minamikawa, S. Fujii, and M. Ikeda, *Synthesis*, 196 (1974).
7. H. C. Brown, "Organic Synthesis *via* Boranes," Wiley, New York, (1975).
8. C. F. Lane, *J. Org. Chem.*, **39**, 1437 (1974).
9. L. I. Krimen and D. J. Cota, *Org. React.*, **17**, 213 (1969).
10. H. C. Brown and J. T. Kurek, *J. Am. Chem. Soc.*, **91**, 5647 (1969).

ALLYLIC OXIDATION WITH HYDROGEN PEROXIDE–SELENIUM DIOXIDE:
trans-PINOCARVEOL

[Bicyclo[3.1.1]heptan-3-ol, 6,6-dimethyl-2-methylene-, (1α,3α,5α)-]

$$\xrightarrow[\substack{(CH_3)_3COH, H_2O \\ 40-50°}]{H_2O_2, SeO_2}$$

Submitted by J. M. Coxon, E. Dansted,
and M. P. Hartshorn[1]
Checked by D. W. Brooks and S. Masamune

1. Procedure

Caution! Selenium compounds are exceedingly toxic (Note 1). Hydrogen peroxide attacks the skin and may decompose violently (Note 2). The reaction should be carried out behind a safety screen and in an efficient fume hood, and the operator should wear safety glasses and rubber gloves.

Benzene has been identified as a carcinogen; OSHA has issued emergency standards on its use. All procedures involving benzene should be carried out in a well-ventilated hood, and glove protection is required.

A 500-ml., three-necked, round-bottomed flask is fitted with a mechanical stirrer, a thermometer, a dropping funnel, and a reflux condenser. A solution of 0.74 g. (0.0067 mole) of selenium dioxide in 150 ml. of *tert*-butyl alcohol is introduced into the flask, followed by 68 g. (0.50 mole) of β-pinene (Note 3). The resulting mixture is warmed to 40° with a hot water bath before 35 ml. (0.62 mole) of 50% aqueous hydrogen peroxide (Note 2) is added dropwise over 90 minutes, during which time the mixture is maintained at 40–50° by occasional immersion in a cold water bath. After stirring for an additional 2 hours, the reaction mixture is diluted with 50 ml. of benzene, washed with three 50-ml. portions of saturated aqueous ammonium sulfate, and dried over sodium sulfate. A small amount of hydroquinone is added (Note 4), and the solvents are removed on a rotary evaporator. *trans*-Pinocarveol is isolated by simple distillation under reduced pressure, yielding 37–42 g. (49–55%), b.p. 60–70° (1 mm.), n_D^{22} 1.4972, $[\alpha]_D^{20}$ + 53.5 to + 60.0° (*c* 2.5, methanol) (Note 5).

2. Notes

1. The physiological properties of selenium compounds are similar to those of arsenic compounds. Any selenium dioxide solid or solution spilled on the skin should be removed immediately by washing under running water.

2. Aqueous 50% hydrogen peroxide causes immediate blistering if allowed to come into contact with the skin. The presence of metal salts may cause decomposition of the hydrogen peroxide.

3. The checkers purchased β-pinene, $[\alpha]_D^{20}$ − 16.6° (*c* 1.9, methanol), from Aldrich Chemical Company, Inc.

4. Hydroquinone stabilizes the product during distillation by reducing traces of peroxide present in the reaction product.

5. GC analysis (capillary column coated with polypropylene glycol, 60.9 m., 100°) indicated that the product was *ca.* 95% pure (submitters). The checkers found the once-distilled material to be analytically pure. Analysis calculated for $C_{10}H_{16}O$: C, 78.90; H, 10.59. Found: C, 78.71; H, 10.55. IR (CCl₄) cm.$^{-1}$: 3600 medium, 3460 broad, medium, 1645 medium; ¹H NMR (CCl₄), δ (multiplicity, coupling constant *J* in Hz., number of protons): 0.63 (s, 3H), 1.26 (s, 3H), 1.6–2.5 (m, 6H), 2.88 (s, 1H, O*H*), 4.33 (approx. d, *J* = 7, 1H), 4.74 (approx. s, 1H), 4.96 (approx. s, 1H).

3. Discussion

trans-Pinocarveol is an important intermediate in the preparation of substituted pinane systems. It has been prepared by oxidation of β-pinene with lead tetraäcetate and hydrolysis of the corresponding ester (32%);[2] by photosensitized oxidation of α-pinene, followed by reduction of the corresponding hydroperoxide (35%);[3] by oxidation of β-pinene with molar quantities of selenium dioxide (53–64%);[4] and by epoxidation of α-pinene followed by isomerization with a variety of bases, of which lithium diethylamide (74–80% yield over the two steps) is best.[5]

The present procedure is a convenient, one-step method of preparing optically active *trans*-pinocarveol. Although lower in yield than the lithium diethylamide procedure, it is more readily adaptable to large-scale work. Moreover, the two methods are complimentary in the conditions required (neutral *vs.* basic) and in the overall transformation accomplished:

Since only catalytic quantities of selenium dioxide are required, the danger of handling large quantities of this material (Note 1) is avoided. Furthermore, the problems associated with the formation of selenium and organoselenides, which commonly arise in oxidations using molar quantities of selenium dioxide, are not encountered.

1. Department of Chemistry, University of Canterbury, Christchurch 1, New Zealand.
2. M. P. Hartshorn and A. F. A. Wallis, *J. Chem. Soc.*, 5254 (1964).
3. G. O. Schenck, H. Eggert, and W. Denk, *Justus Liebigs Ann. Chem.*, **584**, 177 (1953).
4. J. M. Quinn, *J. Chem. Eng. Data*, **9**, 389 (1964).
5. J. K. Crandall and L. C. Crawley, *Org. Synth.*, **Coll. Vol. 6**, 948 (1988) and references cited therein.

BASE-INDUCED REARRANGEMENT OF EPOXIDES TO ALLYLIC ALCOHOLS: *trans*-PINOCARVEOL

[Bicyclo[3.3.1]heptan-3-ol, 6,6-dimethyl-2-methylene-, (1α,3α,5α-]

Submitted by J. K. CRANDALL[1] and L. C. CRAWLEY
Checked by SHOICHIRO UYEO and WATARU NAGATA

1. Procedure

A dry, 300-ml., three-necked, round-bottomed flask is fitted with an effective reflux condenser, a 50-ml. pressure-equalizing dropping funnel, a rubber septum, a magnetic stirring bar, and a nitrogen inlet tube on the top of the condenser to maintain a static nitrogen atmosphere in the reaction vessel throughout the reaction. The flask is flushed with dry nitrogen, then charged with 2.40 g. (0.0329 mole) of diethylamine (Note 1) and 100 ml. of anhydrous diethyl ether (Note 2). The flask is immersed in an ice bath, the stirrer is started, and 25 ml. (0.035 mole) of 1.4 M n-butyllithium in hexane (Notes 3 and 4) is added carefully through the rubber septum with a syringe. After stirring for 10 minutes, the ice bath is removed and 5.00 g (0.0329 mole) of α-pinene oxide (Note 5) in 20 ml. of anhydrous ether is added dropwise over a 10-minute period. The resulting mixture is heated to reflux with stirring for 6 hours (Note 11). After the clear homogeneous mixture is cooled in an ice bath, it is stirred vigorously while 100 ml. of water is added. The ether phase is separated and washed successively with 100 ml. portions of 1 N hydrochloric acid, water, saturated aqueous sodium hydrogen carbonate, and water. The aqueous phase and each washing are extracted twice with 50 ml. portions of ether, and the ethereal extracts are combined and dried over anhydrous magnesium sulfate. Evaporation of the solvent on a rotary evaporator yields a light-yellow, oily residue which is distilled through a short-path distillation head, giving 4.50–4.75 g. (90–95%) of *trans*-pinocarveol as a colorless oil, b.p. 92–93° (8 mm.) n_D^{25} 1.4955 (Notes 12 and 13).

2. Notes

1. Commercial diethylamine, b.p. 55–58°, purchased from Fisher Scientific Company, was distilled from calcium hydride before use. The checkers used material purchased from Kanto Chemical Company, Inc. (Japan) and distilled it from sodium hydride.

2. The checkers used anhydrous ether, distilled from sodium hydride before use.

3. The *n*-butyllithium in hexane solution was purchased from Foote Mineral Company. The checkers obtained their material from Wako Pure Chemical Industries Ltd. (Japan) and titrated it with 0.80 M 2-butanol in xylene using 1,10-phenanthroline as indicator.[2] Care should be exercised in handling *n*-butyllithium solutions.

4. The submitters used about three molar equivalents of lithium diethylamide in about twice as much solvent. The checkers found that an amount of base slightly in excess of one molar equivalent was sufficient to convert the epoxide to exocyclic methylene alcohol of superior purity.

5. The submitters purchased α-pinene oxide from F.M.C. Corporation. However, since the compound is no longer available, the checkers prepared it from α-pinene as follows. A three-necked, round-bottomed flask fitted with a 50-ml. dropping funnel, a thermometer, and a magnetic stirring bar is charged with 22.0 g. (0.102 mole) of *m*-chloroperbenzoic acid (Note 6), 11.0 g. (0.131 mole) of sodium hydrogen carbonate, and 250 ml. of dichloromethane. The suspension is stirred with a powerful stirrer while being cooled with an ice–salt bath. To this mixture is added dropwise a solution of 13.6 g. (0.986 mole) of α-pinene (Note 7) in 20 ml. of dichloromethane at a rate such that the inner temperature is kept between 5–10° (Note 8). During the addition, sodium *m*-chlorobenzoate begins to crystallize indicating that the reaction is proceeding. After completion of the addition, stirring is continued for 1 hour longer at the same temperature (Note 9). A solution of 5 g. of sodium sulfite in 50 ml. of water is added, and the mixture is stirred vigorously at room temperature for 30 minutes. Water (50 ml.) is added, and the dichloromethane phase is separated and washed with 100 ml. of 5% aqueous sodium carbonate. The two aqueous washings are extracted with 50 ml. of dichloromethane, and the organic solutions are combined and dried over anhydrous magnesium sulfate. Evaporation of the solvent on a rotary evaporator gives an oily residue that is distilled through a vacuum-jacketed column, yielding 12.5–12.8 g. (82–85%) of α-pinene oxide as a colorless oil, b.p. 89–90° (28 mm.) (Note 10).

6. *m*-Chloroperbenzoic acid was obtained from F.M.C. Corporation. It was shown to be 80% pure by titration.

7. Technical grade α-pinene, purchased from Wako Pure Chemical Industries Ltd. (Japan), was used without purification.

8. A more efficient cooling system, such as an acetone–dry ice bath, is necessary to shorten the addition time in large-scale preparations.

9. Completion of the reaction may be checked by GC.

10. GC of this product using a 1-m. column containing 5% KF-54 on Chromosorb W at 100° gave a single peak. The material gave the following ^{1}H NMR spectrum (CDCl$_3$): δ 0.95, 1.30, and 1.33 (3 s, 9H, 3CH$_3$), 1.53–2.20 (m, 6H, 2CH$_2$ and 2CH), 3.03 (m, 1H, CHOC). The boiling point is reported to be 70–71° (12 mm.).[3]

11. Completion of the reaction may be checked by GC analysis. Refluxing for prolonged periods can give saturated ketone as an impurity if excess base is used.

12. The reported[4] value is n_D^{20} 1.4993.

13. The spectral properties are: IR (neat) cm.$^{-1}$ 3360 ms (OH), 1644 vw (C=C), 893 ms (C=CH$_2$); ^{1}H NMR (CDCl$_3$): δ 0.65 and 1.28 (2 s, 6H, 2CH$_3$), 1.63–2.55 (m, 6H, 2CH$_2$ and 2CH), 4.42 (d, J = 7 Hz., 1H, CHOH), 4.82 and 5.00 (2 m, 2H, C=CH$_2$). Purity of the product is greater than 98% as determined by GC using a Carbowax 20 M on 60–80 Chromosorb W column or a 1-m. column containing 5% KF-54 on Chromosorb W at 100°.

3. Discussion

Pinocarveol has been prepared by the autoxidation of α-pinene,[5] by the oxidation of β-pinene with lead tetraäcetate,[6] and by isomerization of α-pinene oxide with di-isobutylaluminum,[7] lithium aluminum hydride,[8] activated alumina,[9] potassium *tert*-butoxide in dimethyl sulfoxide,[10] and lithium diethylamide.[11] The present method is preferred for the preparation of pinocarveol, since the others give mixtures of products. It also illustrates a general method for converting 1-methylcycloalkene oxides into the corresponding exocyclic methylene alcohols.[11] The reaction is easy to perform, and the yields are generally high.

In general, the strong base isomerization of epoxides to allylic alcohols constitutes a useful synthetic reaction. Since the rearrangement is a highly specific process, it should be of value in organic synthesis. For example, there is a very high propensity for Hofmann-type eliminations to yield the least substituted double bond from unsymmetrically substituted epoxides.[12] There is also a large conformational effect arising from the operation of a *syn*-elimination mechanism which leads to specificity in eliminations of cyclic epoxides.

1. Department of Chemistry, Indiana University, Bloomington, Indiana 47401.
2. R. A. Ellison, R. Griffin, and F. N. Kotsonis, *J. Organomet. Chem.*, **36**, 209 (1972).
3. J. J. Ritter and K. L. Russell, *J. Am. Chem. Soc.*, **58**, 291 (1936).
4. H. Schmidt, *Ber. Dtsch. Chem. Ges.*, **63**, 1129 (1930).
5. R. N. Moore, C. Golumbic, and G. S. Fisher, *J. Am. Chem. Soc.*, **78**, 1173 (1956).
6. T. Sato, *Nippon Kagaku Zasshi*, **86**, 252 (1965).
7. P. Teisseire, A. Galfre, M. Plattier, and B. Corbier, *Recherches*, **15**, 52 (1966).
8. Y. Chrétien-Bessière, H. Desalbres, and J. P. Monthéard, *Bull. Soc. Chim. Fr.*, 2546 (1963).
9. V. S. Joshi, N. P. Damodaran, and S. Dev, *Tetrahedron*, **24**, 5817 (1968).
10. J. P. Monthéard and Y. Chrétien-Bessière, *Bull. Soc. Chim. Fr.*, 336 (1968).
11. J. K. Crandall and L. H. Chang, *J. Org. Chem.*, **32**, 435 (1967); J. K. Crandall and L. H. C. Lin, *J. Org. Chem.*, **33**, 2375 (1968).
12. R. P. Thummel and B. Rickborn, *J. Org. Chem.*, **36**, 1365 (1971); R. P. Thummel and B. Rickborn, *J. Am. Chem. Soc.*, **92**, 2064 (1970); B. Rickborn and R. P. Thummel, *J. Org. Chem.*, **34**, 3583 (1969).

POLYMERIC CARBODIIMIDE. PREPARATION

**[Benzene, diethenyl-, polymer with ethenylbenzene,
[[[[(1-methylethyl)imino]methylene]amino]methyl]deriv.]**

(P) = styrene–divinylbenzene copolymer

Submitted by NED M. WEINSHENKER,[1] CHAH M. SHEN,
and JACK Y. WONG
Checked by A. FUKUZAWA and S. MASAMUNE

1. Procedure

Caution! Benzene has been identified as a carcinogen; OSHA has issued emergency standards on its use. All procedures involving benzene should be carried out in a well-ventilated hood, and glove protection is required.

A. *Polymeric benzylamine.* A 300-ml., one-necked, round-bottomed flask equipped with a reflux condenser and a magnetic stirrer is charged with 125 ml. of N,N-dimethylformamide (Note 1) and 10.0 g. of chloromethylated polystyrene beads (0.0106 mole of active chloride) (Notes 2 and 3). A gas-inlet is attached to the top of the condenser, and the system is maintained under a slight positive pressure of nitrogen. The temperature is then raised to 100° with an oil bath, and 2.95 g. (0.0159 mole) of potassium phthalimide (Notes 4 and 5) is added while the mixture is stirred. After stirring at 100° overnight, the mixture is cooled and filtered. The polymer beads are washed with 200 ml. each of distilled water and methanol and dried under reduced pressure, giving 11.70 g. of phthalimido polymer.

The beads prepared above (11.58 g.) are suspended in 175 ml. of boiling absolute ethanol, and 0.94 g. (0.016 mole) of 85% aqueous hydrazine monohydrate is added with stirring. The resulting mixture is refluxed for 10 hours, after which the polymer is collected by filtration and washed with 150-ml. portions of ethanol, aqueous 0.2 N sodium hydroxide, distilled water, and anhydrous methanol. After vacuum drying at 60° for four hours, the yield of polymeric benzylamine is 10.38 g.

B. *Polymeric urea.*   A 10.0-g. portion of the benzylamine polymer beads prepared in Part A and 125 ml. of tetrahydrofuran (Note 6) are combined in a 300-ml., three-necked, round-bottomed flask equipped with a magnetic stirrer, a dropping funnel, and a condenser fitted with a gas-inlet tube. A nitrogen atmosphere is established in the system, and the slurry is stirred while 1.35 g. (0.0159 mole) of 2-isocyanatopropane is added. An exothermic reaction ensues and subsides after about 20 minutes. The mixture is then stirred at room temperature for 22 hours, and at reflux for an additional 4 hours. The beads are collected by filtration, washed with 150-ml. portions of tetrahydrofuran (Note 6) and methanol, and dried under reduced pressure over calcium chloride, yielding 9.09 g. of the 2-propyl urea polymer.

C. *Polymeric carbodiimide.*   The polymeric urea prepared above (9.09 g.) is combined with 100 ml. of dichloromethane in a 300-ml., three-necked, round-bottomed flask equipped with a magnetic stirrer, a condenser fitted with a gas-inlet tube, and a stopper. Under a blanket of nitrogen, 5.76 g. (0.0570 mole) of triethylamine and 2.75 g. (0.0145 mole) of p-toluenesulfonyl chloride (Note 7) are added to the stirred reaction mixture. The resulting slurry is refluxed with stirring for 50 hours, cooled to room temperature, and filtered. The polymer beads are washed successively with 100-ml. portions of dichloromethane, ice water, 3 : 1 dioxane–water, dioxane, and anhydrous diethyl ether. Vacuum drying yields 8.61 g. of polymeric carbodiimide, containing 0.98–1.01 millimoles of carbodiimide per gram (Note 8).

2. Notes

1. *N,N*-Dimethylformamide was dried overnight over Linde type 4A molecular sieves.

2. The checkers used beads of chloromethylated polymer available from Bio. Rad. Laboratories, Richmond, California (Bio Beads S·X2). Chlorine analysis (Note 3) showed that the resin contained 1.06 milliequivalents of chlorine per gram, as specified by the manufacturer.

The submitters prepared the polymer as follows. *Caution! Chloromethyl methyl ether is a carcinogen and is listed as such on the OSHA list. Therefore, preparation of the chloromethylated resin must be performed in a fume hood, the operator must wear gloves, and the reagent must be disposed of in an appropriate manner.* A slurry of 200 g. (1.93 moles) of polystyrene crosslinked with 2% divinylbenzene (Amberlite XE-305, obtained from the Rohm and Haas Company, Philadelphia) and 2.5 l. of chloroform was prepared in a 3-l., three-necked, round-bottomed flask equipped with a dropping funnel, a condenser, and a mechanical stirrer. After stirring for 0.5 hour at room temperature, the mixture was cooled in an ice-water bath, and a mixture of chloromethyl methyl ether (430 ml., 5.69 moles) and anhydrous tin(IV) chloride (45 ml., 0.39 mole) was added dropwise with continuous stirring. After the addition was completed, the ice bath was removed, and the mixture was stirred for an additional 4 hours at room temperature. The beads were collected by filtration and washed successively with 2 l. of 3 : 1 dioxane–water, 2 l. of 3 : 1 dioxane–3 N hydrochloric acid, 1 l. of dioxane, 1 l. of water, and 1 l. of methanol. It is desirable to allow each of the solvents used in the washing procedure to be in contact with the beads for 5–10 minutes before filtration to ensure complete penetration. Drying over calcium chloride under reduced pressure yields 252 g. of the chloromethylated polymer. The chlorine content was 15.50%, equivalent to 4.29 milliequivalents of chlorine per gram of polymer (Note 3).

3. The chlorine content can be determined by either chlorine elemental analysis or a potentiometric titration using a chloride-ion electrode. For titration, about 0.2 g. of polymer is heated in 3 ml. of pyridine at 100° for 2 hours. This suspension is then transferred to a 50-ml. beaker containing 30 ml. of 50% acetic acid and 5 ml. of concentrated nitric acid, and the resulting mixture is titrated against aqueous 0.1 N silver nitrate.

4. All chemicals used were reagent grade unless otherwise specified. A 50% molar excess of reagents was employed throughout the synthesis in order to drive the reactions to completion.

5. Potassium phthalimide was washed with acetone prior to use according to the procedure in *Org. Synth.*, **Coll. Vol. 4,** 810 (1963).

6. Tetrahydrofuran was dried and distilled from lithium aluminum hydride prior to use. For a warning concerning potential hazards of this procedure, see *Org. Synth.*, **Coll. Vol. 5,** 976 (1973).

7. *p*-Toluenesulfonyl chloride was recrystallized from hexane prior to use.

8. The *maximum* content of active carbodiimide groups can be determined by a nitrogen elemental analysis. The submitters determined the *minimum* carbodiimide content by treating the reagent with excess acetic acid: *ca.* 1 g. of accurately weighed polymeric carbodiimide was suspended in a mixture of 7 ml. of benzene, 3 ml. of ether, and 1.2 g. of acetic acid. After 20 hours of stirring, the conversion to acetic anhydride was determined by GC using a Carbowax 20M column operated at 160°. Triglyme was used as an internal standard. The final, deactivated polymer still showed a very strong absorption at 2140 cm.$^{-1}$ (KBr) in its IR spectrum. The checkers modified the above procedure slightly by using glutaric acid instead of acetic acid.

3. Discussion

The general procedure described here was originally published by the submitters,[2] who have used this insoluble reagent to prepare aldehydes and ketones under Moffat oxidation conditions.[3] A polymeric reagent offers two advantages: (a) when an oxidation is complete, the urea by-product is cleanly separated from the products by a simple filtration; and (b) the deactivated urea form of the polymer can be recycled efficiently to the carbodiimide form, as outlined in Part C of the present procedure. The use of polyhexamethylenecarbodiimide in peptide syntheses has been mentioned previously.[4]

1. Dynapol, 1454 Page Mill Road, Palo Alto, California 94304.
2. N. M. Weinshenker and C. M. Shen, *Tetrahedron Lett.*, 3281 (1972).
3. N. M. Weinshenker and C. M. Shen, *Tetrahedron Lett.*, 3285 (1972); N. M. Weinshenker, C. M. Shen, and J. Y. Wong, *Org. Synth.*, **Coll. Vol. 6,** 218 (1988).
4. Y. Wolman, S. Kivity, and M. Frankel, *Chem. Commun.*, 629 (1967).

ALKYNES *via* PHASE TRANSFER–CATALYZED DEHYDROHALOGENATION: PROPIOLALDEHYDE DIETHYL ACETAL

(1-Propyne, 3,3-diethoxy-)

Submitted by A. Le Coq[1] and A. Gorgues
Checked by G. Saucy and P. S. Manchand

1. Procedure

A.   *2,3-Dibromopropionaldehyde diethyl acetal.*   A 500-ml., three-necked, round-bottomed flask is equipped with a mechanical stirrer, a pressure-equalizing dropping funnel fitted with a calcium chloride drying tube, and a thermometer. The flask and dropping funnel are charged with 28.0 g. (0.500 mole) of freshly distilled acrolein and 80.0 g. (0.500 mole) of bromine, respectively. The acrolein is stirred rapidly and cooled to 0° in an ice–salt bath, then bromine is added at a rate such that the temperature is kept at 0–5°, until a permanent red color indicates a slight excess of bromine in the flask. A total of 78–79 g. of bromine is added over a 1-hour period. The crude 2,3-dibromopropionaldehyde is stirred while a solution of 80 g. (0.54 mole) of freshly distilled triethyl orthoformate in 65 ml. of absolute ethanol (Note 1) is added over 15 minutes. The solution warms to 45° and is stirred for 3 hours, after which ethyl formate, ethanol, and triethyl orthoformate are removed on a rotary evaporator. Distillation of the residual liquid through a 15-cm. Vigreux column affords 107–112 g. (74–77%) of 2,3-dibromopropionaldehyde diethyl acetal, b.p. 113–115° (11 mm.), as a pale-yellow liquid (Note 2).

B. *Propiolaldehyde diethyl acetal.* A 500-ml., three-necked, round-bottomed flask equipped with a mechanical stirrer (Note 3), a double-walled condenser, and a pressure-equalizing dropping funnel is charged with 100 g. (0.295 mole) (Note 4) of tetrabutylammonium hydrogen sulfate (Note 5) and 20 ml. of water. The mixture is stirred, forming a thick paste to which a solution of 29 g. (0.10 mole) of 2,3-dibromopropionaldehyde diethyl acetal in 75 ml. of pentane is added. The resulting mixture is stirred rapidly and cooled to 10–15° as a cold (10–15°) solution of 60 g. (1.5 moles) of sodium hydroxide in 60 ml. of water is added over 10 minutes. About 5 minutes later the pentane begins to boil and continues to reflux for another 10–20 minutes. The mixture is stirred for 2 hours at room temperature, cooled to 5°, and made slightly acidic (Note 6) by adding *ca.* 120 ml. of cold (*ca.* 5°) 25% sulfuric acid. Stirring is stopped, the layers are allowed to separate for 30 minutes, and the upper organic layer is carefully decanted (Note 7). The lower, aqueous layer is filtered, removing sodium sulfate, extracted with three 50-ml. portions of pentane, and, if desired, processed to recover the tetrabutylammonium salt (Note 8). The pentane solutions are combined, dried over anhydrous sodium sulfate, and evaporated. The colorless concentrate is distilled, giving 7.8–8.6 g. (61–67%) of propiolaldehyde diethyl acetal as a colorless liquid, b.p. 138–139° (760 mm.), 95–96° (170 mm.) (Note 9).

2. Notes

1. Absolute ethanol from a commercial supplier was used without further treatment.
2. The submitters report a yield of 113–122 g. (78–84%), b.p. 113–115° (11 mm.) (lit.,[2] b.p. 108–110°, 10 mm.). The product obtained by the checkers was analyzed. Analysis calculated for $C_7H_{14}Br_2O_2$: C, 28.99; H, 4.87; Br, 55.11. Found: C, 28.81; H, 4.88; Br, 55.37. The ^{1}H NMR spectrum (CDCl₃), δ (multiplicity, coupling constant *J* in Hz., number of protons, assignment): 1.27 (t, $J = 7$, 6H, 2 OCH₂CH₃), 3.6–3.9 (m, 6H, 2 OCH₂CH₃ and CH₂Br), 4.22 (apparent d of t, $J = 4$ and 7, 1H, CHBr), 4.72 (d, $J = 4$, 1H, CH).
3. The submitters used a 1-l. Erlenmeyer flask and a magnetic stirrer. The Erlenmeyer flask was recommended to minimize splattering of the pasty mixture into the condenser. The checkers preferred a 500-ml., round-bottomed flask equipped as described above.
4. The submitters found that the yield of product was reduced to *ca.* 50% when only 1 equivalent (0.2 mole) of tetrabutylammonium hydrogen sulfate was used.
5. The submitters purchased tetrabutylammonium hydrogen sulfate (97% pure) from Fluka AG, Buchs, Switzerland; it was obtained by the checkers from Aldrich Chemical Company, Inc.
6. Care must be exercised during acidification, since excess sulfuric acid lowers the yield, presumably through hydrolysis of the acetal.
7. The checkers often obtained a thick emulsion which separated into three layers after standing for *ca.* 2 hours. When this occurred, the mixture was poured into 500 ml. of water, and the product was extracted with three 150-ml. portions of pentane.
8. The following unchecked procedure has been provided by the submitters for the purpose of recovering the tetrabutylammonium salt. The aqueous layer, which contains 12 g. of sodium bromide, is extracted with two 100-ml. portions of dichloromethane. The solution is dried and evaporated, giving 91–93 g. (96–98%) of crude tetrabutylammonium bromide which can be recrystallized from ethyl acetate or employed directly for regenerat-

ing the hydrogen sulfate salt. The submitters recommend that the bromide be accumulated from several runs and then converted to the hydrogen sulfate by the procedure of Brandström.[3] A two-necked, round-bottomed flask fitted with a short distillation column and a dropping funnel is charged with 196 g. (0.609 mole) of recovered tetrabutylammonium bromide and 300 ml. of chlorobenzene. The contents of the flask are heated, and 92 g. (0.73 mole) of dimethyl sulfate is then added dropwise to the hot solution. The methyl bromide formed distills from the flask and is collected in a trap cooled with acetone–dry ice. As the rate of production of methyl bromide decreases, the heating is increased until the temperature at the top of the distillation column starts to rise rapidly. A solution of 1.5 ml. of concentrated sulfuric acid in 600 ml. of water is then cautiously added. The mixture is heated at reflux for 48 hours and evaporated to dryness under reduced pressure. After the residue has been dissolved in 500 ml. of dichloromethane, the resulting solution is washed with two 50-ml. portions of water and dried with anhydrous sodium sulfate. Evaporation of the solvent provides 202.5 g. of almost pure tetrabutylammonium hydrogen sulfate which can be recrystallized from isobutyl methyl ketone.

9. The submitters report a yield of 8.6–9.5 g. (67–74%), b.p. 95–96° (170 mm.) (lit.,[2] b.p. 138–139.5°, 760 mm.). The submitters recommend that the product be distilled under reduced pressure. The spectral characteristics of the product are as follows: IR (liquid film) cm.$^{-1}$: 3260 ($\equiv$CH), 2125 (C$\equiv$C); ^{1}H NMR (CDCl$_3$), δ (multiplicity, coupling constant J in Hz., number of protons, assignment): 1.24 (t, $J = 7$, 6H, 2CH_3), 2.58 (d, $J = 2$, 1H, $\equiv$CH), 3.71 (apparent q of d, $J = 7$ and 2, 4H, 2CH_AH_BCH$_3$), 5.21 (d, $J = 2$, 1H, CH(OC$_2$H$_5$)$_2$].

3. Discussion

The preparation of 2,3-dibromopropionaldehyde diethyl acetal described here is based on the procedure of Grard.[2,4] The dehydrobromination of the dibromide to propiolaldehyde diethyl acetal has previously been carried out with potassium hydroxide in ethanol[2,4] and with sodium amide in liquid ammonia.[5] In the present procedure the elimination is effected with aqueous sodium hydroxide in the presence of the phase-transfer agent, tetrabutylammonium hydrogen sulfate.[6,7] The principal advantage of the phase-transfer procedure is its operational simplicity. The method has been used to prepare diphenylacetylene (75%), phenylacetylene (87%), p-tolylacetylene (77%), and 3-chloropropiolaldehyde diethyl acetal (70%).[8] The halide reactants were the corresponding 1,2-dibromides in the first two examples and vinyl chlorides in the second two cases. The yields obtained with this method are better than those from traditional procedures, and the conditions are generally milder. In addition, the extent of substitution and dehalogenation, side reactions that frequently complicate the synthesis of acetylenes by elimination with alkoxide or amide bases, is diminished.[9] The ability to recover efficiently the tetrabutylammonium salt enhances the practicality of this procedure.[3]

Propiolaldehyde diethyl acetal has found numerous synthetic applications in the literature. The compound has been utilized in the synthesis of unsaturated and polyunsaturated acetals and aldehydes by alkylation of metallated derivatives,[5,10–13] by Cadiot-Chodkiewicz coupling with halo acetylenes,[13–14] and by reaction with organocuprates.[15] Syntheses of heterocyclic compounds including pyrazoles,[16] isoxazoles,[16] triazoles,[2] and pyrimidines[17,18] have employed this three-carbon building block. Propiolaldehyde diethyl acetal has also been utilized in the synthesis of naturally occurring polyacetylenes[19–23] and steroids.[24]

1. Laboratoire de Synthèse Organique, Université de Rennes, Rennes, France.
2. J. C. Sheehan and C. A. Robinson, *J. Am. Chem. Soc.*, **71**, 1436 (1949).
3. A. Brandström, "Preparative Ion Pair Extraction," Apotekarsocieteten/Hässle Läkemedel, Sweden, 1974, pp. 141–142.
4. M. Grard, *Justus Liebigs Ann. Chem.*, **13**, 336 (1930).
5. J. P. Ward and D. A. Van Dorp, *Recl. Trav. Chim. Pays-Bas*, **85**, 117 (1966).
6. For reviews, see E. V. Dehmlow, *Angew. Chem. Int. Ed. Engl.*, **16**, 493 (1977); W. P. Weber and G. W. Gokel, "Phase Transfer Catalysis in Organic Synthesis," Springer-Verlag, Berlin, 1977.
7. For other procedures using phase-transfer catalysis in this series, see M. Makosza and A. Jończyk, *Org. Synth.*, **Coll. Vol. 6**, 897 (1988); G. W. Gokel, R. P. Widera, and W. P. Weber, *Org. Synth.*, **Coll. Vol. 6**, 232 (1988).
8. A. Gorgues and A. Le Coq, *Tetrahedron Lett.*, 4723 (1976).
9. G. Köbrich and P. Buck, "Synthesis of Acetylenes and Polyacetylenes by Elimination Reactions," in H. G. Viehe, Ed., "Chemistry of Acetylenes," Dekker, New York, 1969, pp. 99–168.
10. Unilever N. V., Neth. Pat. 296,925 (1965) [*Chem. Abstr.*, **65**, 13547c (1966)].
11. E. K. Raunio and H. A. Schroeder, *J. Org. Chem.*, **22**, 570 (1957).
12. P. H. M. Schreurs, W. G. Galesloot, and L. Brandsma, *Recl. Trav. Chim. Pays-Bas*, **94**, 70 (1975).
13. J. P. Ward and D. A. Van Dorp, *Recl. Trav. Chim. Pays-Bas*, **86**, 545 (1967).
14. A. Gorgues, *Justus Liebigs Am. Chem.*, **7**, 211, 373 (1972).
15. A. Alexakis, A. Commercon, J. Villieras, and J. F. Normant, *Tetrahedron Lett.*, 2313 (1976).
16. L. Claisen, *Ber. Dtsch. Chem. Ges.*, **31**, 1021 (1898); L. Claisen, *Ber. Dtsch. Chem. Ges.*, **36**, 3664 (1903).
17. M. L. A. Fluchaire and G. L. A. Bost, U.S. Pat. 2,497,163 (1950) [*Chem. Abstr.*, **44**, 5908a (1950)].
18. G. W. Hearne, T. W. Evans, and H. L. Yale, U.S. Pat., 2,455,172 (1948) [*Chem. Abstr.*, **43**, 1813f (1949)].
19. F. Bohlmann and H. Bornowski, *Chem. Ber.*, **94**, 3189 (1961).
20. S. Prévost, J. Meier, W. Chodkiewicz, P. Cadiot, and A. Willemart, *Bull. Soc. Chim. Fr.*, 2171 (1961).
21. A. G. Fallis, M. T. W. Hearn, E. R. H. Jones, V. Thaller, and J. L. Turner, *J. Chem. Soc., Perkin Trans. 1*, 743 (1973).
22. C. A. Higham, E. R. H. Jones, J. W. Keeping, and V. Thaller, *J. Chem. Soc., Perkin Trans. 1*, 1991 (1974).
23. E. R. H. Jones, V. Thaller, and J. L. Turner, *J. Chem. Soc., Perkin Trans. 1*, 424 (1975).
24. M. Rosenberger, A. J. Duggan, R. Borer, R. Muller, and G. Saucy, *Helv. Chim. Acta*, **55**, 2663 (1972).

REARRANGEMENT OF BRIDGEHEAD ALCOHOLS TO POLYCYCLIC KETONES BY FRAGMENTATION-CYCLIZATION: 4-PROTOADAMANTANONE (TRICYCLO-[4.3.1.0³,⁸]DECAN-4-ONE)

[2,5-Methano-1H-inden-7(4H)-one, hexahydro]

Submitted by Zdenko Majerski[1] and Zdenko Hameršak[1]
Checked by Thomas P. Demuth and Andrew S. Kende

1. Procedure

Caution! Benzene has been identified as a carcinogen; OSHA has issued emergency standards on its use. All procedures involving benzene should be carried out in a well-ventilated hood, and glove protection is required.

A. *endo-7-Iodomethylbicyclo[3.3.1]nonan-3-one.* A 2-1., three-necked, round-bottomed flask equipped with an efficient mechanical stirrer and a reflux condenser is charged with 600 ml. of dry benzene (Note 1). The flask is immersed in a water bath, stirring is initiated, and 58.3 g. (0.132 mole) of lead tetraäcetate (Note 2), 37.4 g. (0.147 mole) of iodine, and 10.0 g. (0.0654 mole) of 1-adamantanol (Note 3) are added (Note 4). The bath temperature is gradually raised to 80° over a 20-minute period, then allowed to cool to 70–75°. Stirring is continued for 2 hours at 70–75° (Note 5) and for an additional hour while the mixture is cooled to room temperature. The inorganic salts are filtered and carefully washed with five 50-ml. portions of diethyl ether. The benzene filtrate and ether washings are combined in a 2-1. separatory funnel and shaken with 500 ml. of saturated aqueous sodium bisulfite (Note 6) until the dark red color disappears. The layers are *not* separated. If the color reappears within 10–15 minutes, the mixture is shaken again until colorless. This procedure is repeated as many times as necessary. The layers are then separated, and the organic layer is washed with 500 ml. of water and 250 ml. of saturated aqueous sodium hydrogen carbonate. The benzene–ether solution is dried over anhydrous magnesium sulfate for 1 hour and concentrated in a 500-ml., round-bottomed flask with a rotary evaporator (Note 7). The resulting crude, oily iodo ketone weighs 14–16 g. (Note 8) and is used immediately in Part B.

B. *4-Protoadamantanone.* The flask containing the crude iodo ketone is equipped with a magnetic stirring bar and a reflux condenser. A solution of 7 g. (0.1 mole) of potassium hydroxide in 150 ml. of methanol is added, and the mixture is stirred and heated at reflux for 3 hours. The contents of the flask are allowed to cool to room temperature and poured into 300 ml. of ice water. The resulting mixture is extracted with five 100-ml. portions of ether. The combined extracts are dried over anhydrous magnesium sulfate and evaporated under reduced pressure, leaving 8.6–9.1 g. of a yellow solid (Note 9). A solution of this crude product in 3 ml. of chloroform is allowed to percolate onto a chromatography column packed with 200 g. of activity III, neutral alumina in pentane (Note 10). The column is eluted first with 100 ml. of pentane, then with 500 ml. of 3:7 (v/v) ether–pentane, as 25-ml. fractions are collected and analyzed by GC (Note 11). Those fractions containing product whose purity is judged to be 98% or greater are combined and evaporated, affording 7.0–8.1 g. (71–82% based on 1-adamantanol) of 4-protoadamantanone as a colorless or pale-yellow solid, m.p. 202–204° (Note 12).

2. Notes

1. Solvent grade benzene was dried over sodium wire prior to use. If the benzene is wet, a considerable amount of starting 1-adamantanol remains unreacted owing to hydrolysis of lead tetraäcetate.
2. Lead tetraäcetate, both purchased from Fluka AG, Buchs, Switzerland, and prepared according to a literature procedure,[2] was used by the submitters without any noticeable difference. Lead tetraäcetate was dried prior to use for at least 12 hours over potassium hydroxide and phosphorus pentoxide in an evacuated desiccator (12 mm.), protected from direct light. If well protected from moisture, lead tetraäcetate can be kept for weeks. However, after exposure to moisture in the air it usually turns brown due to hydrolysis to lead hydroxide. The reactivity of such lead tetraäcetate is diminished somewhat, but it can still be used. If it has turned black, the reagent should be recrystallized from glacial acetic acid and dried prior to use, as described above.
3. 1-Adamantanol is available from the following three suppliers: Aldrich Chemical Company, Inc.; Fluka AG, Buchs, Switzerland; E. Merck, Darmstadt, Germany. It may also be prepared from adamantane by bromination to 1-bromoadamantane and hydrolysis.[3] Adamantane is sold by the same three suppliers.
4. The resulting solution is dark red in color.
5. The temperature of the bath should be *carefully* maintained in this range. At temperatures below 70° the reaction is much slower and increased amounts of unreacted 1-adamantanol will contaminate the product. At temperatures above 75° the amount of tar in the product is increased.
6. Other reducing agents such as sodium thiosulfate or sodium metabisulfite may be used as well.
7. Most of the solvent was evaporated with a bath temperature of 40–50°. The last 40–50 ml. was removed without heating. *endo*-7-Iodomethylbicyclo[3.3.1]nonan-3-one should be handled as quickly as possible, since this iodo ketone is thermally unstable. In the absence of solvent, decomposition may be rapid even at room temperature.
8. The crude iodo ketone usually contains up to 10% benzene, which does not interfere with the cyclization step (Part B). Complete removal of the benzene takes time, during which a considerable proportion of the iodo ketone may decompose.
9. A GC analysis on the crude product was carried out by the submitters using a 1.5 m. × 3.2 mm. column packed with 10% diethylene glycol succinate supported on 60/80

mesh Chromosorb W and heated at 140°. The chromatogram showed peaks for product, 1–3% unreacted 1-adamantanol, and a total of 1–2% of several other minor by-products.

10. Activity III alumina is prepared by adding 6% (w/w) of water to neutral alumina of activity grade I. The submitters used a 50 × 3 cm. glass column for the chromatography.

11. The conditions for GC are given in Note 9. The product was found mainly in fractions 2–20 by the submitters. The first 25-ml. fraction contained considerable amounts of by-products, while fractions 21 and higher contained 1-adamantanol. The checkers collected 10-ml. fractions with an automatic fraction collector.

12. Recrystallization from aqueous methanol raised the melting point to 207–210° (lit.,[4] m.p. 210–212°). The product obtained by the checkers was analytically pure. Analysis calculated for $C_{10}H_{14}O$: C, 79.95; H, 9.39. Found: C, 80.19; H, 9.31. The spectral characteristics of 4-protoadamantanone are as follows: IR (KBr) cm.$^{-1}$: 2920, 2860, 1710 (C=O), 1322, 1235; ^1H NMR (CDCl$_3$), δ (multiplicity, number of protons): 1.0–2.0 (m, ca. 7H), 2.0–3.0 (m, ca. 7H); ^{13}C NMR (CDCl$_3$) δ (assignment): 216.2 (C=O), 51.1 (CH), 45.0 (CH$_2$), 41.4 (CH$_2$), 38.2 (CH$_2$), 37.3 (CH$_2$ and CH), 37.2 (CH), 34.9 (CH$_2$), 29.6 (CH); mass spectrum m/e (relative intensity): 150 (M$^+$, 100), 95 (63), 93 (23), 81 (24), 80 (40), 79 (46), 67 (30), 66 (40).

3. Discussion

4-Protoadamantanone is a versatile intermediate for the synthesis of not only protoadamantane derivatives,[5–7] but also 1,2- and 2,4-disubstituted adamantanes,[8–10] 2-substituted noradamantanes,[11] and 4(5)-substituted 4-homoprotoadamantanes.[12]

4-Protoadamantanone has been prepared by the nitrous acid deamination of 2-amino-1-adamantanol (77%),[5] by aprotic diazotization of endo-7-aminomethylbicyclo[3.3.1]nonan-3-one in benzene with an equivalent amount of acetic acid (67%),[13] and by thermolysis of 1-adamantyl hypohalites followed by base-promoted cyclization of the resulting halo ketones (32–37%).[4,14,15] In spite of low and erratic yields, the last reaction sequence has provided the most convenient route to the protoadamantanes, since the other two approaches require lengthy syntheses of the starting materials.

The procedure described here is a modification of one involving the thermal fragmentation of 1-adamantyl hypoiodite and cyclization of the resulting iodo ketone.[4,14,16] With this procedure, 4-protoadamantanone is obtained from 1-adamantanol with consistent yields in the range of 71 to 82% and a purity greater than 98%. This method is also applicable to the preparation of other polycyclic ketones from the related bridgehead alcohols with α-bridges of zero, one, or two carbon atoms (see Table I).

With unsymmetrical bridgehead alcohols the structure of the product depends on the regioselectivity of both the fragmentation and intramolecular alkylation reactions. The position of the bond cleavage in the fragmentation step appears to be controlled by the relative thermodynamic stability of the keto free radical intermediates which subsequently react with iodine to produce the iodo ketones. In most cases this can be approximated simply by combination of the relative strain energies of the corresponding hydrocarbons and the relative stabilities of the free radical centers. The course of the cyclization is controlled by the balance of at least three factors: preferential enolization toward one

TABLE I
REARRANGED POLYCYCLIC KETONES PREPARED BY FRAGMENTATION AND RECYCLIZATION
OF BRIDGEHEAD ALCOHOLS

Alcohol	Product(s)	Ratio	Yield (%)	Reference
		—	74	16
		2:3	78	17
		1:1	69	18
		2:1[a]	30	18

[a]The base-catalyzed cyclization was carried out in aqueous 70% dioxane at reflux.

α-methylene group, the size of the smallest ring to be formed, and the relative degree of distortion of the preferred collinear arrangement of the two enolate α-carbon atoms and the carbon-leaving group bond.

1. Department of Organic Chemistry and Biochemistry, Rudjer Bošković Institute, 41001 Zagreb, Croatia, Yugoslavia.

2. L. F. Fieser and M. Fieser, "Reagents for Organic Synthesis," Vol. 1, Wiley, New York, 1967, p. 537.
3. H. Stetter, M. Schwarz, and A. Hirschhorn, *Chem. Ber.*, **92**, 1629 (1959); H. W. Geluk and J. L. M. A. Schlatmann, *Tetrahedron*, **24**, 5361 (1968); E. Osawa, *Tetrahedron Lett.*, 115 (1974).
4. W. H. W. Lunn, *J. Chem. Soc. C*, 2124 (1970).
5. D. Lenoir, R. E. Hall, and P. von R. Schleyer, *J. Am. Chem. Soc.*, **96**, 2138 (1974).
6. J. Boyd and K. H. Overton, *J. Chem. Soc., Perkin Trans. 1*, 2533 (1972); K. Mlinarić-Majerski and Z. Majerski, unpublished results.
7. A. Karim, M. A. McKervey, E. M. Engler, and P. von R. Schleyer, *Tetrahedron Lett.*, 3987 (1971); A. Karim and M. A. McKervey, *J. Chem. Soc., Perkin Trans. 1*, 2475 (1974).
8. D. Lenoir, R. Glaser, P. Mison, and P. von R. Schleyer, *J. Org. Chem.*, **36**, 1821 (1971).
9. B. D. Cuddy, D. Grant, and M. A. McKervey, *J. Chem. Soc. C*, 3173 (1971).
10. J. K. Chakrabarti, T. M. Hotten, D. M. Rackham, and D. E. Tupper, *J. Chem. Soc., Perkin Trans. 1*, 1893 (1976).
11. Z. Majerski, R. Šarac-Arneri, D. Škare, and B. Lončar, *Synthesis*, 74 (1980).
12. N. Takaishi, Y. Inamoto, K. Aigami, Y. Fujikura, E. Osawa, M. Kawanisi, and T. Katsushima, *J. Org. Chem.*, **42**, 2041 (1977).
13. J.-H. Liu and P. Kovacic, *J. Org. Chem.*, **38**, 3462 (1973).
14. R. M. Black and G. B. Gill, *J. Chem. Soc. D*, 972 (1970).
15. V. Boido and O. E. Edwards, *Can. J. Chem.*, **49**, 2664 (1971).
16. Z. Majerski, Z. Hameršak, and D. Škare, *Tetrahedron Lett.*, 3943 (1977).
17. Z. Hameršak, D. Škare, and Z. Majerski, *J. Chem. Soc. Chem. Commun.*, 478 (1977).
18. Z. Majerski and J. Janjatović, *Tetrahedron Lett.*, 3977 (1979); J. Janjatović and Z. Majerski, *J. Org. Chem.*, **45**, 4892 (1980).

QUADRICYCLANE

(Tetracyclo[3.2.0.0^{2,7}.0^{4,6}]heptane)

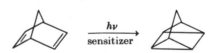

Submitted by CLAIBOURNE D. SMITH[1]
Checked by A. J. TAGGI and J. MEINWALD

1. Procedure

A Hanovia, 550-watt, immersion photochemical reactor (Note 1) equipped with a magnetic stirrer and water condenser (Note 2) is charged with 1 l. of diethyl ether, 180 g. (1.96 moles) of bicyclo[2.2.1]hepta-2,5-diene (2,5-norbornadiene, Note 3), and 8 g. of acetophenone. The system is flushed briefly with a stream of nitrogen, then irradiated for about 36–48 hours (Note 4). After irradiation, the ether is removed by distillation through a 20-cm. Vigreux column (Note 5). The residue, a clear liquid weighing about 185 g., is distilled through a spinning-band column under reduced pressure (Note 6), yielding

126–145 g. (70–80%) of quadricyclane as a colorless liquid, b.p. 70° (200 mm.) (Note 7).

2. Notes

1. The reactor, manufactured by the Hanovia Division of Engelhard Industries, consists of a water-jacketed Pyrex well through which a stream of water is continuously passed. The well is placed in an appropriately shaped flask containing the solution to be irradiated [*Org. Synth.*, **Coll. Vol. 5,** 528 (1973)]. The essentially cylindrical flask is equipped with a sidearm near the top which is connected to a water-cooled condenser. There should be sufficient clearance between the bottom of the well and the flask for a magnetic stirring bar. The flask is so designed that the liquid level is above the top of the lamp.

2. A source of nitrogen is attached to the top of the condenser, protecting the system from oxygen. The condenser serves as a safeguard in case the temperature of the system exceeds the boiling point of the solvent.

3. Bicyclo[2.2.1]hepta-2,5-diene was obtained from Shell Chemical Company and can be used as supplied. However, if the diene is distilled it should be used at once. The uninhibited diene may form a white, insoluble polymer or peroxide if allowed to stand in the presence of air and light. The use of undistilled diene results in a slightly lower yield of quadricyclane. The checkers used bicyclo[2.2.1]hepta-2,5-diene supplied by the Aldrich Chemical Company, Inc., which was distilled prior to use, giving a pure sample, b.p. 89–90.5°, $n_D^{26.5}$ 1.4680.

4. Other sensitizers (acetone, benzophenone) can be used, but with slightly reduced yields. The reaction may be monitored by removing aliquots and analyzing by GC or ^{1}H NMR. For GC analysis a 2-m. column containing 20% by weight of 1,2,3-tris(2-cyanoethoxy)propane suspended on Gas Chrome R (60–80 mesh) is used at a temperature of 75° with a flow rate of 85 ml./minute of helium. The retention time for quadricyclane is 5.2 minutes. For ^{1}H NMR analysis the disappearance of the absorption of the olefinic protons at δ 6.75 is monitored.

Lower wattage lamps can be used, although the irradiation time would be somewhat longer. The checkers found this reaction to be almost complete in about half this time; lamp age and other factors will cause appreciable variation in the irradiation time required.

5. Rapid distillation of the solvent may slightly reduce the yield of product.

6. The only volatile impurity at this point is bicyclo[2.2.1]hepta-2,5-diene. If the irradiation has been carried out for a sufficient period of time, the amount of diene present is less than 2%. Distillation through an efficient column will remove most of the diene, b.p. 91° (760 mm.) or 51° (200 mm.) Traces of acid[2] or noble metal ions and complexes[3] may cause quadricyclane to isomerize to the diene. The checkers used a 60-cm., Teflon-coated, spinning-band column available from Nester-Faust Corporation. The submitter used a similar 43-cm. column.

7. The checkers found n_D^{26} 1.4830 (lit.,[5] $n_D^{26.5}$ 1.4830) for the distillate. The ^{1}H NMR spectrum[2] shows peaks at δ 1.41 (6H) and δ 2.00 (2H). No olefinic absorption was detectable. The IR spectrum (CCl$_4$) shows three unusually well-resolved bands in the C–H stretching region at 3069, 2929, and 2852 cm.$^{-1}$.

3. Discussion

Quadricyclane may be prepared by direct irradiation of bicyclo[2.2.1]hepta-2,5-diene[2] and 2,3-diazatetracyclo[4.3.0.0^{4,8}.0^{7,9}]non-2-ene,[5] or by photosensitized isomerization of bicyclo[2.2.1]hepta-2,5-diene.[4,6] Several substituted quadricyclanes have been prepared by direct irradiation[7-10] and by photosensitization.[11-13] The procedure described above can be used to isomerize substituted bicycloheptadienes to the corresponding quadricyclanes when traces of sensitizers can be conveniently removed or their presence does not interfere with further use of the quadricyclane.

Quadricyclane is a highly strained and reactive compound. It reacts readily with acetic acid, giving a mixture of nortricyclyl acetate and *exo*-norbornyl acetate and with bromine, yielding a mixture of 2,6-dibromonortricyclene and *exo*-5-*anti*-7-dibromonorbornene.[2] Quadricyclane undergoes cycloaddition reactions with a variety of dienophiles, giving 1:1 adducts.[14]

1. Contribution No. 1222 from the Central Research Department, E. I. du Pont de Nemours & Co. (Inc.), Experimental Station, Wilmington, Delaware 19898. [Present address: Fabrics and Finishes Department, E. I. du Pont de Nemours & Co. (Inc.), Marshall Laboratory, Philadelphia, Pennsylvania 19146.]
2. W. G. Dauben and R. L. Cargill, *Tetrahedron*, **15**, 197 (1961).
3. H. Hogeveen and H. C. Volger, *J. Am. Chem. Soc.*, **89**, 2486 (1967).
4. G. S. Hammond, N. J. Turro, and A. Fischer, *J. Am. Chem. Soc.*, **83**, 4674 (1961).
5. R. M. Moriarty, *J. Org. Chem.*, **28**, 2385 (1963).
6. G. S. Hammond, P. Wyatt, C. D. DeBoer, and N. J. Turro, *J. Am. Chem. Soc.*, **86**, 2532 (1964).
7. S. J. Cristol and R. L. Snell, *J. Am. Chem. Soc.*, **76**, 5000 (1954).
8. H. G. Richey, Jr. and N. C. Buckley, *J. Am. Chem. Soc.*, **85**, 3057 (1963).
9. J. A. Claisse, D. I. Davies, and C. K. Alden, *J. Chem. Soc. C*, 1498 (1966).
10. D. I. Davies and P. J. Rowley, *J. Chem. Soc. C*, 2245 (1967).
11. P. R. Story and S. R. Fahrenholtz, *J. Am. Chem. Soc.*, **86**, 527 (1964).
12. P. G. Gassman, D. H. Aue, and D. S. Patton, *J. Am. Chem. Soc.*, **86**, 4211 (1964).
13. H. Prinzbach and J. Rivier, *Tetrahedron Lett.*, 3713 (1967).
14. C. D. Smith, *J. Am. Chem. Soc.*, **88**, 4273 (1966).

1,2,3,4-TETRAHYDRO-β-CARBOLINE

(1H-Pyrido[3,4-b]indole, 2,3,4,9-tetrahydro-)

Submitted by BENG T. HO and K. E. WALKER[1]
Checked by S. TEITEL, J. O'BRIEN, and A. BROSSI

1. Procedure

In a 1-l. Erlenmeyer flask, 25 g. (0.13 mole) of tryptamine hydrochloride (Note 1) is dissolved in 400 ml. of water by stirring and warming on a steam bath to approximately 45°. After cooling to room temperature, a solution of 13.2 g. (0.143 mole) of glyoxylic acid monohydrate (Note 2) in 30 ml. of water is added followed by the slow addition (about 3 minutes) of a cooled solution of 7.05 g. (0.126 mole) of potassium hydroxide in 35 ml. of water (Note 3). Precipitation of tetrahydro-β-carboline-1-carboxylic acid takes place during the addition of the potassium hydroxide solution or soon thereafter. After stirring at ambient temperature for 1 hour, the solid is collected on a filter and washed thoroughly with 100 ml. of water. The damp filter cake is transferred to a 1-l. beaker and suspended in 240 ml. of water; 34 ml. of concentrated hydrochloric acid is slowly added (Note 4) with stirring. The mixture is boiled on a hot plate for 30 minutes before an additional 35 ml. of concentrated hydrochloric acid is added. Heating is continued for another 15 minutes, and the resulting solution is allowed to cool to room temperature. The precipitated hydrochloride salt is collected on a filter and washed with 30 ml. of water.

The product is dissolved in 400 ml. of water by stirring and warming on a steam bath to approximately 55°, and the solution is adjusted to pH 12 with 20% aqueous potassium hydroxide (approximately 50 ml. is required). After cooling to room temperature, the product is collected by suction filtration, washed with 400 ml. of water, and dried in a vacuum desiccator over phosphorus pentoxide, yielding 17.0–17.6 g. (78–80%) of 1,2,3,4-tetrahydro-β-carboline, m.p. 204–205° (Note 5); ^{1}H NMR spectrum (dimethyl sulfoxide-d_6): δ 2.70 (t, 2H, CH_2CH$_2$N), 2.72 (s, 1H, NH), 3.00 (t, 2H, CH$_2$CH_2N), 3.85 (s, 2H, CCH_2N), 6.80–7.50 (m, 4H, C$_6$$H_4$), and 10.53 (s, 1H, N$H$).

2. Notes

1. The checkers used tryptamine hydrochloride (m.p. 253–255°) purchased from Regis Chemical Company.

2. Glyoxylic acid monohydrate is available from Pierce Chemical Company.

3. The resulting solution should have a pH between 3.5 and 4.0; if not, it should be adjusted with either potassium hydroxide or hydrochloric acid solution.

4. If all the hydrochloric acid is added at once, foaming makes the reaction unmanageable.

5. The melting point agrees with that of the literature[3] and is unchanged on recrystallization of the product from ethanol.

3. Discussion

1,2,3,4-Tetrahydro-β-carboline has been prepared by the condensation of tryptamine with formaldehyde in the presence of sulfuric acid[2] and has also been obtained as a by-product in the acid-catalyzed esterification of 1,2,3,4-tetrahydro-β-carboline-1-carboxylic acid.[3]

The described two-step procedure is uncomplicated and can be carried out in 1 day, giving in good yield a product that does not require further purification. This procedure has been used for the preparation of 3-methyl-, 9-methyl-, and 6-methoxy-1,2,3,4-tetrahydro-β-carboline[4] and has been modified for 9-phenyl-1,2,3,4-tetrahydro-β-carboline.[5] The method is generally applicable to the preparation of other 1-unsubstituted tetrahydro-β-carbolines providing the 1-carboxylic acid precursor is soluble in the hot acid used to effect decarboxylation.

Reviews of the chemistry of the carbolines have been published.[6,7]

1. Texas Research Institute of Mental Sciences, Texas Medical Center, Houston, Texas 77025.
2. E. Späth and E. Lederer, *Ber. Dtsch. Chem. Ges.,* **63,** 2102 (1930).
3. Z. J. Vejdělek, V., Trčka, and M. Protiva, *J. Med. Pharm. Chem.,* **3,** 427 (1961).
4. B. T. Ho, W. M. McIsaac, K. E. Walker, and V. Esteves, *J. Pharm. Sci.,* **57,** 269 (1968).
5. B. T. Ho, W. M. McIsaac, and K. E. Walker, *J. Pharm. Sci.,* **57,** 1364 (1968).
6. W. O. Kermack and J. E. McKail, in R. C. Elderfield, "Heterocyclic Compounds," Vol. 7, Wiley, New York, 1961, p. 237.
7. R. A. Abramovitch and I. D. Spenser, *Adv. Heterocycl. Chem.,* **3,** 79 (1964).

AZIRIDINES FROM β-IODOCARBAMATES: 1,2,3,4-TETRAHYDRONAPHTHALENE(1,2)IMINE

(1*H*-Naphth[1,2-*b*]azirine, 1a,2,3,7b-tetrahydro-)

Submitted by C. H. HEATHCOCK[1] and A. HASSNER[2]
Checked by WILLIAM G. KENYON and RICHARD E. BENSON

1. Procedure

A 500-ml., round-bottomed flask equipped with a reflux condenser is charged with a solution of 25 g. of potassium hydroxide in 250 ml. of 95% ethanol, to which is added 16.6 g. (0.0498 mole) of methyl (*trans*-2-iodo-1-tetralin)carbamate (Note 1). The resulting mixture is heated under reflux on a stream bath for 2 hours, cooled, and added to 500 ml. of water. The clear, yellow solution is shaken three times with 100-ml. portions of diethyl ether. The ether layers are combined, washed three times with 125-ml. portions of water and once with 125 ml. of a saturated sodium chloride, dried over 5 g. of anhydrous potassium carbonate, and filtered. The ether is removed by distillation on a steam bath, giving the crude imine as a yellow-brown oil (Note 2). The oil is transferred to a small flask, the container is rinsed with ether, and the rinse is added to the distillation flask. The product is collected by distillation through a small Vigreux column with warm water circulating through the condenser to prevent crystallization of the product. The fraction boiling at 80–82° (0.15–0.25 mm.) is collected as a solid that forms in the receiver, yielding 4.9–5.1 g. (68–70%) of the imine, m.p. 54–56° (Note 2); the IR spectrum has a band at 3205 cm.$^{-1}$ (NH) (Note 3).

2. Notes

1. The methylcarbamate may be prepared by the procedure in *Org. Synth.*, **Coll. Vol. 6**, 795 (1988).
2. The submitters state that product, m.p. 49–51°, can be obtained by direct crystallization of the oil. The oil from a run conducted on a scale twice that described above is cooled to −15° and 30 ml. of pentane is added. Upon scratching the flask, the product crystallizes, is collected by filtration, and washed with a little cold pentane, yielding 9–10 g. (62–69%), m.p. 49–51°.
3. The ^{1}H NMR spectrum (CCl$_4$) shows a broad singlet centered at δ 0.7 (1H) and complex multiplets at 1.1–3.05 (6H) and 6.76–7.30 (4H).

3. Discussion

The procedure reported here, that of Hassner and Heathcock,[3] is more convenient than the Wenker synthesis of aziridines[4] and appears to be more general.[5] It represents a simple route from olefins to aziridines (via β-iodocarbamates).[3,5,6] Aziridines are also useful as intermediates in the synthesis of amino alcohols and heterocyclic systems.[5,7–9]

1. Department of Chemistry, University of California, Berkeley, California 94720.
2. Present address: Department of Chemistry, State University of New York, Binghamton, New York 13901.
3. A. Hassner and C. Heathcock, Tetrahedron, 20, 1037 (1964).
4. O. E. Paris and P. E. Fanta, J. Am. Chem. Soc., 74, 3007 (1952).
5. A. Hassner and C. Heathcock, J. Org. Chem., 30, 1748 (1965).
6. G. Drefahl and K. Ponsold, Chem. Ber., 93, 519 (1960).
7. H. W. Heine, Angew. Chem., 74, 772 (1962) [Angew. Chem. Int. Ed. Engl., 1, 528 (1962)].
8. A. Hassner, M. E. Lorber, and C. Heathcock, J. Org. Chem., 32, 540 (1967).
9. L. A. Paquette and D. E. Kuhla, Tetrahedron Lett., 4517 (1967).

2,3,4,5-TETRAHYDROPYRIDINE TRIMER

(Pyridine, 2,3,4,5-tetrahydro-, trimer)

Submitted by George P. Claxton, Lloyd Allen, and J. Martin Grisar[1]
Checked by Henry F. Russell, Richard J. Sundberg, and Carl R. Johnson

1. Procedure

A. N-*Chloropiperidine*. A 500-ml., three-necked flask fitted with a mechanical stirrer, a dropping funnel, and a thermometer is charged with 170 g. (2.00 moles) of piperidine (Note 1). The flask is cooled in an acetone–ice bath, the piperidine is stirred, and 120 g. (2.00 moles) of glacial acetic acid is added dropwise at such a rate that the temperature does not exceed 10° (Note 2).

A 3-l., three-necked flask fitted with a mechanical stirrer, a dropping funnel, and a thermometer is charged with an aqueous solution of 2.2 moles of calcium hypochlorite (Note 3), and the piperidine acetate prepared above is placed in the dropping funnel. The hypochlorite solution is stirred and cooled to 0° to −5° with a methanol–ice bath, and the piperidine acetate is added dropwise over a period of 1.25 hours while the temperature is maintained below 0°. After an additional 15 minutes of stirring, equal portions of the mixture are placed in two 2-l. separatory funnels and extracted three times with a total of about 1300 ml. of diethyl ether. The ether extract is placed in a 2-l. flask, dried overnight over anhydrous sodium sulfate in a cold room at 4°, and filtered. The bulk of the ether is removed with a water bath maintained below 60° (Note 4).

B. *2,3,4,5-Tetrahydropyridine trimer.* A 3-l., three-necked, round-bottomed flask equipped with a sealed mechanical stirrer, a dropping funnel, and a reflux condenser fitted with a calcium chloride drying tube is charged with 264 g. (4.71 moles) of potassium hydroxide and 1250 ml. of absolute ethanol. This mixture is stirred with a Teflon paddle (Note 5) and heated to reflux, effecting solution. The *N*-chloropiperidine solution prepared in Part A is filtered through glass wool directly into the dropping funnel and added dropwise to the well-stirred, boiling reaction mixture over a period of *ca.* 2.5 hours (Note 6). The resulting mixture is stirred for an additional 2 hours without heating and allowed to stand at least 24 hours at room temperature, during which time tetrahydropyridine trimerizes. Precipitated potassium chloride is removed by filtration, washed with two 150-ml. portions of absolute ethanol, and set aside for later use. The washes are combined with the filtrate, ethanol is distilled off on a steam bath under reduced pressure, and the distillate saved for further processing (Note 7). The residue remaining after distillation and the recovered potassium chloride are then combined in 750 ml. of water, and the resulting solution is extracted four times with a total of 500 ml. of ether. After standing over anhydrous magnesium sulfate for 4 hours, the extract is filtered and concentrated on a rotary evaporator with gentle warming. The resulting oily residue is dissolved in 75 ml. of acetone, and the solution is cooled to −20° overnight. If no seed crystals are available, the walls of the flask are scratched with a glass rod, inducing crystallization. The precipitate is collected by vacuum filtration and washed twice with 20-ml. portions of cold (−20°) acetone, giving 64−80 g. (39−48%) of tetrahydropyridine trimer, m.p. 58−61° (Note 8).

2. Notes

1. Freshly distilled piperidine was used.
2. Ice was added to the mixture as required to dissolve any precipitated material and keep the viscous solution clear.
3. MC and B Manufacturing Chemists HTH grade 70% calcium hypochlorite was used. The solution required for the procedure was prepared by placing 680 g. of 70% calcium hypochlorite and 3 l. of water in a 5-l. flask and stirring overnight. The mixture was allowed to settle for several hours, and the supernatant was vacuum-filtered through Celite and glass wool, giving enough hypochlorite solution for a single run. By centrifugation of the suspension remaining after decanting, enough hypochlorite solution for another run could be obtained. The molarity of the solution was determined by iodometric titration: 1 g. of potassium iodide was dissolved in 25 ml. of water, and 15 ml. of 10% sulfuric acid and 1 ml. of the hypochlorite solution were added. The red solution was titrated with 0.10 *N* sodium thiosulfate. When the color changed to faint yellow, 1 ml. of starch solution was added, and the titration was continued to the colorless end point. The concentration was then determined according to the following formula:

$$¼[(ml. \ Na_2S_2O_3 \ reagent)(normality \ Na_2S_2O_3)] = molarity \ of \ Ca(OCl)_2$$

4. *Caution! To avoid a rapid, spontaneous decomposition that results in complete loss of N-chloropiperidine, the ether should not all be boiled off, nor should the temperature exceed 60°. The crude product should be used immediately in Part B.*

5. Use of a wire stirrer caused darkening of the ethanolic potassium hydroxide solution.

6. Isotripiperidein (m.p. 97–98°) is obtained if insufficient potassium hydroxide is used or if stirring is not sufficiently vigorous[2] (see Discussion).

7. It is recommended that this solvent be reused in later runs.[2,3] The submitters found that substantial amounts of product can be recovered from this distillate after a few days of standing. Thus, it appears that some tetrahydropyridine distils as monomer with the ethanol and trimerizes in the distillate. The checkers found that when the reaction was worked up after 24 hours of standing, the majority of the product was in the ethanol distillate. Therefore, they allowed the distillate to stand for several days, concentrated it by rotary evaporation, and crystallized the residue from acetone. The resulting tetrahydropyridine trimer was combined with that otherwise obtained.

8. Two trimers are known: α- (m.p. 60–62°) and β- (m.p. 70–73°) tripiperidein.[2] The β-form, usually obtained as a crude, white solid (m.p. 40–68°), may be converted to the more stable α-isomer by recrystallizing from acetone containing 2% water.[2] α-Tripiperidein is best stored in a closed container over potassium hydroxide and may be kept for over a year in this manner.

3. Discussion

There is one standard procedure for preparing 2,3,4,5-tetrahydropyridine.[2,3] No acceptable alternative method is available, except that N-chlorosuccinimide may be substituted for calcium hypochlorite.[4] Similar reaction sequences have been used to prepare substituted 2,3,4,5-tetrahydropyridines,[4,5] pyrroline,[6] and substituted pyrrolines.[7]

2,3,4,5-Tetrahydropyridine is useful for condensation with pyrroles and indoles,[8–10] β-ketoacids,[11–14] β-ketoesters,[15] and for a novel and very general reaction with magnesium chelates formed by reaction of methyl ketones with magnesium methyl carbonate.[16,17] The highly reactive monomer trimerizes on standing to tripiperidein, which exists in two interconvertible crystalline forms designated as α and β:

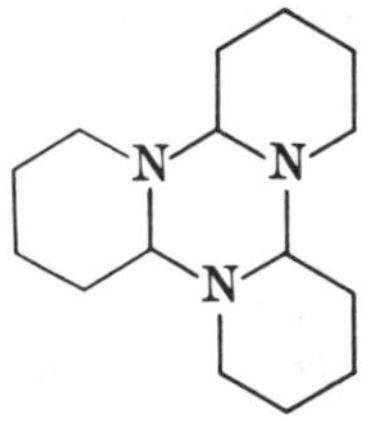

In the absence of base, the trimer can rearrange to iso-tripiperidein, a product of self-condensation.[11,18] In the presence of base, however, α-tripiperidein is stable for over a year. In solution, tripiperidein readily detrimerizes to the monomer, which is in equilibrium with δ-aminovaleraldehyde (5-aminopentanal).[9]

1. Organic Chemistry Department, Merrell Research Center, Merrell-National Laboratories, Division of Richardson-Merrell Inc., Cincinnati, Ohio 45215 [Present address: Centre de Recherche Merrell International, 16 rue d'Aukara, 67084 Strasbourg Cedex, France].
2. C. Schöpf, A. Komzak, F. Braun, and E. Jacobi, *Justus Liebigs Ann. Chem.*, **559**, 1 (1948).
3. C. Schöpf, H. Arm, and H. Krimm, *Chem. Ber.*, **84**, 690 (1951).
4. M. F. Grundon and B. E. Reynolds, *J. Chem. Soc.*, 2445 (1964).
5. M. F. Grundon and B. E. Reynolds, *J. Chem. Soc.*, 3898 (1963).
6. H. Poisel, *Monatsh. Chem.*, **109**, 925–928 (1978).
7. R. Bonnet, V. M. Clark, A. Giddey, and A. Todd, *J. Chem. Soc.*, 2087 (1959).
8. D. W. Fuhlhage and C. A. VanderWerf, *J. Am. Chem. Soc.*, **80**, 6249 (1958).
9. E. E. van Tamelen and G. G. Knapp, *J. Am. Chem. Soc.*, **77**, 1860 (1955).
10. J. Thesing, S. Klüssendorf, P. Ballach, and H. Mayer, *Chem. Ber.*, **88**, 1295 (1955).
11. J. H. Wisse, H. de Klonia, and B. J. Visser, *Recl. Trav. Chim. Pays-Bas*, **85**, 865 (1966).
12. C. Schöpf, F. Braun, K. Burkhardt, G. Dummer, and H. Müller, *Justus Liebigs Ann. Chem.*, **626**, 123 (1959).
13. J. van Noordwijk, J. J. Mellink, B. J. Visser, and J. H. Wisse, *Recl. Trav. Chim. Pays-Bas*, **82**, 763 (1963).
14. J. H. Wisse, H. de Klonia, and B. J. Visser, *Recl. Trav. Chim. Pays-Bas*, **83**, 1265 (1964).
15. J. P. Rosazza, J. M. Bobbitt, and A. E. Schwarting, *J. Org. Chem.*, **35**, 2564 (1970).
16. G. P. Claxton, J. M. Grisar, E. M. Roberts, and R. W. Fleming, *J. Med. Chem.*, **15**, 500 (1972).
17. J. M. Grisar, G. P. Claxton, and K. T. Stewart, *Synthesis*, 284 (1974).
18. C. Schöpf, F. Braun, and A. Komzak, *Chem. Ber.*, **89**, 1821 (1956).

3,3,6,6-TETRAMETHOXY-1,4-CYCLOHEXADIENE

[1,4-Cyclohexadiene, 3,3,6,6-tetramethoxy-]

Submitted by PAUL MARGARETHA[1] and PAUL TISSOT[2]
Checked by RONALD F. SIELOFF and CARL R. JOHNSON

1. Procedure

A 600-ml., tall-form beaker is equipped with a thermometer, a magnetic stirring bar, and two electrodes, a 45-mesh, cylindrical platinum anode (Note 1) surrounded by a cylindrical nickel cathode (Note 2). The electrodes are held in place (distance between anode and cathode: 0.75 cm.) and suspended in the beaker with a clamp formed from Delrin rods (Note 3). The electrodes are connected to an adjustable DC power supply (Notes 4, 5).

A solution of 27.6 g. (0.200 mole) of 1,4-dimethoxybenzene (Note 6), 4.0 g. of potassium hydroxide, and 400 ml. of methanol is placed in the apparatus. The beaker and contents are cooled with a 0° bath. The magnetically stirred solution is electrolyzed for 6 hours at a current intensity maintained at 2.0 amp. (Notes 5, 7). The temperature of the solution varies between 8° and 14°. During this time small amounts of methanol are added from time to time to compensate for evaporation.

After electrolysis the solution is reduced to a volume of 100 ml. with a rotary evaporator and extracted 10 times with 100-ml. portions of hexane. The hexane fractions are dried over anhydrous magnesium sulfate and the hexane evaporated with a rotary evaporator, yielding white crystals. Recrystallization from 75 ml. of pentane (Note 8) affords 27.8–28.2 g. (70–71%) of pure 3,3,6,6-tetramethoxy-1,4-cyclohexadiene, m.p. 40–43° (Note 9).

2. Notes

1. The platinum anode used was Model 611 obtained from Engelhard Industries. This electrode has a height of 5.6 cm. and a diameter of 5.1 cm. The total surface area claimed by the supplier is 200 cm.^{2}

2. The nickel cathode, fashioned from 22 gauge nickel sheet obtained from Huntington Alloys, Inc., Huntington, W. Va. 25720, had a height of 5.6 cm. and a diameter of 6.6 cm. When the sheet was rolled into a cylinder a small gap was left at the seam.

3. The clamp consisted of a 30 cm. vertical Delrin rod (0.5 cm. in diameter) threaded through the center of two 7-cm. horizontal Delrin rods (1.25 cm. in diameter). The horizontal rods were notched to hold the electrodes between them.

4. The power source used by the checkers was a 30-volt, 3-amp. adjustable DC supply.

5. The submitters used a cathode of nickel foil (140 × 71 × 0.5 mm.) rolled into a cylinder 3.5 cm. in diameter surrounded by three curved platinum anodes each having the dimensions 70 × 30 × 1 mm. (total surface area 130 cm.^{2}) with a distance of 0.5–1 cm. between the cathode and the anodes. The submitters electrolyzed for 6 hours at a current maintained at 3.25 amp. This corresponds to a total of 19.5 amp.-hours and an anodic current density of 0.025 amp./cm.^{2} Under these conditions the submitters report yields of 81–84%.

6. 1,4-Dimethoxybenzene was obtained from Aldrich Chemical Company, Inc.

7. This corresponds to 12 amp.-hours (theoretical value is 10.6 amp.-hours). Longer electrolysis times did not significantly increase the yield of product.

8. Cooling is necessary to complete crystallization.

9. ^{1}H NMR (CDCl$_3$), δ (multiplicity, number of protons, assignment): 3.30 (s, 12H, 4OCH$_3$), 6.10 (s, 4H, 4CH).

3. Discussion

The procedure described is essentially that of Belleau and Weinberg[3] and represents the only known way of obtaining the title compound. One other quinone ketal, 1,4,9,12-tetraoxadispiro[4.2.4.2]tetradeca-6,13-diene, has been synthesized by a conventional method (reaction of 1,4-cyclohexanedione with ethylene glycol followed by bromination and dehydrobromination[4]) as well as by an electrochemical method [anodic oxidation of

2,2-(1,4-phenylenedioxy)diethanol[5]]. Quinone ketals have been used as intermediates in the synthesis of 4,4-dimethoxy-2,5-cyclohexadienone, syn-bishomoquinone,[4,6] and compounds related to natural products.[7]

Aromatic diethyl ethers and furans undergo alkoxylation by addition upon electrolysis in alcohol containing a suitable electrolyte.[8-12] Other compounds such as aromatic hydrocarbons, alkenes, N-alkyl amides, and ethers lead to alkoxylated products by substitution. Two mechanisms for these electrochemical alkoxylations are currently discussed. The first one consists of direct oxidation of the substrate, giving the radical cation which reacts with the alcohol, followed by reoxidation of the intermediate radical, and either alcoholysis or elimination of a proton to the final product. In the second mechanism the primary step is the oxidation of the alcoholate, giving an alkoxyl radical which reacts with the substrate, the consequent steps then being the same as above. The formation of quinone ketals in particular seems to proceed via the second mechanism.[5]

1. Département de Chimie Organique, Université de Genève, Switzerland. [Present address: Institut für Organische Chemie and Biochemie, Universität Hamburg, Martin-Luther-King-Platz 6, 2000 Hamburg 13, Germany.]
2. Département de Chimie Minerale, Analytique et Appliquée, Université de Genève, Switzerland.
3. B. Bellau and N. L. Weinberg, J. Am. Chem. Soc., 85, 2525 (1963).
4. J. E. Heller, A. S. Dreiding, B. R. O'Connor, H. E. Simmons, G. L. Buchanan, R. A. Raphael, and R. Taylor, Helv. Chim. Acta, 56, 272 (1973).
5. P. Margaretha and P. Tissot, Helv. Chim. Acta, 58, 933 (1975).
6. G. L. Buchanan, R. A. Raphael, and R. Taylor, J. Chem. Soc. Perkin Trans. I, 373 (1973).
7. M. J. Manning, P. W. Raynolds, and J. S. Swenton, J. Am. Chem. Soc., 98, 5008 (1976).
8. L. Eberson, "Organic Electrochemistry," M. M. Baizer, Ed., Dekker, New York, 1973, p. 785.
9. N. L. Weinberg, "Techniques of Organic Chemistry," Vol. 5, A. Weissberger, Ed., Wiley, New York, 1974, p. 259.
10. A. J. Fry, "Synthetic Organic Electrochemistry," Harper & Row, New York, 1972, p. 298.
11. F. Beck, "Elektroorganische Chemie," Verlag Chemie, Weinheim, Germany, 1974, p. 247.
12. D. R. Henton, R. L. McCreery, and J. S. Swenton, J. Org. Chem., 45, 369 (1980).

2,2,3,3-TETRAMETHYLIODOCYCLOPROPANE

(Cyclopropane, 3-iodo-1,1,2,2-tetramethyl)

$$I_2 + 2 NaOH \rightarrow NaI + NaOI + H_2O$$

Submitted by T. A. Marolewski and N. C. Yang[1]
Checked by T. Nakahira and K. B. Wiberg

1. Procedure

Caution! The intense emission from the light source should be shielded from visibility in order not to damage the eyesight of the experimentalist.

Each of three 250-ml., round-bottomed Pyrex flasks is charged with 8.4 g. (0.10 mole) of 2,3-dimethyl-2-butene (Note 1), 175 ml. of dichloromethane, and 50 ml. of an aqueous 5 M sodium hydroxide solution. The flasks are kept rather full, making more efficient use of the incident light. A Teflon-covered magnetic stirring bar 2.5 cm. in length is added to each flask. Three 170 cm. by 90 cm. Pyrex crystallization dishes are partially filled with an ice–water mixture (Note 2), each dish is placed above a Mag-Mix magnetic stirrer, and each flask is immersed in the ice–water bath and held in place with a clamp. The three assemblies are arranged symmetrically around a Hanovia quartz immersion well (Note 3) cooled with running tap water, containing a Hanovia 450-watt, medium pressure, mercury lamp. The edge of each flask is placed approximately 1 cm. from the wall of the well. After 2.0 g. of iodoform is added to each flask, the mixtures are irradiated with stirring until the yellow color of the iodoform disappears. This process is continued until 39.4 g. (0.100 mole) of iodoform, equally distributed between the flasks, has been consumed (Note 4). After the reaction is complete, the reaction mixtures are combined and the organic layer is separated, washed once with water, and dried over anhydrous sodium sulfate. The solvent is removed with a rotary evaporator and a water pump. The residue is transferred to a 50-ml. flask, and 1.0 g. of sodium methoxide is added (Note 5). The mixture is distilled under reduced pressure in an apparatus with a 5-cm. Vigreux sidearm. The receiver is cooled in an ice–water bath and the first fraction, which boils at 45–48° (5 mm.), n_D^{25} 1.5087, is collected, yielding 14.0–15.0 g. (63–67%) of a clear distillate which should be stored in a refrigerator (Notes 6 and 7).

2. Notes

1. 2,3-Dimethyl-2-butene (99%) was purchased from the Chemical Samples Co.
2. At the beginning of irradiation, the mixture is mostly ice and contains just enough

water to make efficient contact with the flask. The ratio of ice to water will vary during the course of irradiation, and ice is added to replace excess water from time to time.

3. Vycor or Pyrex wells are also satisfactory since the irradiation is carried out in Pyrex flasks.

4. The total period of irradiation was about 8 hours; however, this may vary with the equipment used.

5. The presence of sodium methoxide prevents decomposition of the product during the distillation.

6. The checkers also carried out the reaction using equimolar quantities of 2,3-dimethyl-2-butene and iodoform (0.1 mole each) and obtained 12.6–13.0 g. (56–58%) of the product. The submitters made the same observation. They found that the yield increased slightly as the mole ratio of olefin to iodoform was increased from 1 : 1 to 3 : 1. Use of a larger excess of olefin resulted in no further increase in yield.

7. 1-Iodo-cis,trans-2,3-dimethylcyclopropane, b.p. 25–27° (8–10 mm.), n_D^{25} 1.5105, may be prepared in 56% yield from trans-2-butene (Matheson Gas Products) with this procedure. Both the 2,2,3,3-tetramethyliodocyclopropane and the 1-iodo-cis,trans-2,3-dimethylcyclopropane prepared by this procedure give only one peak on GC. The retention times are 272 and 114 seconds, respectively, on a 60-cm. 20% SE-30 on Chromosorb W column at a temperature of 81° and a helium flow rate of 41 ml. per minute.

3. Discussion

Bromo- and iodocyclopropanes cannot be prepared by the direct halogenation of cyclopropanes. Substituted chloro- and bromocyclopropanes have been synthesized by the photochemical decomposition of α-halodiazomethanes in the presence of olefins;[2] iodocyclopropanes have been prepared by the reaction of an olefin with iodoform and potassium tert-butoxide, followed by the reduction of the diiodocyclopropane formed with tri-n-butyltin hydride.[3] The method described employs a readily available light source and common laboratory equipment, and is relatively safe to carry out. The method can be modified for the preparation of cyclopropanes and halocyclopropanes as well, by using diiodomethane and halodiiodomethanes instead of iodoform.[4,5] If the olefin used gives two isomeric halocyclopropanes, the isomers are usually separable by chromatography.[4]

1. Department of Chemistry, University of Chicago, Chicago, Illinois 60637.
2. G. L. Closs and J. J. Coyle, J. Am. Chem. Soc., 87, 4270 (1965).
3. J. P. Oliver and U. V. Rao, J. Org. Chem., 31, 2696 (1966).
4. N. C. Yang and T. A. Marolewski, J. Am. Chem. Soc., 90, 5644 (1968).
5. N. J. Pienta and P. J. Kropp, J. Am. Chem. Soc., 100, 655 (1978).

4H-1,4-THIAZINE 1,1-DIOXIDE

Submitted by Wayland E. Noland[1] and Robert D. DeMaster[2]
Checked by H. Gurien, G. Kaplan, and A. Brossi

1. Procedure

Caution! Benzene has been identified as a carcinogen; OSHA has issued emergency standards on its use. All procedures involving benzene should be carried out in a well-ventilated hood, and glove protection is required.

A. cis *and* trans-2,6-*Diethoxy*-1,4-*oxathiane* 4,4-*dioxide.* Ozone (Note 1) is passed into a solution of 2,5-dihydrothiophene 1,1-dioxide (30.0 g., 0.254 mole) (Note 2) in 50 ml. of absolute ethanol (Note 3) and 250 ml. of dichloromethane contained in a 1-l., three-necked, round-bottomed flask fitted with a straight glass-inlet tube, a calcium chloride drying tube, and a glass stopper. The solution is cooled in a methanol–dry ice bath and magnetically stirred while the ozone is added. When the solution becomes blue (Note 4), the addition of ozone is stopped and liquid sulfur dioxide (35 ml., 0.78 mole) (Note 5) is added in portions over a period of 10–15 seconds. After 2 minutes, the cold bath is removed and the reaction solution is allowed to warm to room temperature over a period of 8–16 hours. The resulting dark-brown solution is poured into a 4-l. beaker containing a rapidly stirred mixture of aqueous sodium carbonate (120 g. in 1 l. of cold water) and 200 g. of ice. The reaction flask is rinsed with 50 ml. of water, which is added to the basic mixture. After being stirred for 5 minutes, the basic mixture is poured into a 2-l. separatory funnel and the lower dichloromethane layer is separated and saved. The beaker is rinsed with 200 ml. of dichloromethane and 100 ml. of water, which are then added to the separatory funnel. The contents of the separatory funnel are shaken, and the lower, dichloromethane layer is separated and saved. The aqueous layer is extracted with two more 150-ml. portions of dichloromethane. All of the dichloromethane layers and extracts are combined, and washed with 300 ml. of water and 300 ml. of saturated aqueous sodium chloride. The solution is dried over 3–6 g. of anhydrous magnesium sulfate, filtered, and evaporated with a rotary evaporator at 50–60° in a water bath under aspirator pressure. The residual, cream-colored solid (50–52 g., 88–91%), m.p. 76–118°, is dissolved with magnetic stirring in 850–950 ml. of boiling heptane (Note 6) containing 1–2 g. of activated carbon and filtered hot.

The filtrate is cooled to 0° in a refrigerator overnight. The resulting precipitate is filtered, giving *cis*- and *trans*-2,6-diethoxy-1,4-oxathiane 4,4-dioxide as a white solid (42–46 g., 74–81%), m.p. 83–117° (Note 7).

B. 4H-1,4-*Thiazine* 1,1-*dioxide. Caution! This step should be carried out in a hood to avoid exposure to hydrogen chloride gas.* A mixture of *cis*- and *trans*-2,6-

diethoxy-1,4-oxathiane 4,4-dioxide (15.0 g., 0.0669 mole), 3.8 g. (0.071 mole) of ammonium chloride (Note 8), and 300 ml. of glacial acetic acid is placed in a 500-ml., one-necked, round-bottomed flask fitted with a reflux condenser and a magnetic stirring bar. The mixture is placed in an oil bath preheated to 125–130° and refluxed, with magnetic stirring, for 25–35 minutes, during which the ammonium chloride dissolves, hydrogen chloride is evolved, and the solution becomes brownish yellow in color (Note 9). The acetic acid is evaporated with a rotary evaporator at 70–80° in a water bath under aspirator pressure. The residual yellow solid is magnetically stirred with a solution of 75 ml. of diethyl ether containing 10 ml. of 2-propanol for 10 minutes (Note 10). The resulting suspension is filtered, then sucked dry on a Büchner funnel. The yellow solid (8.7–9.2 g.), m.p. 208–212°, is boiled with 225–250 ml. of 2-propanol and filtered hot, removing the residual, greenish-black, insoluble material (0.5–1 g.). The filtrate is cooled to −10° to −5° in a freezer overnight, causing separation of 4.6–5.3 g. (52–60%) of 4H-1,4-thiazine 1,1-dioxide as small yellow needles, m.p. 237–240° (Note 11), which are filtered. Concentration of the filtrate to 50 ml., followed by filtration and cooling, causes separation of an additional 1.5–2.0 g. (17–23%) of crude yellow solid, m.p. 234–240°.

2. Notes

1. A Welsbach Corporation Ozonator, style T-23, was used, with the voltage set at 120 volts and the oxygen pressure at 8 p.s.i. to give a 4–5% ozone concentration. The checkers used a Welsbach Corporation Ozonator, style T-408, to give a 1–2% ozone concentration. The input oxygen was dried by being passed through a tower of color-indicating Hammond Drierite.

2. 2,5-Dihydrothiophene 1,1-dioxide (butadiene sulfone, or 3-sulfolene) was purchased from the Aldrich Chemical Company, Inc.

3. Use of larger amounts of absolute ethanol causes formation of more of the acyclic 3-thiapentane-1,5-dial bis(diethyl acetal) 3,3-dioxide, with a corresponding reduction in yield of the cyclic product.

4. Appearance of the blue color of ozone signals complete cleavage of the double bond. Further addition of ozone could cause undesirable oxidation.

5. Sulfur dioxide was purchased in lecture-size bottles from the City Chemical Corporation. The gas was condensed into a precalibrated, 50-ml. Erlenmeyer flask cooled in the methanol–dry ice bath used for cooling the ozonolysis reaction.

6. Eastman Organic Chemicals Technical Grade "Heptanes," b.p. 96–100°, containing 70% heptanes and the rest octanes, was used.

7. In one instance the submitters obtained an 85% yield when the reaction mixture was stirred with sulfur dioxide for 18 hours, followed by crystallization of the resulting crude material (32 g. per l.) without the use of charcoal.

The product is obtained as an approximately 55:45 mixture of cis- and trans-isomers, as indicated by ^{1}H NMR absorption (CDCl$_3$) at δ 1.27 (t, J = 7 Hz., 5.9H, 2 OCH$_2$CH$_3$), 2.77–3.47 (m, 4.0H, CH$_2$SO$_2$CH$_2$), 3.47–4.27 (m, 4.1H, 2 OCH$_2$CH$_3$), 4.95 (d of d, $J_{a,a}$ = 8 Hz., $J_{a,e}$ = 2 Hz., 1.1H, CH proton of the cis-isomer), and 5.33 (t, J = 4 Hz., 0.9H, CH proton of the trans-isomer). The IR spectrum (Nujol) has strong bands at 1312, 1118, 1029, and 972 cm.$^{-1}$, which are attributed to the SO$_2$ and CO groups. The cis-isomer, m.p. 103–105°, can be separated from the mixture by three or four fractional crystallizations from methanol, while the trans-isomer, m.p. 136–137°, can be separated from the mixture (or from the residue obtained by evaporation of the methanol mother liquors from which the *cis*-isomer was crystallized) by two or three fractional crystallizations from benzene-petroleum ether (b.p. 60–68°).

8. "Baker Analyzed" Reagent Grade ammonium chloride was purchased from the J. T. Baker Chemical Company.

9. Refluxing for longer times causes formation of increased amounts of a dark, greenish-brown by-product, which complicates purification by crystallization. If the acetic acid becomes black-brown, the residue (which is sometimes tarry) obtained on evaporation can be purified by rapid chromatography through a 3.8-cm.-deep column of activated alumina, using acetone as a transfer agent and eluent.

10. The purpose of the wash with ether and 2-propanol is to remove the remaining acetic acid and any residual hydrogen chloride, which may cause decomposition during the subsequent crystallization.

11. The analytical sample melted at 240–241.5°. The IR spectrum (Nujol) has a strong NH band at 3360, a strong band in the double bond region at 1645 and another at 1511, and a group of bands at 1265 and 1255 (medium strong) and 1238, 1226, 1102, and 1093 (all strong), some of which are attributable to the sulfonyl group, and a strong band at 692 cm.$^{-1}$. The ^{1}H NMR spectrum (dimethyl sulfoxide-d_6) has an AA'BB' pattern with major peaks at δ 7.12 and 6.99 (2.0H) and 6.02 and 5.88 (2.0H), attributed to the 4 CH protons. The UV spectrum has maxima (95% C$_2$H$_5$OH) nm at (log ϵ) 226 (3.75), 230, inflection (3.72), 237, inflection (3.47), 277 (3.52), and 287 (3.55).

3. Discussion

This procedure represents the first reported synthesis of *cis*- and *trans*-2,6-diethoxy-1,4-oxathiane 4,4-dioxide[3] and of its further reaction product, 4H-1,4-thiazine 1,1-dioxide.[3] A derivative of the latter, 3,5-diphenyl-4H-1,4-thiazine 1,1-dioxide, has been prepared previously by reaction of phenacyl sulfone with ammonia.[4,5] Primary amines, in addition to ammonia, can be converted to the corresponding 4-substituted 4H-1,4-thiazine 1,1-dioxides by condensation with 2,6-diethoxy-1,4-oxathiane 4,4-dioxide, using the procedure described above. For example, 4-aminobenzoic acid hydrochloride gave 4-(4-carboxyphenyl)-4H-1,4-thiazine 1,1-dioxide in 83% yield.[3] The submitters have also observed,[3] as have others,[4] that the 4H-1,4-thiazine 1,1-dioxide system may be *N*-alkylated with an alkyl halide using potassium carbonate in anhydrous acetone.

The ozonolysis reaction, followed by reductive workup with sulfur dioxide, as described in Part A of the present procedure, illustrates a general method which has been developed for the preparation of acetals.[3] Application of the procedure is illustrated by conversion of the following olefins in alcoholic solution to the corresponding acetals:[3] (1) 1-chloro-4-(2-nitrophenyl)-2-butene to 2-nitrophenylacetaldehyde dimethyl acetal in 84% yield; (2) 1,4-dibromo-2-butene to bromoacetaldehyde dimethyl acetal in 67% yield; (3) 3-butenoic acid to malonaldehydic acid diethyl acetal ethyl ester in 61% yield; (4) cyclopentadiene to malonaldehyde bis(diethyl acetal) in 48% yield; and (5) 1,4-dinitro-2-butene (produced *in situ* from 1,3-butadiene and dinitrogen tetroxide) to nitroacetaldehyde diethyl acetal in 21% yield.

1. School of Chemistry, University of Minnesota, Minneapolis, Minnesota 55455.

2. Safety and Security Systems Laboratory, 3M Company, St. Paul, Minnesota 55101.
3. Robert D. DeMaster, Ph.D. Dissertation, University of Minnesota, Minneapolis, Minnesota, June 1970 [*Diss. Abstr. Int. B*, **31**, 5871 (1971)].
4. C. R. Johnson and I. Sataty, *J. Med. Chem.*, **10**, 501 (1967).
5. I. Sataty, *J. Org. Chem.*, **34**, 250 (1969).

2-THIOPHENETHIOL

Submitted by E. Jones[1] and I. M. Moodie
Checked by E. J. Corey and Joel I. Shulman

1. Procedure

A 3-l. three-necked flask fitted with a mechanical stirrer, a 600-ml. dropping funnel, and filled with dry nitrogen is charged with 500 ml. of tetrahydrofuran (distilled from lithium aluminum hydride; see *Org. Synth.*, **Coll. Vol. 5**, 976 (1973) for warning concerning the purification of tetrahydrofuran) and 56 g. (53 ml., 0.67 mole) of thiophene. This mixture is stirred under nitrogen and cooled to −40° with an acetone–dry ice bath while 490 ml. (0.662 mole) of 1.35 M *n*-butyllithium in pentane (Notes 1 and 2) is added over a 5-minute period *via* the dropping funnel. The temperature of the mixture is held between −30° and −20° for 1 hour, then lowered to −70° by the addition of dry ice to the bath. Powdered sulfur crystals (20.4 g., 0.638 g.-atom) are added in one aliquot to the stirred mixture. After 30 minutes the temperature is allowed to rise to −10°, whereupon the yellow solution is carefully poured into 1 l. of rapidly stirred ice water, dissolving the lithium thiolate and destroying any unreacted 2-thienyllithium. The pentane layer is extracted with three 100-ml. portions of water. These aqueous extracts are combined with the aqueous layer, and the whole is chilled and carefully acidified with 4 N sulfuric acid (Note 3). This aqueous phase is immediately extracted with three 200-ml. portions of diethyl ether (Note 4). The combined ether extracts are washed twice with 100-ml. portions of water to remove acid and remaining tetrahydrofuran and dried over anhydrous sodium sulfate. After removal of ether, the residual, golden-brown oil is purified by distillation at reduced pressure. The portion boiling at 53–56° (5 mm.) is collected, yielding 49.5–53.5 g. (65–70%) of 2-thiophenethiol as a yellow oil, n_D^{25} 1.6110.

2. Notes

1. The checkers obtained *n*-butyllithium from the Foote Mineral Co., Exton, Pennsylvania. The concentration of *n*-butyllithium was determined by the method of Gilman and Haubein.[2] This reagent is conveniently transferred to a precalibrated addition funnel under nitrogen pressure, through a short length of inert (*e.g.*, Teflon) tubing.

2. The submitters employed *n*-butyllithium prepared by the method of Jones and Gilman.[3] The *n*-butyllithium solution so prepared was filtered to remove finely divided lithium, using an apparatus previously described.[4]

3. The general procedure is similar to that described by Gronowitz[5] in the preparation of 3-thiophenethiol, the principal differences being the use of tetrahydrofuran–pentane solvent and the omission of a 10% potassium hydroxide extraction before acidification with sulfuric acid. This omission leads to higher yields of thiol.

4. Undue delay in the ether extraction of the thiol has been found to result in reduced yields.

3. Discussion

Houff and Schuetz[6] have prepared 2-thiophenethiol by two different routes. One involves the sulfurization of 2-thienylmagnesium bromide followed by acidification, giving the thiol; the other method is an *in situ* reduction of 2-thienylsulfonyl chloride with zinc dust and sulfuric acid. Gronowitz[5] has prepared the isomeric 3-thiophenethiol by sulfurization of 3-thienyllithium, which was obtained by metalation of 3-bromothiophene with *n*-butyllithium.

This method is based on the known reactivity of the 2-position of thiophene; the desired 2-thiophenethiol may be prepared in good yield by direct substitution of thiophene. 2-Chloro-5-thiophenethiol may also be prepared by this method in 59% yield from 2-chlorothiophene.[7]

Direct substitution in the 3-position of the thiophene ring is difficult and can be achieved only by activation of this reaction site. Thus, the isomeric 3-thiophenethiol may be prepared by this general method starting with a 3-halogenothiophene. For example, 3-thiophenethiol may be obtained from 3-bromothiophene in 63% yield.[5]

The procedure outlined above also offers a general method for the synthesis of alkyl and aryl thiols starting from the appropriate halides. Thus, thiophenol may be obtained in 62% yield by lithiation and sulfurization of bromobenzene.[8]

1. Work done at the former Arthur D. Little Research Inst., Inveresk Gate, Musselburgh, Midlothian, Scotland.
2. H. Gilman and A. H. Haubein, *J. Am. Chem. Soc.*, **66**, 1515 (1944).
3. R. G. Jones and H. Gilman, *Org. React.*, **6**, 339–366 (1951).
4. H. Gilman, W. Langham, and F. W. Moore, *J. Am. Chem. Soc.*, **62**, 2327 (1940).
5. S. Gronowitz and R. Hakansson, *Ark. Kemi*, **16**, 309 (1960).
6. W. H. Houff and R. D. Schuetz, *J. Am. Chem. Soc.*, **75**, 6316 (1953).
7. E. Jones and I. M. Moodie, *Tetrahedron*, **21**, 2413 (1965).
8. H. Gilman and L. Fullhart, *J. Am. Chem. Soc.*, **71**, 1478 (1949).

p-TOLYLSULFONYLDIAZOMETHANE

[Benzene, 1-[(diazomethyl)sulfonyl]-4-methyl-]

$$CH_3-\langle\ \rangle-SO_2Na + CH_2O + H_2NCO_2C_2H_5 \xrightarrow[\text{water, 70-75°}]{\text{HCO}_2\text{H}}$$

$$CH_3-\langle\ \rangle-SO_2CH_2NHCO_2C_2H_5 + H_2O + HCO_2Na$$

$$CH_3-\langle\ \rangle-SO_2CH_2NHCO_2C_2H_5 + ClNO \xrightarrow[0°]{\text{pyridine}}$$

$$CH_3-\langle\ \rangle-SO_2CH_2\underset{N=O}{NCO_2C_2H_5} + C_5H_5N\cdot HCl$$

$$CH_3-\langle\ \rangle-SO_2CH_2\underset{N=O}{NCO_2C_2H_5} \xrightarrow[\substack{\text{diethyl ether,}\\10-15°}]{\text{alumina}}$$

$$CH_3-\langle\ \rangle-SO_2CHN_2 + CO_2 + C_2H_5OH$$

Submitted by A. M. van Leusen[1] and J. Strating
Checked by Bruce A. Carlson and William A. Sheppard

1. Procedure

Caution! Part B must be conducted in an efficient hood to avoid exposure to toxic nitrosyl chloride.

A. *Ethyl N-(p-tolylsulfonylmethyl)carbamate.* A solution of 178 g. (1.00 mole) of sodium *p*-toluenesulfinate (Note 1) in 1 l. of water is placed in a 3-l., three-necked flask equipped with a condenser, an efficient mechanical stirrer, and a thermometer. After addition of 100 ml. (108 g.) of an aqueous 34–37% solution of formaldehyde (*ca.* 1.2–1.4 moles) (Note 2), 107 g. (1.20 moles) of ethyl carbamate (Note 3), and 250 ml. of formic acid (Note 4), the stirred solution is heated to 70°. Soon after this temperature is reached, the reaction mixture becomes turbid due to separation of the product as oily droplets, which are kept dispersed with vigorous stirring. After heating for 2 hours at 70–75°, the heating mantle is replaced with an ice bath, while stirring is continued. At about 60° the product begins to solidify. Under continued stirring the mixture is cooled further and kept in the ice bath for 2 hours after a temperature of 5° is reached (Note 5). The precipitate is collected by suction filtration and washed three times by stirring efficiently with 400-ml. portions of cold water. After drying at 70° to constant weight,

214–232 g. (83–90%) (see Note 1) of white, microcrystalline ethyl N-(p-toluenesulfonylmethyl)carbamate, m.p. 108–110°, is obtained; it is sufficiently pure for use in the next step of the reaction. Recrystallization from 95% ethanol provides colorless flakes, m.p. 109–111°.

B. *Ethyl* N-*nitroso*-N-(p-*tolylsulfonylmethyl*)*carbamate*. A solution of 154 g. (0.599 mole) of the ethyl N-(p-tolylsulfonylmethyl)carbamate in 600 ml. of pyridine (Note 6) is placed in a 1-l., four-necked flask equipped with a thermometer, a mechanical stirrer, a gas-inlet tube leading into the solution, and a gas outlet leading to the exhaust. The weight of the flask together with its contents is determined, preferably on a balance placed in the same hood. The solution is cooled to 0° in an ice–salt mixture. Gaseous nitrosyl chloride (Note 7) is introduced, *via* a mineral oil bubbler and a trap, into the stirred solution at such a rate that the temperature is kept between 0° and 5°. After 52–65 g. (0.79–0.99 mole) of nitrosyl chloride has been taken up (Note 8), the reaction is completed by stirring for 30 minutes at about 0°, at which point the reaction mixture is poured in a thin stream into a hand-stirred mixture of 4 l. of ice and water, giving a pale-yellow oil that readily solidifies (Note 9). The solid is collected by suction filtration after standing for 1 hour at 0°. Any lumps present are pulverized, and the solid is washed thoroughly in a beaker with four 500-ml. portions of cold water, removing pyridine. The moist product is dissolved in sufficient dichloromethane (*ca*. 1.5 l.), and the water layer is removed. The dichloromethane solution is dried over anhydrous magnesium sulfate and concentrated to dryness under reduced pressure, giving 157–171 g. (92–100%) of crude ethyl N-nitroso-N-(p-tolylsulfonylmethyl)carbamate, m.p. 86–89° (slight dec.). The crude nitroso compound can be used without purification in the next step of the reaction, provided that it is free of starting material (Note 10).

If the nitrosocarbamate is to be stored (preferably at −20°) for periods longer than a month, it should be recrystallized once from 1 : 2 dichloromethane-diethyl ether (Note 11), m.p. 87–89°. In this purified form it can be stored for several months at −20° without noticeable decomposition.

C. p-*Tolylsulfonyldiazomethane*. *Warning!* α-*Diazosulfones slowly decompose under the influence of light. Exposure to light should therefore be kept at a minimum in all stages of the reaction.* A 3-l., three-necked, round-bottomed flask equipped with a condenser, an efficient mechanical stirrer, and a stopper, is charged with 570 g. of alumina (Note 12). The flask is wrapped with aluminum foil or covered with dark cloth or black paper. With stirring, 1.5 l. of ether (Note 13) is added. The mixture is cooled to 10–15° with a water–ice bath, and a solution of 57.3 g. (0.200 mole) of ethyl N-nitroso-N-(p-tolylsulfonylmethyl)carbamate in 150 ml. of dichloromethane (Note 14) is added in one portion. Stirring is continued for 2 hours at 10–15° (Note 15); during this time the ether solution soon develops the bright-yellow color of p-tolylsulfonyldiazomethane. The ether solution is decanted from the alumina. The alumina is extracted thoroughly by stirring for periods of 5 minutes with two portions of 500 ml. and three portions of 250 ml. of ether. The combined ether solutions are filtered through coarse filter paper and concentrated in a vacuum rotary flash evaporator. The temperature of the water bath must not exceed 25°. When the volume is reduced to about 200 ml., the water bath is removed entirely and concentration is continued until the cold residue crystallizes spontaneously

(Note 16). The crystal mass is stirred for about 2 minutes in 50 ml. of ice-cold petroleum ether (40–60°). The crystals are collected on a sintered-glass filter and washed on the filter with two 25-ml. portions of cold petroleum ether. Drying overnight at 0° in a vacuum desiccator over anhydrous calcium chloride, yields 26–30 g. (66–76%) of yellow *p*-tolylsulfonyldiazomethane, m.p. 35–38° (slight dec.) (Note 17). Recrystallization from anhydrous ether-pentane (Note 18) will raise the melting point to 36–38° (slight dec.) at the expense of 5–10% of material. *p*-Tolylsulfonyldiazomethane should be stored at or below 0° in the absence of light in an *unsealed* container (Note 19).

2. Notes

1. Commercially available anhydrous sodium *p*-toluenesulfinate, purum, *ca.* 97% (Fluka A G, Busch S. G., Switzerland) was used. Sodium *p*-toluenesulfinate dihydrate can be used equally well. The checkers used anhydrous sodium *p*-toluenesulfinate from Aldrich Chemical Company, Inc., which was determined by titration to be 87% pure and gave lower yields. The yield stated was obtained by using stoichiometric amounts based on calculated purity. Sodium *p*-toluenesulfinate from other suppliers was found less pure and gave considerably lower yields.

Alternatively, sulfinates can be synthesized conveniently by the method of Truce and Roberts,[2] or by that of Oxley and colleagues.[3]

2. Commercial aqueous formaldehyde solution, containing about 8% of methanol, was used.

3. Ethyl carbamate (J. T. Baker Chemical Company), with a reported melting point of 48–50° (46–49° was found), was used without purification. The solid was added to the reaction mixture.

4. Formic acid (97%) from J. T. Baker Chemical Company was used.

5. By cooling and stirring as described, the product is obtained in a finely divided form, which can be removed from the flask easily and washed efficiently.

6. Commercial pyridine (J. T. Baker Chemical Company) was used without purification.

7. Nitrosyl chloride (Matheson Gas Products) with a purity specified as >97% was used. Occasionally, the needle valve of the nitrosyl chloride tank clogs. After closing the tank, the valve is disconnected and flushed with acetone until the acetone remains colorless. The needle valve is reconnected after being dried with compressed air.

Nitrosyl chloride also can be prepared conveniently from hydrochloric acid and sodium nitrite.[4] Alternatively, the nitrosation can be carried out conveniently with nitrosonium tetrafluoroborate (Aldrich Chemical Company, Inc.). Ethyl *N*-nitroso-*N*-(*p*-tolylsulfonylmethyl)carbamate (99%) was obtained when 0.06 mole of nitrosonium tetrafluoroborate, a hygroscopic solid, was added over 45 minutes, from an Erlenmeyer flask connected to the reaction flask with a piece of Tygon® tubing, to a −10 to 0° solution of ethyl *N*-(*p*-tolylsulfonylmethyl)carbamate (0.05 mole) in 50 ml. of pyridine.

8. The color of the pyridine solution changes rapidly from blue and green to yellow. After roughly 1 equivalent (0.6 mole) of nitrosyl chloride has been taken up (*ca.* 1 hour), the color of the solution changes to dark red-brown. During the reaction a precipitate of pyridinium chloride is formed; however, it will disappear during the workup with water as

described. A larger excess of nitrosyl chloride (up to 3 equivalents) has been used occasionally without any disadvantages.

9. Preferably, a few drops of the pyridine solution are rubbed first with a little water, providing seed crystals so that the product will solidify immediately, giving a finely divided material which can be washed more easily.

10. The presence of ethyl N-(p-tolylsulfonylmethyl)carbamate in the reaction product is most readily detected by the N–H IR absorption band at 3370 cm.$^{-1}$. If the nitrosation is incomplete, the reaction with nitrosyl chloride should be repeated on the mixture of compounds, rather than to try to purify the product by recrystallization.

11. Ether (ca. 1.4 l.) is added to a warm, filtered solution of 100 g. of the crude nitroso compound in ca. 0.7 l. of dichloromethane until the solution becomes slightly turbid. The nitrosocarbamate is collected after cooling overnight at −20° and dried under reduced pressure at 0°, providing 88–93 g. of yellow crystals with a pink luster, m.p. 87–89.

12. Alumina Number 1076, "aktiv basisch," for chromatography (E. Merck, Darmstadt) was usually employed. Occasionally, when alumina Number 1077, "aktiv neutral," from the same company, was used, a longer reaction time was required (compare Note 15).

13. Commercial ether was stored over potassium hydroxide pellets and used without distillation. Because some heat is evolved when the ether is added to the alumina, the flask is equipped with a condenser.

14. Commercial dichloromethane was used without purification.

15. The time necessary for completion of the reaction may vary from 0.5 to 4 hours, depending on the actual activity of the alumina. The progress of conversion should be monitored by IR analysis of a concentrated sample of the solution. Stirring should be continued for 15 minutes after the nitroso band at 1540 cm.$^{-1}$ has disappeared. A strong diazo band at about 2100 cm.$^{-1}$ will then be present. The carbonyl band at 1750 cm.$^{-1}$, initially due to nitrosocarbamate, will usually not disappear completely during the reaction, because some diethyl carbonate is formed in addition to carbon dioxide and ethanol. Diethyl carbonate is removed during the workup procedure.

During the reaction the alumina usually attains a pink color, due to some decomposition of p-tolylsulfonyldiazomethane. However, the colored decomposition products adhere strongly to the alumina and will not, therefore, contaminate the final product. If the alumina becomes reddish rather than pink, the type of the alumina in use may be too basic, causing more extensive decomposition of the p-tolylsulfonyldiazomethane; the reaction time should then be reduced as much as possible to prevent a considerable decrease in yield.

16. If crystallization does not occur, it can be induced readily by scratching.

17. Spectral data of p-tolylsulfonyldiazomethane: IR (Nujol) cm.$^{-1}$: 2125 (C=N=N), 1330 (SO$_2$), 1150 (SO$_2$); ^{1}H NMR (CDCl$_3$), δ (multiplicity, number of protons, assignment): 2.46 (s, 3H, CH$_3$), 5.36 (s, 1H, CH), 7.30, 7.43, 7.73, 7.87 (AB q, 4H, C$_6$H$_4$); visible (40% dioxane-water) nm. max. (log ϵ): 394 (1.8).

18. p-Tolylsulfonyldiazomethane tends to separate as an oil when pentane is in excess of an ether to pentane ratio of 2 : 1. The p-tolylsulfonyldiazomethane is dissolved at room temperature in anhydrous ether (about 20 ml. per 10 g. of p-tolylsulfonyldiazomethane),

and pentane (about 7 ml. per 10 g.) is added, followed by seeding. The solution is cooled first at 0°, then at −20°, before the crystals are collected.

19. *p*-Tolylsulfonyldiazomethane is insensitive to impact detonation; however, it decomposes on warming, evolving significant quantities of nitrogen at temperatures as low as 32°. It should never be stored for any length of time in a sealed container and should be stored at or below 0° if not used immediately.

3. Discussion

p-Tolylsulfonyldiazomethane represents a class of compounds called α-diazosulfones, which was discovered in 1961 by the submitters. Besides being useful synthetic intermediates, α-diazosulfones are the only known source of α-sulfonylcarbenes.[5]

Several new trisubstituted methane derivatives have been prepared by replacing the diazo group of α-diazosulfones.[5] For example, *p*-tolylsulfonyldiazomethane (RSO$_2$CHN$_2$, R = *p*-tolyl throughout) reacts with *p*-toluenesulfenyl chloride,[6] giving chloro-*p*-tolylsulfonyl-*p*-tolylthiomethane [RSO$_2$CH(Cl)SR] in 83% yield;[7] with nitrosyl chloride, yielding the previously unknown *N*-hydroxy-1-(*p*-tolylsulfonyl)methanimidoyl chloride [RSO$_2$C(Cl)=NOH] in 28% yield,[8] with 70% perchloric acid in dichloromethane, giving the isolable covalent perchlorate (RSO$_2$CH$_2$OClO$_3$) in 49% yield;[9,5] and with *tert*-butyl hypochlorite in *tert*-butyl alcohol, giving 1-(1-*tert*-butoxy-1-chloromethylsulfonyl)-4-methylbenzene [RSO$_2$CH(Cl)OC(CH$_3$)$_3$] in 62% yield or 1-(1-chloro-1-ethoxymethylsulfonyl)-4-methylbenzene [RSO$_2$CH(Cl)OC$_2$H$_5$] in 84% yield when carried out in ethanol.[10]

The photolysis of α-diazosulfones dissolved in alkenes provides sulfonyl-substituted cyclopropanes in high yields.[5] This is exemplified by the preparation of 1-(4-methoxyphenylsulfonyl)-2,2,3,3-tetramethylcyclopropane in 75% yield from 4-methoxybenzenesulfonyldiazomethane and 2,3-dimethyl-2-butene. A similar addition to *trans*-2-butene gives (*d,l*)-1-(4-methoxyphenylsulfonyl)-*trans*-2,3-dimethylcyclopropane in 79% yield, resulting from a stereospecific *cis*- addition, indicating a singlet sulfonylcarbene intermediate.[11]

Originally, *p*-tolylsulfonyldiazomethane was prepared by passing an ethereal solution of its precursor, ethyl *N*-nitroso-*N*-(*p*-tolylsulfonylmethyl)carbamate, slowly through a column of alumina.[12] This procedure, which results in yields about 10% higher, is convenient only for small-scale preparations, up to a maximum of 5 g. of *p*-tolylsulfonyldiazomethane. The present modification is that of Middelbos.[13]

The conversion of nitrosocarbamates into α-diazosulfones is also effected with certain bases, notably with aqueous potassium hydroxide.[12] Potassium hydroxide, however, causes rapid decomposition of *p*-tolylsulfonyldiazomethane. Alumina is thought to act as a solid base and does not cause significant decomposition.

Other syntheses of *p*-tolylsulfonyldiazomethane have been worked out. Reaction of 4-carboxybenzenesulfonyl azide and ammonia with *p*-tolylsulfonylacetaldehyde hemihydrate or *p*-tolylsulfonylacetaldehyde enol acetate gives *p*-tolylsulfonyldiazomethane in yields of 73 and 58%, respectively.[14] Furthermore, *p*-tolylsulfonyldiazomethane is obtained in 60% yield by reaction of *p*-tolylsulfonylmethylenetriphenylphosphorane and either *p*-tolylsulfonyl azide or 4-carboxybenzenesulfonyl azide.[14] A method similar to the first of these syntheses has been used for preparation of (alkyl- or arylsulfonyl)phenyl-diazomethanes; however, the present procedure has the advantages of being simple and easily scaled-up, and uses readily available, inexpensive starting materials.

An alternative to the synthesis of arylsulfonylmethylcarbamates by the Mannich condensation as described here,[15] is the Curtius rearrangement of the hydrazides of arylsulfonylacetic acids.[16]

1. Department of Organic Chemistry, Groningen University, Nijenborgh 16, 9747 AG Groningen, The Netherlands.
2. W. E. Truce and F. E. Roberts, *J. Org. Chem.*, **28**, 593 (1963).
3. P. Oxley, M. W. Partridge, T. D. Robson, and W. F. Short, *J. Chem. Soc.*, 763 (1946).
4. J. R. Morton and H. W. Wilcox, *Inorg. Synth.*, **4**, 48 (1953).
5. A. M. van Leusen and J. Strating, *Q. Rep. Sulfur Chem.*, **5**, 67 (1970).
6. F. Kurzer and J. R. Powell, *Org. Synth.*, **Coll. Vol. 4**, 934 (1963).
7. J. Strating and J. Reitsema, *Recl. Trav. Chim. Pays-Bas*, **85**, 421 (1966).
8. J. C. Jagt, I. van Buuren, J. Strating, and A. M. van Leusen, *Synth. Commun.*, **4**, 311 (1974).
9. J. B. F. N. Engberts and B. Zwanenburg, *Tetrahedron Lett.*, 831 (1967).
10. B. Zwanenburg, W. Middelbos, G. J. K. Hemke, and J. Strating, *Recl. Trav. Chim. Pays-Bas*, **90**, 429 (1971); *cf.* W. Middelbos, B. Zwanenburg, and J. Strating, *Recl. Trav. Chim. Pays-Bas*, **90**, 435 (1971).
11. A. M. van Leusen, R. J. Mulder, and J. Strating, *Recl. Trav. Chim. Pays-Bas*, **86**, 225 (1967).
12. A. M. van Leusen and J. Strating, *Recl. Trav. Chim. Pays-Bas*, **84**, 151 (1965).
13. W. Middelbos, Ph.D. thesis, Groningen University, 1971.
14. A. M. van Leusen, B. A. Reith, and D. van Leusen, *Tetrahedron*, **31**, 597 (1975).
15. J. B. F. N. Engberts and J. Strating, *Recl. Trav. Chim. Pays-Bas*, **84**, 942 (1965).
16. A. M. van Leusen and J. Strating, *Recl. Trav. Chim. Pays-Bas*, **84**, 140 (1965).

p-TOLYLSULFONYLMETHYL ISOCYANIDE

[Benzene, 1-((isocyanomethyl)sulfonyl)-4-methyl-]

$$CH_3-\!\!\langle\ \rangle\!\!-SO_2Na + CH_2O + H_2NCHO \xrightarrow[90-95°]{H_2O,\ HCOOH}$$

$$CH_3-\!\!\langle\ \rangle\!\!-SO_2CH_2NHCHO + H_2O$$

$$CH_3-\!\!\langle\ \rangle\!\!-SO_2CH_2NHCHO + POCl_3 \xrightarrow[\substack{dimethoxyethane,\\diethyl\ ether,\\-5°\ to\ 0°}]{(C_2H_5)_3N}$$

$$CH_3-\!\!\langle\ \rangle\!\!-SO_2CH_2N\!\!=\!\!C + HOPOCl_2 + HCl$$

Submitted by B. E. HOOGENBOOM, O. H. OLDENZIEL, and A. M. VAN LEUSEN[1]
Checked by TERESA Y. L. CHAN and S. MASAMUNE

1. Procedure

Caution! The reaction should be conducted in a well-ventilated fume hood.

Benzene has been identified as a carcinogen; OSHA has issued emergency standards on its use. All procedures involving benzene should be carried out in a well-ventilated hood, and glove protection is required.

A. *N-(p-Tolylsulfonylmethyl)formamide.*[2] A 3-l., three-necked, round-bottomed flask equipped with a mechanical stirrer, a condenser, and a thermometer is charged with 267 g. (1.50 moles) of sodium p-toluenesulfinate (Note 1). After addition of 750 ml. of water, 350 ml. (378 g.) of an aqueous 34–37% solution of formaldehyde (*ca.* 4.4 moles) (Note 2), 680 g. (600 ml., 15.5 moles) of formamide (Note 3), and 244 g. (200 ml., 5.30 moles) of formic acid (Note 4), the stirred reaction mixture is heated at 90°. The sodium p-toluenesulfinate dissolves during heating, and the clear solution is kept at 90–95° for 2 hours (Note 5). The reaction mixture is cooled to room temperature in an ice–salt bath with continued stirring, then further cooled overnight in a freezer at −20°. The white solid (Note 6) is collected by suction filtration and washed thoroughly in a beaker by stirring with three 250-ml. portions of ice water. The product is dried under reduced pressure over phosphorus pentoxide at 70° (Note 7), yielding 134–150 g. (42–47%) of crude *N-(p-tolylsulfonylmethyl)formamide*, m.p. 106–110° (Note 8), which is sufficiently pure for use in the next step.

B. *p-Tolylsulfonylmethyl isocyanide.* A 3-l., four-necked, round-bottomed flask equipped with a mechanical stirrer, a thermometer, a 250-ml. dropping funnel, and a drying tube is charged with 107 g. (0.502 mole) of crude *N*-(*p*-tolylsulfonylmethyl)formamide, 250 ml. of 1,2-dimethoxyethane, 100 ml. of anhydrous diethyl ether, and 255 g. (350 ml., 2.52 moles) of triethylamine (Note 9). The stirred suspension is cooled in an ice–salt bath to $-5°$. A solution of 84 g. (50 ml., 0.55 mole) of phosphorus oxychloride (Note 10) in 60 ml. of 1,2-dimethoxyethane is added from the dropping funnel at such a rate that the temperature is kept between $-5°$ and $0°$ (Note 11). During the reaction, the *N*-(*p*-tolylsulfonylmethyl)formamide gradually dissolves and triethylamine salts precipitate. Near the completion of the reaction the white suspension slowly turns brown (Note 12). After stirring for another 30 minutes at $0°$, 1.5 l. of ice water is added with continued stirring. The solid material dissolves, giving a clear, dark-brown solution before the product begins to separate as a fine, brown, crystalline solid. After stirring for an additional 30 minutes at $0°$, the precipitate is collected by suction filtration and washed with 250 ml. of cold water. The wet product is dissolved in 400 ml. of warm benzene $(40-60°)$, the aqueous layer is removed with a separatory funnel, and the dark-brown benzene solution is dried over anhydrous magnesium sulfate. After removal of the magnesium sulfate, 2 g. of activated carbon (Note 13) is added, and the mixture is heated at about $60°$ for 5 minutes and filtered (Note 14). One liter of petroleum ether (b.p. $40-60°$) is added to the filtrate with thorough swirling. After 30 minutes the precipitate is collected by suction filtration and dried in a vacuum desiccator, yielding 74–82 g. (76–84%) of crude *p*-tolylsulfonylmethyl isocyanide as a light-brown, odorless solid, m.p. 111–114° (dec.) (Note 15), which can be used for synthetic purposes without further purification.

Completely white material is obtained by rapid chromatography through alumina (Note 16). An analytically pure product, m.p. 116–117° (dec.), is obtained after one recrystallization from methanol.

2. Notes

1. The submitters used anhydrous sodium *p*-toluenesulfinate ("purum" quality, *ca.* 97% from Fluka A G), and the checkers purchased the reagent from Aldrich Chemical Company, Inc. Compare also Note 1 in the preparation of *p*-tolylsulfonyldiazomethane [*Org. Synth.*, **Coll. Vol. 6,** 981 (1988)].

2. Commercial aqueous formaldehyde solution containing 8% methanol was used. Formaldehyde is needed in excess; otherwise the yield is considerably diminished.

3. Commercial formamide (E. Merck, Darmstadt) was used. The use of a large excess of formamide with respect to the sulfinate is required to obtain the yield specified.[2]

4. Commercial 97% formic acid (J. T. Baker Chemical Company) was used.

5. Prolonged heating lowers the yield considerably.

6. When no solid is formed overnight, crystallization may be induced by scratching. In this case the solution should be kept for another 4 hours at $-20°$ before the product is collected.

7. The drying process can be speeded up by dissolving the wet product in dichloromethane, removing the water layer in a separatory funnel, drying the dichloromethane solution over anhydrous magnesium sulfate, and removing the solvent on a rotary evaporator.

8. Recrystallization from 95% ethanol or benzene raised the m.p. to 108–110°.

9. 1,2-Dimethoxyethane and triethylamine, both in "zur Synthese" quality, were purchased from E. Merck, Darmstadt. Ether was distilled from phosphorus pentoxide and stored over sodium wire.

10. Commercial phosphorus oxychloride ("tout pur" from UCB, Belgium) was used without purification.

11. The addition requires about 1 hour. At the beginning of the reaction much heat is evolved, and, therefore, the phosphorus oxychloride solution should initially be added very slowly.

12. The development of a brown color indicates that sufficient phosphorus oxychloride has been added. If the mixture remains colorless, the final product is likely to be contaminated with unreacted *N*-(*p*-tolylsulfonylmethyl)formamide. It is therefore advantageous to add more phosphorus oxychloride and continue stirring until the brown color is obtained.

13. Activated carbon was purchased from J. T. Baker Chemical Company.

14. The color of the solution is lightened only slightly by treatment with activated carbon, but eventually a purer product is obtained.

15. The product has the following spectral properties; IR (Nujol) cm.$^{-1}$: 2150 (N=C), 1320 and 1155 (SO$_2$); ^{1}H NMR (CDCl$_3$), δ (multiplicity, number of protons, assignment): 2.5 (s, 3H, CH$_3$), 4.6 (s, 2H, CH$_2$), 7.7 (q, 4H, C$_6$H$_4$).

16. A solution of 50 g. of *p*-tolylsulfonylmethyl isocyanide in 150 ml. of dichloromethane is placed on a 40 × 3 cm. column containing about 100 g. of neutral alumina slurried in dichloromethane. A nearly colorless solution (*ca.* 700 ml.) is collected over about 1 hour. This solution is evaporated to dryness on a rotary evaporator, providing 42–47 g. of white *p*-tolylsulfonylmethyl isocyanide, m.p. 113–114° (dec.).

3. Discussion

p-Tolylsulfonylmethyl isocyanide was originally obtained by irradiation of *p*-tolylsulfonyldiazomethane[3] in liquid hydrogen cyanide.[4] This isocyanide represents a group of sulfonylmethyl isocyanides, most of which have been prepared, as in the present procedure, by dehydration of the corresponding formamides.[4,5] *p*-Tolylsulfonylmethyl isocyanide has also been prepared by reaction of *p*-tolylsulfonyl fluoride with isocyanomethyllithium.[4,6] The advantages of the present dehydration method are twofold: (1) it is a simple procedure using readily available and inexpensive starting materials and (2) the use of the foul-smelling methyl isocyanide is avoided.

p-Tolylsulfonylmethyl isocyanide is a useful and versatile reagent.[7] It has been used for the synthesis of several azole ring systems by base-induced addition of its C—N=C moiety to various C=O, C=N, C=S, C=C, and N=N containing substrates. Thus, oxazoles,[7-10] imidazoles,[7,8,11,12] thiazoles,[7,13] pyrroles,[7,8] and 1,2,4-triazoles,[7] have been prepared, respectively. Furthermore, *p*-tolylsulfonylmethyl isocyanide has found use in a one-step conversion of ketones and aldehydes to cyanides containing one more carbon atom.[7,14]

Another application is based on the umpolung principle. In this sense *p*-tolylsulfonylmethyl isocyanide is a formaldehyde anion or dianion equivalent,[7,10] which has been used in the synthesis of symmetrical and unsymmetrical ketones,[7,15,16] α-diketones,[7,17] α-hydroxy aldehydes,[7] and α-hydroxy ketones.[7] Under reducing reaction conditions amino, methylamino or β-hydroxy methylamino compounds have been prepared from the same intermediates.[7]

Finally, p-tolylsulfonylmethyl isocyanide can be transformed into 1-isocyano-1-tosylalkenes,[7] which are useful synthons in their own right, as appears from their use in the synthesis of imidazoles[12] and pyrroles.[7]

1. Department of Organic Chemistry, Groningen University, Nijenborgh 16, 9747 AG Groningen, the Netherlands.
2. T. Olijnsma, J. B. F. N. Engberts, and J. Strating, *Recl. Trav. Chim. Pays-Bas*, **91**, 209 (1972).
3. A. M. van Leusen and J. Strating, *Org. Synth.*, **Coll. Vol. 6**, 981 (1988). Review: A. M. van Leusen and J. Strating, *Q. Rep. Sulfur Chem.*, **5**, 67 (1970).
4. A. M. van Leusen, G. J. M. Boerma, R. B. Helmholdt, H. Siderius, and J. Strating, *Tetrahedron Lett.*, 2367 (1972); A. M. van Leusen, R. J. Bouma, and O. Possel, *Tetrahedron Lett.*, 3487 (1975).
5. H. Böhme and G. Fuchs, *Chem. Ber.*, **103**, 2775 (1970).
6. U. Schöllkopf, R. Schröder, and E. Blume, *Justus Liebigs Ann. Chem.*, **766**, 130 (1972).
7. Brief review: A. M. van Leusen, *Lect. Heterocycl. Chem.*, **5**, S111 (1980) (a supplementary issue of volume 17 of *J. Heterocycl. Chem.*).
8. O. Possel and A. M. van Leusen, *Heterocycles*, **7**, 77 (1977).
9. H. Saikachi, T. Kitagawa, H. Sasaki, and A. M. van Leusen, *Chem. Pharm. Bull.*, **27**, 793 (1979), and **29**, in press.
10. S. P. J. M. van Nispen, C. Mensink, and A. M. van Leusen, *Tetrahedron Lett.*, 3723 (1980).
11. A. M. van Leusen, J. Wildeman and O. H. Oldenziel, *J. Org. Chem.*, **42**, 1153 (1977).
12. A. M. van Leusen, F. J. Schaart, and D. van Leusen, *Recl. Trav. Chim. Pays-Bas*, **98**, 258 (1979).
13. A. M. van Leusen and J. Wildeman, *Synthesis*, 501 (1977).
14. O. H. Oldenziel, J. Wildeman, and A. M. van Leusen, *Org. Synth.*, **Coll. Vol. 6**, 41 (1988); O. H. Oldenziel, D. van Leusen, and A. M. van Leusen, *J. Org. Chem.*, **42**, 3114 (1977); A. M. van Leusen and P. G. Oomkes, *Synth. Commun.*, **10**, 399 (1980).
15. O. Possel and A. M. van Leusen, *Tetrahedron Lett.*, 4229 (1977).
16. D. van Leusen and A. M. van Leusen, *Synthesis*, 325 (1980).
17. D. van Leusen and A. M. van Leusen, *Tetrahedron Lett.*, 4233 (1977).

TRI-*tert*-BUTYLCYCLOPROPENYL TETRAFLUOROBORATE

[Cyclopropenylium, 1,2,3-tris(1,1-dimethylethyl)-, tetrafluoroborate(1-)]

$$(CH_3)_3CCH_2MgCl + (CH_3)_3CCH_2COCl \rightarrow$$
$$(CH_3)_3CCH_2COCH_2C(CH_3)_3 + MgCl_2$$

$$(CH_3)_3CCH_2COCH_2C(CH_3)_3 + 2\,Br_2 \rightarrow$$
$$(CH_3)_3CCHBrCOCHBrC(CH_3)_3 + 2\,HBr$$

$$(CH_3)_3CCHBrCOCHBrC(CH_3)_3 + 2\,(CH_3)_3COK \rightarrow$$

$$+ 2\,(CH_3)_3COH + 2\,KBr$$

Submitted by J. Clabattoni,[1] E. C. Nathan, A. E. Feiring, and P. J. Kocienski
Checked by L. M. Leichter and S. Masamune

1. Procedure

A. *Dineopentyl ketone.* A dry, 2-1., three-necked flask is fitted with a reflux condenser, a precision, pressure-equalizing addition funnel, and a mechanical stirrer. A gas-inlet tube at the top of the condenser is used to maintain a static nitrogen atmosphere in the reaction vessel throughout the reaction. The flask is charged with 900 ml. of anhydrous diethyl ether, 45 g. (1.85 g.-atoms) of magnesium turnings, and 74.6 g. (0.700 mole) of 1-chloro-2,2-dimethylpropane (Note 1). The vigorously stirred mixture is heated to gentle reflux before 156 g. (0.839 mole) of 1,2 dibromoethane in 150 ml. of dry ether is added over a 12-hour period (Note 2). After addition is complete, the reaction mixture is refluxed for an additional 2 hours. The mixture is then cooled to 0–5° in an ice bath, and 71.7 g. (0.533 mole) of *tert*-butylacetyl chloride (Note 3) in 150 ml. of dry ether is added dropwise to the rapidly stirred Grignard reagent over a period of 1.5 hours (Note 4). The mixture is then stirred at 0–5° for an additional 1.5 hours and poured with stirring onto a mixture of 800 g. of cracked ice and 150 ml. of concentrated hydrochloric acid. The ether layer is separated and washed consecutively with 100 ml. each of water, 5% aqueous sodium carbonate solution, and finally saturated aqueous sodium chloride. After drying over anhydrous magnesium sulfate, filtration, and removal of solvent with a rotary evaporator, the yellow residual oil is distilled under reduced pressure, yielding 81.2 g. (90%) of dineopentyl ketone as a colorless oil, b.p. 86–90° (22 mm.) (Notes 5 and 6).

B. *α,α'-Dibromodineopentyl ketone. Caution! The dibromoketone, a highly volatile compound with lachrymatory properties, is a skin irritant which may induce allergic effects. Therefore, steps B and C should be performed in a well-ventilated hood. Rubber gloves should be worn.* A 1-l., three-necked flask fitted with a thermometer, an addition funnel, a magnetic stirring bar, and a gas-exit tube, which is connected with Tygon tubing to a funnel inverted over a beaker of water for trapping hydrogen bromide, is charged with 82 g. (0.48 mole) of dineopentyl ketone in 500 ml. of dichloromethane, and the solution is cooled to $0-5°$ in an ice bath. Over a period of 5 hours, 160 g. (1.00 mole) of bromine is added dropwise (Note 7), and the mixture is stirred an additional hour at $0-5°$. The reaction mixture is carefully transferred to a 1-l. separatory funnel, and the excess bromine is destroyed by extraction with 100 ml. of saturated aqueous sodium sulfite (Note 8). After washing with 100 ml. of 5% aqueous sodium hydrogen carbonate solution and 100 ml. of aqueous saturated sodium chloride, the organic layer is dried over magnesium sulfate, filtered, and evaporated with a rotary evaporator. The yellow, crystalline residue is dissolved in 350 ml. of hot hexane and cooled in ice, giving the dibromoketone as white needles, which are collected by suction filtration, washed with 100 ml. of cold hexane (Note 9), and air dried in a well-ventilated hood, yielding 83 g. of product. Concentration of the mother liquors provides two additional crops of crystals (40 g. and 12 g.) for a total yield of 135 g. (85%) of dibromoketone, m.p. $69-72°$ (Note 10).

C. *Di-*tert-*butylcyclopropenone. Caution! The same precautions described in Part B should be exercised in this step.* A dry, 1-l., three-necked flask fitted with an efficient mechanical stirrer, a low temperature thermometer, and a solid addition assembly (Note 11) is charged with 97 g. (0.30 mole) of the dibromoketone and 700 ml. of anhydrous tetrahydrofuran (Note 12). After the reaction vessel has been flushed with nitrogen, a static nitrogen atmosphere is maintained in the reaction vessel throughout the remainder of the reaction. The vigorously stirred solution is cooled to $-70°$ in an acetone–dry ice bath before 80 g. (0.71 mole) of powdered potassium *tert*-butoxide (Note 13) is added over a 2-hour period. The addition is completed, the mixture is stirred an additional hour at $-70°$, and 50 ml. of 10% hydrochloric acid is added dropwise. The cooling bath is then removed, and the mixture is allowed to warm to room temperature. The precipitated salts are filtered and washed with 100 ml. of tetrahydrofuran, and the filtrate and washing are combined. Most of the tetrahydrofuran is removed under reduced pressure, the nearly colorless residue is dissolved in 450 ml. of hexane (Note 14), and this solution is extracted twice with 100 ml. of water and once with 100 ml. of saturated aqueous sodium chloride. The organic layer is dried over magnesium sulfate, filtered, and concentrated with a rotary evaporator, leaving a pale-yellow oil which crystallizes upon standing. Sublimation of the crude product at 55° (1 mm.) provides 39–41 g. (79–83%) of the cyclopropenone, m.p. $61-63°$ (Notes 15 and 16).

D. *Tri-*tert-*butylcyclopropenyl tetrafluoroborate.* A dry, 500-ml., three-necked flask equipped with a magnetic stirring bar, a pressure-equalizing addition funnel, and a nitrogen inlet system is charged with 56 ml. of a 2.34 *M* commercial solution of *tert*-butyllithium (0.126 mole) in pentane (Note 17) and cooled in an ice bath to 0°. A solution of 20.0 g. (0.120 mole) of di-*tert*-butylcyclopropenone in 200 ml. of pentane is added over 30 minutes, and then the ice bath is removed. The reaction mixture is stirred an additional 30 minutes, and the nearly colorless mixture is poured, with vigorous stirring, into 150 ml. of water. The pentane layer is separated, washed with two 75-ml. portions of water, dried over magnesium sulfate, filtered, and concentrated using a rotary evaporator, leaving a pale-yellow oil which is transferred to a 1-l. Erlenmeyer flask and diluted with 600 ml. of ether. After cooling the ether solution to 0° with an ice bath, a freshly prepared 10% solution of fluoroboric acid in acetic anhydride (Note 18) is added with rapid magnetic stirring. After the resulting suspension is stirred for 20 minutes the white precipitate is collected on a sintered glass funnel (medium porosity) under vacuum and washed thoroughly with 75-ml. portions of ether. The product is dissolved in a minimal amount of boiling acetone (*ca.* 300 ml.), cooled in a freezer (*ca.* −25°), filtered, and washed with ether, yielding 15.3 g. of tri-*tert*-butylcyclopropenyl tetrafluoroborate as white needles. Concentration of the mother liquors gives two additional crops (6.6 and 2.1 g.) of pure product, for a total yield of 24–28 g. (68–79%). When heated on a hot stage, the material darkens with decomposition above 300° (Note 19).

2. Notes

1. Reagent grade anhydrous ether is employed in all operations without prior purification. The magnesium turnings were available from the J. T. Baker Chemical Company, and the 1-chloro-2,2-dimethylpropane (b.p. 84–85°), obtained from Matheson, Coleman and Bell, should be distilled before use.

2. The 1,2-dibromoethane (Eastman Organic Chemicals) was used without prior purification.

3. The *tert*-butylacetyl chloride (b.p. 126–129°) was purchased from Aldrich Chemical Company, Inc., and used without prior purification.

4. Inverse addition of the Grignard reagent to an ethereal solution of the *tert*-butylacetyl chloride does not improve the yield.

5. Fractional distillation is not necessary since the only volatile components are ether and dineopentyl ketone. A dark, high-boiling residue remains in the pot.

6. The ^{1}H NMR spectrum (CDCl$_3$) shows two singlets at δ 1.03 (18H) and 2.29 (4H). The IR spectrum (CCl$_4$) exhibits bands at 2950 (s), 2900 (s), 2865 (s), 1715 (s), 1480 (s), 1470 (s), 1370 (s), and 1355 cm^{-1} (s).

7. Irradiation with a sun lamp or addition of one drop of concentrated hydrochloric acid may be necessary to initiate the bromination.

8. *Caution! The extraction with sodium sulfite should be performed with caution since a considerable amount of heat is evolved.*

9. The dibromoketone is quite soluble in hexane; therefore, filtration should be conducted as rapidly as possible. The hexane should be precooled to at least 0°.

10. Very pure dibromoketone (m.p. 70–71°) may be obtained by sublimation at 25° (1 mm.); however, the product obtained by recrystallization is sufficiently pure for the next step. The ^{1}H NMR spectrum (CDCl$_3$) shows singlets at δ 1.18 (18H) and 4.42 (2H). The IR spectrum (CCl$_4$) shows peaks at 2960 (s), 2935 (m), 2910 (m), 2870 (m), 1740 (s), 1720 (m), 1715 (m), 1485 (s), 1475 (s), 1400 (m), 1375 (s), and 1340 cm.$^{-1}$ (s).

11. A convenient apparatus for the addition of potassium *tert*-butoxide consists of a 250-ml. filter flask connected to the reaction vessel with Gooch tubing. The sidearm of the filter flask serves as a nitrogen inlet.

12. The tetrahydrofuran should be freshly distilled from lithium aluminum hydride. See *Org. Synth.*, **Coll. Vol. 5,** 926 (1973) for a warning concerning the purification of tetrahydrofuran.

13. Alcohol-free potassium *tert*-butoxide, obtained from the MSA Research Corporation, Callery, Pennsylvania, should be weighed and transferred under anhydrous conditions.

14. Hexane is most effective in permitting extraction of residual tetrahydrofuran into water. Failure to remove the tetrahydrofuran can delay the crystallization of the cyclopropenone.

15. The checkers noted the presence of an oil which also distilled onto the cold-finger during the sublimation. The easiest way to remove this oil is to press the sublimed cyclopropenone between two sheets of filter paper. The melting point is recorded after having removed the oil in this manner. This oil poses no hindrance in the next step. Alternatively, the submitters report that the pure cyclopropenone can be obtained more rapidly, but in slightly reduced yield, by recrystallization of the crude product from pentane at low temperature (*ca.* −70°).

16. The ^{1}H NMR spectrum (CDCl$_3$) shows a singlet at δ 1.34. The IR spectrum (CCl$_4$) shows bands at 2980 (s), 2945 (m), 2920 (m), 2880 (m), 1875 (m), 1855 (s), 1820 (s), 1640 (s), 1485 (m), 1465 (m), and 1375 cm.$^{-1}$ (m). The UV spectrum exhibits a maximum at 260 nm (log ε 1.66) in 95% C$_2$H$_5$OH and at 285 nm (log ε 1.79) in cyclohexane. The mass spectrum shows peaks at *m/e* 166, 138, 123, 95, 81, and 67.

17. The *tert*-butyllithium was obtained from Alfa Inorganics, Inc.

18. *Caution! The reaction of fluoroboric acid with acetic anhydride is exothermic and should be conducted with caution.* Under an atmosphere of nitrogen, 102 g. (1.00 mole) of acetic anhydride (J. T. Baker Chemical Company) is cooled to −40° in an acetone–dry ice bath. With magnetic stirring, 20.4 g. (0.116 mole) of 50% fluoroboric acid (J. T. Baker Chemical Company) is added over 10 minutes. After carefully warming to 0°, the freshly prepared solution is used immediately.

19. The ^{1}H NMR spectrum (CDCl$_3$) shows a singlet at δ 1.58. The IR spectrum (KBr) exhibits bands at 2980 (s), 1485 (s), 1465 (m), 1425 (m), 1370 (s), 1225 (s), 1197 (m), 1070 (s), 940 (m), 860 (m), and 525 cm.$^{-1}$ (m).

3. Discussion

The preparation of tri-*tert*-butylcyclopropenyl tetrafluoroborate involves a modification of the method originally employed in the synthesis of triphenylcyclopropenyl perchlorate[2] from diphenylcyclopropenone.[2,3] The overall yields of tri-*tert*-butylcyclopropenyl tetrafluoroborate and di-*tert*-butylcyclopropenone are 40% and 52%, respectively. Other substituted di-*tert*-butylcyclopropenyl cations have also been prepared by the above procedure.[4] It should be pointed out, however, that the synthesis of cyclopropenyl cations using the reaction of cyclopropenones with organometallic reagents is not general.[2,4] Possible competing side reactions include conjugate addition[5] and proton abstraction (if the cyclopropenone contains an acidic hydrogen).

1. Department of Chemistry, Brown University, Providence, Rhode Island 02912.
2. R. Breslow, T. Eicher, A. Krebs, R. A. Peterson, and J. Posner, *J. Am. Chem. Soc.*, **87**, 1320 (1965).
3. R. Breslow and J. Posner, *Org. Synth.*, **Coll. Vol. 5**, 514 (1973).
4. J. Ciabattoni and E. C. Nathan, III, *J. Am. Chem. Soc.*, **91**, 4766 (1969).
5. J. Ciabattoni, P. J. Kocienski, and G. Melloni, *Tetrahedron Lett.*, 1883 (1969).

TRICARBONYL[(2,3,4,5-η)-2,4-CYCLOHEXADIEN-1-ONE]IRON AND TRICARBONYL[(1,2,3,4,5-η)-2-METHOXY-2,4-CYCLOHEXADIEN-1-YL]IRON(1+) HEXAFLUOROPHOSPHATE(1−) FROM ANISOLE

[Iron, tricarbonyl[(2,3,4,5-η)-2,4-cyclohexadien-1-one] and Iron(1+), tricarbonyl[(1,2,3,4,5-η)-2-methoxy-2,4-cyclohexadien-1-yl] hexafluorophosphate (1−)]

Submitted by A. J. Birch[1] and K. B. Chamberlain
Checked by Susumu Kamata, Tsutomu Aoki,
and Wataru Nagata

1. Procedure

Caution! Parts A and B must be conducted in an efficient hood to prevent exposure to ammonia, iron pentacarbonyl, and carbon monoxide.

A. *1-Methoxy-1,4-cyclohexadiene.* A 3-l., three-necked, round-bottomed flask equipped with an inlet tube, mechanical stirrer, and an acetone–dry ice condenser fitted with a drying tube is charged with 150 ml. of tetrahydrofuran, 250 ml. of *tert*-butyl alcohol, and 50 g. (0.46 mole) of anisole (Note 1). About 1.5 l. of dried liquid ammonia (Note 2) is distilled into the reaction vessel from a steam bath. Lithium (11.5 g., 1.66 g.-atoms) (Notes 3, 4) is added cautiously with stirring and, when the addition is complete, the stirring is continued for 1 hour with refluxing. The blue color is discharged by cautiously adding methanol dropwise (about 100 ml. is required); 750 ml. of water is then added carefully. The excess ammonia is allowed to evaporate overnight, more water is added, dissolving the lithium salts, and the mixture is extracted three times with 100-ml. portions of petroleum ether (b.p. 30–40°) (Note 5). The combined extracts are washed four times with 75-ml. portions of water, removing *tert*-butyl alcohol and methanol, and dried over anhydrous magnesium sulfate, and the solvent is removed through a 30-cm. Vigreux column (Note 6) under reduced pressure (20 mm.). Distillation of the residue yields 1-methoxy-1,4-cyclohexadiene (38–40 g., about 75%) (Notes 7, 8), b.p. 40° (20 mm.).

B. *Tricarbonyl[(1,2,3,4-η)-1- and 2-methoxy-1,3-cyclohexadiene]iron.* A 500-ml., three-necked, round-bottomed flask equipped with a nitrogen-inlet tube, a condenser provided with a gas bubbler, and a stopper is flushed with nitrogen and charged with 39 g. (0.35 mole) of 1-methoxy-1,4-cyclohexadiene, 320 ml. of dibutyl ether (Note 9), and 95 g. (65 ml., 0.49 mole) of filtered iron pentacarbonyl (Notes 10–12). Using a heating mantle, the mixture is refluxed for 18 hours (Note 13) under a slow nitrogen stream. After cooling, the reaction mixture is filtered by suction through Celite, removing iron particles (Note 14), the Celite is washed twice with 15-ml. portions of dibutyl ether, and the washings and filtrate are combined. The crude product (Note 15) is obtained by evaporating excess iron pentacarbonyl, unreacted diene, and the dibutyl ether using a rotary evaporator (in a fume hood), with a hot-water bath and ice cooling of the receiver. The distillate is again refluxed for 18 hours under nitrogen as before and worked up in the same manner. This procedure is then repeated again. Distillation of the combined residues using a nitrogen leak (Note 16) yields 54 g. of the product as a yellow oil, b.p. 66–68° (0.1 mm.) (Note 17). The distillation residue, after elution through a short acidic alumina column with light petroleum ether and solvent evaporation, yields an additional 5 g. of product, giving a total yield of 59–68 g. (67–78%) (Notes 18, 19).

C. *Tricarbonyl[(1,2,3,4,5-η)-1- and 2-methoxy-2,4-cyclohexadien-1-yl]iron*(1+) *tetrafluoroborate*(1−). Triphenylmethyl tetrafluoroborate (34 g., 0.10 mole) (Note 20) is dissolved in a minimum volume of dichloromethane, and 18 g. (0.072 mole) of tricarbonyl (1- and 2-methoxy-1,3-cyclohexadiene)iron dissolved in a like volume of dichloromethane is added. The resulting dark solution is left for 20–30 minutes and added with stirring to three times its volume of diethyl ether (Note 21). The precipitate is collected and washed with ether, yielding 21–22 g. (87–91%) of product as a yellow solid (Note 19).

D. *Tricarbonyl[(2,3,4,5-η)-2,4-cyclohexadien-1-one]iron.* The tetrafluoroborate mixture from Part C (21 g., 0.062 mole) is heated on a steam bath for 1 hour in 450 ml. of water, during which time orange crystals separate. After cooling, the mixture is extracted three times with 100-ml. portions of ether, into which most of the solid dissolves. (The aqueous layer is used in Part E.) The extracts are dried over anhydrous magnesium sulfate, and the ether is evaporated, yielding 7–7.5 g. (47–51%) of the yellow crystalline dienone complex (Note 22).

E. *Tricarbonyl[(1,2,3,4,5-η)-2-methoxy-2,4-cyclohexadien-1-yl]iron(1+) hexafluorophosphate(1−).* To the aqueous layer from Part D is added with swirling 7.1 g. (0.044 mole) of ammonium hexafluorophosphate (Note 23) in 30 ml. water. After 30 minutes, the light-yellow product is filtered, washed with water, and air dried, yielding 9–10 g. (35–44%) (Notes 19, 24).

2. Notes

1. Anisole (500 g.) was purified[2] by washing twice with 50 ml. of 2 N sodium hydroxide, twice with 50 ml. of water, drying over anhydrous magnesium sulfate, and distillation, b.p. 43–46° (20 mm.). The checkers used anisole obtained from Kanto Chemical Co., Ltd., Japan.

2. Liquid ammonia, from a cylinder, is purified by addition of 2–3 g. of sodium cut into small pieces and distillation into the reaction vessel.

3. The submitters used lithium wire (Merck & Company, Inc.) (12 in. = 1 g.) cut into small pieces. The checkers used a block of lithium cut into small pieces.

4. The lithium pieces must be small and added to the ammonia solution cautiously. If too much is added at one time, the reaction becomes violent and froths.

5. Frequently not all of the ammonia evaporates; the first extraction should be by swirling in a separatory funnel without a stopper, and subsequent extractions should be done with frequent pressure release.

6. The solvent must be carefully removed; use of a rotary evaporator results in considerable loss of the product.

7. The IR spectrum of the 1-methoxy-1,4-cyclohexadiene shows the absence of strong aromatic absorption at 1600 cm.$^{-1}$; the UV spectrum shows absence of absorption at 270 nm., indicating absence of the conjugated isomer.

8. ^{1}H NMR spectrum, δ (number of protons): δ 2.5–2.9 (4H), 3.48 (3H), 3.50 (1H), 5.60 (2H).

9. Dibutyl ether must be dry and peroxide-free. This can be achieved by filtering it through a large column of basic alumina, or by leaving overnight over sodium wire and distillation. If these precautions are not observed, low yields result. The checkers purified dibutyl ether by distillation from a sodium hydride dispersion.

10. During the reaction the hood must be operating at all times, as carbon monoxide is evolved.

11. Iron pentacarbonyl is toxic and volatile; consequently, it should only be handled in a good hood, while wearing gloves.

12. The submitters used iron pentacarbonyl "pract." grade obtained from Fluka A G. The checkers used iron pentacarbonyl obtained from Merck, Germany.

13. When the reaction was followed by ^{1}H NMR spectroscopy, it was found that the yield reached a maximum after 18 hours; longer refluxing resulted in decomposition and lower yields. At that point as well, the maximum proportion of the 1-methoxy isomer was produced; this isomer is converted into the dienone complex.

14. Care should be exercised in filtering the reaction mixture. The solid collected is largely finely divided iron and pyrophoric; it should not be allowed to dry.

15. The crude product is unstable and should be stored under nitrogen with refrigeration.

16. Alternatively, rapid magnetic stirring will prevent bumping and allow distillation without a gas bleed.

17. For smaller batches, purification by elution of the product through a short column of acidic alumina with light petroleum ether and evaporation of the solvent is satisfactory.

18. In one of the checker's experiments almost all the material could be distilled, giving 68.14 g. (77.9%) of the product, b.p. 66–67° (0.3–0.4 mm.).

19. For ^{1}H NMR and IR spectra, see Birch and co-workers.[3]

20. Triphenylmethyl fluoroborate is prepared by dissolving 27 g. (0.10 mole) of triphenylmethanol ("purum," Fluka A G) in 260 ml. of propionic anhydride by warming on a steam bath. With an acetone–dry ice bath the solution is cooled to 10° and maintained between 10° and 20° while 31 ml. of 43% (w/w) fluoroboric acid is added portionwise with swirling. The yellow solid is collected, washed well with dry ether, and dried in a desiccator under vacuum, yielding 34 g. (90–99%). The product is very hygroscopic, taking up water with hydrolysis. It is desirable to prepare this reagent immediately before use.

21. The ether should be reagent grade but not sodium-dried. The traces of water present destroy excess reagent, leading to a cleaner product.

22. The product at this stage is sufficiently pure for most purposes. It can be recrystallized in small batches from water,[3] m.p. 104–104.5°; it can be chromatographed on silica or acidic alumina or sublimed at 80–90° (0.2 mm.). The checkers obtained the purified material (m.p. 104–105°) by recrystallization from dichloromethane-ether. The purified material has the following IR spectrum (CHCl$_3$) cm.$^{-1}$: 2070 strong, 2000 strong, 1665 strong; ^{1}H NMR spectra (CDCl$_3$), δ (multiplicity, number of protons): 2.28–2.45 (m, 2H), 3.1–3.46 (m, 2H), 5.6–6.1 (m, 2H).

23. Obtained from Ozark-Mahoning Chemical Co.

24. The complex can be stored for long periods under nitrogen in a refrigerator with only slight darkening. The purified material has the following IR spectrum (Nujol) cm^{-1}: 2110 strong, 2060 strong, 1515 weak, 1494 weak, 1253 medium strong.

3. Discussion

Anisole is reduced using the solvent system of Dryden and colleagues.[4] Iron pentacarbonyl and 1-methoxy-1,4-cyclohexadiene react as shown by Birch and co-workers,[3] although dibutyl ether has been found to be a superior solvent.[5] The tricarbonyl(methoxy-1,3-cyclohexadiene)iron isomers undergo hydride abstraction[3,6] with triphenylmethyl tetrafluoroborate, forming the dienyl salt mixture, of which the 1-methoxy isomer is hydrolyzed by water to the cyclohexadienone complex. The 2-methoxy isomer can be recovered by precipitation as the hexafluorophosphate salt. By this method the 3-methyl-substituted dienone complex has also been prepared[3] from 1-methoxy-3-methylbenzene. The use of the conjugated 1-methoxy-1,3-cyclohexadiene in Part B led to no increase in yield or rate and resulted chiefly in another product of higher molecular weight. An alternative route to the dienone is the reaction of tricarbonyl(1,4-dimethoxycyclohexadiene)iron with sulfuric acid.[7]

The dienone complex is an effective phenylating agent for aromatic amines; *e.g.,* aniline and tricarbonylcyclohexadienoneiron react in glacial acetic acid at 75° overnight, giving diphenylamine in 95% yield.[8] Under appropriate experimental conditions cyclohexadienone complexes react with lithium alkyls and, upon removal of iron tricarbonyl, the alkylbenzene is formed.[9] The 2-methoxy cation has now been resolved, *via* the menthoxy derivative.[10] This cation is equivalent to sterically directed 2-cyclohexen-1-one 4-cation.[9]

1. Research School of Chemistry, Australian National University, P.O. Box 4, Canberra, ACT 2600, Australia.
2. D. D. Perrin, W. L. F. Armarego, and D. R. Perrin, "Purification of Organic Compounds," Pergamon Press, Oxford, 1966.
3. A. J. Birch, P. E. Cross, J. Lewis, D. A. White, and S. B. Wild, *J. Chem. Soc. A,* 332 (1968).
4. H. L. Dryden, Jr., G. M. Webber, R. R. Burtner, and J. A. Cella, *J. Org. Chem.,* **26,** 3237 (1961).
5. M. Cais and N. Maoz, *J. Organomet. Chem.,* **5,** 370 (1966).
6. E. O. Fischer and R. D. Fischer, *Angew. Chem.,* **72,** 919 (1960).
7. A. J. Birch, L. F. Kelly, and D. J. Thompson, *J. Chem. Soc., Perkin Trans. 1,* (1980).
8. A. J. Birch and I. D. Jenkins, *Tetrahedron Lett.,* 119 (1975).
9. A. J. Birch, B. M. R. Bandara, K. Chamberlain, B. Chauncy, P. Dahler, A. I. Day, I. D. Jenkins, L. F. Kelly, T.-C. Khor, G. Kretschmer, A. J. Liepa, A. S. Narula, W. D. Raverty, E. Rizzardo, C. Sell, G. R. Stephenson, D. J. Thompson, and D. H. Williamson, *Tetrahedron,* **37,** Suppl. 9, 289 (1981).
10. A. J. Birch and T.-C. Khor, unpublished work.

ALKYLATION OF DIMEDONE
WITH A TRICARBONYL(DIENE)IRON COMPLEX:
TRICARBONYL[2-[(2,3,4,5-η)-4-METHOXY-2,4-CYCLOHEXADIEN-1-YL]-5,5-DIMETHYL-1,3-CYCLOHEXANEDIONE]IRON

[Iron, tricarbonyl[2-[(2,3,4,5-η)-4-methoxy-2,4-cyclohexadien-1-yl]-5,5-dimethyl-1,3-cyclohexanedione]-]

Submitted by A. J. Birch and K. B. Chamberlain[1]
Checked by T. Aoki, S. Kamata and W. Nagata

1. Procedure

A 500-ml., round-bottomed flask equipped with a condenser is charged with 5 g. (0.013 mole) of tricarbonyl[(1,2,3,4,5-η)-2-methoxy-2,4-cyclohexadien-1-yl]iron(1+) hexafluorophosphate(1−) (Note 1), 150 ml. of water, and 50 ml. of ethanol and heated on the steam bath with occasional swirling until the salt is dissolved. Dimedone (1,3-Cyclohexanedione, 5,5-dimethyl-) (2.5 g., 0.018 mole) is dissolved in 50 ml. of ethanol by warming. The two solutions are mixed and refluxed for 15 minutes. After cooling to about 25°, the mixture is poured into 500 ml. of water with stirring, and the precipitate is collected, washed with water, and air-dried, yielding 4.4 g. of crude product. The product is recrystallized by first dissolving in a minimum volume of boiling ethanol and adding water until the first sign of turbidity. On standing under refrigeration overnight, crystallization occurs. After collection by filtration, the product is washed with water and air-dried, yielding 3.7 g. of small, white to buff-colored crystals (Note 2), which darken above 140° but do not melt. On addition of water to the filtrate a crop of 0.6 g. is obtained, giving a total yield of 4.3 g. (87%).

2. Notes

1. The preparation of this salt is described in *Org. Synth.*, **Coll. Vol. 6, 996** (1988).
2. ^{1}H NMR (100 MHz, acetone-d_6), δ (multiplicity, number of protons): δ 1.02 (s, 6H), 1.88 (q, 1H), 2.23 (s, 2.52 (quintet), 2.48–3.52 (m)—the preceding four signals account for 10 protons—3.66 (s, 3H), 5.24 (q, 1H); IR (Nujol) cm.$^{-1}$: 2040, 1980, 1950, 1560; mass spectrum m/e: 388 (M+).

3. Discussion

Hexafluorophosphate and tetrafluoroborate dienyl salts react with many nucleophiles.[2] The tetrafluoroborate salts are preferred, being more soluble in organic solvents than the hexafluorophosphates.

In all cases so far investigated, including a useful direct reaction with ketones,[3] the methoxydienyl salt is substituted at the 5-position and not at the alternative 1-position. The iron tricarbonyl group can conveniently be removed from many of these adducts by the action of iron(III) chloride in ethanol.[3]

1. Research School of Chemistry, Australian National University, P.O. Box 4, Canberra, A.C.T. 2600, Australia.
2. A. J. Birch, P. E. Cross, J. Lewis, D. A. White, and S. B. Wild, *J. Chem. Soc. A*, 332 (1968).
3. A. J. Birch, K. B. Chamberlain, M. A. Haas, and D. J. Thompson, *J. Chem. Soc. Perkin Trans. 1*, 1882 (1973); A. J. Birch, K. B. Chamberlain, and D. J. Thompson, *J. Chem. Soc. Perkin Trans. 1*, 1900 (1973).

CYCLOBUTADIENE IN SYNTHESIS: endo-TRICYCLO[4.4.0.0²,⁵]DECA-3,8-DIENE-7,10-DIONE

Submitted by L. Brener, J. S. McKennis, and R. Pettit[1,2]
Checked by R. E. Ireland and G. Brown

1. Procedure

Caution! Because of the evolution of carbon monoxide, this procedure should be carried out in a well-ventilated hood.

A 500-ml., three-necked, round-bottomed flask fitted with a sealed mechanical stirrer and an outlet leading to a gas bubbler, is charged with a solution containing 4.0 g. (0.021 mole) of cyclobutadieneiron tricarbonyl[2] and 2.0 g. (0.018 mole) of freshly sublimed p-benzoquinone (Note 1) in 72 ml. of acetone and 8 ml. of water. To the vigorously stirred, ice-cold solution, approximately 40–42 g. of ceric ammonium nitrate (Note 2) is added portionwise over a period of 10–12 minutes (Note 3), until the carbon monoxide evolution has ceased. The reaction mixture is then poured into 600 ml. of cold brine, and the resulting mixture is extracted with five 150-ml. portions of diethyl ether. The combined extracts are washed with four 250-ml. portions of water and dried over anhydrous magnesium sulfate.

Removal of the solvent under reduced pressure affords 1.9–2.1 g. of the crude yellow adduct (Note 4). This crude material is dissolved in 8 ml. of hot dibutyl ether (70–80°) and rapidly percolated through approximately 2.0 g. of Florisil (Note 5). Cooling and filtering the eluent affords 1.2–1.3 g. (40–44%) of yellow crystals, m.p. 77–80°. An additional recrystallization from dibutyl ether or ethyl acetate–hexane yields pale yellow crystals, m.p. 78.5–80° (Note 6).

2. Notes

1. In most syntheses using cyclobutadiene, it is advantageous to use an excess of the trapping agent, but here excess p-benzoquinone hampers isolation of the pure adduct.

2. Other oxidizing agents may be used to degrade cyclobutadieneiron tricarbonyl; in those cases in which the reactants or products are sensitive to the acidic ceric ammonium nitrate solutions, lead tetraäcetate in pyridine can be used.

3. Slower addition results in a diminished yield.

4. The ^{1}H NMR spectrum (C_6H_6-d_6) indicated the presence of less than 5% p-benzoquinone. This material darkens upon standing, even in a refrigerator; recrystallization should be performed as soon as possible.

5. It is imperative to use a hot narrow column to prevent crystallization and to avoid passage of dark material through the column. The column was made in a Liebig condenser (6 mm. dia.). The Florisil filled 14 cm. of the condenser. A temperature of 70–80° is sufficient to melt even pure product but avoid passage of dark material.

6. The pure adduct had the following ^{1}H NMR spectrum ($CDCl_3$), δ (multiplicity, number of protons, assignment): 3.5 (broad m, 4H, cyclobutane protons), 6.20 (m, 2H, cyclobutene vinyl protons), and 6.75 (s, 2H, cyclohexene vinyl protons).

3. Discussion

This procedure is illustrative of the synthetic use of cyclobutadieneiron tricarbonyl[3] as a source of highly reactive cyclobutadiene. Cyclobutadiene has been employed, for example, in the synthesis of cubane, Dewar benzenes, and a variety of other systems.[3,4]

The synthesis of endo-tricyclo[4.4.0.0^{2,5}]deca-3,8-dien-7,10-dione and verification of its endo-configuration has been reported earlier.[3] This adduct is a useful starting material for the syntheses of tetracyclo[5.3.0.0^{2,6}.0^{3,10}]deca-4,8-diene,[5] tricyclo[4.4.0.0^{2,5}]deca-3,7,9-triene,[6] cis, syn, cis-tricyclo[5.3.0.0^{2,6}]deca-4,8-dien-3,10-dione,[7] and 4-oxahexacyclo[5.4.0.0^{2,6}.0^{3,10}.0^{5,9}.0^{8,11}]undecane.[8]

1. Deceased December 10, 1981; work done at Department of Chemistry, University of Texas, Austin, Texas 78712.

2. J. Henery and R. Pettit, Org. Synth., Coll. Vol. 6, 422 (1988); R. Pettit and J. Henery, Org. Synth., Coll. Vol. 6, 310 (1988); R. H. Grubbs, J. Am. Chem. Soc., 92, 6693 (1970).

3. J. C. Barborak, L. Watts, and R. Pettit, J. Am. Chem. Soc., 88, 1328 (1966).

4. R. Pettit, Pure Appl. Chem., 17, 253 (1968); J. C. Barborak and R. Pettit, J. Am. Chem. Soc., 89, 3080 (1967); G. D. Burt and R. Pettit, Chem. Commun., 517 (1965).

5. J. S. McKennis, L. Brener, J. S. Ward, and R. Pettit, J. Am. Chem. Soc., 93, 4957 (1971).

6. E. Vedejs, J. Chem. Soc. D, 536 (1971).

7. P. E. Eaton and S. A. Cerefice, J. Chem. Soc. D. 1494 (1970).

8. J. S. Ward and R. Pettit, unpublished results.

TRIFLUOROACETYLATION OF AMINES AND AMINO ACIDS UNDER NEUTRAL, MILD CONDITIONS: N-TRIFLUOROACETANILIDE AND N-TRIFLUOROACETYL-L-TYROSINE

[Acetamide, 2,2,2-trifluoro-N-phenyl- and L-Tyrosine, N-(trifluoroacetyl)-]

$$F_3CCCCl_3 + C_6H_5NH_2 \xrightarrow[25-35°]{(CH_3)_2SO} C_6H_5NHCCF_3 + CHCl_3$$

$$F_3CCCCl_3 + p\text{-HO}-C_6H_4CH_2\underset{\underset{NH_2}{|}}{C}HCO_2H \xrightarrow[25-35°]{(CH_3)_2SO}$$

$$p\text{-HO}-C_6H_4CH_2\underset{\underset{\underset{O}{\|}}{\underset{NHCCF_3}{|}}}{C}HCO_2H + CHCl_3$$

Submitted by C. A. PANETTA[1]
Checked by R. M. FREIDINGER and G. BÜCHI

1. Procedure

Caution! Benzene has been identified as a carcinogen; OSHA has issued emergency standards on its use. All procedures involving benzene should be carried out in a well-ventilated hood, and glove protection is required.

A. *N-Trifluoroacetanilide.* A two-necked, round-bottomed flask fitted with a thermometer, a Drierite tube, and a magnetic stirring bar is charged with 4.66 g. (4.56 ml., 0.0501 mole) of aniline (Note 1) and 15 ml. of dimethyl sulfoxide (Note 2). The resulting solution is stirred and cooled in an ice water bath, and when the internal temperature has dropped to 10–15°, 21.5 g. (0.0998 mole) of 1,1,1-trichloro-3,3,3-trifluoroacetone (Note 3) is added in portions through the condenser. A mild exotherm results, and the addition is extended over *ca.* 5 minutes, maintaining a reaction temperature below 40°. When the addition is complete, the ice bath is removed, and the amber solution is stirred at room temperature for 22 hours. The reaction mixture is poured into 750 ml. of water. A crystalline solid separates, and the resulting slurry is stirred for 1 hour before filtration. After being washed with water, the crystals are dried in a vacuum oven at 50° for 40 minutes, giving 6.49–6.53 g. (69%) of N-trifluoroacetanilide, m.p. 86.0–86.5° (cor.) (Notes 4–6).

B. *N-Trifluoroacetyl-L-tyrosine.* A two-necked, round-bottomed flask fitted with a thermometer, a condenser protected with a Drierite tube, and a magnetic stirrer is charged with 18.12 g. (0.1001 mole) of L-(-)-tyrosine (Note 7) and 130 ml. of dimethyl sulfoxide (Note 2). The suspension is stirred and cooled in an ice water bath. When the internal temperature reaches 10–15°, 64.62 g. (0.2999 mole) of 1,1,1-trichloro-3,3,3-trifluoroacetone (Note 3) is added through the condenser at a rate such that the temperature of the reaction temperature does not exceed 35°. The cooling bath is removed, and the reaction mixture is stirred at room temperature for 22 hours, during which time the suspension becomes a solution. This solution is poured into 660 ml. of ice water, and the resulting mixture is extracted with two portions (660 ml. and 400 ml.) of 1-butanol. The organic extracts are concentrated, first on the rotary evaporator and then at 40° (0.1 mm.), giving a red-orange semisolid, which is dissolved in a minimum amount of acetone and placed on a column of silica gel (Note 8). Elution with benzene–acetone mixtures (Note 9) provides 20.0–22.2 g. (72–80%) of N-trifluoroacetyl-L-tyrosine as a colorless to light yellow solid. Recrystallization from either benzene–acetone or water gives white needles, m.p. 192.5–193.5° (cor.) (Notes 6 and 10).

2. Notes

1. Commercial aniline from Fisher Scientific Company (purified grade) was used as supplied.
2. Dimethyl sulfoxide from the J. T. Baker Chemical Company (reagent grade) was used as supplied.
3. 1,1,1-Trichloro-3,3,3-trifluoroacetone is available from PCR, Inc., P.O. Box 14318, Gainesville, Florida. It may also be prepared easily by the following procedure. Fresh, anhydrous aluminum chloride (18.5 g., 0.139 mole) and 35.0 g. (0.192 mole) of chloropentafluoroacetone (b.p. 7.8°; available from PCR, Inc., or Allied Chemical Corp.) are combined in a flask fitted with a dry ice condenser and a magnetic stirring bar. The refluxing mixture is stirred for 4–6 hours and allowed to warm gradually to room temperature. The contents of the flask are extracted three times with anhydrous ether, and the combined extracts are distilled at atmospheric pressure. After the ether has been removed, continued distillation gives 22.8–28.5 g. (55–69%) of 1,1,1-trichloro-3,3,3-trifluoroacetone, b.p. 83.5–84.5°; IR (film) 1790 cm.$^{-1}$. This compound is stored at room temperature in a tightly stoppered bottle. In the absence of reliable toxicity data, it should be handled with normal precautions.
4. IR (CH_2Cl_2) cm.$^{-1}$: 3401 and 3049 (NH, CH), 1740 (C=O), 1235 (C—F), (lit.,[2] m.p. 87.6°).
5. The checkers suspected that some product was lost during the drying process. Therefore, they purified the crude product by sublimation at 55° (0.15 mm.), which gave 3.76–3.84 g. (80–81%) of N-trifluoroacetanilide in half-scale runs (Note 6).
6. The checkers ran Parts A and B at half the submitter's scale, and the yields were comparable or higher in all cases.
7. Reagent-grade L-(-)-tyrosine was obtained from Fisher Scientific Company.
8. Silica gel 60 (70–230 mesh) was purchased from E. Merck, Darmstadt, Germany.
9. The checkers, working at one half the submitter's scale, obtained 17.5 g. of crude product and used 175 g. of silica gel in their column. They eluted as follows:

Fraction	Eluent (Benzene : Acetone Ratio, ml.)	Eluate
1	9:1, 425	1.2 g., yellow oil
2	17:3, 100	1.7 g., pale yellow solid, m.p. 85–92°
3	3:1, 850	9.4 g., colorless solid, m.p. 194–196°
4	3:2, 600	0.4 g., colorless solid, m.p. 209–212°

Fraction 2 was recrystallized from benzene–acetone to give 1.2 g. of colorless solid, m.p. 194–196° (uncor.), which was combined with fraction 3 to give 10.6 g. (76%) of product.

10. IR (Nujol) cm.$^{-1}$: 1695 (C=O), 1180 (C—F), (lit.,[3] m.p. 192.5–193.5°).

3. Discussion

The original procedure for the trifluoroacetylation of amino acids used trifluoroacetic anhydride,[4] which, although inexpensive and readily available, has certain disadvantages: it is a highly reactive compound and has caused undesired reactions such as the cleavage of amide or peptide bonds;[5] unsymmetrical anhydrides are formed between the newly formed N-trifluoroacetylamino acids and the by-product trifluoroacetic acid; and excess trifluoroacetic anhydride has caused racemization of asymmetric centers.

Thus, other trifluoroacetylation reagents have been investigated. S-Ethyl trifluorothio-acetate[6] has none of the above disadvantages. It does require, however, weakly basic

TABLE I

TRIFLUOROACETYLAMINO COMPOUNDS PREPARED WITH CF_3COCCl_3[a]

Product	Yield (%)
TFA-Aniline	69
TFA-L-Valine	94
TFA-DL-Phenylalanine	52
TFA-L-Phenylalanine	57
TFA-L-Leucine	100
TFA-L-Tyrosine	80
TFA-L-Proline	100
TFA-DL-Alanine	20
TFA-Glycylglycine	43
TFA-L-Prolyglycine, ethyl ester	23
TFA-L-Asparagine	26
TFA-Dehydroabietylamine	11

[a]Except for the first entry, all the compounds listed were prepared by the procedure of Part B.

conditions (pH 8–9) and an aqueous medium. Phenyl trifluoroacetate[7] effects trifluoroacetylation of amino acids under essentially neutral conditions. Its main disadvantages are high cost and the elevated temperatures (120–150°) required.

1,1,1-Trichloro-3,3,3-trifluoroacetone is a relatively unreactive compound that is volatile and easily handled. It may be obtained either commercially or by a simple laboratory preparation (Note 3), and it trifluoroacetylates amino groups in amino acids and other compounds under neutral and extremely mild conditions.[8] Table I lists some compounds that have been prepared with this reagent.

1. Department of Chemistry, University of Mississippi, University, Mississippi 38677.
2. *Beilstein,* **12,** 2nd Suppl., 141 (1950).
3. H. J. Shine and C. Niemann, *J. Am. Chem. Soc.,* **74,** 97 (1952).
4. F. Weygand and E. Leising, *Chem. Ber.,* **87,** 248 (1954).
5. F. Weygand, R. Geiger, and U. Glocker, *Chem. Ber.,* **89,** 1543 (1956).
6. E. E. Schallenberg and M. Calvin, *J. Am. Chem. Soc.,* **77,** 2779 (1955).
7. F. Weygand and A. Röpsch, *Chem. Ber.,* **92,** 2095 (1959).
8. C. A. Panetta and T. G. Casanova, *J. Org. Chem.,* **35,** 4275 (1970).

ALDEHYDES FROM ACID CHLORIDES BY MODIFIED ROSENMUND REDUCTION: 3,4,5-TRIMETHOXYBENZALDEHYDE

(Benzaldehyde, 3,4,5-trimethoxy-)

Submitted by A. I. RACHLIN, H. GURIEN, and D. P. WAGNER[1]
Checked by JAMES H. SHERMAN and RICHARD E. BENSON

1. Procedure

A pressure vessel (Note 1) is charged in order with 600 ml. of dry toluene (Note 2), 25 g. (0.30 mole) of anhydrous sodium acetate (Note 3), 3 g. of dry, 10% palladium-on-carbon catalyst (Note 4), 23 g. (0.10 mole) of 3,4,5-trimethoxybenzoyl chloride (Note 5), and 1 ml. of Quinoline S (Note 6). The pressure vessel is flushed with nitrogen, sealed, evacuated briefly, and pressured to 50 p.s.i. with hydrogen. The mixture is shaken with 50 p.s.i. of hydrogen for 1 hour at room temperature (Note 7), then heated at 35–40° for 2 hours. Agitation is continued overnight while the reaction mixture cools to room temperature. The pressure on the vessel is released, the vessel is opened, and the mixture is filtered through 10 g. of Celite filter aid, and the insoluble material is washed with 25 ml.

of toluene. The combined filtrates are washed successively with 25 ml. of 5% sodium carbonate solution and 25 ml. of water. The toluene solution is dried over 5 g. of anhydrous sodium sulfate and filtered. The filtrate is concentrated by distillation at reduced pressure using a water aspirator. The residue (Note 8) is distilled through a 10-cm. Vigreux column with warm water circulating through the condenser, to prevent crystallization of the distillate, yielding 12.5–16.2 g. (64–83%) of 3,4,5-trimethoxybenzaldehyde, b.p. 158–161° (7–8 mm.), m.p. 74–75° (Notes 9 and 10).

2. Notes

1. Both glass-lined and stainless-steel autoclaves have been used successfully. The checkers used a 1.2-l., Hastelloy autoclave.

2. Reagent grade toluene was heated at reflux to remove a small forerun, then allowed to cool.

3. Anhydrous sodium acetate was dried in a vacuum oven at 115° for 48 hours. The use of less than 3 moles of sodium acetate per mole of acid chloride results in a lower yield of product.

4. A catalyst available from Engelhard Industries was used after being dried in a vacuum oven at 115° for 48 hours. *Caution! Palladium-on-carbon is pyrophoric, and vacuum drying increases this hazard. Catalysts kept in the oven for longer periods of time were extremely pyrophoric.*

5. The acid chloride or the acid may be purchased from Aldrich Chemical Company, Inc. The acid chloride must be pure (99% minimum by GC analysis) whether purchased or prepared. Purification was effected by recrystallization from Skellysolve B.

6. Quinoline S was prepared according to the procedure in *Org. Synth.*, **Coll. Vol. 3,** 629 (1955).

7. Repressuring with hydrogen is required during this period. The amount of repressuring required is dependent upon the free space of the pressure vessel. The submitters report lower yields if the pressure falls below 30 p.s.i. No further repressuring is made at the end of 1 hour.

8. The crude aldehyde (prior to distillation) is sufficiently pure for most purposes. Isolation of the aldehyde may also be achieved *via* the bisulfite-addition compound.[2]

9. The product shows a strong IR band (KBr) at 1690 cm.$^{-1}$ (C=O). The ^{1}H NMR spectrum (CCl$_4$) has peaks at δ 3.84 (s, 3H), 3.87 (s, 6H), 7.03 (s, 2H), and 9.76 (s, 1H).

10. The submitters state that the aldehyde is obtained in 78–84% yield when the reaction is conducted on a scale 5 times that described. The amount of catalyst and Quinoline S need not be increased proportionately. The pressure vessel is charged with 3 l. of dry toluene, 123 g. of anhydrous sodium acetate, 10 g. of dry, 10% palladium-on-carbon catalyst, 115 g. of 3,4,5-trimethoxybenzoyl chloride, and 4 ml. of Quinoline S.

3. Discussion

3,4,5-Trimethoxybenzaldehyde has been prepared by the classical Rosenmund[3–5] reduction, by methylation of 5-hydroxyvanillin,[6] and by oxidation of 3,4,5-trimethoxybenzyl alcohol.[7]

The normal Rosenmund reduction has often been used for small-scale reactions, but for large preparations it has the following disadvantages: long reaction cycles at elevated temperatures, inefficient use of hydrogen, the hazard of passing hydrogen through and away from a hot reaction, the use of relatively high catalyst to substrate ratios, and the necessity of monitoring evolved hydrogen chloride as a means of following the reaction. These shortcomings have been eliminated by carrying out the reaction in a closed system at low pressure in the presence of a hydrogen chloride acceptor.

The reaction has been carried out on large- and small-scale batches (Note 10). This modification[8] has been applied by the submitters to the preparation of other aldehydes,[9] such as 3,4-dimethylbenzaldehyde[10] (90% yield), 3-benzyloxy-4,5-dimethoxybenzaldehyde[11] (88% yield, with retention of the benzyl group), and 3-methoxy-4-nitrobenzaldehyde[12] (62% yield, with retention of the nitro group).

1. Chemical Research Department, Hoffmann La-Roche Inc., Nutley, New Jersey 07110.
2. A. I. Vogel, "A Textbook of Practical Organic Chemistry," 3rd ed., Wiley, New York, 1956, p. 322.
3. K. W. Rosenmund, *Ber. Dtsch. Chem. Ges.*, **51**, 591 (1918).
4. K. H. Slotta and H. Heller, *Ber. Dtsch. Chem. Ges.*, **63**, 3029 (1930).
5. F. Benington and R. D. Morin, *J. Am. Chem. Soc.*, **73**, 1353 (1951).
6. I. A. Pearl and D. L. Beyer, *J. Am. Chem. Soc.*, **74**, 4262 (1952).
7. A. Heffter and R. Capellmann, *Ber. Dtsch. Chem. Ges.*, **38**, 3636 (1905).
8. H. Gurien, D. P. Wagner, and A. I. Rachlin, U.S. Pat. 3,517,066 (1970).
9. D. P. Wagner, H. Gurien, and A. I. Rachlin, *Ann. N. Y. Acad. Sci.*, 172 (9), 186 (1970).
10. L. F. Hinkel, E. E. Ayling, and W. H. Morgan, *J. Chem. Soc.*, 2707 (1932).
11. E. Spath and H. Röder, *Monatsh. Chem.*, **43**, 93 (1922).
12. M. Ulrich, *Ber. Dtsch. Chem. Ges.*, **18**, 2572 (1885).

OXIDATION WITH THE NITROSODISULFONATE RADICAL. I. PREPARATION AND USE OF DISODIUM NITROSODISULFONATE: TRIMETHYL-*p*-BENZOQUINONE

(Nitrosodisulfonic acid, disodium salt and 2,5-Cyclohexadiene-1,4-dione, 2,3,5-trimethyl-)

$$NaNO_2 + 2 SO_2 + NaHCO_3 \longrightarrow HON(SO_3Na)_2 + CO_2$$

$$HON(SO_3Na)_2 + OH^- \xrightarrow[\substack{\text{stainless}\\\text{steel}\\\text{anode}}]{-e^-} :\overset{\cdot}{O}-\overset{\cdot\cdot}{N}(SO_3Na)_2 + H_2O$$

$$+ 2:\overset{\cdot}{O}-\overset{\cdot\cdot}{N}(SO_3Na)_2 \xrightarrow[C_7H_{16},12°]{H_2O}$$

$$+ HON(SO_3Na)_2 + HN(SO_3Na)_2$$

Submitted by Pius A. Wehrli and Foster Pigott[1]
Checked by Don Koepsell and Herbert O. House

1. Procedure

A. *Disodium nitrosodisulfonate.* A 1-l., resin kettle equipped with a mechanical stirrer, a thermometer, a gas-inlet tube suspended about 0.5 cm. above the bottom of the vessel, and an ice-cooling bath is charged with 15.0 g. (0.217 mole) of sodium nitrite (Note 1), 16.8 g. (0.200 mole) of sodium hydrogen carbonate (Note 1), and 400 g. of ice. Sulfur dioxide (25.6 g., 0.400 mole, Note 2) is passed into the cold, initially heterogeneous mixture with stirring over a period of 40 minutes. Near the end of the sulfur dioxide addition, the light brown color of the reaction mixture fades almost completely. The resulting colorless to pale-yellow solution of disodium hydroxylaminedisulfonate (Note 3), which has an approximate pH of 4, is stirred for 10 minutes before 59.5 g. (0.480 mole) of sodium carbonate monohydrate (Note 1) is added, giving a solution of pH 11. The gas-inlet tube is removed from the reaction vessel and replaced with a rectangular anode constructed from a 3.5 cm. by 4.7 cm. piece of stainless-steel mesh (about 16 mesh/cm.^{2}) with a stainless-steel wire as an electrical lead. The cathode is a cylindrical coil formed from a 1.5-mm. by 40-cm. piece of stainless-steel wire suspended in a 5-cm. by 10-cm. porous porcelain thimble filled with aqueous 10% sodium carbonate. The porcelain thimble containing the cathode is suspended in the reaction vessel so that the liquid levels in the anode and cathode compartments are the same. The cathode-anode resistance of the electrolysis cell should be in the range of 5–10 ohms. While the reaction solution is continuously stirred and maintained at a temperature of 12° with an ice bath, the electrolysis is started by applying a sufficient potential (approximately 10 volts, Note 4) to the anode and cathode leads, giving a cell current of 2.0 amp. As the electrolysis proceeds, the potential applied to the cell is adjusted to maintain a cell current of 2.0 amp. The formation of the nitrosodisulfonate radical is evidenced by the appearance of a deep purple color (Note 5). The electrolysis is continued with stirring and cooling until quantitative measurement of the optical density of the reaction solution (Note 5), indicates the concentration of disodium nitrosodisulfonate to be 0.42–0.47 M (84–94% yield). The typical reaction time is 4 hours; the amount of electricity passed through the cell totals approximately 28,800 coulombs or 8 amp-hours (theoretically, 19,300 coulombs or 5.4 amp-hours). This solution of the nitrosodisulfonate radical is removed from the anode compartment of the electrolysis cell and used directly in the next step (Note 6).

B. *Trimethyl-p-benzoquinone.* The aqueous solution containing approximately 0.17 mole of disodium nitrosodisulfonate is placed in a 1-l., round-bottomed flask fitted with a mechanical stirrer, a thermometer, and an ice bath. A solution of 10.0 g. (0.0734 mole) of 2,3,6-trimethylphenol (Note 7) in 100 ml. of heptane is added to the reaction flask, and the resulting mixture is stirred vigorously for 4 hours with continuous cooling, maintaining the reaction temperature below 12°. The yellow heptane layer is separated, and the brown aqueous phase is extracted with two 100-ml. portions of heptane. The combined heptane solutions are quickly (Note 8) washed with three 50-ml. portions of cold (0–5°), 4 M aqueous sodium hydroxide, followed by two 100-ml. portions of saturated aqueous sodium chloride. The organic solution is dried over anhydrous magnesium sulfate and concentrated at 40° with a rotary evaporator, yielding 10.0–10.9 g. (91–99%) of crude trimethyl-*p*-benzoquinone as a yellow liquid, which crystallizes when cooled below room temperature. Further purification may be accomplished by distillation under reduced pressure, yielding 8.5–8.7 g. (77–79%) of the quinone, b.p. 53° (0.4 mm.), which crystallizes on standing as yellow needles, m.p. 28–29.5° (Note 9).

2. Notes

1. Reagent grades of these inorganic reagents were employed.

2. Sulfur dioxide was purchased from Matheson Gas Products. It is convenient to use sulfur dioxide contained in a lecture bottle so that the small cylinder can be mounted on a balance allowing continuous measurement of the weight of sulfur dioxide added.

3. The following alternative procedure may be used to prepare a solution of disodium hydroxylaminedisulfonate. Sodium nitrite (15 g., 0.22 mole) and 41.6 g. (0.400 mole) of sodium bisulfite are added to 250 g. of ice. With stirring, 22.5 ml. (0.393 mole) of acetic acid is added in one portion, and the mixture is stirred for 90 minutes in an ice bath. At the end of the stirring period the reaction solution is pH 5 and a potassium iodide-starch test is negative. A solution of 50 g. (0.47 mole) of sodium carbonate in water (total volume 250 ml.) is added. This buffered solution of disodium hydroxylaminedisulfonate may be used for electrolytic oxidation.

4. Since accurate control of the anode potential is not required in this oxidation, a variety of direct current sources may be employed, provided they are able to supply continuously a current of about 2 amp. at a potential of about 12 volts. One of the simplest direct current sources is an unfiltered rectifier of the type used to charge automobile batteries.

5. In 1 M aqueous potassium hydroxide solution, the nitrosodisulfonate radical has a maximum in the visible at 544 nm (ϵ 14.5).

6. The submitters report that approximately half of the nitrosodisulfonate radical had decomposed after the solution was stored at 0° for 2 weeks. They report the following procedure for the isolation of Fremy's salt (dipotassium nitrosodisulfonate).

Caution! Fremy's salt may decompose spontaneously in the solid state.[2]

To the cold (12°), purple solution of disodium nitrosodisulfonate was added, dropwise and with stirring, a solution of 37.3 g. (0.501 mole) of potassium chloride in 100 ml. of water. The resulting mixture, from which the orange-yellow dipotassium nitrosodisulfonate crystals precipitated, was allowed to stand overnight in the refrigerator, and the crystals were filtered with suction and washed with 100 ml. of 1 M aqueous potassium-hydroxide. The damp crystals weighed 55 g. A 1-g. aliquot of the wet material was dried at room temperature in a desiccator over Drierite, leaving 0.76 g. of orange crystals (72% yield based on sodium nitrite). On two occasions small samples of dried material decomposed spontaneously. *It is again stressed that if the electrolyzed solution is not used directly, any isolated Fremy's salt should be stored as a slurry in 0.5 M potassium carbonate at 0°.*

7. Crude 2,3,6-trimethylphenol, purchased from Aldrich Chemical Company, Inc., was purified by recrystallization from either hexane or chlorobenzene, m.p. 62–63°.

8. This extraction with aqueous sodium hydroxide removes phenolic by-products and starting material and must be performed quickly because p-quinones are unstable with strong bases.

9. The product has IR absorption (CHCl$_3$) at 1645 (conjugated C=O) and 1619 cm^{-1} (C=C) with UV maxima (95% C$_2$H$_5$OH) nm (ϵ) at 256 (16,700), 341 (3880), and 430 (shoulder, 30). The material has ^{1}H NMR peaks (CDCl$_3$) at 1.9–2.2 (partially resolved m, 9H, 3CH_3) and 6.5–6.7 (m, 1H, CH), with the following abundant peaks in its mass spectrum: m/e (rel. int.), 150 (100, M$^+$), 122 (24), 121 (14), 107 (30), 79 (18), 68 (20), 54 (14), 40 (18), and 39 (16).

3. Discussion

The nitrosodisulfonate salts, particularly the dipotassium salt (Fremy's salt), are useful reagents for the selective oxidation of phenols and aromatic amines to quinones (the Teuber reaction).[3,5] Dipotassium nitrosodisulfonate, commercially available, has been prepared by the oxidation of a hydroxylaminedisulfonate salt with potassium permanganate,[3-5] with lead dioxide,[6] or by electrolysis.[2,7] The present procedure illustrates the electrolytic oxidation, forming an alkaline, aqueous solution of the relatively soluble disodium nitrosodisulfonate and avoids a preliminary filtration, which is required to remove manganese dioxide formed when potassium permanganate is used as the oxidant.[3-5]

Solutions of nitrosodisulfonate salts are most stable in weakly alkaline solutions (pH 10) and decompose rapidly when the solution is acidic or strongly alkaline.[3] Dipotassium nitrosodisulfonate has been reported to decompose spontaneously,[2,3] suggesting that procedures involving the use of substantial quantities of the dry solid salt may be hazardous. In the present procedure, separation and use of the solid salt is avoided since the disodium nitrosodisulfonate is formed and used in aqueous solution. In this procedure, two moles of the preformed nitrosodisulfonate salt are consumed in the oxidation of one mole of the phenol to the benzoquinone derivative.[3] The submitters report that only one molar equivalent of the nitrosodisulfonate salt is required if the electrochemical oxidation is carried out in a heptane solution of the phenol.

Trimethyl-*p*-benzoquinone has been prepared by the oxidation of 2,3,5-trimethyl-1,4-benzenediamine with iron(III) chloride[8] and by the oxidation of 2,3,5-trimethylphenol with dipotassium nitrosodisulfonate.[9]

1. Chemical Research Department, Hoffman-La Roche Inc., Nutley, N.J. 07110.
2. P. A. Wehrli and F. Pigott, *Inorg. Chem.* **9**, 2614 (1970).
3. H. Zimmer, D. C. Lankin, and S. W. Horgan, *Chem. Rev.*, **71**, 229 (1971).
4. G. Brauer, "Handbuch der Präparativen Anorganischen Chemie," Vol. 1, Ferdinand Encke Verlag, Stuttgart, 1960, p. 452.
5. H.-J, Teuber and G. Jellinek, *Chem. Ber.*, **85**, 95 (1952) and subsequent publications.
6. G. Harvey and R. G. W. Hollingshead, *Chem. Ind. (London)*, 244 (1953).
7. W. R. T. Cottrell and J. Farrar, *J. Chem. Soc. A*, 1418 (1970).
8. L. I. Smith, *J. Am. Chem. Soc.*, **56**, 472 (1934).
9. H.-J. Teuber and W. Rau, *Chem. Ber.*, **86**, 1036 (1953).

2,2-(TRIMETHYLENEDITHIO)CYCLOHEXANONE

(1,5-Dithiaspiro[5.5]undecan-7-one)

Submitted by R. B. WOODWARD,[1] I. J. PACHTER,[2] and M. L. SCHEINBAUM[3]
Checked by G. S. BATES and S. MASAMUNE

1. Procedure

Caution! Benzene has been identified as a carcinogen; OSHA has issued emergency standards on its use. All procedures involving benzene should be carried out in a well-ventilated hood, and glove protection is required.

A. *1-Pyrrolidinocyclohexene.*[4]   A solution of 29.4 g. (0.300 mole) of cyclohexanone and 28.4 g. (0.394 mole) of pyrrolidine in 150 ml. of benzene is placed in a 500-ml., one-necked flask attached to a Dean-Stark trap. The solution is refluxed under a nitrogen atmosphere until the separation of water ceases (Note 1). The excess pyrrolidine and benzene are removed from the reaction mixture on a rotary evaporator. The resulting residue is stored under refrigeration and distilled just before use in the next step, yielding 44.6 g. (98%) of 1-pyrrolidinocyclohexene, b.p. 76–77° (0.5 mm.), 105–106° (13 mm.).

B. *2,2-(Trimethylenedithio)cyclohexanone.*   A solution of 3.02 g. (0.0200 mole) of freshly distilled 1-pyrrolidinocyclohexene, 8.32 g. (0.0200 mole) of trimethylene dithiotosylate[4] (Note 2), and 5 ml. of triethylamine (Note 3) in 40 ml. of anhydrous acetonitrile (Note 4), is refluxed for 12 hours in a 100-ml., round-bottom flask under a nitrogen atmosphere. The solvent is removed with a rotary evaporator, and the residue is treated with 100 ml. of 0.1 N hydrochloric acid for 30 minutes at 50° (Note 5). The mixture is cooled to ambient temperature and extracted with three 50-ml. portions of diethyl ether. The combined ether extracts are washed with 10% aqueous potassium hydrogen carbonate solution (Note 6), until the aqueous layer remains basic to litmus, and with saturated sodium chloride solution. The ethereal solution is dried over anhydrous sodium sulfate, filtered, and concentrated on a rotary evaporator. The resulting oily residue is diluted with 1 ml. of benzene then with 3 ml. of cyclohexane. The solution is poured into a chromatographic column (13 × 2.5 cm.), prepared with 50 g. of alumina (Note 7) and 3 : 1 cyclohexane–benzene. With this solvent system, the desired product moves with the solvent front; the first 250 ml. of eluent contains 95% of the total product. Elution with an additional 175 ml. of solvent removes the remainder. The combined fractions are evaporated, and the pale-yellow, oily residue crystallizes readily on standing. Recrystallization of this material from pentane gives 1.82 g. (45% yield) of white, crystalline 2,2-(trimethylenedithio)cyclohexanone, m.p. 52–55° (Note 8).

2. Notes

1. The time required for this operation generally is 3.5–5 hours.

2. Trimethylene dithiotosylate, m.p. 66–67°,[4] as described in *Org. Synth.*, **Coll. Vol. 6,** 1016 (1988) was employed.

3. Eastman white label triethylamine was distilled from sodium hydroxide.

4. Fisher Reagent acetonitrile was distilled from phosphorus pentoxide.

5. Treatment with the dilute acid effects aqueous extraction of pyrrolidine and hydrolysis of unreacted dithiotosylate and enamine starting materials.

6. Hydrogen carbonate washing ensures removal of the sulfonic and sulfinic acids.

7. The checkers used "Aluminum Oxide" purchased from J. T. Baker Chemical Company.

8. The ^{1}H NMR spectrum of the product ($CDCl_3$) exhibits multiplets in the region δ 1.65–2.45. The IR spectrum ($CHCl_3$) shows peaks at 2980 (m), 2940 (s), 2870 (m), 1690 (s), 1445 (m), 1420 (m), 1120 (m), 1110 (m), and 910 (s) cm^{-1}.

3. Discussion

The preparation of dithianes from enamines by reaction with trimethylene dithiotosylate has been applied with enamines derived from cholestan-3-one, acetoacetic ester, and phenylacetone.[5] Reactions of trimethylene dithiotosylate with hydroxymethylene derivatives of ketones also give rise to dithianes; thus, the hydroxymethylene derivative of cholest-4-en-3-one can be converted to 2,2-(trimethylenedithio)cholest-4-en-3-one.[6] 1,3-Dithiolanes are obtained in a similar manner by reaction of ethylene dithiotosylate[7] with the appropriately activated substrate.[5,8]

1. 1965 Novel Laureate in Chemistry; deceased July 8, 1979; formerly at the Department of Chemistry, Harvard University, Cambridge, Massachusetts 02138.

2. Present address: Bristol Laboratories, Division of Bristol-Myers Company, Syracuse, New York 13201.

3. Present address: Sterling-Winthrop Research Institute, Rensselaer, New York 12144.

4. L. A. Cohen and B. Witkop, *J. Am. Chem. Soc.,* **77,** 6595 (1955); G. Stork, A. Brizzolara, H. Landesman, J. Szmuszkovicz, and R. Terrell, *J. Am. Chem. Soc.,* **85,** 207 (1963).

5. R. B. Woodward, I. J. Pachter, and M. L. Scheinbaum, *J. Org. Chem.,* **36,** 1137 (1971).

6. R. B. Woodward, A. A. Patchett, D. H. R. Barton, D. A. J. Ives, and R. B. Kelley, *J. Chem. Soc.,* 1131 (1957).

7. R. B. Woodward, I. J. Pachter, and M. L. Scheinbaum, *Org. Synth.,* **Coll. Vol. 6,** 1016 (1988).

8. R. B. Woodward, I. J. Pachter, and M. L. Scheinbaum, *Org. Synth.,* **Coll. Vol. 6,** 590 (1988).

TRIMETHYLENE DITHIOTOSYLATE AND ETHYLENE DITHIOTOSYLATE

(Benzenesulfonothioic acid, 4-methyl-, S,S'-1,3-propanediyl and S,S'-1,2-ethanediyl esters)

$$KOH + H_2S \xrightarrow[0°]{H_2O} KHS + H_2O$$

$$CH_3C_6H_4SO_2Cl + 2\ KHS \xrightarrow[55-60°]{H_2O} CH_3C_6H_4SO_2SK + KCl + H_2S$$

$$2\ CH_3C_6H_4SO_2SK + Br(CH_2)_3Br \xrightarrow[reflux]{C_2H_5OH,\ KI}$$

$$CH_3C_6H_4SO_2S(CH_2)_3SSO_2C_6H_4CH_3 + 2\ KBr$$

$$2\ CH_3C_6H_4SO_2SK + Br(CH_2)_2Br \xrightarrow[reflux]{C_2H_5OH,\ KI}$$

$$CH_3C_6H_4SO_2S(CH_2)_2SSO_2C_6H_4CH_3 + 2\ KBr$$

Submitted by R. B. Woodward,[1] I. J. Pachter,[2] and Monte L. Scheinbaum[3]
Checked by P. A. Rossy and S. Masamune

1. Procedure

Caution! This procedure should be carried out in a hood to avoid inhalation of hydrogen sulfide.

A. *Potassium thiotosylate.* A solution of 64.9 g. (1.00 mole) of 86.5% potassium hydroxide (Note 1) in 28 ml. of water is cooled in an ice bath, saturated with hydrogen sulfide, and flushed with nitrogen to ensure complete removal of excess hydrogen sulfide (Notes 2 and 3). The freshly prepared potassium hydrosulfide solution is diluted with 117 ml. of water and stirred under nitrogen at 55–60° before 95.3 g. (0.500 mole) of finely ground p-toluenesulfonyl chloride (Note 3) is introduced in small portions at a uniform rate so that the reaction temperature is maintained at 55–60° (Note 2). A mildly exothermic reaction ensues, and the solution becomes intensely yellow. After about 90 g. of the tosyl chloride has been introduced, the yellow color disappears, and the dissolution of the chloride ceases. The reaction mixture is rapidly filtered with suction through a warmed funnel, and the filtrate is cooled several hours at 0–5°. The crystals of potassium thiotosylate are filtered, dissolved in 200 ml. of hot 80% ethanol, filtered hot to remove traces of sulfur, and cooled several hours at 0–5°. The recrystallized salt is filtered and air-dried, providing 48–55 g. (42–49%) of white crystals.

B. *Trimethylene dithiotosylate.* To 150 ml. of 95% ethanol containing 10–20 mg. of potassium iodide is added 40 g. (0.18 mole) of potassium thiotosylate and 20 g. (0.10

mole) of 1,3-dibromopropane (trimethylene dibromide) (Note 4). The mixture is refluxed with stirring for 8 hours in the dark and under nitrogen. The reaction mixture is cooled to ambient temperature, diluted with an equal volume of cold water, and agitated. After decantation of the supernatant liquid, the residual honeylike layer of product is washed with three 200-ml. portions of cold water, once with 100 ml. of cold 95% ethanol, and once with 100 ml. of cold absolute ethanol. The crude product (Note 5) is dissolved in 10 ml. of acetone, diluted with 80 ml. of hot absolute ethanol, and stirred under nitrogen at 0°. The oil which separates is redissolved by the addition of a minimum amount (ca. 5 ml.) of acetone. Seed crystals are introduced (Note 6), and the mixture is stirred for 1 hour at 0° under nitrogen and stored at −30° for several hours. The microcrystalline product is collected by filtration and weighs 20.2 g., m.p. 63.5–65.0°. Three recrystallizations from nine parts (180 ml.) of ethanol give 17.2 g. (41%) of white needles, m.p. 66–67° (Note 7). During the recrystallizations some of the material oils out when the solution is cooled to room temperature. The supernatant liquid is decanted, seeded, and stored at −30° for several hours. The oil is not further purified. The recrystallized material is chemically pure for further use [*Org. Synth.*, **Coll. Vol. 6,** 590 (1988)].

C. *Ethylene dithiotosylate*. To 200 ml. of ethanol containing 10–20 mg. of potassium iodide is added 45.3 g. (0.200 mole) of potassium thiotosylate and 18.8 g. (0.100 mole) of 1,2-dibromopropane. The mixture is refluxed with stirring for 8 hours in the dark and under a nitrogen atmosphere. The solvent is removed, and the resulting white solid is washed with a mixture of 80 ml. of ethanol and 150 ml. of water. After decantation, the solid is washed three times with 50-ml. portions of water and recrystallized from approximately 150 ml. of ethanol, yielding 28.7 g. of crude product, m.p. 72–75°. Three recrystallizations from a mixture of ethyl acetate and ethanol afford 24 g. (60%) of white crystals, m.p. 75–76° (Note 8).

2. Notes

1. Potassium hydroxide pellets of reagent grade commonly available, such as that from Fisher Scientific Company, contain 10–15% water. The checkers used the amount calculated on the basis of 86.5%, as specified.

2. Hydrogen sulfide is undesirable because its presence can lead to the formation of potassium p-toluenesulfinate. The latter can be formed by the desulfurization of thiotosylate by hydrogen sulfide generated in the reaction of potassium hydrosulfide with the tosyl chloride. Attention should be directed toward control of the reaction temperature so that hydrogen sulfide is rapidly removed, thereby ensuing survival of the S—S bond of the thiotosylate. p-Toluenesulfinate ion can displace bromide to form stable sulfones which are less soluble in common solvents, such as benzene, than trimethylene dithiotosylate. Therefore, purification of the dithiotosylate contaminated with the sulfones is difficult to achieve by fractional recrystallization.

3. The p-toluenesulfonyl chloride should be free of p-toluenesulfonic acid, otherwise potassium p-toluenesulfonate will be formed and result in the formation of tosylates, rather than thiotosylates. The reagent used by the checkers was obtained from British Drug Houses Ltd. and was purified by washing benzene solution of the tosyl chloride with

5% aqueous sodium hydroxide, drying with magnesium sulfate, and distilling under reduced pressure, b.p. 146° (15 mm.).[4]

4. Trimethylene dibromide, available from Eastman Organic Chemicals, was distilled prior to use (b.p. 167–168°).

5. The checkers found that the crude oily product crystallizes after storage for a few days under nitrogen at −30°. Some of this solid was saved and used as seed crystals.

6. The submitters reported that seed crystals were obtained by column chromatography, using 40 parts by weight of Woelm neutral alumina (activity grade one) and benzene elution. The center cuts of m.p. 65° or higher were combined and recrystallized from nine parts of ethanol to give white needles, m.p. 67°. Two recrystallizations of chromatographed gave trimethylene dithiotosylate, m.p. 67.5°.

7. The purified trimethylene dithiotosylate exhibits IR bands (CHCl₃) at 3030 (w), 2930 (w), 1590 (w), 1490 (w), 1440 (w), 1410 (w), 1325 (s), 1300 (m), 1180 (w), 1140 (s), 1075 (s), 1015 (w), and 810 (m) cm^{-1}. The ^{1}H NMR spectrum (CDCl₃ included signals at δ 1.98 (quinlet, J = 7 Hz., 2H, CH₂CH_2CH₂), 2.43 (s, 6H, 2 CH_3), 2.97 (t, J = 7 Hz., 4H, CH_2CH₂CH_2), 7.30 (d, J = 9 Hz., 4H), and 7.75 (d, J = 9 Hz., 4H). Analysis calculated for C₁₇H₂₀O₄S₄: C, 49.01; H, 4.84; S, 30.79. Found: C, 49.13; H, 4.81; S, 30.51 (submitters). Found: C, 48.71; H, H, 4.64; S, 30.45 (checkers). The checkers found that the product exhibited a single peak on a 3 ft. × ⅛ in. Waters Associates Analytical Liquid Chromatographic column, packed with Durapak-Carbowax 400/Poracil C. Chloroform was used as the eluting solvent.

8. The ^{1}H NMR spectrum (CDCl₃) has absorptions at δ 2.47 (s, 6H, 2CH_3), 3.31 (s, 4H, 2CH_2), 7.48 (complex in, J = 9 Hz., 4H) and 7.97 (d, J = 9 Hz., 4H). Analysis calculated for C₁₆H₁₈S₄O₄: C, 47.73; H, 4.51; S, 31.86. Found: C, 47.89; H, 4.44; S, 32.22.

3. Discussion

Although it has been long known that trimethylene dithiotosylate can be prepared by the reaction of thiotosylate ion with trimethylene dibromide,[5] various difficulties are associated with the preparation. These problems are to a considerable extent related to the mode of preparation and the resultant purity of potassium thiotosylate. The thiotosylate salt must be free of tosylate and p-toluenesulfinate impurities, otherwise side products such as tosylates or sulfones will form. One such by-product, tosyltrimethylene thiotosylate, CH₃C₆H₄SO₂(CH₂)₃SSO₂C₆H₄CH₃, m.p. 92°, was isolated from contaminated samples of trimethylene dithiotosylate. It is products such as these, that make crystallization of the dithiotosylate difficult. The procedure described herein serves as a reliable technique for minimizing these experimental difficulties. More recently, it has been shown[6] that trimethylene dithiotosylate can be prepared easily by the reaction of tosyl chloride and 1,3-propanedithiol in pyridine. This procedure, however, is unchecked.

Trimethylene dithiotosylate can react with activated methylene groups, enamines, or hydroxyethylene derivatives of carbonyl compounds to form dithiane derivatives. In this context, trimethylene dithiotosylate has been employed in the preparation[7] and modification[8] of several steroids. It has also been used in the synthesis of alkaloids,[9] 10-membered ring lactones,[10] and vernolepin analogues.[11] Ethylene dithiotosylate undergoes similar reactions, forming dithiolanes.[3,12]

1. 1965 Nobel Laureate in Chemistry; deceased July 8, 1979; formerly at the Department of Chemistry, Harvard University, Cambridge, Massachusetts 02138.
2. Bristol Laboratories, Division of Bristol-Myers Company, Syracuse, New York 13201.
3. Sterling-Winthrop Research Institute, Rensselaer, New York 12144.
4. D. D. Perrin, W. L. F. Armarego, and D. R. Perrin, "Purification of Laboratory Chemicals," Pergamon Press, London, 1966, p. 268.
5. (a) J. C. A. Chivers and S. Smiles, *J. Chem. Soc.*, 697 (1928); (b) L. G. S. Brooker and S. Smiles, *J. Chem. Soc.*, 1723 (1926).
6. P. R. Heaton, J. M. Midgley, and W. B. Walley, *J. Chem. Soc., Perkin Trans. 1*, 1011 (1978).
7. T. Kametani, H. Matsumoto, H. Nemoto, and K. Fukumoto, *J. Am. Chem. Soc.*, **100**, 6218 (1978).
8. (a) P. J. Hylands, J. M. Midgley, C. Smith, A. F. A. Whallis, and W. B. Whalley, *J. Chem. Soc., Perkin Trans. 1*, 817 (1977); (b) J. M. Midgley, W. B. Whalley, P. A. Dodson, G. F. Katekar, and B. A. Lodge, *J. Chem. Soc., Perkin Trans. 1*, 823 (1977).
9. (a) S. Takano, M. Sasaki, H. Kanno, K. Shisido, and K. Ogasawara, *Heterocycles*, **7**, 143 (1977); (b) S. Takano, M. Sasaki, H. Kanno, K. Shisido, and K. Ogasawara, *J. Org. Chem.*, **43**, 4169 (1978); (c) T. Kametani, T. Ohsawa, and M. Ihara, *Heterocycles*, **12**, 913 (1979).
10. T. Wakamatsu, K. Akasaka, Y. Ban, *J. Org. Chem.*, **44**, 2008 (1979).
11. J. A. Marshall and D. E. Seitz, *J. Org. Chem.*, **40**, 534 (1975).
12. R. B. Woodward, I. J. Pachter, and M. L. Scheinbaum, *J. Org. Chem.*, **36**, 1137 (1971).

TRIMETHYLOXONIUM TETRAFLUOROBORATE

[Oxonium, trimethyl- tetrafluoroborate(1-)]

$$4(C_2H_5)_2O \cdot BF_3 + 6(CH_3)_2O + 3ClCH_2CH\!-\!CH_2$$
$$\underset{O}{\diagdown\diagup}$$

$$\downarrow$$

$$3(CH_3)_3O^+BF_4^- + 4(C_2H_5)_2O + \quad B(OCHCH_2OCH_3)_3$$
$$\underset{CH_2Cl}{|}$$

Submitted by T. J. Curphey[1]
Checked by A. Eschenmoser, R. Keese, and A. Daniel

1. Procedure

A 500-ml., three-necked flask fitted with a mechanical stirrer, a Dewar condenser (Note 1) connected by a T-tube to a mineral oil bubbler and a source of dry nitrogen, and a gas-inlet tube connected to a source of dry dimethyl ether (Note 2) is charged with 80 ml. of dichloromethane and 38.4 g. (33.3 ml., 0.271 mole) of boron trifluoride diethyl etherate (Note 3). After establishing a nitrogen atmosphere in the flask, the condenser is filled with an acetone – dry ice mixture. With gentle stirring, dimethyl ether is passed into the solution until approximately 75 ml. has collected (Note 4). The gas-inlet tube is replaced with a pressure-equalizing dropping funnel containing 28.4 g. (24.1 ml., 0.307 mole) of epichlorohydrin, which is added dropwise with vigorous stirring over a 15-minute period. The mixture is stirred overnight under an atmosphere of nitrogen (Note 5). The stirrer is replaced by a filter stick, and the supernatant liquid is drawn off from the crystalline trimethyloxonium tetrafluoroborate, while keeping the mixture under nitrogen. The oxonium salt is washed with two 100-ml. portions of anhydrous dichloromethane and two 100-ml. portions of sodium-dried diethyl ether (Note 6), and dried by passing a stream of nitrogen over the salt until the odor of ether is no longer detected, yielding 28 – 29 g. (92.5 – 96.5%) of a white crystalline solid, m.p. (sealed tube) 179.6 – 180.0° (dec.), (Notes 7 and 8).

2. Notes

1. A Kontes K-45750 condenser was used.

2. Dimethyl ether and nitrogen were dried by passage through columns of Drierite. Boron trifluoride etherate (Eastman Practical Grade) was redistilled. Epichlorohydrin (Eastman Organic Chemicals) and dichloromethane (Fisher Scientific Company) were used as received.

3. According to ^{1}H NMR analysis the use of boron trifluoride etherate does not cause any detectable introduction of ethyl groups into the product.

4. This may conveniently be done by placing, prior to conducting the reaction, a mark on the reaction flask at a level of 190 ml., and collecting dimethyl ether up to the mark. The exact amount of dimethyl ether used is not critical.

5. After 2 – 3 hours of stirring the reaction appears to be over, and the dry ice in the condenser need no longer be renewed. The reaction mixture may be worked up at this point without appreciable reduction in the product yield or purity.

6. According to analysis by ^{1}H NMR the use of diethyl ether at this point does not cause any detectable exchange of methyl by ethyl groups in the oxonium salt. A user[2] has reported obtaining the best samples of oxonium salt by using boron fluoride dimethyl etherate instead of the diethyl etherate and omitting the diethyl ether washing of the product. Oxonium salt prepared in this way was used to prepare methyl esters (from the corresponding amides) with no detectable (by GC analysis) ethyl esters.

7. The melting point of trimethyloxonium tetrafluoroborate apparently depends upon the procedure by which it is prepared and the method of melting-point determination. It has, for example, been reported to melt at 124.5°,[3] 141 – 143° [*Org. Synth.*, **Coll. Vol. 5**, 1096 (1973)], and 175°.[4] The ^{1}H NMR spectrum, determined (liquid SO_2, purissimum Fluka AG) in a sealed tube at room temperature shows a single methyl resonance at δ 4.54; a trace of impurity is discernible as a singlet at δ 3.39.

8. When prepared as described, the oxonium salt is stable and nonhygroscopic, and may readily be handled in the air for short periods of time. A sample kept in a desiccator over Drierite for 1 month at −20° showed no change in melting point, and batches stored in this manner for over a year have been successfully used for alkylations.

3. Discussion

Trialkyloxonium salts were first discovered by Meerwein,[3] who also investigated much of their chemistry. A discussion of the literature prior to 1963 has been published.[5] Simple trialkyloxonium cations which have been prepared, other than trimethyl, include triethyl,[6] tri-n-propyl,[7] and tri-n-butyl,[8] with tetrafluoroborate or hexachloroantimonate anions, in most cases. Methods used to prepare trimethyloxonium tetrafluoroborate, which are typical of the class as a whole, include the reaction of boron trifluoride with epichlorohydrin in the presence of dimethyl ether,[3,4,9] the reaction of dimethyloxonium tetrafluoroborate with diazomethane or diazoacetic ester,[10] and the alkylation of dimethyl ether by triethyloxonium tetrafluoroborate[11] or dimethoxycarbonium tetrafluoroborate.[12] Several of these reactions involve the initial formation of a mixed oxonium ion $[R_1R_2OCH_3]^+$, which then methylates dimethyl ether, providing R_1R_2O and the trimethyloxonium ion. Of the available procedures, the one described here is probably the most convenient, involving as it does a single-step preparation from inexpensive, commercially available, and nonhazardous reagents. Under the proper conditions (Note 8), the resulting product has storage properties comparable to those of the less-accessible trimethyloxonium 2,4,6-trinitrobenzenesulfonate.[13]

The trialkyloxonium salts are powerful alkylating agents. Trimethyl- and triethyloxonium tetrafluoroborates, in particular, have been widely employed for methylation and ethylation of sensitive or weakly nucleophilic functional groups. Alkylations of over 50 such functional groups have been reported in the literature. Examples include amides,[4,7,14-16] lactams,[16-19] sulfides,[3,20] nitro compounds,[9] enols and enolates,[3,21] ethers,[7,11,22] phenols,[3] sulfoxides,[3,7] amine oxides,[3,7,23] carboxylic acids,[3] lactones,[3,4] ketones,[3,16] metal carbonyls,[12,24] thiophenes,[25] and phosphonitriles.[26]. Oxonium salts have also been advantageously employed as quarternizing agents for a variety of heterocyclic amines.[27-34] In this way the first diquarternary salts of several heterocyclic diazines have been prepared,[30,31] as have reagents for peptide synthesis,[33,34] for the synthesis of polycyclic ketones,[32] and for cyanine dyes.[28]

One of the major advantages of oxonium salts is that alkylations can be effected under reaction conditions that are generally much milder than those necessary with the more conventional alkyl halides or sulfonates. Triethyloxonium tetrafluoroborate, for example, has usually been employed at room temperature in dichloromethane or dichloroethane solution. Occasionally chloroform[17,23] or no solvent at all[4,21] is used. Difficult alkylations can be effected in refluxing dichloroethane.[30,31] The less soluble trimethyloxonium tetrafluoroborate has been used as a suspension in dichloromethane or dichloroethane, or as a solution in nitromethane or liquid sulfur dioxide. Reports of alkylations in water[24] and trifluoroacetic acid[22] have also appeared. Direct fusion with trimethyloxonium tetrafluoroborate has succeeded in cases where other conditions have failed.[26,31]

Alkylations by oxonium salts have added several new weapons to the synthetic chemist's armamentarium. For example, the O-alkylated products from amides $[R_1C(OR)=NR_2R_3]^+$ ($R=CH_3$ or C_2H_5) may be hydrolyzed under mild conditions to amines and esters,[15,35] reduced to the amines $R_1CH_2NR_2R_3$ by sodium borohydride,[14] converted to amide acetals $R_1C(OR)_2NR_2R_3$ by alkoxides,[4,16] and (for $R_3=H$) deprotonated to the imino esters $R_1C(OR)=NR_2$.[17-19] Amide acetals and imino esters are themselves in turn useful synthetic intermediates. Indeed, oxonium salts transform the rather intractable amide group into a highly reactive and versatile functionality, a fact elegantly exploited in recent work on the synthesis of corrins.[35]

Other reagents which approach or exceed the oxonium salts in alkylating ability include dialkoxycarbonium ions,[36] alkyl trifluoromethanesulfonates,[37] alkyl fluorosulfonates,[38] dialkylhalonium ions,[39] and alkyl halides in the presence of silver salts.[25,37,40] In terms of availability, stability, and freedom from hazards,[25] oxonium salts often appear to be the reagents of choice. When either methylation or ethylation is acceptable, methylation may be preferable. Triethyloxonium tetrafluoroborate must be stored under ether and handled in a dry box,[6] whereas the trimethyl salt can be stored solvent-free in the freezing compartment of a refrigerator and dispensed in the open atmosphere. Moreover, while information on the relative alkylating ability of the oxonium salts is not extensive, a few cases have been reported in which trimethyloxonium tetrafluoroborate effected alkylations which the triethyl analog did not.[20,31] The trimethyloxonium salt, therefore, appears to be the more potent alkylating agent.

1. Department of Chemistry, St. Louis University, St. Louis, Missouri 63156 [Present address: Department of Pathology, Dartmouth Medical School, Hanover, New Hampshire 03755].
2. Robert F. Myers (with William S. Johnson), Department of Chemistry, Stanford University, Stanford, California 94305, private communication to the editor-in-chief, May 5, 1970.
3. H. Meerwein, G. Hinz, P. Hofmann, E. Kroning, and E. Pfeil, *J. Prakt. Chem.*, [2], **147**, 257 (1937).
4. H. Meerwein, P. Borner, O. Fuchs, H. J. Sasse, H. Schrodt, and J. Spille, *Chem. Ber.*, **89**, 2060 (1956).
5. H. Meerwein, in "Methoden der Organischen Chemie" (Houben-Weyl), Vol. 6/3, Georg Thieme Verlag, Stuttgart, 1965, p. 325.
6. H. Meerwein, *Org. Synth.*, **Coll. Vol. 5**, 1080 (1973).
7. H. Meerwein, E. Battenberg, H. Gold, E. Pfeil, and G. Willfang, *J. Prakt. Chem.*, [2], **154**, 83 (1939).
8. G. Hilgetag and H. Teichmann, *Chem. Ber.*, **96**, 1446 (1963).
9. N. Kornblum and R. A. Brown, *J. Am. Chem. Soc.*, **86**, 2681 (1964).
10. F. Klages, H. Meuresch, and W. Steppich, *Justus Liebigs Ann. Chem.*, **592**, 81 (1955).
11. H. Meerwein, *Org. Synth.*, **Coll. Vol. 5**, 1096 (1973).
12. R. B. Silverman and R. A. Olofson, *Chem. Commun.*, 1313 (1968).
13. G. K. Helmkamp and D. J. Pettitt, *Org. Synth.*, **Coll. Vol. 5**, 1099 (1973).
14. R. F. Borch, *Tetrahedron Lett.*, 61 (1968).
15. H. Muxfeldt, J. Behling, G. Grethe, and W. Rogalski, *J. Am. Chem. Soc.*, **89**, 4991 (1967).
16. H. Meerwein, W. Florian, N. Schön, and G. Stopp, *Justus Liebigs Ann. Chem.*, **641**, 1 (1961).
17. S. Petersen and E. Tietze, *Justus Liebigs Ann. Chem.*, **623**, 166 (1959).
18. E. Vogel, R. Erb, G. Lenz, and A. A. Bothner-By, *Justus Liebigs Ann. Chem.*, **682**, 1 (1965).
19. L. A. Paquette, T. Kakihana, J. F. Hansen, and J. C. Philips, *J. Am. Chem. Soc.*, **93**, 152 (1971).
20. J. E. Baldwin, R. E. Hackler, and D. P. Kelly, *J. Am. Chem. Soc.*, **90**, 4768 (1968).
21. G. Hesse, H. Broll, and W. Rupp, *Justus Liebigs Ann. Chem.*, **697**, 62 (1966).
22. P. E. Peterson and F. J. Slama, *J. Am. Chem. Soc.*, **90**, 6516 (1968).
23. C. Reichardt, *Chem. Ber.*, **99**, 1769 (1966).
24. R. Aumann and E. O. Fischer, *Chem. Ber.*, **101**, 954 (1968).
25. R. M. Acheson and D. R. Harrison, *J. Chem. Soc. C*, 1764 (1970).
26. J. N. Rapko and G. Feistel, *Inorg. Chem.*, **9**, 1401 (1970).

27. H. Balli and F. Kersting, *Justus Liebigs Ann. Chem.*, **647**, 1 (1961).
28. C. Reichardt, *Justus Liebigs Ann. Chem.*, **715**, 74 (1968).
29. H. Quast and S. Hünig, *Chem. Ber.*, **99**, 2017 (1966); *Chem. Ber.* **101**, 435 (1968); H. Quast and E. Schmitt, *Chem. Ber.*, **101**, 1137 (1968).
30. T. J. Curphey, *J. Am. Chem. Soc.*, **87**, 2063 (1965).
31. T. J. Curphey and K. S. Prasad, *J. Org. Chem.*, **37**, 2259 (1972).
32. G. Stork, S. Danishefsky, and M. Ohashi, *J. Am. Chem. Soc.*, **89**, 5459 (1967).
33. R. B. Woodward, R. A. Olofson, and H. Mayer, *Tetrahedron*, Suppl. 8, 321 (1966).
34. R. A. Olofson and Y. L. Marino, *Tetrahedron*, **26**, 1779 (1970).
35. A. Eschenmoser, *Q. Rev. Chem. Soc.*, **24**, 366 (1970).
36. S. Kabuss, *Angew. Chem.*, **78**, 714 (1966) [*Angew. Chem. Int. Ed. Engl.*, **5**, 675 (1966)] and references therein.
37. A. J. Boulton, A. C. G. Gray, and A. R. Katritzky, *J. Chem. Soc. B*, 911 (1967).
38. M. G. Ahmed, R. W. Alder, G. H. James, M. L. Sinnott, and M. C. Whiting, *Chem. Commun.*, 1533 (1968).
39. G. A. Olah and J. R. DeMember, *J. Am. Chem. Soc.*, **92**, 2562 (1970).
40. H. Meerwein, V. Hederich, and K. Wunderlich, *Arch. Pharm. Weinheim, Ger.*, **291**, 541 (1958).

3,5,5-TRIMETHYL-2-(2-OXOPROPYL)-2-CYCLOHEXEN-1-ONE

[2-Cyclohexen-1-one, 3,5,5-trimethyl-2-(2-oxopropyl)-]

Submitted by Z. Valenta[1] and H. J. Liu[2]
Checked by T. H. O'Neill, W. Thompson, D. H. Hawke, and R. E. Ireland

1. Procedure

Caution! Benzene has been identified as a carcinogen; OSHA has issued emergency standards on its use. All procedures involving benzene should be carried out in a well-ventilated hood, and glove protection is required.

A. *7-Acetoxy-4,4,6,7-tetramethylbicyclo[4.2.0]octan-2-one.* The apparatus used for the photocycloaddition reaction is shown in Figure 1. In the reaction vessel is placed a solution of 34.5 g. (0.250 mole) of 3,5,5-trimethyl-2-cyclohexen-1-one (isophorone) (Note 1) and 500 g. (5.00 moles) (Note 2) of 1-propen-2-yl acetate (Note 3) in 625 ml. of benzene. A constant and moderate flow of argon (Note 4) is maintained, agitating the solution throughout the reaction period. The trap is filled with 2-propanol and dry ice (Note 5). The solution is irradiated with a 450-watt, Hanovia high-pressure, quartz, mercury vapor lamp, using a Pyrex filter, for 96 hours (Note 6). Concentration of the resulting solution under reduced pressure (water aspirator) gives 65–80 g. of crude 7-acetoxy-4,4,6,7-tetramethylbicyclo[4.2.0]octan-2-one (Note 7).

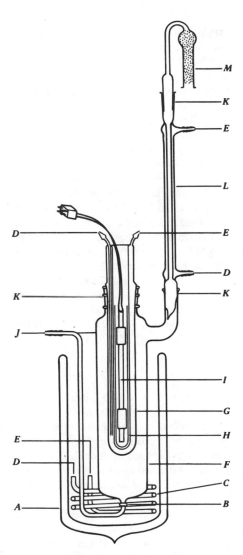

Figure 1. *A*, Dewar flask; *B*, sintered-glass filter; *C*, metal cooling coil; *D*, water inlet; *E*, water outlet; *F*, reaction vessel; *G*, quartz immersion well; *H*, Pyrex filter; *I*, lamp; *J*, nitrogen gas inlet; *K*, ground glass joint; *L*, condenser; *M*, calcium chloride drying tube.

B.   7-*Hydroxy*-4,4,6,7-*tetramethylbicyclo*[4.2.0]*octan*-2-*one*.   The preceding crude photo-adduct is dissolved in 250 ml. of methanol and transferred to a 1-l., three-necked, round-bottomed flask fitted with overhead stirrer with Teflon blade, addition funnel, and argon inlet. The solution is cooled with an ice bath, and 500 ml. of 4 *M* aqueous sodium hydroxide is added, with stirring, over a period of 20 minutes. Upon completion of the addition, the ice bath is removed, and stirring is continued for 16 hours. The brown solution is extracted with four 500-ml. portions of chloroform. The organic extract is washed with saturated sodium chloride solution, dried over magnesium sulfate, filtered, and concentrated under reduced pressure. Distillation of the residue through a 6-cm. Vigreux column affords, after a small forerun, 23.0–27.9 g. (47–57%) of 7-hydroxy-4,4,6,7-tetramethylbicyclo[4.2.0]octan-2-one (Note 8), collected at 92–101° (0.2 mm.).

C. *3,5,5-Trimethyl-2-(2-oxopropyl)-2-cyclohexen-1-one.* A 2-l., three-necked, round-bottomed flask, fitted with an overhead stirrer with Teflon blade, a Fries condenser, and a stopper, is charged with a solution of 9.8 g. (0.050 mole) of 7-hydroxy-4,4,6,7-tetramethylbicyclo[4.2.0]octan-2-one in 600 ml. of 50% (by volume) aqueous acetonitrile, and 82 g. (0.15 mole) (Note 9) of ceric ammonium nitrate (Note 10) is added in one portion with stirring. Immediately after completion of the addition, the flask is immersed in an oil bath preheated to 170°. Refluxing occurs in about 10 minutes and is continued for 5 minutes. During this period the color of the solution changes from light brown to pale yellow. At the end of this time the reaction mixture is immediately poured onto crushed ice and extracted with four 600-ml. portions of chloroform. The combined extracts are washed with saturated sodium hydrogen carbonate and saturated sodium chloride, dried over magnesium sulfate, and filtered. The solvent is removed under reduced pressure, and the residue distilled, using a short-path distillation apparatus. All material boiling at 70–100° (0.25 mm.) is collected. Fractionation of the yellow oil through a 6-cm. Vigreux column gives 4.68–4.71 g. (48–50%) of 3,5,5-trimethyl-2-(2-oxopropyl)-2-cyclohexene-1-one, b.p. 81–85° (0.4 mm.) (Note 11).

2. Notes

1. Isophorone obtained from M C and B Manufacturing Chemists was freshly distilled, b.p. 73° (4.5 mm.).

2. To minimize the formation of cyclohexenone dimer and achieve a cleaner photoadduct, it is essential to use a large excess of olefin.[3]

3. 1-Propen-2-yl acetate was supplied by Aldrich Chemical Company, Inc.

4. Submitters used nitrogen purified by passing it through a set of gas wash bottles containing Fieser's solution,[4] concentrated sulfuric acid, sodium hydroxide, and calcium chloride.

5. The submitters filled the Dewar flask with ice and water. After 2 hours the ice had melted and water was left in the flask for cooling.

6. The progress of the reaction was monitored by injecting, after each 24-hour period, an aliquot into a GC and checking the peak corresponding to isophorone. Alternatively, TLC (E. Merck 0.25-mm. silica gel plates developed with ethyl acetate) can be used.

7. This crude product is contaminated mainly by polymeric compounds. An attempted distillation of this material was unsuccessful; partial decomposition occurred at 110–125° (0.3 mm.). If it is desirable, purification can be achieved by extensive silica gel column chromatography with 5% ether in benzene.

8. The product is a mixture of at least two diasteriomers as indicated by its ^{1}H NMR spectrum (CCl₄), showing eight singlets at δ 0.9–1.22, for a total of twelve methyl

protons. Its IR spectrum (neat) exhibits absorption bands at 3440 and 1695 cm.$^{-1}$. A molecular ion peak at 196.1447 (calcd. for $C_{12}H_{20}O_2$: 196.1463) is displayed in its mass spectrum.

9. Use of less of the reagent resulted in partial recovery of the starting material.

10. Ceric ammonium nitrate was supplied by Fisher Scientific Company.

11. IR (neat) cm.$^{-1}$: 1720, 1670, 1645; ^{1}H NMR (CDCl$_3$), δ: 1.03, 1.87, 2.13, 2.20, 2.25, 3.43 (all singlets); mass spectrum m/e 194.1299 (M$^+$).

3. Discussion

Recently, the application of photocycloaddition reactions to organic synthesis has been gaining importance.[5,6] The procedure described is illustrative of a general method,[3] based on a photocycloaddition reaction, for the introduction of an activated alkyl group specifically to the α-carbon atom of an α,β-unsaturated cyclohexenone. Especially significant is the fact that the method is also applicable to α,β-unsaturated cyclohexenones which do not possess enolizable γ-hydrogen atoms and to which normal alkylation reactions[7] cannot be applied. A closely related procedure involving the photocycloaddition of vinyl acetate to 2-cyclohexenones (in which enolization toward the 6-position is forbidden) followed by bromination and fragmentation of the adduct has been reported.[3,8] It has also been observed[9] that photoadducts of cycloalkenones and vinylene carbonate undergo fragmentation upon alkali treatment, giving 2-(2-oxoethyl)-2-cycloalken-1-ones.

1. Department of Chemistry, University of New Brunswick, Fredericton, N.B., Canada E3B 6E2.
2. Department of Chemistry, University of Alberta, Edmonton, Alberta, Canada T6G 2G2.
3. N. R. Hunter, G. A. MacAlpine, H. J. Liu, and Z. Valenta, *Can. J. Chem.,* **48,** 1436 (1970).
4. L. F. Fieser, *J. Am. Chem. Soc.,* **46,** 2639 (1924).
5. References cited in Ref. 3.
6. For reviews, see P. E. Eaton, *Acc. Chem. Res.,* **1,** 50 (1968); P. de Mayo, *Acc. Chem. Res.,* **4,** 41 (1971); P. G. Bauslaugh, *Synthesis,* 287 (1970).
7. H. O. House, "Modern Synthetic Reactions," 2nd ed, W. A. Benjamin, Menlo Park, Calif., Chap. 9.
8. H.-J. Liu, Z. Valenta, J. S. Wilson, and T. T. J. Yu, *Can. J. Chem.,* **47,** 509 (1969).
9. P. T. Ho, S. F. Lee, D. Chang, and K. Wiesner, *Experientia,* **27,** 1377 (1971).

ALDEHYDES BY OXIDATION OF TERMINAL OLEFINS WITH CHROMYL CHLORIDE: 2,4,4-TRIMETHYLPENTANAL

(Pentanal, 2,4,4-trimethyl-)

$$
\underset{\underset{CH_3}{|}}{\overset{\overset{CH_3}{|}}{CH_3-C-CH_2-C=CH_2}} \quad \xrightarrow[\text{2. Zn, H}_2\text{O}]{\text{1. CrO}_2\text{Cl}_2} \quad \underset{\underset{CH_3}{|}\quad\underset{CH_3}{|}}{\overset{\overset{CH_3}{|}}{CH_3-C-CH_2-CH-CHO}}
$$

Submitted by Fillmore Freeman,[1] Richard H. DuBois, and Thomas G. McLaughlin
Checked by Graham Hagens and Peter Yates

1. Procedure

A 5-l., three-necked flask fitted with a mechanical stirrer, a thermometer, and a dropping funnel equipped with a calcium chloride drying tube is charged with 112.2 g. (1.002 mole) of freshly distilled 2,4,4-trimethyl-1-pentene (Note 1) and 1 l. of dichloromethane (Note 2). The flask is immersed in an ice–salt bath, and the stirred solution is cooled to 0–5°. A solution of 158 g. (1.02 moles) (Note 3) of freshly distilled chromyl chloride (Note 4) in 200 ml. of dichloromethane (Note 5) is added dropwise with stirring from the dropping funnel while the temperature is maintained at 0–5° (Note 6). The reaction mixture is stirred for 15 minutes, and 184 g. of 90–95% technical grade zinc dust (Note 7) is added. The mixture is stirred for 5 minutes, 1 l. of ice water and 400 g. of ice are added as rapidly as possible (Note 8), and the mixture is stirred for an additional 15 minutes. The ice–salt bath is replaced with a heating mantle, and the flask is fitted for steam distillation. After distillation of the dichloromethane the residue is steam distilled (Note 9). The distillate is transferred to a separatory funnel, the organic layer is separated, and the aqueous layer is washed with three 50-ml. portions of dichloromethane. The combined organic phases are distilled (Note 10) through a 56-cm., vacuum-jacketed, Vigreux column, removing the solvent. The product is transferred to a 250-ml. round-bottomed flask and distilled. After removal of a small amount of dichloromethane the fraction boiling at 45–52° (15 mm.) is collected, giving 90–100 g. (70–78%) (Note 11) of 2,4,4-trimethylpentanal.

2. Notes

1. The alkene is available from Aldrich Chemical Company, Inc., or Phillips Petroleum Company, and can be used without distillation.

2. The material available from Matheson, Coleman and Bell or Eastman Organic Chemicals is satisfactory except as explained in Note 5.

3. Since chromyl chloride is easily hydrolyzed, a slight excess is used.

4. The fraction, b.p. 115.5–116.5°, is used. Chromyl chloride is available from Alfa Inorganics, Inc.

5. Chromyl chloride tends to react slowly with the commercially available dichloromethane. This can be avoided with a slight increase in yield if the dichloromethane used to dissolve the chromyl chloride is distilled through a 15–20 cm. Vigreux column immediately before use.

6. The time required for the addition is about 60 minutes.

7. This approximate fivefold excess is necessary to reduce the higher valence chromium salts, thereby eliminating overoxidation and double bond cleavage. The zinc dust used was obtained from Allied Chemical Corporation.

8. The temperature usually increases to 8–10°.

9. Steam distillation is discontinued when the distillate gives a negative test with 2,4-dinitrophenylhydrazine.

10. It is not necessary to dry the organic phase.

11. The checkers, working at two-thirds scale, obtained the product in 70–71% yield.

3. Discussion

2,4,4-Trimethylpentanal has been prepared by the catalytic isomerization of 1,2-epoxy-2,4,4-trimethylpentane in both the liquid and gas phases (77–92%),[2] and by the oxidation of 2,4,4-trimethyl-1-pentene with chromium trioxide in acetic anhydride.[3] Although the catalytic isomerization of the epoxide[2] gives 2,4,4-trimethylpentanal in good yield, this requires epoxidation of the alkene as the first step. The chromyl acetate and chromic acid oxidative methods give unsatisfactory yields.[3] In the preparation described here, 2,4,4-trimethylpentanal is obtained from the alkene in good yield, in one step. Also, this preparation illustrates a general and convenient procedure for the direct oxidation of 2,2-disubstituted-1-alkenes (Table I) to unstable and reactive aldehydes.[4] The reaction is very fast and the aldehyde is the major product.

4,4-Dimethyl-2-neopentylpentanal,[6] 2-phenylpropanal,[7] and Diphenylacetaldehyde[8] are generally prepared by isomerization of the corresponding epoxide and/or by multistep syntheses.

In contrast to the relative simplicity of the chromyl chloride oxidation of 2,2-disubstituted-1-alkenes to aldehydes, the chromyl acetate and chromic acid oxidations generally lead to epoxides, acids, and carbon–carbon double bond cleavage. For ex-

TABLE I
ALDEHYDES BY OXIDATION OF TERMINAL OLEFINS
WITH CHROMYL CHLORIDE

Alkene	Aldehyde	Yield, %
4,4-Dimethyl-2-neopentyl-1-pentene	4,4-Dimethyl-2-neopentylpentanal	80.8[a]
2-Phenylpropene	2-Phenylpropanal	60.0[b]
1,1-Diphenylethylene	Diphenylacetaldehyde	62.7[b]

[a]Reference 4.
[b]Reference 5.

ample, chromyl acetate oxidizes 4,4-dimethyl-2-neopentyl-1-pentene primarily to 1,2-epoxy-4,4-dimethyl-2-neopentylpentane in low yield,[9] and chromic acid oxidizes the alkene principally to 4,4-dimethyl-2-neopentylpentanoic acid.[6,10]

1. Department of Chemistry, California State College, Long Beach, California 90801. This investigation was supported by the Long Beach California State College Foundation, the Research Corporation, and the Petroleum Research Fund administered by the American Chemical Society. [Present address: Department of Chemistry, University of California, Irvine, California 92717].
2. E. J. Gasson, A. R. Graham, A. F. Millidge, I. K. M. Robson, W. Webster, A. M. Wild, and D. P. Young, *J. Chem. Soc.*, 2170 (1954).
3. A. Byers and W. J. Hickinbottom, *J. Chem. Soc.*, 1334 (1948).
4. F. Freeman, P. J. Cameron, and R. H. DuBois, *J. Org. Chem.*, **33**, 3970 (1968).
5. F. Freeman, R. H. DuBois, and N. J. Yamachika, *Tetrahedron*, **25**, 3441 (1969).
6. P. D. Bartlett, G. L. Fraser, and R. B. Woodward, *J. Am. Chem. Soc.*, **63**, 495 (1941).
7. C. F. H. Allen and J. VanAllan, *Org. Synth.*, **Coll. Vol. 3**, 733 (1955).
8. D. J. Reif and H. O. House, *Org. Synth.*, **Coll. Vol. 4**, 375 (1963).
9. W. J. Hickinbottom and D. G. M. Wood, *J. Chem. Soc.*, 1600 (1951).
10. F. C. Whitmore and J. D. Surmatis, *J. Am. Chem. Soc.*, **63**, 2200 (1941).

TRIMETHYLSILYL AZIDE

(Silane, azidotrimethyl-)

$$(CH_3)_3SiCl + NaN_3 \rightarrow (CH_3)_3SiN_3 + NaCl$$

Submitted by L. BIRKOFER[1] and P. WEGNER
Checked by R. F. MERRITT and W. D. EMMONS

1. Procedure

Caution! This reaction should be conducted behind a safety screen in a hood. If the system is not completely dry, the presence of toxic hydrazoic acid is probable.

A 1-l., three-necked flask fitted with a stirrer, reflux condenser equipped with a drying tube, and addition funnel provided with a pressure-equalizer arm is dried in a 100° oven and assembled while warm. The warm apparatus is immediately purged with dry nitrogen, introducing the nitrogen at the top of the addition funnel. The flask is charged with 81 g. (1.2 moles) of sodium azide (Note 1) and 500 ml. of freshly distilled diethylene glycol dimethyl ether (Note 2). A simple distillation apparatus is then dried in the oven and assembled while warm under a slow nitrogen purge. The distillation flask is charged with 112 g. of chlorotrimethylsilane (Note 3), and after a forerun of approximately 2 g. the remaining material is distilled (b.p. 57–58°) directly into the addition funnel of the reaction flask. During this distillation it is convenient to disconnect the nitrogen stream from the top of the addition funnel and introduce it into the distillation flask. After the distillation is complete, the distillation apparatus is disconnected and the nitrogen stream is again introduced at the top of the addition funnel. The chlorotrimethylsilane (108.6 g.,

1.000 mole) (Note 4) is then added rapidly to the sodium azide slurry, and this mixture is stirred at 70° for 60 hours. During this period the nitrogen flow is terminated (Note 5).

After the heating period is complete, the nitrogen stream is again initiated, and the mixture is cooled to 30°. The addition funnel and reflux condenser are replaced with two gas-inlet tubes with stopcocks. One inlet tube is connected to the nitrogen source and the other to a standard vacuum trap, of at least 150 ml. capacity. A vacuum (15–20 mm.) is applied to the trap after the latter is cooled to −78°, and the product is then distilled at 30° (15 mm.) into the trap. Slight heating is necessary to maintain 30°, and rapid stirring should be continued throughout. Removal of volatile product is complete within 5 hours under these conditions. The entire system is then slowly pressurized to atmospheric pressure with nitrogen, and the product is redistilled through a 5-cm. Vigreux column. From 121 g. of crude flash distillate are obtained 4.0 g. of forerun and 98 g. (85%) of pure trimethylsilyl azide, b.p. 95–99°. During the distillation the pot temperature is maintained at 135–140° with a thermostated oil bath. The pot residue contains 19 g. of diethylene glycol dimethyl ether with traces of trimethylsilyl azide. The purity of the product cut as established by ^{1}H NMR (CCl$_4$) is 98%. A single peak at 13 cps. downfield from tetramethylsilane is observed, the only impurity being siloxane hydrolysis products. Chlorotrimethylsilane is conspicuous by its absence.

2. Notes

1. Sodium azide was obtained from Alpha Inorganics, Inc., Beverly, Massachusetts, and the freshly opened material was used without further purification or drying.

2. Diethylene glycol dimethyl ether from Aldrich Chemical Co. was distilled under a nitrogen atmosphere, and the fraction boiling at 161–162° was used.

3. The chlorotrimethylsilane was obtained from Pennisula Chem-research Corp., Gainesville, Florida.

4. It is undesirable to reweigh the chlorotrimethylsilane in the addition funnel because moisture contamination is possible. An excess of sodium azide is used in this preparation, and the exact amount of the silane used is not critical.

5. If the nitrogen flow is maintained during the heating period, the volatile materials will be swept out and the yield will be reduced.

3. Discussion

Trimethylsilyl azide has been prepared by the thermolysis of 1-trimethylsilyl-5-trimethylsilylaminotetrazole, by reaction of hydrazoic acid with hexamethyl-disilazane, and by reaction of chlorotrimethylsilane with sodium azide.[2] With a suitable solvent and anhydrous conditions the last procedure is the method of choice and has been extended to other trialkyl and triarylsilyl azides.[3]

Unlike hydrazoic acid, trimethylsilyl azide is thermally quite stable. Even at 200° it decomposes slowly and without explosive violence. Accordingly, it is a very convenient and safe substitute for hydrazoic acid in many reactions. A notable example is the cycloaddition of hydrazoic acid to acetylenes, a general route to substituted triazoles.[4] The reaction of trimethylsilyl azide with acetylenes is also a general reaction, from which

2-trimethylsilyl-1,2,3-triazoles may be obtained in good yield.[5] These adducts are hydrolyzed under mild conditions to the parent alkyl 1,2,3-triazoles.[5]

$$RC{\equiv}CR + (CH_3)_3SiN_3 \longrightarrow \left[\begin{array}{c} R \\ R \end{array} \diagup\hspace{-0.2em}\begin{array}{c} N \\ N \\ N \\ | \\ Si(CH_3)_3 \end{array} \right]$$

$$(CH_3)_3SiOH + \underset{R \quad H}{\underset{}{\diagdown N \diagup N}} \xleftarrow{H_2O} \underset{R}{\overset{R}{\diagdown}} N{-}Si(CH_3)_3$$

Another interesting application of trimethylsilyl azide is as a convenient preparation of trialkyl- or triarylphosphinimines, first prepared by Appel and Hauss using chloramine.[6]

$$R_3P + (CH_3)_3SiN_3 \xrightarrow{-N_2} R_3P{=}NSi(CH_3)_3 \xrightarrow{ROH} R_3P{=}NH$$

This synthesis is quite simple and its success lies in the facile cleavage of the Si—N bond.[7,8] Trimethylsilyl azide also reacts with aldehydes, giving the stable adducts,

$$(CH_3)_3SiN_3 + RCHO \xrightarrow{ZnCl_2} \underset{N_2}{RCHOSi(CH_3)_3} \xrightarrow[-N_2]{\Delta} \underset{O}{RCNHSi(CH_3)_3}$$

1-trimethylsiloxyalkyl azides, which on thermolysis yield N-trimethylsilyl amides.[9]

1. Institut für Organische Chemie der Universität Düsseldorf, 4 Düsseldorf, Germany.
2. L. Birkofer, A. Ritter, and P. Richter, *Chem. Ber.*, **96**, 2750 (1963).
3. N. Wiberg and B. Neruda, *Chem. Ber.*, **99**, 740 (1966).
4. O. Dimroth and G. Fester, *Ber. Dtsch. Chem. Ges.*, **43**, 2219 (1910).
5. L. Birkofer and P. Wegner, *Chem. Ber.*, **99**, 2512 (1966).
6. R. Appel and A. Hauss, *Chem. Ber.*, **93**, 405 (1960).
7. L. Birkofer, A. Ritter, and S. M. Kim, *Chem. Ber.*, **96**, 3099 (1963).
8. L. Birkofer and S. M. Kim, *Chem. Ber.*, **97**, 2100 (1964).
9. L. Birkofer, F. Muller, and W. Kaiser, *Tetrahedron Lett.*, **29**, 2781 (1967).

3-TRIMETHYLSILYL-3-BUTEN-2-ONE: A MICHAEL ACCEPTOR

[3-Buten-2-one, 3-(trimethylsilyl)-]

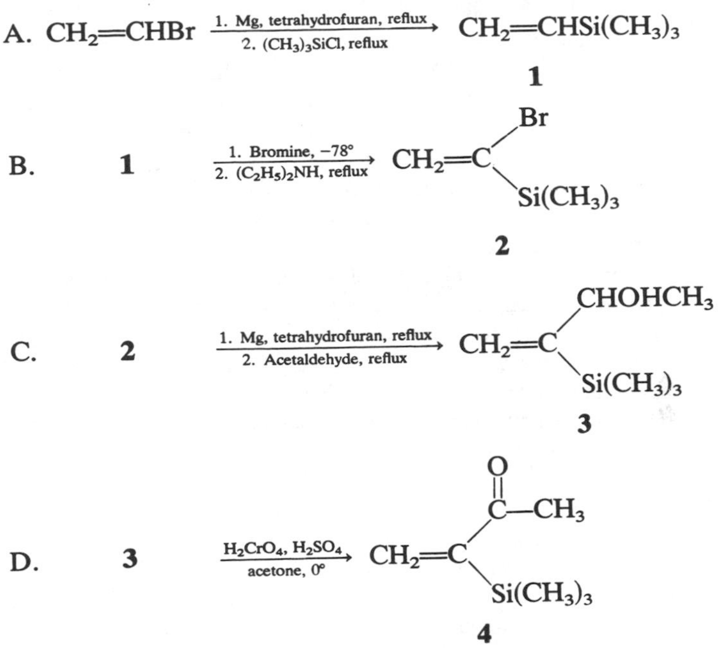

A. $CH_2{=}CHBr$ $\xrightarrow[\text{2. }(CH_3)_3SiCl,\ \text{reflux}]{\text{1. Mg, tetrahydrofuran, reflux}}$ $CH_2{=}CHSi(CH_3)_3$

1

B.   **1** $\xrightarrow[\text{2. }(C_2H_5)_2NH,\ \text{reflux}]{\text{1. Bromine, }-78°}$ $CH_2{=}C\begin{smallmatrix}Br\\ \\Si(CH_3)_3\end{smallmatrix}$

2

C.   **2** $\xrightarrow[\text{2. Acetaldehyde, reflux}]{\text{1. Mg, tetrahydrofuran, reflux}}$ $CH_2{=}C\begin{smallmatrix}CHOHCH_3\\ \\Si(CH_3)_3\end{smallmatrix}$

3

D.   **3** $\xrightarrow[\text{acetone, 0°}]{H_2CrO_4,\ H_2SO_4}$ $CH_2{=}C\begin{smallmatrix}O\\ \|\\ C{-}CH_3\\ \\Si(CH_3)_3\end{smallmatrix}$

4

Submitted by ROBERT K. BOECKMAN, JR.,[1,2] DAVID M. BLUM,[1] BRUCE GANEM,[3] and NEIL HALVEY[3]
Checked by WILLIAM R. BAKER and ROBERT M. COATES

1. Procedure

A. *Vinyltrimethylsilane* (**1**). A 2-l., three-necked, round-bottomed flask fitted with a mechanical stirrer, a reflux condenser, and a 500-ml., pressure-equalizing dropping funnel (Note 1) is charged with 26.4 g. (1.09 g.-atom) of magnesium turnings and 800 ml. of dry tetrahydrofuran (Note 2). A solution of 107 g. (70.5 ml., 1.00 mole) of vinyl bromide (Note 3) in 200 ml. of tetrahydrofuran is placed in the addition funnel and slowly added dropwise to the reaction vessel. After the reaction has begun (Note 4), the addition rate is regulated, maintaining a gentle reflux during the remainder of the addition period. The mixture is heated at reflux for an additional hour, and a solution of 108 g. (0.995 mole) of chlorotrimethylsilane (Note 5) in 100 ml. of tetrahydrofuran is added dropwise while the reaction is maintained at reflux with continued heating and stirring (Note 6). The suspension is stirred for another 2 hours under reflux, then cooled to room temperature and stirred overnight.

The condenser and dropping funnel are removed, and the flask is equipped for distillation with a 30.5-cm. Vigreux column. The distillate (b.p. 60–65°) is collected, transferred to a separatory funnel, and washed with 10–20 100-ml. portions of water (Note 7), yielding 67–78 g. (67–78%) of silane **1** as a colorless liquid containing small amounts of tetrahydrofuran (Note 8).

B. (1-Bromoethynl)trimethylsilane (**2**). A 1-l., three-necked, round-bottomed flask equipped with a mechanical stirrer and a 250-ml. dropping funnel is charged with 89.8 g. (approximately 0.90 mole, Note 8) of silane **1**. The contents of the flask are stirred and cooled to −78°, and 168 g. (1.06 mole) of bromine is added dropwise over *ca*. 1 hour. The cooling bath is removed, and the red viscous mixture is warmed to room temperature. The flask is fitted with an efficient, water-cooled condenser, and 425 g. (600 ml., 5.82 moles) of diethylamine (Note 9) is cautiously (Note 10) added with continued stirring. After the addition is complete, the reaction mixture is heated at reflux for 12 hours, during which time a precipitate of diethylamine hydrochloride forms. The salts are separated from the cooled suspension by filtration and washed with several 300-ml. portions of diethyl ether. The ether filtrate is carefully washed, first with 100-ml. portions of 10% hydrochloric acid until the aqueous layer remains acidic (pH *ca*. 2), then with 100 ml. of water and 200 ml. of saturated aqueous sodium chloride. The ether solution is dried with anhydrous magnesium sulfate, concentrated with a rotary evaporator, and distilled under reduced pressure through a 20.3-cm. Vigreux column, affording 104 – 110 g. (65 – 68%) of silane **2,** b.p. 72 – 75° (120 mm.) (Note 11).

C. 3-*Trimethylsilyl*-3-*buten*-2-*ol* (**3**). A 500-ml., three-necked, round-bottomed flask equipped with two 30.5-cm. Liebig condensers connected in series, a pressure-equalizing dropping funnel, and a magnetic stirrer is charged with 9.2 g. (0.38 g.-atom) of magnesium turnings and 100 ml. of tetrahydrofuran (Notes 1 and 2). About 2 g. of 1,2-dibromoethane is added, initiating the formation of the Grignard reagent. When the supernatant solution becomes warm and begins to reflux from reduction of 1,2-dibromoethane, a solution of 50 g. (0.28 mole) of silane **2** in 75 ml. of tetrahydrofuran is added dropwise to the stirred mixture at a rate that maintains gentle reflux. After the addition is complete, the reaction mixture is kept at reflux for an additional hour before freshly distilled acetaldehyde (25.0 g., 0.568 mole) is introduced. The temperature is maintained at reflux, and stirring is continued throughout the addition and for an additional hour. The flask is then fitted with a distillation head and heated until *ca*. 100 ml. of distillate has been collected. The reaction mixture is cooled (ice-water bath) and stirred, diluted with 100 ml. of ether, and hydrolyzed by addition of enough saturated ammonium chloride (approximately 50 ml.) to dissolve the thick, sticky precipitate. The salts are filtered and washed with ether, and the aqueous layer of the filtrate is extracted with three 150-ml. portions of ether. The combined ether layers are washed with saturated aqueous sodium chloride, dried over anhydrous magnesium sulfate, and concentrated by distillation at atmospheric pressure (Note 12), giving 48 – 55 g. of crude butenol **3** as a liquid that is used in the next step without further purification.

D. 3-*Trimethylsilyl*-3-*buten*-2-*one* (**4**). A solution of 55 g. of crude butenol **3** in 100 ml. of acetone is placed in a 500-ml., three-necked, round-bottomed flask equipped with a mechanical stirrer and a 250-ml. dropping funnel. The reaction vessel is immersed in an ice-water bath, and 95 ml. of an aqueous solution containing chromium trioxide and sulfuric acid (Note 13) is added to the stirred acetone solution. After completion of the addition, 2-propanol is added to the reaction mixture until a green endpoint is reached, indicating consumption of excess oxidant. The contents are poured into 450 ml. of ether, 300 ml. of water are added, and the aqueous layer is saturated with sodium chloride. The layers are separated, and the aqueous solution is extracted with five 150-ml. portions of ether. The combined ether solutions are washed with two 150-ml. portions of saturated aqueous sodium chloride, dried with anhydrous magnesium sulfate, and concentrated by distillation at atmospheric pressure through a 30.5-cm. Vigreux column. Continued distillation under reduced pressure gives, after separation of a low boiling forerun, 14.7 – 15 g. (37 – 38%) of butenone **4** as a pale yellow liquid, b.p. 98 – 103° (100 mm.) (Notes 14 and 15).

2. Notes

1. The apparatus is flamed dry under an argon atmosphere and maintained under argon during the reaction.

2. The submitters purified the tetrahydrofuran by distillation from lithium aluminum hydride. The checkers used tetrahydrofuran that had been distilled from the sodium ketyl of benzophenone. (*Caution!* See *Org. Synth.*, **Coll. Vol. 5**, 976 (1973), *for a warning regarding the purification of tetrahydrofuran.*)

3. Gaseous vinyl bromide is condensed in a 500-ml. flask cooled in an acetone – dry ice bath and diluted with 200 ml. of tetrahydrofuran. The submitters used vinyl bromide supplied by J. T. Baker Chemical Company; the checkers purchased this reagent from Linde Specialty Gases. Vinyl bromide is also available from Aldrich Chemical Company, Inc.

4. Approximately 70 ml. is added over a 20-minute period before formation of the Grignard reagent begins. The total addition time is *ca.* 1 hour.

5. The submitters used practical grade chlorotrimethylsilane purchased from PCR, Inc., which was distilled before use. Chlorotrimethylsilane from Aldrich Chemical Company, Inc., was employed by the checkers, both with and without prior distillation. Approximately the same yield was obtained in either case.

6. A white precipitate, presumed to be magnesium salts, is deposited as the solution of chlorotrimethylsilane is added.

7. The layers must be allowed to separate completely to avoid sizable mechanical losses. The submitters used 10 water extractions, whereas the checkers continued the extractions until the organic layer reached an approximately constant weight.

8. The ^{1}H NMR spectrum of the product obtained by the checkers revealed the presence of 6 ± 2% of tetrahydrofuran.

9. Diethylamine is available from Aldrich Chemical Company, Inc.

10. The reaction of the excess bromine with diethylamine is exothermic; consequently it may be necessary to moderate the reaction by cooling with an ice-water bath during the early stages of the addition.

11. ^{1}H NMR spectrum (CDCl$_3$), δ (multiplicity, coupling constant J in Hz., number of protons, assignment): 0.16 [s, 9H, Si(CH_3)$_3$], 6.12 (d, J = 2, 1H, CH), 6.21 (d, J = 2, 1H, CH).

12. The checkers terminated the distillation when the head temperature reached 44°.

13. The oxidizing reagent is prepared as described in *Org. Synth.*, **Coll. Vol. 5**, 866 (1973).

14. GC analysis of the product by the submitters on a 1.85-m. 3% silicone gum rubber (SE-30) column at 25° gave a single peak. Butenone **4** has the following ^{1}H NMR spectrum (CCl$_4$), δ (multiplicity, coupling constant J in Hz., number of protons, assignment): 0.14 [s, 9H, Si(CH_3)$_3$], 2.23 (s, 3H, CH_3), 6.18 (d, $J = 2$, 1H, CH), 6.53 (d, $J = 2$, 1H, CH).

15. This material showed no tendency to deteriorate when stored under an argon atmosphere at $-20°$.

3. Discussion

Butenone **4** has been obtained by Brook and Duff[4] from the reaction of 1-trimethylsilylvinylmagnesium bromide and acetic anhydride at $-120°$. However, the product was a mixture of butenone **4** and a dimeric substance which apparently resulted from subsequent conjugate addition of the Grignard reagent to the ketone. The procedures in the present reaction sequence for the preparation of silane **1**, silane **2**, and 1-trimethylsilylvinylmagnesium bromide are based on those reported by Ottolenghi, Fridkin, and Zilkha.[5] Similarly, other 1-trimethylsilylvinyl ketones may be prepared by reaction of the appropriate aldehyde with 1-trimethylsilylvinylmagnesium bromide and subsequent oxidation with chromic acid. The 1-trimethylsilylvinyl ketones are remarkably stable and useful in a variety of conjugate addition reactions.[6]

1. Department of Chemistry, Wayne State University, Detroit, Michigan 48202.
2. Present address: Department of Chemistry, University of Rochester, Rochester, New York 14627.
3. Department of Chemistry, Cornell University, Ithaca, New York 14853.
4. A. G. Brook and J. M. Duff, *Can. J. Chem.*, **51**, 2024 (1973).
5. A. Ottolenghi, M. Fridkin, and A. Zilkha, *Can. J. Chem.*, **41**, 2977 (1963).
6. R. K. Boeckman, Jr., D. M. Blum, and B. Ganem, *Org. Synth.*, **Coll. Vol. 6**, 666 (1988).

TROPOLONE

[2,4,6-Cycloheptatrien-1-one, 2-hydroxy-]

Submitted by RICHARD A. MINNS[1]
Checked by ARTHUR J. ELLIOTT
and WILLIAM A. SHEPPARD

1. Procedure

Caution! Benzene has been identified as a carcinogen; OSHA has issued emergency standards on its use. All procedures involving benzene should be carried out in a well-ventilated hood, and glove protection is required.

A. *7,7-Dichlorobicyclo[3.2.0]hept-2-en-6-one.* A 2-l., three-necked, round-bottomed flask fitted with an addition funnel, a reflux condenser, and a mechanical stirrer is charged with 100 g. (0.678 mole) of dichloroacetyl chloride (Note 1), 170 ml. (2 moles) of cyclopentadiene (Note 2), and 700 ml. of pentane (Note 3). The solution is heated to reflux under nitrogen and rapidly stirred while a solution of 70.8 g. (0.701 mole) of triethylamine (Note 4) in 300 ml. of pentane is added over a period of 4 hours (Note 5). After the cream-colored mixture has been refluxed for an additional 2 hours, 250 ml. of distilled water is added, dissolving the triethylamine hydrochloride; the layers are separated in a 2-l. separatory funnel. After extraction of the aqueous layer with two 100-ml. portions of pentane, the combined organic layers are filtered and dried by passage through absorbent cotton. Pentane and excess cyclopentadiene are then removed by rapid distillation. The resulting viscous, orange liquid is fractionally distilled under reduced pressure through a 30-cm. Vigreux column. Heat is supplied from an oil bath held at 105°. During collection of the first fraction, which consists mainly of dicyclopentadiene (Note 6), b.p. 61 – 62° (9 mm.), the cold finger and take-off tube must be warmed periodically with a heat gun to prevent plugging. The 7,7-dichlorobicyclo[3.2.0]hept-2-en-6-one, 101 – 102 g. (84 – 85%), is collected as a colorless liquid, b.p. 66 – 68° (2 mm.), n_D^{25} 1.5129, having a purity >99% as determined by GC analysis (Notes 6 and 7).

B. *Tropolone.* A 1-l., three-necked, round-bottomed flask equipped with a mechanical stirrer, addition funnel, and a reflux condenser is charged with 500 ml. of glacial acetic acid and then, *cautiously,* 100 g. of sodium hydroxide pellets. After the pellets have dissolved, 100 g. (0.565 mole) of 7,7-dichlorobicyclo[3.2.0]hept-2-en-6-one is added and the solution is maintained at reflux under nitrogen for 8 hours. Concentrated hydrochloric acid is added until the mixture is about pH 1; approximately 125 ml. of acid is required. After the addition of 1 l. of benzene, the mixture is filtered and the solid sodium chloride is washed with three 100-ml. portions of benzene. The two phases of the filtrate are separated, and the aqueous phase is transferred to a magnetically stirred, 1-l., continuous extractor (Note 8). The combined benzene phase is transferred to a 2-l. pot connected to the extractor, and the aqueous phase is extracted for 13 hours. Following distillation of the benzene, the remaining orange liquid is distilled under reduced pressure through a 30-cm. Vigreux column, removing acetic acid. When tropolone begins to distill into the column, the condenser is replaced with a two-necked flask immersed in ice water. With vacuum applied through one neck of this receiver, tropolone distills at 60° (0.1 mm.) and is collected as a crude yellow solid, 66.4 g. (96%). A solution of the impure product in 150 ml. of dichloromethane is diluted with 600 ml. of pentane, 4 g. of activated carbon is added, and the mixture is heated to boiling. After removal of the carbon by filtration, the solution is maintained at −20° until crystallization is complete. Tropolone, 53 g. (77%) (Note 9), is collected as white needles, m.p. 50 – 51°, by filtration. Evaporation of the filtrate to dryness, dissolution of the residue in 800 ml. of pentane, treatment with activated carbon, and cooling to −20° yields an additional 8 g. (12%) of tropolone as pale-yellow crystals, m.p. 49.5 – 51°.

2. Notes

1. Freshly opened bottles of dichloroacetyl chloride from Aldrich Chemical Company, Inc., were used. The acid chloride can also be prepared by the dropwise addition of 1 volume of dichloroacetic acid to 2.5 volumes of phthaloyl chloride heated to 140°. After the addition is complete, the solution is vigorously heated and dichloroacetyl chloride, b.p. 106 – 108°, is distilled through a 30-cm. column packed with glass beads; the yield is 85%.

2. Cyclopentadiene was prepared by cracking dicyclopentadiene [*Org. Synth.,* **Coll. Vol. 4,** 475 (1963)] of 95% purity purchased from Aldrich Chemical Company, Inc.

3. Technical grade pentane from Fisher Scientific Company was used.

4. Triethylamine from Eastman Organic Chemicals was used without further purification.

5. Faster addition results in some polymerization of the dichloroketene and darkens the precipitate.

6. Fractions were analyzed by GC (column: 0.3 × 120 cm., 20% SE-52 on Chromosorb P 60/80, 130°, helium flow rate of 60 ml./min.). Retention times of 1.9 minutes for dicyclopentadiene and 4.6 minutes for the 7,7-dichlorobicyclo[3.2.0]hept-2-en-6-one were found.

7. 7,7-Dichlorobicyclo[3.2.0]hept-2-en-6-one has the following spectral characteristics: IR (neat) cm.$^{-1}$: 1806 (C=O), 1608 (C=C); ^{1}H NMR (CCl$_4$), δ (multiplicity,

number of protons, assignment): 2.70 (m, 2H, CH_2), 4.10 (m, 2H, 2CH), 5.90 (m, 2H, $CH{=}CH$).

8. A continuous extractor has been described in *Org. Synth.*, **Coll. Vol. 5**, 630 Note 10 (1973).

9. Tropolone has the following spectral characteristics: IR (KBr pellet) cm.$^{-1}$: 3210 (OH), 1613 (C=O), 1548 (C=C); ^{1}H NMR (CDCl$_3$), δ (multiplicity, number of protons, assignment): 7.33 (m, 5H, 5CH), 8.76 (s, 1H, OH).

3. Discussion

Tropolone has been made from 1,2-cycloheptanedione by bromination and reduction,[2] and by reaction with *N*-bromosuccinimide;[3] from cycloheptanone by bromination, hydrolysis, and reduction;[4] from diethyl pimelate by acyloin condensation and bromination;[5] from cycloheptatriene by permanganate oxidation;[6] from 3,5-dihydroxybenzoic acid by a multistep synthesis;[7] from 2,3-dimethoxybenzoic acid by a multistep synthesis;[8] from tropone by chlorination and hydrolysis,[9] by amination with hydrazine and hydrolysis,[10] or by photoöxidation followed by reduction with thiourea;[11] from cyclopentadiene and tetrafluoroethylene;[12] and from cyclopentadiene and dichloroketene.[13,14]

The present procedure, based on the last method, is relatively simple and uses inexpensive starting materials. Step A exemplifies the 2 + 2 cycloaddition of dichloroketene to an olefin,[15-17] and the specific cycloadduct obtained has proved to be a useful intermediate in other syntheses.[18-20] Step B has been the subject of several mechanistic studies,[21-24] and its yield has been greatly improved by the isolation technique described above. This synthesis has also been extended to the preparation of various tropolone derivatives.[14,21,22,25-28]

1. Converse Memorial Laboratory, Harvard University, Cambridge, Mass. 02138. [Present address: Polaroid Corporation, 730 Main Street, Cambridge, Massachusetts 02139.]
2. J. W. Cook, A. R. Gibb, R. A. Raphael, and A. R. Somerville, *J. Chem. Soc.*, 503 (1951).
3. T. Nozoe, S. Seto, Y. Kitahara, M. Kunori, and Y. Nakayama, *Proc. Jpn. Acad.*, **26** (7), 38 (1950).
4. T. Nozoe, Y. Kitahara, T. Ando, and S. Masamune, *Proc. Jpn. Acad.*, **27** (8), 415 (1951); T. Nozoe, Y. Kitahara, T. Ando, S. Masamune, and H. Abe, *Sci. Rep. Tohoku Univ.*, *Ser. 1*, **36**, 166 (1952) [*Chem. Abstr.*, **49**, 11615e (1955)].
5. J. D. Knight and D. J. Cram, *J. Am. Chem. Soc.*, **73**, 4136 (1951).
6. W. von E. Doering and L. H. Knox, *J. Am. Chem. Soc.*, **73**, 828 (1951).
7. E. E. van Tamelen and G. T. Hildahl, *J. Am. Chem. Soc.*, **78**, 4405 (1956).
8. O. L. Chapman and P. Fitton, *J. Am. Chem. Soc.*, **85**, 41 (1963).
9. A. P. ter Borg, R. van Helden, and A. F. Bickel, *Recl. Trav. Chim. Pays-Bas*, **81**, 177 (1962).
10. T. Nozoe, T. Mukai, and K. Takase, *Sci. Rep. Tohoku Univ.*, *Ser. 1*, **39**, 164 (1955) [*Chem. Abstr.*, **51**, 7316c (1957)].
11. M. Oda and Y. Kitahara, *Tetrahedron Lett.*, 3295 (1969).
12. J. J. Drysdale, W. W. Gilbert, H. K. Sinclair, and W. H. Sharkey, *J. Am. Chem. Soc.*, **80**, 245, 3672 (1958).
13. H. C. Stevens, D. A. Reich, D. R. Brandt, K. R. Fountain, and E. J. Gaughan, *J. Am. Chem. Soc.*, **87**, 5257 (1965); L. Ghosez, R. Montaigne, and P. Mollet, *Tetrahedron Lett.*, 135 (1966); P. D. Bartlett, U.S. Pat. 3,448,155 (1969) [*Chem. Abstr.*, **71**, 91130x (1969)].

14. H. C. Stevens, J. K. Rinehart, J. M. Lavanish, and G. M. Trenta, *J. Org. Chem.*, **36**, 2780 (1971).
15. L. Ghosez, R. Montaigne, A. Roussel, H. Vanlierde, and P. Mollet, *Tetrahedron*, **27**, 615 (1971), and references therein.
16. P. R. Brook and J. G. Griffiths, *J. Chem. Soc. D*, 1344 (1970).
17. R. E. Harmon, W. D. Barta, S. K. Gupta, and G. Slomp, *J. Chem. Soc. C*, 3645 (1971).
18. M. Rey, U. A. Huber, and A. S. Dreiding, *Tetrahedron Lett.*, 3583 (1968).
19. P. R. Brook, *Chem. Commun.*, 565 (1968).
20. P. A. Grieco, *J. Org. Chem.*, **37**, 2363 (1972).
21. T. Asao, T. Machiguchi, T. Kitamura, and Y. Kitahara, *J. Chem. Soc. D*, 89 (1970).
22. P. D. Bartlett and T. Ando, *J. Am. Chem. Soc.*, **92**, 7518 (1970).
23. T. Asao, T. Machiguchi, and Y. Kitahara, *Bull. Chem. Soc. Jpn.*, **43**, 2662 (1970).
24. W. T. Brady and J. P. Hieble, *J. Am. Chem. Soc.*, **94**, 4278 (1972).
25. R. W. Turner and T. Seden, *Chem. Commun.*, 399 (1966).
26. T. R. Potts and R. E. Harmon, *J. Org. Chem.*, **34**, 2792 (1969).
27. J. H. Shim, *Taehan Hwahak Hoechi*, **13**, 75, 83 (1969) [*Chem. Abstr.*, **72**, 3194p, 3195q (1970)].
28. K. Tanaka and A. Yoshikoshi, *Tetrahedron*, **27**, 4889 (1971).

PREPARATION OF ALKENES BY REACTION OF LITHIUM DIPROPENYLCUPRATES WITH ALKYL HALIDES: (*E*)-2-UNDECENE

This procedure [*Org. Synth.*, **55**, 103 (1976)] has been deleted because the required starting material, lithium dispersion containing 1% sodium, is no longer commercially available.

GENERAL INDEX

This comprehensive, alphabetical index includes references to the reagents, catalysts, less common solvents, distinctive apparatus, isolable intermediates, products, and selected compound and reaction types involved in the preparations. For emphasis, the isolable intermediates and products are named in capital letters often followed by the Chemical Abstracts Service (CAS) Registry Numbers in brackets. An effort has been made to also include in the index all organic compounds mentioned in the notes or discussion sections, even if shown only as structural formulas in the tables. Besides the compound names which appear in the text, alternate or better systematic names are also frequently given. All names which appear in the Formula Index are included. Under headings for compound types appearing in capital letters, to avoid unnecessary repetition of names with prefixes to the headings, in most cases only compound names are listed which do not already appear nearby in alphabetical order. The page number listed for each entry is the point of first appearance in the preparation, whether it is as a name or a structural formula.

1041

FORMULA INDEX

All organic compounds (except solvents) mentioned or shown as structures in the preparation and discussion sections are included in this comprehensive index. The system of indexing is similar to that used by *Chemical Abstracts* and in previous volumes of this series. The essential principles are that: (1) The arrangement of symbols in formulas is alphabetical except C always comes first followed immediately by H if present. (2) The arrangement of formulas is also alphabetical except that the number of atoms of one kind influences the order of compounds. For example, all formulas with one carbon atom precede those with two carbon atoms; thus, the order CH_2I_2, CH_3NO_2, CH_5N, $C_2H_2O_3$, etc. (3) The arrangement of entries under any formula is alphabetical according to the names of the isomers, except that italicized prefixes used as locants or stereochemical descriptors are disregarded. (4) In most cases inorganic salts are listed under the formulas of the parent organic compounds from which they are derived, but organometallic compounds are listed with the metal in the formula. In some cases alternate or better names are given after the name which appears in the text. The same names and additional alternate names are given in the General Index.

$CClNO_3S$ Chlorosulfonyl isocyanate, 304, 465, 788

CCl_2O Phosgene, 199, 282, 418

CCl_4 Carbon tetrachloride, 634

CD_2N_2 Dideuteriodiazomethane, 432

$CHBr_3$ Bromoform, 187

$CHCl_3$ Chloroform, 87, 232, 731

$CHFO$ Formyl fluoride, 9

CHF_3O_3S Trifluoromethanesulfonic acid, 757

CHI_3 Iodoform, 974,

CHN Hydrogen cyanide, 14

CH_2I_2 Diiodomethane, 327

CH_2N_2 Diazomethane, 386, 432, 613

CH_2O Formaldehyde, 981, 987

CH_2O_2 Formic acid, 8, 631

CH_3Br Methyl bromide, 675

CH_3ClO_2S Methanesulfonyl chloride, 56, 482, 652, 727

CH_3I Methyl iodide, 64, 442, 501, 552, 683, 704, 830

CH_3NO Formamide, 987

CH_3NO_2 Methyl nitrite, 199
Nitromethane, 797

CH_4O Methanol, 788

CH_4O_3S Methanesulfonic acid, 312

CH_4S Methanethiol, 601, 683

CH_5N Methylamine, 818

CH_5NO *N*-Methylhydroxylamine, 670

CO_2 Carbon dioxide, 207

C_2Cl_3N Trichloroacetonitrile, 507

C_2Cl_4O Trichloroacetyl chloride, 618

$C_2Cl_4O_2$ Trichloromethyl chloroformate, 715

$C_2F_6O_5S$ Trifluoromethanesulfonic anhydride, 757

$C_2F_6O_5S_2$ Trifluoromethanesulfonic acid anhydride, 324

C_2HCl_3O Dichloroacetyl chloride, 1037

$C_2HF_3O_2$ Trifluoroacetic acid, 273, 324
Thallium(III) trifluoracetate, 709

C_2H_2 Acetylene, 940

$C_2H_2O_3$ Glyoxylic acid, 965

$C_2H_2O_4$ Oxalic acid, 64, 359, 754, 901

C_2H_3Br Vinyl bromide, 1033

C_2H_3ClO Acetyl chloride, 8, 34, 883

$C_2H_3ClO_2$ Methyl chloroformate, 715

C_2H_3N Methyl isocyanide or (Carbylamino)methane, 232

$C_2H_3NO_2S$ Methanesulfonyl cyanide, 727

$C_2H_3NO_4$ Dipotassium salt of nitroacetic acid, 797

C_2H_4 Ethylene, 744

C_2H_4BrF 1-Bromo-2-fluoroethane, 836

$C_2H_4Br_2$ 1,2-Dibromoethane, 991, 1016, 1034

$C_2H_4ClNO_4S$ Methyl (chlorosulfonyl)-carbamate, 788

$C_2H_4Cl_2O$ Dichloromethyl methyl ether, 138

$C_2H_4F_2$ 1,2-Difluoroethane, 836

C_2H_4O Acetaldehyde, 901, 1033

$C_2H_4O_2$ Acetic acid, 56, 181, 210, 246, 299, 461, 968
Methyl formate, 620
Thallium(I) acetate, 348

$C_2H_4O_3$ Peracetic acid, 862

HAZARD INDEX

AUTHOR INDEX

CONCORDANCE INDEX

The annual volume (in bold-faced type) and page number where each procedure first appeared is given on the left; the page number for the revised procedure is given on the right. For a concordance listing by title of procedure, see the Table of Contents.

Annual Volume		This Volume		Annual Volume		This Volume
Number	Page	Page		Number	Page	Page
50	1	8		20	631
	3	10		24	905
	6	161		31	751
	9	207		39	869
	13	215		44	830
	15	276		48	910
	18	304		53	967
	21	310		55	133
	24	196		60	142
	27	392		66	172
	31	403		70	268
	36	422		73	271
	38	442		76	316
	43	454		82	150
	50	461		86	389
	52	465		90	245
	56	501		94	700
	58	517		96	259
	62	531		100	618
	65	555		103	371
	66	901		106	179
	72	556		109	744
	75	558		112	795
	77	613		115	883
	81	664		121	936
	84	837		128	928
	88	852		133	962
	94	913		136	965
	97	925		139	824
	102	12		142	1019
	104	979			
	107	1030	**52**	1	28
				5	644
51	1	21		11	78
	4	1028		16	101
	8	1007		19	240
	11	312		22	153
	17	815		33	401

Annual Volume		This Volume	Annual Volume		This Volume
Number	Page	Page	Number	Page	Page
55 *(continued)*				72	835
	114	56		74	856
	122	248		78	412
	127	776		80	873
				83	893
56	1	39		88	727
	3	1		92	971
	8	109		95	981
	15	581		102	987
	19	115		107	996
	25	946		113	1024
	28	625		117	1037
	32	187			
	36	648	**58**	1	595
	40	788		4	507
	44	567		12	887
	49	462		17	512
	52	51		24	137
	59	576		32	943
	65	Replaced by 615		37	571
	68	890		43	342
	72	601		52	36
	77	704		56	520
	83	747		64	424
	88	263		67	427
	95	951		75	628
	99	218		79	615
	101	769		83	774
	107	368		86	652
	112	560		98	698
	118	968		101	334
	122	1004		106	670
				113	542
57	1	167		122	220
	8	41		127	407
	11	338		134	451
	16	1001		138	130
	18	468		143	833
	22	163		147	235
	26	273		152	1033
	30	301		158	666
	33	662		163	445
	36	320			
	41	361	**59**	1	95
	45	418		10	954
	50	440		16	184
	53	482		20	90
	60	503		26	282
	62	505		35	353
	65	564		42	293
	69	242		49	492

Annual Volume		This Volume		Annual Volume		This Volume
Number	Page	Page		Number	Page	Page
59 *(continued)*				132	226	
	53	866		141	533	
	58	23		147	958	
	66	414		153	474	
	71	82		159	252	
	79	640		169	348	
	85	845		176	43	
	95	199		183	620	
	102	807		190	5	
	113	327		195	715	
	122	875		202	737	